Machine Learning for Civil & Environmental Engineers

Machine Learning for Civil & Environmental Engineers

A Practical Approach to Data-driven Analysis, Explainability, and Causality

M. Z. Naser, PhD, PE

School of Civil and Environmental Engineering & Earth Sciences (SCEEES) and
Artificial Intelligence Research Institute for Science and Engineering (AIRISE),
Clemson University, Clemson, SC, USA

Published by John Wiley & Sons, Inc., Hoboken, New Jersey.

Published simultaneously in Canada.

Library of Congress Cataloging-in-Publication Data

Names: Naser, M. Z., author. | John Wiley & Sons, publisher.
Title: Machine learning for civil & environmental engineers : a practical approach to data-driven analysis, explainability, and causality / M. Z. Naser.
Description: Hoboken, New Jersey : Wiley, [2023] | Includes bibliographical references and index. | Summary: "Synopsis: The theme of this textbook revolves around how machine learning (ML) can help civil and environmental engineers transform their domain. This textbook hopes to deliver the knowledge and information necessary to educate engineering students and practitioners on the principles of ML and how to integrate these into our field. This textbook is about navigating the realm of data-driven ML, explainable ML, and causal ML from the context of education, research, and practice. In hindsight, this textbook augments ML into the heart of engineering. Together, we will go over the big ideas behind ML. We will ask and answer questions such as, what is ML? Why is ML needed? How does ML differ from statistics, physical testing, and numerical simulation? Can we trust ML? And how can we benefit from ML, adapt to it, adopt it, wield it, and leverage it to overcome many, many of the problems that we may face? This book is also about showing you, my dear reader, how to amplify your engineering knowledge with a new tool. A tool that is yet to be formally taught in our curriculum. A tool that many civil and environmental engineering departments and schools may not fully appreciate ; yet are eager to know more about!"-- Provided by publisher.
Identifiers: LCCN 2023003262 (print) | LCCN 2023003263 (ebook) | ISBN 9781119897606 (hardback) | ISBN 9781119897620 (pdf)
Subjects: LCSH: Machine learning. | Civil engineering--Data processing. | Environmental engineering--Data processing.
Classification: LCC Q325.5 .N37 2023 (print) | LCC Q325.5 (ebook) | DDC 006.3/1--dc23/eng20230506
LC record available at https://lccn.loc.gov/2023003262
LC ebook record available at https://lccn.loc.gov/2023003263

Cover Image: © Dall•E 2/OpenAI
Cover Design: Wiley

Set in 9.5/12.5pt STIXTwoText by Integra Software Services Pvt. Ltd, Pondicherry, India

SKY10050279_070523

To the future of civil and environmental engineering

Contents

Preface

Hello, friend.

Elliot Alderson (Mr. Robot)[1]

Synopsis

The theme of this textbook revolves around how machine learning (ML) can help civil and environmental engineers transform their domain. This textbook hopes to deliver the knowledge and information necessary to educate engineering students and practitioners on the principles of ML and how to integrate these into our field. This textbook is about navigating the realm of data-driven ML, explainable ML, and causal ML from the context of education, research, and practice. In hindsight, this textbook augments ML into the heart of engineering.

Together, we will go over the big ideas behind ML. We will ask and answer questions such as, what is ML? Why is ML needed? How does ML differ from statistics, physical testing, and numerical simulation? Can we trust ML? And how can we benefit from ML, adapt to it, adopt it, wield it, and leverage it to overcome many, many of the problems that we may face?

This textbook is also about showing you, my dear reader, how to amplify your engineering knowledge with a new tool. A tool that is yet to be formally taught in our curriculum. A tool that many civil and environmental engineering departments and schools may not fully appreciate;[2] yet are eager to know more about!

Tone and Rhythm

I have come to the conclusion that ML is one of those ideas that we constantly hear but do not *really* seem to fully understand, know what it is, what it does, or why we need it. Here is a prime example, our tablets and phones have plenty of ML built into them. More often than not, you will hear someone say, *Oh. I hate the new update. They changed the algorithm. Again!*[3] I have always struggled[4] with the following question, *how do you write about something [ML] that your audience "thinks" that they know but, in reality, do not seem to grasp?* This made writing this book an equally challenging and unique experience.

By now, you might have noticed that I am breaking free from the typical and expected formality as well as high technicality associated with engineering writing. Frankly speaking, my goal was not to write a heavy *technical* textbook[5] *per se.* I felt a smoother tone might be more inviting for an engineering resource with a particular theme on ML that is designed to

1 *eps1.0_hellofriend.mov* is the title to the pilot episode of Mr. Robot.

2 At this moment in time, there is a handful of departments/schools that cover ML in a civil and environmental engineering curriculum. I do not know why. I can only speculate! However, in a few years from now, I am expecting that ML will be a cornerstone in our curriculum.

3 This reminds me of the following: "*I have a cue light I can use to show you when I'm joking, if you like.*" – TARS (Interstellar, 2014).

4 More often than not, especially whenever I work on research proposals with a ML component.

5 I have had the pleasure of writing some and reading/collecting many myself.

be a textbook for its audience. While I admittedly do so, I still maintain an above average degree of formality, as I believe that such formality helps convey key ideas and discussions.

A technical tone is also great when describing facts, methodologies, and results. Such a tone sounds assuring – even in the muddy waters of ML. Personally, I wanted the experience of navigating the pages of this textbook to *feel* more like we are learning this material together and, for the first time – fostering an atmosphere of trust and positivity.

At some points, you may find that I have realized a balanced tone and rhythm, while in others, my *alter ego* may prevail, and I will try to walk you through my thinking process and then answer any questions that I imagine might be lurking in the shadows of your subconscious.[6] In the latter, you will likely see my reliance on footnotes.[7] I like footnotes as I feel they resemble a place where I can extend the conversation[8] with a bit less formality.

I also like quotes,[9] and I have purposefully used some across my chapters to help deliver my thoughts or as anchors to introduce and solidify concepts. I will be adding some short storylines that I believe will come in handy to explain some notions and ideas. You might find these at the beginning of chapters to help reinforce their themes.[10] These are true stories that I have the pleasure of experiencing personally or seeing firsthand.

I would like this textbook to be a journey. I have designed it as one that more or less mimics the way I learned and researched ML.[11] I believe that we need to learn ML as a *methodology* and *philosophy* instead of *applied* science or one that revolves around *coding*. The journey I am anticipating is conversational, spanning intriguing questions and covering the quest to answer them.

Start of Our Conversation

This seems like a good spot to start this conversation. I am M.Z. Naser,[12] an assistant professor at the School of Civil and Environmental Engineering & Earth Sciences and Artificial Intelligence Research Institute for Science and Engineering (AIRISE) at the College of Engineering, Computing and Applied Sciences at Clemson University.[13] I was fortunate enough to complete my undergraduate[14] and MSc degrees at the American University of Sharjah, Sharjah, United Arab Emirate, and my PhD and Post-doc at Michigan State University, East Lansing, USA. Surprisingly, my research area started with structural fire engineering[15] and then propagated into ML.[16] I had the pleasure of authoring/co-authoring over 150 papers and four books – most of them revolve around structural fire engineering and ML.[17] I actively serve on a number of international editorial boards and conferences, as well as codal building committees (in American Society of Civil Engineers, American Concrete Institute, Precast, Prestressed Concrete Institute, and International Federation for Structural Concrete,[18] to name a few). At the time of this writing, I am humbled to the chair of the ASCE Advances in Technology committee.

---

6 I was new to ML not too long ago! I still have many burning questions that I would like to answer.

7 Hello!

8 Shout out to Nick Huntington-Klein, the author of *The Effect: An Introduction to Research Design and Causality* [1].

9 Those extracted from scenes from movies/shows/songs/poems (especially those in Arabic). I did not know that these could be added into textbooks until I read *Causal Inference: The Mixtape* by Scott Cunningham [2].

10 Some might call this, *learning by storytelling* [3]!

11 Of course, with much less focus on failures.

12 I mainly go by my last name (thanks to the good folks at Michigan State University who started this trend). M.Z. is for Mohannad Zeyad. Note the presence of the two N's in my first name.

13 In Clemson, South Carolina, 29634.

14 I usually start every one of my courses with a short introduction and then add that I graduated with a GPA of 2.69 and had many semesters where my GPA dropped below 2.20 (I was repeatedly ranked at the bottom of my class). If you happen to be an above C student, then congratulations on your hard work! For those of you who happen to share my fortune, I truly understand. Do not give up!

15 A niche area of structural engineering where we focus on how fire (thermal effects) affects materials and structures. A real conversation starter and icebreaker!

16 I must acknowledge Prof. Ghassan Abu Lebdah for introducing me to this area.

17 As always, I am thankful for all of my collaborators and students, who without them, many of these works would not have seen the light.

18 Or simply, Fib (Fédération internationale du béton) – thanks to Thomas Gernay!

Elemental Questions

Around 2018, I was working on developing a new and simple method to predict the fire resistance of reinforced concrete (RC) columns.[19] I was hoping to develop a unified method that, unlike existing methods that do not seem to converge,[20] works in a one-step approach.[21] The objective was simple: to create a method that does not suffer from the same symptoms and limitations of existing methods.

I started exploring and, of course, crossed paths with ML. I built a ML model, and upon examination, this model outperformed every existing method (whether codal, accepted, or published) I have tested against. This was intriguing. How does the ML model perform so well? Why does it perform so well?

It made sense to pursue and investigate *how* and *why* the model works well.[22] My rationale is simple. Knowing the *how* and *why* allows us to see what the model sees, and if we knew the answers to these two questions, then it is plausible that we could come up with a new theory to predict the fire resistance of RC columns! Or perhaps to distill new knowledge that we have not seen before. I was excited.

Answers to Elemental Questions

The answer to the *how* is simple. ML models are nonparametric and are designed to identify hidden patterns in data.[23] Those patterns are the *something* that the ML model sees that we, as well as our traditional methods, may not see well.

The answer to the *why* is not so simple. It turns out that there is a whole field dedicated to explaining how ML models predict the way they do. So, if we could learn the reasoning behind model predictions, we could start to better grasp the hidden patterns.[24] Even better, start to see such patterns without the need for ML! To get closer to the essence[25] of why the phenomena we are interested in occur!

Surprisingly, explanations from ML models may or may not agree with our domain knowledge or physics! This is both concerning and problematic! Thus, we ought to know if a model's *how* and *why* are correct and if so, then it is possible that the model understands the true *cause* and *effect* of the patterns it sees.[26] Guess what? There is a whole domain about establishing causality too. We will visit this domain as well.

By now, I hope you can appreciate the overarching goal behind this textbook. This book is about convergence to building a practical approach to the above elemental questions.

Anatomy of This Textbook

This textbook comprises a series of chapters. Each chapter covers one *big idea* about ML. The chapters are organized in a manner that resembles the most commonly thought process in adopting ML into a problem within civil and environmental engineering. These chapters also contain examples and tutorials. Some of the examples are short, while others may span a couple of chapters. You will find that I tend to favor illustrative examples on simple components (i.e., beams and columns) as these are easy to visualize and can be thought of as metaphors that are likely to be extended to other sub-fields as well.[27] Collectively, this book draws inspiration and examples from most of the sub-areas belonging to civil and environmental engineering.

---

19 At that point, a few tools existed – as you will see in an example covered in a later chapter.

20 We know why this happens. It is because each method is built on certain assumptions and has a list of specific conditions. Therefore, a convergence across existing methods is not guaranteed, nor expected.

21 One-step approach is a procedure that allows the prediction/evaluation of an outcome in one step.

22 To be frank, the question I had in mind then was, *what* does the ML model see that engineers/other methods do not see?

23 This is the *data driven* part to the title.

24 This is the *explainability* part to the title.

25 Formally, this is called the data generating process.

26 Here is the *causality* part to the title.

27 Two more reasons, 1) I am a visual learner and value visual examples, and 2) some examples were part of my research work which, as we have learned, stems from structural fire engineering.

Since this is a textbook, it made sense to include sections on *definitions*, *blueprints*, as well as *questions* and *problems* at the end of each chapter. The definitions section summarizes the main terms highlighted in each chapter, and the blueprints lay out a visual representation of each chapter's concepts.

The *questions* can be answered by going over the discussion presented in each corresponding chapter.[28] On the other hand, the *problems* may require you to *build* ML models via coding-free/coding-less platforms or via simple *coding*. For these, I have provided you with ample codes (scripts) and databases that can ease you into building models, programming, and coding. These can be conveniently accessed through my website.[29] My goal from these questions and problems is to invite you to learn and practice ML a bit more.

Speaking of programming and coding. I did not intend for this book to be another source on *learning ML by coding*.[30] As such, I did not dedicate much landscape to specifically teach you how to program/code. My thinking process is that 1) I would like you to learn about ML as a methodology – not as a coding exercise, and 2) there are plenty of resources on teaching coding out there. This book is about ML in civil and environmental engineering, and it is <u>not</u> about coding civil and environmental engineering problems using ML. A distinction, while subtle, is necessary and warranted to make clear.

Now, let us talk about each chapter.

Chapter 1 describes a series of strategies, approaches, and pathways to a few teaching methods I envision/propose for this textbook. This chapter also hopes to navigate the persisting challenge of modernizing the civil and environmental engineering curriculum. I am setting this chapter as a foundation for students and faculty who are interested in ML but may not be proficient in ML. This chapter could be interesting to advanced ML users as well.

Chapter 2 formally introduces us to ML. This chapter starts with a brief history of ML from computer science as well as civil and environmental engineering points of view. We will also be going over a systematic framework (pipeline) to build and tailor proper ML models.

Chapter 3 is where we discuss the role of data and statistics in ML. Here, we will chat a bit about where to get data from, what to do when we collect such data, what are some of the data handling techniques that a typical ML requires, and so on.

Chapter 4 is all about the different ML algorithms. These are the main tools that will help us apply and create ML models to solve many of our problems. This chapter presents key theoretical and practical concepts to supervised and unsupervised algorithms.

Chapter 5 is perhaps the last theoretical chapter before we get to practice[31] ML. In this chapter, we cover principles and notions behind ML performance fitness indicators and error metrics needed to quantify the adequacy of our models.

Chapter 6 is where we practice ML! We will be introduced to the realm of coding-free and coding-based ML. We will also explore a few exciting platforms, problems, and analyses. You will get to see and enjoy plenty of examples and tutorials.

Chapter 7 goes beyond the traditional blackbox ML modeling to showcase the role of explainable and interpretable ML. In this chapter, we first dive into explainability methods and then continue with examples and tutorials we have seen and practiced in Chapter 6 but adding the necessary tools (such as SHAP and partial dependence plots, etc.) to unbox the blackbox.

Chapter 8 introduces causality, along with its principles from the lens of ML and civil and environmental engineering. This chapter also goes over methods and algorithms to identify the data generating process, the process of creating causal ML models, and causally inferring knowledge from such models.

Chapter 9 covers advanced topics. These topics span a multitude of concepts that could extend a bit over the duration of a course designed for beginners. I can see these topics as more applicable for research projects and a possible guest lecture to take place toward the end of a typical course to possibly attract and retain students interested in ML to further explore *other big ideas*.

Chapter 10 presents recommendations, suggestions, and best practices that can be of aid in ML-based investigations. I find these informative and applicable to many, many of our problems. The same can also give you a glimpse into the state of our domain and how our colleagues are utilizing and adopting ML.

---

28 In some rare instances, I may refer you to a scholarly article or a website.

29 www.mznaser.com, as well as the companion website to this book www.wiley.com/go/Naser/MachineLearningforCivilandEnvironmentalEngineers

30 More on this in the first chapter.

31 Here is the *practical approach* part to the title.

The main body of the textbook concludes with Chapter 11. This chapter is designed to display some of the possible future trends that could potentially benefit from ML. Some of these ideas may be trimmed and tuned to be tackled in future research problems and projects.

The Appendix is divided into sub-sections where I list and share what I feel and believe will be some valuable information, sources, techniques, and additional examples that I did not get to cover within the main chapters of this book.

A few section titles were assigned the following snowflake[32] symbol *. These sections are advanced and may be beneficial for well-versed users. I opted to keep these here to push the envelope a bit. Perhaps intrigue you a bit. Feel free to skim through these to have a feeling of what is available *out there*.

Things I Would like You to Try

Here are a few things that I would like you to try.

Please feel free to reach out to me with your thoughts on this book, ideas for future examples/problems, or chapters. Do not be afraid to share some of the typos or inconsistencies you might come across, and do share the codes you have developed.

Advancements in ML are consistently rapid. It is amazing to see the number of publications that hypothesize, create, or apply new ML methods. Follow up with the authors of the cited references – one thing about academics is that many would love to chat about their research. Many ML enthusiasts and developers have Twitter and LinkedIn accounts.[33] Ask them questions too. Follow trending topics on Google Scholar and social media. There are many YouTube channels that stay up to date on this front.[34]

Subscribing to scholarly journal updates is another effective method to stay in touch with the latest ML-based paper in one's field, especially if you happen to be a student where your school's library offers access to publishers of such journals.[35] In parallel, seek to attend conferences and stay up-to-date with society updates.

While you do not need to become an advanced coder/programmer, remember that it is never too late to learn something new. Wondering where to start? Start at any point! Pick a programming language.[36] There are plenty of free resources or paid services that you can use. These might include webbooks, forums, or MOOCs.[37]

Some Additional Thoughts

I fought the temptations to present ML from the context of Natural Language Processing (NLP)[38] or heavy-duty computer vision in this edition.[39] I am confident that these topics, along with many others, will find a home in future editions, as well as other books dedicated to civil and environmental engineering.

To keep our conversation going, I may recall a few of my dialogues, events, or correspondents and use these to navigate some of my current discussions. I may even recall a few examples of good and poor use of ML I came across throughout the past few years.

---

32 The snowflake is a throwback to Michigan!

33 They are there to follow them. I do!

34 This is 2023, so things might change a bit by the time you get your hands on this textbook. We are barely recovering from a couple of years of pandemic.

35 I cannot offer an exhaustive list of recommended journals. The following are some that come to mind: Computer-Aided Civil and Infrastructure Engineering, Automation in Construction, Water Research, and ASCE Journal of Computing in Civil Engineering, Engineering Structures, Journal of Building Engineering, Fire Technology, and many more.

36 If you are looking for a sign, here is one. Pick *R* or *Python*.

37 Massive open online courses. I have a list of recommended reads in the Appendix.

38 Can come in very handy in many sub-fields, especially in construction management and planning.

39 If you are wondering how this book came to be, then let me walk you through the following short story. Kalli Schultea reached out to me regarding a seminar I was presenting at the 2021 ASCE convention titled, *Modernizing Civil Engineering through Machine Intelligence*. She was wondering if I was interested in writing a book on ML, and I said, absolutely yes (since at the same time, I was actually working on developing a course on ML at Clemson University to be delivered in the Spring of 2023). The timing made perfect sense! The actual project took off from there.

Many individuals[40] and organizations[41] were kind enough to keep me motivated, knowingly or unknowingly. You will find those cited throughout the pages of this book. I thank you.

Much of my pursuit hopes to follow the steps of my mentors,[42] Rami Hawileh, Jamal Abdalla, Hayder Rasheed, and of course, the guru, Venkatesh Kodur.[43] I wish I could be their student again. Life was much easier and simpler back then. If you happen to cross their path one day, my advice is to get to learn from them as much as you can.

To my students. Haley Hostetter, Deanna Craig, Mohammad (Moe) Al-Bashiti, Arash Tapeh, Aditya Daware, Alireza Ghasemi, and Mohammad Abedi. Their hard work, support, dedication, publications, and help in putting some of the scripts and charts together made this journey brighter and is evident in this book. You are brilliant, and I thank you!

To Clemson University and my home and faculty (many of which are mentioned within the pages of this book). Thank you for your endless support!

To John Wiley & Sons, Inc. This book would not have seen the light if it was not for believing in this project and assistance along the way. My thanks go to Kalli Schultea, Amy Odum, Sarah Lemore, Britta Ramaraj, and Isabella Proietti.

Last but not least. My family, parents, siblings, nieces, and nephews – thank you for your unconditional love (despite my many, many shortcomings). My thanks also go to Pirate and Snow, my pets, for keeping me company during this journey. Thanks to Ena for making life a lil bit more interesting (and, possibly, less dramatic)! Special thanks go to my wife, Jasmine. You are my anchor and gym partner. She constantly reminds me that life is not just about research or writing papers. She says, "*life is a lil bit more than that!*". She is always supportive. I continue to learn from you. You are the source of my inspiration and motivation!

And to those forgotten, yet intentionally or unintentionally not listed here. I thank you.

Finally, this book is not an individual effort. It is a community effort. We learn together. We rise together. We advance together.

References

1 Huntington-Klein, N. (2021). *The Effect: An Introduction to Research Design and Causality*. Boca Raton: Chapman and Hall/CRC. https://doi.org/10.1201/9781003226055.

2 Cunningham, S. (2021). *Causal Inference: The Mixtape*. Yale University Press. ISBN-13: 978-0300251685.

3 Alterio, M. and McDrury, J. (2003). *Learning Through Storytelling Higher Education: Using Reflection & Experience to Improve Learning*. https://doi.org/10.4324/9780203416655.

---

40 Talk about a healthy competition! Whether by reading their publications, or answering my/their questions. A special thanks go to those that invited me to present about ML (e.g., Wojciech Węgrzyński, Nan Hu, Negar Elhami Khorasani, Barbara Lane, Ting Lin etc.). Also, thanks to the creators of many of the tools that I have showcased throughout this book. They were kind enough enough to lend me their tools, including but not limited to: Eduardo García Portugués, Victor Powell, Lewis Lehe, Minsuk Kahng, Naftali Harris, Adam Harley, Jonas Greitemann, Stephanie Yee, Sasha Luccioni, etc.

41 American Concrete Institute, American Society of Civil Engineers (on many occasions!), National Fire Protection Association, International Association for Fire Safety Science, BigML, DataRobot, Dataiku, Exploratory, Clarifai, and others.

42 It is not surprising to note that we are still in direct touch after all those years!

43 Remember my GPA story? Well, I was rejected from the graduate program at AUS and then again at MSU, and Prof. Hawileh and Prof. Kodur were kind enough to override the school's decision. Thank you!

About the Companion Website

This book is accompanied by a companion website:

www.wiley.com/go/Naser/MachineLearningforCivilandEnvironmentalEngineers

This website includes

- Appendix
- Datasets and codes
- PowerPoint slides
- Lecture notes
- Homework answers

1

Teaching Methods for This Textbook

Deciding is for machines because its computational. Choosing is very human.

Joseph Weizenbaum

Synopsis

This brief chapter outlines some proposed suggestions for adopting this textbook into a civil and environmental engineering curriculum. I try to remain flexible with describing my suggestions[1] as I understand the complexity of integrating new courses and/or modules into a dense curriculum – just like ours. I hope these can be of aid.[2]

1.1 Education in Civil and Environmental Engineering

There is a rich body of literature that tackles a variety of teaching and learning methods targeting engineering students [1–3]. If we are to distill this literature into its fundamental facets, we arrive at the elemental conclusion that many engineering students can be described as *active learners*. Such learners are likely to be experimentalists and tend to value cause-and-effect demonstrations. This presents an exciting challenge to drafting this book – and to be quite honest, it remains a challenge that I try to overcome every time I present ML to a new or junior audience. Before jumping into some possible means of delivery, please allow me to walk you through the following.

Foundational methods within our domain are developed systematically and with methodical consistency. In the majority of the time, such methods have a clear approach. We start from the first principles and work our way up to a goal (i.e., a solution). Suppose we think of designing a system (say, a load bearing element or a transportation network). This design exercise requires knowing the first principles associated with this particular system. Such principles often entail a set of provisions/guidelines, formulae, or tables/charts that can be applied to realize a *proper* design.[3]

We value first principles because they are authentic and transparent. They describe the logic that governs a problem (say, the design of aforenoted systems). They are compact and easy to use – for the most part. They could also, and perhaps most importantly, be verified via laboratory testing or some other means [4–6]. All of these qualities are essential to establish the merit and validity of adopting new methods into our domain.[4]

1 *Quick story* : I was a student of Prof. Venkatesh Kodur for about 8 years during my stay at Michigan State University. One of the eye-opening lessons I have learned from him is that writing and teaching are best when there is an impeccable flow – *like a plot in a movie!* The more intuitive and smooth the delivery, the more attention the audience ought to display.

2 I believe that this is a community effort. As such, I will be on the lookout for your suggestions, thoughts, and experiences. Please feel free to share yours with me.

3 Thanks to all of the great engineers that worked hard on creating, verifying, and publishing such tools.

4 Think of ML!

Machine Learning for Civil & Environmental Engineers: A Practical Approach to Data-driven Analysis, Explainability, and Causality, First Edition. M. Z. Naser.

Companion Website: www.wiley.com/go/Naser/MachineLearningforCivilandEnvironmentalEngineers

Take a look at Equation 1.1. This is a fundamental equation that ties the moment capacity (M, or simply *resistance*) of a W-shaped steel beam to its primary factors (i.e., the plastic section modulus, Z, and the yield strength, f_y, of the grade of the structural steel used in the same beam).

$$M = Z \times f_y \tag{1.1}$$

Equation 1.1 *visually* represents the main parameters governing the sectional capacity of a W-shaped steel beam under bending. Equation 1.1 also *expresses* the functional relationship between M, Z, and f_y (i.e., the *resistance* equals the multiplication of Z and f_y). More importantly, this question shows a perceived *causal* look into the sectional capacity; that is, an increase in either Z and/or f_y is expected to increase the *resistance* of steel beams.

While the discussion on Equation 1.1 may seem trivial – especially for many of the advanced readers of this book, extending the same discussion to more complex problems is not as straightforward.[5]

Let us think about the collapse of buildings as a phenomenon. At this point in time, we do not have an elegant equation/formula that can represent such a phenomenon. First of all, this is a hefty problem with a vast number of involved parameters (of which some we know and many, many others that we do not know). If we are to speculate on a form of an equation that can represent this phenomenon, then opting for a formula with a highly nonlinear form would not be a far stretch. Evidently, complex problems are often represented with complex procedures. I hope you can now appreciate my emphasis on visualization.[6] It is much easier to deal with problems once you visualize/describe them elegantly.

In the large scheme of our domain, we lack similar equations to Equation 1.1 on countless problems. On a more positive note, we have other means to *visualize* and tackle complex phenomena.

When I was a graduate student, I was introduced to numerical simulations. I was fortunate enough to learn ANSYS [7], from which I was able to model and simulate complex problems. If you take a minute to think about it, finite element (FE) models are similar to Equation 1.1. Such models visually present and describe the logic that governs phenomena. They are compact, intuitive, and can also be verified via laboratory testing, as well as differential equations and other means of verifications.

Numerical models are a form of visualization that can be applied to simple problems (e.g., moment capacity of a beam) and to more complex problems (i.e., the collapse of buildings). These models implicitly retain our preference for first principles and the notion of transparent and verifiable methods. These are some of the qualities that are elemental to the success of ML adoption and education in our domain.

1.2 Machine Learning as an Educational Material

When developing a new course, or educational material, a necessary question often arises, *what are the learning objectives for such a course?* Overall, the learning objectives are statements of what we intend to cover and instill during an education experience.

I believe that the objectives[7] of an accompanying ML-themed course to this textbook can be grouped under three components; 1) introduce the principles of ML and contrast those against traditional methods favored by civil and environmental engineers (i.e., scientific method, statistical, numerical, and empirical analysis, etc), 2) present case studies to pinpoint the merit space of where ML can be most impactful,[8] and 3) provide a platform for our students to practice, collaborate, develop, and create ML solutions for civil and environmental engineering problems.

A deep dive into the above shows that I did not mention the terms *coding* or *programming* as I do not see ML as a coding/programming subject or exercise. In my eyes, ML is both an *instrument* and a *philosophy* and hence needs to be taught as one.

---

5 Given my background and research interest, I "may" tend to gravitate a bit and borrow examples from structural engineering and/or structural fire engineering. From now on, I promise to do my best to maintain some balance between the main civil engineering areas.

6 The same discussion can be extended to the process of creating a new building material derivative with self-healing properties, or planning mass evacuation due to natural disasters.

7 You will notice that I am following Bloom's [8] taxonomy – a key component of ASCE's ExCEEd Teaching Workshop. Hello David Saftner and Fethiye Ozis! If you happen to be a faculty member, I would highly suggest you attend this workshop.

8 We also need to acknowledge that ML, just like other methods, may not be fully compatible/applicable to all of our problems.

I believe that ML is best taught in a manner similar to our foundational courses, such as statics or mechanics. Thus, for the most part, I do not support the notion that the majority of our students or engineers are expected to become programmers – just like I do not also support the push aimed to convert our students into becoming FE experts or modelers. On the contrary, I believe that the majority of our engineers are expected to be familiar with ML, just as much as they are familiar with setting up physical tests, statistical methods, and numerical models.

I once heard a wise person saying *that a good indication to forecasting where one area is trending is to see its analog in different domains.* A look into the recent rise and interest in *big* and *small* data research, and the success of ML in parallel engineering fields that often grow in line with civil and environmental engineering (i.e., mechanical engineering, industrial engineering, etc.), implies that ML is soon to find a permanent residence in our domain. As such, providing elementary and advanced educational material covering ML to civil and environmental engineering undergraduates and graduate students, as well as practicing engineers, will be essential to the growth of our domain.

1.3 Possible Pathways for Course/Material Delivery

From a practical perspective, adding a new course or a series of specialized courses to an already heavy civil engineering curriculum can be challenging.[9] However, if such a course is an elective course, it would ease the burden on faculty, students, and curriculum development committees.

Thus, the path of least resistance to creating a simple means of delivery for ML-based education material is to introduce a new course. This course can build upon a faculty's experiences with ML. In such a pathway, a faculty can tailor the layout of a course plan with modules and chapters, each of which is dedicated to a specific concept on ML. The table of contents and chapters of this textbook could potentially serve as an initial guide for such a course.[10]

At its surface, this textbook is designed to be a companion for a course on ML. The early chapters of this book cover introductory ML themes, and later chapters go further to outline specifics of how ML can be adopted and applied as a solution to some of the common problems faced in our domain.[11]

In lieu of the above simple solution, I also believe there are multiple ways to deliver a ML-themed course.[12] I will cover three herein. The first is through what I refer to as *teaching ML by programming*, the second is by *teaching ML by application*, and the third is a hybrid of the two.

Teaching ML by programming heavily relies on teaching students how to code and program ML algorithms and scripts. You will notice that I am intentionally avoiding this practice of solely *teaching ML by programming*. This decision stems from two viewpoints; 1) my hope to change the inertia against ML as a coding experience as opposed to an engineering experience,[13] and 2) the rise of new platforms that provide *coding-free* ML solutions (i.e., *Big ML, DataRobot, Dataiku*, and *Exploratory*).

Such platforms allow users to *create, manipulate*, and *apply* data and ML algorithms via friendly interfaces (i.e., without any programming experience).[14] These platforms are driven by the desire to *teaching*[15] *ML by application* and provide free access to students/faculty. Hence, they are valuable for the many of us who struggle with the lack of programming in our curriculums. I foresee these platforms as the future of our domain, so I am inclined to present them in this book.[16] The good news is that we can also create our own platform as I will show you in a later chapter.

---

9 Some of the presented ideas build upon my findings from my recent paper [9] on ML and civil engineering education. A quick review of this paper might shed some light into the origin of some of the proposed modalities.

10 As always, please feel free to edit/revise and share your thoughts with me.

11 I have always valued straight-to-the-point solutions and hence my preference for this path of least resistance approach.

12 I am adding a few possible course outlines that you may wish to follow/build upon at the end of this chapter (see Table 1.1).

13 In other words, untie the need for someone to be a coder for them to be a ML user/developer.

14 One can think of a FE model (say, ANSYS or ABAQUS), where the user can build models through the graphical user interface as opposed to building models from matrices and differential equations.

15 From a business point of view; *adopting ML by application*.

16 More recently, the creation of *AlphaCode* [10] and *InCoder* [11], two ML algorithms that can code, are opening the frontier for a new form ML- one that requires little operation by users. Similarly, OpenAI's ChatGPT can do the same with ease. These algorithms can be handy for users from our domain since they minimize the need for the coding component, yet keep the need to stay up-to-date on the theory component.

Keep in mind that I am not opposed to coding, but rather I am trying to present an alternative to overcome the coding obstacle that continues to shed shades of grey in introducing ML to our engineers.[17] I will still present *some* degree of programming throughout the pages of this book. I feel that this is necessary to give the reader a glimpse of what happens *behind-the-scenes* with respect to ML.[18] Thus, instead of delivering another textbook on *learning to code* with a specific programming language, I thought it might be a better option for this edition to focus on ML and provide accompanying success stories in terms of case studies that cover common problems pertaining to our domain.

As you will see in the next few chapters, this book provides fully solved examples and tutorials based on *Python* and *R*. Personally; I do not have a preference for a specific programming language. I happen to be fortunate enough to have two of my PhD students, *Moe AlBashiti* and *Arash Tapeh*, who are proficient in these two languages. They were kind enough to offer their support in building many of the presented tutorials. We thought that presenting our examples in two languages might cater to readers of various backgrounds and possibly initiate interest in a cross-over between the two languages.[19]

1.3.1 Undergraduate Students

In lieu of the path of least resistance, another pathway to presenting the concepts homed in this book is by offering *bits & pieces* of ML in existing civil and environmental engineering courses. For example, courses with heavy software or analytical backgrounds such as *Computational* or *Numerical methods*, a common junior/senior course, may provide an organic pathway to covering introductory concepts of data-driven ML techniques that can complement or augment statistical methods.[20]

The same *bits & pieces* approach may also be designed to span several courses that maximize the expertise of resident faculty, wherein those with strong programming backgrounds can lead the *teaching by programming* side, and those with application backgrounds can lead the *teaching by application* side.

The following comes to mind; students enrolled in a *statics* course can be tasked with *programming* a new script or *applying* a pre-established script that decomposes a vector into its three basic components. The same exercise is then re-introduced at a *mechanics* course to arrive at internal stresses within a body (i.e., by programming/applying the following stress = force/area). In all cases, hand calculations can be used to verify the output of ML scripts.

Then, higher-level undergraduate courses (say, within the structures track such as *theory of structures* and *structural design*) can build on students' exposures to ML to solve stability problems or optimize the design of load bearing elements. Such courses often involve a class project which presents a smooth opportunity for interested students to explore ML (say, to develop an ML model through a platform that selects the lightest element for a set of loading conditions vs. attempting to arrive at such a section through the iterative and commonly used procedure often introduced to students). Finally, a cornerstone project could potentially be structured to have the option to utilize some form of ML as a primary or secondary component.

In the presented approaches, interested students are likely to experience a consistent dose of ML knowledge across multiple courses within the same semester or parallel semesters. In this particular example, a series of courses at the sophomore, junior, and senior level could include modules on ML programming or the use of ML platforms. As such, the chapters of this book would need to be split into modules that can fit existing courses.

---

17 If you happen to serve at an institution where coding is not a hurdle, then I envy you!

18 Not to mention the limited duration of a typical course (~16 weeks) which makes it a challenge to deliver a full coding-based ML experience. However, if you happen to be proficient in coding, then kudos to you! I trust that you will find re-creating the presented examples from scratch interesting and the discussion on coding-free platforms exciting.

19 As we will see in the next few chapters, each programming language and algorithm has its positives and negatives.

20 Say for regression or optimization.

1.3.2 Graduate Students and Post-docs

Oftentimes, graduate courses are designed to place distinct emphasis on specialized topics and to spark the interest of students and/or complement research projects/theses of graduate students of attending faculty members. Thus, the graduate level presents a natural fit for integrating a new and dedicated course on ML.[21]

At the graduate level, ML can be thought of as a method to guide engineering intuition to discover new knowledge, test hypotheses, analyze data and observations, or be adopted in the traditional sense for prediction. As such, I see the implementation of ML here as one with the highest potential simply because graduate students and post-doctoral associates can be a bit more independent and possibly more curious than their undergraduate counterparts and/or mentors.

A graduate student can, for example, develop a *physics-informed*[22] *ML model* or a *causal*[23] *ML model* to address research questions pertaining to a particular problem/thesis/project. The same student could explore new algorithms and techniques, collect data, etc., to push the boundary of the comfort zone of their research area. Some of the problems that could be of interest include developing ML models for structural health monitoring applications, tracking the safety of workers in sites, forecasting traffic detours in the wake of an accident, etc. For graduate students and post-doctoral associates, the possibilities for ML can be endless.

In a graduate course (i.e., CE4000/6000 or 6000/8000 level), a faculty member can task one group of students to develop a ML model to predict the failure point (i.e., sectional capacity) of a load bearing member and then task another group to develop a model to predict the stress/strain distribution of the same member at failure. Results from both groups can then be compared against a laboratory test (or FE simulation) of an identical specimen/FE model. An exercise of this magnitude is perhaps feasible to conduct at each course offering and, if appropriately designed, may be able to mesh all three aspects of investigations, namely, physical testing, FE simulation, and ML modeling, to our students.

The above example could be extended to other disciplines of civil and environmental engineering with some modification. For example,

- Construction materials:
  - Undergraduates – create a ML model to predict the compressive strength of a given concrete mixture,
  - Graduates – develop a ML model to predict the stress-strain curve of such a mixture,
  - Verification – compare predictions from ML via testing cylinders of the same mixture in a laboratory.
- Water resources:
  - Undergraduates – develop a ML model to predict head loss due to friction in a circular pipe,
  - Graduates – create a ML model to predict the fluid's Reynolds' number for the same pipe and conditions,
  - Verification – validate the above ML models through an experiment.
- Transportation:
  - Undergraduates – build a ML model to predict the average daily traffic across a highway section,
  - Graduates – create a ML model to predict the influence of traffic delays on the same segment during a holiday season or post a traffic disturbance,
  - Verification – compare models' predictions against historical or simulated data.

In all of the above examples, data from previously collected laboratory tests can also be used to develop ML models.[24]

Another possible and logical solution is to offer a minor degree in ML. A department[25] can set-up a plan that incorporates a few courses on various ML and civil and environmental engineering tracks. Such a plan could be of interest to the next generation of our students. The same program can be extended to practicing engineers.

---

21 This textbook was inspired through my development of such a course at Clemson University.

22 A model that is constrained to satisfy physics and first principles.

23 A model that examines the cause-and-effect as opposed to trying to associate the occurrence of the phenomenon from a data-driven perspective.

24 Please note that my goal is get our students to exercise developing simple ML models as opposed to creating full-scale deployable models. If such data is not available, then please feel free to use some of the datasets used throughout this book.

25 With or without a collaboration with a computer science department.

1.3.3 Engineers and Practitioners

The aforementioned modalities are primarily intended for students who have access to school or university resources such as testing facilities or computer laboratories. In hindsight, having access to such facilities has its advantages. However, ML research and education do not require laboratory facilities such as those often needed in civil and environmental engineering departments. In contrast, ML models can be developed via handheld devices (i.e., PC, laptops[26]) and through free/open-source packages (i.e., *Python*, *R*) or online coding-free platforms (see [12]). This brings an unprecedented path for affordability and inclusivity.

The same also reduces the logistical and monetary resources necessary for learning and utilizing ML. As such, summer courses, Massive Open Online Courses (MOOCs), and public seminars can also be intelligently stitched to introduce civil engineers to the concepts and principles of ML. For example, most of my ML experience comes from watching online tutorials on *YouTube*, and reading pre-prints posted on *Arxiv.org*.[27]

More recently, I was humbled to chair and participate in seminars hosted by the American Society of Civil Engineering (ASCE), the American Concrete Institute (ACI), the American Association for Wind Engineering (AAWE), and the International Association for Fire Safety Science (IAFSS),[28] which focused on the role of ML in infrastructure. Many, if not all, of these seminars can be found online and are available for engineers. Such events could be further supplemented by certificates or participant awards to encourage attendance.[29]

In parallel, the rise of engineering podcasts presents an exciting opportunity for engineers to *learn-on-the-go*. These podcasts cover trending topics in civil engineering and environmental, such as ML. For example, I had the pleasure of participating in the *Civil Engineering Vibes* podcast and the *Fire Science Show* to chat about ML and how we can maximize our adoption of ML.[30] I also had the pleasure of presenting many case studies to leading industry giants such as ARUP, Warrington Fire, and universities around the world.

1.3.4 A Note

I must admit that at the time of this edition, the majority of civil engineers involved with ML are considered *appliers* as opposed to *creators* of ML. This is as expected as it is natural. A series of questions then arise:

Is it enough to be an applier of ML? Do we need our engineers to be creators of ML? If so, how and when will we get there?

I doubt that we have answers to the above question now, let alone years from now. However, a few years ago, I learned the following mental exercise. Take a complex question and paraphrase it into something you are familiar with. So, let me rephrase the above questions into something that we are a bit more familiar with.

Is it enough to be an applier of FE modeling? Do we need our engineers to be creators of FE programs? If so, how and when will we get there?

An honest look into these questions shows that, despite the long history of the FE method in engineering, the majority of engineers remain not only appliers of FE but with basic knowledge of FE modeling. For example, many graduate courses and textbooks on the FE method still cover 1D and 2D elements and briefly mention 3D elements.[31] In fact, a typical help manual of a FE package can contain explicit details that are rarely captured in graduate courses/textbooks. Despite the above, the FE industry in civil and environmental engineering is thriving.[32]

---

26 Remember our discussion on coding-free platforms. Well, for these all one needs is access to the internet as the computational power needed to create ML models resides in the cloud!

27 Thanks to many YouTubers that spend many, many hours creating interactive content. Shout out to Google and Cornell University.

28 Let us not forget other seminars and workshops led by my colleagues such as Ting Li, Xilei Zhao, Henry Burton, Seymour Spence, and Xinyan Huang, to name a few.

29 For example, my ACI session "*The Concrete Industry in the Era of Artificial Intelligence*" had Professional Development Hours (PDH) awarded to all participants. I am also aware of two more seminars by ASCE on ML to take place at the 2022 ASCE convention and 2023 ASCE Structures Congress – I can see the future holding more of these too.

30 Thanks to the hosts, Abdel-Rahman Atif and Ghanim Kashwani, as well as Wojciech Węgrzyński, who keep us up-to-date with the recent advances in our domains.

31 For fairness, deriving 3D elements and accompanying matrices/differential equations is a hefty exercise.

32 Without such software, it is unclear to know if we have made it this far.

I doubt that we will have clear-cut answers to many of the above questions, as well as many of those I have not listed above. Let us wait and see what the future holds. Personally, I do see merit in partnering with our computer science colleagues to learn from their expertise (i.e., on the educational front as in how to teach programming, as well as ML creation front). The same partnership can go a long way in terms of establishing certifications and future events that mesh and converge engineering and computing students via competitions. This exercise can help engage students from cross disciplines and can also be viewed as an excellent team-building workout.

Table 1.1 Possible course outline.

	Undergraduate		Graduate	
Chapter no.	**Material covered**	**Timeline**	**Material covered**	**Timeline**
Chapter 01. Teaching Methods for This Textbook	–	–	–	–
Chapter 02. Introduction to Machine Learning	✓	Wk01–Wk02	✓	Wk01–Wk02
Chapter 03. Data and Statistics	✓	Wk03–Wk04	✓	Wk03
Chapter 04. Machine Learning Algorithms	✓ [up to unsupervised learning]	Wk05–Wk06	✓	Wk04–Wk05
Chapter 05. Performance Fitness Indicators and Error Metrics	✓ [up to unsupervised learning]	Wk07–Wk08	✓	Wk06
Chapter 06. Coding-free and Coding-based Approaches to Machine Learning	✓	Wk08–Wk12	✓	Wk07–Wk09
Chapter 07. Explainability and Interpretability	✓ [covering the big ideas]	Wk13–Wk14	✓	Wk10–Wk12
Chapter 08. Causal Discovery and Causal Inference	–	Possible seminar	✓	Wk13–Wk14
Chapter 09. Advanced Topics	–	–	✓ [Covering GreenML, Ensembles, and AutoML. The rest of the topics can be covered in a seminar]	Wk15–Wk16
Chapter 10. Recommendations, Suggestions, and Best Practices	✓ [a few of each]	Wk15	✓	Wk16
Chapter 11. Final Thoughts and Future Directions	–	–	✓	Wk16

1.4 Typical Outline for Possible Means of Delivery

These outlines are built having in mind a full course (over 16 weeks) or a summer course (over 6 weeks) of delivery. The bolded lines indicate possible timelines for midterms (please note that I am not anticipating a final exam for the graduate students. I believe a final project might be a better suit). A summer course layout will consist of constraining the material covered in each midterm to be delivered in 2.5–3 weeks.

Chapter Blueprint

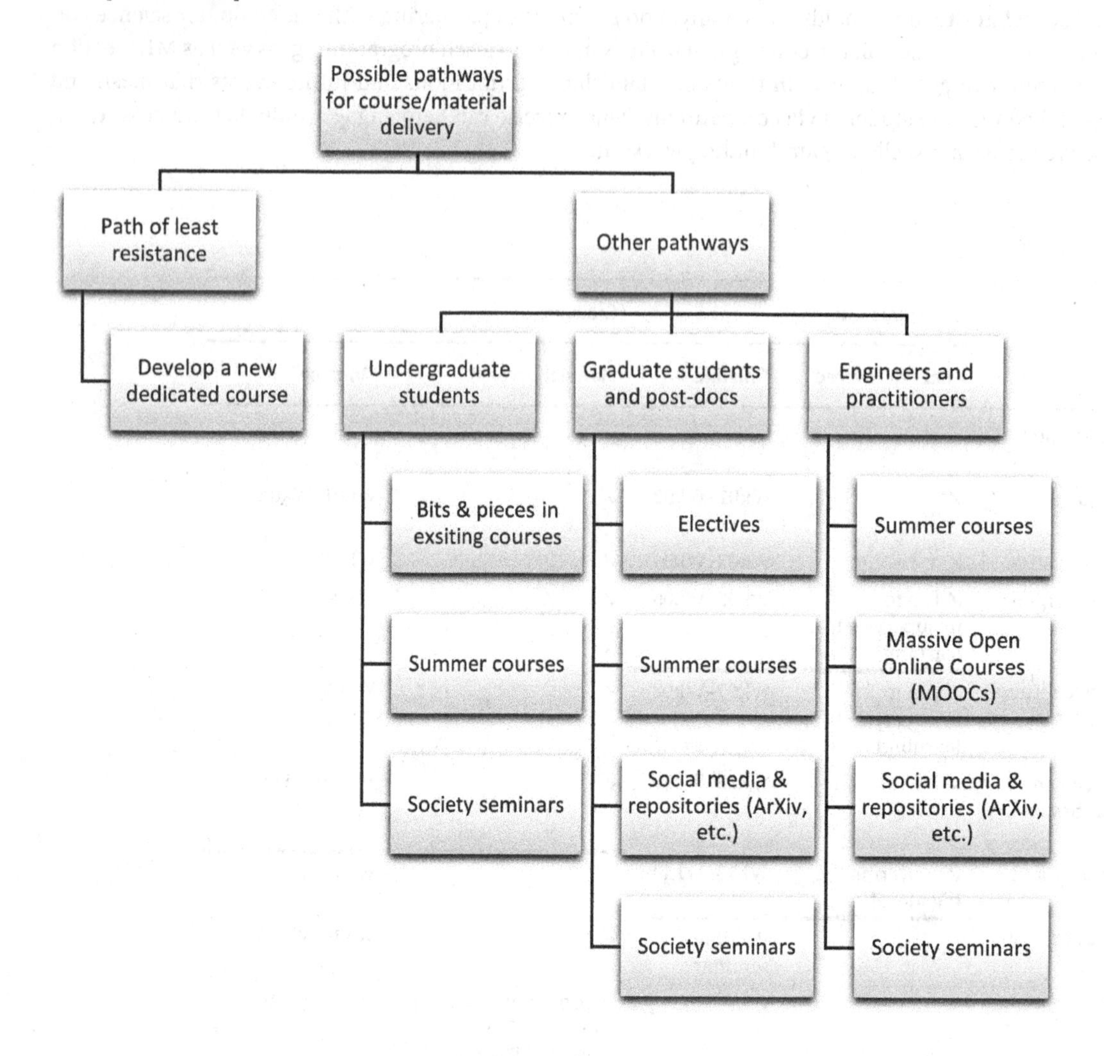

Questions and Problems

1.1 List two possible pathways to ML education in our domain.

1.2 Design a bits & pieces exercise that incorporates ML.

1.3 What are some of the learning objectives for an ML-themed course in civil and environmental engineering?

1.4 Summarize the main points discussed at the ACI session "*The Concrete Industry in the Era of Artificial Intelligence*" [https://www.mznaser.com/media-mentions].

1.5 Contrast *teaching ML by programming* to *teaching ML by application.*

References

1 Dunn, R. and Carbo, M. (1981). Modalities: An Open Letter to Walter Barbe, Michael Milone, and Raymond Swassing. *Educ. Leadersh.* 5: 381–382.

2 Felder, R. and Silverman, L. (1988). Learning and Teaching Styles in Engineering Education. *Eng. Educ.* 78(7): 674–681.

3 Harvey, D., Ling, C., and Shehab, R. (2010). Comparison of Student's Learning Style in STEM Disciplines. Proceedings of the 2010 Industrial Engineering Research Conference, Cancun, Mexico, June 5–9.

4 Sacks, R. and Barak, R. (2010). Teaching Building Information Modeling as an Integral Part of Freshman Year Civil Engineering Education. *J. Prof. Issues Eng. Educ. Pract.* 136(1): 30–38. https://doi.org/10.1061/(ASCE)EI.1943-5541.0000003.

5 Cox, M.F., London, J.S., Ahn, B. Zhu, J., Torres-Ayala, A. T., Frazier, S., Cekic, O., Chavela Guerra, R. C., Cawthorne, J. E.(2011). *Attributes of Success for Engineering Ph.D.s: Perspectives from Academia and Industry.* Paper presented at 2011 ASEE Annual Conference & Exposition, Vancouver, BC. 10.18260/1-2--17548.

6 Shaeiwitz, J.A. (1996). Outcomes Assessment in Engineering Education. *J. Eng. Educ.* https://doi.org/10.1002/j.2168-9830.1996.tb00239.x.

7 Kohnke, P.C. (2013). ANSYS. © ANSYS, Inc.

8 Armstrong, P. (2017). Bloom's Taxonomy, Center for Teaching. Vanderbilt University.

9 Naser, M. (2022). A Faculty's Perspective on Infusing Artificial Intelligence into Civil Engineering Education. *J. Civ. Eng. Educ.* 148. https://doi.org/10.1061/(ASCE)EI.2643-9115.0000065.

10 Li, Y., Choi, D., Chung, J. et al. (2022). Competition-Level Code Generation with AlphaCode. 2022–2025. https://doi.org/10.48550/arxiv.2203.07814.

11 Fried, D., Aghajanyan, A., Lin, J. et al. (n.d.). InCoder: A Generative Model for Code Infilling and Synthesis. https://sites.google.com/view/incoder-code-models (accessed 23 May 2022).

12 Naser, M.Z. (2023). Machine learning for all! Benchmarking automated, explainable, and coding-free platforms on civil and environmental engineering problems. *J. Infra. Intell. and Resi.* 2(1): 100028. https://doi.org/10.1016/j.iintel.2023.100028. (accessed 01 Mar 2023).

2

Introduction to Machine Learning

We are tellers of story.

Roland Blum (The Good Fight)

Synopsis

This chapter starts our journey into machine learning. We trace the modern[1] origins of this technology and outline some of its key milestones. This chapter sets the stage for defining most of the terminology we will be adopting in this textbook.[2] In parallel, this chapter will answer questions such as, what is machine learning? How does machine learning relate to artificial intelligence? What is some of the common procedure(s) to create and/or apply machine learning? How does machine learning differ from some of our traditional methods (i.e., experiments, analytical/numerical simulation, and statistical methods)? [3]

2.1 A Brief History of Machine Learning

As this textbook is about machine learning (ML), let us start by defining what ML is before we visit its history. Machine learning[4] creates specific types of computer programs[5] that mimic the way humans learn to gradually improve their performance.[6] Simply, a machine is trained[7] on a task, and the more it learns about it, the more it becomes fluent and capable of accurately performing it. Recently, AI and ML have often been used interchangeably.[8]

As you might have noticed, the definitions of ML and AI do not mention algorithmic details or how these technologies practically relate to our domain. The same also infers that ML and AI are primarily and highly driven by data.

1 The history of computing arguably starts by the invention of the abacus (~2400 BCE), if not earlier [26].

2 At the moment, the ML field, at least from a perspective from civil and environmental engineering, remains fluid with conflicting definitions that can be used to describe similar, if not identical, concepts. As such, it is necessary to articulate our own set of definitions to clearly convey the present chapter, as well as those upcoming.

3 This chapter continues and builds upon [8].

4 Is a branch of artificial intelligence (AI). **Artificial intelligence** as defined by John McCarthy [27]: *It is the science and engineering of making intelligent machines, especially intelligent computer programs. It is related to the similar task of using computers to understand human intelligence, but AI does not have to confine itself to methods that are biologically observable.*

5 Aka. **Algorithms**: Methods or processes in which a ML system handles data (often optimized and well established). Codes are often used instead of algorithms. However, a **code** (or script) is defined as a list of instructions for machines to execute. Another term that you are likely to have heard is **Deep learning (DL)**: a specific type of AI that deals with neural networks with deep and fluid architectures.

6 This definition builds upon that adopted by IBM [28].

7 Keep in mind that a machine may not have been explicitly programmed to carry this task! i.e., learns as it goes.

8 Yet, in reality. Data science → AI → ML → DL.

Machine Learning for Civil & Environmental Engineers: A Practical Approach to Data-driven Analysis, Explainability, and Causality, First Edition. M. Z. Naser.

Companion Website: www.wiley.com/go/Naser/MachineLearningforCivilandEnvironmentalEngineers

The first mention[9] of the term AI as a tactic to resolve problems that defy solution via traditional techniques was noted at the 1956 Dartmouth Summer Research Project on Artificial Intelligence (DSRPAI) (which was hosted by John McCarthy and Marvin Minsky) – see Figure 2.1 for a series of exciting milestones [1, 2]. Nowadays, AI, ML, and DL have formed a well-established and unique discipline that generates thousands of publications per year. A glance at Figure 2.2 shows the publication trends in AI, ML, and DL and compares that to our discipline. As you can see, the publications in AI, ML, and DL are multiples of that in civil and environmental engineering.[10]

2.2 Types of Learning

Traditional statistical methods are often used to carry out behavioral modeling, wherein such methods may require the idealization of complex processes and approximation of varying conditions. More importantly, statistical methods frequently assume linear[11] relationships between predictors and output(s). Yet, these assumptions do not always hold true – especially in the context of real engineering data. As such, statistical methods often fall short of delivering high-quality solutions to our problems.[12]

As you might have noticed by now, ML has the term *learning* embedded in it. This brings us to the concept of learning.[13] We tend to learn from our experiences[14] because we have the ability to communicate, reason, and understand. From the context of ML, this term was coined by Arthur Samuel[15] [3] after he designed a computer program for playing checkers. The more the program played, the more it *learned* to make better predictions for future moves from its past experiences.

In addition to the aforenoted simple definition, machine learning can formally be defined as techniques that are designed to enable machines to learn/understand/identify meaningful patterns and trends from data [4]. According to Michalski et al. [5], a *"computer program is said to learn from experience E with respect to some class of tasks T and performance measure P, if its performance at tasks in T, as measured by P, improves with experience E."* Simply, ML is the process of training machines to carry out pre-identified tasks with the goal of improving the performance of such machines (with time and new knowledge[16]) in the hope of understanding/explaining[17] the data on hand.

In general, there are three types of learning that ML can fall under; *supervised learning*, *unsupervised learning*, and *semi-supervised learning* [6].

Supervised learning is a learning process that involves inputs (x_1, x_2, ... x_n) and an output[18] variable(s) (Y); however, a function, $f(x)$, that represents the relation between inputs, as well as inputs and outputs is not known nor is easily derivable (see Equation 2.1). Hence, the main goal behind supervised learning is to map the inputs[19] to the output. Supervised learning can be divided into: *regression* (when the output variable is numeric, i.e., moment capacity of a beam, number of

---

9 Let us pause for a minute to acknowledge the fact that AI/ML history is intertwined. While some sources date ML to the early 1940s with the publication of Walter Pitts and Warren McCulloch paper, *A logical calculus of the ideas immanent in nervous activity* [29] and Donald Hebb's book, *The Organization of Behavior* [30], others seem to have adopted the workshop held in the Dartmouth College version. In all cases, this looks like a good project for research paper homework.
There are a few equally interesting points in time, known as the AI winter(s). There are periods of time (most notably, 1973–1980 and 1987–1993) where interest in AI plummets due to technological challenges, immature approaches, etc. The current moment in time is known as the AI spring (i.e., increased interest, funding, notable success, etc.). For more information on similar milestones, please visit: https://achievements.ai/

10 It is worth noting that this figure does not compare publications in computer science (as a discipline) to civil and environmental engineering. The publications in the computer science are much larger (and faster) than that in ours.

11 And in some cases, nonlinear.

12 Remember, the intention is not that statistical methods do not perform well nor that they do not deliver. The intention is that these methods do work, and have been working into domain for a long while, however, in some/many scenarios, predictions from such methods remain of low quality (given their inherent limitations). As you will see later on, every known method, including ML, has its positives and negatives.

13 **Learning**: The process of seeking knowledge [31].

14 One can think of daily interactions as an example, etc.

15 A pioneer computer scientist at IBM.

16 A closer look at the quoted definition infers that machines do not learn by reasoning but rather by algorithms.

17 Glennan [32] argues that scientific explanations have (at least) two notions: 1) explain genuinely singular events, or 2) explain general regularities.

18 Note that I am using the following terms interchangeably: inputs, features, independent variables. I am also using output, outcome, response, target, and dependent variable interchangeably as well.

19 In the majority of cases, the result of a ML analysis is an implicit function, with only a few being capable of generating an explicit function. Such a function may not be easily formulated using traditional methods, if at all.

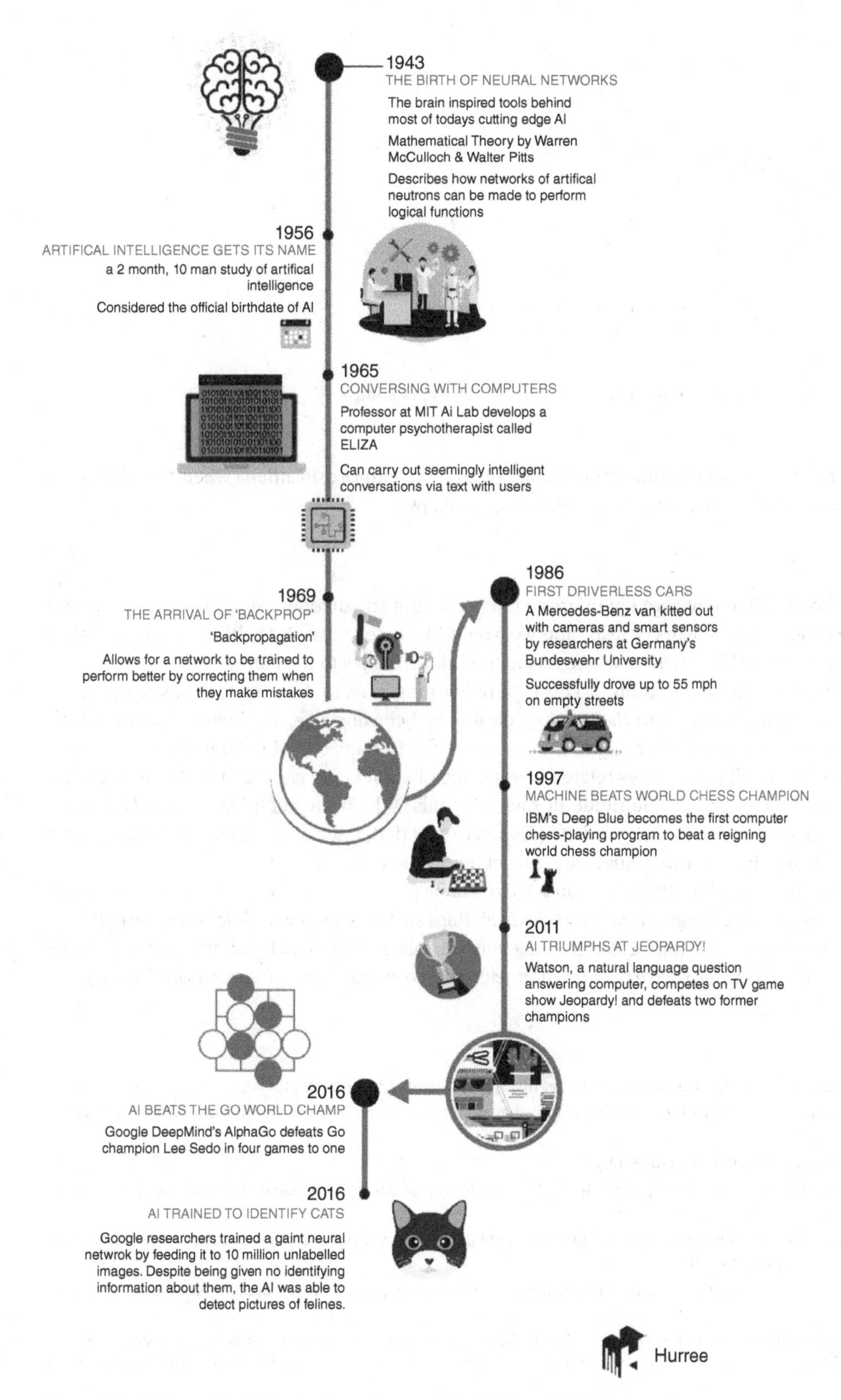

The History of Artifical Intelligence
1943
THE BIRTH OF NEURAL NETWORKS
The brain inspired tools behind most of todays cutting edge AI
Mathematical Theory by Warren McCulloch & Walter Pitts
Describes how networks of artifical neutrons can be made to perform logical functions
1956
ARTIFICAL INTELLIGENCE GETS ITS NAME
a 2 month, 10 man study of artifical intelligence
Considered the official birthdate of AI
1965
CONVERSING WITH COMPUTERS
Professor at MIT Ai Lab develops a computer psychotherapist called ELIZA
Can carry out seemingly intelligent conversations via text with users
1969
THE ARRIVAL OF 'BACKPROP'
'Backpropagation'
Allows for a network to be trained to perform better by correcting them when they make mistakes
1986
FIRST DRIVERLESS CARS
A Mercedes-Benz van kitted out with cameras and smart sensors by researchers at Germany's Bundeswehr University
Successfully drove up to 55 mph on empty streets
1997
MACHINE BEATS WORLD CHESS CHAMPION
IBM's Deep Blue becomes the first computer chess-playing program to beat a reigning world chess champion
2011
AI TRIUMPHS AT JEOPARDY!
Watson, a natural language question answering computer, competes on TV game show Jeopardy! and defeats two former champions
2016
AI BEATS THE GO WORLD CHAMP
Google DeepMind's AlphaGo defeats Go champion Lee Sedo in four games to one
2016
AI TRAINED TO IDENTIFY CATS
Google researchers trained a gaint neural netwrok by feeding it to 10 million unlabelled images. Despite being given no identifying information about them, the AI was able to detect pictures of felines.
Hurree

Figure 2.1 Infographic displaying key milestones in AI, ML, and DL [hurree.co].

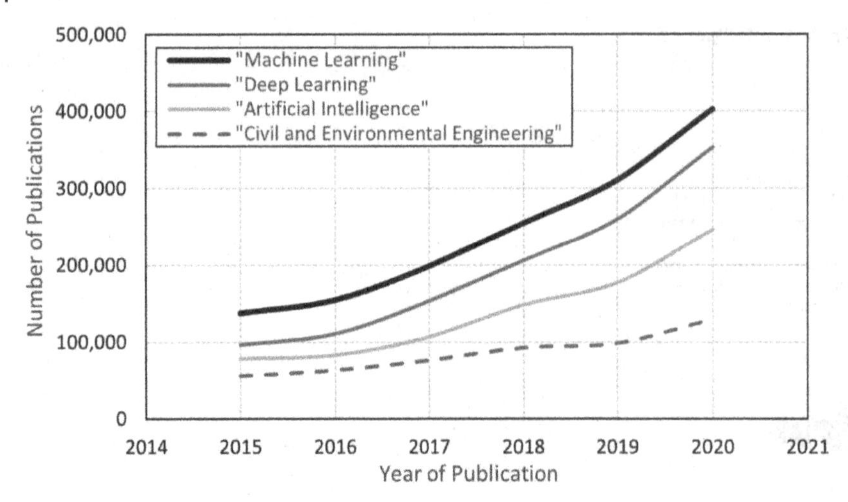

Figure 2.2 Publication trends in AI, ML, and DL, as well as civil and environmental engineering.

vehicles passing through an intersection, concentration of pollutants in air, etc.) and *classification* (when the output is a categorical variable, i.e., failure/no failure, minor damage/major damage/collapse).

$$Y = f(x) = f(x_1, x_2, ..x_n) \tag{2.1}$$

Now, what if the output does not exist? What if we are not interested in predicting an outcome per se? For example, what if we have plenty of data that we collect from sensors or tests, and we would like to see what lies behind such data? Is/are there something that ties the data together? Well, here is where unsupervised ML comes to the aid.

Unsupervised learning can answer the above questions.[20] This type of learning aims to discover the underlying structure/distribution of the data and is often grouped into *clustering* (grouping by behavior[21] i.e., concretes of dense microstructure tend to have a relatively high compressive strength, drivers under the influence tend to be prone to accidents, etc.), *dimensionality reduction* (e.g., out of all the features related to a specific phenomenon, can we identify the main features? For example, concrete mixtures comprise a multitude of raw materials. Which are such key materials that are directly related to the strength property?) and *association rules*[22] (discover rules that describe the data, i.e., bridges made from green construction material tend to have a lower damaging impact on the environment, etc.).

The third type of learning is that known as *semi-supervised* learning. This learning process is suited where only some of the outputs are labeled (i.e., in a dataset of images, some images showing cracked buildings are labeled as such while others are not[23]). As such, an algorithm tries to learn from the smaller (labeled) data to identify the labels of the unlabeled data correctly. Semi-supervised learning includes other types of learning such as *active*,[24] *transfer*,[25] *reinforcement*,[26] and *self-supervised*[27] learning.

---

20 I hope it is clear that, unlike supervised learning, there is no specific target this type of learning is trying to predict.

21 As you will see in the next chapter, clusters may have different sizes, shapes, densities. Clusters can be formed with a hierarchal order and can overlap.

22 Association and clustering can often be combined under clustering.

23 In this example, a semi-supervised model learns the labels from the limited labeled images and is then applied to label the rest of the images. You can think of this as a merge between classification and clustering.

24 **Active learning**: A process in which a ML algorithm could potentially reach a notable accuracy, despite using a smaller number of training data, if it were allowed to choose its own training data [33].

25 **Transfer learning**: A process where a ML model developed for a task is reused as the starting point for a model on a second task. This type of learning ensures efficiency [34].

26 **Reinforcement learning**: A ML system that is programmed to teach itself following prespecified reward mechanisms. For instance, this type of learning can be used to train robots how to navigate their environments (i.e., when a robot does well, it receives a reward to reinforce the good behavior).

27 **Self-supervised learning**: A hybrid between supervised and unsupervised learning that was developed by Facebook. This type of learning aims to predict unobserved inputs from any observed input and is often used in computer vision and speech recognition [35].

2.3 A Look into ML from the Lens of Civil and Environmental Engineering

Unlike statistical methods, ML adaptively learns from its experiences. In fact, one of the major advantages of ML over traditional statistical methods is its ability to derive a relationship(s) between inputs and outputs without assuming prior forms or existing relationships.[28]

As you have seen in the previous section, we are in the midst of a computational revolution that brings exciting opportunities to our domain (with the potential to overcome the limitations of traditional approaches [7]). Nevertheless, this revolution remains underutilized despite the early works[29] on ML in our area [8–10].

A good chunk as to why this underutilization persists can be due to the opaqueness[30] of ML.[31] This opaqueness is an idea we are familiar with, but we do not quite like – especially when compared to other transparent[32] methods (e.g., experimental, numerical, and analytical methods).

Let us deep dive into the above.

Most of us (i.e., engineers) have been trained, accustomed, and more or less primed to carrying out experiments or some form of numerical/analytical investigations[33] [11]. In reality, we rarely teach experimental methods or design of experiments, but rather we conduct experiments. We also rarely develop finite element software nor teach three dimensional elements; but rather we cover one/two dimensional elements in our lectures and then provide a glimpse into higher-order elements.[34]

If you feel that there is a bit of contradiction hidden between the above few lines in which our curriculum does not prepare our engineers for what they are expecting to actually face in the field (i.e., for example, most problems tackled by us are inherently complex and do require means of advanced computations), then let me assure you there is not one. A curriculum is designed to convey first principles and provide students with the arsenal to tackle the real world!

This can be reflected in *practice, need*, and *continuing education* that enables an engineer to get familiar with the innings of a methodology of investigation [12].[35] Similarly, I firmly believe that with continued practice, an engineer can also adapt to and be convinced of the potential of ML.[36]

Here are some of the reasons why we need ML.[37]

- Learning a new method of investigation is likely to positively expand and improve our understanding and arsenal. This may allow us to realize new knowledge that we did not have before, as well as explore solutions that we have not thought about! ML can help us on this front.[38]

---

28 For completion, there is an area of statistics that fall under non-parametric methods. Unlike parametric models (i.e., the number of parameters is fixed with respect to the sample size), the number of parameters in non-parametric models can grow with the data size.

29 It is worth noting that some of the earliest works that mention the use of AI (a more prominent term at the time) in civil engineering date back to the 1980s and early 1990s [36–38]. It is also worth noting that one of the first reviews on the use of AI in civil engineering applications was conducted by Prof. Hojjat Adeli [39] over 20 years ago. A side note, the term *expert systems* was more common at the beginning of that era.

30 Meaning, *difficult to see through.*

31 I wish I can share panel reviews on many of my submitted proposals that did not see the light. While generally helpful, such reviews would also reveal a dark, and sad, side to how panelists perceive ML.

32 Here is my intention, I am using *transparent* to showcase that we are more likely to be comfortable in adopting a method of investigation that we are familiar with than other methods that we are not – in hindsight, this is not only expected but also makes a lot of sense. The same also sounds like a standard case of *cognitive bias* and a good idea for a research article – stay tuned!

33 Whether during our schooling or in practice.

34 In some cases, software help manuals do contain more information on finite elements than standard textbooks. This is making a lot of sense, since textbooks are not primarily designed to cover proprietary details on elements. If this sounds familiar, then it is because it is likely to be true! What I am trying to get at is that we have accepted the above. However, we have not accepted the opaque nature of ML. Perhaps a bit more time, and a bit more practice might get us there!

35 Simply, not every single engineer needs to be an expert on FE modeling! Yet, most engineers are expected to be, at a minimum, familiar with such modeling. The same goes to experimentalists, and theorists, etc. One can think of Statics, it is an early undergraduate course that is taught to all civil and environmental engineers – yet only a portion of these students get to actually use it in their specialties! *Principles! Principles!*

36 Let me be crystal clear. ML is a method and a tool, and just like other tools and methods, it has its positives and negatives. Similarly, ML is not to be abused but rather to be used responsibly – we will chat more on this in the following chapters.

37 Or perhaps, in need of ML.

38 A key goal of all sciences is to advance our knowledge!

- Domains experience technological shifts every so often. What we are experiencing at the moment is likely to be one such shift – a computational transformation.[39] We ought to learn and maximize its potential!
- Many of our engineering tasks can be described as repetitive[40] and hence can be automated for simplicity and productivity [13].[41]
- A considerable volume of computation and time is needed to overcome the complexity of modern engineering projects. In fact, analyzing project-sized numerical models requires high-capacity workstations, access to proper software and license, as well as the availability of expert users, and many, many iterations. This makes developing and solving such models through traditional methods challenging and, in some instances, impractical[42] (due to cost or time constraints) [14].
- From a logistical point of view, civil and environmental engineering projects often amount to procedures which have been, historically, yet seamlessly and systematically developed by engineering authorities (i.e., regulatory committees). Provisions of such procedures[43] were arrived at from years of research and practical experience. Creating new provisions/updating existing ones to tackle the significant need of our ever-demanding domain is likely to be costly and timely.[44]
- Safety checks, risk management, compliance, and decision making are key aspects of civil and environmental engineering projects. The fact that ML is designed to process various data types and sizes, positions it as one of the most attractive methods to help us conduct safety checks and provide us with valuable insights to deliver optimal decisions.[45]
- Many other ideas and items, such as those pertaining to the perspectives of energy, surveying, construction, education, community resilience, etc., are likely to make use of ML and to positively benefit.

A question worth asking is, what is the status of ML in our domain at the moment?

Well, one way to answer the above question is by looking at Figure 2.3. This figure demonstrates the number of publications generated between 2015 and 2020 for three series, *Civil And Environmental Engineering, Machine Learning* AND *Civil And Environmental Engineering,* as well as *Deep Learning* AND *Civil And Environmental Engineering, and Artificial Intelligence* AND *Civil And Environmental Engineering.* Two observations[46] can be seen from this figure. The first is that all trends are positive and rising. The second is that AI/ML/DL-based publications consistently rank about 20–30% of the total number of publications in our domain.

Another way to also answer the same question is to look at the current interest of civil and environmental engineering organizations and institutions in ML. For example, in 2021, two leading bodies in our domain (namely, the American Concrete Institute (ACI) and the American Society of Civil Engineers (ASCE)) have sponsored ML-based sessions and initiatives, including the 2021 American Concrete Institute (ACI) session "The Concrete Industry in the Era of AI" and the 2021 American Society of Civil Engineers (ASCE) Convention "Modernizing Civil Engineering through Machine Intelligence." These sessions aimed at introducing ML to civil and environmental engineers from an industry/practice point of view. The ACI session has resulted in the publication of SP-350 [16] with the same title as the session. At the moment,[47] I am also aware of the planning of similar future sessions to be taken in 2022 and 2023. Other bodies and organizations[48] within our domain are also sponsoring similar sessions.

---

39 "*People fear what they don't understand and hate what they can't conquer.*" – Andrew Smith.

40 Can you guess the average number of shear tab connections a design firm go over a year? Over a decade? While not all connections are identical, we can still acknowledge the similarity of this, as well as other design tasks.

41 Time is money!

42 A simple example would be carrying out a performance-based design structural fire design for a building. Ask my good friend, Kevin LaMalva [40]! Here is another example, evaluating the structural health monitoring of a typical structure (i.e., bridge, or a dam) is associated with a series of sensor networks that continually measure and record valuable information spanning the service life of such a structure. Analyzing such data in real-time, or near real-time, mechanically or by means of legacy software can be infeasible.

43 Say an equation or a chart – presumably, homed in a code or standard.

44 There is more than one way to do something!

45 Ever think about how people evacuate under extreme conditions – here is a good read on the role of ML in mass evacuation from Xilei Zhao's lab [41]. Here is another on forecasting risk [42]. Imagine going through 100s of contractual pages, ML can help skim through these with ease [43] (Tuyen Le's office is two doors down from my office!).

46 A third, but minor, observation of publications adopting machine ranking ahead of deep learning or artificial intelligence can also be deduced.

47 July 2022.

48 Such as the Structures in Fire conference series, etc.

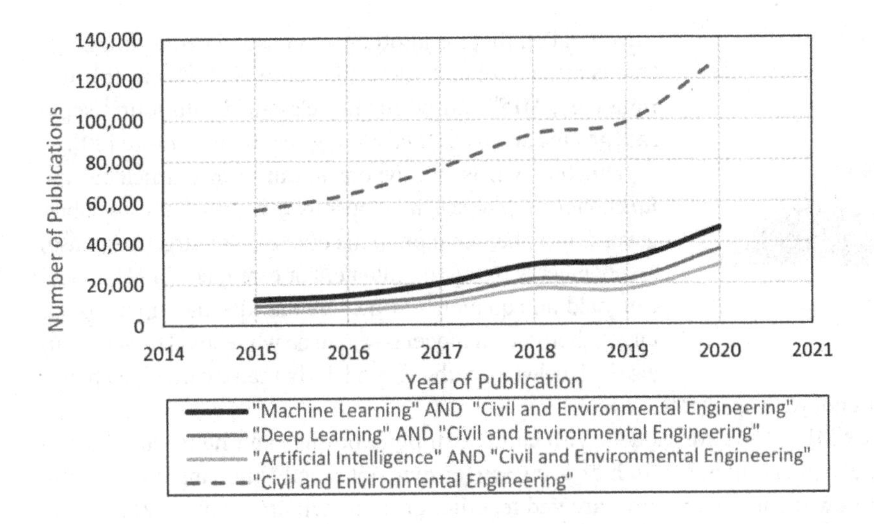

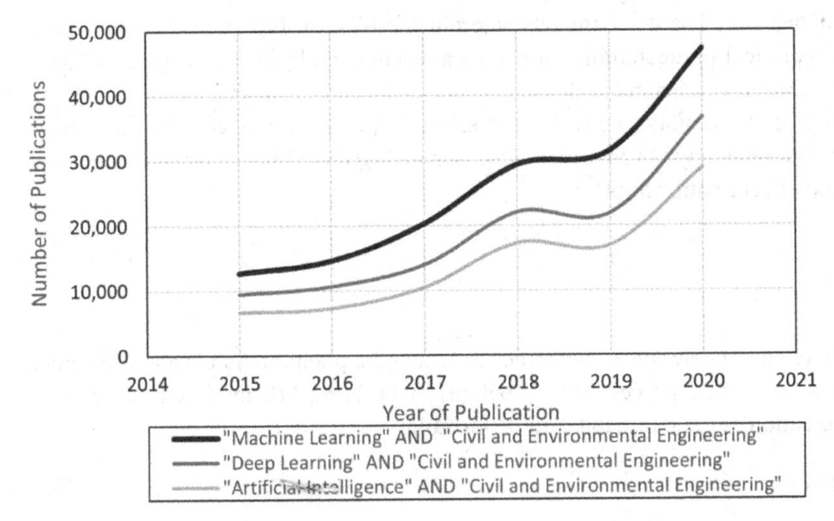

Figure 2.3 Publication trends (2000–2020) [using the *Dimensions* [15] / Digital Science & Research Solutions Inc.].

A third, and perhaps least known look, stems from recent news broadcasted by leading industry firms noting new initiatives to integrate ML into their office. Some of these initiatives are being practiced by leading firms such as Thornton Tomasetti [17], Giatec [18], and Keck & Wood [19], to name a few.[49]

2.4 Let Us Talk a Bit More about ML

So far, we have established that ML is, in a way, a computational method. Collectively, computational methods can be divided into two groups, *soft computing*, and *hard computing*. Hard computing methods are those that integrate a high degree of certainty into solving problems, while their soft computing counterparts are done via approximate means [20]. ML falls under the latter.

49 One more item before I end this section. There is a rising interest from funding agencies in ML and civil and environmental engineering. For example, The National Science Foundation (NSF)'s programs *Engineering for Civil Infrastructure (ECI)* as well as *Civil Infrastructure Systems (CIS)* now welcome ML-based research proposal. The Transportation Research Board (TRB) also welcomes the same.

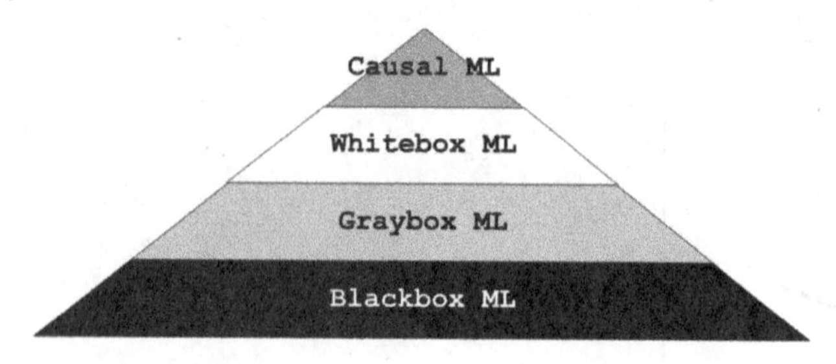

Figure 2.4 Pyramid of ML.

In parallel, there is another set of associated terminology that is often applied to describe features of ML. In this terminology, ML[50] can be further classified into four[51] separate groups: *whitebox*, *blackbox*, *graybox*, and *causal* [20].

Whitebox ML is one where it can clearly articulate the *functional* or *physical* relationship(s) between the variables governing a phenomenon to the outcome (or target/response) an engineer is trying to evaluate. For example, if a ML model can yield an equation[52] or a chart that ties the inputs to the target, then such a model is said to be whitebox. The same can also be labeled to methods/models that can clearly show how a prediction is arrived at even if the model does not yield an expressive tool.

The opposite of whitebox ML is blackbox ML. These are models with complex inner workings and hence are hard to interpret[53] yet can still deliver an outcome of a phenomenon with high accuracy (despite not exactly knowing how, or why such a model attains high predictivity nor how the inputs/features are tied together or to the target/response [21][54]). The same classification can also be said to ML models that are purely built on data, or data-driven models.[55]

Graybox ML is somewhere between the above two. These are models wherein the relationship and/or mathematical model is partially built on some degree of theoretical understanding or prior knowledge while also making use of data-driven-like approaches [22]. Purely empirical models can belong to this class.

Causal models, on the other hand, are those that articulate the cause and effect between features and the outcome of interest. This class of ML is at the top of the pyramid shown in Figure 2.4. Not surprisingly, traditional ML does not often yield, nor is it capable of identifying cause and effect relationships![56]

2.5 ML Pipeline

The truth of the matter is that leveraging ML to adequately solve/overcome/understand a phenomenon is an art as much as it is science.[57] Fortunately, there exists a systematic process (i.e., pipeline) of building ML models to examine a phenomenon. Such a process has led to the creation of the following generic formula:

$$Machine\ Learning = Algorithms + Data \tag{2.2}$$

---

50 As well as computational methods.

51 Some sources might opt to use two groups, *causal* and *non causal*. Others may use *whitebox* and *blackbox*. Other terms such as *transparent* or *glassbox* may be used to refer to variants of whitebox models. There is also a mention of physics-based or physics-guided models. These incorporate some form of physics constraints or differential equations. The point to take home is that there are many ways to classify models and such classification seems to be fitted into specialized area/subfields. To avoid all confusions, I will stick to the four groups as outlined above (where models other than causal models are assumed to be non causal).

52 This can be a bit sketchy, but bear with me. Arriving at an equation can be as simple as a regression exercise. Now, this equation can simply represent the data (as in best fits the data) or it can be one that matches our physics and domain knowledge. The latter is an entirely different story than the former. For example, an empirical equation is selected only because they fit experimental data and without a theoretical justification – this is what many ML models can get us (whether whitebox, blackbox or graybox). Empirical and theoretical equations both have a room and a place. A good read by Ribarič and Šušteršičis [44] which they list many success stories of how empirical equations have actually led to new theories will be worthy of your time. We will cover both in this textbook.

53 If at all possible!

54 The majority of ML may fall under blackbox ML.

55 **Data-driven model**: a model driven by the data – change the data and the model will change. Unlike, for example, the model for bending in beams, no matter how the beams differ, the model remains the same to a large extent.

56 Unfortunately, very few know this fact. A side note, knowing the cause and effect may not be needed for every single problem! However, knowing which is which is definitely needed!

57 The same can also be said for conducting experiments or building and solving numerical models. Tweaking an experimental set-up or finetuning an FE model have a large creative component.

When the emphasis is placed on the algorithm, then such a practice is called a model-centric[58] approach to ML. On the other hand, if the emphasis is placed on the data,[59] then the approach switches to what has been commonly known as a data-centric approach to ML.

Irrespective of the approach,[60] a typical systematic process to ML is summarized in Figure 2.5 and starts with the need to create a ML-based tool or a solution to address a hypothesis or answer intriguing questions that existing methods do not seem to address or could simply stem from curiosity for exploring the space of available data.[61] Let us now dive into each process of ML in more detail in the next few sections.

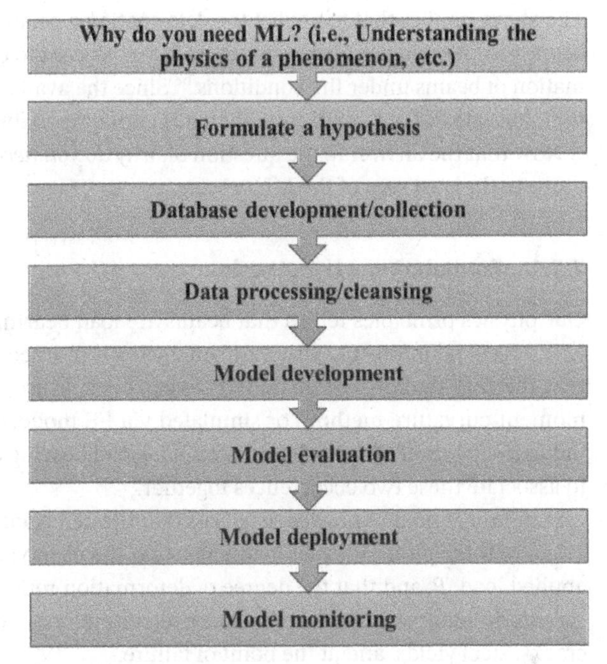

Figure 2.5 Typical ML pipeline.

Answering the question of why do you need ML to tackle a problem/phenomenon?

I would like you to think of a problem or phenomenon X that you find intriguing. While there could be plenty of reasons as to why such a phenomenon intrigues you, suppose that one of them is that you do not know how to predict this phenomenon.

I promised to[62] share my experiences of how I adopt ML in my research. This could then be a good way to share one of my early attempts at developing a test-free and simulation-free method for predicting/forecasting the deformation of beams under fire conditions. My goal was to attempt to create a predictive method that is quick and intuitive. To know how a phenomenon occurs requires a form of investigation. So, what can I do?

In civil and environmental engineering,[63] one could carry out a few types of investigations.

- The first, we could design an experimental campaign aimed at exploring how beams deflect under fire. As you might know, experiments are costly, and fire tests are not only costly,[64] but they require specialized facilities[65] and months of preparations as well.
- The second, re-visiting first principles and theoretical aspects of fire- and stress-induced deformations could be another solution. However, and just like most theoretical investigations, there will be a need to confine such investigations to a series of idealized assumptions and have a stronghold of mathematics.
- The third and perhaps the most affordable solution, developing and validating FE simulations of fire-exposed beams. However, such models can be complex,[66] and we are yet to embrace a systematic procedure to ensure properly validating FE models.

58 This comprises the majority of the practiced ML.

59 Especially to label data by experts.

60 If you really step back and think about Equation 2.2, you will come to the realization that finding a suitable algorithm that best fits a set of data can be automated! More importantly, you will also realize that what really governs the performance of the algorithm is the quality of data! Thus, to improve the performance of a typical algorithm, it could be far more superior to finetune the data to ensure properly treating outliers, wrong labels, insufficient and missing values!

61 A reviewer once commented on one of my papers by saying that the presented ML analysis *felt* like a *fishing expedition*. That was a proper assessment for that particular paper. On a more positive note, that expedition resulted in a new look into the spalling phenomenon. Sometimes an expedition can yield new insights, and in others it fails to do so. My advice is to use your time *wisely*.

62 [try to].

63 And most other engineering areas as well.

64 As in, +$20 000 a test.

65 On occasion, I do miss MSU's fire laboratory.

66 Shout out to Jean-Marc Franssen and Thomas Gernay – the developers of SAFIR [45] (a brilliant FE software for structural fire simulations).

The above implies that to predict the deformation a beam undergoes under fire; an engineer is bound to test or model such a beam. Simply, and as you can see, there is a clear *need* to create a test-free and simulation-free method[67] to predict the deformation of beams under fire conditions.[68] Since the available methods of investigation do not seem to provide us with a solution that satisfies our conditions, then it is perhaps possible to try to seek such a solution via a different approach (i.e., ML).

Now that the answer to the question of, *why do you need ML to tackle a problem/phenomenon?* has been answered. Let us move to the next step of the ML process.

2.5.1 Formulating a Hypothesis

Our physics principles tell us that beams are load bearing members designed to satisfy strength and serviceability criteria [23]. A central interest to structural engineers is to trace the load-deformation history of beams. From a practical perspective, the deformation a beam undergoes can be measured in tests or calculated following mechanics principles (such as the moment-curvature method) or simulated via FE models. Given that the extent of deformation a beam undergoes largely indicates the level of degradation within its load bearing capacity arising from the applied loading, then it is quite possible to associate these two occurrences together.

Similarly, domain knowledge tells us that the deformation history of a loaded beam in bending results in a parabolic-like or curved-like plot. This plot insinuates that the observed deformations result from stresses generated from a continually applied load, *P*, and that the degree of deformation reflects the degradation of the beam's sectional capacity (a function of geometric features, material properties, restraints, etc.). In RC beams, such a plot can have distinct points at which concrete cracks, steel yields, and at the point of failure.

Figure 2.6 demonstrates the deformation history of two idealized beams, *Beam 1* and *Beam 2*. The only difference between these two beams is that *Beam 1* is deeper than *Beam 2*. Thus, under identical loading conditions, *Beam 2* will fail before *Beam 1* given its larger size. Arriving at the above conclusion is trivial since only one parameter (*depth of beam*) is varied between the two beams. Now, if other parameters were to be varied, then the problem substantially grows as the linkage between physics principles and domain knowledge becomes harder to read.

I can, then, argue that load-deformation history plots contain the essence or DNA of a beam as they distill the interaction between the geometrical and material features, as well as loading and restraint conditions. This load-deformation history is what we seek to obtain when we design our tests and remains a primary outcome of our investigations. The same history gives us valuable insights into a beam's response (i.e., cracking, yielding, etc.), performance (ductile, brittle, etc.), and failure (via bending, shear, etc.). Thus, it is essential to be able to obtain such a history. Hence, an opportunity to create a

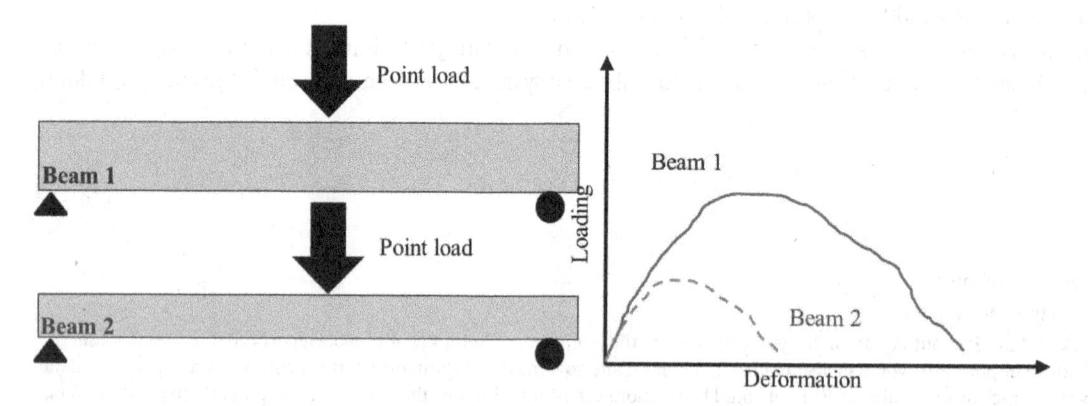

Figure 2.6 Illustration of load-deformation history under ambient conditions.

67 Remember: repeat the above mental exercise with a problem of your own that you can relate best to.

68 One other thing. In structural fire engineering, we declare failure of beams by primarily tracing their deformation under fire (unlike in ambient testing, where failure is simply declared once a beam *breaks*, meaning its sectional capacity has been exceeded). Under fire, such capacity may not need to be reached for a beam to fail! Simply, going under large deformation can be as detrimental (if not more) as reaching the sectional capacity.

new and instantaneous approach to trace the deformation history of beams (or any load bearing member for that matter) presents itself – especially if such an approach can minimize the need for tests or simulations.

Now, let us re-visit the above example by adding the presence of thermal effects.[69] The addition of thermal effects to the above beams introduces thermal deformation and temperature-induced degradation to the beam's sectional capacity (a function of temperature rise, geometric features, material properties, restraints etc.).[70]

Unlike ambient testing, the magnitude of P remains constant during fire testing.[71] Thus, the extent of mid-span deflection in beams is a mere reflection of the degradation in material properties (i.e., how much mechanical properties have degraded due to temperature rise) as well as associated losses in cross section size[72] (if any).

To better visualize the above example, let us compare two identical beams (see Figure 2.7). Since these beams are identical and are being subjected to the same fire conditions, both beams will undergo identical temperature rise and degradation.

In the event that Beam 1 is loaded with P_1 (where $P_1 > P_2$ and P_2 is applied to the second beam[73]), then this beam will experience a higher degree of stress which reflects into larger deformations and fail at an early point in time (see Figure 2.7).

Figure 2.7 is simple and its simplicity stems from the fact that we only varied one parameter (i.e., magnitude of load) throughout our discussion. This triviality is deemed necessary to establish a logical understanding of how beams deflect

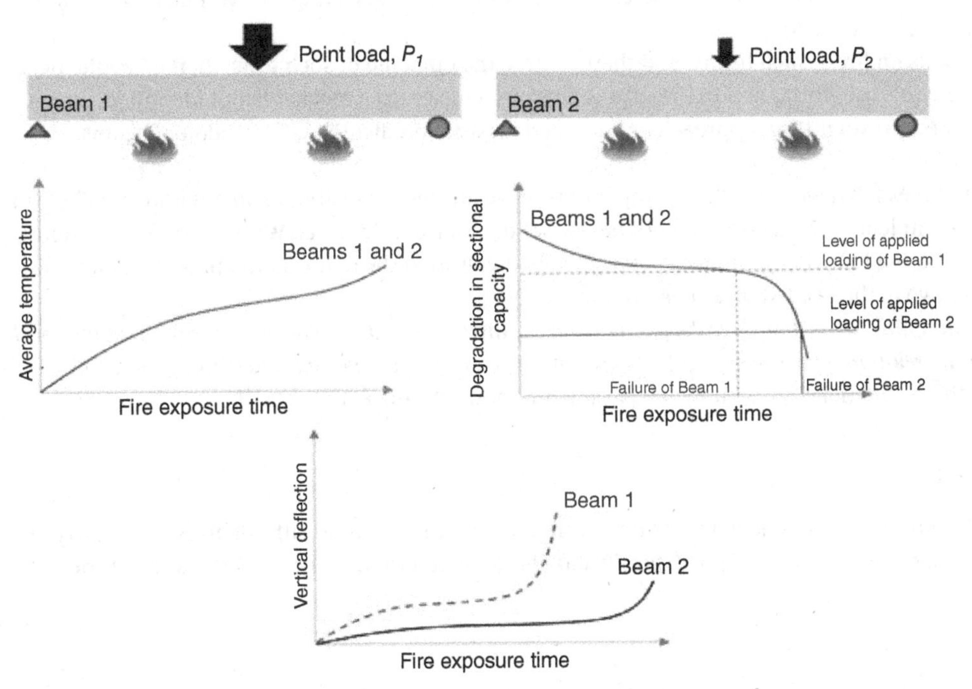

Figure 2.7 Typical response of beams under fire conditions. [Note: failure occurs once the beams deformations exceed a predefined limit and/or once the degradation in sectional capacity falls below the level of applied loading].

69 After all, it is the deformation of beams *under* fire that we are looking for!

70 Bear with me for a minute as this loaded statement might be a bit confusing at first. Let us simplify. **Temperature-induced degradation**: Represent degradation (or loss) in property (i.e., strength) due to the presence of elevated temperatures (which can trigger physio-chemical reactions that weakens materials). In addition, the rise in temperature may lead to the formation of additional effects (say cracking) that *would not have been* witnessed if it was not for the fire. We will re-visit the italics component in the chapter on causality!

71 I hope you can appreciate the complexity of fire testing and hence the selection of this example!

72 One way to think of sectional losses is by visualizing a beam made from timber being exposed to fire. With prolonged exposure, such a beam tends to get smaller and smaller in size as its layer burn/char away. In this scenario, a timber beam loses its capacity on two fronts due to the degradation in material strength and reduction to sectional size. Remember: sectional capacity primarily comprises two factors, material properties and geometry!

73 Remember what we said about testing under fire conditions. The applied loads remain constant!

under fire.[74] As you can probably assume, the more parameters we explore, the trickier the exercise of understanding when and how beams will deform under fire conditions.

While we may not know how varying all parameters, as well as the interaction of all parameters, can affect the deformation behavior,[75] we can form a hypothesis by leaning on our understanding of physics and engineering knowledge. Thus, in order to obtain the deformation history in fire-exposed beams, an engineer ought to identify the *key parameters*, in addition to the *governing relation* that connects these parameters to the phenomenon of deformation in fire-exposed beams. In other words,

$$\textit{Deformation in beams is a function of key parameters} + \textit{governing relation(s)} \quad (2.3)^{76}$$

We know that the *key parameters* revolve around beams': geometry (i.e., width, height, span, etc.), material properties (strength, modulus, etc.), loading (type, magnitude, etc.), and boundary conditions (viz., simply supported, etc.), as well as fire conditions (intensity, heating rate, duration etc.) and so on. On the other hand, the missing component that we do not have is the *governing relation*[77] that ties these parameters together [24]. Hence, our pursuit of arriving at such a relation via a method of investigation such as ML.

Now, if our hypothesis proves correct, then engineers will be able to predict the deformation history of fire-exposed beams and, by extension, other members with ease (and possibly without the need for tests or complex simulation!).

How can we address such a hypothesis, especially since we cannot carry out our own testing, derivation, or simulation? Here is where the role *data*[78] becomes apparent.

If we can observe a series of tests on fire-loaded beams and then analyze the outcome of such tests, then it is quite plausible that we extract the mechanism[79] of deformation in fire-exposed beams. Once such a mechanism is identified, then we can transform/convert such a mechanism into an approach or a method to easily predict the deformation of beams.

Allow me to simplify this further.

Looking at the above formula shows that we have three components. If we are confident about two, then we are likely to realize the third! We are fairly confident of the deformation of beams we obtain from fire tests. We are also fairly comfortable with our physics and engineering knowledge of the key parameters involved in the deformation process. Thus, effectively, we have a good grasp on two of the needed three components.

Simply, *the rationale behind adopting ML is that since the phenomenon of interest (i.e., deformation history) can be observed (say in fire tests), then a governing relation tying such effect to its known key features could be deduced (through ML[80]).*

In other words, we observe the phenomenon, we know the parameters, and we can get the data[81]!

2.5.2 Database Development

To get the data, we can conduct our own comprehensive testing/simulation program, or more affordably, we can carry out a detailed[82] literature review. Collecting *healthy* and *good quality* data is elemental to the success of the to-be-carried out

74 A similar logic can also be used to assess many problems and phenomenon in our domain. Use your imagination!

75 If we knew, then we would not need to showcase this example!

76 The key parameters are likely to be known (to some extent), and the governing relation(s) are likely to be unknown (to some extent). Also, this equation can be generalized into: $\textit{Phenomenon} = \textit{Key parameters} + \textit{Governing relation}(s)$

77 In my eyes, such a relation can be explicit (whitebox or causal) or implicit (blackbox or graybox).

78 **Data**: information, facts, quantities, and statistics collected for the purpose of reference or analysis.

79 Or mechanisms!

80 The same can be true for adopting other methods of investigations! As you will see in later chapters, the governing relation is often highly nonlinear and hence traditional methods (i.e., statistical analysis) might fail to properly capture such relation. What if traditional methods can properly capture such relation? Then, you may use these methods with or without ML! You are the designer!

81 This may sound overly ambitious or optimistic! Let us pause for a minute. I would not want you to think that this is a simple data collection exercise (in some problems, it is in fact a simple process – however this is not the case in most problems). As in, let us collect some data to run a couple of algorithms to see what we can get. There is more to data that can be fitted into this footnote – hence the next chapter!

82 As in, take your time and collect quality and notable sources. Aim for both *quality* and *quantity*.

ML analysis. Such data will be compiled into a dataset and is to home a consistent spread of observations that cover a variety of conditions and information on possible features (parameters) that govern the phenomena on hand.[83]

Datasets can vary in type and size. The most common datasets are those of tabular nature (i.e., homes numeric values and/or characters). These datasets can come in handy in regression and/or classification analysis. The same can also be true for clustering, anomaly detection, and dimensionality reduction. In the case of a ML analysis that examines footage, images, videos, or audio, a second type of dataset can be more appropriate wherein files of the same format can be compiled. We will use many types of datasets in this book.[84]

2.5.3 Processing Observations

By now, a dataset will be compiled and ready and we can comfortably start our ML analysis. To start such analysis requires us to identify/adopt a ML algorithm[85] or a set of algorithms.[86] At this moment in time, I would like you to think of algorithms as *tools/software* that you can pick and choose from. Some work well on specific problems (i.e., clustering algorithms are designed to cluster data and hence are primarily used in clustering tasks). Others work well for many problems (e.g., neural networks perform equally on a variety of tasks).

It is quite common for an engineer to prefer to use one software over another. This could be due to experience,[87] preference,[88] or simply availability.[89] The same is also true in the case of selecting a ML algorithm. In all cases, the selected algorithm needs to be trained and then validated.

Here is why I have asked you to think of algorithms as software (for now). It is because by doing so, we can realize that these algorithms can accommodate any data! For instance, a civil engineer can use algorithm A by inputting civil engineering data. An environmental engineer can also use the same exact algorithm by inputting environmental data. Algorithms do not understand our data nor our problems.[90]

2.5.4 Model Development

We train algorithms by using our datasets. The goal of this training is to familiarize an algorithm with the first two components of Equation 2.2 (i.e., the observations (or response/target we seek to predict) and the parameters we hypothesize to be elemental to the phenomenon on hand).

There are several ways to *train* a ML algorithm. Two of the most widely used methods are referred to as variants of *ratio sampling* and *cross-validation*.

In the first method, the dataset is shuffled and then *randomly* split into two parts: a larger training subset and a likely smaller testing subset.[91] Common split ratios used in this method vary between 50 and 80% for training, with the remaining samples used for validation/testing. This method works well for healthy and balanced datasets.[92] In the case where the

83 For completeness, the space of possible features may or may not be identical for phenomena. Further, there is no guarantee that two independent research groups hoping to tackle the same phenomenon will arrive at the same features, number of features, or types of features, etc. As an exercise, identify five research articles that developed ML models to predict the compressive strength of concrete given mixture proportion and see how many of these articles contain the same features, and cite the same sources.

84 For now, I would like you to know that we will be collecting data for ML. I will go over the data aspect in-depth in the next chapter. In case you are wondering why do we need such data, we will be using this data to teach ML algorithms.

85 **Algorithm**: A method or process (set of steps/rules) in which a ML system tackles the data. Chapter 4 covers algorithms in depth and hence I will not be diving into these right now.

86 One can combine a group of algorithms together. Such a group is called **ensemble**. One can also use one algorithm to tune another algorithm. Other combinations will be discussed in future chapters. For now, let us focus on the big picture!

87 Say an engineer was trained to use *software A*, but then got introduced to the more intuitive *software B*. It is plausible that the same engineer will gravitate toward the latter.

88 For example, my preferred FE software was, and remains, ANSYS. Thanks to the many working hours during my grad school!

89 Accessible/free software can be more affordable than those hiding behind a paywall.

90 Saying this aloud reveals some of the dangers of ML!

91 Which can also be further split into a validation and testing subsets. Also note that the training and testing subsets may or may not be equal.

92 **Balanced dataset**: A dataset where each output class is represented by a similar number of observations.

dataset is imbalanced, then adopting this training method may lead to bias.[93] To overcome the bias issue, a *stratified ratio sampling* method can be adopted. In this method, the distribution of classes/representations in all subsets is preserved.[94] In this process, the performance of the algorithm is documented at the training and testing subsets.

A more modern approach to training ML algorithms is to use the cross-validation method (and its derivatives). In this approach, the dataset is also randomly split into two subsets; a training set and a testing set. However, unlike the other traditional methods, the training set is further divided into k number of parts. The model is then trained using the k-1 parts and is then validated on k part.[95] This process is repeated k times until each unique part has been used as the validation subset (see Figure 2.8) [25]. As you can see, the performance of the algorithm is documented on the training, validation, and testing[96] subsets.

Indeed, the cross-validation[97] method is an improvement over the ratio sampling method. However, left unchecked, the cross-validation may also suffer from bias. Thus, the use of the *stratified k-fold cross-validation* method is often suggested.

Our overarching goal of incorporating the above methods is to ensure proper training of the algorithm to prevent *fitting*[98] issues, and *lucky guessing*,[99] as to improve the accuracy and confidence of the algorithm.

2.5.5 Model Evaluation

You might be asking a question, say that we have adopted the stratified k-fold cross-validation method to train an algorithm; how can we monitor and evaluate model performance throughout its development process? How do we know if the training process is indeed successful?

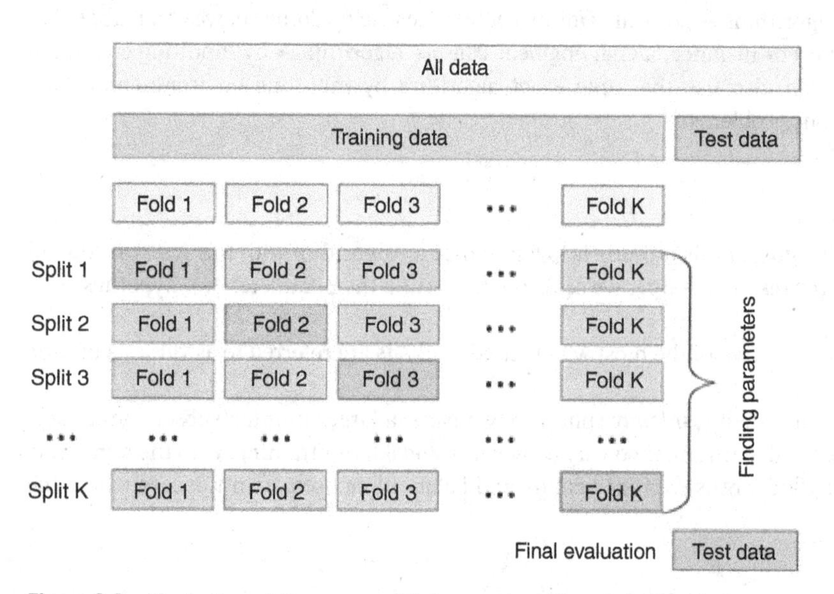

Figure 2.8 Illustration of the cross-validation method (Copyright Elsevier).

93 Think of it this way, while we are randomly splitting the data, there is no guarantee that we are representing all outputs/classes equally (i.e., the training subset may contain large number of observations pertaining to a specific class or features, etc.). On a regression front, bias may arise from lumping many observations on pH (say from *site A*, while representing some samples from other sites).

94 For example, say we have a dataset of 1000 observations of which 700 belong to *class A* and the rest of the observations belong to *class B*. Then, the training subset will contain 70% (700/1000) observations from *class A* and 30% (300/1000) from *class B*.

95 Here is an example. Suppose $k = 5$, then we are to perform a 5-fold cross-validation. In this process, we split our datasets into 5 equal parts (labeled as Parts 1, 2, 3, 4, and 5) that contain no overlapping observations. We will now consider each part as a new dataset (i.e., for the first dataset, use Part 1 as the validation set, and the rest of the parts for training). Given that $k = 5$ implies that each new dataset forms 20% of the original dataset. Thus, each dataset has an 80%-20% train-validation split ratio. The k-cross validation method can be generalized as:

$$\textit{Percentage of data in the validation subset} = (100 / k)\%$$

96 In case a number of algorithms are applied, it is not uncommon that only the performance of the best algorithm is reported on the testing subset (i.e., the testing subset is only used once the best algorithm is identified).

97 For completion, there are other variants to the cross validation than those outlined herein (such as Leave-one-out cross validation).

98 **Overfitting**: A modeling error that occurs when predictions from a ML model align very closely to the data used in training. For example, when model performance on training data significantly outperforms that on testing data. Counter to overfitting is **underfitting**: A modeling error where model is unable to capture the essence of the training data properly. More on these errors in the next chapter.

99 **Lucky guessing**: When a ML model accurately predicts the outcome through sheer luck.

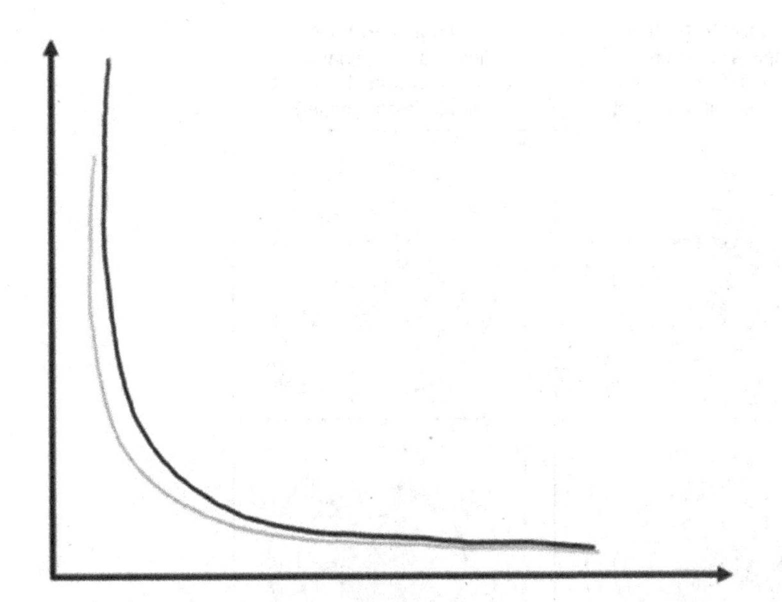

Figure 2.9 Sample of a proper learning curve. [Note: Dark line represents training loss and gray line represents validation loss].

One way to monitor the performance of algorithms is by viewing their learning curves.[100] Figure 2.9 provides a sample of such a learning curve. As you can see, both the training and validation loss decrease in parallel and then stabilize (with a minimal gap between the two, especially toward the end of the analysis[101]).

The above discussion and comparison between the training and validation learning curves bring a discussion on the bias[102] and variance[103] trade-off.[104] Figure 2.10 shows a visual representation of the bias-variance trade-off. In general, ML models with low variance tend to have a simple[105] architecture and hence can be stable (consistent but inaccurate) to training data. In other words, such models lack the needed flexibility to learn the true pattern of the data and hence suffer from underfitting.[106] On the other hand, ML models of low bias tend to be more complex[107] in structure and, as such, are flexible enough to accommodate nonlinearities. These models, when unmonitored, can overfit the data (i.e., memorize training data and/or capture noise) and mistakenly provide poor generalization of the data.

---

100 **Learning curve**: A line plot that displays learning over experience or amount of data used.

101 If the stabilized portion happen to be excessively long, this may indicate a case where the algorithm is learning too fast and hence a slower learning rate can be of aid. Further examples on possible learning curves, techniques (i.e., early stopping, etc.) and how to diagnose such curves are shown in the Appendix.

102 **Bias**: The difference between the average prediction and the ground truth (i.e., true or correct value). In other words, bias measures the distance between the best we can predict to the ground truth.

103 **Variance**: The variability of model prediction for a given data instance (i.e., a measure of spread). In other words, the variance indicates how well we can get close to the best we can predict given different training data.

104 This is a trade-off because increasing one will decrease the other and hence a balance is needed. Overall, the error in a ML model equals to that due to the bias, variance, and irreducible error (e.g., a type of error that cannot be reduced through generation of models as this is a measure of the noise in our data and/or formulation of our problem). Mathematically, $Error = Bias^2 + Variance + Irreducible\ error$ (Note that both the bias and variance be negative terms, this effectively imply the persistence of some error that we cannot simply/fully overcome through modeling). In case you are wondering why the bias is squared, then I will invite to see the proof in the *Appendix* (well, the error is in terms of the mean squared error and variance is already squared!).

105 Such as parametric models. One can think of linear regression where the model is already pre-built on the linear assumption between variables. Such a model, as you might have guessed may limit the algorithm's ability to learn the true underlying structure of the data. Thus, such a model is essentially *biased* to linearity! A sidenote on biased models, say that we are using a model that properly accounts for some degree of nonlinearity. Yet, the same model continues to underperform despite acconting for sure nonlinearity. Then, this could be another sign of underfitting in which the model is not using enough number of features to train.

106 Models with high bias pay little attention to the data and hence, underfit the data on hand.

107 Such as non-parametric models.

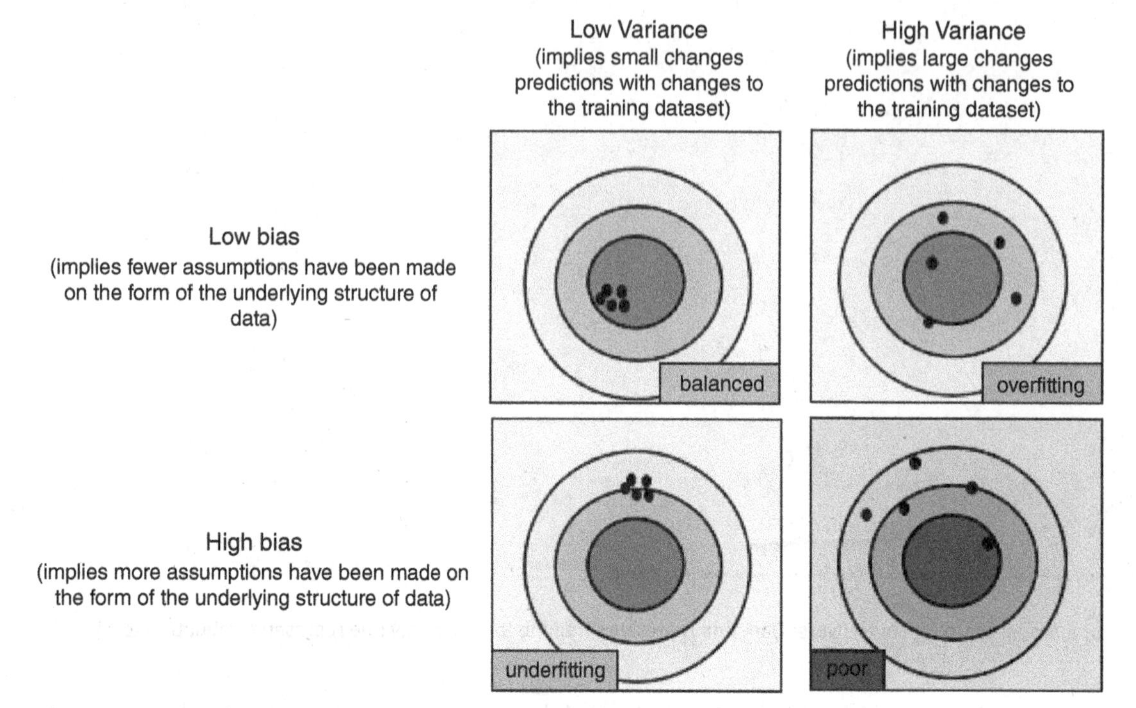

Figure 2.10 Illustration of bias-variance trade-off. [Note: high bias + low variance → underfitting. Low bias + high variance → overfitting].[108]

When the algorithm on hand completes its training and validation, the performance of this algorithm is also evaluated through error metrics and performance indicators.[109] These are mathematical or logical constructs that examine how predictions converge or diverge from the real observations (or ground truth). When algorithmic predictions match the ground truth well[110], then the algorithm is deemed fit, and this shifts our terminology use from *algorithm* to *model*.[111]

There exist a variety of metrics and indicators which, for the most part, can be quite similar to those often used in statistics. The simplest form of such metrics in a regression problem is that given by deducting a predicted value from its corresponding actual/observed value. For example, civil and environmental engineers are accustomed to regression-based metrics such as average error, correlation coefficient, etc. In reality, one, a multiple, or a combination of metrics and indicators can be used to examine the performance of a ML model.

Nevertheless, not all metrics and indicators are equal[112] or applicable. For instance, using a regression metric (such as the correlation coefficient) is not adequate in evaluating a classification model. In such classification problems, the simplest form of evaluation is to compare how well the model predicts the class of interest. Other common classification-based metrics include the area under the receiver operating characteristic, accuracy, etc. Up to this point in time, we are yet to identify a set of standard performance metrics.[113]

---

108 Additional notes on the bias and variance: A model is said to be biased when it systematically generates incorrect predictions (implying that the model cannot capture the phenomenon on hand). High bias (or under-fitting) can be handled by incorporating features to capture the unmodeled complexity, or by using a different model type capable of capturing the complexity on hand. A model is said to have high variance when it does not generalize well when fed with new data (the model has essentially learned the data that was used in its training vs. learning the phenomenon). High variance (or overfitting) can be handled by decreasing the complexity of the model.

109 We have chapter 5 to cover these.

110 I acknowledge the qualitative nature of this term. As I said, we will visit this point in more detail in a few chapters from now.

111 On the other hand, if algorithmic predictions do not match that of the ground truth, then additional finetuning could be applied. This tuning can involve adopting different training strategies, inclusion of additional data, or tuning model hyperparameters. **Hyperparameter**: A model configuration that is independent of the data. Unlike hyperparameters, model **parameters** can be estimated from the data. An example of such parameters include: *weights* in a neural network or *support vectors* in support vector machines. Terms in italics are covered in Chapter 4.

112 As in their evaluation power and applicability to specific/general problems. Some can also be dependent on others (the coefficient of determination, R^2, is the square of the correlation coefficient, R!).

113 Hence my advice is to apply a combination of metrics and of different formulations (i.e., error-based, objective-based, correlation-based etc.) to overcome some of the limitations associated with individual metrics (i.e., biasness, oversimplifications, etc.).

2.5.6 Model Optimization

Once the ML model is properly validated, this model can be further tuned (optimized) by manipulating the size and/or content of data used in training/validation/testing or via adjusting the architecture of the algorithm (e.g., number of layers/ neurons in a neural network, etc.), hyperparameters (for metaheuristics), etc.

2.5.7 Model Deployment

In this stage, the ML model is ready to be deployed[114] and shared with other users (i.e., in offices/design firms) or be put forth for use as a service to address real-world problems. The same models can also be converted into software and web applications with intuitive graphical user interfaces.[115] To ensure seamless integration, ML models must be easy to utilize.

2.5.8 Model Management (Monitoring, Updating, Etc.)

With time, the performance of deployed models can degrade as new data[116] and trends become available, etc. As such, it is a good practice to continue to monitor and tune deployed ML models.

2.6 Conclusions

This chapter presents an introductory look into the history of ML and the big ideas behind this technology. This chapter also covers the primary types of learning available at the moment and covers the complete process of developing and creating ML models – from a civil and environmental engineering perspective. Finally, various definitions were given and articulated to set the stage for the coming chapters.

Definitions

Active learning A process in which a ML algorithm could potentially reach a notable accuracy, despite using a smaller number of training data, if it were allowed to choose its own training data.
Algorithm A method or process (set of steps/rules) in which a ML system tackles the data (often optimized and well established).
Artificial intelligence Intelligence displayed by machines in which a system gathers information from its surroundings (i.e., data) to take actions.
Association A type of unsupervised learning that discovers rules with the data.
Balanced dataset A dataset where each output class is represented by a similar number of observations.
Bias The difference between the average prediction and the ground truth (i.e., true or correct value).
Blackbox ML A ML approach that covers models with complex inner workings and hence are hard to interpret.
Classification A type of supervised learning that involves an output variable of a categorical nature.
Clustering A type of unsupervised learning that groups by behavior.
Code A list of instructions for machines to execute.
Data jump and **data drift** Expected, unexpected, and/or undocumented changes to data structure as a result of model architecture, or environment or practices. These are symptoms of degradation of model prediction that occurs over time.
Data-driven model A model driven by the data – change the data and the model will change. Unlike, for example, the model for bending in beams, no matter how the beams differ, the model remains the same to a large extent.
Data Information, facts, quantities, and statistics collected for the purpose of reference or analysis.
Data-centric ML A ML approach that emphasizes data aspects.[117]

114 **Model deployment**: The process of applying the model toward predictions and/or new data.
115 Visit the Appendix for more information on this! I have two examples that you can follow (*Python* and *R*).
116 Possible examples include *data jump* and *data drift* and that the use of commonly used construction material becomes less (or more) common due to policy, inflation, availability, or changes to practice, respectively.
117 Answers questions like, how can we change data labels to improve model performance?

Deep learning Neural networks with deep and fluid architectures.
Dimensionality reduction A type of unsupervised learning that reduces the number of parameters into those that can explain the data the most.
Ensemble A group of algorithms combined into one.
Graybox ML ML models wherein the relationship and/or mathematical model is partially built on some degree of theoretical understanding or prior knowledge while also making use of data-driven-like approaches.
Hyperparameter A model configuration that is independent of the data.
Inputs/features/independent variables A set of attributes, often represented as a vector and are associated with an output.
Learning Curve A line plot that displays learning over experience or amount of data used.
Learning The process of seeking knowledge.
Lucky guessing When a ML model accurately predicts the outcome through sheer luck.
Machine learning A branch of artificial intelligence (AI) that creates computer programs that mimic the way humans learn and gradually improve their accuracy. They are defined as techniques that are designed to enable machines to learn/understand/identify meaningful patterns and trends from data.
Model deployment The process of applying the model toward predictions and/or new data.
Model parameters can be estimated from the data. An example of such parameters includes: *weights* in a neural network or *support vectors* in support vector machines.
Model-centric ML A ML approach that emphasizes algorithmic aspects.[118]
Output/outcome/response/target/dependent variable: Observations or labels we are interested in.
Overfitting A modeling error that occurs when predictions from a ML model aligns very closely to the data used in training. For example, when model performance on training data significantly outperforms that on testing data.
Regression A type of supervised learning that involves an output variable of a numeric nature.
Reinforcement learning A ML system that is programmed to teach itself following prespecified reward mechanisms.
Self-supervised learning This type of learning aims to predict unobserved inputs from any observed input and is often used in computer vision and speech recognition.
Semi-supervised learning A ML process that combines a small amount of labeled data with larger (unlabeled) data.
Testing subset Examples or observations used to test and evaluate the performance of a ML algorithm.
Training subset Examples or observations used to train a ML algorithm.
Transfer learning A process where a ML model developed for a task is reused as the starting point for a model on a second task. This type of learning ensures efficiency.
Underfitting A modeling error where a model is unable to capture the essence of the training data properly. More on these errors in the next chapter.
Unsupervised learning A ML process that hopes to uncover the underlying structure/distribution of the data.
Validation subset Examples or observations used to validate a ML algorithm.
Variance The variability of model prediction for a given data instance (i.e., a measure of spread).
Whitebox ML (also, **transparent** or **glassbox ML**): A ML approach that articulates the *functional* or *physical* relationship(s) between the variables governing a phenomenon to the outcome (or target/response).

118 Answers questions like, how can we change algorithmic settings to improve model performance?

Chapter Blueprint

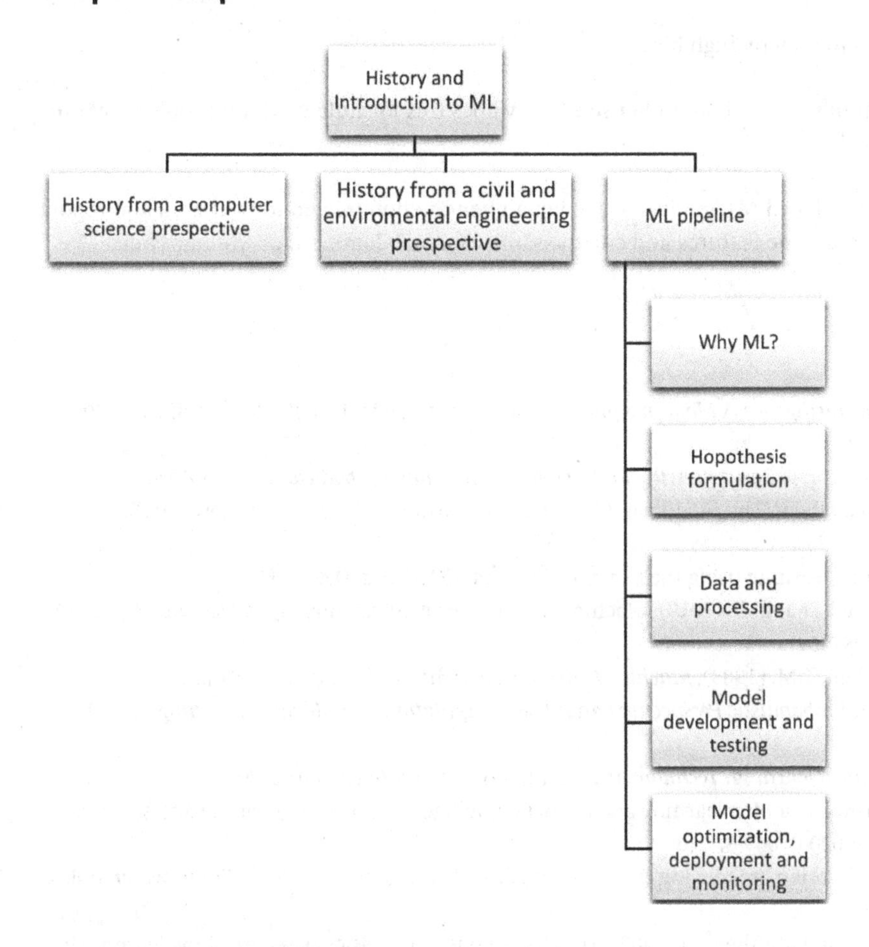

Questions and Problems[119]

2.1 Write a two-page paper on the history of AI/ML and DL.

2.2 Discuss the importance of ML to civil and environmental engineers.[120]

2.3 What are algorithms and ensembles?

2.4 Define whitebox, graybox, blackbox, and causal ML.

2.5 What is supervised learning? What is unsupervised learning?

2.6 Explain the difference between active, transfer, and reinforcement learning. Supplement your answers with examples.

---

119 These questions could be considered teasers to faculty members.

120 Bonus points to discussions that go beyond that provided in this chapter.

2.7 Describe the overall ideas behind ML pipeline.

2.8 How to remedy high variance? How to remedy high bias?

2.9 In your eyes, do you think that formula no. 3 is comprehensive? How does this formula work for problems within your field?

2.10 Identify five research articles that developed ML models to predict a phenomenon related to your area and report how many of these articles contain the same features and cite the same sources. Comment on your answers.

References

1 Russell, S. and Norvig, P. (2009). *Artificial Intelligence A Modern Approach*, 3e. Cambridge University Press. doi: 10.1017/S0269888900007724.

2 Steels, L. and Brooks, R. (2018). *The Artificial Life Route to Artificial Intelligence: Building Embodied, Situated Agents.* Routledge. https://books.google.com/books/about/The_Artificial_Life_Route_to_Artificial.html?id=7b5aDwAAQBAJ (accessed 9 July 2022).

3 Samuel, A. (1959). Some studies in machine learning using the game of checkers. *IBM J. Res. Dev.* 44 (1.2).

4 Fu, G.-S., Levin-Schwartz, Y., Lin, Q.-H., and Zhang, D. (2019). Machine learning for medical imaging. *J. Healthc. Eng.* 2019, Article ID 9874591: 2. doi: 10.1155/2019/9874591.

5 Michalski, R.S., Carbonell, J.G., and Mitchell, T.M. (1983). *Machine Learning: An Artificial Intelligence Approach.* Springer.

6 Murphy, K.P. (2012). *Machine Learning: A Probabilistic Perspective (Adaptive Computation and Machine Learning Series).* The MIT Press.

7 Ziegel, E.R. (2003). The elements of statistical learning. *Technometrics* 45 (3). doi: 10.1198/tech.2003.s770.

8 Naser, M.Z. (2021). Mechanistically informed machine learning and artificial intelligence in fire engineering and sciences. *Fire Technol.* 57: 1–44. doi: 10.1007/s10694-020-01069-8.

9 Shen, C., Chen, X., and Laloy, E. (2021). Editorial: broadening the use of machine learning in hydrology. *Front. Water.* doi: 10.3389/frwa.2021.681023.

10 Vadyala, S.R., Betgeri, S.N., Matthews, J.C., and Matthews, E. (2022 March). A review of physics-based machine learning in civil engineering. *Results Eng.* 13. doi: 10.1016/j.rineng.2021.100316.

11 Mills, H. and Treagust, D. (2003). Engineering education. Is Problem-Based or Project-Based Learning the Answer? *Australas. J. Eng. Educ.*

12 Stevens, R., O'Connor, K., Garrison, L. et al. (2013). Becoming an engineer: toward a three dimensional view of engineering learning. *J. Eng. Educ.* doi: 10.1002/j.2168-9830.2008.tb00984.x.

13 CSI. (2016). SAP2000. Analysis reference manual. CSI Berkeley (CA, USA). Comput. Struct. INC.

14 Behnood, A. and Golafshani, E.M. (2020). Machine learning study of the mechanical properties of concretes containing waste foundry sand. *Constr. Build. Mater.* 243. doi: 10.1016/j.conbuildmat.2020.118152.

15 Dimensions (2021). Dimensions.ai. https://www.dimensions.ai (accessed January 2022).

16 Naser, M. and Mulluer, K. (2021). SP-350: the concrete industry in the era of artificial intelligence. American Concrete Institute. https://www.concrete.org/store/productdetail.aspx?ItemID=SP350&Format=PROTECTED_PDF&Language=English&Units=US_Units (accessed 6 March 2022).

17 Hiriyur, B. (2020). Thornton Tomasetti Launches T2D2, an AI Solution Company to detect, classify and monitor deterioration | Thornton Tomasetti. Thornton Tomasetti. https://www.thorntontomasetti.com/news/thornton-tomasetti-launches-t2d2-ai-solution-company-detect-classify-and-monitor-deterioration (accessed 6 March 2022).

18 Hearns, A. (2019). Giatec unveils first artificial intelligence solution for concrete | Giatec. Giatec. https://www.giatecscientific.com/press/giatec-unveils-first-artificial-intelligence-solution-for-concrete (accessed 6 March 2022).

19 Keck, Wood. (2021). Uses of artificial intelligence in civil engineering – Keck & Wood civil engineers. Duluth, Fayetteville GA, Rock Hill, North Charleston SC. https://keckwood.com/news-updates/uses-of-artificial-intelligence-in-civil-engineering (accessed 6 March 2022).

20 Magdalena, L. (2010). What is soft computing? revisiting possible answers. *Int. J. Comput. Intell. Syst.* 3: 148–159. doi: 10.1080/18756891.2010.9727686.

21 Zarringol, M., Thai, H.T., Thai, S., and Patel, V. (2020). Application of ANN to the design of CFST columns. *Structures* 28. doi: 10.1016/j.istruc.2020.10.048.

22 Bohlin, T.P. (2013). *Practical Grey-box Process Identification: Theory and Applications*. American Society of Civil Engineers.

23 ASCE. (2016). *Minimum Design Loads for Buildings and Other Structures*. American Society of Civil Engineers. doi: 10.1061/9780872629042.

24 Naser, M., Hostetter, H., and Daware, A. (2020). AI modelling & mapping functions: a cognitive, physics-guided, simulation-free and instantaneous approach to fire evaluation. *11th International Conference on Structures in Fire (SiF2020)*. The University of Queensland, Brisbane, Australia.

25 Anguita, D., Ghelardoni, L., Ghio, A. et al. (2012). The 'K' in K-fold Cross Validation. *ESANN 2012 proceedings, 20th European Symposium on Artificial Neural Networks, Computational Intelligence and Machine Learning. Bruges (Belgium), 25-27 April 2012*, i6doc.com publ., ISBN 978-2-87419-049-0.

26 Flegg, G. (1989). Numbers through the ages. Macmillan Education; Open University. Distributed in the USA by Sheridan House, Houndmills Basingstoke Hampshire. Milton Keynes [England]; Dobbs Ferry NY.

27 Mccarthy, J. (2004). What is Artificial Intelligence? http://www-formal.stanford.edu/jmc (accessed 13 July 2022).

28 IBM Cloud Education. (2020). What is Artificial Intelligence (AI)? | IBM. IBM Cloud Education.

29 McCulloch, W.S. and Pitts, W. (1943). A logical calculus of the ideas immanent in nervous activity. *Bull. Math. Biophys.* 5. doi: 10.1007/BF02478259.

30 Hebb, D.O. (2005). The organization of behavior : a neuropsychological theory. *Organ. Behav.* 378. doi: 10.4324/9781410612403.

31 Mahdavi, S., Rahnamayan, S., and Deb, K. (2018). Opposition based learning: a literature review. *Swarm Evol. Comput.* 39. doi: 10.1016/j.swevo.2017.09.010.

32 Glennan, S. (2022). Rethinking mechanistic explanation. *Philos. Sci.* 69. doi: 10.1086/341857.

33 Settles, B. (2012). Active Learning. *Synth. Lect. Artif. Intell. Mach. Learn.* 100. doi: 10.2200/S00429ED1V01Y201207AIM018.

34 Goodfellow, I., Bengio, Y., and Courville, A. (2016). *Deep Learning (Adaptive Computation and Machine Learning series)*. Ian Goodfellow, Yoshua Bengio, Aaron Courville: 9780262035613. Amazon.com. Books. MIT Press.

35 Facebook (2021). Self-supervised learning: the dark matter of intelligence. https://ai.facebook.com/blog/self-supervised-learning-the-dark-matter-of-intelligence (accessed 11 July 2022).

36 Blockley, D.I. (1979). The role of fuzzy sets in civil engineering. *Fuzzy Sets Syst.* 2. doi: 10.1016/0165-0114(79)90001-0.

37 Flood, I. (1989). A neural network approach to the sequencing of construction tasks. *Proceedings of the 6th International Symposium on Automation & Robotics in Construction*. doi: 10.22260/isarc1989/0026.

38 Vanluchene, R.D. and Sun, R. (1990). Neural networks in structural engineering. *Comput. Civ. Infrastruct. Eng.* 5. doi: 10.1111/j.1467-8667.1990.tb00377.x.

39 Adeli, H. (2002). Neural networks in civil engineering: 1989–2000. *Comput. Civ. Infrastruct. Eng.* 16: 126–142. doi: 10.1111/0885-9507.00219.

40 LaMalva, K.J. and Medina, R.A. (2022). Building codes and the fire regulatory context of smart and autonomous infrastructure. In: *Handbook of Cognitive and Autonomous Systems for Fire Resilient Infrastructures* (ed. M.Z. Naser), 117–138. doi: 10.1007/978-3-030-98685-8_5.

41 Zhao, X., Lovreglio, R., and Nilsson, D. (2020). Modelling and interpreting pre-evacuation decision-making using machine learning. *Autom. Constr.* 113. doi: 10.1016/j.autcon.2020.103140.

42 Wagenaar, D., Curran, A., Balbi, M. et al. (2020). Invited perspectives: how machine learning will change flood risk and impact assessment. *Nat. Hazards Earth Syst. Sci.* 200. doi: 10.5194/nhess-20-1149-2020.

43 Ul Hassan, F. and Le, T. (2020). Automated requirements identification from construction contract documents using natural language processing. *J. Leg. Aff. Disput. Resolut. Eng. Constr.* 12. doi: 10.1061/(asce)la.1943-4170.0000379.

44 Ribarič, M., Šušteršič, L., and Stefan, J. (n.d.). Empirical relationship as a stepping-stone to theory.

45 Franssen, J.M. and Gernay, T. (2017). Modeling structures in fire with SAFIR®: theoretical background and capabilities. *J. Struct. Fire Eng.* 8. doi: 10.1108/JSFE-07-2016-0010.

3

Data and Statistics

Data have no meaning in themselves; they are meaningful only in relation to a conceptual model of the phenomenon studied.

G. Box, W. Hunter, and J. Hunter[1]

Synopsis

This chapter covers the principles and techniques of data collection, handling, and manipulation needed for a variety of machine learning (ML) investigations. In addition, this chapter reviews fundamentals of statistics and serves as a refresher and/or a reinforcer to some of the essential principles behind statistical analyses. First, we define data and then lay out strategies for visualizing and plotting the different types of data we are likely to encounter in our field. Then, we spend some time exploring and diagnosing data, and for this, a series of methods will be presented and showcased. I will be sharing techniques to carry out fundamental[2] data analysis and visualization using *Excel*, *Python*, *R*, and *Exploratory*.

3.1 Data and Data Science

We have previously defined data as pieces (or units) of information, facts, quantities, and statistics collected for the purpose of reference or analysis. Data is a big element of ML,[3] especially one that is *data*-driven (or driven by the data[4]). The data has two main components: an explanatory portion (a numerical value or a category) and a label describing such a unit. Numerical data are made of numeric or numbers, and categorical data are made from categories or classes. We also have footage data, audio data, video data, text data, time series data, etc.

---

1 From Chapter 9 in *Statistics for Experimenters: An Introduction to Design, Data Analysis, and Model Building* (the 1978 edition) [12].

2 It must be noted that we will briefly re-visit this concept in Chapter 6. I will show how to easily automate this process using a collection of coding-free software.

3 Actually, data is a huge component of the architecture, engineering, and construction (AEC) industry. The average firm with 100 projects is adding 1 petabyte (PB) of data every year. 1 PB is equivalent to 1024 terabytes (TB). The average adult human brain has the ability to store the equivalent of 2.5 PB of memory [18]. Here is another surprising statistic: According to Klaus Nyengaard [19], co-founder and chairman of the construction technology company *Geniebelt*, "In 2017, *as much as 95% of all data in the construction industry was either being thrown away – or not even collected in the first place*."

4 As in, if the data changes, the model that is built on this data is likely to change as well (whether; slightly – as expected, or largely – which may turn problematic). Aside from being used as input into a ML model, the data is also necessary, as I hope you remember, to continuously improve the performance of ML models over time. Thus, ensuring that we have high-quality data is essential for monitoring and creating self-improving models. This brings me to this *Quick story*: one of my students pointed out this scene from the movie Sherlock Holmes (2009), *"Never theorize before you have data. Invariably, you end up twisting facts to suit theories, instead of theories to suit facts"* – said Sherlock Holmes to Dr. Watson.

Machine Learning for Civil & Environmental Engineers: A Practical Approach to Data-driven Analysis, Explainability, and Causality, First Edition. M. Z. Naser.

Companion Website: www.wiley.com/go/Naser/MachineLearningforCivilandEnvironmentalEngineers

Let us say that we have acquired the data needed for a specific ML analysis. What do we do with such data? Well, we can use the data in three primary ways:

- To *describe* the phenomenon the data was collected on.
  - For example, say that we have built 20 reinforced concrete beams and then tested these beams under flexure.[5] The results from these tests can be used to *describe* a few observations, such as the patterns of cracks in beams before failure; some beams failed with large deformations while others were under small deformations. A few beams failed at larger loads than others, etc. Such descriptions can be obtained via data exploration and/or unsupervised ML.
- To *predict* the phenomenon the data was collected on.
  - For example, we can use the data we collected on the above beams to build a model that can forecast/estimate how similar beams would behave. Structural engineers can be interested in knowing the bending capacity[6] of beams. Thus, we make use of the data we collected on failure loads to develop a model[7] that can predict the failure load in beams.[8] This is where the majority of supervised ML applications fall.
- To *causally* understand or infer cause and effect relations.
  - For example, out of the 20 failures we have seen, what has/have caused the failures in each beam? Can we identify a causal mechanism that uncovers how failure takes place? Answering such a question is where causal ML comes into the picture.

Thus, to use the data in any of the three above ways to arrive at knowledge,[9] we need to follow systematic and well-established approaches. Hence, *data science* is the subject that uses scientific methods, processes, algorithms, and systems to extract knowledge from data.[10] This chapter will primarily focus on data.

From this view, it is of merit to list the properties of data. These fall under:

- Scale and size.
- Variety of types.
- Rate of generation and recording.
- Meaningfulness.
- Degree of confidence.

3.2 Types of Data

There are numerous types of data, and five are presented herein.[11]

3.2.1 Numerical Data

This type represents quantitative and measurable data such as forces, weights, deformations, age, etc. Exact and whole numbers[12] are called discrete numbers. Continuous numbers are those that fall under a range (i.e., the axial capacity is 1234.2 kN). Continuous data can be further divided into two groups: *interval* and *ratio* data. Interval data can be measured on a known scale where each point is placed at an equal distance from its neighbors.[13] This type of data can

---

5 Say that we have set up the beams in a simply supported manner and applied a point load at each beam's mid-span.

6 The load at which beams fail or break. More specifically, under bending conditions, a beam will fail once the bending effect (i.e., the effect of applying bending forces) exceeds the beam's bending capacity (i.e., the resistance of the beam to bending effects).

7 Mathematical or empirical or numerical, etc.

8 Preferably of similar features and conditions to the tested beams used to build the model. This statement is critical to note! Keep it in mind when building models (ML, or otherwise).

9 Is that not one of the goals of collecting and analyzing data?

10 Or simply, "*Data science* is the study of the generalizable extraction of knowledge from data" [20].

11 Programming languages also have other types of data. For example, *Python* has *integer* (for whole numbers), *float* (for real numbers with decimals), etc.

12 Such as *Bridge A* has 5 girders.

13 You can think of temperature. Temperature has a scale where each point is spaced equally from its neighbors. Keep in mind interval data do not have zeros.

accommodate addition and subtraction operations. Unlike interval data, ratio data can have zero values. Overall, numerical data can be statistically and arithmetically calculated (i.e., averaged) and/or arranged in an ascending or descending order.

I would like you to try a mental exercise and read the following numbers: 33, 35, 48, 86, 88, 91, and 97 out loud. Now, I would like you to answer the following two questions, what are these numbers? What do they convey?

Go back to this chapter's quote. As you can see, "*data have no meaning in themselves*"! Now, let us assign units to these numbers, say, MPa. These numbers now refer to pressure or stress.[14] In the context of construction materials, the compressive strength of concrete is given in terms of MPa.[15] If we use different units, then these numbers would represent the air quality in a certain region (in $\mu g/m^3$), the length of highways (in km), depth of footing for a structure (in feet).

Knowing the context allows us to build a story. Our story is that we have seven concrete cylinders that we have tested and recorded their compressive strength. Knowing the context also allows us to effectively name all these values under one title. Now, these are no longer random values. These are data points that describe the measured compressive strength of concrete of each tested cylinder. In Chapter 2, we discussed how most supervised ML revolves around building regression and classification models. Thus, these data points can be readily used in a regression model. If this model is to predict the compressive strength of concrete, then these values are now our target (or response).

What if we would like to run a classification model on this data? To do so, we will have to use categorical data. Let us then discuss categorical data.

3.2.2 Categorical Data

Categorical data are non-numerical and are sorted by their labels or characteristics such as construction type (low-rise, high-rise, etc.), failure mode (flexures, shear, etc.), etc. Given their labels, such data cannot be added or averaged. Categorical data can be further divided into *nominal*[16] data and *ordinal* data.

Nominal data contain names or titles (such as color or title). This type of data can be collected from surveys or questionnaires. On the other hand, ordinal data comes with a preconceived order/rating scale. While this type of data can be ranked, it cannot be measured (i.e., minor damage, major damage, collapse). A simple way to remember numerical and categorical data is coded in Figure 3.1.

Let us now revisit our aforenoted question. By definition, a classification model classifies categorical data. However, the data we collected from testing the concrete cylinders is numerical. To convert numerical data into categorical data, we will have to create categories that our numerical data can be fitted into. One such example is to create two classes, namely, normal strength concrete and high strength concrete.

The cylinders with a compressive strength of concrete under 55 MPa can fall into the normal strength concrete category,[17] and those of higher strength must then fall under the high strength concrete category. Thus, our target now is no longer a numerical value but rather a class.

From this context, we can now build regression-based or classification-based ML models to predict our target based on some input. This input could be numerical, categorical, or a mixture of the two.[18] I would like to emphasize this point. The target in a tabular format is more likely to be homogeneous than the inputs. As in, the target will be numerical values (in case of regression) or classes (in case of classification) – see Table 3.1 for an example that lists the results from a series of tests conducted on structural elements.

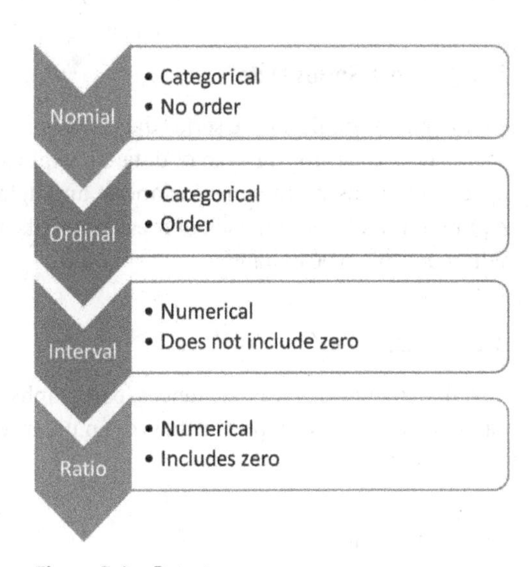

Figure 3.1 Data types.

14 Pa is pascal (named after Blaise Pascal). A pascal describes a force over an area (such as N/mm^2). Force over an area is pressure or stress.

15 Equivalently, ksi.

16 From the Latin nomenclature *Nomen* (meaning name).

17 I am following the recommendation of the American Concrete Institute (ACI).

18 We will get to these in a few sections from now. Look under the *manipulation* section.

Table 3.1 Example of various data types.

	Feature 1	Feature 2	Target for regression	Target for classification
No.	Specimen type (nominal)	Strength level (ordinal)	Applied load (numeric)	Failed? (class)
1	Beam	High	450 kN.m	Yes
2	Column	Low	200 kN	No
3	Joint	Medium	321 kN	Yes
...	...	...	...	...

3.2.3 Footage

Unlike tabular data, data for computer vision comprises images/footage. In this case, the images themselves are used as both the input and output. The images are used as input since we are usually interested in something specific about these images. For example, visual inspection of damage (say, cracks in pavements) is a lengthy and time-consuming process. Here is an example.

The state of Michigan has a total of 120,256 miles[19] of paved roadway. Now, imagine the number of resources needed to visually inspect those roadways! What about instead of asking state pavement engineers to manually check for cracks across state roadways, we attach cameras to a fleet of autonomous vehicles we program to drive across the roadways within Michigan. We take this footage and ask our engineers to watch and identify roads in need of repair. This is a feasible solution that, at the very least, will not only improve their quality of life but will reduce the needed resources to carry out such an inspection.

By now, you might be wondering, could not we automate this process using ML? The answer is yes!

All it takes is to use this footage to train a computer vision model on how to identify cracks.[20] In this scenario, the footage is used as input to such a model where the same engineers can label (or annotate) the different concepts they would like the model to learn (i.e., fatigue cracks, slippage cracks, etc.). Train the model, and then once fully validated, these engineers can then use this model to autonomously inspect visual footage of paved roadways.[21]

3.2.4 Time Series Data*

As the name implies, time series data covers a collection of observations or recordings taken through repeated measurements over time. In this type of data, an engineer may opt to record such data in a consistent manner (i.e., over equally spaced intervals).[22] In some instances, an engineer may not have access to consistently recorded data whether due to equipment malfunction, lack of measurements, funding limitations, etc.[23] In all cases, data indexed in time[24] order often fall under this type of data.

3.2.5 Text Data*

Text data includes words, sentences, paragraphs, or documents that can be collected and organized. Unlike numerical and categorical data, this type of data is often unstructured.[25]

19 ~193,533.7 km.

20 As well as any other concepts related to pavement inspection (rotting, etc.).

21 In reality, the engineers will annotate a portion of the footage. Then, a model will be created to annotate the rest of the footage. Another model will then be trained to either classify footage that display poor pavement conditions, or detect portions of the footage that show poor quality (pavement). In some advanced version of the above, a ML model can flag cases where its confidence in its own predictions fall below a specific threshold.

22 Think of measuring daily air quality, or seasonal changes in temperature.

23 Perhaps an engineer may not have access to consistent historical data! It is possible that some of such data is lost or misplaced.

24 This is also the key difference between numerical data and time series data, the former is not indexed in time (i.e., does not have start and end points).

25 **Unstructured data**: data that cannot be organized into rows or columns. **Structured data**: data that can be organized into rows or columns. **Semi-structured data**: data that does not fit into rows or columns but has some form of a fixed structure.

3.3 Dataset Development

Tabulated data can be structured into a matrix. This matrix has rows (r) that represent the collected samples (or measurements or instances/observations) and columns (c) that represent the characteristics or features of the measured samples. Hence, the dimension of a typical tabular dataset is equal to ($r \times c$) – see Table 3.1. If supervised ML is sought, a designated column is reserved for the target (or dependent variable or response);[26] otherwise, such a column is unnecessary.

The engineer is advised to create a *healthy* and *balanced* database.[27] The intentional use of the descriptive term healthy aims to describe a database that contains a uniform spread of observations to cover a variety of features and conditions.[28] Say an engineer is building a dataset to examine the response of beams under bending, then this engineer is to be mindful of collecting data on:

- Varying geometric and material features.[29]
  - To increase the predictive capability of the to-be-developed ML model.
- Tests from different eras.
  - To reflect upon changes in materials and design detailing.
- Beams tested under similar conditions.
  - To control the dimensionality of the problem and/or normalize the effect of testing regimes.
- From different groups/laboratories.
  - To account for minor changes due to facilities and conditions.
- Complete response.
  - Since any missing data will add to the uncertainty and complexity of the future model.
- Duplicated specimens.
  - To ensure the reliability of measurements and minimize outliers.

In case of footage data:

- Properly tag and identify the regions of interest.
- Include a collection of texture, colors, quality, etc., displaying a varying magnitude of the phenomenon on hand (i.e., damage).

Once a dataset is compiled, this database can be labeled as the *master dataset*.

3.4 Diagnosing and Handling Data

At this stage, the integrity of the master dataset demands evaluation. For example, it is common for a dataset to contain noise, outliers, missing points, or contain erroneous data. Thus, the master dataset is to be processed to minimize the effects of missing or erroneous data, noise, and outliners.

This process is often referred to as diagnosing and handling.[30] This process is essential to guarantee that our dataset is appropriately organized, informative, consistent, and hence can yield proper ML models. This process seeks to make it easier for us (and our ML models) to identify true trends and insights on data distribution, involved features, etc. In

---

26 Note the variety of terminology used.

27 The Appendix contains repositories that you can explore to build datasets.

28 Frank and Todeschini [21] recommended a general rule of thumb of maintaining a ratio of three and five between the number of observations to features. Another, independent, and perhaps outdated, rule of thumb is called the *10-times rule*. In this rule, the amount of observations should be ten times more than the number of model's parameters. A more systematic approach is to train the ML model on a portion of the dataset and then evaluate the performance of the model on each subset to try to identify the gain in performance with the addition of data. There are a couple of studies that noted success with using 10–30 observations to train ML models. I am not citing these herein as I am not quite sure that I would like to entertain this practice without further examination. I keep wondering, why do we need to build a ML for such small data? Can we wait a bit to collect more data? Would not a statistical analysis be sufficient? I recognize that answers to these questions can vary and extend beyond this footnote. Hence, my advice is to do your diligence.

29 It is often preferable for a dataset to be deeper than wider, i.e., containing a much larger number of observations as opposed to features.

30 Others may call this *data handling* or *data cleansing*.

addition, by cleaning the dataset, we ensure that only needed and high-quality information is left for us to analyze – which cuts down on unnecessary cycles of iterations. Removing duplicated, missing, and incorrect data enables the model to better capture[31] the phenomena on hand.

Here, we will go over some of the main techniques for handling missing data and outliers.

Data can be *missing* due to countless reasons.[32] However, the bottom line is that some data are not present in our dataset. Remedies to missing data can vary in design and complexity. In its simplest form, missing data can be deleted or ignored.[33] A more flexible approach is to fill missing data with some other data. For example, the average[34] value of a feature can be used to substitute an instance of missing data. Interpolating between neighboring points can be used to impute a missing data point. Let us take a look at our data from concrete cylinders:

Original data = 33, 35, 48, 86, 88, 91, and 97 with an average of 68.28 MPa.

Say that the fourth point (i.e., 86 MPa) was missing. Simply substituting this data with the average may alter the group's new average a bit (new average = 65.71 MPa) despite the significant deviation between the two points (86 − 68.28 = 17.7 MPa).[35] Hence, in a large enough dataset with a few missing points, this can be an attractive solution.[36]

More complex methods may try to predict the missing data point based on the known information from the neighboring observations with similar or close enough features. Such methods work well if there is some relationship between the observation.[37] For instance, a ML model[38] can be built to predict such missing data! A probabilistic model can also be built to draw a replacement for missing data from the distribution of the parent feature.

Outliers are rare and exceptional data that do not conform[39] to the norm/observed trend in a feature. Outliers could occur due to measurement or human error, undocumented behavior, or anomalies.[40] A very large outlier may shift the average[41] of data or may cause an engineer to mistakenly believe/assume faulty behavior. The remedies for outliers are similar to those for missing data.[42]

Both missing data and outliers can be present in their parent dataset or come to light when merging different datasets.[43]

3.5 Visualizing Data

Now that the dataset has been compiled, the next step is to understand our data. I will start by showcasing simple methods to visualize a dataset. Our goal is to try to identify *general* trends or insights pertaining to our data. Such insights might help us set forth assumptions or point out any needs for refining our dataset.[44]

---

31 By avoiding bias.

32 There are three, or more, such reasons: *missing by design* (when data is missing deliberately, i.e., data on small cracks are not recorded since they are not needed), *missing completely at random* (when the data is missing due to a random factor such as being damaged or simply lost), *missing at random* (when the data is missing is not related to the missing data but rather to some of the observed data – an example is that we do not report crack width since the test specimen is not made from crackable material), and *missing not at random* (when the probability of data being missed varies for reasons unknown to us).

33 As in, delete the row in the dataset where there is a missing data. You may also delete an entire feature column if it contains a missing data point or a number of missing data (you can assign a threshold to do so!). A reminder that dropping a feature or even an observation due to one single missing data point may lead to losing too much information. One may suggest to only drop the observations that actually hurt the performance of the model.

34 You may also use the median or mode to do the same. The weighted average seems to work well in the case of time-based data.

35 Please recognize that this is but one example. This procedure is sensitive to the available number of data and their scale.

36 This is a good time to remember to read the review on statistics at the end of this chapter.

37 To circle back to the cylinders example. The cylinder of 86 MPa strength clearly belong to the high strength concrete class. Thus, we could use the average of the cylinders from that particular mixture (i.e., 88, 91, and 97) vs. including those that belong to normal strength concrete. The average of the cylinders of high strength is 90.5 MPa. Substituting this average into this group (*90.5*, 88, 91, and 97) returns 91.63 MPa! Thus, this is a more accurate representation of the missing data point.

38 Such as K-Nearest Neighbors (KNN) or tree-models. For simple datasets, linear models can do the trick.

39 They can be too large or too small, etc.

40 In some instance, a data point can be thought of to be an outlier, just to turn out that it is not. These may occur due to the examining of a space that has not been explored before, or due to change/shift that occurs naturally, etc.

41 We will chat more on this in the second half of this chapter.

42 I have also seen some examples where a new column is added to the master dataset to keep track of which observations had missing data, and/or outliers, and to add comments on which remedies were used on those observations.

43 Additional notes: Tree-based and Naive Bayes algorithms seem to be a bit robust when handling missing data.

44 I highly advise visiting Kuhn and Johnson's book, *Feature Engineering and Selection: A Practical Approach for Predictive Models* [6] and online version [22].

The first appropriate step is to assess this dataset by means of a simple statistical analysis. In this analysis, we are trying to describe the properties of the data. Here is a list of such properties.[45] The formulas/commands for these properties in *Excel*, *Python*, and *R* are shown in the companion table.

- *Range*: shows the spread of data.
- *Min/max values*: lists the smallest and largest value.
- *Mean*: the average of values and is calculated as $\mu = \frac{1}{n}\sum_{i=1}^{n} x_i$
- *Variance*:[46] measures the spread of data and is calculated as $\sigma^2 = \frac{1}{n}\sum_{i=1}^{n}(x_i - \mu)^2$
- *Standard deviation*:[47] measures the dispersion of a dataset relative to its mean and is calculated as $\sigma = \sqrt{\frac{1}{n}\sum_{i=1}^{n}(x_i - \mu)^2}$
- *Median*: identifies the middle number of the data by ordering all data points.[48]
- *Mode*: the most common data value.
- *Skewness*:[49] describes the degree of asymmetry and is calculated as $\mu_3 = \frac{\sum_{i=1}^{n}(x_i - \mu)^3}{(n)\sigma^3}$
- *Kurtosis*: describes the tailedness[50] of the data and is calculated as $\mu_4 = \frac{\sum_{i=1}^{n}(x_i - \mu)^4}{(n)\sigma^4}$
- *Percentiles*: values indicating the percent of a distribution that is equal to or below it.[51]

Companion table for formulas and commands.

Property	*Excel*	*Python*	*R*
Range	`=MAX(Cells)-MIN(Cells)`	-	`range()`
Min	`=MIN(Cells)`	`min()`	`min()`
Max	`=MAX(Cells)`	`max()`	`max()`
Mean	`=AVERAGE(Cells)`	`from statistics import mean`	`mean()`
Variance	`=VAR(Cells)` or `=VARP(Cells)`	`from statistics import variance`	`var()`
Standard deviation	`=STDEV.P(Cells)` or `= STDEV.S(Cells)`	`from statistics import stdev`	`sd()`
Median	`=MEDIAN(Cells)`	`from statistics import median`	`median()`
Mode	`=MODE(Cells)`	`from statistics import mode`	-
Skewness	`=SKEW(Cells)`	`from scipy.stats import skew`	`skewness()`
Kurtosis	`=KURT(Cells)`	`from scipy.stats import kurtosis`	`kurtosis()`

45 Side note, you might have seen some of the following formulas using *n-1* as opposed to *n* as the divisor. This goes back to a more in-depth discussion on unbiased estimators – a talk for a different day. I advise you to review [23].

46 Tells us how data points differ from the mean. This is also referred to as the second standardized moment.

47 Tells us, *on average*, how far each score lies from the mean.

48 In the event that there are two middle numbers, the median would be the mean of those two numbers.

49 Skewness >1.0 or < −1.0 indicates a high skewedness. A value between 0.5 and 1 or −0.5 and −1.0 indicates moderate skewedness. A value between −0.5 and 0.5 indicates light skewedness or fairly symmetrical. Skewness is also referred to as the third standardized moment.

50 If the data is heavily or lightly tailed. The kurtosis for a normal distribution is 3 (the distribution for a kurtosis greater than +1.0 describes the peakdness while for −1.0 indicates flatness). In other words, this property measures the deviation from the peak in a Gaussian distribution due to outliers. Kurtosis is also referred to as the fourth standardized moment.

51 For example, 25th, 50th, and 75th percentile (interquartile range (IQR)).

Table 3.2 Statistical insights.

Property	Diameter (mm)	Tube thickness (mm)	Effective length (mm)	Yield strength (MPa)	Compressive strength of concrete (MPa)	Axial capacity (kN)
Min	44.45	0.52	152.35	178.28	9.17	45.20
Max	1020.00	16.54	5560.00	853.00	193.30	46,000.00
Mean	158.52	4.31	1060.07	336.53	50.21	2349.58
Standard deviation	105.42	2.45	1005.28	90.89	31.57	4042.29
Median	127.30	4.00	662.00	325.00	41.00	1244.60
Skew	3.71	1.59	1.98	2.19	2.07	5.40
Kurtosis	20.46	3.69	3.84	8.47	5.17	38.63

Allow me to demonstrate some of the above properties on a more realistic dataset. In this example, data on concrete-filled steel tubes[52] are collected into a dataset. If we apply some of the above properties to this dataset, we will arrive at Table 3.2 which will help us visualize this dataset. This table also indicates the range of applicability we expect our to-be-developed ML model to be applied to.[53]

While Table 3.2 is indeed helpful, perhaps we can visualize our dataset via other means. For example, we can plot our dataset. There are many, many ways to do so. Let us go over a few.

Histograms: display the frequency distribution for a numerical feature. By default, a histogram also shows the range of a feature as well as intervals of equal length (or bins).[54] For example, a look in Figure 3.2a shows the distribution of the tube thickness in our dataset as plotted in *Microsoft Excel*.[55] As you can see, this distribution is a bit skewed to the left and seems to have a good number of observations on its skewed side. On the other hand, the number of observations heavily drops on the right side. This reflects the presence of a limited number of tubes (or columns) with large thicknesses.[56] Figures 3.2b and 3.2c plot different styles of histograms for other features using *Python* and *R*.[57] Histograms are a great tool to show individual features.[58]

Boxplots: displays a summary of the main characteristics of numerical features through the mean, median, IQR, and possible outliers. These plots can be used to display one or multiple features.[59] As you can see, Figure 3.3 shows a boxplot for two features (yield strength of tube and compressive strength of concrete filling). This boxplot marks the mean with an × and the median with a line that crosses the box.[60] The lines of this box represent the 25th and 75th percentile. The farthest horizontal lines[61] from the box are calculated as 1.5 × IQR. Possible outliers are also plotted as dots.

---

52 These are columns made of steel tubes. Steel tubes have small thickness and are hollow. To improve their axial strength (capacity), structural engineers fill these hollow tubes with concrete. In this configuration, the tubes act as formwork to host the concrete and the filled concrete delays buckling of the thin tubes.

53 Which also translates to the limits to which we can apply such a model (with some degree of confidence).

54 Bins are not always of equal size. You may use Sturge's rule [24] to choose the number of bins given the size of data, number of bins $= \log 2(n) + 1$. You may also use Scott's rule [25], binwidth $= 3.5\sigma n^{-1.3}$.

55 https://www.microsoft.com.

56 This can be a factor in the accuracy of how the model predicts such columns. If the model is not exposed to many examples, then its predictivity could be affected.

57 Made through *Python*'s *Seaborn* [26] and *R*'s *ggplot2* [27] libraries. You can find the source files on my website.

58 They can also come in handy to compare multiple features. In all cases, avoid intentionally manipulating the horizontal or vertical axes to hide a weak spot. A good practice is to fix the vertical axis of all features in a dataset to allow for a uniform comparison.

59 However, for multiple features the vertical axis can be of different units (unless the features are normalized/standardized). In this case, comparing two features can give you insights on the range of features (as you can see in Figure 3.3, the yield strength of steel tubes is much larger and of a much wider rangethan the compressive strength of concrete), and tidiness of data between the features.

60 The *box* homes 50% of the data.

61 Also called, *whiskers*.

(a) Using *Excel*

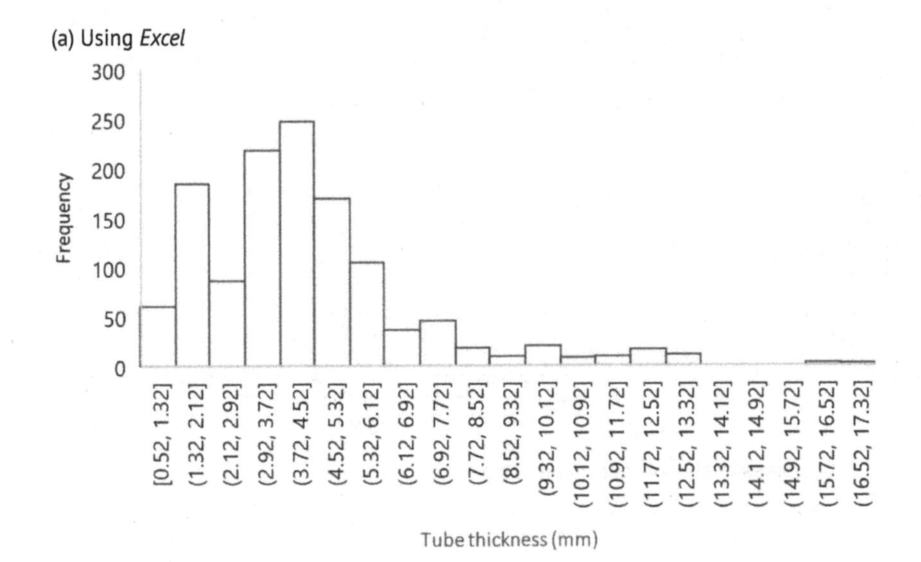

(b) Using *Python*

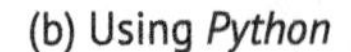

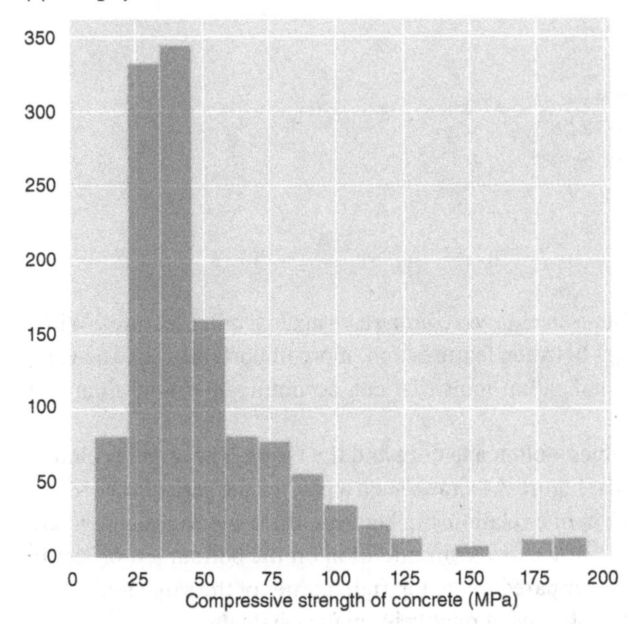


```
import matplotlib as plt
import seaborn as sns

df = pd.read_excel("Chapter 3 [Examples].
xlsx")
sns.set(rc = {'figure.figsize':(6,6)})

df

sns.distplot(df['Compressive strength of
concrete (MPa)'], bins=15, kde=False)
```

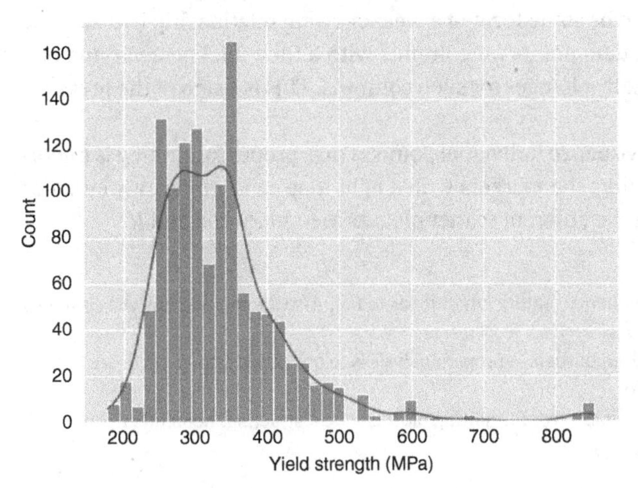


```
import matplotlib as plt
import seaborn as sns

df = pd.read_excel("Chapter 3 [Examples].
xlsx")

sns.set(rc = {'figure.figsize':(6,6)})

df

sns.histplot(data = df, x = "Yield strength
(MPa)", kde = True)
```

(Continued)

(c) Using *R*

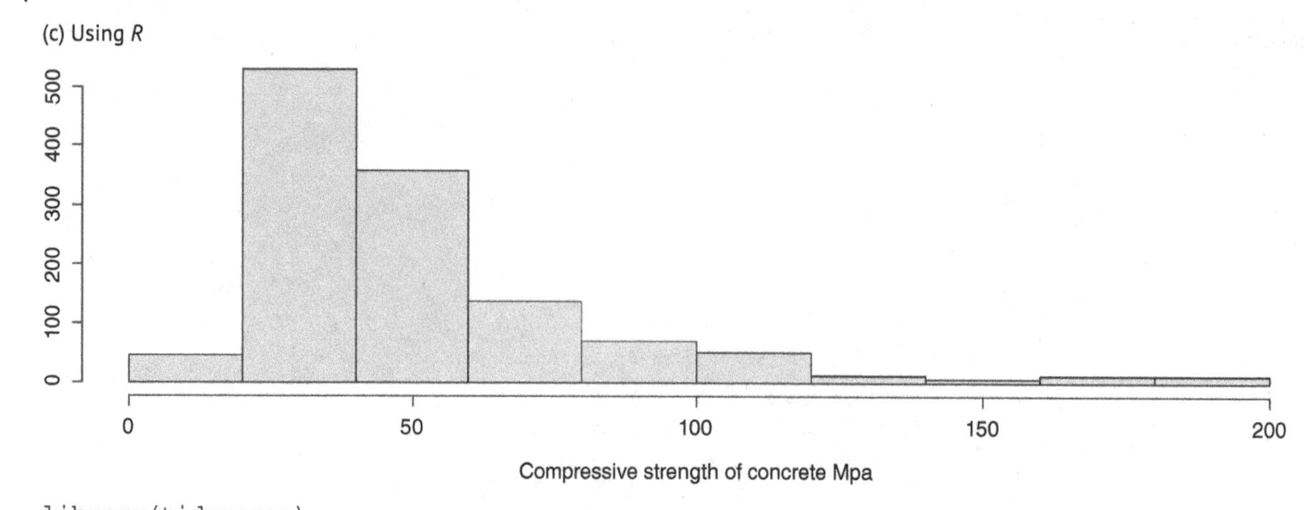


```
library(tidyverse)
library(ggplot2)

data=read.csv('C:/Chapter 3.csv')
data
str(data)

hist(data$Compressive.strength.of.concrete..MPa.,
     main = " Compressive strength of concrete Mpa",
     xlab = " Compressive strength of concrete Mpa",
                          ylab = "count")
```

Figure 3.2 (Cont'd) Demonstration of histograms.

Up to this point, we were interested in visualizing features by themselves or via comparison against other features. While such insights are helpful, we are also interested in the relationships between features and, more importantly, between features and the target of interest. Here is where we go toward visualization tools that can accommodate more than one variable.

Scatter plot: this plot depicts the relationship between two variables – often a feature and the target. The target is plotted on the vertical axis, and the feature is plotted on the horizontal axis. Figure 3.4 shows such a plot for the case of tube diameter and axial capacity. As you can see, there seems to be some form of a relationship between these two variables, where columns with larger diameters tend to have larger axial capacity.[62] If we focus our attention on the bottom left of Figure 3.4a, we can clearly note the presence of a dense group of data (as compared to the top right corner of the same plot). This effectively implies that the dataset contains plenty of examples on columns of relatively smaller diameter.

The bottom of Figure 3.4a also shows a similar scatter plot but for the case of tube thickness. This plot has some similarities with the plot on tube diameter. For example, there seems to be some form of a weaker rising relationship between the tube diameter and the axial capacity. In addition, if we look at data points for columns with a tube thickness varying between 8 and 14 mm, we will notice that the axial capacity drastically changes for such columns. This is a sign of the presence of possible outliers.[63]

Figure 3.4 shows that the data clutters and it becomes hard to visualize individual points. Thus, proper formatting is important to ensure that we gain the most of this plot. For example, filling the markers with a light tone can significantly enhance the quality of this plot.[64] Similar observations can be made from the different scatter plots as seen in *Python* and *R*.

---

62 This makes sense since larger columns may have larger thicknesses and hence require larger loads to fail. Also, larger columns can contain larger amounts of concrete.

63 This is a visual inspection. To ensure that such possible outliers are indeed outliers, a proper analysis is to be carried out. You will see in the next few chapters that we do have methods and algorithms that can help us!

64 Side note: aim to design the plots to be as simple and informative as possible. Over formatting plots can be distracting. Remember your audience!

(a) Using *Excel*

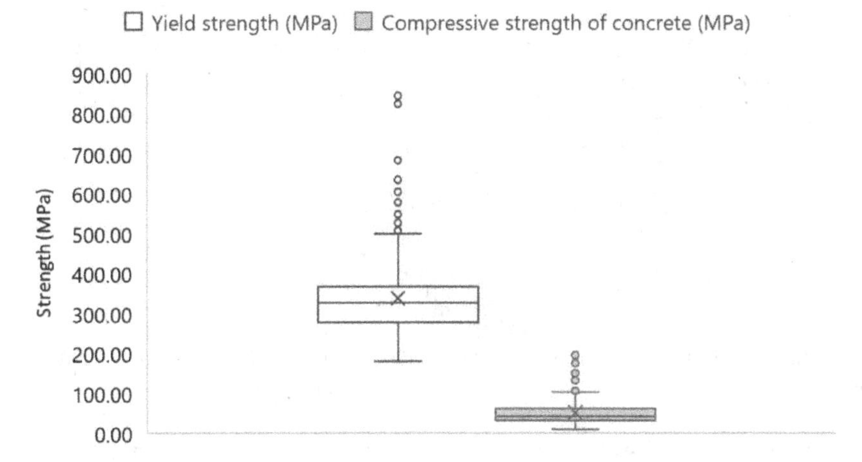

(b) Using *Python*

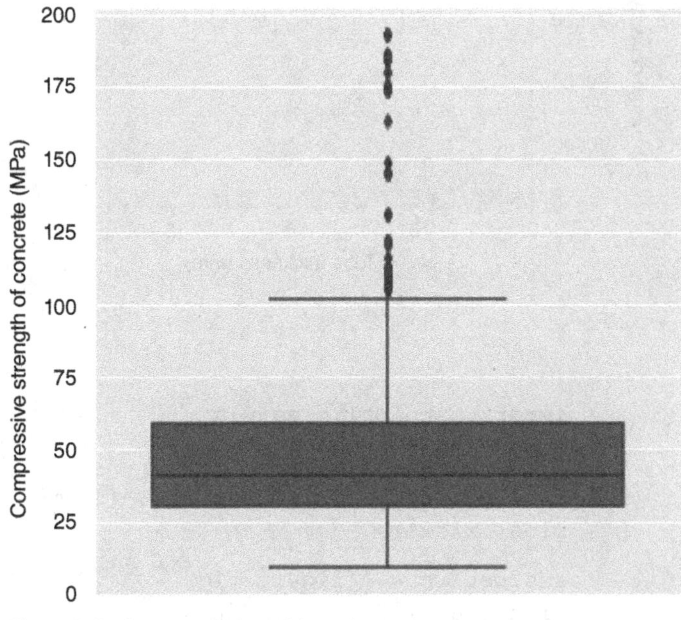


```
import matplotlib as plt
import seaborn as sns

df = pd.read_excel("Chapter 3
[Examples].xlsx")

sns.set(rc = {'figure.fig-
size':(6,6)})

df

   sns.boxplot( y=df['Compressive
   strength of concrete (MPa)'] );
```

Figure 3.3 Demonstration of box plots.

As mentioned above, scatter plots can accommodate multiple features[65] (see Figure 3.5). Looking at Figure 3.5 shows how adding multiple features to a scatter plot can be inefficient as a visualization tool. This could be improved a bit by formatting the plot. Another approach is to reduce the number of features involved.[66] A more practical approach is to use a different visualization tool.

For completion, there are other visualization tools that can be used and applied, such as violin plots, point plots, funnel plots, time series plots, etc. You will see these sprinkled throughout the examples of this book. For now, I would like you to remember that the goal of using such tools is to try to gain insights into your dataset.

65 Depending on the features – these plots may work well with two features. For visualizing high dimensional datasets, the Principal component analysis (PCA) algorithm can be used (to be discussed in a later chapter).

66 For example, we can divide the compressive strength by the yield strength. We can also do the same for tube thickness and diameter. The first ratio can be thought of as a measure of material strength and the second ratio can be thought of as a measure for the sectional properties of the tube.

(a) Using *Excel*

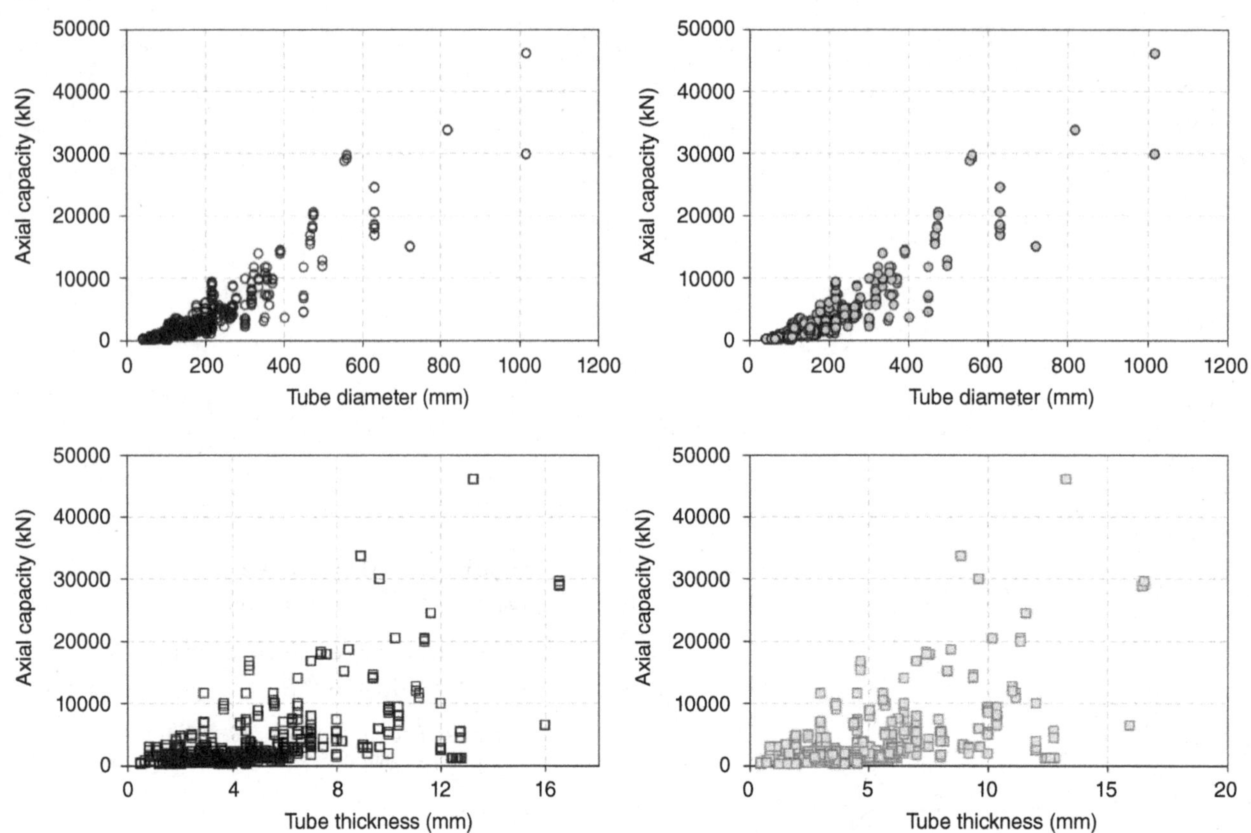

(b) Using *Python*

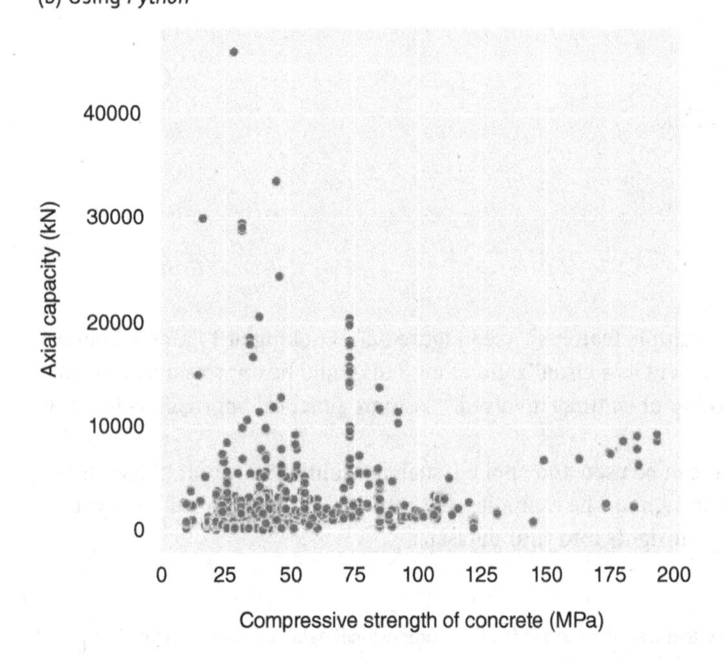


```
import matplotlib as plt
import seaborn as sns

df = pd.read_excel("Chapter 3 [Exam-
ples].xlsx")

sns.set(rc = {'figure.fig-
size':(6,6)})

df

sns.scatterplot(x='Compressive
strength of concrete (MPa)', y='Axial
capacity (kN)', data=df)
```

(Continued)

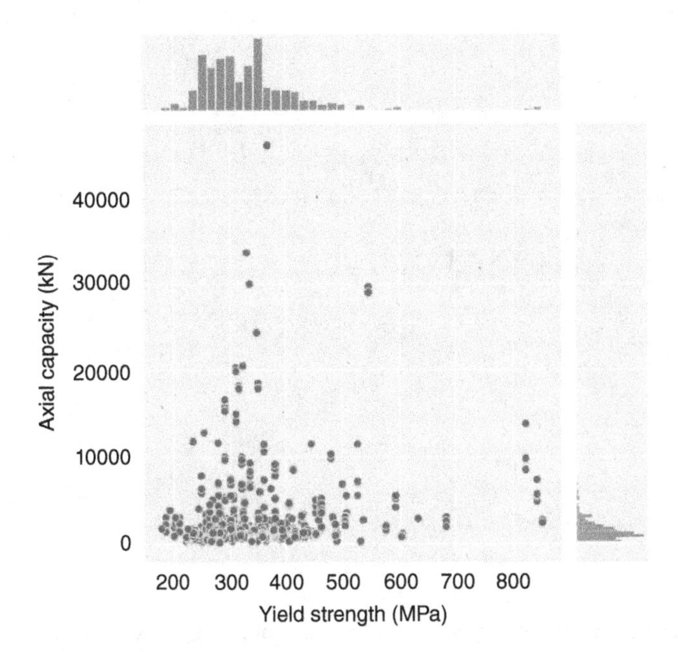


```
import matplotlib as plt
import seaborn as sns

df = pd.read_excel("Chapter 3 [Exam-
ples].xlsx")

sns.set(rc = {'figure.figsize':(6,6)})

df

sns.jointplot('Yield strength (MPa)',
'Axial capacity (kN)', data=df);
```

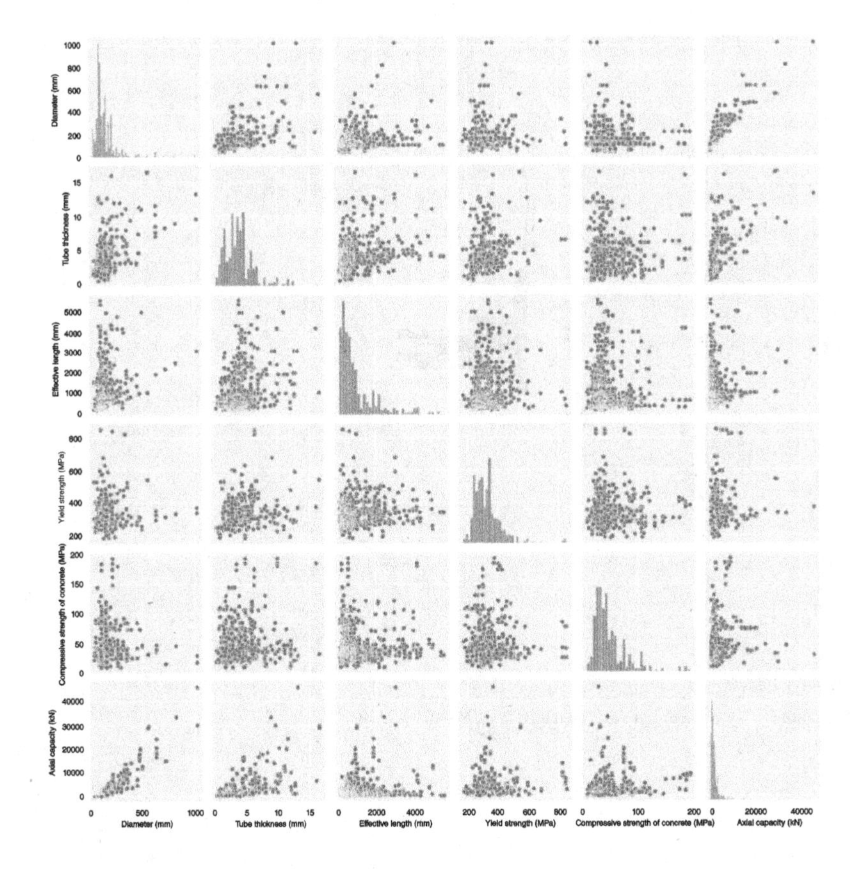


```
import matplotlib as plt
import seaborn as sns

df = pd.read_excel("Chapter 3
[Examples].xlsx")

sns.set(rc = {'figure.fig-
size':(6,6)})

df

sns.pairplot(df)
```

(Continued)

```
library(tidyverse)
library(ggplot2)

data=read.csv('C:/Chapter 3.csv ')
data
str(data)

#scatter plot
p <- ggplot(data, aes(x=Compressive.strength.of.concrete..MPa., y=Axial.capacity..kN.)) +
geom_point() +

theme(legend.position="none")
```

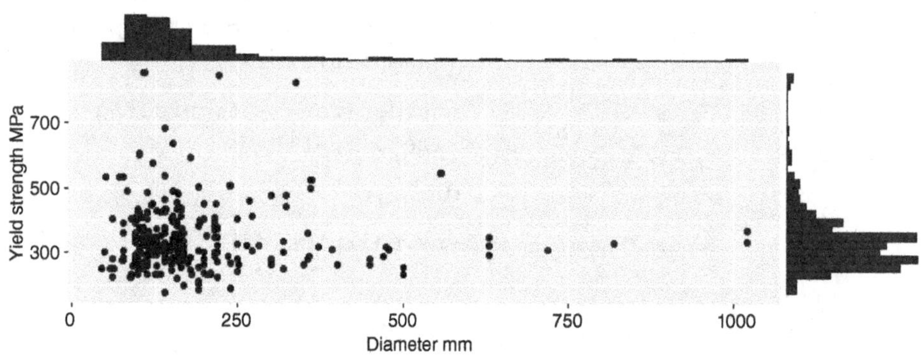


```
library(tidyverse)
library(ggplot2)

data=read.csv('C:/Chapter 3.csv ')
data
str(data)

p2<-lines(density(data$Diameter..mm.))
p <- ggplot(data, aes(x=Diameter..mm., y=Yield.strength..MPa.)) +
geom_point() +

theme(legend.position="none")

#marginal histoplot
p3 <- ggMarginal(p, type="histogram")
```

(Continued)

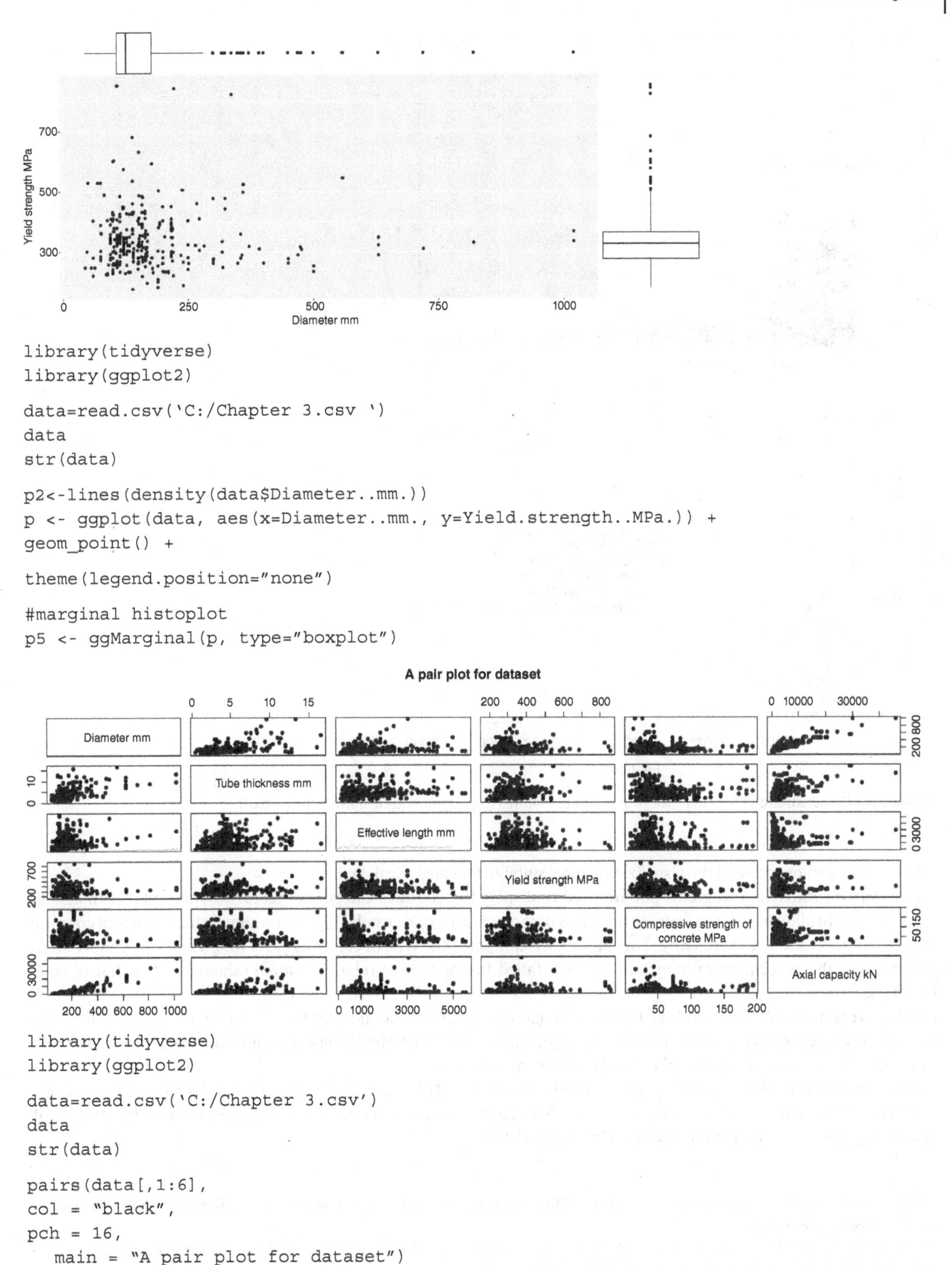


```
library(tidyverse)
library(ggplot2)

data=read.csv('C:/Chapter 3.csv ')
data
str(data)

p2<-lines(density(data$Diameter..mm.))
p <- ggplot(data, aes(x=Diameter..mm., y=Yield.strength..MPa.)) +
geom_point() +

theme(legend.position="none")

#marginal histoplot
p5 <- ggMarginal(p, type="boxplot")
```

```
library(tidyverse)
library(ggplot2)

data=read.csv('C:/Chapter 3.csv')
data
str(data)

pairs(data[,1:6],
col = "black",
pch = 16,
  main = "A pair plot for dataset")
```

Figure 3.4 (Cont'd) Demonstration of a scatter plot (left: unformatted, right: formatted, top: unformatted, bottom: formatted).

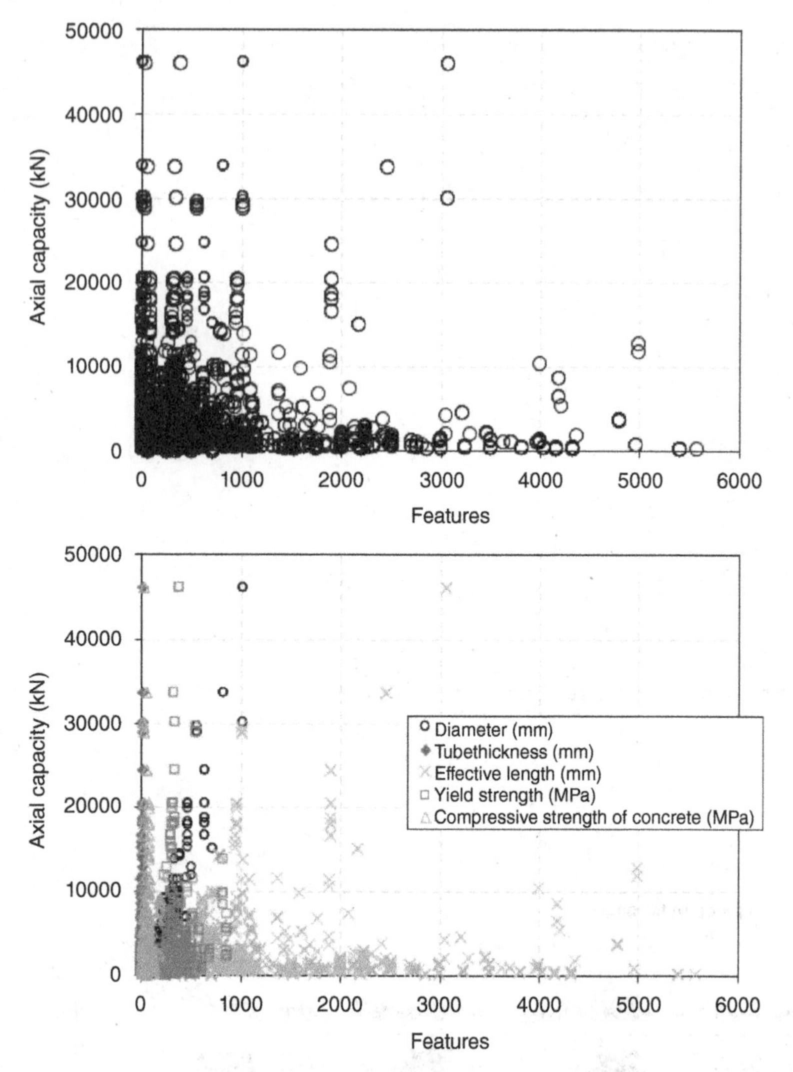

Figure 3.5 Demonstration of scatter plot with multiple features (top: unformatted, bottom: formatted).

Let us now explore some of the visualization tools suitable for categorical data.

Bar charts: similar to histograms but are used to display the count of categorical features. These plots can also be formatted to display the ratio of features (in terms of their class). For example, Figure 3.6 shows the number of columns that failed in buckling and crushing. This figure also compares the type of columns (very long vs. long vs. short[67]). As you can see, the percentage of long columns that failed in crushing is relatively small (about 20%) of that of short columns.

Other types of visualizations include boxplots,[68] contingency tables, and mosaic plots. Examples of a contingency table (which shows frequencies for combinations of categorical variables) and Mosaic plot (which shows the whole-to-part relationships for a number of categorical features) are shown in Figure 3.7.

For completion, allow me to showcase the software *Exploratory*. This is a *R*-based software that can be used to create ML models and one that we will be using in the next few chapters. *Exploratory* has full capabilities to visualize data. I will showcase a few examples of numerical and categorical data.[69]

67 Quick mechanics review. Long columns are likely to fail in buckling than in crushing and short columns are likely to crush than buckle.

68 Which we have seen earlier.

69 The data is available for download from my website and *Exploratory* from here [28]. You can download this example from https://exploratory.io/git/XfH4zib6KA/Exploratory__Examples__lkW8pVg3.git

(a) Using *Excel* (top: traditional, bottom: stacked).

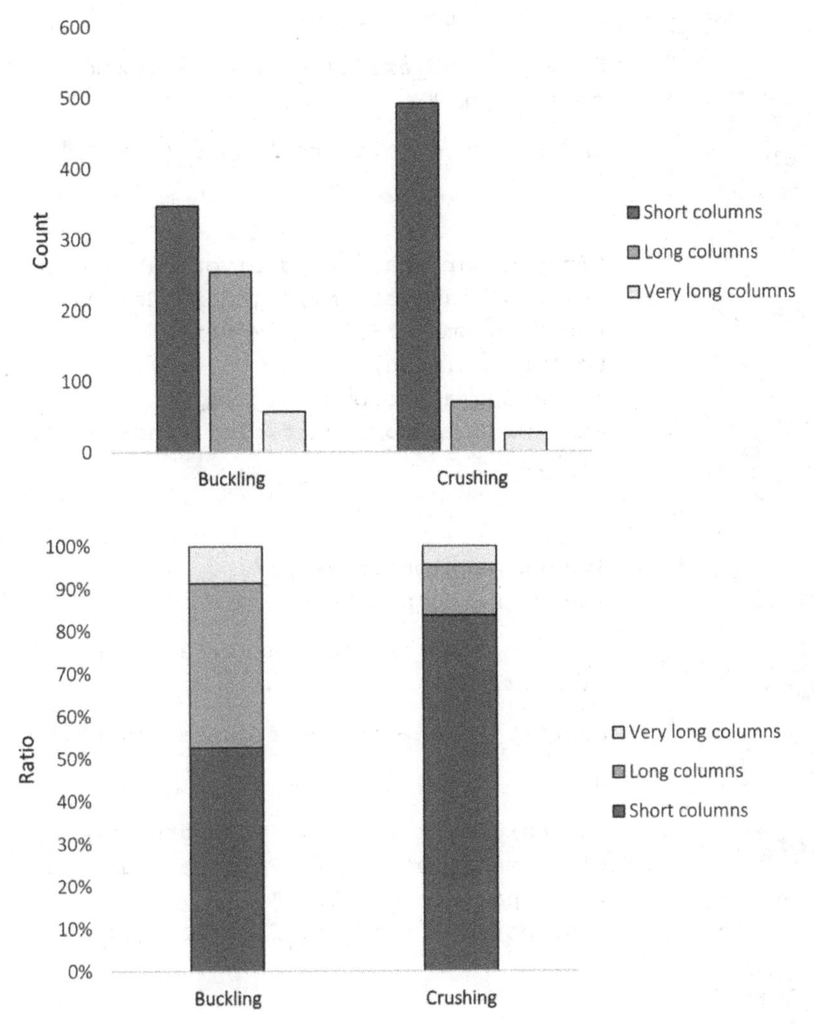

b)Using *Python*

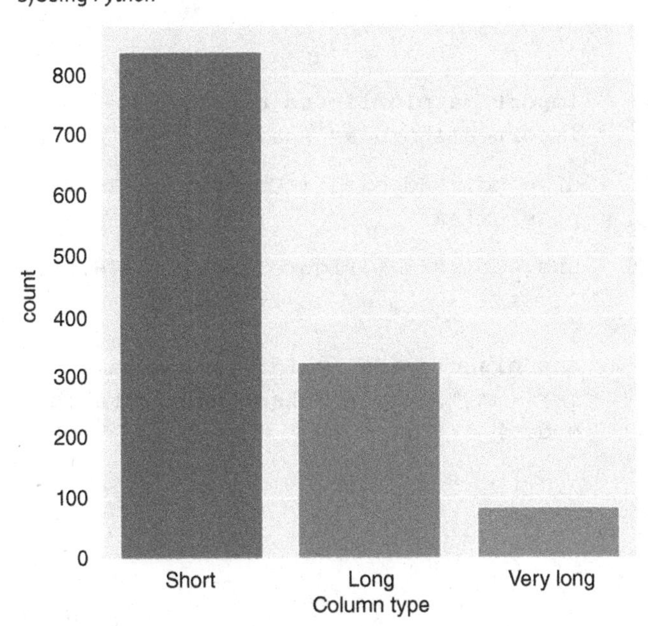


```
import matplotlib as plt
import seaborn as sns

df = pd.read_excel("Chapter 3 [Exam-
ples].xlsx")

sns.set(rc = {'figure.figsize':(6,6)})

df

sns.countplot(df['Column type'],)
```

(Continued)

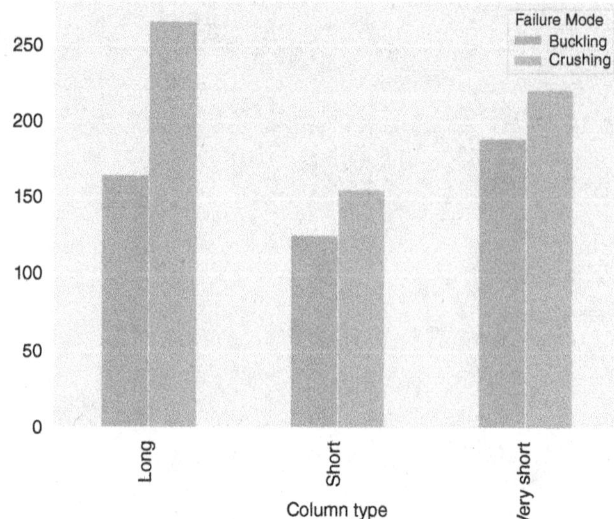

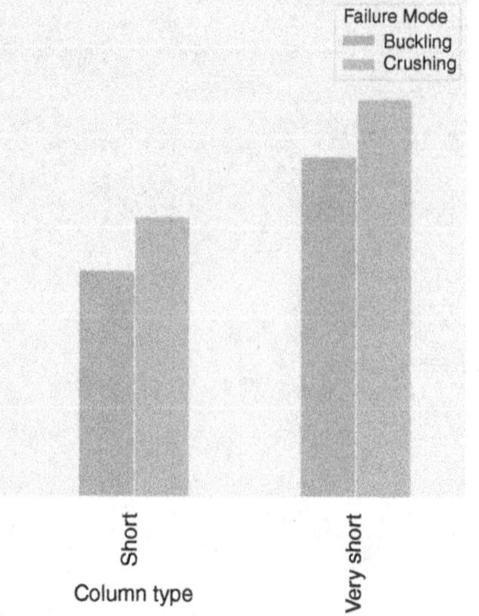

```
import matplotlib as plt
import seaborn as sns

df = pd.read_excel("Chapter 3 [Exam-
ples].xlsx")

sns.set(rc = {'figure.figsize':(6,6)})

df

Grouped_bar_chart = pd.pivot_table(df,
values="Diameter (mm)",index="Column
type",columns="Failure Mode",
aggfunc=np.mean)
#Creating a grouped bar chart
ax = Grouped_bar_chart.plot(kind="bar",
alpha=0.5)
```

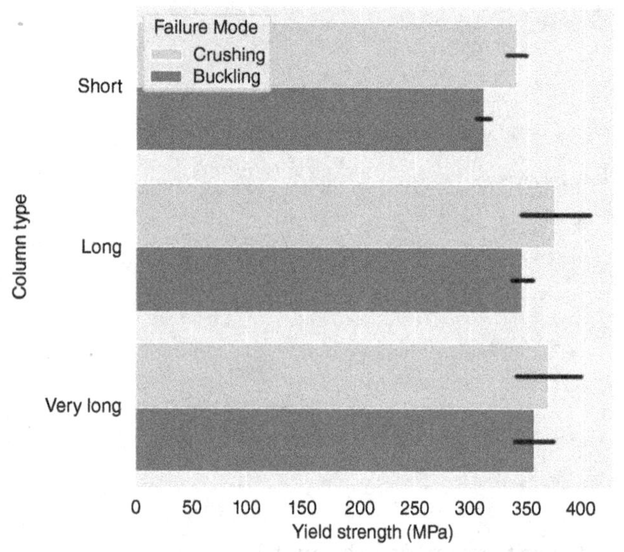


```
import matplotlib as plt
import seaborn as sns

df = pd.read_excel("Chapter 3 [Exam-
ples].xlsx")

sns.set(rc = {'figure.figsize':(6,6)})

df

sns.barplot(x = 'Yield strength (MPa)',y
='Column type',hue = 'Failure Mode',data
= df, palette = "Blues")
sns.set(rc = {'figure.figsize':(14,6)})
```

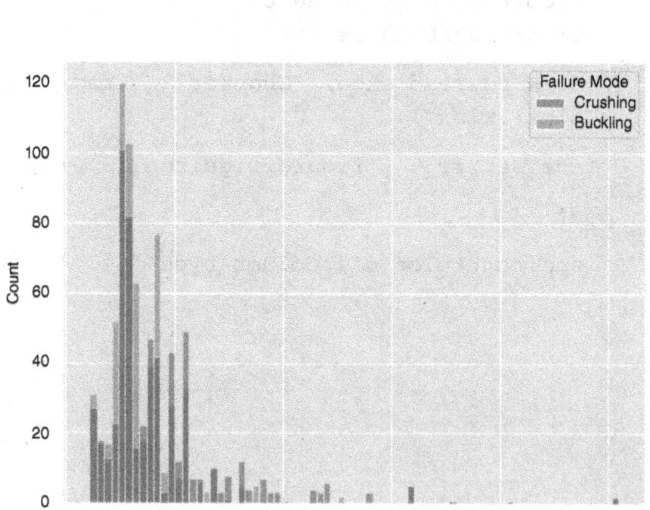


```
import matplotlib as plt
import seaborn as sns

df = pd.read_excel("Chapter 3 [Exam-
ples].xlsx")

sns.set(rc = {'figure.figsize':(6,6)})

df

sns.histplot(data = df, x = "Diame-
ter (mm)", kde = False, hue = "Failure
Mode")
```

(Continued)

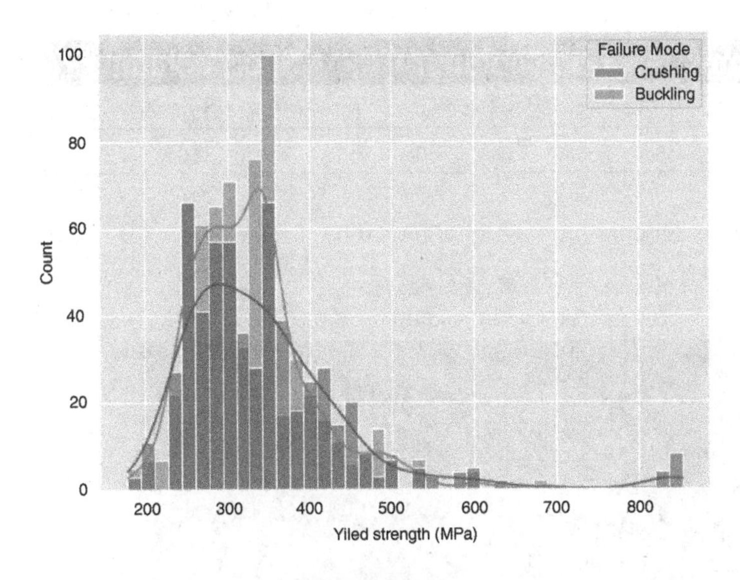


```
import matplotlib as plt
import seaborn as sns

df = pd.read_excel("Chapter 3 [Exam-
ples].xlsx")

sns.set(rc = {'figure.fig-
size':(6,6)})

df

sns.histplot(data = df, x = "Yield
strength (MPa)", kde = True, hue =
"Failure Mode")
```

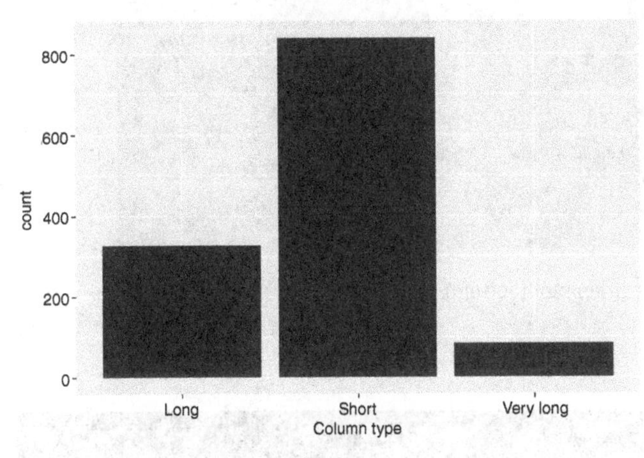


```
library(tidyverse)
library(ggplot2)

data=read.csv('C:/ Chapter 3 [Exam-
ples].xlsx')

data
str(data)

ggplot(data = data, aes(x =Column.
type)) +
  geom_bar()
```

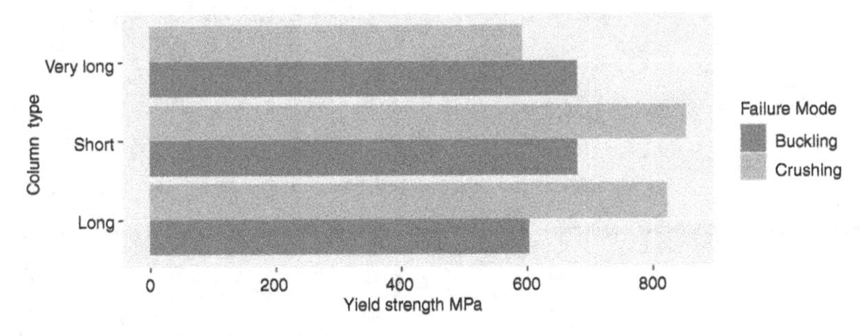


```
library(tidyverse)
library(ggplot2)

data=read.csv(Chapter 3 [Examples].xlsx')
data
str(data)

# Barplot
ggplot(data, aes(y=Column.type, x=Yield.strength..MPa.,

fill=Failure.Mode)) +
  geom_bar(stat = "identity",position = "dodge")
```

Figure 3.6 (Cont'd) Demonstration of bar charts.

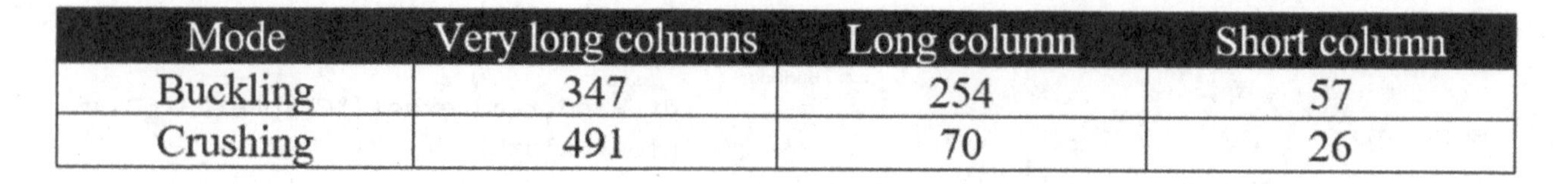

Mode	Very long columns	Long column	Short column
Buckling	347	254	57
Crushing	491	70	26

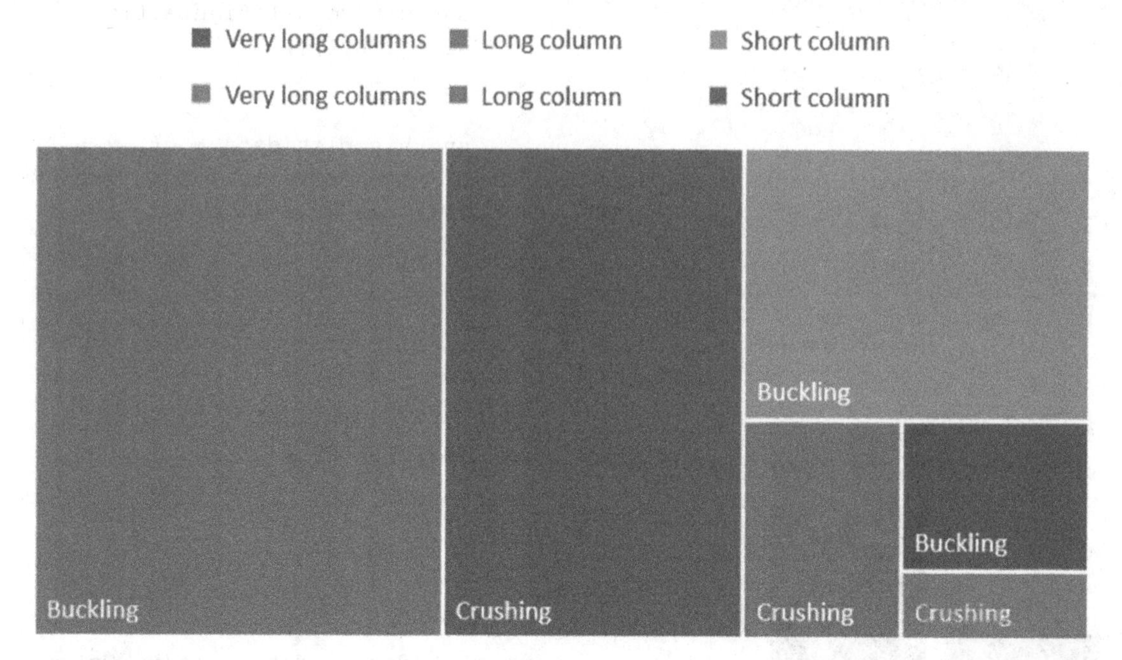

Figure 3.7 Demonstration of a contingency table (top) and a mosaic (tree) plot (bottom).

Step 1: Upload the dataset. Go to *Data Frames*, then *File data*.

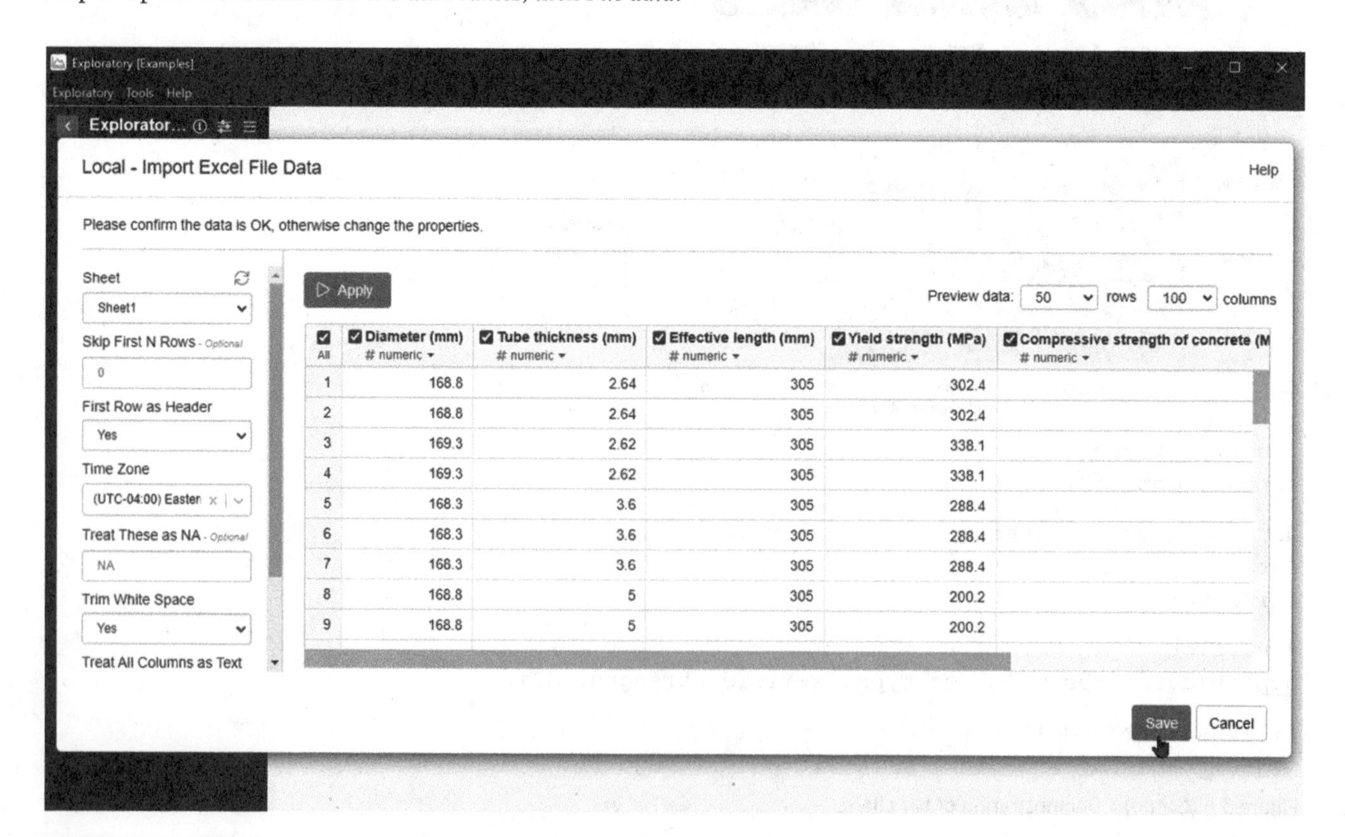

Step 2: Visit the *Summary* to examine statistical insights of each variable tab and then navigate to the *Charts* tab.

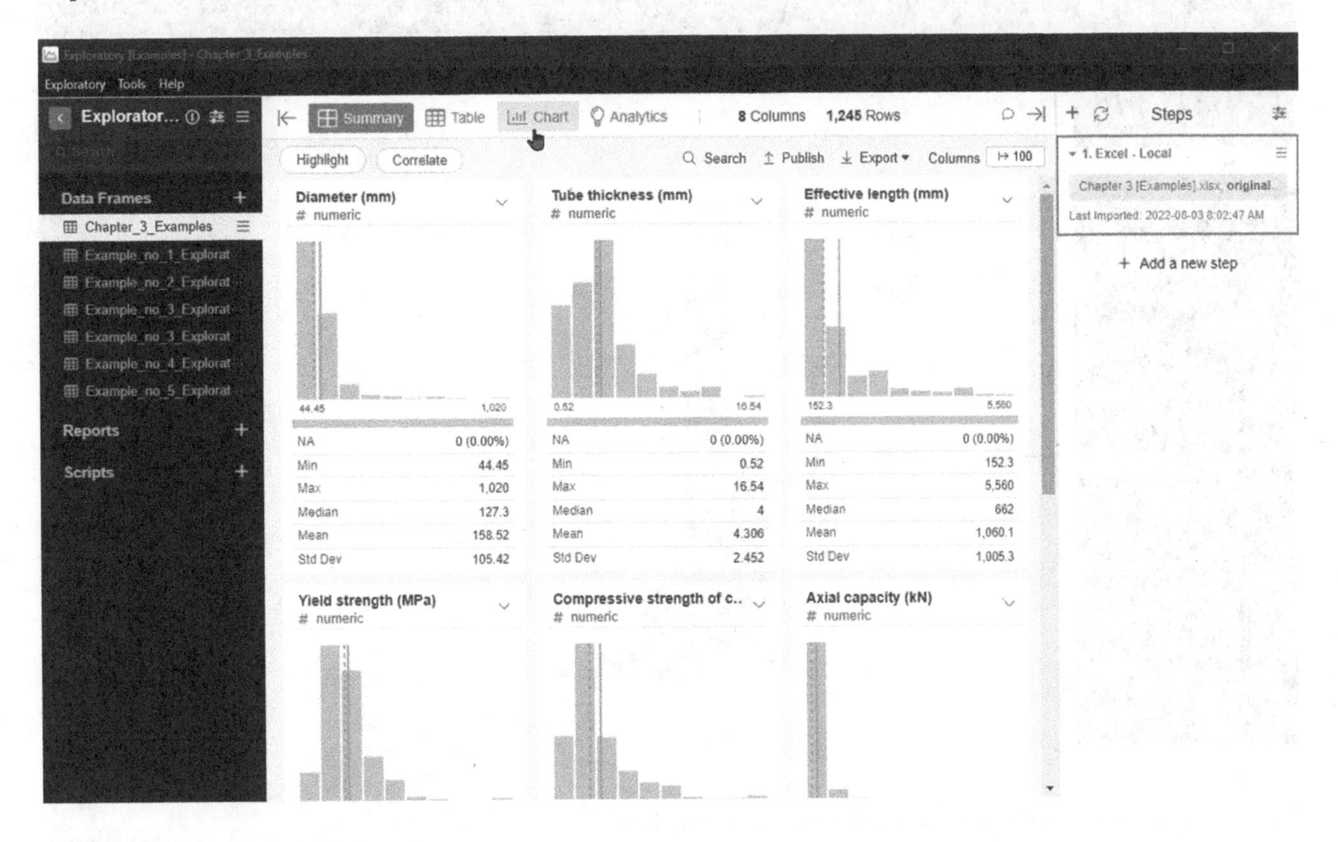

Step 3: Select *Histogram* and plot a feature.

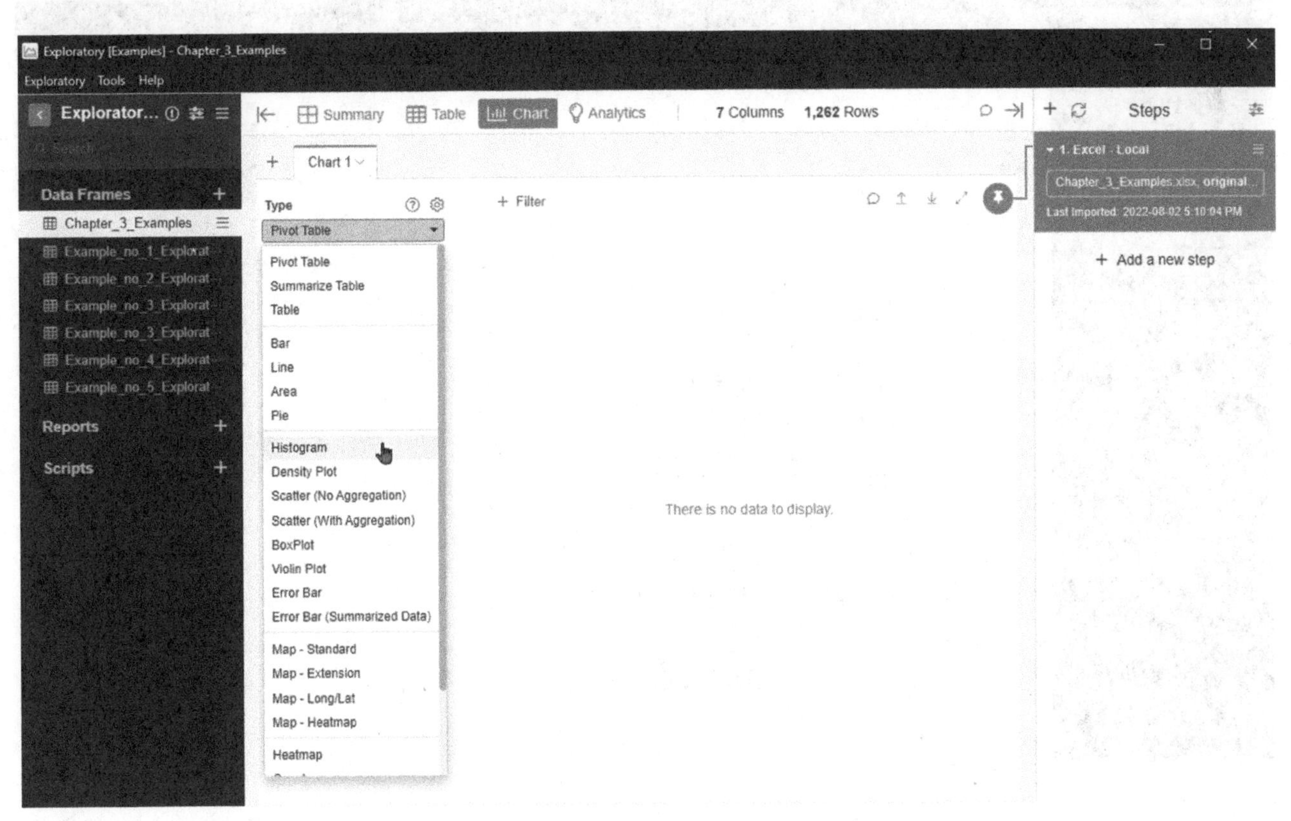

(Continued)

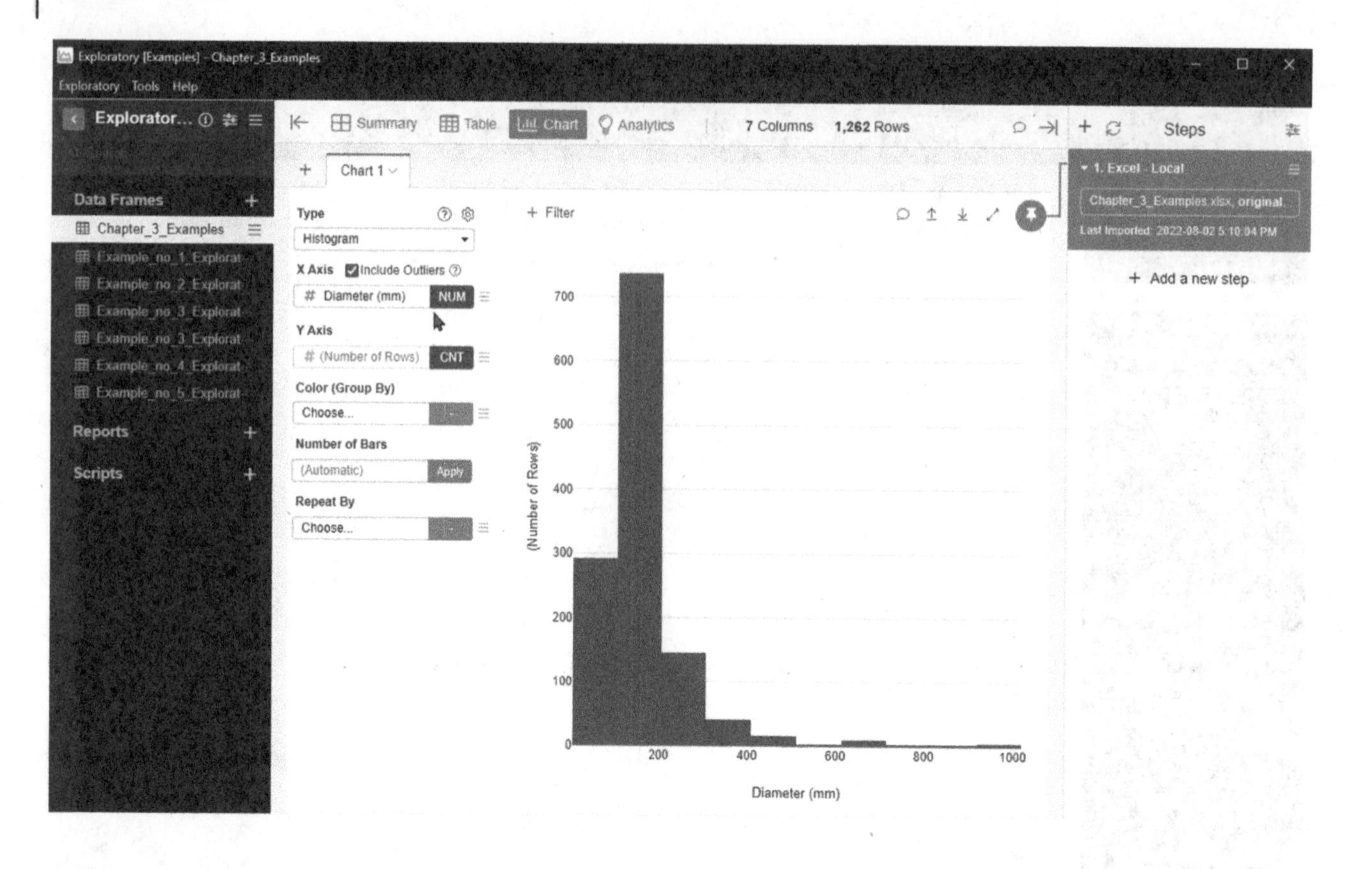

Step 4: You can select to color the histogram with the failure *Mode*.

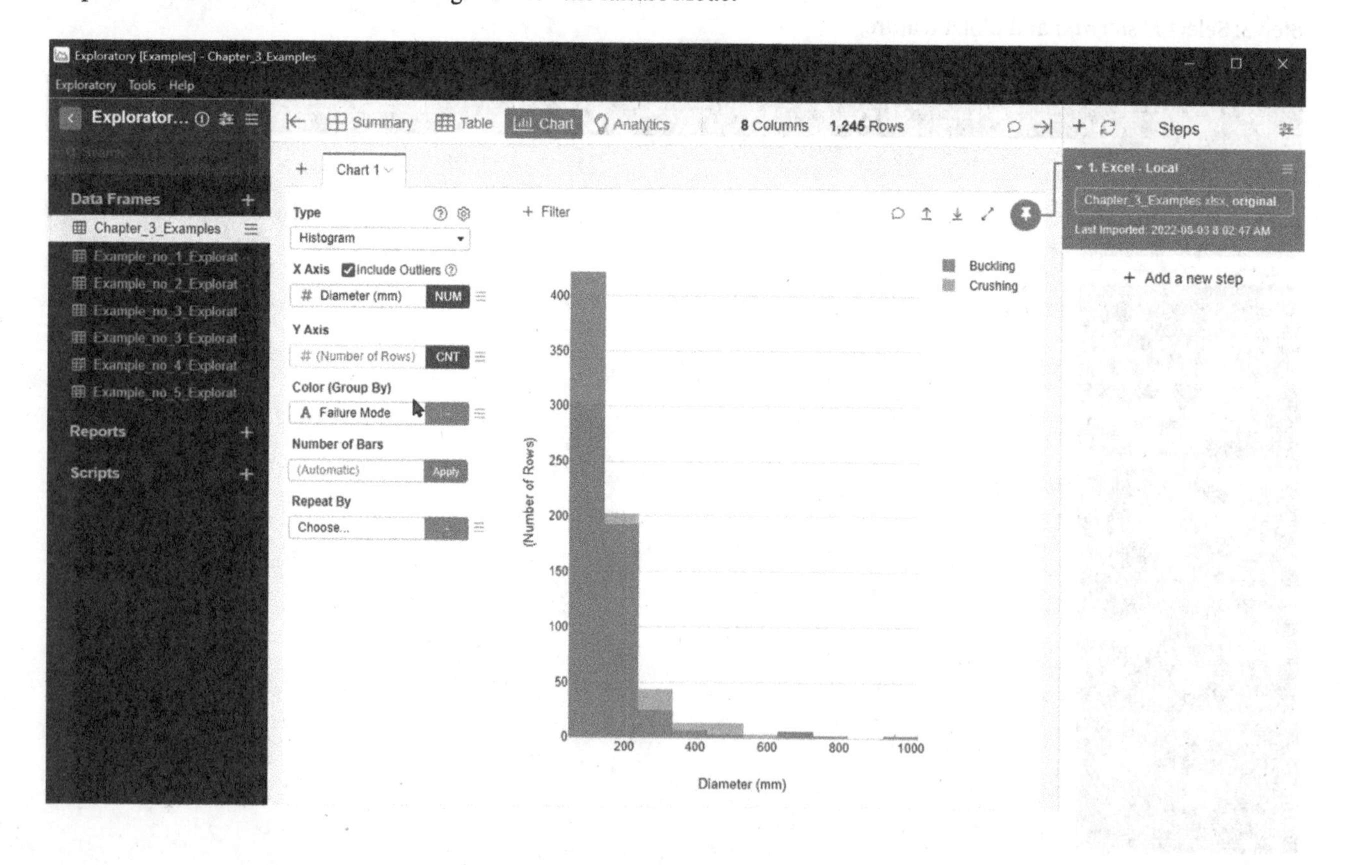

Step 5: Change the chart type into *Scatter*. Select a feature to plot on the horizontal axis and plot the axial capacity on the vertical axis.

Step 6: Format your chart as you can see below. Notice the *Size* effect.

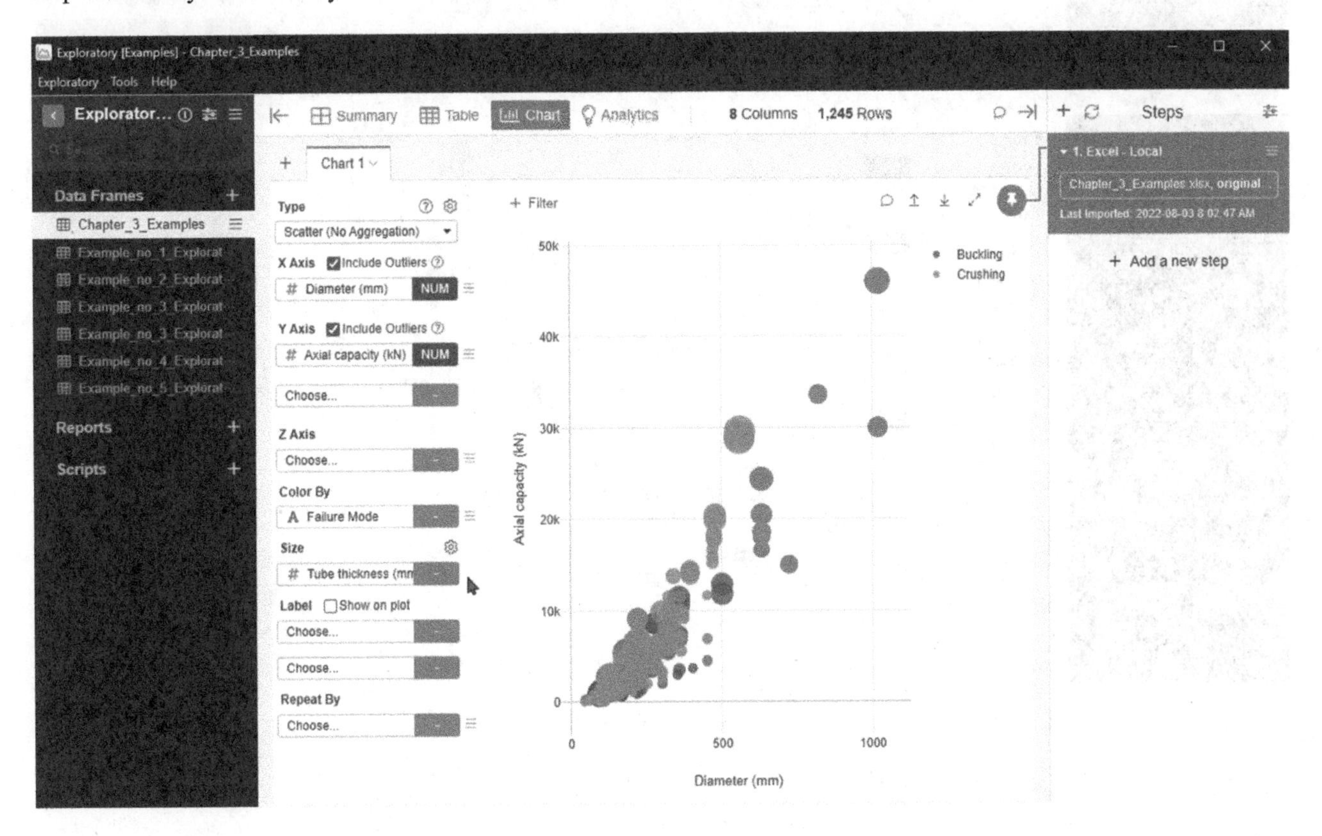

Step 7: You can repeat the plot for multiple cases.

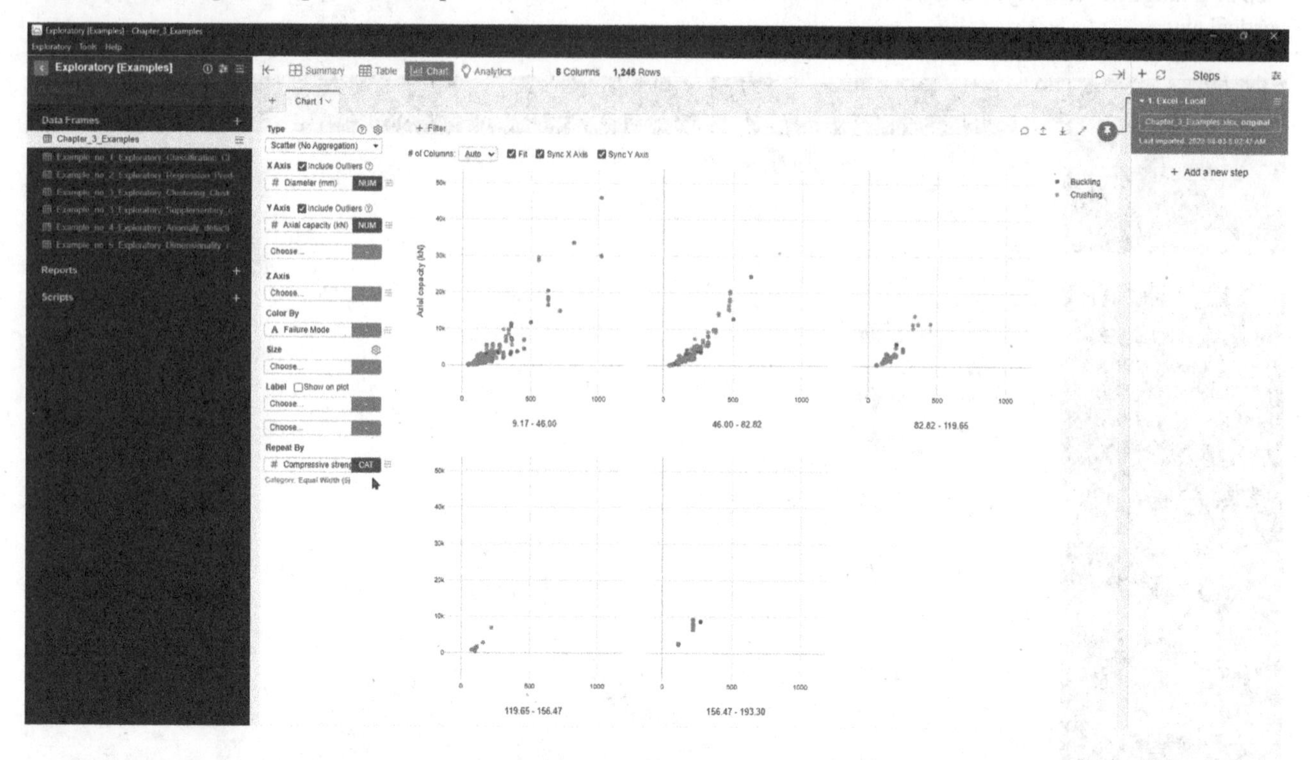

Step 8: Please try other plots. A variety is provided below.

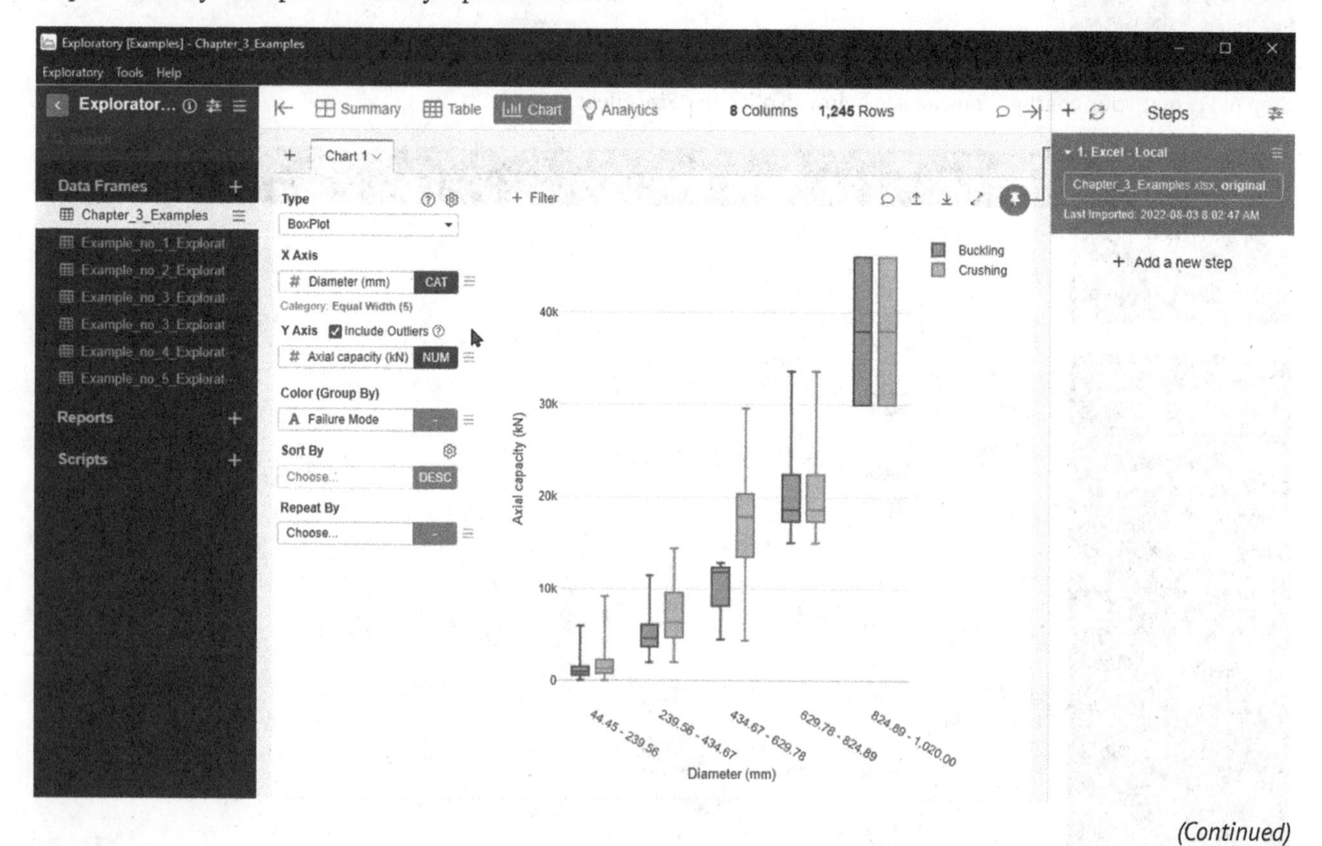

(Continued)

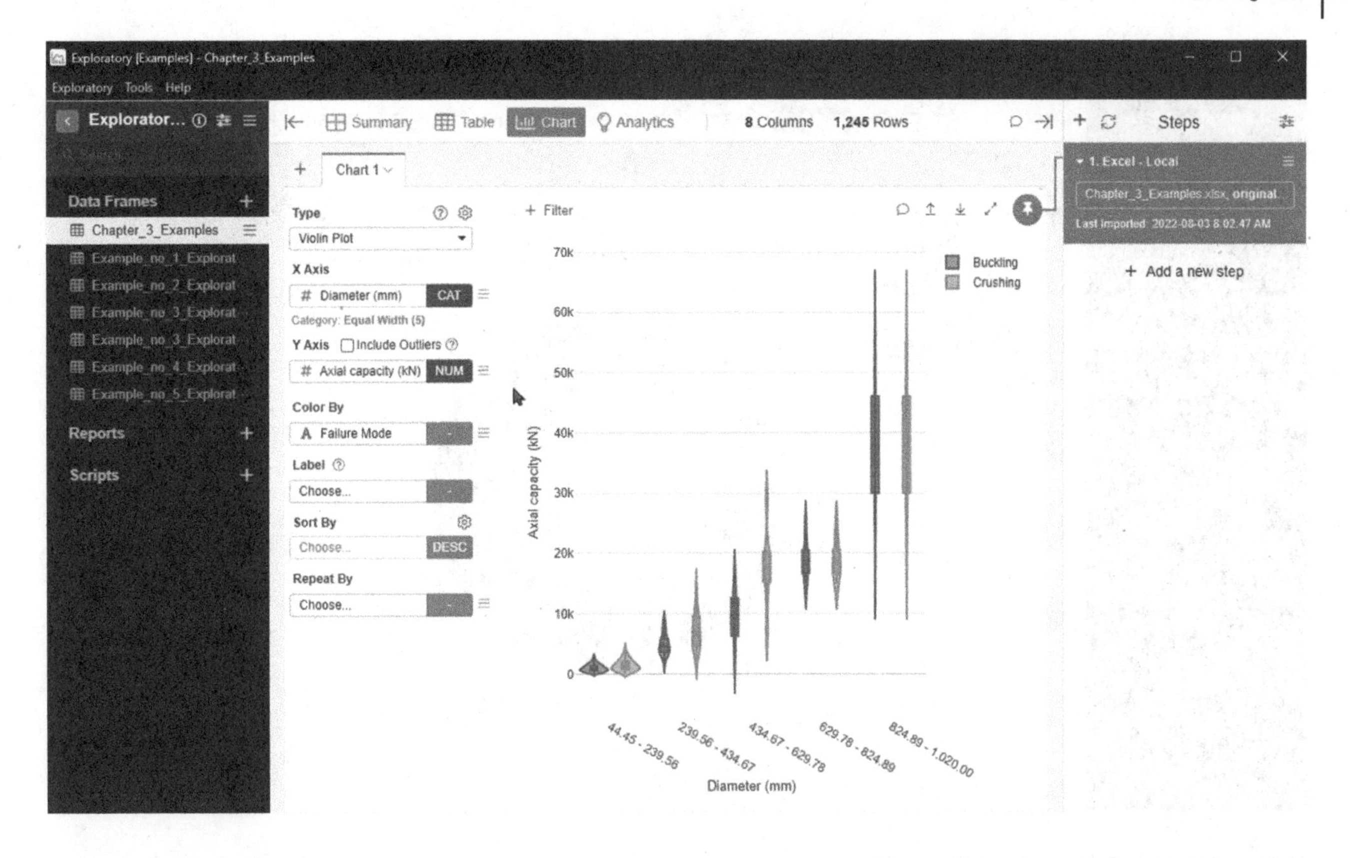

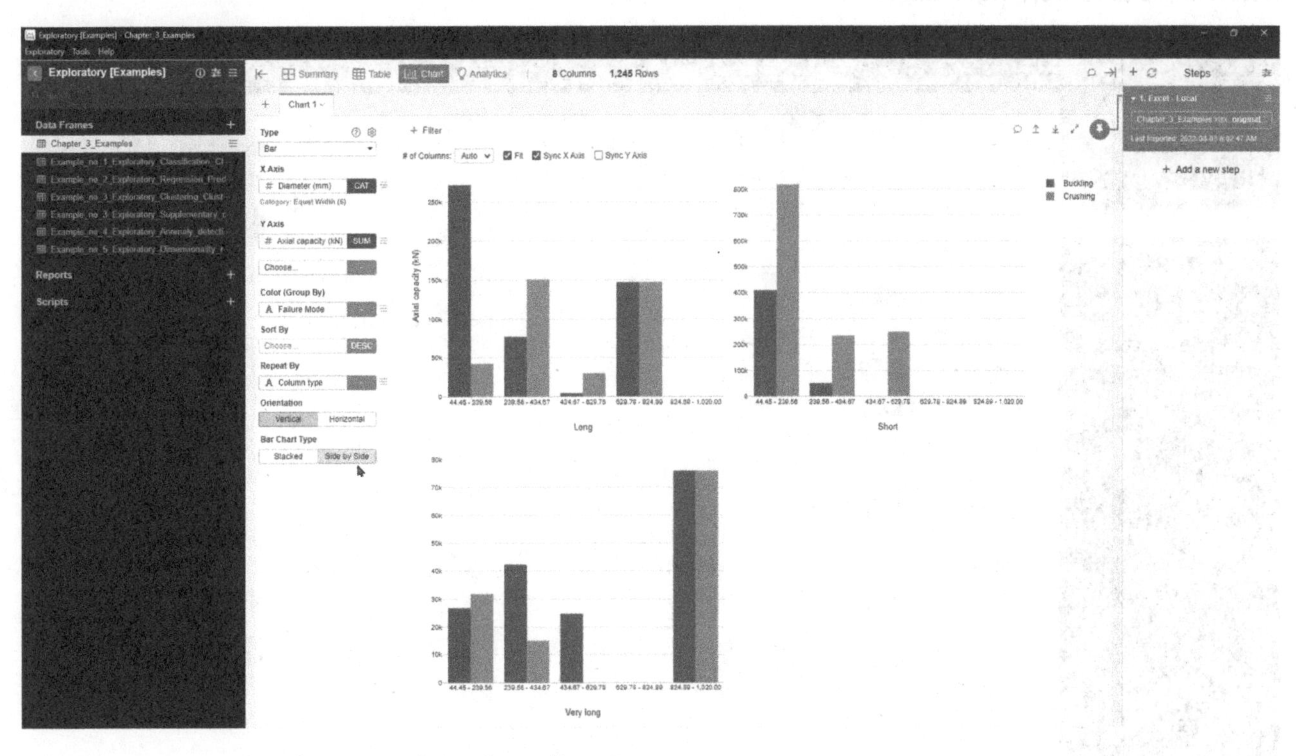

Note that you can update the *Orientation* and *Bar Chart Type*.

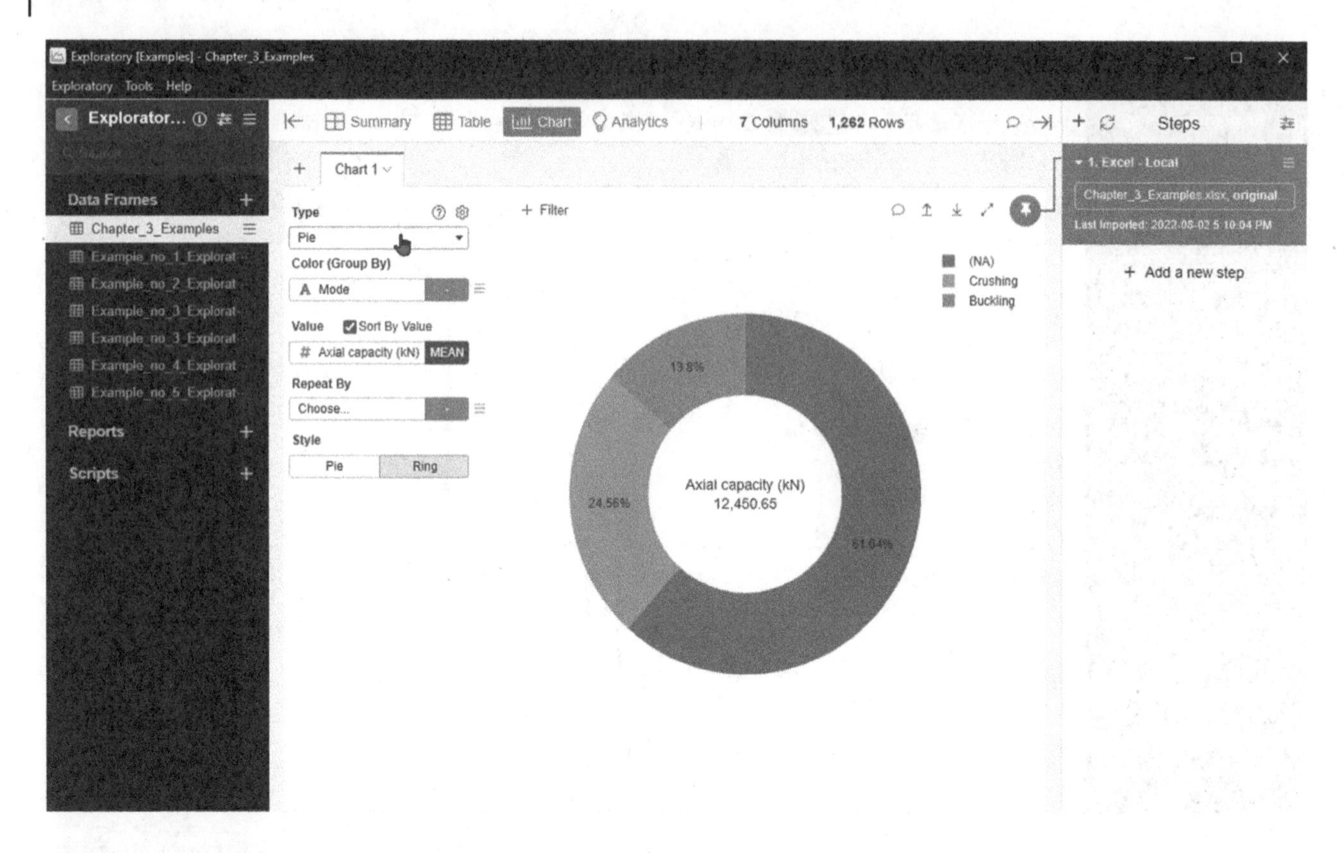

Note that you can select a *Pie* or *Ring* style.

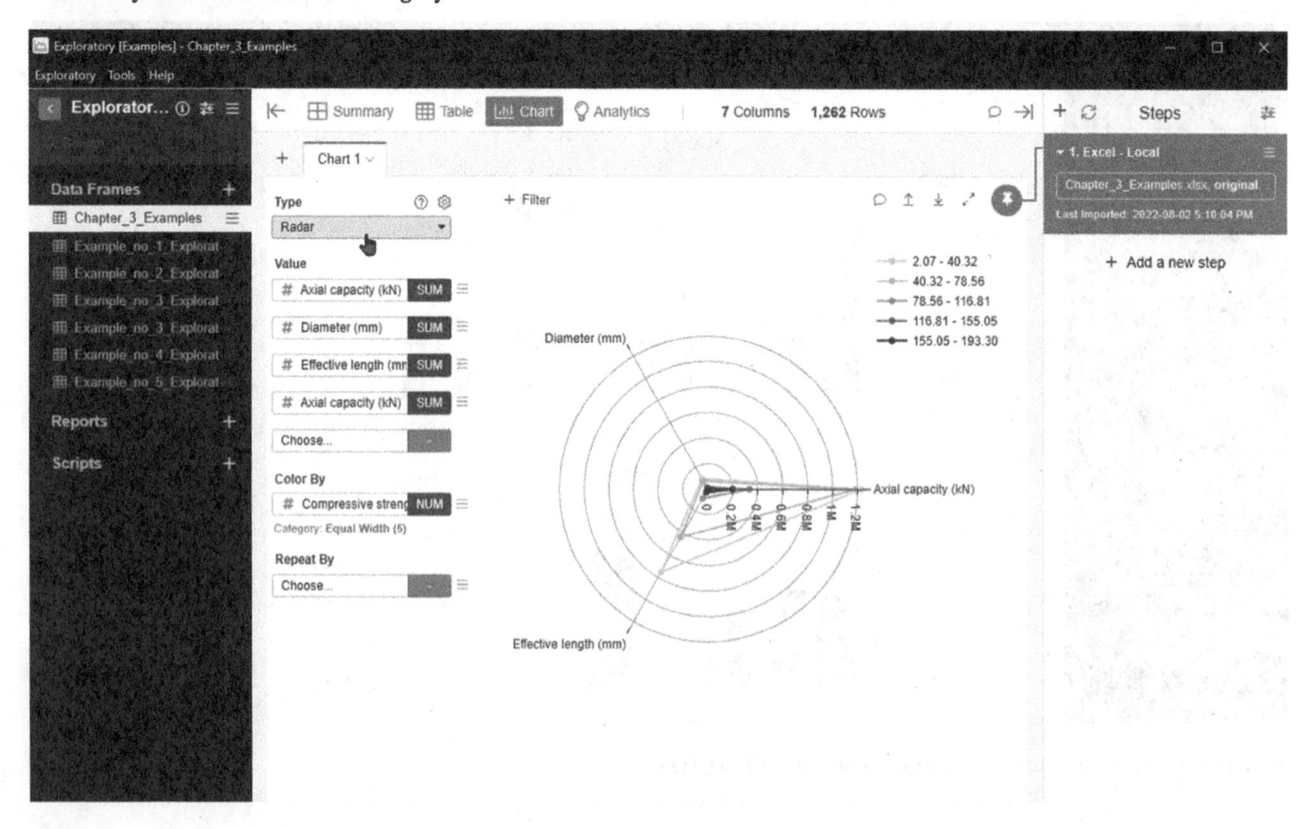

3.6 Exploring Data[70]

Once the data is visualized, we then turn gears to further explore this data. Our goal is to go past the above simple visualization tools and toward quantifying relationships[71] between features, as well as features and the target. There are a few techniques that we can use to explore relationships.

3.6.1 Correlation-based and Information-based Methods

Let us start with Pearson's correlation coefficient (r_{xy}).[72] This coefficient measures the *linear*[73] relationship between two numerical variables (say, x and y). This linear relationship is described in terms of strength and direction by:

$$r_{xy} = \frac{\sum_{i=1}^{n} (x_i - \mu_x)(y_i - \mu_y)}{\sigma_x \sigma_y} \tag{3.1}$$

This coefficient has a range from −1.0 to +1.0. The larger the absolute value of the Pearson correlation coefficient, the stronger the linear relationship between the two variables. A value of zero indicates no linear relationship. Positive and negative values to this coefficient indicate a linear relationship with a positive and negative slope, respectively.[74] For example, a Pearson correlation coefficient of +0.9 implies a positive correlation with a strength magnitude of 0.9.

This correlation can be calculated for any two variables as well as multiple variables. In the latter, the results can be fitted into a matrix. You can easily calculate Pearson's correlation coefficient using *Excel*, as you can see in Figure 3.8.

Step 1: Select the data. Navigate to the *Data* tab and then pick *Data Analysis*[75] and *Correlation*.

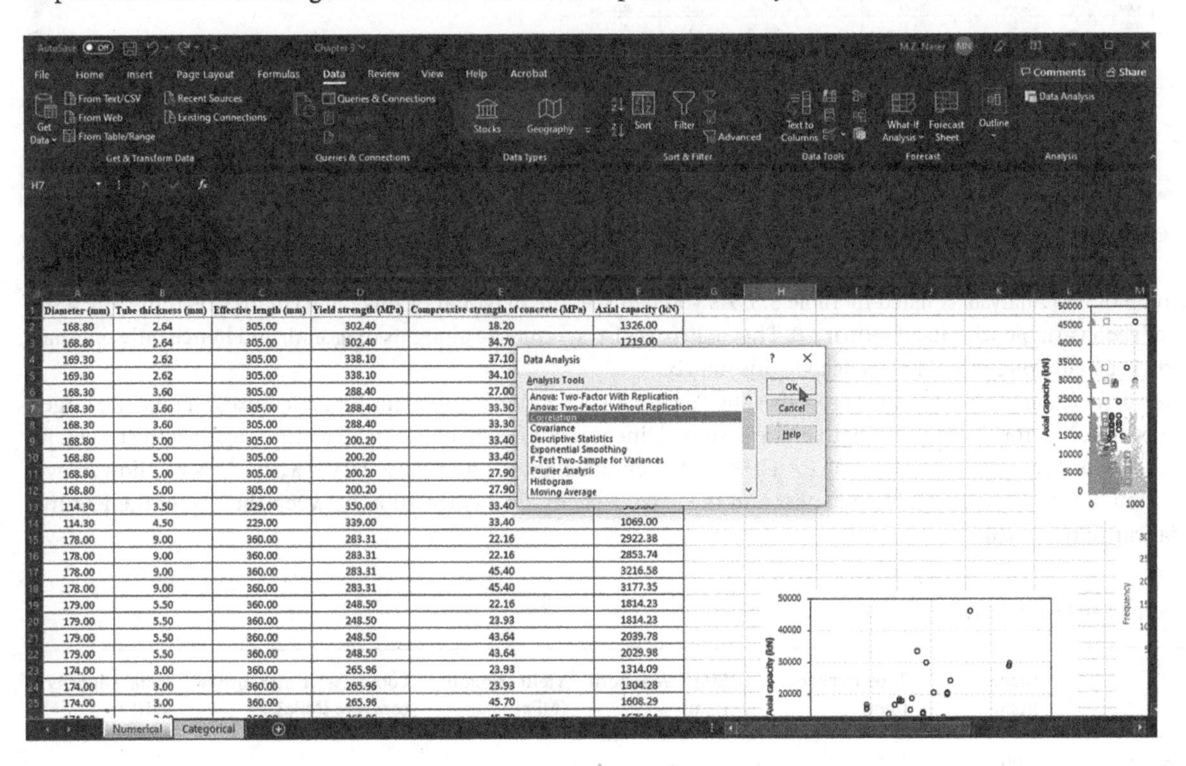

---

70 In some sources, this step could be merged with visualizing the data. I chose to present this as a separate section since I believe *visualizing* the data is a *qualitative* exercise and *exploring* the data is a *quantitative* exercise.

71 For example, if a scatter chart indicates the presence of a linear relationship between a feature and the target, here we are trying to estimate how linear is this relationship.

72 Or simply, R (not to be confused with the programming language, *R*).

73 While this may seem self-explanatory, I am afraid that it can be overlooked. This coefficient assumes that the relationship is linear and quantifies such a relationship. This does not mean that the relationship is indeed linear, nor that there is a linear relationship at all. Further, this does not also mean that one of the variables causes the other (this may or may not be true). In reality, a good portion of relationships are likely to have some form of nonlinearity. Be careful when interpreting such coefficients. *Additional note*: we will re-visit this coefficient and its usefulness as a performance metric in Chapter 5.

74 Up to 0.19 can be considered a very weak correlation. 0.2–0.39 weak correlation. 0.4–0.59 moderate correlation. 0.6–0.79 strong correlation and 0.8–1.0 as a very strong correlation.

75 If you cannot access this tool, then you may add it from the *Analysis ToolPak* under the *Excel Add-ins* within the *Manage* feature in *Excel's Options*.

Step 2: Choose the *Input Range*. Check the *Labels in first row box*. Choose a cell to *Output* the correlation matrix.

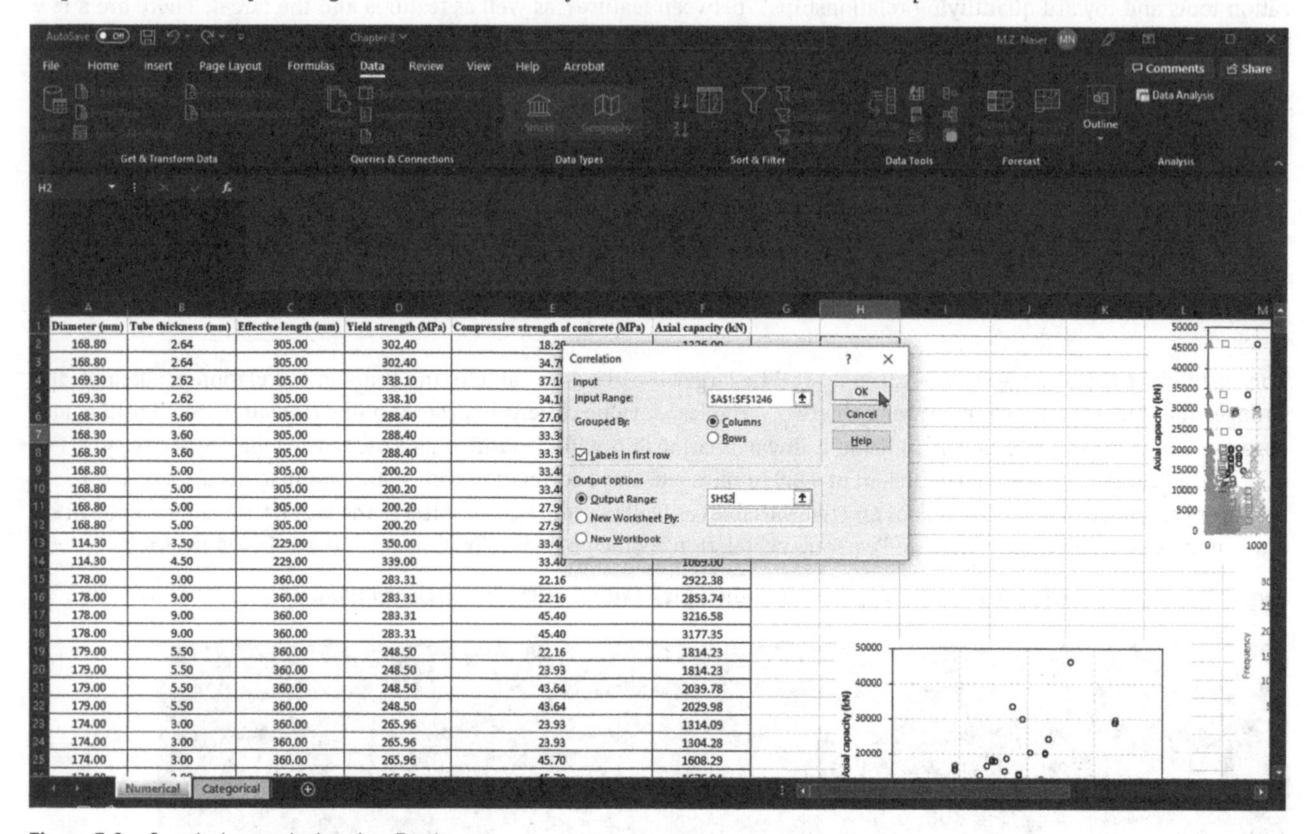

Figure 3.8 Correlation analysis using *Excel*.

The outcome of this analysis is listed in Table 3.3. As you can see, the diagonal of this table (matrix) is unity since the correlation of each variable with itself return 1.0. This matrix is also symmetric along its diagonal, and hence only one-half of this matrix is usually presented. Let us decode this matrix.

I would like to focus on the last row. This row lists the Pearson's correlation coefficients for all features against the target. To be more specific, this row lists the strength of the assumed linear relationship between all features against the target. As you can see, there is a very strong correlation (> +0.90) between the diameter and the axial capacity. This is followed by a moderate correlation between the tube thickness and the axial capacity. The correlation between the rest of the features and the axial capacity is weak.[76]

Table 3.3 Results of Pearson's correlation analysis.

Variable	Diameter (mm)	Tube thickness (mm)	Effective length (mm)	Yield strength (MPa)	Compressive strength of concrete (MPa)	Axial capacity (kN)
Diameter (mm)	1.00					
Tube thickness (mm)	0.48	1.00				
Effective length (mm)	0.20	0.22	1.00			
Yield strength (MPa)	0.07	0.24	0.08	1.00		
Compressive strength of concrete (MPa)	0.00	−0.02	−0.15	0.03	1.00	
Axial capacity (kN)	0.91	0.55	0.11	0.15	0.13	1.00

76 This information is helpful as it tells us that the linear correlation is weak. However, domain knowledge and mechanics note that there is some form of strong relationship between these features and axial strength. If this was a purely data-based problem where these features represent data (without any physical essence), then we might be able to justify dropping the features with low correlation.

Once you have examined the last row, now is a good time to explore the correlation between the features themselves. As you can see, the linear correlation between the features varies between weak and moderate. There seems to be a negative correlation between tube thickness and effective length and the compressive strength of concrete. In all cases, please remember that correlation (in any strength or magnitude) may not imply causation.

You can use *Exploratory*[77] to calculate and plot the Pearson correlation coefficient. Here is how:

Step 1: Navigate to the *Analytics* tab and then select *Correlation*.

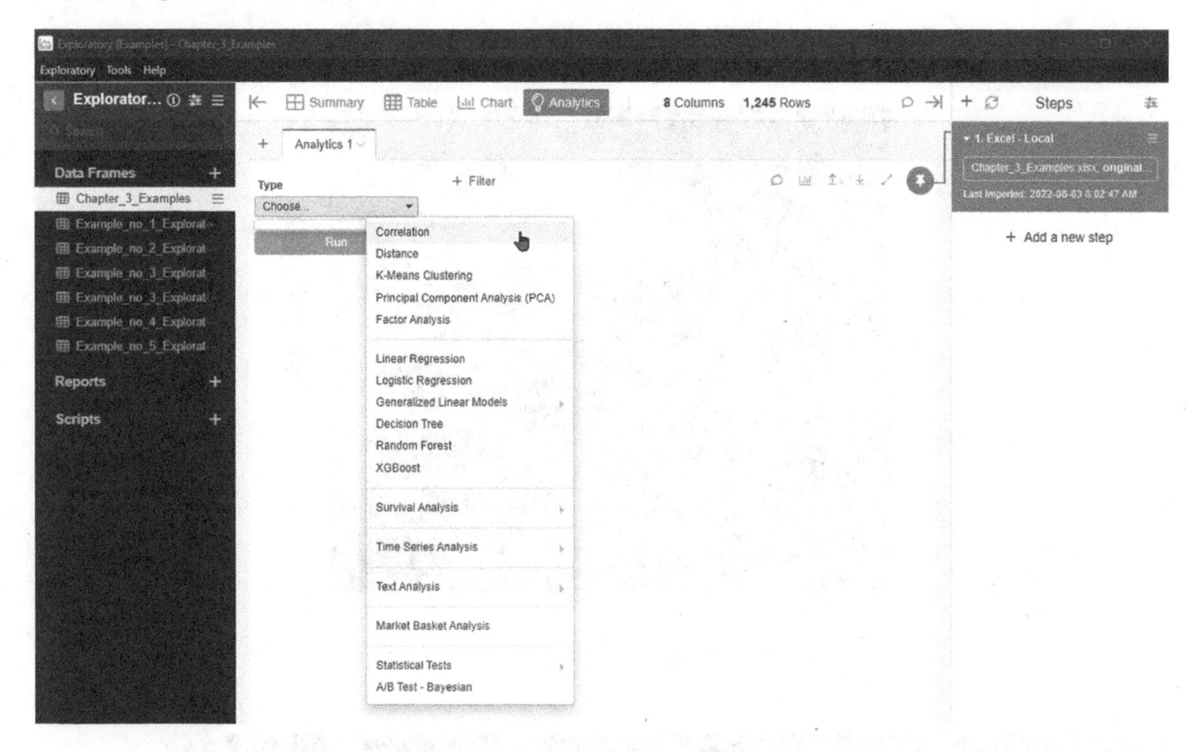

Step 2: Select the features and target. *Run* the analysis.

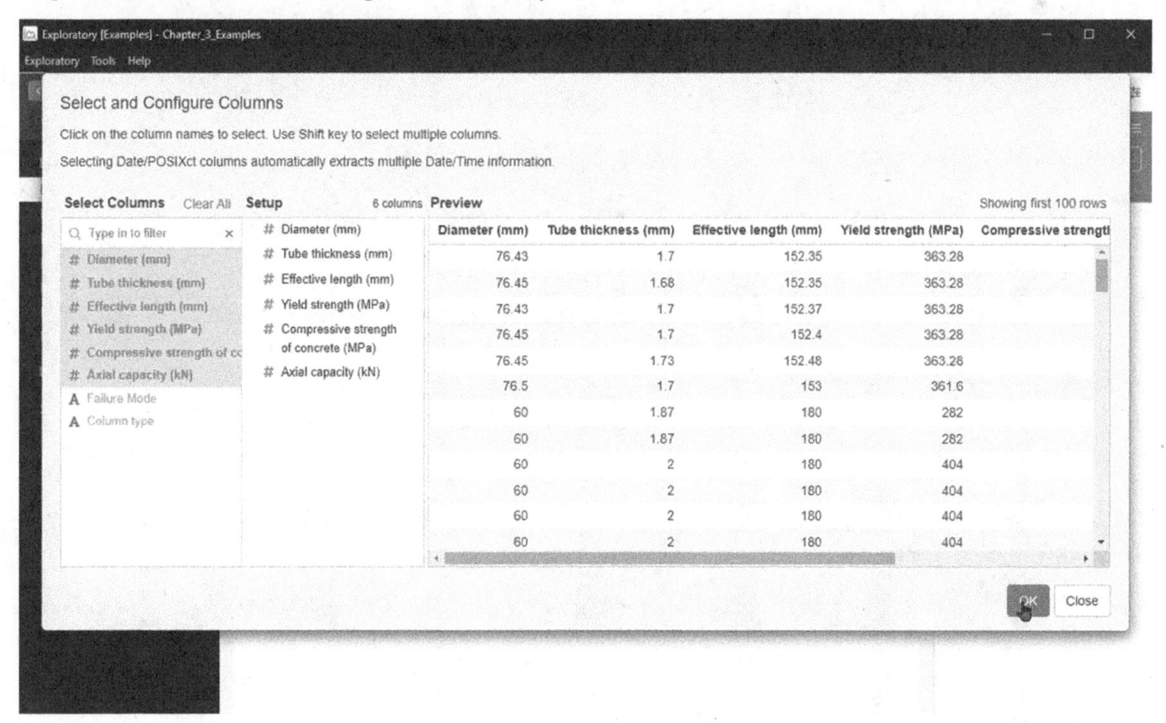

77 As well as other software such as *Excel*, or *Python*, *R*, etc.

Step 3: Visualize the results.

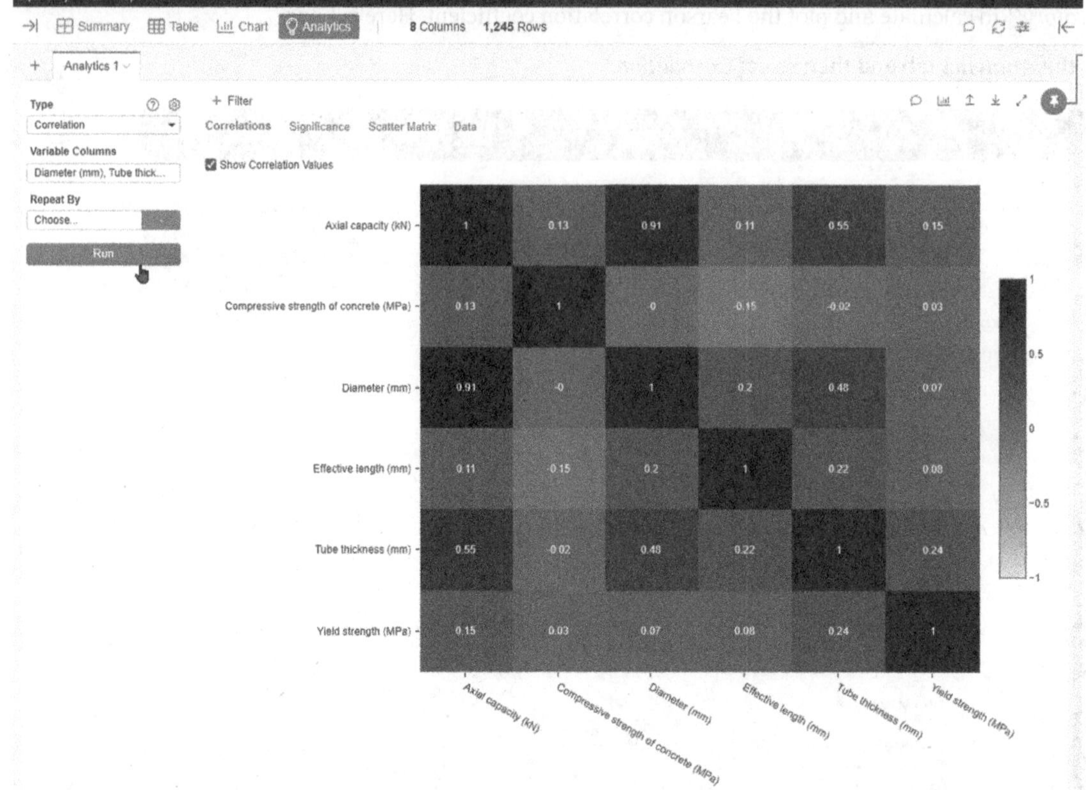

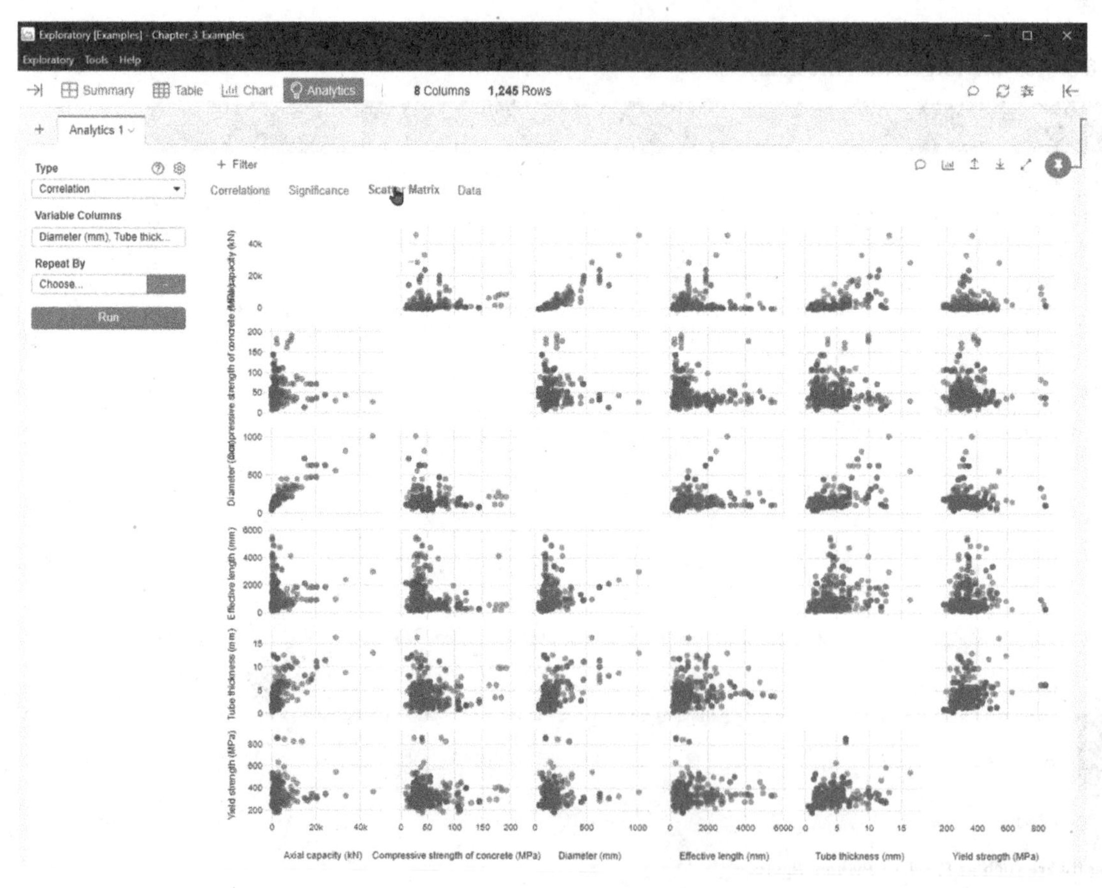

Table 3.4 Results of Spearman's correlation analysis.

Variable	Diameter (mm)	Tube thickness (mm)	Effective length (mm)	Yield strength (MPa)	Compressive strength of concrete (MPa)	Axial capacity (kN)
Diameter (mm)	1.00					
Tube thickness (mm)	0.53	1.00				
Effective length (mm)	−0.07	0.01	1.00			
Yield strength (MPa)	−0.40	−0.04	−0.20	1.00		
Compressive strength of concrete (MPa)	−0.35	−0.55	−0.63	−0.25	1.00	
Axial capacity (kN)	0.57	0.57	0.24	−0.34	−0.24	1.00

Pearson correlation coefficient is a neat and simple technique; however, it is limited to describing linear relationships. To overcome this limitation, we can use other measures, such as Spearman's rank correlation coefficient, which describes the *monotonic*[78] relationship between variables.

Similar to the Pearson correlation coefficient, Spearman's coefficient (ρ_{xy}) also ranges from −1.0 to +1.0. A positive coefficient means that as one variable increases, the other tends to increase (and vice versa for a negative coefficient). Unfortunately, the procedure to calculate Spearman's coefficient using *Excel* is lengthy, and hence I am providing the outcome in Table 3.4. The formula to calculate this coefficient is:

$$\rho_{xy} = 1 - 6\frac{\sum_{i=1}^{n}\left(\text{rank}(x_i) - \text{rank}(y_i)\right)^2}{n\left(n^2 - 1\right)} \tag{3.2}$$

where rank: ranking/ordering the data in ascending order.[79]

A look at Table 3.4 shows that the calculated coefficients are different from those in Table 3.3. However, it is clear that the diameter and tube thickness remain the largest two variables.

3.6.1.1 A Note on Pearson and Spearman Correlation Coefficients

It is worth noting to remember that Pearson correlation assumes features to be normally distributed and to satisfy homoscedasticity (i.e., equally distributed about the regression line). On the other hand, the Spearman correlation does not carry any assumptions with regard to the distribution of the data [1]. You can carry out a correlation analysis in *Python* or *R* using the following packages (`(x.corr(y, method='spearman')` and `x.corr(y, method='spearman'))` and `(cor(x, y, "pearson"), cor(x, y, "spearman"))`, respectively.

The third technique, that is independent of correlation, that we can use to explore the relationship within our dataset is that known as *mutual information* (MI).[80] Mutual information measures the nonlinear relations between two variables of any type and tells us the amount of information[81] one variable reveals about another variable.[82] This measure ranges from 0 to ∞ and does not depend on the functional form of the true relationship between variables. The mutual information between two variables can be calculated as:

78 Meaning that as one variable increases, the other variable either increases or decreases (not both). For example, data plotted in a pattern that resembles a happy or sad smile is not monotonic (nor linear). Data plotted in a pattern that resembles a straight lines at 45 degrees is monotonic (with strong linear correlation).

79 Additional notes: another difference between Pearson's coefficient and Spearman's coefficient is that the former works with raw data whereas the latter works with ranked data. Other forms of correlations include, Chatterjee correlation coefficient, etc.

80 In *Python*, from `sklearn.feature_selection import mutual_info_regression (from sklearn.feature_selection import mutual_info_classif as MIC)`. In *R*, `(mutinformation(X, Y, method="emp")` other "methods" can be found under [29]).

81 You can think of mutual information as the reduction in uncertainty about one variable given our knowledge of another variable. For example, there is a high mutual information between timber construction and corrosion; since timber is not known to corrode!

82 Mutual information between two discrete variables is calculated through their joint probabilities as estimated from the frequency of the number of observed samples in each combination of variable categories [30].

Table 3.5 Results of Spearman's correlation analysis.

Variable	Diameter (mm)	Tube thickness (mm)	Effective length (mm)	Yield strength (MPa)	Compressive strength of concrete (MPa)	Axial capacity (kN)
Diameter (mm)	0.84					
Tube thickness (mm)	0.36	0.43				
Effective length (mm)	0.32	0.22	0.03			
Yield strength (MPa)	0.36	0.36	0.24	0.10		
Compressive strength of concrete (MPa)	0.84	0.20	0.20	0.26	0.01	
Axial capacity (kN)	0.35	0.24	0.22	0.23	0.17	1.00

$$\mathrm{MI}(X,Y) = E\left(\ln \frac{p\,(x,y)}{p\,(x)\,p\,(y)}\right) = \sum_{x,y} \mathrm{p}(x,y)\left[\ln p(x,y) - \ln p(x)p(y)\right] \tag{3.3}$$

The mutual information creates a weighted sum of the joint probabilities of variables. Table 3.5 presents the scores for our example.[83] Unlike the other correlation-based coefficients, the diagonal of this matrix may return a value other than unity since mutual information of a variable with itself equals the entropy[84] within this variable.

The scores listed in Table 3.5 are quite different from those in Tables 3.3 and 3.4. This is a testament to the expected non-linearity of the relationship between the variables involved. Not surprisingly, the last row indicates that knowing information on the features may not reveal much information about the axial capacity of CFSTs.

3.6.2 Feature Selection and Extraction Methods

Another key goal of exploring the data is to identify the features of high merit/fidelity.[85] This goal hopes to minimize the number of features to be incorporated into a ML analysis without compromising the essence of the problem on hand or the predictive performance of the ML model.

Feature selection builds upon the Law of Parsimony of *Occam's Razor*, which often translates to "*Entities should not be multiplied beyond necessity*." This law strives for simplicity and establishes that the best explanations are those that involve the fewest possible assumptions [2]. Therefore, a ML model avoids overfitting, provides faster and more cost-effective models, allows debugging, and allows a deeper insight into the underlying processes that generated the observations – all of which indicate an improved performance.

Practically, in the space of a dataset, some features can be classified as *relevant* (i.e., strongly, weakly, or irrelevant) and/or can be classified per their *redundancy* as redundant/not redundant. Overall, engineers may opt to adopt feature selection and feature extraction techniques. Features[86] can be selected through three methods, *filter*, *wrapper*, or *embedded methods* [3]. Let us go over each of these methods herein.

83 Please note that we will seek to use coding-free and coding-based to calculate all of the above techniques in the next few chapters. For now, our goal is to see how we can best use the outcome of such techniques in exploring our dataset.

84 **Entropy**: A characterization of the unpredictability of a variable.

85 I would like to pause here for a minute. The next sub-section will cover feature selection techniques. In a pure ML pipeline features are often selected once a process called *feature engineering* is completed. In this process, the features are manipulated such that the ML model yield high predictivity. I am of the opinion that feature engineering fits best under the next section which is titled *manipulating data* for three reasons: 1) engineering features is a manipulation process in which we alter features, 2) we ought to select the core features first before we engineer the input features, and 3) when features are heavily manipulated/engineered, they may lose some of the physics behind them as we are turning the to-be-developed ML into a prediction box with little insights into the actual physics. In all cases, you can opt to examine all collected features, engineer some and then select the final features, or you may directly select the features first. My advice is to try both procedures and see which one best fits your problem and data. The pipeline to ML is flexible.

86 Additional information with regard to the history and the background of feature selection methods can be found in [5, 9, 31–33].

Table 3.6 Common filtering methods.

	Features	
Target	**Continuous**	**Categorial**
Continuous	• Pearson's correlation coefficient • Spearman's rank coefficient • Mutual information • F-test • Neighborhood component analysis • Sequential feature selection	• Mutual information • ANOVA correlation coefficient • Kendall's rank coefficient • Linear discriminant analysis • F-test • Sequential feature selection
Categorial	• Mutual information • One Way ANOVA • Kendall's rank coefficient • Minimum Redundancy Maximum Relevance • Neighborhood component analysis	• Contingency table • Chi-Squared test • Mutual Information • Minimum Redundancy Maximum Relevance • Neighborhood component analysis

3.6.2.1 Filter Methods

Filter methods select features according to their relationship with the target by means of statistical analysis, feature importance methods, and prior domain knowledge.[87] These methods operate prior to the ML analysis and hence reduce (or filter) the number of features to be used in the analysis. In most cases, a relevance[88] score is calculated for all features, and features with low scores are removed.

Table 3.6 lists a number of filtering techniques. These techniques often follow the types of inputs and targets (i.e., whether numerical or categorial). For regression problems, correlation-based methods can be applied, such as Pearson's correlation coefficient, Spearman's rank coefficient, etc. On the other hand, for classification problems, the following techniques can be used: Kendall's rank coefficient and Chi-Squared test. Some methods, such as mutual information, can be used for regression or classification problems.

The advantages of filter methods include 1) computational simplicity and efficiency, 2) independence of the ML analysis and algorithm, 3) performed only once and prior to the start of the ML analysis, and 4) effective at capturing global trends. On the other hand, the same methods are associated with user preference[89] and tend to disregard feature dependencies and interaction with the target.[90]

3.6.2.2 Wrapper Methods

Wrappers, in contrast to filters, look for features that allow a model to perform well by comparing every conceivable feature combination to an assessment criterion within a specific ML algorithm [4]. In order to find a mix of features that maximizes the performance of ML, wrapper search algorithms add or remove features.[91] Despite their excellence and feature dependency accounting, their wide search space may be expensive.

There are three primary techniques for wrapper methods: forward selection, backward selection, and stepwise selection. In the first technique, the analysis starts with a null set of features which expands by iteratively adding relative features if the model's performance continues to improve. On the contrary, backward selection starts with all the features and then

87 Here is where your engineering knowledge/practice and codal provision becomes elemental!

88 For example, an engineer may decide that a mutual information of less than 0.1 to be of low relevance. Hence, features with such value (and lower) are removed from the pool of features. An engineer may also opt to pick one of two highly correlated features (instead of including both features in their model).

89 An engineer must select a confidence level or relevance metric to be applied during the filtering process. Since such metrics are not standardized, then the ML analysis can suffer from user bias.

90 For completion, solutions to the aforenoted problems exist (see [5, 9]).

91 Also, wrappers may work on smaller sub-sets of data vs. full dataset.

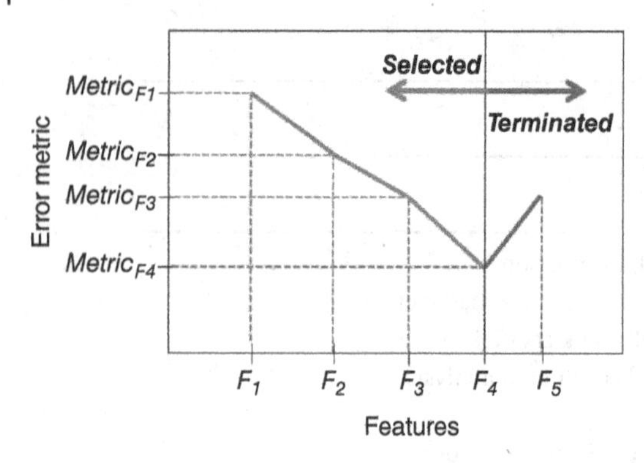

Figure 3.9 Demonstration of RFE algorithm.

removes the worst-performing features. This selection procedure ends once no improvement is observed by the removal of new features [5].[92] If a characteristic's relevance is determined to be minimal during a stepwise selection, the newly added feature is eliminated similarly to how it would be in a backward selection.

Wrapper techniques can choose features using greedy or non-greedy algorithms. In the former, a feature search path always goes in the direction that at the time of the iteration appears to be conducive to implementing a solution. If it seems to have a possible benefit, non-greedy techniques are willing to dive into an undesirable path for space search [6].

Regression and classification problems can be solved with popular wrapper techniques like Simulated annealing (SA) [7] and Recursive feature elimination (RFE) [8]. Figure 3.9 demonstrates the RFE algorithm, where F_1 is the first appended feature with the corresponding $Metric_{F1}$, F_2 is the second appended feature with the corresponding $Metric_{F2}$, and so on. During the analysis, F_1 is randomly selected, and its corresponding $Metric_{F1}$ is calculated. Then, F_2 is added to the model such that a new metric ($Metric_{F2}$) is calculated. This process continues until a feature (F_5) is added in which its corresponding $Metric_{F5}$ is larger than the previous metric $Metric_{F4}$. At this point, only the features that caused a decrease in errors are selected.

3.6.2.3 Embedded (Intrinsic) Methods

Embedded methods select features during the ML model training process, and hence they have a built-in capability to identify features of merit via the implementation of regularizers (L1, L2, etc.), constraints, or objective functions. These methods account for feature interactions, and since they are embedded within the model development process, they may require less computational resources. However, embedded methods are specific to the used algorithm, which may cause algorithmic bias. Commonly used embedded methods include tree-based models and regularized regressions (which include a penalty: to reduce overfitting or to features that do not contribute to the target variable, etc.) [9].

3.6.3 Dimensionality Reduction

When a significantly large number of features exist, then it is possible to condense their space and eliminate redundant features or those with minimal influence[93] [10]. This can be completed via adopting feature extraction techniques that rely on applying feature reduction (or dimensionality), such as Linear Discriminant Analysis (LDA) or Principal Component Analysis (PCA) [94] [11].

3.7 Manipulating Data

In this stage, we look at the trimmed[95] dataset we have and try to further[96] tune it (if needed/applicable). Tuning our dataset, specifically, the features within this set, is what is referred to as *feature engineering*. Feature engineering is the process of manipulating the features to realize new features that happen to be *numerically* (and may not be physically meaningful) more effective in predicting the phenomenon on hand than the original features. We may also require such data manipulation as a precursor to applying some ML algorithms. We have two main approaches to manipulating data, pending a regression or classification analysis.

---

92 Forward selection, backward selection fall under sequential feature selection methods.

93 This is important since larger numbers of features consume processing time and clutter our model. Most notably, visualization becomes challenging.

94 Examples on those are provided in our discussion of PCA in Chapter 6. Here is the big picture, feature selection selects or excludes features without changing them. On the other hand, dimensionality reduction transforms the features into a lower dimension (i.e., changes features).

95 Our dataset can be trimmed due to the removal of missing data, or features of low relevance.

96 As such, this stage is often optional.

Table 3.7 Normalization vs. standardization.

Normalization	Standardization
Applicable when features do not follow a Gaussian/normal distribution.	Applicable when features distribution is Gaussian/normal.
Sensitive to outliers.	Less sensitive to outliers.
Bounded by [0, +1] or [−1, +1]	Not bounded.
Python package: `scaler = MinMaxScaler()`	*Python* package: `scaler = StandardScaler()`
R package: `normalr`	*R* package: `standardize`

3.7.1 Manipulating Numerical Data

Numerical data comes with varying units and scales. For example, the length of CFST columns listed in Table 3.2 is in millimeters. This length could have been listed in terms of meters.[97] If so, the value of this particular feature would be small as compared to the diameter.[98] Comparatively, a ML model may deduce that this feature has little impact, if at all, on the target value.[99] It is then advisable to take the units and physics of the phenomena into account when manipulating data.

Looking again at Table 3.2 shows that the features have two different units, mm and MPa. These are units that represent length and stress. These units are not complementary. A value of 10 mm cannot be compared with 10 MPa. Thus, one workaround is to scale (or normalize) the data.[100] Normalization transforms features to be on a similar scale between 0 and +1.0 or −1.0 and +1.0. A simple approach to normalizing data is to use the following equation:

$$\text{Feature}_x = \frac{x - x_{min}}{x_{max} - x_{min}} \tag{3.4}$$

where x_{min} and x_{max} are the minimum and the maximum values of the feature, respectively.[101]
Another workaround is to standardize[102] the data to center the data around the mean of zero with a unit standard deviation.

$$\text{Feature}_x = \frac{x - \text{Mean}}{\text{Standard deviation}} \tag{3.5}$$

A comparison between normalization and standardization is listed in Table 3.7 for completion.

We can also combine features together. For example, we could merge two features into one through a mathematical procedure (multiplication, division, etc.). This new feature replaces the parent features. By doing so, we hope that the new feature can deliver additional[103] information to the ML model. This new feature may or may not be representative of a real or physical quantity.[104]

97 I already knew about scaling beforehand and hence used mm for all geometric features in Table 3.2!

98 Keep in mind, we do not input units into ML models.

99 Imagine this, for a ML user/engineer without domain knowledge, removing this feature can be devastating since the axial capacity of columns is influenced by the length – here is where domain knowledge can come in handy and here is also where ML may not prove as smart as we would like it to be!

100 Tree-based algorithms can be insensitive to normalization. Neural network and distance-based (i.e., KNN, SVM, PCA) algorithms can be. Neural networks work via gradient descent which require data to be scaled to find a local minimum of a model's differentiable function. Distance-based algorithms use distance between data points to assign similarity and hence are influenced by the scale of data.

101 Additional note: In the event where the data conform to power law (a handful of features have many points, and most other features have few points), then you might want to try log scaling = log(x). A simpler approach that needs to be approached with *caution* is to remove extreme data.

102 Standardization is also called Z-Score normalization.

103 Or at least, maintain the same level of information the parents hold. Later on, you will see that we can trade off a slight reduction in such amount of information on the premise of having leaner models and improved performance.

104 Here is where the line of applied ML and civil and environmental engineering becomes blurry. Do we need all features to be physically representative? Do we have to have these? Can we use non-dimensional features? These are questions with a philosophical component that need to be discussed in open forum. Read rule no. 20 in Google's Rules of Machine Learning [17].

3.7.2 Manipulating Categorical Data

There are a few means to manipulate data to conform with a classification analysis. For example, categorical features can be transformed into numerical values. Say that we have a categorical feature with three items, namely, freeway, highway, and expressway. These items can be transformed into 0, 1, and 2, respectively.[105] This is a common manipulation procedure and can also be used to turn a numerical target into a categorical one.[106] For example, a target that assigns a numerical value to describe the level of damage (where zero describes a state of no damage) can be converted such that all numerical values returns 1.0 while the rest is zero.

Similarly, numerical values can be transformed into categorical ones. For instance, it is common to label load bearing elements according to their fire resistance rating (1 hour, 2 hours, 3 hours, and 4 hours). However, this assignment takes place once such elements are tested. While testing, the rating is measured in terms of minutes. In this case, if an element fails in 70 min, then this 70 min can then be transformed into 1 hour. In this example, all measured ratings can be converted into hourly ratings that fall under the aforementioned four classes. This is called discretization.

3.7.3 General Manipulation

The target in a dataset can be heavily skewed toward a specific class or value. As a result of this skewness, the ML model may fail to correctly predict observations belonging to the smaller class/or those of fewer representations. To cover this issue, a dataset can be handled via under-sampling and/or over-sampling techniques. Such techniques produce a new dataset with equally distributed classes.

Under-sampling may randomly remove samples from the larger class until this class contains as many observations as the minority class.[107] On the contrary, over-sampling adds new copies of observations belonging to the minority class data to the datasets until the dataset is balanced. Both of these techniques can be carried out manually or through a coding script.[108] A more advanced technique that I will cover in a later chapter is where we create synthetic observations that we draw from the statistical distribution of the minority class.[109]

3.8 Manipulation for Computer Vision

In the case of footage and imagery, data manipulation aims to minimize issues related to limited images, quality of images, etc. Data augmentation can be used to enlarge the size of a dataset by duplicating existing images and adding filters, textures, rotations, and other edits (contrast, white/gray/black) etc. In addition, some images can be split into new images to showcase different views.

Once the dataset[110] has been compiled, diagnosed, visualized, explored, and manipulated, then the dataset can be fed into a, or a series of, ML algorithms. A discussion on how to do so starts with the next chapter.

3.9 A Brief Review of Statistics

3.9.1 Statistical Concepts

Statistics revolves around methods and procedures for collecting and analyzing data.[111] Oftentimes, engineers do not have access to the full population[112] of data.[113] We do, however, have access to a sample[114] of this population. Any property[115] that describes a population is referred to as a parameter, and the same property is called a statistic when it describes a sample.

---

105 For a binary item, the use of 0 and 1 can suffice. Do you still remember our example of concrete cylinders belonging to normal strength or high strength class?

106 Some may call this, **Dummy variables**: Artificial numerics to capture some aspect of categorical variables.

107 Just like deleting features with missing data, this practice can also discard potentially useful information from the dataset.

108 Also, these techniques may add/remove data with or without replacing the original observations.

109 Also referred to as Synthetic Minority Oversampling Technique (SMOTE) – see Chapter 9.

110 Up to this point in time, we are short of establishing benchmarked datasets suitable to our domain. On a more positive light, a few datasets have been published and benchmarked (see [34]). I have added a description of these in the Appendix.

111 **Statistical inference**: Drawing valid conclusions or insights from a data sample that represents a population.

112 **Population**: Any set of data of interest.

113 Imagine dealing with 1 petabyte (PB) of construction data every year!

114 **Sample**: A subset of the population. You can think of a sample as a dataset we use to explore a phenomenon.

115 We have covered such properties and used them to describe the datasets.

A descriptive statistic illustrates central tendency and dispersion (variability). For example, the central tendency of a sample is the average value of a variable being observed. This tendency is also called the mean. The mean is calculated using all points in a dataset (unlike the mode or median), and hence it can be sensitive to outliers or extreme values.[116] This brings us to the *Central Limit Theorem*, which roughly states that whenever the data is moderately large, then the mean can be approximated with a normal distribution.[117] Describing the data by the mean can be helpful, but there are many examples where the means of two samples can be the same, but the data within the samples are dispersed differently (see Figure 3.10).

The variability in the data can be expressed by a statistic suitable for illustrating such a description. Let us consider the variation (or how far) a particular point is from the mean. More formally, this is called the *deviation*, and here is where the variance and standard deviation come to play. The variance describes the variability in terms of squared deviation from the mean and has squared units.[118] As such, we prefer to use the standard deviation (which is calculated as the square root of the variance and is hence noted as a measure of spread). Generally, we use the mean and standard deviation to describe data and, more importantly, their distributions.

A distribution is a description of how often the numerical data occur.[119] It is also a description of the percentage/ratio of how often categorical data appear. Distributions can be plotted in histograms or chart bars or can be listed in tables. When bins in a histogram are merged, they return a smooth curve that we call a density plot. The area under this curve is the probability and is equal to unity. Thus, such a plot can organically return the probability of a data point being in a given range.

The normal distribution is one that looks like those plotted in Figure 3.10. It is continuous, symmetric, and has a bell shape.[120] Since it has a known shape, then all that it takes to describe the normal distribution is to know its mean and standard deviation.

There are many other concepts that can be reviewed. For example, classification of variables, error types, statistical estimation and hypothesis testing, sampling, association, correlation, and controlling. For more information on these and many more, please refer to standard textbooks on statistics. I find Box et al.'s [12] as well as Kleinbaum et al.'s [13] textbooks

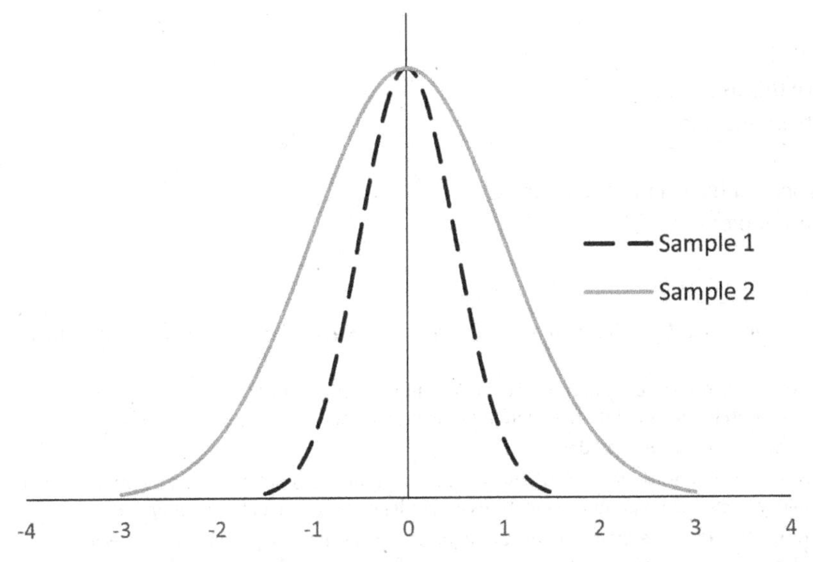

Figure 3.10 Illustration of two samples with the same mean but different variation.

116 This makes it easy to associate the mean with a line of balance that tells us at which point the data distribution is balanced.

117 Regardless of the distribution of the underlying variable (or feature) – I hope the rationale for the substitution exercise we carried out when describing missing data is clearer now. I will circle back to the normal distribution in a minute.

118 Which may not be very helpful in some scenarios (years^2, etc.). More specifically, the variance is the mean value of the deviation.

119 For example, Figure 3.2a shows the distribution of the tube thickness.

120 Categorical data do not belong here. They fall under the discrete distribution.

to be invaluable. For a more focused review, I find the first few chapters of Nick Huntington-Klein [14] and Scott Cunningham [15] books[121] to be informative and brilliant.

3.9.2 Regression

Regression[122] is perhaps one of the first things that come to an engineer's mind when they are tasked with analyzing data. A common goal[123] is to create a model that can be used to either predict this data or possibly explain the underlying mechanism that has generated such data. In reality, regression can be used for many other applications. In lieu of the above two applications, regression can also be used to:

- Characterize relationships between dependent variables and the independent variable.[124]
- Determine the importance of dependent variables with regard to the independent variable.
- Compare possible combinations of dependent variables (i.e., form of regression, need for interaction terms).

There are many forms of regressions, most of which are a function of the availability and types of the present data. I will start with the simplest case of linear regression.[125] The simplest form of regression is one where we fit a line on the data to describe the relationship in the data. Such a line is likely to have a slope β_1 and an intercept β_0, and hence the following equation turns suitable to describe such a line:

$$Y = \beta_0 + \beta_1 X + \epsilon \tag{3.6}$$

In Equation 3.6, the Y is the dependent variable, the X is the independent variable, and ϵ is the error term. Thus, this is a simple[126] linear[127] regression for a single independent variable and one dependent variable. There is a clear interpretation of the slope and intercept. Simply, β_0 is the mean of Y when $X = 0$, and β_1 is the change in Y due to one unit change in X.[128]

There are five assumptions to linear regression. These include:[129]

- Existence of a linear relationship.
  - The dependent and independent variables should be linearly related
    - Can be determined via scatter plots.
  - If this assumption is violated, then:
    - Transforming the variables can restore this assumption.
    - Add new variables to replace the problematic ones.
- Multivariate normality.
  - Any linear combination of the variables should have a normal distribution.
    - Can be checked by inspecting distribution plots or QQ-plots.

---

121 These two are great sources for causal discovery and inference. We will get there in a few chapters. What I like about these sources is that they are delivered in a simple and not-so-formal style.

122 Sir Francis Galton (1885) introduced the concept of *regression* while examining the relationship observed for the heights of fathers and their sons [35, 36]. Sir Galton noted that sons' height *regress* to the mean of the population of the father's height.

123 A subconscious one, that I still get to this day, whenever I see a bunch of data.

124 I am sticking to the common language used in regression. For example, dependent variables → inputs (or predictors) and the independent variable → output (or target/response). In ML terminology, dependent variables → features and independent variable → phenomenon.

125 Please keep in mind that my goal of including a discussion on regression is not meant to be exhaustive but rather to be treated as a refresher and a reinforcer. For more details on regression, as well as other statistical tools, please refer to the cited resources above, as well as the PennState STAT501 online course [37].

126 Because we have one independent variable. The addition of more independent variables transforms this regression into a multi-linear regression.

127 The term, *linear*, refers to the relationship between model parameters (e.g., β_0 and β_1) and the dependent variable. Please note that this does not imply that the relationship between the data and dependent variable is linear!

128 For the case of one independent variable, the slope can be estimated by dividing the co-variance of X and Y over the variance of X. The **co-variance** is a measure of the joint variability of two variables.

129 As you will see, much of our real data is likely to violate one or a few of the above assumptions – hence our pursuit of new modeling techniques. Keep in mind that this does not imply that linear regression (or regression in general) is not suited for our problems! It simply implies that regression can be an effective technique to use for problems that satisfy its assumptions. When such assumptions are met, regression becomes a tremendous tool given its simplicity and interpretability! Keep in mind that, just like other methods, regression has pros and cons. ML also has the same! The discussion on more modern regression methods continues in Chapter 4 once we discuss generalized linear models (GLM) and generalized additive models (GAM).

  - If this assumption is violated, then:
    - Transforming the variables can restore this assumption.
- No or little multicollinearity.
  - There is no, or very little, correlation between the independent variables.
    - Can be checked by examining the correlation matrix or via the variance inflation factor (VIF) [130] method.
  - If this assumption is violated, then:
    - You may remove the problematic observations.
    - You may use a different regression approach.
- Independence
  - The observations of the independent variable are statistically independent of one another.
    - Can be checked through hypothesis testing (i.e., Durbin-Watson Test, etc.).
  - If this assumption is violated, then:
    - You may remove the problematic observations.
    - Select a different regression approach.
- Homoscedasticity
  - The variance of residuals is the same for any value of the independent variable.
    - Can be checked through plotting predictions vs. actuals.
  - If this assumption is violated, then:
    - You may transform the dependent variables.
    - You may use weighted regression.[131]

Now, say that we have data that satisfies the above assumptions, and we are looking to fit a straight line into such data. How do we place such a line?[132]

We can do so by using the least squares method.[133] The *least* in the least squares denotes that this is a minimization approach in which we are hoping to minimize the sum of the squares of vertical[134] lines drawn from the observed data to our fitted line. For example, have a look at the visualization tool created by Eduardo García-Portugués [16]. The dotted lines are those vertical[135] lines we are hoping to minimize the sum of their squares. As you can see, the sum of these lines equals 163.29. By adding a slope of 1.25, this sum reduces to 61.11. Now, adding an intercept of −0.5, further reduces this sum. The sum minimizes at a slope of 1.395 and intercept of −0.615, as noted by García-Portugués.

The minimization of the Sum of Squares due to Error (SSE) can be calculated as:

$$\mathrm{SSE} = \sum_{i=1}^{n}(Y_i - \beta_0 - \beta_1 X_i)^2 \tag{3.7}$$

And, here are the mathematical equations to solve for model parameters:

$$\hat{\beta}_1 = \frac{\text{Sample covariance}}{\text{Sample variance}} = \frac{\frac{1}{n}\sum_{i=1}^{n}(X_i - \bar{X})(Y_i - \bar{Y})}{\frac{1}{n}\sum_{i=1}^{n}(X_i - \bar{X})^2} \tag{3.8}$$

$$\hat{\beta}_0 = \bar{Y} - \hat{b}_1\bar{X} \tag{3.9}$$

where the bars denoted sample means, and the hats denote sample estimates.

---

130 A value of 5 or 10 can be used as a cut-off value.

131 By assigning a relative weight to data points based on their variance. In this procedure, more weight is given to the observations with smaller variance (as they are tied to being more reliable).

132 A simple and manual method is to place such a line by inspection. However, this can be subjective and imprecise.

133 Other methods can be used, such as the minimum sum of absolute errors or the minimum-variance method. The famous 19th century mathematician, Carl Gauss, proved that the least squares method coincide with maximum-likelihood estimate (a method of estimating model parameters of an assumed probability distribution) for independent and normally distributed errors with zero mean and equal variances.

134 Not perpendicular!

135 Initially, these lines are perpendicular as the fitted line is horizontal.

Figure 3.12 repeats the above analysis in *Excel* as follows to return a comparable result[136] to that seen in Figure 3.11. Note the influence of intercept.

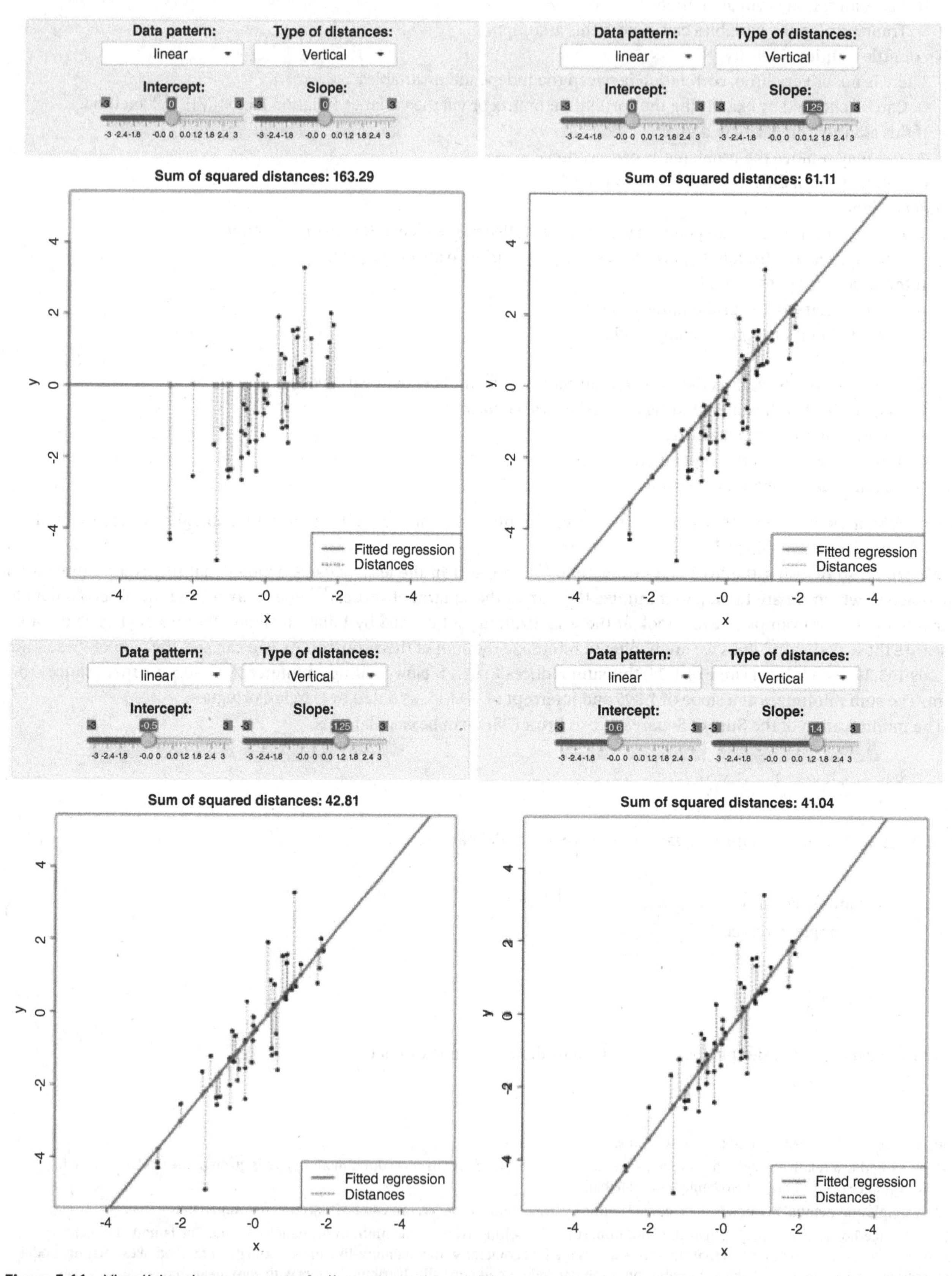

Figure 3.11 Visualizing the parameters of a linear regression model.

136 I must admit that I did not have the exact values as those in Figure 3.11 and hence the minor differences. In a perfect world, both methods would lead to identical results.

Step 1: Select the data series and right-click. Choose *Add Trendline*.

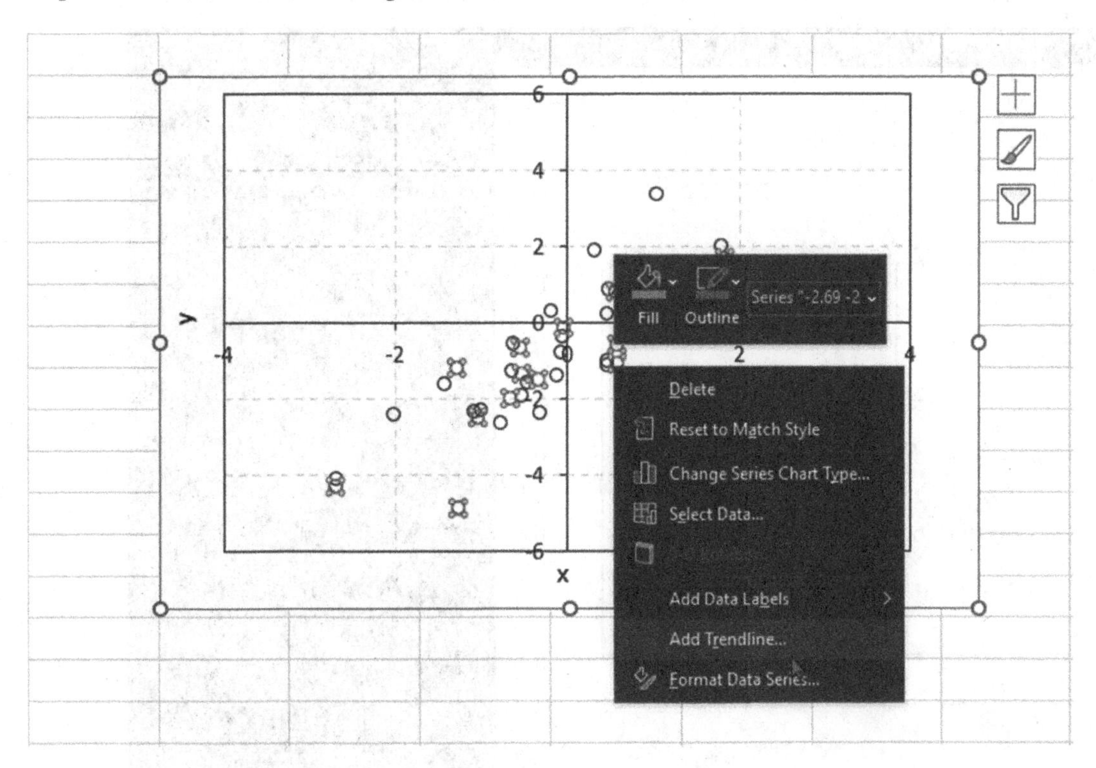

Step 2: Select *Display Equation* and *Display R-squared value on chart.*

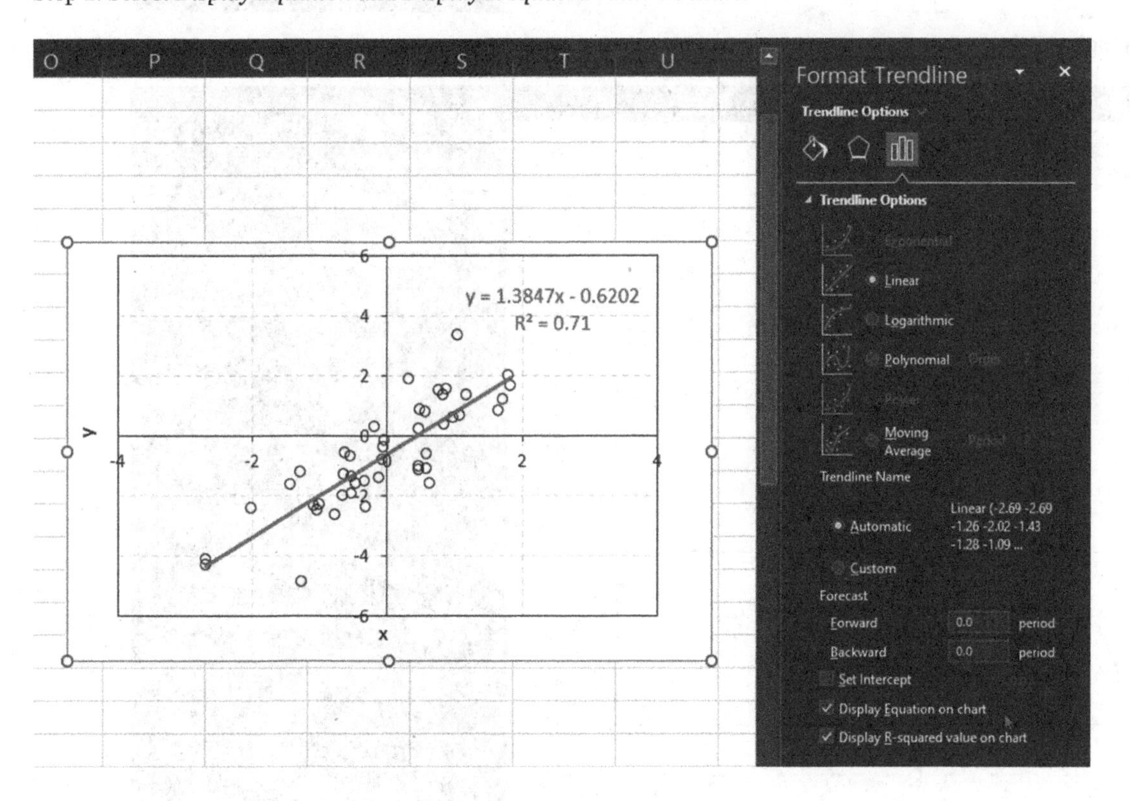

Step 3: Select *Set Intercept*.

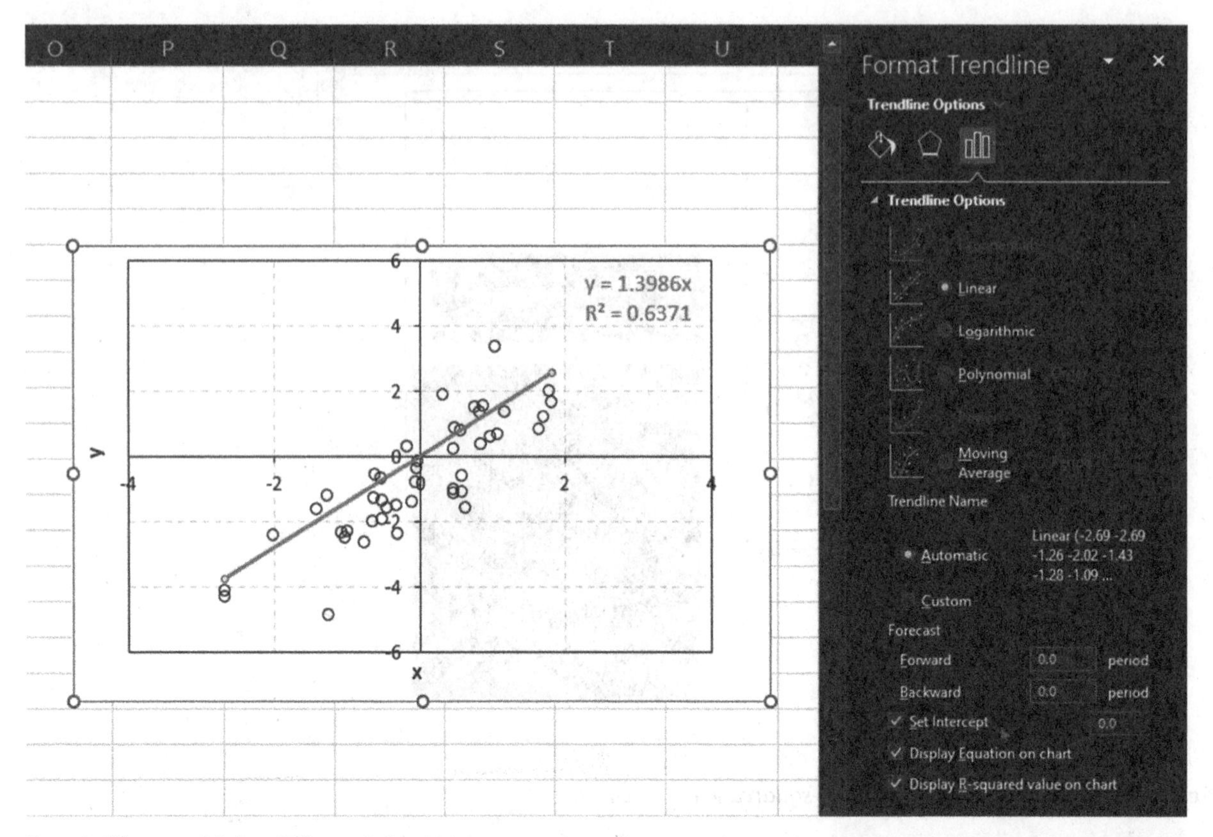

Step 4: Choose a higher *Polynomial* to compare your results.

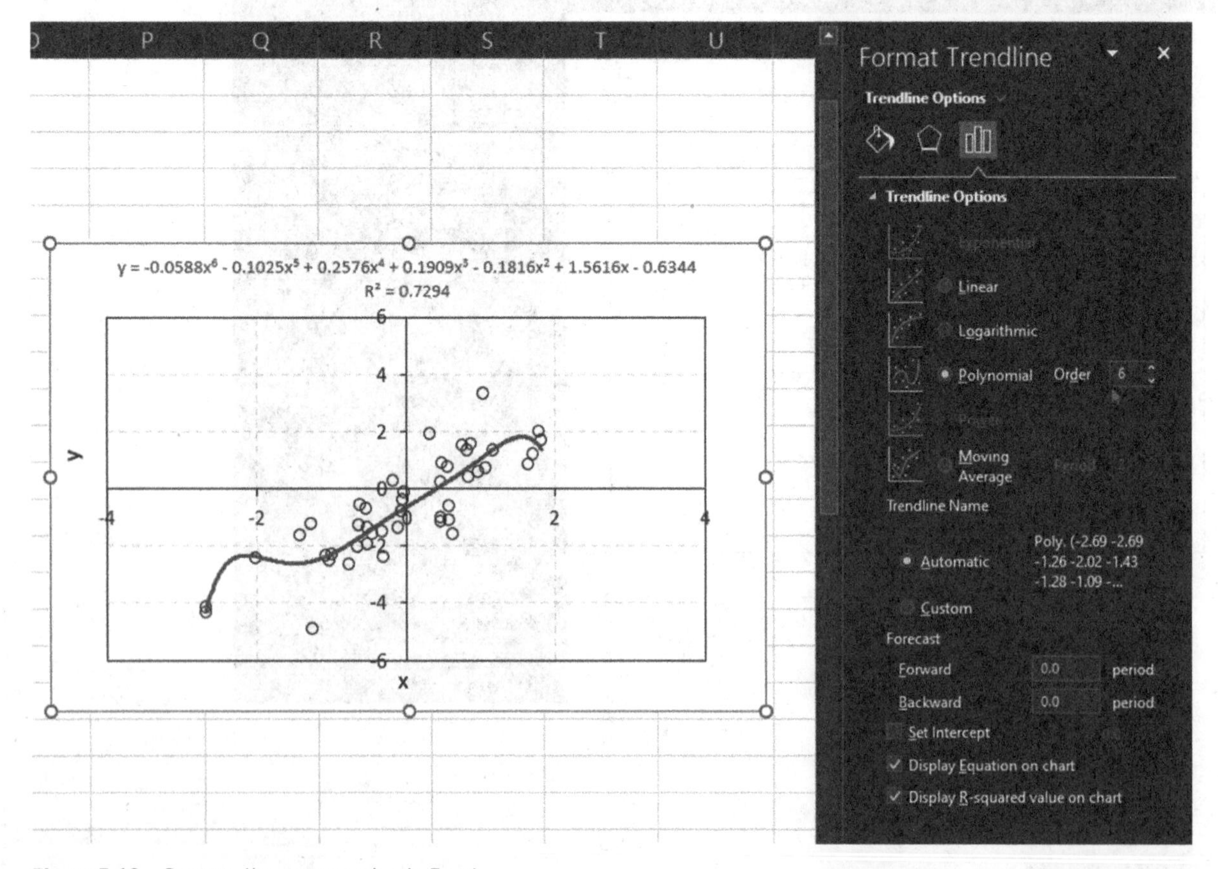

Figure 3.12 Steps to linear regression in Excel.

If a straight line is not suitable for the data we have, then we can use other forms of linear regression. The use of polynomials can be of aid (see Step 4 of Figure 3.12). Polynomials use the same variable with powers applied to themselves. These fit non-strength lines through the data in search of a better fit.[137] The interpretation of coefficients in this form turns tricky[138] since all coefficients are to be interpreted together.

$$Y = \beta_0 + \beta_1 X + \beta_2 X^2 + \ldots + \beta_n X^n \quad (3.10)$$

In the case there are additional independent variables, then the approach can be extended to *multiple linear regression* to accommodate the new variables. In this event, Equation 3.6 transforms to

$$Y = \beta_0 + \beta_1 X_1 + \cdots + \beta_n X_n + \varepsilon \quad (3.11)$$[139]

Up to this point, the presented regression models assume that the majority of the variation in the dependent variable is explained by the cumulative effect of individual independent variables. However, this variation can also be a function of two, or more, variables working in conjunction.[140] In this case, our regression model becomes:

$$Y = \beta_0 + \beta_1 X_1 + \beta_2 X_2 + \beta_3 X_1 X_2 + \varepsilon \quad (3.12)$$

In Equation 3.12, β_0 represents the overall average response, β_1 and β_2 represent the average rate of change due to X_1 and X_2, respectively. β_3 represents the incremental rate of change due to the combined effect of X_1 and X_2 where this incremental change cannot be explained by X_1 and X_2 alone. Kuhn and Johnson [6] identified four cases for such regression:

1) If β_3 is not significantly different from zero, then the interaction between X_1 and X_2 is not useful for explaining the variation in the dependent variable. This also implies that the relationship between these two independent variables is *additive*.
2) If β_3 is negative and significantly different from zero, while X_1 and X_2 alone also affect the dependent variable, then the interaction is *antagonistic*.
3) If β_3 is positive and significantly different from zero while X_1 and X_2 alone also affect the dependent variable, then the interaction is *synergistic*.
4) The final scenario occurs when β_3 is significantly different from zero, but either one or both of X_1 and X_2 do not affect the dependent variable. In this case, the average response of X_1 across the values of X_2 (or vice versa) has a rate of change that is essentially zero, while the average value of X_1 at each value of X_2 is different from zero. This interaction is rare and hence is referred to as *atypical*.

The regression realm grows past the above two models into nonlinear[141] and nonparametric[142] regression. A generalization of nonparametric models is shown in Equation 3.13, and an example is shown in Equation 3.14.

$$Y = m(X_i) + \epsilon \quad (3.13)$$

$$Y = \beta_1 x_1^{\beta_2} + \tan(x_2^{\beta_3} x_3) + \epsilon \quad (3.14)$$[143]

The simplicity of logistic regression to classification resembles that of what the linear model presents to regression. Essentially, logistic regression predicts the outcome when the dependent variable is a class. Let us visit the case of a binary outcome. Unlike the linear regression model, the logistic does not require a linear relationship between variables. The

137 Keep in mind that sometimes such fits can be minor (as seen in Figure 3.12) or substantial. Remember that complexity and interpretation can get blurry with complex forms of regression.

138 Say, X = *strength*. Then, X^2 equals *strength*^{2} and X^3 equals *strength*^{3}. How do we interpret such terms? What does *strength*^{2} and *strength*^{3} mean?

139 This is the case for many ML problems that you are likely to face. Note that the coefficients retain their interpretations noting that β_i is the change in Y due to one unit change in X_i provided that the other independent variables remain constant. We will re-visit regression in Chapter 8.

140 For example, webs and transverse stiffeners in steel girders work hand in hand to resist shear loading.

141 A form that does not obey the assumptions of linear regression.

142 A model that does not have a pre-determined form.

143 Notice the absence of linearity assumption between Y and model parameter.

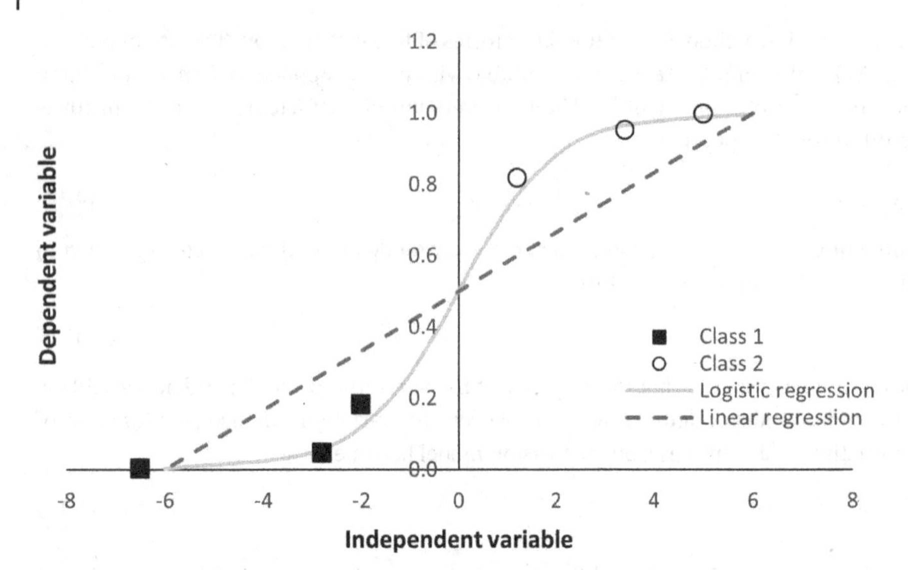

Figure 3.13 Logistic function (and a comparison against linear regression, which returns a value of zero and unity).

logistic function makes use of the logistic function shown in Equation 3.15. This type of regression does not predict values but rather generates a probability for a given class between zero and unity[144] (see Figure 3.13).

$$\text{Logistic function} = \frac{1}{1+e^{-z}} \tag{3.15}$$

where $z = \beta_0 + \sum_{i=1}^{n} \beta_i X_i$

There are three main types of logistic regression: binary, multinomial, and ordinal. In the first type (depicted in Figure 3.13), there are two possible values. On the other hand, multinomial logistic regression deals with three or more values, and ordinal logistic regression can capture three or more classes with a natural order.

3.10 Conclusions

This chapter covers some of the main and key principles of data and data science. We covered the different types of data we are likely to encounter and learned how to build datasets. We looked at methods to diagnose and handle data, visualize data, explore data, and manipulate data in pursuit of creating suitable ML models in a variety of ways, using *Microsoft Excel*, *Python*, *R*, and *Exploratory*. We reviewed statistical concepts with a focus on regression.

Definitions

Co-variance is a measure of the joint variability of two variables.
Dummy variables Artificial numeric to capture some aspect of categorical variables.
Entropy A characterization of the unpredictability of a variable.
Population Any set of data of interest.
Sample A subset of the population.
Semi-structured data Data that does not fit into rows or columns but has some form of a fixed structure.

144 Despite its simplicity, this technique has proved very effective in tackling complex problems (see some of the success stories my group had with applying logistic regression [38, 39]).

Statistical inference Drawing valid conclusions or insights from a data sample that represents a population.
Structured data Data that can be organized into rows or columns.
Unstructured data Data that cannot be organized into rows or columns.

Chapter Blueprint

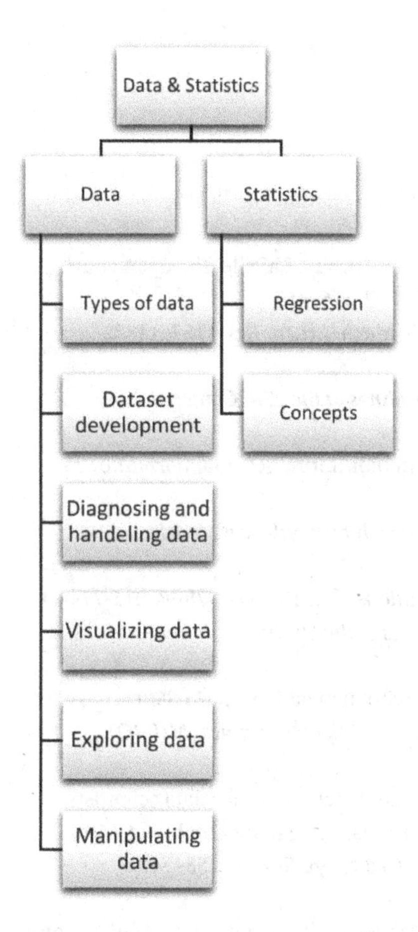

Questions and Problems[145]

3.1 Find how much data your field generates annually. Cite your references.

3.2 List and explain three different types of data.

3.3 Plot histograms for all the features shown in the dataset on CFSTs.

3.4 Identify and define two methods to handle missing data.

3.5 What are two differences between correlation and mutual information?

145 Note that additional problems on this chapter will be provided in Chapter 6.

3.6 Compare filter, wrapper, and embedded feature selection methods.

3.7 What is over- and under-sampling? Contrast these two approaches.

3.8 Read and summarize Google's Rules of Machine Learning [17] into one page.

3.9 How does linear regression differ than logistic regression?

References

1 Hauke, J. and Kossowski, T. (2011). Comparison of values of Pearson's and Spearman's correlation coefficients on the same sets of data. *Quaest. Geogr.* 30. doi: 10.2478/v10117-011-0021-1.

2 Hildebrand, C.D.W. (1938 December). Ockham, studies and selections by Stephen Chak Tornay. LaSalle, Ill.: open court publishing company, 1938. viii, 207 pages. $1.75. *Church Hist.* 7 (4): 385–387. doi: 10.2307/3160457.

3 Guyon, I. and Elisseeff, A. (2003). An introduction to variable and feature selection. *J. Mach. Learn. Res.* 1157–1182. doi: 10.5555/944919.944968.

4 Gu, Q., Li, Z., and Han, J. (2011). Generalized fisher score for feature selection. *Proceedings of the 27th Conference on Uncertainty in Artificial Intelligence, UAI 2011.*

5 Saeys, Y., Inza, I., and Larrañaga, P. (2007). A review of feature selection techniques in bioinformatics. *Bioinformatics* 23. doi: 10.1093/bioinformatics/btm344.

6 Kuhn, M. and Johnson, K. (2019). *Feature Engineering and Selection: A Practical Approach for Predictive Models.* doi: 10.1201/9781315108230.

7 Laarhoven, P.J.M. and Aarts, E.H.L. (1987). *Simulated Annealing: Theory and Applications.* doi: 10.1007/978-94-015-7744-1.

8 Chen, X.W. and Jeong, J.C. (2007). Enhanced recursive feature elimination. *Proceedings of the 6th International Conference on Machine Learning and Applications ICMLA 2007.* doi: 10.1109/ICMLA.2007.35.

9 Jović, A., Brkić, K., and Bogunović, N. (2015). A review of feature selection methods with applications. *2015 38th International Convention on Information and Communication Technology, Electronics and Microelectronics (MIPRO).* doi: 10.1109/MIPRO.2015.7160458.

10 Zheng, A. and Casari, A. (2018). Feature engineering for machine learning: principles and techniques for data scientists.

11 Unglert, K., Radić, V., and Jellinek, A.M. (2016). Principal component analysis vs. self-organizing maps combined with hierarchical clustering for pattern recognition in volcano seismic spectra. *J. Volcanol. Geotherm. Res.* 320: 58–74. doi: 10.1016/J.JVOLGEORES.2016.04.014.

12 Box, G.E.P., Hunter, J.S., and Hunter, W.G. (1978). *Statistics for Experimenters: An Introduction to Design, Data Analysis, and Model Building.* John Wiley & Sons.

13 Klienbaum, D.G., Kupper, L.L., Nizam, A., and Rosenberg, E.S. (2014). *Applied Regression Analysis and Other Multivariable Methods.* Cengage Learning.

14 Huntington-Klein, N. (2021). *The Effect : An Introduction to Research Design and Causality.* Boca Raton: Chapman and Hall/CRC. doi: 10.1201/9781003226055.

15 Cunningham, S. (2021). Causal inference: the mixtape.

16 García-Portugués, E. (2022). *Notes for Predictive Modeling.* https://bookdown.org/egarpor/PM-UC3M (accessed 3 August 2022).

17 Zinkevich, M. (2022). Rules of machine learning: best Practices for ML Engineering | Google Developers. https://developers.google.com/machine-learning/guides/rules-of-ml (accessed 4 August 2022).

18 Reber, P. (2010). What is the memory capacity of the human brain? Scientific American. https://www.scientificamerican.com/article/what-is-the-memory-capacity (accessed 1 August 2022).

19 Shapiro, I. (2018). What to do with the 95% of data construction companies throw away. BDO. https://www.bdo.global/en-gb/blogs/tech-media-watch-blog/december-2018/what-to-do-with-the-95-of-data-construction-companies-throw-away (accessed 1 August 2022).

20 Dhar, V. (2013). Data science and prediction. *Commun. ACM* 56. doi: 10.1145/2500499.

21 Frank, I. and Todeschini, R. (1994). *The Data Analysis Handbook*. Elsevier. https://books.google.com/books?hl=en&lr=&id=SXEpB0H6L3YC&oi=fnd&pg=PP1&ots=zfmIRO_XO5&sig=dSX6KJdkuav5zRNxaUdcftGSn2k (accessed 21 June 2019).

22 Kuhn, M. and Johnson, K. (2019). *Feature Engineering and Selection: A Practical Approach for Predictive Models*. Chapman and Hall/CRC. https://bookdown.org/max/FES (accessed 1 August 2022).

23 Radziwill, N.M. (2019). Statistics (the easier way) with R.

24 Scott, D.W. (2009). Sturges' rule. *Wiley Interdiscip. Rev. Comput. Stat.* 1. doi: 10.1002/wics.35.

25 Scott, D.W. (2010). Scott's rule. *Wiley Interdiscip. Rev. Comput. Stat.* 6. doi: 10.1002/wics.103.

26 Waskom, M. (2021). seaborn: statistical data visualization. *J. Open Source Softw.* 6: 3021. doi: 10.21105/JOSS.03021.

27 ggplot2 (2022). Create elegant data visualisations using the grammar of graphics • ggplot2. https://ggplot2.tidyverse.org (accessed 7 July 2022).

28 Exploratory (2022). About. https://exploratory.io/about (accessed 6 July 2022).

29 Meyer, P. (2008). R: mutual information computation. https://search.r-project.org/CRAN/refmans/infotheo/html/mutinformation.html (accessed 2 August 2022).

30 Seok, J. and Kang, Y.S. (2015). Mutual information between discrete variables with many categories using recursive adaptive partitioning. *Sci. Rep.* 5. doi: 10.1038/srep10981.

31 Chandrashekar, G. and Sahin, F. (2014). A survey on feature selection methods. *Comput. Electr. Eng.* 40. doi: 10.1016/j.compeleceng.2013.11.024.

32 Liu, H. and Yu, L. (2005). Toward integrating feature selection algorithms for classification and clustering. *IEEE Trans. Knowl. Data Eng.* 17. doi: 10.1109/TKDE.2005.66.

33 Kumar, V. and Minz, S (2014). Feature selection: a literature review. *Smart Comput. Rev.* 44. https://faculty.cc.gatech.edu/~hic/CS7616/Papers/Kumar-Minz-2014.pdf.

34 Naser, M.Z., Kodur, V., Thai, H.-T. et al. (2021). StructuresNet and fireNet: benchmarking databases and machine learning algorithms in structural and fire engineering domains. *J. Build. Eng.* 102977. doi: 10.1016/J.JOBE.2021.102977.

35 Ethington, C.A., Thomas, S.L., and Pike, G.R. (2002). Back to the basics: regression as it should be. doi: 10.1007/978-94-010-0245-5_6.

36 Senn, S. (2011). Francis Galton and regression to the mean. *Significance* 44. doi: 10.1111/j.1740-9713.2011.00509.x.

37 Pardoe, I., Simon, L., and Young, D. (2022). Welcome to STAT 501! | STAT 501. https://online.stat.psu.edu/stat501 (accessed 3 August 2022).

38 Naser, M.Z. (2019). Heuristic machine cognition to predict fire-induced spalling and fire resistance of concrete structures. *Autom. Constr.* 106: 102916. doi: 10.1016/J.AUTCON.2019.102916.

39 Tapeh, A. and Naser, M.Z. (2022). Discovering graphical heuristics on fire-induced spalling of concrete through explainable Artificial Intelligence. *Fire Technol.* 58. doi: 10.1007/s10694-022-01290-7.

4

Machine Learning Algorithms

Where others saw chaos, I saw a pattern.

Hari Seldon (Foundation, 2021)

Synopsis

This chapter presents the tools we will use to create, develop, and/or explore ML models and solutions. These tools are referred to as algorithms. There are numerous ML algorithms as well as algorithmic families.[1] Much of them are appropriate to engineering problems. Most[2] of which are applicable to civil and environmental engineering. Many are easy to use, and some may require a bit of work.[3] I will primarily focus on algorithms that utilize tabular data and then showcase the use of deep learning and computer vision techniques. I will also show visualization tools that you can try on your own to better visualize how the presented algorithms handle and process data (i.e., features) and then arrive at their predictions. Finally, I will be attaching several code libraries[4] and already-built scripts for each discussed algorithm, as well as in the Appendix, for you to try and apply.

4.1 An Overview of Algorithms[5]

This section provides an overview of some of the current and commonly used ML algorithms that have been successful in solving or overcoming problems and case studies belonging to the domain of civil and environmental engineering (see Figure 4.1). For simplicity, I will be presenting such algorithms in alphabetical order. The majority of the algorithms that

1 In reality, developing algorithms may not be as hard as you think. Creating universal algorithms (i.e., that can be applied to various problems and return proper results) is as hard as you think, if not harder!

2 Also, most of the presented algorithms can be found within the coding-free platforms we will be exploring such as *BigML*, *DataRobot*, *Dataiku*, *Exploratory*, and *Clarifai* I will present in Chapter 6.

3 *Quick story*: This brings me to this chapter's story. Ena is a first-year resident in a general surgery program. A part of the training Ena has to go through is to practice a number of surgery cases (~150) per year. These surgeries cover different procedures and require the use of various tools. From the context of where we stand, this chapter aims to provide you with the necessary algorithms (aka. tools) to successfully tackle the various problems that you might face!

4 Thanks to the many contributors from online forums that were kind enough to share their codes online. Admittedly, tracing the original contributors was an ambitious endeavor given that many used avatars, etc. To those that were not explicitly credited, I thank you for sharing your codes and I hope that your path crosses this book. If so, please reach out to me and I will be happy to update the list of references and citations.

5 One thing to note. Each algorithm has its own set of parameters which makes it challenging to cover all the parameters of each algorithm, let alone those that can be tweaked in different programming languages. So, instead of listing all the parameters of each algorithm, I will discuss the theory and big ideas behind each algorithm and briefly discuss their specialized parameters. Luckily, a full discussion on such parameters can be found in the original sources pertaining to each algorithm. I have cited these for your reference. The same can also be found in *Python*- and *R*-related help manuals. My advice is to always start with the default settings and then adjust, if necessary. A side note – you can download much of the discussed algorithms and datasets from my website: www.mznaser.com.

Machine Learning for Civil & Environmental Engineers: A Practical Approach to Data-driven Analysis, Explainability, and Causality, First Edition. M. Z. Naser.

Companion Website: www.wiley.com/go/Naser/MachineLearningforCivilandEnvironmentalEngineers

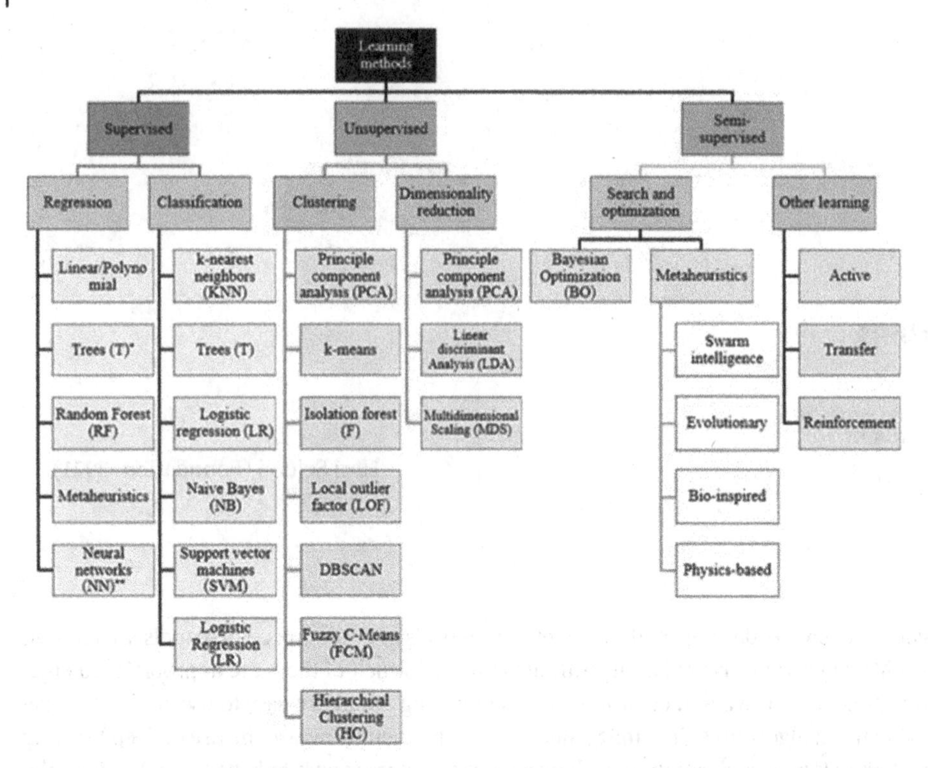

Figure 4.1 A snippet of notable ML algorithms [1,2].[6] *Decision trees (DT) and gradient boosted trees (GBTs).

fall under supervised learning can accommodate regression and classification problems. On the other hand, algorithms belonging to unsupervised learning are traditionally reserved for this particular type of learning, and these too are presented in the second half of this chapter.

4.1.1 Supervised Learning

As you remember, ML problems comprising datasets with labeled[7] targets fall under supervised learning. These problems encompass the majority of the ML realm and our domain.[8] Let us go over some of the most notable algorithms that fall under supervised learning and work well for regression and classification problems.

4.1.1.1 Adaptive Neuro-fuzzy Interface System (ANFIS)

The *adaptive neuro-fuzzy interface system (ANFIS)* is a multilayer adaptive network-based fuzzy inference system. This system contains a number of nodes connected through directional links. Each node is identified by a function with a fixed or adjustable parameter.

The ANFIS system can accommodate numerical, verbal, logical, or conditional terms of "if-then" rules to capture imprecise knowledge of fuzzy variables[9] [3]. This process involves mapping through membership functions to numerically define the partial belonging (i.e., fuzziness) of a statement by assigning values on a scale between zero and unity.

How does ANFIS work?

One approach to apply ANFIS is via the first order Takagi-Sugeno model [4], which includes two inputs, x and y (see Figure 4.2). This model is set with two *if-then* rules, as shown by the following formulations, and yields a weighted average of each output rule:

6 This chapter builds on my group's two review papers [1, 2].

7 Please remember that semi-supervised learning is applied when the labels are only available for a small percentage of targets.

8 This makes sense! Especially when we recognize that we heavily use ML to predict phenomena (which implies that we know the target and we can label it).

9 A **fuzzy variable** is one that can be an adjective, i.e., strong and weak, hot and cold. It is worth noting that fuzzy logic was created by Prof. Lotfi Zadeh. Fuzzy variables can be attractive to capture the uncertainty of features.

$$\text{Rule 1: If } x \text{ is } A_1 \text{ and } y \text{ is } B_1, \text{ then } f_1 = a_1 x + b_1 y + r_1 \tag{4.1}$$

$$\text{Rule 2: If } x \text{ is } A_2 \text{ and } y \text{ is } B_2, \text{ then } f_2 = a_2 x + b_2 y + r_2 \tag{4.2}$$

where x, y = input arguments, A, B = the linguistic labels, a, b, r = output function parameters.

A typical fuzzy inference contains five layers [6]:

Layer 1: every node i in this layer is an adaptive node characterized by a bell-like function, e.g.:

$$\mu_i(x) = \frac{1}{1 + \left|\frac{x - \delta_1}{\alpha_1}\right|^{2\beta_1}} \tag{4.3}$$

where $\mu_i(x)$ = i^{th} resultant of 1st layer, x = node input, α_1, β_1, δ_1 = parameter set. Parameters of the first layer are usually denoted as the premise set. The resultants of the first layer are the membership values of the premise part.

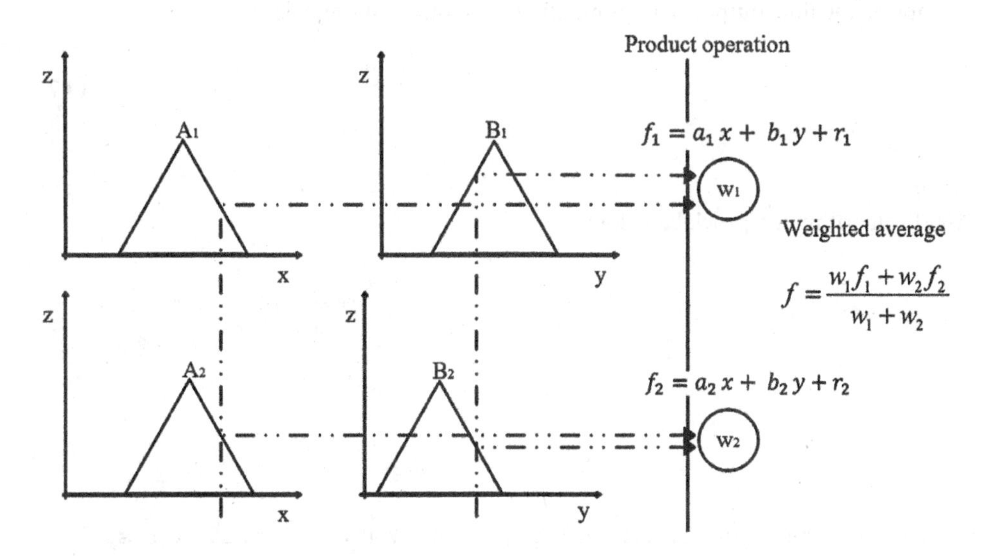

(a) First order Takagi-Sugeno fuzzy model with two rules and two inputs

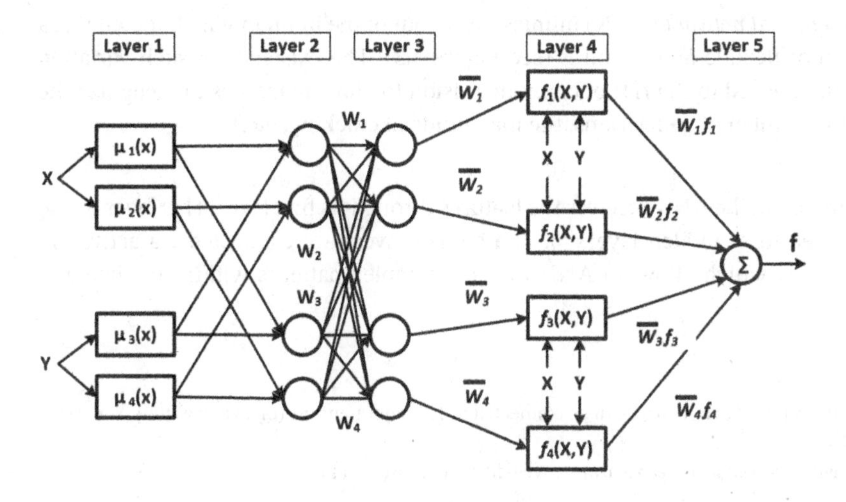

(b) Equivalent ANFIS architecture

Figure 4.2 Layout of a typical ANFIS model [5].

Layer 2: comprises nodes that multiply incoming signals and send the product of this multiplication to the third layer. The output of each node indicates the firing strength of a given rule.

$$W_i = \mu_{ki}(x)\mu_{li}(y), i = 1, 2, \ldots \tag{4.4}$$

where W_i = 2nd layer i^{th} output, $\mu_{ki}(x)$, $\mu_{li}(y)$ = signals coming from the 1st layer.

Layer 3: is made of nodes that compute the ratio of the ith rule's firing strength to the summed value of all rules' firing strengths. The resultant of this layer is known as the normalized firing strength:

$$\bar{W}_i = \frac{W_i}{W_1 + W_2 + W_3 + W_4}, \; i = 1, 2, \ldots \tag{4.5}$$

Layer 4: includes functional adaptive nodes such that:

$$\bar{W}_i f_i = \bar{W}_i(a_i x + b_i y + r_i), \; i = 1, 2, \ldots \tag{4.6}$$

where $\bar{W}_i$ = 3rd layer i[th] output, a_i, b_i, c_i = parameter set.

Layer 5: holds a single node that computes the final output. The summation of all incoming signals is:

$$f = \sum_{i=1}^{n} W_i f_i \tag{4.7}$$

4.1.1.1.1 Programming This Algorithm

The *Python* and *R* codes for the ANFIS algorithm are provided below:

Python Package

```
import anfis
model=myanfis.ANFIS()
```

R Package:

```
>install.packages("'frbs'")
>library('frbs')
>frbs.learn(data.train, range.data = NULL, method.type = c("WM"), control = list())
```

4.1.1.2 Artificial Neural Networks (ANN)

First introduced in the 1940s [7], the *Artificial neural network* (ANN) mimics the layout of the brain in which it contains a series of layers[10] – which allows the creation of flexible architectures (see Figure 4.3). The existence of such variation allows ANNs to be used in many problems and has led to the rise of computer vision to allow machines to recognize the content of images and videos (i.e., estimate the number of pedestrians, the magnitude of cracking, etc.) [8].

How does ANN work?

Each layer houses processing units (or neurons[11]). The ANN receives the features through its first layer. This layer is also connected to the second set of layer(s) (referred to as hidden layers[12]). The hidden layers are connected via activation functions,[13] e.g., Step, Logistic, ReLu, TanH, etc., which allow an ANN to learn complex patterns within the data and

10 Traditional ANNs have a minimum of three layers.

11 The human brain has an average of 86 billion neurons [78]. Each neuron can be connected to 10^4 other neurons (on average) [79]. You can think of a neuron as a switch (on/off) with a threshold.

12 An ANN algorithm with +2 hidden layers is formally referred to as *deep learning*. We will get to those in a bit.

13 Step $= 0$ for $x < 0$ and 1 for $x \geq 0$. Logistic $= 1/(1+e^{-x})$. ReLu(Recitified Linear unit) $= \begin{cases} 0 \text{ for } x < 0 \\ x \text{ for } x \geq 0 \end{cases}$. $\tan H = \frac{2}{1+e^{-2x}} - 1$.

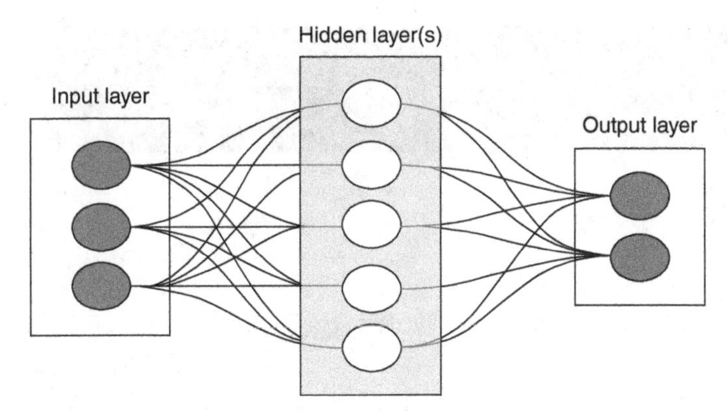

Figure 4.3 Layout of a typical ANN.

generate an approximation form that permits gradient-based optimization,[14] etc. This process is called feed-forward[15] and continues until the selected performance metric is satisfied [9]. Other approaches to feeding an ANN also exist, such as multilayer perceptron network, carpenter network, Hopfield network, Keras, and back-propagation,[16] to name a few.

$$\text{net}_j = \sum_{i=1}^{n} In_i w_{ij} + b_j \tag{4.8}$$

$$Y = f(\text{net}_j) \tag{4.9}$$

where In_i and b_j are the ith input signal and the bias value of jth neuron, respectively, w_{ij} is the connecting weight between ith input signal and jth neuron and f is an activation function.

In addition, an ANN comprises neurons. These are mathematical functions that calculate the weighted average of the data supplied and then send the data through an activation function. The neurons in each layer (as well as other layers) are linked via connections. The strength of the relationship between these is represented by a weight[17] that is hoped to minimize the chances of losing error. The error is propagated backward to adjust the ANN to yield better predictions (which are propagated forward). A key setting to ANN is the learning rate (which determines how quickly or slowly the weights are updated) and the derivative of the loss function (which is calculated in reference to each weight value using back-propagation and then subtracted from that weight).[18]

4.1.1.2.1 Visualizing ANN

There is a Google TensorFlow visualization[19] tool [10] that I usually like to showcase in my courses (see Figure 4.4). I have a couple of snippets from this tool below to present regression and classification visualizations. Please feel free to try this visualization tool. Please also note that this tool can accommodate up to six hidden layers, and hence we will be re-visiting this tool when describing DL.

---

14 These functions also decide if a neuron is to be active or not.

15 This is the most common type of ANN architecture, where a neuron is only connected to neurons in the next layer. An equally important term is, *fully-connected*, which refers to the state where each neuron in one layer is connected to all neurons in the next layers.

16 Where we update the error from the previous training stage by back propagating the error from output nodes to the input nodes.

17 Weights can be updated after the training when all training patterns have been evaluated (*batch training*), or after each training pattern (*online training*). Can we automatically update these weights? Yes, through the gradient descent, which finds a local minimum of a differentiable function (thus, an activation function mush be differentiable). Simply, the gradient descent finds the coefficients of a function's parameters to minimize the cost function as much as possible.

18 Additional notes: ANN are simple and flexible neural architectures. They do have a lot of parameters to tune.

19 https://playground.tensorflow.org.

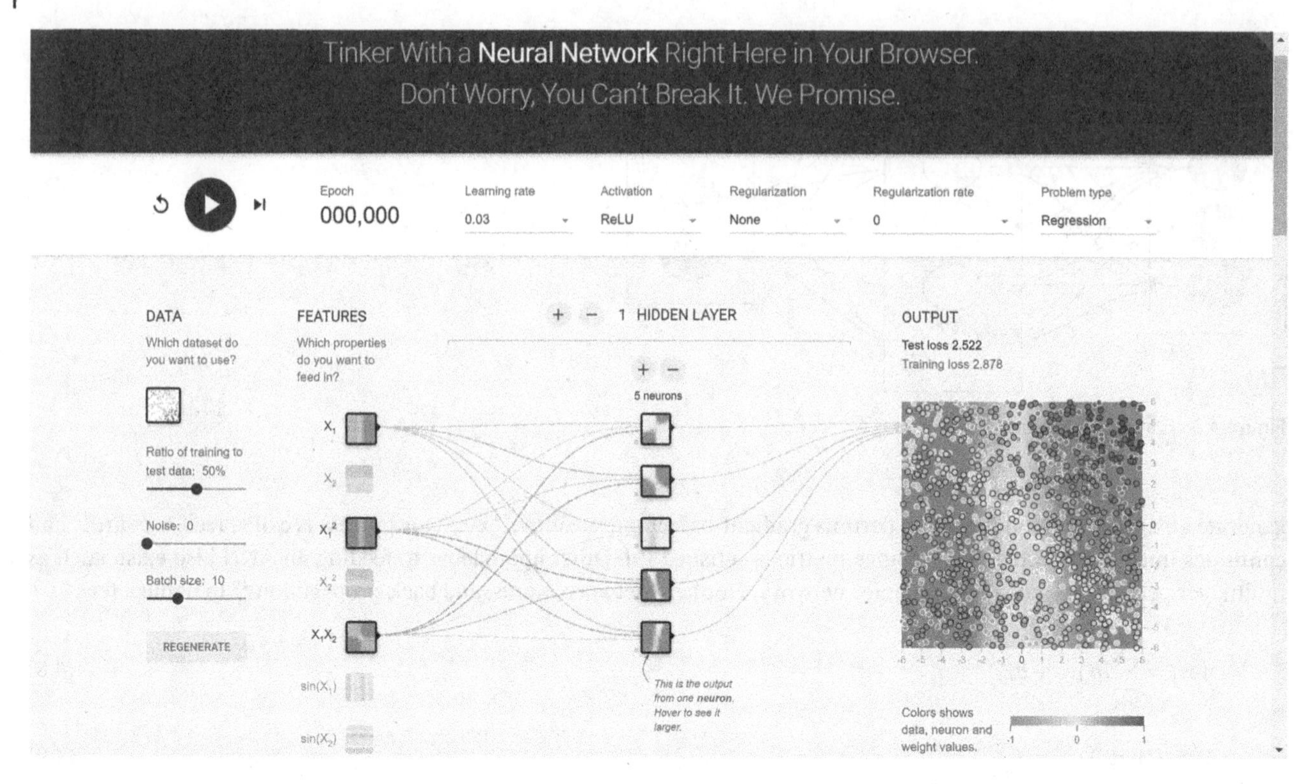

(a) Regression (pre-analysis)

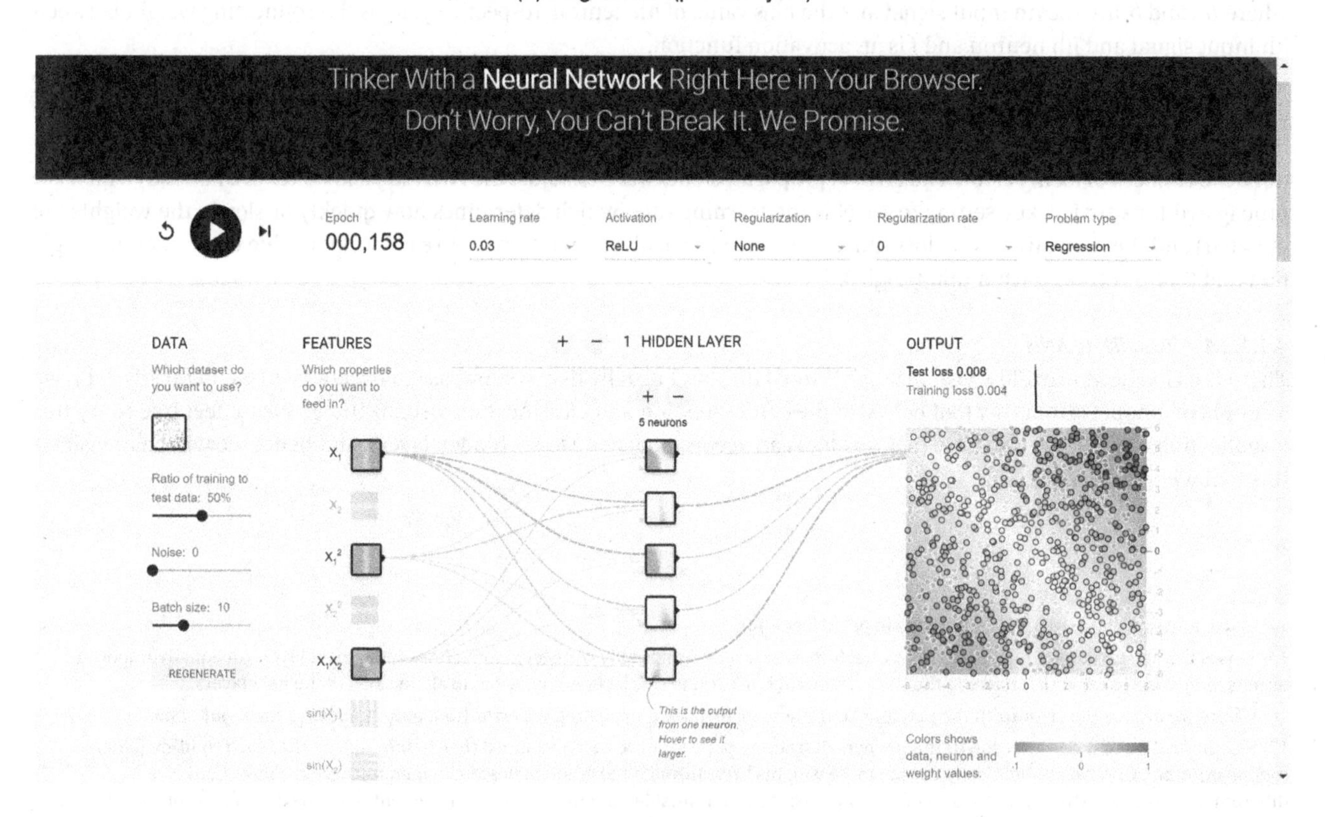

(a) Regression (post-analysis)

(Continued)

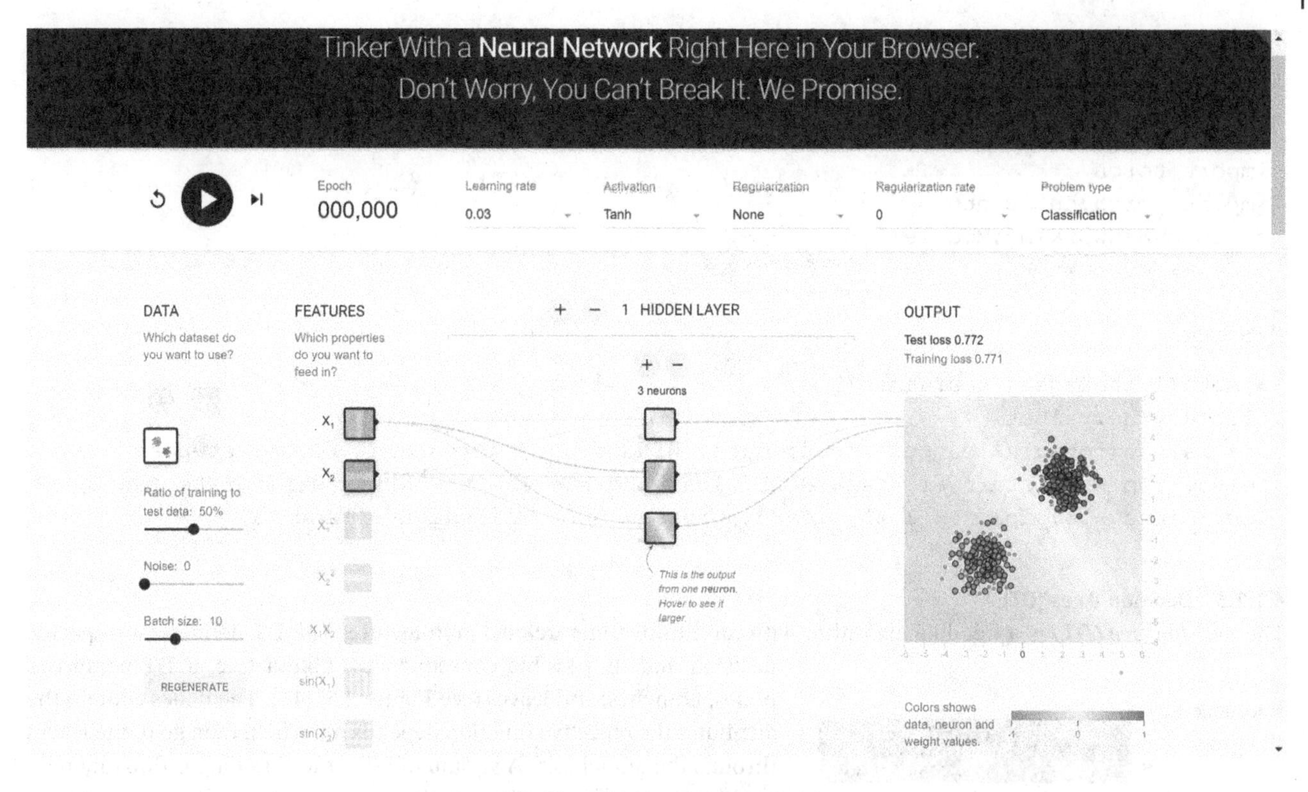

(b) Classification (pre-analysis)

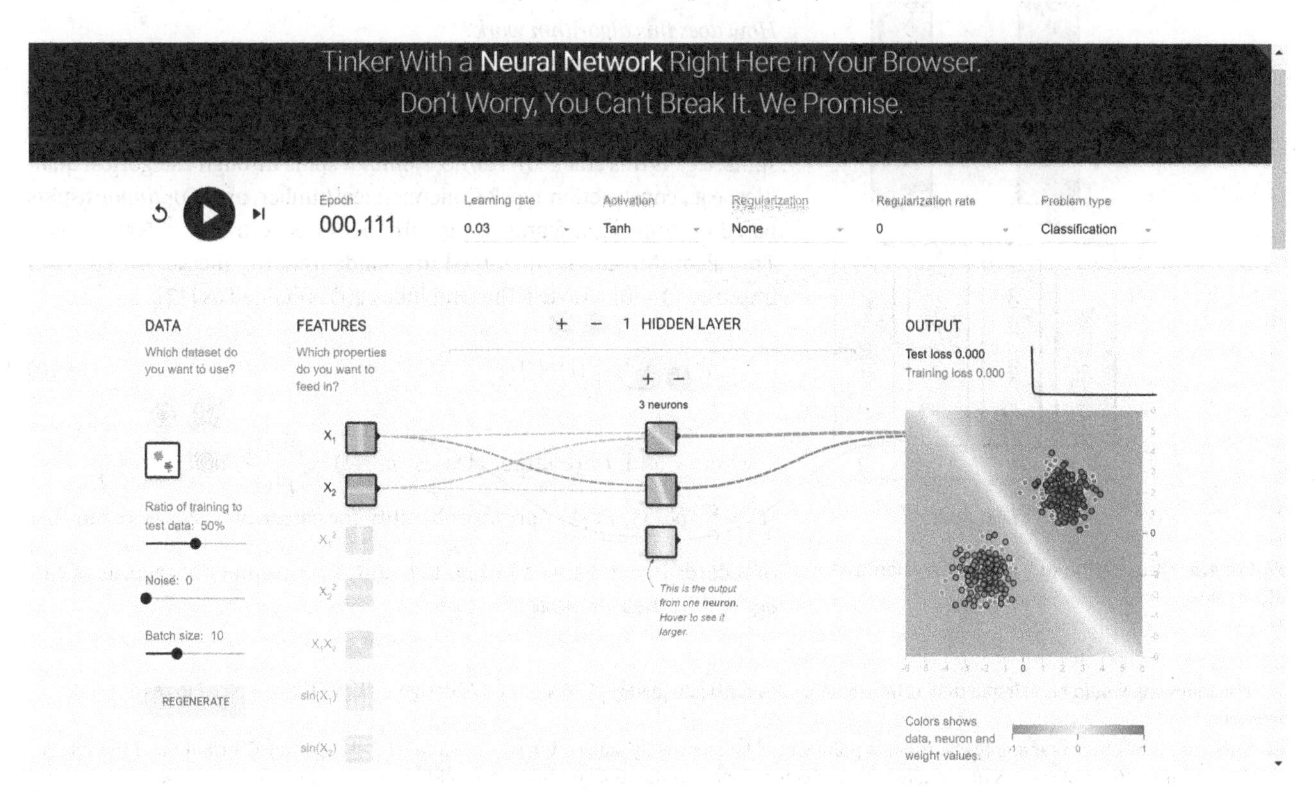

(b) Classification (post-analysis)

Figure 4.4 (Cont'd) Demonstration of a typical ANN (from TensorFlow neural network playground [10]).

4.1.1.2.2 Programming This Algorithm

The *Python* and *R* codes for a typical ANN are provided below:

Python Package

```
import torch
import torch.nn as nn
class MyNetwork(nn.Module):
model = MyNetwork()
```

R Package:

```
>install.packages("neuralnet")
>library(neuralnet)
>neuralnet(formula= output~input, data, hidden = 1, threshold = 0.01, stepmax =
1e+05, rep = 1, startweights = NULL, learningrate.limit = NULL, arningrate.factor =
list(minus = 0.5,plus = 1.2), learningrate = NULL, lifesign = "none"…)
```

4.1.1.3 Decision Trees (DT)

The *decision tree (DT)* is a tree-like algorithm with an intuitive hierarchical architecture. The DT defines each specific decision and its possible consequences. Like a tree, a DT comprises nodes, branches, and leaves (see Figure 4.5) [11]. The nodes contain the attributes the objective function depends on, which then go to the leaves through the branches.[20] A variant of DT is the *CART* algorithm (abbreviated for classification and regression tree).

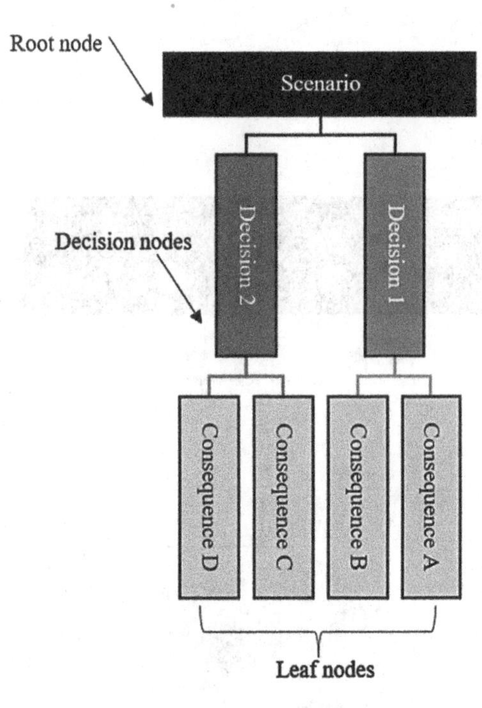

Figure 4.5 Illustration of a typical decision tree algorithm.

How does this algorithm work?

DT is a simple and easy-to-use algorithm and can be applied in regression and classification problems. This ease stems from its graphical representation of its thought process. DT splits the data via three options (*Boolean* splits: e.g., is this cracked? Yes/no, *Nominal* splits through categorical qualities: e.g., construction type? Concrete/steel/timber, and *Continuous* splits based on numerical features: e.g., number of pedestrians? $\leq 500/>500$). This algorithm can be finetuned to include impurity measures (e.g., Gini impurity[21]) – for a node t, the Gini index $g(t)$ is defined as [12]:

$$g(t)=\sum_{j\neq i} p(j\,|\,t)p(i\,|\,t) \tag{4.10}$$

where i and j are target classes, $p(j\,|\,t)=\frac{p(j.t)}{p(t)}$, $p(jt)=\frac{\pi(j).N_j(t)}{N_j}$, $(t)=\sum_j p(j.t)$, $P(j)$ = prior probability for category j, $N_j(t)$ = number of records in category of j of node t, and N_j = number of records of category j in the root node.[22]

20 The following would be a simple tree. Is this member designed adequately? If Yes → End of design process. If No → Need to re-design the member.

21 Measures how often a randomly chosen data point would be incorrectly labeled if it was randomly labeled per the distribution of the subset. Other measures also exist such as *Information Gain* which rationalizes that the best feature to split on is the one producing the largest reduction in entropy from the original set, and *Gain Ratio* which normalizes the information gain of a feature against how much entropy it has.

22 Additional notes: DT can be useful for data with features of unknown importance and tend to be insensitive to normalization issues or noise. Given their hierarchical structure, DT may reveal unexpected dependencies in the data. DT stops the analysis once a node contains one pattern or once all features are used. Strive to prune trees to avoid overgrown trees, complexity, and overfitting. There are two methods to prune DT, *pre-pruning* which stops the construction of the tree during training by assigning a minimum size for the tree leaves, and *post-pruning* which reduces the size of an overgrown tree by replacing poor branches by leaves, or a subtree by its largest branch without jeopardizing model performance. Techniques for pruning include [80], *reduced error pruning*, *pessimistic pruning*, etc.

4.1.1.3.1 Visualizing This Algorithm

An intuitive visualization[23] I found on DT is that developed by Stephanie Yee and Tony Chu (from R2D3) [13]. I have taken a snippet on an example problem that classifies homes in New York from homes in San Francisco based on elevation, year built, prices, among other features. This DT is plotted in Figure 4.6.

4.1.1.3.2 Programming This Algorithm

The *Python* and *R* codes for the DT algorithm are provided below:

Python Package

```
from sklearn import tree
from sklearn.tree import DecisionTreeClassifier
clf = tree.DecisionTreeClassifier()
```

R Package:

```
>install.packages("rpart")
>library(rpart)
rpart(formula, data, weights, subset, na.action = na.rpart, method, model = FALSE, x
= FALSE, y = TRUE, parms, control, cost, ...)
```

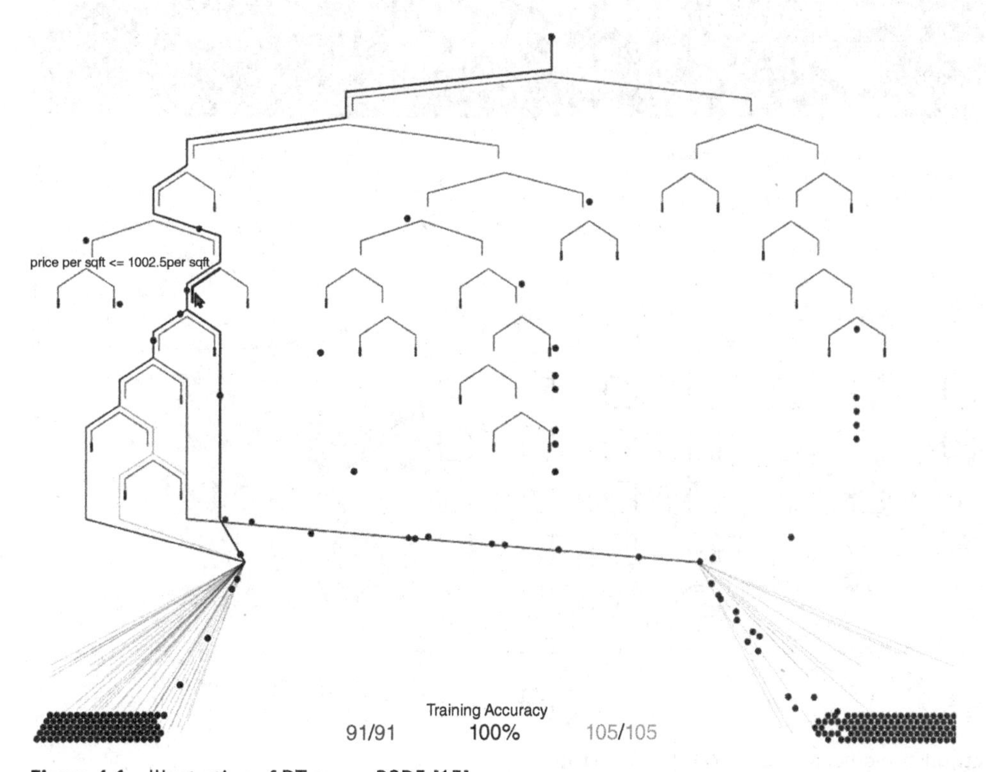

Figure 4.6 Illustration of DT as per R2D3 [13].

4.1.1.4 Deep Learning (DL)

Deep learning (DL) is a form of ANN that contain more than one[24] hidden layer and hence is a natural progression of ANN.[25] Unlike ANN, in DL, each layer of nodes is trained on a distinct set of features based on the previous layer's output.

---

23 http://www.r2d3.us/visual-intro-to-machine-learning-part-1.

24 Hence, the term *deep*!

25 Notable pioneers of DL are Yoshua Bengio, Geoffrey Hinton, and Yann LeCun [81].

The most common architectures for DL include Recursive Neural Networks (RNN) [26] and Convolutional Neural Networks (CNN). RNN are primarily designed to handle text data, while CNN are better at handling images and videos.[27]

How do RNN and CNN work?
An RNN is trained through a sequence of inputs where the output of some layers is fed back to the previous layer, effectively allowing for a feedback loop. Such looping enables learning contextual concepts, and hence RNN can remember important details from the input they receive. This allows RNN to precisely predict what's coming next.[28]

CNN is built on the concept of convolution[29] operations, defined using the three spatial dimensions: length, width, and depth. Unlike layers in ANN, neurons in convolutional layers do not connect to each and every neuron in the following layer.[30] To simplify the data between the convolution layers, pooling layers are often added between convolutional layers. Fully connected layers appear at the end. In such layers, neurons are fully connected, looking for what features most accurately describe the task (say, an image).

4.1.1.4.1 Visualizing RNN and CNN

A typical visualization tool for DL is the same as that we have seen in the ANN example (from the minds in TensorFlow [10]), as shown in Figure 4.7. All that I did was extend the former ANN example by adding new hidden layers and randomly adjusting the number of neurons.

To give a glimpse into computer vision applications, allow me to present the next visualization tool. Adam Harley [14] created an interactive tool[31] that can help us visualize how a CNN predicts numbers[32] written manually (see Figure 4.8).

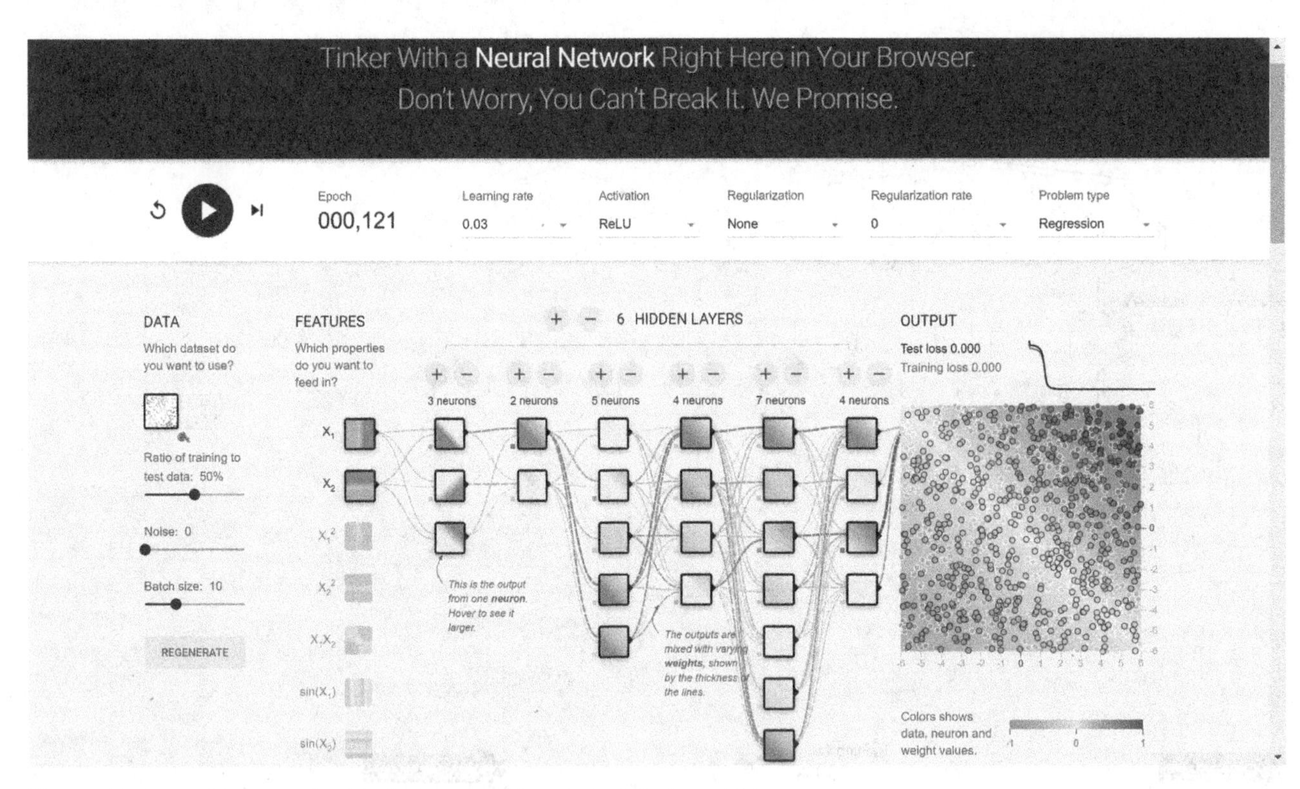

Figure 4.7 TensorFlow neural network playground for a typical DL (from [10]).

---

26 Side note: *Apple's Siri* and *Google's voice search* are based on RNN.

27 We can combine RNN and CNN together. The result is called *DanQ* architecture [82].

28 Think of auto-correct, or auto-search!

29 **Convolution operation**: a grouping function that occurs between two matrices. This operation combines two functions to make a third and hence merges information.

30 You can visualize this by imagining a sliding window (which is unlike that of connecting every neuron to others).

31 https://adamharley.com/nn_vis.

32 A similar approach can be used to predict cracking or other phenomena that we are interested in.

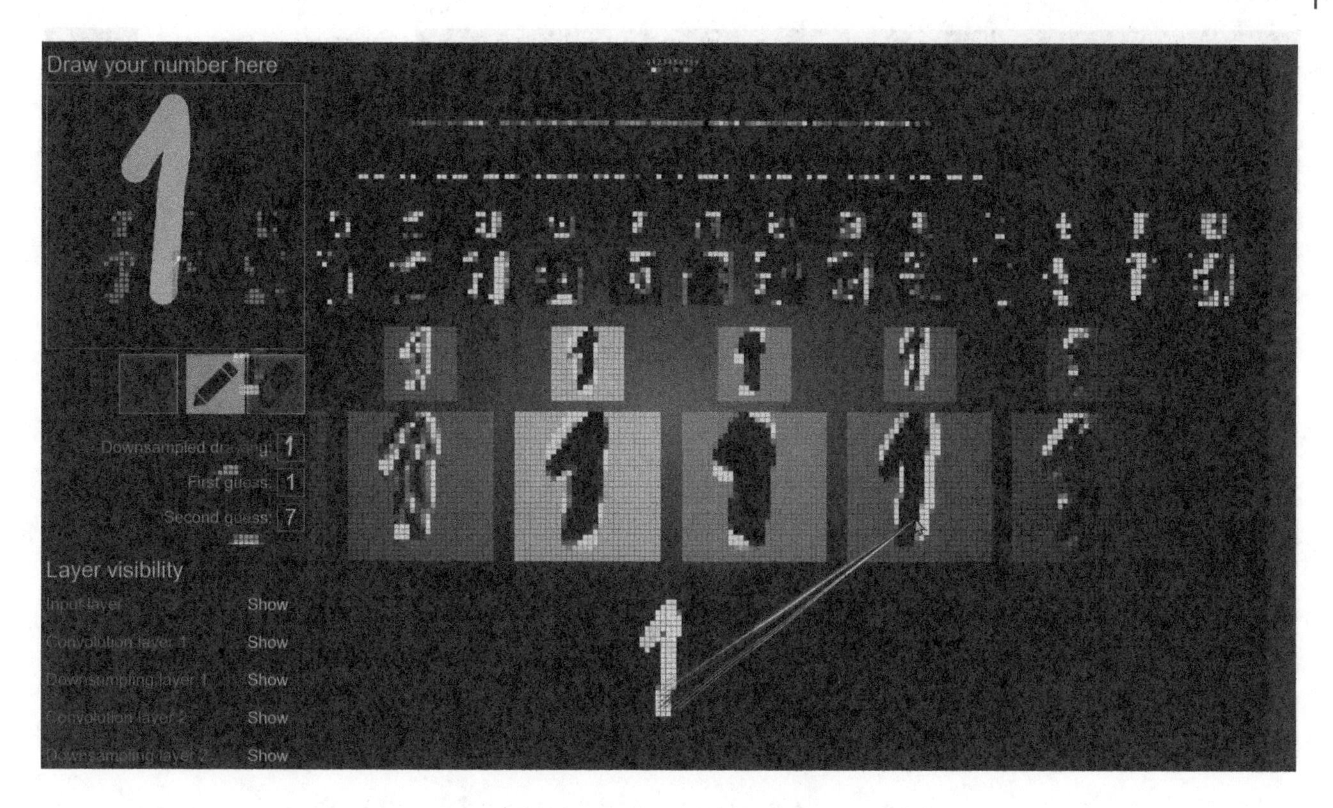

(a) First step: handwrite number one

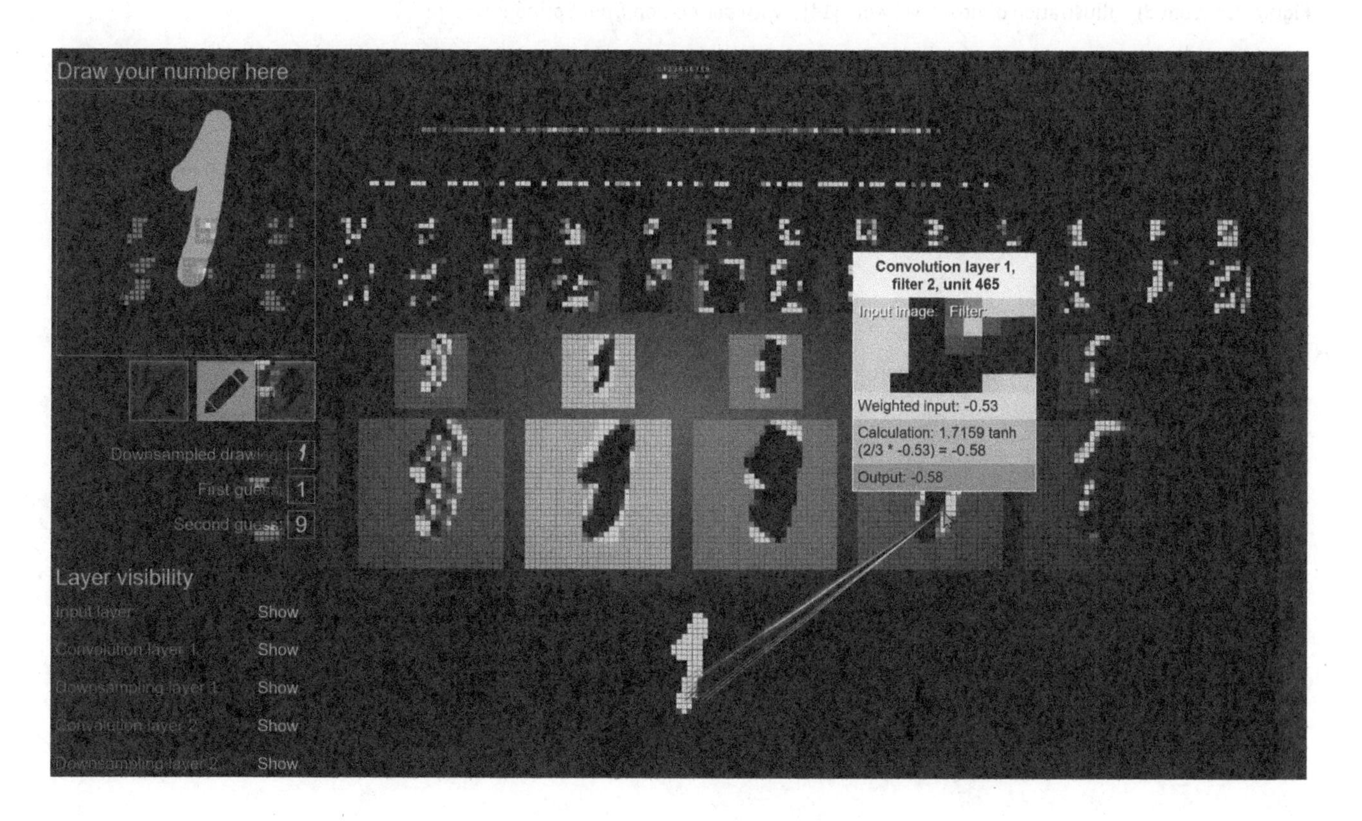

(b) Demonstration of the first convolution layer

(Continued)

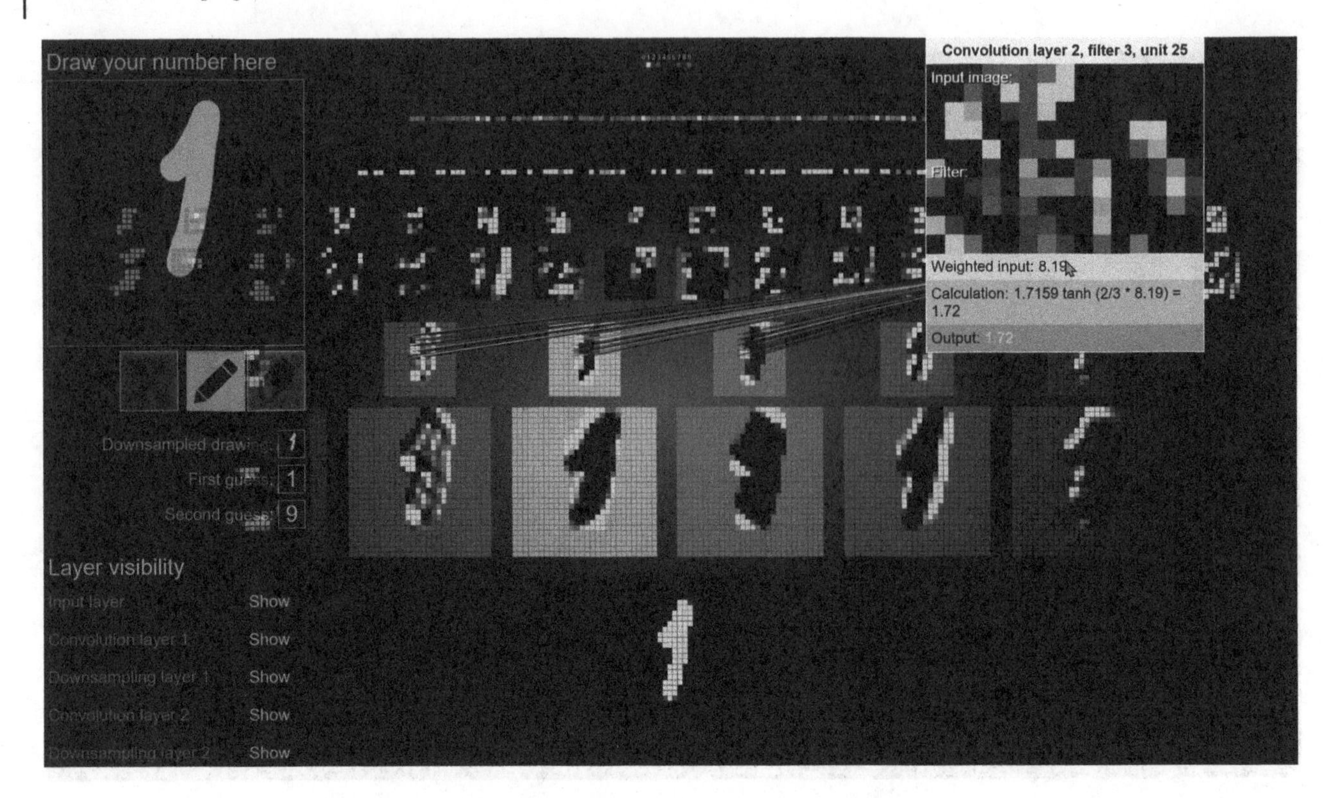

(c) Demonstration of the second convolution layer

Figure 4.8 (Cont'd) Illustration of how CNN work [14] / with permission from Springer Nature.

I am presenting the following illustrations of my handwriting of the number one. As you can see, this tool showcases each process taken to arrive at the first guess of one and second guess of nine.

The final visualization tool under DL is that created by Andreas Madsen [15]. As you can see, Figure 4.9 creates a neat visualization[33] for RNN that is based on textual data for an autocomplete task. This tool shows the vanishing gradient problem where the contribution from the earlier steps becomes insignificant at advanced steps (i.e., text input early on to later inputs).

4.1.1.4.2 Programming CNN and RNN

The *Python* and *R* codes for typical DL using the TensorFlow deep learning library:

Python Package

```
from tensorflow.keras.models import Sequential
from tensorflow.keras.layers import Dense
model = Sequential()
model.add(Dense(12, input_shape=(8,), activation='relu'))
model.add(Dense(8, activation='relu'))
model.add(Dense(1, activation='sigmoid'))
```

R Package:

```
library(keras)
install_tensorflow()
```

33 https://distill.pub/2019/memorization-in-rnns.

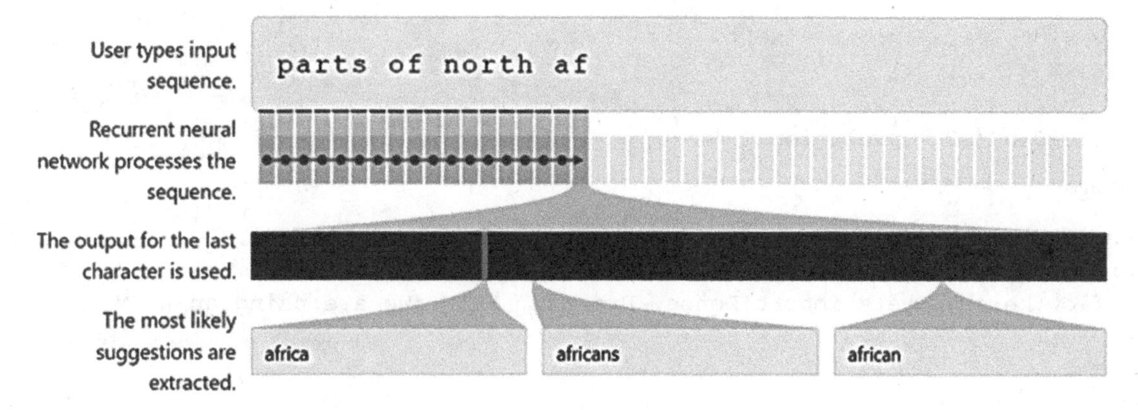
User types input sequence.
parts of north af
Recurrent neural network processes the sequence.
The output for the last character is used.
The most likely suggestions are extracted.
africa
africans
african
Autocomplete: An application that has a humanly interpretable output, while depending on both short and long-term contextual understanding. In this case, the network uses past information and understands the next word should be a country.

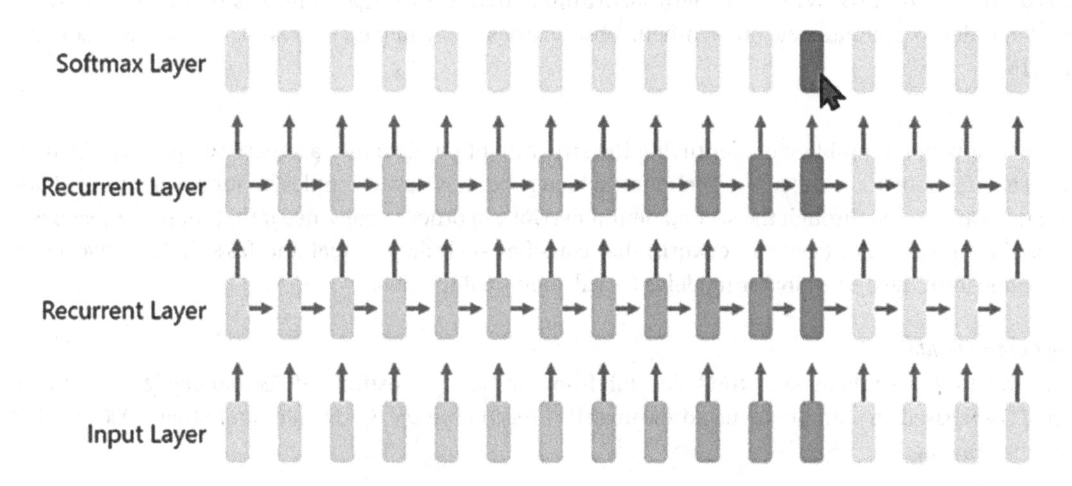
Softmax Layer
Recurrent Layer
Recurrent Layer
Input Layer
Vanishing Gradient: where the contribution from the earlier steps becomes insignificant in the gradient for the vanilla RNN unit.

Figure 4.9 Demonstration of RNN (from [15]).

The *Python* and *R* codes for typical CNN and RNN are provided below:

CNN Python Package

```
import tensorflow as tf
from keras.models import Sequential
from keras.layers import convolutional
from convolutional import Conv2D
model = Sequential()
model.add(Conv2D)
```

CNN R Package:

```
>install.packages("keras")
>library(keras)
>k_convNd(x, kernel, strides = c(1, 1, 1),
```

```
padding = "valid", data_format = NULL,
dilation_rate = c(1, 1, 1)
)
N=[1,2,3] for 1D/2D/3D convolutional
```

RNN Python Package

```
import tensorflow as tf
from tensorflow.keras.models import Sequential
from tensorflow.keras.layers import Dense, Dropout, LSTM #We are using an LSTM
variant here
```

RNN R Package:

```
install.packages('rnn')
```

4.1.1.5 Generative Adversarial Networks (GAN)

The *Generative Adversarial Networks (GAN)* is a modern algorithmic architecture that contrasts two neural networks against one another [34] in order to generate new but synthetic observations that can pass for real data. GAN is credited to Ian Goodfellow et al. [16].

How does GAN work?
GANs re-formulate the unsupervised problem of identifying the structure of the data into a supervised problem by using two neural networks. The first network is called the generator, which generates new examples of our observations. These new synthetic observations are passed through the second neural network in order to convince it that these are real observations. However, it is the discriminator (second network) that classifies such data as real and fake. Both networks are trained in a zero-sum game until the discriminator model is fooled about half the time.[35]

4.1.1.5.1 Visualizing GLM and GAM

A great visualization[36] tool for GAN is created by the Polo Club from the Georgia Institute of Technology [17, 18]. Figure 4.10 is an illustration of GAN based on a simple shape I drew myself.[37] As you can see, GAN takes more than 7000 epochs[38] to be able to replicate the drawing.

4.1.1.6 Generalized Models (GM)

There are two *generalized*[39] *models* (GM) that we will go over herein, namely, the *generalized linear model* (GLM) and the *generalized additive model* (GAM). For completion, we will compare both with traditional linear regression.

The GLM was formulated by John Nelder and Robert Wedderburn in 1972 and fits traditional linear models by maximizing the log-likelihood [19]. This algorithm is an extension of linear models and hence is quick, interpretable, and efficient. In GLM, the distribution of the response (or target) can be explicitly non-normal and can be non-continuous (categorical). This response is arrived at via a linear combination of the features involved connected through a link function.[40] There are variations to GLM, such as Poisson generalized linear model (PGLM), negative binomial generalized linear model (NBGLM), generalized linear mixed model (GLMM), etc.

A special case of GLM is the *generalized additive model (GAM)*. GAM was created by Trevor Hastie and Robert Tibshirani in 1986 [20]. GAM aims to minimize prediction errors of the target by defining the model in terms of smooth function to

---

34 And hence the *adversarial*.

35 This means that generator network is actually generating plausible observations!

36 https://poloclub.github.io/ganlab.

37 Excuse the lack of imagination. I did try a couple of more complex shapes and convergence took a while...

38 **Epoch**: one complete pass of the training data through the algorithm.

39 Remember that the term *general linear model* refers to the traditional linear regression model.

40 Typical link functions include: Log link : $f(z)=\log(z)$. Inverse link : $f(z)=1/z$. Identity link : $f(z)=z$. Logit link : $f(z)=\log(z/(1-z))$.

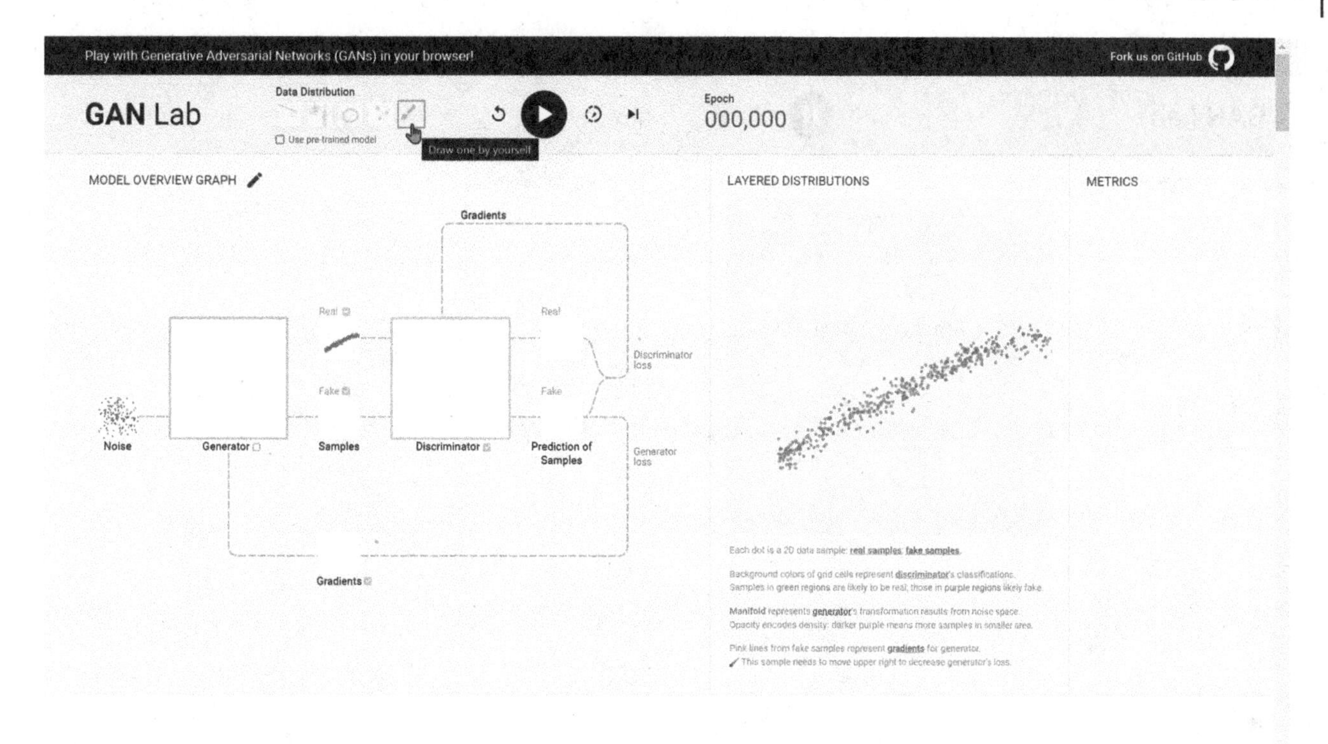

(a) Select *Draw one by yourself* and draw a shape

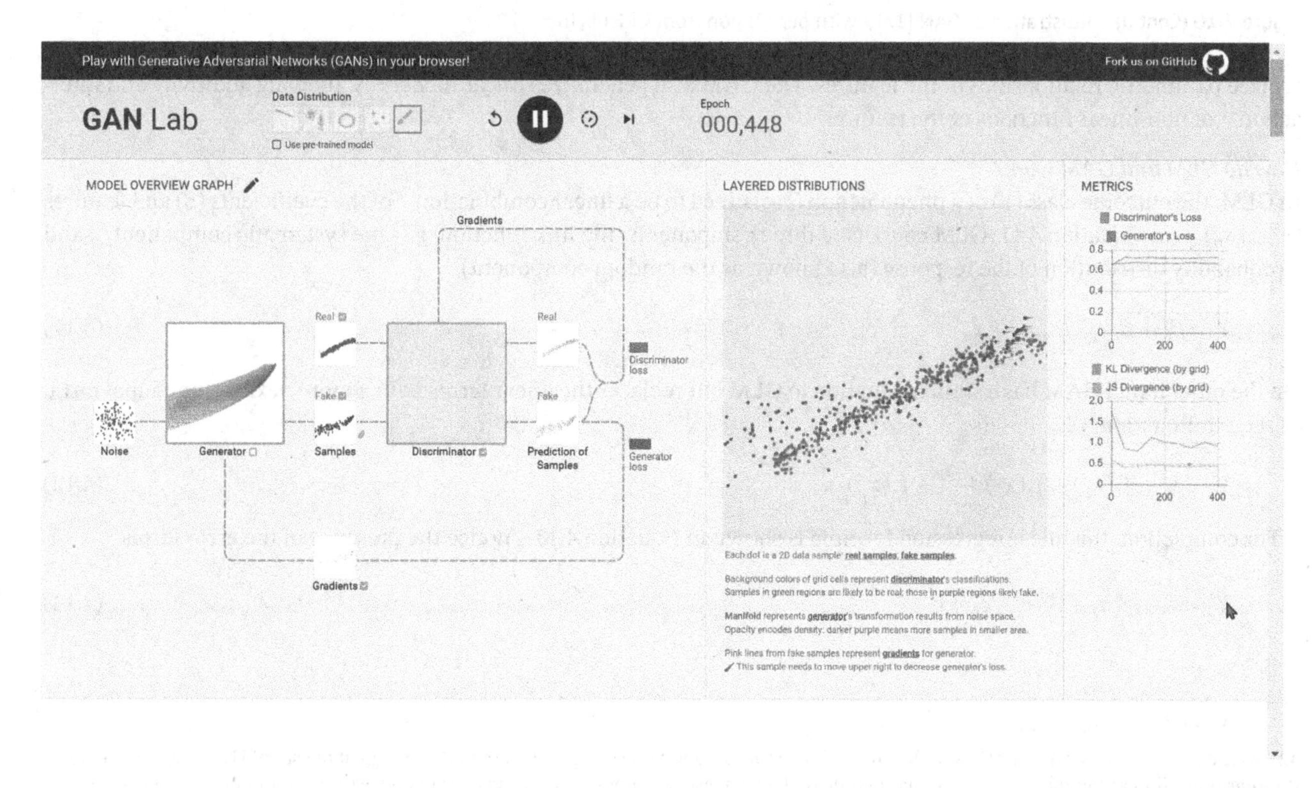

(Continued)

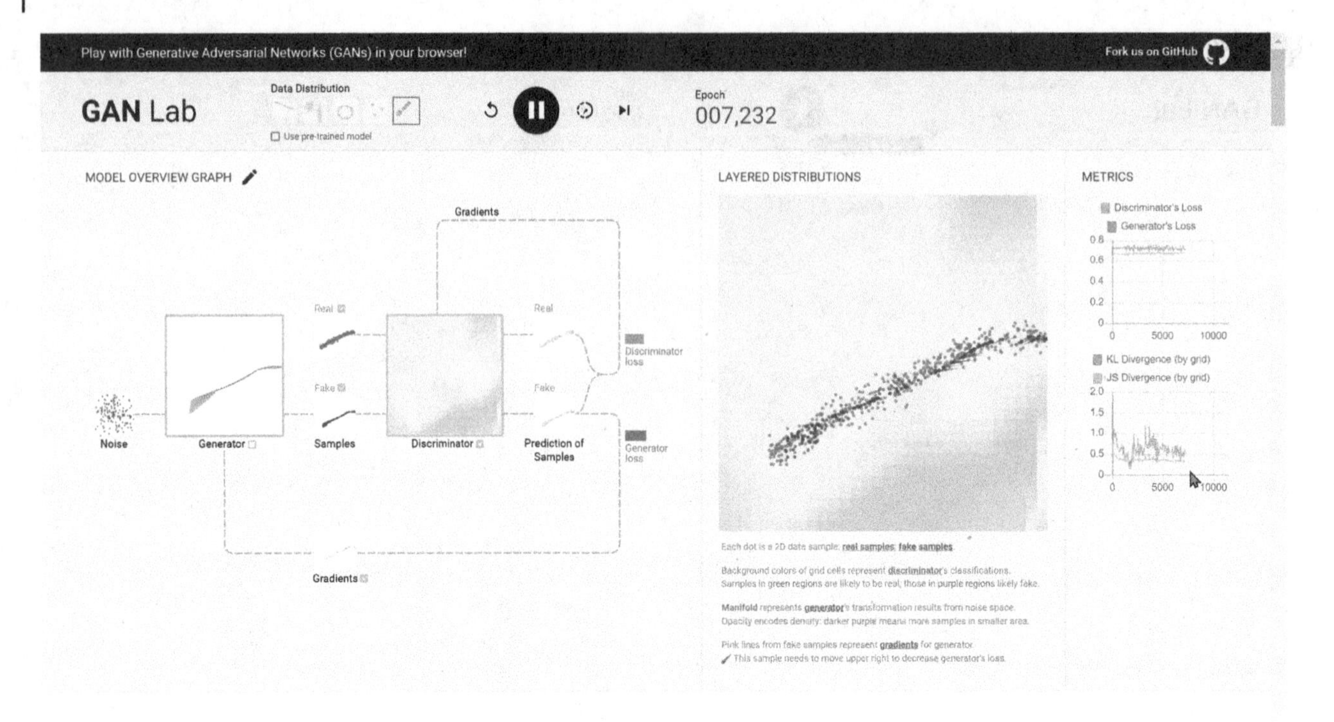

(b) Processing in GAN

Figure 4.10 (Cont'd) Illustration of GAN [17] / with permission from GitHub, Inc.

replace parametric relationships of the features. The GAM also generalizes linear models by allowing additivity and interaction[41] of non-linear functions of the features.

How do GLM and GAM work?

In GLM, the outcome class (*Y*) of a phenomenon is assumed to be a linear combination[42] of the coefficients (β) and features (x_1, ..., x_n) – see Equation 4.11. GLM consists of three components: the link function, g,[43] the systematic component,[44] and a probability distribution of the response (also known as the random component).

$$Y = \beta_0 + \beta_1 x_{1,i} + \ldots + \beta_n x_{n,i} \quad (4.11)$$

On the other hand, GAM has a similar equation to GLM but replaces the linear terms with a more flexible functional form, as seen in Equation 4.12.

$$Y = \beta_0 + f_1(x_1) + f_2(x_2) + \ldots + f_p(x_p) \quad (4.12)$$

For completion, the linear regression formula is shown in Equation 4.13 – notice the presence of the error terms.

$$Y = \beta_0 + \beta_1 x_1 + \ldots + \beta_p x_p + \epsilon \quad (4.13)$$

41 GAM with interaction → GA2M!

42 In case you are wondering how GLM differs from simple (or multiple) linear regression, then this is a great question! Here is an answer to this question. GLM accommodates various distribution of the response (i.e., non-normal) and does not have to be continuous, i.e., it can be binomial, multinomial, or ordinal multinomial. GLM also predicts the response from a linear combination of features that are connected via a link function (e.g., a function that maps nonlinear relationship into a linear relationship). The linear regression assumes a linear relationship between the response and features, while GLM does not.

43 g is the link function that defines how the expected value of the target (*E*(*Y*)) is related to the linear predictor of its features.

44 Refers to the linear combination of features. $g(E[y] = \beta_0 + \beta_1 x_{1,i} + \ldots + \beta_n x_{n,i}$ where the red, green, and blue coloring schemes represent the link function, random, and systematic components, respectively.

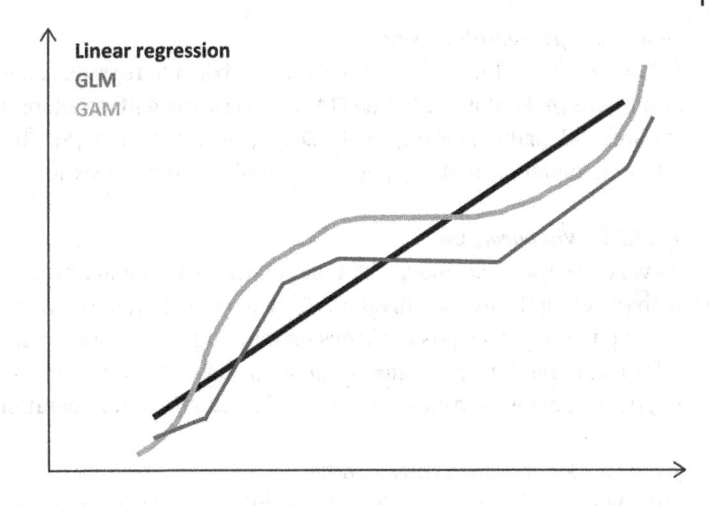

Figure 4.11 Illustration of linear regression, GLM, and GAM into the same set of hypothetical data.

4.1.1.6.1 *Visualizing GLM and GAM*

A look into Figure 4.11 shows a glimpse into a comparison between a hypothetical fitting of linear regression, GLM, and GAM models into the same data. I hope you can appreciate how GLM can be a bit smoother than GAM (which can be clearly seen by comparing Equations 4.11 and 4.12.

4.1.1.6.2 *Programming This Algorithm*

The *Python* and *R* codes for the GLM algorithm are provided below:

Python Package

```
#For GLM
from statsmodels.formula.api import glm
Model=glm()
```

```
#For GAM
from pygam import GAM, s, te
```

or,

```
from pygam import LinearGAM
```

R Package

```
#For GLM
>install.packages("stats")
>library("stats")
>glm(formula, family = gaussian, data, weights, subset, na.action, start = NULL,
etastart, mustart, offset, control = list(...), model = TRUE, method = "glm.fit", x =
FALSE, y = TRUE, singular.ok = TRUE, contrasts = NULL, ...)
```

```
#For GAM
library(mgcv)
gam_y <- gam(y ~ s(x), method = "REML")
```

4.1.1.7 Genetic Algorithm (GA)

The *genetic algorithm (GA)* is a population-based metaheuristic[45] algorithm that imitates the Darwinian evolutionary principle of survival of the fittest as introduced by John Holland [21].

45 **Heuristic**: Act of discovering. A heuristic is an approach to solve a specific method. **Metaheuristic**: A high-level problem-independent algorithmic framework (or simply search strategy) that develops heuristic optimization algorithms. Such approaches stem inspiration from nature and the biological evolutionary process. These can become very attractive in searching for surrogate models – a topic we will cover in upcoming chapters. More on Heuristics and metaheuristics in a bit.

How does this algorithm work?

In GA, possible solutions[46] in the population begin to reproduce and undergo genetic evolutions in their structure and pass it into future generations [22]. The GA analysis starts with a randomly selected population of possible solutions (i.e., comprise features and mathematical symbols (i.e., exp, log, ×, +, etc.)) [5]. The GA continues to evolve with evolutionary operations (i.e., selection, mutation, and cross-over) taking place to enhance the created solution and to arrive at the fittest solution.[47]

4.1.1.7.1 Visualizing GA

As we have discussed above, GA relies on the evolutionary process and the process of natural selection. This process starts with selecting the fittest individuals from a population to produce offsprings (which will be added to the next generation of the population). This process keeps on iterating to produce a generation with the fittest individuals. In terms of mutation, this process randomly mutates the genes in a chromosome to help maintain diversity and prevent premature convergence. Figure 4.12 helps us visualize GA as well as evolutionary operations.

4.1.1.7.2 Programming This Algorithm

The *Python* and *R* codes for the GA algorithm are provided below:

Python Package

```
import pygad
model = pygad.GA()
```

OR

```
from nia.algorithms import GeneticAlgorithm
model = GeneticAlgorithm()
```

R Package:

```
>install.packages("GA")
>library("GA")
>ga(type = c("binary", "real-valued", "permutation"), fitness, …, lower, upper,
nBits,…)
```

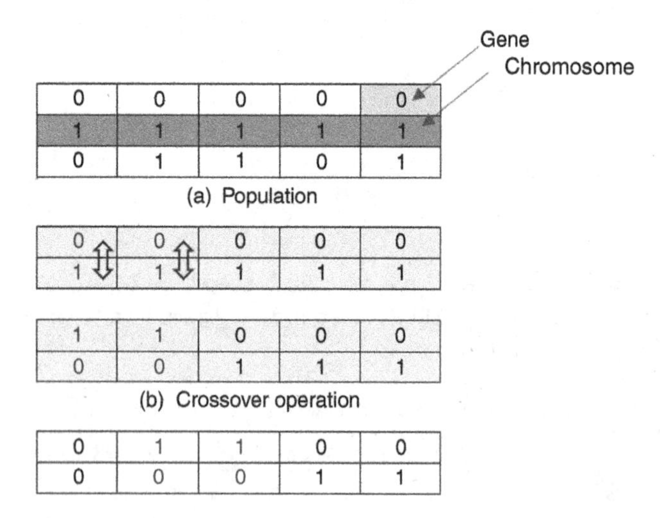

(c) Mutation operation (Top: before mutation, Bottom: after mutation)

Figure 4.12 Illustration of GA.

46 For now, let us say that the *possible solutions* resemble *expressions* or *equations* that describe the phenomenon on hand. Thus, a possible solution (equation) would start with a certain structure layout, and then evolves into a different structure as the search process progresses.

47 One that maximizes predefined metrics.

4.1.1.8 Genetic Programming (GP)*

Genetic programming (GP) is a modern variant and a special case of GA and can be further grouped under *linear genetic programming (LGP)* and *genetic expression programming (GEP)* [23, 24]. Unlike GA, which is made of expressions (or simple strings of formulas), GP creates computer programs in the form of a tree-like structure (as in codes[48]) [25]. John Koza is credited for his invention of GP in 1988.[49]

4.1.1.8.1 Programming This Algorithm

The *Python* and *R* codes for the GP algorithm are provided below:

Python Package:

```
#Use package DEAP from https://pypi.org/project/deap/ [26]
pip install deap
import random
from deap import creator, base, tools, algorithms
```

R Package:

```
>install.packages("grp")
>library("grp")
>geneticProgramming(fitnessFunction, stopCondition=makeTimeStopCondition(5),
population, populationSize=100, eliteSize=ceiling(0.1 * populationSize),
elite=list(), functionSet=mathFunctionSet
```

4.1.1.9 Gradient Boosted Trees (GBTs)

GBTs are simple tree-like models. In their simplest form, they use the first model's error as a feature to build successive models. Focusing on errors obtained from a feature when building successive models reduces the overall error in the model. There are variants to GBT, most notably the *Extreme Gradient Boosting (XGBoost)*, the *Light Gradient Boosting Machine (LightGBM)*, and the *CatBoost*[50] *algorithm.*

How do these algorithms work?

The XGboost brings two techniques to improve the performance of a GBT; a weighted quantile sketch by means of an approximation algorithm that can determine how to make splits candidates in a tree and the sparsity-aware split finding technique. The XGboost uses a pre-sorted algorithm and a histogram-based algorithm for computing the best split. This algorithm was created by Chen and Guestrin [27] and can be found at [28, 29]. This algorithm does not use any weighted sampling techniques.

$$Y = \sum_{k=1}^{M} f_k(x_i),\ f_k \in F = \left\{ f_x = w_{q(x)}, q:\ R^p \rightarrow T, w \in R^T \right\} \tag{4.14}$$

where M is additive functions, T is the number of leaves in the tree, w is a leaf weights vector, w_i is a score on i-th leaf, and $q(x)$ represents the structure of each tree that maps an observation to the corresponding leaf index [30].

On the other hand, the LightGBM algorithm, created by Microsoft [31], also introduces two new techniques to improve the performance of a GBT. These techniques are gradient-based one-side sampling[51] and exclusive feature bundling[52] to accelerate the processing time within this algorithm. Finally, the CatBoost algorithm is the latest of the discussed algorithms and was developed by Yandex in 2017. This algorithm adopts the minimal variance sampling techniques in which it weighs sampling at the tree-level to maximize the accuracy of split scoring.

---

48 To re-iterate, GA searches for expressions and symbols to fit into an expression. On the other hand, GP searches for software components to fit into a code.

49 The first record of the proposal to evolve programs is attributed to Alan Turing in the 1950s.

50 Multiple Categories of data + Boosting = CatBoost.

51 Which identifies the most informative observations and skips those less informative.

52 To group features in a near-lossless way.

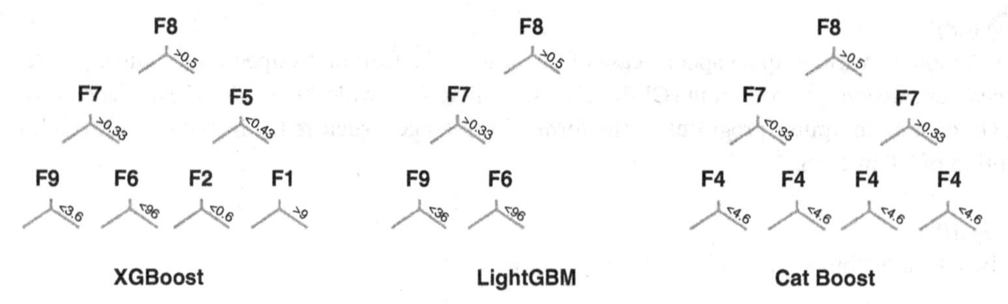

Figure 4.13 Comparison between XGboost, LightGBM, and CatBoost (inspired by [32]).

4.1.1.9.1 Visualizing GBTs

Figure 4.13 shows how XGboost, LightGBM, and CatBoost differ in the way they build and grow their trees. As you can see, the XGboost splits up to the specified maximum depth set by the user and then prunes the tree backward to remove splits beyond which there is no positive gain. On the other hand, the LightGBM uses leaf-wise tree growth as it chooses to grow the leaf that minimizes the algorithmic loss (which may allow the growth of imbalanced trees).[53] Finally, the CatBoost grows balanced trees where at each level, the feature-split pair brings the lowest loss.

4.1.1.9.2 Programming This Algorithm

The *Python* and *R* codes for the XGBoost and LGBM algorithms are provided below:

Python Package

```
import xgboost as xgb
```

```
#For classification
from xgboost import XGBClassifier
model2 = XGBClassifier()
```

```
#For regression
from xgboost import XGBRegressor
model = XGBRegressor()
```

```
#For classification
import lightgbm as ltb
model = ltb.LGBMClassifier()
```

```
#For regression
import lightgbm as ltb
model = ltb. LGBMRegressor()
```

```
#For regression
from catboost import CatBoostRegressor
```

```
#For classification
from catboost import CatBoostClassifier
```

R Package:

```
#For XGBoost
>install.packages("'xgboost'")
>library("'xgboost'")
```

53 This procedure may lead to overfitting when the data is small.

```
> xgb.train(params = list(),data,nrounds, watchlist = list(),obj = NULL,feval =
NULL, verbose = 1,print_every_n = 1L, early_stopping_rounds = NULL,
maximize = NULL, save_period = NULL,save_name = "xgboost.model",xgb_model =
NULL,callbacks = list(),…)
```

```
#For LGBM
>install.packages("'lightgbm'")
>library('lightgbm')
>lightgbm( data, label = NULL, weight = NULL, params = list(), nrounds = 100L, ver-
bose = 1L, eval_freq = 1L, early_stopping_rounds = NULL, save_name = "lightgbm.
model", init_model = NULL, callbacks = list(), … )
```

```
#For CatBoost
> library(catboost)
```

4.1.1.10 Graph Neural Networks (GNN)*

Graph[54] neural networks (GNN[55]) are a class of deep learning methods designed to process data described by graphs. GNN was initially proposed by Scarselli and Gori [33]. Examples of a GNN would be city layouts (i.e., addresses are nodes and highways are edges), construction projects (e.g., workers are nodes, roles/tasks between them are edges), etc.

How does it work?

Every node in a graph is associated with a label, and the GNN aims to predict the label of such nodes (i.e., in a similar manner to a classification problem). In a way, a GNN uses the known labels of nodes and information from their neighborhood to predict the labels of unknown nodes.

4.1.1.10.1 Visualizing the RF Algorithm

Benjamin Sanchez-Lengeling et al. [34] developed a great graphical interface[56] that allows users to visualize GNNs. I have added Figure 4.14 from such an interface as it builds on the Leffingwell Odor dataset to predict the properties of molecules.

4.1.1.10.2 Programming This Algorithm

The *Python* code for a typical GNN is provided here:

Python Package

```
from torch_geometric.nn import GCNConv
model = GCN()
```

4.1.1.11 Heuristics (HS) and Metaheuristics (MHS)

Heuristics (HS) and *metaheuristics* (MHS) are algorithms designed to find solutions for optimization problems. The main difference between heuristics and metaheuristics is that the heuristics are problem-dependent, wherein they adapt to a specific problem to take full advantage of its particularities. Thus, it is common for heuristics to be trapped in local optima.[57] On the contrary, metaheuristics are problem-independent and can be used in many problems. Metaheuristics may accept a temporary decline in the fitness of a solution to allow a more thorough exploration within the solution space in pursuit of the global optimum.[58]

---

54 **Graph**: A data structure that consists of vertices (nodes) and envelopes.

55 Let us not confuse GNN with GANs (Generative adversarial networks).

56 https://distill.pub/2021/gnn-intro.

57 Hence the term, *greedy* heuristics.

58 If this reminds you of the different feature selection techniques, then great!

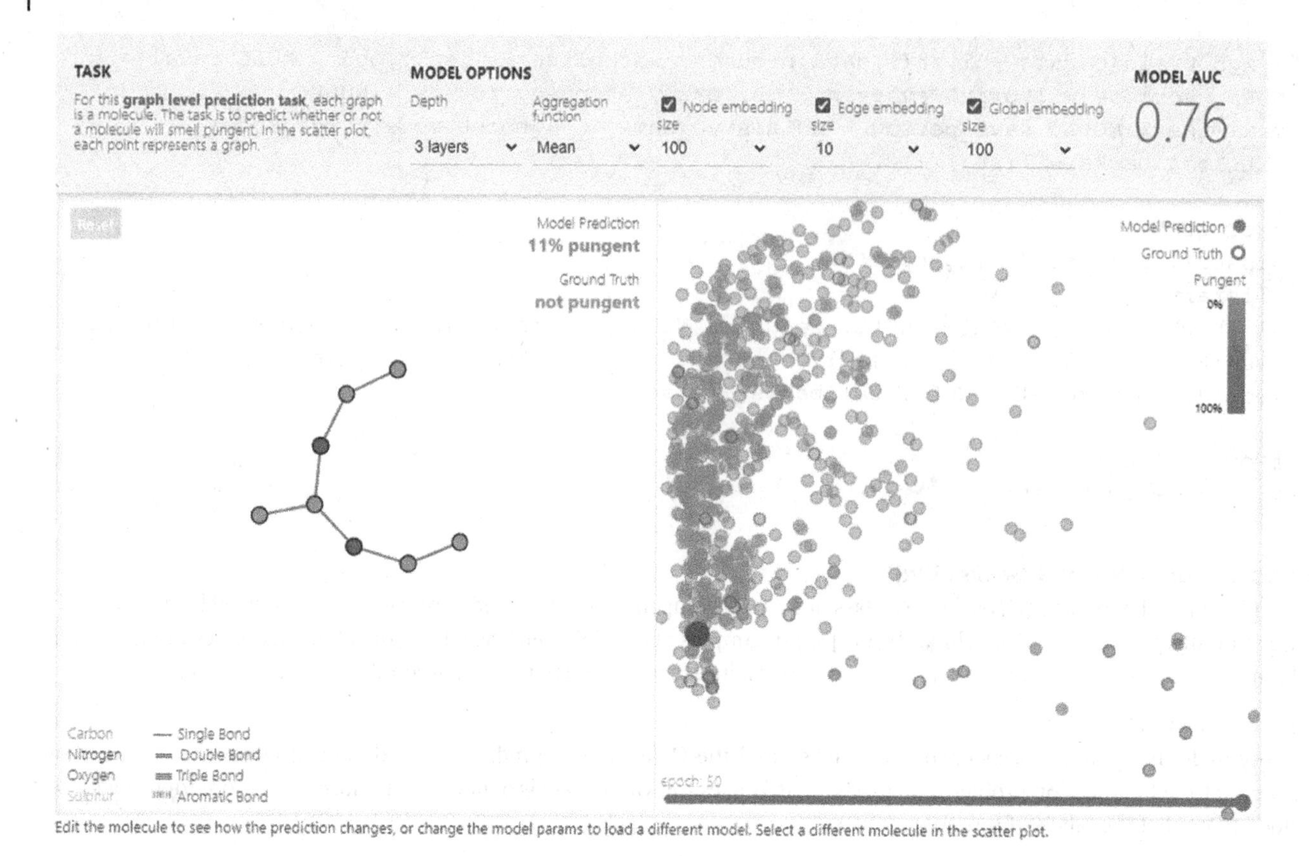

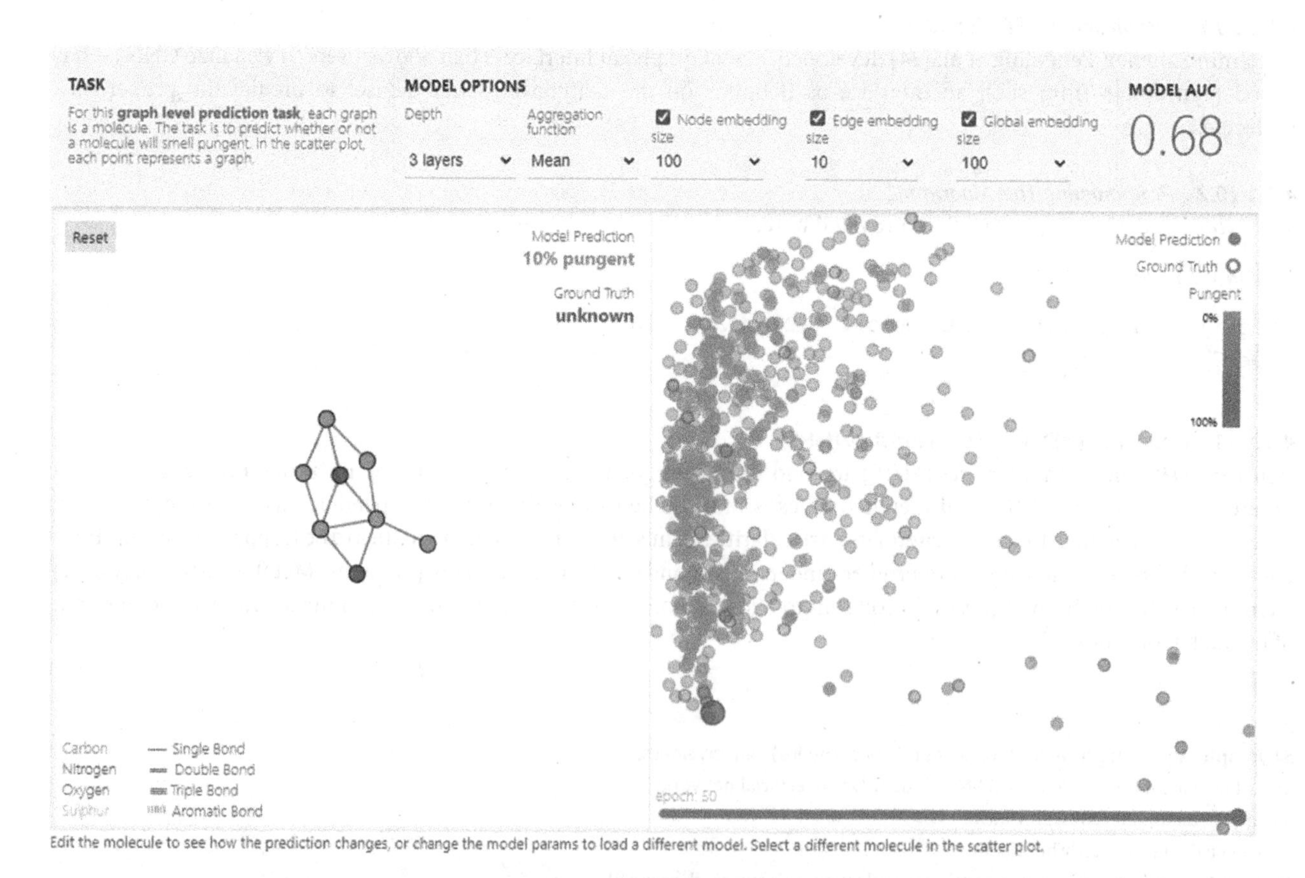

Figure 4.14 Illustration of two GNNs [34] / with permission from Distill.com.

There are various MHS. For the sake of simplicity, I will be broadly grouping these under four[59] classes (see Figure 4.15). Overall, MHS ranges from simple search procedures to complex learning processes. Thus, it is easy to split them into two groups: trajectory-based and population-based methods.

A trajectory-based method starts with an initial solution that is then substituted by an alternative (often the best) solution found at each iteration. On the other hand, population-based algorithms use a population of randomly generated solutions that are improved through a series of evolutionary operations (i.e., replacement by new solutions with superior properties). In some instances, evolutionary algorithms[60] can take advantage of past search experiences to guide the search on hand [35].

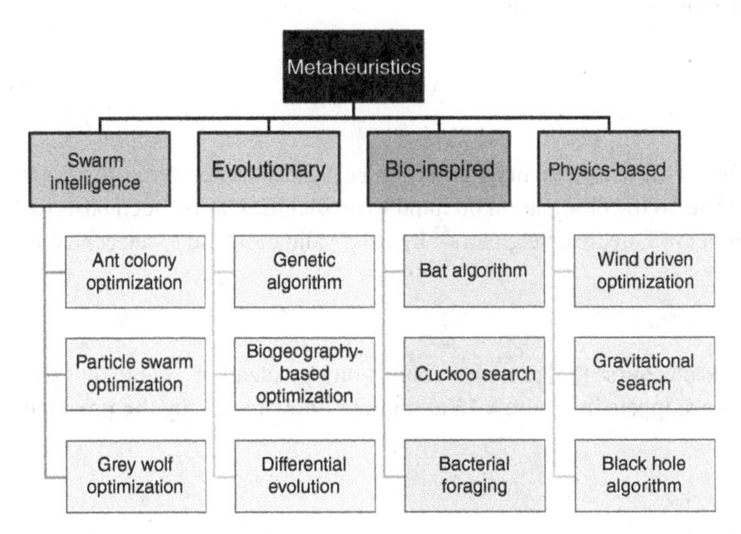

Figure 4.15 Algorithms falling under metaheuristics.

4.1.1.11.1 Programming This Algorithm

The *R* code for the HS and MHS algorithm is provided below. Please note that the algorithms shown in Figure 4.15 have very specific codes that are directly related to each algorithm. These were not added herein for brevity.

R Package:

```
>install.packages("'metaheuristicOpt'")
>library('metaheuristicOpt')
>metaOpt(FUN, optimType = "MIN", algorithm = "PSO", numVar, rangeVar,
  control = list(), seed = NULL)
```

4.1.1.12 K-nearest Neighbor (KNN)

The *K-nearest neighbor* (KNN) was created by Fix and Hodge in 1951 [36]. This algorithm employs a distance metric, *d*, to find *K* data points near the case data.[61] For this purpose, KNN often uses two distance-based metrics to determine the length between data points, namely, Euclidean distance and Manhattan distance.

How does it work?

For example, if we consider two points, $i = (y_{i1}, y_{i2}, \ldots y_{in})$ and $j = (y_{j1}, y_{j2}, \ldots y_{jn})$ with *n*- numeric attributes, respectively, their distance equals to:

---

59 One should note that the number of classes and classifications of heuristics and metaheuristics significantly varies across different works/disciplines.

60 For more information on these, please check out the many works of Amir Alavi [83], Ali Behnood, and Emadaldin Golafshani [84].

61 In general, the nearest neighbor algorithms fall under instance-based learning where, instead of building a model to generalize beyond its training, the model stores all training data as examples. Then, the model uses the stored data to derive predictions on new data (i.e., use the target of the closest neighbor to the observation (or case) on hand in the training set to predict the outcome of the observation on hand). As you can guess, prediction on a large dataset can be demanding.

$$\text{Euclidean}\, d(i.j) = \sqrt{(y_{1i} - y_{1j})^2 + (y_{2i} - y_{2j})^2 + \ldots + (y_{ni} - y_{nj})^2} \tag{4.15}$$

$$\text{Manhattan}\, d(i.j) = |y_{1i} - y_{1j}| + |y_{2i} - y_{2j}| + \ldots + |y_{ni} - y_{nj}| \tag{4.16}$$

If satisfied, then these equations are subject to the following four conditions: 1) $d(i.j) \gg 0$, 2) $d(i.j) = 0$, 3) $d(i.j) = d(j.i)$ and 4) $d(i.j) \ll d(i.k) + d(k.j)$.

Also, for the positive number, *K*, and the observed data, *x*, the number of data close to *x* is equal to the conditional probability for *x* in class *K* as estimated by [37], and can be found online at [38]:

$$P_k(X) = P_r(Y = k.X = x) = \frac{1}{k} \sum_{i \in N_k} I(y_i = k) \tag{4.17}$$

In the case of regression, the model uses the mean value of the k nearest neighbors. For a classification problem, the KNN model selects the majority class of the k nearest neighbors to the observation on hand to predict the class of such observation. In all cases, it is recommended to choose the number of k nearest neighbors[62] by cross-validation and to select an odd number (especially in classification, to avoid ties).

4.1.1.12.1 Visualizing the KNN Algorithm

The Stanford's CS231n: Deep Learning for Computer Vision course [39] provides a hands-on visualization[63] demo that you can use to try out the KNN algorithm. I have taken two snippets in Figure 4.16 to highlight how to classify the provided data (by moving some data points around).

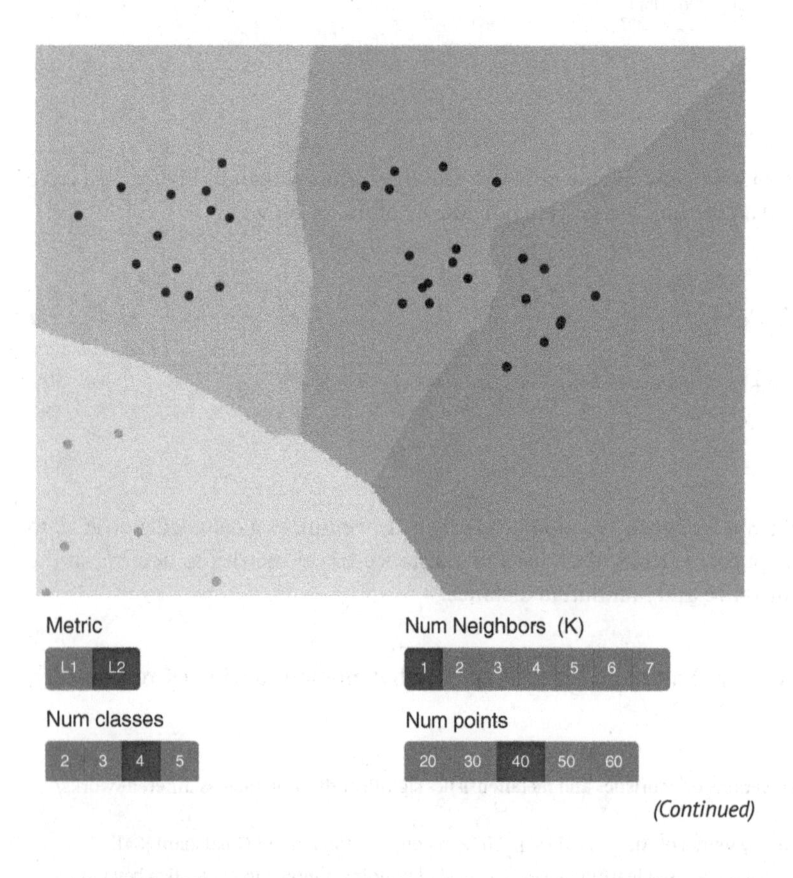

(Continued)

62 In theory, k can vary from 1 to all data points.

63 http://vision.stanford.edu/teaching/cs231n-demos/knn.

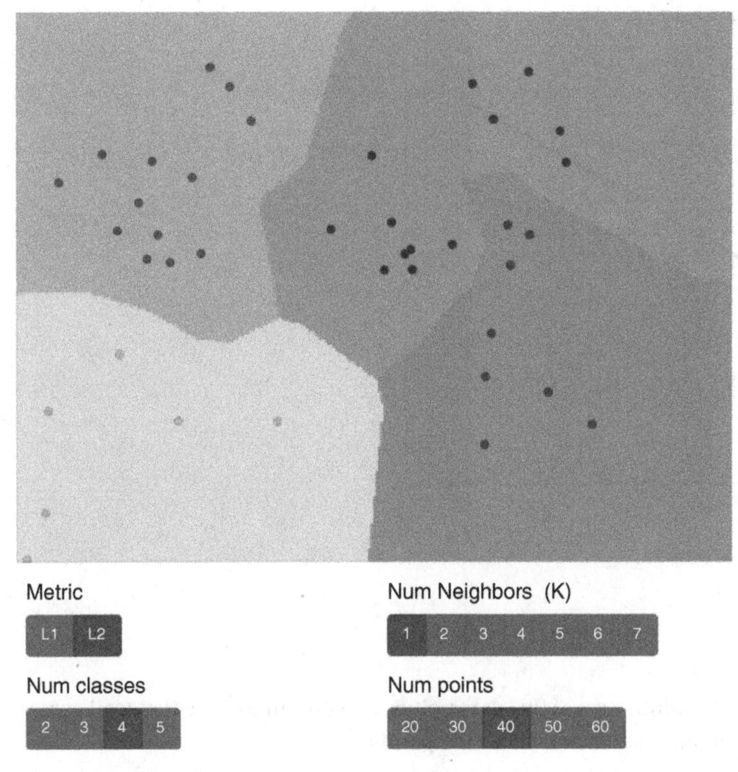

Figure 4.16 (Cont'd) Illustration of the KNN (from [39]).

4.1.1.12.2 Programming This Algorithm

The *Python* and *R* codes for the KNN algorithm are provided below:

Python Package

```
import sklearn
from sklearn.neighbors import NearestNeighbors
model = NearestNeighbors()
```

R Package:

```
>install.packages("class")
>library(class)
>knn(train, test, cl, k = 1, l = 0, prob = FALSE, use.all = TRUE)
```

4.1.1.13 Linear Regression (LiR)

This is perhaps one of the simplest supervised ML algorithms. *Linear regression (LiR)* is applied in regression problems to tie a target with one or more independent variables.[64]

How does it work?

In LiR, the coefficients of the linear form are estimated using the least-squares criterion, which minimizes the sum of squared residuals. In this process, the LiR tries to find the weights (namely, slope and y-intercept) that lead to the best-fitting line for the input data.

64 When one feature is used the LiR is called simple, and is called multiple LiR when more than one feature are used.

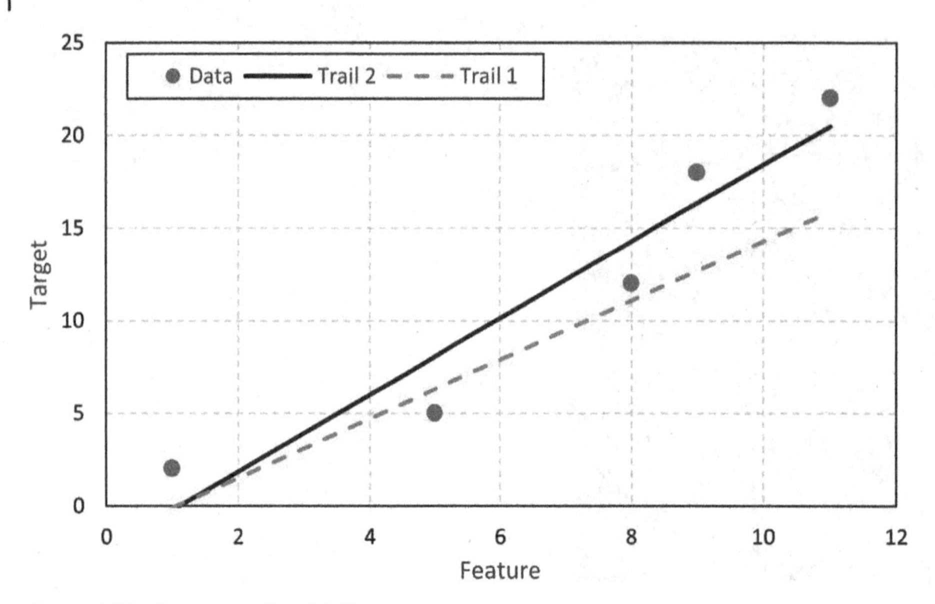

Figure 4.17 Demonstration of LiR.

4.1.1.13.1 *Visualizing the LiR Algorithm*

Figure 4.17 shows two attempts at fitting the LiR algorithm into a set of five data points. As you can see, the first trail comes close to fitting the outer points. The second trail then manages to better fit all the points.

4.1.1.13.2 *Programming This Algorithm*

The *Python* and *R* codes for the LR algorithm are provided below:

Python Package

```
from sklearn.linear_model import LinearRegression
model= LinearRegression()
```

R Package

```
lm(Y ~ model)
```

4.1.1.14 Logistic Regression (LR)

The *logistic regression*[65] *(LR)* maximizes the likelihood of observing a continuous or a categorical phenomenon. LR is a generalized linear model that applies a binomial distribution to fit regression models to binary response variables.

How does it work?

The LR is often used in classification problems wherein the outcome (target) falls under two or more labels (i.e., fatal/non-fatal, collapse/major damage/minor damage, etc.) [40]. LR approximates the multi-linear regression function as shown in Equation 4.18:

$$\text{logit}(p) = \beta_0 + \beta_1 x_1 + \ldots + \beta_n x_n \quad (4.18)$$

where p is the probability of the presence of a phenomenon. The logit transformation is defined as the logged odds:

$$\text{odds} = \frac{p}{1-p} \quad (4.19)$$

and,

65 Created by Joseph Berkson [85].

$$\mathrm{logit}(p) = \ln\left(\frac{p}{1-p}\right) \tag{4.20}$$

4.1.1.14.1 *Visualizing the LR Algorithm*

To better visualize the LR algorithm, a look into Figure 4.18 can be of aid.[66] The analysis starts with an attempt at horizontal fitting and then graduates into a logistic fitting.

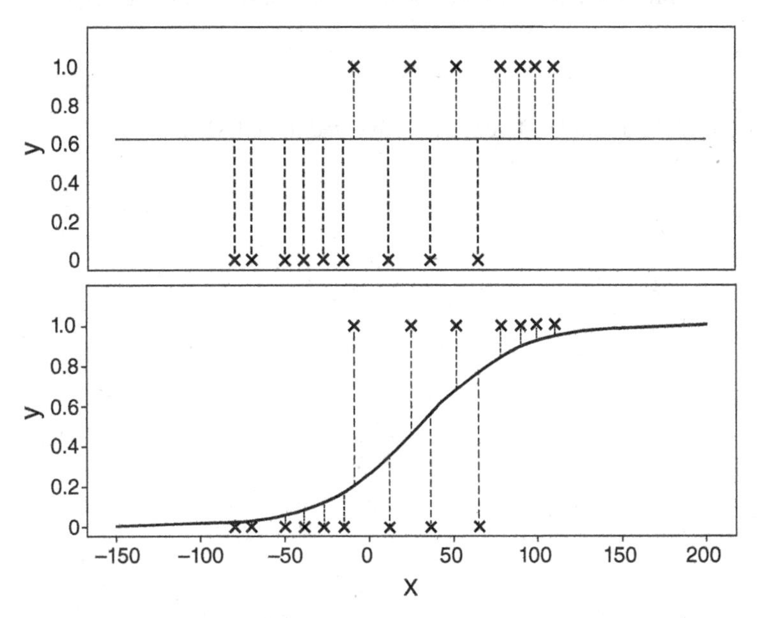

Figure 4.18 Illustration of LR (inspired by [41]) [top: initial state, and bottom: final state].

4.1.1.14.2 *Programming This Algorithm*

The *Python* and *R* codes for the LR algorithm are provided below:

Python Package

```
import sklearn
from sklearn.linear_model import LogisticRegression
model = LogisticRegression()
```

R Package:

```
>install.packages("rms")
>library(rms)
>lrm(formula, data=environment(formula),
subset, na.action=na.delete, method="lrm.fit", model=FALSE, x=FALSE, y=FALSE,
linear.predictors=TRUE, se.fit=FALSE, penalty=0, penalty.matrix, tol=1e-7, strata.
penalty=0, var.penalty=c('simple','sandwich'),
weights, normwt, scale=FALSE, ...)
```

4.1.1.15 Long Short-Term Memory (LSTM)*

The *Long Short-Term Memory* (LSTM) network is a special type of RNN capable of learning order dependence and was introduced by Hochreiter and Schmidhuber in 1997 [42]. LSTM has feedback connections and can process single data points (i.e., image) as well as sequences of data (such as speech or video).

66 This figure was inspired by the blog of Tobias Roeschl [41].

How does it work?

In an LSTM, the input gate determines which of the input values should be used to change the memory. In this process, the sigmoid function determines whether to allow 0 or 1 values through, and the tanh function assigns weight to the importance of the data on a scale of −1 to 1. Next to the input gate is the forget gate, and this gate identifies which part of the data should be removed (also via a sigmoid function). For each number in the cell state C_{t-1}, it looks at the preceding state (h_{t-1}) and the content input (X_t) and produces a number between zero (omit) and one (keep). Finally, the output gate with a sigmoid function determines whether to allow 0 or 1 values through. This gate also homes a tanh function that assigns weight to the values provided, determining their relevance on a scale of −1 to 1 and multiplying it with the sigmoid output.

4.1.1.15.1 Visualizing the RF Algorithm

Figure 4.19 shows the anatomy of an LSTM and outlines the cells and different gates involved in an LSTM.

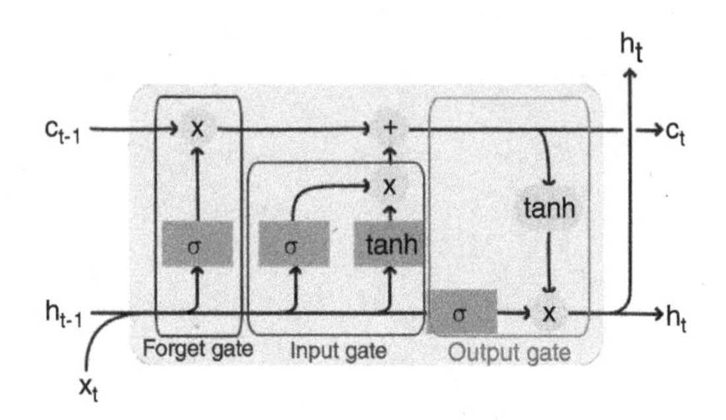

Figure 4.19 Sketch of LSTM.

4.1.1.15.2 Programming This Algorithm

The *Python* and *R* codes for LSTM are provided below:

Python Package

```
from keras.layers import LSTM
model = Sequential()
model.add(LSTM())
```

R Package

```
>library(keras)
>library(tensorflow)
>install_keras()
>install_tensorflow

>lstm_model %>%
  layer_lstm(units = 50,
       batch_input_shape = c(1, 12, 1),
       return_sequences = TRUE,
       stateful = TRUE) %>%
  layer_dropout(rate = 0.5) %>%
  layer_lstm(units = 50,
        return_sequences = TRUE,
        stateful = TRUE) %>%
  layer_dropout(rate = 0.5) %>%
  time_distributed(keras::layer_dense(units = 1))
```

4.1.1.16 Naïve Bayes (NB)

Naïve Bayes (NB) [67] is a probabilistic algorithm based on the Bayes Theorem.[68] Thus, this is an iterative algorithm that evaluates conditional probabilities to verify pre-set rules in order to predict discrete or continuous observations.

How does it work?

In a classification setting with a set of features, i.e., x_1, ..., x_n categorized under a class observation, Y, the NB algorithm starts the analysis to maximize the posterior probability of the class variable given the set of features through the following relations [43]:

$$\arg\max P_{c\in C}(Y \mid x_1,\ldots, x_n) \tag{4.21}$$

The application of Bayes rule in classification is formulated as:

$$P_{c\in C}(Y \mid x_1,\ldots, x_n) = \frac{P(Y)P(x_1,\ldots, x_n \mid Y)}{P(x_1,\ldots, x_n \mid Y)} \tag{4.22}$$

and further simplifies to:

$$\arg\max P_{c\in C}(Y)\prod_{i=1}^{n} P(x_i \mid Y) \tag{4.23}$$

4.1.1.16.1 *Programming This Algorithm*

The *Python* and *R* code scripts for the NB algorithm are provided below:

Python Package

```
import sklearn
from sklearn.naive_bayes import GaussianNB
Model = GaussianNB()
```

R Package:

```
>install.packages("naivebayes")
>library(naivebayes)
>naive_bayes(formula, data, prior = NULL, laplace = 0, usekernel = FALSE, usepoisson
= FALSE, subset, na.action = stats::na.pass, …)
```

4.1.1.17 Random Forest (RF)

Random forest (RF) is an algorithm that integrates multiple DTs[69] by ensemble learning methods to form a more powerful prediction model [44]. Similar to other algorithms, RF can be used in regression and classification and is defined as a non-parametric[70] classifier. The RF algorithm was created initially by Tin Kam Ho and then by Leo Breiman [45].

How does it work?

In an RF-based analysis, all individual DTs reach an outcome. In a regression problem, the average result of all individuals is calculated to find the mean, which also comes to be the final outcome. In a classification problem, the majority voting method is used to arrive at the most common vote, which comes to be the output of RF analysis.[71] A typical formulation of RF is presented herein:

---

67 The term *Naïve* reflects the emphases on the independency of the assumptions used in this algorithm and the term *Bayes* implies the belonging to Bayes' theorem to evaluate the probability of an observation to belong to a specific class.

68 Bayes Theorem describes the probability of an event given our prior knowledge of conditions that might be related to the event. More specifically, it states that the conditional probability of an event, based on the occurrence of another event, is equal to the likelihood of the second event given the first event multiplied by the probability of the first event.

69 Many trees = forest!

70 Does not require assumptions on the form of relationship between the features and target.

71 It is worthwhile to compare RF and GBT. For example, RF builds trees independently, while GBT builds trees one at a time. In addition, and as we have seen, RF combines the outcome of the ensemble learning only at the end of the RF analysis, while GBT combines results of the ensemble learning throughout the analysis process. Unlike RF, the GBT iteratively corrects developed ensembles by comparing predictions against true observations throughout the analysis (which allows next iterations to help correct for previous mistakes).

$$Y = \frac{1}{J}\sum_{j=1}^{J} C_{j,\text{full}} + \sum_{k=1}^{K}\left(\frac{1}{J}\sum_{j=1}^{J}\text{contribution}_j(x,k)\right) \tag{4.24}$$

where J is the number of trees in the forest, k represents a feature in the observation, K is the total number of features, c_{full} is the average of the entire dataset (initial node).

4.1.1.17.1 *Visualizing the RF Algorithm*

Figure 4.20 shows a schematic of the RF algorithm. In this figure, there are three DTs where each tree generates one prediction. Voting takes place, and the majority of votes are used to arrive at the RF's prediction.

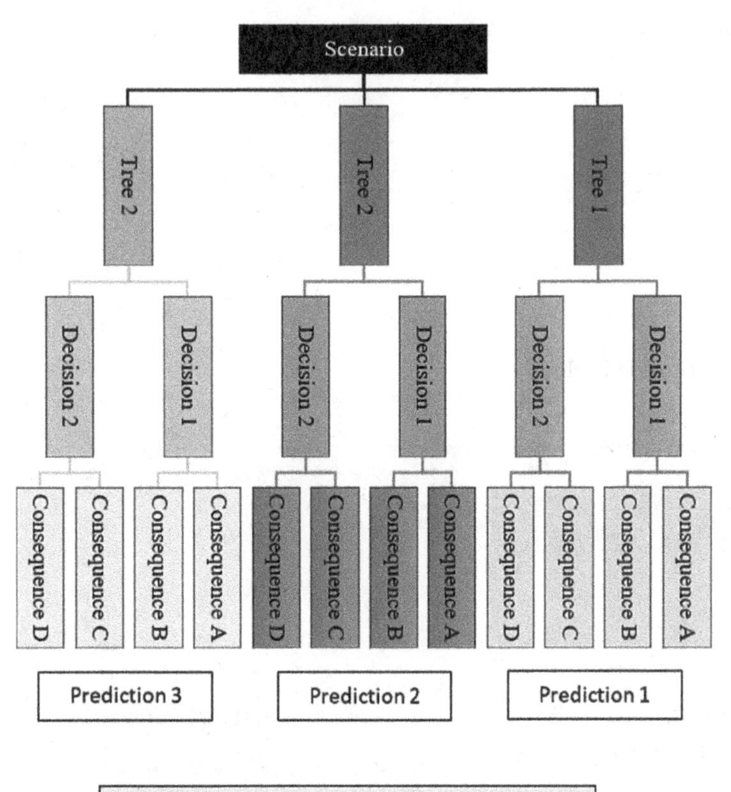

Figure 4.20 Illustration of random forest tree algorithm in a classification problem.

4.1.1.17.2 *Programming This Algorithm*

The *Python* and *R* codes for the RF algorithm are provided below:

Python Package

```
import sklearn
from sklearn.ensemble import RandomForestRegressor
model = RandomForestRegressor()
```

R Package:

```
>install.packages("randomForest")
>library(randomForest)
>randomForest(formula, data=NULL, …, subset, na.action=na.fail)
```

4.1.1.18 Support Vector Machine (SVM)

The *support vector machine* (SVM) is an algorithm suited for multi-dimensional input regression and classification problems. SVM can linearly and/or nonlinearly identify key features to solve binary or multi-class phenomena by separating hyperplanes of a distance d.[72] This algorithm was initially developed by Vladimir Vapnik and Alexey Ya [46, 47].

How does it work?
In SVM, operators aim to identify a line or a boundary (referred to as a hyperplane or decision boundary) in an n-dimensional space (where n represents the number of features) that can be used to classify data into separate classes. Such a plane maximizes the distance between data points of each class to allow for confident classification [48]. The term *support vectors* refer to points in close proximity to the hyperplane, and hence these can significantly influence the position and orientation of the plane (see Figure 4.21). SVM uses a special form of mathematical functions defined as kernels (k) which transforms inputs into the required form. Typical kernel (k) functions for classification are as follows:

$$\text{Linear: } k(x_i.y_i) = x_i^T x_j \tag{4.25}$$

$$\text{Polynomial: } k(x_i.y_i) = (\gamma x_i^T x_j + r)^d . \gamma > 0 \tag{4.26}$$

$$\text{Radial basis function(RBF): } k(x_i.y_i) = \exp(-\gamma x_i^T x_j x_i^T x_j^d) . \gamma > 0 \tag{4.27}$$

$$\text{Sigmoid: } k(x_i.y_i) = \tanh\,(\gamma x_i^T x_j + r) \tag{4.28}$$

While SVM was initially designed for classification, this technique can also accommodate regression [49]. When used for regression, the SVM algorithm employs an intense loss function to maintain a maximum margin. The linear model of the intense loss function is as follows:

$$L^{\in}(x.y.f) = |y - f(x)|_{\in} = \begin{cases} 0 & \text{if } |y - f(x)| < \in \\ |y - f(x)| - \in & \text{otherwise} \end{cases} \tag{4.29}$$

To fit a model of the form:

$$f(x) = \sum_{i=1}^{n} c_i k(x.x_i) \tag{4.30}$$

where c_i refers to a choice of coefficient and $k(x.x_i)$ is the Gaussian kernel function.

Similar to the classification version, the regression SVM also requires the use of regression-based kernel functions.

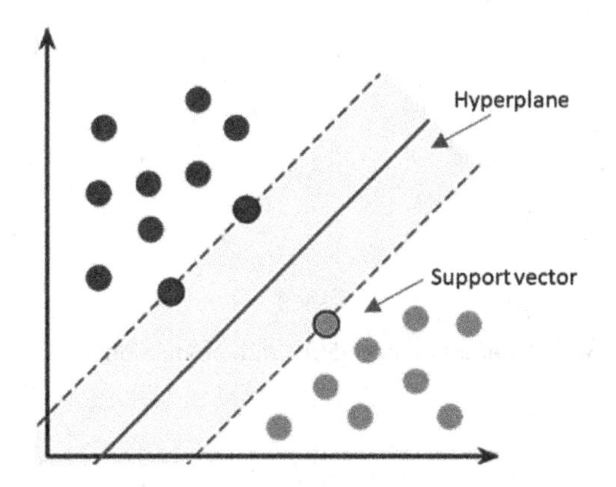

Figure 4.21 Illustration of SVM.

72 Note that KNN is a form of SVM. Also note that sometimes such planes may allow soft margins (to allow minor violations) or hard margins (does not allow violations or misclassifications).

4.1.1.18.1 Visualizing this Algorithm

Jonas Greitemann [50] has developed a neat online interface[73] and tool[74] that showcases the working of the SVM algorithm. Figure 4.22 presents two snippets from this interface. You can start by adding random points and selecting a kernel function. I am showing the generated support vectors for a linear and a sigmoid function.

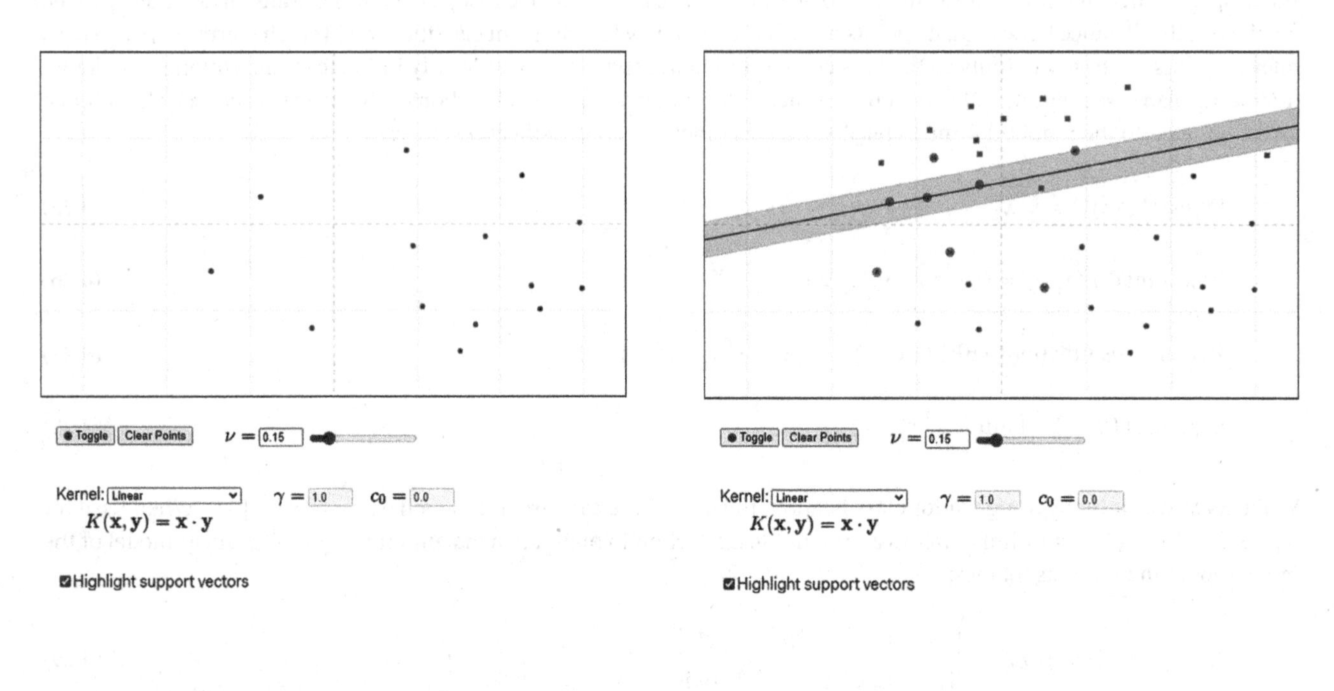

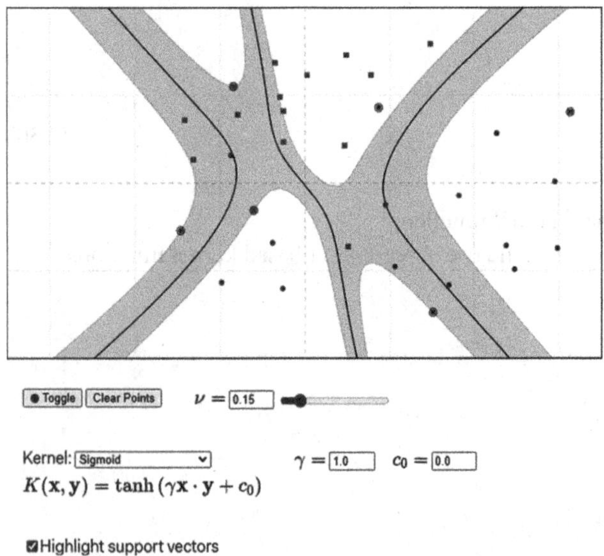

Figure 4.22 Illustrations of the SVM algorithm using Jonas Greitemann's interface with different kernels [50] / with permission.

73 https://jgreitemann.github.io/svm-demo.

74 https://jgreitemann.github.io/messing-with-wasm/svm-demo.

4.1.1.18.2 Programming This Algorithm

The *Python* and *R* codes for the SVM algorithm are provided below:

Python Package

```
import sklearn
from sklearn import svm
model = svm.SVC()
```

R Package:

```
>install.packages("e1071")
>library(e1071)
>svm(formula, data = NULL, …, subset, na.action = na.omit, scale = TRUE)
```

4.1.1.19 TensorFlow (TF)*

The *TensorFlow* (TF) is an open-source and free neural network-based model that uses deep learning and has been hosted by Google since 2015 (and revamped in 2019) [51]. This model creates dataflow graphs that describe how data moves through a series of processing nodes, where each node represents a mathematical operation, and each connection between nodes is a multi-dimensional data array (tensor) [51].

4.1.1.19.1 Visualizing TF

Computational processes are described as graphs in TF. One such graph is shown in Figure 4.23.

4.1.1.19.2 Programming This Algorithm

The *Python* and *R* codes to use TF are provided below:

Python Package

```
import tensorflow as tf
from keras.models import Sequential
tf.keras.Sequential()
```

R Package:

```
>install.packages("tensorflow")
>library(tensorflow)
>install_tensorflow()
```

4.1.1.20 Transformers (T)*

A *transformer (T)* is a deep learning model that was introduced by Google Brain [52]. Transformers adopt a self-attention mechanism to identify the significance of parts of the input data. This mechanism relates different positions of a single sequence in order to compute a representation of the sequence (without using sequence-aligned recurrent architecture). In a way, this mechanism calculates a weighted average of feature representations with the weight proportional to a similarity score between pairs of representations. This algorithm is primarily used in problems related to natural language processing[75] (NLP) and computer vision.[76]

4.1.1.20.1 Programming This Algorithm

A sample *Python* code script for this algorithm is provided below:

---

75 Where can NLP be helpful in civil and environmental engineering? In fields for extracting and exchanging information, management, and decision-making. For example, NLP can be used in construction management (contracts, legality), etc.

76 Transformers can be valuable in problems dealing with image and footage analysis (i.e., identify cracking, poor quality road conditions, etc.).

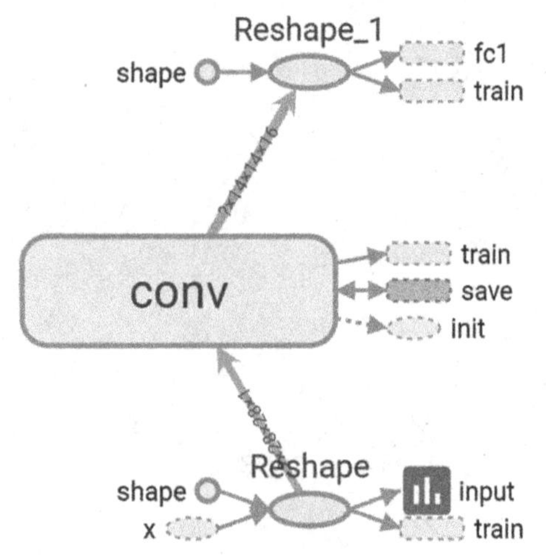

Figure 4.23 Sample of dataflow in TF.

Python Package (from PyTorch)

```
#Import required libraries
import torch
from pytorch_transformers import GPT2Tokenizer,
GPT2LMHeadModel
```

4.1.1.21 Vowpal Wabbit (VW)*

Vowpal Wabbit (VW) is an open-source fast and online interactive machine learning system library that started at Yahoo add then moved to Microsoft [53]. The VW is an out-of-core algorithm learner capable of streaming data which comes in handy in large datasets that cannot be supported by existing hardware.[77]

4.1.1.21.1 Programming This Algorithm

The *Python* and *R* codes for the VW algorithm are provided below:

Python Package

```
import vowpalwabbit
model = vowpalwabbit.Workspace(quiet=True)
```

R Package

```
>install.packages("RVowpalWabbit")
>library(RVowpalWabbit)
>vw(args, quiet=TRUE)
```

4.1.2 Unsupervised Learning

In this section, we shift our attention to problems tackling datasets without labels for the target (response) or those without a target variable. I will cover anomaly detection[78] algorithms first and then dive into clustering and dimensional reduction algorithms.[79]

4.1.2.1 Isolation Forest (IF) Anomaly Detection Algorithm

The *isolation forest* (IF) *anomaly detection* algorithm[80] entertains the idea that outliers, unlike inliers, are easy to isolate.[81] This algorithm has a tree structure, and hence the number of splits required to isolate a sample is equivalent to the *path length.*[82]

How does it work?

The IF algorithm isolates observations in a given dataset by randomly choosing a feature and then arbitrarily selecting a split value that lies somewhere between the maximum and minimum of the same selected feature. This algorithm uses 100 trees to start a random forest, with an expected outlier fraction of 10% as recommended by its developers [54, 55].

4.1.2.1.1 Visualizing this Algorithm

Figure 4.24 depicts a visual look into the IF algorithm where the anomaly is the side of the tree with the shortest path length.

---

77 May not be an issue in most of problems. However, we do not know what the future holds!

78 Anomaly detection (also called outlier detection) identifies unusual patterns/deviations in data as compared to the normal data.

79 Admittedly, I do not cover *association rules* as they are adequate for data mining than ML. For more information on these, please visit [86].

80 It is worth noting that there is an extension to this algorithm called Extended Isolation Forest [87].

81 Think of it like this, outliers are few and different; hence they are easy to isolate!

82 **Path length**: A measure of normality where anomalies are linked to having shorter paths.

4.1.2.1.2 Programming This Algorithm

The *Python* and *R* code scripts for the IF algorithm are provided below.

Python Package

```
import sklearn
from sklearn.ensemble import IsolationForest
model=IsolationForest()
```

R Package

```
>install.packages("solitude")
> library("solitude")
> isolationForest$new( sample_size = 256,
num_trees = 100, replace = FALSE,seed = 101,nproc =
NULL, respect_unordered_factors = NULL, max_depth =
ceiling(log2(sample_size)) )
```

Figure 4.24 Demonstration of a tree from the isolation forest (IF) anomaly detection algorithm [Note: the shallow depth of the anomaly at the far left].

4.1.2.2 Local Outlier Factor (LOF) Anomaly Detection Algorithm

The *local outlier factor* (LOF) *anomaly detection* algorithm measures the density[83] deviation of a particular observation with respect to its neighbors. In this mechanism, the algorithm can identify outliers as those with a substantially lower density than their neighbors. An anomaly is considered when its anomaly score returns a value that is larger than 0.5 [54]. In lieu of the above algorithms, other anomaly algorithms can also be used, such as Mahalanobis distance [56].

How does it work?

The local density for a particular data point is calculated by estimating distances between data points that are neighboring data points. The estimated distances fall under the *k*-nearest neighbors[84] and are used to calculate the reachability distance.[85] All reachability distances are calculated to determine the local reachability density[86] of that point. When the points are close, their distance is lesser, resulting in a higher density. In other words, low values of local reachability density indicate that the closest cluster is far from the point on hand.

4.1.2.2.1 Visualizing this Algorithm

A look into Figure 4.25 shows a demonstration of the LOF process where the hollow point (outside of the green host) is labeled as an outlier.

4.1.2.2.2 Programming This Algorithm

The *Python* and *R* code scripts for the LOF algorithm are provided below. This algorithm can also be found within the *DataRobot* platform.

Python Package

```
import sklearn
from sklearn.neighbors import LocalOutlierFactor
model = LocalOutlierFactor()
```

R Package

```
>install.packages("'DMwR'")
>library('DMwR')
>lofactor(data, k)
```

---

83 An observation is considered an outlier if its neighborhood does not have enough other points (hence the low density).

84 In this instance, the *k* value determines how many neighbors we are considering. For example, when $k = 4$, we are looking at the distance to the nearest fourth neighbor.

85 **Reachability distance**: The maximum of (the distance between two points) and the (k-distance of that particular point).

86 **Local reachability density**: measures the density of k-nearest points around a point. This measure is calculated by taking the inverse of the sum of all of the reachability distances of all the k-nearest neighboring points.

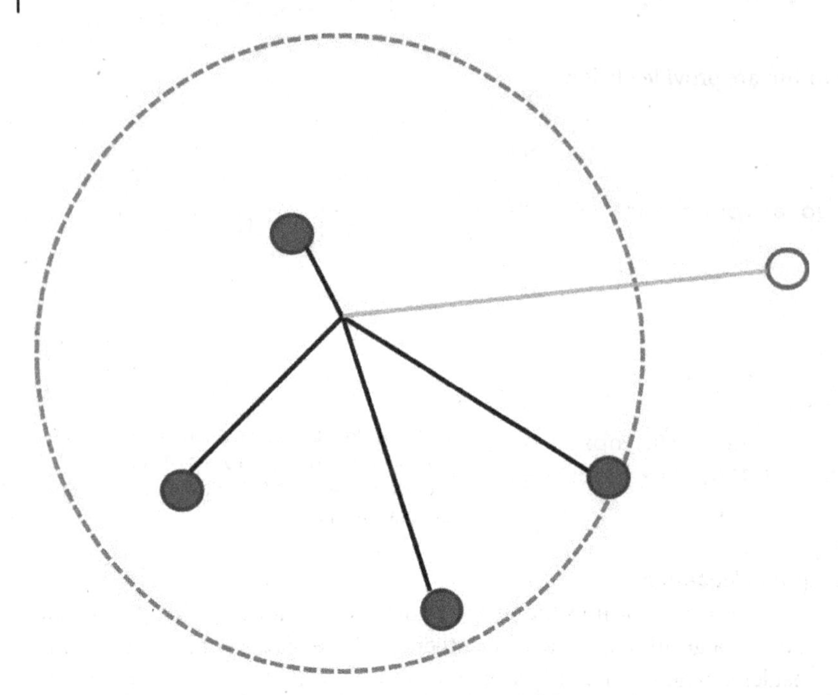

Figure 4.25 Illustration of the local outlier factor (LOF) anomaly detection process [Note: dark lines represent a $k = 4$ (all within reachability). The light line represents a situation of out of reachability].

4.1.2.3 Seasonal Hybrid Extreme Studentized Deviate (S-H-ESD) Algorithm

The Seasonal Hybrid Extreme Studentized Deviate (S-H-ESD) is the third anomaly detection algorithm that I will highlight. Unlike the others, this algorithm detects anomaly in *time series* and has the capability to detect such anomalies at the global and local (i.e., seasonality trends) levels, as well as identify positive and negative anomalies.[87] This algorithm was developed by Twitter [57].[88]

How does it work?

The S-H-ESD employs time series seasonal and trend decomposition to detect anomalies. S-H-ESD model is an extension of a generalized extreme studentized deviate test, where the ESD sample mean and standard deviation are used to detect anomalies in time series, while the S-H-ESD model uses the median to minimize the number of false positives [58]. The S-H-ESD algorithm also employs piecewise approximation for long time series.

4.1.2.3.1 Programming This Algorithm

The *Python* and *R* codes for the S-H-ESD algorithm are provided below. This algorithm can also be found within the *Exploratory* platform.

Python Package

```
import sesd
model = sesd.seasonal_esd()
```

R Package

```
>import sesd
>model = sesd.seasonal_esd()
```

87 Global and local anomalies refer to anomalies extending above or below expected seasonality (i.e., not subject to seasonality and underlying trends), or anomalies which occur inside seasonal patterns, respectively. Positive and negative anomalies are those in the positive or negative direction, respectively.

88 Please go over the cited blog to learn how and why Twitter needed to develop this algorithm. Hint: it is Christmas time!

```
>install.packages("devtools")
>devtools::install_github("twitter/AnomalyDetection")
>library(AnomalyDetection)
```

4.1.2.4 Density-based Spatial Clustering of Application with Noise (DBSCAN) Clustering Algorithm

The DBSCAN (density-based spatial clustering of application with noise) algorithm considers the region with dense data points as a cluster [59]. Density-based clustering tends to perform the best on numerical data.

How does it work?
This algorithm separates clusters based on differences in their density [60, 61]. The clustering analysis starts with checking whether a randomly selected data point in a dataset is a core point or not. If this point turns into a core point, then this point and its neighbors[89] are added to a new cluster. If a data point is not a core point, then a check for border or noise points is carried out. The analysis terminates when all data points are clustered [62].

4.1.2.4.1 Visualizing this Algorithm

As you can see in Figure 4.26,[90] there are two parameters that govern the DBSCAN analysis, namely, epsilon (a distance measure) and *minPoints*. An arbitrary point is first picked for analysis, and if there are more than *minPoints* points within a distance of epsilon from that point, then a cluster is formed. The analysis continues on this cluster to expand it. This process checks all of the new points to see if they too are surrounded by more than *minPoints* points within a distance of epsilon. If so, the cluster grows and vice versa.

4.1.2.4.2 Programming This Algorithm

The *Python* and *R* codes for the DBSCAN algorithm are provided below:

Python Package:

```
import sklearn
from sklearn.cluster import DBSCAN
model = DBSCAN()
```

R Package:

```
>install.packages("dbscan")
>library(dbscan)
>dbscan(x, eps, minPts = 5, weights = NULL, borderPoints = TRUE, …)
```

4.1.2.5 Fuzzy C-Means (FCM) Clustering Algorithm

The Fuzzy C-Means (FCM) clustering algorithm calculates the membership of a data point by determining the degree to which it belongs to clusters [64, 65]. Given its fuzziness, the FCM calculates the *probability* that a point belonging to any cluster is between zero and one.[91] Unlike other algorithms, the FCM can assign a data point to more than one cluster at the same time. Thus, the FCM turns effective for data points that do not clearly fall into specific clusters [66].

How does it work?
The FCM starts the analysis by randomly selecting cluster centroids. This algorithm randomly initializes memberships, U_{ij}, and computes centroid vectors v_j. Each membership is updated via a minimization objective function (see Equation 4.31), at which point the analysis terminates [67].

$$J_{\text{FCM}} = \sum_{i=1}^{N}\sum_{j=1}^{C} U_{ij}^{m}(p_i - v_j)^2 \tag{4.31}$$

89 Now to be called, *minPoints*, are the minimum number of data points inside the largest radius of the neighborhood.
90 A neat visualization tool from Naftali Harris [63] (https://www.naftaliharris.com/blog/visualizing-dbscan-clustering).
91 And the sum of the membership probabilities for each data point is always 1.0.

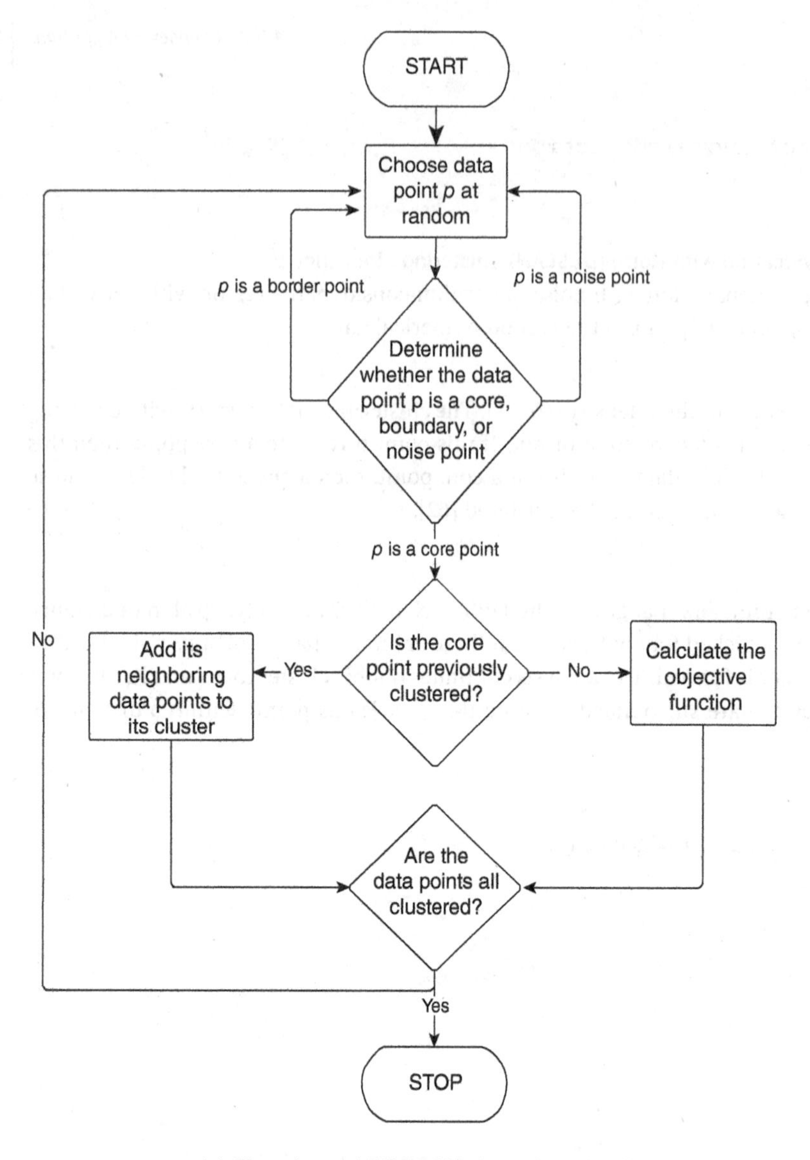

(a) Flowchart of DBSCAN clustering algorithm

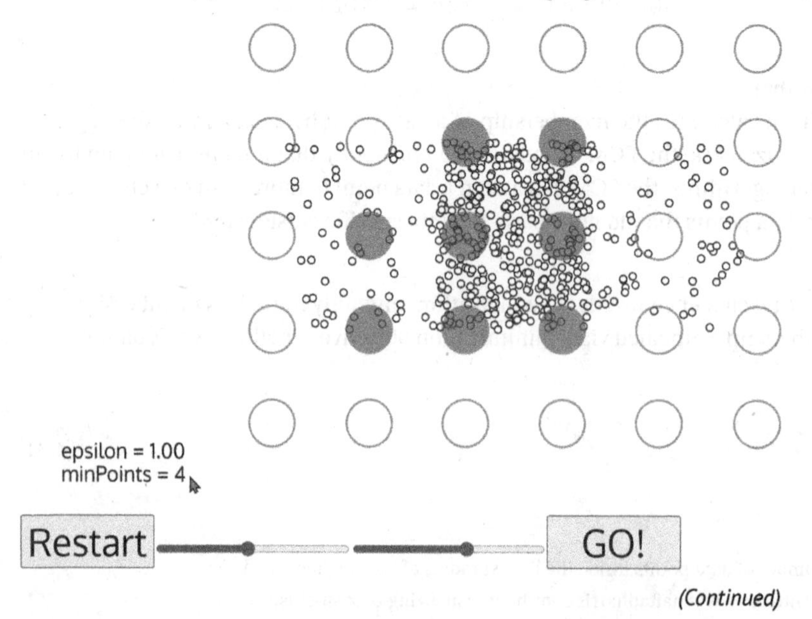

(Continued)

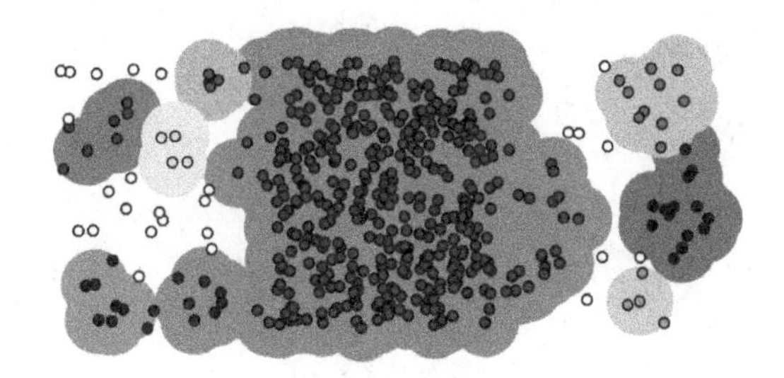

(b) Visualization of clustering

Figure 4.26 (Cont'd) Demonstration of the DBSCAN (from [63]).

where p_i is the ith data point, v_j is the centroid of the cluster j, C represents the number of clusters, and N denotes the total number of data points in the dataset. Moreover, U_{ij} is the degree of membership of p_i data point in the jth cluster, and under the $1 \leq m < \infty$ constraint, m denotes the degree of fuzziness.[92]

Also, note that the membership matrix U_{ij} is randomly assigned as:

$$U_{ij} = \frac{1}{\sum_{k=1}^{c} \left(\frac{|p_i - v_j|}{|p_i - v_k|} \right)^{\frac{2}{m-1}}} \tag{4.32}$$

And, the cluster centroid is calculated as:

$$v_j = \frac{\sum_{i=1}^{N} U_{ij}^{m} \cdot p_i}{\sum_{i=1}^{N} U_{ij}^{m}} \tag{4.33}$$

4.1.2.5.1 *Visualizing this Algorithm*

Figure 4.27 shows a flowchart of the FCM algorithm.

4.1.2.5.2 *Programming This Algorithm*

The *Python* and *R* code scripts for the FCM algorithm are provided below:

Python Package

```
from fcmeans import FCM
model = FCM()
```

R Package:

```
>install.packages("ppclust")
>library(ppclust)
> fcm(x, centers=N)
```

92 *m* infinity results in complete fuzziness.

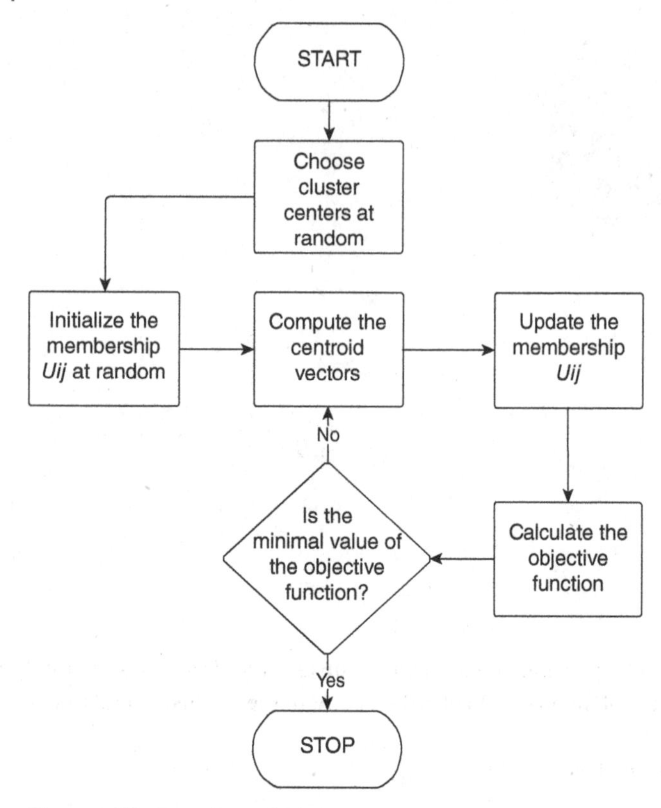

Figure 4.27 Flowchart of FCM clustering algorithm.

4.1.2.6 Hierarchical Clustering Algorithm

The *hierarchical clustering* (HC) algorithm is another clustering tool that agglomerates or divides data points into groups/ clusters [68]. The *agglomerative* approach employs bottom-up clustering, wherein each point in the dataset is initially considered a cluster, and the most similar points are then clustered together. On the other hand, the *divisive* approach initially groups all the observations into one cluster and then recursively splits the most appropriate clusters until one of two conditions are met, namely, reaching the identified number of clusters, or until each observation in the dataset creates its own cluster (see Figure 4.28)[93] [68]. This algorithm was proposed by Ward [69].

How does it work?

The HC starts by forming *N* number of clusters by making each data point an individual cluster. The HS then determines the two closest data points in terms of the Euclidean Distance as calculated by Equation 4.34. This process combines the data points into a single cluster (which forms *N*-1 clusters), and this process is repeated until there is only one large cluster [70].

$$\text{distance}(P_1, P_2) = \sqrt{(x_2 - x_1)^2 + (y_2 - y_1)^2} \qquad (4.34)$$

where P_1 *and* P_2 are the two closest data points.

4.1.2.6.1 *Visualizing this Algorithm*

Figure 4.28 shows a schematic of how the HC algorithm works in terms of its two approaches, namely, the agglomerative and divisive approaches. Note the presence of a *dendogram*, which is a common means to represent a diagram of HC analysis. The same figure also shows a flowchart of the HC algorithm.

4.1.2.6.2 *Programming This Algorithm*

The *Python* and *R* codes for the HC algorithm are provided below.

93 Note: the agglomerative approach is used more widely than the divisive approach.

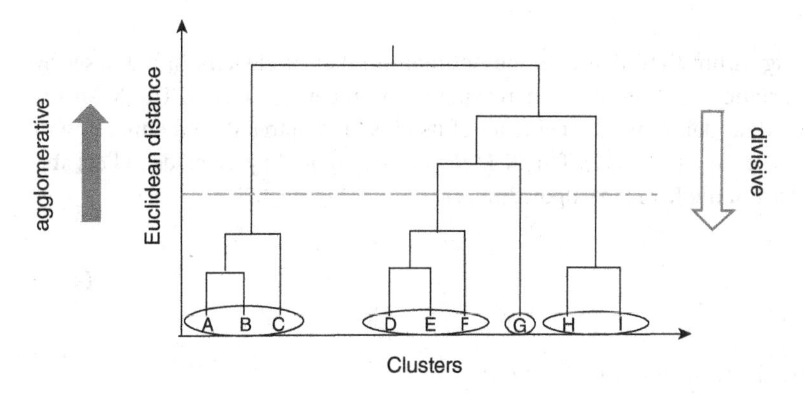

(a) A dendrogram example for HC algorithm

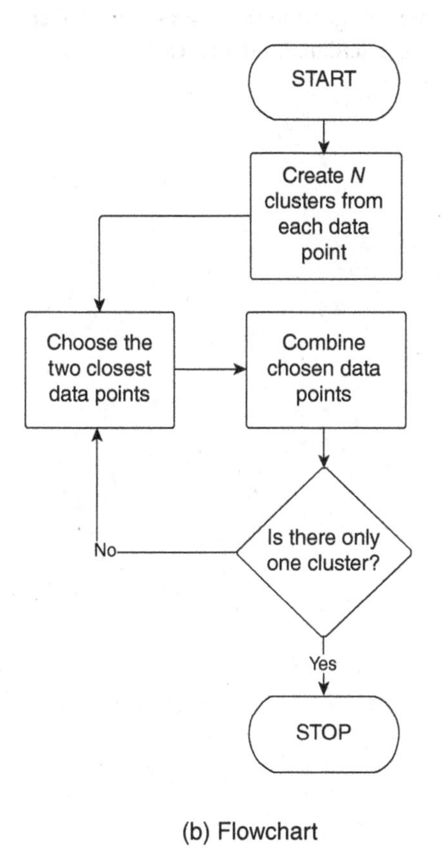

(b) Flowchart

Figure 4.28 Flowchart of HC algorithm.

Python Package:

```
import scipy.cluster.hierarchy as sch
model = sch.dendrogram()
```

R Package:

```
>install.packages("'fastcluster'")
>library(fastcluster)
>hclust(d, method="complete", members=NULL)
```

4.1.2.7 K-Means Clustering Algorithm

K-Means is another commonly used clustering algorithm that identifies the optimal number of clusters in a dataset by examining the total decrease in intra-cluster[94] variation attained from increasing the number of clusters. The K-Means minimizes the Euclidean distances between each data point and the centroid of its cluster, squares it, and sums it with others (aka. the sum of the squared errors, SSE) – see Equation 4.35). This K-Means was proposed by MacQueen [71] and finetuned by Hartigan and Wong[95] [72]. A flowchart of the K-Means algorithm is shown in Figure 4.29.

$$SSE = \sum_{p=1}^{k} \sum_{x_i \in C_p} p_i - \mu_p^2 \tag{4.35}$$

where p_i is the data point, μ_p is the centroid of the clusters, and C_p *is* p^{th} cluster.

How does this algorithm work?
A typical K-Means analysis starts by arriving at the number of clusters[96] k (see Figure 4.29). This algorithm then selects random points in the data space[97] as the centroid of clusters (μ_p). Data points are then assigned to the nearest centroid so that k clusters are formed. Now that the clusters are formed, the new centroids can be calculated. Finally, each data point

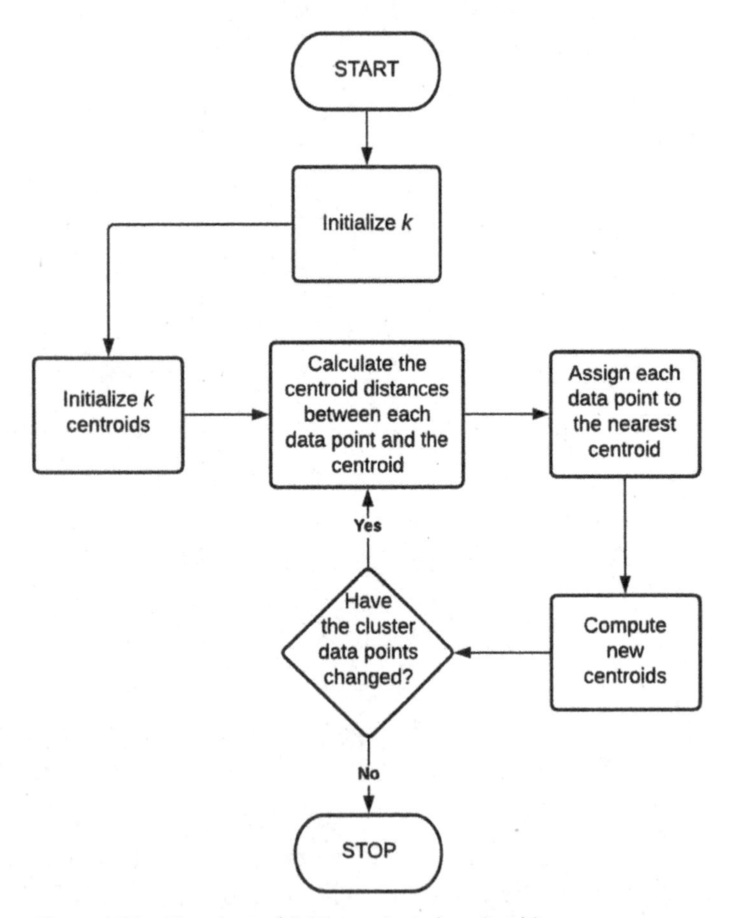

Figure 4.29 Flowchart of K-Means clustering algorithm.

94 The distance between data point with cluster center.

95 There are two ideas to the K-Means algorithm that I would like to briefly mention. The first is that this algorithm may cluster loosely related observations together (even for scattered observations) and since clusters depend on the mean value of cluster elements, each data's weight counts in forming the clusters. Another challenge is that you need to specify the number of clusters to apply it.

96 Via metrics such as elbow method, etc. as will be discussed in the next chapter.

97 Hence, these points do not necessarily have to be from the dataset.

whose nearest centroid has changed (depending on the displacement of the centroids) to the new nearest centroid. All other data points whose centroid did not change are left intact, and the analysis terminates.[98]

4.1.2.7.1 Visualizing this Algorithm

Akif AKKUŞ at the Computer Engineering Department in the Middle East Technical University in Turkey has developed an accessible visualization[99] tool that we can use to trace how the K-Means algorithm work [73]. A look into Figure 4.30 illustrates a series of the interaction of a clustering process where we assume that we have three clusters. The larger and hollow circles are the assumed centroids for this problem, while the filled dots represent the data points. To maximize this exercise, please keep an eye on how the centroids update their location as the analysis progresses throughout the presented seven iterations.

4.1.2.7.2 Programming This Algorithm

The *Python* and *R* code scripts for the K-Means algorithm are provided below.

Python Package:

```
import sklearn
from sklearn.cluster import KMeans
model = KMeans()
```

R Package:

```
>install.packages("cluster")
>install.packages("factoextra")
>library(cluster)
>library(factoextra)
> kmeans(df, centers = 2, nstart = 25)
```

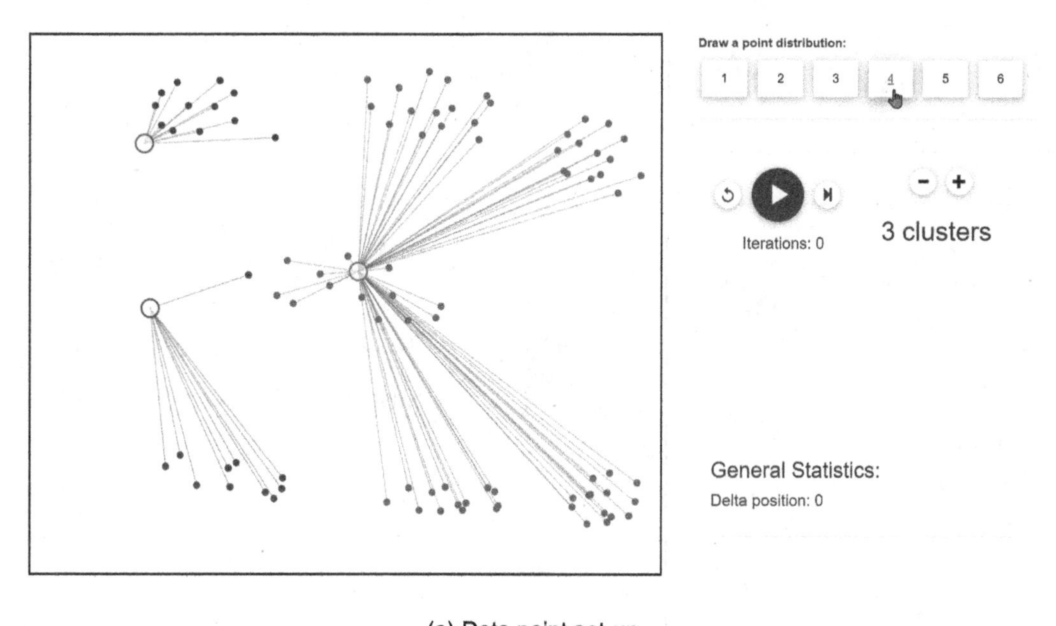

(a) Data point set-up

(Continued)

98 Additional notes: Instead of relying on the mean value as a reference point in a cluster, *medoids*, which are the most centrally located points, can be used.

99 https://user.ceng.metu.edu.tr/~akifakkus/courses/ceng574/k-means.

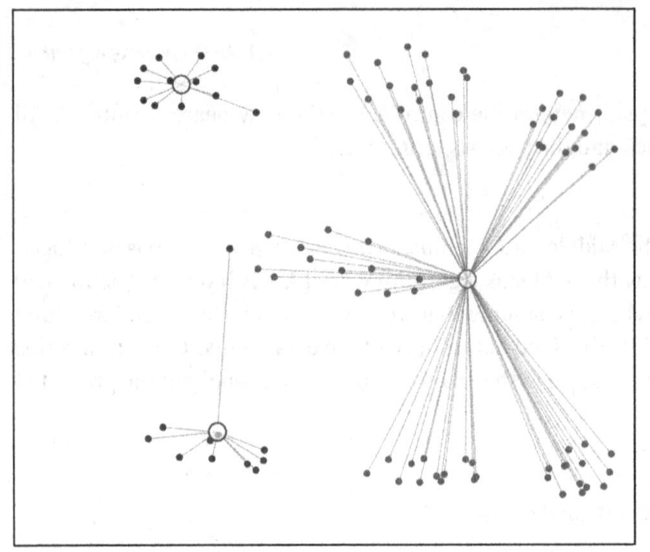

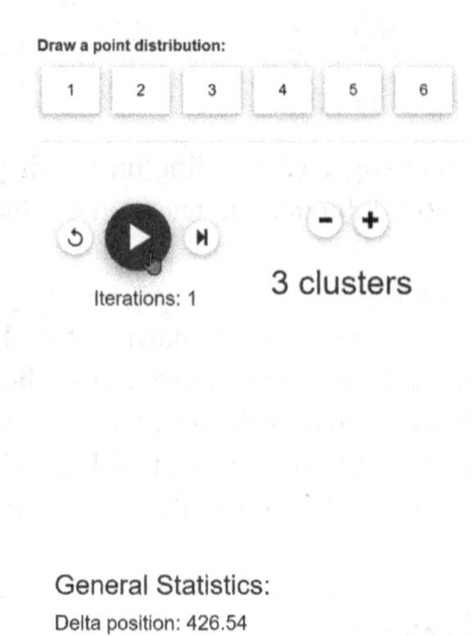

Draw a point distribution:
1 2 3 4 5 6
Iterations: 1
3 clusters
General Statistics:
Delta position: 426.54

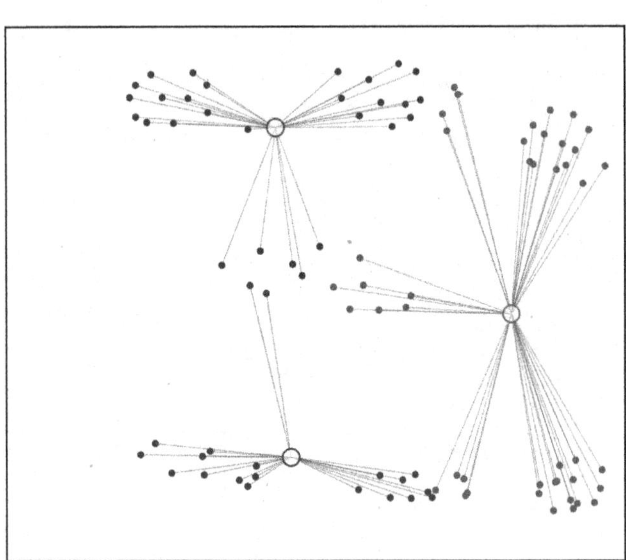

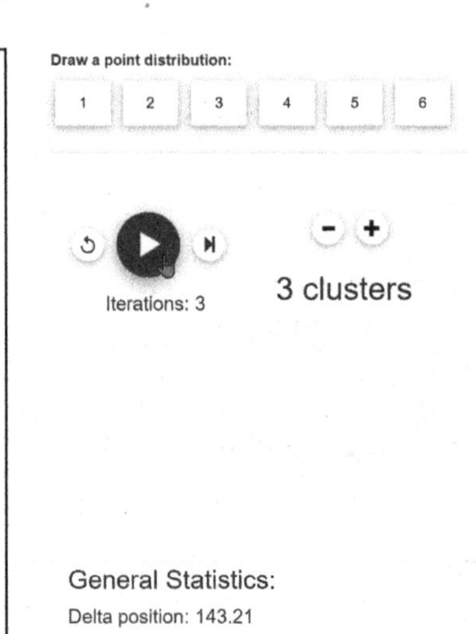

Draw a point distribution:
1 2 3 4 5 6
Iterations: 3
3 clusters
General Statistics:
Delta position: 143.21

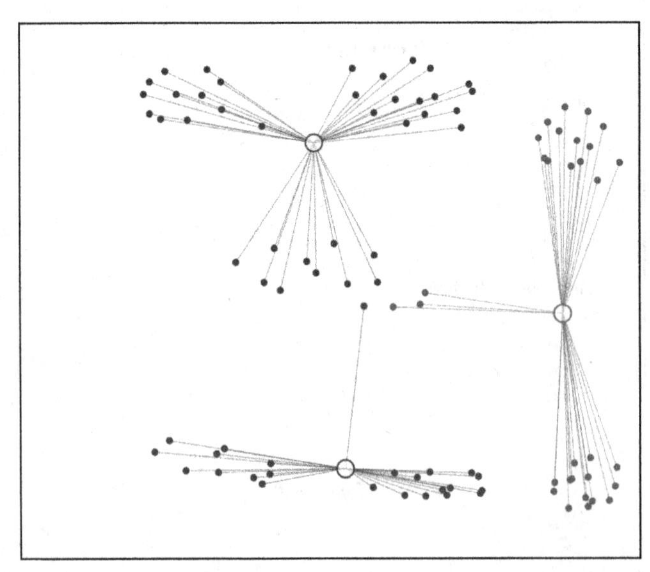

Draw a point distribution:
1 2 3 4 5 6
Iterations: 5
3 clusters
General Statistics:
Delta position: 48.29

(Continued)

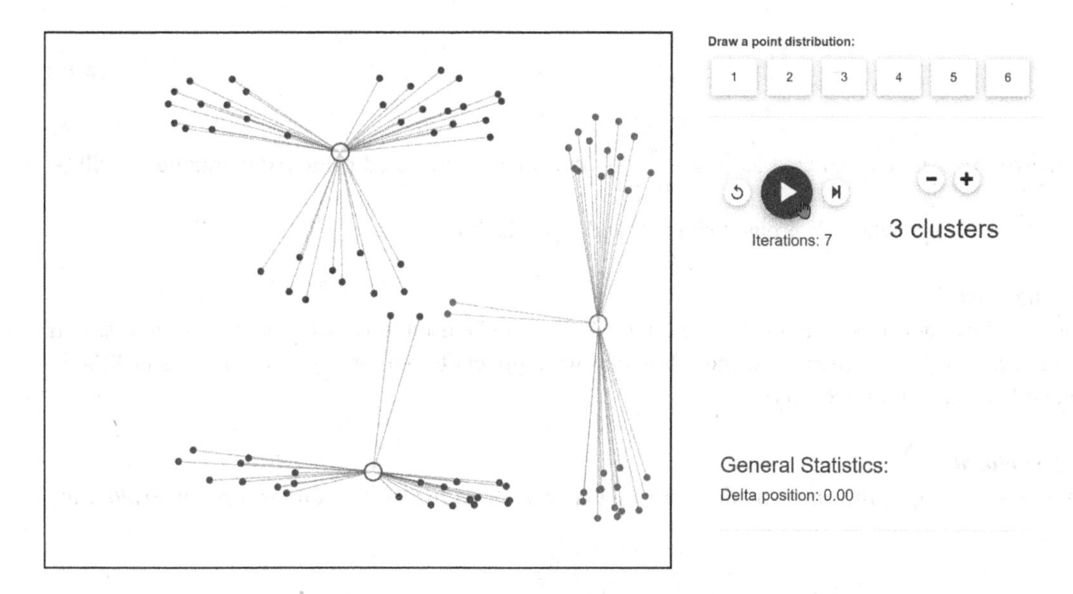

(b) Iterations of the K-Means analysis

Figure 4.30 (Cont'd) A visual demonstration of a typical K-Means analysis.

4.1.2.8 Principal Component Analysis (PCA) for Dimensionality Reduction

The *principal component analysis* (PCA) is a dimension reduction[100] algorithm that applies a linear static method to convert multi-dimensional inputs to leaner feature space data [74]. Such a conversion process identifies patterns in data in order to distill the observations down to their most important features.[101] It is worth noting that the PCA was developed by Karl Pearson [75]. The outcome of the PCA analysis is a list of ordered principal components (PCs), wherein PC1 is the most significant component, and hence it stores the maximum possible information from the data.[102]

How does it work?

The PCA creates new features by transforming the original features into a new set of features (dimensions) and into a new set of coordinates, called *principal components*, using eigenvectors and eigenvalues[103] calculated from a covariance matrix of the original variables[104] [76]. The PCA algorithms can consider a dataset, x_{ij}, such that $(i = 1, 2, \ldots m)$ ands_j, where m and k are equal to the feature dimension and number of data points, respectively. First, we calculate the mean $\tilde{x}_j$ and the standard deviation s_j for data jth column which is equal to:

$$\tilde{x}_j = \frac{1}{m}\sum_{i=1}^{m} x_{ij} \tag{4.36}$$

$$s_j = \sqrt{\frac{\sum_{i=1}^{m}(x_{ij} - \overline{x}j)^2}{m}} \tag{4.37}$$

Then, we obtain $[\,\tilde{x}\,]$ from the $[x]$ transformation matrix. The normalized elements $\tilde{x}_{ij}$ is obtained as follows.

---

100 While only PCA is covered here, I would still like to know that other methods are available such as *Linear Discriminant Analysis (LDA)*, which preserves the separation among classes, and *Multidimensional Scaling (MDS)*, which preserves the distances between data objects.

101 A fancy way of saying that the PCA simplifies a dataset without losing valuable attributes. PCA preserves the structure of the data by using its variance.

102 PC1 is followed by PC2, etc. In an *x-y* plane, two principal components will be perpendicular to each other because they should be able to span the whole *x-y* area. The first principal component (PC1) is computed so that it explains the largest amount of variance in the original features and the second component (PC1) is orthogonal to PC1 to explain the largest amount of variance left.

103 An **Eigenvector** of a linear transformation is a *nonzero* vector that changes at most by a scalar factor when that linear transformation is applied to it (and thus does not change direction). This factor is called the **Eigenvalue**. Simply, the Eigenvalue is a number that tells us how the dataset is spread out on the direction of an Eigenvector.

104 The PCA often employs data standardization/normalization.

$$\tilde{x}_{ij} = \frac{x_{ij} - \overline{x}j}{s_j} \tag{4.38}$$

Then the covariance matrix $[c]$ can be calculated as $[c] = \frac{[\tilde{x}]^T [\tilde{x}]}{m-1}$ and finally, the stated principal components will be obtained in the form of $[c]\{p_i\} = \lambda_i \{p_i\}$, where λ_i is eigenvalue i^{th} and $\{p_i\}$ is its related vector.[105]

4.1.2.8.1 Visualizing this Algorithm

Victor Powell and Lewis Lehe have developed an interactive interface that can be used to visualize how PCA works. This interface can be accessed here[106] [77]. I have created the following two snippets to demonstrate the process of PCA in Figure 4.31. Have a look and give this interface a try!

4.1.2.8.2 Programming This Algorithm

The *Python* and *R* codes for the PCA algorithm are provided below. This algorithm can also be found within the *Exploratory* platform.

Python Package:

```
import sklearn
from sklearn.decomposition import PCA
model = PCA()
```

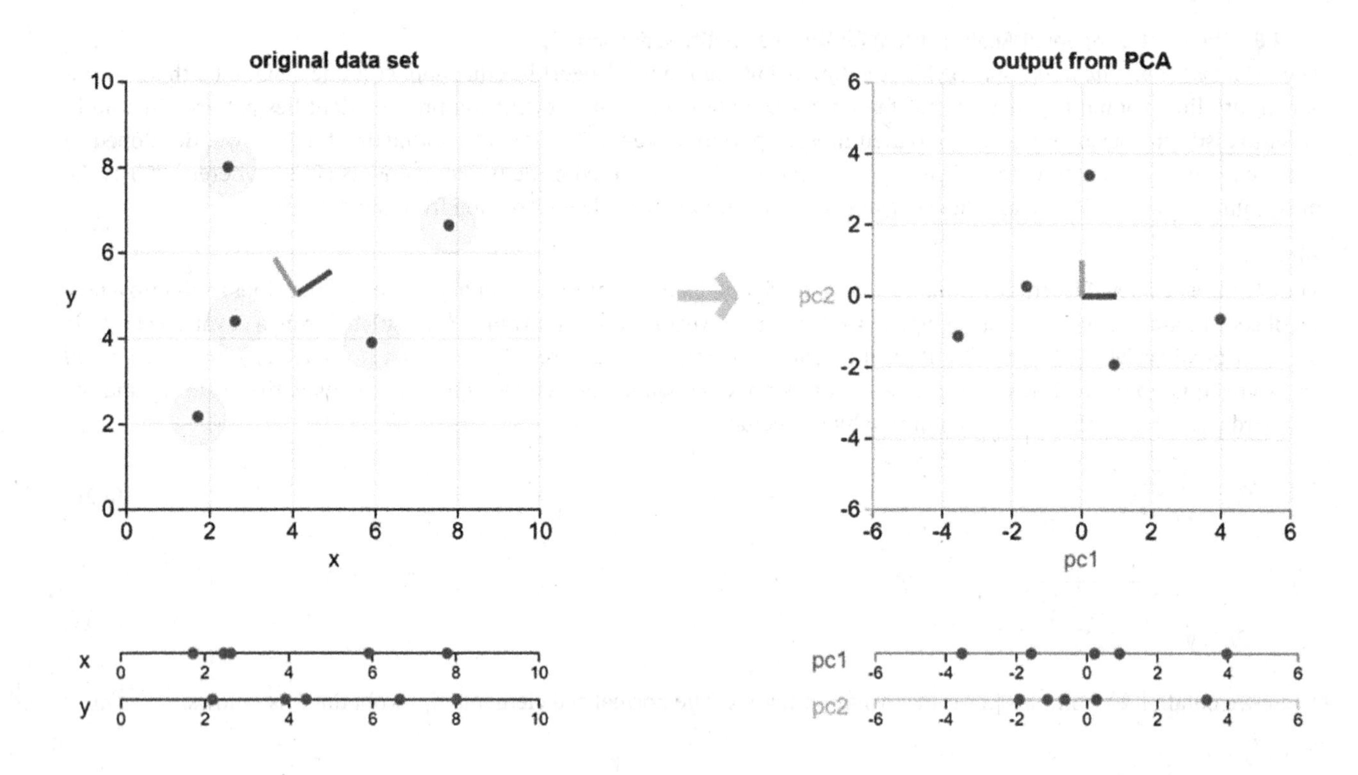

(a) Example 1

(Continued)

105 Additional notes: the results of a PCA can be difficult to interpret and can be effective on numeric features.
106 https://setosa.io/ev/principal-component-analysis.

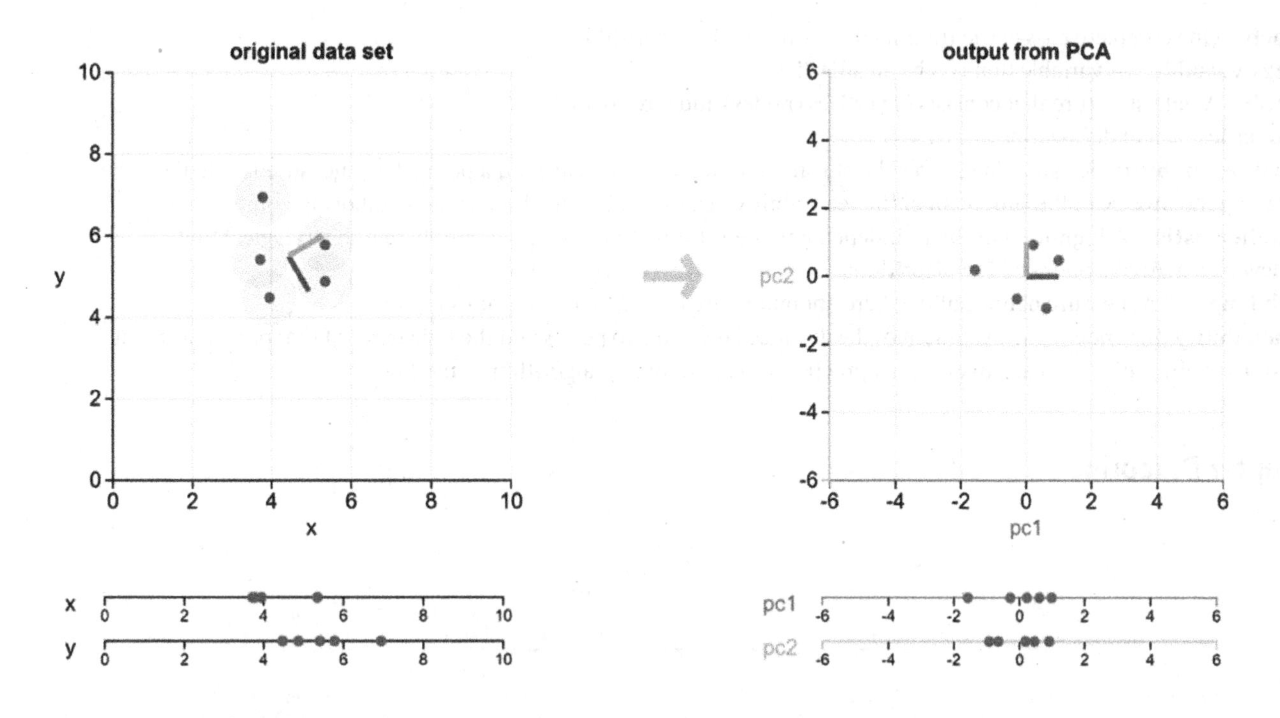

Figure 4.31 (Cont'd) Illustration of the PCA algorithm [77].

R Package:

```
>install.packages("FactoMineR")
>library(FactoMineR)
> PCA(X, scale.unit = TRUE, ncp = 5, ind.sup = NULL, quanti.sup = NULL, quali.sup =
NULL, row.w = NULL, col.w = NULL, graph = TRUE, axes = c(1,2))
```

4.2 Conclusions

This chapter covered a wide range of algorithms that fall under supervised and unsupervised learning methods. Our goal is twofolds to 1) get familiar with some of the most commonly used algorithms in our domain and 2) try to explore such algorithms by means to enhance and update our arsenal of algorithms. *Python* and *R* packages were also presented. A number of visualization tools created were also cited. I invite you to try these to get a closer and personal look into ML algorithms.[107]

Definitions

Convolution operation A grouping function that combines two functions to make a third and hence merges information.

Eigenvalue A number that tells us how the dataset is spread out on the direction of an Eigenvector.

107 Please note that the Appendix contains a series of benchmarked algorithms on some of the notable datasets from our domain.

Epoch One complete pass of the training data through the algorithm.
Fuzzy variable A variable that can be an adjective.
Graph A data structure that consists of vertices (nodes) and envelopes.
Heuristic Act of discovering.
Local reachability density measures the density of k-nearest points around a point. This measure is calculated by taking the inverse of the sum of all of the reachability distances of all the k-nearest neighboring points.
Metaheuristic A high-level problem-independent algorithmic framework (or simply search strategies) that develops heuristic optimization algorithms.
Path length A measure of normality where anomalies are linked to having shorter paths.
Reachability distance The maximum of the distance between two points and the k-distance of that particular point.
Visualization tool A means to examine, practice, or experience an algorithm or method.

Chapter Blueprint

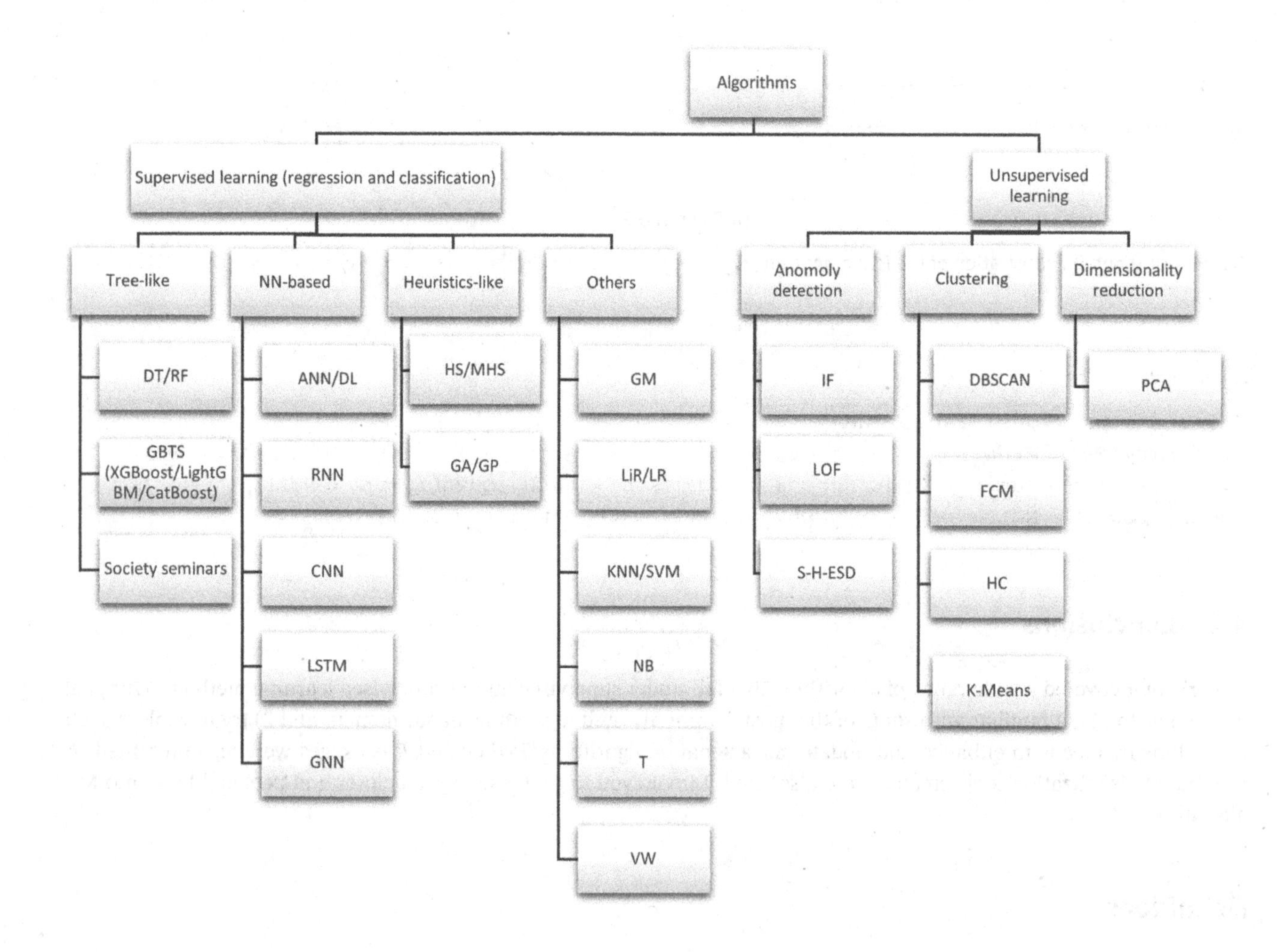

Questions and Problems

4.1 How many algorithms were covered in this chapter? Why are there many types of algorithms?

4.2 List three supervised and unsupervised learning algorithms.

4.3 What are activation functions? Why are they needed?

4.4 List two applications of CNN and RNN.

4.5 What are the main differences between GA and GP?

4.6 What is the reason for calling the RF algorithm?

4.7 How are GLM and GAM different than traditional linear regression?

4.8 How do heuristics differ from metaheuristics?

4.9 Discuss the application of graph neural networks and transformers from the context of your area.*

4.10 Who created the XGBoost, LightGBM, and CatBoost algorithms? List the main differences between these algorithms.

4.11 Sketch the prediction process of a regression and classification RF.

4.12 Select two clustering algorithms and compare their processes and flowcharts.

4.13 Explain the difference between clustering and dimensionality reduction.

References

1 Naser, M.Z. (2021). Mechanistically informed machine learning and artificial intelligence in fire engineering and sciences. *Fire Technol.* 57: 1–44. doi: 10.1007/s10694-020-01069-8.

2 Tapeh, A. and Naser, M.Z. (2022). Artificial intelligence, machine learning, and deep learning in structural engineering: a scientometrics review of trends and best practices. *Arch. Comput. Methods Eng.* doi: 10.1007/s11831-022-09793-w.

3 Zadeh, L.A. (1995). Discussion: probability theory and fuzzy logic are complementary rather than competitive. *Technometrics* 37: 271–276. doi: 10.1080/00401706.1995.10484330.

4 Takagi, T. and Sugeno, M. (1985). Fuzzy identification of systems and its applications to modeling and control. *IEEE Trans. Syst. Man.* 15. doi: 10.1109/TSMC.1985.6313399.

5 Seitlllari, A. and Naser, M.Z.Z. (2019). Leveraging artificial intelligence to assess explosive spalling in fire-exposed RC columns. *Comput. Concr.* 24. doi: 10.12989/cac.2019.24.3.271.

6 Jang, J.S.R., Sun, C.T., and Mizutani, E. (1997). Neuro-fuzzy and soft computing-a computational approach to learning and machine intelligence. *Autom. Control. IEEE.* 42. doi: 10.1109/TAC.1997.633847.

7 McCulloch, W.S. and Pitts, W. (1943). A logical calculus of the ideas immanent in nervous activity. *Bull. Math. Biophys.* 5. doi: 10.1007/BF02478259.

8 Naser, M.Z. (2020). Autonomous fire resistance evaluation. *ASCE Jounral Struct. Eng.* 146. doi: 10.1061/(ASCE)ST.1943-541X.0002641.

9 Cobaner, M., Unal, B., and Kisi, O. (2009). Suspended sediment concentration estimation by an adaptive neuro-fuzzy and neural network approaches using hydro-meteorological data. *J. Hydrol* 367: 52–61. doi: 10.1016/J.JHYDROL.2008.12.024.

10 TensorFlow (2022). A neural network playground. https://playground.tensorflow.org/#activation=relu&batchSize=10&dataset=gauss®Dataset=reg-plane&learningRate=0.03®ularizationRate=0&noise=0&networkShape=3,2,5,4,7,4&seed=0.62778&showTestData=true&discretize=false&percTrainData=50&x=true&y=true&xTimesY=false&xSquared=false&ySquared=false&cosX=false&sinX=false&cosY=false&sinY=false&collectStats=false&problem=regression&initZero=false&hideText=false (accessed 4 July 2022).

11 Musharraf, M., Khan, F., and Veitch, B. (2019). Validating human behavior representation model of general personnel during offshore emergency situations. *Fire Technol.* 55. doi: 10.1007/s10694-018-0784-1.

12 Chou, J.-S.-S., Tsai, C.-F.-F., Pham, A.-D.-D., and Lu, Y.-H.-H. (2014). Machine learning in concrete strength simulations: multi-nation data analytics. *Constr. Build. Mater.* 73: 771–780. doi: 10.1016/j.conbuildmat.2014.09.054.
13 Yee, S. and Chu, T. (2015). A visual introduction to machine learning. http://www.r2d3.us/visual-intro-to-machine-learning-part-1 (accessed 1 July 2022).
14 Harley, A.W. (2015). An interactive node-link visualization of convolutional neural networks. In: *Lecture Notes Computer Science* (Including Subser. Lect. Notes Artif. Intell. Lect. Notes Bioinformatics). (ed. G. Bebis, R. Boyle, B. Parvin, et al.) Springer. doi: 10.1007/978-3-319-27857-5_77.
15 Madsen, A. (2019). Visualizing memorization in RNNs. *Distill.* doi: 10.23915/distill.00016.
16 Goodfellow, I., Pouget-Abadie, J., Mirza, M. et al. (2020). Generative adversarial networks. *Commun. ACM* 63. doi: 10.1145/3422622.
17 Polo Club (2022). Polo Club of data science @ georgia tech: human-centered AI. Deep Learning Interpretation & Visualization, Cybersecurity, Large Graph Visualization and Mining | Georgia Tech | Atlanta, GA 30332, United States. https://poloclub.github.io (accessed 4 July 2022).
18 Kahng, M., Thorat, N., Chau, D.H.P. et al. (2018). GAN Lab: understanding complex deep generative models using interactive visual experimentation. *IEEE Trans. Vis. Comput. Graph.* 25. doi: 10.1109/TVCG.2018.2864500.
19 Nelder, J.A. and Wedderburn, R.W.M. (1972). Generalized linear models. *J. R. Stat. Soc.* 135.
20 Hastie, T. and Tibshirani, R. (1986). Generalized additive models. *Stat. Rev.* 1: 297–310. doi: 10.1214/SS/1177013604.
21 Holland, J.H. (1975). *Adaptation in Natural and Artificial Systems : An Introductory Analysis with Applications to Biology, Control, and Artificial Intelligence.* MIT Press.
22 Alavi, A.H., Gandomi, A.H., Sahab, M.G., and Gandomi, M. (2010). Multi expression programming: a new approach to formulation of soil classification. *Eng. Comput.* 26: 111–118. doi: 10.1007/s00366-009-0140-7.
23 Willis, M.-J. (2005). Genetic programming: an introduction and survey of applications. doi: 10.1049/cp:19971199.
24 Aslam, F., Farooq, F., Amin, M.N. et al. (2020). Applications of gene expression programming for estimating compressive strength of high-strength concrete. *Adv. Civ. Eng.* 2020. doi: 10.1155/2020/8850535.
25 Zhang, Q., Barri, K., Jiao, P. et al. (2020). Genetic programming in civil engineering: advent, applications and future trends. *Artif. Intell. Rev.* 54. doi: 10.1007/s10462-020-09894-7.
26 Fortin, F.A., De Rainville, F.M., Gardner, M.A. et al. (2012). DEAP: evolutionary algorithms made easy. *J. Mach. Learn. Res.* 13.
27 Chen, T. and Guestrin, C. (2016). XGBoost: a scalable tree boosting system. Proceedings of the 22nd ACM SIGKDD International Conference on Knowledge Discovery and Data Mining. doi: 10.1145/2939672.2939785.
28 Scikit (2020). sklearn.ensemble. GradientBoostingRegressor — scikit-learn 0.24.1 documentation. https://scikit-learn.org/stable/modules/generated/sklearn.ensemble.GradientBoostingRegressor.html (accessed 9 February 2021).
29 XGBoost Python Package (2020). Python package introduction — xgboost 1.4.0-SNAPSHOT documentation. https://xgboost.readthedocs.io/en/latest/python/python_intro.html#early-stopping (accessed 10 February 2021).
30 Gradient boosted tree (GBT). (2019).
31 Ke, G., Meng, Q., Finley, T. et al. (2017). LightGBM: a highly efficient gradient boosting decision tree. *Advances in Neural Information Processing Systems 30 (NIPS 2017).*
32 Nahon, A. (2019). XGBoost, LightGBM or CatBoost — which boosting algorithm should I use? https://medium.com/riskified-technology/xgboost-lightgbm-or-catboost-which-boosting-algorithm-should-i-use-e7fda7bb36bc (accessed 1 July 2022).
33 Scarselli, F., Hagenbuchner, M., Yong, S.L. et al. (2005). Graph neural networks for ranking Web pages. *IEEE WIC ACM International Conference on Web Intelligence (WI).* doi: 10.1109/WI.2005.67.
34 Sanchez-Lengeling, B., Reif, E., Pearce, A., and Wiltschko, A. (2021). A gentle introduction to graph neural networks. *Distill.* doi: 10.23915/distill.00033.
35 De Leon-Aldaco, S.E., Calleja, H., and Aguayo Alquicira, J. (2015). Metaheuristic optimization methods applied to power converters: a review. *IEEE Trans. Power Electron.* 30. doi: 10.1109/TPEL.2015.2397311.
36 Fix, E. and Hodge, J. (1951). Discriminatory analysis nonparametric discrimination: consistency properties. https://apps.dtic.mil/dtic/tr/fulltext/u2/a800276.pdf (accessed 1 July 2022).
37 Mangalathu, S. and Jeon, J.S. (2018). Classification of failure mode and prediction of shear strength for reinforced concrete beam-column joints using machine learning techniques. *Eng. Struct.* 160. doi: 10.1016/j.engstruct.2018.01.008.
38 Scikit (2021). sklearn.neighbors. NearestNeighbors — scikit-learn 0.24.1 documentation. https://scikit-learn.org/stable/modules/generated/sklearn.neighbors.NearestNeighbors.html?highlight=knearest#sklearn.neighbors.NearestNeighbors.kneighbors (accessed 5 April 2021).

39 K-Nearest Neighbors Demo (n.d.) http://vision.stanford.edu/teaching/cs231n-demos/knn (accessed 1 July 2022).
40 Jafari Goldarag, Y., Mohammadzadeh, A., and Ardakani, A.S. (2016). Fire risk assessment using neural network and logistic regression. *J. Indian Soc. Remote Sens.* 44. doi: 10.1007/s12524-016-0557-6.
41 Roeschl, T. (2020). Animations of logistic regression with. https://towardsdatascience.com/animations-of-logistic-regression-with-python-31f8c9cb420 (accessed 1 July 2022).
42 Hochreiter, S. and Schmidhuber, J. (1997). Long short-term memory. *Neural Comput.* 9. doi: 10.1162/neco.1997.9.8.1735.
43 Shiri Harzevili, N. and Alizadeh, S.H. (2018). Mixture of latent multinomial naive Bayes classifier. *Appl. Soft Comput.* 69: 516–527. doi: 10.1016/J.ASOC.2018.04.020.
44 Liaw, A. and Wiener, M. (2001). Classification and regression by RandomForest. https://www.researchgate.net/publication/228451484 (accessed 8 April 2019).
45 Breiman, L. (2001). 4_method_random_forest. *Mach. Learn.* 44: 5–32.
46 Boser, B.E., Guyon, I.M., and Vapnik, V.N. (1992). A training algorithm for optimal margin classifiers. COLT'92: Proceedings of the Fifth Annual Workshop on Computational Learning Theory. doi: 10.1145/130385.130401.
47 Solhmirzaei, R., Salehi, H., Kodur, V., and Naser, M.Z. (2020). Machine learning framework for predicting failure mode and shear capacity of ultra high performance concrete beams. *Eng. Struct.* 224. doi: 10.1016/j.engstruct.2020.111221.
48 Çevik, A., Kurtoğlu, A.E., Bilgehan, M. et al. (2015). Support vector machines in structural engineering: a review. doi: 10.3846/13923730.2015.1005021.
49 Cherkassky, V. and Ma, Y. (2004). Practical selection of SVM parameters and noise estimation for SVM regression. *Neural Netw.* 17. doi: 10.1016/S0893-6080(03)00169-2.
50 Greitemann, J. (2018). Interactive demo of Support Vector Machines (SVM). https://jgreitemann.github.io/svm-demo (accessed 1 July 2022).
51 Abadi, M. (2016). TensorFlow: learning functions at scale. ACM SIGPLAN Not. doi: 10.1145/3022670.2976746.
52 Vaswani, A., Shazeer, N., Parmar, N. et al. (2017). Attention is all you need. *Advances in Neural Information Processing Systems.*
53 VowpalWabbit (2021). vowpalwabbit.sklearn — vowpalWabbit 8.11.0 documentation. https://vowpalwabbit.org/docs/vowpal_wabbit/python/latest/vowpalwabbit.sklearn.html (accessed 26 April 2021).
54 Liu, F.T., Ting, K.M., and Zhou, Z.H. (2008). Isolation Forest. *Advances in Neural Information Processing Systems 30 (NIPS 2017).* doi: 10.1109/ICDM.2008.17.
55 Scikit (2021). sklearn.ensemble. IsolationForest — scikit-learn 0.24.1 documentation. Scikit. https://scikit-learn.org/stable/modules/generated/sklearn.ensemble.IsolationForest.html (accessed 22 March 2021).
56 De Maesschalck, R., Jouan-Rimbaud, D., and Massart, D.L. (2000). The Mahalanobis distance. *Chemom. Intell. Lab. Syst.* 50. doi: 10.1016/S0169-7439(99)00047-7.
57 Kejariwal, A. (2015). Introducing practical and robust anomaly detection in a time series. https://blog.twitter.com/engineering/en_us/a/2015/introducing-practical-and-robust-anomaly-detection-in-a-time-series (accessed 28 June 2022).
58 Ray, S., McEvoy, D.S., Aaron, S. et al. (2018). Using statistical anomaly detection models to find clinical decision support malfunctions. *J. Am. Med. Informatics Assoc.* 25. doi: 10.1093/jamia/ocy041.
59 Ester, M., Kriegel, H.-P., Sander, J., and Xu, X. (1996). A density-based algorithm for discovering clusters in large spatial databases with noise. AAAI Press. 226–231.
60 Tran, T.N., Wehrens, R., and Buydens, L.M.C. (2006). KNN-kernel density-based clustering for high-dimensional multivariate data. *Comput. Stat. Data Anal.* 51: 513–525. doi: 10.1016/j.csda.2005.10.001.
61 Neto, A.C.A., Sander, J., Campello, R.J.G.B., and Nascimento, M.A. (2019). Efficient computation and visualization of multiple density-based clustering hierarchies. *IEEE Trans. Knowl. Data Eng.* 33: 3075–3089. doi: 10.1109/TKDE.2019.2962412.
62 Zhu, Q., Tang, X., and Elahi, A. (2021). Application of the novel harmony search optimization algorithm for DBSCAN clustering. *Expert Syst. Appl.* 178: 115054. doi: 10.1016/j.eswa.2021.115054.
63 Harris, N. (2022). Visualizing DBSCAN clustering. https://www.naftaliharris.com/blog/visualizing-dbscan-clustering (accessed 27 June 2022).
64 Haralick, R.M. and Shapiro, L.G. (1985). Image segmentation techniques. *Comput. Vision, Graph. Image Process* 29: 100–132. doi: 10.1016/S0734-189X(85)90153-7.
65 Dias, M.L.D. (2019). Fuzzy-c-means: an implementation of Fuzzy c-means clustering algorithm. https://scholar.google.com/scholar?hl=en&as_sdt=0%2C41&q=Dias%2C+M.L.D.+%282019%29.+Fuzzy-c-means%3A+an+implementation+of+Fuzzy+c-means+clustering+algorithm&btnG=.

66 Adhikari, S.K., Sing, J.K., Basu, D.K., and Nasipuri, M. (2015). Conditional spatial fuzzy C-means clustering algorithm for segmentation of MRI images. *Appl. Soft Comput. J.* 34: 758–769. doi: 10.1016/j.asoc.2015.05.038.

67 Alruwaili, M., Siddiqi, M.H., and Javed, M.A. (2020). A robust clustering algorithm using spatial fuzzy C-means for brain MR images. *Egypt. Inform. J.* 21: 51–66. doi: 10.1016/j.eij.2019.10.005.

68 Hexmoor, H. (2015). Diffusion and contagion. In: *Computational Network Science: An Algorithmic Approach* (ed. H. Hexmoor), 45–64. Elsevier.

69 Ward, J.H. (1963). Hierarchical grouping to optimize an objective function. *J. Am. Stat. Assoc.* 58. doi: 10.1080/01621459.1963.10500845.

70 Yeturu, K. (2020). Chapter 3 - Machine learning algorithms, applications, and practices in data science. In: *Handbook of Statistics* (ed. A.S.R. Srinivasa Rao and C.R. Rao), 81–206. Elsevier. doi: 10.1016/bs.host.2020.01.002.

71 MacQueen, J. (1967). Some methods for classification and analysis of multivariate observations. *Proceedings of the Fifth Berkeley Symposium on Mathematical Statistics and Probability.*

72 Hartigan, J.A. and Wong, M.A. (1979). Algorithm AS 136: A K-means clustering algorithm. *Appl. Stat.* 28. doi: 10.2307/2346830.

73 Akkuş, A. (2022). K-Means clustering demo. https://user.ceng.metu.edu.tr/~akifakkus/courses/ceng574/k-means (accessed 27 June 2022).

74 Jollife, I.T. and Cadima, J. (2016). Principal component analysis: a review and recent developments. *Philos. Trans. R. Soc. A Math. Phys. Eng. Sci.* 28. doi: 10.1098/rsta.2015.0202.

75 Pearson, K. (1901). LIII. On lines and planes of closest fit to systems of points in space. *London, Edinburgh, Dublin Philos. Mag. J. Sci.* 374. doi: 10.1080/14786440109462720.

76 Li, J., Dackermann, U., Xu, Y.-L., and Samali, B. (2011). Damage identification in civil engineering structures utilizing PCA-compressed residual frequency response functions and neural network ensembles. *Struct. Control Heal. Monit.* 18: 207–226. doi: 10.1002/stc.369.

77 Powell, V. and Lehe, L. (2022). Principal component analysis explained visually. https://setosa.io/ev/principal-component-analysis (accessed 27 June 2022).

78 Herculano-Houzel, S. (2012). The remarkable, yet not extraordinary, human brain as a scaled-up primate brain and its associated cost. *Proc. Natl. Acad. Sci. U. S. A.* 109. doi: 10.1073/pnas.1201895109.

79 Zimmer, C. (2011). 100 trillion connections: new efforts probe and map the brain's detailed architecture. *Sci. Am.* 304: 58–63. doi: 10.1038/SCIENTIFICAMERICAN0111-58.

80 Esposito, F., Malerba, D., and Semeraro, G. (1997). A comparative analysis of methods for pruning decision trees. *IEEE Trans. Pattern Anal. Mach. Intell.* 19. doi: 10.1109/34.589207.

81 LeCun, Y., Bengio, Y., and Hinton, G. (2015). Deep learning review. *Nature* 521.

82 Quang, D. and Xie, X. (2016). DanQ: a hybrid convolutional and recurrent deep neural network for quantifying the function of DNA sequences. *Nucleic Acids Res.* 44. doi: 10.1093/nar/gkw226.

83 Gandomi, A.H. and Alavi, A.H. (2012). Krill herd: a new bio-inspired optimization algorithm. *Commun. Nonlinear Sci. Numer. Simul.* 17. doi: 10.1016/j.cnsns.2012.05.010.

84 Golafshani, E.M. and Behnood, A. (2019). Estimating the optimal mix design of silica fume concrete using biogeography-based programming. *Cem. Concr. Compos.* 96: 95–105. doi: 10.1016/J.CEMCONCOMP.2018.11.005.

85 Cramer, J.S. (2002). The origins of logistic regression. *SSRN Electron. J.* doi: 10.2139/ssrn.360300.

86 Kotsiantis, S. and Kanellopoulos, D. (2006). Association rules mining: a recent overview. *Science (80-.)*.32.

87 Hariri, S., Kind, M.C., and Brunner, R.J. (2019). Extended isolation forest. *IEEE Trans. Knowl. Data Eng.* 33. doi: 10.1109/TKDE.2019.2947676.

5

Performance Fitness Indicators and Error Metrics

But fairness is squishy and hard to quantify.

Cathy O'Neil[1]

Synopsis

In the past few chapters, we learned about data, various ML algorithms, and how we can turn these into ML models. A ML model is declared *proper* once it satisfies a rigorous training/validation/testing process. We quantify such a process through metrics and indicators. What are these? Why are they needed? And, what makes them a suitable judge of our models? This chapter not only hopes to shed light on the above three questions but aims to go one step further toward showcasing and recommending a series of *metrics* and *indicators*[2] that could be of interest to some of the problems you will be exploring.

5.1 Introduction

Machine learning, by far, is a data-driven approach. It looks for patterns in data in search of a model[3] that can describe the patterns in our data. This implies that a good ML model is one that *performs optimally and best describes the phenomenon on hand*. The first component (i.e., *performs optimally*) can be addressed by properly developing and tuning the model. The second component (i.e., *best describes the phenomenon on hand*) is where we need some form of comparison that contrasts model predictions against a known ground truth. The objective is simple; if the predictions fall in line with an acceptable level of the ground truth, then our model can be declared to have *good fitness*.[4]

At first glance, the notion of using metrics to quantify the goodness of a model may sound foreign. However, this practice is not uncommon in civil and environmental engineering. Here is an example.

Suppose we are interested in exploring the effect of using a new type of construction material on the deformation[5] of a beam that we happen to have tested during an earlier experiment. In order to carry out this investigation, and instead of fabricating

1 Cathy O'Neil, the author of *Weapons of Math Destruction* [66].
2 **Indicator**: is a measure that comprises a set of metrics.
3 Or, models.
4 Quick story: Jasmine and I have two dogs, *Pirate* and *Snow*. Whenever we play catch, we noticed that *Snow* consistently excels at catching toys much more than *Pirate* (despite *Snow*'s smaller frame and the fact that she is a three-legged dog!). So far, *Snow* owns the record to the number of successful catches in a row (18/20 vs. 12/20)! Does this tell us that *Snow* has a better accuracy catching toys than *Pirate*? This chapter is about establishing means to evaluate performance comparisons.
5 For the sake of this example, let us assume that we would like to see how the new material would affect the load-deformation history of the beam on hand.

Machine Learning for Civil & Environmental Engineers: A Practical Approach to Data-driven Analysis, Explainability, and Causality, First Edition. M. Z. Naser.

Companion Website: www.wiley.com/go/Naser/MachineLearningforCivilandEnvironmentalEngineers

and testing a new beam, we will opt to develop an FE model that replicates the same characteristics and conditions of the tested beam. We build our FE model and then run a simulation to extract the load-deformation history of the simulated beam.[6]

It is commonly accepted for engineers to compare the predicted load-deformation history to the actual load-deformation history obtained from the test. A qualitative assessment is then established wherein if the predicted load-deformation history *matches* the actual one, then the FE model is deemed suitable. This process is what is referred to as *validating* the FE model.

Figure 5.1 depicts our example by plotting the actual load-deformation history against three predicted load-deformation histories (obtained from three different FE models). Here is a question, *which of these three FE models would you consider suitable?* Let us walk through each model.

By inspection, we can clearly see that *Model 1* seems to provide the poorest predicted load-deformation history. The predicted history is deemed poor since the model does not seem to match the same trend observed in the actual test and throughout the loading history[7] (despite the good agreement initially). Furthermore, this model predicts a drop in the applied loading with increasing deformation once the beam reaches its maximum load capacity. This behavior is not present in the actual test. In other words, *Model 1* does not match our expected history at the global level (throughout the load-deformation history) as well as the local level (post-peak).

In contrast to *Model 1*, predictions from *Model 2* and *Model 3* better align with the observed load-deformation history. However, between *Model 2* and *Model 3*, we can see that *Model 2*'s predictions fluctuate between a stiffer response (initially) and a smoother response toward the end of the history plot. Unlike *Model 2*, *Model 3*'s predictions are much more consistent in capturing the full response of the tested beam. Hence, we can declare that *Model 3* as validated and can now be used to explore how the use of a new type of material can influence the deformation of the tested beam.

As you might have noticed, the above discussion is primarily *qualitative* and does not provide us with a *quantitative* scale to establish the validity of FE models. This is because we do not have established quantitative criteria (yet) to declare the validity of an FE model.[8] Let us now bring this discussion back to the case of ML.

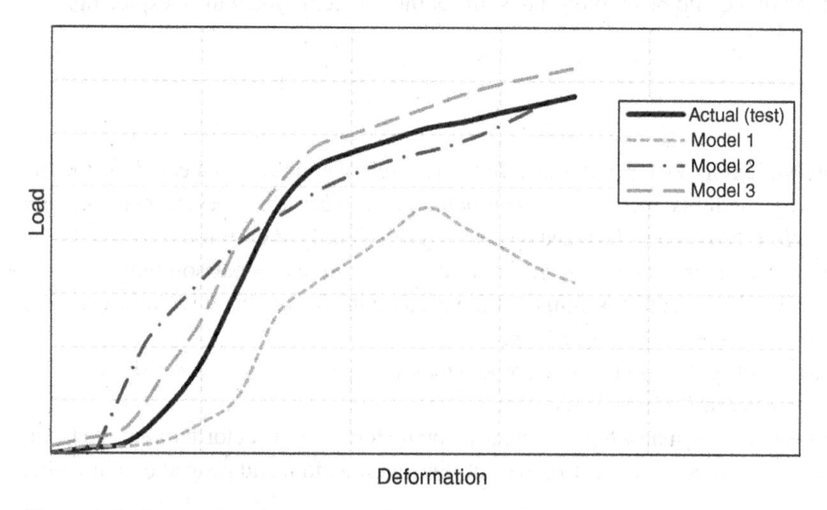

Figure 5.1 Comparison between predicted and actual load-deformation history.

5.2 The Need for Metrics and Indicators

Unlike FE simulations, we do have a series of tools and measures that we can use to establish the validity of ML models. These are referred to as metrics and indicators, and these are necessary to *quantify* how good/poor a model is in predicting the phenomenon on hand. There are other reasons why we need to learn about such measures.

6 This simulated beam is a replica of the tested beam, i.e., we are not exploring the effect of the new construction material yet.

7 This is called a history, as it documents the response (i.e., deformation) of the beam under various load levels. Since we start with smaller loads first and then increase such loads, we can plot the response as a function of the increasing loads. It is a successive and chronological process.

8 From this perspective, the validity of an FE model is established if the variation between predicted results and measured observations is between 5 and 15% [67]. Please note that the validation of an FE model is also governed by satisfying convergence criteria input in the FE software. My good friend Thomas Gernay, and his colleagues, has a nice paper of establishing validation via FE simulations [68]. More on this can be found elsewhere [69, 70].

As you remember, a major component of ML revolves around data. In some problems, the data could be limited or expensive/hard to collect (i.e., need specialized instrumentation/facilities [1]). As such, if one is to develop a ML model, then such a model could be ill-posed or biased.[9] In a more serious case, an engineer with little experience with ML could apply *improper* metrics or *misinterpret* the outcome of metrics [2, 3]. The aforenoted scenarios are examples of why we need to learn about metrics and indicators.

The bottom line is that we would like to assess the ML model to ensure: 1) the validity of the developed model in predicting a phenomenon given our dataset, and 2) the proper extension of the same model toward new/future datasets.

Let us now define what the *performance fitness indicators* and *error metrics (PFIEM)* are. Metrics are defined as logical and/or mathematical constructs intended to measure the closeness of actual observations to that expected (or predicted) by ML models. Often, PFIEM are derived with regard to specific assumptions and limitations, and hence these must be complied with.

Simply, PFIEM are used to establish an understanding of how the predictions from a ML model compare to the real (or measured) observations by explicitly stating such variation [4–6]. In this textbook, I will visit PFIEM, which can be used in supervised learning[10] (i.e., regression and classification) and unsupervised learning (clustering).

Personally, I prefer metrics that are easy to apply and interpret. I also prefer to select 3–7 independent metrics[11] to be applied to a particular problem. I will be providing some of these, together with code/packages for both *Python* and *R*, toward the end of this chapter.

5.3 Regression Metrics and Indicators

From the lens of regression, PFIEM can provide a quantitative description (i.e., some form of deviation or distance) of how model predictions fair against the ground truth. A simple metric results from subtracting a predicted value from its corresponding actual/observed value. Such a metric is simple, easy to interpret, and yields the magnitude of error (or difference) in the same units as those measured. Hence, it is intuitive to indicate if the model overestimates or underestimates observations (by analyzing the sign of the remainder).[12]

This metric can be supplemented with absoluteness or squareness. For example, an absolute error[13] $|A\text{-}P|$ can yield non-negative values. Analogous to traditional error, the absolute error also maintains the same units of predictions (and observations) and hence is easily relatable. However, due to its nature, the bias in absolute errors cannot be determined.[14]

Similar to the same concept of the absolute error, the squared error $(A\text{-}P)^2$ also tries to mitigate mutual cancellation. This metric is continuously differentiable, which comes in handy in optimization problems. Unlike the absolute error, this metric results in relatively large errors and could be susceptible to outliers.[15] The fact that the units of squared error are also squared can complicate the interpretation of errors (e.g., squared degrees or years, or as we saw earlier, strength).

Another simple metric divides the predicted value by its actual counterpart (e.g., P/A). Such a metric can also be supplemented with other operators to yield other error types. For example, these may include logarithmic quotient error (i.e., $\ln(P/A)$) as well as absolute logarithmic quotient error (i.e., $|\ln(P/A)|$).

For completion, Table 5.1 presents a list of relevant metrics that can be used in regression problems, together with some of their limitations/shortcomings and properties. These metrics are general – mostly field independent, and hence can be used in many problems. Metrics that include n are applicable to the general/full dataset as their calculation may require incorporating the number of data points in some shape or form.[16] The same metrics can also be applied to the training, validation, and testing sets. This can provide the engineer with a look into the performance of the ML within the specific bounds of the training, validation, and testing sets or any other subsets that comprise the aforenoted sets (e.g., a hybrid subset[17] that comprises 30% observations from the training set, and 40% from the validation set and 30% from the testing set).

9 Since it was not developed/tested across a comprehensive dataset.

10 Botchkarev [71] has a great review on preferred metrics reported by researchers during the 1980–2007 era. Some of the provided metrics also appear in my review [72].

11 For example, R and R^2 are related and hence dependent.

12 A remainder of zero implies a perfect prediction.

13 Throughout this chapter, note that A: actual measurements, P: predictions, n: number of data points.

14 Since the absoluteness always result in positive remainders.

15 **Outliers**: observations that lie at an abnormal distance from other values (in a random sample from a population).

16 In parallel, the first metric, E, in Table 1 is applicable at individual points. On the other hand, the metric ME, averages this error across the whole dataset.

17 More on this in a few sections.

Table 5.1 List of commonly used regression metrics.

No.	Metric	Definition/remarks	Formula	Notes
1	Error (E)	The amount by which a prediction differs from its actual counterpart.	$E = A - P$	• Intuitive. • Easy to apply.
2	Mean error (ME)	The average of all errors in a set.	$ME = \frac{\sum_{i=1}^{n} E_i}{n}$	• May not be helpful in cases where positive and negative predictions cancel each other out.
3	Mean Normalized Bias (MNB)	Associated with an observation-based minimum threshold.	$MNB = \frac{\sum_{i=1}^{n} E_i / A_i}{n}$	• Biased toward overestimations.
4	Mean Percentage Error (MPE)	Computed average of percentage errors.	$MPE = \frac{\sum_{i=1}^{n} E_i / A_i}{n / 100}$	• Undefined whenever a single actual value is zero.
5	Mean Absolute Error (MAE)*	Measures the difference between two continuous variables.	$MAE = \frac{\sum_{i=1}^{n} \lvert E_i \rvert}{n}$	• Uses a similar scale to input data [7]. • Compares different scales.
6	Mean Absolute Percentage Error (MAPE)*	Measures the extent of error in percentage terms.	$MAPE = \frac{100}{n} \sum_{i=1}^{n} \lvert E_i \rvert / \lvert A_i \rvert$	• Commonly used as a loss function [8]. • It cannot be used if there are zero values. • Percentage error cannot exceed 1.0 for small predictions. • No upper limit. • Non-symmetrical (adversely affected if a predicted value is larger or smaller than the corresponding actual value) [8].
7	Relative Absolute Error (RAE)	Expressed as a ratio comparing the mean error to errors produced by a trivial model.	$RAE = \sum_{i=1}^{n} \lvert E_i \rvert / \lvert A_i - A_{mean} \rvert$	• E_i ranges from zero (ideal) to infinity.
8	Mean Absolute Relative Error (MARE)	Measures the average ratio of absolute error to random error.	$MARE = \frac{1}{n} \sum_{i=1}^{n} \lvert E_i \rvert / \lvert A_i \rvert$	• Sensitive to outliers. • Undefined if division by zero.
9	Mean Relative Absolute Error (MRAE)	The ratio of accumulation of errors to the cumulative error of random error.	$MRAE = \frac{\sum_{i=1}^{n} \lvert E_i \rvert / \lvert A_i - A_{mean} \rvert}{n}$	• For a perfect fit, the numerator equals zero [9].

10	Geometric Mean Absolute Error (GMAE)*	Defined as the n-th root of the product of error values.	$GMAE = \sqrt[n]{\prod_{i=1}^{n} \lvert E_i \rvert}$	• It can be more appropriate for averaging relative quantities vs. arithmetic mean [10]. • It can be dominated by large outliers and minor errors.
11	Fractional Absolute Error (FAE)	Evaluates the absolute fractional error.	$FAE = \frac{1}{n} \sum_{i=1}^{n} \frac{2 \times \lvert E_i \rvert}{\lvert A_i \rvert + \lvert P_i \rvert}$	• Works with numeric data.
12	Mean Squared Error (MSE)	Measures the average of the squares of the errors.	$MSE = \frac{\sum_{i=1}^{n} E_i^2}{n}$	• Scale dependent [11]. • Values closer to zero present an adequate state. • Heavily weights outliers. • Highly dependent on the fraction of data used (low reliability) [12].
13	Normalized Mean Squared Error (NMSE)	Estimates the overall deviations between measured values and predictions.	$NMSE = \frac{\frac{\sum_{i=1}^{n} E_i^2}{n}}{variance^2}$ $variance = \frac{\sum (x_i - mean)^2}{n-1}$	• Biased toward over-predictions [13]. • An extension of MSE.
14	Root Mean Squared Error (RMSE)	Root square of average squared error.	$RMSE = \sqrt{\frac{\sum_{i=1}^{n} E_i^2}{n}}$	• Scale-dependent. • A lower value for RMSE is favorable. • Sensitive to outliers. • Highly dependent on the fraction of data used [12]. • An extension of MSE.
15	Root Mean Square Percentage Error (RMSPE)*	Evaluates the mean of squared errors in percentages.	$RMSPE = \sqrt{\frac{\sum_{i=1}^{n} (\lvert E_i \rvert / \lvert A_i \rvert)^2}{n/100}}$	• Scale independent. • It can be used to compare predictions from different datasets. • Non-symmetrical [8]. • An extension of RMSE.
16	Normalized Root Mean Squared Error (NRMSE)**	Normalizes the root mean squared error.	$NRMSE = \frac{\sqrt{\frac{\sum_{i=1}^{n} E_i^2}{n}}}{A_{mean}}$	• It can be used to compare predictions from different datasets [14]. • Works with numeric data • An extension of RMSE.

(Continued)

Table 5.1 (Continued)

No.	Metric	Definition/remarks	Formula	Notes
17	Geometric Root Mean Squared Error (GRMSE)	Evaluates the geometric root squared errors.	$GRMSE = \sqrt[2n]{\prod_{i=1}^{n} E_i^2}$	• Scale-dependent. • Less sensitive to outliers than RMSE. [11]. • An extension of RMSE.
18	Relative root mean squared error (RRMSE) [15]	Present percentage variation in accuracy.	$RRMSE = \sqrt{\frac{1}{n}\sum (A-P)^2}$	• Lower RRMSE values are favorable.
19	Sum of Squared Error (SSE)	Sums the squared differences between each observation and its mean.	$SSE = \sum_{i=1}^{n} E_i^2$	• A small SSE indicates a good fit [16].
20	Relative Squared Error (RSE)	Normalizes total squared error by dividing by the total squared error.	$RSE = \sum_{i=1}^{n} E_i^2 / (A_i - A_{mean})^2$	• A perfect fit is achieved when the numerator equals zero [9].
21	Root Relative Squared Error (RRSE)	Evaluates the root relative squared error between two vectors.	$RRSE = \sqrt{\sum_{i=1}^{n} E_i^2 / (A_i - A_{mean})^2}$	• Ranges between zero and 1.0, with zero being ideal [9].
22	Mean Square Percentage Error (MSPE)*	Evaluates the mean of square percentage errors.	$MSPE = \frac{\sum_{i=1}^{n} \left(\lvert E_i\rvert / \lvert A_i\rvert\right)^2}{n/100}$	• Non-symmetrical [8].
23	Correlation coefficient (R)	Measures the strength of association between variables.	$R = \frac{\sum_{i=1}^{n} (A_i - \bar{A}_i)(P_i - \bar{P}_i)}{\sqrt{\sum_{i=1}^{n} (A_i - \bar{A}_i)^2 \sum_{i=1}^{n} (P_i - \bar{P}_i)^2}}$	• R>0.8 implies strong correlation [17]. • Does not change by equal scaling. • Cannot distinguish between independent and dependent variables.
24	Coefficient of Determination (R^2)	The square of correlation.	$R^2 = 1 - \sum_{i=1}^{n} (P_i - A_i)^2 / \sum_{i=1}^{n} (A_i - A_{mean})^2$	• R^2 values close to 1.0 indicate a strong correlation. • Related to R. • The addition of predictors increases R^2 (even if due to chance alone).
25	Mean Absolute Scaled Error (MASE)	Mean absolute errors divided by the mean absolute error.	$\frac{\sum_{i=1}^{n} \frac{E_i}{A_i}}{n/100} / \left(\frac{1}{n} - 1\right) \sum_{i=1}^{n} \lvert A_i - A_{i-1}\rvert$	• Scale independent. • Stable near zero [11].

26	Golbraikh and Tropsha's [18] criterion	–	*At least one slope of regression lines (k or k') between the regressions of actual (A_i) against predicted output (P_i) or P_i against A_i through the origin, i.e., $A_i = k \times P_i$ and $P_i = k' A_i$, respectively.* $k = \frac{\sum_{i=1}^{n}(A_i \times P_i)}{A_i^2}$ $k' = \frac{\sum_{i=1}^{n}(A_i \times P_i)}{P_i^2}$ $m = \frac{R^2 - R_o^2}{R^2}$ $n = \frac{R^2 - R_{o'}^2}{R^2}$	• k and k' need to be close to 1.0 or at least within the range of 0.85 and 1.15. • m and n are performance indexes and their absolute value should be lower than 0.1.
27	Q.S.A.R. model by Roy and Roy [19]	–	$R_m = R^2 \times \left(1 - \sqrt{\left\|R^2 - R_o^2\right\|}\right)$ *where* $-\frac{\sum_{i=1}^{n}(P_i - A_i^o)^2}{\sum_{i=1}^{n}(P_i - P_{mean})^2}, A_i^o = k \times P_i R_0'2$ $= 1 - \frac{\sum_{i=1}^{n}(A_i - P_i^o)^2}{\sum_{i=1}^{n}(A_i - A_{mean})^2}, P_i^o = k' \times A_i$	• R_m is an external predictability indicator. $R_m > 0.5$ implies a good fit.
28	Frank and Todeschini [20]	–	*Recommend maintaining a ratio of 3–5 between the number of observations and input parameters.*	–

(Continued)

Table 5.1 (Continued)

No.	Metric	Definition/remarks	Formula	Notes
29	Objective function by Gandomi et al. [21]	A multi-criteria metric.	$\text{Function} = \left(\frac{No._{Training} - No._{Validation}}{No._{Training} + No._{Validation}}\right)\frac{RMSE_{Training} + MAE_{Learning}}{R_{Learning} + 1} + \frac{2No._{Validation}}{No._{Training} + No._{Validation}} \frac{RMSE_{Validation} + MAE_{Validation}}{R_{Training} + 1}$ *where, $No._{Training}$ and $No._{Validation}$ are the number of training and validation data, respectively.*	• This function needs to be minimized to yield the highest fitness.
30	Reference index (RI) by Cheng et al. [22]	A multi-criteria metric that uniformly accounts for RMSE, MAE, and MAPE.	$RI = \frac{RMSE + MAE + MAPE}{3}$	• Each fitness metric is normalized to achieve the best performance. • An extension of RMSE, MAE, and MAPE.
31	Scatter index (SI) [23]	Applied to examine whether RMSE is good or not.	$SI = \frac{\sqrt{\frac{\sum_{i=1}^{n}\left(P_{max(A)} - P_{max(p)}\right)^2}{n}}}{P'_{max(p)}}$ *where n = number of datasets used during the training phase. $P'_{max(p)}$ = mean actual observations data*	• SI is RMSE normalized to the measured data mean. • If SI is less than one, then estimations are acceptable. • "excellent performance" when SI < 0.1, a "good performance" when 0.1 < SI < 0.2, a "fair performance" when 0.2 < SI < 0.3, and a "poor performance" when SI > 0.3.
32	Synthesis index (SyI) [24]	Comprehensive performance measure based on MAE, RMSE, and MAPE.	$\text{SyI} = \frac{1}{n}\sum_{i=1}^{n}\left(\frac{P_i - P_{min,i}}{P_{max,i} - P_{min,i}}\right)$ *where, n = number of performance measures; and P_i = ith performance measure.*	• The SI ranged from 0.0 to 1.0; an SI value close to 0 indicated a highly accurate predictive model.
33	Performance index (PI) [15]	Performance index to evaluate predictivity of a model.	$PI = \frac{RRMSE}{1 + R}$	• Lower PI values result in more accurate model predictions.

34	$a_{20\text{-index}}$ [25]	Performance index to evaluate predictivity of a model within 20% variation.	$a_{20\text{-index}} = \frac{m_{20}}{M}$ *where m_{20} is the number of samples with the ratio of experimental value over predicted value falling from 0.8 to 1.2 and M is the number of samples in the dataset.*	• Presents the number of samples with the difference between the predicted value and the ground truth within ±20%.
35	Fractional bias (FB) [26]	A measure of shift between the observed and predicted values.	$FB = \frac{2\sum_{i=1}^{n}(A-P)}{\sum_{i=1}^{n}(A+P)}$	• Dimensionless metric, which is convenient for comparing the results from studies involving different scales. • Symmetrical and bounded; values for the fractional bias range between −2.0 (extreme underprediction) to +2.0 (extreme overprediction) • A perfect model has FB of zero.
36	Relative index of agreement (RD) [27]	A standardized measure of the degree of model prediction error.	$\text{RD}=1-\frac{\sum_{i=1}^{N}\left(\frac{A-P}{A}\right)}{\sum_{i=1}^{N}\left(\frac{(\lvert P-\bar{A}\rvert+\lvert A-\bar{A}\rvert)}{\bar{A}}\right)^2}$	• A value of 1.0 indicates a perfect match, and zero indicates no agreement. • Overly sensitive to extreme values.
37	Nash–Sutcliffe coefficient (NSE) [28]	A metric often used in flow predictions.	$NSE = 1 - \left[\frac{\sum_{i=1}^{N}(A-P)^2}{\sum_{i=1}^{N}(A-\bar{A})^2}\right]$	• NSE = 1.0 indicates perfect correspondence. • NSE = 0.0 indicates that the model simulations have the same explanatory power as the mean of the observations. • NSE < 0.0 indicates that the model is a worse predictor than the mean of the observations.
38	Kling–Gupta efficiency (KGE) [29]	A metric often used in flow predictions.	$\text{KGE} = 1-\sqrt{(r-1)^2+(\alpha-1)^2+(\beta-1)^2}$, where r is the linear correlation between the predicted and actuals. α is the magnitude of the variability calculated as the standard deviation in predictions divided by the standard deviation in actuals. β is the bias term calculated as the prediction means divided by the actual mean. N is the number of the dataset over the training and testing phases.	• KGE = 1.0 indicates perfect agreement between actuals and predictions. • KGE< 0.0 indicates that the mean of actuals provides a better estimate than predictions. • For other values of KGE, please refer to [30].

*Has a median derivative.

**Can be normalized by the standard deviation of actual observations.

***The reader is encouraged to review the cited references for full details on specific metrics.

5.4 Classification Metrics and Indicators

In classification problems, we hope to categorize the output/target/independent variable into two classes (i.e., the beam fails in bending, the beam does not fail in bending[18]) or more classes (e.g., the beam fails in bending, shear, or instability). Therefore, an intuitive way to check if our ML classifier is proper, and just like the case of regression, is to compare its predicted class against the ground truth. If the predicted class matches the actual class, then the model properly predicted this *instance* and vice versa.

Now, *what about if we would like to gauge performance of the classifier in terms of the whole class?* Well, instead of only comparing instances against their possible classes, we will also need to compare the predicted classes for such instances against *all other possible* classes. As you can guess, this would require us to use a matrix format. This matrix is referred to as the *confusion matrix*.[19] This matrix homes statistics about actual and predicted classifications undertaken by a specific classifier. Each column in this matrix signifies *predicted* instances, while each row represents *actual* instances. Figure 5.2 shows a sample confusion matrix.

As you can see, the main diagonal of the confusion matrix represents the number of instances correctly identified by the ML classifier. On the other hand, the second diagonal represents the *confused* (or mistakenly) identified classes. The same matrix can also be used to arrive at supplementary metrics such as accuracy, sensitivity, specificity, etc. These metrics, along with others belonging under classification, are listed in Table 5.2.

		Prediction	
	Total population Positive + Negative	Positive	Negative
Actual	Positive (P)	True positive (TP) Say, the model arrives at 4/5 → 4 out 5 are correctly classified	False negative (FN) Say, the model arrives at 1/5 → 1 out 5 are falsely classified
	Negative (N)	False positive (FP) Say, the model arrives at 1/3 → 1 out 3 is falsely classified	True negative (TN) Say, the model arrives at 2/3 → 2 out 3 are correctly classified

Figure 5.2 Anatomy of a confusion matrix [20] [Note: In this table, *P (denotes number of actual positives), N (denotes number of actual negatives), TP (denotes true positives), TN (denotes true negatives), FP (denotes false positives), and FN (denotes false negatives)*].

5.5 Clustering Metrics and Indicators

In unsupervised learning, the observations do not have labels, and hence it is up to the ML algorithm to identify the hidden structures in unlabeled observations. In such an instance, a ML algorithm seeks to cluster the data on the basis of how the data are spread out according to its centroid, density, etc.

The absence of labels means that we do not know the ground truth! Without knowing such truth, how could a metric describe any variation between model predictions and their actual counterparts? Luckily, the open literature provides us with tools and indexes that can be applied to examine the *quality* of clusters, namely, the *Elbow method*, the *Silhouette Coefficient*,[21] and the *Dunn Index* (see Table 5.3) [54].

---

18 Other examples, the project will be delayed/not delayed, a wildfire will breakout/will not breakout, etc.

19 **Confusion matrix**: A table used to describe the performance of a classifier (i.e., how confused is the classifier in identifying the correct classes?)

20 If we assume Positive to be Class A and Negative to be Class B, the top left corner will read as 4 out of 5 are correctly classified as Class A. In parallel, the top right corner will read as Say, the model arrives at 1 out of 5 are falsely classified as Class B.

21 Other methods exist, such as the Gap statistic [73].

Table 5.2 List of commonly used classification metrics.

No.	Metric	Definition/remarks	Formula	Notes
1	True Positive Rate (TPR)	Measures the proportion of actual positives that are correctly identified as positives.	$TPR = \frac{TP}{P} = \frac{TP}{TP + FN} = 1 - FNR$	• Also known as Sensitivity or Recall. • Does not account for indeterminate results.
2	True Negative Rate (TNR)	Measures the proportion of actual negatives that are correctly identified negatives.	$TNR = \frac{TN}{N} = \frac{TN}{TN + FP} = 1 - FPR$	• Also known as Specificity or Selectivity.
3	Positive Predictive Value (PPV) or Precision	The proportions of positive observations that are true positives.	$PPV = \frac{TP}{TP + FP} = 1 - FDR$	• Ranges from zero to unity (ideal).
4	Negative Predictive Value (NPV)	The proportions of negative observations that are true positives.	$NPV = \frac{TN}{TN + FN} = 1 - FOR$	• Ranges from zero to unity (ideal).
5	False Positive Rate (FPR)	Measures the proportion of positive cases that are correctly identified as positives.	$FPR = \frac{FP}{N} = \frac{FP}{FP + TN} = 1 - TNR$	• Also known as Fall-out.
6	False Discovery Rate (FDR)	Describes the expected proportion of false observations.	$FDR = \frac{FP}{FP + TP} = 1 - PPV$	–
7	False Omission Rate (FOR)	Measures the proportion of false negatives that are incorrectly rejected.	$FDR = \frac{FN}{FN + TPN} = 1 - NPV$	–
8	Positive likelihood ratio (LR+)	Evaluates the change in the odds of having a diagnosis with a positive test.	$LR+ = \frac{TPR}{FPR}$	• Measures the ratio of TPR to FPR. • Presents the likelihood ratio for increasing certainty about a positive diagnosis.
9	Negative likelihood ratio (LR-)	Evaluates the change in the odds of having a diagnosis with a negative test.	$LR- = \frac{FNR}{TNR}$	• Describes the ratio of FNR to TNR.
10	Diagnostic odds ratio (DOR)	Measures the effectiveness of a (diagnostic) test.	$DOR = \frac{LR+}{LR-} = \frac{TP / FP}{FN / TN}$	• Often used in binary classification.
11	Accuracy (ACC)	Evaluates the ratio of the number of correct predictions to the total number of samples.	$ACC = \frac{TP + TN}{P + N} = \frac{TP + TN}{TP + TN + FP + FN}$	• Presents performance in a single class. • Assumes equal cost for errors [31].
12	F_1 score	Harmonic mean of the precision and recall.	$F_1 = \frac{2PPV \times TPR}{PPV + TPR} = \frac{2TP}{2TP + FP + FN}$	• Describes the harmonic mean of precision and sensitivity. • Focuses on one class only. • Biased to the majority class [32].

(Continued)

Table 5.2 (Continued)

No.	Metric	Definition/remarks	Formula	Notes
13	Matthews Correlation Coefficient (MCC)	Measures the quality of binary classifications analysis.	$MCC = \frac{TP \times TN - FP \times FN}{\sqrt{(TP+FP)(TP+FN)(TN+FP)(TN+PN)}}$	• Measures the quality of binary and multi-class classifications. • Applicable for classes of different sizes. • When MCC equals +1 → perfect prediction, → 0 equivalent to a random prediction and → −1 false prediction. • Considered a balanced measure as it involves values of all the four quadrants of a confusion matrix [33].
14	Bookmaker Informedness (BM) or Youden's J statistic	Evaluates the discriminative power of the test [34].	$BM = TPR + TNR - 1$	• Describes the probability of an informed decision (vs. a random guess). • Has a range between zero and unity (ideal). • Considers both real positives and real negatives. • Takes into account all predictions [35]. • Suitable for imbalanced data. • It does not change concerning the differences between sensitivity and specificity [34].
15	Markedness (MK)	Measures trustworthiness of positive and negative predictions.	$MK = PPV + NPV - 1$	• Measures trustworthiness of positive and negative predictions by a model [36]. • Considers both predicted positives and predicted negatives. • Specifies the probability that a condition is marked by the predictor (as opposed to luck/chance) [37]. • Sensitive to data changes (not suitable for imbalanced data) [34].
16	Average Class Accuracy (ACA)	Measures the average accuracy of predictions in a class.	$ACA = W\left(\frac{TP}{TP+FP}\right) + (1-W)\left(\frac{TN}{TN+FP}\right)$ $where\ 0 < W < 1$	• Used with unbalanced data. • Choosing a good weighting factor a priori [32]. • When $W > 0.5$, minority class accuracy contributes more than majority class. • Presents performance at a single class threshold.

17	Receiver Operating Characteristic (ROC)	Plots the diagnostic ability of a binary classifier system as its discrimination threshold is varied.	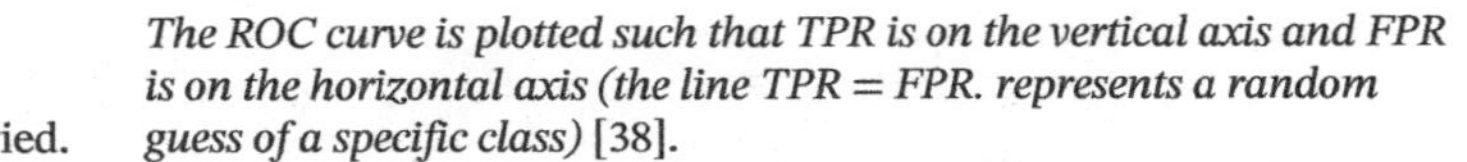 *The ROC curve is plotted such that TPR is on the vertical axis and FPR is on the horizontal axis (the line TPR = FPR. represents a random guess of a specific class)* [38]. 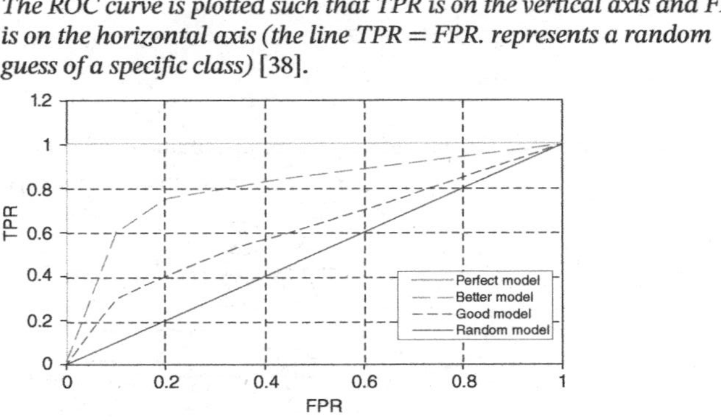	• Characterizes tradeoff between TPR and FPR. [39]. • Takes a value between zero and 1 to relate the probability distribution to a single state [40]. • A threshold of zero ensures the highest sensitivity, and 1 ensures the best specificity. • Estimate the cost ratio (slope of the line tangent to ROC. curve). • It should be used in datasets with roughly equal numbers of observations for each class [41, 42].
18	Area under the ROC curve (AUC)	Measures the two-dimensional area underneath the entire ROC curve.	$AUC=\sum_{i=1}^{N-1}\frac{1}{2}(FP_{i+1}-FP_i)(TP_{i+1}-TP_i)$ *or* $AUC=\frac{1}{2}w(h+h'),$ *where, w = width, and h and h' = heights of the sides of a trapezoid histogram*	• Not dependent on a single class threshold. • Associated with increased training times. • Higher areas represent better performance.
19	Precision-Recall curve	Plots the tradeoff between precision and recall for different thresholds.	*Plots precision (in the vertical axis) and recall (in the horizontal axis) for different thresholds.*	• Applicable in cases of moderate to large class imbalance [41]. • Used in binary classification.
20	Log Loss Error (LLE)	Measures where the prediction input is a probability value.	$LLE=-\sum_{c=1}^{M}A_i logP,$ *where M: number of classes, c: class label, y: binary indicator (0 or 1) if c is the correct classification for a given observation.*	• Measures the uncertainty of the probabilities by comparing predictions to the ground truth. • Penalizes for being too confident in the wrong prediction. • Has probability between zero and 1. • A log loss of zero indicates a perfect model.
21	Hinge Loss Error (HLE)	–	$HLE=\max(0,1-q\cdot y)$ *where q= ±1 and y: classifier score*	• Linearly penalize incorrect predictions. • Primarily used in support vector machine.
22	Wilcoxon–Mann–Whitney (WMW) test [32]	–	$WMW=\frac{\sum_{i\in Minorclass}\sum_{i\in Majorclass} I_{wmw}\left(P_i,P_j\right)}{\lvert Minor\ class\rvert\times\lvert Major\ class\rvert},$ *where P_i and P_j: outputs when evaluated on an example from the minority and majority classes, respectively*	• Used in scenarios with unbalanced data. • The indicator function I_{wmw} returns 1 if $_{Pi} > P_j$ and $P_i \geq 0$ or 0 if otherwise.

(Continued)

Table 5.2 (Continued)

No.	Metric	Definition/remarks	Formula	Notes
23	Fitness Function *Amse* (FFA) [32]	Measures pattern difference between input and output.	$FFA = \frac{1}{K}\sum_{c=1}^{K}\left(1 - \frac{\sum_{i=1}^{N_c}\left(1 - sig\left(P_{ci}\right) - T_c\right)}{N_c \times 2}\right)^2,$ $sig(x) = \frac{2}{1+e^{-x}} + 1$ *where P_{ci}: output of a classifier evaluated on the ith example, N_c: number of examples, K: number of classes, T_c: target values (equals to -0.5 and 0.5 for majority and minority classes, respectively)*	• Used in scenarios with unbalanced data. • Needs to be scaled to a range of [−1, 1] and hence the need for sigmoid function. • FFA = 1 presents an ideal scenario.
24	Fitness Function *Incr* (FFI) [32]	–	$Incr = \frac{1}{K}\sum_{c=1}^{K}\left(\frac{\sum_{j=1}^{M_c}\left[I_{zt}\left(j, D_{cj}, c\right).\sum_{i=1}^{N_c} Eq\left(D_{cj}, P_{ci}\right)\right]}{\frac{1}{2}N_c(N_c+1)}\right)$ $I_{zt}\left(r,k,c\right) = \begin{cases} r, & \text{if } k \geq 0 \text{ and } c \in \textit{Minority class} \\ & \text{or if } k < 0 \text{ and } c \in \textit{Majority class} \\ 0, & \text{otherwise} \end{cases}$ $Eq\left(p,q\right) = \begin{cases} 1, & \text{if } p = q \\ 0, \textit{otherwise} \end{cases}$	• Used in scenarios with unbalanced data. • Assigns incremental rewards to predictions that fall further away from the class boundary. • Ranges between zero and unity (ideal).
25	Fitness Function Correlation (FFC)	–	$FFC = \frac{1}{K}(r + I_{zt}(1, \mu_{minor}, \mu_{major}),$ $r = \sqrt{\frac{\sum_{c=1}^{K} N_c\left(\mu_c - \bar{\mu}\right)^2}{\sum_{c=1}^{K}\sum_{i=1}^{N_c}\left(P_{ci} - \bar{\mu}\right)^2}}$ $\mu_c = \frac{\sum_{i=1}^{N_c} P_{ci}}{N_c}, \bar{\mu} = \frac{\sum_{c=1}^{K} N_c \mu_c}{\sum_{c=1}^{K} N_c}.$ *where r: correlation ratio, μ_{minor} and μ_{major}: mean for minor and major classes, respectively*	• Used in scenarios with unbalanced data.

26	Fitness Function Distribution (FFD)	Measures the distance between class distributions as a function of class separability.	$FFD = \frac{\lvert \mu_{\min} - \mu_{\mathrm{maj}} \rvert}{\sigma_{\min} + \sigma_{\mathrm{maj}}} \times I_{zt}(2, \mu_{\min}, \mu_{\mathrm{maj}})$ $\mu_c = \frac{\sum_{i=1}^{N_c} P_{ci}}{N_c}, \sigma_c = \sqrt{\frac{1}{N_c}\sum_{i=1}^{N_c}(P_{ci} - \mu_c)^2}.$ *where μ_c and σ_c: mean and standard deviation of the class distribution, respectively,*	• Used in scenarios with unbalanced data. • Treats predictions as independent distributions. • Measures separability (i.e., the distance between class distributions) [43] – high separability (no overlap), and this distance turn large (go to +∞). • Uses I_{zt} to enforce zero class threshold.
27	Canberra Metric (CM)	Measures the distance between pairs of points in a vector space.	$CM = \sum_{i=1}^{n} \frac{\lvert E_i \rvert}{A_i + P_i}$	-
28	Wave Hedges Distance (WHD)	–	$WHD = \sum_{i=1}^{n} \frac{\lvert E_i \rvert}{max(A_i, P_i)}$	• Normalizes the difference of each pair of coefficients with its maximum [44–46].
29	Lift [47]	Measures the performance of a model at predicting or classifying cases.	$LIFT = \frac{\%of\ true\ positives\ above the threshold}{\%of\ dataset\ above the threshold}$	• Measures betterness of a classifier than a baseline classifier that randomly predicts positives. • A threshold is set as a static fraction of the positive dataset. • Lift and Accuracy do not always correlate well.
30	Mean Cross Entropy (MXE)	Measures the performance of a model where the output is a probability between zero and one.	$MXE = -\frac{1}{N}\sum True \times \ln(Predicted) + (1 - True) \times \ln(1 - Predicted)$ *(The assumptions are that Predicted ∈ [0, 1] and True ∈ {0, 1})*	• Minimizing MXE gives the maximum likelihood [35].
31	Probability Calibration (CAL)	–	*Order cases 1–100 by their predicted in the same bin.* 1) *Evaluate the percentage of true positives.* 2) *Calculate the mean prediction for true positives.* 3) *Calculate the mean prediction calibration error for this bin (using the absolute value of the difference between the observed frequency and the mean).* 4) *Repeat steps 1–4 for cases 2–101, 3–102, etc.* 5) *CAL is calculated as the mean of these binned calibration errors* [35].	• Lengthy.
32	Precision-recall break-even point	The point at which the precision-recall-curve intersects the bisecting line.	*Precision = Recall*	• Defines the point when precision and recall are equal.

(Continued)

Table 5.2 (Continued)

No.	Metric	Definition/remarks	Formula	Notes
33	Average precision (AP)	Combines recall and precision for ranking.	$AP = \sum_n (Recall_n - Recall_{n-1}) Precision_n$	• Describes the weighted mean of precision in each threshold with the increase in recall from the previous threshold used.
34	Balanced accuracy [48]	Calculates the average of the correctly identified proportion of individual classes.	*Defined as the average of recall obtained in each class.*	• Used in binary and multi-class classification problems. • Accommodates imbalanced datasets.
35	Brier score (BS)	Measures the accuracy of probabilistic-based predictions.	$BS = \frac{1}{N}\sum_{i=1}^{N}(f_i - A_i)^2$ *in which {\displaystyle f_{t}}f_i is the probability that was forecast, {\displaystyle o_{t}} A_i the actual outcome of the event at instance {\displaystyle t}i*	• Measures the mean squared difference between the predicted probability and the actual outcome. • Takes on a value between zero and 1.0 (the lower the score is, the better the predictions). • Composed of refinement loss and calibration loss. • Appropriate for binary outcomes. • Inappropriate for ordinal variables.
36	Cohen's kappa (CK) [49]	Measures interrater (agreement) reliability.	$\kappa = (p_o - p_e)/(1 - p_e)$ *where p_o: empirical probability of agreement on the label assigned to any sample, p_e: expected agreement when both annotators assign labels randomly and this is estimated using a per-annotator empirical prior over the class labels.*	• Measures inter-annotator agreement. • Expresses the level of agreement between two annotators [50]. • Ranges between -1.0 and 1.0 (ideal).
37	Hamming loss (HL)	Fraction of the wrongly identified labels.	$HL = \frac{1}{m}\sum_{i=1}^{m} 1_{\widehat{P_{i \neq A_i}}}$	• Describes the fraction of labels that are incorrectly predicted. • Optimal value is zero [51].
38	Fitness (T) [52]	-	$Fitness(T) = Q(T) + \alpha * R(T) + \beta * Cost(T)$ *where Q(T): accuracy, R(T): sum of $R(T_i)$ in all multi-tests of the T tree, Cost(T): sum of the costs of attributes constituting multi-tests. The default parameters values are: α=1.0 and β=−0.5,* $R(T_i) = \frac{\lvert X_i \rvert}{\lvert X \rvert} * \sum_{j=1}^{\lvert mt_i \rvert - 1} r_{ij}$ *where X: learning set, X_i: instances in i-th node, and $\lvert mt_i \rvert$: size of a multi-test.* $Cost(T_i) = \frac{\lvert X \rvert}{\lvert X_i \rvert} * C(a_{ij})$ *where: a_{ij}: j-th attribute of the i-th multi-test, $C(a_{ij})$: cost of the a_{ij} attribute.*	• Used for fitting decision trees. • This function needs to be maximized to achieve high performance.

39	F2 score [53]	Measured as the weighted average of precision and recall.	$F_\beta = 1 + \beta 2 \times \frac{precision \times recall}{(\beta 2 \times precision) + recall}$ *where:* $\beta = 2$.	• Computes a weighted harmonic mean of Precision and Recall. • Learning about the minority class.
40	Distance score (D score) [53]	–	$D_{sc} = \frac{2 \times C1 \times C2}{C1 + C2}$ *where:* $C1 = \frac{\sum_{i=0}^{N_{maj}} sig(P_{Maj}) \times \lvert T - sig(P_{Maj}) \rvert}{N_{maj}} \times func(1, P_{Maj})$ $sig(x) = \frac{2}{1+e-x} - 1$ $C2 = \frac{\sum_{i=0}^{N_{min}} sig(P_{Min}) \times \lvert T - sig(P_{Min}) \rvert}{N_{min}} \times func(1, P_{Min})$ $func(1,k) = \begin{cases} 1, & \text{if } k \leq 0 \text{formajorityclassinstance} \\ 1, & \text{if } k > 0 \text{forminorityclassinstance} \\ 0, & \text{otherwise} \end{cases}$ *C1* for the majority class and *C2* for the minority class.	• Used in genetic programming and medical fields. • Distance score (D score) which learns about both the classes by giving them equal importance and being unbiased. • The range of both C1 and C2 is 0.0 to 1.0 (ideal).
41	Predictivity indicator	–	$P = \sum_{i}^{n} \frac{ACC \times AUC}{LLE}$ where i = training, validation, and testing. Higher values of this metric are favorable; for example, for a hypothetical case of ACC and AUC of unity, and LLE = 0.001, this metric yields 1000.	• Combines ACC, AUC, and LLE • When ACC and AUC equals unity, and LLE = 0.001, this metric yields a hypothetical value of 1000.

* The reader is encouraged to review the cited references for full details on specific metrics.
[Note: the first 16 metrics are derived from the confusion matrix].

Table 5.3 List of commonly used clustering metrics.

No.	Metric	Definition/remarks	Formula/illustration	Notes
1	Elbow method [55, 56]	A visual method is expressed by the Within-Cluster Sum of Square (WCSS).	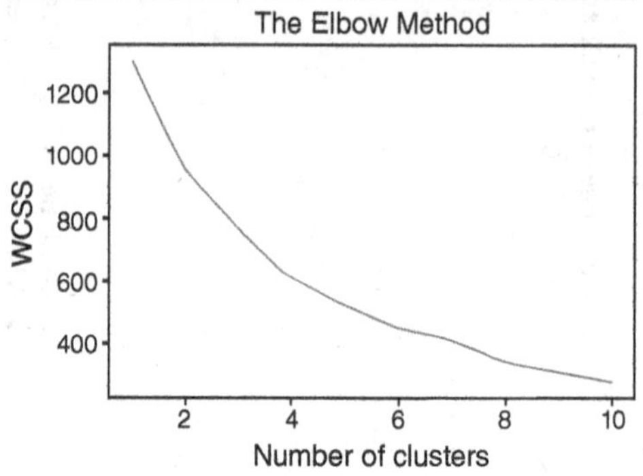 The Elbow Method (WCSS vs. Number of clusters) $$WCSS = \sum_{i=1}^{n_c} \sum_{x \in c} d(x, x_c)^2$$ where n_c is number of clusters, x_c is cluster centroid. WCSS represents the sum, over each cluster, of the squared errors between each point belonging to a particular cluster and the cluster's centroid.	• Choose a number of clusters such that adding another cluster does not improve total WCSS by much.
2	Silhouette Coefficient	Measures clustering quality by looking at how far data belonging to different clusters are from each other and how close data belonging to the same cluster are closer to each other.	$$SIL(p) = \frac{b(p) - a(p)}{max\{a(p), b(p)\}}$$ *where a(p) is the average of the distances of data point p to all data points in its own cluster, and b(p) is the minimum value of the average distances of data point p to all data points in the other clusters* [57]. 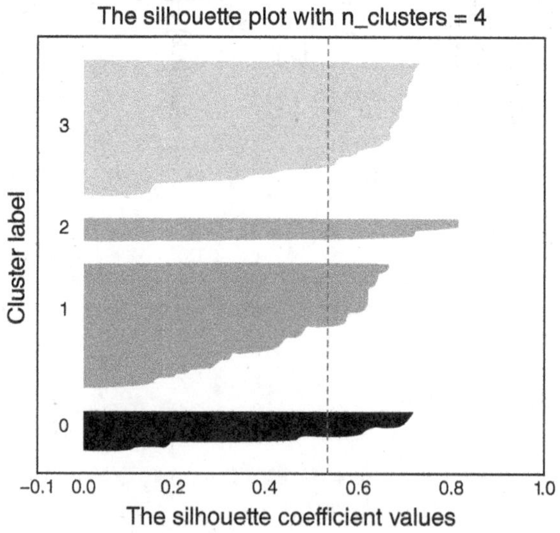 The silhouette plot with n_clusters = 4 (Cluster label vs. The silhouette coefficient values)	• The larger the value of Silhouette coefficient near +1, the more likely a sample is far away from neighboring clusters. • Good clusters have a high silhouette coefficient.
3	Dunn Index	Measures the ratio of the smallest distance between observations not in the same cluster to the largest intra-cluster distance.	$D = min.separation / max.diameter$	• Ranges between zero and infinity. • It should be maximized.

5.6 Functional Metrics and Indicators*

In more recent years, researchers started to create new sets of metrics/indicators aimed at describing various model behaviors. Given that many of such measures were developed with a specific nature, such metrics and indicators can be thought of as *functional*.[22] For example, in some instances, a ML model could be large and complex and require a tremendous amount of energy and resources to operate. From this view, such a model may not be worthy of deployment even if it achieves high predictivity metrics. An equivalent model of lower energy expenditure[23] could be more attractive [58].

Similarly, some problems may require the derivation of specialized metrics that are more suitable for examining the behavior of a ML model from the lens of a particular problem as opposed to traditional metrics (listed in the above tables). This section further dives into the case of special metrics from two perspectives: energy, and domain-specific.

5.6.1 Energy-based Indicators[24]

The performance of ML models can be examined from an energy perspective via a series of measures.[25] Most of such measures relate the accuracy of the model to some aspect(s) of the ML energy consumption (i.e., training time or energy, model size, and/or inference time or energy). I will cover a few of these here. Table 5.4 lists such indicators for your reference.

The objective of describing such indicators here is to advocate for the idea of favoring ML models with balanced accuracy and energy consumption – especially since such energy consumption can be traced back to carbon emissions.[26] In a way, incorporating energy consumption as a dimension for comparing ML models could allow us to adopt Green ML in civil and environmental engineering problems at an early stage of our embracing of ML. Try the *ImpactTracker Python* tool [59] or the *Machine Learning Emissions Calculator* [60] website[27] to visualize the energy consumption of your models (see Fig. 5.3).

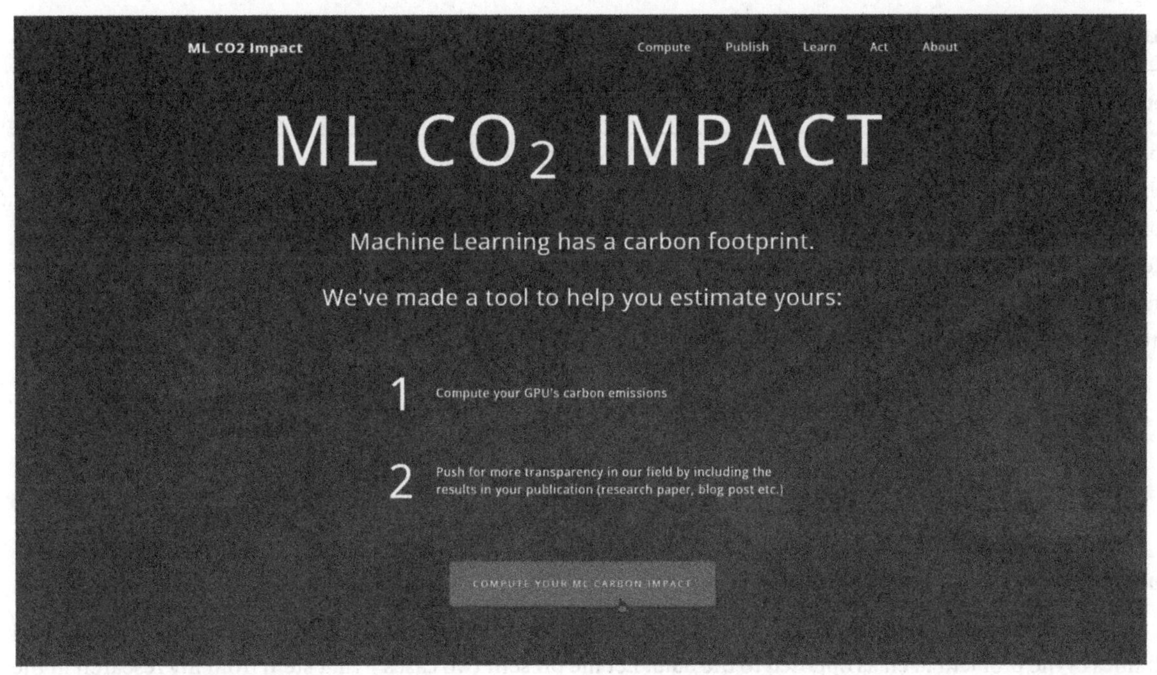

(Continued)

22 They serve a specific function or purpose.

23 Dare I also say, lower predictivity! Suppose we have two models; *Model A* (accuracy = 96%) consumes 15% more energy than *Model B* (accuracy = 94%). Both models are designed to predict the average vehicles passing through an intersection at a given hour. For a problem like this, the small difference in accuracy can be minute and may not warrant the deployment of *Model A*. At the same time, let us be cognizant of the following: 1) simple models can also consume large amount of energy, and vice versa, 2) some problems may require maintaining a robustly high predictivity (regardless of energy expenditure), and 3) achieving a balance between energy consumption and accuracy (or other factors) may not always be possible.

24 I will be covering the energy component more in-depth in the chapter on *Advanced Topics*.

25 Please refer to [74–77] for some great ideas and discussions.

26 A not so fun fact, "*computational costs of state-of-the art AI research has increased 300,000 × in recent years*" [78]. According to Strubell et al. [74], one deep learning model has the potential to generate over 280,000 kg of carbon emissions (note that an average American car produces 16,400 kg of emissions annually).

27 https://mlco2.github.io/impact/#compute.

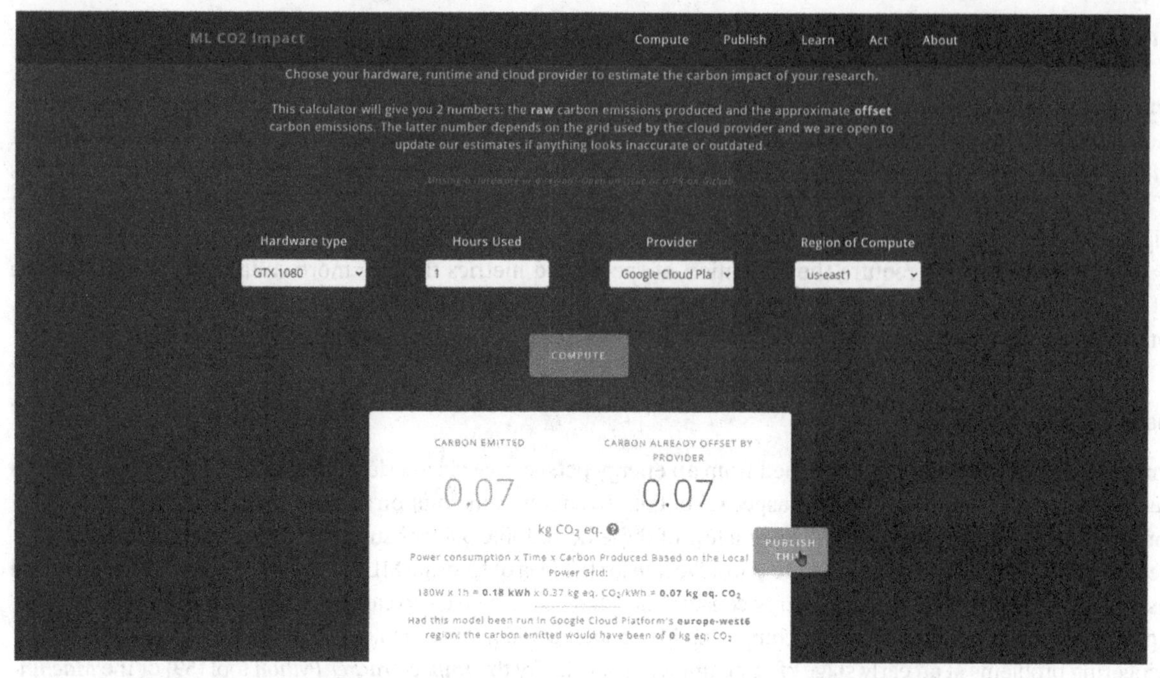

Figure 5.3 (Cont'd) Illustrative example of the ML CO_2 Impact tool [28] / with permission from BigML, Inc.

Table 5.4 List of commonly used energy-based metrics [61, 62].

No.	Metric	Formula	Notes
1	Model Size Efficiency	*Model Size Efficiency = Accuracy / model size*	–
2	Training Time Efficiency	*Training Time Efficiency = Accuracy / training time*	–
3	Training Energy Efficiency	*Training Energy Efficiency = Accuracy / training energy*	–
4	Inference Time Efficiency	*Inference Time Efficiency = Accuracy / inference time*	–
5	Inference Time Energy	*Inference Time Energy = Accuracy / inference energy*	–
6	Estimated Energy	*Estimated Energy* $= MS \times (TT + PT)$ *where MS: model size, TT: training time, and PT: prediction time.*	• Combines model size with training and prediction time. • Smaller values are favorable. • Has a hypothetical minimum value = 1.0 MB × (0.5 + 0.5) sec = 1 MB.sec.

5.6.2 Domain-specific Metrics and Indicators

Functional metrics and indicators can showcase new insights into the performance of ML models that can be more related to the mechanics of the problem itself as opposed to the data. Let me present two cases[29] that stem from my research in the structural fire engineering area [63].

Fire resistance (FR) time of columns is often measured in minutes. This FR is then compared to a range that spans 60 min,[30] i.e., < 60 min, 60–120 min, 120–180 min, and 180–240 min. Suppose a column fails after 179 min under standard heating conditions. Then, FR for this column is 179 min.

Building codes do not recognize FR, but instead adopt fire resistance rating (FRR). FRR returns FR to the closest (safest) full 60 min range. Thus, the same column of FR = 179 min would have a FRR of 120 min.[31]

Unlike other regression or classification metrics, which give statistical insights into the predictivity of the ML model at the global level, a new functional metric could be applied to *individual observations*. This metric, to be referred to as the *Clemson*

---

28 https://mlco2.github.io/impact.

29 I am fairly confident that a similar methodology can be extended to other sub-fields as well.

30 This is a simplification. For more information on fire resistance, please review [79, 80].

31 Since FR is less than 179 < 180 min, then the column cannot be rounded up. It is more conservative to round it down to the nearest full 60 min.

Metric (CM), examines the predicted FRR obtained from a ML model to the observed FRR from fire tests, as seen in Equation 5.1. For example, say that a ML model predicts FRR of a given RC column to be 120 min; however, this column actually failed within 60 min. In this particular example, CM returns a non-zero value (i.e., [120–60]/60 = 1.0), which indicates poor fitness.

$$Clemson\ Metric(CM) = \frac{FRR_{predicted} - FRR_{ground\ truth}}{60} \tag{5.1}$$

where CM is applicable for FRR ranging between 0.0 and 240 min, with a zero as a favorable performance that suggests a correct ML prediction. Negative scores of CM indicate *conservative* predictions, and positive scores indicate *unconservative* predictions.

To extend CM from assessing individual observations to examining the global predictivity of the ML model, the CM can be extended into another functional metric, the *Cumulative Clemson Metric (CCM)* – see Equation 5.2. As you can see, the CMM compares the percentage of all observations that pass the CM metric (i.e., return zero) to the total number of observations examined by the ML model.

$$Cumulative\ Clemson\ Metric(CCM) = \frac{no.of\ observations\ of \left(\frac{FRR_{predicted} - FRR_{ground\ truth}}{60} = zero \right)}{total\ number\ of\ observations} \tag{5.2}$$

To establish the degree of model predictivity, the CMM is supplemented with four hypothetical cut-offs of <25%, 25–50%, 50–75%, and >75%.[32] These cut-offs articulate the expected goodness of a ML model. For instance, if the model predicts the correct FRR for all elements under examination with *87%*[33] accuracy or higher, then this model is deemed of "*excellent*" fitness. Given the nature of their derivation, the CM and CMM metrics can be used in regression and classification problems.[34]

Let us shift our discussion toward a more involved problem where a ML model predicts a continuous response (as opposed to a single value or class). Say we create a ML to help us predict deformation-time history[35] of load bearing members under fire conditions.

A glimpse into the above scenario shows that we are looking at a metric or indicator that also happens to be continuous. Such a metric is called the *Maximum Response Deviation (MRD)* indicator. This MRD evaluates ML predictions within a specified range of observations (say, during the fire exposure time, within the first 30 min of fire exposure, etc.). This indicator builds an upper and lower bound based on the largest observation obtained (i.e., temperature or deformation), then divides its value over equally spaced intervals of the duration of the fire test.[36] ML predictions, if proper, would fit within the upper and lower bounds.

$$Maximum\ Response\ Deviation(MRD) = Observed\ response_{at\ time,t} \pm \frac{Maximum\ observed\ response}{\frac{Largest\ observed\ time}{Selected\ time\ interval}} \tag{5.3}$$

For example, say that the observed temperature was measured at 50 min as 500°C. Then, an engineer may construct the upper bound of the MRD indicator to the nearest 10 minutes such that:

$$\text{MRD at 1 min} = 25°\text{C}_{\text{at 1 min}} + 500/(50/10) = 125°\text{C}_{\text{at 1 min}}$$
$$\text{MRD at 10 min} = 75°\text{C}_{\text{at 10 min}} + 500/(50/10) = 175°\text{C}_{\text{at 10 min}}$$
$$\text{MRD at 20 min} = 153°\text{C}_{\text{at 20 min}} + 500/(50/10) = 253°\text{C}_{\text{at 20 min}}$$

And so on...

The time and corresponding temperatures as calculated from MRD can then be plotted against the observed and ML-predicted responses.

ML model predictions can then be compared across these two bounds. If all predictions lay within the bounds, then the model is said to have a *perfect* match. If < 25%, 25–50%, 50–75%, and > 75% of model predictions fall within these bounds,

32 <25% (poor), 25–50% (fair), 50–75% (strong), and >75% (excellent) fitness.

33 It is worth noting that this value is used herein as an example. Also note that like other traditional metrics, the CCM can be applied for training, validation, and testing stages, as well as to the whole database.

34 I am thinking out loud here. I can see similar metrics to CM and CMM behind derived problems of equivalent structure to that of the fire resistance of columns. Here are some examples: concentration of chemicals in a lake, classification of air quality in urban regions, or transportation volume in a given highway network, etc.

35 **The deformation-time history**: A plot that depicts how the deformation of a load bearing member evolves with fire exposure time. A similar history, the **temperature-time history**, depicts the temperature rise at a specific section in a fire-tested member as a function of heating.

36 I suggest somewhere between 5 and 20 min interval for fire exposures that happen to be larger than 60 min. Further testing might be necessary to establish a more proper bound.

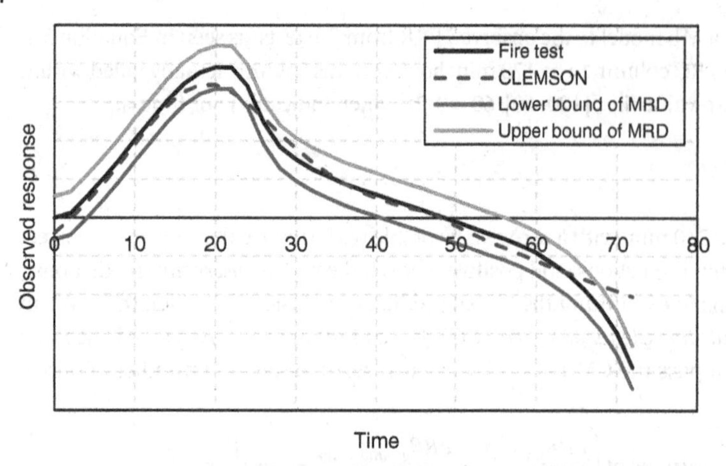

Figure 5.4 Illustration of MRD indicator in tracing ML model prediction for a Concrete Filled Steel Tube (CFST) column exposed to standard fire conditions. [Note: MRD = 89% (excellent agreement)].

then the model is said to be of *poor*, *fair*, *strong*, and *excellent* match with observations from fire tests.[37] An illustrative sample of the MRD indicator is shown in Figure 5.4.

5.6.3 Other Functional Metrics and Indicators

As you can see, functional metrics/indicators can stem from a need to relate model predictions to a useful comparison that may better represent the goodness of a ML model. Much of the available metrics are designed by individual efforts, and hence I would advise trying to create metrics and indicators for your problems.

5.7 Other Techniques (Beyond Metrics and Indicators)

This section goes a little further than purely applying PFIEM to describe four procedures that can improve our assessment of ML models. These procedures are referred to as *spot analysis*, *case-by-case examination*, *drawing and stacking*, *rational vetting*, and incorporating *confidence intervals*. I will discuss these via regression examples here but keep in mind that the described procedures can also be extended to other ML analyses as well.

5.7.1 Spot Analysis

Suppose you have developed a ML model that predicts a continuous response (e.g., mid-span deformation of a beam under fire loading).[38] Naturally, the deformation slowly grows. Thus, we can expect the deformation to continue to increase organically until failure. If we re-organize the ground truth data in an ascending or descending order, then we could arrive at a plot similar to Figure 5.5 which shows that the ordered ground truth points can be represented in a smooth line.[39] The same figure also shows that the predicted data pertaining to the ordered ground truths oscillates around the ground truths.[40]

It is clear from the same figure that there are some *spots* where the oscillation (deviation) is small, while in others, the oscillation is large. It is also evident that the model seems to predict deformations up to 80 mm more adequately than larger deformations. In fact, the model seems to consistently underpredict deformation over 80 mm.[41,42]

---

37 I hope it is clear that MRD is a *dynamic*, *unique*, and *customizable* indicator that relates to the observed performance of the particular member on hand and checks ML predictions against bounds derived for that specific element to measure the quality of the model fitness (vs. an arbitrary bound say at 5% error, or 10% error, etc.).

38 Pause! Here is a mental exercise. Try to visualize the following two sentences and imagine conducting such a test.

39 This makes sense since for each unique loading point, we have a corresponding deformation point. Keep in mind that, sometimes, a loading point may have multiple deformation points (say collected from two devices, etc.).

40 This also makes sense especially for a ML that does not have perfect scores.

41 This slightly under performance could be partly due to the limited number of observations at failure (i.e., could arise from complexity in capturing deformation at failure under fire). Fire tests are complex! And data from fire tests make great examples! I may be a bit biased.

42 Let me add the following. There is something about organizing the deformation ground truth data in the manner shown in Figure 5.5. This data nicely fits a quadratic or cubic polynomial form! I will be providing such data for you to further explore.

The spot analysis can help engineers decide if their model is functioning adequately. For example, the model shown in Figure 5.5 scores well across three performance metrics[43] (MAE = 6.97 mm, RMSE = 9.55 mm, and R^2 = 91.50%). However, practically speaking, the model underpredicts large deformations, which could be problematic as beams tend to fail at larger deformations. In hindsight, this model is not conservative. In contrast, this model can be useful for predicting fire-induced deformations of beams up to 80–120 mm.[44]

While Figure 5.5 showcases an example of a continuous response, the spot analysis can also be used for single value predictions as well. Figure 5.6 shows one such example where a ML model is developed to predict the shear capacity of cold-formed steel channels with slotted webs. I hope you can see how the ML model generally performs well, except for channels with ultimate shear strength exceeding 70 kN. This implies that this model is most helpful at predicting the ultimate capacity of cold-formed steel channels with slotted webs that do not exceed 70 kN.[45]

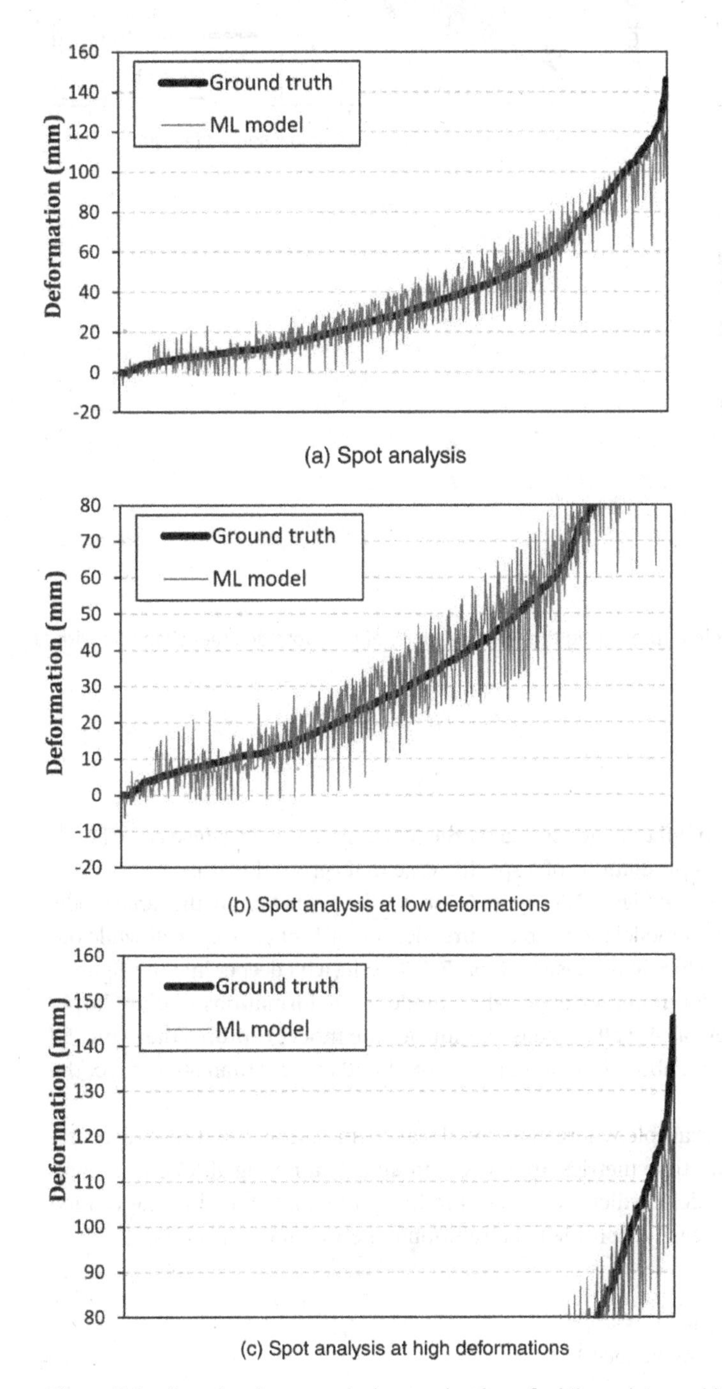

Figure 5.5 Sample of spot analysis examination of a ML model capable of predicting the mid-span deformation of fire-tested beams.

43 Averaged across the training, validation, and testing sets.

44 I can see some engineers arguing that pushing this model is good to go (given its metrics). Others may or may not agree! Hence, the spot analysis.

45 For completion, model metrics are provided herein MAE = 3.10 kN, RMSE = 5.0 kN, and R^2 = 85.67%.

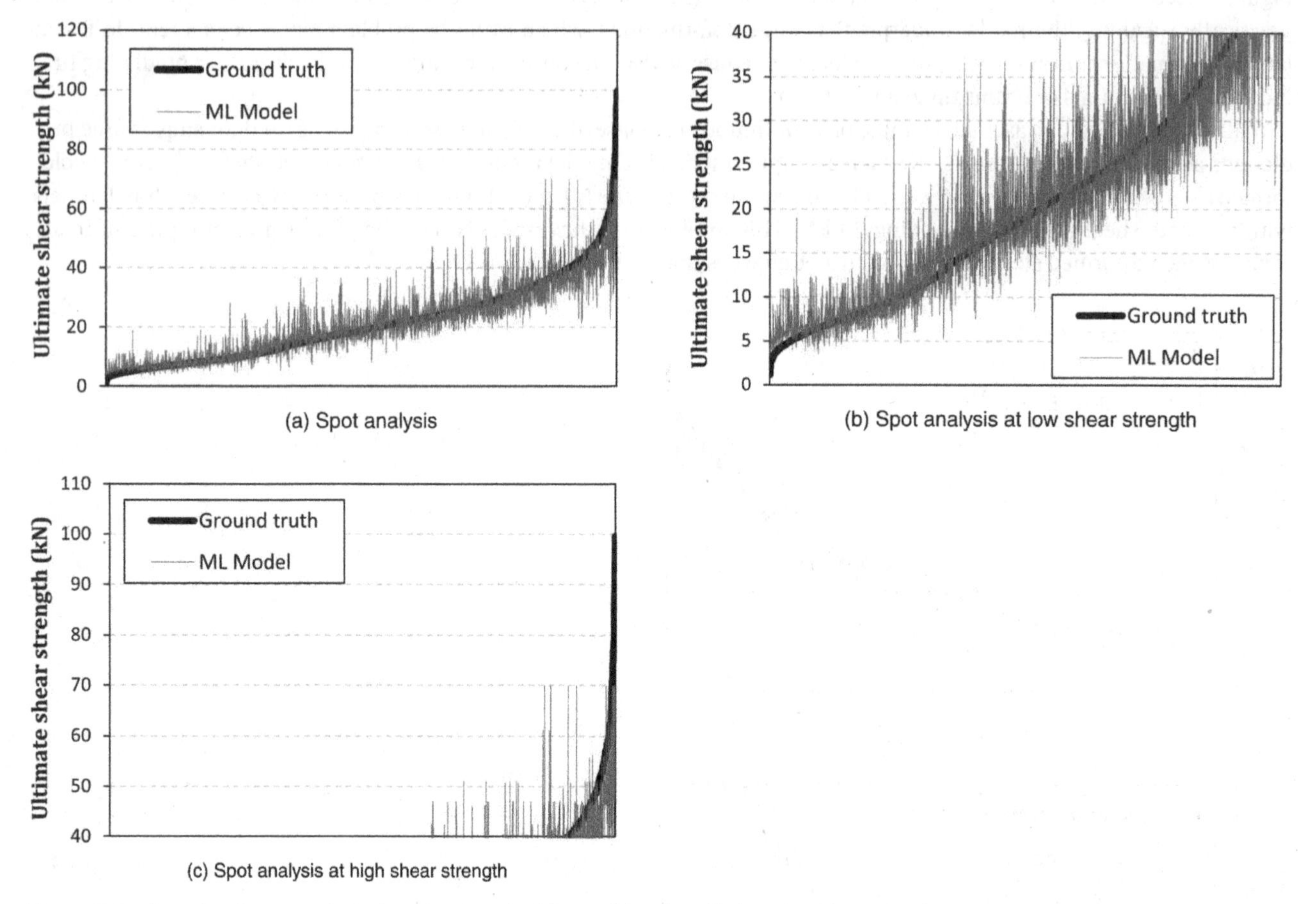

Figure 5.6 Sample of spot analysis examination of a ML capable of predicting the shear capacity of cold-formed steel channels with slotted webs.

5.7.2 Case-by-Case Examination

Unlike the spot analysis, which looks at a holistic picture of all data generated by the ML model, the *case-by-case (CbC)* [46] *examination* only looks at unique cases. The CbC compares predictions of a specific case to its ground truth.

For instance, let us borrow the same ML model[47] we discussed in Figure 5.5. Figure 5.7 shows the predictions of the same model when compared for two different beams.[48] As you can see, this model seems to capture the case of *Beam A* very well while off-shoots toward the initial and final stages of the deformation history in the case of *Beam B*. Let us dig a bit deeper into this figure.

We can clearly infer from the spot analysis that the model underperforms when modeling deformations that are larger than 80 mm. Surprisingly, the mid-span deformation in *Beam A* well exceeds 80 mm, yet the model captures these nicely! On the other hand, the mid-span deformations in *Beam B* are about half of that of *Beam A* and are less than 80 mm, yet the model underperforms in predicting these deformations.

The analysis presented in Figures 5.5 and 5.7 is a prime example where global vs. local examination might reveal contradicting findings![49] This is also a good reminder to adopt multiple metrics. In a way, and upon examining all *CbCs*, an engineer might be able to note persistent issues with their model predictions. These are likely to be attributed to the limited generalizability of the ML model.[50] Such issues may not be easily identified via traditional metrics and indicators.

46 I believe that the *CbC* can be most helpful in continuous phenomena.

47 Remember, MAE = 6.97 mm, RMSE = 9.55 mm, and R^2 = 91.50% for this model.

48 Or cases!

49 Hence, the introduction of Beam A in this example!

50 How can we fix this issue? This is a good question. For a model to be able to generalize better, the addition of more data/examples could be of aid!

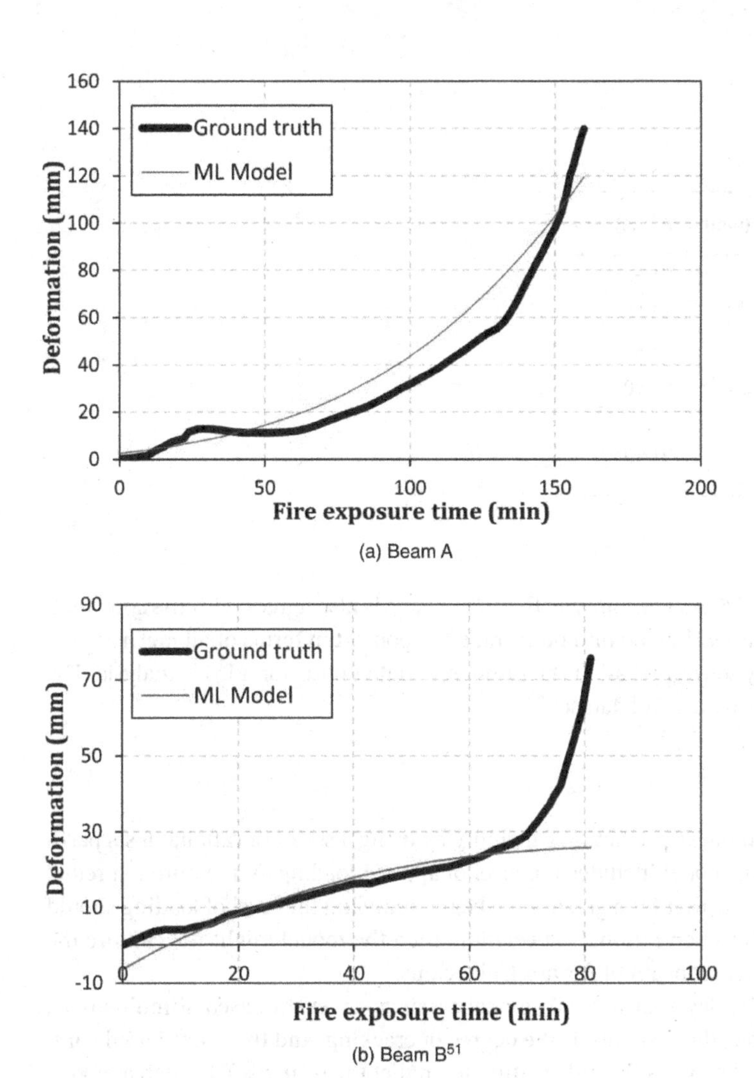

Figure 5.7 Illustration of CbC examination for two cases (*Beam A* and *Beam B*).[51]

5.7.3 Drawing and Stacking

Drawing and stacking (D&S) refer to the practice of examining ML models against sub-sets of data that are either drawn from the full dataset *deliberately* or *randomly* or in a *stacked* manner (i.e., a random sub-set can comprise *T*% from the training sub-set, *V*% from the validation sub-set, and *S*% from the testing sub-set).[52]

Let me illustrate D&S by[53] using the model from Figure 5.5. I will use the simple case of deliberately drawing data points from the full dataset. In this example, I will use the outcome of the *spot analysis* wherein the model seems to capture the deformation well up to a deflection of 80 mm.

Here is what I am going to do. I will split the dataset into two groups,[54] a sub-set with deformation data up to 80 mm, and another for deformations larger than 80 mm. Then, I will re-apply the same metrics (MAE, RMSE, and R^2) to these two

51 I would like to clarify the following. My discussion on model predictions of *Beam B* might be a bit harsh. Let me track back a bit. To be quite honest, model predictions of *Beam B* are not "that bad" since the model captures the deformation history well between 10–65 mm. My concern is with regard to the failure of the model to capture the trend toward the ultimate stage (before collapse), wherein the model predictions stabilize past the 65 mm mark – implying that the beam does not show any signs of failure. My biggest concern is the predicted response shows a concave *downward* trend as opposed to an *upward* concave.

52 If you are wondering, how about examining the model against the full dataset? Then, yes. You can! Also, the summation of *T*, *V*, and *S* should add to 100%.

53 I hope you are frustrated by my continued use of the beam example. I thought that having one example to showcase three metrics/indicators would be beneficial (as opposed to three different examples). See the *Questions and Problems* section.

54 Regardless of how they were used in training, validation, or testing.

Table 5.5 Comparison of metrics.

	MAE (mm)	RMSE (mm)	R^2 (%)
Full dataset	6.97	9.55	91.50
Deformations up to 80 mm	6.55	8.47	85.50
Deformations more than 80 mm	9.57	14.52	45.80
50% of data points were randomly selected from the full dataset (1st trail)	7.26	9.89	91.90
50% of data points were randomly selected from the full dataset (2nd trail)	6.28	8.90	91.40

sub-sets. The results of my analysis are listed in Table 5.5. As you can see, the D&S analysis also agrees with the spot analysis in which the sub-set representing deformations larger than 80 mm performed the poorest in terms of all metrics.

In addition, I also carried out two trials of randomly selecting half of the number of data points for a D&S analysis. The results, not surprisingly, are similar to that obtained from the full dataset.[55]

5.7.4 Rational Vetting*

In the *rational vetting* technique, we hope to test our model's prediction capability by using a series of rational tests pertinent to our domain knowledge or physics. Say that our model includes the level of applied loading as a feature to predict the deformation of beams. A rational vetting would imply that for a given same beam, reducing the level of loading should result in lesser deformation. If the model fails to predict such rational expectation, then the model might not capture the reality of the problem on hand – whether for this particular beam or for most/all beams.

The following could also serve as another example. Say that our ML model predicts an event-based stimulus (e.g., cracking due to elevated temperatures). In this example, the response is the degree of cracking, and the event-based stimulus is the occurrence of heating (i.e., fire – which could serve as the only feature to predict the response). In such a model, cracking is assumed to be solely tied to the occurrence of heating. Thus, a rational vetting would imply that if the value of the elevated temperature feature was to be set at 25°C, then the model should, in theory, predict zero/minimal cracking (due to the absence of stimulus). If the model predicts a considerable degree of cracking at ambient temperature, then the model is likely to be faulty.[56]

5.7.5 Confidence Intervals*

When we develop a ML model, we train, validate, and test this model on the data we have. Therefore, we can easily quantify the fitness of the model by comparing how far[57] its predictions are from the ground truth. In the event wherein the model is deployed and is fed with new data that it has not seen before or input with data that we do not know its ground truth, then how could we quantify the predictivity of the model?

Here is where we can augment our model with confidence intervals.[58] Such intervals are limits that can quantify the uncertainty of a prediction via two bounds, an upper and a lower bound. I would generally advise incorporating confidence

55 Another reminder of the importance of proper validation and influence of chance/luck in ML investigations.

56 Here, my examples are intentionally simple in the hope of conveying this concept. For instance, the cracking is only tied to thermal effects, and the absence of such effects should not result in cracking. In reality, cracking is a much more complex problem that stretches beyond one feature. In the event that cracking is not only tied to thermal effects but to other features as well, then the model may return some degree of cracking (given the relation between all the features). In hindsight, I would encourage you to think about logical and representative tests to carry out rational vettings for your models. Some may say that logical vetting are hypothetical or physics-informed tests – I would agree! These are also called logical counterfactuals. We will visit the concept of counterfactuals in Chapter 8.

57 Or, close!

58 Other intervals also exist such as *prediction interval* and *tolerance interval* [81].

intervals into ML models – however, I am more inclined to leave the discussion on these to a minimum.[59] The general formula for the confidence interval is:

$$\mu \pm z\frac{\sigma}{\sqrt{n}} \tag{5.4}$$

where the value of z (referred to as the confidence level value) for a 90% and 95% confidence interval, we use $z = 1.64$ and 1.96, respectively.

5.8 Conclusions

Performance fitness indicators and error metrics (PFIEM) are mathematical and/or logical constructs that are used to describe the goodness of ML models. Overall, PFIEM provides us with qualitative and quantitative measures as to how model predictions match the ground truth (i.e., measured observations in tests, etc.). It is likely, as we have seen above, that a ML model scores well against traditional PFIEM, only to be later identified that such a model does not properly represent the phenomenon on hand.[60] This can be avoided by adopting rigorous training/validation[61] and examination procedures[62] [64, 65].

Unfortunately, many of the published studies in the area of ML civil and environmental engineering continue to not include multi-criteria/additional validation phases and rely on traditional performance metrics (such as R or R^2). Please remember to try to adopt multi metrics that are also independent of each other. Be diligent in assessing model performance at the global and local levels. There is going to be a need to create functional indicators and metrics. The foundation of such metrics could be statistical, empirical, or may stem from domain knowledge/needs. Therefore, it is likely that the performance of the existing metrics can be further improved through systematic collaboration between researchers within our domain, as well as those in parallel domains.

At the end of this chapter, there are a couple of things I would like you to keep in mind:

- Understand the underlying assumptions pertaining to metrics and indicators.
- For a given ML investigation, select a proper metric, or preferably, a set of metrics.
- Check model performance at the global level and local level.

Definitions

Confusion matrix A table used to describe the performance of a classifier (i.e., how confused is the classifier in identifying the correct classes?)

Drawing and stacking Refer to the practice of examining ML models against sub-sets of data that are either drawn from the full dataset deliberately or randomly or in a stacked manner (i.e., a random sub-set can comprise T% from the training sub-set, V% from the validation sub-set, and S% from the testing sub-set).

Functional metrics or indicators Metrics and indicators which serve specific function or purpose.

Indicator A measure that comprises a set of metrics.

Metrics Defined as logical and/or mathematical constructs intended to measure the closeness of actual observations to that expected (or predicted) by ML models.

Outliers Observations that lie at an abnormal distance from other values (in a random sample from a population).

Spot analysis A procedure that orders ground truth points in a smooth line to reveal oscillation around the ground truths.

---

59 Sebastian Raschka has a good and healthy discussion on confidence intervals which can be seen at his blog and paper [82].

60 Or a portion of the phenomenon!

61 Such as *k*-cross validation.

62 Please refer to metrics and indicators discussed throughout this chapter.

Chapter Blueprint

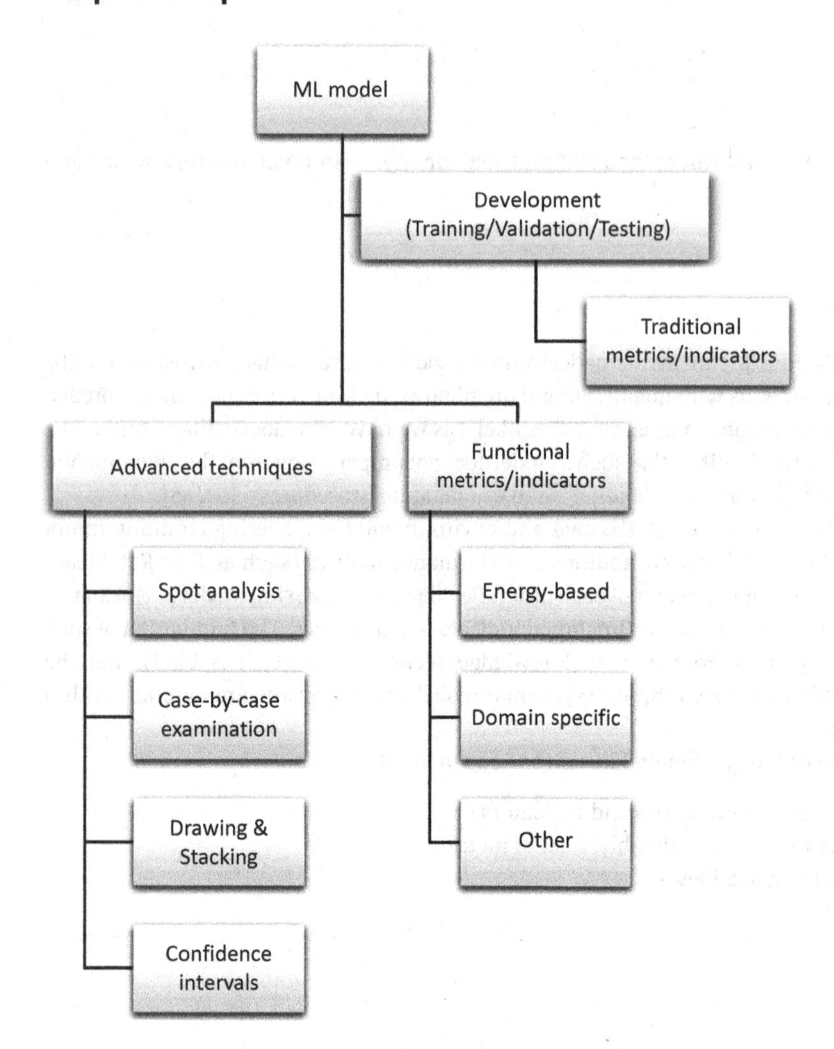

Questions and Problems

5.1 Define performance fitness indicators and error metrics. Why are these constructs needed?

5.2 What is the main difference between indicators and metrics?

5.3 When are functional metrics and indicators needed?

5.4 What is the benefit(s) of examining ML models via *spot analysis* and *CbC*?

5.5 Why is there a need to incorporate energy-based metrics and indicators?

5.6 Build a ML model of your choosing and evaluate its outcome via traditional PFIEM as well as spot analysis, CbC, and D&S. You may use the supplied dataset accompanying Figure 5.5.

5.7 Explore the lack of standardized PFIEM for civil and environmental engineering problems.

5.8 What are the conditions that may require a D&S examination?

5.9 Discuss how and why functional metrics can be useful (i.e., provide additional insights beyond PFIEM).

Suggested Metrics and Packages

Table 5.6 lists a series of ML packages for PFIEM.

Table 5.6 Suggested metrics and packages*.

No.	Metric	*Python* code/package	*R* code/package
Regression			
1	Error (E)	from **sklearn.metrics** import **max_error**	**#Function for error Computation:** e <- function (actual, predicted){ return((actual-predicted))
2	Mean Absolute Error (MAE)*	from **sklearn.metrics** import **mean_absolute_error**	**#Package for error Computation:** >install. Packages("Metrics") >library("Metrics") >mea(actual, predicted)

3	Mean Squared Error (MSE)	from **sklearn.metrics** import **mean_squared_error**	**#Function for MSE Computation:**
			Se<- function(actual, predicted){
			return((actual-predicted)^2)
			}
			mse<- function(actual, predicted){
			return(sum(se(actual, predicted)))
			}
			
			#Package for MSE Calculation:
			>install. Packages("Metrics")
			>library("Metrics")
			>mse(actual, predicted)

4	Root Mean Squared Error (RMSE)	import **math**
		MSE= mean_squared_error (actual, predicted)
		RMSE = math.sqrt(MSE)
			#Function for RMSE Computation:
			rmse<- function(actual, predicted){
			return(sqrt(mse(actual, predicted)))
			}
			#Package for RMSE Calculation:
			>install. Packages("Metrics")
			>library("Metrics")
			>rmse(actual, predicted)

5	Correlation coefficient (R)	import **numpy as np** **R = np.corrcoef**(actual, predicted)	**#Package for Cc (Correlation coefficient) calculation:** >Installed. Packages("tidymodel", dependencies=TRUE) >library(tidymodels) cor(data$var1,data$var2) resource:

(Continued)

Table 5.6 (Continued)

No.	Metric	*Python* code/package	*R* code/package
6	Coefficient of Determination (R^2)	from **sklearn.metrics** import **r2_score**	**#Package for R2 calculation:** >install. Packages("stats") >library("stats") Model<- lm(formula, data, subset, weights, .) summary(Model)
Classification			
1	Confusion matrix	from **sklearn.metrics** import **confusion_matrix**	**#Insatll required packages** install.packages('caret') **#Import required library** library(caret) **#Creating confusion matrix example** confusionMatrix(data=predicted_value, reference = expected_value) **#Display results** example

2	Accuracy (ACC)	from **sklearn.metrics** import **accuracy_score**	**#Function for ACC Computation:**
			accuracy <- function(actual, predicted) {
			return(1 - ce(actual, predicted))
			}
			#Package for ACC Calculation:
			>install. Packages("Metrics")
			>library("Metrics")
			>accuracy(actual, predicted)

3	F_1 score	from **sklearn.metrics** import **f1_score**
		f1_score(actual,predicted, **average='macro'**)
		f1_score(actual,predicted, **average='micro'**
		f1_score(actual,predicted, **average='weighted'**
			#Function for F1 Computation:
			f1 <- function (actual, predicted)
			{
			act <- unique(actual)
			pred <- unique(predicted)
			tp <- length(intersect(act,pred))
			fp <- length(setdiff(pred,act))
			fn <- length(setdiff(act,pred))
			precision <- ifelse ((tp==0 & fp==0), 0,
			tp/(tp+fp))
			recall <- ifelse ((tp==0 & fn==0), 0,
			tp/(tp+fn))
			score <- ifelse ((precision==0 & recall==0),
			0, 2*precision*recall/(precision+recall))
			score
			}
			#Package for F1 Calculation:
			>install. Packages("Metrics")
			>library("Metrics")
			>f1(actual, predicted)

Table 5.6 (Continued)

No.	Metric	*Python* code/package	*R* code/package

4	Balanced accuracy [48]	from **sklearn.metrics** import **balanced_accuracy_score**	**#Function for recall Computation:**
			recall <- function(actual, predicted) {
			return(mean(predicted[actual == 1]))
			}

5	Receiver Operating Characteristic (ROC)	from **sklearn.metrics** import **roc_curve** from **sklearn.metrics** import **roc_auc_score** from **matplotlib** import **pyplot** **ns_fpr, ns_tpr, _ = roc_ curve(y_test, ns_ prediction)** **M_fpr, M_tpr, _ = roc_ curve(y_test, M_ prediction)** **pyplot.plot(ns_fpr, ns_tpr, linestyle='--', label='No Skill')** **pyplot.plot(M_fpr, M_tpr, marker='.', label='regression, classification,etc.')**	**#Package for ROC Calculation:** >install. Packages("pROC") >library("pROC") >roc(predicted class, model. fit$fitted.value, plot=True)

6	Area under the ROC curve (AUC)	from **sklearn.metrics** import **roc_auc_score**
		roc_auc_score(y, model. predict_proba(X)[:, 1])
			#Function for AUC computation:
			auc <- function(actual, predicted)
			{
			r <- rank(predicted)
			n_pos <- sum(actual==1)
			n_neg <- length(actual) - n_pos
			auc
			<- (sum(r[actual==1]) - n_pos*(n_ pos+1)/2)
			/ (n_pos*n_neg)
			auc
			}
			
			#Package for AUC calculation:
			>install. Packages("Metrics")
			>library("Metrics")
			>auc(actual, predicted)

7	Log Loss Error (LLE)	from **sklearn.metrics** import **log_loss**	**#Function for Log Loss Computation:**
			ll <- function(actual, predicted) {
			score <- -(actual*log(predicted) + (1 - actual)*log(1 - predicted))
			score[actual == predicted] <- 0
			score[is.nan(score)] <- Inf
			return(score)
			}
			
			#Package for Log Loss calculation:
			>install. Packages("Metrics")
			>library("Metrics")
			>ll(actual, predicted)

(Continued)

Table 5.6 (Continued)

No.	Metric	*Python* code/package	*R* code/package
Clustering			
1	The Elbow method [55, 56]	**#K-means using Euclidean distance/distortion.** **distortions = []** **#Total number of clusters.** **K = range(1,10)** **#Calcualte distortion per cluster.** **for k in K:** **kmeanModel = KMeans(n_clusters=k).fit(X)** **kmeanModel.fit(X)** **distortions.append(sum(np.min(cdist(X, kmeanModel.cluster_centers_, 'euclidean'), axis=1)) / X.shape[0])** **#Plotting.** **plt.plot(K, distortions, 'bx-')** **plt.xlabel('k-clusters')** **plt.ylabel('Distortion')** **plt.xlim([0, 10])**	**#Package for Plotting Number of Cluster calculation:** >install. Packages("NbClust") >library("NbClust") > set.seed(1234) nc <- NbClust(df, min.nc=2, max.nc=15, method="kmeans")

2	The Silhouette Coefficient	from **sklearn.metrics** import **silhouette_samples, silhouette_score**
		#Compute the silhouette scores for each sample.
		sample_silhouette_values = silhouette_samples (X, cluster_labels)
		#The silhouette_score gives the average value for all the samples.
		silhouette_avg = silhouette_score (X, cluster_labels)
			#Package for Plotting clustering quality in Dataframe:
			>install. Packages("cluster")
			>install. Packages("vegan")
			>library("cluster")
			>library("vegan")
			dis = vegdist(data)
			res = pam(dis,5) # or whatever your choice of
			clustering algorithm is
			sil = silhouette (res$clustering,dis) # or use your cluster vector of yours
			plot(sil)

* Compiled from open sources such as *SciKit-Learn.org*, cran.r-project.org, *rdocumentation.org*, and *https://github.com/mfrasco6/Metrics/blob/master/R/regression.R.* To apply metrics in *Python*, insert *(actual, predicted)* right after the metric. For example, **metric** *(actual, predicted)*. You could also, **print** *("Root Mean Square Error:",* **RMSE**) to print the value of the metric.

References

1 Kodur, V.K.R., Garlock, M., and Iwankiw, N. (2011). Structures in fire: state-of-the-art, research and training needs. *Fire Technol.* 48: 825–839. doi: 10.1007/s10694-011-0247-4.

2 Naser, M.Z. (2019). Fire resistance evaluation through artificial intelligence – a case for timber structures. *Fire Saf. J.* 105: 1–18. doi: 10.1016/j.firesaf.2019.02.002.

3 Domingos, P. (2012). A few useful things to know about machine learning. *Commun. ACM*. doi: 10.1145/2347736.2347755.

4 Schmidt, M.D. and Lipson, H. (2010). Age-fitness pareto optimization. doi: 10.1145/1830483.1830584.

5 Cremonesi, P., Koren, Y., and Turrin, R. (2010). Performance of recommender algorithms on top-n recommendation tasks categories and subject descriptors. *RecSys '10: Proceedings of the Fourth ACM Conference on Recommender Systems.*

6 Laszczyk, M. and Myszkowski, P.B. (2019). Survey of quality measures for multi-objective optimization: construction of complementary set of multi-objective quality measures. *Swarm Evol. Comput.* 48: 109–133. doi: 10.1016/J.SWEVO.2019.04.001.

7 Willmott, C.J. and Matsuura, K. (2005). Advantages of the mean absolute error (MAE) over the root mean square error (RMSE) in assessing average model performance. *Clim. Res.* 30. doi: 10.3354/cr030079.

8 Makridakis, S. (1993). Accuracy measures: theoretical and practical concerns. *Int. J. Forecast.* 9. doi: 10.1016/0169-2070(93)90079-3.

9 Ferreira, C. (2001). Gene expression programming: a new adaptive algorithm for solving problems. *Gene Expr. Program. a New Adapt. Algorithm Solving Probl. Complex Syst.* 13. https://www.semanticscholar.org/paper/Gene-Expression-Programming%3A-a-New-Adaptive-for-Ferreira/3232b2a24c2584ca8e81cb5bf6f55aef34f0aefe (accessed 16 March 2019).

10 D. Pollock (1999). Handbook of Time Series Analysis, Signal Processing, and Dynamics. *Geology, Computer Science.* https://www.elsevier.com/books/handbook-of-time-series-analysis-signal-processing-and-dynamics/pollock/978-0-12-560990-6. doi: 10.1016/b978-0-12-560990-6.x5000-3.

11 Hyndman, R.J. and Koehler, A.B. (2006). Another look at measures of forecast accuracy. *Int. J. Forecast.* 22. doi: 10.1016/j.ijforecast.2006.03.001.

12 Shcherbakov, M.V., Brebels, A., Shcherbakova, N.L. et al. (2013). A survey of forecast error measures. *World Appl. Sci. J.* 24.

13 Poli, A.A. and Cirillo, M.C. (1993). On the use of the normalized mean square error in evaluating dispersion model performance. *Atmos. Environ. A- Gen. Top.* 27. doi: 10.1016/0960-1686(93)90410-Z.

14 Armstrong, J.S. and Collopy, F. (1992). Error measures for generalizing about forecasting methods: empirical comparisons. *Int. J. Forecast.* 8. doi: 10.1016/0169-2070(92)90008-W.

15 Sadat Hosseini, A., Hajikarimi, P., Gandomi, M. et al. (2021). Genetic programming to formulate viscoelastic behavior of modified asphalt binder. *Constr. Build. Mater.* 286. doi: 10.1016/j.conbuildmat.2021.122954.

16 Bain, L.J. (1967). Applied regression analysis. *Technometrics* 9. doi: 10.1080/00401706.1967.10490452.

17 Smith, G. (1986). *Probability and Statistics in Civil Engineering.* London: Collins.

18 Golbraikh, A., Shen, M., Xiao, Z. et al. (2003). Rational selection of training and test sets for the development of validated QSAR models. *J. Comput. Aided. Mol. Des.* 17: 241–253. doi: 10.1023/A:1025386326946.

19 Roy, P.P. and Roy, K. (2008). On some aspects of variable selection for partial least squares regression models. *QSAR Comb. Sci.* 27: 302–313. doi: 10.1002/qsar.200710043.

20 Frank, I. and Todeschini, R. (1994). *The Data Analysis Handbook.* Elsevier. https://books.google.com/books?hl=en&lr=&id=SXEpB0H6L3YC&oi=fnd&pg=PP1&ots=zfmIRO_XO5&sig=dSX6KJdkuav5zRNxaUdcftGSn2k (accessed 21 June 2019).

21 Gandomi, A.H., Yun, G.J., and Alavi, A.H. (2013). An evolutionary approach for modeling of shear strength of RC deep beams. *Mater. Struct. Constr.* 46. doi: 10.1617/s11527-013-0039-z.

22 Cheng, M.Y., Firdausi, P.M., and Prayogo, D. (2014). High-performance concrete compressive strength prediction using Genetic Weighted Pyramid Operation Tree (GWPOT). *Eng. Appl. Artif. Intell.* 29. doi: 10.1016/j.engappai.2013.11.014.

23 Alwanas, A.A.H., Al-Musawi, A.A., Salih, S.Q. et al. (2019). Load-carrying capacity and mode failure simulation of beam-column joint connection: application of self-tuning machine learning model. *Eng. Struct.* 194. 10.1016/j.engstruct.2019.05.048.

24 Chou, J.S., Tsai, C.F., Pham, A.D., and Lu, Y.H. (2014). Machine learning in concrete strength simulations: multi-nation data analytics. *Constr. Build. Mater.* 73. doi: 10.1016/j.conbuildmat.2014.09.054.

25 Nguyen, T.T., Pham Duy, H., Pham Thanh, T., and Vu, H.H. (2020). Compressive strength evaluation of fiber-reinforced high-strength self-compacting concrete with artificial intelligence. *Adv. Civ. Eng.* 2020. doi: 10.1155/2020/3012139.

26 Sultana, N., Zakir Hossain, S.M., Alam, M.S. et al. (2020). Soft computing approaches for comparative prediction of the mechanical properties of jute fiber reinforced concrete. *Adv. Eng. Softw.* 149. doi: 10.1016/j.advengsoft.2020.102887.

27 Willmott, C.J. (1981). On the validation of models. *Phys. Geogr.* 2. doi: 10.1080/02723646.1981.10642213.

28 Nash, J.E. and Sutcliffe, J.V. (1970). River flow forecasting through conceptual models part I – a discussion of principles. *J. Hydrol.* 10. doi: 10.1016/0022-1694(70)90255-6.

29 Gupta, H.V., Kling, H., Yilmaz, K.K., and Martinez, G.F. (2009). Decomposition of the mean squared error and NSE performance criteria: implications for improving hydrological modelling. *J. Hydrol.* 377. doi: 10.1016/j.jhydrol.2009.08.003.

30 Knoben, W.J.M., Freer, J.E., and Woods, R.A. (2019). Technical note: inherent benchmark or not? comparing Nash-Sutcliffe and Kling-Gupta efficiency scores. *Hydrol. Earth Syst. Sci.* 23. doi: 10.5194/hess-23-4323-2019.

31 Huang, H. and Burton, H.V. (2019). Classification of in-plane failure modes for reinforced concrete frames with infills using machine learning. *J. Build. Eng.* 25. doi: 10.1016/j.jobe.2019.100767.

32 Bhowan, U., Johnston, M., and Zhang, M. (2011). Developing new fitness functions in genetic programming for classification with unbalanced data. *IEEE Trans. Syst. Man Cybern B: Cybern.* 42. doi: 10.1109/TSMCB.2011.2167144.

33 Boughorbel, S., Jarray, F., and El-Anbari, M. (2017). Optimal classifier for imbalanced data using Matthews correlation coefficient metric. *PLoS One* 12. doi: 10.1371/journal.pone.0177678.

34 Tharwat, A. (2020). Classification assessment methods. *Appl. Comput. Inform.* 17. doi: 10.1016/j.aci.2018.08.003.

35 Caruana, R. and Niculescu-Mizil, A. (2004). Data mining in metric space: an empirical analysis of supervised learning performance criteria. *KDD '04: Proceedings of the Tenth ACM SIGKDD International Conference on Knowledge Discovery and Data Mining*. doi: 10.1145/1014052.1014063.

36 Jurman, G., Riccadonna, S., and Furlanello, C. (2012). A comparison of MCC and CEN error measures in multi-class prediction. *PLoS One* 7. doi: 10.1371/journal.pone.0041882.

37 Powers, D.M.W. (2011). Evaluation: from precision, recall and F-measure to ROC, informedness, markedness and correlation. *J. Mach. Learn. Technol.* doi: 10.48550/arXiv.2010.16061.

38 Bradley, A.P. (1997). The use of the area under the ROC curve in the evaluation of machine learning algorithms. *Pattern Recognit.* 30. doi: 10.1016/S0031-3203(96)00142-2.

39 Hanley, J.A. and McNeil, B.J. (1982). The meaning and use of the area under a receiver operating characteristic (ROC) curve. *Radiology* 143. doi: 10.1148/radiology.143.1.7063747.

40 Zhang, Y., Burton, H.V., Sun, H., and Shokrabadi, M. (2018). A machine learning framework for assessing post-earthquake structural safety. *Struct. Saf.* 72. doi: 10.1016/j.strusafe.2017.12.001.

41 Davis, J. and Goadrich, M. (2006). The relationship between Precision-Recall and ROC curves. *ICML '06: Proceedings of the 23rd International Conference on Machine Learning*. doi: 10.1145/1143844.1143874.

42 Bi, J. and Bennett, K. (2003). Regression error characteristic curves. *20th International Conference on Machine Learning, 2003.*

43 Zhang, M. and Smart, W. (2006). Using Gaussian distribution to construct fitness functions in genetic programming for multiclass object classification. *Pattern Recognit. Lett.* 27. doi: 10.1016/j.patrec.2005.07.024.

44 Kocher, M. and Savoy, J. (2017). Distance measures in author profiling. *Inf. Process. Manag.* 53. doi: 10.1016/j.ipm.2017.04.004.

45 Patel, B.V. and Meshram, B.B. (2012). Content based video retrieval systems. *Int. J. UbiComp.* 3. doi: 10.5121/iju.2012.3202.

46 Giusti, R. and Batista, G.E.A.P.A. (2013). An empirical comparison of dissimilarity measures for time series classification. *2013 Brazilian Conference on Intelligent Systems*. doi: 10.1109/BRACIS.2013.22.

47 Vuk, M. and Curk, T. (2006). ROC curve, lift chart and calibration plot. Metod. Zv.

48 Brodersen, K.H., Ong, C.S., Stephan, K.E., and Buhmann, J.M. (2010). The balanced accuracy and its posterior distribution. *2010 20th International Conference on Pattern Recognition*. doi: 10.1109/ICPR.2010.764.

49 Cohen, J. (1960). A coefficient of agreement for nominal scales. *Educ. Psychol. Meas.* 20. doi: 10.1177/001316446002000104.

50 Artstein, R. and Poesio, M. (2008). Inter-coder agreement for computational linguistics. *Comput. Linguist.* 34. doi: 10.1162/coli.07-034-R2.

51 Destercke, S. (2014). Multilabel prediction with probability sets: the hamming loss case. *IPMU 2014: Information Processing and Management of Uncertainty in Knowledge-Based Systems*. doi: 10.1007/978-3-319-08855-6_50.

52 Czajkowski, M. and Kretowski, M. (2019). Decision tree underfitting in mining of gene expression data. An evolutionary multi-test tree approach. *Expert Syst. Appl.* 137. doi: 10.1016/J.ESWA.2019.07.019.

53 Devarriya, D., Gulati, C., Mansharamani, V. et al. (2020). Unbalanced breast cancer data classification using novel fitness functions in genetic programming. *Expert Syst. Appl.* 140: 112866. doi: 10.1016/J.ESWA.2019.112866.

54 Rousseeuw, P.J. (1987). Silhouettes: a graphical aid to the interpretation and validation of cluster analysis. *J. Comput. Appl. Math.* 20. doi: 10.1016/0377-0427(87)90125-7.

55 Thorndike, R.L. (1953). Who belongs in the family? *Psychometrika* 18. doi: 10.1007/BF02289263.

56 Nainggolan, R., Perangin-angin, R., Simarmata, E., and Tarigan, A.F. (2018). Improved the performance of the K-means cluster using the Sum of Squared Error (SSE) optimized by using the elbow method. *J. Phys. Conf. Ser.* 1361: 12015. doi: 10.1088/1742-6596/1361/1/012015.

57 Arbelaitz, O., Gurrutxaga, I., Muguerza, J. et al. (2013). An extensive comparative study of cluster validity indices. *Pattern Recognit.* 46: 243–256. doi: 10.1016/j.patcog.2012.07.021.

58 Naser, M.Z. (2023, January 1). Do we need exotic models? Engineering metrics to enable green machine learning from tackling accuracy-energy trade-offs. *J. Clean. Prod.* 382: 135334. doi: 10.1016/j.jclepro.2022.135334.

59 Henderson, P., Hu, J., Romoff, J. et al. (2020). Towards the systematic reporting of the energy and carbon footprints of machine learning. *J. Mach. Learn. Res.* 21.

60 Lacoste, A., Luccioni, A., Schmidt, V., and Dandres, T. (2019). Quantifying the carbon emissions of machine learning. (accessed 10 June 2022) https://mlco2.github.

61 Canziani, A., Paszke, A., and Culurciello, E. (2016). An analysis of deep neural network models for practical applications. doi: 10.48550/arxiv.1605.07678.

62 Dalgren, A. and Lundegård, Y. (2019). GreenML: a methodology for fair evaluation of machine learning algorithms with respect to resource consumption. Linköpings universitet. https://www.diva-portal.org/smash/record.jsf?pid=diva2%3A1345249&dswid=-9453 (accessed 8 June 2022).

63 Naser, M.Z. (2022). CLEMSON: an automated machine learning (AutoML) virtual assistant for accelerated, simulation-free, transparent, reduced-Order and inference-Based reconstruction of fire response of structural members. *ASCE J. Struct. Eng.* 148. doi: 10.1061/(ASCE)ST.1943-541X.0003399.

64 Bhaskar, H., Hoyle, D.C., and Singh, S. (2006). Machine learning in bioinformatics: a brief survey and recommendations for practitioners. *Comput. Biol. Med.* doi: 10.1016/j.compbiomed.2005.09.002.

65 Kohavi, R. (1995). A study of cross-validation and bootstrap for accuracy estimation and model selection. *International Joint Conference on Articial Intelligence IJCAI*. Vol. 2.

66 O'Neil, C. (2016). *Weapons of Math Destruction: How Big Data Increases Inequality and Threatens Democracy*. Crown.

67 Franssen, J.M. and Gernay, T. (2017). Modeling structures in fire with SAFIR®: theoretical background and capabilities. *J. Struct. Fire Eng.* 8. doi: 10.1108/JSFE-07-2016-0010.

68 Ferreira, J., Gernay, T., Franssen, J., and Vassant, O. (2018). Discussion on a systematic approach to validation of software for structures in fire. *SiF 2018: 10th international conference on structures in fire*. http://hdl.handle.net/2268/223208 (accessed 1 April 2020).

69 Kohnke, P.C. (2013). ANSYS. © ANSYS, Inc.

70 Abaqus 6.13, Abaqus 6.13 (2013). Anal. User's Guid. Dassault Syst.

71 Botchkarev, A. (2019). A new typology design of performance metrics to measure errors in machine learning regression algorithms. *Interdiscip. J. Inform. Knowl. Manag.* 14: 045–076. doi: 10.28945/4184.

72 Naser, M.Z. and Alavi, A.H. (2021). Error metrics and performance fitness indicators for artificial intelligence and machine learning in engineering and sciences. *Archit. Struct. Constr.* 1: 1–19. doi: 10.1007/S44150-021-00015-8.

73 Tibshirani, R., Walther, G., and Hastie, T. (2002). Estimating the number of clusters in a data set via the gap statistic. *J. R. Stat. Soc. Ser. B: Stat. Methodol.* 63. doi: 10.1111/1467-9868.00293.

74 Strubell, E., Ganesh, A., and McCallum, A. (2020). Energy and policy considerations for modern deep learning research. *Proceedings of the AAAI Conference on Artificial Intelligence*. doi: 10.1609/aaai.v34i09.7123.

75 Jackson, R. (2019). A roadmap to reducing greenhouse gas emissions 50 percent by 2030. https://earth.stanford.edu/news/roadmap-reducing-greenhouse-gas-emissions-50-percent-2030#gs.zij3zf.

76 García-Martín, E., Rodrigues, C.F., Riley, G., and Grahn, H. (2019). Estimation of energy consumption in machine learning. *J. Parallel. Distrib. Comput.* 134. doi: 10.1016/j.jpdc.2019.07.007.

77 Yang, T.J., Chen, Y.H., Emer, J., and Sze, V. (2017). A method to estimate the energy consumption of deep neural networks. *2017 51st Asilomar Conference on Signals, Systems, and Computers*. doi: 10.1109/ACSSC.2017.8335698.

78 Schwartz, R., Dodge, J., Smith, N.A., and Etzioni, O. (2020). Green AI. *Commun. ACM*. doi: 10.1145/3381831.

79 Kodur, V. and Naser, M.Z. (2020). *Structural Fire Engineering*, 1e. McGraw Hill Professional.

80 Buchanan, A. and Abu, A. (2017). *Structural Design for Fire Safety*. Wiley. 440. **ISBN-10**: 9780470972892.

81 Parker, P.A. (2020). Statistical intervals: a guide for practitioners and researchers. *J. Qual. Technol.* 52. doi: 10.1080/00224065.2019.1584508.

82 Raschka, S. (2018). Model evaluation, model selection, and algorithm selection in machine learning. *ArXiv*. doi: 10.48550/arXiv.1811.12808.

6

Coding-free and Coding-based Approaches to Machine Learning

You take the blue pill, the story ends, you wake up in your bed and believe whatever you want to believe. You take the red pill, you stay in wonderland, and I show you how deep the rabbit hole goes.

Morpheus (The Matrix, 1999)

Synopsis

A few chapters ago, I took the liberty of outlining my thoughts on why a good component of the inertia against ML lies in our lack of coding/programming in our curriculum.[1] It felt more beneficial to find some solutions that do not require us to change our domain abruptly. The simplest of all solutions was to adopt a coding-free approach to ML where engineers with limited knowledge/resources of programming can still leverage the full potential of ML without the need to resort to traditional coding. Thus, this chapter offers a look into the two main approaches to practicing ML, *coding-free* and *coding-based*. In this journey, we will go over a few platforms that I found most suitable for engineers, and especially students, to implement and use. All of the presented platforms offer free services to engineering students and hence are adopted in this chapter.[2] Along the way, we will cover the origins and capabilities of such platforms. Toward the end of this chapter, we will cover the use of *Python* and *R* as representatives of coding-based approaches. In all cases, I have supplemented my discussions with a series of examples and tutorials.[3] I have listed and appended these here for you to practice along the way.

6.1 Coding-free Approach to ML

An instinctive way to start this chapter is by showcasing the traditional approach to ML in which a user codes or programs an executable script. However, I find it to be more exciting to start this chapter with a look into the new approach to ML. In this approach, coding is no longer required to build ML models.[4] Similarly, this approach negates the need for specialized software/hardware infrastructure needed to carry out ML modeling and analysis. It only requires access to an online coding-free platform!

1 Quick story: My good friend, Sami Hasasneh, used to say; *if you are ever faced with a problem, all that you need to do is to find a solution. Most intuitive solutions are simple. Simply, remove the problem!* Re-iterating this story from the context of this chapter implies the following. If one of the main problems to integrate ML in our area arises from the heavy reliance on coding/programming, then let us try to remove this problem (at least for a bit)!

2 Some also offer free services/downloads to any user – including practicing engineers. It is worth noting that other platforms than those explicitly mentioned here are available (i.e., *DataBricks*, *OpenML*, *Google Colab*, *Kaggle*, etc.). Please feel free to try them.

3 Admittedly, most of these examples cover tabular data and two will cover the application of computer vision. In all cases, similar, modified, and extended examples can also be tailored and applied. Use your imagination to extend these beyond the scope of this book and toward your field!

4 This would qualify as a simple solution!

Machine Learning for Civil & Environmental Engineers: A Practical Approach to Data-driven Analysis, Explainability, and Causality, First Edition. M. Z. Naser.

Companion Website: www.wiley.com/go/Naser/MachineLearningforCivilandEnvironmentalEngineers

Imagine this; if there is no need for coding and specialized equipment, then ML becomes a *bit* more accessible to various users and industries.[5] From this end, the user is expected to become familiar with the platform of their choice and general ideas behind ML as opposed to learning ML via learning how to code. The latter is indeed one of the biggest hurdles to embracing ML in a variety of engineering industries.

In the coding-free approach, a civil or environmental engineer can access an online portal with an intuitive graphical interface.[6] It is through this interface that the engineer can upload their data, select the type of analysis they envision to carry out, select one or a series of ML algorithms, run the ML analysis, and examine the outcome of this analysis.[7] In some instances, the engineer could deploy their final model to use it outside of the platform. The engineer could seamlessly share and collaborate with others on projects.

I will be covering five platforms in this chapter. My reasoning for selecting these five stems from the positive experiences I gained while using such platforms, as well as the fact that these platforms offer free access and academic support to students and faculty. These platforms are *BigML*, *DataRobot*, *Dataiku*, *Exploratory*, and *Clarifai*.[8]

6.1.1 BigML

According to *BigML*,[9]

> *BigML is a consumable, programmable, and scalable Machine Learning platform that makes it easy to solve and automate Classification, Regression, Time Series Forecasting, Cluster Analysis, Anomaly Detection, Association Discovery, and Topic Modeling tasks.*

BigML was founded in 2011[10] with the mission of making *ML easy and beautiful for everyone* [1]. Of all the presented platforms in this textbook, *BigML* could arguably be the most general in applications. *BigML* also has a strong academic program that supports ML education and research and allows students to freely access and practice its platform. A unique feature of BigML is that it allows its users to download the completed ML models in scripts[11] (i.e., once a user completes building a model, the same user can download their model's code in a programming language of their choice to share with other users, etc.). *BigML* also supports explainable ML.[12] A screenshot of *BigML*'s homepage is shown in Figure 6.1. We will use this platform to cover four examples,[13] as you will see below. Please feel free to follow my steps.

Example no. 1 [*BigML*, Classification]: Categorizing failure mode in FRP-strengthened RC beams.
In this example, we have a dataset[14] that homes the outcome of over 100 experimental tests on RC beams strengthened with FRP[15] composites. This dataset includes compressive strength of concrete f_c, yield strength of steel reinforcement f_y, the ratio of steel reinforcement r_s, FRP ratio r_f, modulus of FRP E_{fr}, moment arm a, and strengthening type T.[16] The target in this dataset is the failure mode[17] (*FM*) observed in each test (i.e., steel yielding/concrete crushing or failure through the FRP system). Let us use the *BigML* platform to carry out this example.

5 As you can see, there is an entrepreneurship business model behind such services that revolves around simplicity and convenience.

6 Which you will get to see a few examples on shortly.

7 As you will see, some of these steps can be very well automated and all that a user needs to do is to upload their data! In my eyes, the above sounds both *exciting* and *terrifying*!

8 I would highly encourage you to stay up-to-date on coding-free platforms as these services seem to be gaining popularity. In addition, please explore and visit the websites of the presented platforms for up-to-date information on their capabilities. Keep in mind that we will be visiting other software in Chapter 9: *Advanced Topics* that are deemed to have very specific applications (vs. those showcased in this chapter). As always, there are many, many features, platforms, and software that I do not get to cover here, some of which may include RapidMiner [26], Weka [27], KNIME [28], etc.

9 https://bigml.com.

10 Headquartered in Corvallis, Oregon.

11 This ability is referred to as *Exportable ML*.

12 Please visit https://bigml.com/documentation for more information on the capabilities within *BigML*.

13 The first two examples use the same dataset.

14 Full details on this dataset and data sources can be found in [29, 30].

15 Fun fact, my first research article was on FRP. FRP are strengthening/retrofitting systems that we can attach to aging or old structural members (mainly those made of concrete, but some have used them on metallics as well). When attached, the FRP system improves the load bearing capacity of the structural member.

16 There are traditionally two strengthening types, *plated* (e.g., where plates of FRP are bonded to the surface of the member and *mounted* (aka. near surface mounted (NSM) where rebars or strips are embedded via grooves cut into the sides of the member)).

17 Please note that 42 beams failed via steel yielding, and 59 beams failed due to FRP system failure (debonding: 18 beams, concrete cover separation: 18 beams, mixed-mode: 16 beams, and FRP rupture: 7 beams). I will be presenting the case of binary classification (i.e., failure due to steel yielding or due to FRP system failure) in this chapter. Please feel free to extend this example into a multi-classification problem!

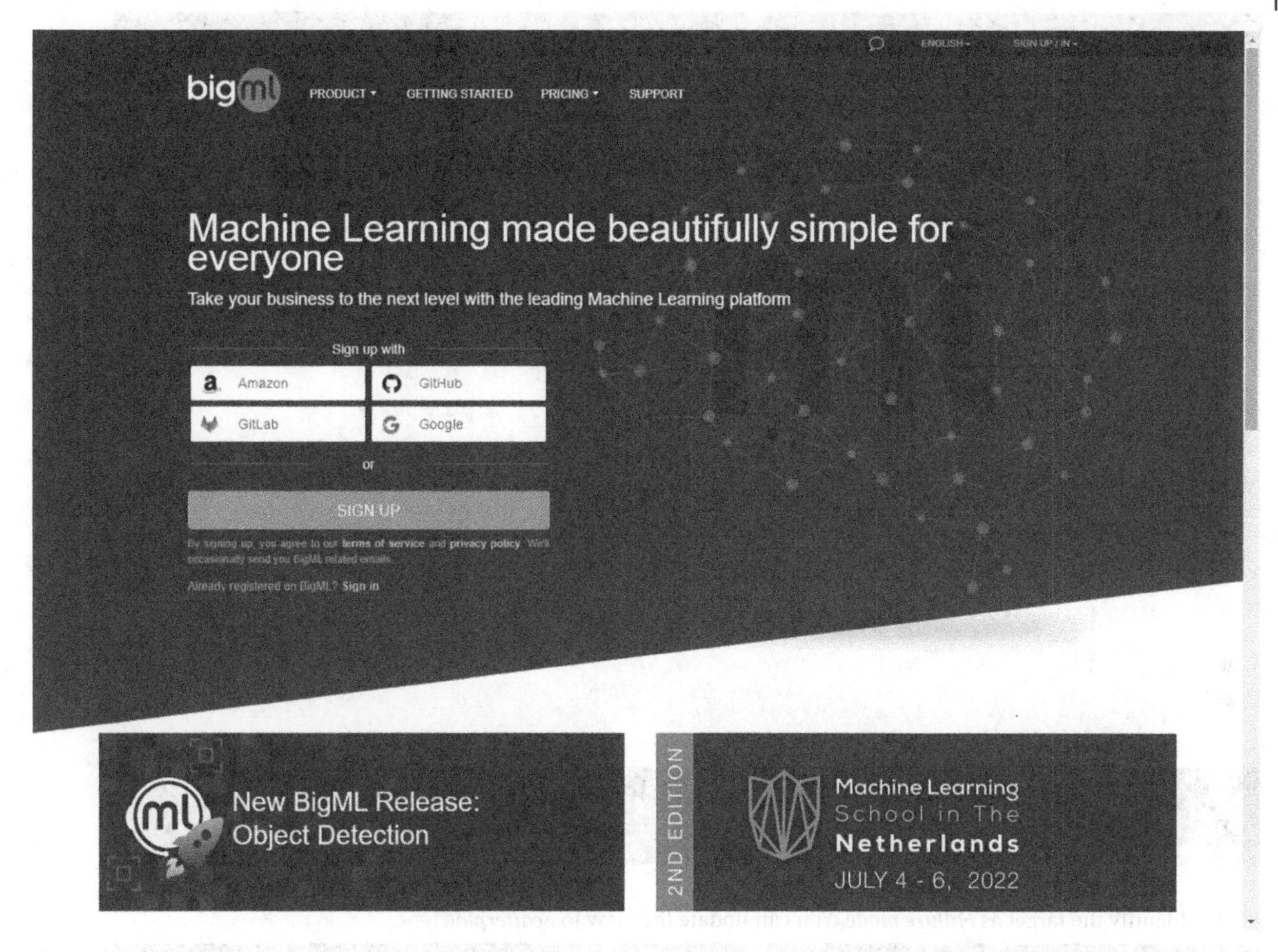

Figure 6.1 Snippet of the *BigML home page* BigML, Inc., https://bigml.com, last accessed under August 17, 2022.

Step 1: Upload your dataset

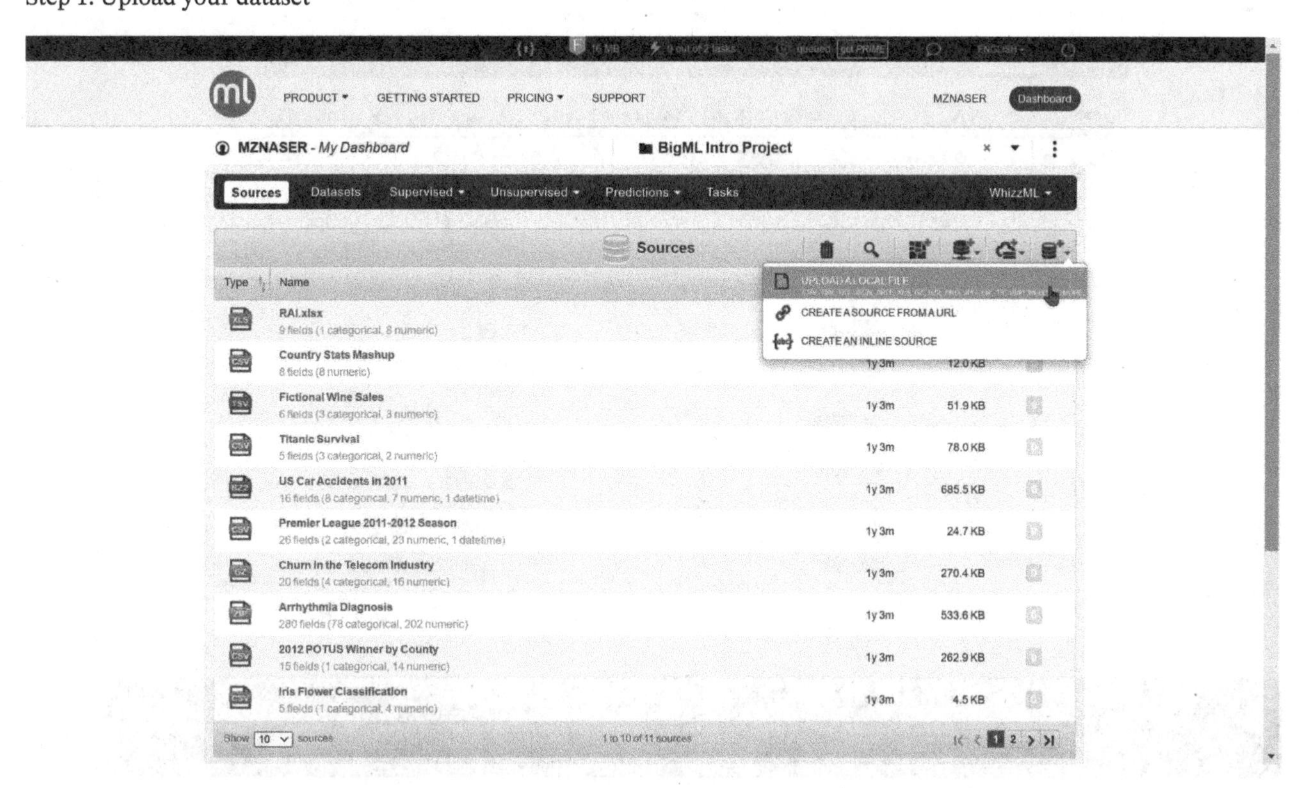

Step 2: Select *1-Click Dataset* and navigate to the *Datasets* tab.

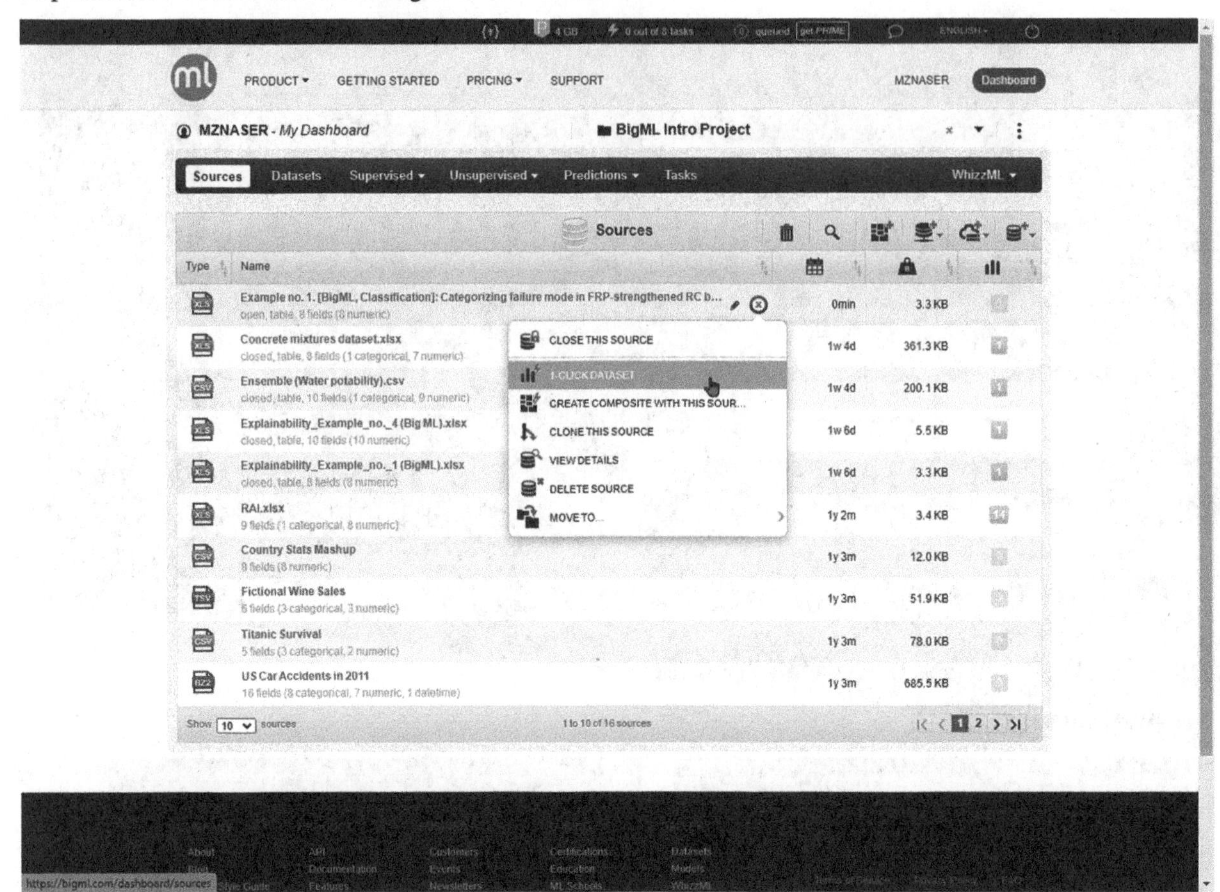

Step 3: Identify the target as *Failure mode*. You can update the view to *Scatterplot*.

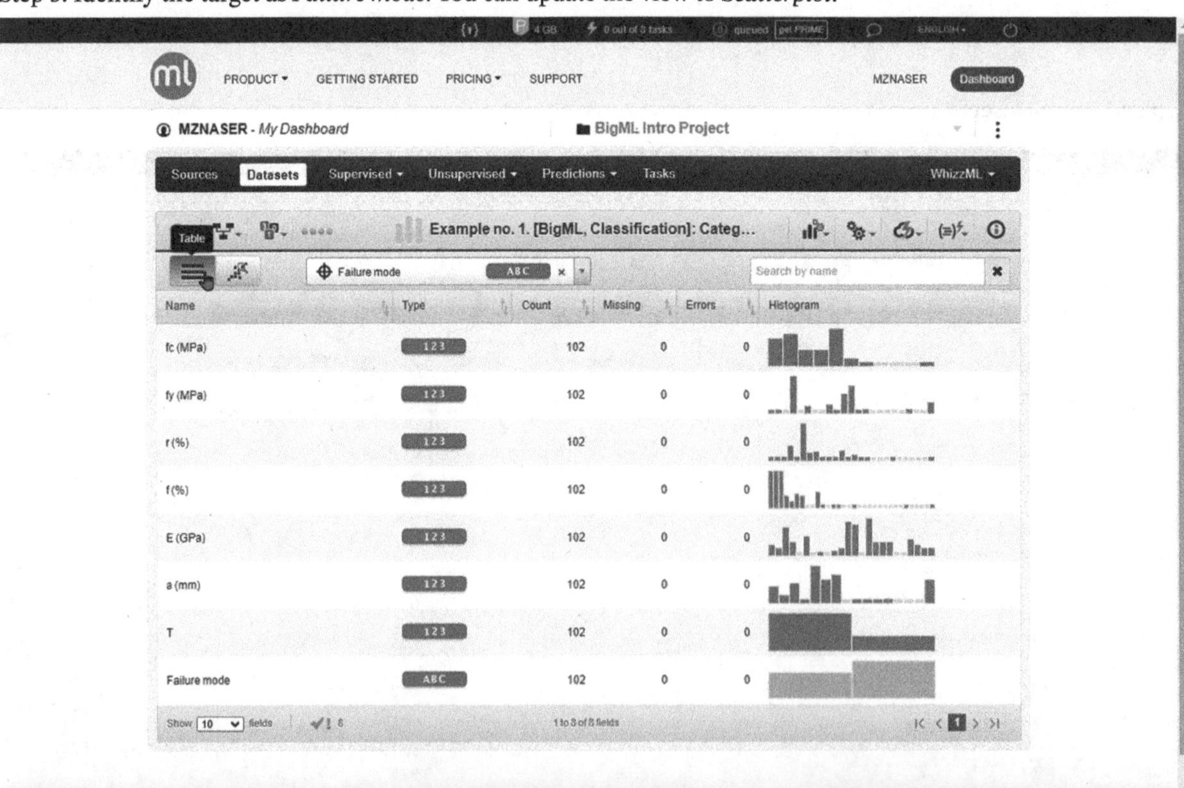

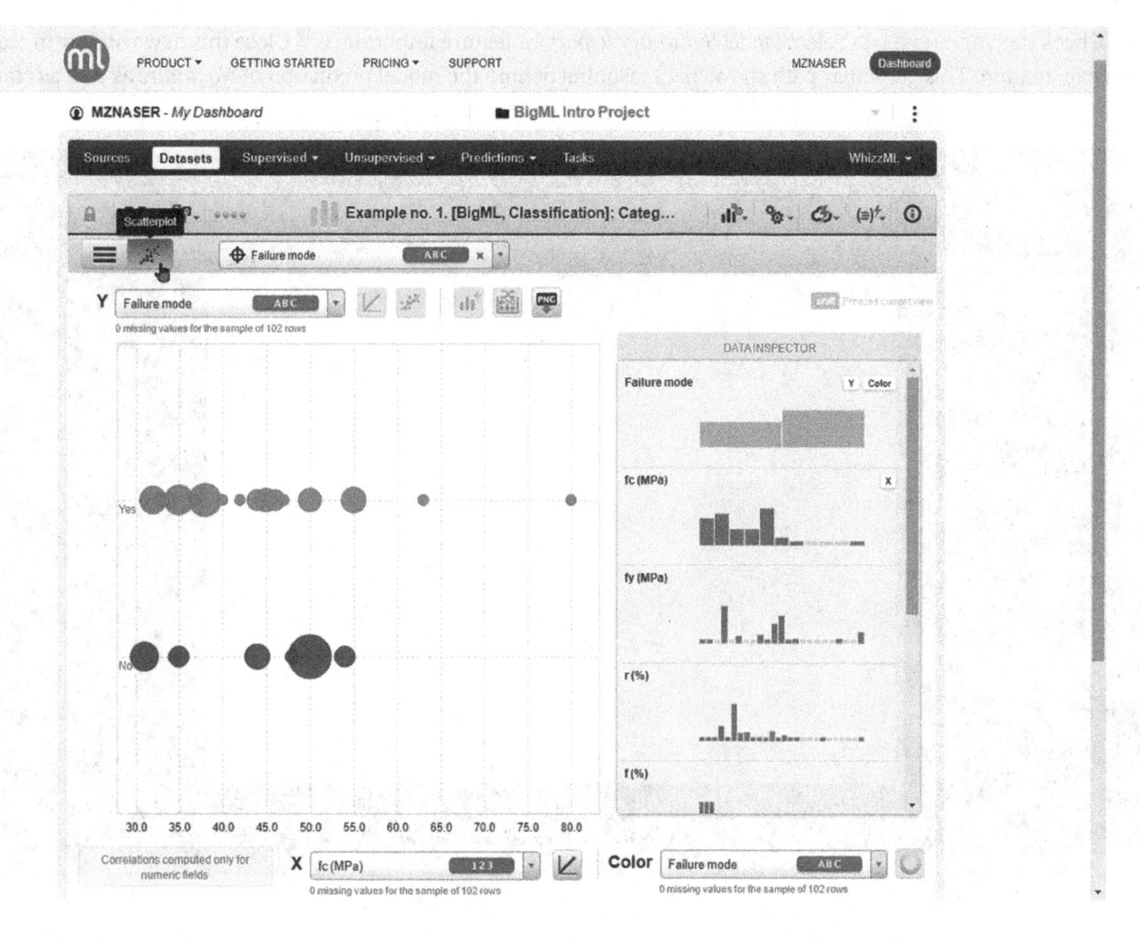

Step 4: Select *Model*

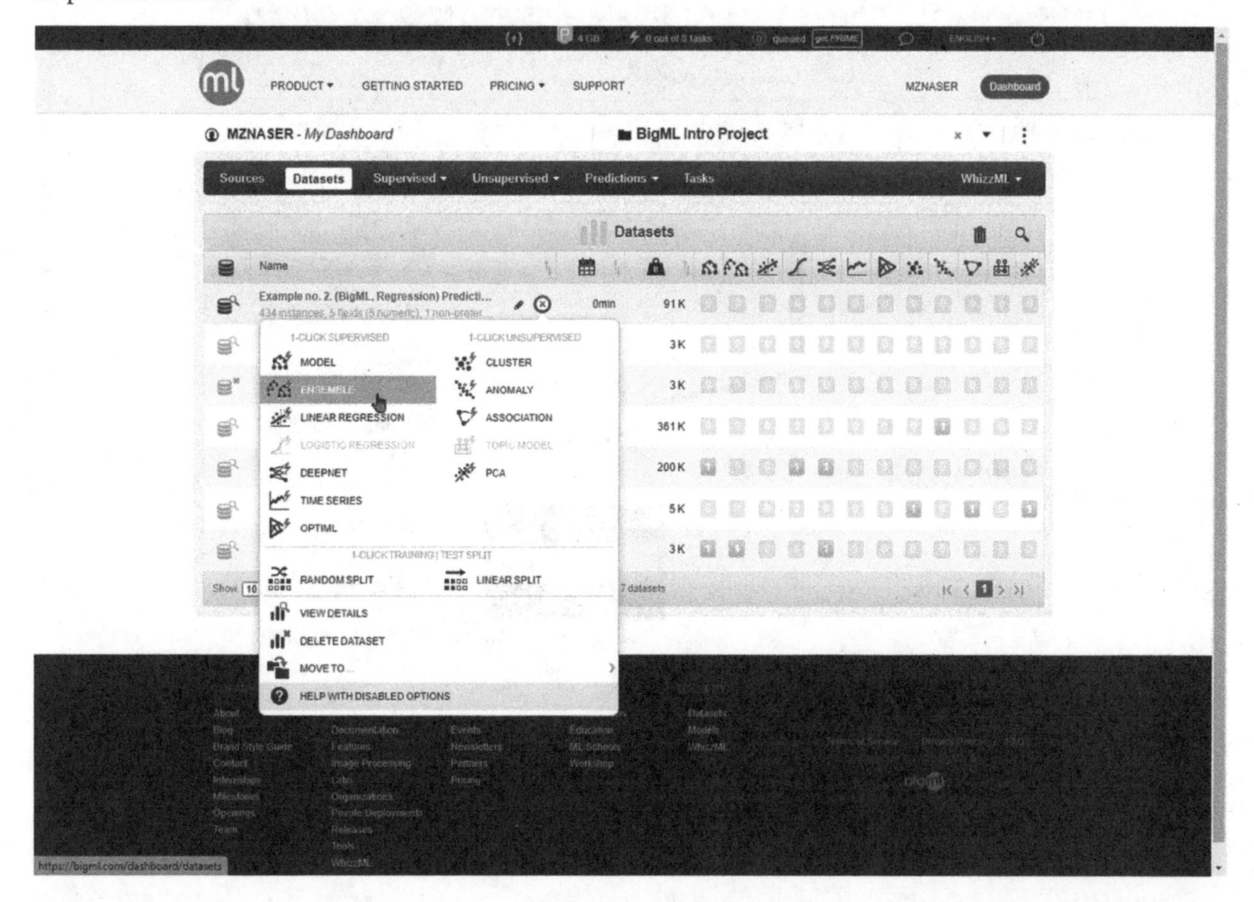

Step 5: Check the *Supervised* tab. Select *Model Summary Report* for feature importance.[18] Close this new window to see the prediction explanation. This particular path shows the reasoning behind the model prediction of *No failure* as well as confidence.

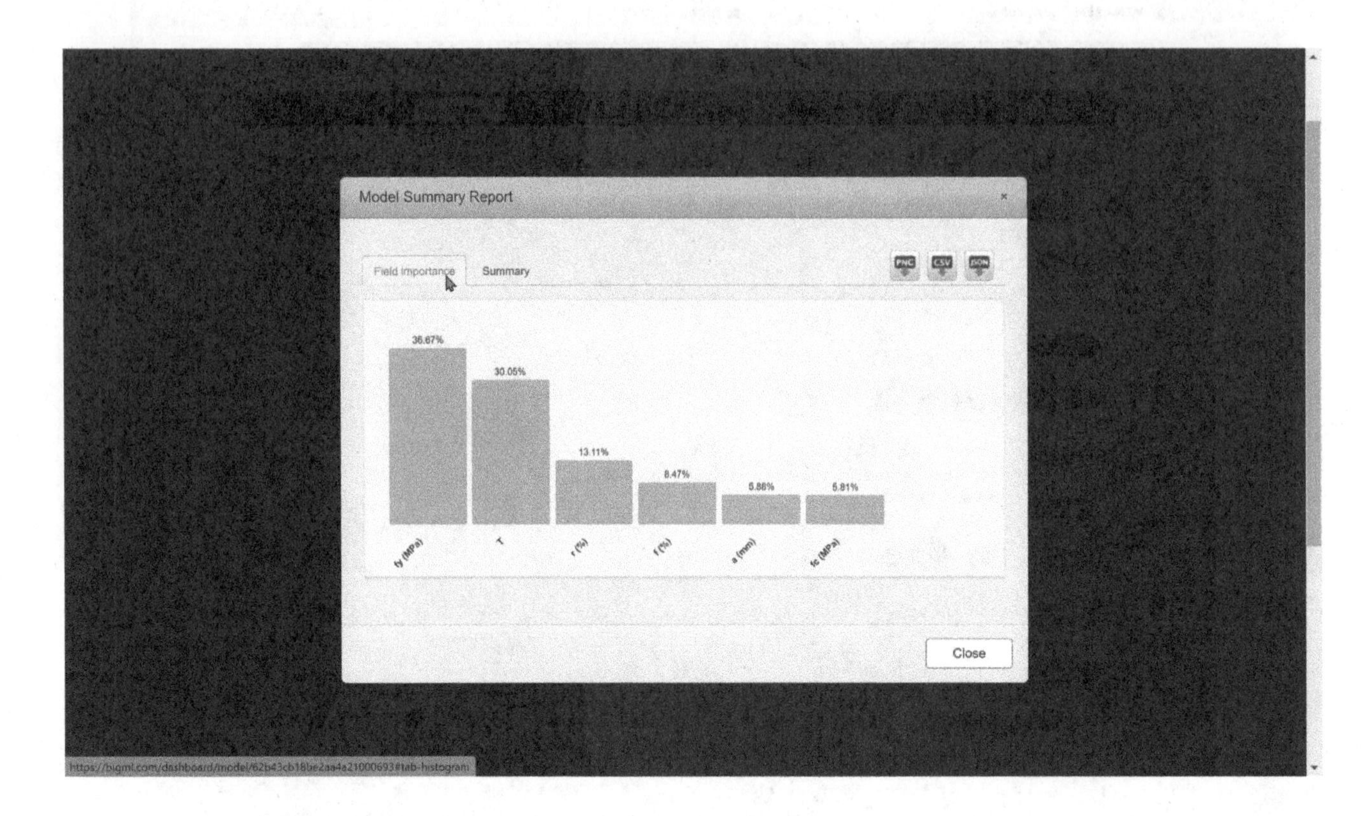

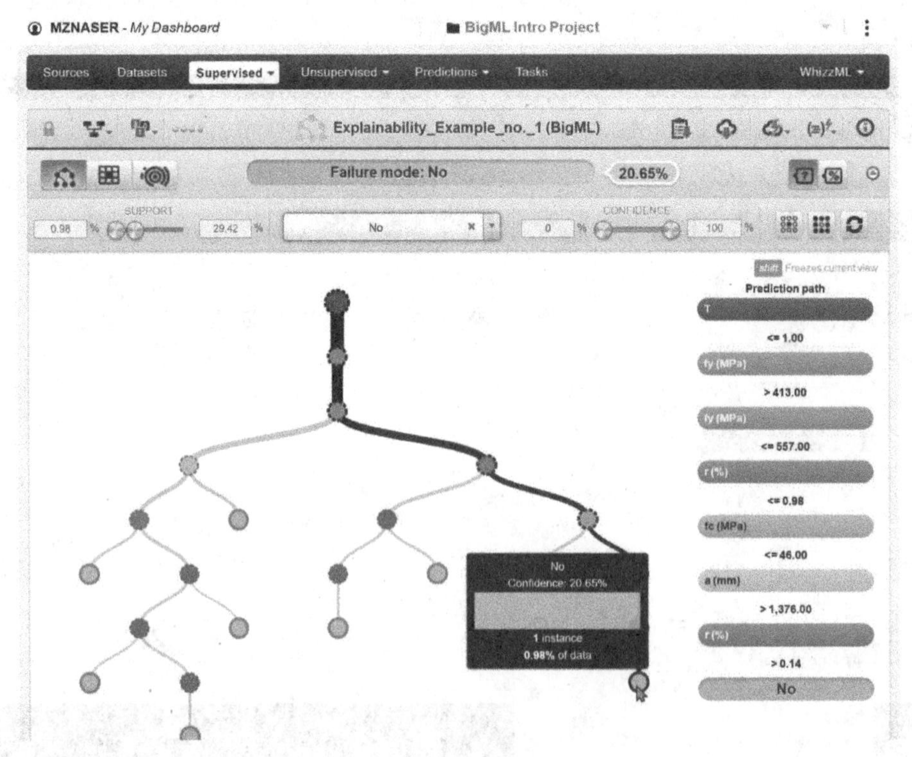

18 I will not go into details on feature importance and prediction explanation now. I have reserved the next chapter to discuss explainability. For now, you can think of these two concepts as means for us to see *how* and *why* the model arrived at its predictions.

Step 6: Go for *Predict Question by Question*. This option allows you to see how predictions change per the addition/manipulation of each feature.

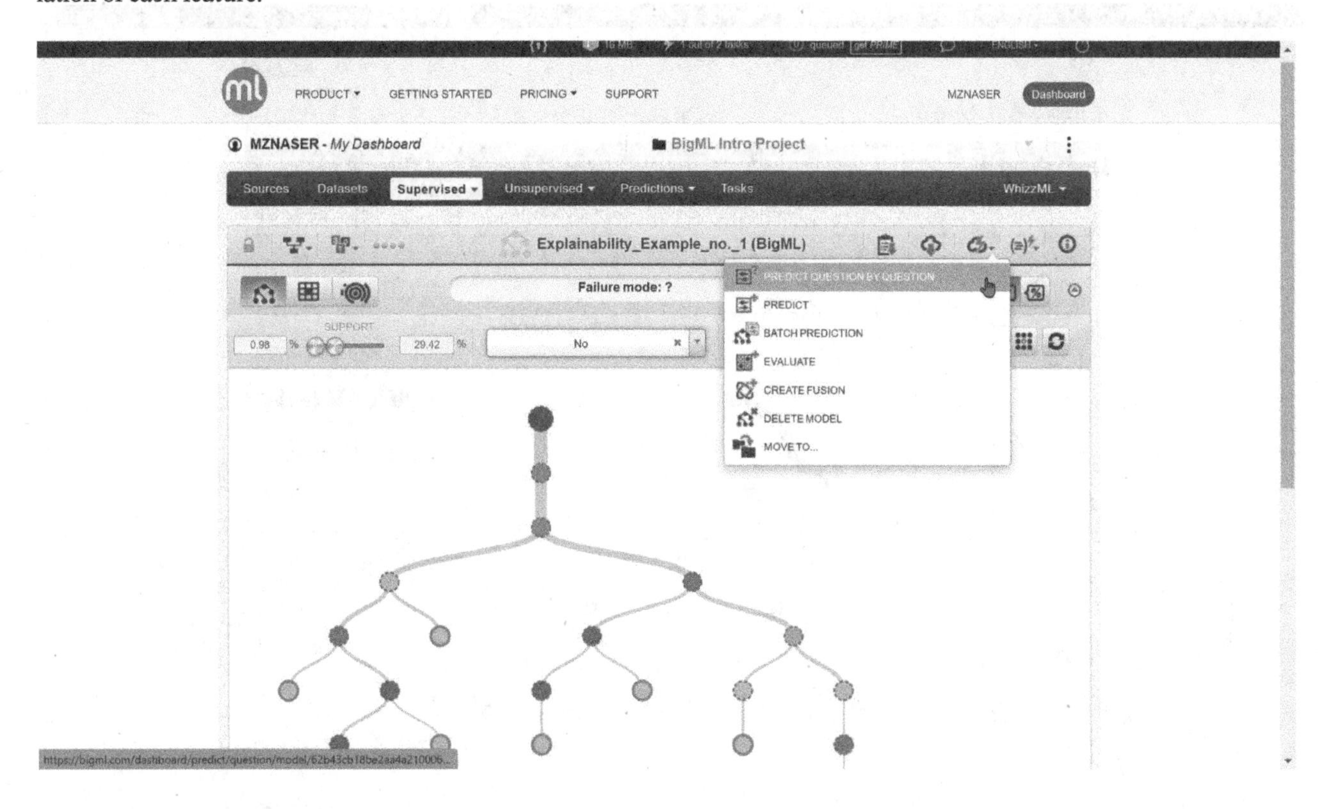

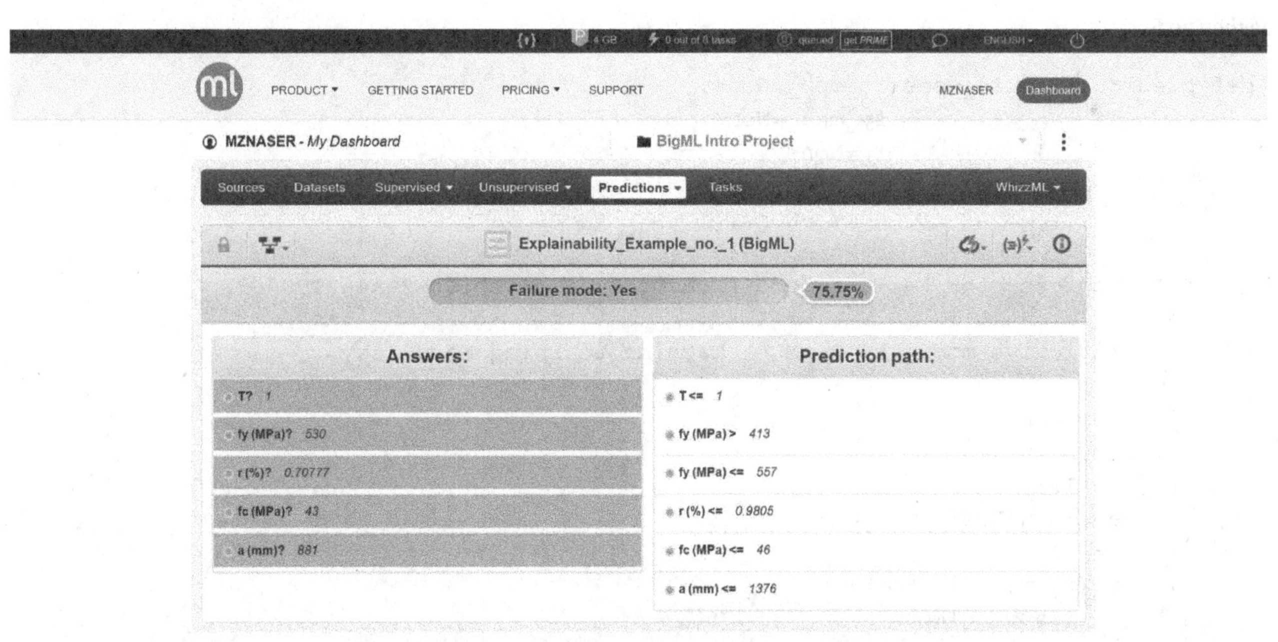

Step 7: Select *Download Actionable Model* to download this model into a programming language of your choosing.

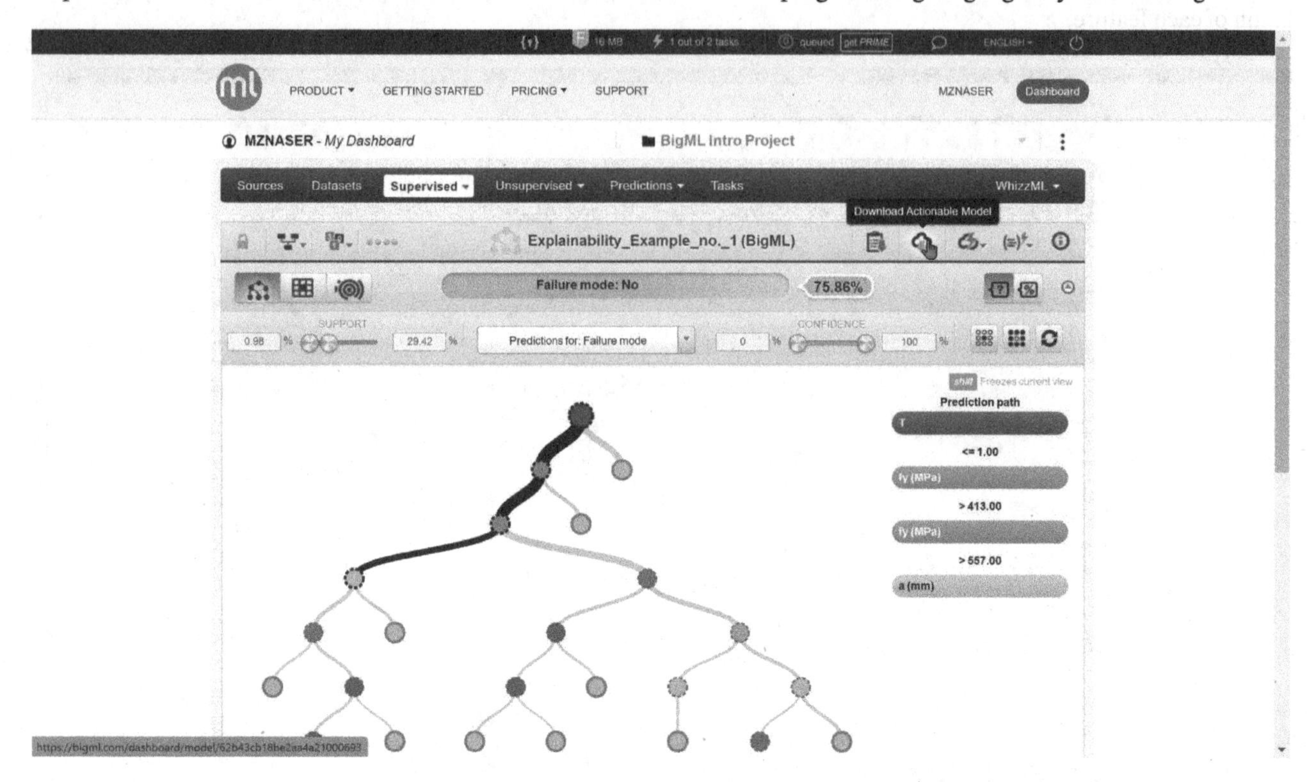

Python code:

```
def predict_failure_mode(fc_mpa_=None,
                         fy_mpa_=None,
                         r_=None,
                         f_=None,
                         a_mm_=None,
                         t=None):
    """ Predictor for Failure mode from model/62b43cb18be2aa4a21000693

        Predictive model by BigML - Machine Learning Made Easy
    """
    if (t is None):
        return u'Yes'
    if (t > 1):
        return u'Yes'
    if (t <= 1):
        if (fy_mpa_ is None):
            return u'No'
        if (fy_mpa_ > 413):
            if (fy_mpa_ > 557):
                if (a_mm_ is None):
                    return u'No'
                if (a_mm_ > 656):
                    if (r_ is None):
                        return u'No'
```

```
                if (r_ > 0.9315):
                    return u'No'
                if (r_ <= 0.9315):
                    if (f_ is None):
                        return u'No'
                    if (f_ > 0.3):
                        if (f_ > 0.75):
                            return u'No'
                        if (f_ <= 0.75):
                            if (f_ > 0.5):
                                return u'No'
                            if (f_ <= 0.5):
                                return u'Yes'
                    if (f_ <= 0.3):
                        return u'No'
            if (a_mm_ <= 656):
                return u'No'
        if (fy_mpa_ <= 557):
            if (r_ is None):
                return u'No'
            if (r_ > 0.9805):
                if (f_ is None):
                    return u'No'
                if (f_ > 0.3025):
                    if (f_ > 0.46):
                        return u'No'
                    if (f_ <= 0.46):
                        return u'Yes'
                if (f_ <= 0.3025):
                    return u'No'
            if (r_ <= 0.9805):
                if (fc_mpa_ is None):
                    return u'Yes'
                if (fc_mpa_ > 46):
                    if (a_mm_ is None):
                        return u'No'
                    if (a_mm_ > 431):
                        return u'No'
                    if (a_mm_ <= 431):
                        return u'Yes'
                if (fc_mpa_ <= 46):
                    if (a_mm_ is None):
                        return u'Yes'
                    if (a_mm_ > 1376):
                        if (r_ > 0.142):
                            return u'No'
                        if (r_ <= 0.142):
                            return u'Yes'
                    if (a_mm_ <= 1376):
                        return u'Yes'
    if (fy_mpa_ <= 413):
        return u'Yes'
```

R code:

```
# Predictor for Failure mode from model/62b43cb18be2aa4a21000693
#
# Predictive model by BigML - Machine Learning Made Easy
#
predictFailureMode <- function(fc_mpa_=NA,
fy_mpa_=NA,
r_=NA,
f_=NA,
a_mm_=NA,
t=NA){
if (is.na(t)){
return("Yes")
}
if (t > 1) {
return("Yes")
}
if (t <= 1) {
if (is.na(fy_mpa_)){
return("No")
}
if (fy_mpa_ > 413) {
if (fy_mpa_ > 557) {
if (is.na(a_mm_)){
return("No")
}
if (a_mm_ > 656) {
if (is.na(r_)){
return("No")
}
if (r_ > 0.9315) {
return("No")
}
if (r_ <= 0.9315) {
if (is.na(f_)){
return("No")
}
if (f_ > 0.3) {
if (f_ > 0.75) {
return("No")
}
if (f_ <= 0.75) {
if (f_ > 0.5) {
return("No")
}
if (f_ <= 0.5) {
return("Yes")
}
}
}
if (f_ <= 0.3) {
return("No")
}
```

```
}
}
if (a_mm_ <= 656) {
return("No")
}
}
if (fy_mpa_ <= 557) {
if (is.na(r_)){
return("No")
}
if (r_ > 0.9805) {
if (is.na(f_)){
return("No")
}
if (f_ > 0.3025) {
if (f_ > 0.46) {
return("No")
}
if (f_ <= 0.46) {
return("Yes")
}
}
if (f_ <= 0.3025) {
return("No")
}
}
if (r_ <= 0.9805) {
if (is.na(fc_mpa_)){
return("Yes")
}
if (fc_mpa_ > 46) {
if (is.na(a_mm_)){
return("No")
}
if (a_mm_ > 431) {
return("No")
}
if (a_mm_ <= 431) {
return("Yes")
}
}
if (fc_mpa_ <= 46) {
if (is.na(a_mm_)){
return("Yes")
}
if (a_mm_ > 1376) {
if (r_ > 0.142) {
return("No")
}
if (r_ <= 0.142) {
return("Yes")
}
}
```

```
if (a_mm_ <= 1376) {
return("Yes")
}
}
}
}
}
if (fy_mpa_ <= 413) {
return("Yes")
}
}
return(NA)
}
```

Example no. 2 [*BigML*, Regression]: Predicting moment capacity of FRP-strengthened RC beams.
This example continues the above on FRP-strengthened RC beams but aims to predict the moment capacity of such beams as opposed to the observed failure mode.

The first few steps are similar to what we did in the first example. I will skip these for now.[19] Let us now build an ensemble.[20]

Step 1: After uploading the dataset and navigating to the *Datasets* tab. Select *Ensemble* under the *Supervised* tab.

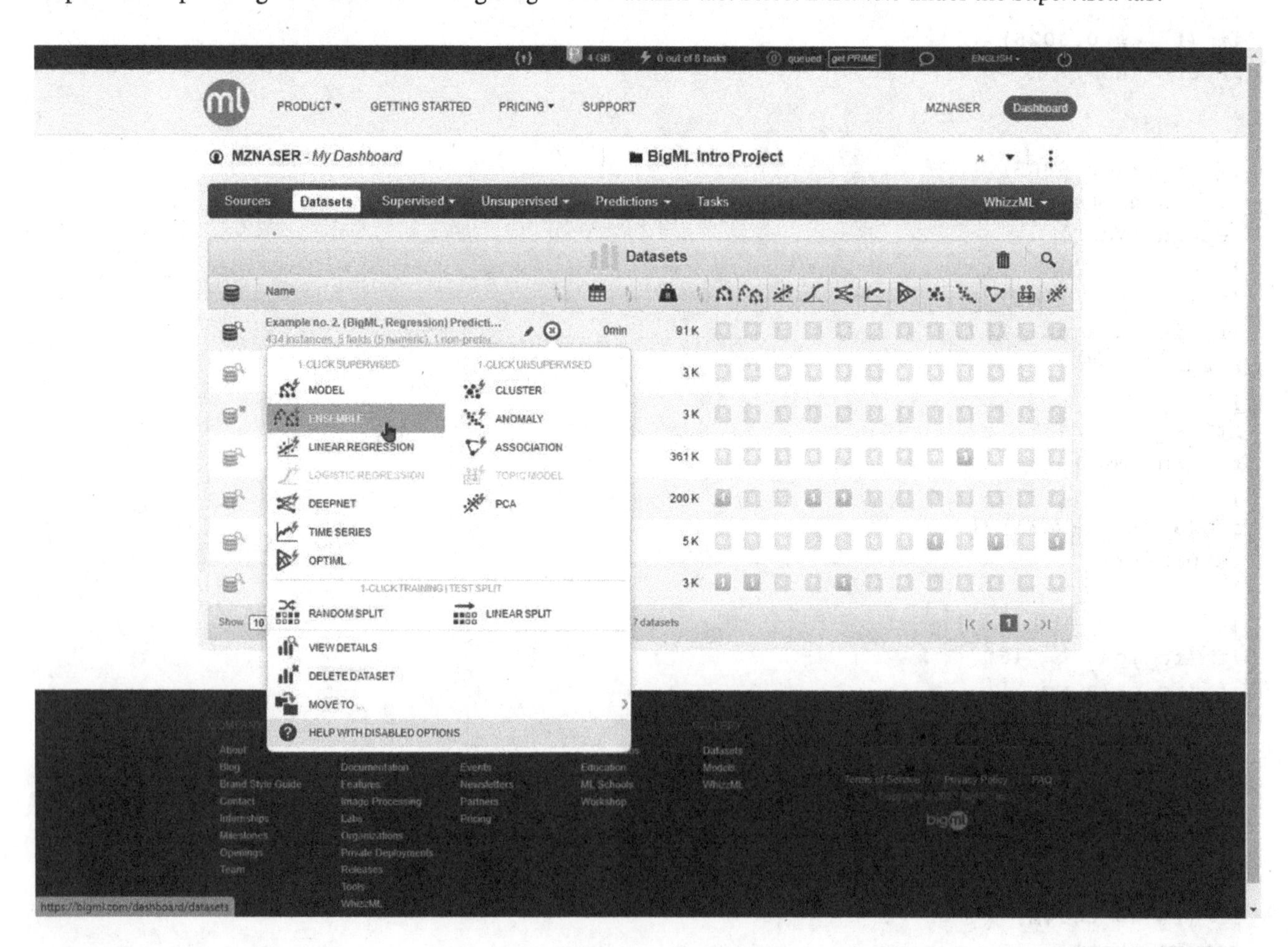

19 I trust you!
20 An **ensemble** is a combination of algorithms – we will discuss these at a later chapter.

Step 2: Navigate to the *Supervised* tab and select *Ensemble Summary Report*. This will allow you to see the importance of mode features.

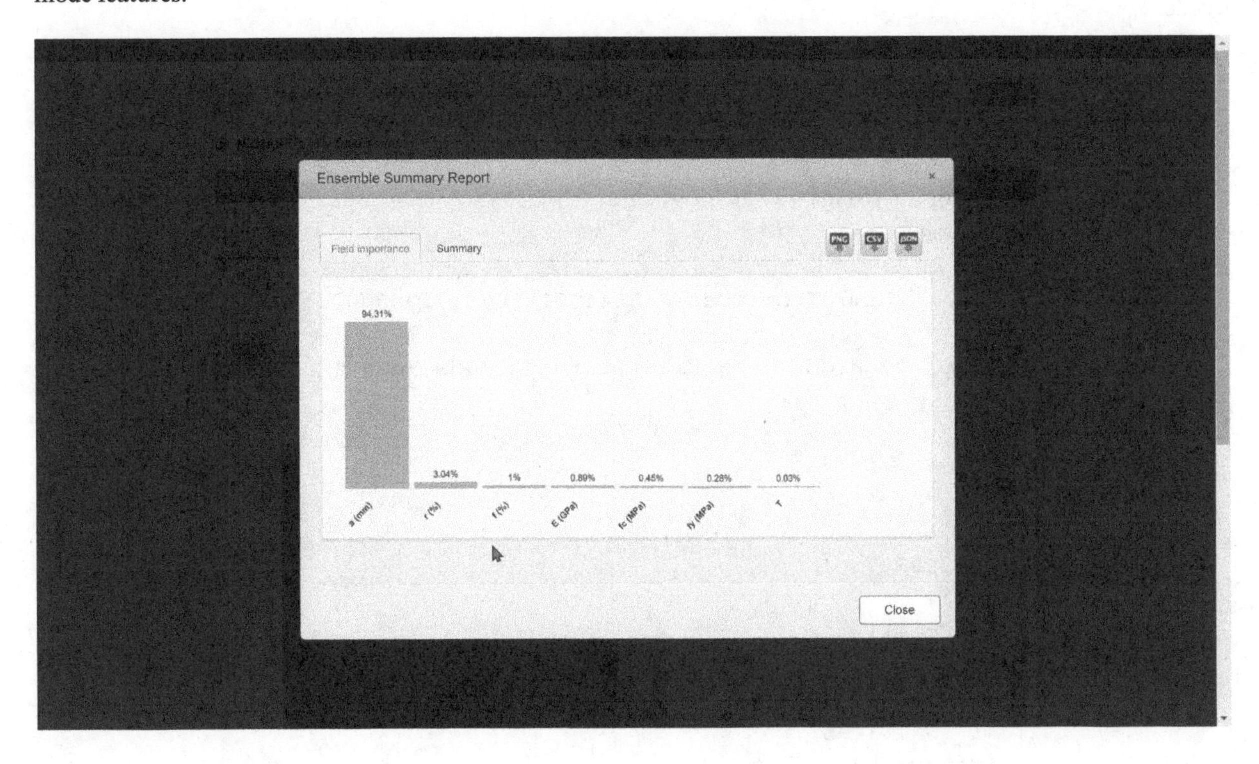

Step 3: Check out the created ensemble tree and the reasoning behind a specific prediction.

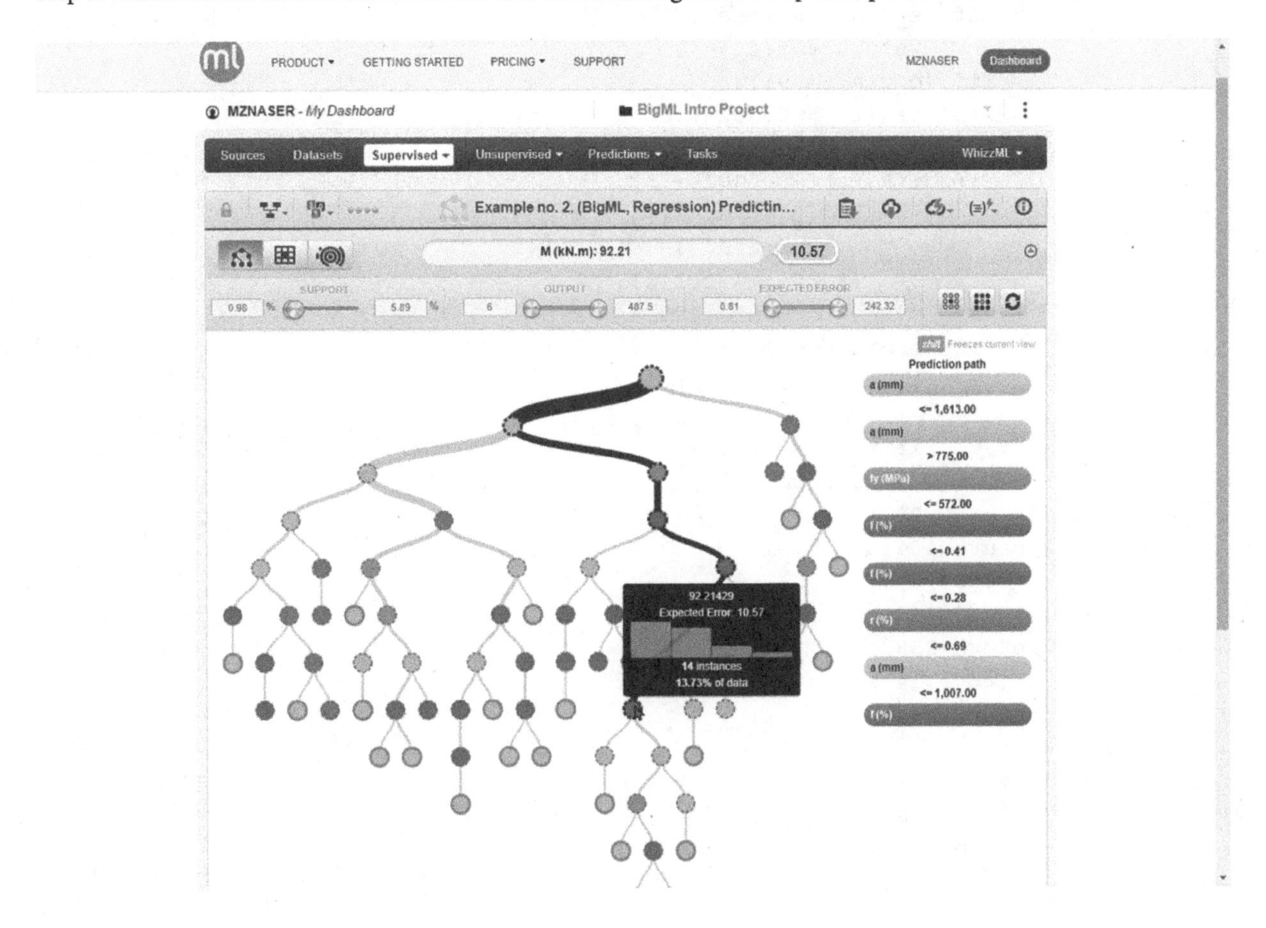

Step 7: Select *Download Actionable Model* to download this model into a programming language of your choosing.

Python code:

```
def predict_m_kn_m_(fc_mpa_=None,
                    fy_mpa_=None,
                    r_=None,
                    f_=None,
                    e_gpa_=None,
                    a_mm_=None,
                    t=None):
    """ Predictor for M (kN.m) from model/62b445018f679a7d9a0008cc

        Predictive model by BigML - Machine Learning Made Easy
    """
    if (a_mm_ is None):
        return 119.92157
    if (a_mm_ > 1613):
        if (r_ is None):
            return 457.92857
        if (r_ > 0.667):
            if (f_ is None):
                return 481.75
            if (f_ > 0.063):
                if (r_ > 1.3495):
                    return 585
                if (r_ <= 1.3495):
                    if (e_gpa_ is None):
                        return 487.44444
                    if (e_gpa_ > 93):
                        if (f_ > 0.3025):
                            return 514
                        if (f_ <= 0.3025):
                            if (f_ > 0.163):
                                return 478.5
                            if (f_ <= 0.163):
                                return 487.5
                    if (e_gpa_ <= 93):
                        if (a_mm_ > 1828):
                            return 468
                        if (a_mm_ <= 1828):
                            return 479.5
            if (f_ <= 0.063):
                return 404.5
        if (r_ <= 0.667):
            if (r_ > 0.142):
                return 338
            if (r_ <= 0.142):
                return 292
    if (a_mm_ <= 1613):
        if (a_mm_ > 790):
            if (fy_mpa_ is None):
                return 96.21622
```

```
    if (fy_mpa_ > 572):
        return 225
    if (fy_mpa_ <= 572):
        if (f_ is None):
            return 92.63889
        if (f_ > 0.407):
            if (a_mm_ > 919):
                if (f_ > 0.825):
                    if (f_ > 1.95):
                        return 98
                    if (f_ <= 1.95):
                        if (f_ > 1.15):
                            return 76
                        if (f_ <= 1.15):
                            return 95
                if (f_ <= 0.825):
                    if (f_ > 0.575):
                        return 65
                    if (f_ <= 0.575):
                        return 76
            if (a_mm_ <= 919):
                if (f_ > 1.096):
                    if (f_ > 2.372):
                        return 61
                    if (f_ <= 2.372):
                        return 53
                if (f_ <= 1.096):
                    return 40
        if (f_ <= 0.407):
            if (r_ is None):
                return 100.66667
            if (r_ > 1.672):
                if (r_ > 2.232):
                    return 155
                if (r_ <= 2.232):
                    return 134
            if (r_ <= 1.672):
                if (f_ > 0.283):
                    if (r_ > 0.6265):
                        return 136
                    if (r_ <= 0.6265):
                        return 125
                if (f_ <= 0.283):
                    if (a_mm_ > 812):
                        if (fc_mpa_ is None):
                            return 99.85714
                        if (fc_mpa_ > 40):
                            if (fc_mpa_ > 43):
                                if (f_ > 0.045):
                                    if (t is None):
                                        return 98.25
                                    if (t > 1):
                                        if (f_ > 0.075):
```

```
                        if (a_mm_ > 1000):
                            if (a_mm_ > 1250):
                                return 106
                            if (a_mm_ <= 1250):
                                return 105
                        if (a_mm_ <= 1000):
                            return 114
                    if (f_ <= 0.075):
                        return 101
                if (t <= 1):
                    if (r_ > 0.4935):
                        if (r_ > 0.806):
                            if (r_ > 1.1945):
                                return 98
                            if (r_ <= 1.1945):
                                return 100
                        if (r_ <= 0.806):
                            return 89
                    if (r_ <= 0.4935):
                        return 73
            if (f_ <= 0.045):
                return 80
        if (fc_mpa_ <= 43):
            return 74
    if (fc_mpa_ <= 40):
        if (r_ > 0.501):
            if (a_mm_ > 1012):
                if (a_mm_ > 1300):
                    return 125
                if (a_mm_ <= 1300):
                    return 120
            if (a_mm_ <= 1012):
                return 114
        if (r_ <= 0.501):
            return 99
if (a_mm_ <= 812):
    if (r_ > 0.884):
        return 112
    if (r_ <= 0.884):
        if (r_ > 0.519):
            return 43
        if (r_ <= 0.519):
            if (e_gpa_ is None):
                return 87.85714
            if (e_gpa_ > 117):
                return 91
            if (e_gpa_ <= 117):
                if (f_ > 0.168):
                    if (f_ > 0.238):
                        return 85
                    if (f_ <= 0.238):
                        return 83.5
                if (f_ <= 0.168):
```

```
                                        if (e_gpa_ > 40.5):
                                            return 94
                                        if (e_gpa_ <= 40.5):
                                            return 87
if (a_mm_ <= 790):
    if (a_mm_ > 531):
        if (r_ is None):
            return 57.5625
        if (r_ > 0.582):
            if (a_mm_ > 687):
                if (t is None):
                    return 66.66667
                if (t > 1):
                    return 47
                if (t <= 1):
                    if (f_ is None):
                        return 68.45455
                    if (f_ > 0.144):
                        if (f_ > 0.4):
                            if (f_ > 0.75):
                                return 65
                            if (f_ <= 0.75):
                                return 76.5
                        if (f_ <= 0.4):
                            if (a_mm_ > 725):
                                return 57
                            if (a_mm_ <= 725):
                                return 67
                    if (f_ <= 0.144):
                        if (f_ > 0.063):
                            if (f_ > 0.1045):
                                return 81
                            if (f_ <= 0.1045):
                                return 82
                        if (f_ <= 0.063):
                            return 59
            if (a_mm_ <= 687):
                return 105
        if (r_ <= 0.582):
            if (fc_mpa_ is None):
                return 46.22222
            if (fc_mpa_ > 52):
                return 36.5
            if (fc_mpa_ <= 52):
                if (a_mm_ > 656):
                    if (f_ is None):
                        return 46.5
                    if (f_ > 0.3):
                        if (f_ > 0.75):
                            return 45
                        if (f_ <= 0.75):
                            if (f_ > 0.5):
                                return 50.5
```

```
                        if (f_ <= 0.5):
                            return 46.5
                    if (f_ <= 0.3):
                        return 40
                if (a_mm_ <= 656):
                    if (f_ is None):
                        return 55.25
                    if (f_ > 0.1985):
                        if (f_ > 0.33):
                            return 65
                        if (f_ <= 0.33):
                            return 61
                    if (f_ <= 0.1985):
                        if (f_ > 0.1315):
                            return 43
                        if (f_ <= 0.1315):
                            return 52
    if (a_mm_ <= 531):
        if (a_mm_ > 478):
            if (f_ is None):
                return 38
            if (f_ > 0.119):
                if (r_ is None):
                    return 42.33333
                if (r_ > 0.7355):
                    return 47
                if (r_ <= 0.7355):
                    if (r_ > 0.4635):
                        return 41
                    if (r_ <= 0.4635):
                        return 39
            if (f_ <= 0.119):
                return 25
        if (a_mm_ <= 478):
            if (a_mm_ > 378):
                if (f_ is None):
                    return 11
                if (f_ > 0.275):
                    if (f_ > 0.8345):
                        return 13
                    if (f_ <= 0.8345):
                        return 14
                if (f_ <= 0.275):
                    if (f_ > 0.1065):
                        return 8
                    if (f_ <= 0.1065):
                        return 6
            if (a_mm_ <= 378):
                if (f_ is None):
                    return 21.2
                if (f_ > 0.0545):
                    if (r_ is None):
                        return 23.375
```

```
                            if (r_ > 0.2585):
                                if (f_ > 0.1085):
                                    if (r_ > 0.699):
                                        return 25
                                    if (r_ <= 0.699):
                                        return 26
                                if (f_ <= 0.1085):
                                    if (r_ > 0.406):
                                        return 24
                                    if (r_ <= 0.406):
                                        return 25
                            if (r_ <= 0.2585):
                                if (f_ > 0.0985):
                                    if (f_ > 0.164):
                                        return 22
                                    if (f_ <= 0.164):
                                        return 15
                                if (f_ <= 0.0985):
                                    return 24
                        if (f_ <= 0.0545):
                            if (r_ is None):
                                return 12.5
                            if (r_ > 0.406):
                                return 13
                            if (r_ <= 0.406):
                                return 12
```

R code:

```
# Predictor for M (kN.m) from model/62b445018f679a7d9a0008cc
#
#       Predictive model by BigML - Machine Learning Made Easy
#
predictMKnM <- function(fc_mpa_=NA,
                        fy_mpa_=NA,
                        r_=NA,
                        f_=NA,
                        e_gpa_=NA,
                        a_mm_=NA,
                        t=NA){
    if (is.na(a_mm_)){
        return(119.92157)
    }
    if (a_mm_ > 1613) {
        if (is.na(r_)){
            return(457.92857)
        }
        if (r_ > 0.667) {
            if (is.na(f_)){
                return(481.75)
            }
            if (f_ > 0.063) {
                if (r_ > 1.3495) {
```

```
                    return(585)
                }
                if (r_ <= 1.3495) {
                    if (is.na(e_gpa_)){
                        return(487.44444)
                    }
                    if (e_gpa_ > 93) {
                        if (f_ > 0.3025) {
                            return(514)
                        }
                        if (f_ <= 0.3025) {
                            if (f_ > 0.163) {
                                return(478.5)
                            }
                            if (f_ <= 0.163) {
                                return(487.5)
                            }
                        }
                    }
                    if (e_gpa_ <= 93) {
                        if (a_mm_ > 1828) {
                            return(468)
                        }
                        if (a_mm_ <= 1828) {
                            return(479.5)
                        }
                    }
                }
            }
            if (f_ <= 0.063) {
                return(404.5)
            }
        }
        if (r_ <= 0.667) {
            if (r_ > 0.142) {
                return(338)
            }
            if (r_ <= 0.142) {
                return(292)
            }
        }
    }
    if (a_mm_ <= 1613) {
        if (a_mm_ > 790) {
            if (is.na(fy_mpa_)){
                return(96.21622)
            }
            if (fy_mpa_ > 572) {
                return(225)
            }
            if (fy_mpa_ <= 572) {
                if (is.na(f_)){
                    return(92.63889)
```

```
        }
        if (f_ > 0.407) {
            if (a_mm_ > 919) {
                if (f_ > 0.825) {
                    if (f_ > 1.95) {
                        return(98)
                    }
                    if (f_ <= 1.95) {
                        if (f_ > 1.15) {
                            return(76)
                        }
                        if (f_ <= 1.15) {
                            return(95)
                        }
                    }
                }
                if (f_ <= 0.825) {
                    if (f_ > 0.575) {
                        return(65)
                    }
                    if (f_ <= 0.575) {
                        return(76)
                    }
                }
            }
            if (a_mm_ <= 919) {
                if (f_ > 1.096) {
                    if (f_ > 2.372) {
                        return(61)
                    }
                    if (f_ <= 2.372) {
                        return(53)
                    }
                }
                if (f_ <= 1.096) {
                    return(40)
                }
            }
        }
        if (f_ <= 0.407) {
            if (is.na(r_)){
                return(100.66667)
            }
            if (r_ > 1.672) {
                if (r_ > 2.232) {
                    return(155)
                }
                if (r_ <= 2.232) {
                    return(134)
                }
            }
            if (r_ <= 1.672) {
                if (f_ > 0.283) {
```

```
        if (r_ > 0.6265) {
            return(136)
        }
        if (r_ <= 0.6265) {
            return(125)
        }
    }
    if (f_ <= 0.283) {
        if (a_mm_ > 812) {
            if (is.na(fc_mpa_)){
                return(99.85714)
            }
            if (fc_mpa_ > 40) {
                if (fc_mpa_ > 43) {
                    if (f_ > 0.045) {
                        if (is.na(t)){
                            return(98.25)
                        }
                        if (t > 1) {
                            if (f_ > 0.075) {
                                if (a_mm_ > 1000) {
                                    if (a_mm_ > 1250) {
                                        return(106)
                                    }
                                    if (a_mm_ <= 1250) {
                                        return(105)
                                    }
                                }
                                if (a_mm_ <= 1000) {
                                    return(114)
                                }
                            }
                            if (f_ <= 0.075) {
                                return(101)
                            }
                        }
                        if (t <= 1) {
                            if (r_ > 0.4935) {
                                if (r_ > 0.806) {
                                    if (r_ > 1.1945) {
                                        return(98)
                                    }
                                    if (r_ <= 1.1945) {
                                        return(100)
                                    }
                                }
                                if (r_ <= 0.806) {
                                    return(89)
                                }
                            }
                            if (r_ <= 0.4935) {
                                return(73)
                            }
```

```
                }
            }
            if (f_ <= 0.045) {
                return(80)
            }
        }
        if (fc_mpa_ <= 43) {
            return(74)
        }
    }
    if (fc_mpa_ <= 40) {
        if (r_ > 0.501) {
            if (a_mm_ > 1012) {
                if (a_mm_ > 1300) {
                    return(125)
                }
                if (a_mm_ <= 1300) {
                    return(120)
                }
            }
            if (a_mm_ <= 1012) {
                return(114)
            }
        }
        if (r_ <= 0.501) {
            return(99)
        }
    }
}
if (a_mm_ <= 812) {
    if (r_ > 0.884) {
        return(112)
    }
    if (r_ <= 0.884) {
        if (r_ > 0.519) {
            return(43)
        }
        if (r_ <= 0.519) {
            if (is.na(e_gpa_)){
                return(87.85714)
            }
            if (e_gpa_ > 117) {
                return(91)
            }
            if (e_gpa_ <= 117) {
                if (f_ > 0.168) {
                    if (f_ > 0.238) {
                        return(85)
                    }
                    if (f_ <= 0.238) {
                        return(83.5)
                    }
                }
```

```
                                        if (f_ <= 0.168) {
                                            if (e_gpa_ > 40.5) {
                                                return(94)
                                            }
                                            if (e_gpa_ <= 40.5) {
                                                return(87)
                                            }
                                        }
                                    }
                                }
                            }
                        }
                    }
                }
            }
        }
    }
    if (a_mm_ <= 790) {
        if (a_mm_ > 531) {
            if (is.na(r_)){
                return(57.5625)
            }
            if (r_ > 0.582) {
                if (a_mm_ > 687) {
                    if (is.na(t)){
                        return(66.66667)
                    }
                    if (t > 1) {
                        return(47)
                    }
                    if (t <= 1) {
                        if (is.na(f_)){
                            return(68.45455)
                        }
                        if (f_ > 0.144) {
                            if (f_ > 0.4) {
                                if (f_ > 0.75) {
                                    return(65)
                                }
                                if (f_ <= 0.75) {
                                    return(76.5)
                                }
                            }
                            if (f_ <= 0.4) {
                                if (a_mm_ > 725) {
                                    return(57)
                                }
                                if (a_mm_ <= 725) {
                                    return(67)
                                }
                            }
                        }
                        if (f_ <= 0.144) {
```

```
                    if (f_ > 0.063) {
                        if (f_ > 0.1045) {
                            return(81)
                        }
                        if (f_ <= 0.1045) {
                            return(82)
                        }
                    }
                    if (f_ <= 0.063) {
                        return(59)
                    }
                }
            }
        }
        if (a_mm_ <= 687) {
            return(105)
        }
    }
    if (r_ <= 0.582) {
        if (is.na(fc_mpa_)){
            return(46.22222)
        }
        if (fc_mpa_ > 52) {
            return(36.5)
        }
        if (fc_mpa_ <= 52) {
            if (a_mm_ > 656) {
                if (is.na(f_)){
                    return(46.5)
                }
                if (f_ > 0.3) {
                    if (f_ > 0.75) {
                        return(45)
                    }
                    if (f_ <= 0.75) {
                        if (f_ > 0.5) {
                            return(50.5)
                        }
                        if (f_ <= 0.5) {
                            return(46.5)
                        }
                    }
                }
                if (f_ <= 0.3) {
                    return(40)
                }
            }
            if (a_mm_ <= 656) {
                if (is.na(f_)){
                    return(55.25)
                }
                if (f_ > 0.1985) {
                    if (f_ > 0.33) {
```

```
                        return(65)
                    }
                    if (f_ <= 0.33) {
                        return(61)
                    }
                }
                if (f_ <= 0.1985) {
                    if (f_ > 0.1315) {
                        return(43)
                    }
                    if (f_ <= 0.1315) {
                        return(52)
                    }
                }
            }
        }
    }
}
if (a_mm_ <= 531) {
    if (a_mm_ > 478) {
        if (is.na(f_)){
            return(38)
        }
        if (f_ > 0.119) {
            if (is.na(r_)){
                return(42.33333)
            }
            if (r_ > 0.7355) {
                return(47)
            }
            if (r_ <= 0.7355) {
                if (r_ > 0.4635) {
                    return(41)
                }
                if (r_ <= 0.4635) {
                    return(39)
                }
            }
        }
        if (f_ <= 0.119) {
            return(25)
        }
    }
    if (a_mm_ <= 478) {
        if (a_mm_ > 378) {
            if (is.na(f_)){
                return(11)
            }
            if (f_ > 0.275) {
                if (f_ > 0.8345) {
                    return(13)
                }
                if (f_ <= 0.8345) {
```

```
                    return(14)
                }
            }
            if (f_ <= 0.275) {
                if (f_ > 0.1065) {
                    return(8)
                }
                if (f_ <= 0.1065) {
                    return(6)
                }
            }
        }
        if (a_mm_ <= 378) {
            if (is.na(f_)){
                return(21.2)
            }
            if (f_ > 0.0545) {
                if (is.na(r_)){
                    return(23.375)
                }
                if (r_ > 0.2585) {
                    if (f_ > 0.1085) {
                        if (r_ > 0.699) {
                            return(25)
                        }
                        if (r_ <= 0.699) {
                            return(26)
                        }
                    }
                    if (f_ <= 0.1085) {
                        if (r_ > 0.406) {
                            return(24)
                        }
                        if (r_ <= 0.406) {
                            return(25)
                        }
                    }
                }
                if (r_ <= 0.2585) {
                    if (f_ > 0.0985) {
                        if (f_ > 0.164) {
                            return(22)
                        }
                        if (f_ <= 0.164) {
                            return(15)
                        }
                    }
                    if (f_ <= 0.0985) {
                        return(24)
                    }
                }
            }
            if (f_ <= 0.0545) {
```

```
                        if (is.na(r_)){
                            return(12.5)
                        }
                        if (r_ > 0.406) {
                            return(13)
                        }
                        if (r_ <= 0.406) {
                            return(12)
                        }
                    }
                }
            }
        }
    }
  }
  return(NA)
}
```

Example no. 3 [*BigML*, Clustering]: Clustering analysis of fire-exposed RC columns.
The compiled database used in this example contains information on 144 RC columns in terms of column width, *W* (mm), steel reinforcement ratio, *r* (mm), column length, *L* (mm), concrete compressive strength, f_c (MPa), steel yield strength, f_y (MPa), restraint conditions by means of effective length factor, concrete cover to reinforcement, *C* (mm), eccentricity in applied loading in the main axes, e_x (mm), the magnitude of applied loading, *P* (kN), and fire resistance, *FR* (min). The goal is to cluster these columns.

Step 1: Let us upload the fire-exposed RC columns dataset, then select *Cluster*.

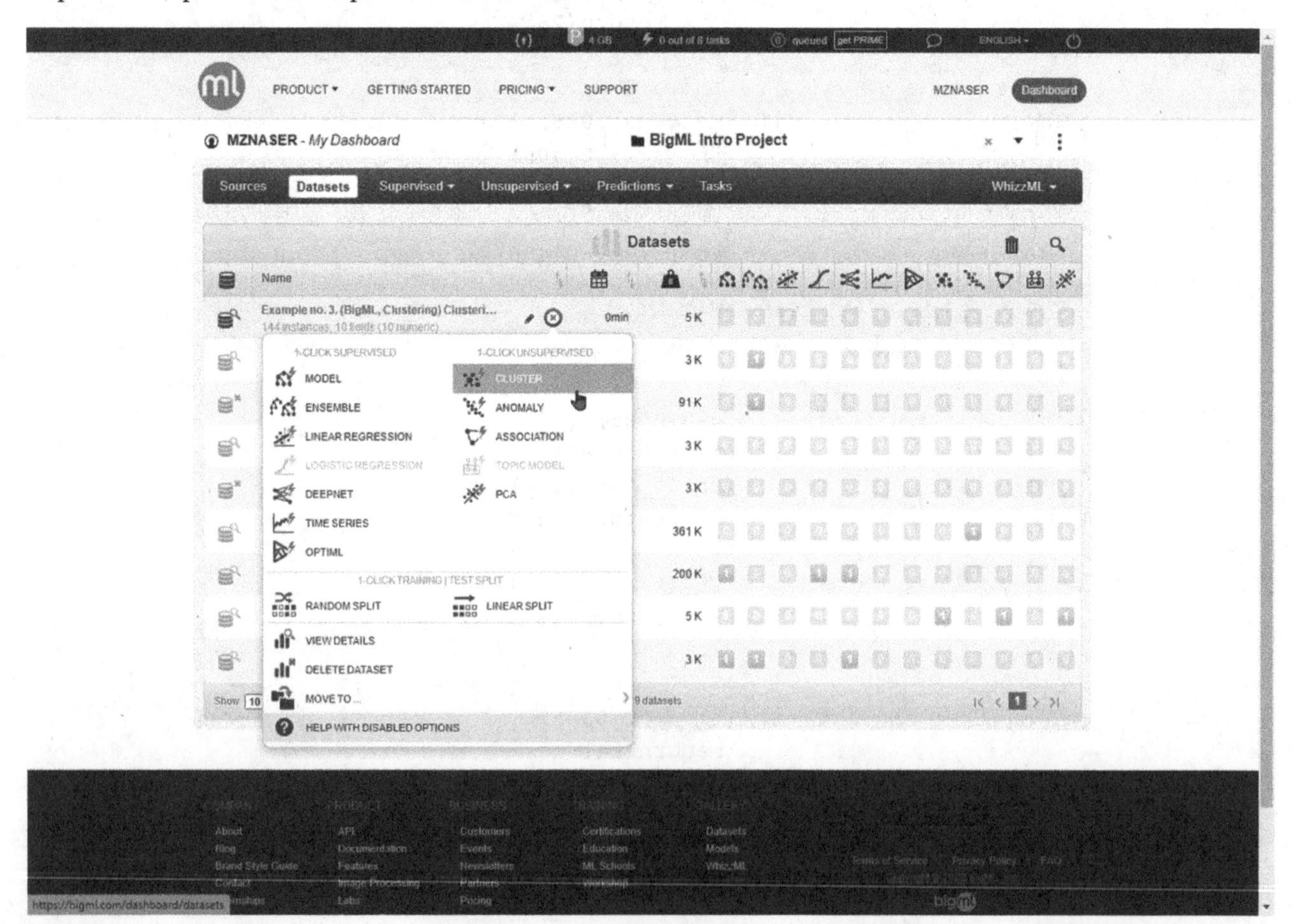

Step 2: Navigate to the *Unsupervised* tab to see the outcome of this analysis. As you can see, the *G-means* algorithm identifies two clusters.

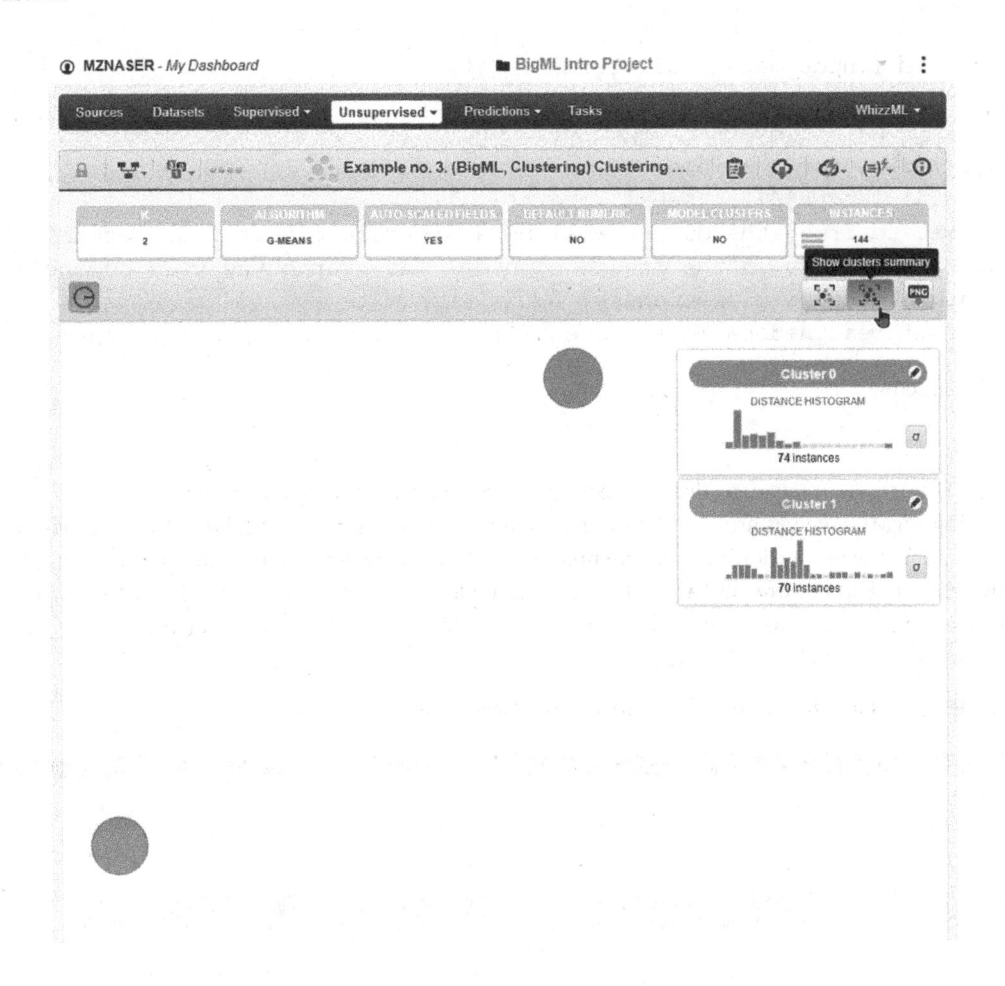

Step 3: Select *Download Actionable Model* to download this model into a programming language of your choosing.

Python code:

```
# Requires BigML Python bindings
#
# Install via: pip install bigml
#
# or clone it:
# git clone https://github.com/bigmlcom/python.git

from bigml.ensemble import Ensemble

# Downloads and generates a local version of the ensemble, if it
# hasn't been downloaded previously.
from bigml.api import BigML

ensemble = Ensemble('ensemble/62b443ceaba2df1b780006ce',
api=BigML("mznaser",
"33dcd88020b98f21a24f68ce3b9c943e6ab4e4b6",
domain="bigml.io"))
```

```
# To make predictions fill the desired input_data in next line.
input_data = {}

ensemble.predict(input_data, full=True)

#
# input_data: dict for the input values
# (e.g. {"petal length": 1, "sepal length": 3})
# full: if set to True, the output will be a dictionary that includes all the
# available information in the predicted node. The attributes vary depending
# on the ensemble type. Please check:
# https://bigml.readthedocs.io/en/latest/#local-ensemble-s-predictions
```

R code is not reproducible.

Example no. 4 [*BigML*, Anomaly detection]: Detecting anomalous poor quality concrete mixtures.
This example using *BigML* revolves around identifying anomalies in a large dataset (about 10,000 concrete mixtures[21]). This dataset documents the proportions of concrete mixtures in terms of water, cement, and fly ash contents (in kg/m^3 of concrete), water-reducing admixture (WRA), and air-entraining admixture contents (AEA in 0.01 kg/kg of cementitious material), coarse and fine aggregate contents (in kg/m^3 of concrete), and fresh air content (in volume %), and the in-situ measured compressive strength for each mixture.

Step 1: Upload the dataset and navigate to the *Dataset* tab. Select *Anomaly*.

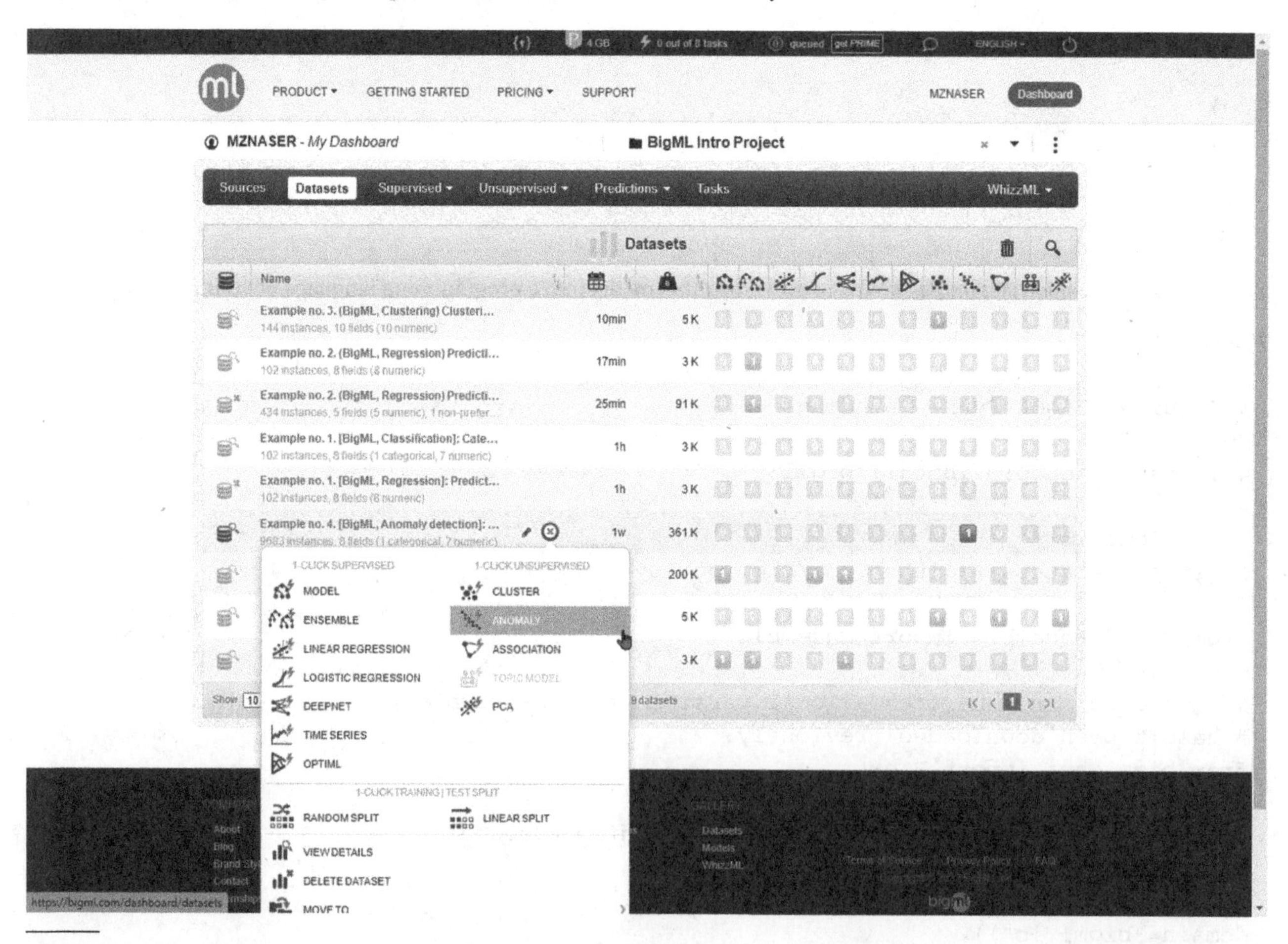

21 I first came across this dataset from the work of Young et al. [31] and thanks to them, this dataset helped me produce the following article [32] – another reason of how sharing data can be helpful and beneficial!

Step 2: Go to the *Unsupervised* tab and then select *View details*.

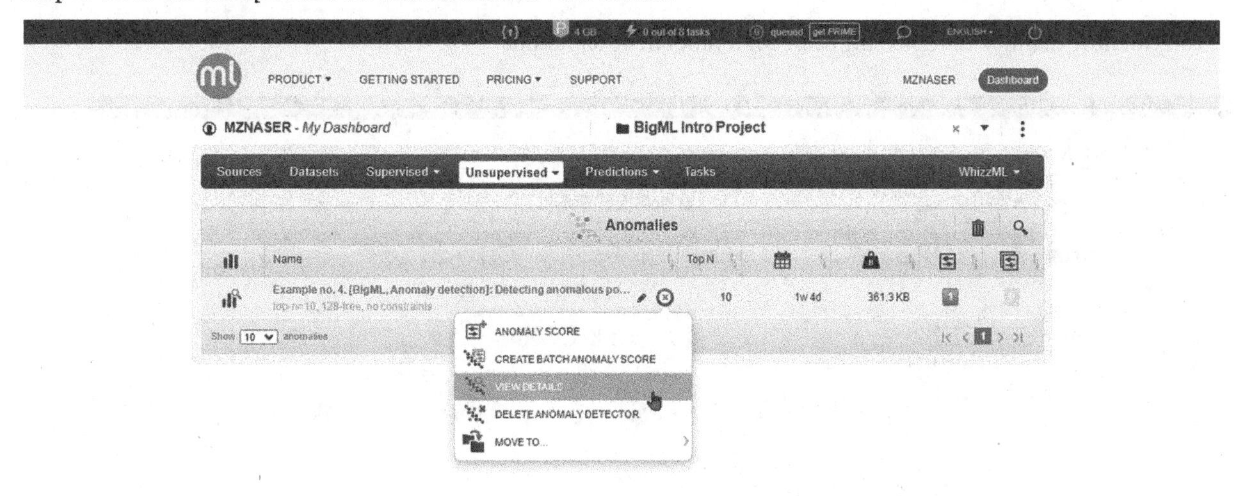

Step 3: You will notice that there are ten anomalies listed in a sample of 1024 observations.

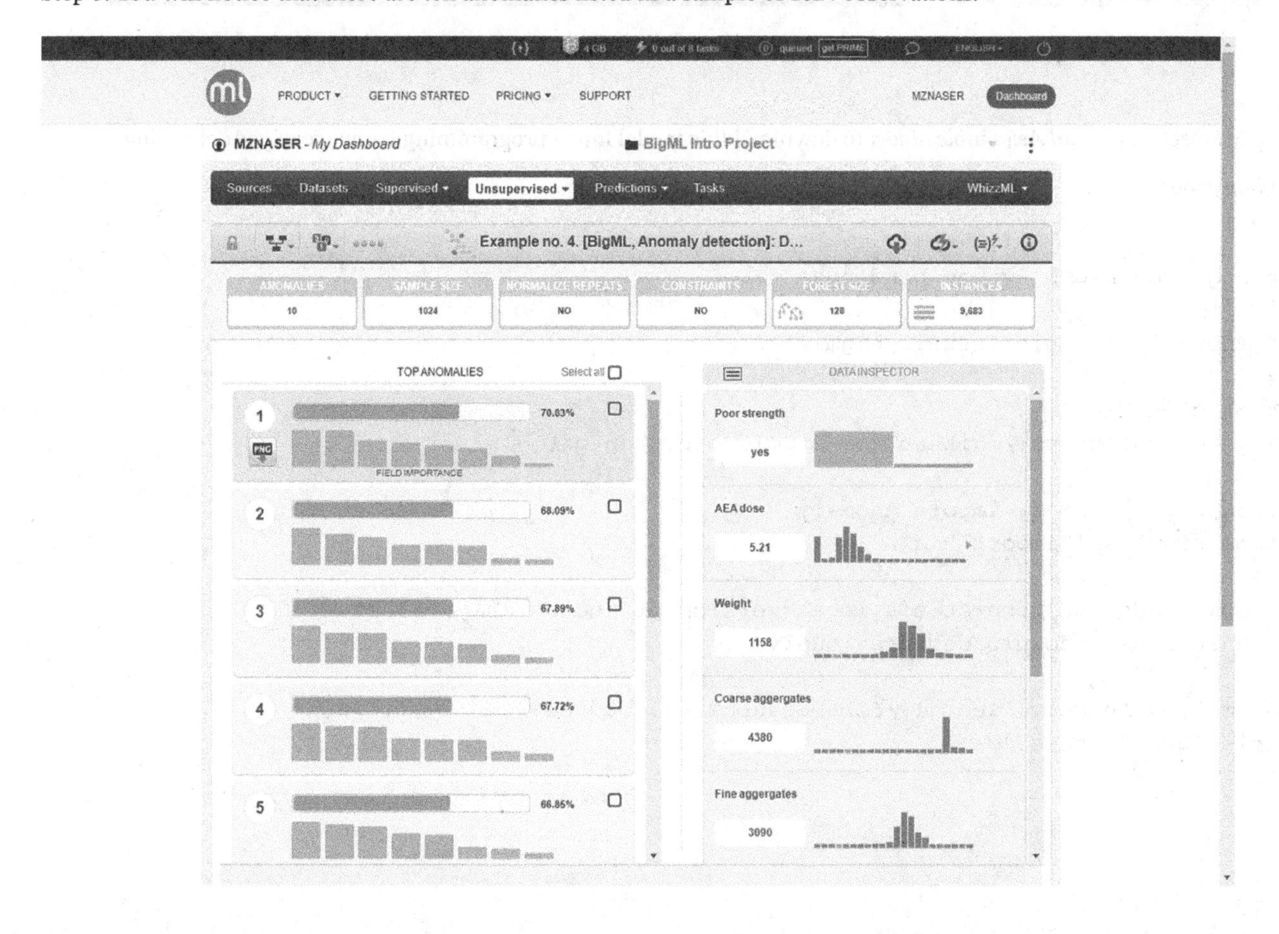

Step 4: You can now examine the *Anomaly Score* for a random observation.

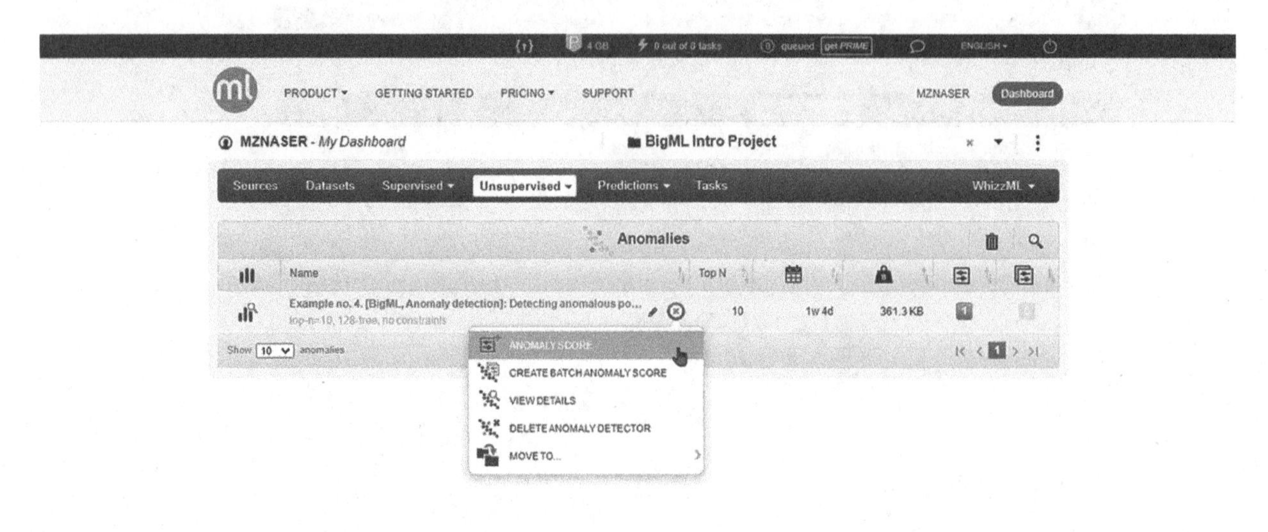

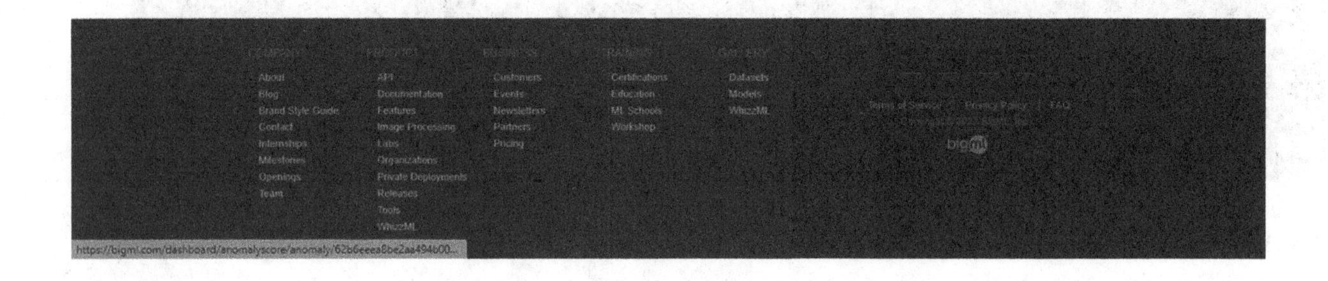

Step 5: Select *Download Actionable Model* to download this model into a programming language of your choosing.

Python code:

```
# Requires BigML Python bindings
#
# Install via: pip install bigml
#
# or clone it:
# git clone https://github.com/bigmlcom/python.git

from bigml.anomaly import Anomaly
from bigml.api import BigML

# Downloads and generates a local version of the anomaly, if it
# hasn't been downloaded previously.

anomaly = Anomaly('anomaly/62b6eeea8be2aa494b000cd7',
api=BigML("mznaser",
```

```
"33dcd88020b98f21a24f68ce3b9c943e6ab4e4b6",
domain="bigml.io"))

# To predict scores fill the desired input_data
# in next line. Numeric fields are compulsory.
input_data = {
"Awra dose": 1
"W/(C+P)": 1
"Fly ash": 1
"AEA dose": 1
"Fine aggergates": 1
"Weight": 1
"Coarse aggergates": 1}
anomaly.anomaly_score(input_data)

# The result is a float number which value is between 0 and 1.
```

R code is not reproducible.

Example no. 5 [*BigML*, Dimensionality reduction]: Reducing the number of features in an expanded spalling dataset. In this example, we will be using an expanded dataset on spalling.[22] This dataset contains the following features: water/binder ratio (%), aggregate/binder ratio (%), sand/binder ratio (%), heating rate (°C/min), moisture content (%), maximum exposure temperature (°C), silica fume/binder ratio (%), aggregate size (mm), GGBS/binder ratio (%), FA/binder ratio (%), PP fibers quantity (kg/m^3), PP fibers diameter (um), PP fibers length (mm), steel fibers quantity (kg/m^3), steel fibers diameter (mm), and steel fibers length (mm), as well as one target (spalling/no spalling). Given the large number of features, let us run a dimensionality reduction using the PCA algorithm.

Step 1: Upload the supplied dataset and select a *PCA* analysis.

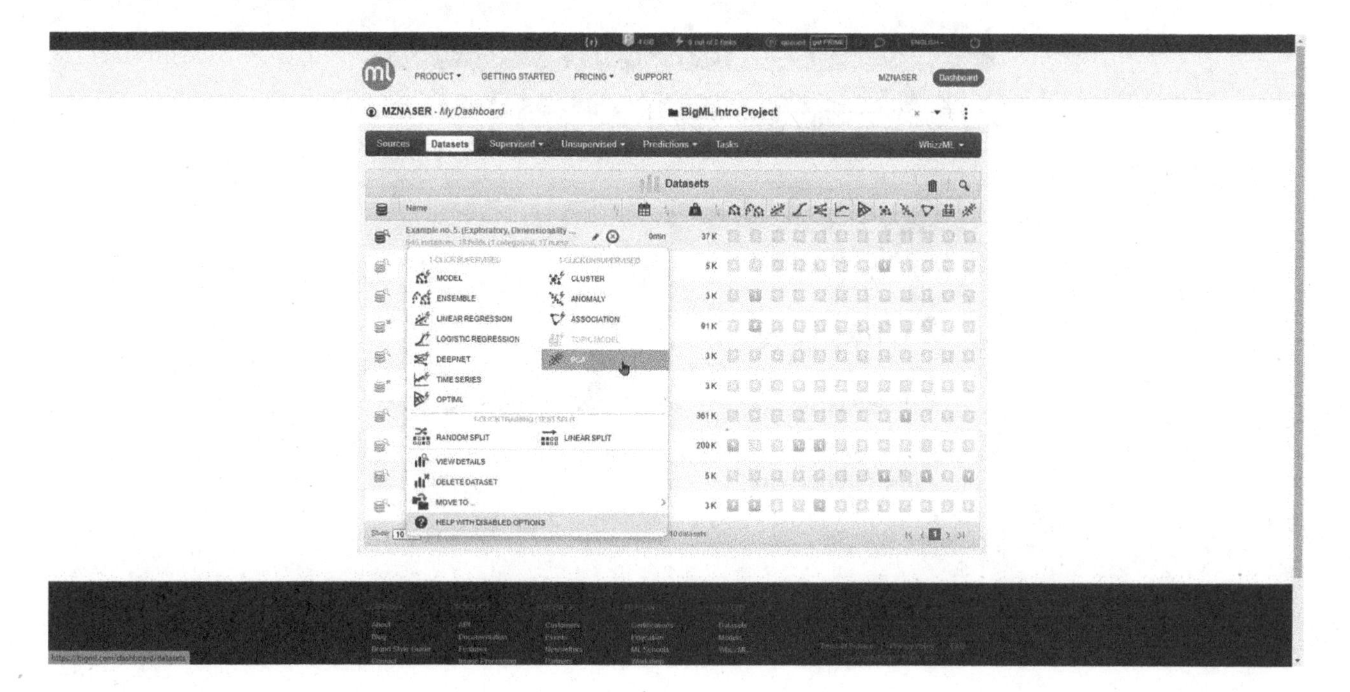

22 Thanks to Prof. Hailong Ye for sharing this dataset.

Step 2: Navigate to the PCA option to view the results of this analysis.

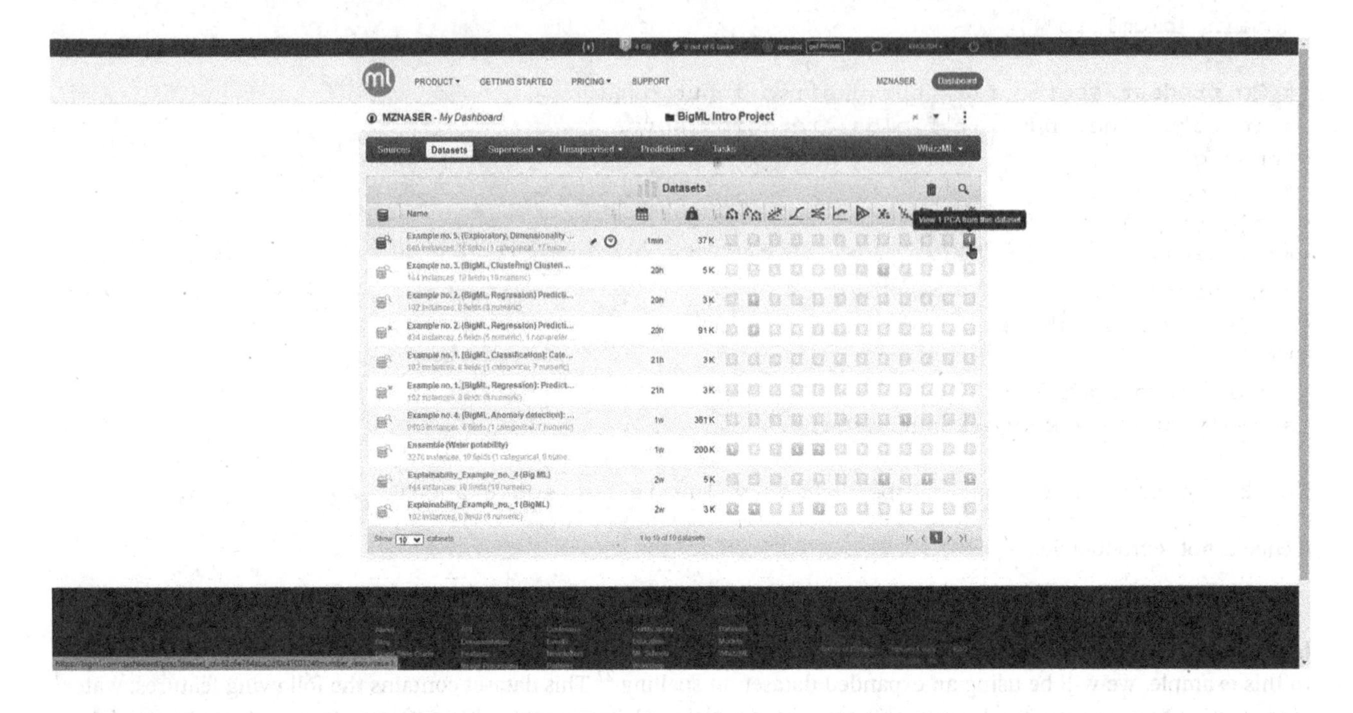

Step 3: Visualize the outcome of this PCA analysis. As you can see, the first two principal components explain about 50%[23] of the information in this dataset.

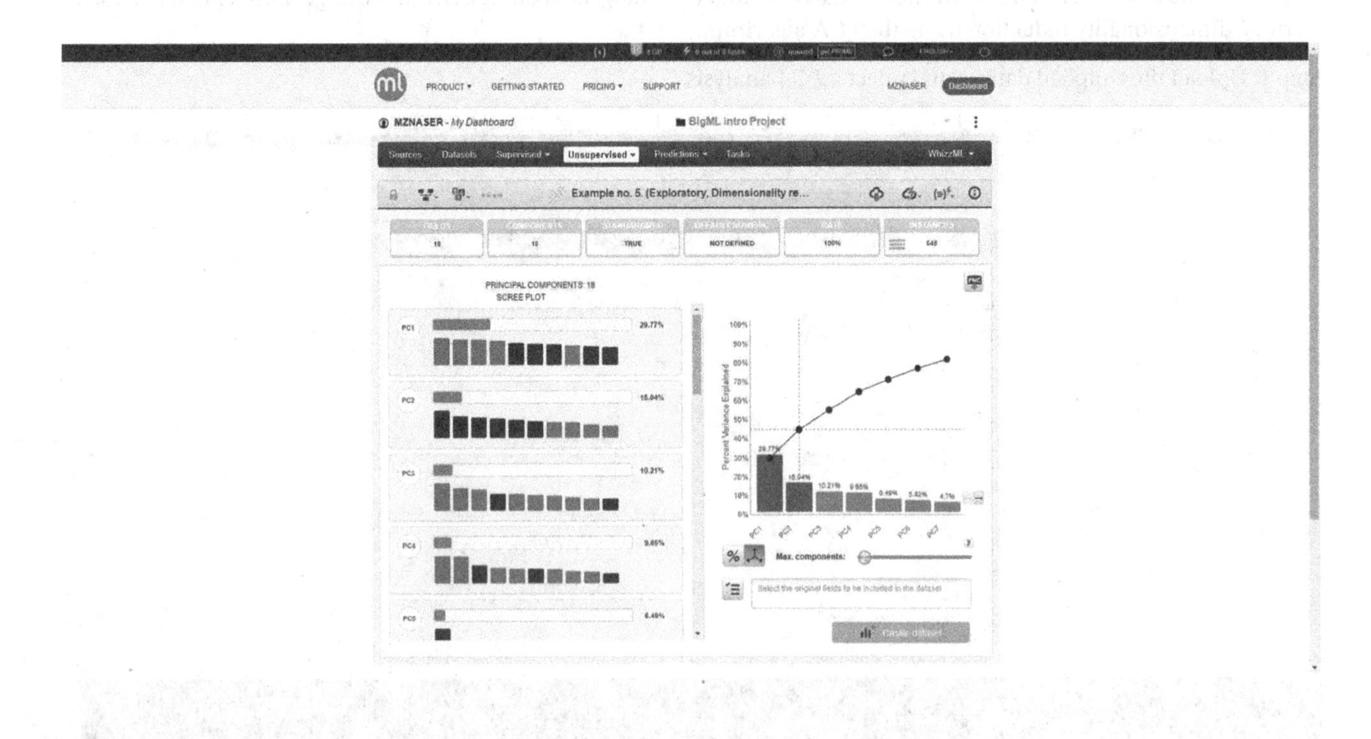

23 A good number for a decent PCA analysis would be about 70%.

Step 4: Select *Download Actionable Model* to download this model into a programming language of your choosing. Here, we use *Python*.

Python code:

```
# Requires BigML Python bindings
# Install via: pip install bigml
# or clone it:
# git clone https://github.com/bigmlcom/python.git

from bigml.pca import PCA
from bigml.api import BigML

# Downloads and generates a local version of the PCA,
# if it hasn't been downloaded previously.

pca = PCA('pca/62c6e79e8f679a6f64001537',
          api=BigML("mznaser",
                    "33dcd88020b98f21a24f68ce3b9c943e6ab4e4b6",
                    domain="bigml.io"))

# To project the PCA fill the desired input_data in the line below.
# Only the ID of the fields to project with are required.
input_data = {
    "maximum exposure temperature": 1,
    "silica fume/binder ratio": 1,
    "moisture content": 1,
    "characteristic distance": 1,
    "sand/binder ratio": 1,
    "heating rate": 1,
    "Water/binder ratio": 1,
    "aggregate/binder ratio": 1,
    "aggregate size": 1,
    "GGBS/binder ratio": 1,
    "s fiber diameter(mm)": 1,
    "pp fiber length (mm)": 1,
    "s fiber quantity (kg/m3)": 1,
    "PP fiber quantity (kg/m3)": 1,
    "pp fiber diameter(um)": 1,
    "FA/binder ratio": 1,
    "s fiber length (mm)": 1
}
pca.projection(input_data)
```

This concludes our examples using the *BigML* platform.

6.1.2 DataRobot

According to *DataRobot*,[24]

"DataRobot was founded in 2012 *to democratize access to AI. Today, DataRobot is the AI Cloud leader, with a vision to deliver a unified platform for all users, all data types, and all environments to accelerate delivery of AI to production for every organization."*

24 https://datarobot.com.

DataRobot is a Boston-based data science company that enables users to build predictive analytics with or without knowledge of ML (or programming). This cloud/online platform[25] uses automated ML to build and deploy accurate predictive models in a short span of time. *DataRobot* provides unique and personalized solutions to various industries, including those related to civil and environmental engineering. What is unique about this platform is that it contains a wide variety of algorithms belonging to different programming languages (i.e., not limited to a specific programming language). *DataRobot* also incorporates explainable ML, deployable ML, fair[26] ML, and trustable ML, among others[27] (see Figure 6.2). Finally, this platform is not limited to tabular data but also provides user access to use computer vision tools.

DataRobot has an in-house academic support program where faculty and students get to access and use its online platform at no cost. They also provide a short course to introduce new users to its platform.[28] Their YouTube channel is also active and continuously updates its content on ML and new additions to the *DataRobot* family.

I will be showcasing three different examples using *DataRobot*.

Example no. 1 [*DataRobot*, Classification]: Predicting fire-induced spalling of concrete.
The used dataset contains 646 test samples collected by the group of Prof. Hailong Ye [2–6]. This dataset comprises 16 features and one target, which describes the occurrence of spalling via two labels: *no spalling* or *spalling*. The 16 independent

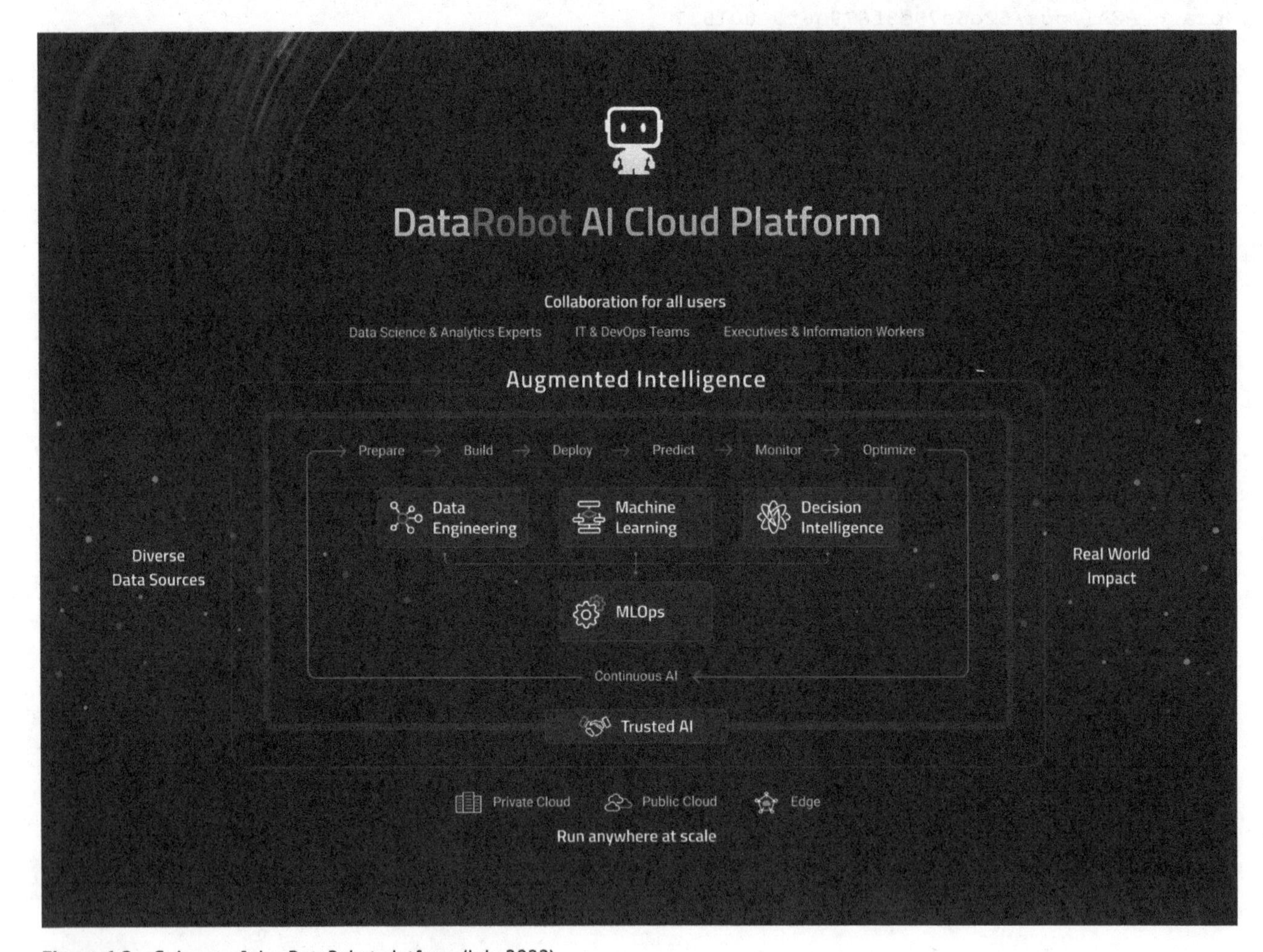

Figure 6.2 Snippet of the *DataRobot* platform (July 2022).

25 A **cloud-based platform** implies that a user only needs to access the online platform to use it (i.e., no need to download specialized software or acquire hardware, etc.).

26 This is one of the coolest features of *DataRobot* – you can find this under *Bias and Fairness* under the *Advanced Option* tab – where a model seeks to mitigate bias and discrimination into its decision making.

27 Such as editable ML model (*Composable ML*).

28 I did enjoy this particular course. Also, shout out to Katherine Kopp, Matt Charoenrath, and William Disch who were kind enough to lend me their platform.

variables are: water/binder ratio (%), aggregate/binder ratio (%), sand/binder ratio (%), heating rate (°C/min), moisture content (%), maximum exposure temperature (°C), silica fume/binder ratio (%), aggregate size (mm), GGBS/binder ratio (%), FA/binder ratio (%), PP fibers quantity (kg/m^3), PP fibers diameter (um), PP fibers length (mm), steel fibers quantity (kg/m^3), steel fibers diameter (mm), steel fibers length (mm).

Here are the steps to carry out this classification analysis. We will be using the XGBoost algorithm.

Step 1: Upload the dataset and select *Output* as *Target*.

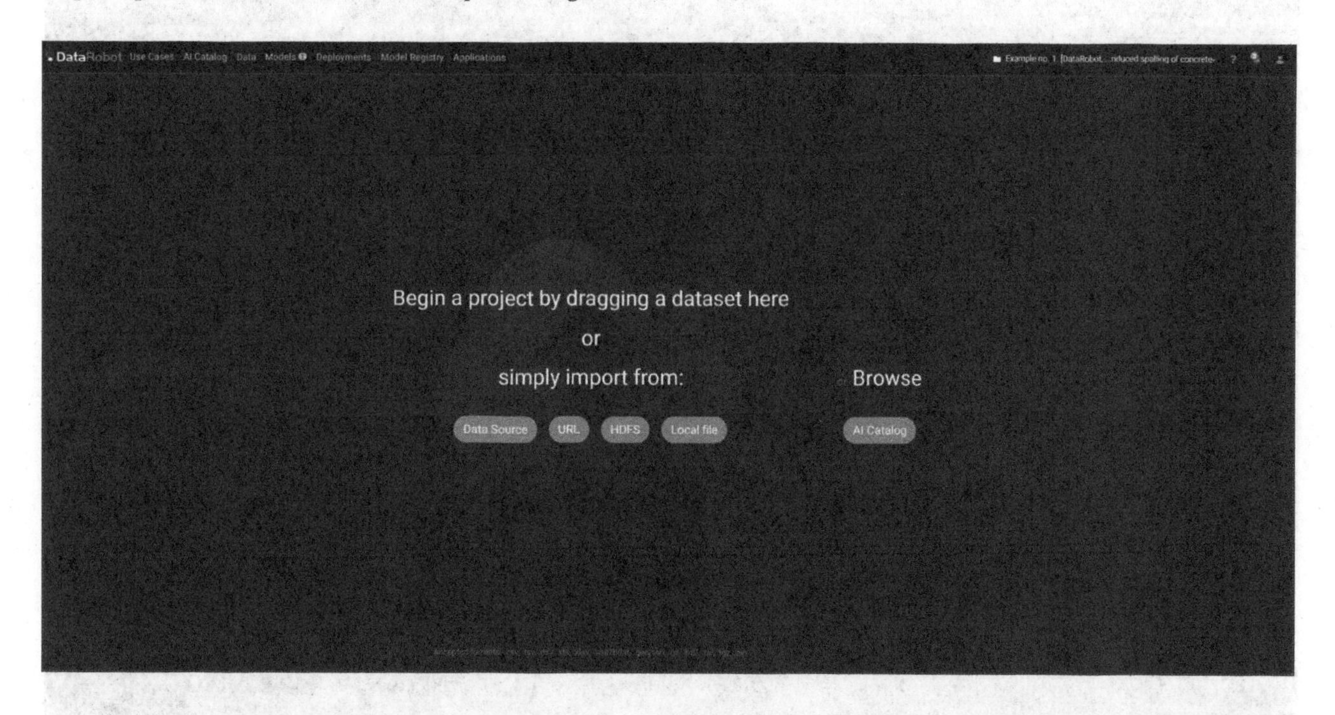

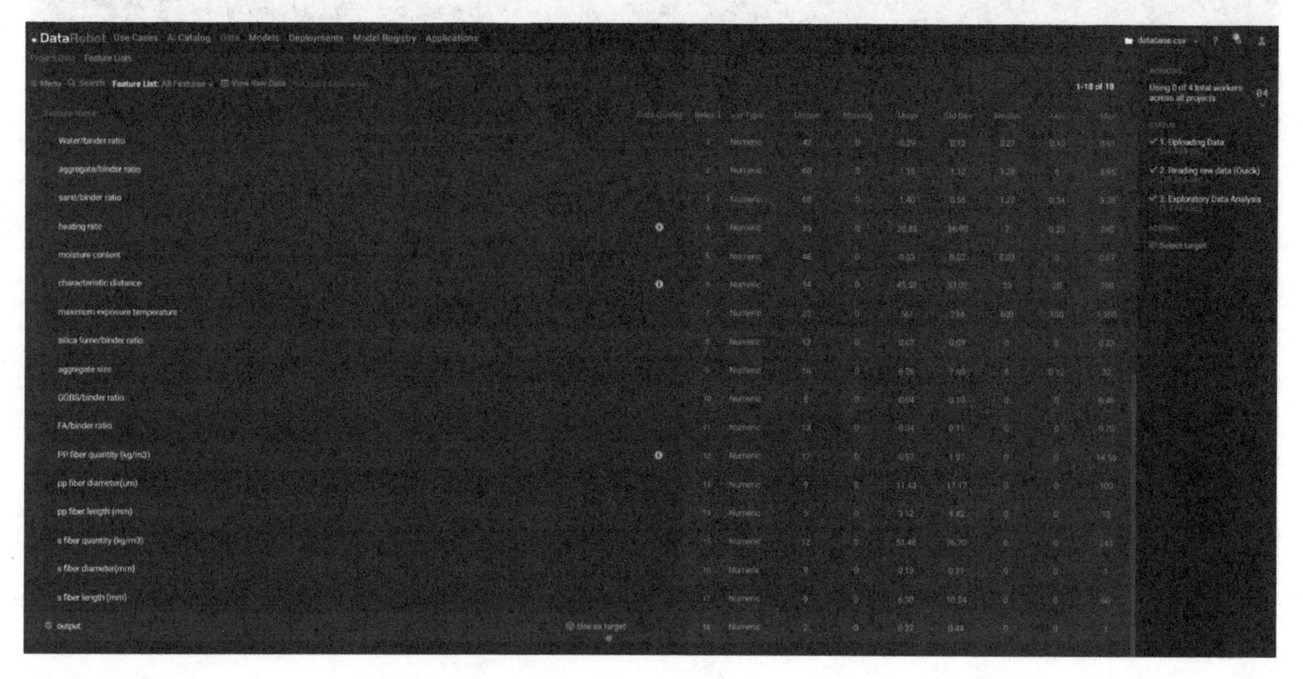

Step 2: You will notice that *DataRobot* changes this analysis into *Classification*. Navigate to *Show advanced options*, then to *Additional* and check *Search for interaction* and uncheck *Create blenders from top models*. Note the 0 = No spalling, and 1 = Spalling.

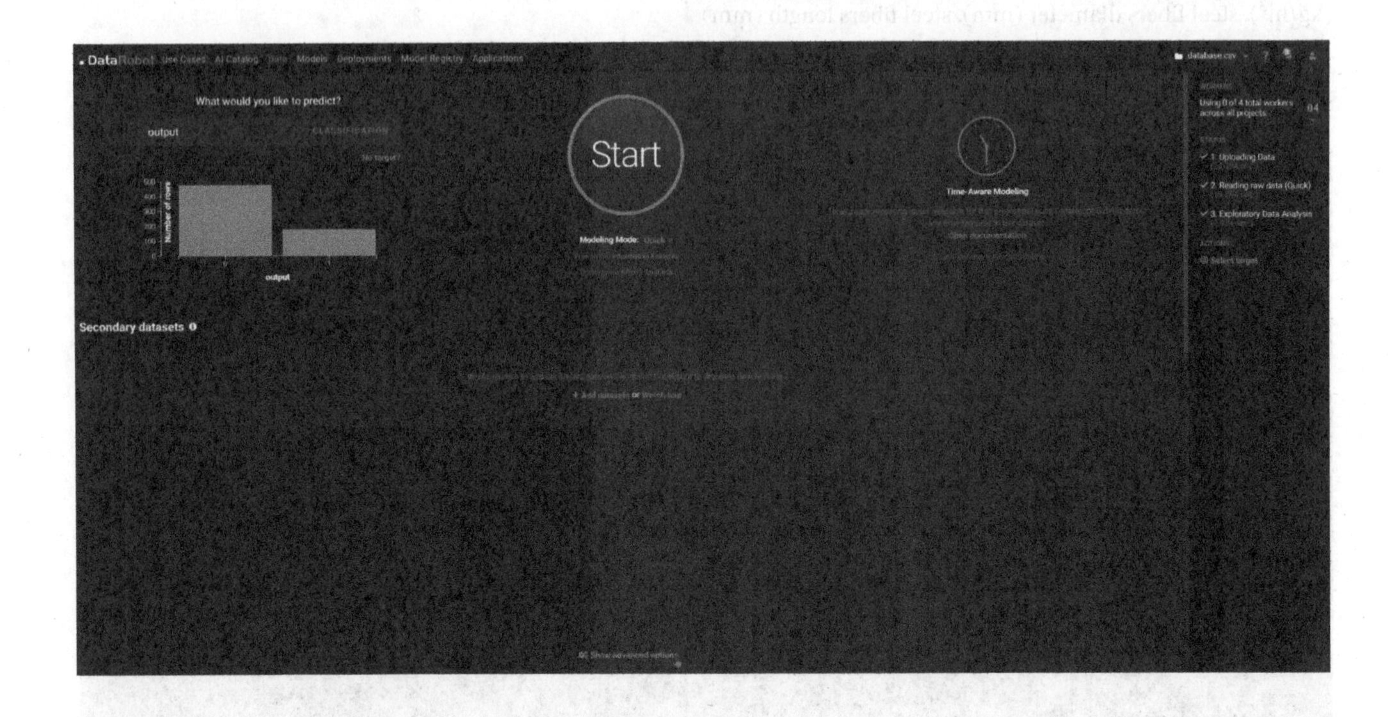

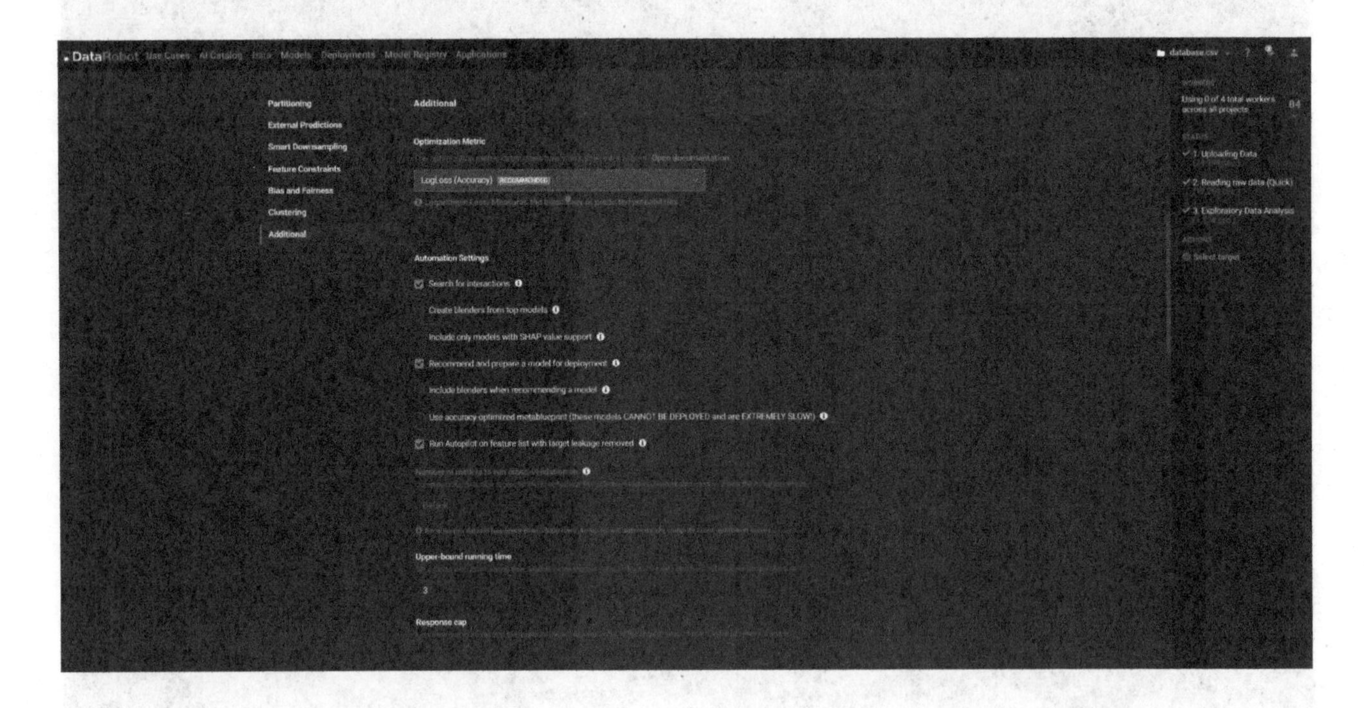

Step 3: Go to *Modeling mode* and select *Manual*. *Start* the analysis.

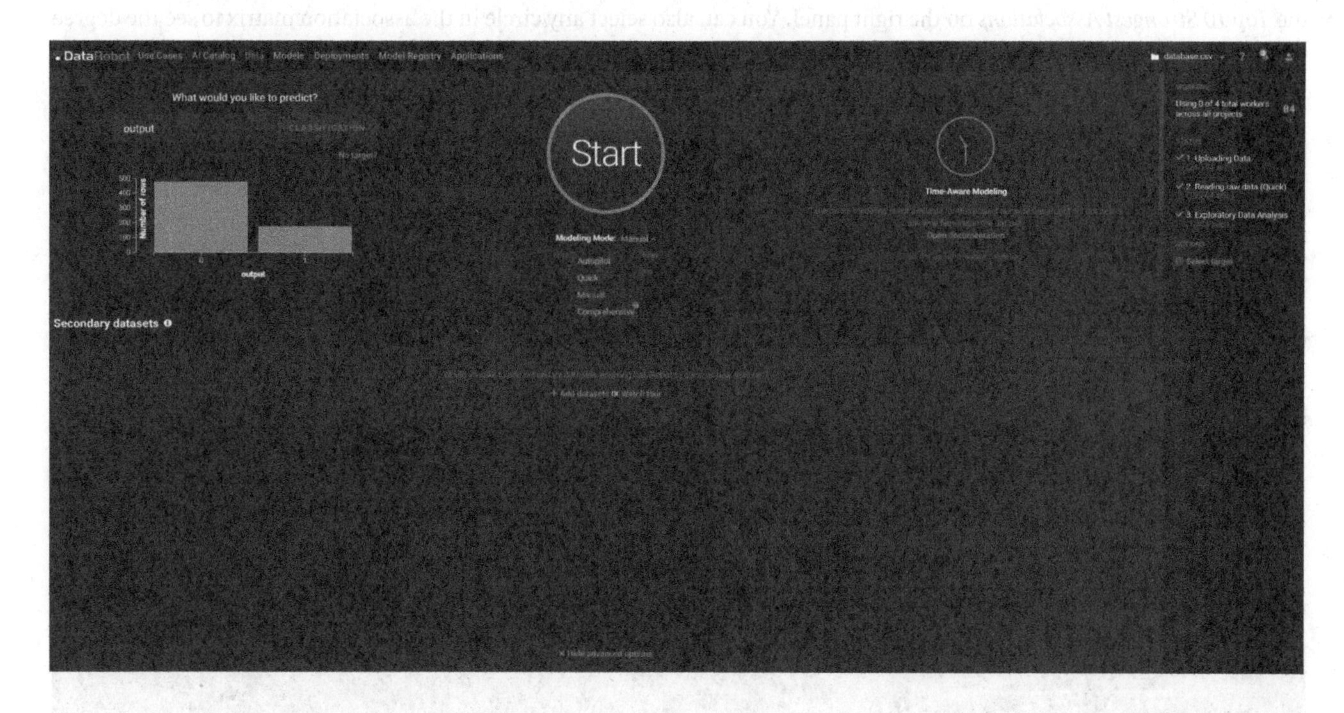

Step 4: To have a visual look into the distribution of each feature. Go to the *Data* tab and select a feature (the output feature is shown here). A histogram will pop up. Additional statistical information on this feature in terms of the mean, standard deviation, median, etc. are also shown to the side.

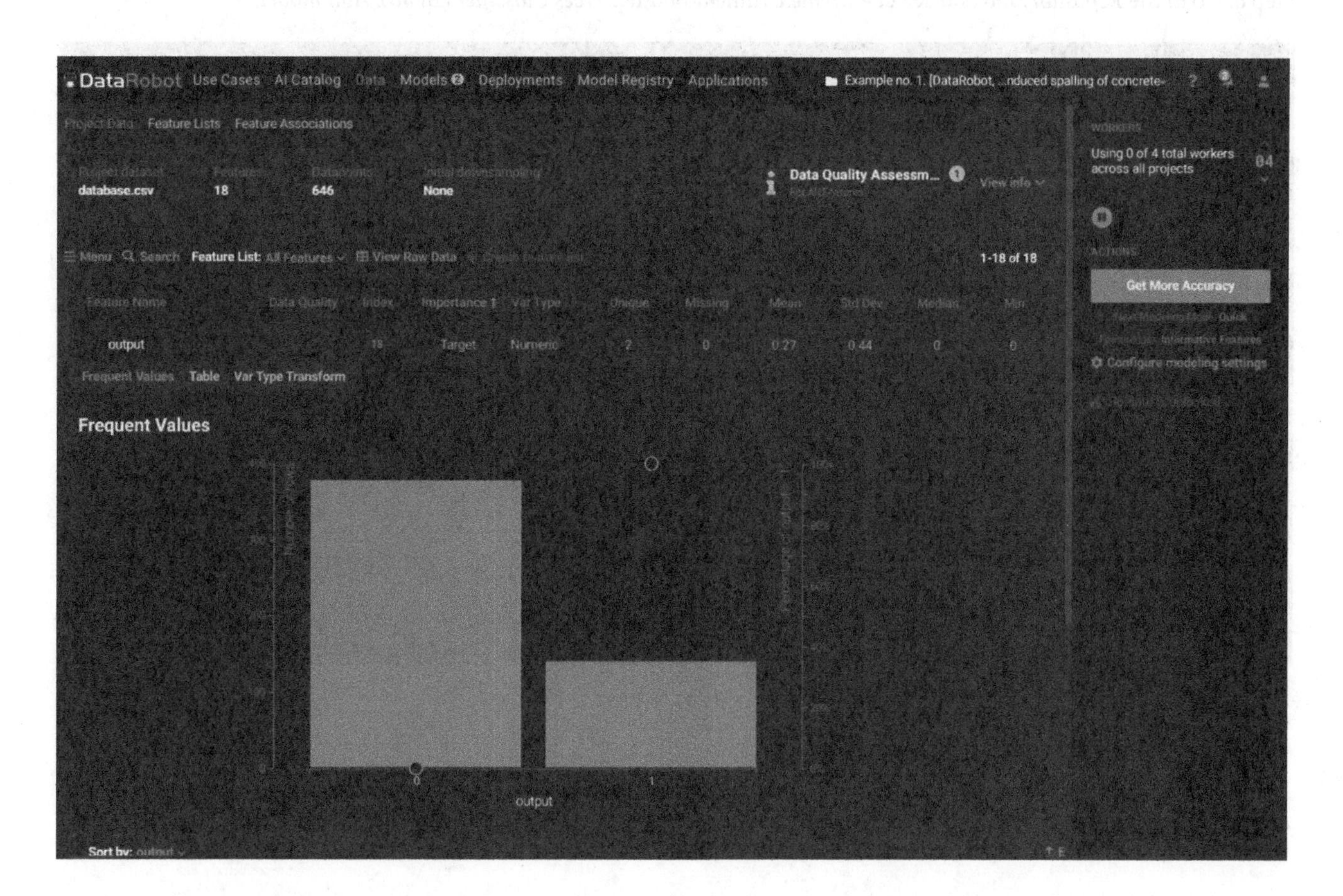

Step 5: Let us see some information about our data. Navigate to the *Feature Associations* under the *Data* tab. You can see the *Top 10 Strongest Associations* on the right panel. You can also select any circle in the association matrix to see the degree of association between two features. As you can see, *DataRobot* clusters the features in colored groups based on the strength of association.[29]

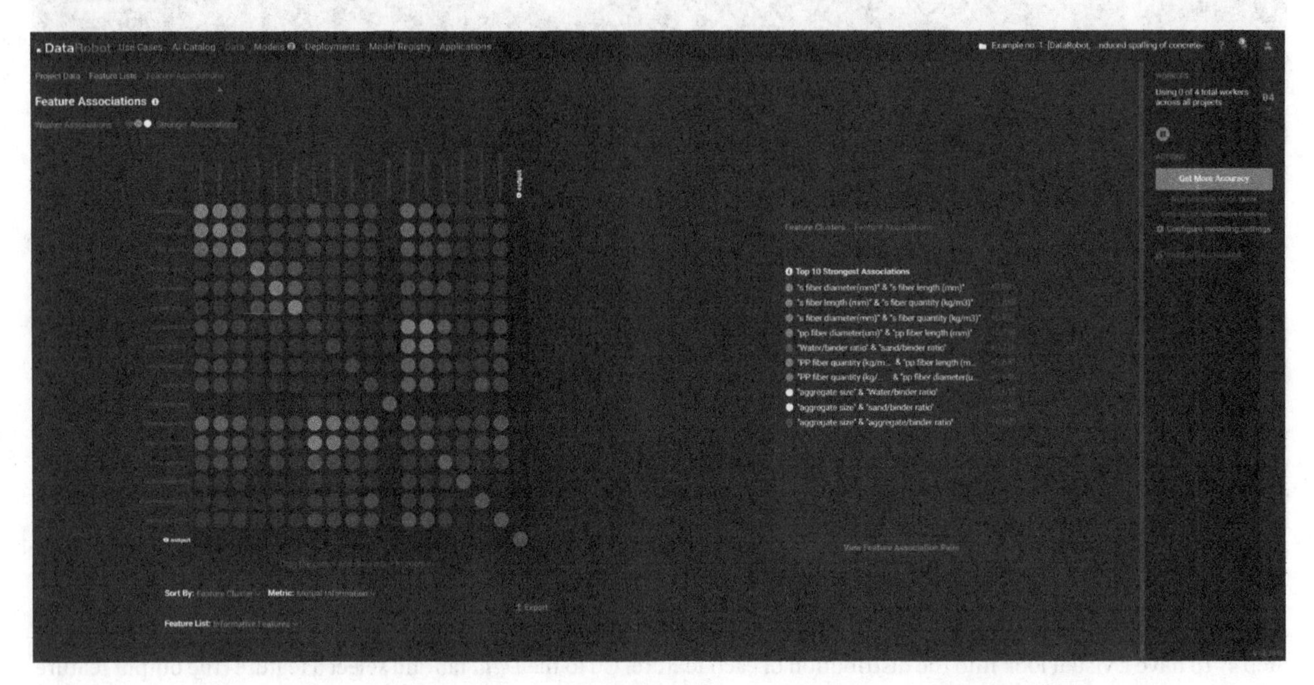

Step 6: Go to the *Repository* tab and select *eXtreme Gradient Boosted Trees Classifier (BP40). Add model.*

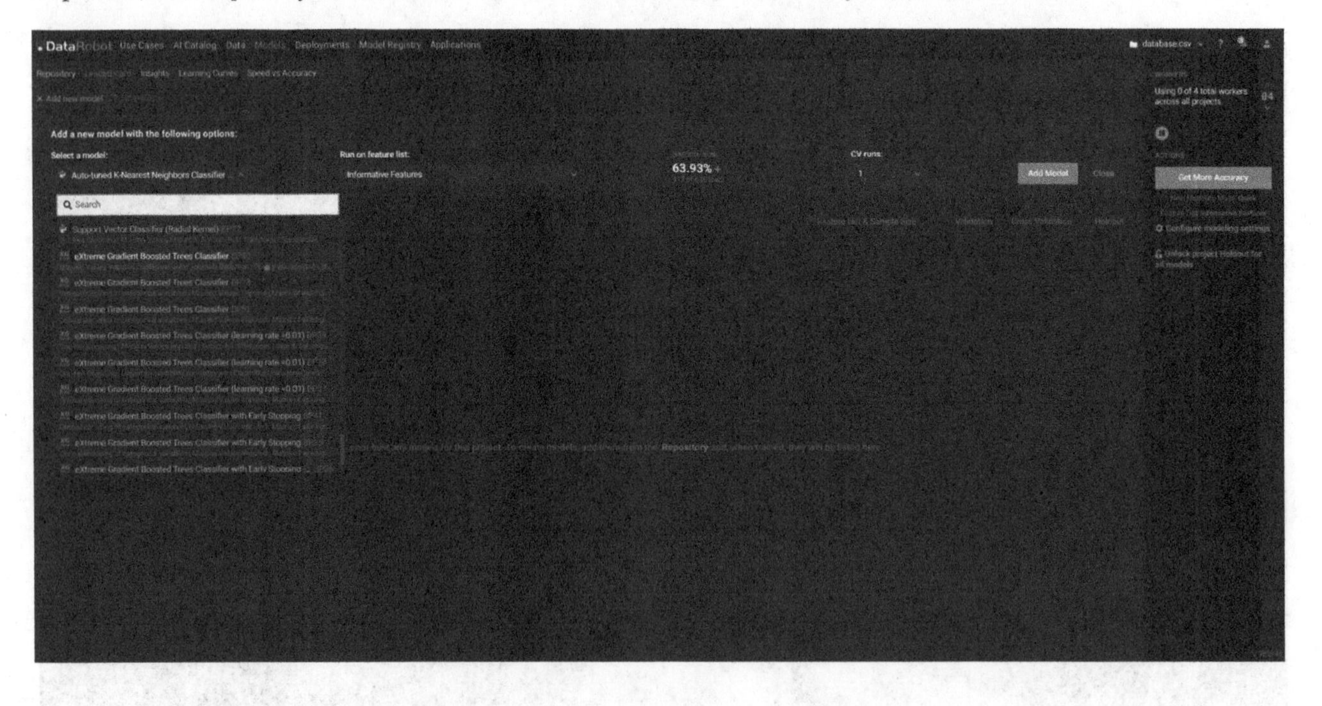

29 You may adjust the *Sort by* option and *Metric* option to arrive at other forms of associations.

Step 7: Navigate to the *Leaderboard* tab. Check the outcome of the analysis. The figure below shows model performance in terms of the AUC metric.

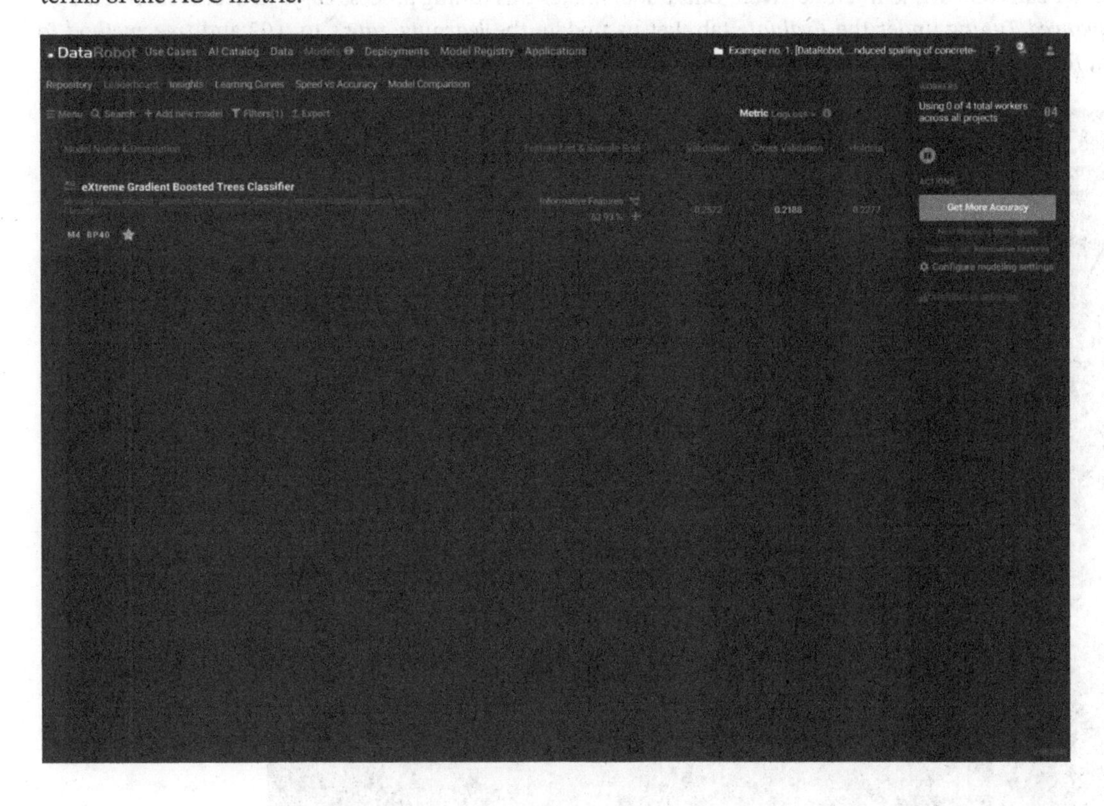

Step 8: Go to the *Evaluate* tab and then *ROC Curve* to show the ROC curve and confusion matrix. Change the *Data Selection* to *Validation*.

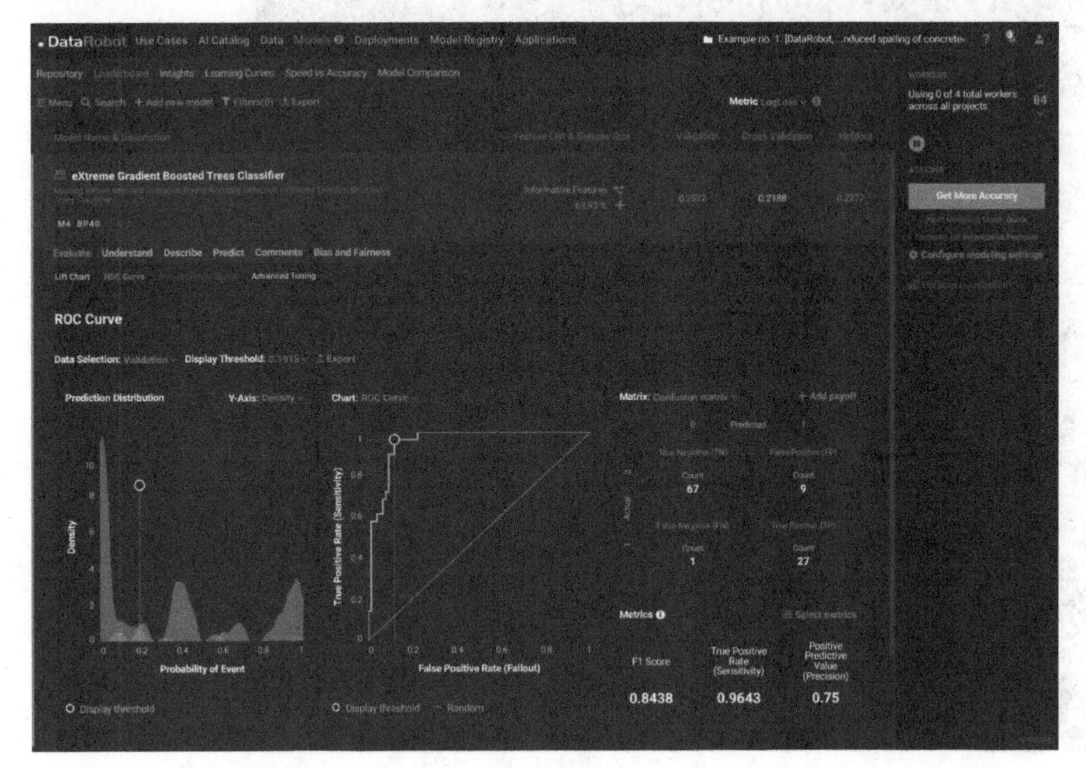

Remember how in one of the footnotes of the chapter on algorithms, I mentioned that each algorithm has a set of settings and parameters that we can control and finetune? Well, DataRobot makes this tuning process easy. To access such settings, navigate to the *Advanced Tuning* under the *Evaluate* tab. Let us update the *learning_rate*[30] to *0.02* and *tree_method* to *exact*.[31] Then, select *Begin Tuning*. Compare the outcome of the default and tuned models.

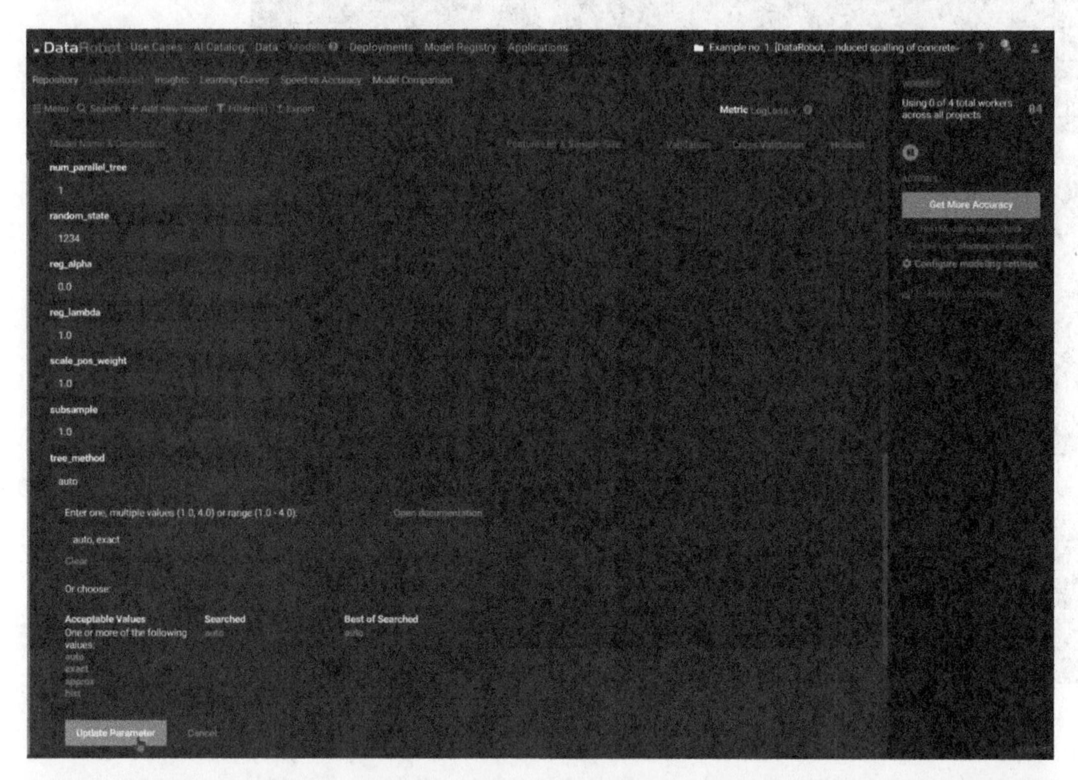

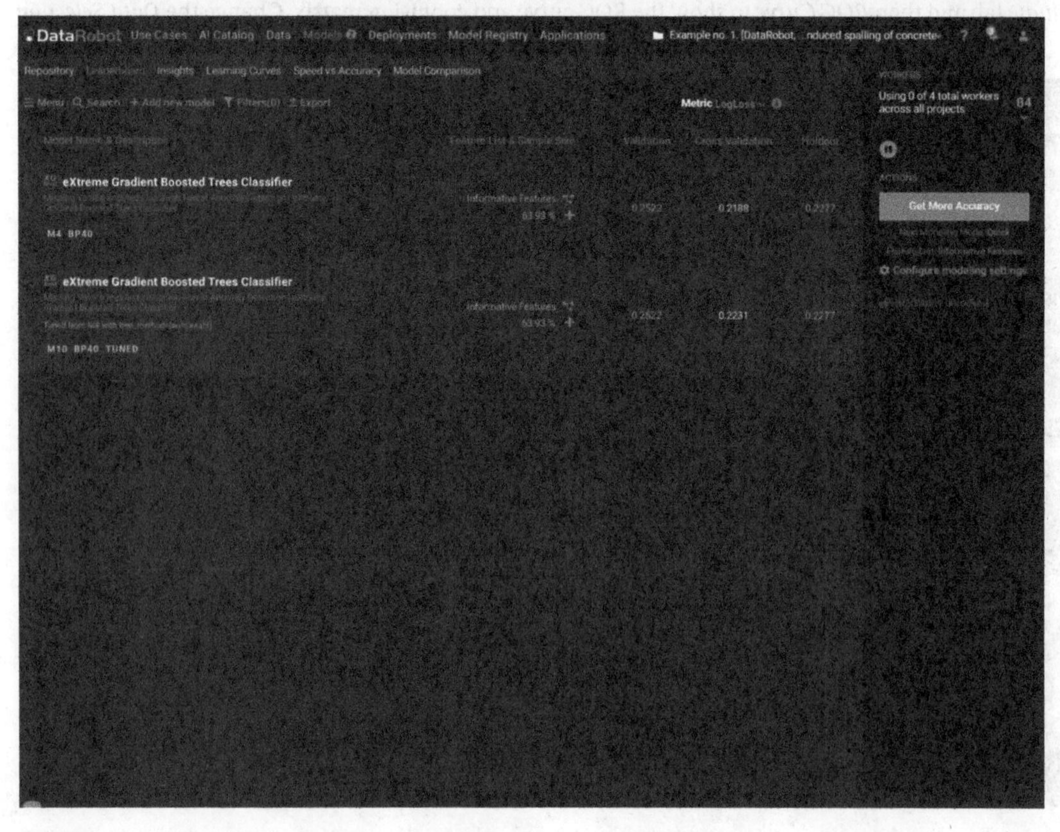

30 The shrinkage at every step to minimize overfitting.
31 Uses exact greedy algorithm.

Example no. 1 [*DataRobot*, Supplementary multi-classification]: Predicting damage to bridges.

In this supplementary example, we will examine a dataset comprising 135 international bridges that underwent some degree of fire damage (minor, major, and collapse). Each incident is documented with regard to eight different features, namely, structural system, *S*, and construction materials, *M*, used in load bearing, span length, *L*, and age, *A*, of the bridge, geographical significance, *G* as well as the number of lanes, *N*, type of fuel, *T*, involved in the fire and closeness of fire break out, *C*. As you can see, the inputs comprise numerical and categorical data.[32]

Let us carry out the analysis using the *DataRobot* platform and upload our dataset. Select *Damage* as the target. We will not be targeting a specific algorithm; instead, we will use the *Autopilot*[33] mode in *DataRobot*.

Step 1: Navigate to the *Start* and Modeling Mode: *Autopilot*.

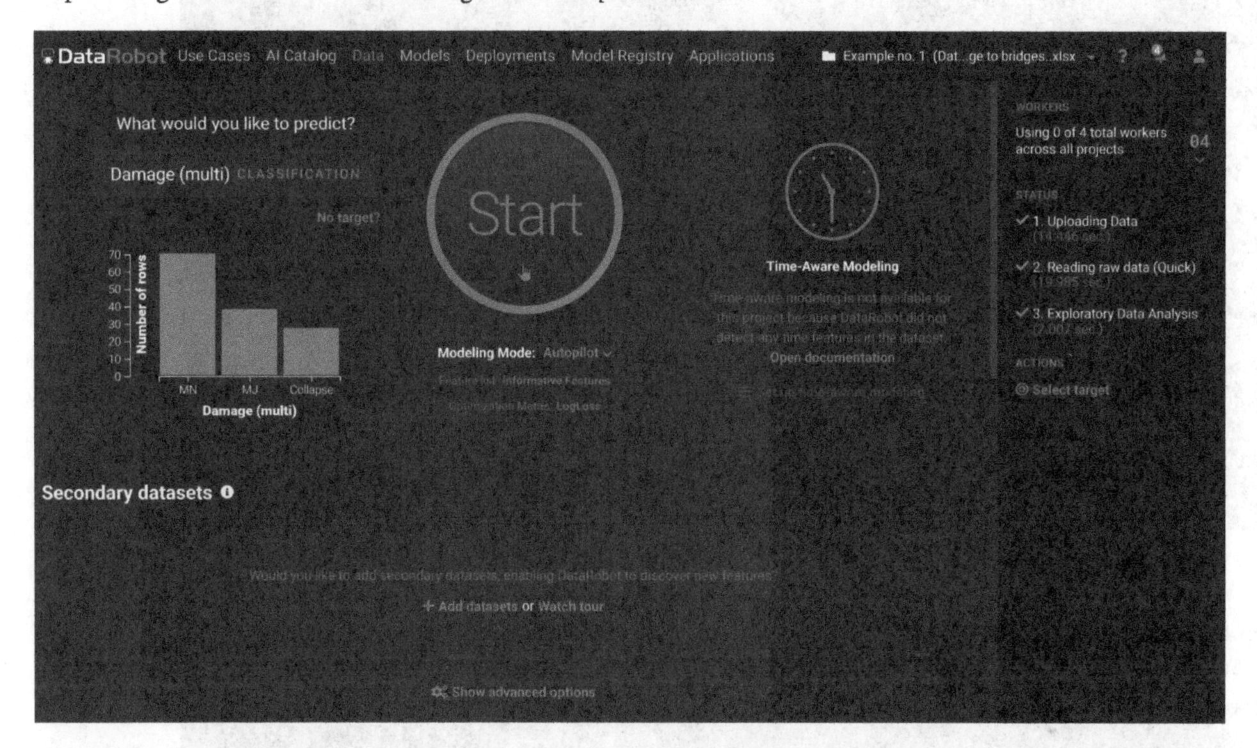

Step 2: Check out the histograms of the features (the target is shown below).

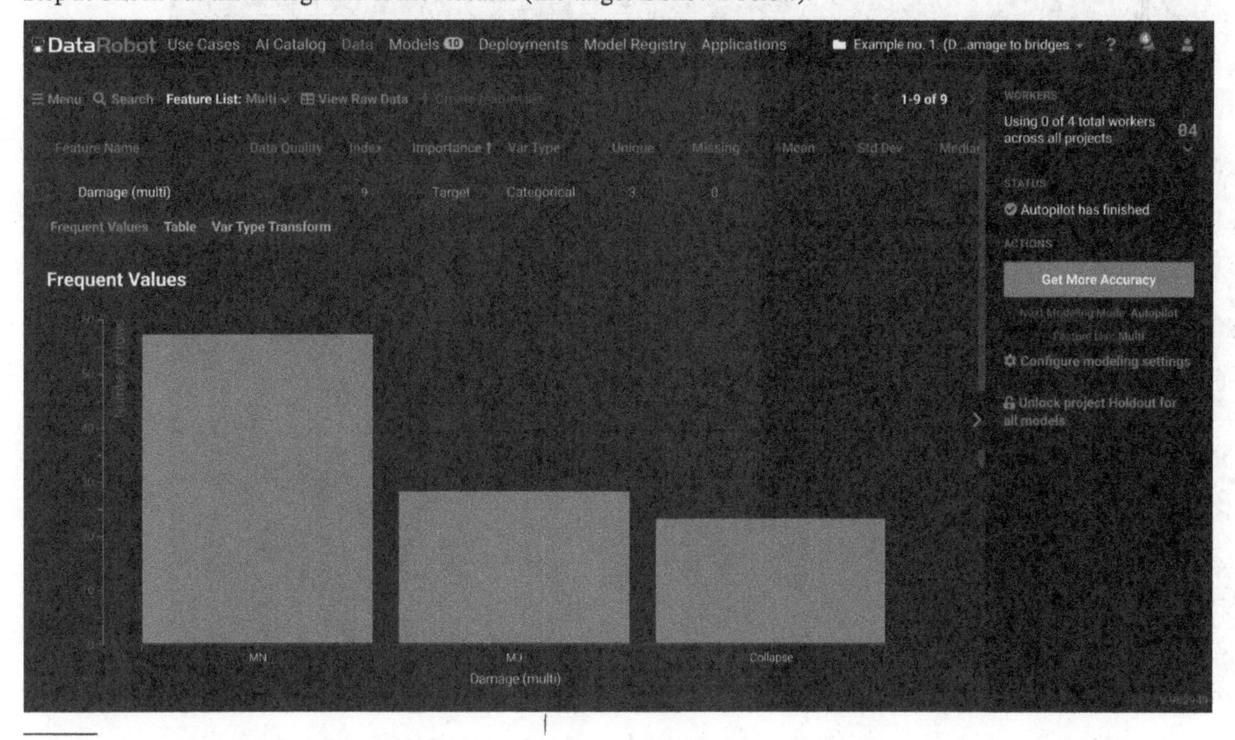

32 We do not need to manipulate the inputs. *DataRobot* will automatically do so.

33 In this mode, *DataRobot* examines a series of algorithms in a competitive manner to arrive and recommend the best performing model.

Step 3: *DataRobot* automatically runs the *Autopilot* mode. The outcome of this analysis can be viewed under the *Leaderboard* tab. As you can see, the *RandomForest Classifier (Gini)* (BP43) scores the best. You can also see that *DataRobot* recommends this model for deployment.

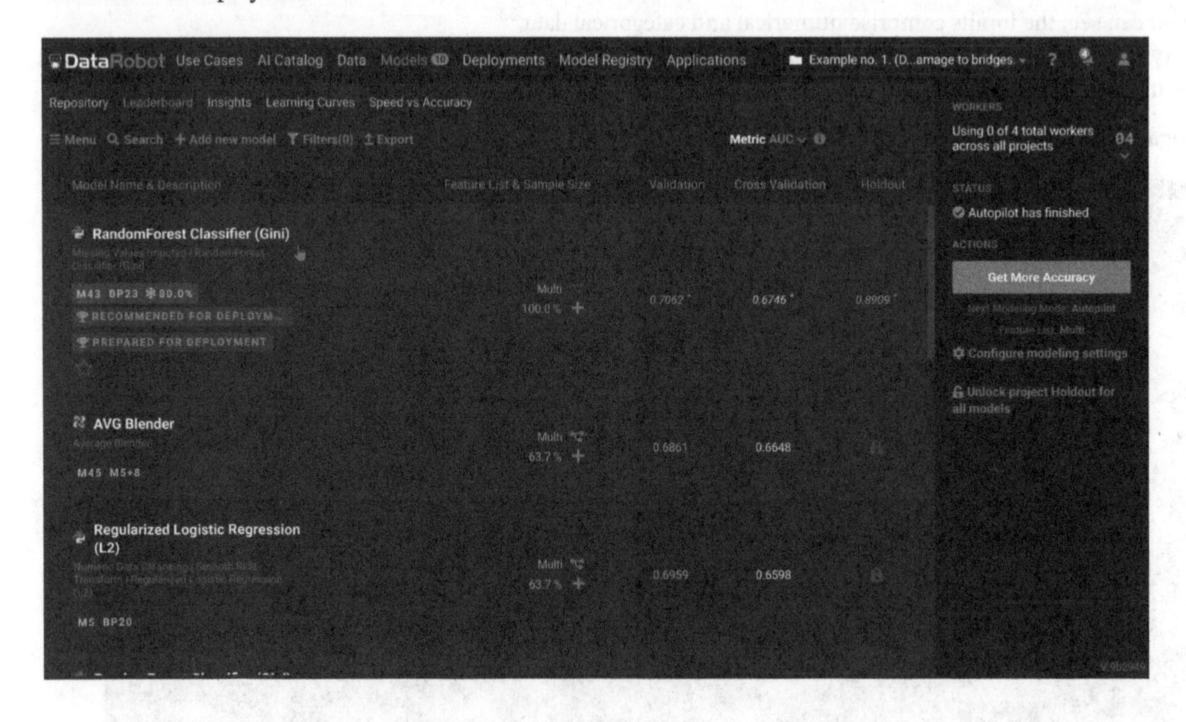

Step 4: Let us now examine the performance of this model in terms of the confusion matrix, which can be accessed through the *Evaluate* tab.

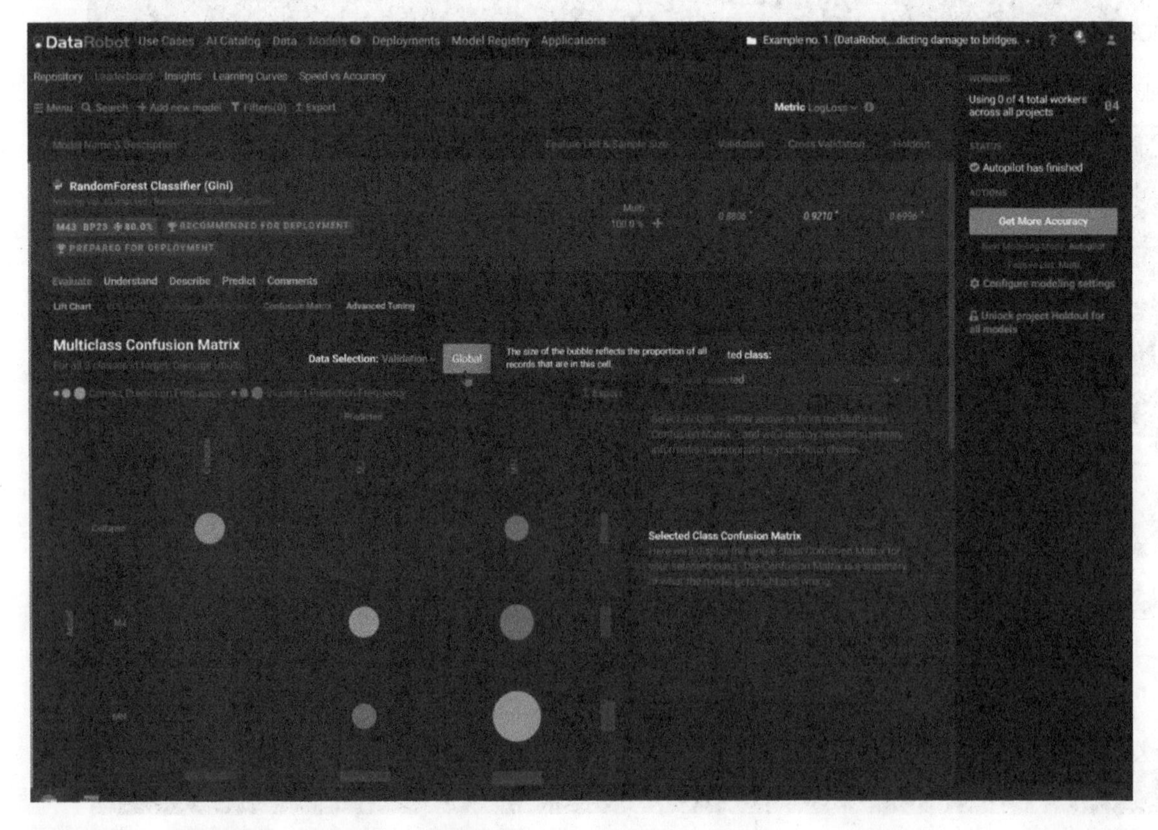

Step 5: Have a look at the importance of each feature as well as the effect of the feature on model predictions.[34] You can navigate the *Select Class* option to visualize the *impact* and *effect* of features upon the whole dataset (*Aggregation*) or a particular class.

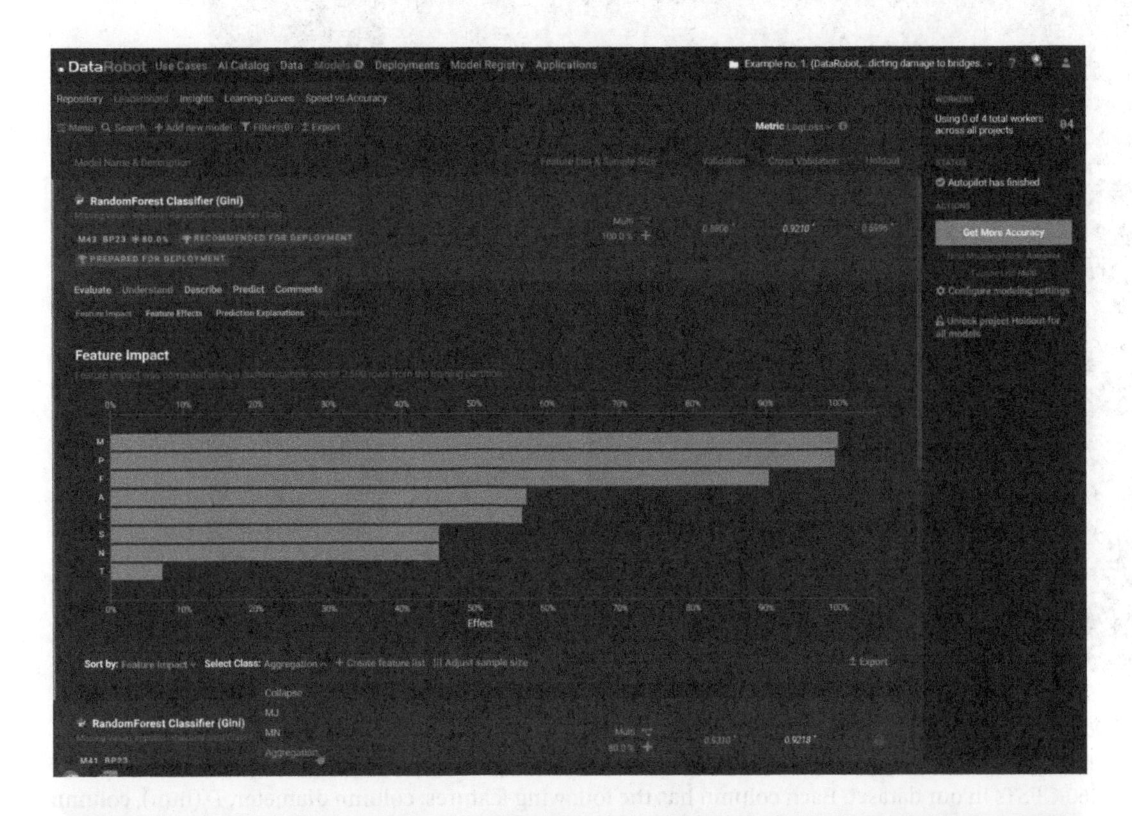

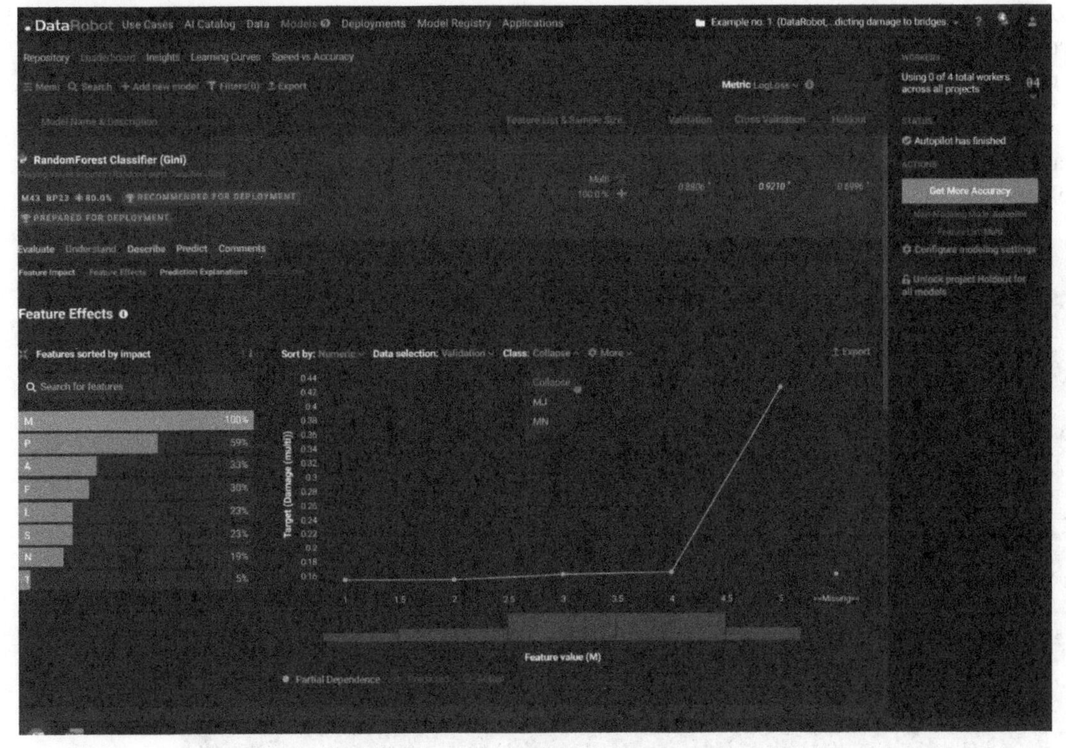

34 I realize that I did not cover these concepts here. I am reserving the discussion until we reach the next chapter. For now, you can think of these as a means to help us understand how the model was able to arrive at its predictions.

Step 6: Finally, you navigate the Data Quality Handling Report to see how the model handled any arising issues within the data/dataset (i.e., missing values, etc.).

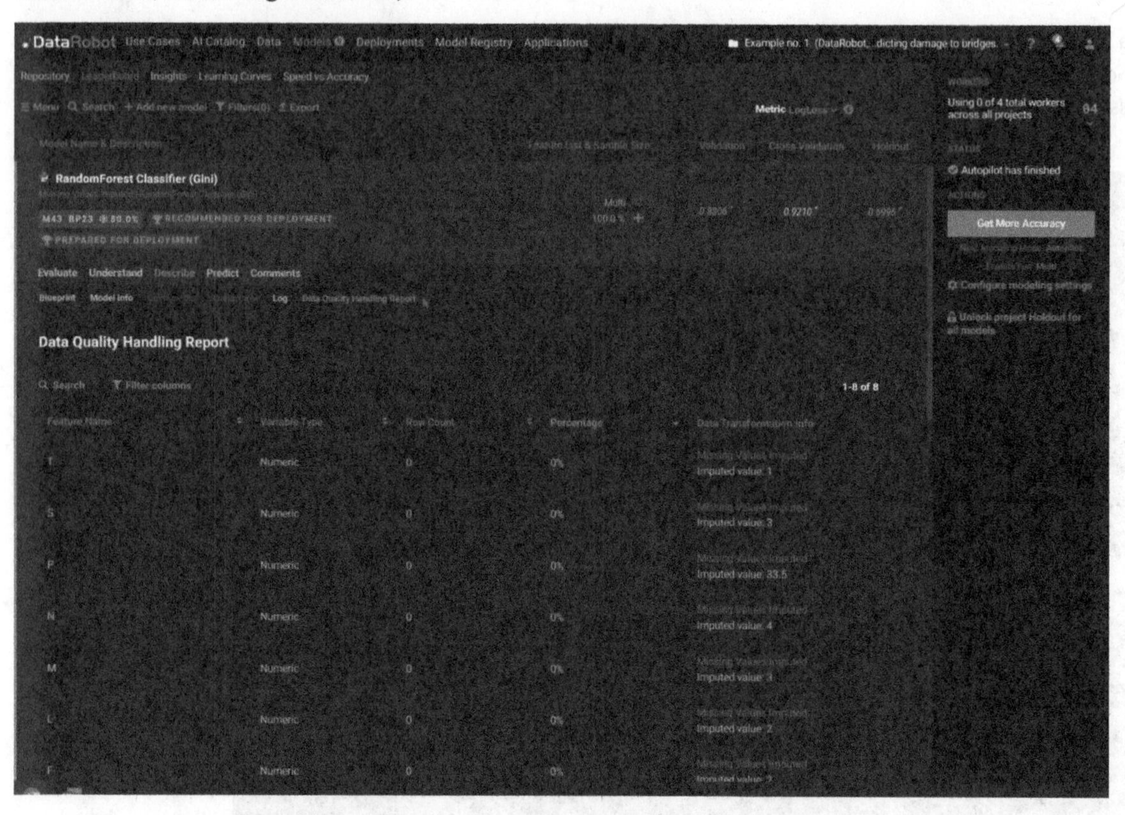

Example no. 2 [*DataRobot*, Regression]: Predicting axial strength of CFSTs.

The second example aims to use ML to predict the axial compressive strength capacity of concrete-filled steel tubes[35] (CFST). There are 1260 CFSTs in our dataset. Each column has the following features: column diameter, D (mm), column thickness, t (mm), column length, L (mm), compressive strength of concrete f_c' (MPa), and steel yielding stress, f_y (MPa). The target for each column was the axial compressive strength N (kN).

Follow me in the next few snippets to use *DataRobot* in this regression example where we will be using the Random Forest algorithm.

Step 1: Once the dataset is uploaded and select N as the *Target*. You will notice that the analysis changes to *Regression*. Select *Manual* modeling mode and *Start* the analysis.

35 This is a type of column that comprises a steel tube that is filled with concrete. In this system, the two materials work synergically and the composite column can develop a significant axial strength capacity.

Step 2: Go to the *Repository* tab and select *Random Forest Regressor (BP 75)* and *Run Task*.

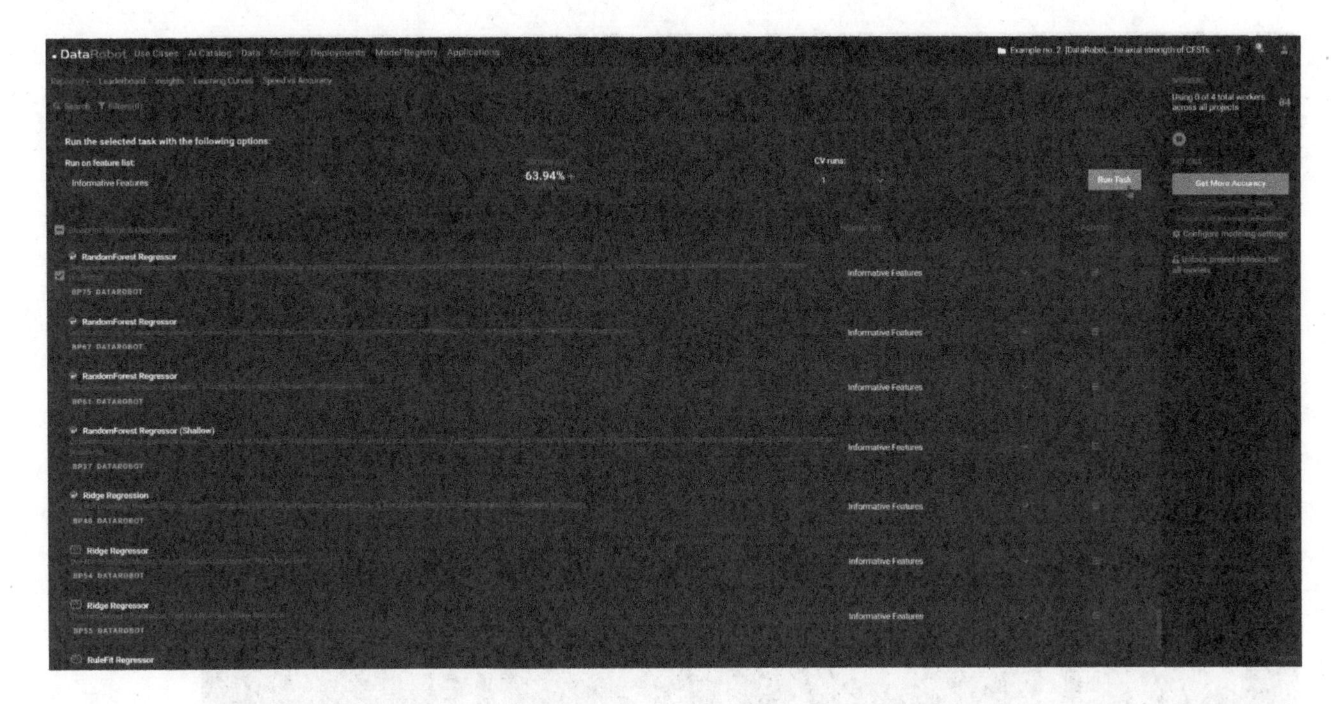

Step 3: Once the analysis completes. Visit the *Evaluate* tab, then *Residuals*. Update the *Data Selection* to Validation (or *Holdout*). You can also check the performance of the model in terms of the various metrics *DataRobot* adopts.

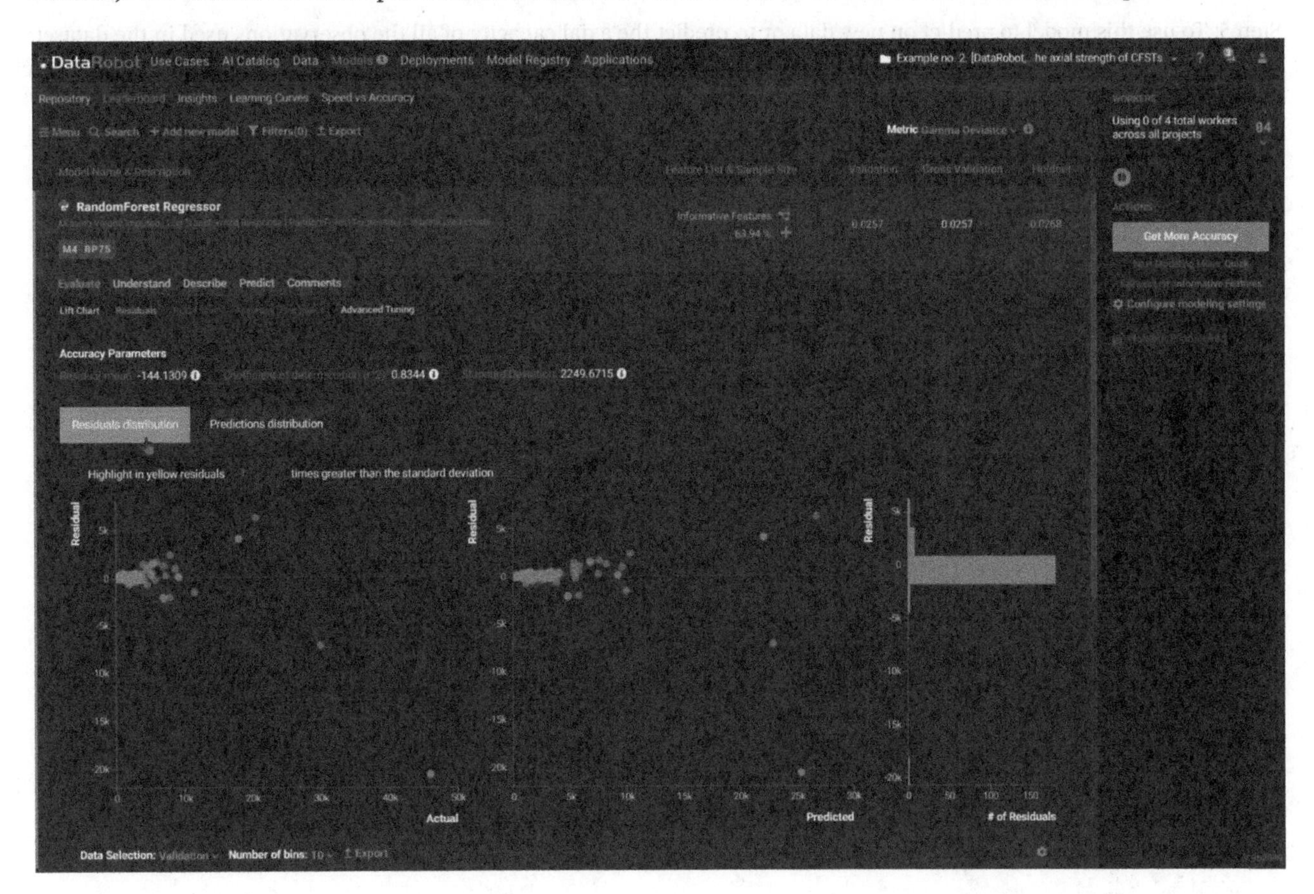

Step 4: Change the view from *Residuals distribution* to *Predictions distribution* and check *Highlight in yellow residuals*.

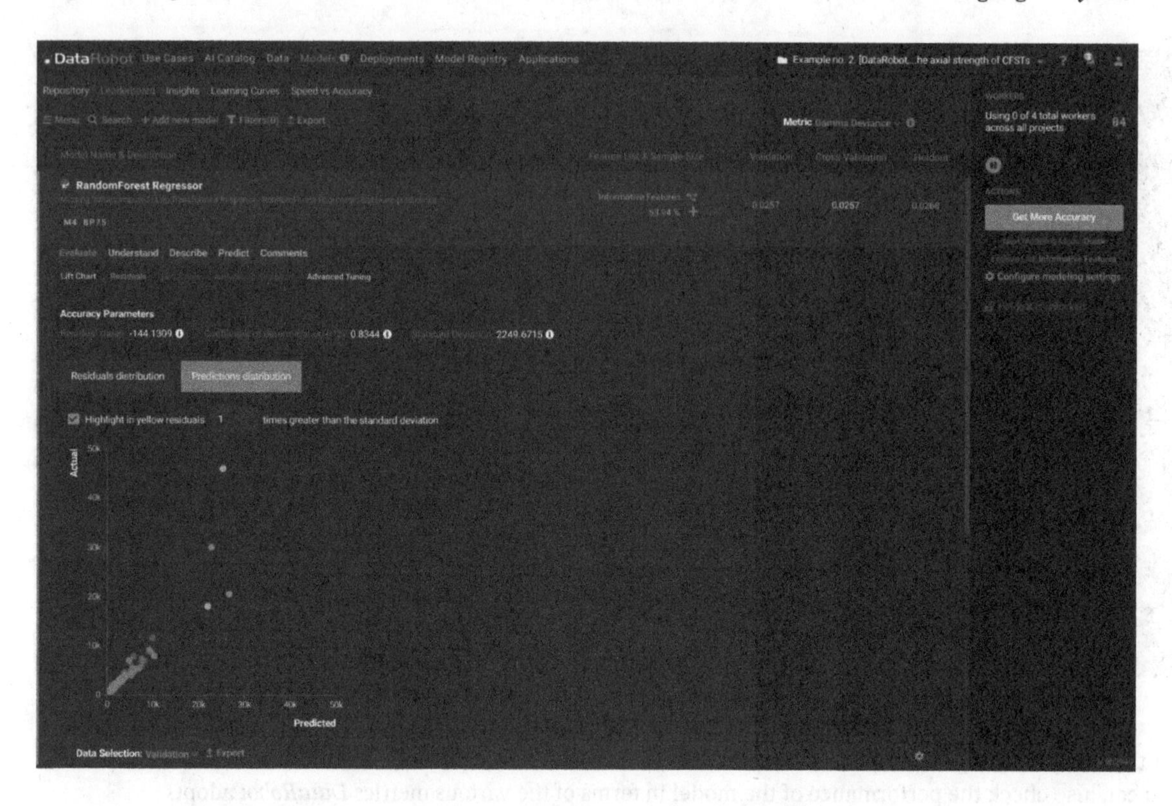

Step 5: To use this model to predict on new data or to predict the axial capacity of all the observations used in the dataset, navigate to the *Predict* tab. Upload a new sheet to predict new observations and simply select compute predictions. The figure below shows the latter for the data used in validation and holdout. Then, *Download predictions*.

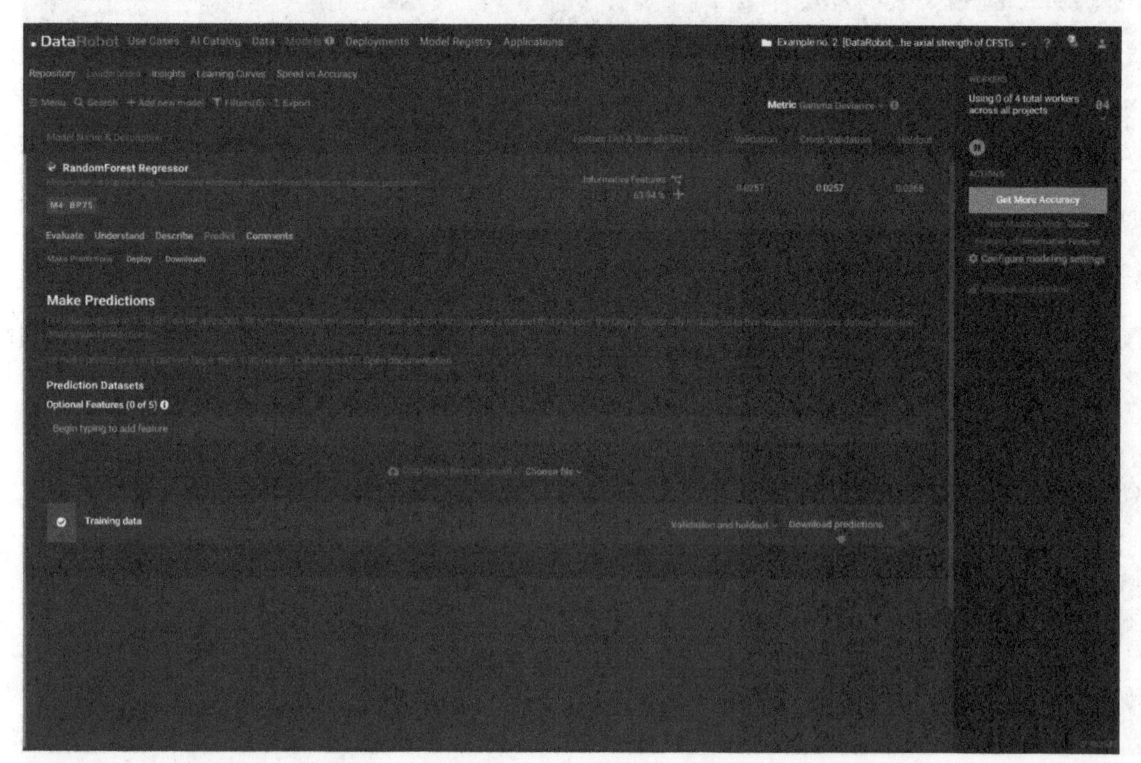

Example no. 3 [*DataRobot*, Anomaly detection[36]]: Predicting anomalies in fire resistance of RC columns.
In this example, we are trying to identify anomalous behavior in terms of the observed fire resistance. For example, of the available RC columns in the dataset, are there any columns that achieved a fire resistance degree that can be considered anomalous?

Here is some information on our dataset, which compiles data on 144 RC columns: column width, W (mm), steel reinforcement ratio, r (mm), column length, L (mm), concrete compressive strength, f_c (MPa), steel yield strength, f_y (MPa), restraint conditions by means of effective length factor, concrete cover to reinforcement, C (mm), eccentricity in applied loading in the main axes, e_x (mm), the magnitude of applied loading, P (kN), and fire resistance, FR (min).

Let us go to the *DataRobot* platform and carry on with the following steps. We will apply the *Local Outlier Factor Anomaly Detection with Calibration* and the *Isolation Forest Anomaly Detection with Calibration* algorithms.

Step 1: Upload the dataset and select *No target?* To switch into an unsupervised analysis. Then, select *Anomalies*. Select *Modeling Mode: Manual.*

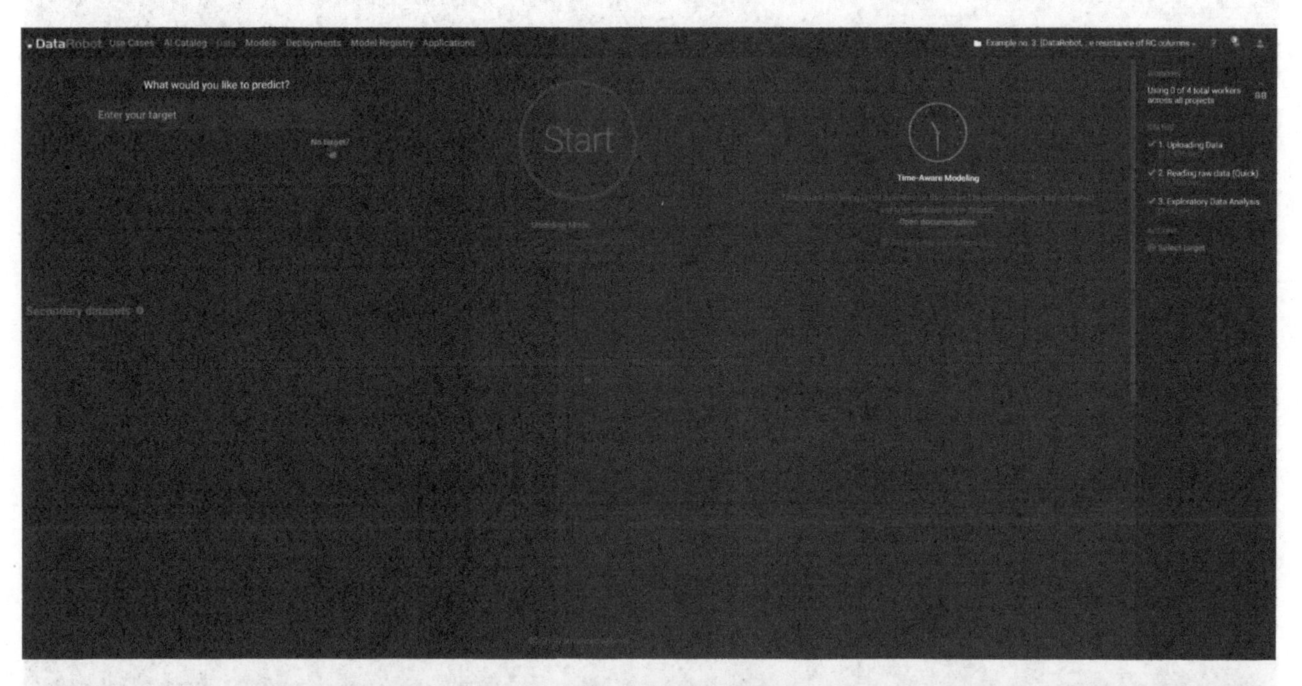

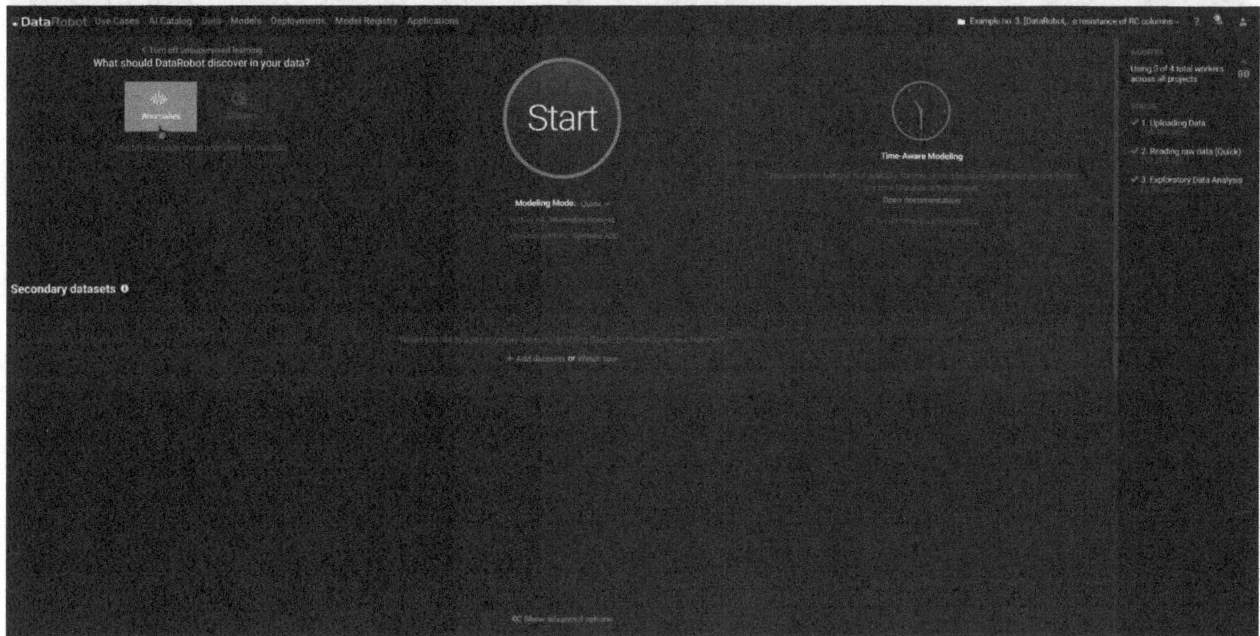

36 At this moment in time, *DataRobot* does not include clustering in its student version. As such, I will be showing this example on anomaly detection to cover unsupervised learning. We will re-visit clustering in *BigML* and *Exploratory*.

Step 2: *Local Outlier Factor Anomaly Detection with Calibration (PB 8)* and *Isolation Forest Anomaly Detection with Calibration (PB7)* algorithms.[37]

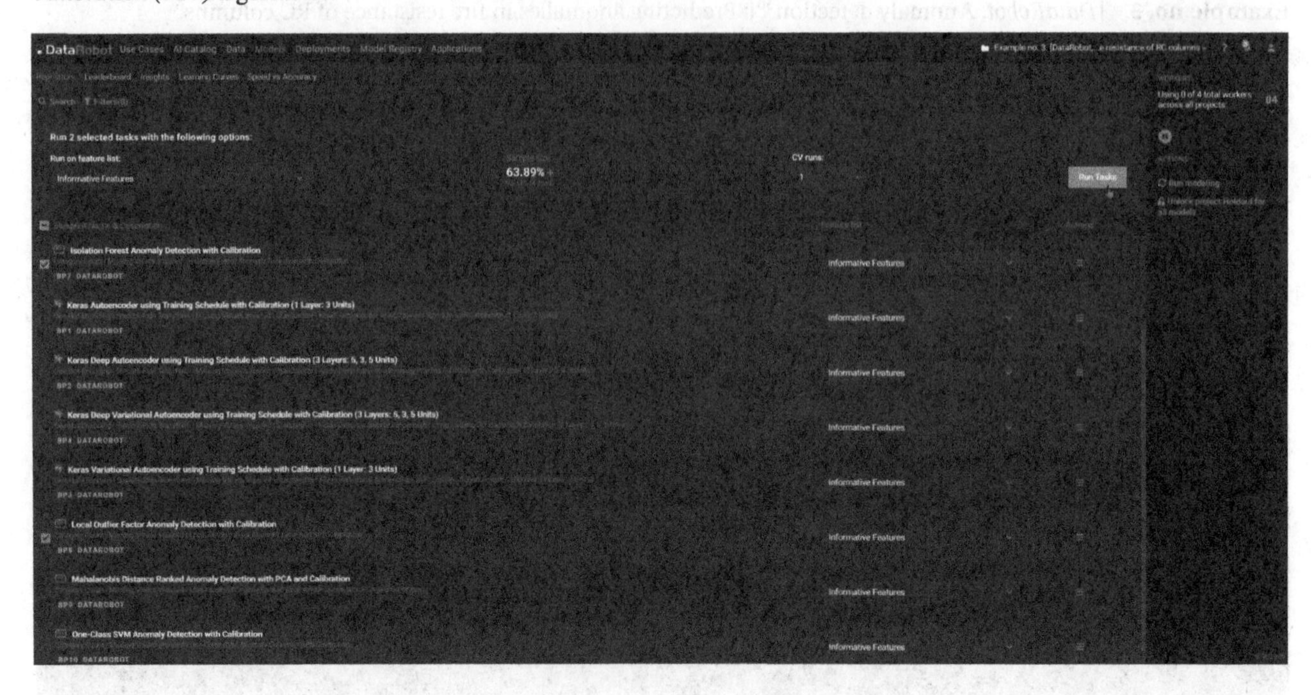

Step 3: Compare the results of your analysis in terms of the Synthetic AUC metric.[38]

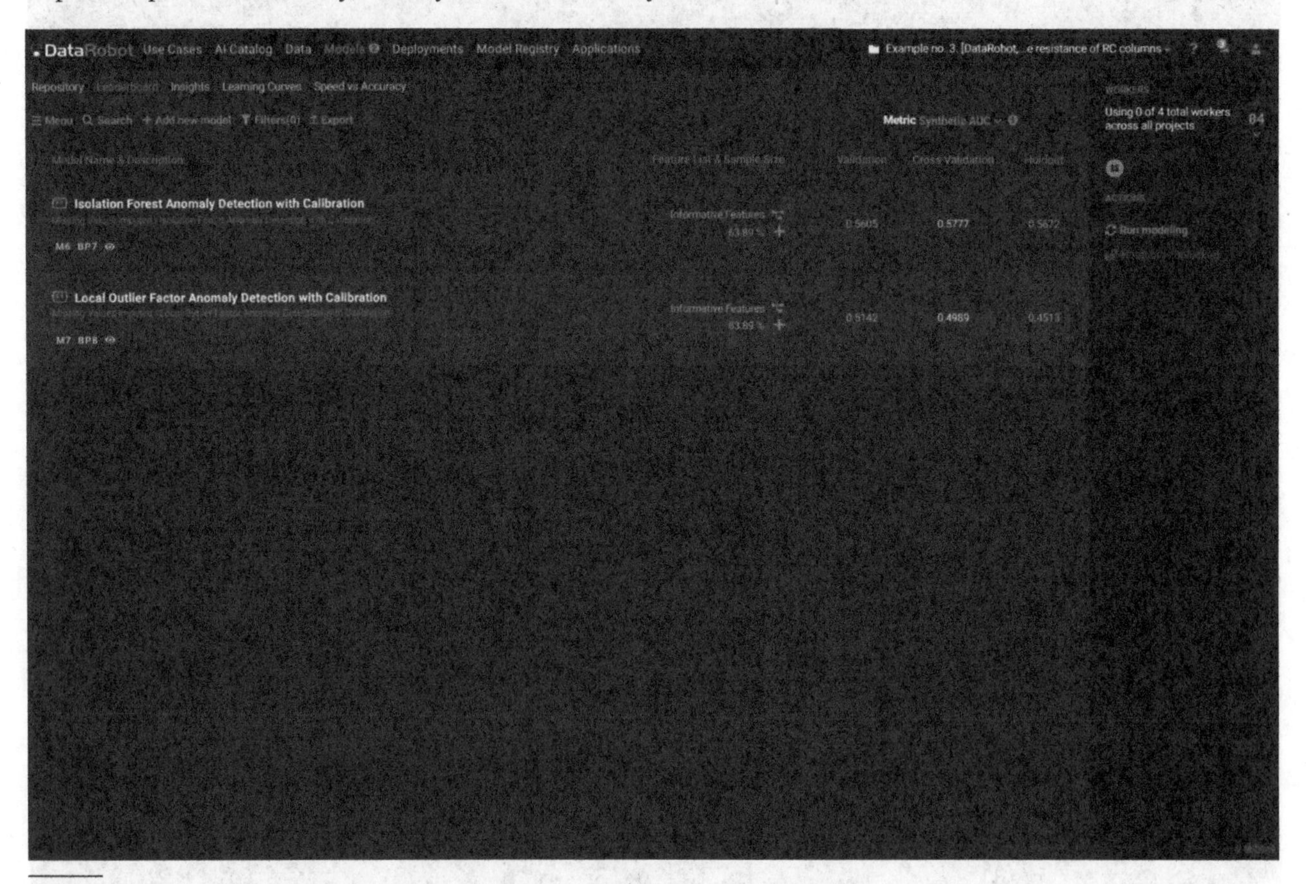

37 Also, feel free to try other algorithms that we did not cover in the algorithm chapter.

38 According to *DataRobot* [33]: This metric generates two synthetic datasets out of the validation sample – one made more normal, one made more anomalous, and the samples are labeled accordingly. The model then calculates anomaly score predictions for both samples using artificial labels as the ground truth. For example, if a model has a Synthetic AUC of 0.9 is likely to outperform another model with a Synthetic AUC score of 0.5.

Step 4: Leave the *Leaderboard* tab and navigate to the *Insights* tab. Check the *Anomaly scores.*[39] Switch between the two models to evaluate each model's scores. You can simply download the scores for the dataset from the *Make Predictions* option under the *Predict* tab.

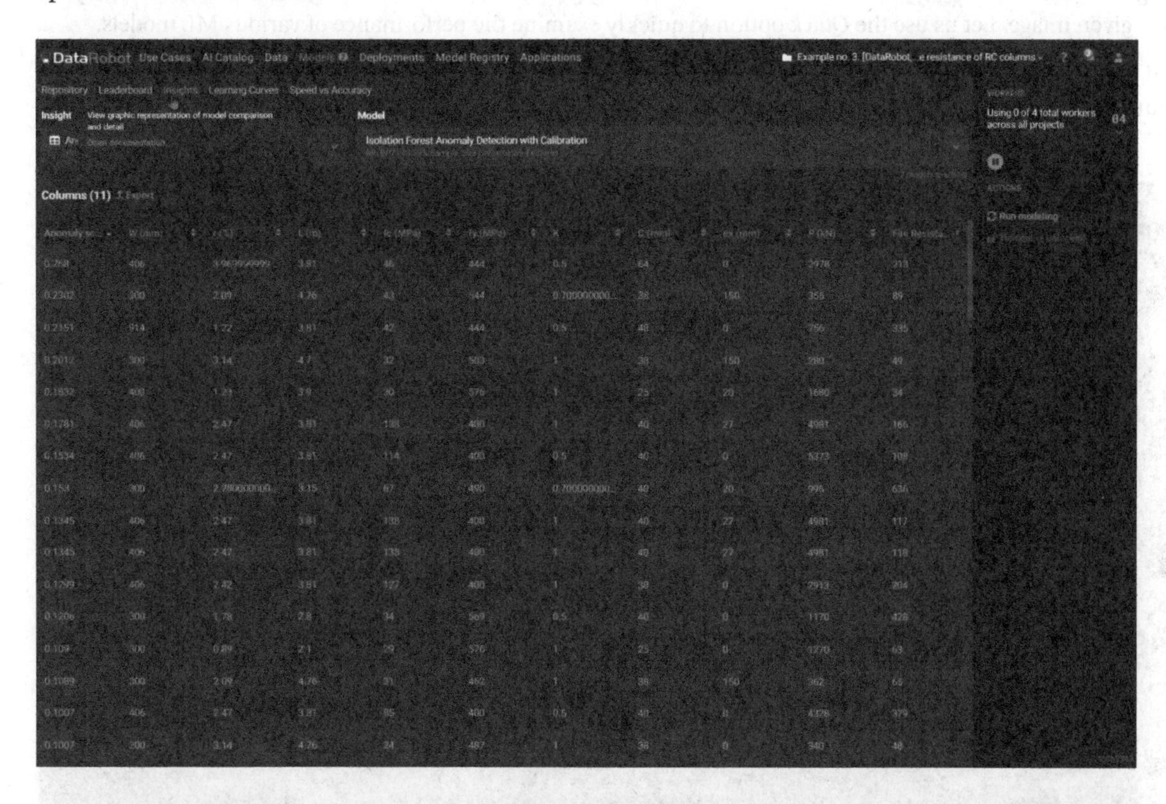

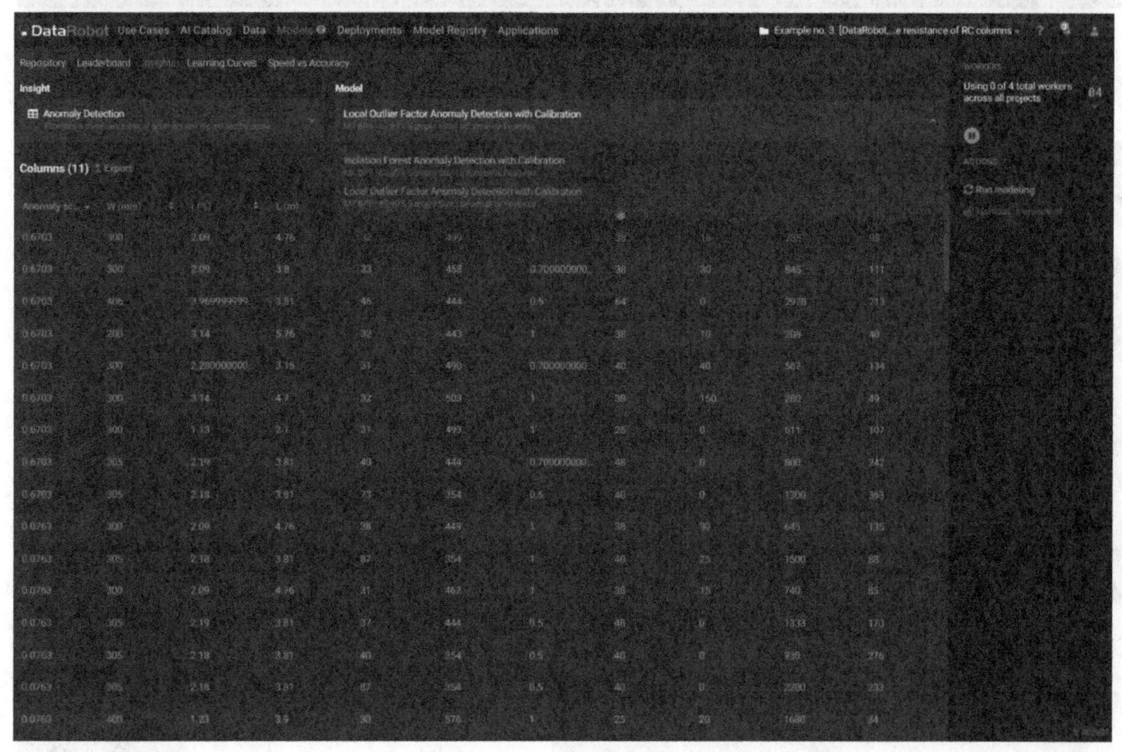

39 Larger scores imply higher tendency for anomalies.

Example no. 4 [*DataRobot*, Classification]: Computer vision for crack detection.
DataRobot provides its users with a simple mean to apply computer vision to footage. In this example, I have a dataset of about 120 images on cracked and uncracked concrete elements. The goal is to build a ML classifier capable of identifying cracks within a given image. Let us use the *Quick* option to quickly examine the performance of various ML models.

Step 1: Upload the provided dataset. Select *Class* as the *Target*. Select *2* for the cross validation option (note, we are selecting *2* as the number of images is small, in reality, a much larger pool of images is needed).

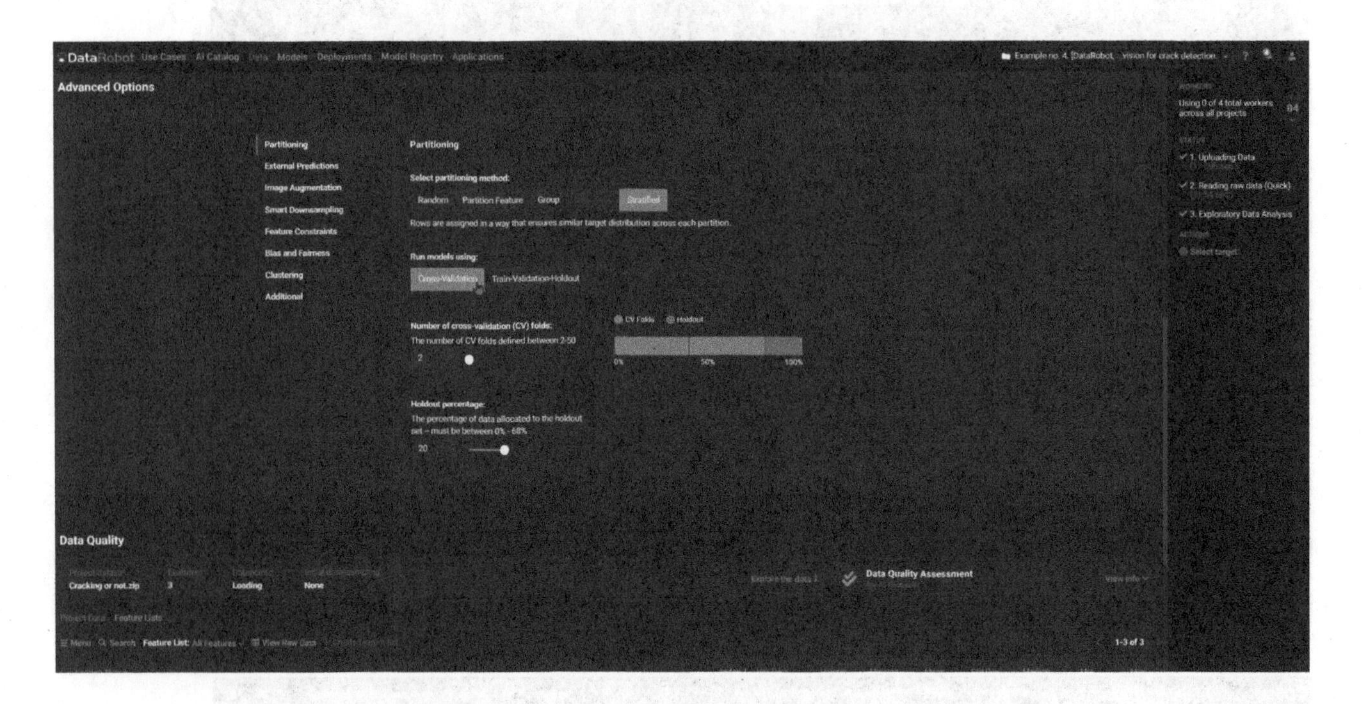

Step 2: Explore the uploaded images by selecting *Images*.

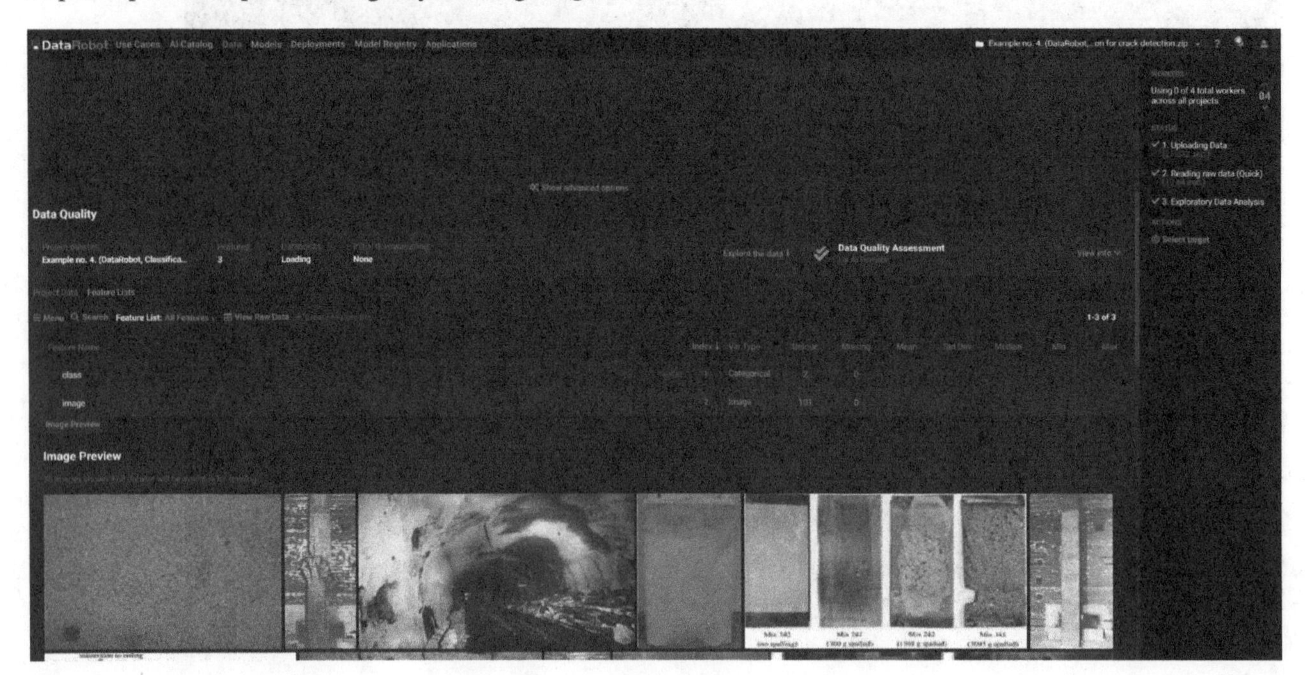

Step 3: *Start* the analysis.

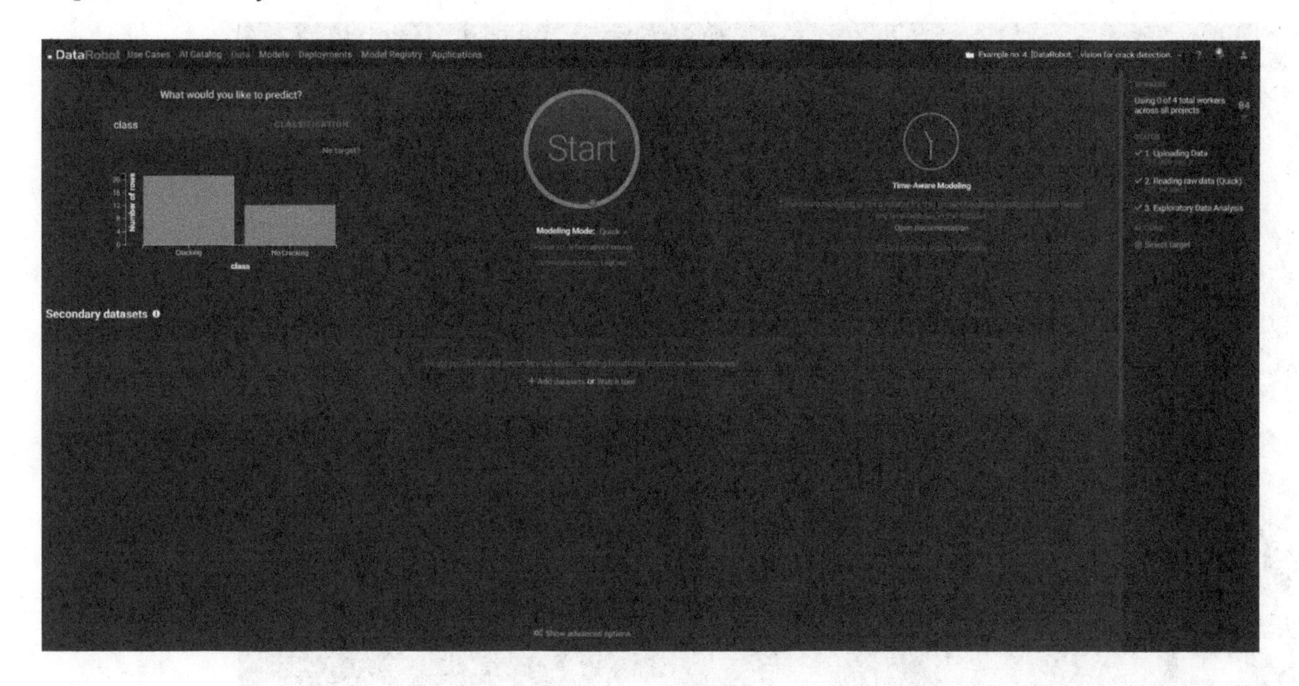

Step 4: Wait until the analysis finishes. As you can see, *DataRobot* recommends the *Elastic-Net Classifier (L2/Binomial Deviance) (BP 35)* model given its good performance in terms of AUC.

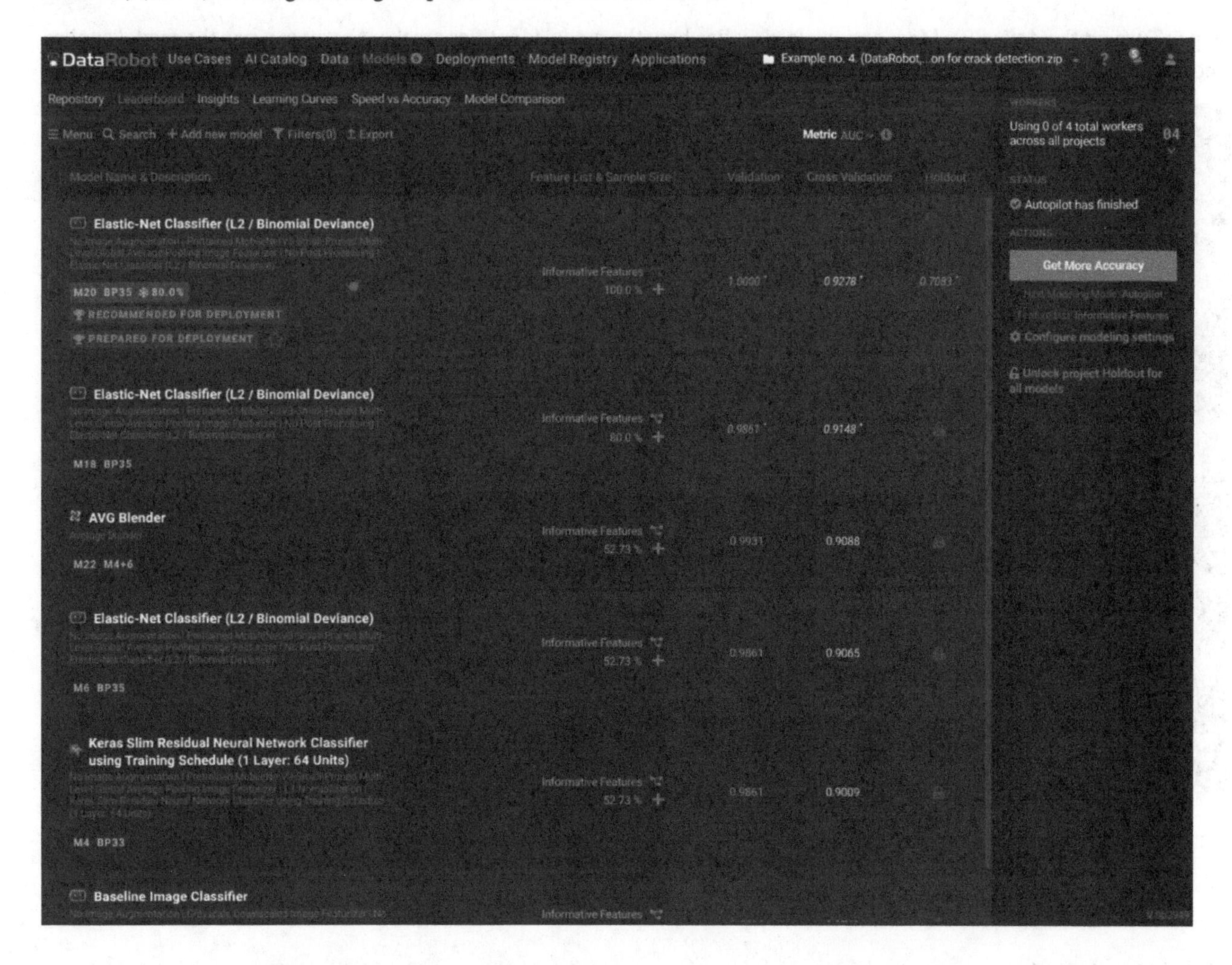

Step 5: Examine model performance which can be found under the *ROC Curve* tab under the *Evaluate* tab.

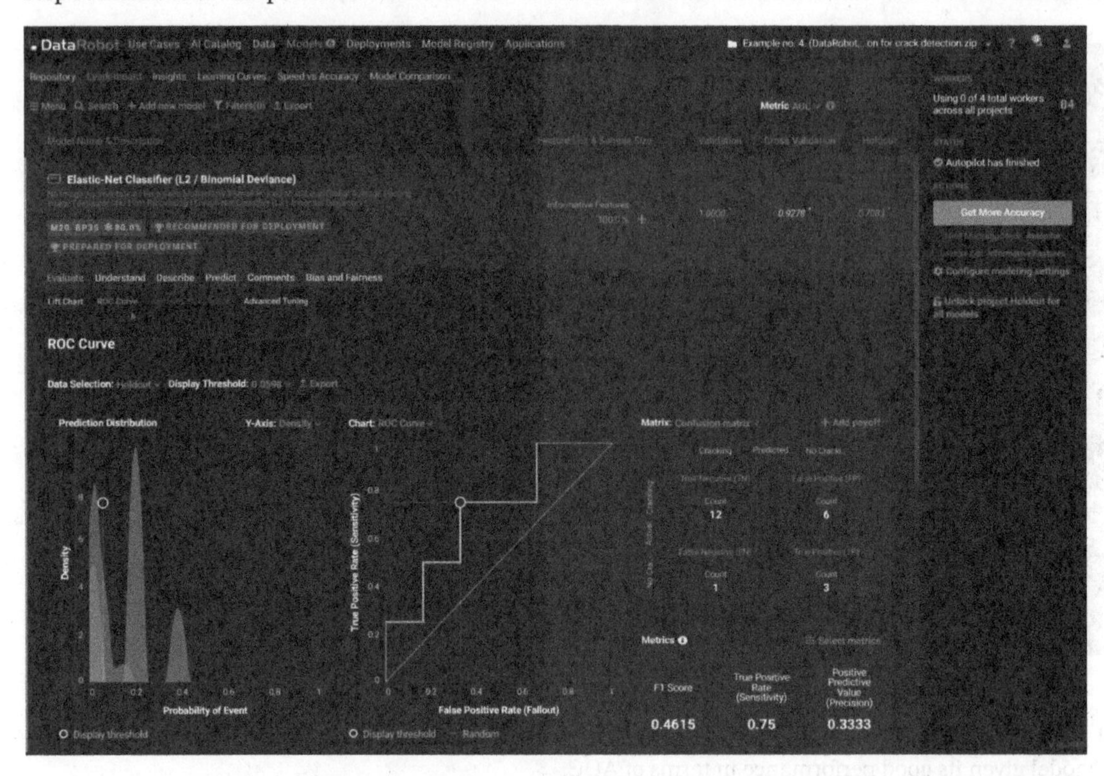

Step 6: Turn on the *Activation Maps* options to visualize how the model identifies cracks. As you can see, the model applies a contour over the image which aligns with cracks (hard warms colors).

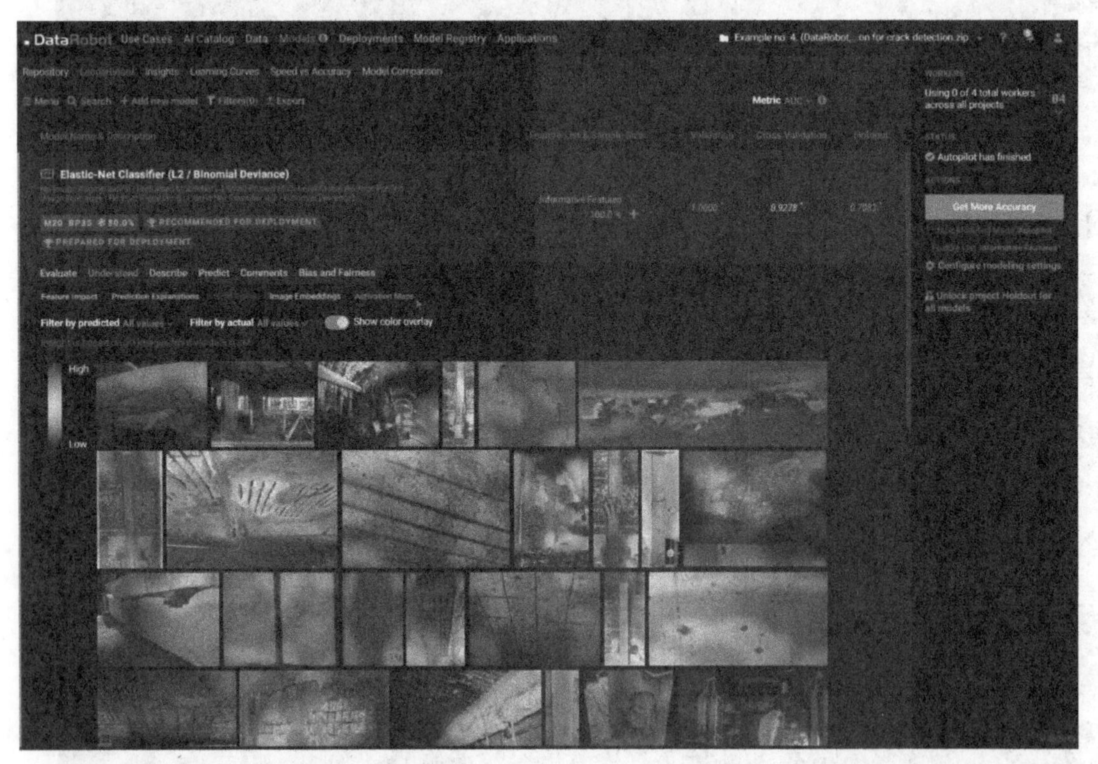

This completes our examples using *DataRobot*.

6.1.3 Dataiku

According to their Dataiku,[40]

> "*Systemizing the use of data. Everyday AI, Extraordinary People. Systemizing the Use of Data. Dataiku is one central solution for the design, deployment, and management of AI applications.*"

In 2013, Florian Douetteau, Clément Stenac, Thomas Cabrol, and Marc Batty founded Dataiku as an AI and ML company. *Dataiku*[41] has offices in New York, Paris, London, Munich, Sydney, Singapore, and Dubai. *Dataiku* also provides a wealth of information, blogs, and ebooks on the use of ML on their website. This platform has a dedicated Academy[42] that students and users can join and learn from. *Dataiku* can accommodate tabular data and image (i.e., computer vision) analysis. The homepage for *Dataiku* is shown in Figure 6.3. We will be going over a few examples herein.

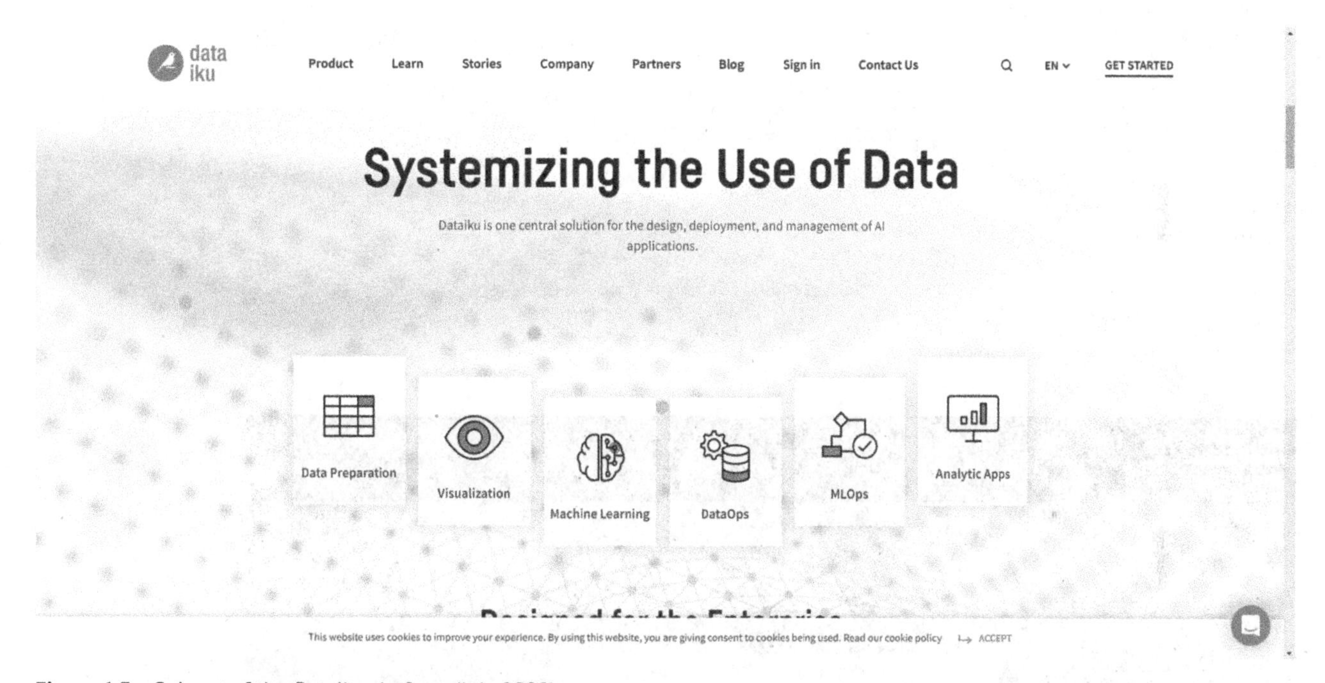

Figure 6.3 Snippet of the *Dataiku* platform (July 2022).

Example no. 1 [*Dataiku*, Classification]: Classifying CO2 rating in vehicles.
In this example, the Canadian fuel consumption rating dataset is used to classify CO2 rating in vehicles produced in 2022 [7]. This dataset contains 945 observations and four features (engine size, number of cylinders, and fuel consumption in the city) and one target, CO2 rating[43] (10 classes[44]).

40 https://www.dataiku.com.

41 Dataiku is a portmanteau for Data and **Haiku:** A short, structured Japanese poem [34].

42 Thanks to Grace Corey for her support!

43 The next example will cover the regression counterpart by assigning the CO2 emissions as the target. This example builds on the same approach we covered in Chapter 3.

44 We will re-do this example by adding explainability measures in Chapter 7.

Step 1: Let us upload our dataset and create a name for this project.

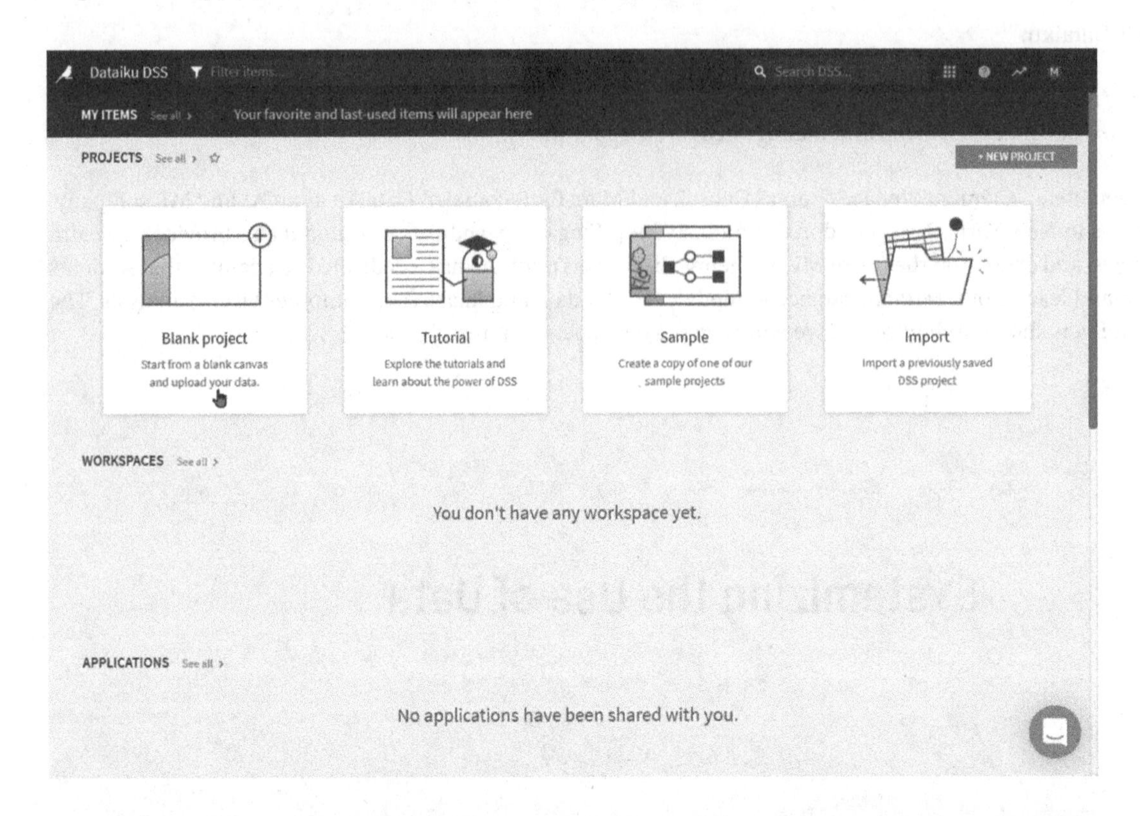

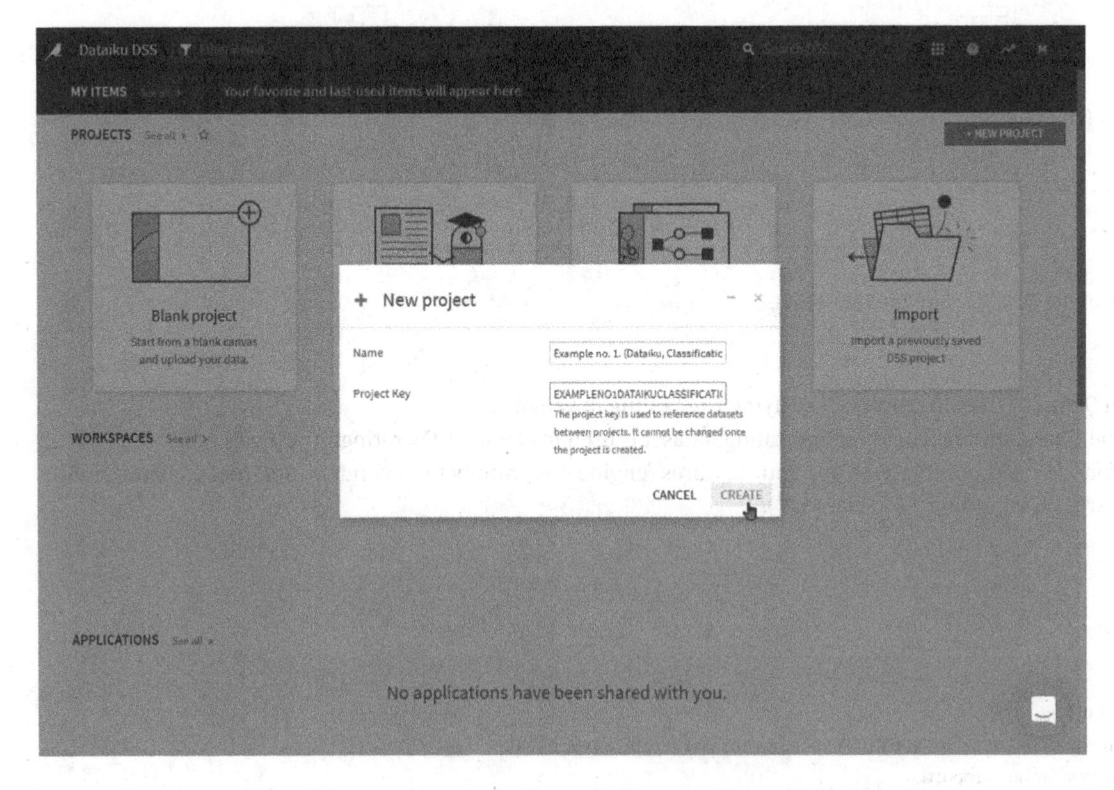

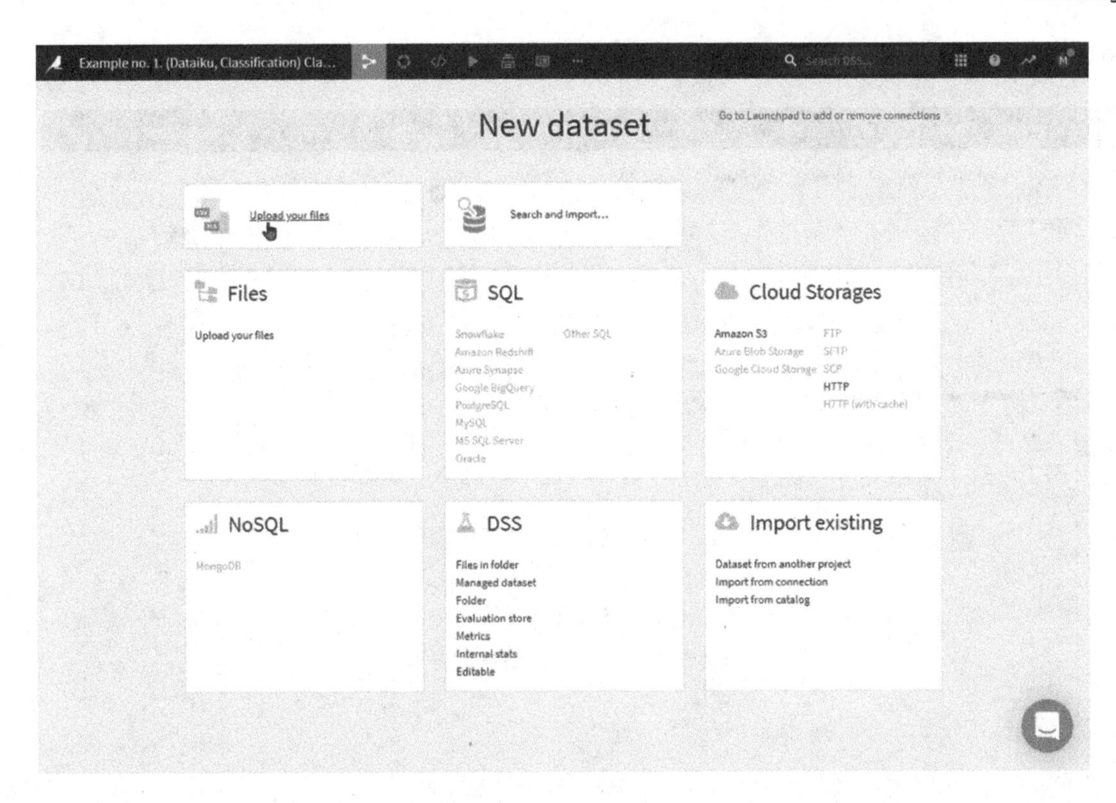
Example no. 1. (Dataiku, Classification) Cla...
New dataset
Go to Launchpad to add or remove connections
Upload your files
Search and import...
Files
Upload your files
SQL
Cloud Storages
Amazon S3
HTTP
NoSQL
DSS
Files in folder
Managed dataset
Folder
Evaluation store
Metrics
Internal stats
Editable
Import existing
Dataset from another project
Import from connection
Import from catalog

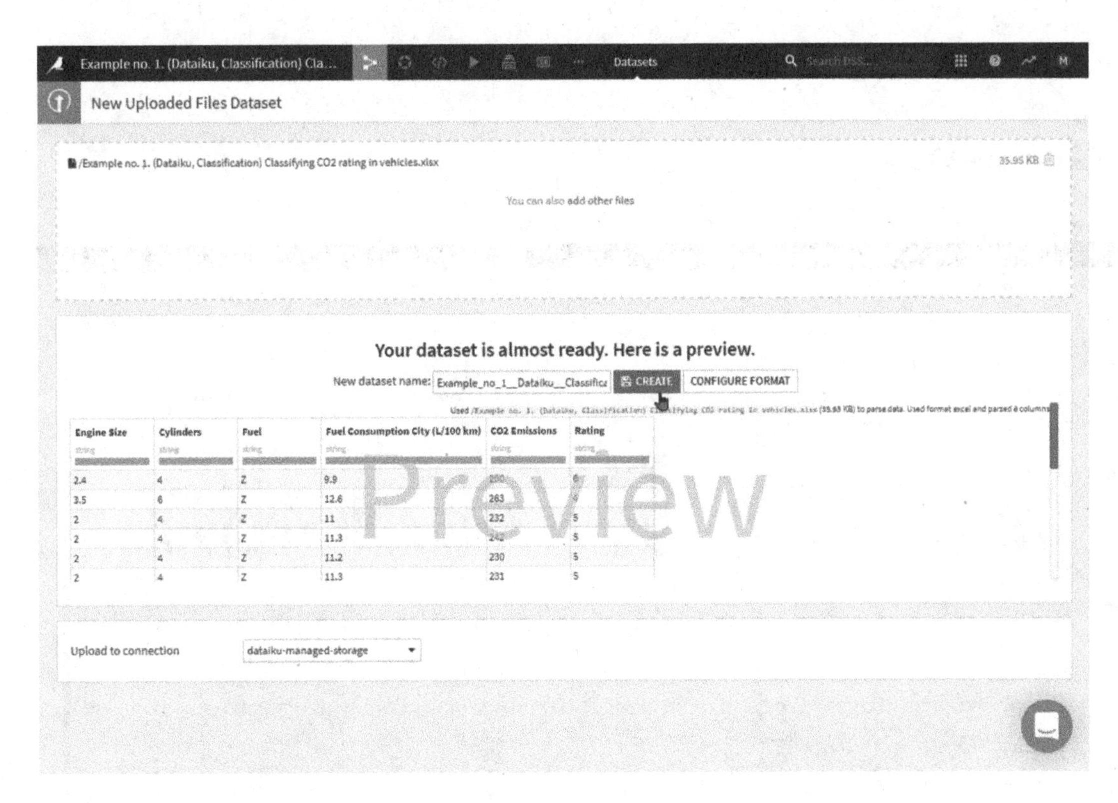
Example no. 1. (Dataiku, Classification) Cla...
Datasets
New Uploaded Files Dataset
/Example no. 1. (Dataiku, Classification) Classifying CO2 rating in vehicles.xlsx
35.95 KB
You can also add other files
Your dataset is almost ready. Here is a preview.
New dataset name:
Example_no_1__Dataiku__Classific
CREATE
CONFIGURE FORMAT
Engine Size
Cylinders
Fuel
Fuel Consumption City (L/100 km)
CO2 Emissions
Rating
Preview
Upload to connection
dataiku-managed-storage

Step 2: Navigate to the *Statistics* tab to view insights on this dataset.

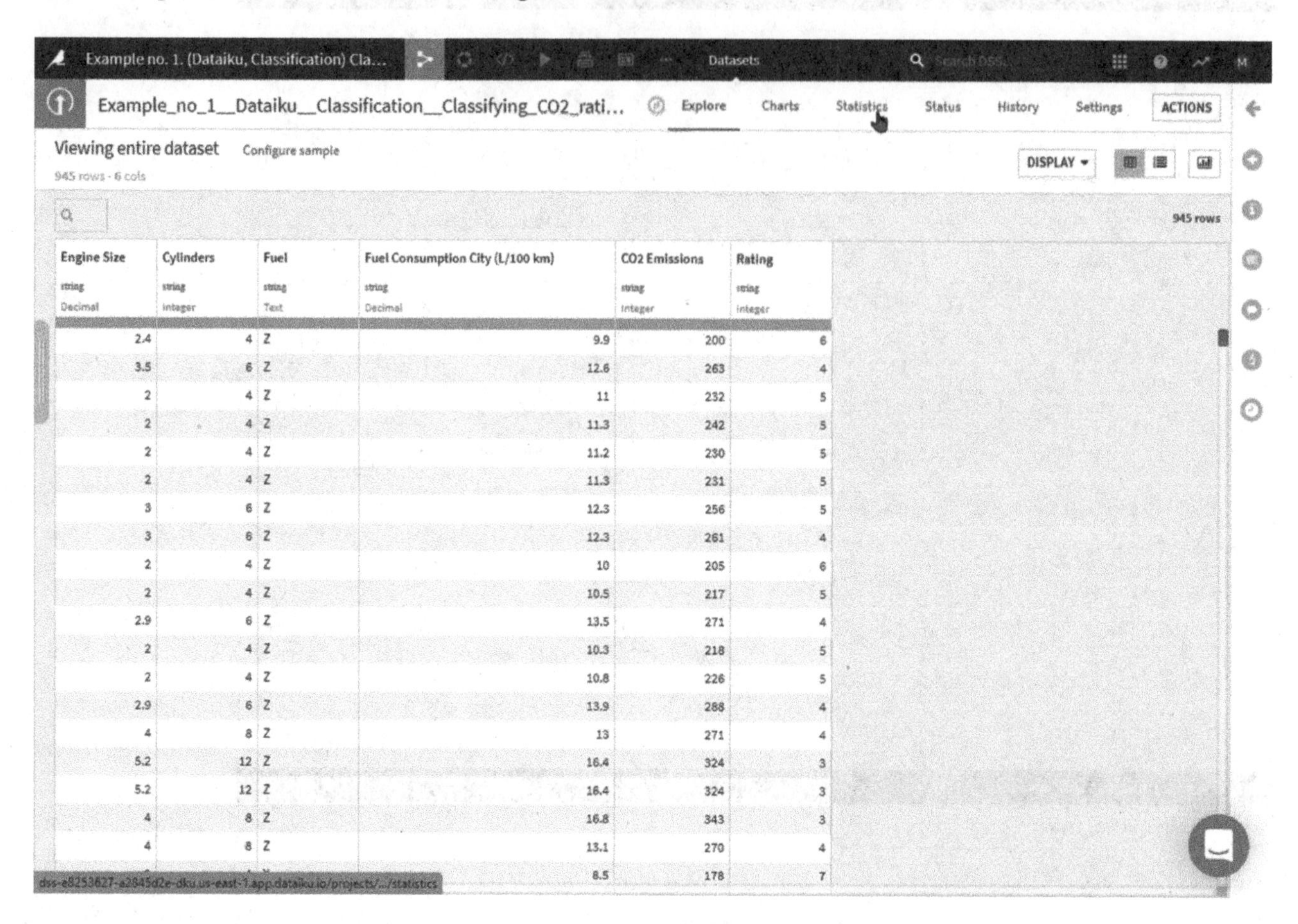

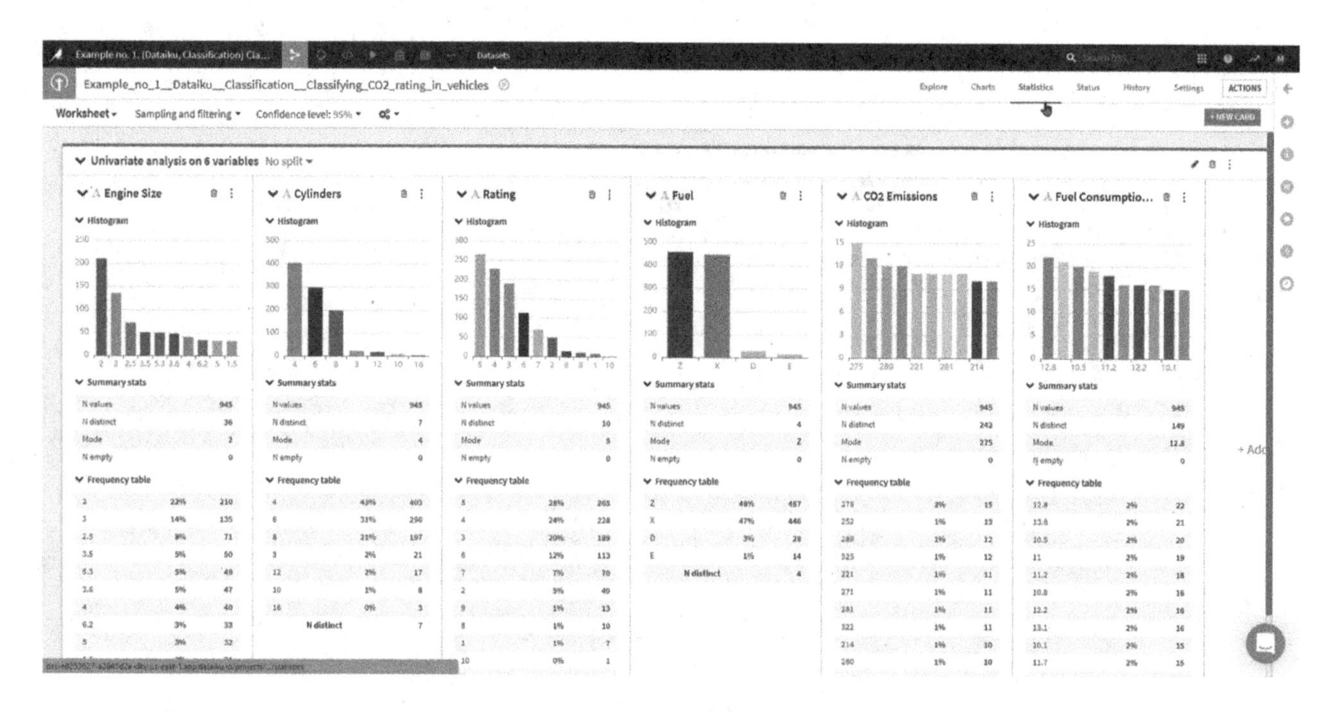

Step 3: Navigate to *Home* and then to *Lab*.

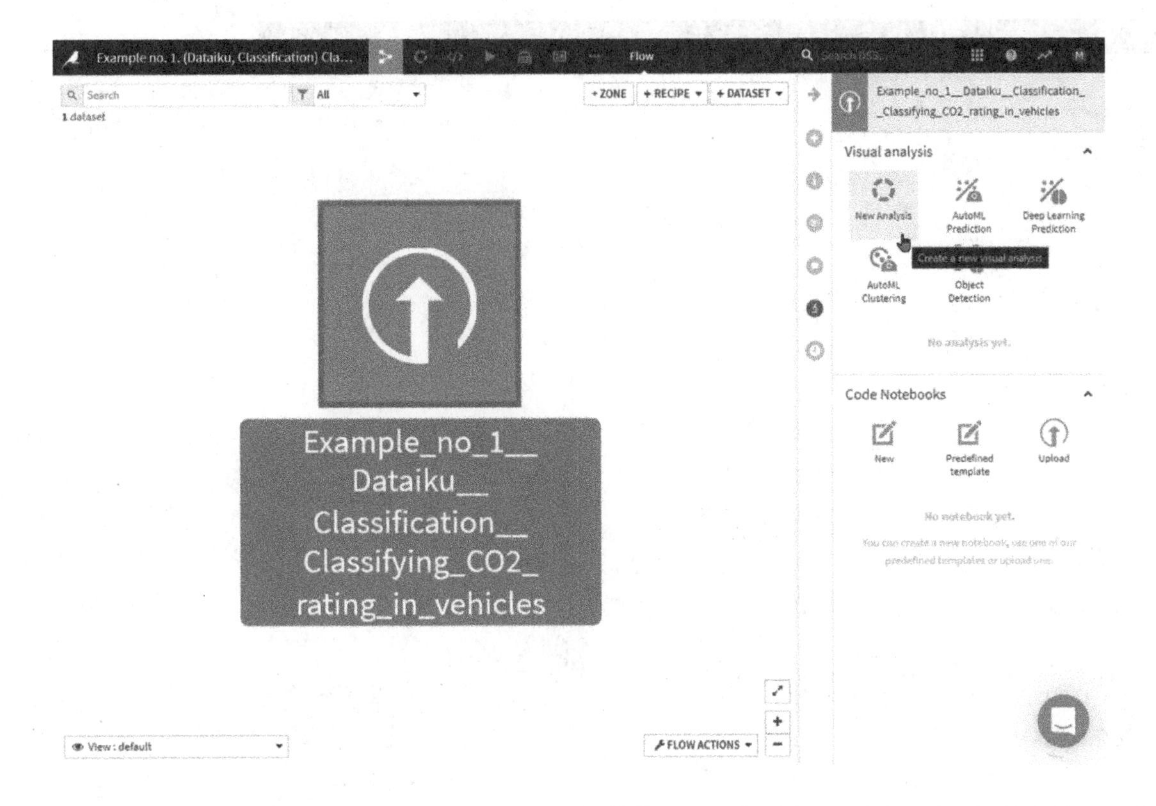

Step 4: Delete *CO2 Emissions*

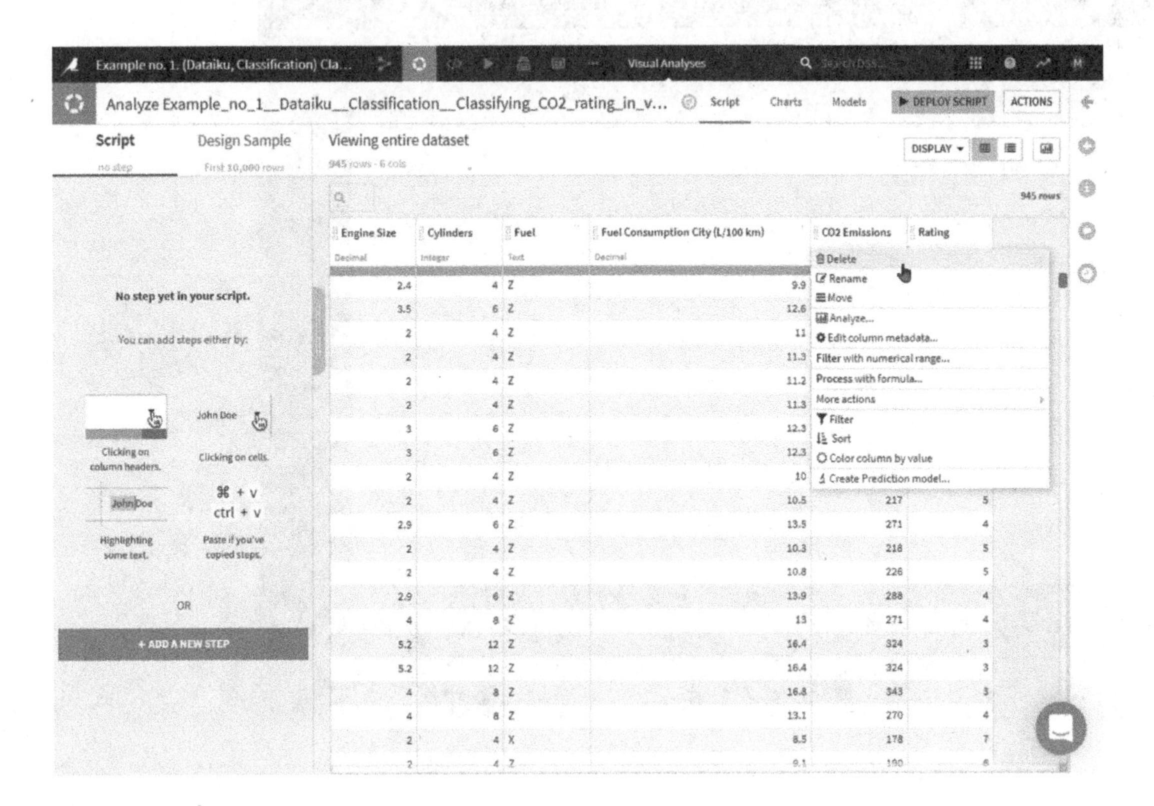

Step 5: Navigate to *Models*.

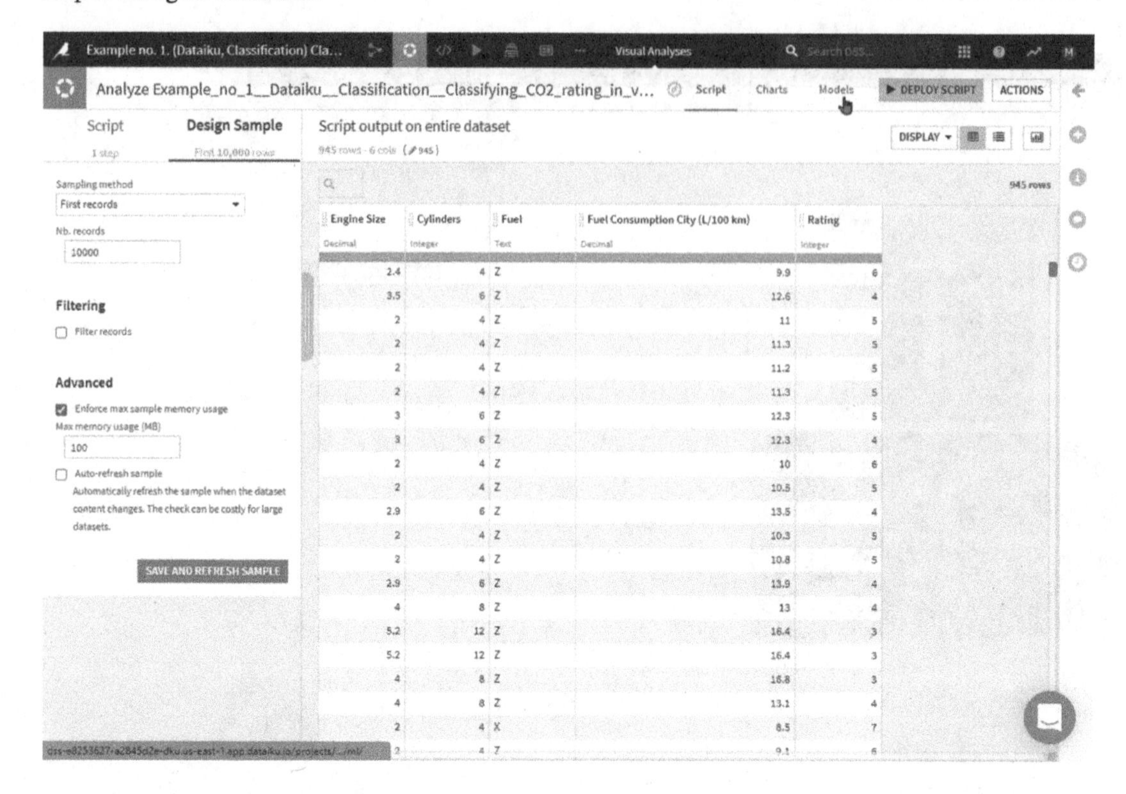

Engine Size	Cylinders	Fuel	Fuel Consumption City (L/100 km)	Rating
2.4	4	Z	9.9	6
3.5	6	Z	12.6	4
2	4	Z	11	5
2	4	Z	11.3	5
2	4	Z	11.2	5
2	4	Z	11.3	5
3	6	Z	12.3	5
3	6	Z	12.3	4
2	4	Z	10	6
2	4	Z	10.5	5
2.9	6	Z	13.5	4
2	4	Z	10.3	5
2	4	Z	10.8	5
2.9	6	Z	13.9	4
4	8	Z	13	4
5.2	12	Z	16.4	3
5.2	12	Z	16.4	3
4	8	Z	16.8	3
4	8	Z	13.1	4
2	4	X	8.5	7

Step 6: Select *Rating* as the target for your model. Choose *Quick Prototypes*.

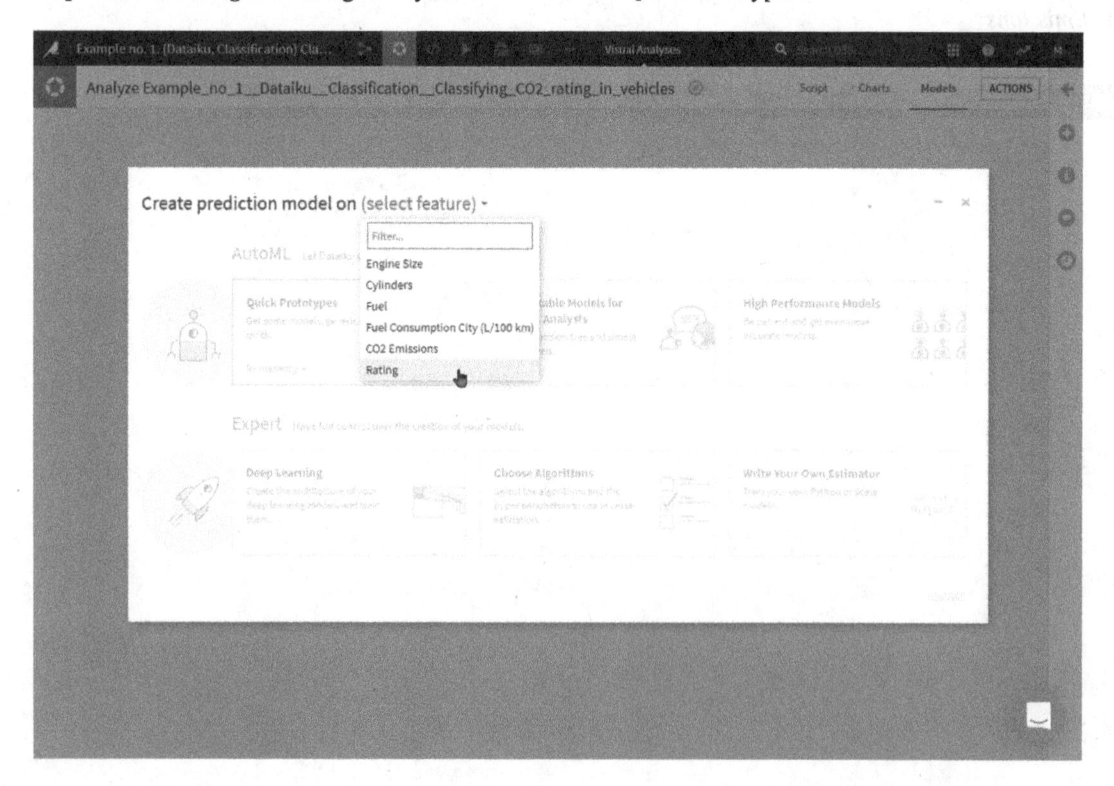

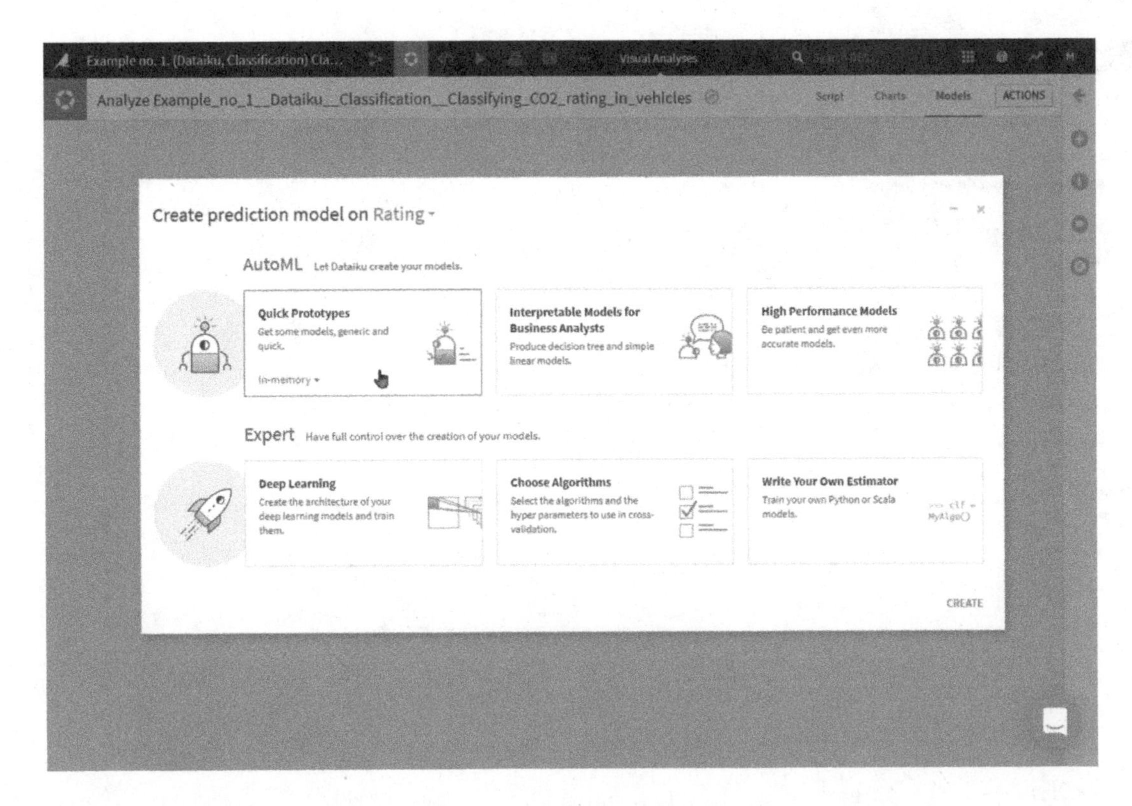

Step 7: Go to *Design* and change the *Prediction type* for *Multiclass classification*. Feel free to update other settings.

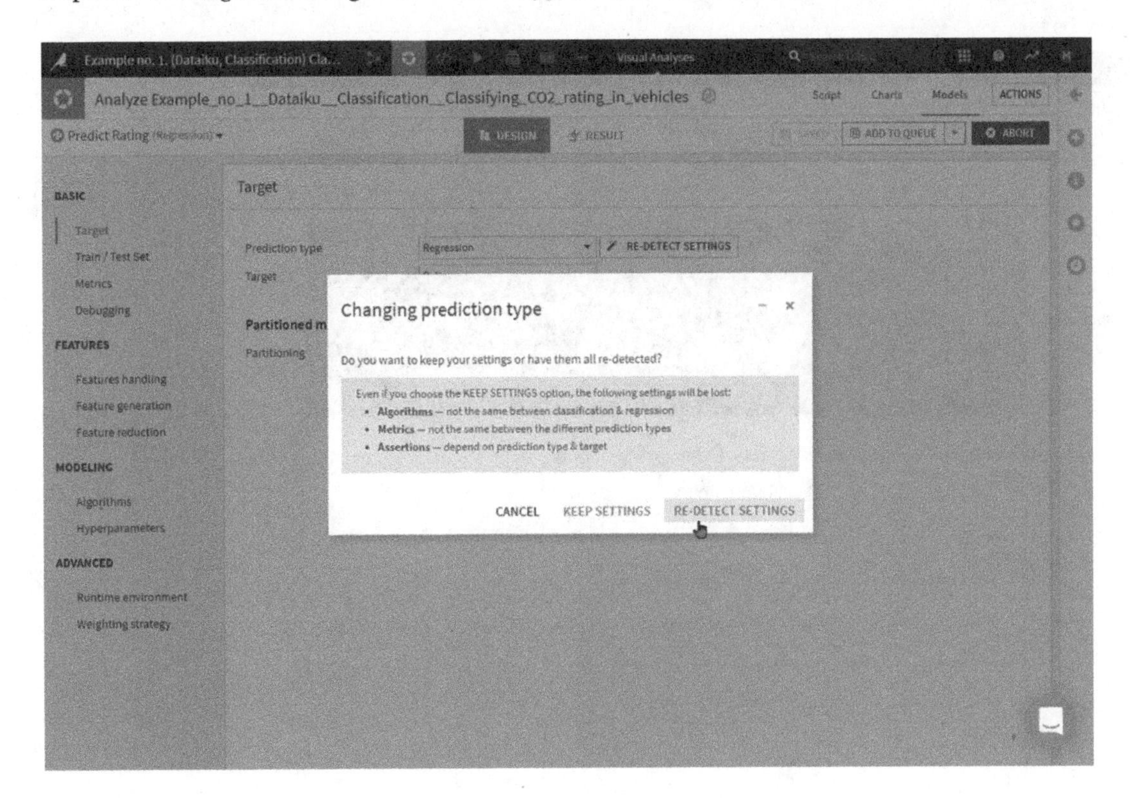

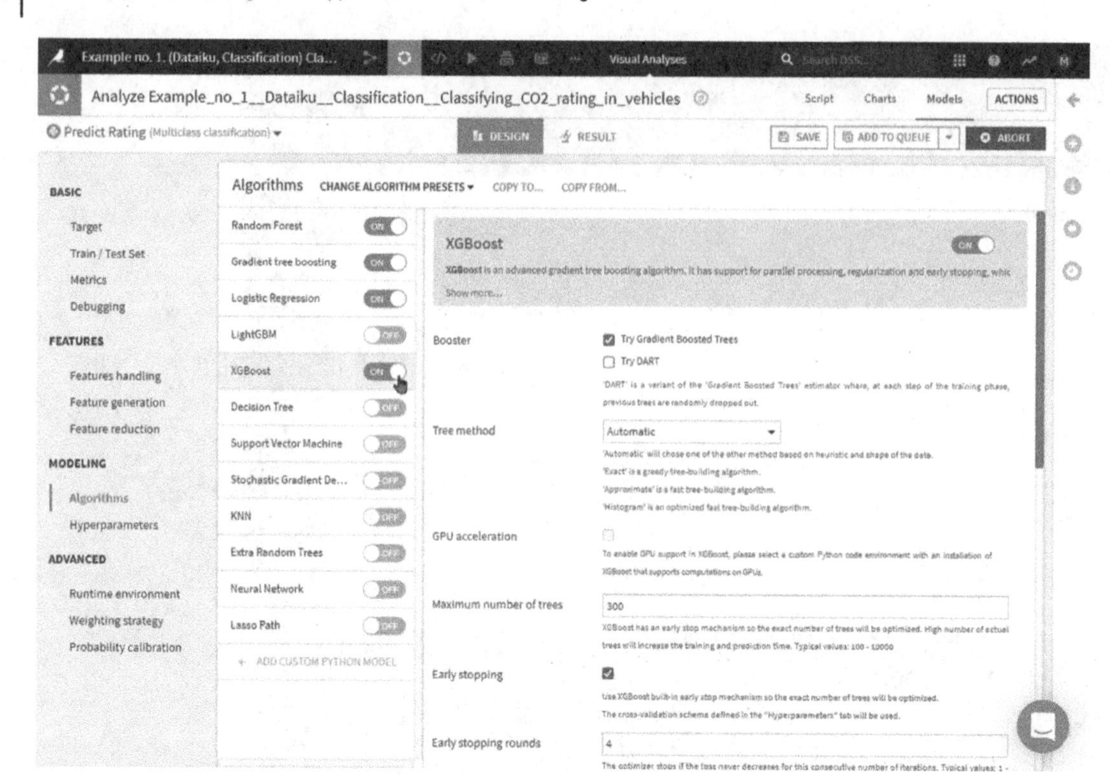

Step 8: *Add to Queue* and wait to evaluate the results of analysis.[45]

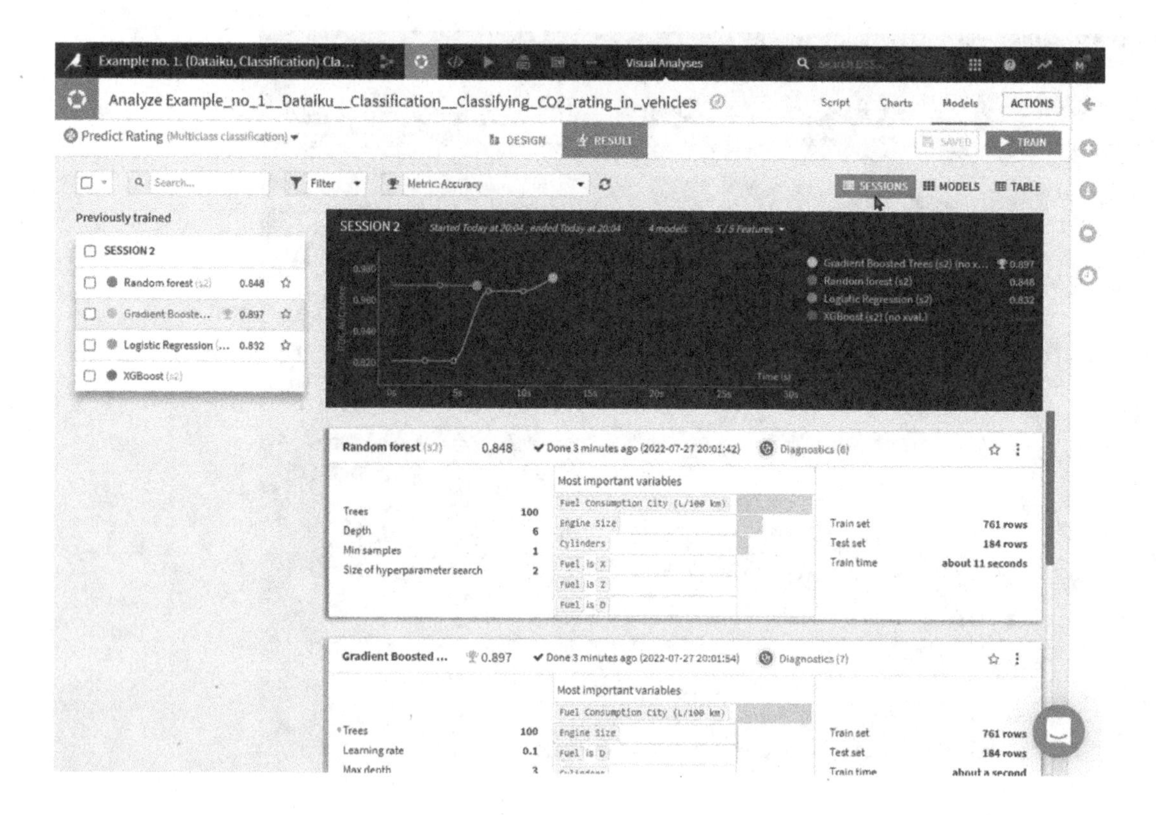

45 You may note that the XGBoost model failed to complete its analysis. This is due to the low count of *Rating 10*. Hover above the *Diagnostics* bar to read dull details of this error.

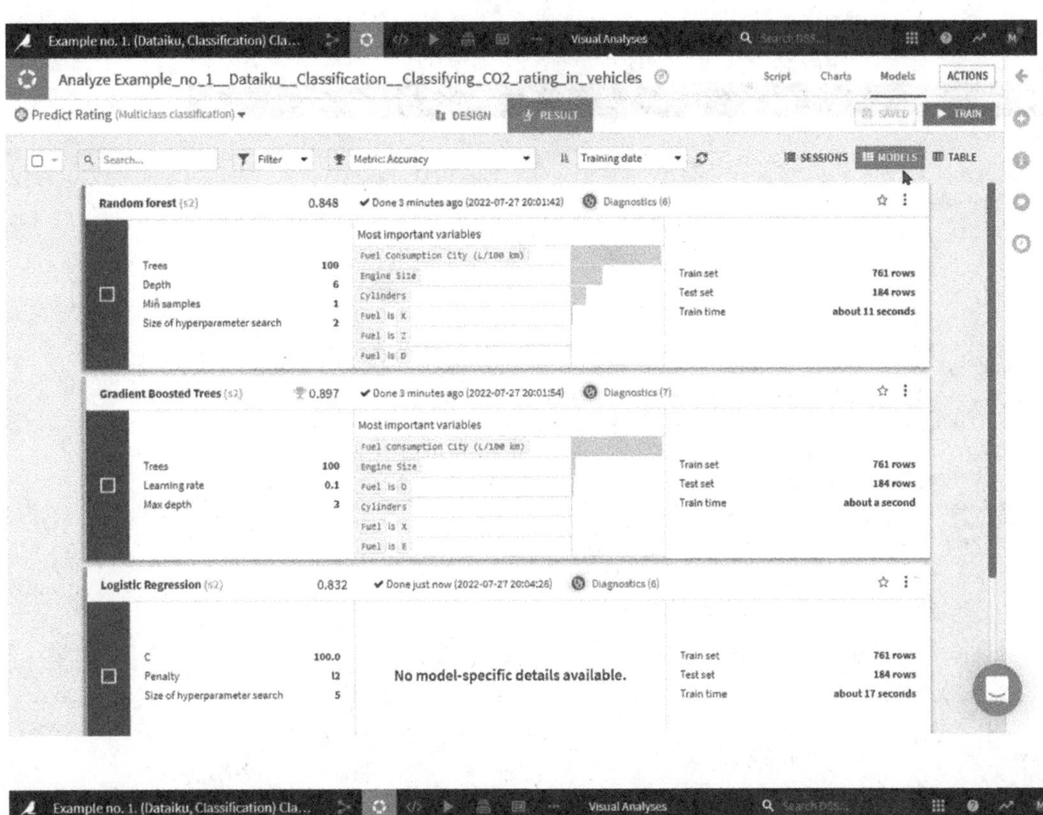

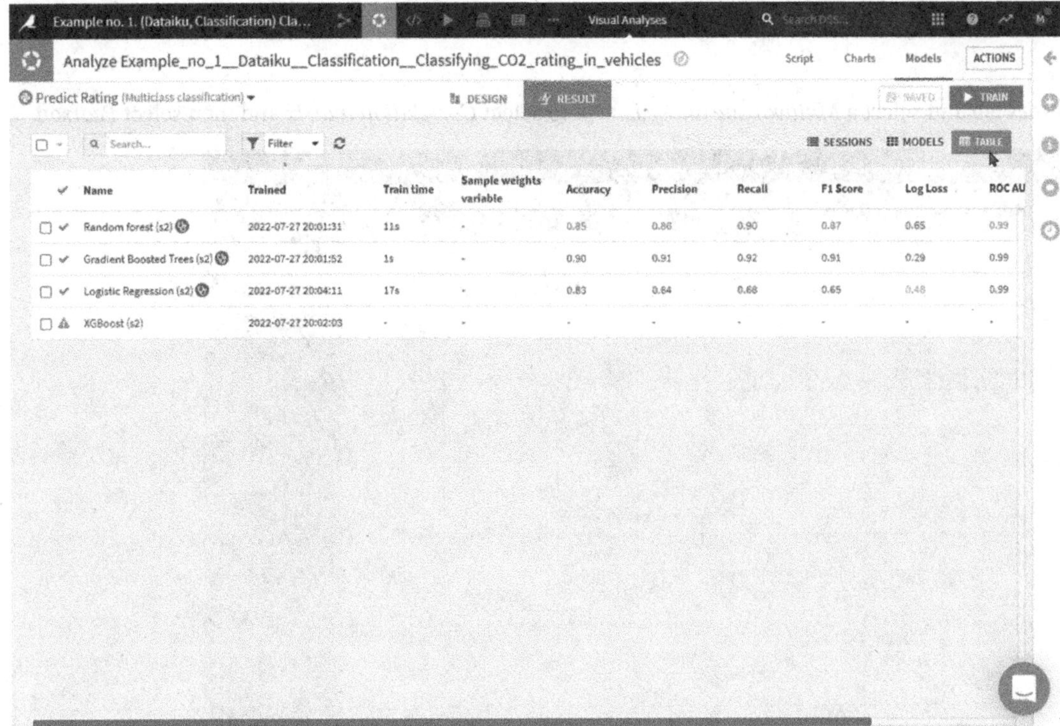

Name	Trained	Train time	Sample weights variable	Accuracy	Precision	Recall	F1 Score	Log Loss	ROC AU
Random forest (s2)	2022-07-27 20:01:31	11s	-	0.85	0.86	0.90	0.87	0.65	0.99
Gradient Boosted Trees (s2)	2022-07-27 20:01:52	1s	-	0.90	0.91	0.92	0.91	0.29	0.99
Logistic Regression (s2)	2022-07-27 20:04:11	17s	-	0.83	0.84	0.88	0.85	0.48	0.99
XGBoost (s2)	2022-07-27 20:02:03	-	-	-	-	-	-	-	-

Example no. 2 [*Dataiku*, Regression]: Predicting CO2 Emissions in vehicles.
Let us repeat Example no. 2 by changing the target to CO2 Emissions (i.e., to carry out a regression analysis).

Step 1: Upload and create your dataset.[46]

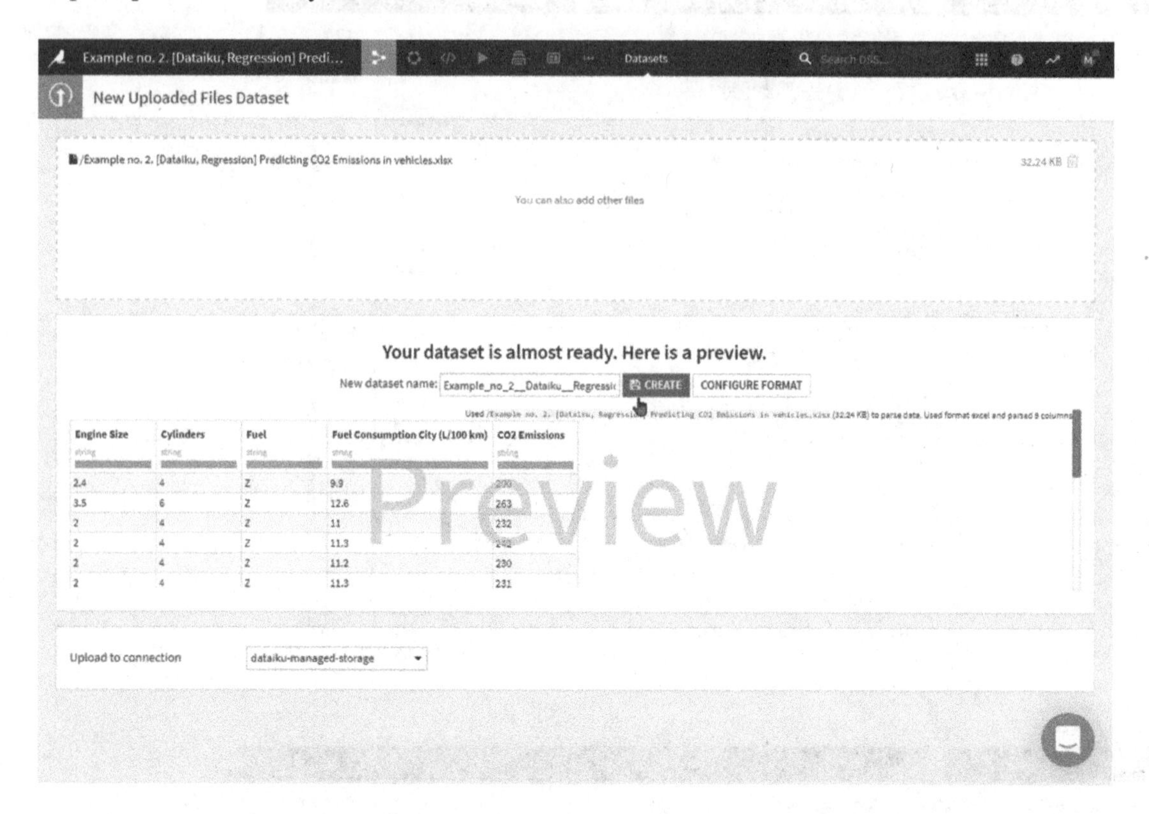

Step 2: Visit the *Statistics* tab and select a *Multivariate analysis*. Opt to select *Correlation matrix* and then select *Pearson*.

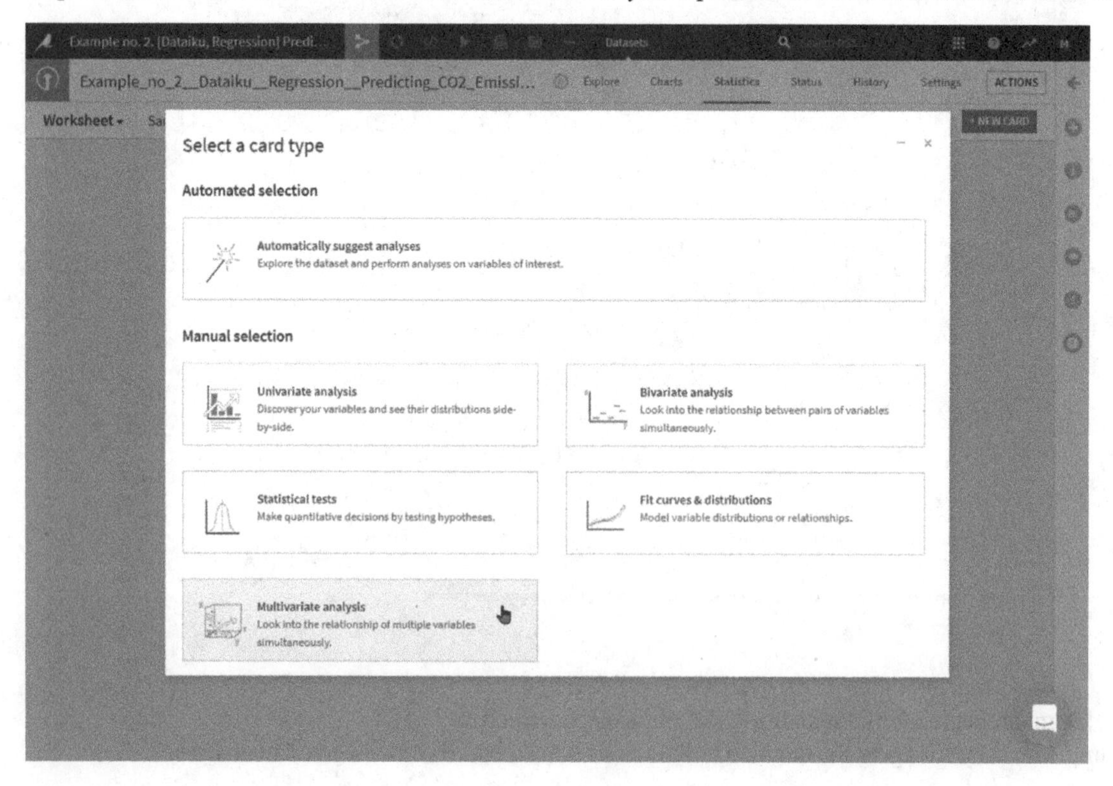

46 You may follow Step 1 from the previous example.

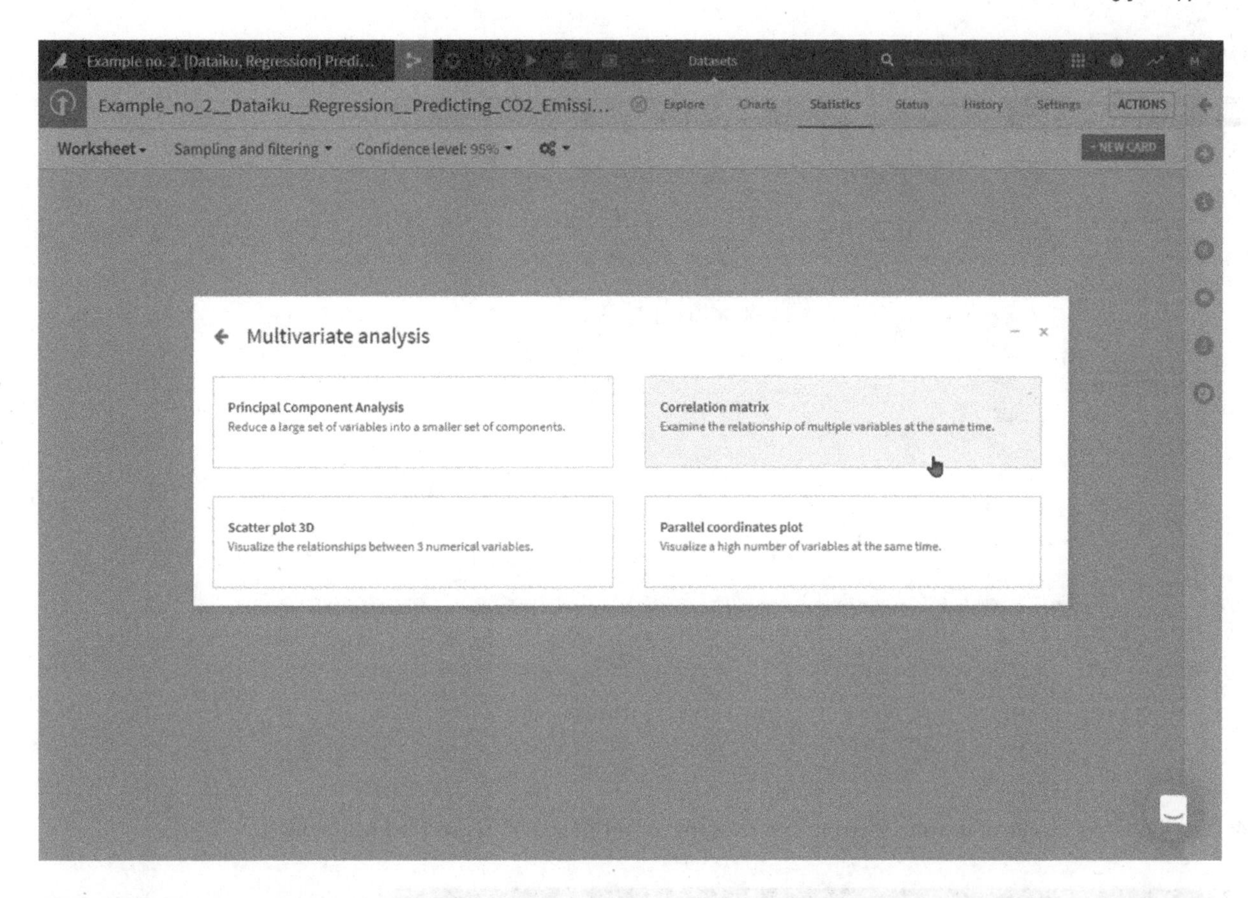
Example_no_2__Dataiku__Regression__Predicting_CO2_Emissi...
Explore
Charts
Statistics
Status
History
Settings
ACTIONS
Worksheet
Sampling and filtering
Confidence level: 95%
Multivariate analysis
Principal Component Analysis
Reduce a large set of variables into a smaller set of components.
Correlation matrix
Examine the relationship of multiple variables at the same time.
Scatter plot 3D
Visualize the relationships between 3 numerical variables.
Parallel coordinates plot
Visualize a high number of variables at the same time.

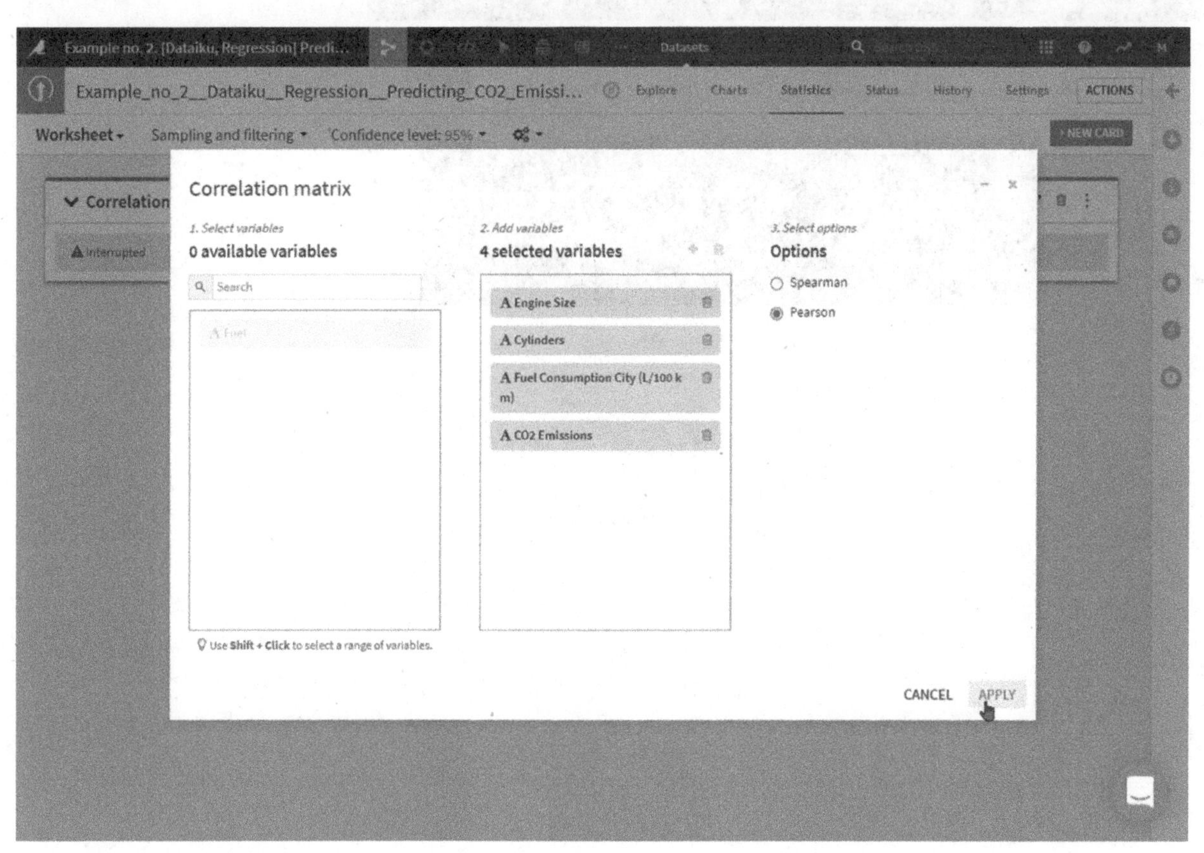
Example_no_2__Dataiku__Regression__Predicting_CO2_Emissi...
Explore
Charts
Statistics
Status
History
Settings
ACTIONS
Worksheet
Sampling and filtering
Confidence level: 95%
Correlation matrix
1. Select variables
0 available variables
Search
2. Add variables
4 selected variables
Engine Size
Cylinders
Fuel Consumption City (L/100 km)
CO2 Emissions
3. Select options
Options
Spearman
Pearson
Use Shift + Click to select a range of variables.
CANCEL
APPLY

Step 3: Create a *quick prototype*. Update analysis settings by adding new ML algorithms.

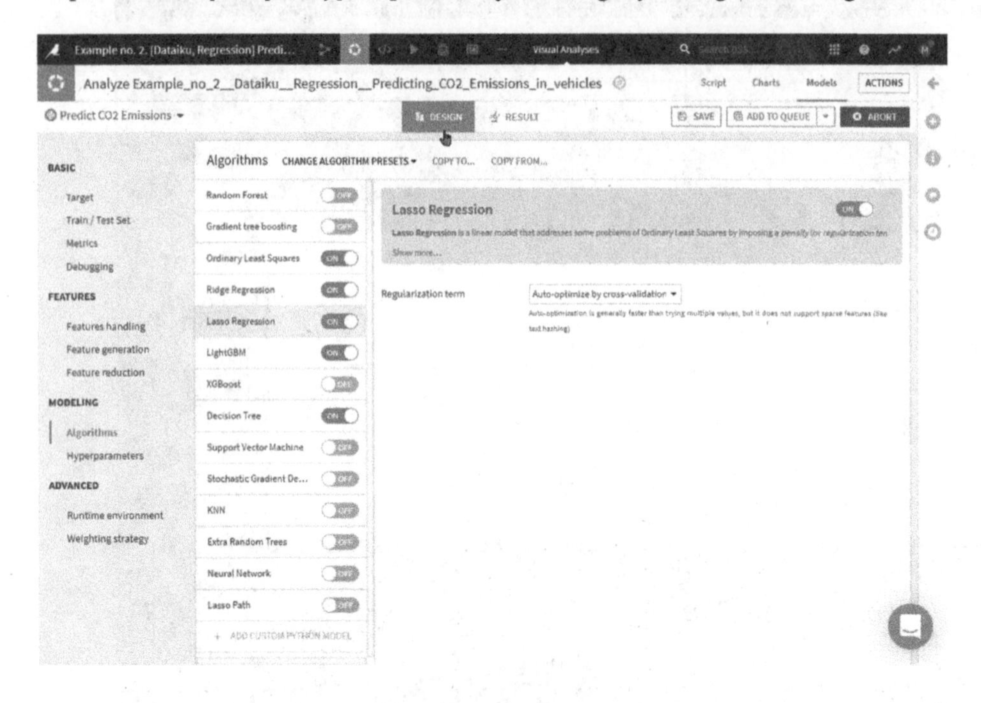

Step 4: Check out the results of the analysis and compare the performance of the created ML models.

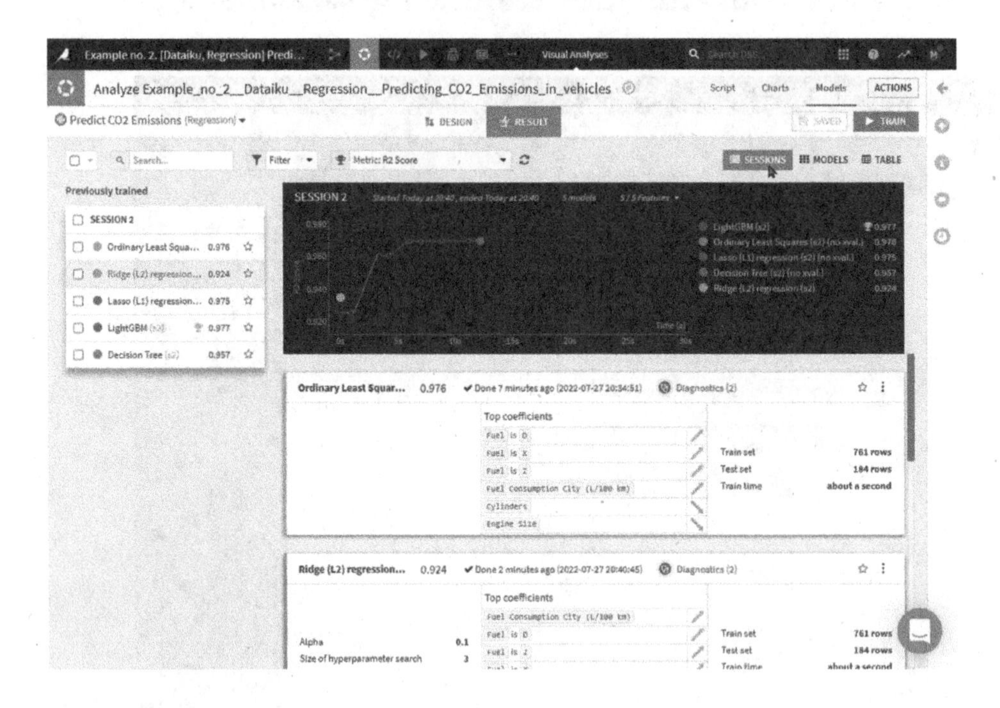

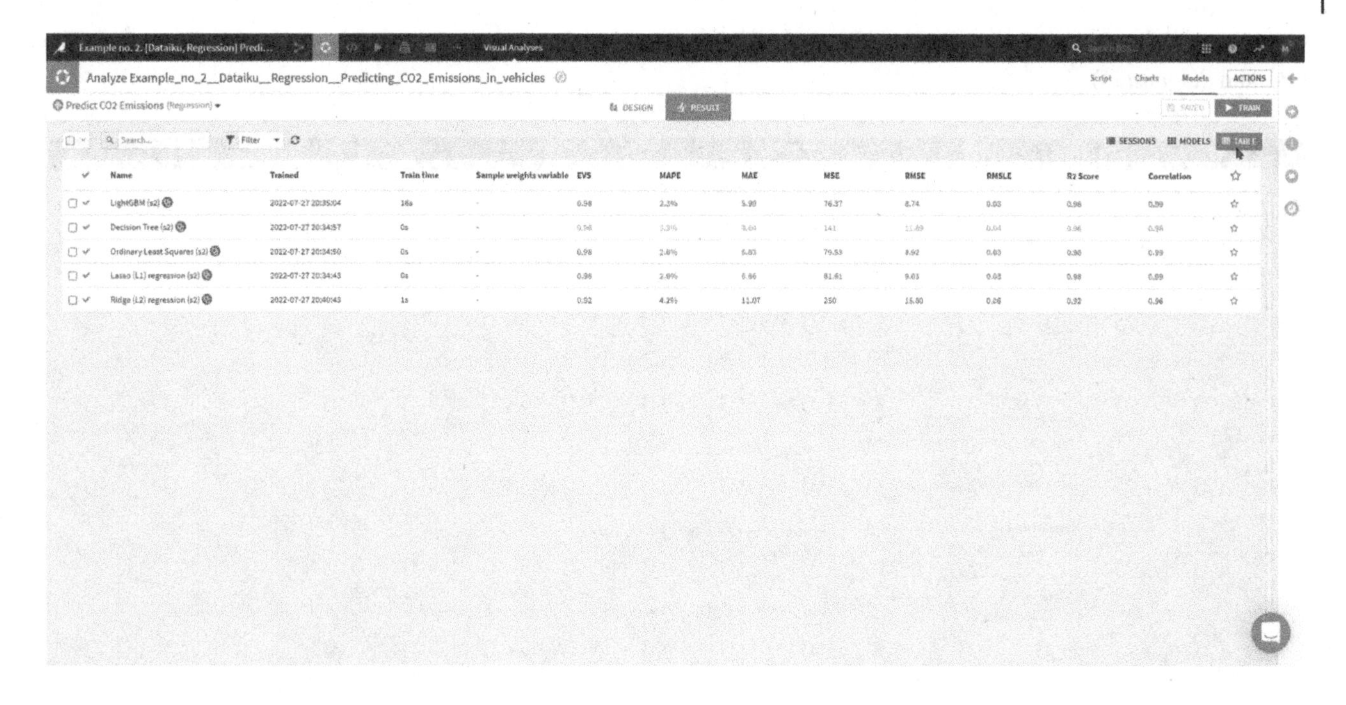

Step 5: Select the LightGBM model (best performance). Have a look at the feature (variable) importance provided by the analysis. It is clear that *Fuel Consumption City* has the highest importance.

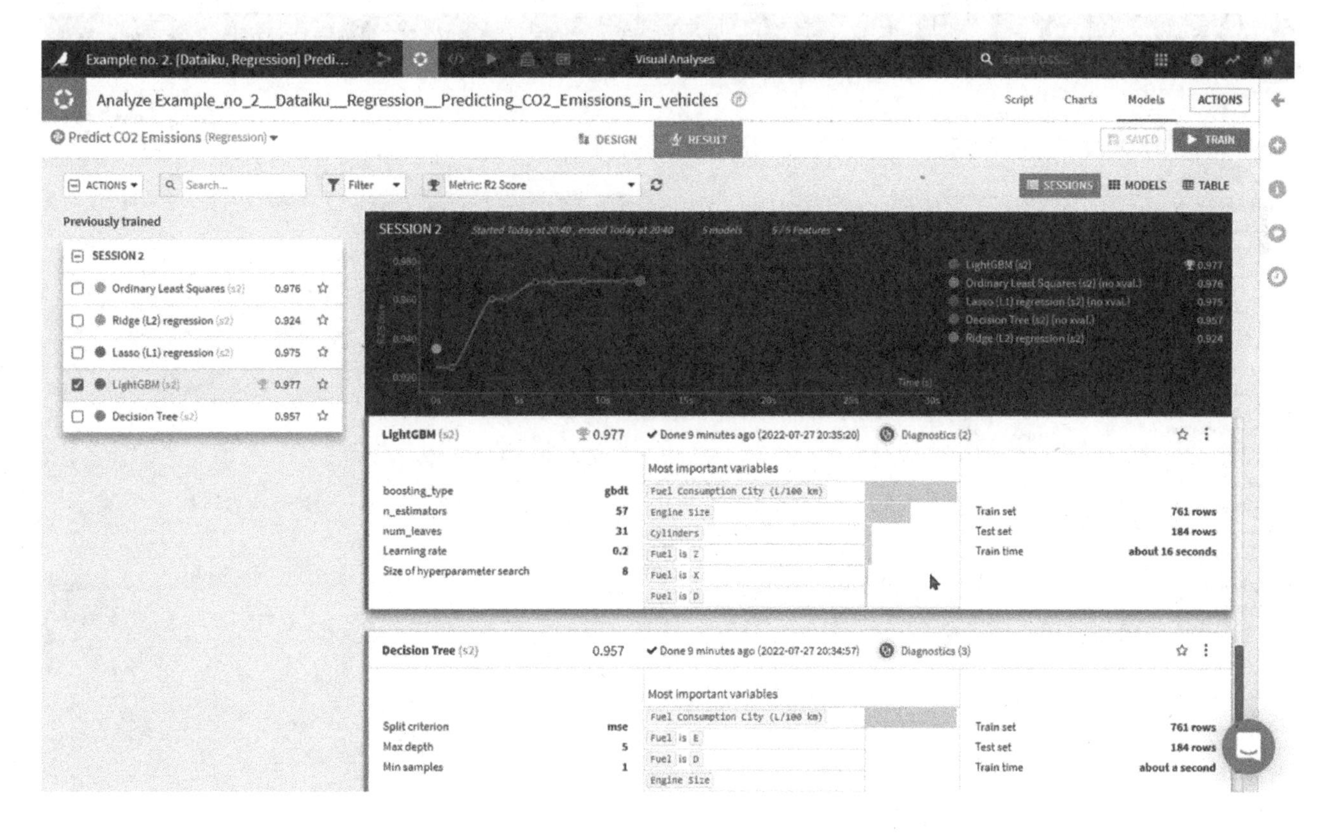

Step 6: Visit the *Model Information* tab to glance at additional details on this model (such as *algorithmic setting*, *learning rate*, and *diagnostics*).

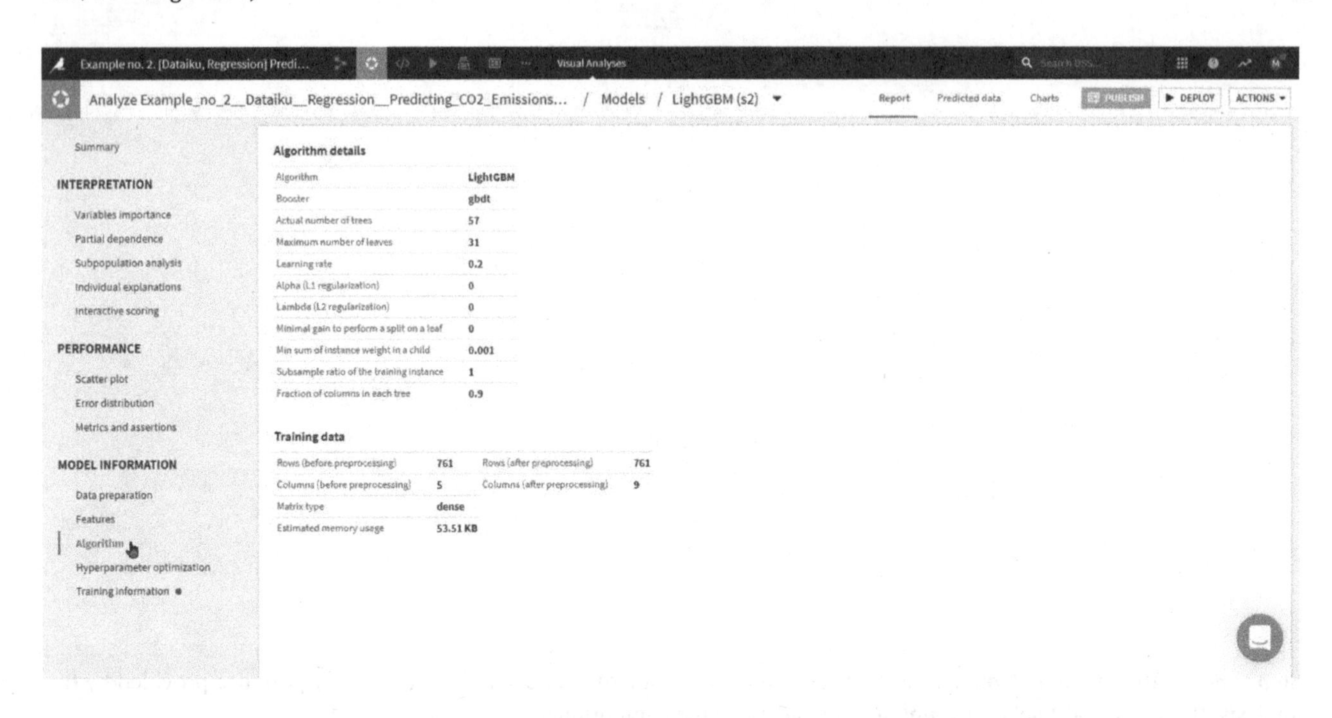

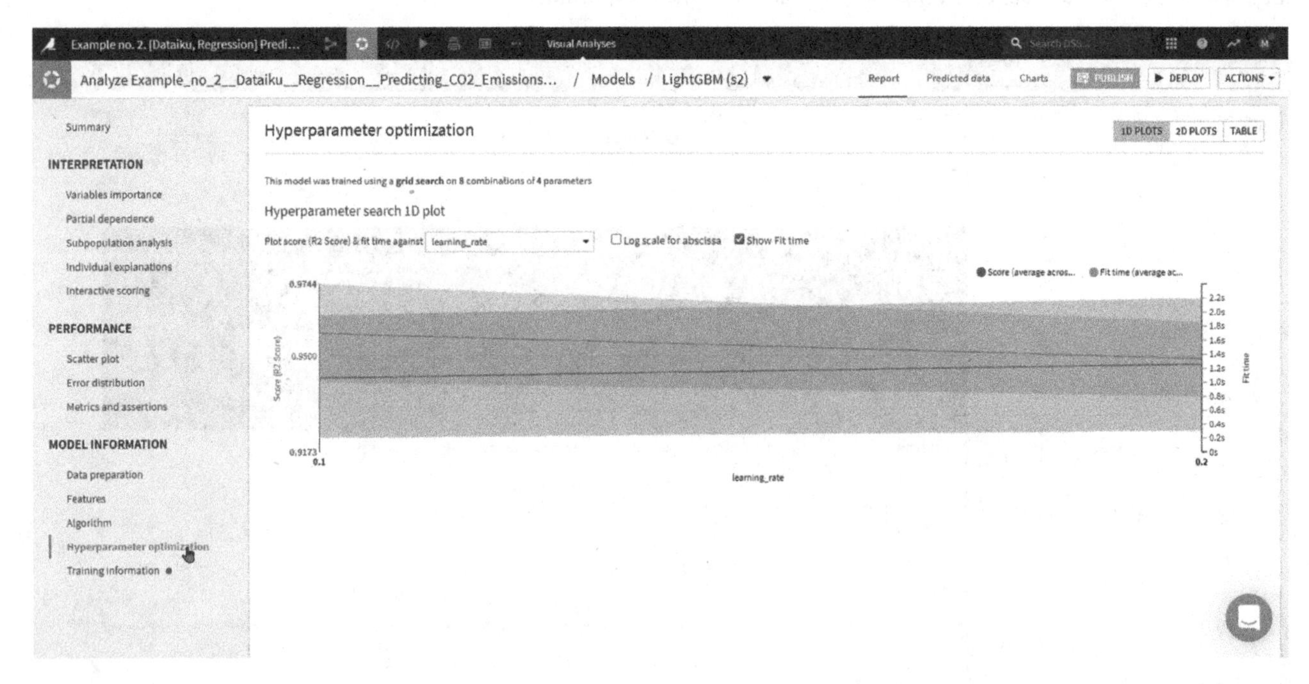

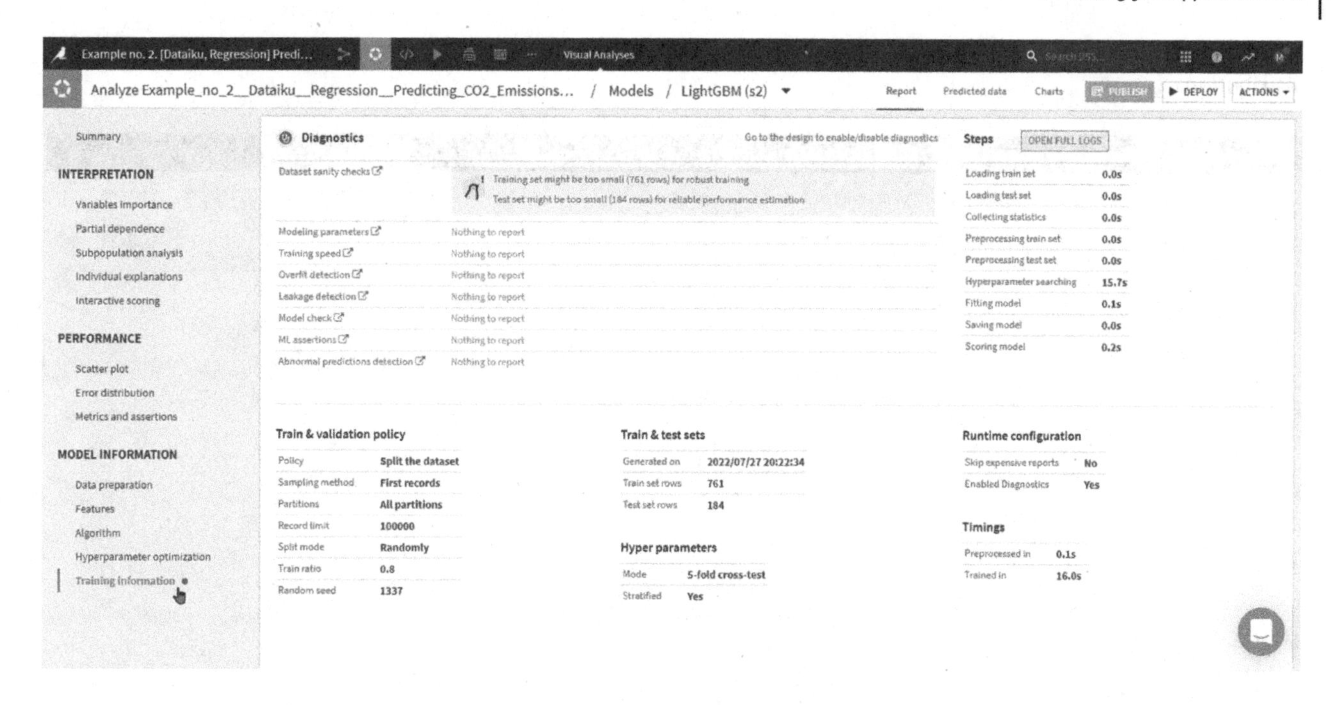

Example no. 3 [*Dataiku*, Clustering]: Clustering bridges based on conditions in Ontario, Canada.
This example uses a large dataset collected from bridges (and culverts) in Ontario, Canada [8]. All bridges are inspected every two years using the Ontario Structure Inspection Manual (OSIM) format. The features are highway name, material, year built, last major rehab, number of span/cells, deck/culverts length, width total, operation status. The target is the Current Bridge Condition Index (BCI).

Step 1: Create your project, upload your dataset and check out the features from the uploaded dataset.

Example no. 3. (Dataiku, Clustering) Cluste... | Datasets

Example_no_3__Dataiku__Clustering__Clustering_bridges_based_on_condition... | Explore | Charts | Statistics | Status | History | Settings | ACTIONS

Viewing entire dataset | Configure sample | DISPLAY

5,053 rows - 10 cols | 5,053 rows

HIGHWAY NA...	CATEGORY	MATERIAL 1	YEAR BUILT	LAST MAJOR REHAB	NUMBER OF SPAN / CELLS	DECK / CULVERTS LENGTH (m)	WIDTH TOTAL (m)	OPERAT
string Integer	string Text	string Text	string Integer	string Integer	string Integer	string Decimal	string Decimal	string Text
403	Culvert	Reinforced Cast-In-Place Concrete	1990	1990		156	4	Open
403	Culvert	Reinforced Cast-In-Place Concrete	1990	1990		55	4	Open to
403	Bridge	Prestressed Precast Concrete	1965	2014	4	66.6	25.4	Open to
403	Bridge	Prestressed Precast Concrete	1963	2014	4	62.47	18.39	Open to
403	Bridge	Prestressed Precast Concrete	1962	2013	4	60.35	17.332	Open to
24	Bridge	Steel	2011	2011	1	41.2	14.5	Open to
403	Bridge	Post-Tensioned Cast-In-Place Concrete	1975	2013	2	64	17.07	Open to
403	Bridge	Reinforced Cast-In-Place Concrete	1977		2	86.13	16.46	Open to
403	Bridge	Steel	1965	2013	4	93.4	15.46	Open to
403	Bridge	Steel	1964	2013	4	93.36	15.456	Open to
403	Bridge	Prestressed Precast Concrete	1965	2014	4	64.99	12.29	Open to
403	Bridge	Prestressed Precast Concrete	1965	2013	4	79.63	11.378	Open to
403	Bridge	Prestressed Precast Concrete	1964	2014	4	63.9	20.22	Open to
403	Bridge	Steel	1963	2013	3	70.409	11.76	Open to
403	Bridge	Steel	1963	2013	3	70.409	11.76	Open to
403	Bridge	Steel	1976		4	223.06	11.23	Open to
403	Bridge	Steel	1976	2014	4	223.06	11.23	Open to
403	Bridge	Other	1983	1984	3	48.96	11.9	Open to
403	Bridge	Prestressed Precast Concrete	1983	1984	3	48.96	11.9	Op
403	Bridge	Post-Tensioned Cast-In-Place Concrete	1983	1993	2	60.9	9.4	O
403	Bridge	Post-Tensioned Cast-In-Place Concrete	1983	2006	2	60.88	9.4	Op

Step 2: Navigate to the *Statistics* tab. Add a new card for *Automated selection*. This gives you insights in feature distribution.

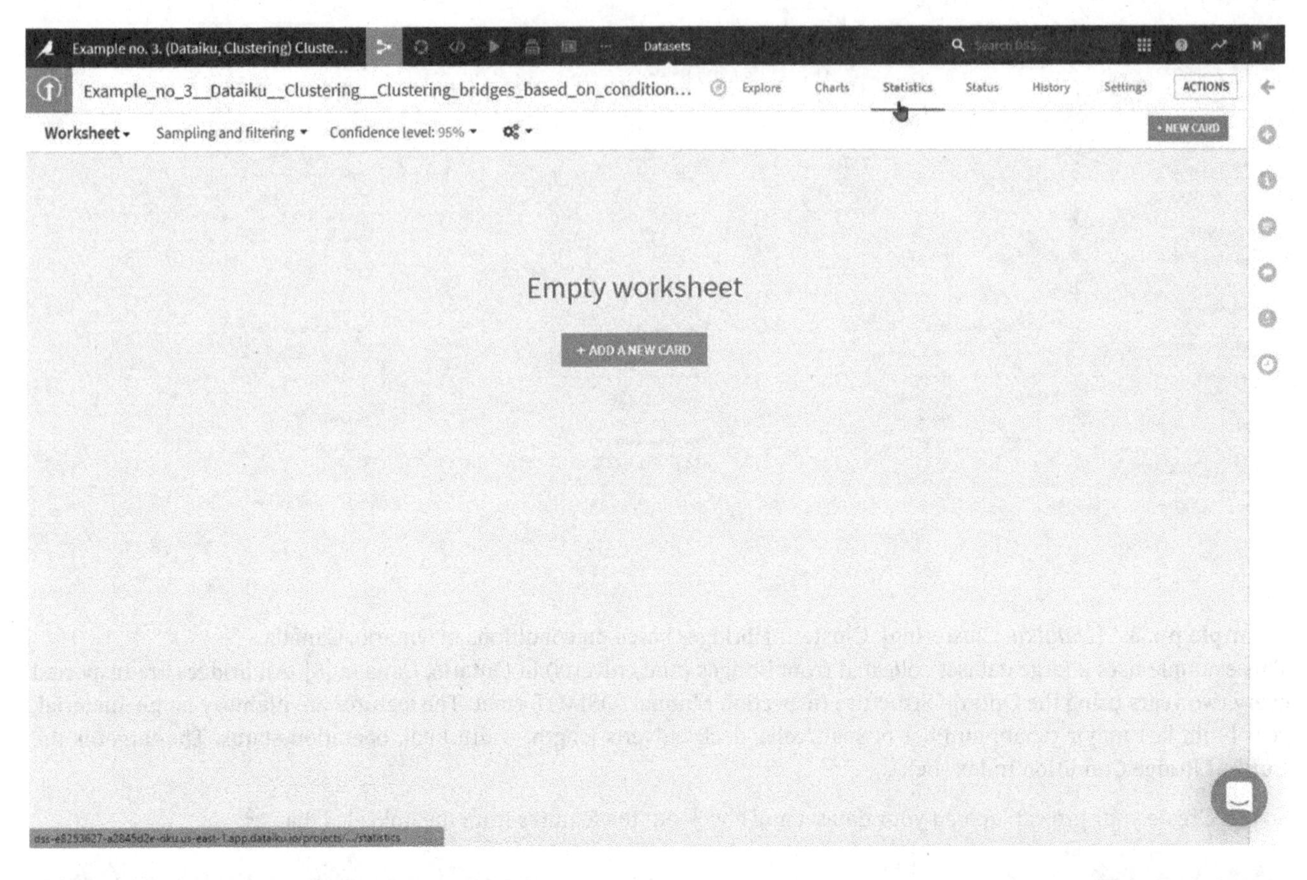

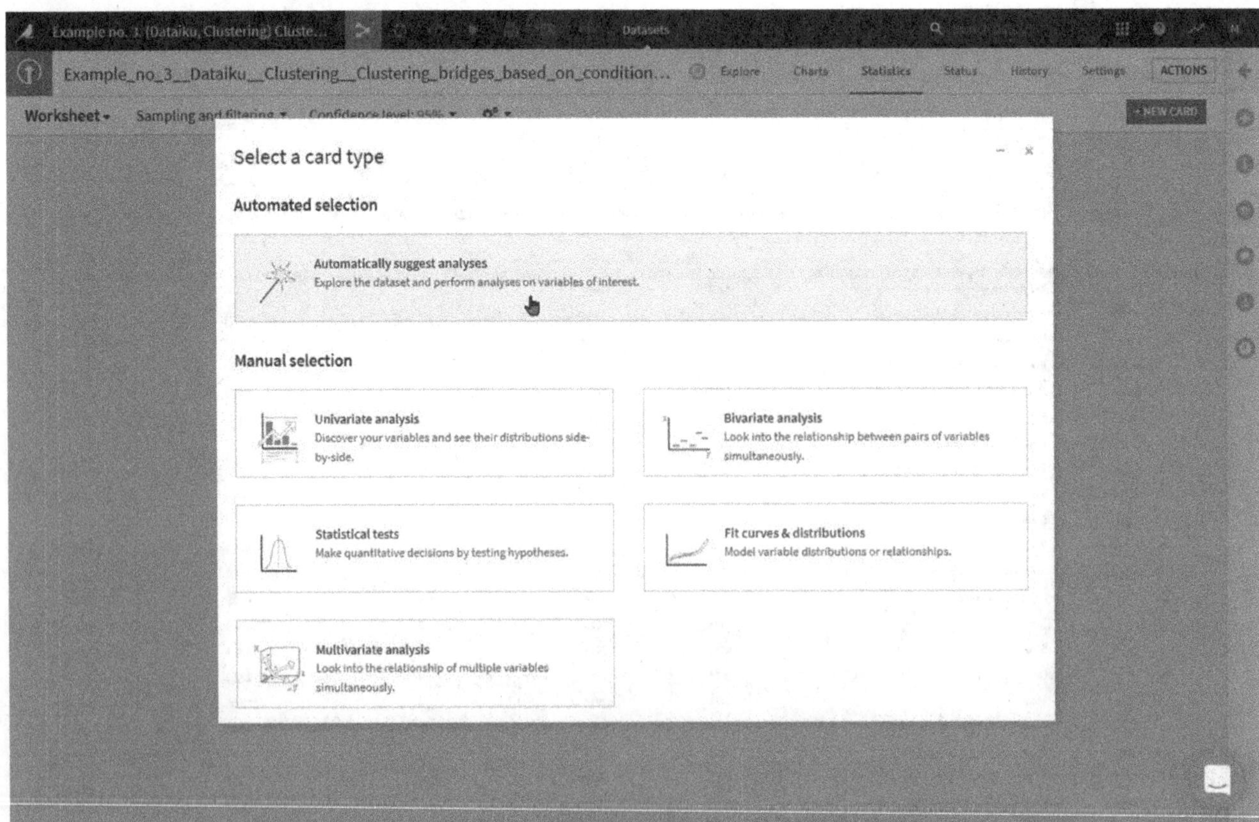

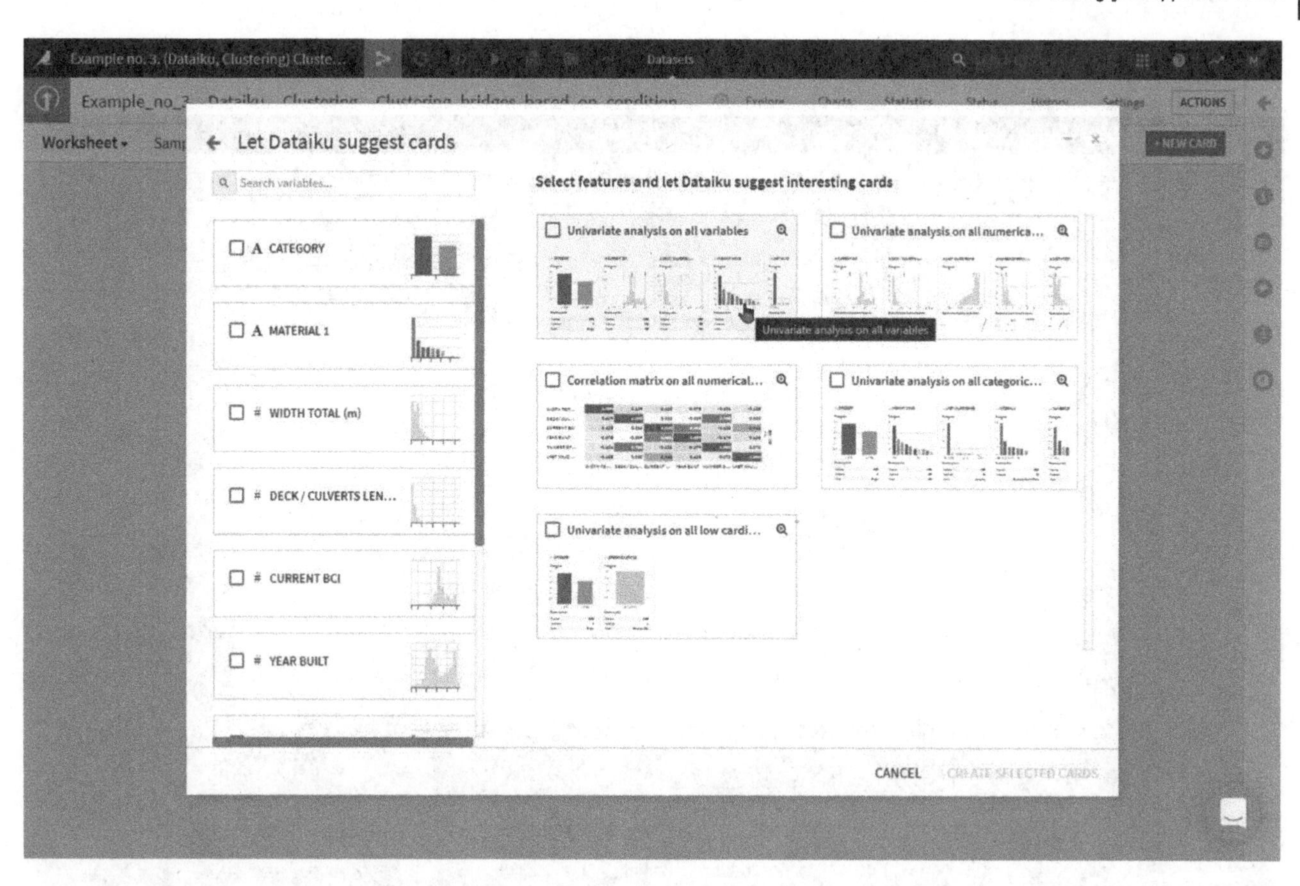
Worksheet
Let Dataiku suggest cards
Search variables...
A CATEGORY
A MATERIAL 1
WIDTH TOTAL (m)
DECK / CULVERTS LEN...
CURRENT BCI
YEAR BUILT
Select features and let Dataiku suggest interesting cards
Univariate analysis on all variables
Univariate analysis on all numerica...
Univariate analysis on all variables
Correlation matrix on all numerical...
Univariate analysis on all categoric...
Univariate analysis on all low cardi...
CANCEL
ACTIONS

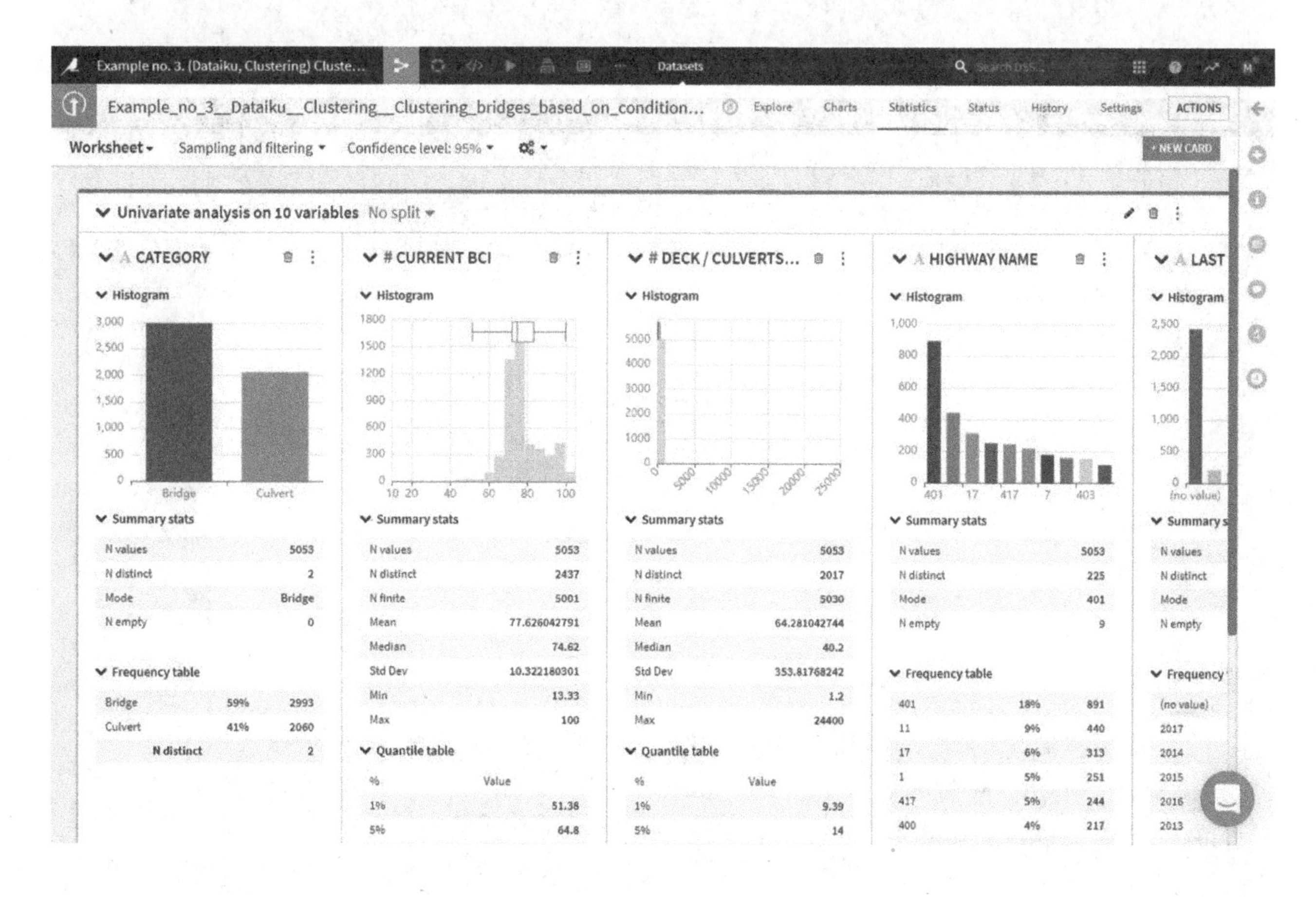
Example no. 3. (Dataiku, Clustering) Cluste...
Datasets
Example_no_3__Dataiku__Clustering__Clustering_bridges_based_on_condition...
Explore
Charts
Statistics
Status
History
Settings
ACTIONS
Worksheet
Sampling and filtering
Confidence level: 95%
+ NEW CARD
Univariate analysis on 10 variables
No split
A CATEGORY
CURRENT BCI
DECK / CULVERTS...
A HIGHWAY NAME
A LAST
Histogram
Bridge
Culvert
Summary stats
N values 5053
N distinct 2
Mode Bridge
N empty 0
Frequency table
Bridge 59% 2993
Culvert 41% 2060
N distinct 2
N values 5053
N distinct 2437
N finite 5001
Mean 77.626042791
Median 74.62
Std Dev 10.322180301
Min 13.33
Max 100
Quantile table
% Value
1% 51.38
5% 64.8
N values 5053
N distinct 2017
N finite 5030
Mean 64.281042744
Median 40.2
Std Dev 353.81768242
Min 1.2
Max 24400
1% 9.39
5% 14
N values 5053
N distinct 225
Mode 401
N empty 9
401 18% 891
11 9% 440
17 6% 313
1 5% 251
417 5% 244
400 4% 217
(no value)
2017
2014
2015
2016
2013

Step 3: Move to the *Analysis* tab and select a *Clustering* analysis.

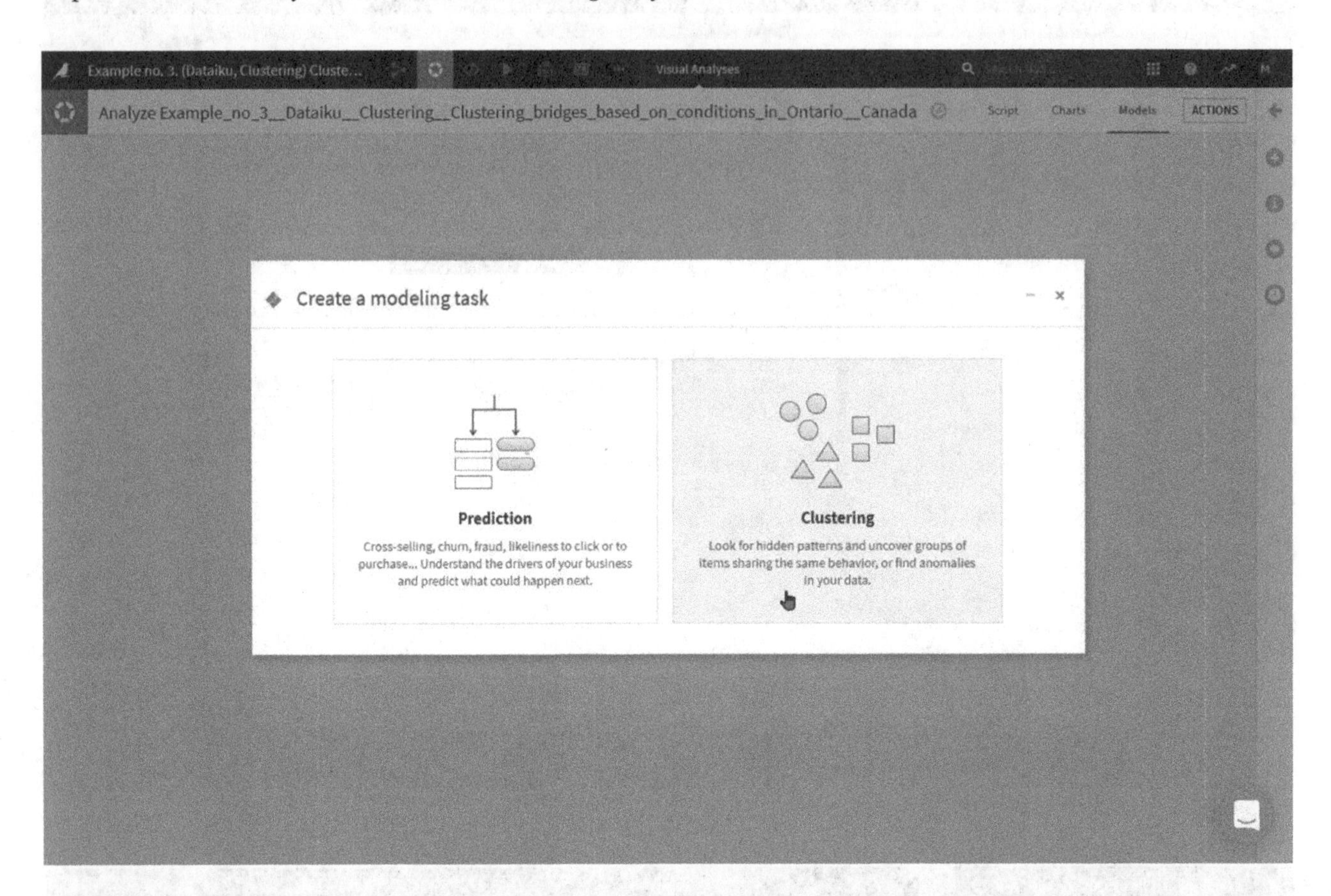

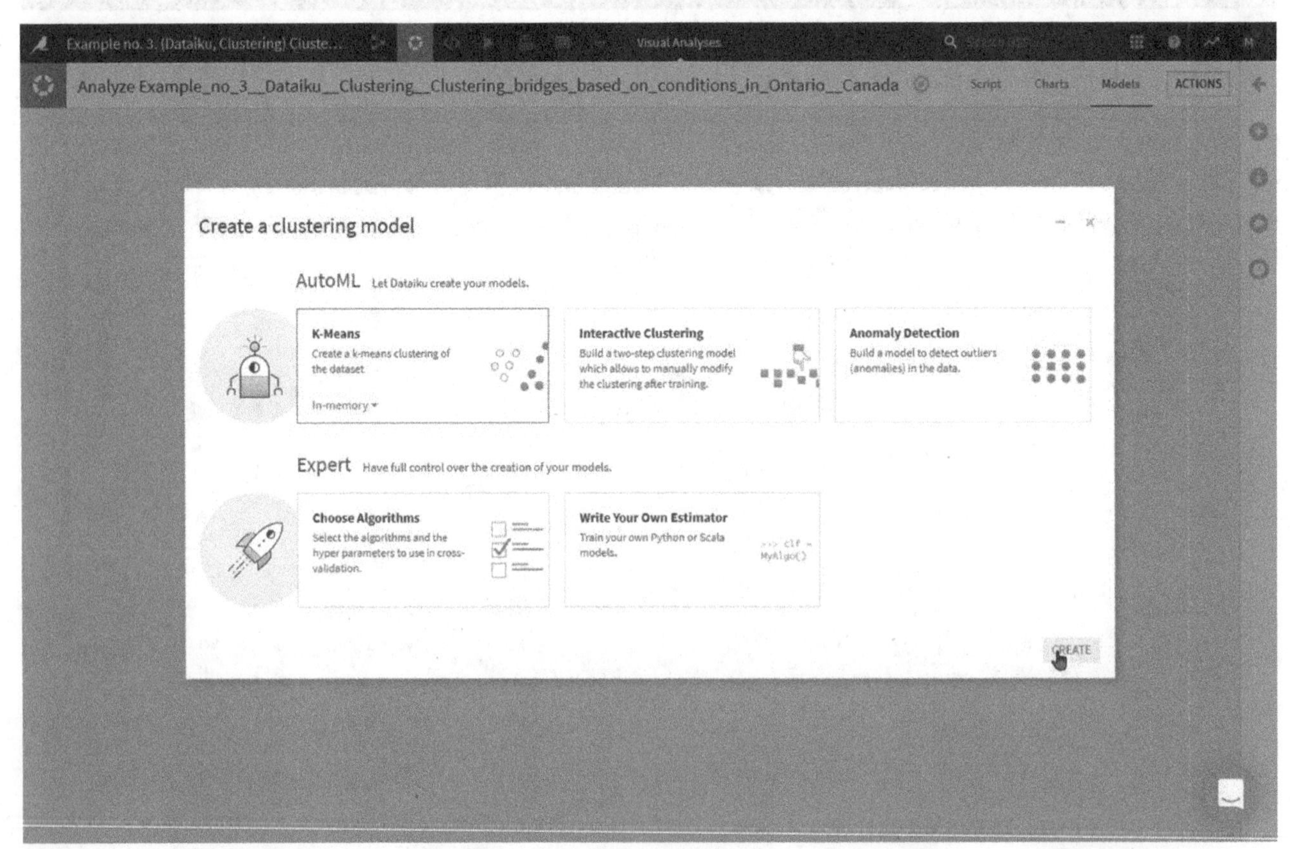

Step 4: Navigate to the *Design* tab and select clustering algorithms for analysis. Start *Training*.

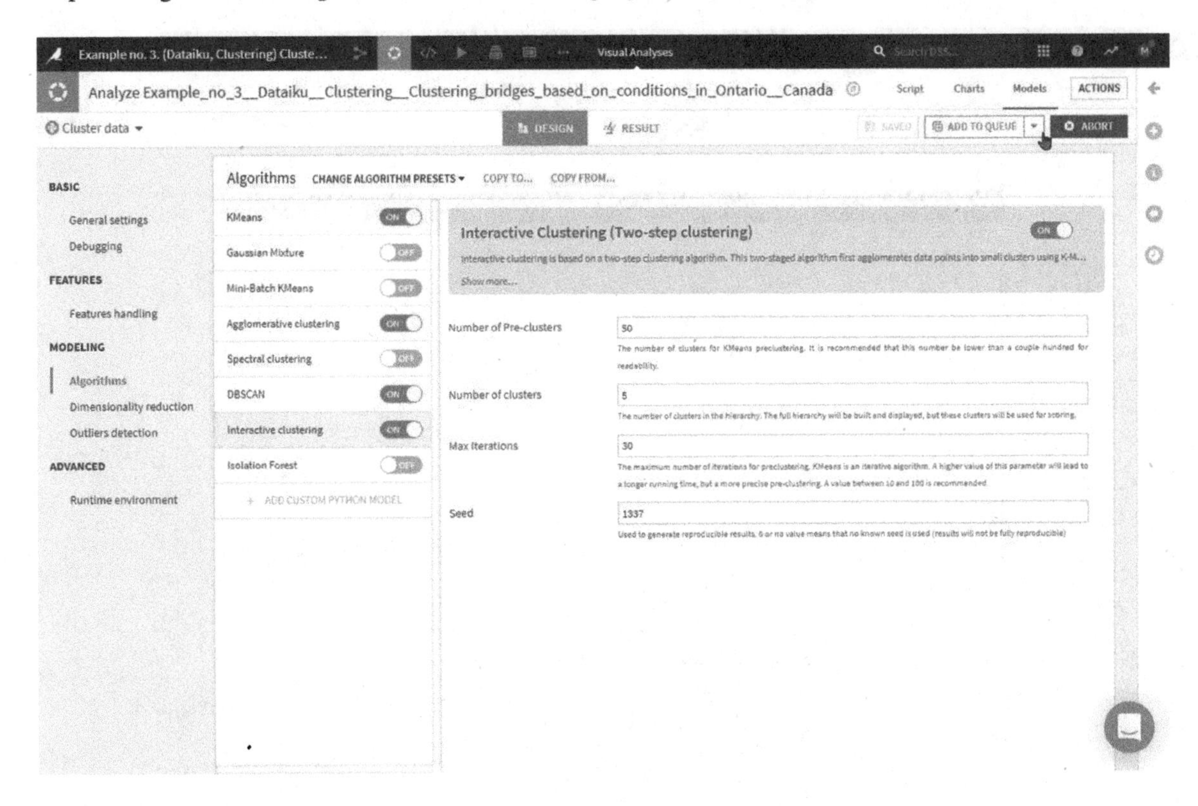

Note that, *Dataiku* by default[47] clusters anomalies into a dedicated cluster.

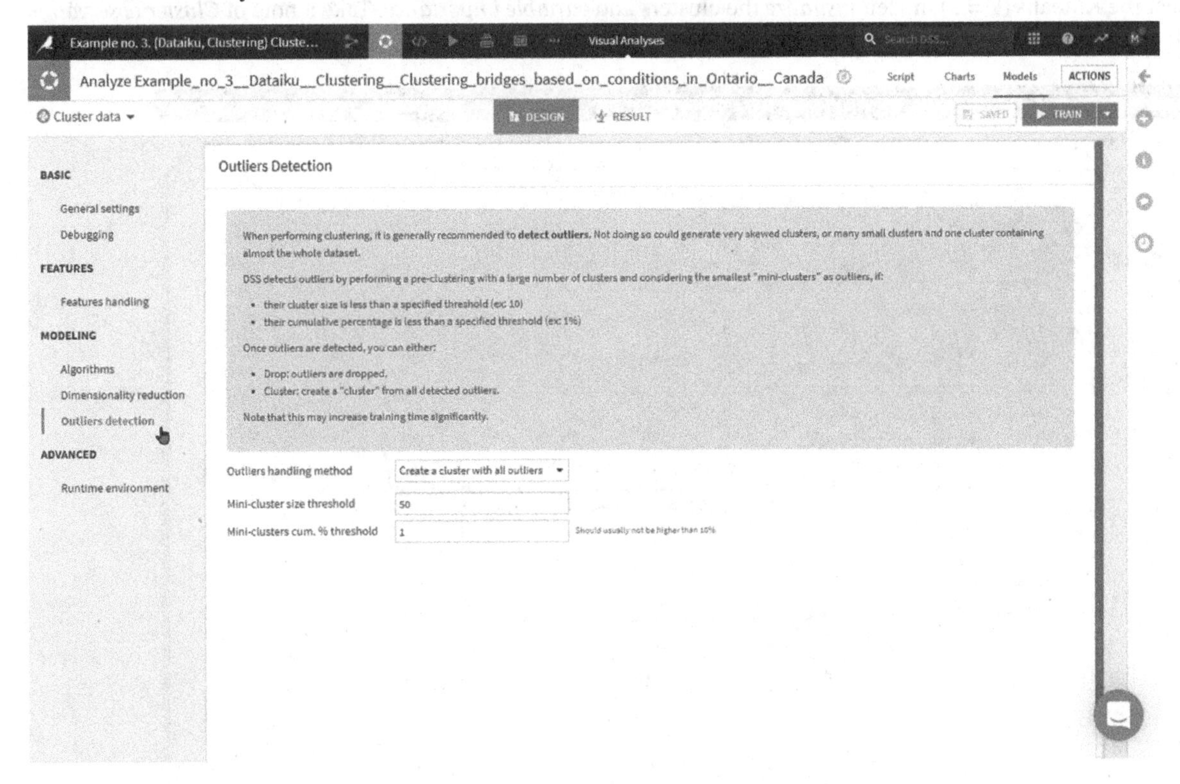

47 Dimensionality reduction could also be used.

Step 5: Explore the results of the analysis. Note the best performing model based on the Silhouette metric.

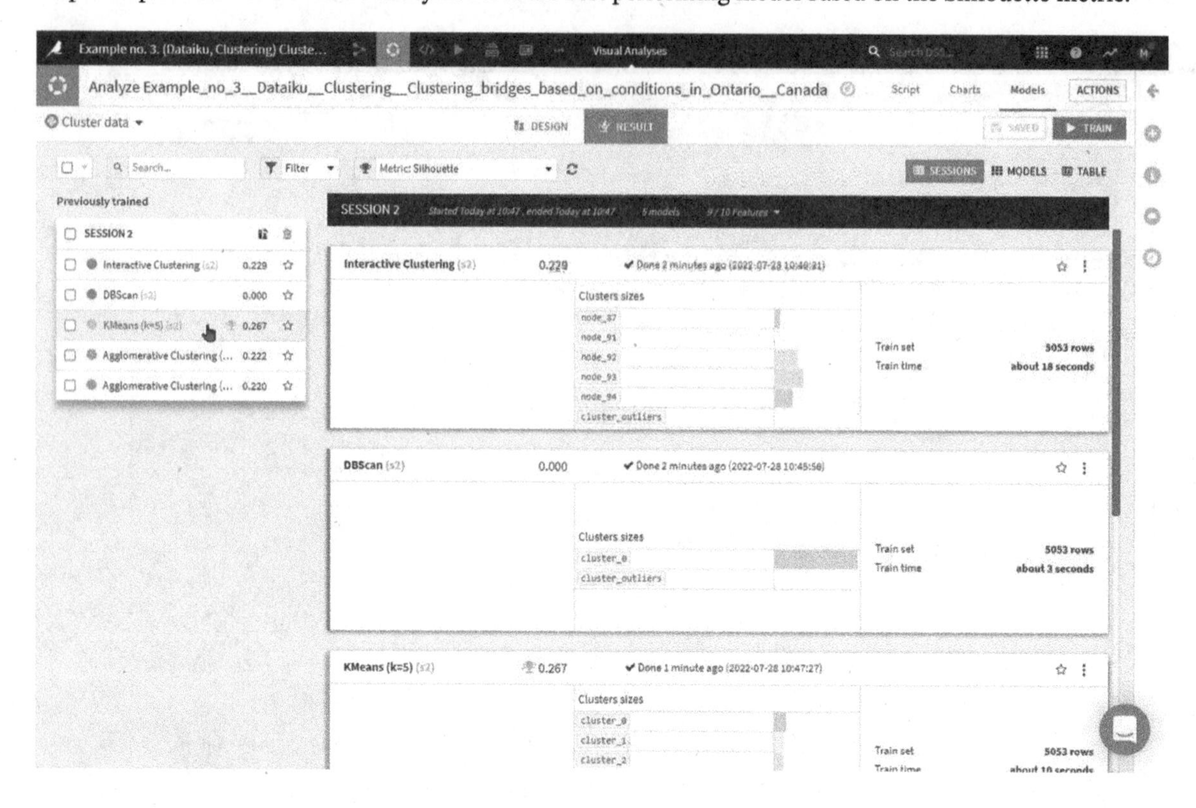

Step 6: Select the KMeans (k = 5) model. *Visualize* the clusters and *variable importance*. Take a note of *Cluster observation* under each cluster.

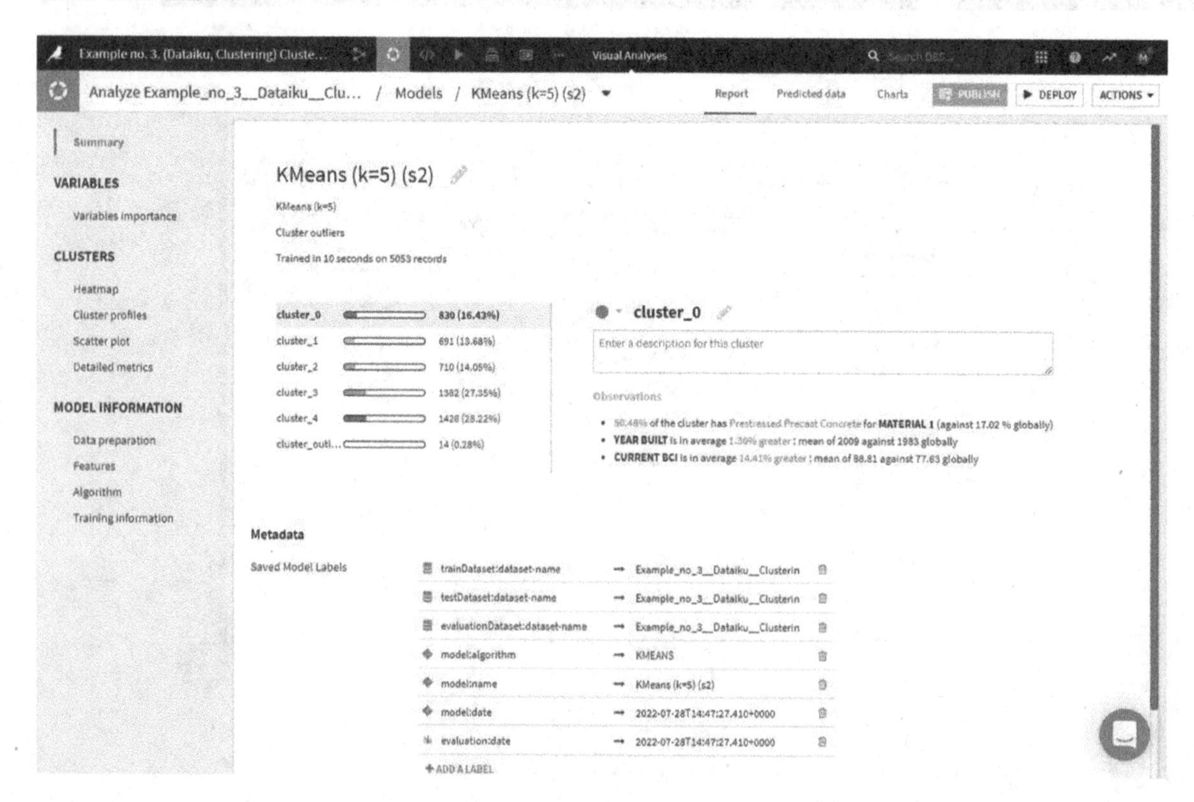

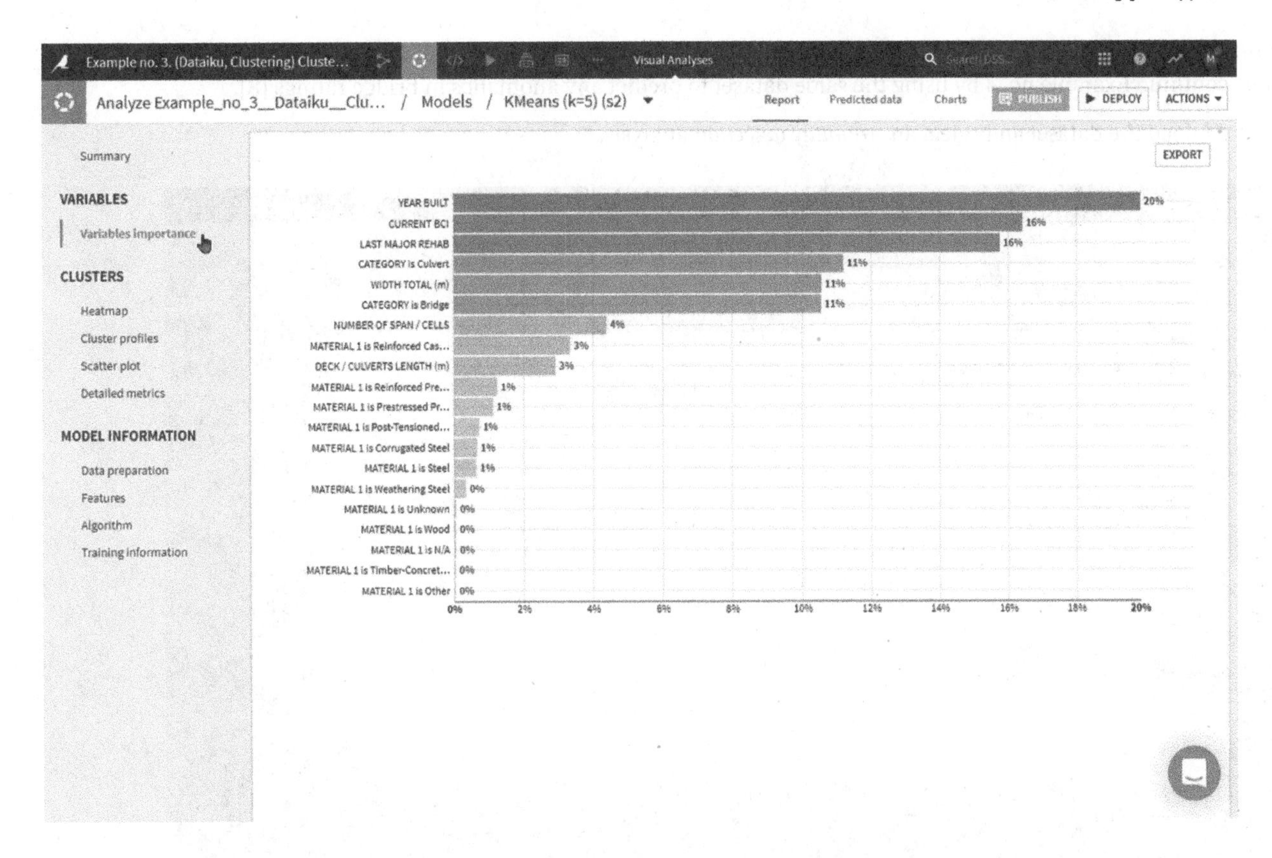

Step 7: A *Scatter plot* allows you to better visualize the distribution of clusters. Feel free to assign the axes.

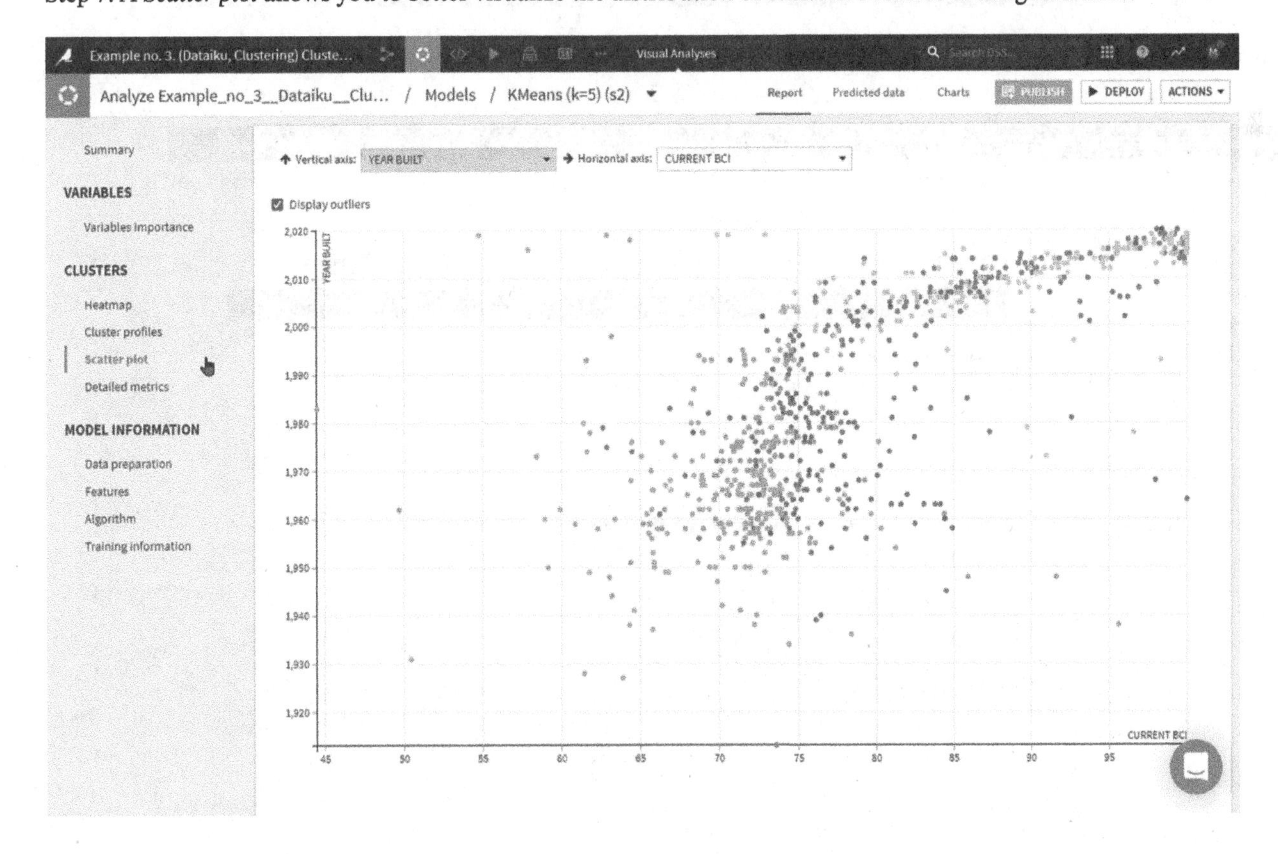

Example no. 4 [*Dataiku*, Anatomy detection]: Predicting anomalies in bridges based on conditions in Ontario, Canada. Let us continue Example no. 4 by using the same dataset to predict any anomalies in bridge ratings [8].

Step 1: Upload the dataset and select an *Anomaly detection* analysis.

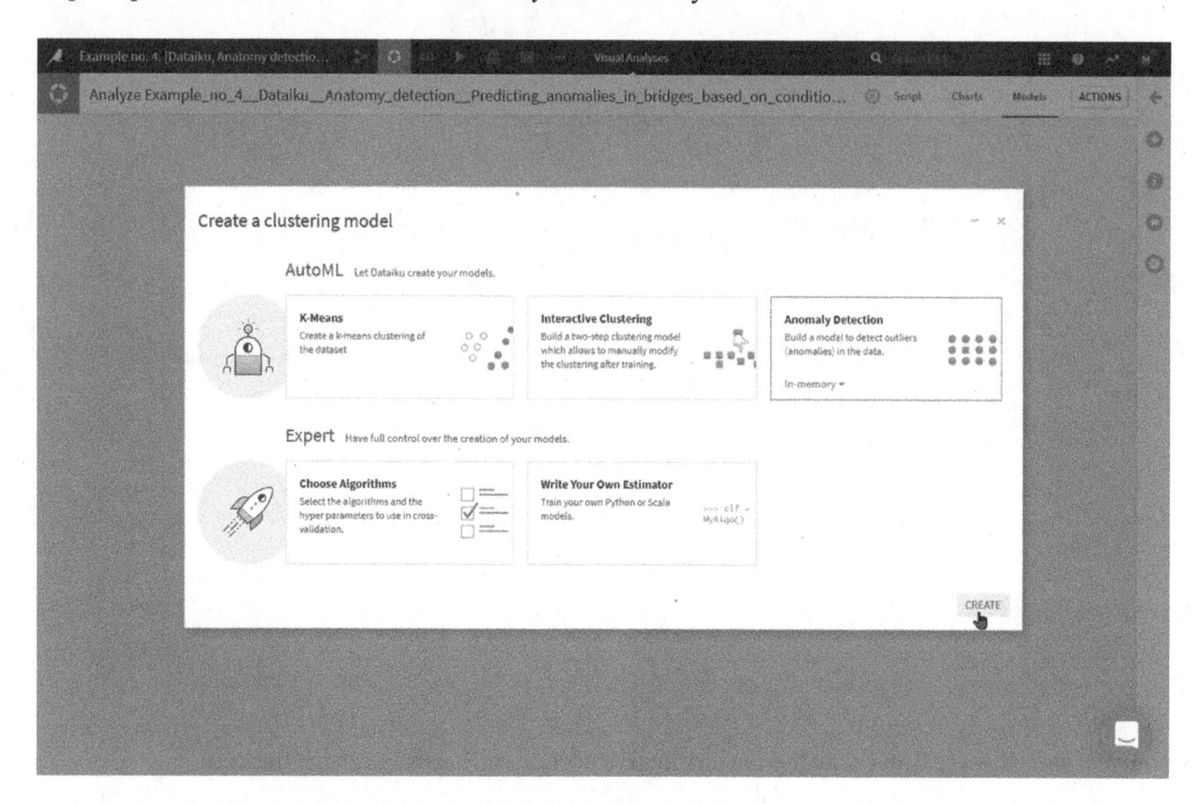

Step 2: The analysis returns two clusters (*regular* and *anomalies*)

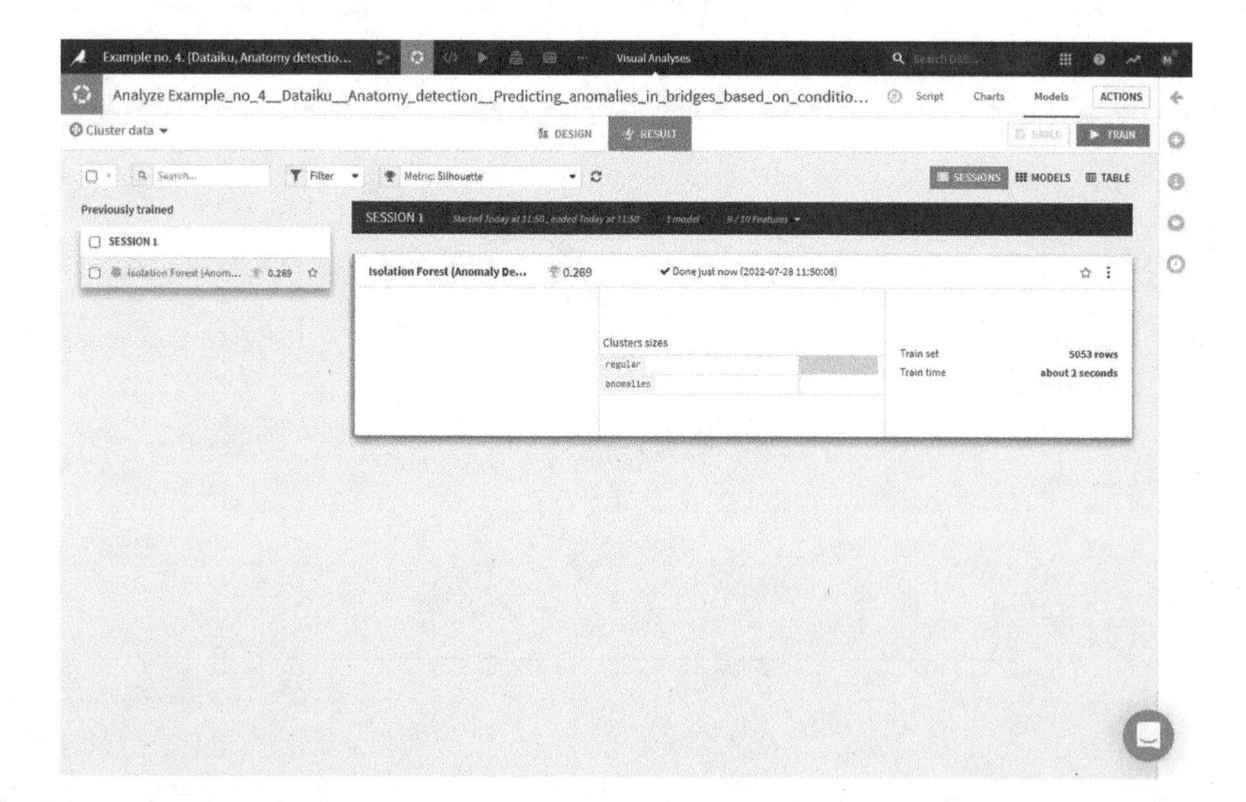

Step 3: Further investigate the results of the analysis by checking the *Summary* and *Variables importance* tabs. Note that about 253 observations are labeled as anomalies.

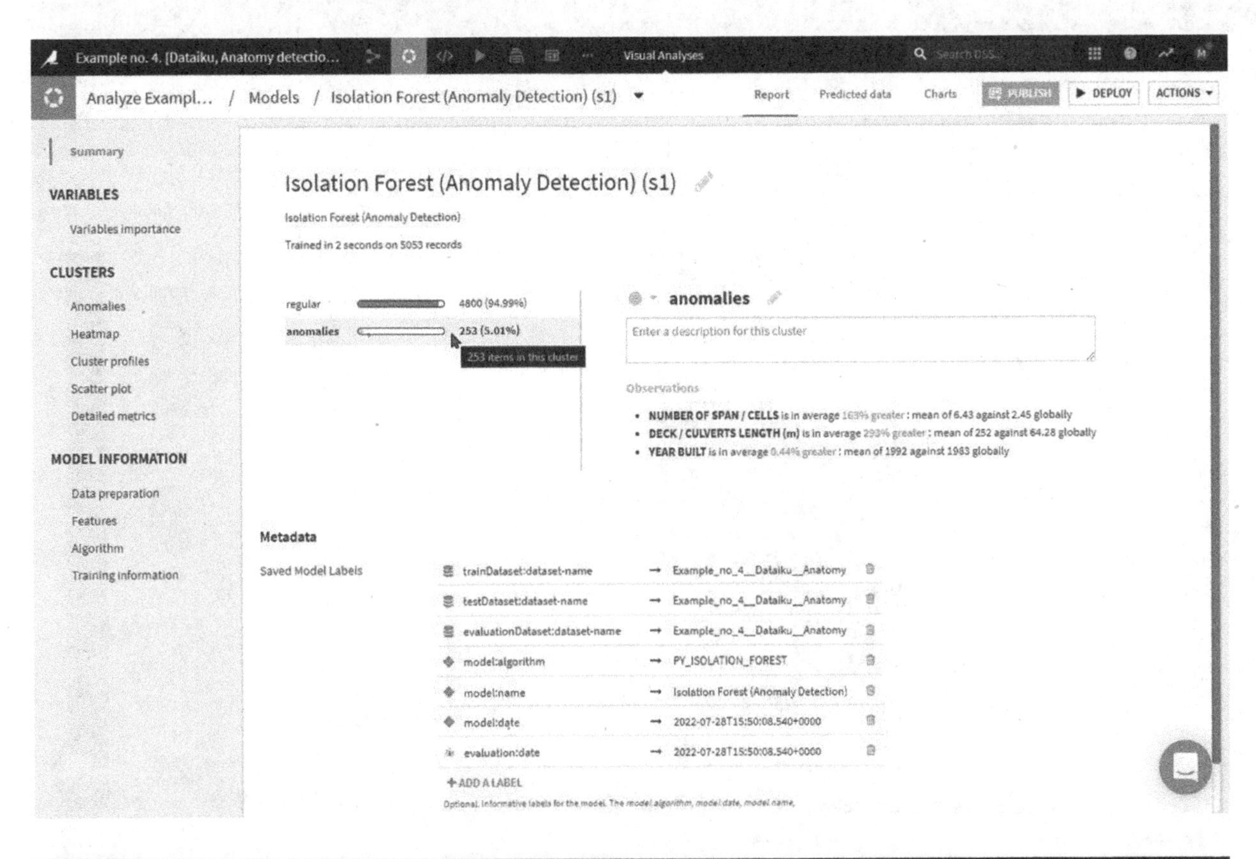

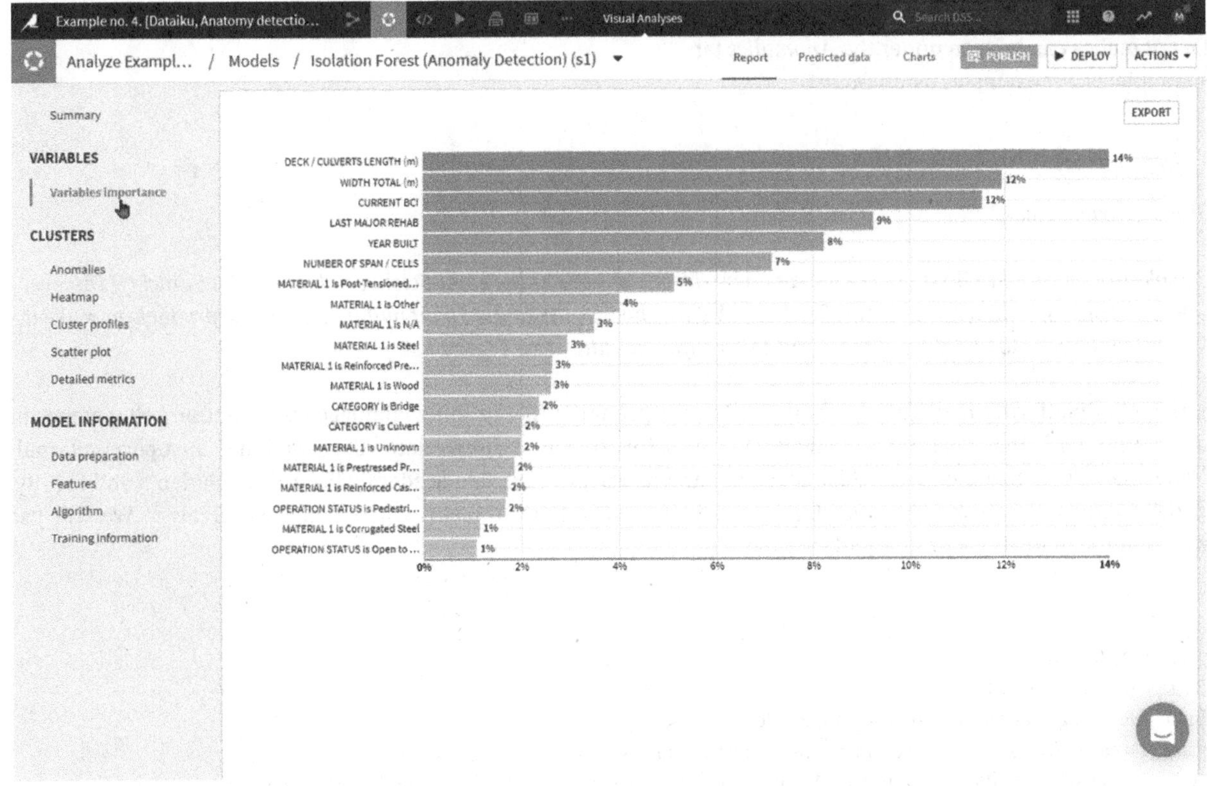

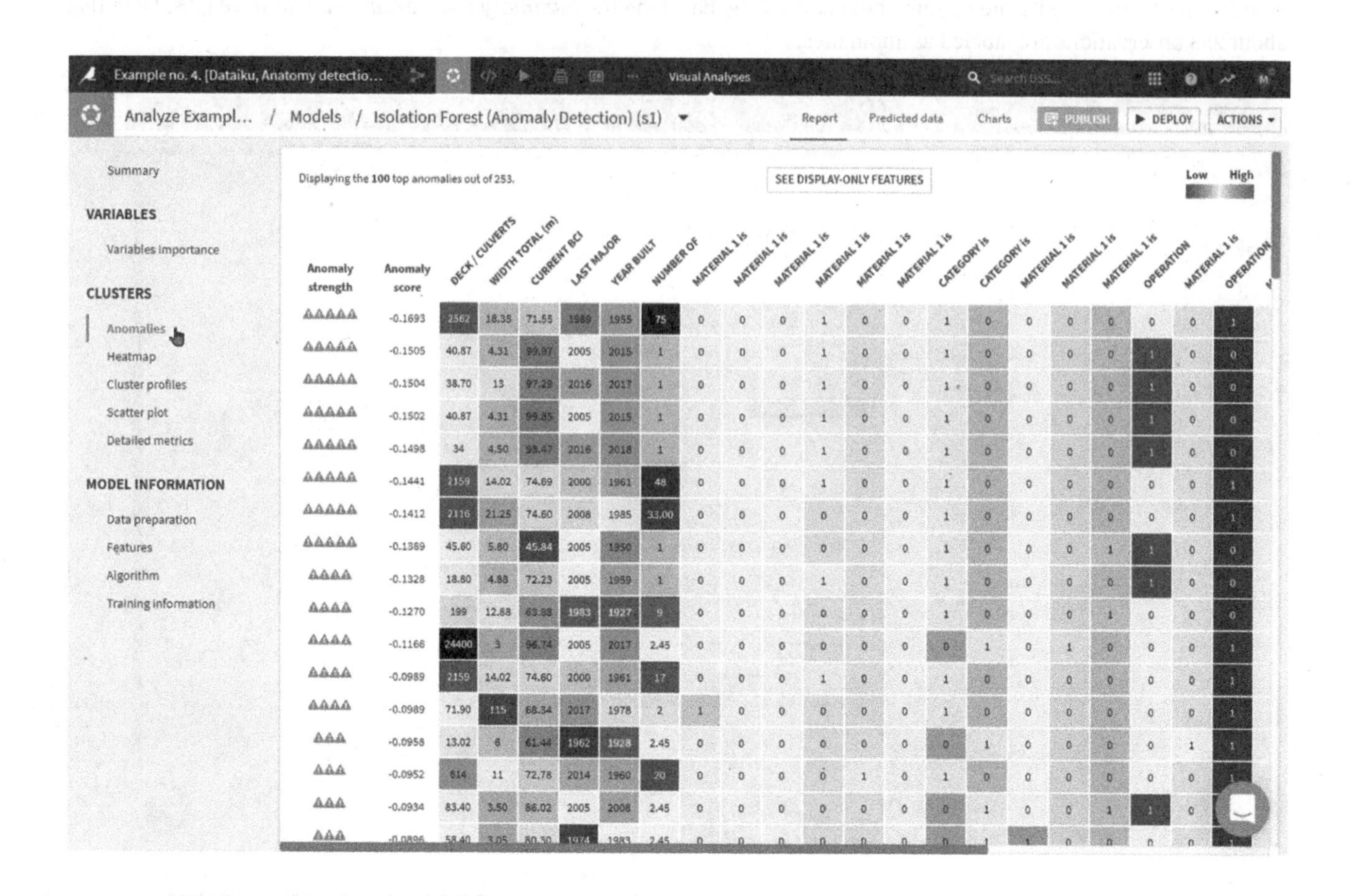

Specific anomalies can be seen under the *Anomalies* tab.

This completes our examples using *Dataiku*.

6.1.4 Exploratory

According to *Exploratory*,[48]

> *Exploratory's Simple UI experience makes it possible for anyone to use Data Science to Explore data quickly, Discover deeper insights, and Communicate effectively. We are always passionate about building awesome products to make it easier for anyone to understand the world better through data.* – see Figure 6.4.

Unlike the other platforms, *Exploratory* is built on *R* as the primary programming language and a platform that comes in software (with GUI[49] and programming support). This software can accommodate supervised and unsupervised analysis.[50] This platform is based in California, USA [9]. *Exploratory* provides free academic licenses to students and faculty. One of the features of *Exploratory* is its ability to download and share projects with other users freely.[51] We will use *Exploratory* in the following four different examples.[52]

48 https://exploratory.io.

49 As in support for coding-free.

50 Including clustering, anomaly detection, and time series analysis.

51 As such, you can download all examples from here. Thanks to Kan Nishida!

52 Which are going to be a bit diverse and tackle other than structural engineering and construction material problems.

Figure 6.4 Snippet of the *Exploratory* platform (July 2022).

Example no. 1 [*Exploratory*, Classification]: Classifying water potability.
In this example, we will be investigating a dataset with regard to water potability.[53] In this effort, we will analyze 3276 instances[54] in terms of pH value,[55] hardness,[56] solids, chloramines,[57] sulfate, conductivity,[58] organic carbon, trihalomethanes,[59] and turbidity[60] to predict the potability of water (i.e., bad quality water or good quality water). We will be adopting the decision tree and random forest algorithms in this example.

Step 1: Let us start by uploading our dataset.

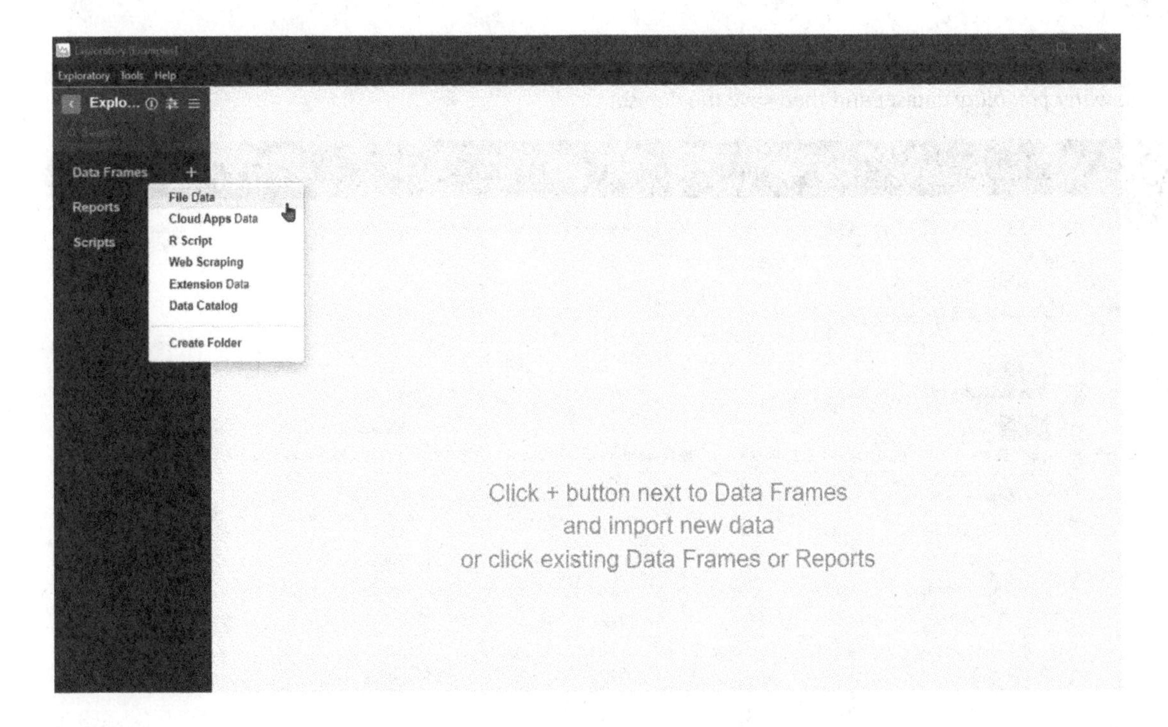

53 Suitability for human consumption.
54 Thanks to Jatin Sadhwani [35] for sharing this dataset.
55 A measure of acid–base balance of water.
56 A measure of mineral content (especially, calcium and magnesium salts).
57 Common disinfectants in public water systems.
58 A measure of transmitting current.
59 Chemicals present in water treated with chlorine.
60 A measure of light emitting properties of water.

Step 2: Select the supplemented dataset.

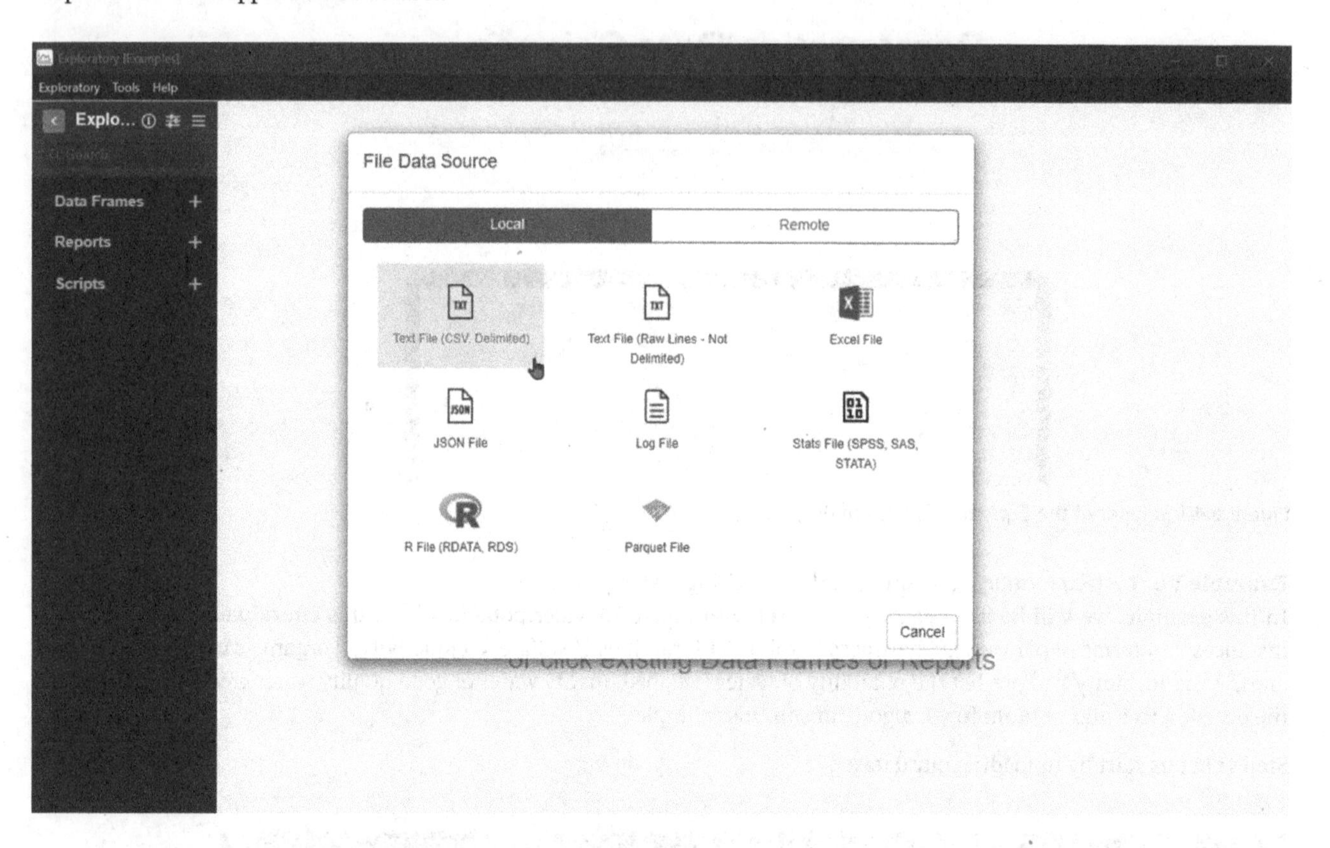

Step 3: Upload the *water potability* dataset and then save the dataset.

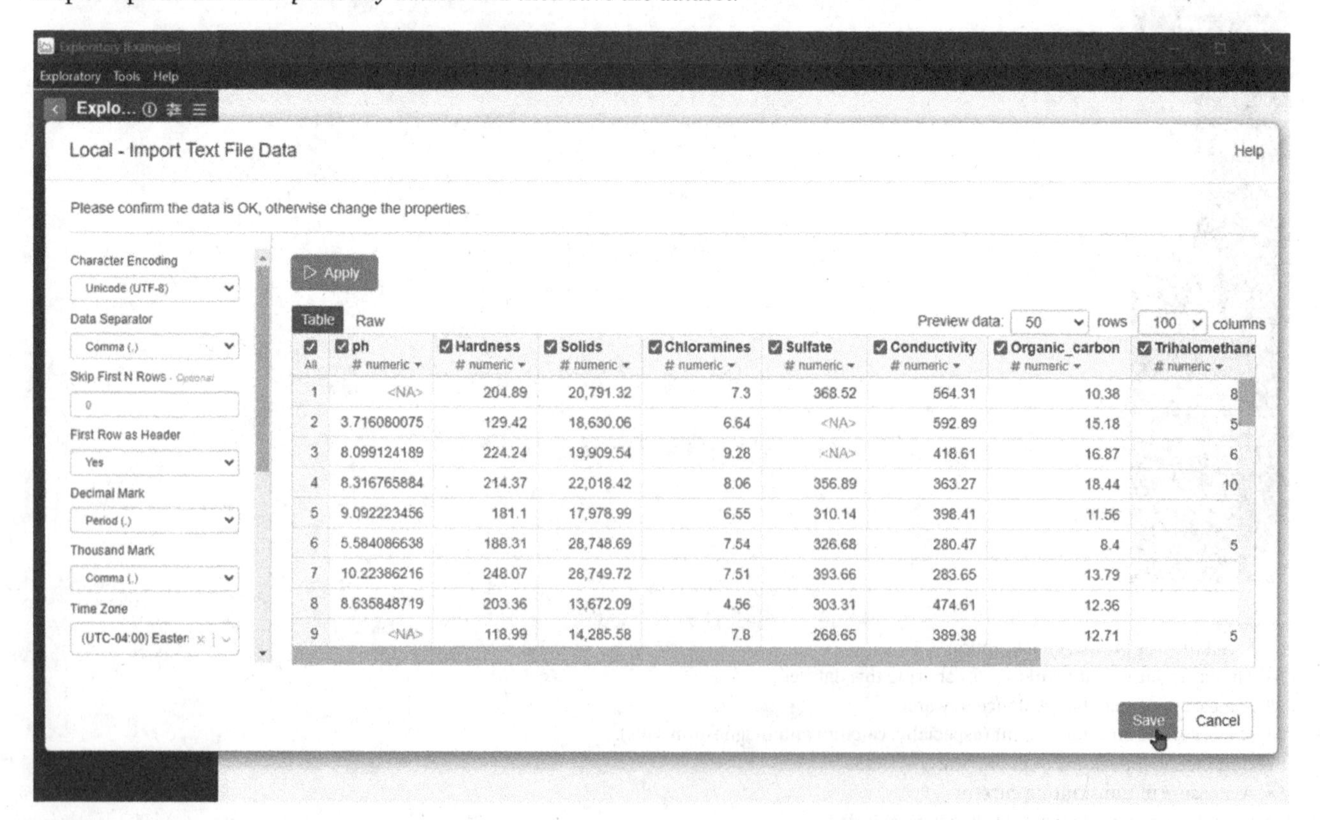

Step 4: Examine the generated histograms and notice the number of missing data in each feature.

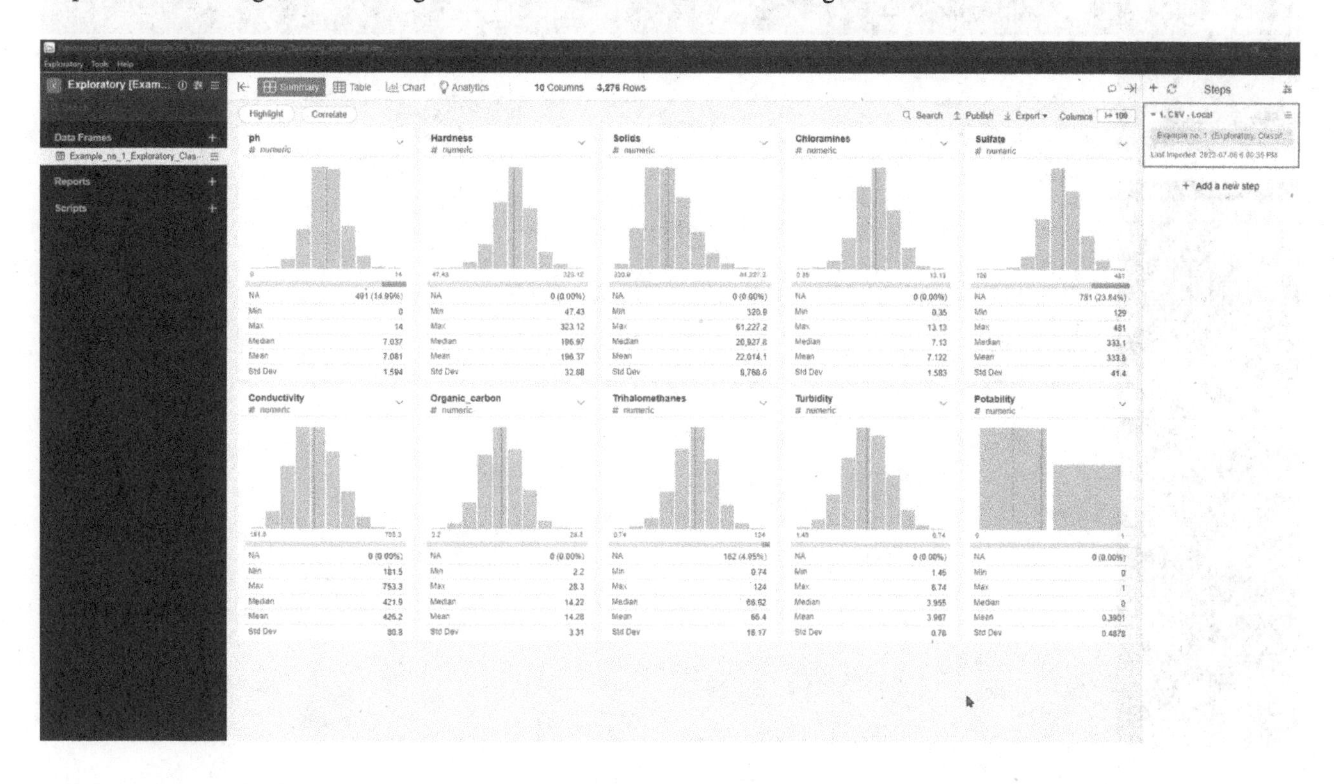

Step 5: Visit the *Table* tab. Go to *Replace values* and then *With new values.* Our goal is to update the characters into terms where 0 means *Bad quality water (Not Potable)* and 1 refers to *Good quality water (Potable).* Re-visiting the *Summary* tab will show the new type of data.

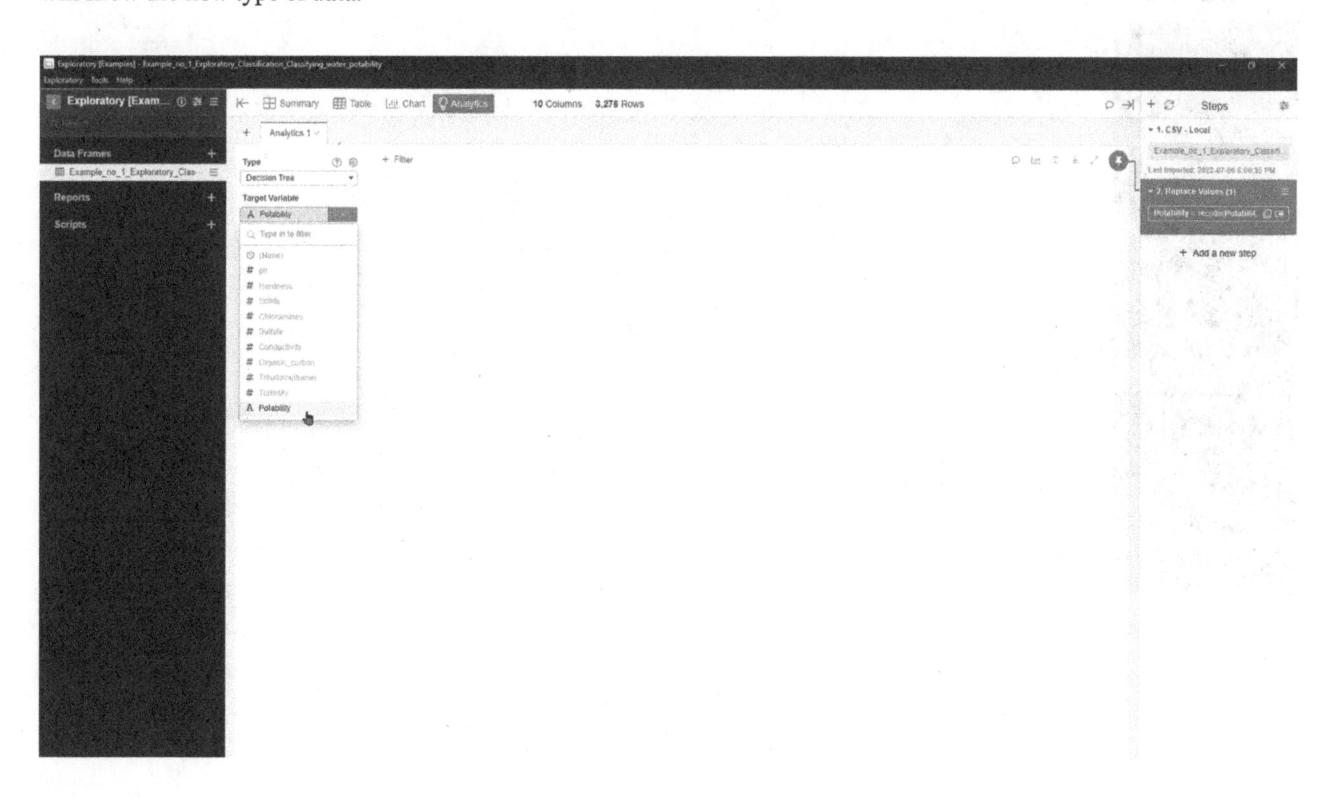

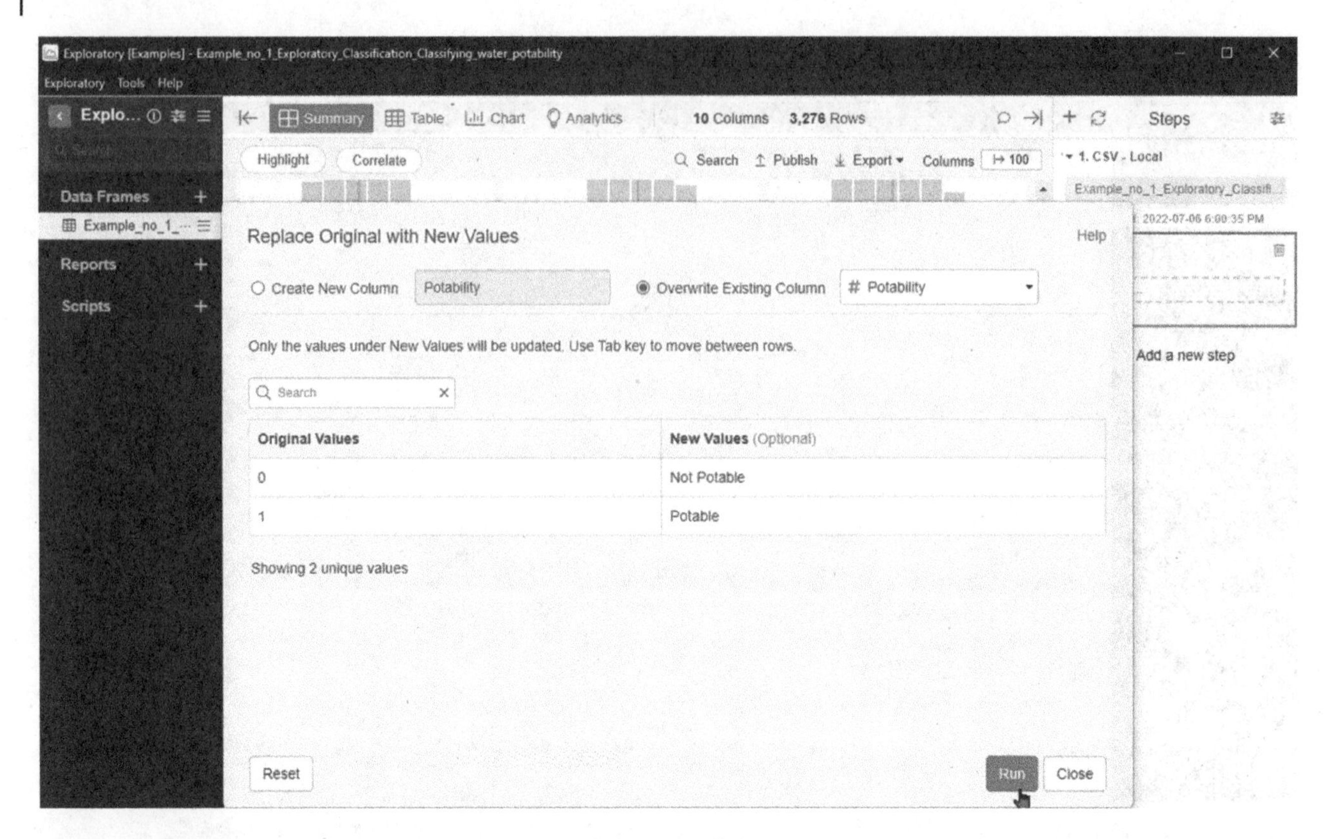
Exploratory [Examples] - Example_no_1_Exploratory_Classification_Classifying_water_potability
Exploratory Tools Help
Explo...
Summary
Table
Chart
Analytics
10 Columns
3,276 Rows
Steps
Highlight
Correlate
Search
Publish
Export
Columns
100
1. CSV - Local
Data Frames
Example_no_1_...
Reports
Scripts
Replace Original with New Values
Help
Create New Column
Potability
Overwrite Existing Column
Potability
Only the values under New Values will be updated. Use Tab key to move between rows.
Search
Original Values
New Values (Optional)
0
Not Potable
1
Potable
Showing 2 unique values
Add a new step
Reset
Run
Close

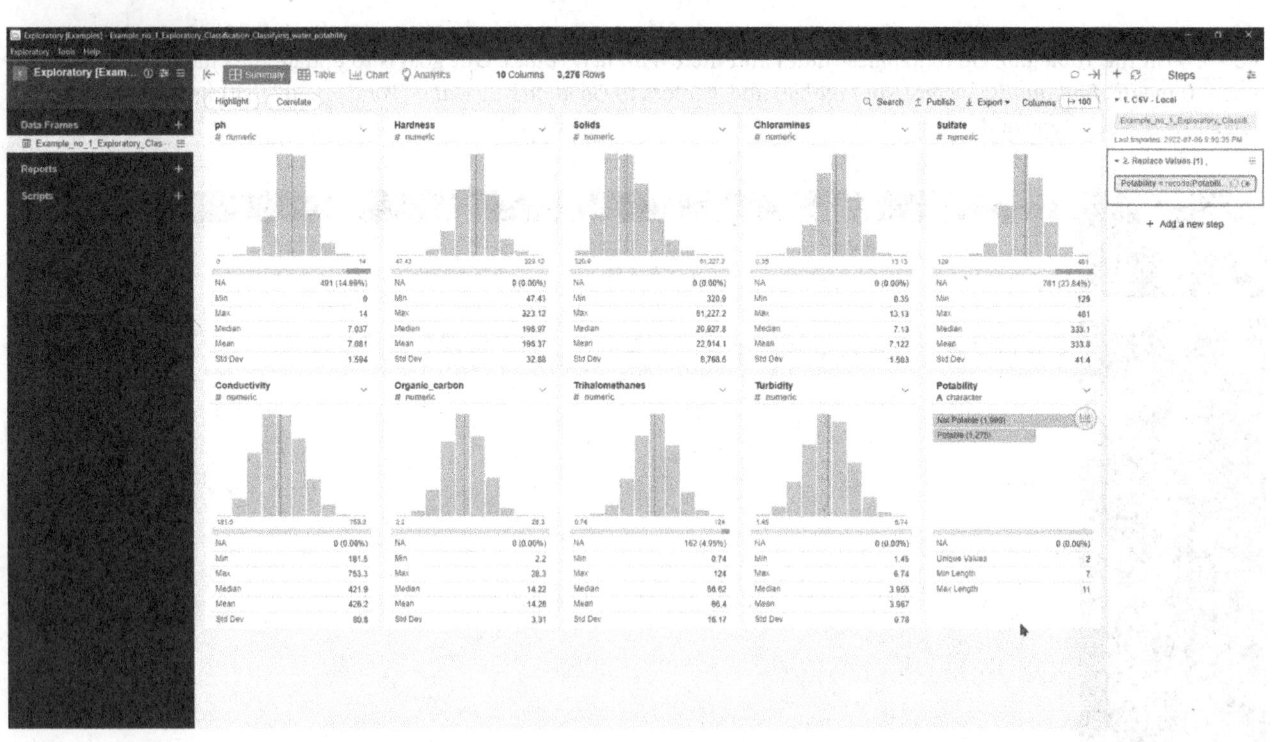
Exploratory [Exam...
Summary
Table
Chart
Analytics
10 Columns
3,276 Rows
Steps
Highlight
Correlate
Data Frames
Reports
Scripts
ph
Hardness
Solids
Chloramines
Sulfate
Conductivity
Organic_carbon
Trihalomethanes
Turbidity
Potability
Add a new step

Step 6: Select *Decision Tree*, Set the target to *Potability* and set the features. *Run* the analysis.

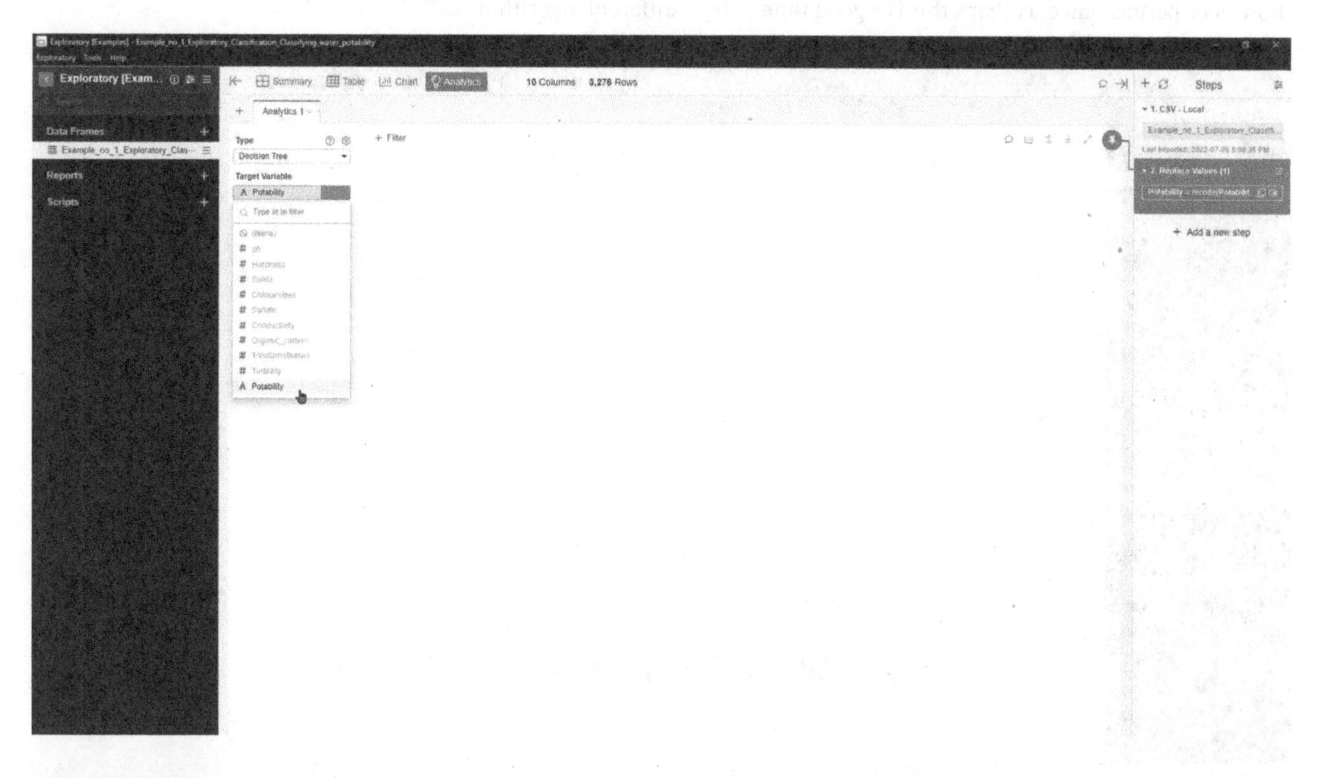

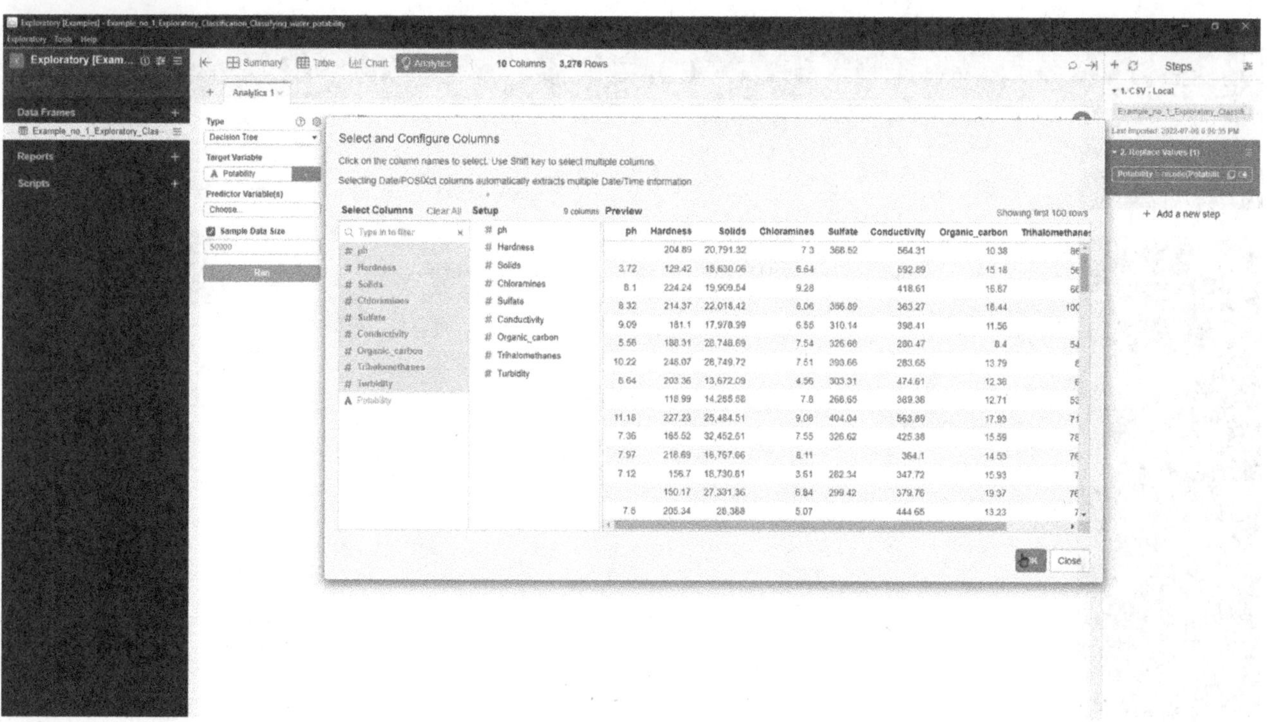

Step 7: Examine the arrived at decision tree and summary of analysis. As you can see, the analysis and confusion matrix show poor performance. Perhaps this is a good time to try a different algorithm.

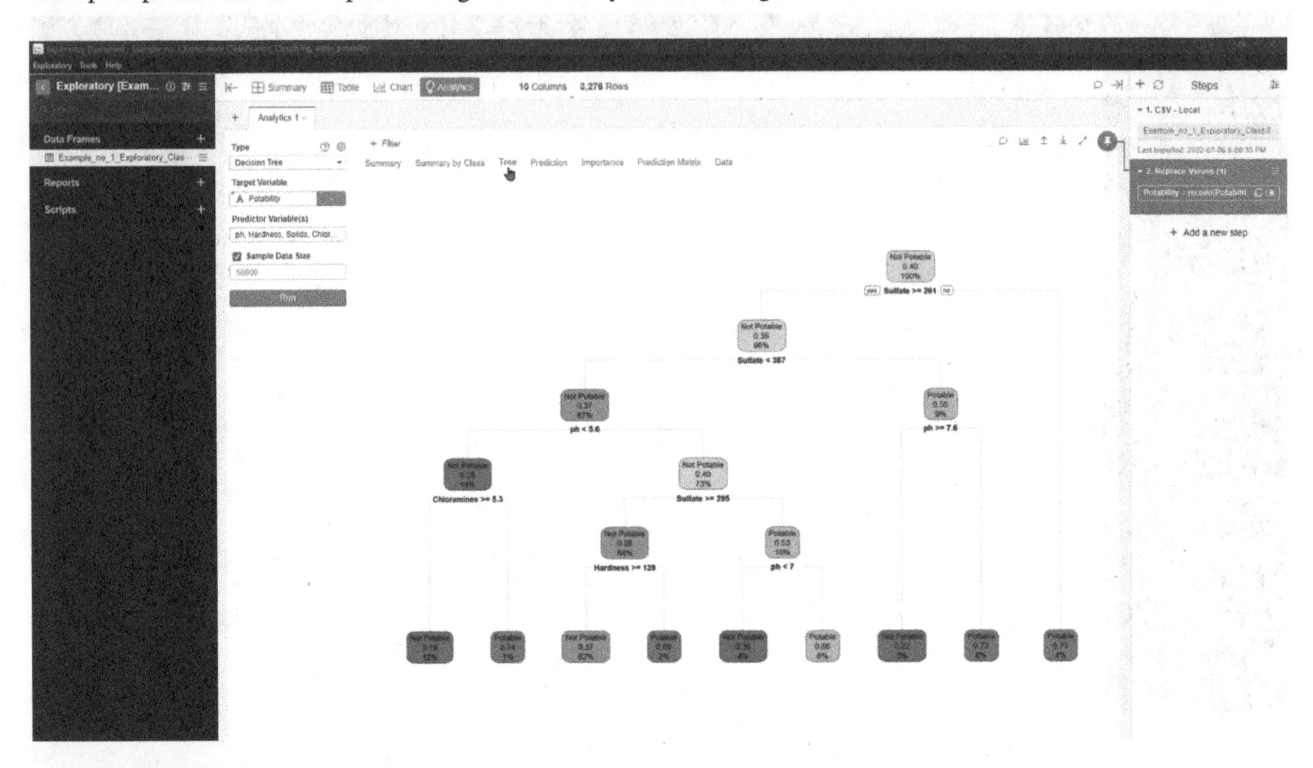

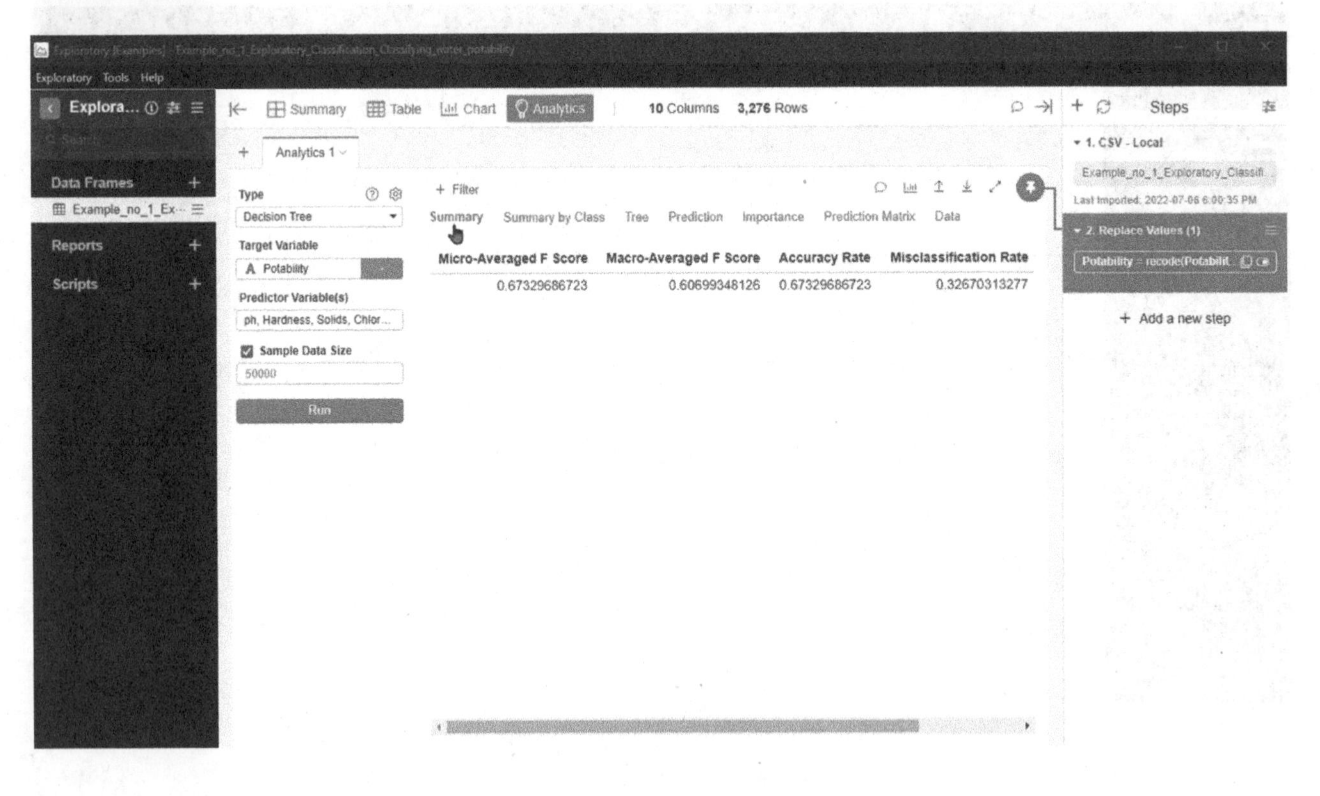

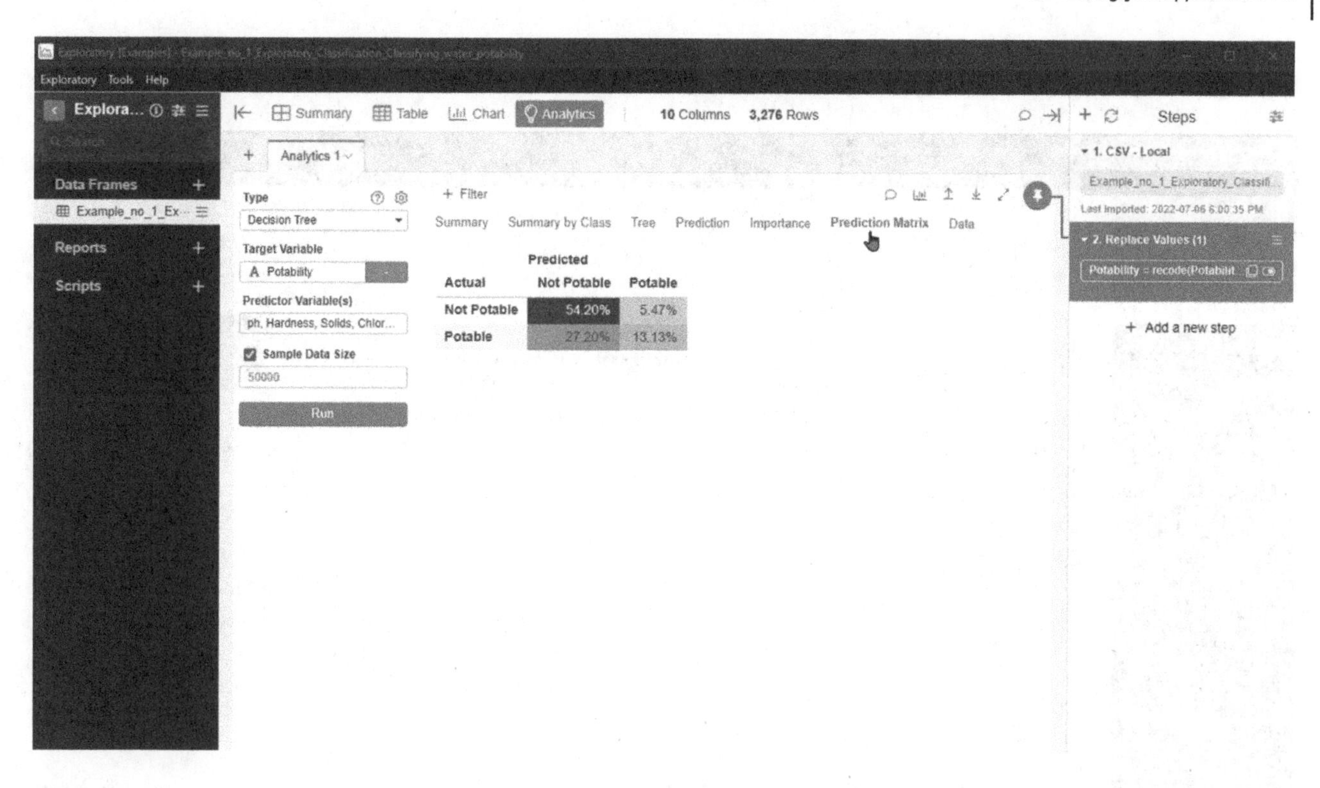

Step 8: Add a new *Analytics* tab. Switch over to the *Random Forest* algorithm in pursuit of improved performance.

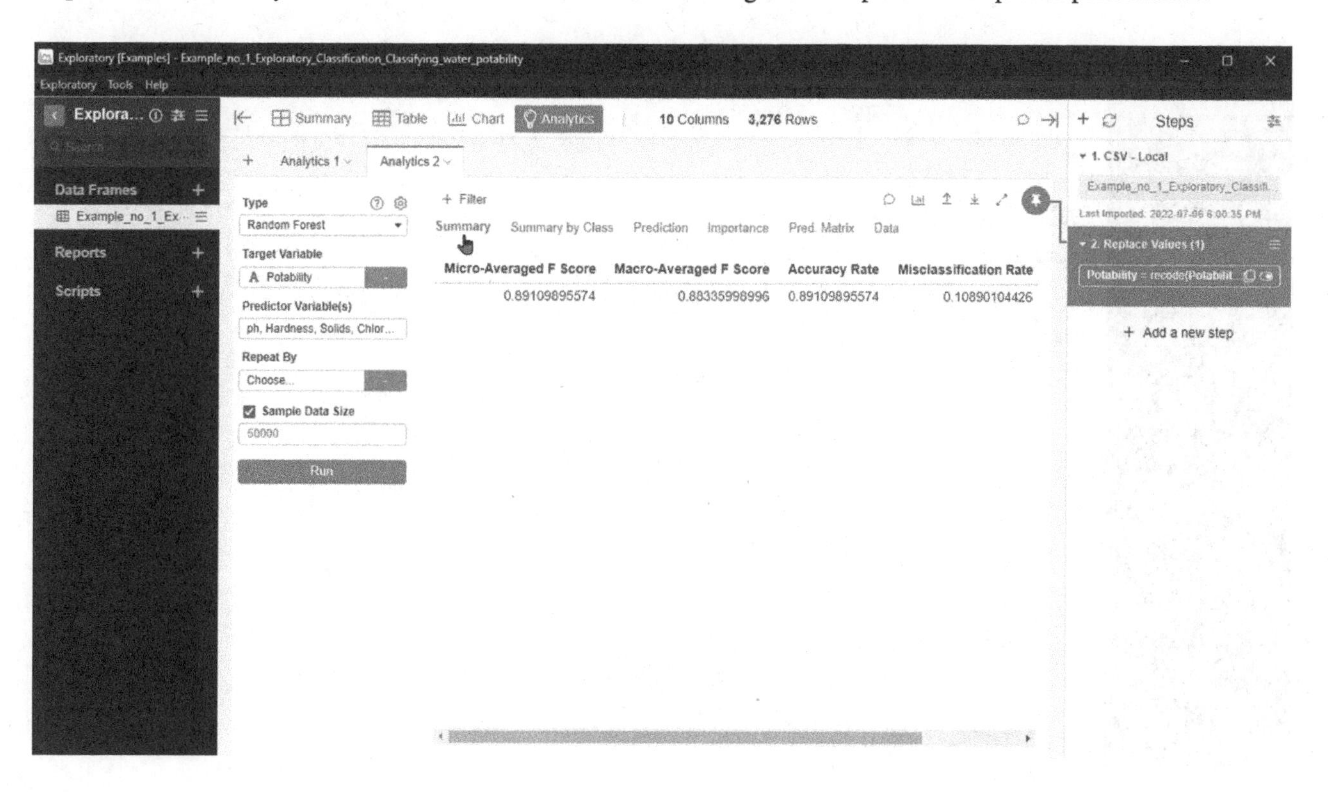

Step 9: Analyze the outcome of the analysis. As you can see, this analysis reveals that the model predicts the target with similar accuracy.

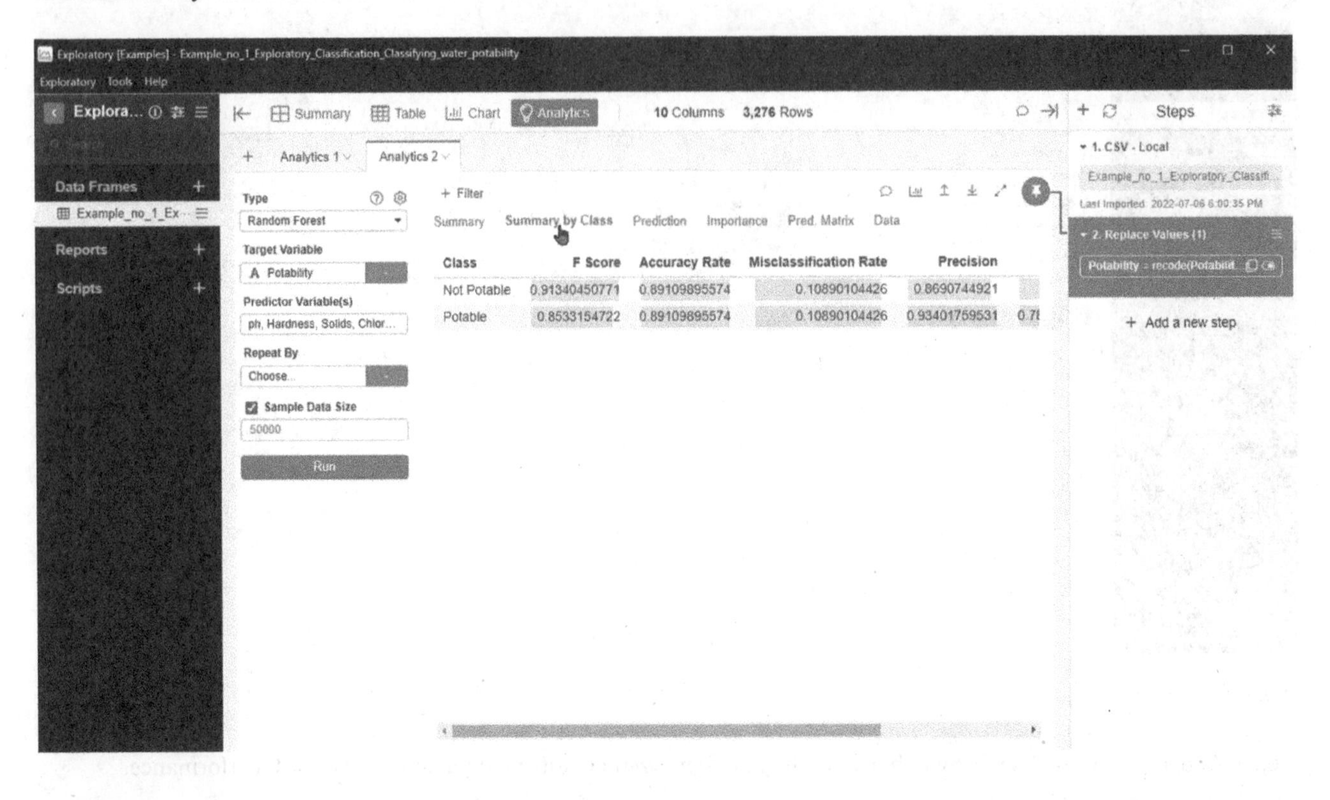

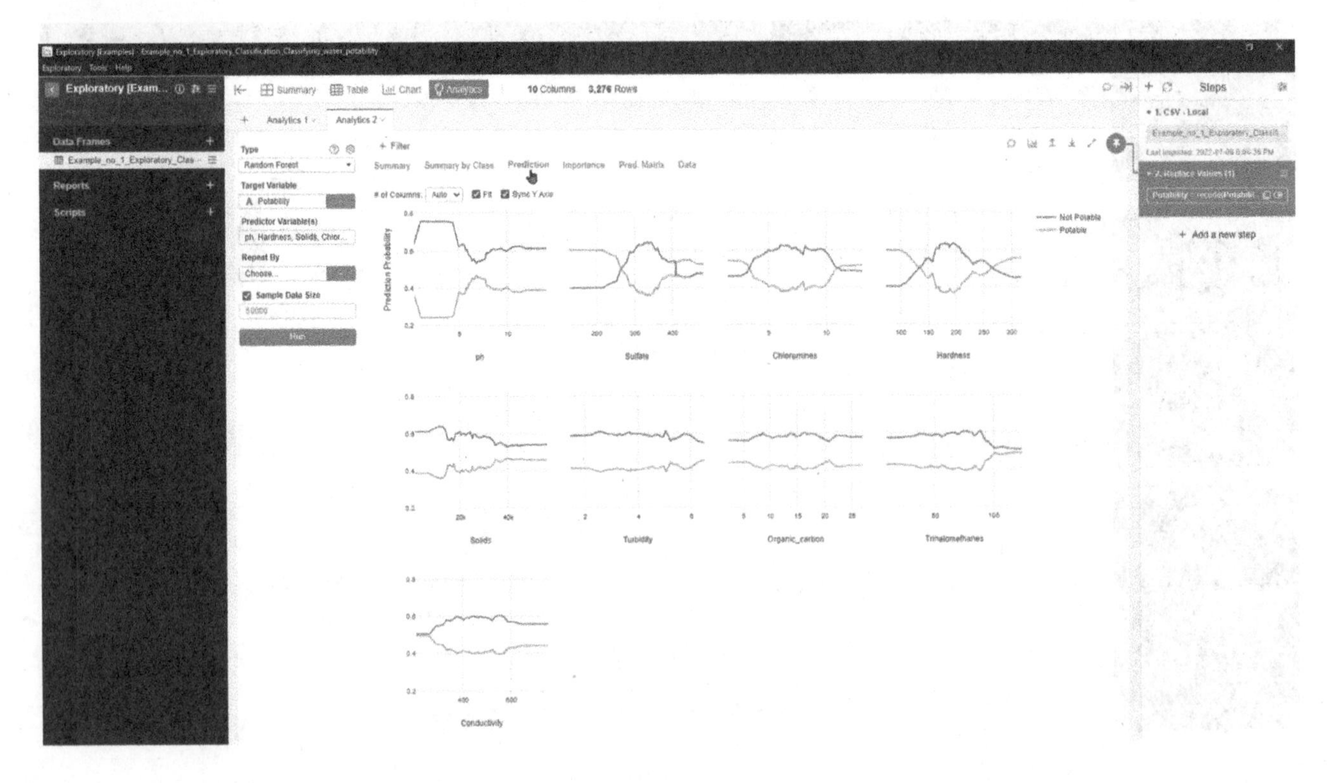

Step 10: Explore the *Importance* of features and confusion matrix.

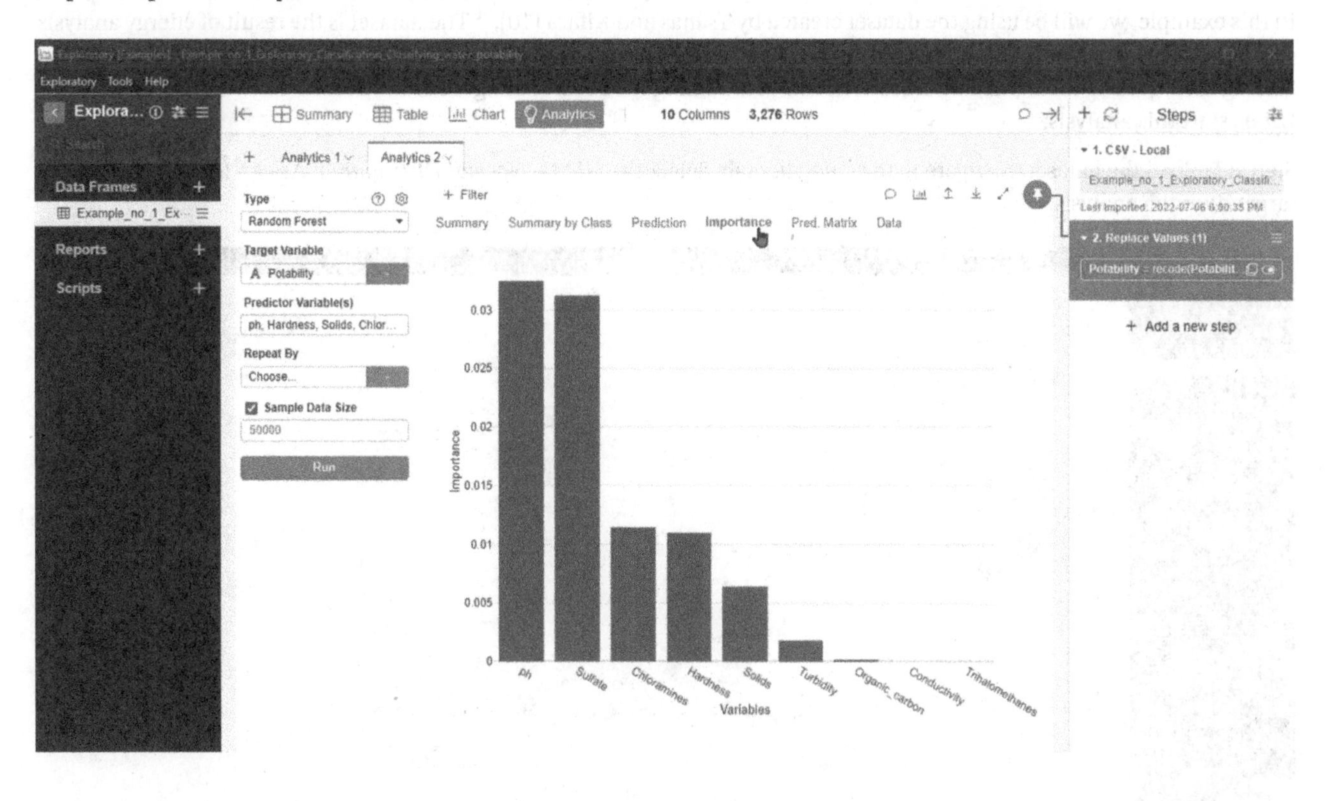

Step 11: Evaluate the confusion matrix under the *Pred. Matrix*.

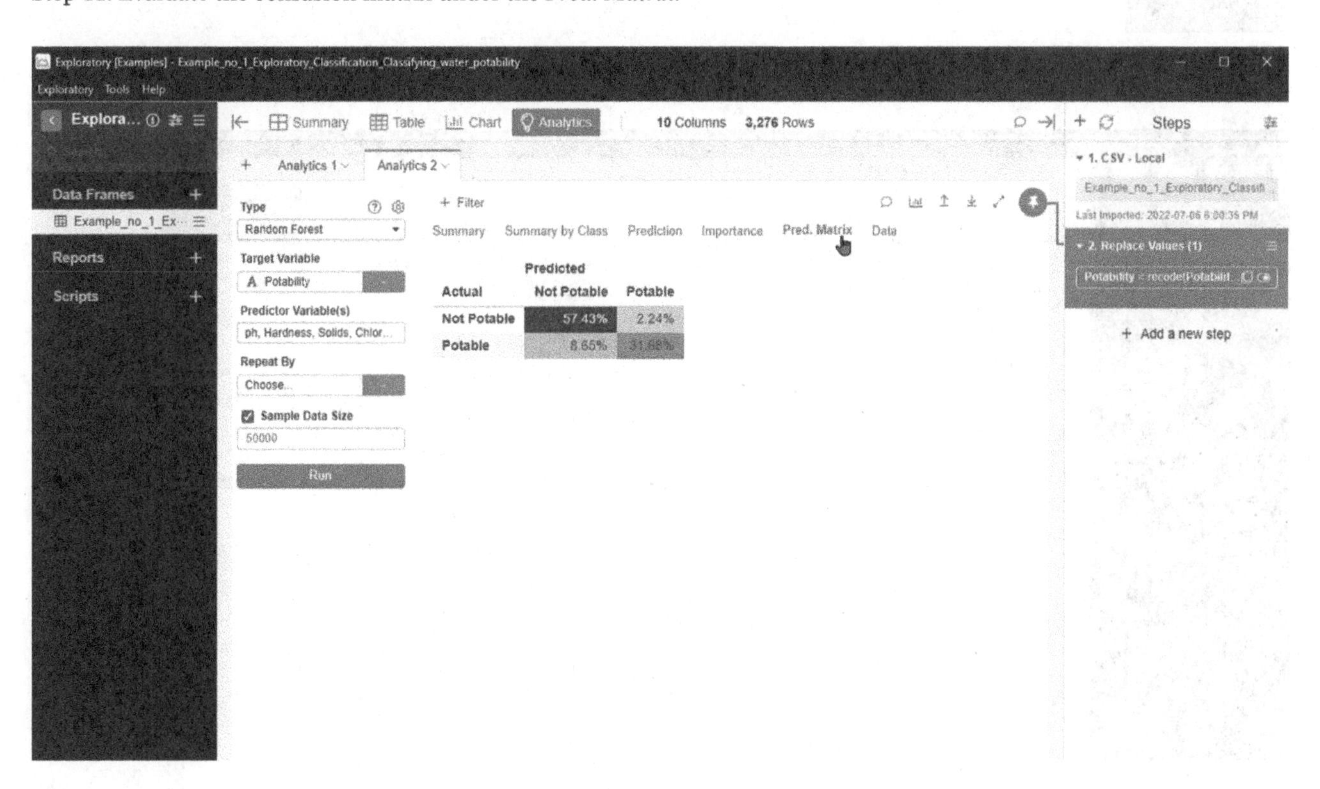

Example no. 2 [*Exploratory*, Regression]: Predicting energy performance of residential buildings.
In this example, we will be using the dataset created by Tsanas and Xifara [10].[61] The dataset is the result of energy analysis on 12 buildings. The features include relative compactness,[62] surface area, wall area, roof area, overall height, orientation, glazing area, glazing area distribution, and two targets heating load and cooling load.[63]
Let us start this analysis:

Step 1: Upload the dataset. Navigate to the *Analytics* tab. Select the *GLM – Normal Distribution*. Assign features and the target. *Run* the analysis.

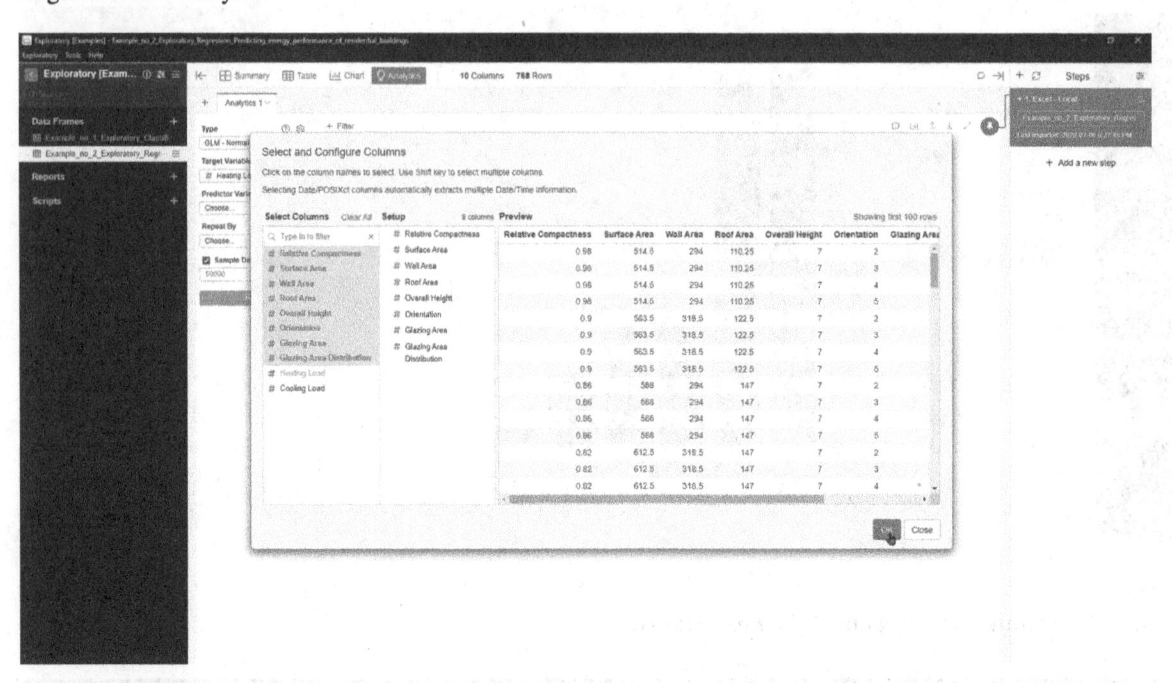

Step 2: Navigate to the *Summary* tab to check the outcome of the analysis. Do the same to the *Prediction* tab and *Importance* tab.

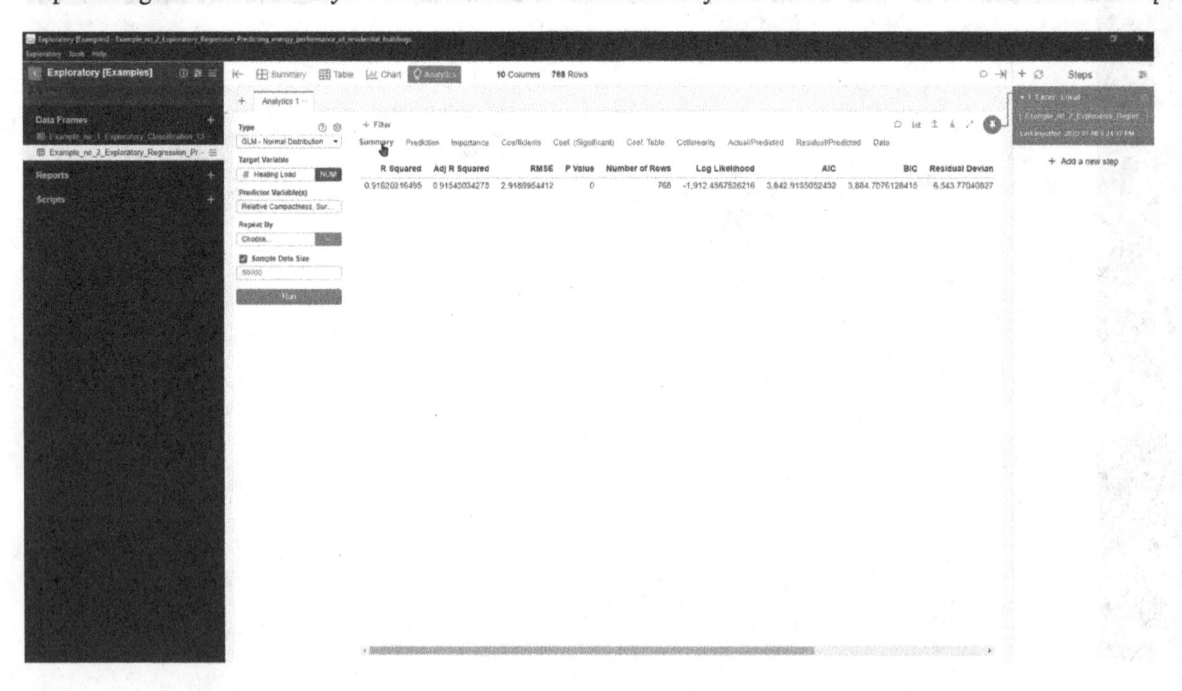

61 The actual dataset can be found here [36].

62 **Relative compactness**: defined as the volume to surface ratio to that of the most compact shape with the same volume.

63 For obvious reasons, I will show this example in terms of heating load as the target and will ask you to repeat it for the cooling load.

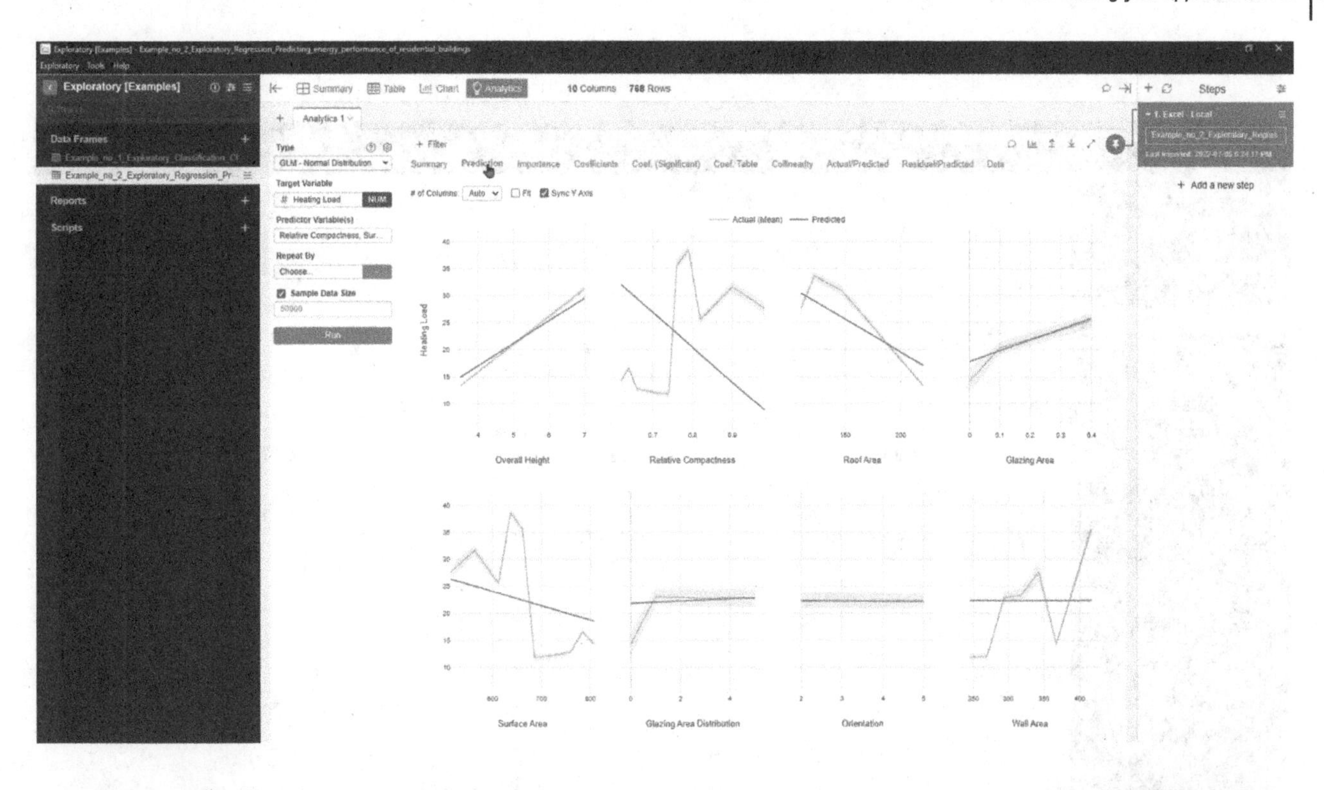
Exploratory [Examples]
Summary
Table
Chart
Analytics
10 Columns
768 Rows
Steps
Analytics 1
Data Frames
Reports
Scripts
Type
GLM - Normal Distribution
Target Variable
Heating Load
NUM
Predictor Variable(s)
Relative Compactness, Sur...
Repeat By
Choose...
Sample Data Size
50000
Run
Filter
Summary
Prediction
Importance
Coefficients
Coef. (Significant)
Coef. Table
Collinearity
Actual/Predicted
Residual/Predicted
Data
Add a new step
Actual (Mean)
Predicted
Heating Load
Overall Height
Relative Compactness
Roof Area
Glazing Area
Surface Area
Glazing Area Distribution
Orientation
Wall Area

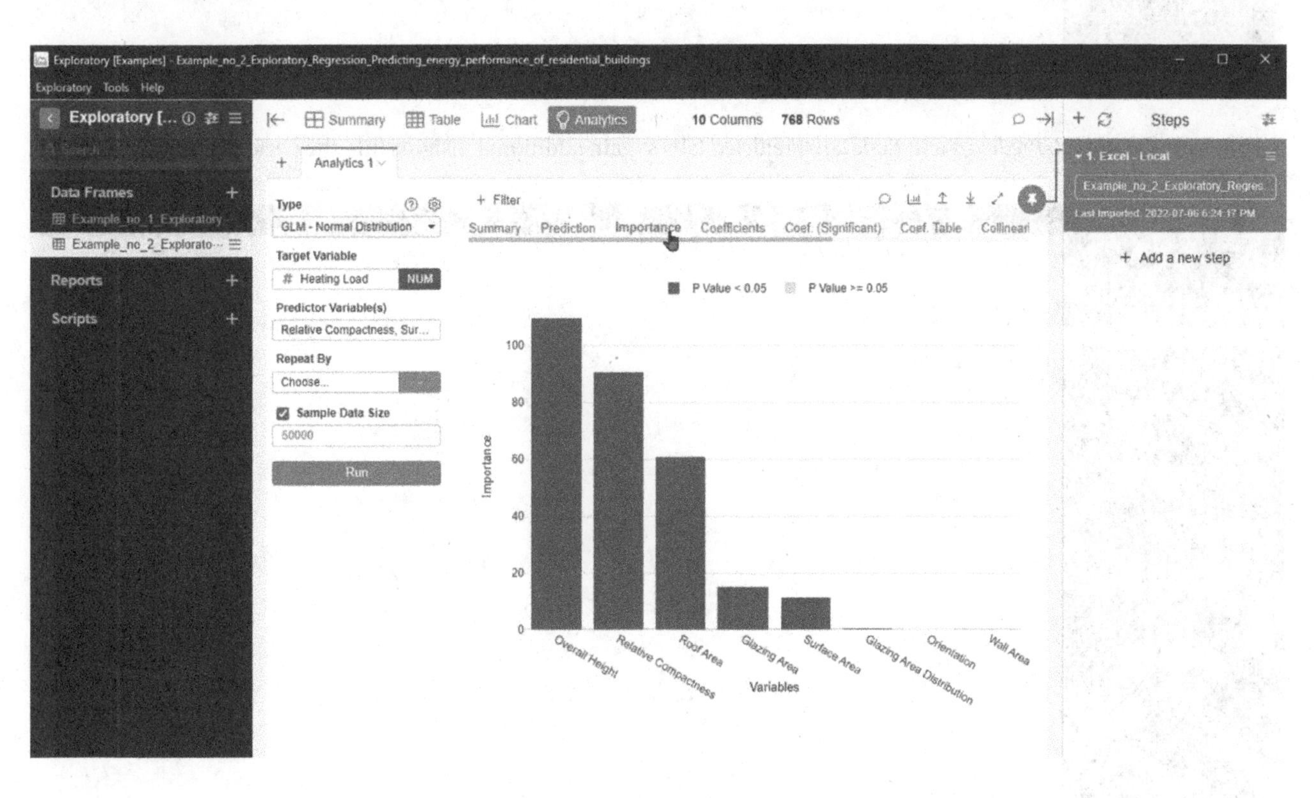
Exploratory [Examples] - Example_no_2_Exploratory_Regression_Predicting_energy_performance_of_residential_buildings
Exploratory
Tools
Help
Exploratory [...
Summary
Table
Chart
Analytics
10 Columns
768 Rows
Steps
1. Excel - Local
Example_no_2_Exploratory_Regres
Last Imported: 2022-07-06 6:24:17 PM
Add a new step
Analytics 1
Data Frames
Example_no_1_Exploratory
Example_no_2_Explorato...
Reports
Scripts
Type
GLM - Normal Distribution
Target Variable
Heating Load
NUM
Predictor Variable(s)
Relative Compactness, Sur...
Repeat By
Choose...
Sample Data Size
50000
Run
Filter
Summary
Prediction
Importance
Coefficients
Coef. (Significant)
Coef. Table
Collinear
P Value < 0.05
P Value >= 0.05
Importance
100
80
60
40
20
0
Overall Height
Relative Compactness
Roof Area
Glazing Area
Surface Area
Glazing Area Distribution
Orientation
Wall Area
Variables

Step 3: Visit the *Coef. Table* tab. These are the coefficients of the LM fitted model.

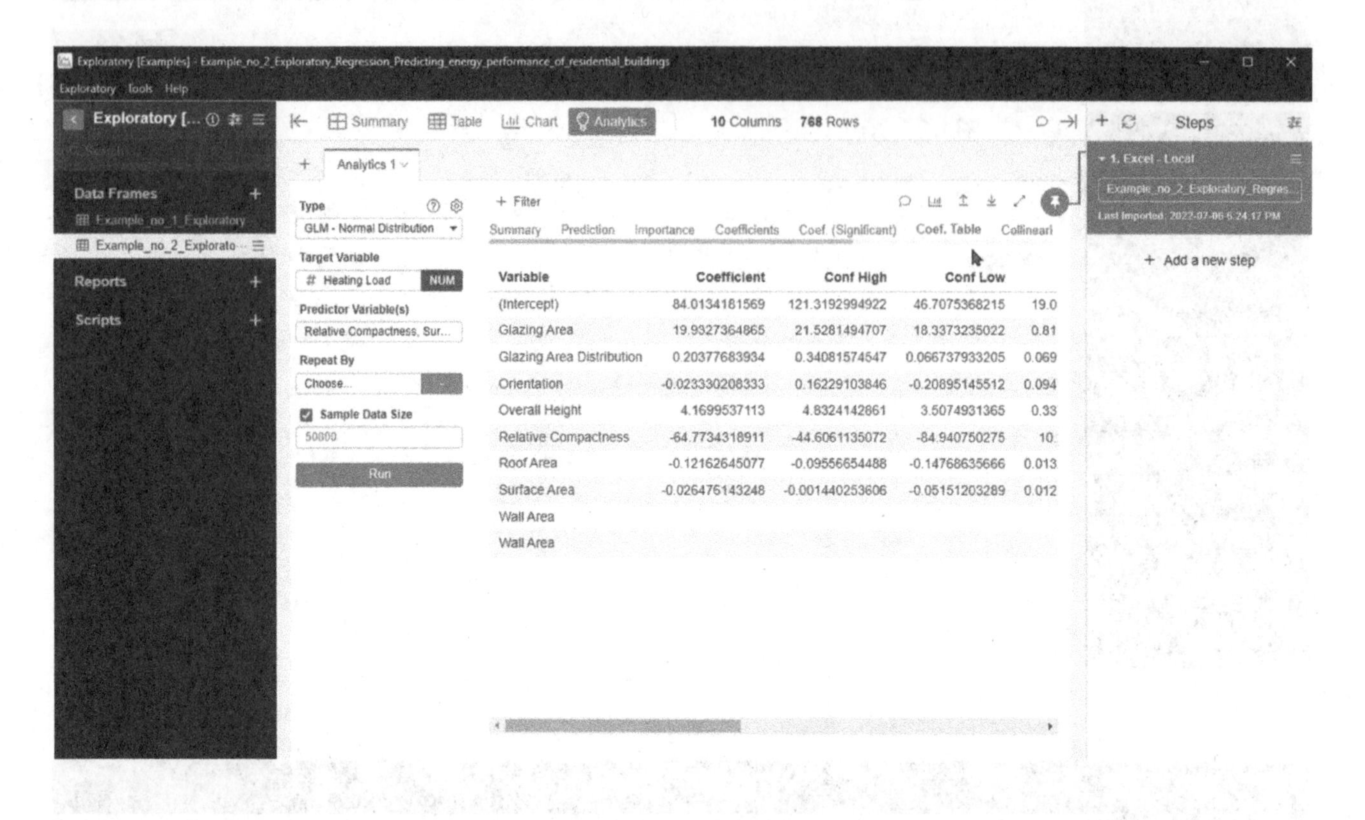

Variable	Coefficient	Conf High	Conf Low	
(Intercept)	84.0134181569	121.3192994922	46.7075368215	19.0
Glazing Area	19.9327364865	21.5281494707	18.3373235022	0.81
Glazing Area Distribution	0.20377683934	0.34081574547	0.066737933205	0.069
Orientation	-0.023330208333	0.16229103846	-0.20895145512	0.094
Overall Height	4.1699537113	4.8324142861	3.5074931365	0.33
Relative Compactness	-64.7734318911	-44.6061135072	-84.940750275	10.
Roof Area	-0.12162645077	-0.09556654488	-0.14768635666	0.013
Surface Area	-0.026476143248	-0.001440253606	-0.05151203289	0.012
Wall Area				
Wall Area				

Step 4: Visit the *Actual/Predicted* and *Residual/Predicted* tabs to gain additional insights into the performance of the model.

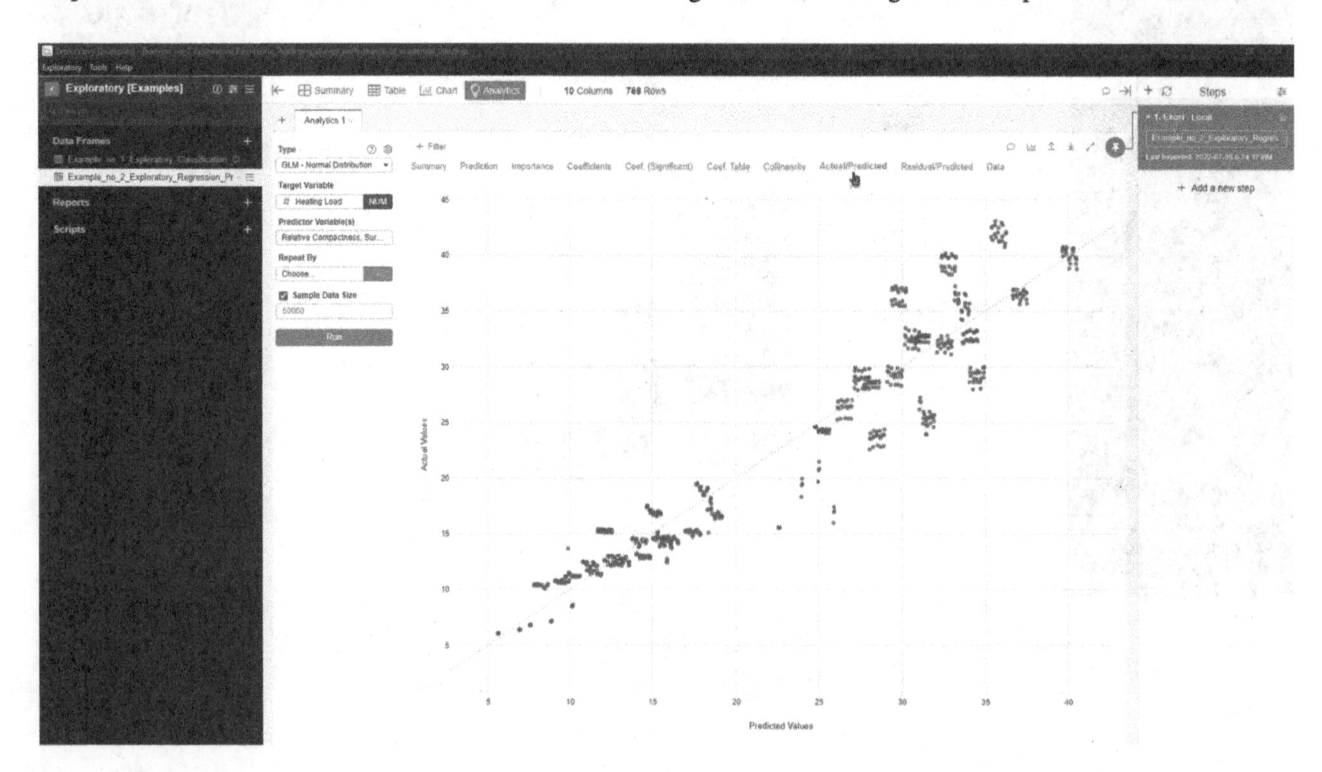

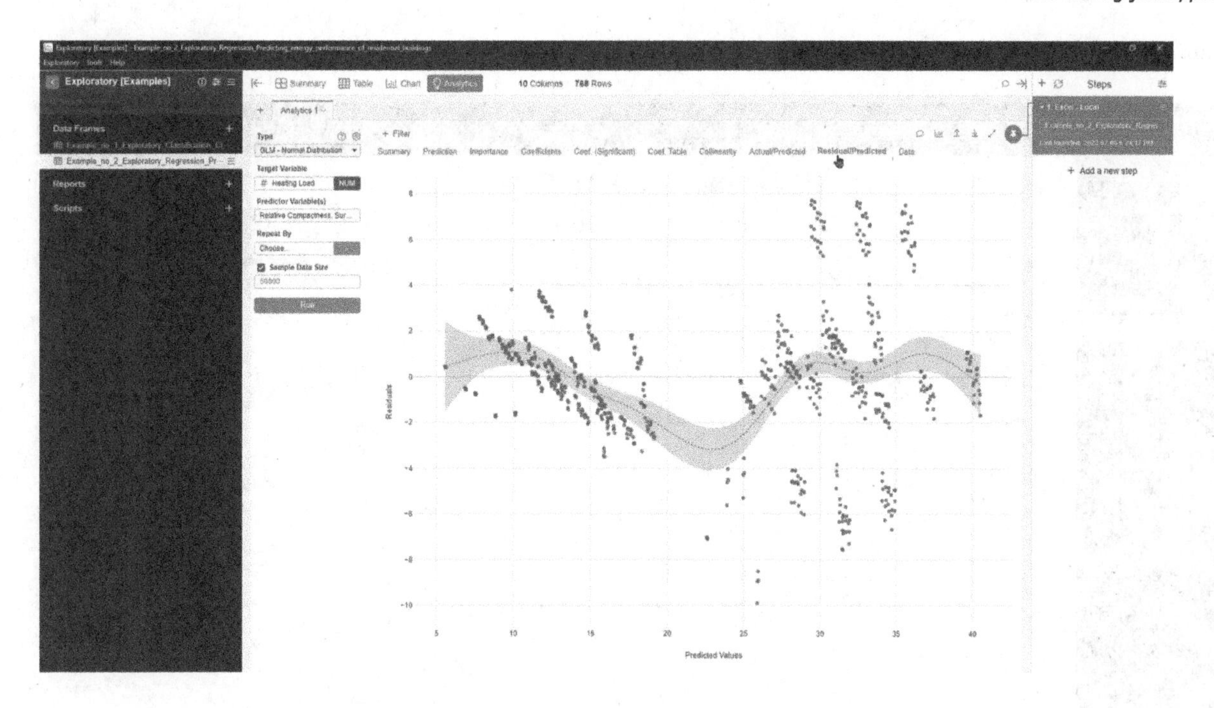

Example no. 3 [*Exploratory*, Clustering]: Clustering of cities in terms of carbon emissions.
In this example, a dataset created by Aditya Chandavale[64] sums up information on carbon emissions in 28 Indian cities. This dataset contains three features: carbon monoxide, CO, Carbon dioxide, CO2, Methane, CH4, in kg per capita.
To carry out this analysis, upload this dataset.

Step 1: Select the *K-Means clustering* analysis. Navigate to *Property*[65] and then *Find Optimal Number of Clusters* option and select *TRUE*. This allows Exploratory to find the optimal number of clusters via the elbow method.

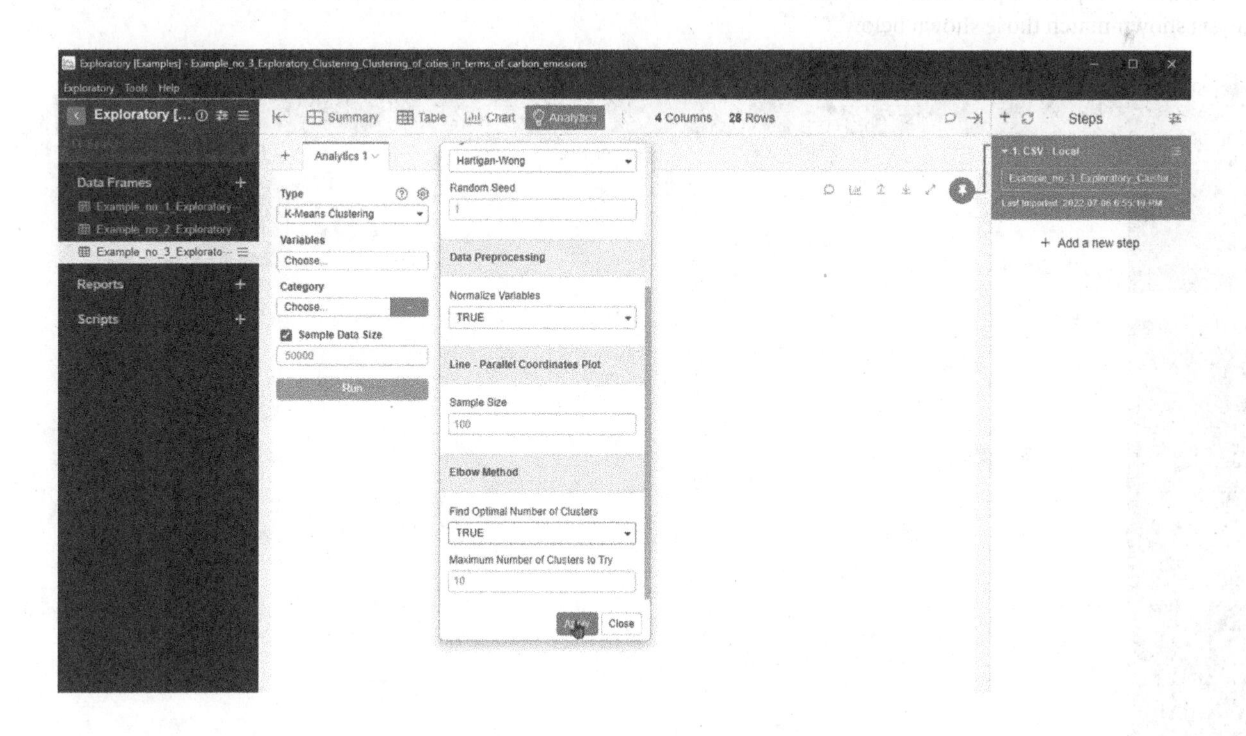

64 This dataset can be found here [37].
65 The *gear* symbol next to the *Type* tab.

Step 2: Run the analysis and notice that there are four possible clusters.

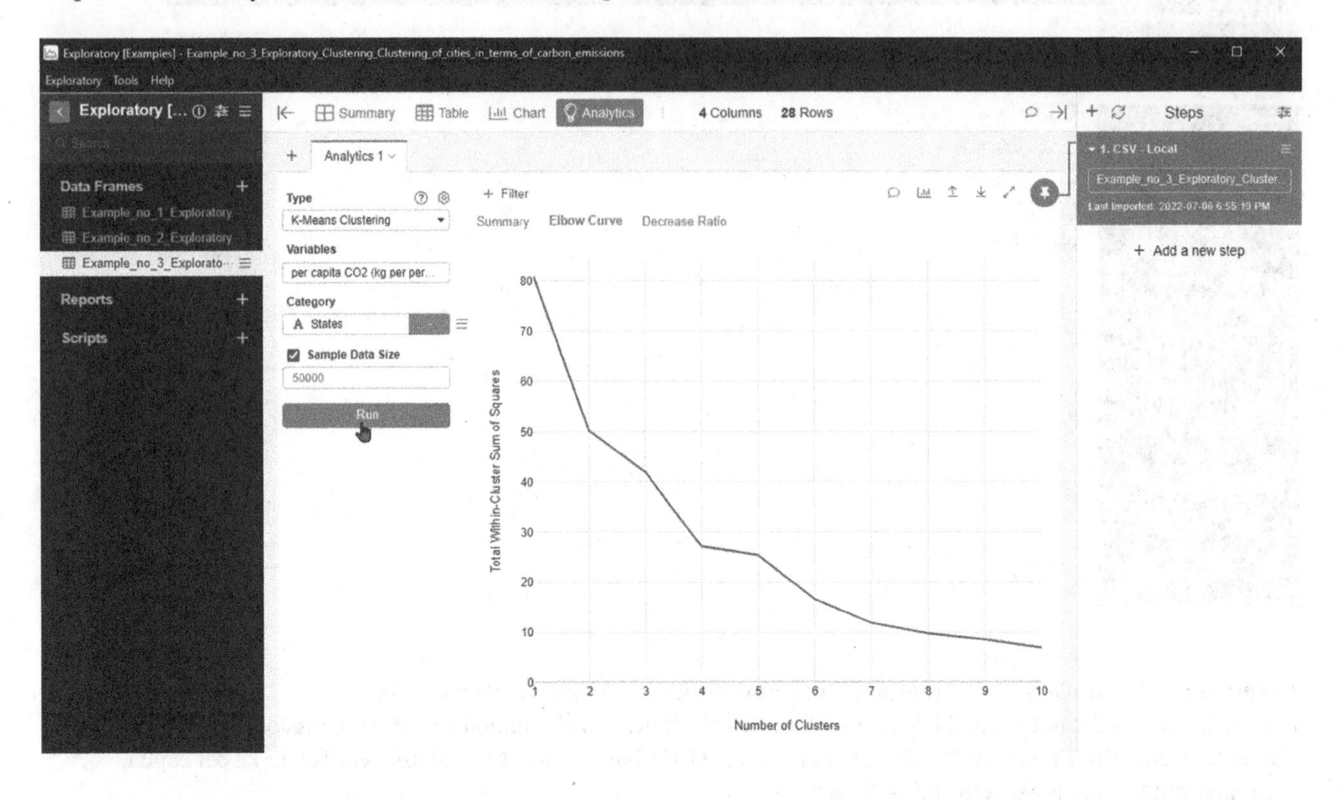

Step 3: Navigate to *Property*. Update the *Find Optimal Number of Clusters* option to FALSE. *Run* the analysis. The results of the analysis shown match those shown below.

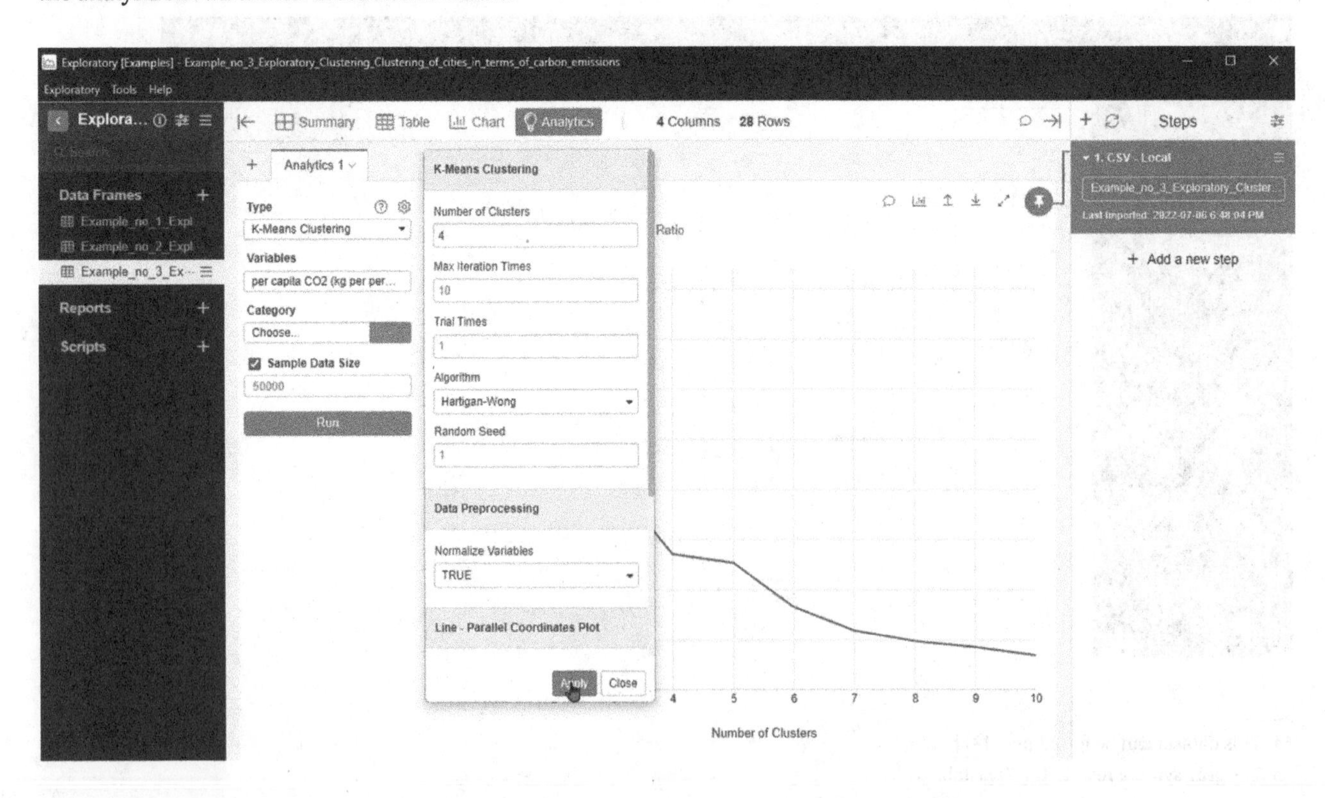

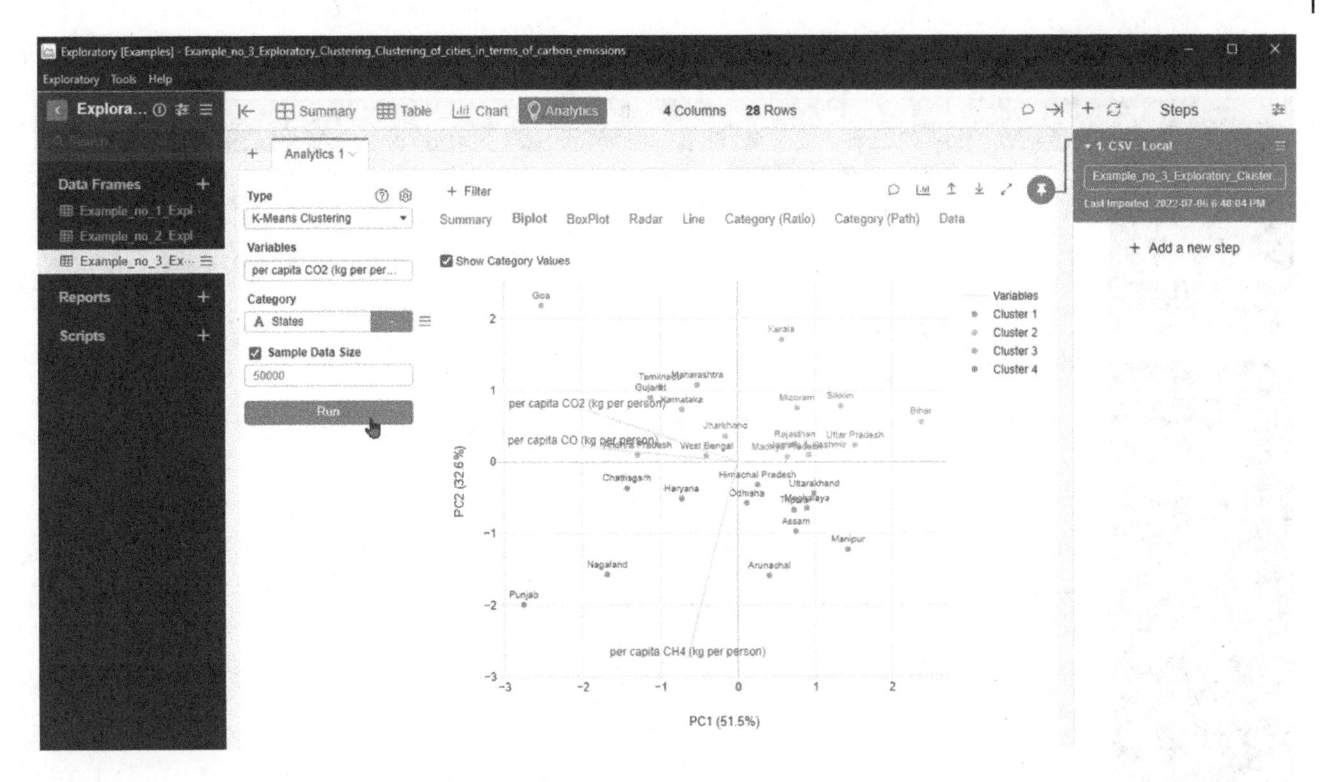

Example no. 3 [*Exploratory*, Supplementary clustering]: Clustering of RC columns against fire.
For completion, I will carry out the same clustering example we did in *BigML* as we will re-visit this example in the following section.[66]

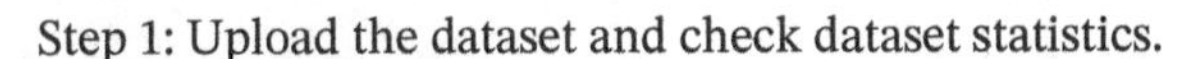

Step 1: Upload the dataset and check dataset statistics.

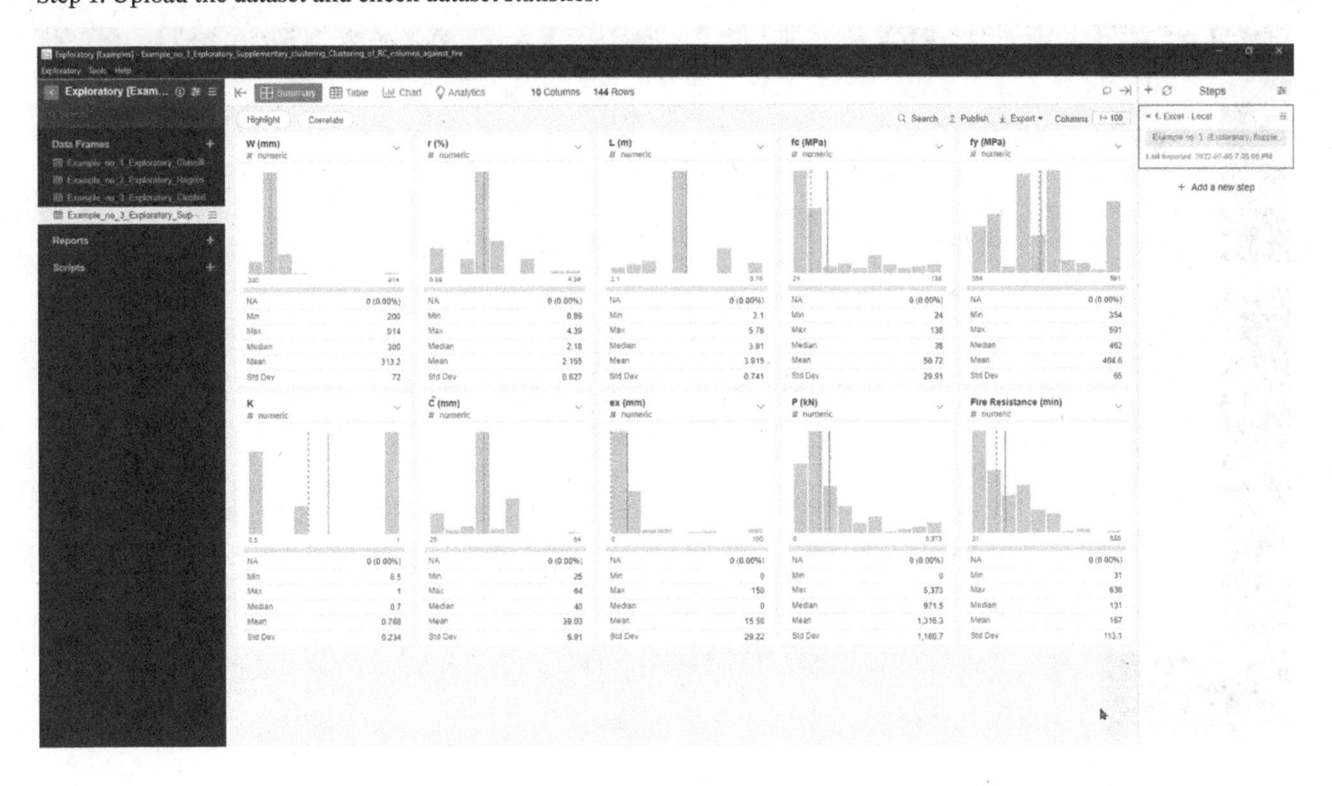

66 Practice makes perfect!

Step 2: Navigate toward the *Analytics* tab and select *K-Means Clustering*.

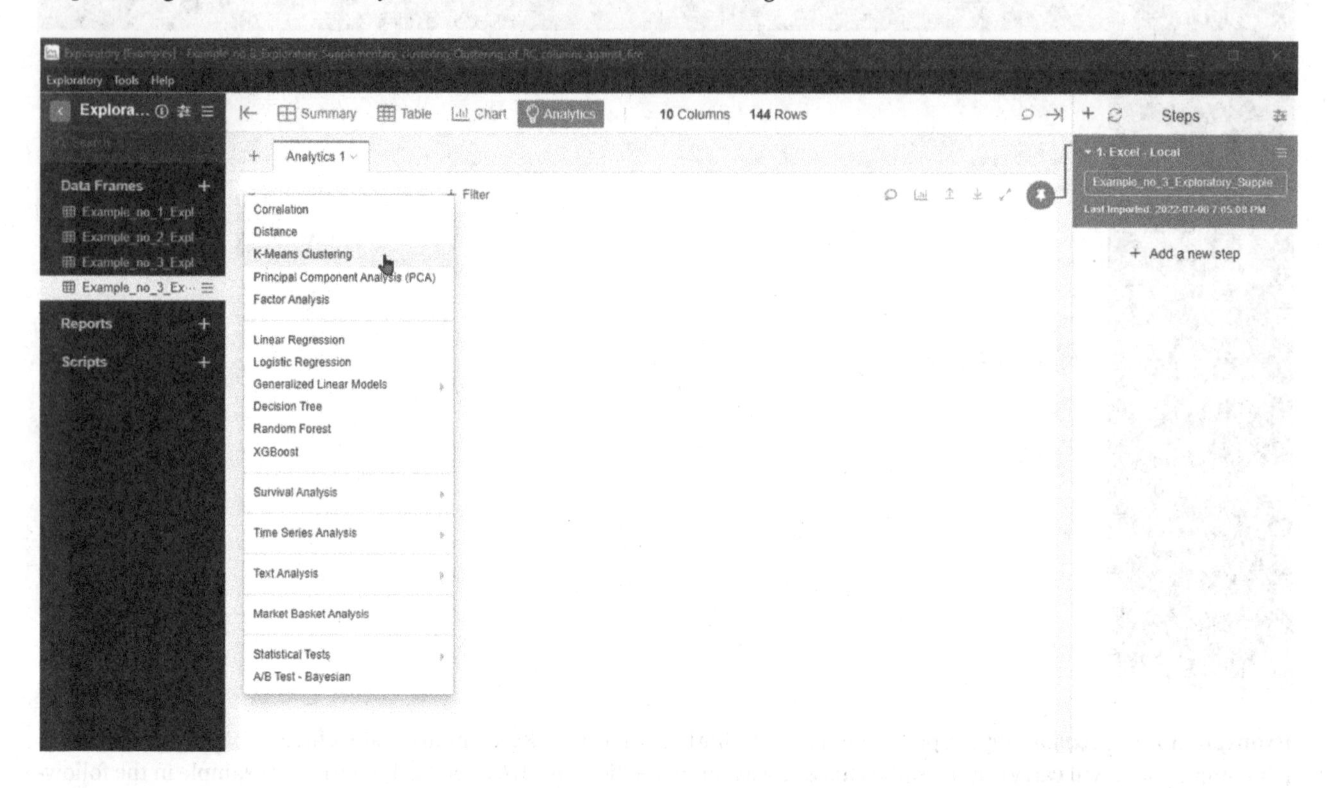

Step 3: Select all features and then *OK*.

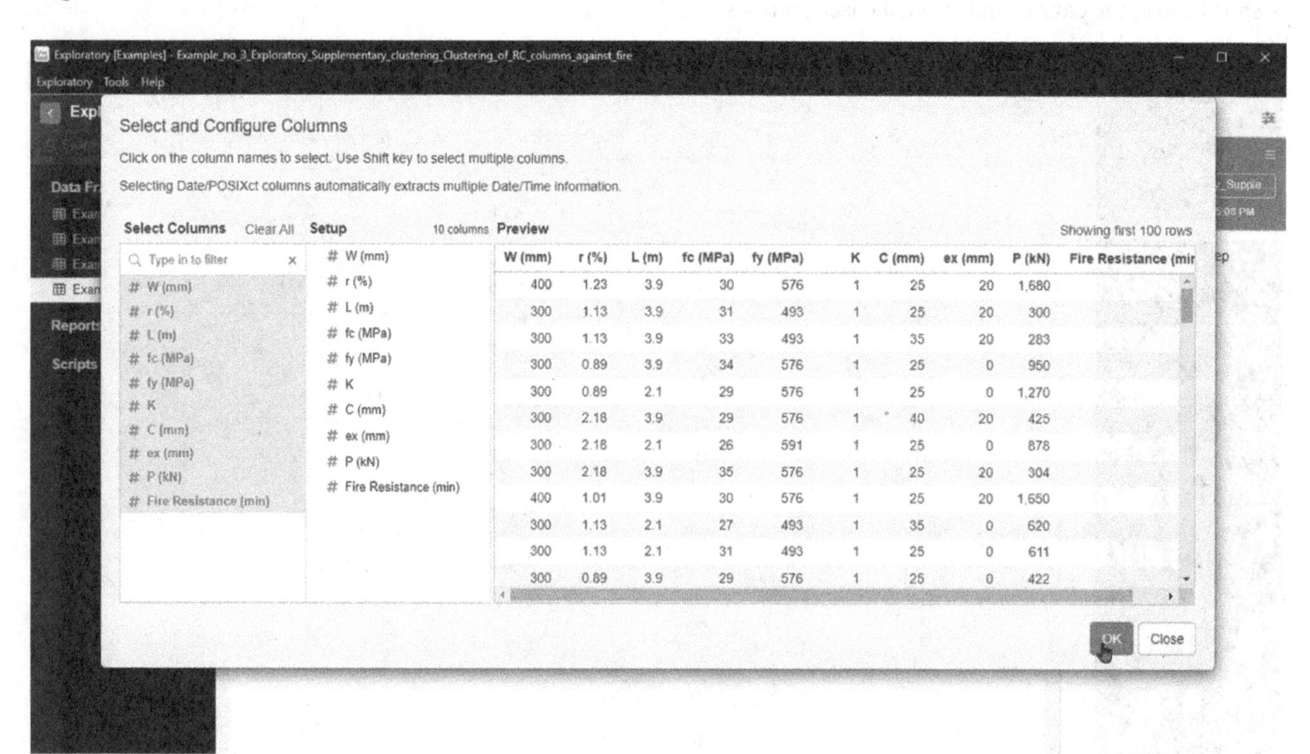

Step 4: Go to *Property* → *Elbow Method* → *TRUE*. You can clearly see the elbow at 4 clusters.

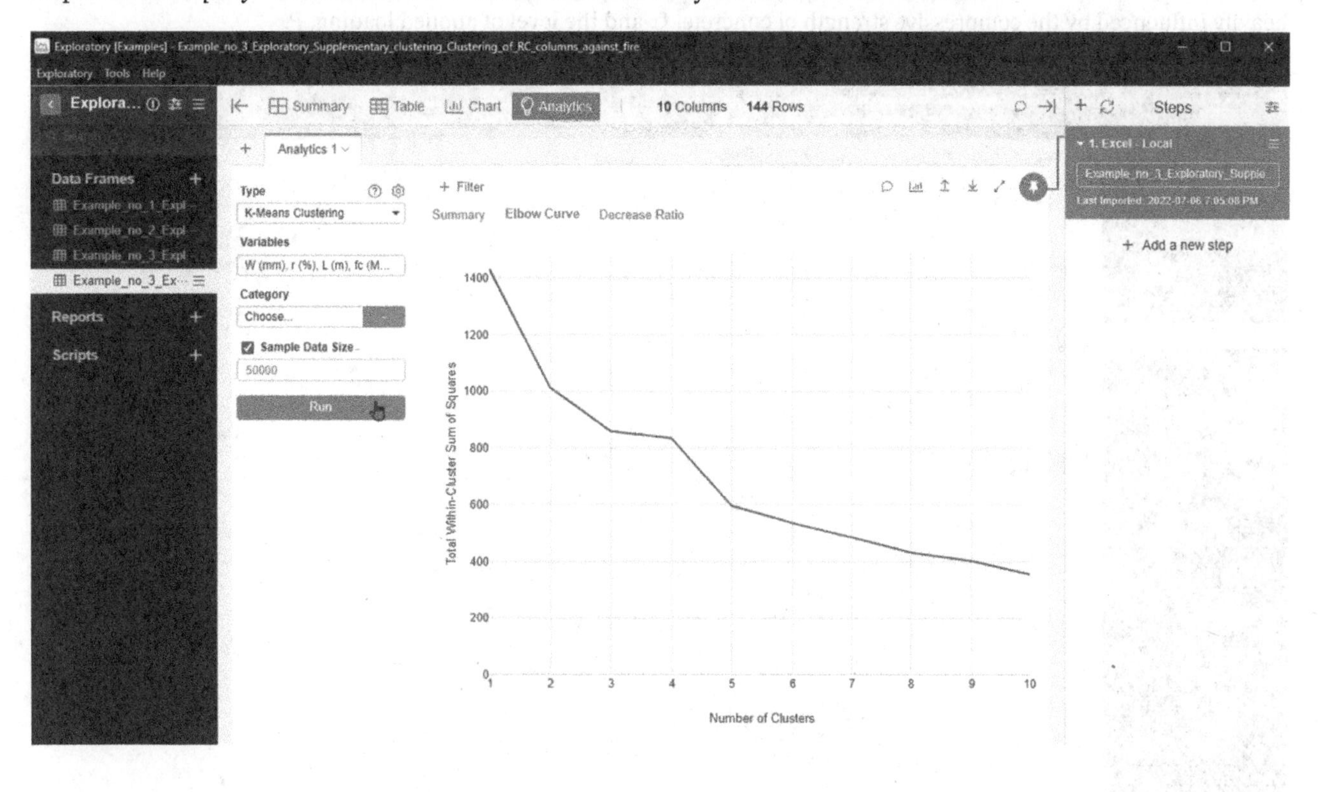

Step 5: Go back to *Property* → *Elbow Method* → *FALSE*. Also, change *Number of Clusters* to *4*. You should be able to see the following plot.

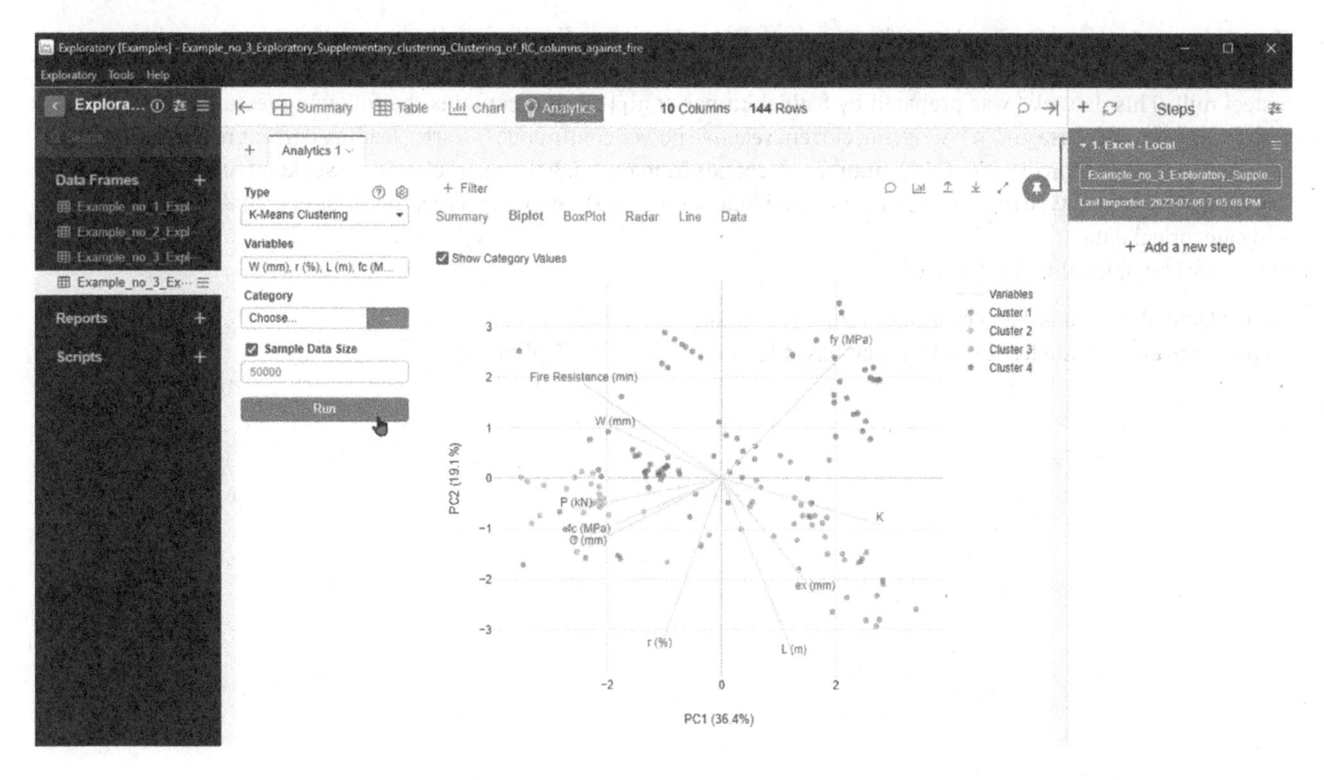

Step 6: Navigate to the *Radar* tab to visualize how each cluster compares to the other clusters. For example, Cluster 2 is heavily influenced by the compressive strength of concrete, f_c, and the level of applied loading, P.

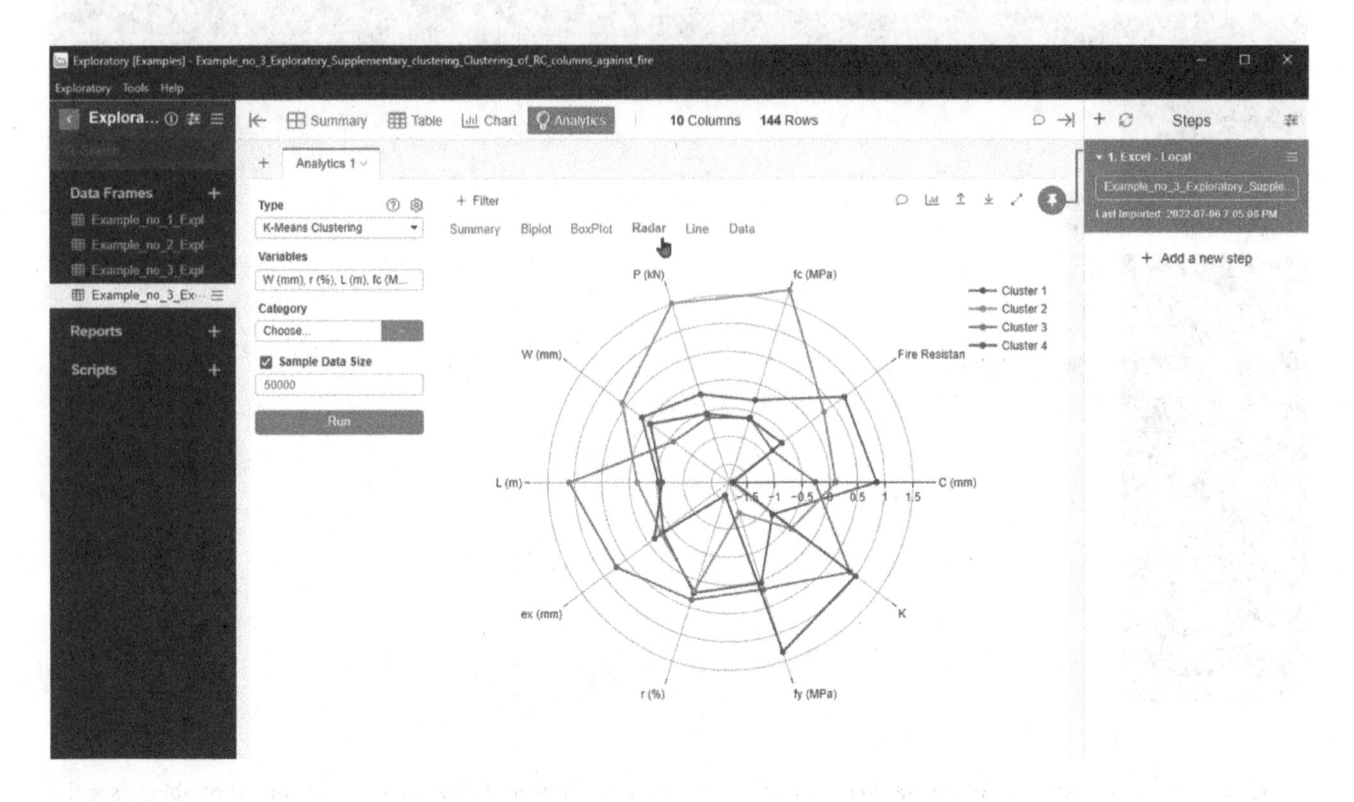

Example no. 4 [*Exploratory*, Anomaly detection]: Detecting anomalies in energy consumption in a steel mill.

In the final example, we will be using the anomaly detection option within Exploratory to analyze energy consumption of a steel mill. This dataset[67] was prepared by Sathishkumar et al. [11] and contains the following features, date, industry energy consumption usage (kWh), lagging current reactive power continuous (kVarh), leading current reactive power continuous (kVarh), carbon dioxide (CO2), number of seconds from midnight (sec), week status (weekend (0) or a weekday(1)), day of the week, load type (light load, medium load, maximum load). As you can see, this dataset contains both numerical and categorical data.

Let us start by uploading the dataset.

Step 1: Upload the dataset and examine feature histograms.
Step 2: Convert the date feature into a time-based feature readable by *Exploratory*.

67 Which can be accessed herein [38].

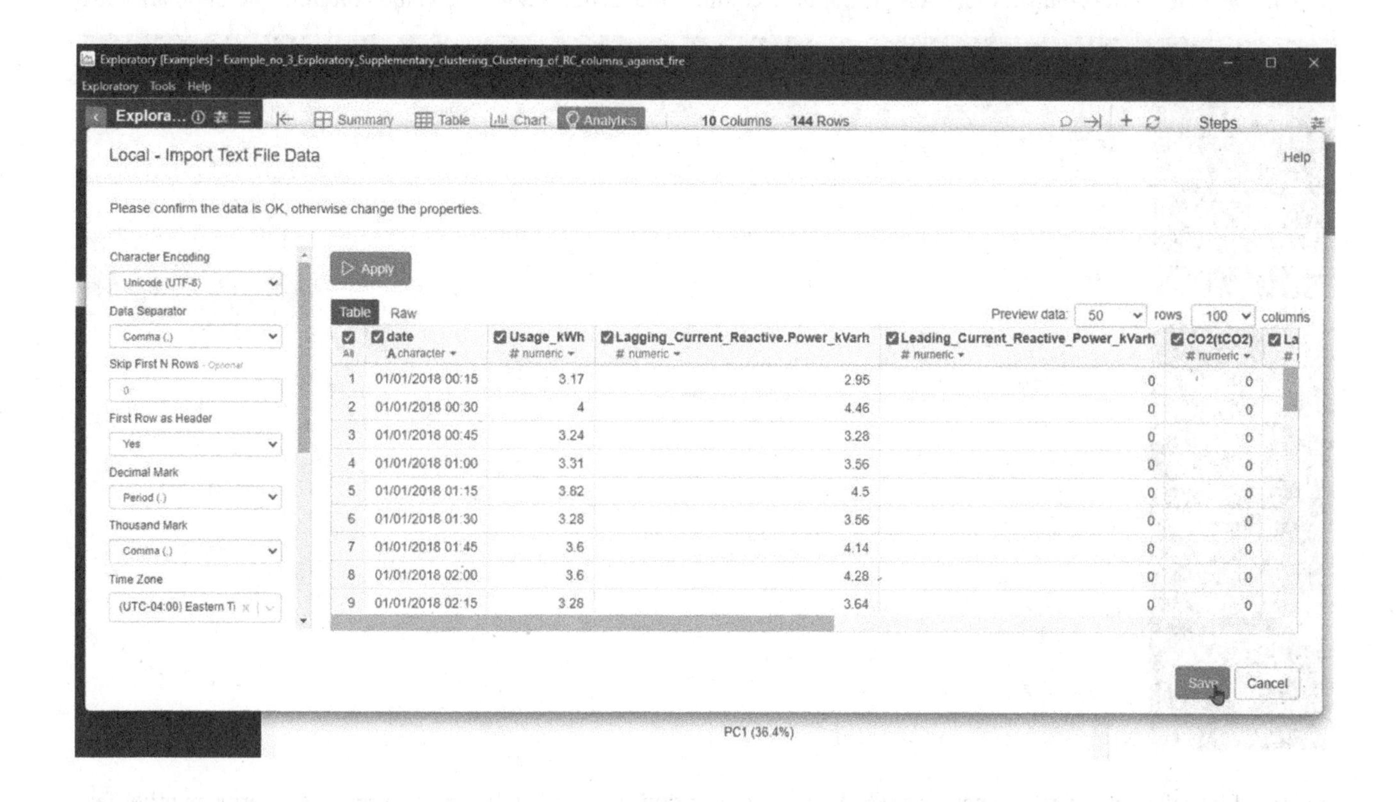

Step 3: Move to the *Analytics* tab and select *Anomaly Detection*

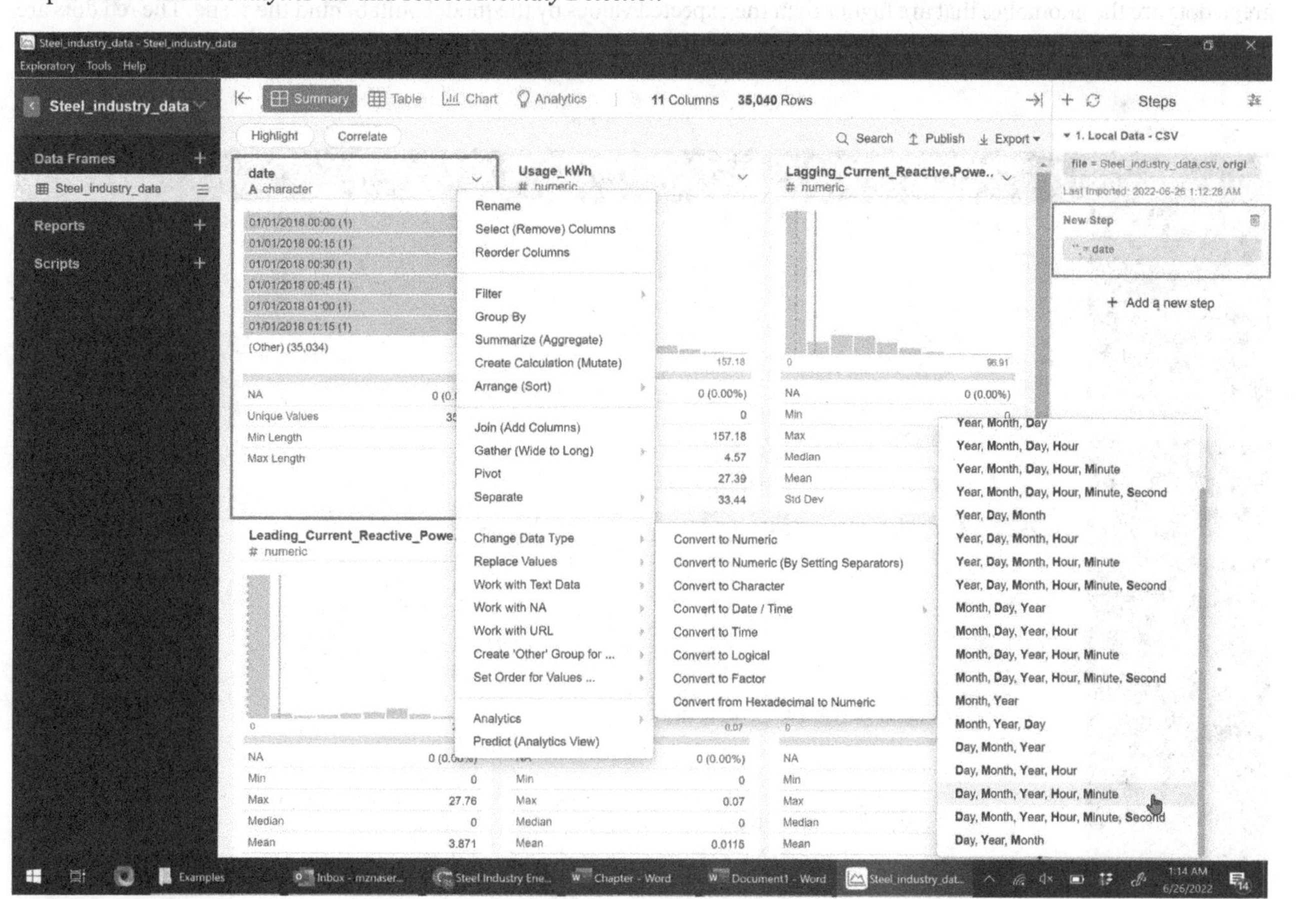

Step 4: Navigate to the *Analytics* tab. Assign the date column and *Usage_kWh* as a Value column. The anomalies are

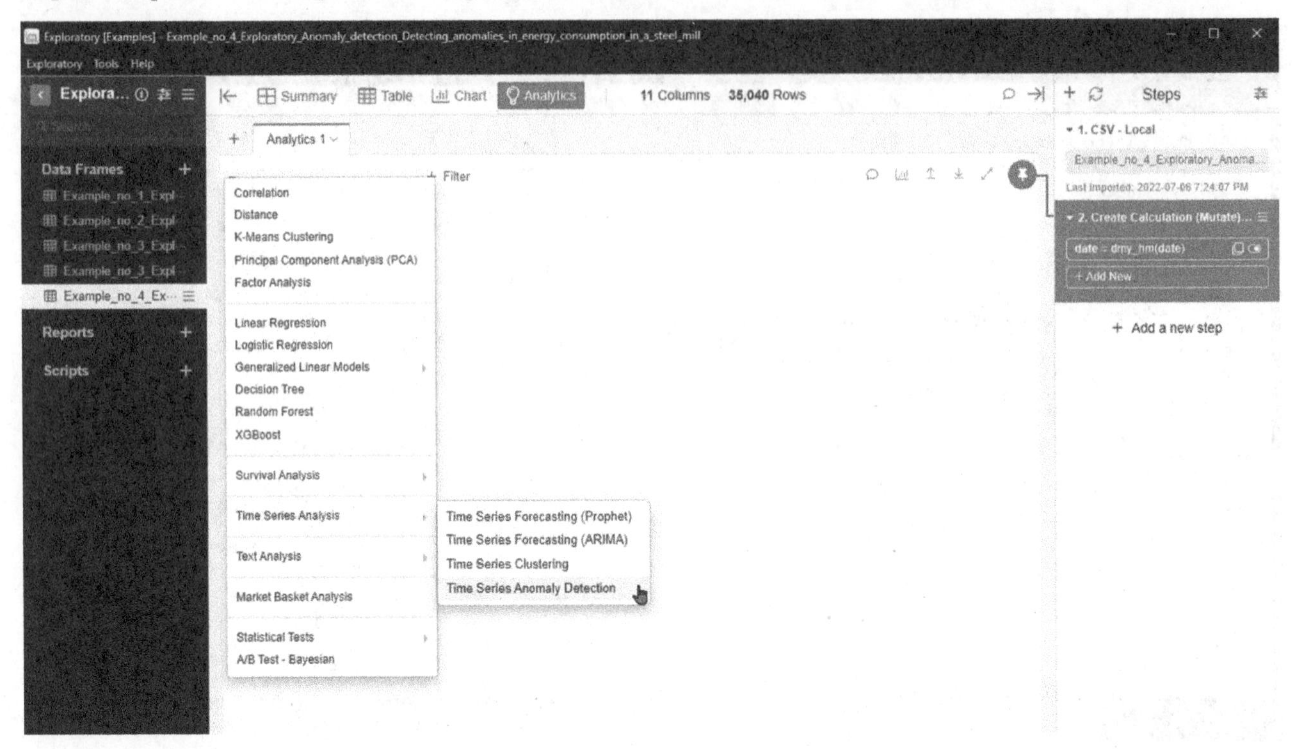

presented by colorful circles. You can also use the *Repeat By* option to visualize the same anomalies in terms of other features (such as *Day of week*). The blue line is the original data, in this case, that is the number of unique page views. The green dots are the anomalies that are higher than the expected values by the model built behind the scene. The red dots are the opposite, which means that they are lower than expected.

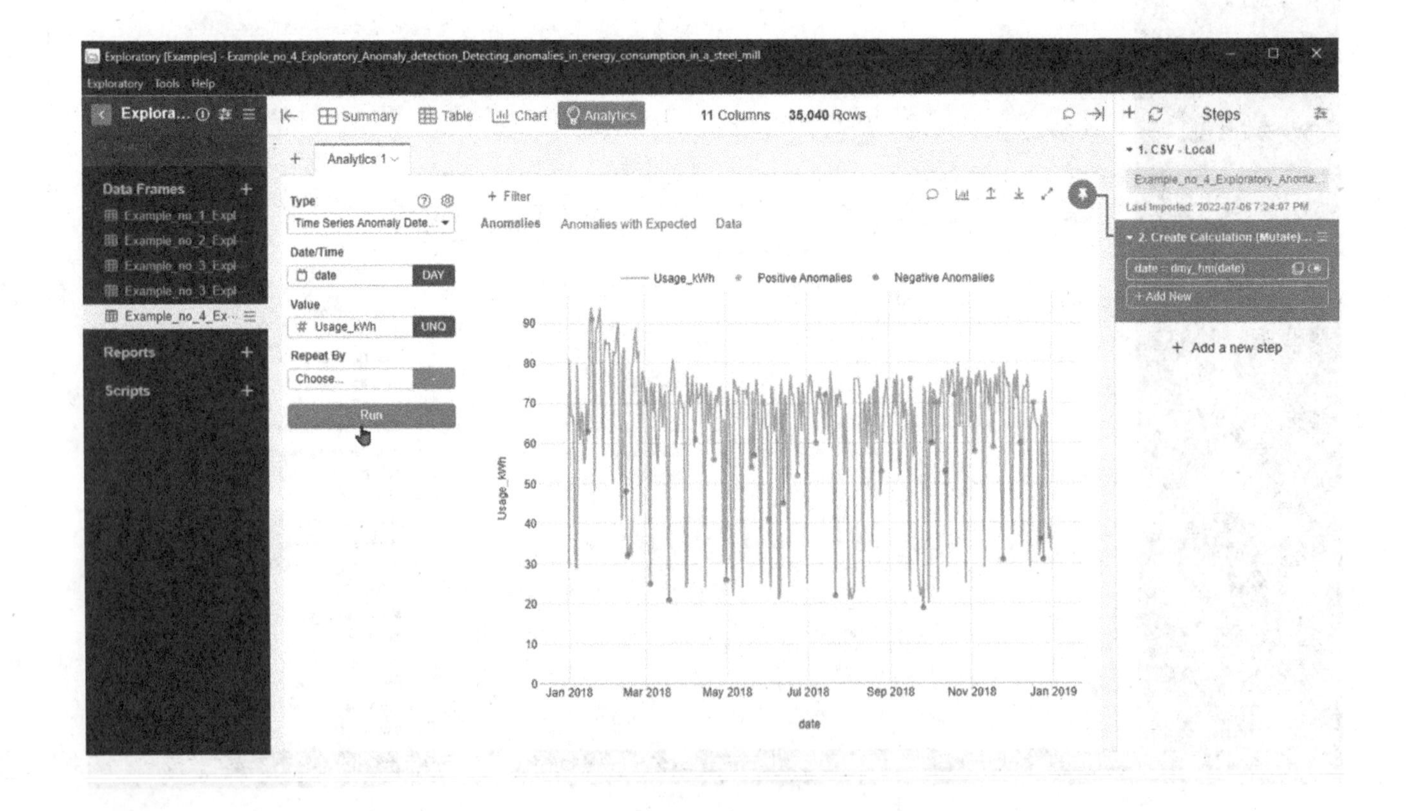

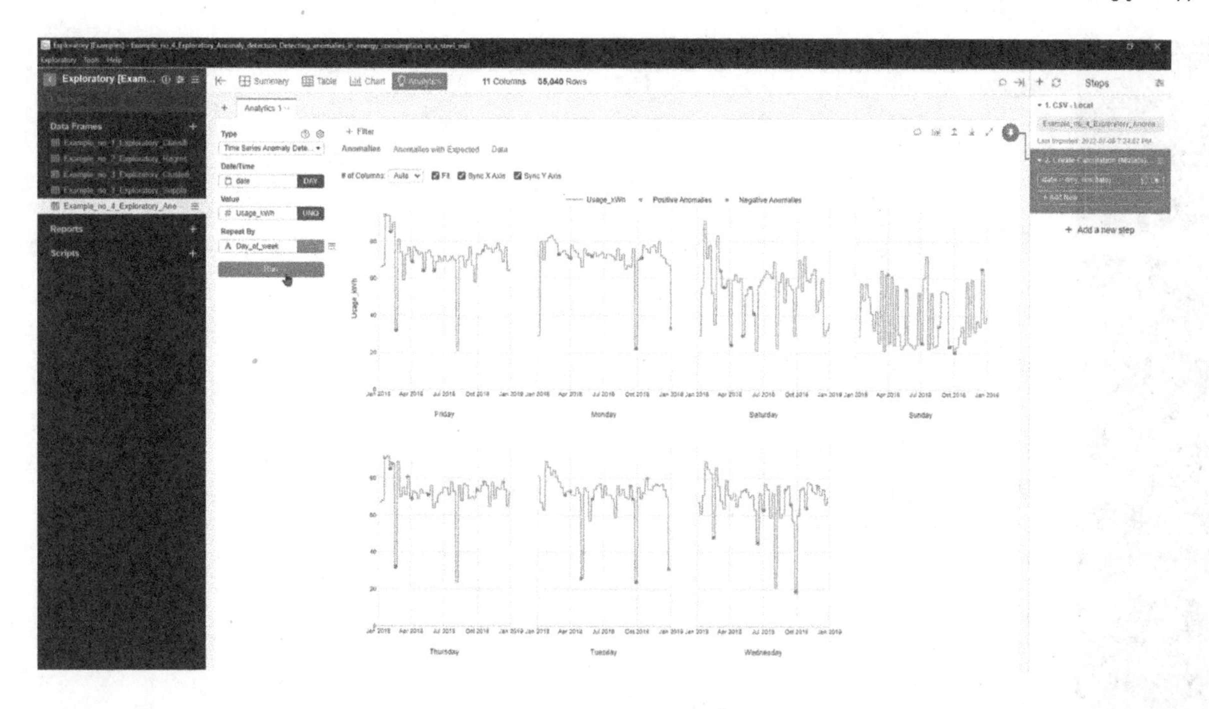

Example no. 5 [*Exploratory*, Dimensionality reduction]: Reducing the number of features in an expanded spalling dataset.

To follow up on Example no. 5 in *BigML*, let us re-run the same dataset using the *PCA algorithm* in *Exploratory*. As you remember, this dataset contains the following features: water/binder ratio (%), aggregate/binder ratio (%), sand/binder ratio (%), heating rate (°C/min), moisture content (%), maximum exposure temperature (°C), silica fume/binder ratio (%), aggregate size (mm), GGBS/binder ratio (%), FA/binder ratio (%), PP fibers quantity (kg/m^3), PP fibers diameter (um), PP fibers length (mm), steel fibers quantity (kg/m^3), steel fibers diameter (mm), and steel fibers length (mm), as well as mode (spalling/no spalling).

Step 1: Upload the dataset and update the output (target) into *No spalling* and *Spalling*, as shown below.[68]

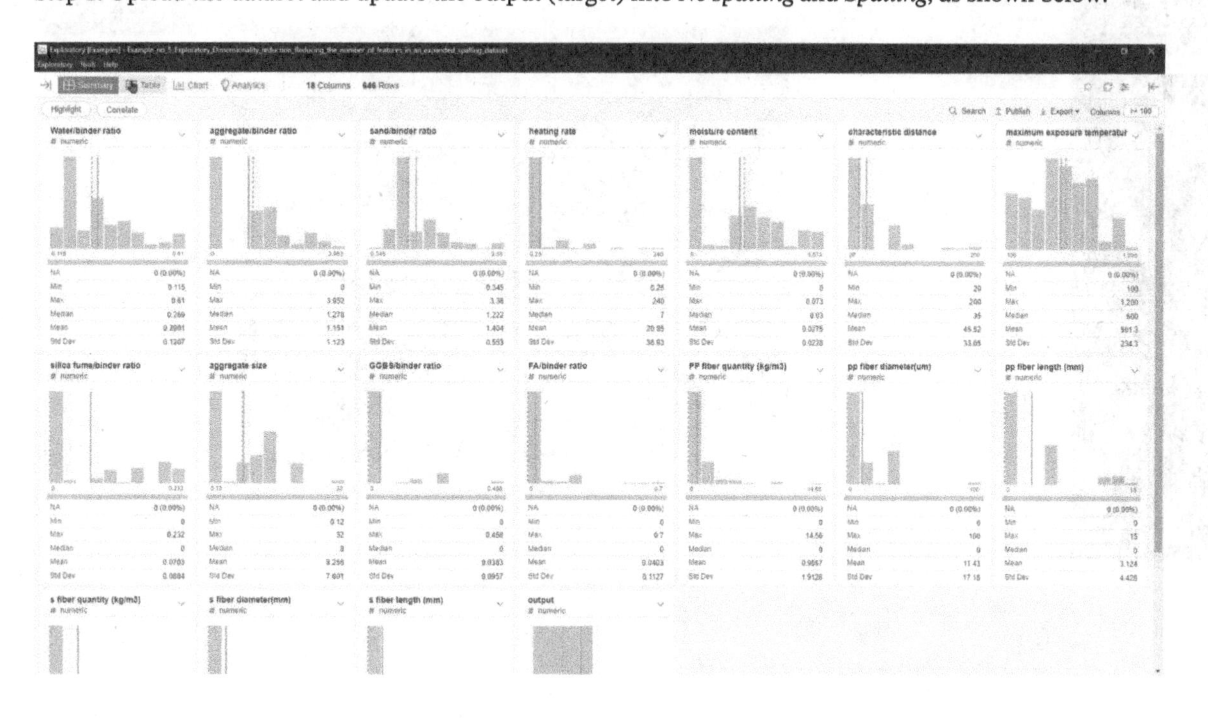

68 You could re-run the analysis without updating the target and using it as a feature!

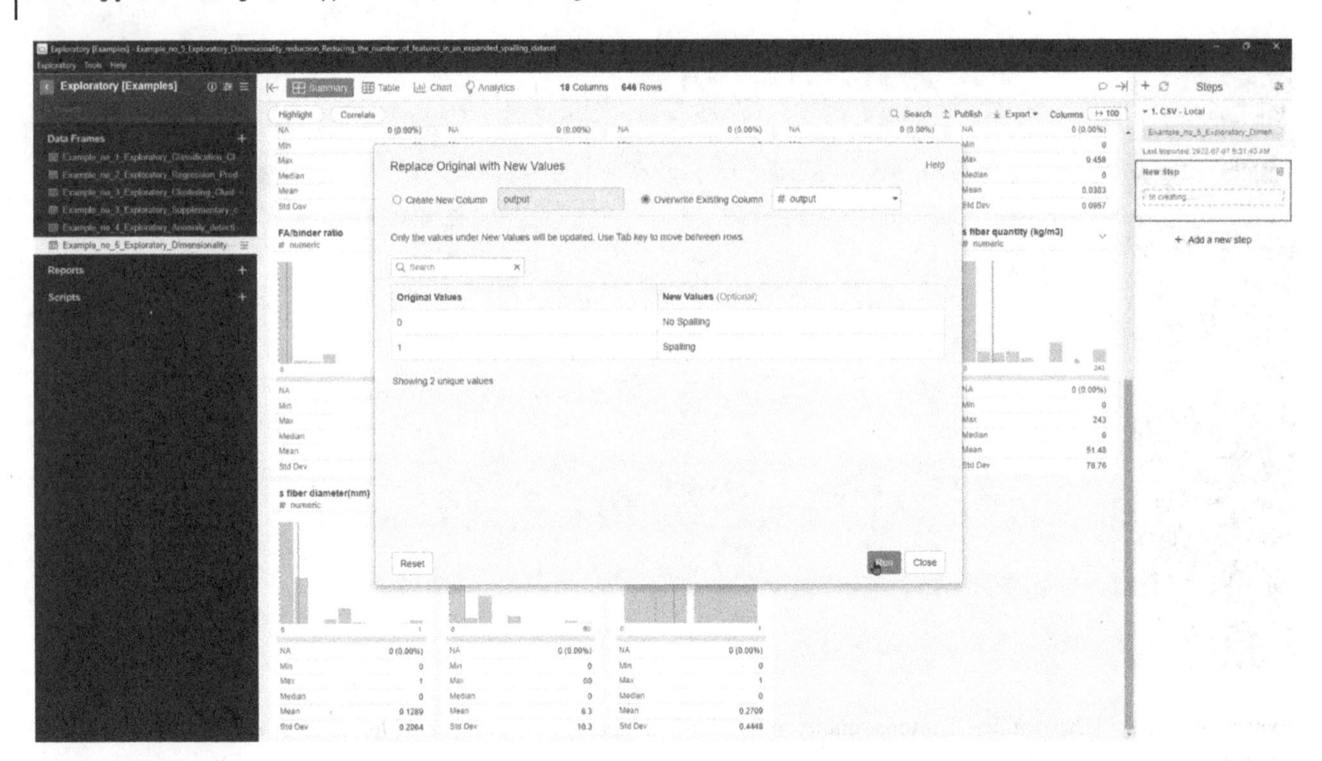

Step 2: Go to the *Analytics* tab and select Correlation to arrive at the correlation matrix.

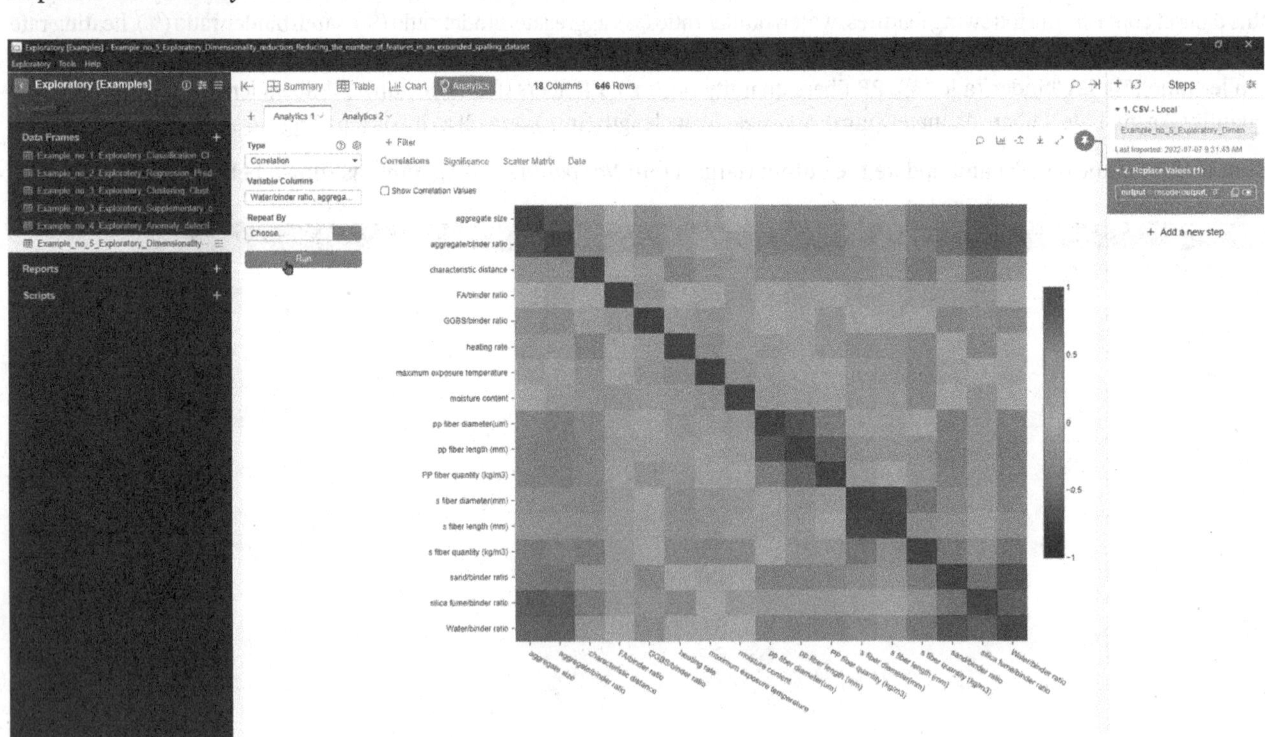

Step 3: Change the analysis type and select *Principal Component Analysis (PCA).*

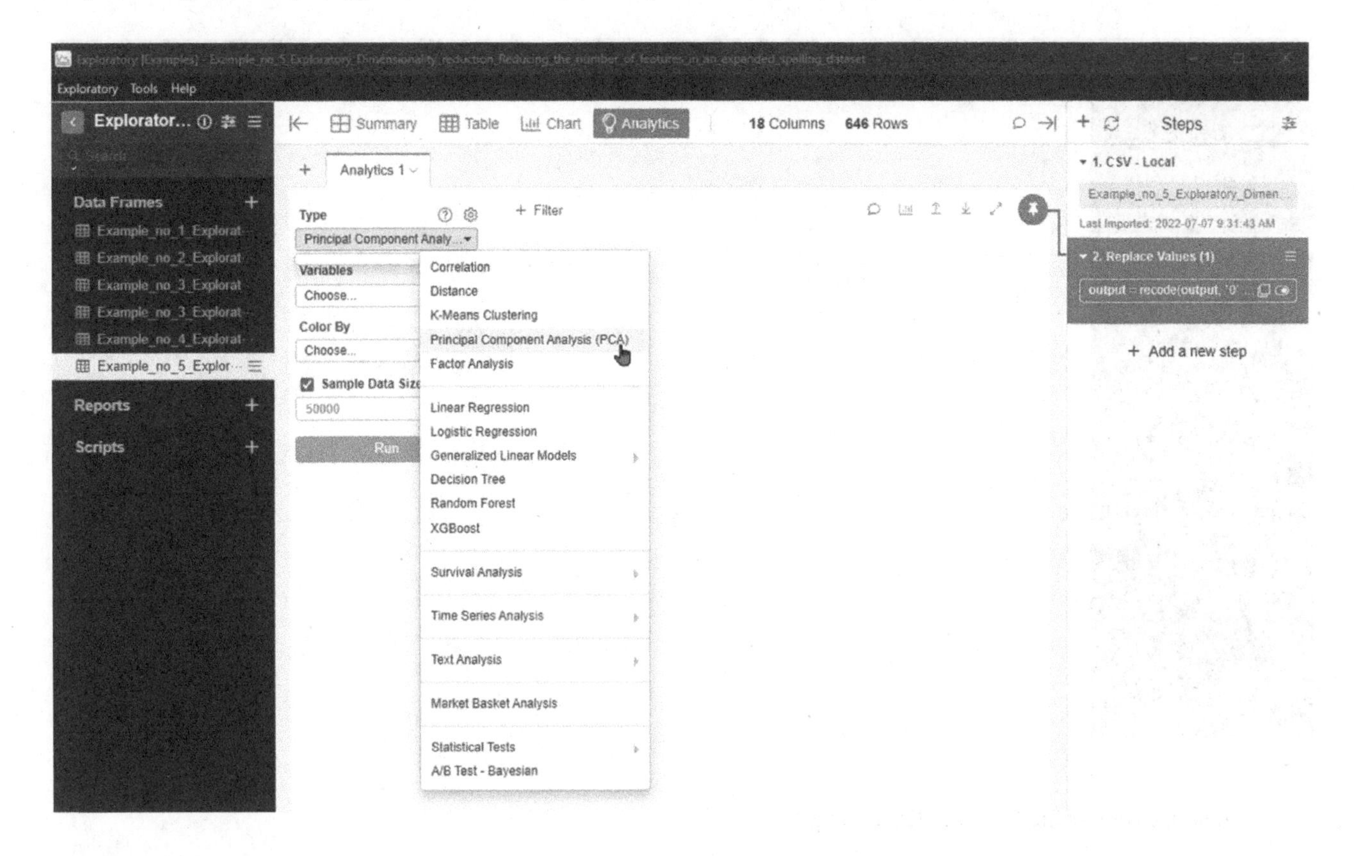

Step 4: Select all features except the output.

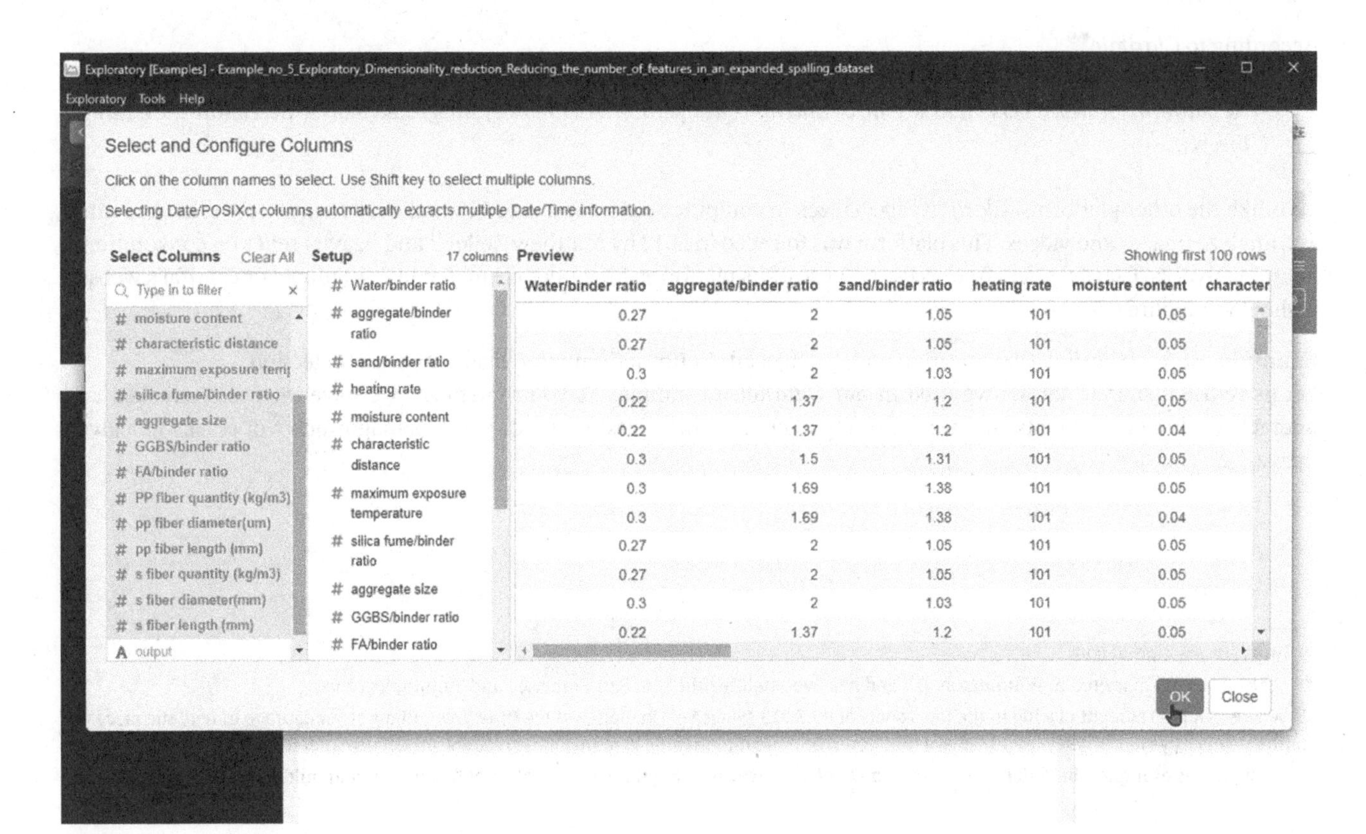

Step 5: *Run* the analysis and note that PC1 and PC2 sum to about 46% of the information in this dataset.

This concludes our examples using *Exploratory*.

6.1.5 Clarifai

According to *Clarifaie*,[69]

> *AI platform for developers, data scientists, and no-code operators. Seamlessly integrated tools to accelerate the entire AI lifecycle.*

Unlike the other platforms, *Clarifai*[70] specializes in computer vision and uses ML and deep neural networks to identify and analyze images and videos. This platform was founded in 2013 by Matthew Zeiler[71] and heavily relies on convolutional neural networks. Clarifai offers free access and use of its platforms for students and faculty. A snippet of Clarifai's website is shown in Figure 6.5.

Example no. 1 [*Clarifai*, Classification and/or Object detection]: Computer vision for crack detection.
Let us re-use a similar dataset we used in our *DataRobot* example. As you remember, we have images of cracked and uncracked concrete elements, and we are trying to create a ML classifier to identify the magnitude[72] of cracks in a given image.

---

69 https://www.clarifai.com.

70 *Clarifai* is headquartered in Wilmington, DE and has two satellite offices in San Francisco and Tallinn, Estonia.

71 Who was a PhD student placing in the top 5 spots of the 2013 *ImageNet* Challenge at the time. I would highly encourage to read the piece on Matthew Zeiler by Aaron Tilley [39]. I believe this would be a great motivator to many of the readers, especially students, of this book!

72 To make this example a bit different than that in *DataRobot*, we are using three classes of cracking, minor, and mild.

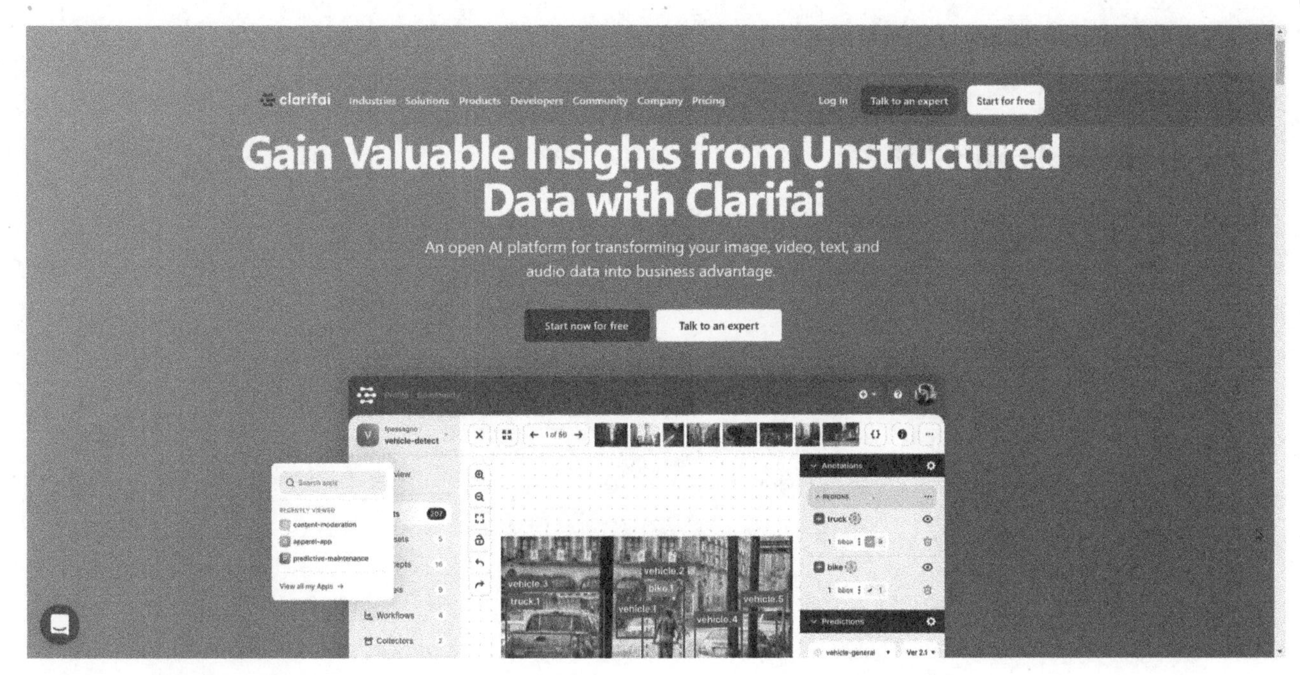

Figure 6.5 Clarifai (July 2022).

Step 1: Create a new *Application*

Step 2: Fill in details on the newly created *Application*

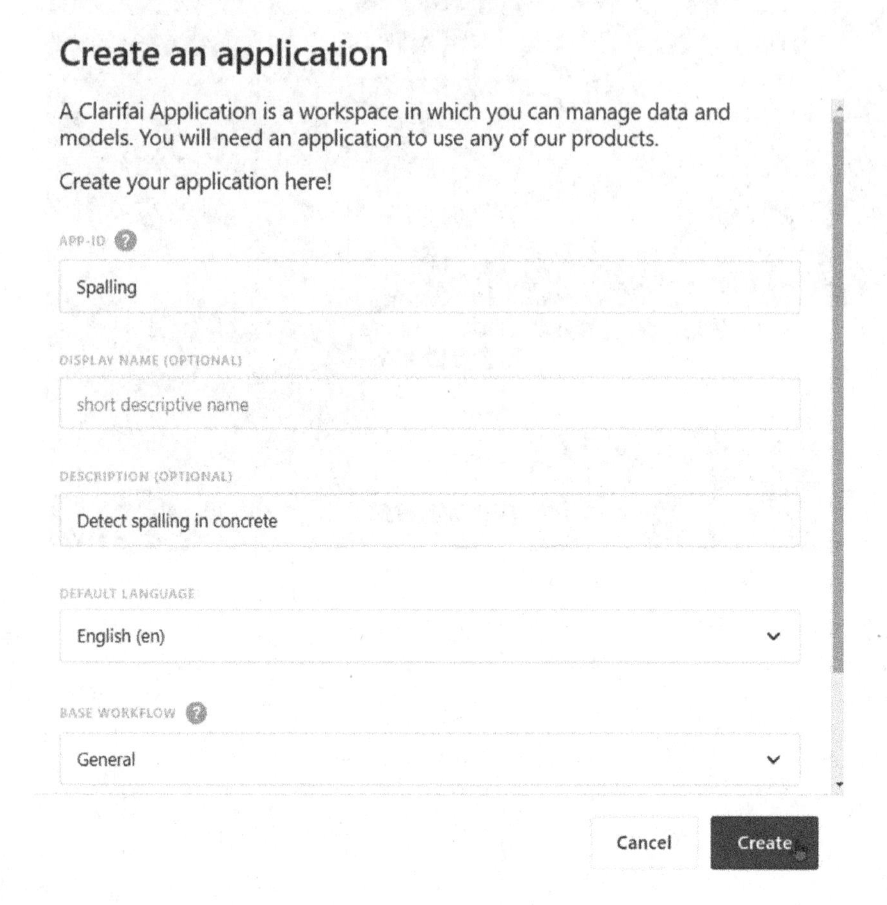

Step 3: Input the images through the *Data Mode* option.

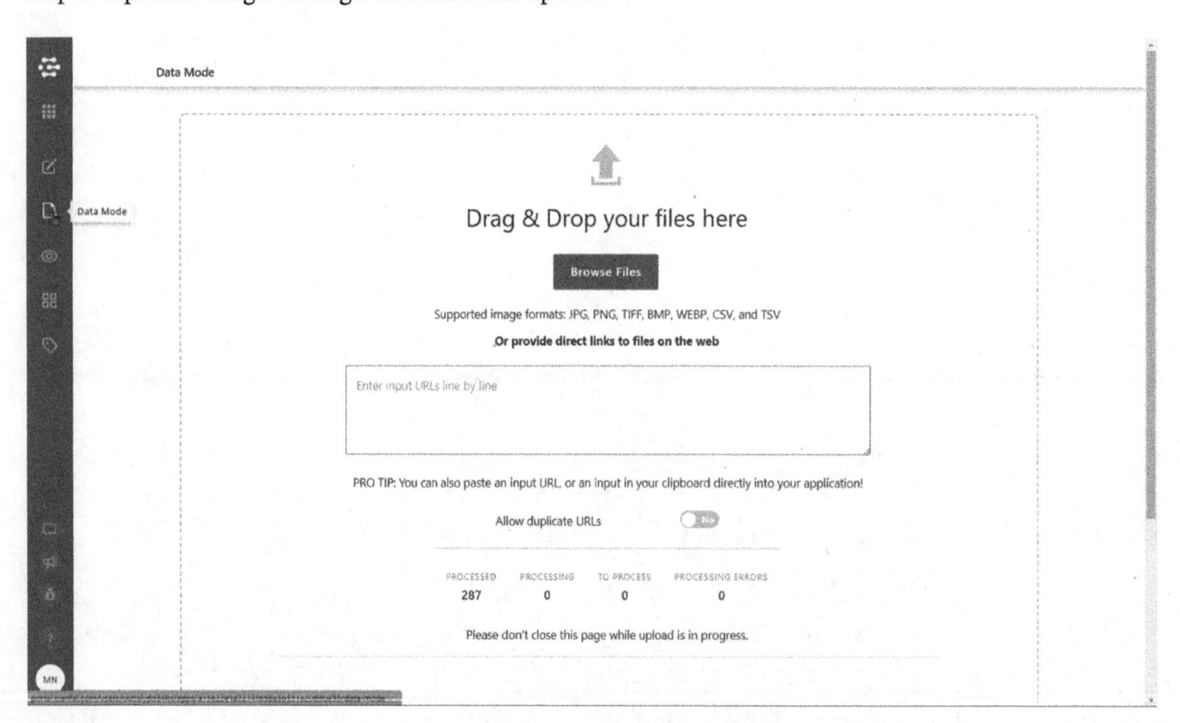

Step 4: Navigate to *Create a Model*. Select *Visual Detector* (or *Visual Classifier*)

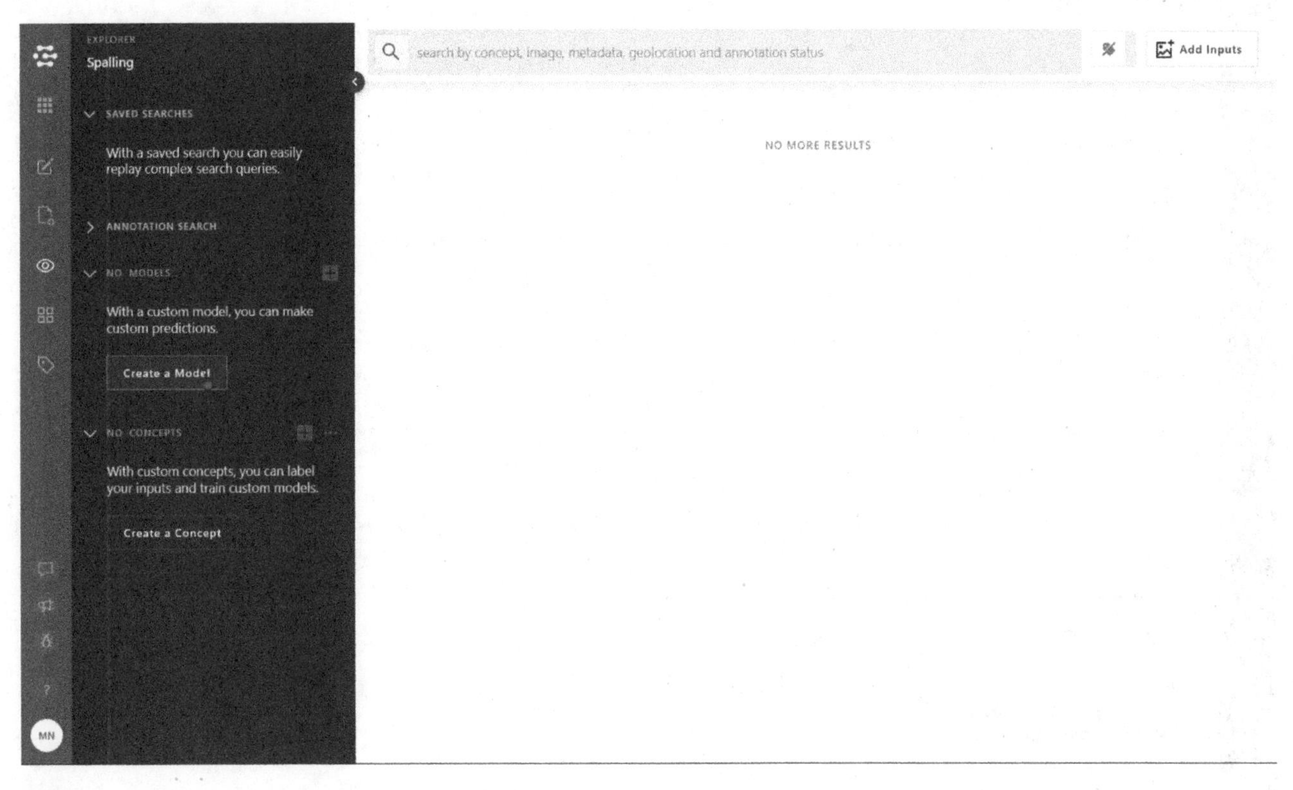

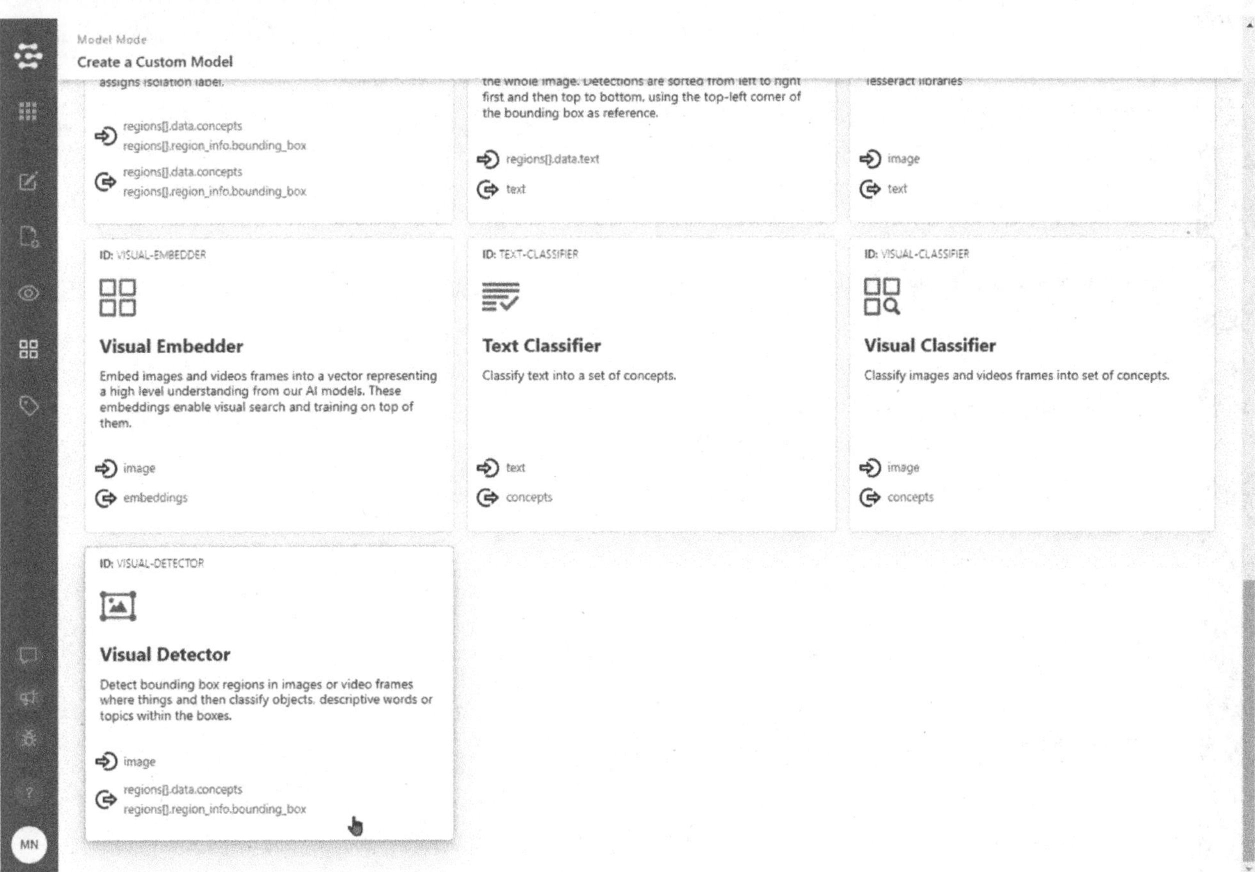

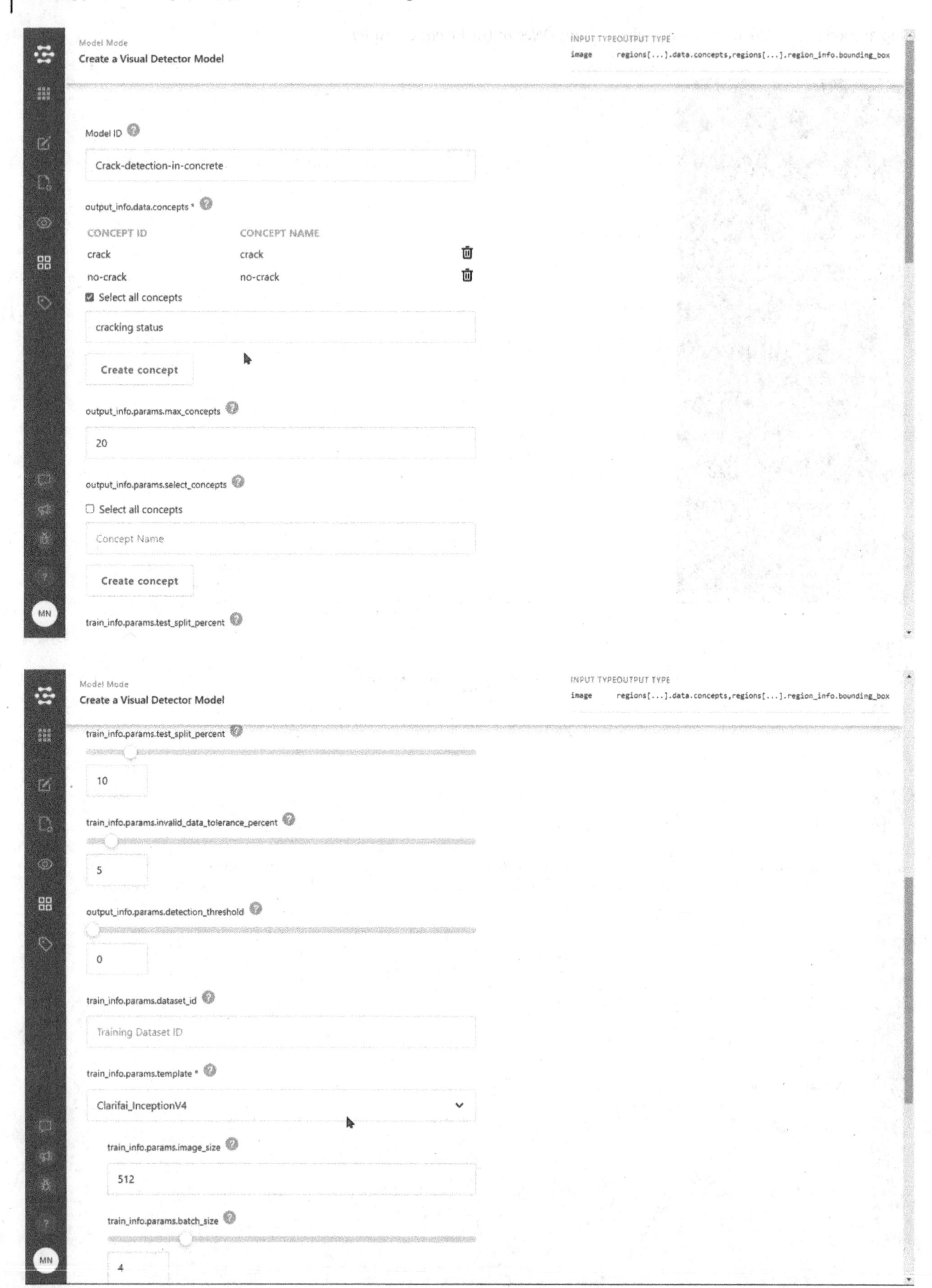

Model Mode
Create a Visual Detector Model
INPUT TYPE OUTPUT TYPE
image regions[...].data.concepts,regions[...].region_info.bounding_box
Model ID
Crack-detection-in-concrete
output_info.data.concepts *
CONCEPT ID
CONCEPT NAME
crack
crack
no-crack
no-crack
Select all concepts
cracking status
Create concept
output_info.params.max_concepts
20
output_info.params.select_concepts
Select all concepts
Concept Name
Create concept
train_info.params.test_split_percent
MN
Model Mode
Create a Visual Detector Model
INPUT TYPE OUTPUT TYPE
image regions[...].data.concepts,regions[...].region_info.bounding_box
train_info.params.test_split_percent
10
train_info.params.invalid_data_tolerance_percent
5
output_info.params.detection_threshold
0
train_info.params.dataset_id
Training Dataset ID
train_info.params.template *
Clarifai_InceptionV4
train_info.params.image_size
512
train_info.params.batch_size
4
MN

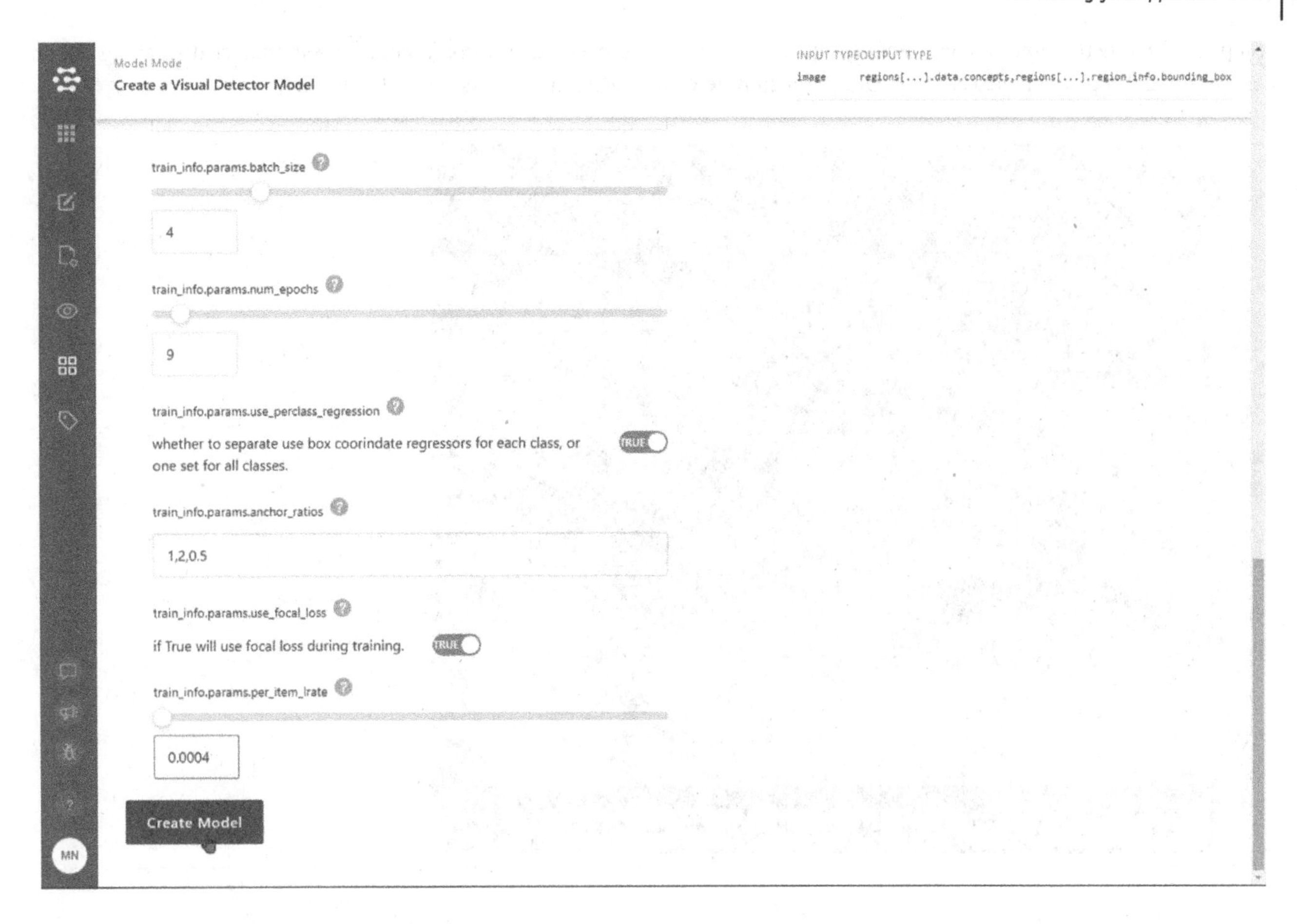

Step 5: Navigate to the *Explorer* tab.

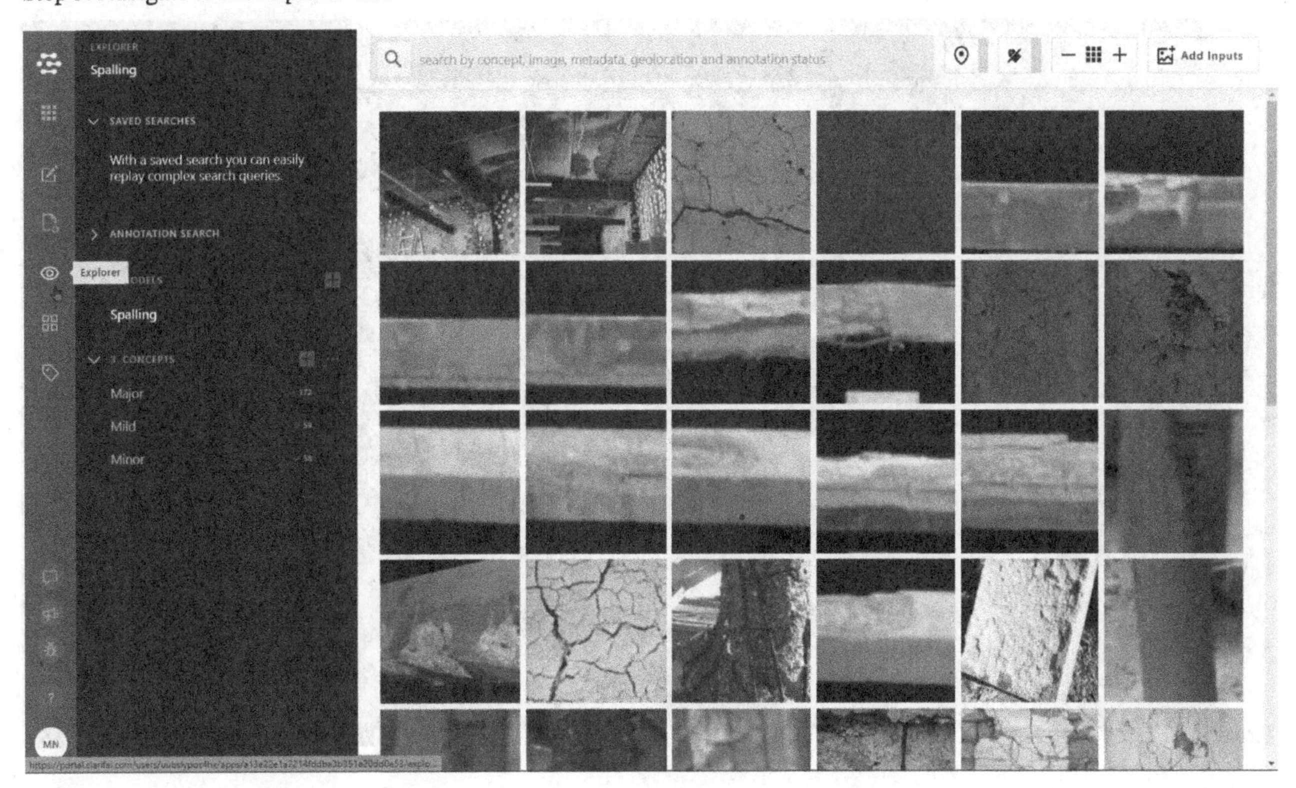

Step 6: *Annotate* the input images. For example, in the case where a *Classification* model is used, images displaying major cracking/spalling are labeled as such. Once you complete annotating the inputs, *Train* the model.

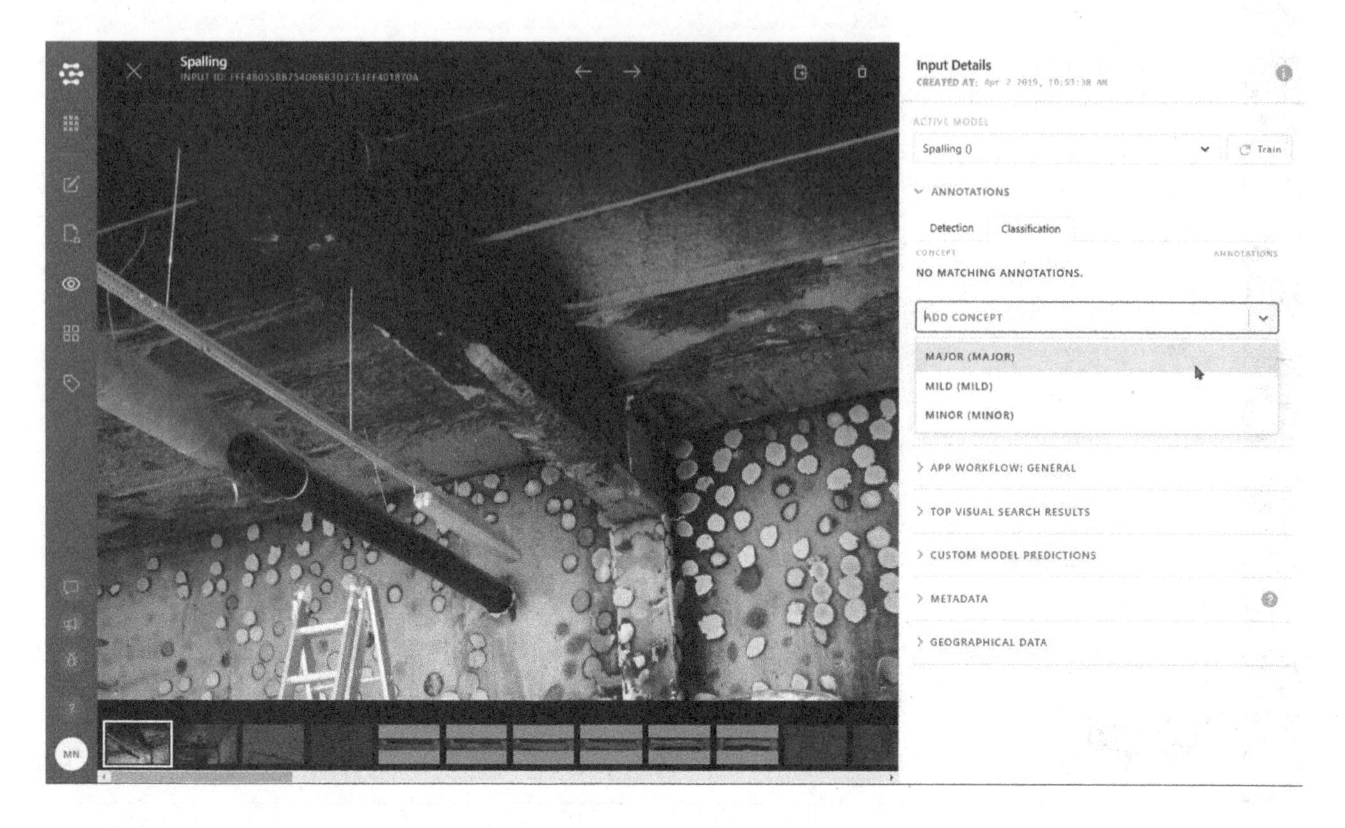

On the other hand, when a *Detection* model is used, please use bounding boxes to specifically bound the areas of interest.

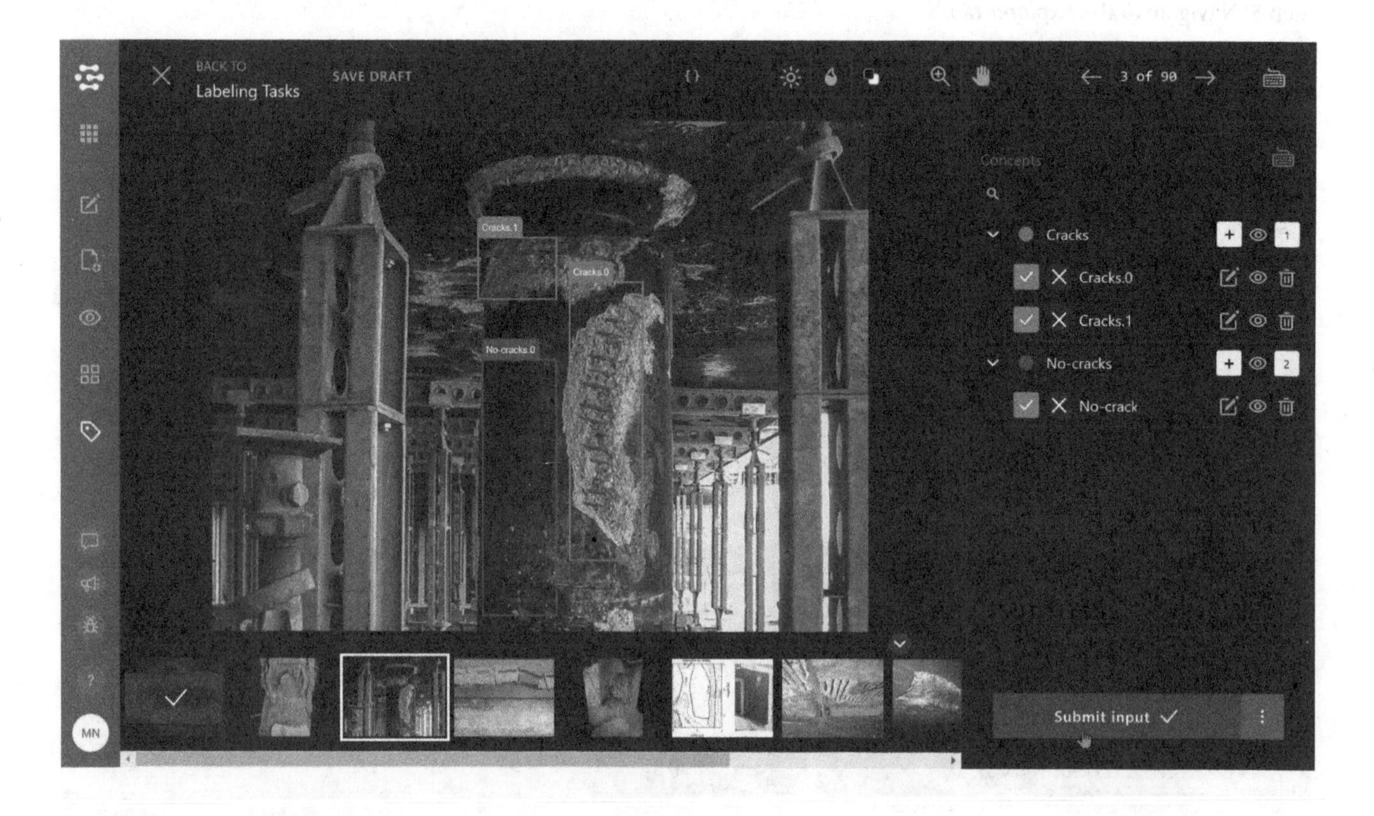

To annotate, you may need to submit a *Task* to label the images (follow the following steps before being able to apply bounding boxes)

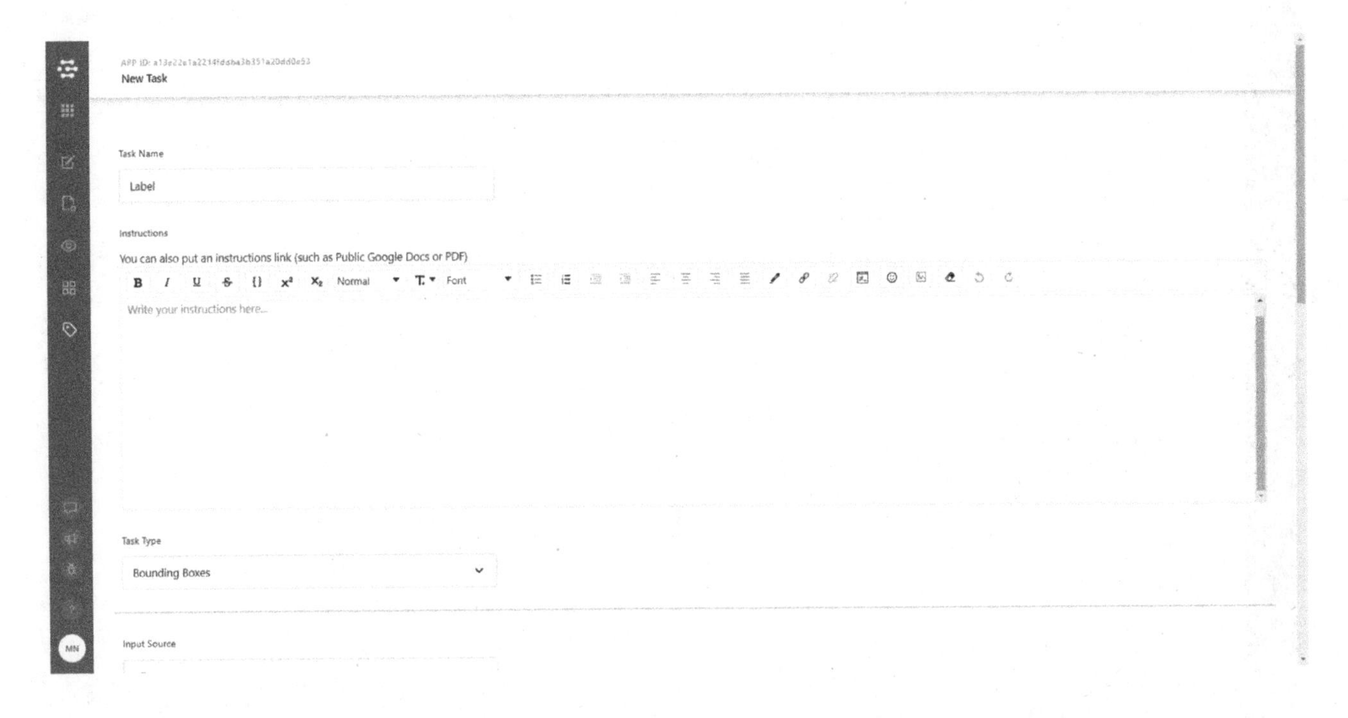

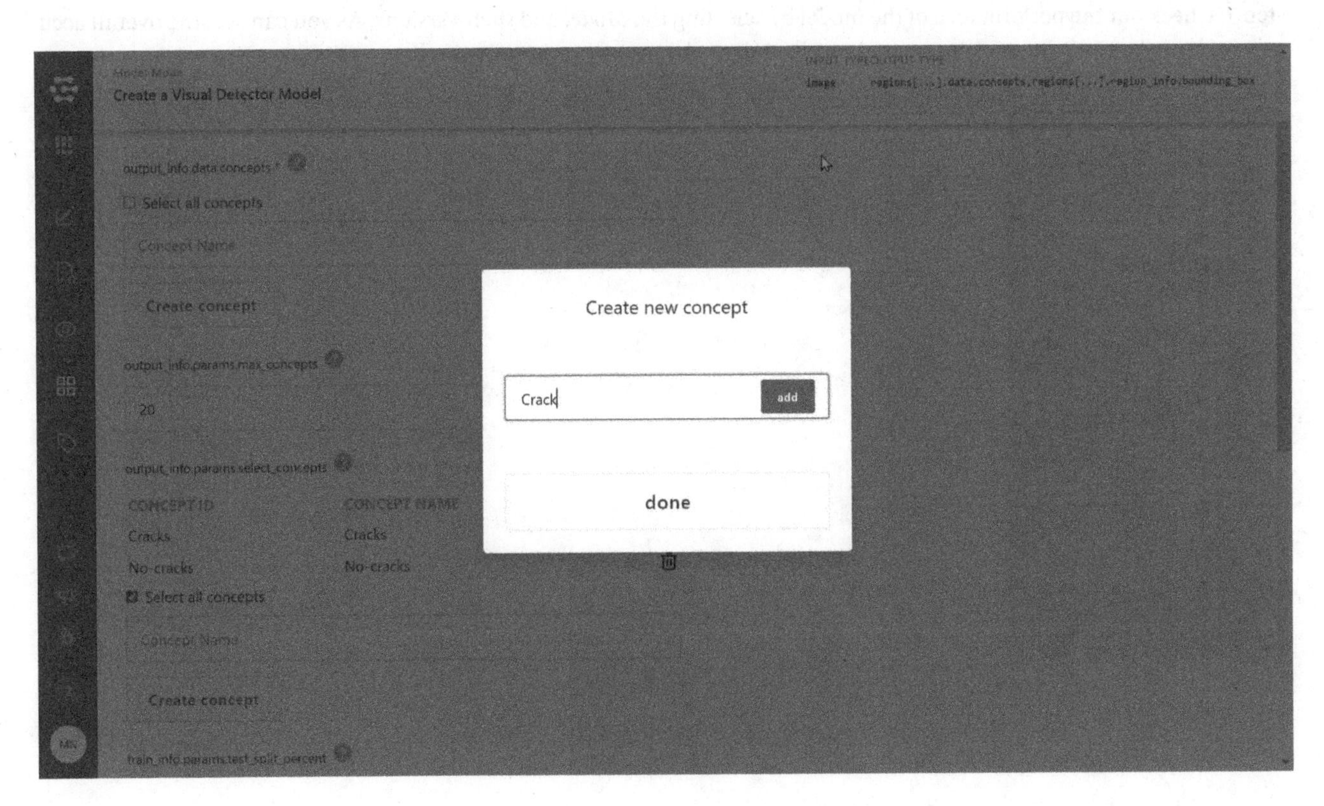

Step 7: Check out the performance of the model by selecting the *Model* and then *Versions*. As you can see, the overall accuracy is about 88.8%. Other metrics such as recall, and precision can be viewed as well.

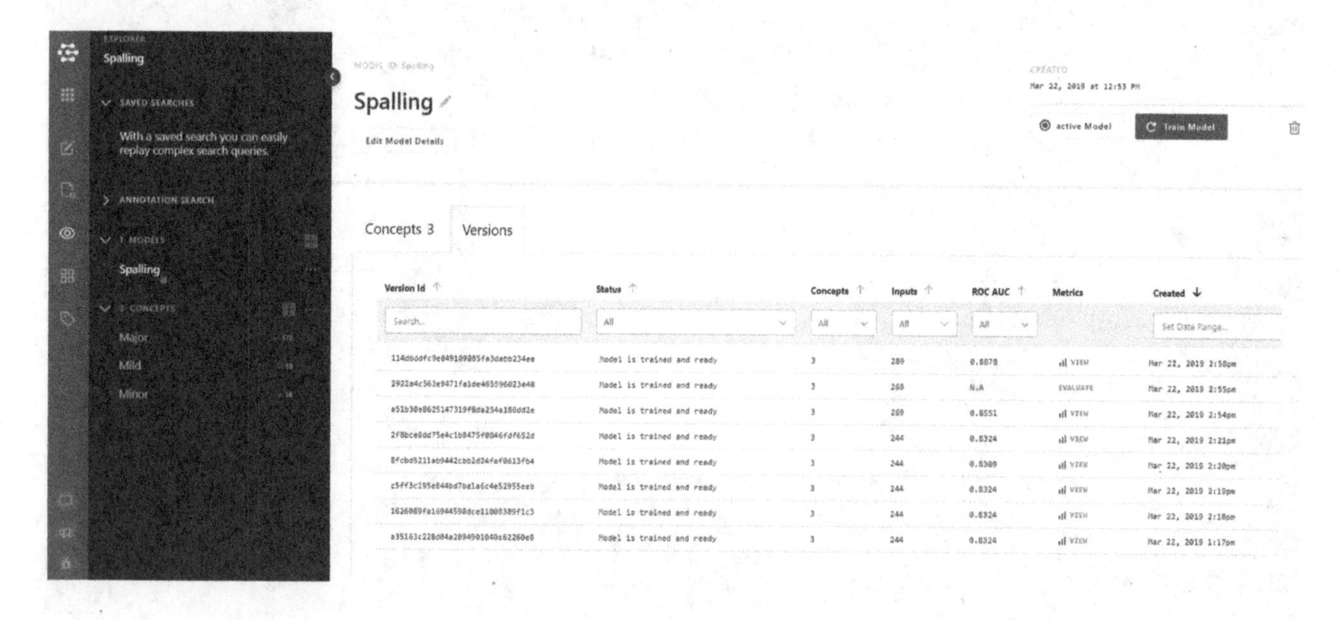

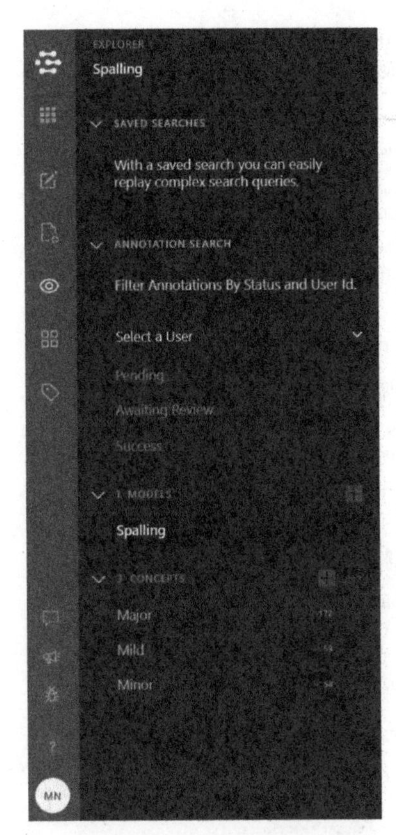

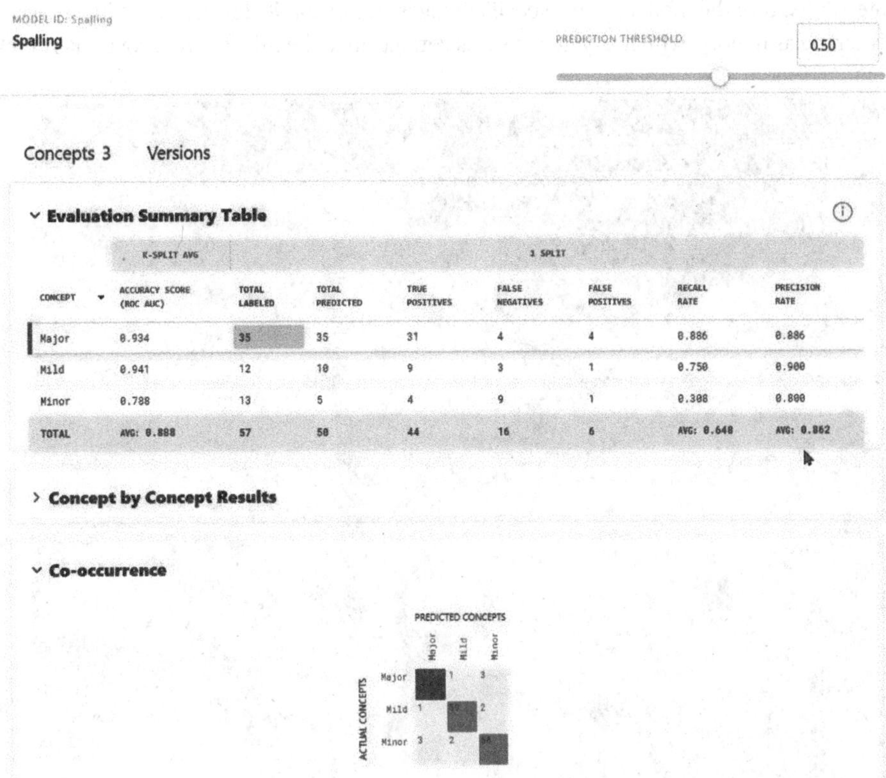

CONCEPT	ACCURACY SCORE (ROC AUC)	TOTAL LABELED	TOTAL PREDICTED	TRUE POSITIVES	FALSE NEGATIVES	FALSE POSITIVES	RECALL RATE	PRECISION RATE
Major	0.934	35	35	31	4	4	0.886	0.886
Mild	0.941	12	10	9	3	1	0.750	0.900
Minor	0.788	13	5	4	9	1	0.308	0.800
TOTAL	AVG: 0.888	57	50	44	16	6	AVG: 0.648	AVG: 0.862

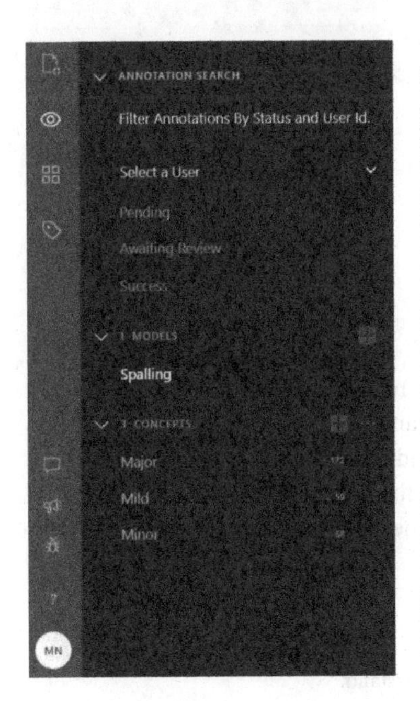

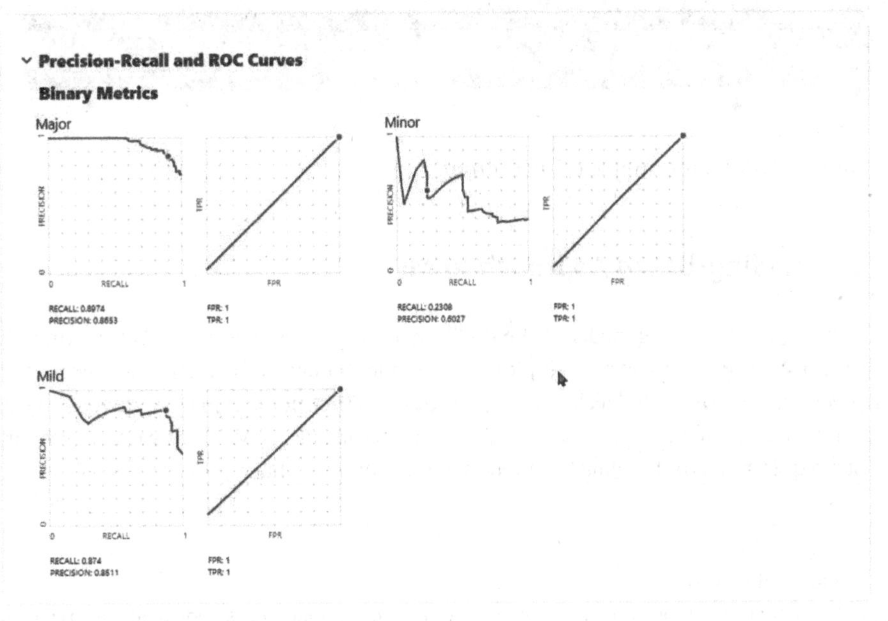

Step 8: Explore the model against specific images. For example, for the following image, the model identifies the magnitude of cracking as major – which as you can see is correct since the full concrete cover has cracked thus exposing the reinforcement.

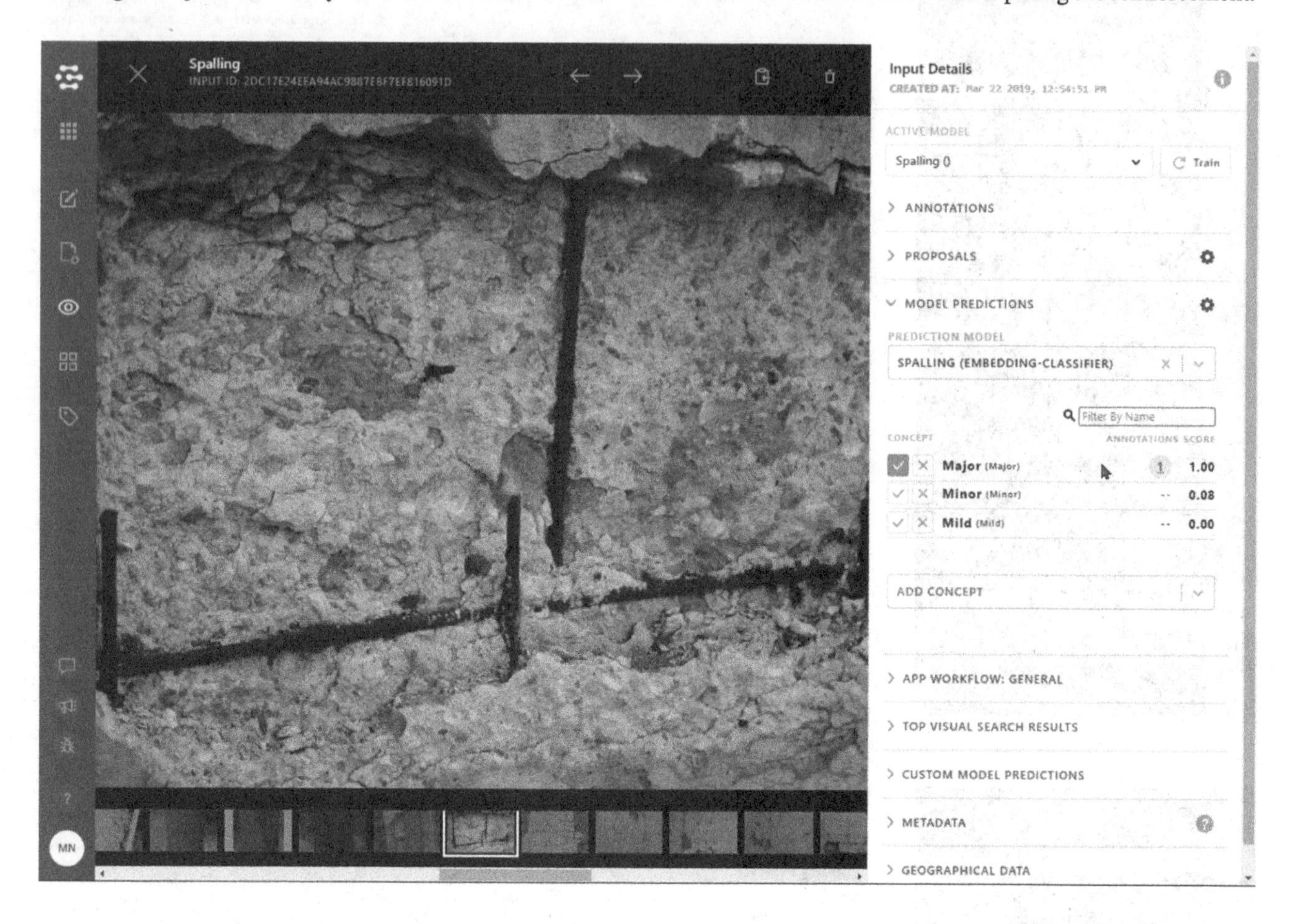

This concludes our example using *Clarifai*.

6.2 Coding-based Approach to ML

Coding or programming models is the traditional[73] approach to building and creating ML models and tools. In this approach, a user must subscribe to a (or multiple) programming language wherein the user writes and executes scripts of procedures to convey[74] an algorithm or data handling procedure. At this moment in time, many[75] civil and environmental engineering courses do not have tailored courses to teach ML coding specifically.[76] While, as we have seen in the first half of this chapter, coding is not a requirement for building ML models, it is my opinion that our engineers will still benefit from learning how to code.[77]

73 Some might say *old school!*

74 We are trying to script our thoughts into a procedure that a computing station (computer) can understand.

75 If not, most.

76 Which explains a good component of the inertia and difficulty in adopting ML.

77 Personally, I often think of coding/programming as a second (or maybe third/fourth) language. Hence, the second half of this chapter. As you can imagine, adding a full discussion on how to code in *Python* and *R* into this textbook might dilute its content and, quite honestly, there are plenty of comparable commands and means that make it infeasible to cover in one chapter (as such I would prefer to direct you to the plenty of available and notable resources that are specifically designed to educate new users on the principles of coding, such as [40, 41] which has an accompanying free website for the *R for Data Science* book}). Remember that you can also learn coding via interactive *YouTube* channels or *GitHub* sites – two of my favorites. While I won't be covering such aspects in detail, the provided scripts of codes will contain annotations and descriptions to the adopted commands and libraries. These will be tremendously helpful from two aspects; explaining procedural steps in each script, and form an anchor for you to search for additional commands/libraries. A good thread I have come across is that created by Microsoft, which covers *Python* and *R*, and is freely hosted on GitHub: *Data-Science-For-Beginners* [42] – give it a shot!

Let us now visit two coding-based programming languages,[78] *Python* and *R*. I believe it would be a good exercise to repeat some of the above-mentioned examples solved in the online platforms by using *Python* and *R* to contrast the results from the coding-free platforms.

6.2.1 Python

Python was invented by Guido van Rossum[79] in the 1980s and released in 1991. *Python* has a high level of abstraction, incorporates comprehensive libraries,[80] and can accommodate a multitude of problems. *Python* is open-source and free to use. The code source can be downloaded from here [12].[81] *Python* is a universal language found in a variety of different applications. Many *Python* packages can be found online as well as on the *Python Package Index (PyPI)* repository [13]. Figure 6.6 shows a snippet of the *Python* website.

Some of the most commonly used *Python* packages include:

- *Keras*: A simple package to express and build neural networks with ease [14].
- *Matplotlib*: A library for creating static, animated, and interactive visualizations [15]. *Seaborn* is a subset of this library.
- *NumPy*: A package for scientific computing and large dimensional arrays [16].
- *Pandas*: provides fast, flexible, and expressive data structures for working with labeled data. This package makes data manipulation and indexing easier. Pandas can be found here [17].
- *Scikit-learn*: A package for ML and statistical modeling tools that includes classification, regression, clustering, and dimensionality reduction [18]. *Sklearn.metrics* is another package that contains performance indicators and error metrics.
- *TensorFlow*: A package for the TensorFlow framework [19].

Let us use *Python* in the following three examples to cover binary classification, regression, and clustering problems.[82]

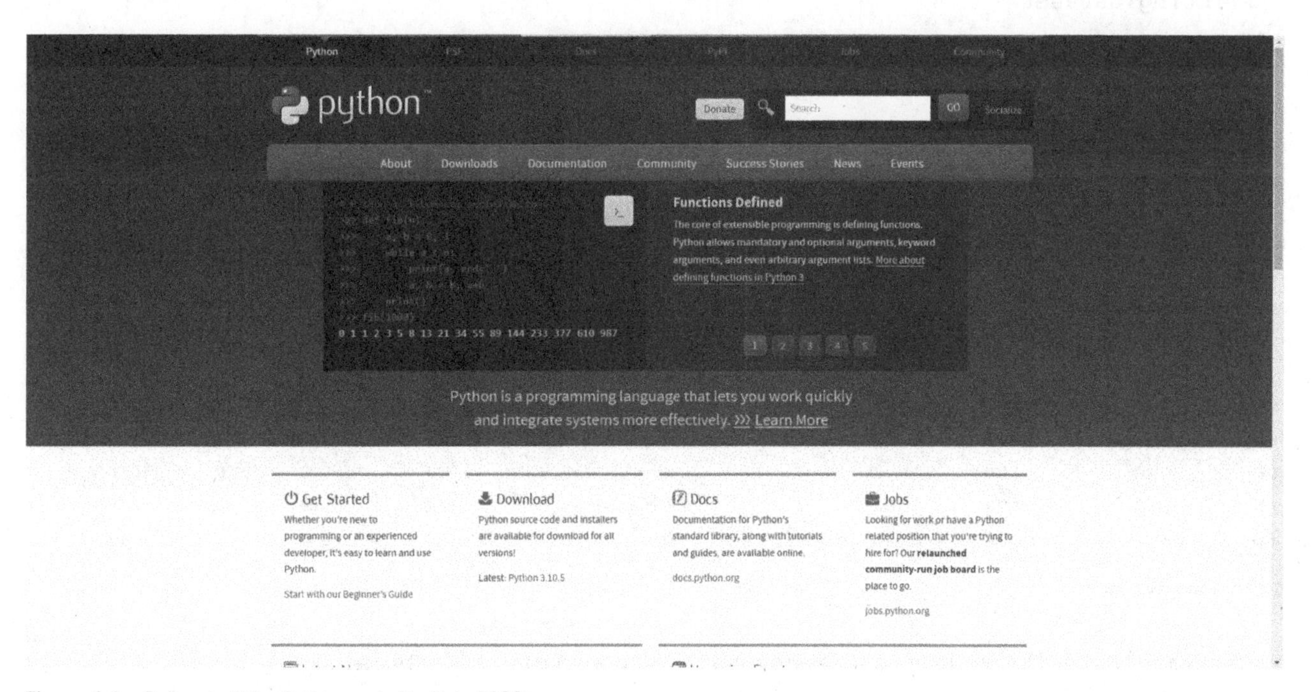

Figure 6.6 Snippet of the *Python* website (July 2022).

---

78 As you know, other programming languages also exist such as *JavaScript*, *C++*, *Go*, and others. I choose *Python* and *R* due to my experience with them (as well as Moe and Arash's experiences) and the fact that these two languages consistently rank among the highest cited coding languages on an annual basis and are regularly updated.

79 The name Python comes from the published scripts from "*Monty Python's Flying Circus*", a BBC comedy series from the 1970s [43].

80 **Code library**: A collection of pre-built modules and codes that can be used repeatedly in different problems.

81 You might also want to download and install the GUI interface, *Anaconda Navigator* [44] to help in launching *Python* packages and applications.

82 You may use Jupyter Notebook [20] or Google Collaboratory [45] to run these examples – both are free to use! All codes are also available for you to use and download from my website.

Example no. 1 [Python, Classification]: Predicting fire-induced spalling of concrete mixtures.
This example repeats that carried out in Example no. 1 in *DataRobot*. As you remember, we used a dataset of 646 test samples comprising 16 features and one target (with two labels: *no spalling* or *spalling*). Our goal is to build a ML algorithm that we can input our dataset into to arrive at a ML model that can precisely predict if a concrete mixture is going to spall or not. We will use the XGBoost algorithm to build such a model, as you can see in the script below. We will also build a web application of this model (visit the Appendix for the complete example on how to do so).

```
#Importing libraries
from xgboost import XGBClassifier
import xgboost as xgb
import seaborn as sns
from matplotlib import pyplot
from sklearn.model_selection import train_test_split
import pandas as pd
import numpy as np
from sklearn.metrics import confusion_matrix,classification_report
from sklearn.metrics import plot_confusion_matrix
```

```
#Importing dataset
from pandas import MultiIndex, Int64Index
fire=pd.read_csv('Classification.csv')
```

```
#Printing dataset
fire
```

	Water/binder ratio (%)	Aggregate/binder ratio (%)	Sand/binder ratio (%)	Heating rate (C/min)	Maximum exposure temperature (C)	Silica fume/binder ratio (%)	Aggregate size (mm)	GGBS/binder ratio (%)	FA/binder ratio (%)	PP fiber quantity (kg/m3)	Output
0	0.273	2.000	1.049	101.0	1034	0.0	7.0	0.25	0.0	0.0	1
1	0.273	2.000	1.049	101.0	1034	0.0	14.0	0.25	0.0	0.0	1
2	0.300	2.000	1.033	101.0	1034	0.0	20.0	0.25	0.0	0.0	1
3	0.218	1.374	1.199	101.0	1034	0.0	7.0	0.25	0.0	0.0	1
4	0.218	1.374	1.199	101.0	1034	0.0	7.0	0.25	0.0	0.0	1
...	...	...	...	...	...	...	...	...	...	...	...
641	0.200	0.000	1.040	15.0	800	0.2	0.6	0.00	0.0	1.0	1
642	0.200	0.000	1.040	15.0	800	0.2	0.6	0.00	0.0	1.5	1
643	0.200	0.000	1.040	15.0	800	0.2	0.6	0.00	0.0	2.0	1
644	0.200	0.000	1.040	15.0	800	0.2	0.6	0.00	0.0	2.5	1
645	0.200	0.000	1.040	15.0	800	0.2	0.6	0.00	0.0	3.0	0

646 rows × 11 columns

```
#Generate data frame [a two-dimensional data structure (tabular)]
x=fire.drop(['Output'],axis=1)
y=fire['Output']
x_train,x_test,y_train,y_test=train_test_split(x,y,test_size=0.30,random_state=111)
```

```
wb=fire['Output']
pyplot.hist(wb,edgecolor='black')
pyplot.title('Water/binder ratio')
pyplot.xlabel('Ratio')
pyplot.ylabel('No. of samples')
pyplot.show()
```

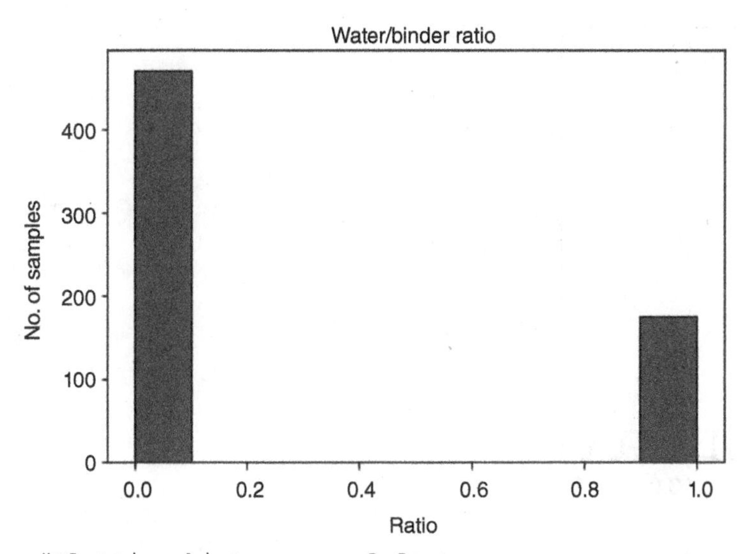


```
#Plotting histograms of features
for i, col in enumerate(fire.columns):
    pyplot.figure(i)
    sns.histplot(fire[col])
```

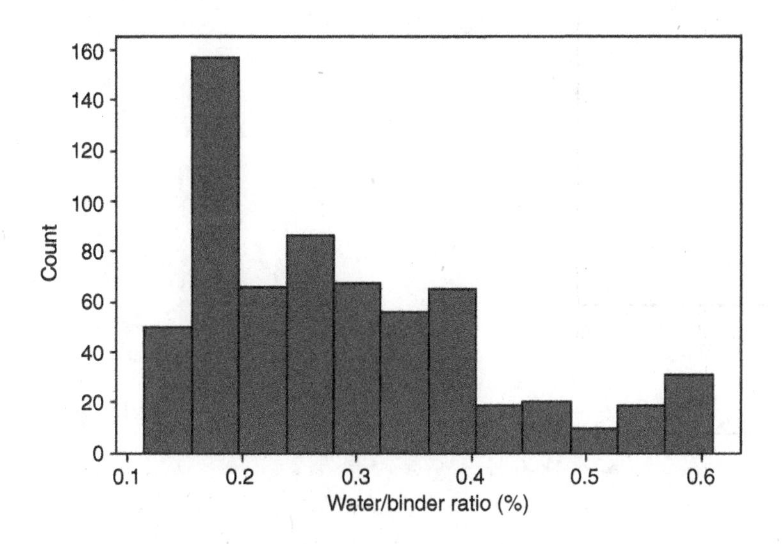

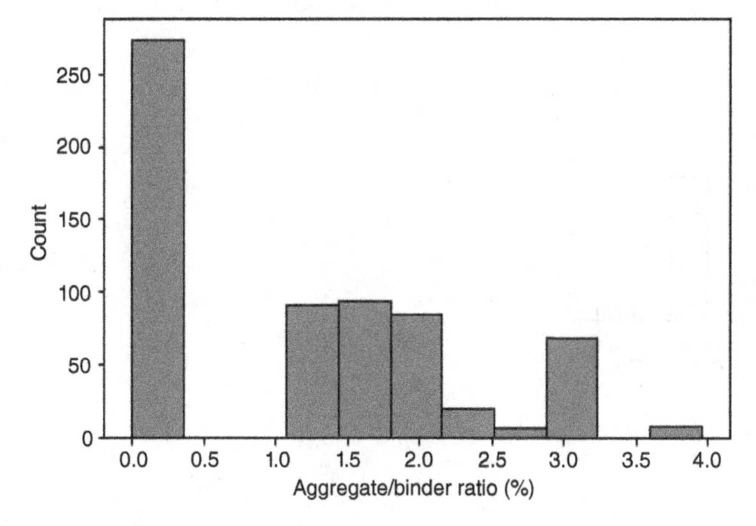

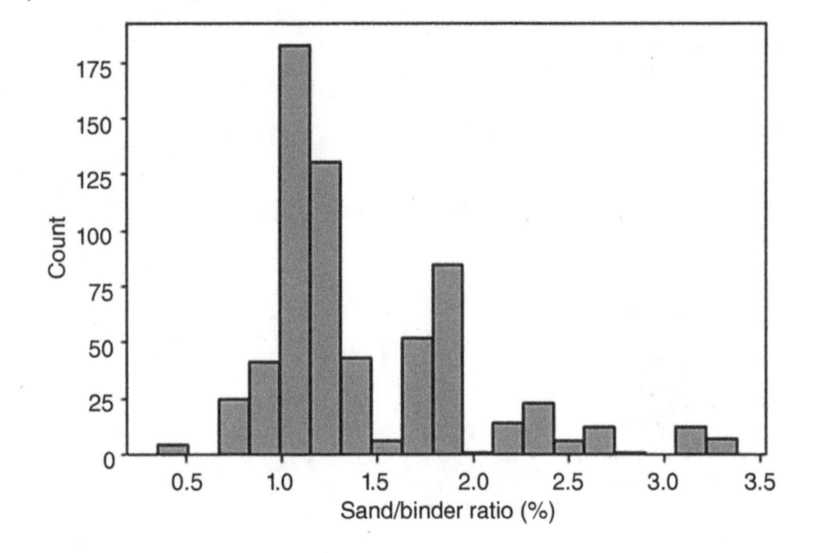
Count
175
150
125
100
75
50
25
0
0.5
1.0
1.5
2.0
2.5
3.0
3.5
Sand/binder ratio (%)

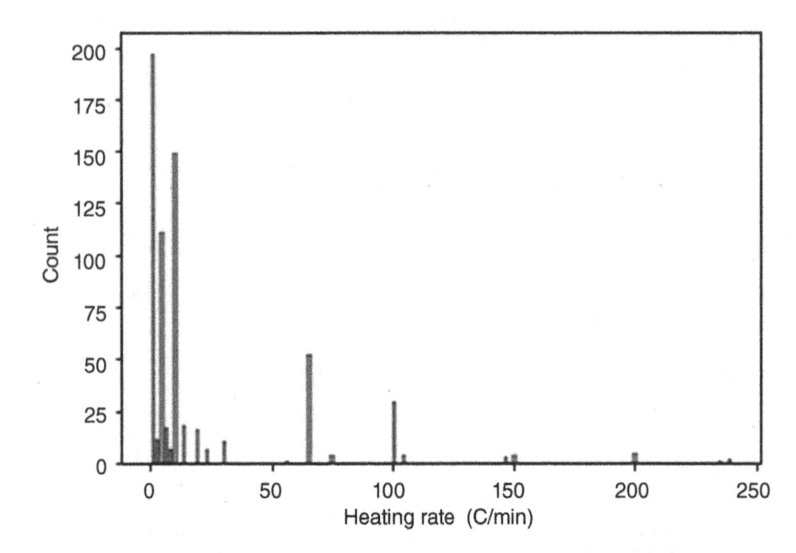
Count
200
175
150
125
100
75
50
25
0
0
50
100
150
200
250
Heating rate (C/min)

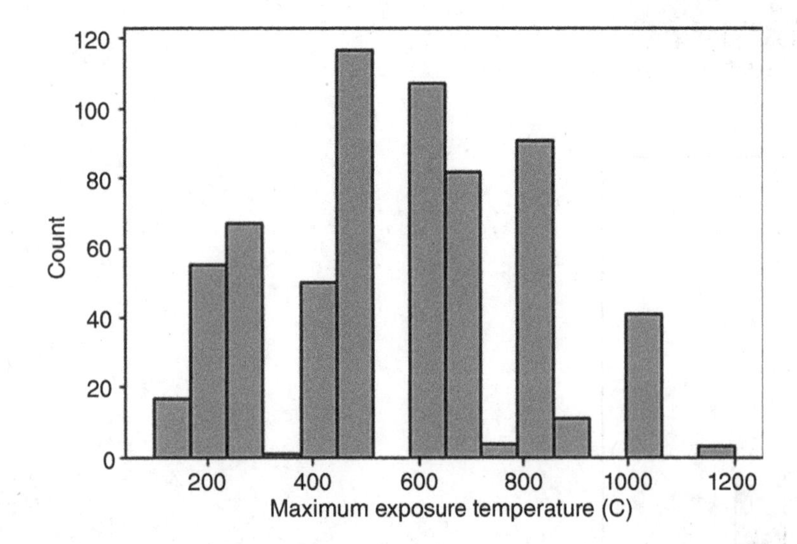
Count
120
100
80
60
40
20
0
200
400
600
800
1000
1200
Maximum exposure temperature (C)

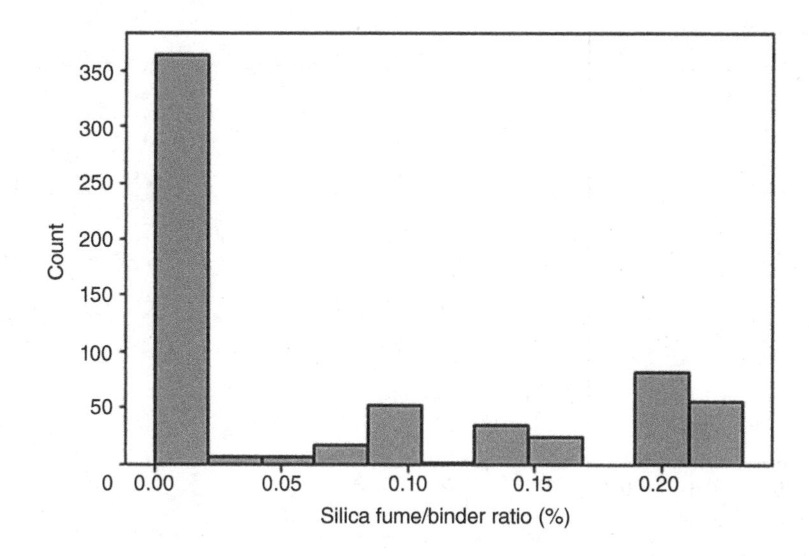
Count
350
300
250
200
150
100
50
0
0.00
0.05
0.10
0.15
0.20
Silica fume/binder ratio (%)

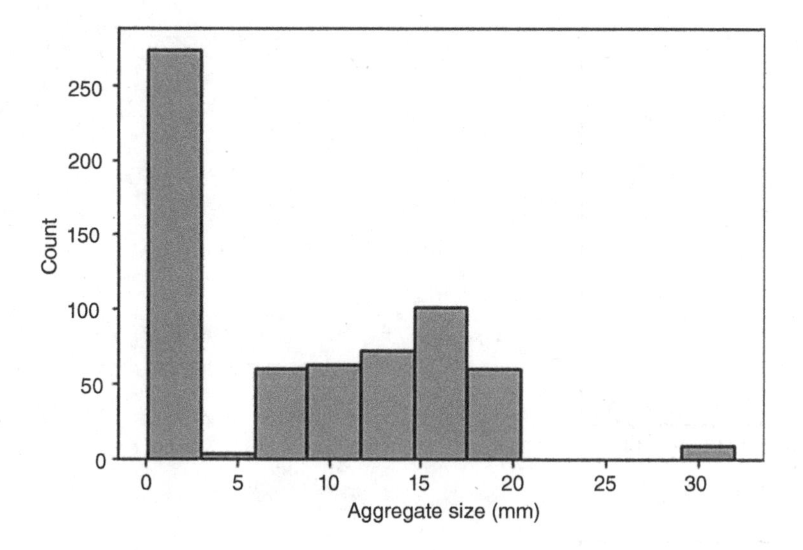
Count
250
200
150
100
50
0
0
5
10
15
20
25
30
Aggregate size (mm)

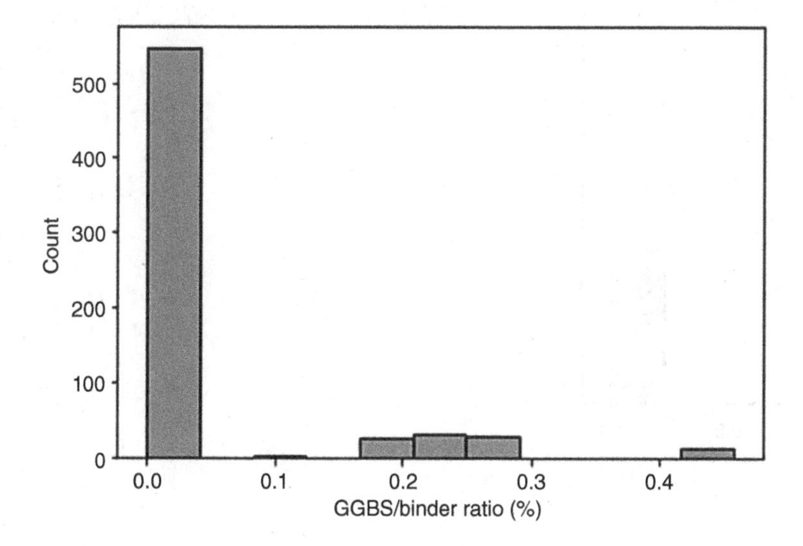
Count
500
400
300
200
100
0
0.0
0.1
0.2
0.3
0.4
GGBS/binder ratio (%)

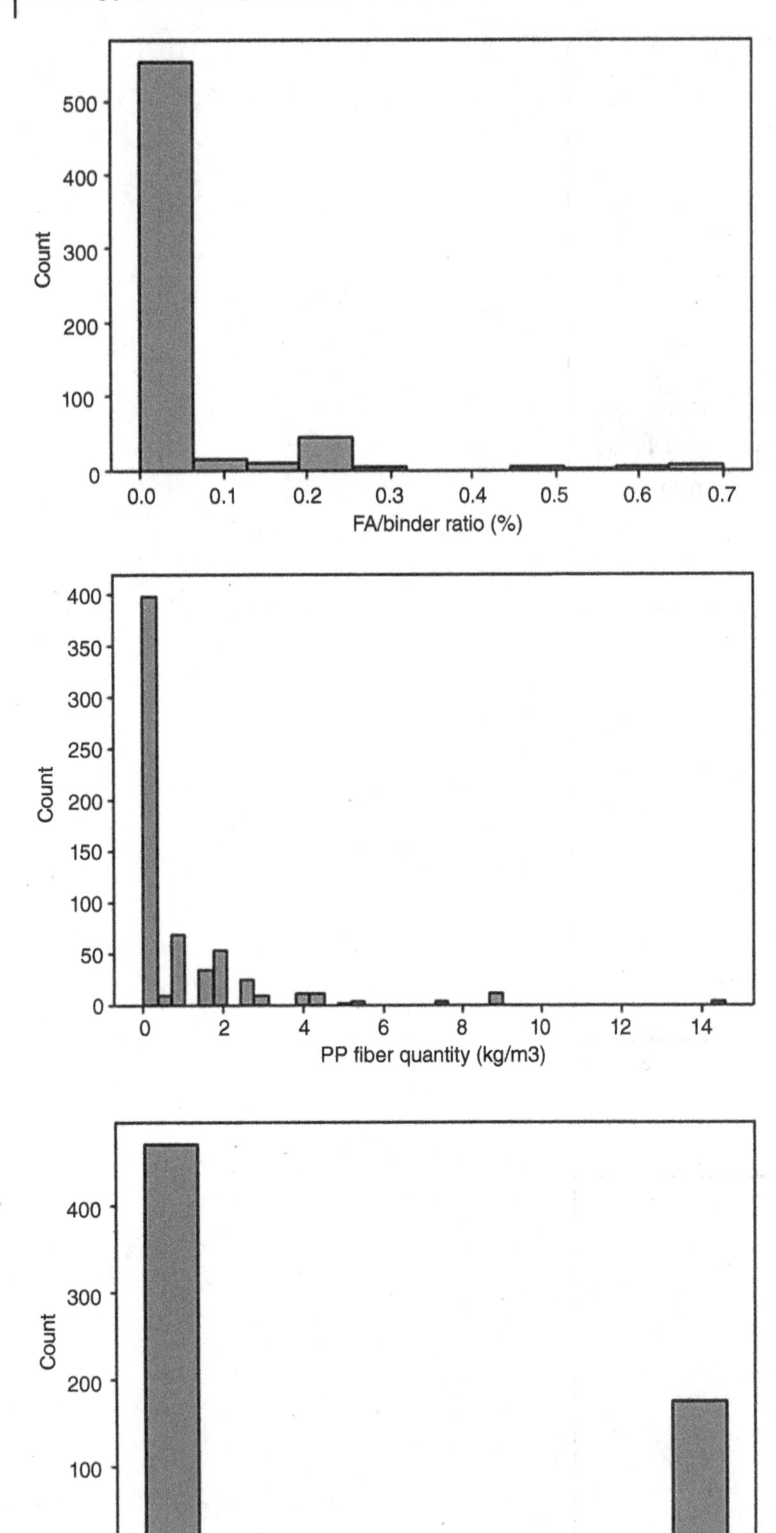

Count
500
400
300
200
100
0
0.0
0.1
0.2
0.3
0.4
0.5
0.6
0.7
FA/binder ratio (%)
Count
400
350
300
250
200
150
100
50
0
0
2
4
6
8
10
12
14
PP fiber quantity (kg/m3)
Count
400
300
200
100
0
0.0
0.2
0.4
0.6
0.8
1.0
Output

```
#Excute and fit the algorithm.
```

```
xgbc=xgb.XGBClassifier(objective ='binary:logistic',missing=1,seed=42,learning_rate
= 0.05, max_depth = 3)
xgbc.fit(x_train,y_train,verbose=True,early_stopping_rounds=50,eval_
metric='aucpr',eval_set=[(x_test,y_test)])
predictions = xgbc.predict(x_test)
```

```
[0]validation_0-aucpr:0.67211
[1]validation_0-aucpr:0.67211
[2]validation_0-aucpr:0.74269
...
...

[97]validation_0-aucpr:0.89769
[98]validation_0-aucpr:0.89683
[99]validation_0-aucpr:0.89600
```

```
class_names = ['no spalling ', 'spalling']
disp = plot_confusion_matrix(xgbc, x_train, y_train, display_labels=class_
names,colorbar=True, cmap=pyplot.cm.Blues, xticks_rotation='vertical')
pyplot.title('Training set')
```

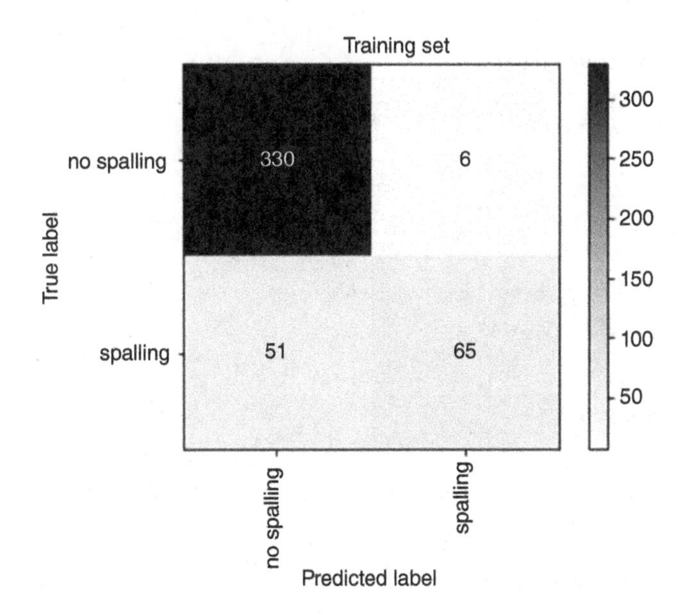


```
class_names = ['no spalling ', 'spalling']
disp = plot_confusion_matrix(xgbc, x_test, y_test, display_labels=class_names,
cmap=pyplot.cm.Blues, xticks_rotation='vertical')
pyplot.title('Testing set')
print (classification_report(y_test,predictions))
```

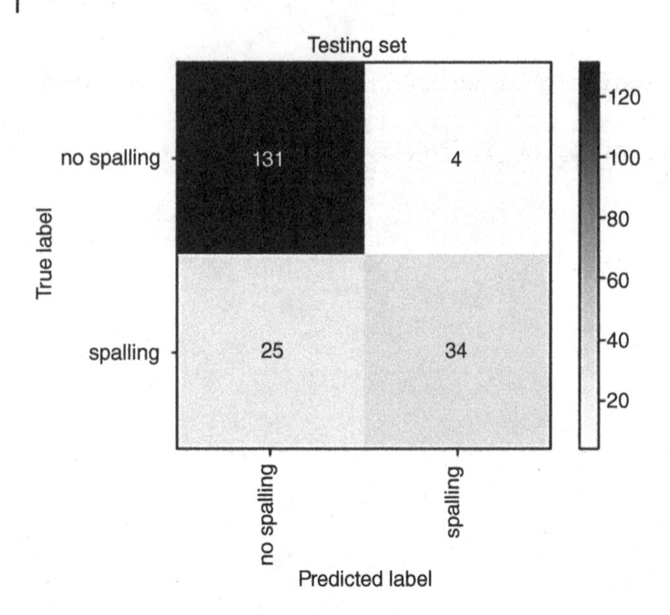


```
r_auc=roc_auc_score(y_test,predictions)
r_auc
```

0.773320778405524

```
yhat = xgbc.predict_proba(x_train)
```

```
#Retrieve just the probabilities for the positive class
pos_probs = yhat[:, 1]

#Plotting the ROC curve
pyplot.plot([0, 1], [0, 1], linestyle='--')

#Calculating the ROC curve
fpr, tpr, _ = roc_curve(y_train, pos_probs)

#Plotting the ROC curve
pyplot.plot(fpr, tpr, marker='.', label='AUC=0.907')

#Adding axis labels
pyplot.xlabel('False Positive Rate')
pyplot.ylabel('True Positive Rate')

#Showing legend
pyplot.legend()
pyplot.title('Training set')

#Showing the plot
pyplot.show()
```

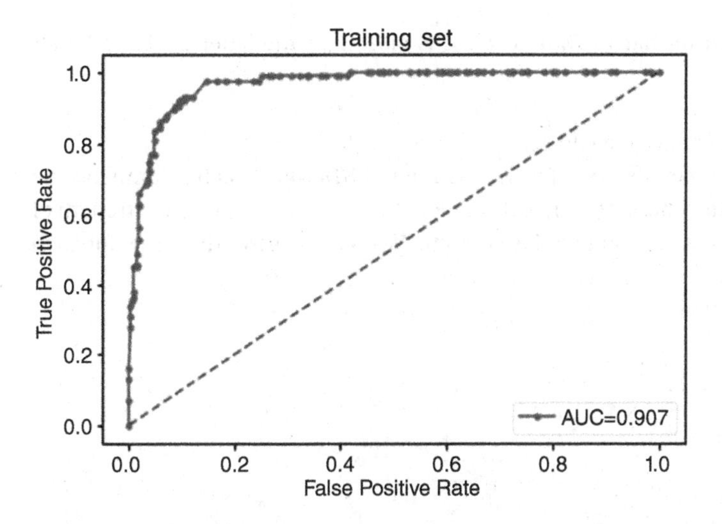


```
yhat = xgbc.predict_proba(x_test)

#Retrieve just the probabilities for the positive class
pos_probs = yhat[:, 1]

#Plotting the ROC curve
pyplot.plot([0, 1], [0, 1], linestyle='--')

#Calculating the ROC curve
fpr, tpr, _ = roc_curve(y_test, pos_probs)

#Plotting the ROC curve
pyplot.plot(fpr, tpr, marker='.', label='AUC=0.866')

#Adding axis labels
pyplot.xlabel('False Positive Rate')
pyplot.ylabel('True Positive Rate')

#Showing legend
pyplot.legend()
pyplot.title('Testing set')

#Showing the plot
pyplot.show()
```

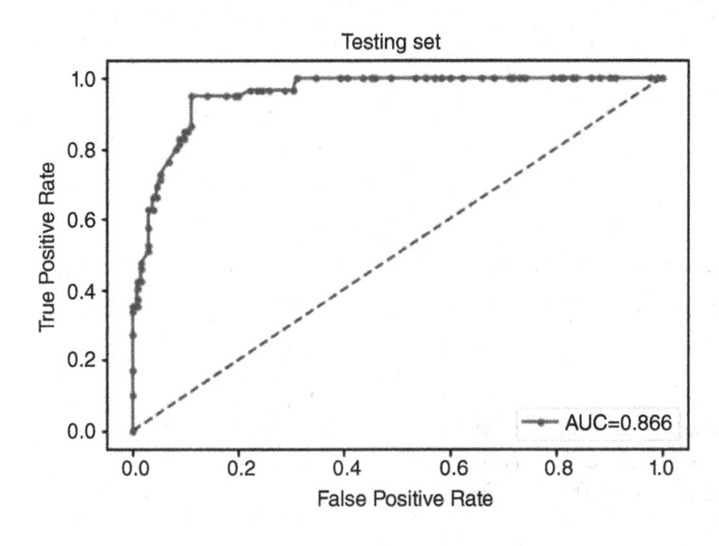

Comparing the performance of the model from *Python* to that in *DataRobot*, it is clear that the latter achieves better performance in terms of the AUC metric.

Example no. 2 [*Python*, Regression]: Predicting axial strength of CFSTs.
Let us repeat the second example of *DataRobot*. As you remember, our dataset has 1260 CFSTs where each column has the following features: column diameter, *D* (mm), column thickness, *t* (mm), column length, *L* (mm), compressive strength of concrete f_c' (MPa), and steel yielding stress, f_y (MPa). The axial compressive strength *N* (kN) comprises the target for each column. We will be using the Random Forest algorithm.

The *Python* code script for this example is provided below.

```
#Importing libraries
import pandas as pd
from pandas import MultiIndex, Int64Index
import numpy as np
import xgboost as xgb
from sklearn.ensemble import RandomForestRegressor
from sklearn.model_selection import train_test_split
from matplotlib import pyplot
import seaborn as sns

#Importing error metrics
from sklearn.metrics import mean_squared_error
from sklearn.metrics import accuracy_score
from sklearn.metrics import r2_score

#Importing dataset
CFST=pd.read_csv('CFST.csv')
CFST.head(5)  #Printing the first five rows of the data
```

	D (mm)	t (mm)	L (mm)	fy (MPa)	fc (Mpa)	N Test (kN)
0	168.8	2.64	305.0	302.4	18.2	1326.0
1	168.8	2.64	305.0	302.4	34.7	1219.0
2	169.3	2.62	305.0	338.1	37.1	1308.0
3	169.3	2.62	305.0	338.1	34.1	1330.0
4	168.3	3.60	305.0	288.4	27.0	1557.0

```
CFST.dtypes
D (mm)          float64
t (mm)          float64
L (mm)          float64
fy (MPa)        float64
fc (Mpa)        float64
N Test (kN)     float64
dtype: object

#Try the agg function to aggregate data into a summary
#of statistical insights for the parameters of the dataset.

Statistical_insights=CFST.agg(
    {
"D (mm)": ["min", "max", "median", "skew", "std", "mean"],
"t (mm)": ["min", "max", "median", "skew", "std", "mean"],
"L (mm)": ["min", "max", "median", "skew", "std", "mean"],
```

```
"fy (MPa)": ["min", "max", "median", "skew", "std", "mean"],
"fc (Mpa)": ["min", "max", "median", "skew", "std", "mean"]
    }
        )
print(Statistical_insights)
```

```
              D (mm)      t (mm)       L (mm)     fy (MPa)     fc (Mpa)
min        44.450000    0.520000   152.350000   178.280000     7.590000
max      1020.000000   16.540000  5560.000000   853.000000   185.940000
median    129.000000    4.000000   662.000000   324.700000    37.400000
skew        3.720495    1.582544     2.004010     2.167322     2.151266
std       104.836319    2.455314  1000.293269    90.956430    30.665549
mean      158.832135    4.274690  1055.142127   335.675413    47.504040
```

```
#Assigning the 'X' variable to use the features as a data frame f#or X.
#Assigning the target label for y variable as a target)
x=CFST.drop(['N Test (kN)'], axis=1)
y=CFST['N Test (kN)']

#Splitting the dataset into a 80% training set and 20% as a testing set.
x_train,x_test,y_train,y_test=train_test_split(x,y
                                               ,test_size=0.2
                                               ,random_state=200
                                              )
#Getting the Random Forest algorithm
RF=RandomForestRegressor(max_depth = 20,
                         random_state = 0,
                         n_estimators = 10000
                        )
RF.fit(x_train,y_train)

#Fitting our split data into training and testing datasets
RandomForestRegressor(max_depth=20, n_estimators=10000, random_state=0)

#Assigning prediction functions
Predictions = RF.predict(x_test)

#Prediction score for the training set
Score =RF.score(x_train,y_train)

#Evaluating model performance
print ("score : % f" %(Score))
print ('Accuracy',100-(np.mean(np.abs((y_test - Predictions) / y_test)) * 100))
print ('r2_score' (y_test, Predictions))
print ('MAE',np.mean(abs(y_test-Predictions)))
print ('RMSE',np.sqrt(np.mean(np.square(y_test-Predictions))))
score :  0.994561
Accuracy 91.62039170328033
r2_score 0.9767946127523274
MAE 190.27976293716608
RMSE 553.1462150693716
```

As you can see, the R^2 of this model is much improved over that obtained from the *Random Forest Regressor (BP 75)* in *DataRobot*.

Example no. 3 [Python, Clustering]: Grouping fire-exposed RC columns.
This example replicates *Example no. 3* from the *BigML* section.

To recap, the compiled database contains the following features from 144 RC columns: column width, W (mm), steel reinforcement ratio, r (mm), column length, L (mm), concrete compressive strength, f_c (MPa), steel yield strength, f_y (MPa), restraint conditions by means of effective length factor, concrete cover to reinforcement, C (mm), eccentricity in applied loading in the main axes, e_x (mm), the magnitude of applied loading, P (kN), and fire resistance, FR (min).

Here is the *Python* code[83] for you to try. I am using the K-Means algorithm here, which I am importing from the *sklearn.cluster* package. I will also provide scripts for the FCM and DBSCAN algorithms.

```
#Importing libraries
import numpy as np
import matplotlib.pyplot as plt
import pandas as pd
from sklearn.preprocessing import StandardScaler
import seaborn as sns
```

```
#Importing dataset
dataset = pd.read_excel('Fire-exposed RC columns.xlsx')
X = dataset.iloc[:, :9].values
pd.plotting.scatter_matrix(dataset, alpha = 0.3, figsize = (26,20), diagonal =
'kde');
```

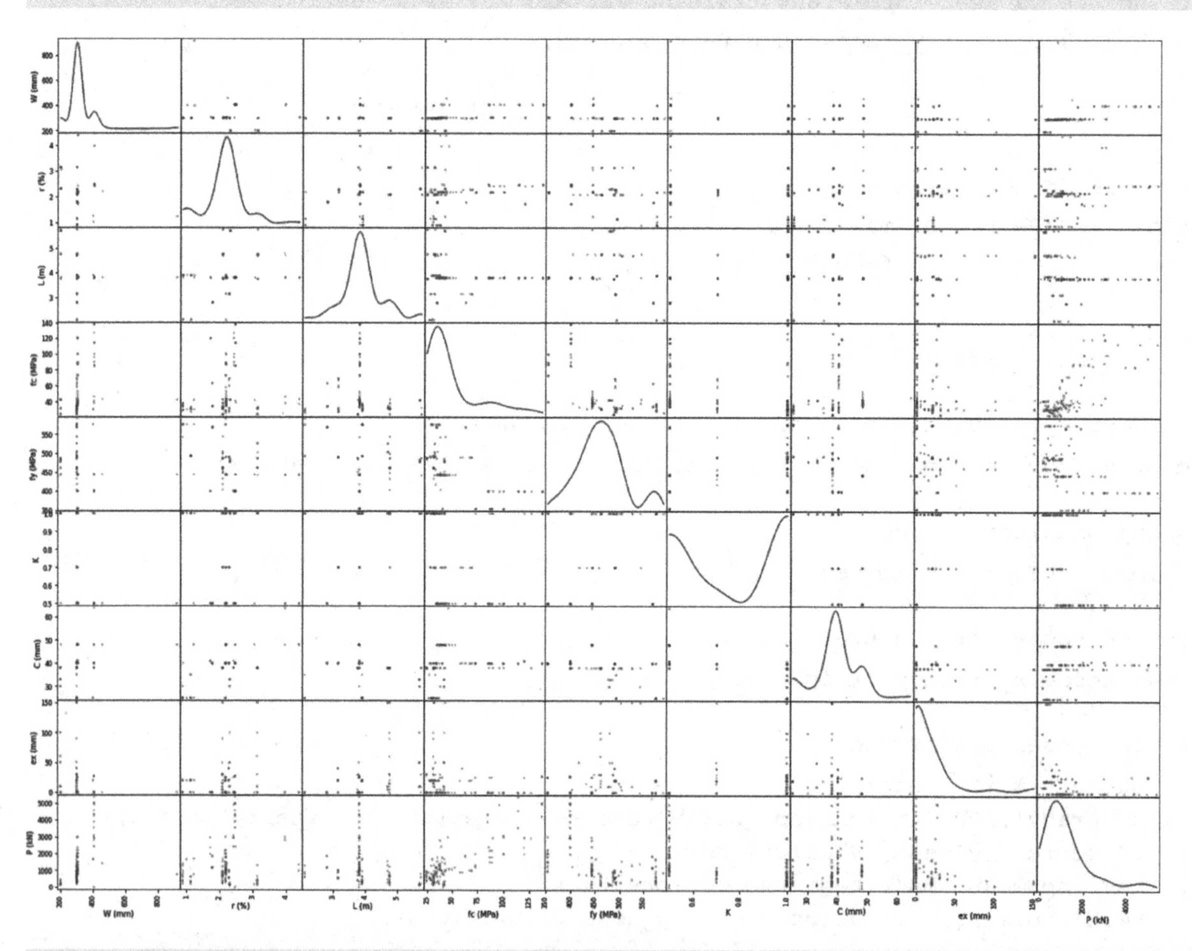


```
from sklearn.cluster import KMeans
from sklearn.metrics import silhouette_samples, silhouette_score
import matplotlib.cm as cm
range_n_clusters = range( 2, 6 )
for n_clusters in range_n_clusters:
```

83 Thanks to Aybike for building this code [46]!

```
#Create a subplot with 1 row and 2 columns
    fig, (ax1, ax2) = plt.subplots(1, 2)
    fig.set_size_inches(18, 7)

#The 1st subplot is the silhouette plot
#The silhouette coefficient can range from -1, 1 but in this
#example lies within [-0.1, 1]
    ax1.set_xlim([-0.1, 1])

#The (n_clusters+1)*10 is for inserting blank space between
#silhouette
#Plotting individual clusters to demarcate them clearly.
    ax1.set_ylim([0, len(X) + (n_clusters + 1) * 10])

#Initialize the clusterer with n_clusters value and a random
#generator seed of 10 for reproducibility.
    clusterer = KMeans(n_clusters=n_clusters, random_state=10)
    cluster_labels = clusterer.fit_predict(X)

#The silhouette_score gives the average value for all the samples
#This gives a perspective into the density of the formed clusters
    silhouette_avg = silhouette_score(X, cluster_labels)
    print("For n_clusters =", n_clusters,
          "The average silhouette_score is :", silhouette_avg)

#Compute the silhouette scores for each sample.
    sample_silhouette_values = silhouette_samples(X, cluster_labels)

    y_lower = 10

    for i in range(n_clusters):
#Aggregate the silhouette scores for samples belonging to cluster #i, and sort them
        ith_cluster_silhouette_values = \
            sample_silhouette_values[cluster_labels == i]

        ith_cluster_silhouette_values.sort()

        size_cluster_i = ith_cluster_silhouette_values.shape[0]
        y_upper = y_lower + size_cluster_i

        color = cm.nipy_spectral(float(i) / n_clusters)
        ax1.fill_betweenx(np.arange(y_lower, y_upper),
                          0, ith_cluster_silhouette_values,
                          facecolor=color, edgecolor=color, alpha=0.7)

#Label the silhouette plots with their cluster numbers
        ax1.text(-0.05, y_lower + 0.5 * size_cluster_i, str(i))

#Compute the new y_lower for next plot
y_lower = y_upper + 10  # 10 for the 0 samples

    ax1.set_title(("The silhouette plot with n_clusters = %d" % n_
clusters),fontsize=23)
    ax1.set_xlabel(("The silhouette coefficient values"),fontsize=23)
    ax1.set_ylabel(("Cluster label"),fontsize=23)
```

```
#The vertical line for average silhouette score of all the values
    ax1.axvline(x=silhouette_avg, color="red", linestyle="--")
    ax1.set_yticks([])  #Clear the y_axis labels / ticks
    ax1.set_xticks([-0.1, 0, 0.2, 0.4, 0.6, 0.8, 1])
    ax1.tick_params(axis='both', which='major', labelsize=16)

#2nd Plot showing the formed clusters
    colors = cm.nipy_spectral(cluster_labels.astype(float) / n_clusters)
    ax2.scatter(X[:, 0], X[:, 1], marker='.', s=30, lw=0, alpha=0.7,
                c=colors, edgecolor='k')

    '''# Labeling the clusters
    centers = clusterer.cluster_centers_
    # Draw white circles at cluster centers
    ax2.scatter(centers[:, 0], centers[:, 1], marker='o',
                c="white", alpha=1, s=200, edgecolor='k')

    for i, c in enumerate(centers):
        ax2.scatter(c[0], c[1], marker='$%d$' % i, alpha=1,
                    s=50, edgecolor='k') '''

    ax2.set_title(("The visualization of the clustered data."),fontsize=22)
    ax2.set_xlabel(("Feature space for the 1st feature"),fontsize=22)
    ax2.set_ylabel(("Feature space for the 2nd feature"),fontsize=22)

    plt.suptitle(("Silhouette analysis for KMeans clustering on sample data "
                  "with n_clusters = %d" % n_clusters),
                 fontsize=14, fontweight='bold')
plt.show()
For n_clusters = 2 The average silhouette_score is : 0.7237647056094885
For n_clusters = 3 The average silhouette_score is : 0.6267411875523311
For n_clusters = 4 The average silhouette_score is : 0.5363610626675143
For n_clusters = 5 The average silhouette_score is : 0.5499181214709731
```

Silhouette analysis for KMeans clustering on sample data with n_clusters = 2

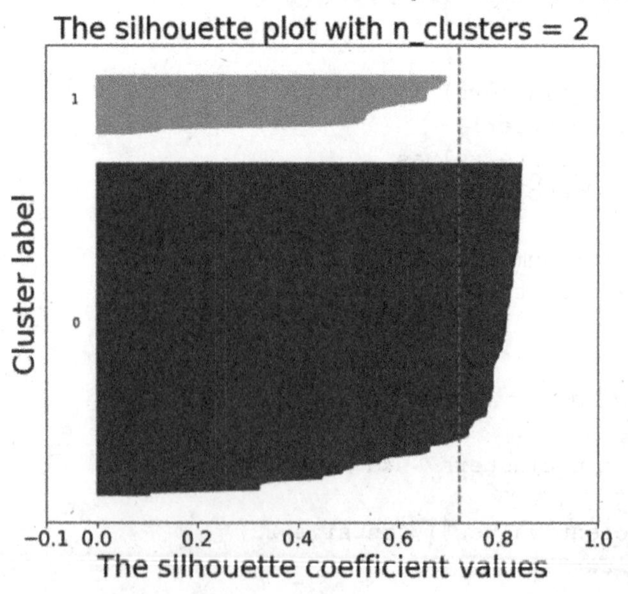

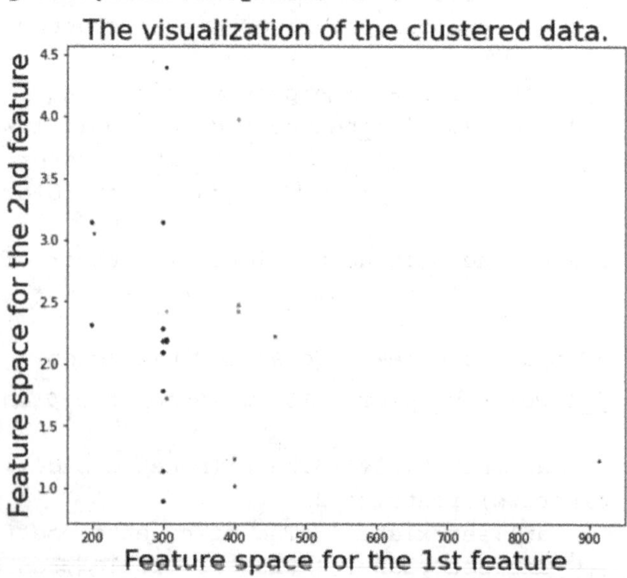

Silhouette analysis for KMeans clustering on sample data with n_clusters = 3

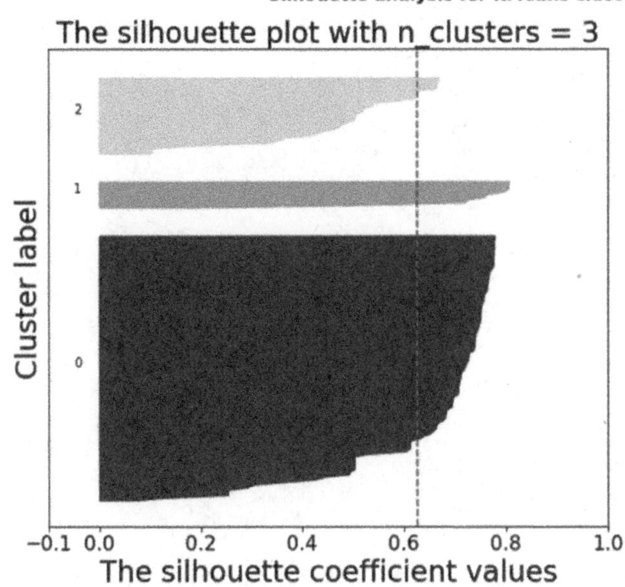

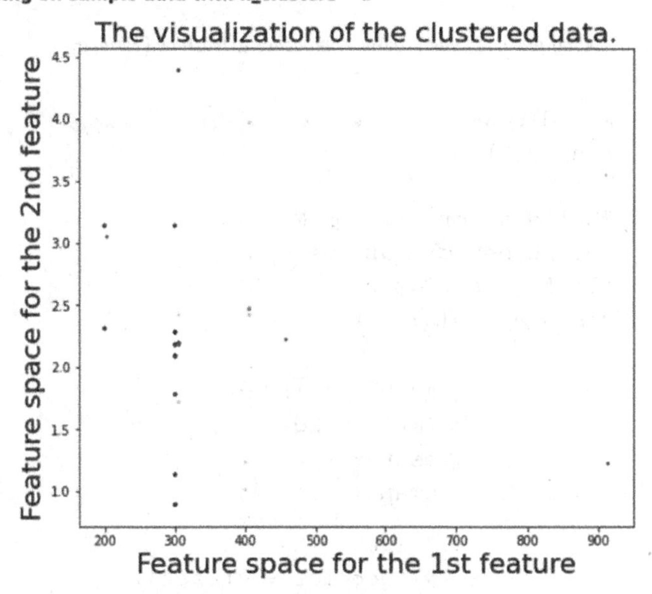

Silhouette analysis for KMeans clustering on sample data with n_clusters = 4

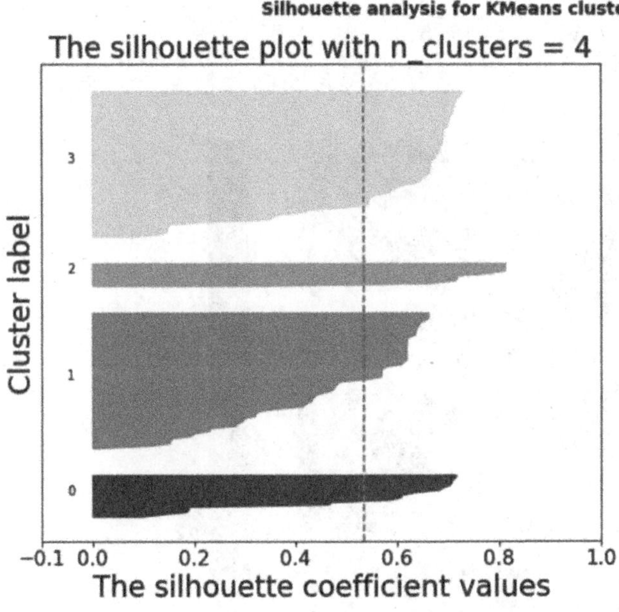

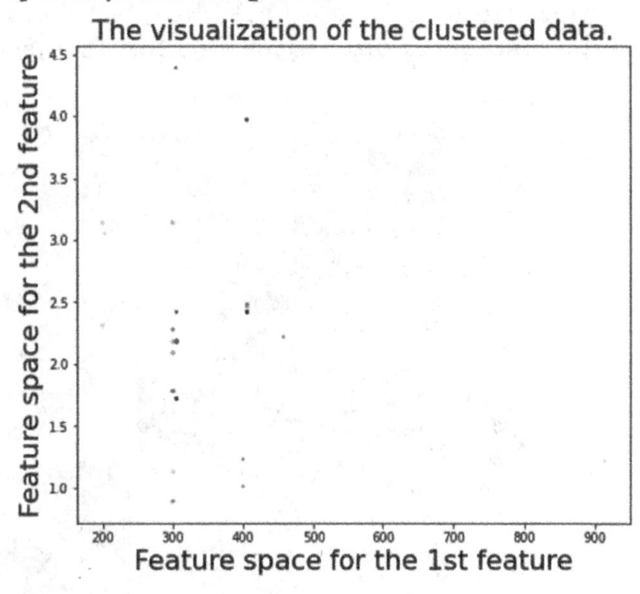

Silhouette analysis for KMeans clustering on sample data with n_clusters = 5

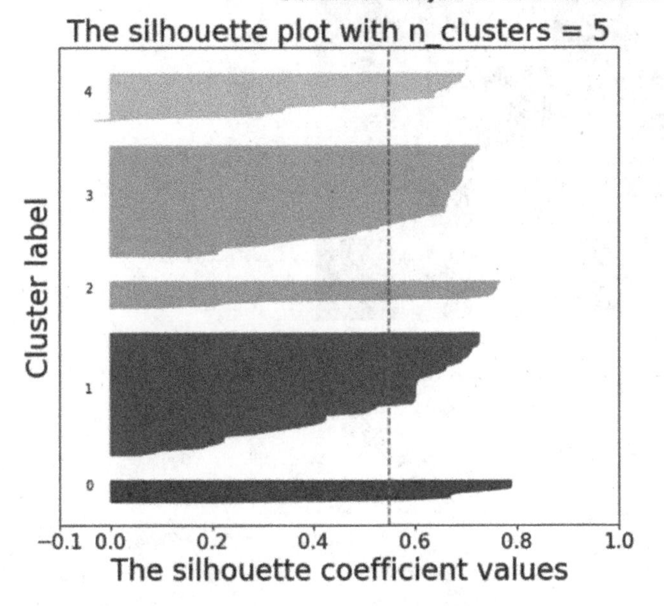

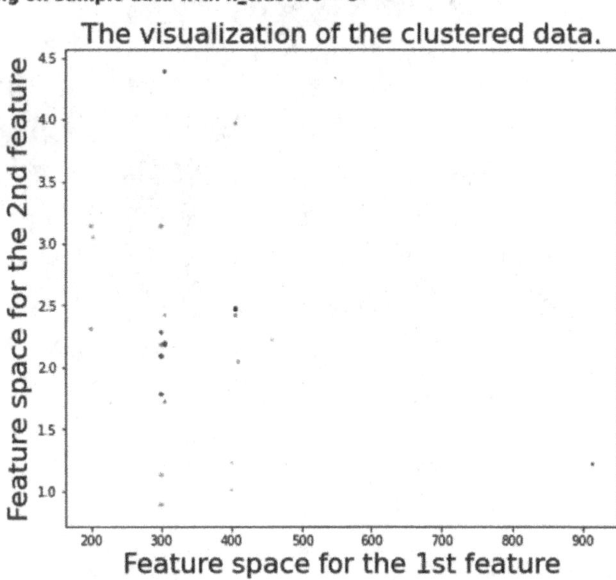

```
X = dataset
X.head(1)

#Build a correlation Matrix
import seaborn as sns
plt.figure(figsize=(15,13))
sns.heatmap(X.corr(),

            #annot = True,
            fmt = '.1g',
            center = 0,
            cmap = 'GnBu')
            #linewidths = 1)
            #linecolor = 'black')
plt.xticks(rotation=45,fontsize=18)
plt.yticks(rotation=45,fontsize=18)
(array([0.5, 1.5, 2.5, 3.5, 4.5, 5.5, 6.5, 7.5, 8.5, 9.5]),
 <a list of 10 Text major ticklabel objects>)
```

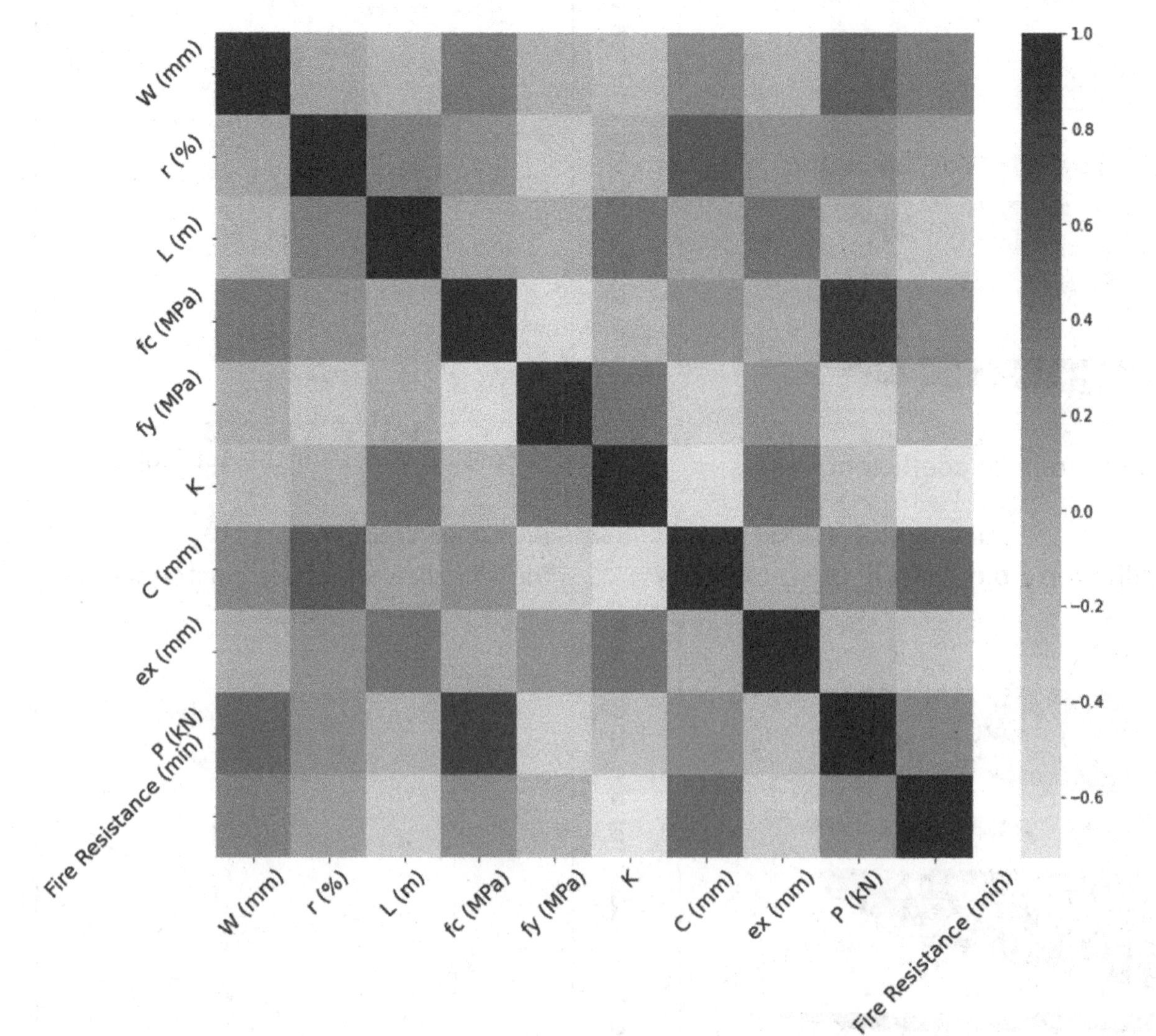

```
#Feature Scaling
X = dataset.iloc[:, :9].values
from sklearn.preprocessing import StandardScaler
sc = StandardScaler()
X = sc.fit_transform(X)
```

```
#X=pd.DataFrame(data=X)
#X.to_excel('Xscaled.xlsx', index=False)
#Using the elbow method to find the optimal number of clusters
from sklearn.cluster import KMeans
wcss = []
for i in range(1, 11):
    kmeans = KMeans(n_clusters = i, init = 'k-means++', random_state = 42)
    kmeans.fit(X)
    wcss.append(kmeans.inertia_)
plt.plot(range(1, 11), wcss)
plt.title(('The Elbow Method'),fontsize=20)
plt.xlabel(('Number of clusters'),fontsize=20)
plt.ylabel(('WCSS'),fontsize=20)
plt.xticks(fontsize = 14)
plt.yticks(fontsize = 14)
plt.show()
```

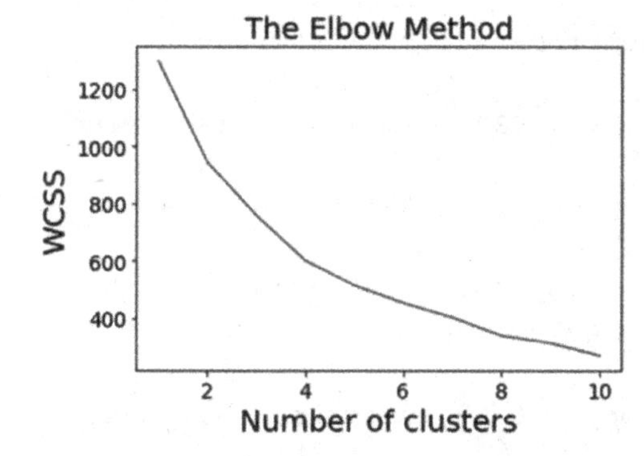


```
#Training the K-Means model on the dataset
from sklearn.cluster import KMeans
kmeans = KMeans(
        n_clusters=4, init="k-means++",
        n_init=10,
        tol=1e-04, random_state=42)
kmeans.fit(X)
y = kmeans.fit_predict(X)
mapping = {0:1, 1:2, 2:3, 3:4}
y = [mapping[i] for i in y]
```

```
#Describing features with a new figure
import plotly.express as px
clusters=pd.DataFrame(X,columns=dataset.columns)
clusters['cluster']=y
polar=clusters.groupby("cluster").mean().reset_index()
polar=pd.melt(polar,id_vars=["cluster"])
```

```
#Fig4 = px.line_polar(polar, r="value", theta="variable", color="cluster",line_clos
e=True,height=800,width=1000)
fig4 = px.line_polar(polar, r="value", theta="variable", color="cluster",line_
close=True,height=600)
fig4.update_layout(
    polar = dict(
      radialaxis_tickfont_size = 21,
      angularaxis = dict(
        tickfont_size=26,
      )))

fig4.update_layout(legend_font_size=26)
fig4.show()
```

```
#Classification and regression trees (CART)
```

```
X_normal = dataset.iloc[:, :9].values
from sklearn.tree import DecisionTreeClassifier
classifier = DecisionTreeClassifier(criterion = 'entropy', random_state = 0)
classifier.fit(X_normal, y)
DecisionTreeClassifier(criterion='entropy', random_state=0)
from sklearn import tree
fig, axes = plt.subplots(nrows = 1, ncols = 1, figsize = (8,8), dpi=200)
tree.plot_tree(classifier, feature_names = dataset.columns,filled = True, class_nam
es=['1','2','3','4'],fontsize=9)
[Text(0.625, 0.9285714285714286, 'fc (MPa) <= 71.0\nentropy = 1.913\nsamples = 144\
nvalue = [32, 21, 56, 35]\nclass = 3'),
 Text(0.5, 0.7857142857142857, 'L (m) <= 4.3\nentropy = 1.537\nsamples = 113\nvalue
= [1, 21, 56, 35]\nclass = 3'),
 Text(0.375, 0.6428571428571429, 'C (mm) <= 36.5\nentropy = 1.169\nsamples = 82\
nvalue = [1, 21, 56, 4]\nclass = 3'),
 Text(0.25, 0.5, 'entropy = 0.0\nsamples = 20\nvalue = [0, 20, 0, 0]\nclass = 2'),
 Text(0.5, 0.5, 'K <= 0.85\nentropy = 0.58\nsamples = 62\nvalue = [1, 1, 56, 4]\
nclass = 3'),
 Text(0.25, 0.35714285714285715, 'r (%) <= 1.5\nentropy = 0.137\nsamples = 52\nval-
ue = [1, 0, 51, 0]\nclass = 3'),
 Text(0.125, 0.21428571428571427, 'entropy = 0.0\nsamples = 1\nvalue = [1, 0, 0,
0]\nclass = 1'),
 Text(0.375, 0.21428571428571427, 'entropy = 0.0\nsamples = 51\nvalue = [0, 0, 51,
0]\nclass = 3'),
 Text(0.75, 0.35714285714285715, 'P (kN) <= 820.0\nentropy = 1.361\nsamples = 10\
nvalue = [0, 1, 5, 4]\nclass = 3'),
 Text(0.625, 0.21428571428571427, 'P (kN) <= 384.5\nentropy = 0.722\nsamples = 5\
nvalue = [0, 1, 0, 4]\nclass = 4'),
 Text(0.5, 0.07142857142857142, 'entropy = 0.0\nsamples = 1\nvalue = [0, 1, 0, 0]\
nclass = 2'),
 Text(0.75, 0.07142857142857142, 'entropy = 0.0\nsamples = 4\nvalue = [0, 0, 0, 4]\
nclass = 4'),
 Text(0.875, 0.21428571428571427, 'entropy = 0.0\nsamples = 5\nvalue = [0, 0, 5,
0]\nclass = 3'),
 Text(0.625, 0.6428571428571429, 'entropy = 0.0\nsamples = 31\nvalue = [0, 0, 0,
31]\nclass = 4'),
 Text(0.75, 0.7857142857142857, 'entropy = 0.0\nsamples = 31\nvalue = [31, 0, 0,
0]\nclass = 1')]
```

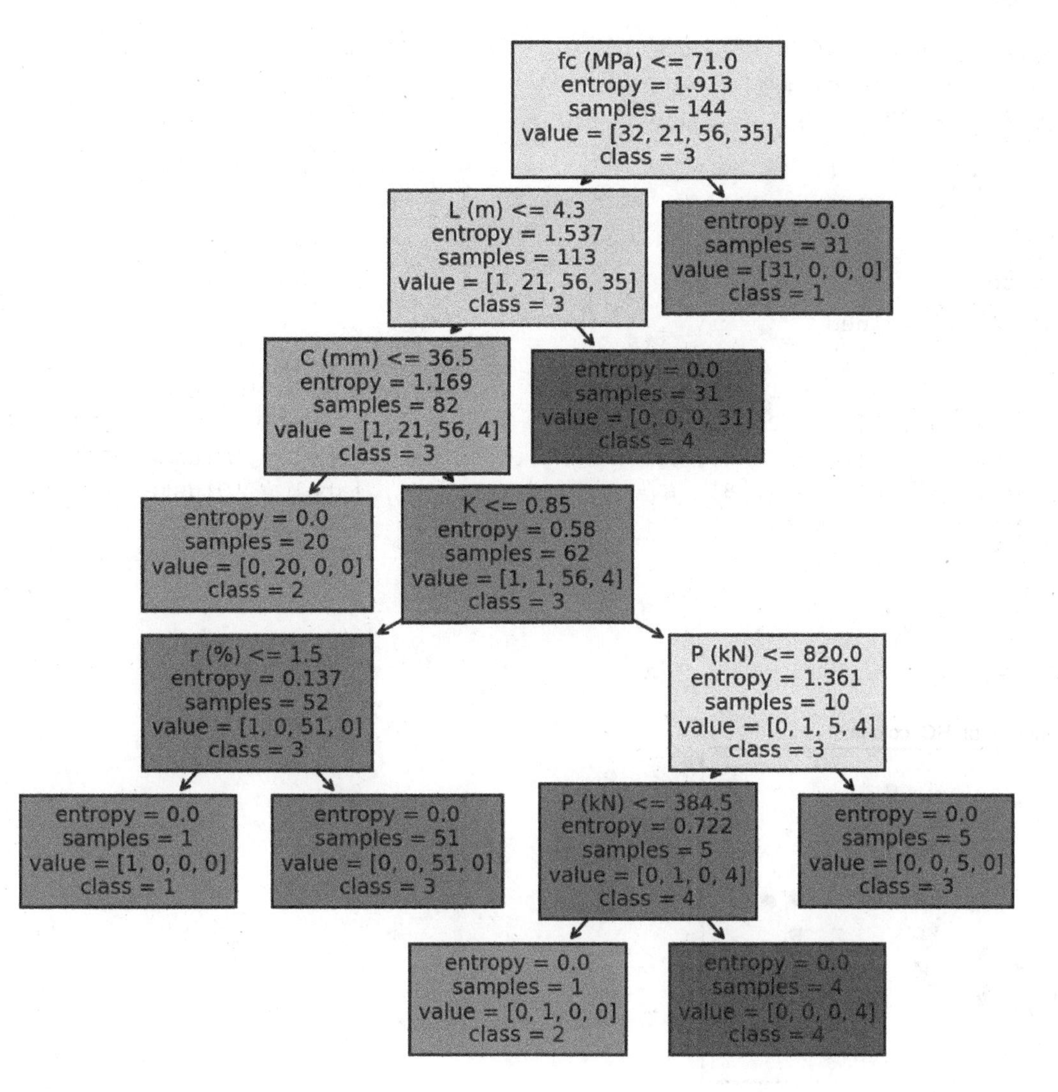


```
text_representation = tree.export_text(classifier)
print(text_representation)
|--- feature_3 <= 71.00
|   |--- feature_2 <= 4.30
|   |   |--- feature_6 <= 36.50
|   |   |   |--- class: 4
|   |   |--- feature_6 >  36.50
|   |   |   |--- feature_5 <= 0.85
|   |   |   |   |--- feature_1 <= 1.50
|   |   |   |   |   |--- class: 2
|   |   |   |   |--- feature_1 >  1.50
|   |   |   |   |   |--- class: 3
|   |   |   |--- feature_5 >  0.85
|   |   |   |   |--- feature_8 <= 820.00
|   |   |   |   |   |--- feature_8 <= 384.50
|   |   |   |   |   |   |--- class: 4
|   |   |   |   |   |--- feature_8 >  384.50
|   |   |   |   |   |   |--- class: 1
```

```
|   |   |   |   |--- feature_8 >  820.00
|   |   |   |   |   |--- class: 3
|   |--- feature_2 >  4.30
|   |   |--- class: 1
|--- feature_3 >  71.00
|   |--- class: 2
```

```
#Visualizing clusters
X = dataset.iloc[:, :9].values
y=np.array(y)
plt.scatter(X[y == 1, 3], X[y == 1, 8], s = 100, c = 'red', label = 'Cluster 1')
plt.scatter(X[y == 2, 3], X[y == 2, 8], s = 100, c = 'green', label = 'Cluster 2')
plt.scatter(X[y == 3, 3], X[y == 3, 8], s = 100, c = 'blue', label = 'Cluster 3')
plt.scatter(X[y == 4, 3], X[y == 4, 8], s = 100, c = 'cyan', label = 'Cluster 4')
plt.title('Clusters of RC columns')
plt.xlabel('fc')
plt.ylabel('P')
plt.legend()
plt.show()
```

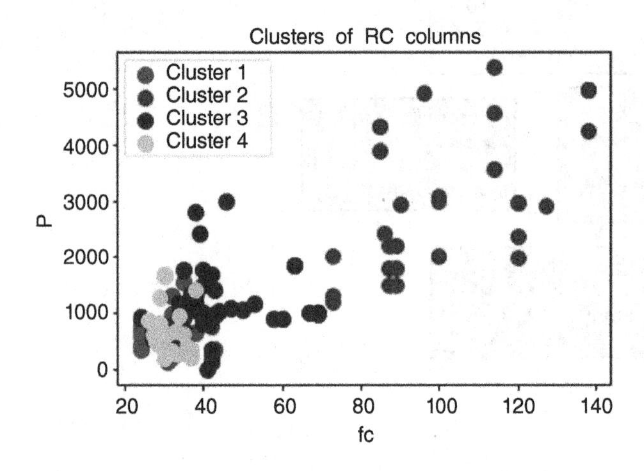


```
#FCM algorithm
%matplotlib inline
import pandas as pd
import numpy as np
from fcmeans import FCM
from matplotlib import pyplot as plt
```

```
dataset = pd.read_excel('Fire-exposed RC columns.xlsx')
X = dataset.iloc[:, :9].values
```

```
X = dataset.iloc[:, :9].values
from sklearn.preprocessing import StandardScaler
sc = StandardScaler()
X = sc.fit_transform(X)
```

```
from sklearn.metrics import silhouette_samples, silhouette_score
import matplotlib.cm as cm
range_n_clusters = range( 2, 11 )
for n_clusters in range_n_clusters:

    fig, (ax1, ax2) = plt.subplots(1, 2)
    fig.set_size_inches(18, 7)

    ax1.set_xlim([-0.1, 1])
    ax1.set_ylim([0, len(X) + (n_clusters + 1) * 10])

    fcm = FCM(n_clusters=n_clusters)
    fcm.fit(X)
    y_fcm =fcm.predict(X)
    cluster_labels = y_fcm
    silhouette_avg = silhouette_score(X, cluster_labels)
    print("For n_clusters =", n_clusters,
          "The average silhouette_score is :", silhouette_avg)

    sample_silhouette_values = silhouette_samples(X, cluster_labels)

    y_lower = 10
    for i in range(n_clusters):
        ith_cluster_silhouette_values = \
            sample_silhouette_values[cluster_labels == i]

        ith_cluster_silhouette_values.sort()

        size_cluster_i = ith_cluster_silhouette_values.shape[0]
        y_upper = y_lower + size_cluster_i

        color = cm.nipy_spectral(float(i) / n_clusters)
        ax1.fill_betweenx(np.arange(y_lower, y_upper),
                          0, ith_cluster_silhouette_values,
                          facecolor=color, edgecolor=color, alpha=0.7)

        ax1.text(-0.05, y_lower + 0.5 * size_cluster_i, str(i))

        y_lower = y_upper + 10  # 10 for the 0 samples

    ax1.set_title("The silhouette plot for the various clusters.")
    ax1.set_xlabel("The silhouette coefficient values")
    ax1.set_ylabel("Cluster label")

    ax1.axvline(x=silhouette_avg, color="red", linestyle="--")

    ax1.set_yticks([])  # Clear the yaxis labels / ticks
    ax1.set_xticks([-0.1, 0, 0.2, 0.4, 0.6, 0.8, 1])

    colors = cm.nipy_spectral(cluster_labels.astype(float) / n_clusters)
    ax2.scatter(X[:, 0], X[:, 1], marker='.', s=30, lw=0, alpha=0.7,
                c=colors, edgecolor='k')
```

```
    '''# Labeling the clusters
    centers = clusterer.cluster_centers_
    # Draw white circles at cluster centers
    ax2.scatter(centers[:, 0], centers[:, 1], marker='o',
                c="white", alpha=1, s=200, edgecolor='k')

    for i, c in enumerate(centers):
        ax2.scatter(c[0], c[1], marker='$%d$' % i, alpha=1,
                    s=50, edgecolor='k') '''

    ax2.set_title("The visualization of the clustered data.")
    ax2.set_xlabel("Feature space for the 1st feature")
    ax2.set_ylabel("Feature space for the 2nd feature")

    plt.suptitle(("Silhouette analysis for KMeans clustering on sample data "
                  "with n_clusters = %d" % n_clusters),
                 fontsize=14, fontweight='bold')

plt.show()
```

```
For n_clusters = 2 The average silhouette_score is : 0.7237647056094885
For n_clusters = 3 The average silhouette_score is : 0.5953875977051863
For n_clusters = 4 The average silhouette_score is : 0.5334878502078647
For n_clusters = 5 The average silhouette_score is : 0.5511675243934873
For n_clusters = 6 The average silhouette_score is : 0.5053391540138767
For n_clusters = 7 The average silhouette_score is : 0.4862022318993492
For n_clusters = 8 The average silhouette_score is : 0.4951367315588422
For n_clusters = 9 The average silhouette_score is : 0.4637781703297825
For n_clusters = 10 The average silhouette_score is : 0.4906067627985123
```

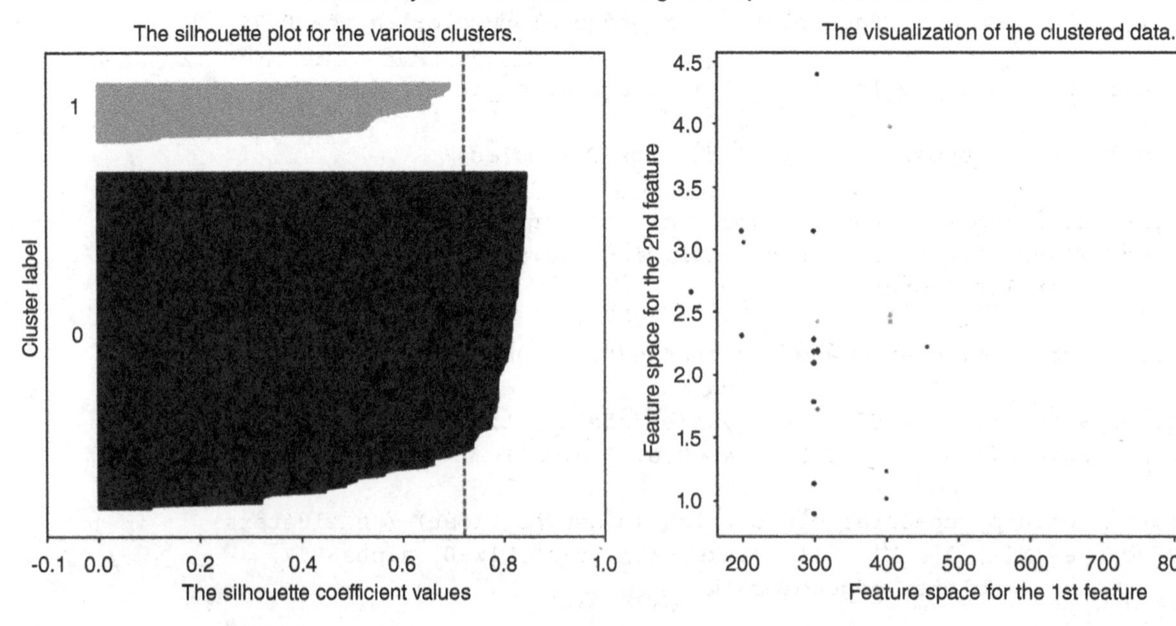

Silhouette analysis for KMeans clustering on sample data with n_clusters = 3

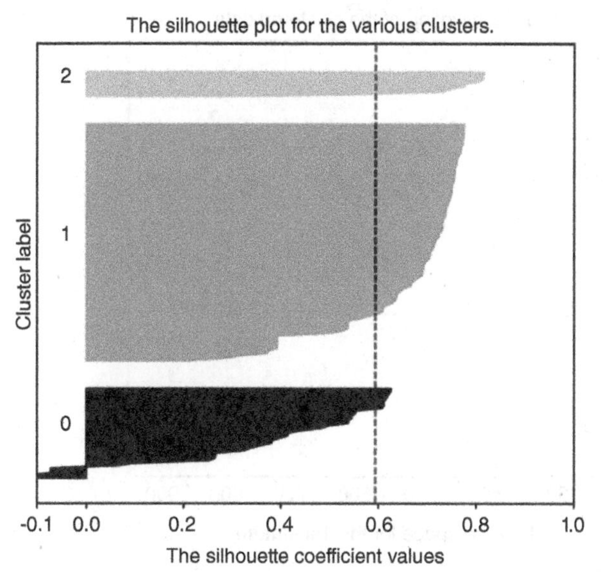

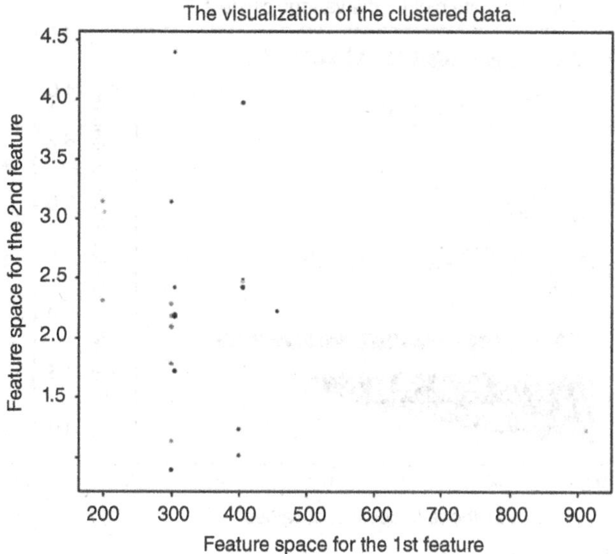

Silhouette analysis for KMeans clustering on sample data with n_clusters = 4

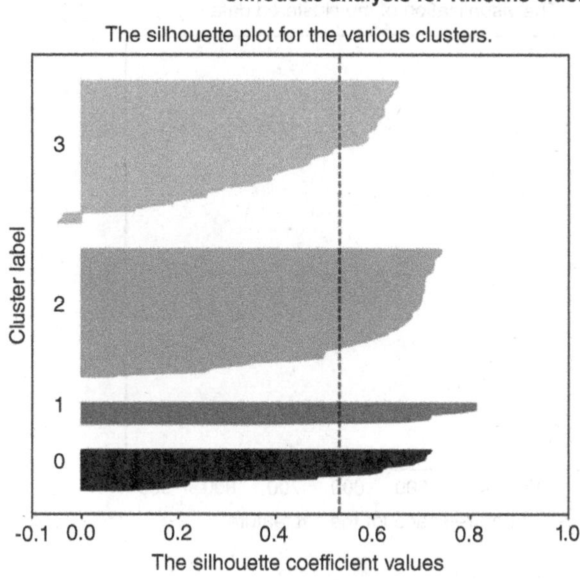

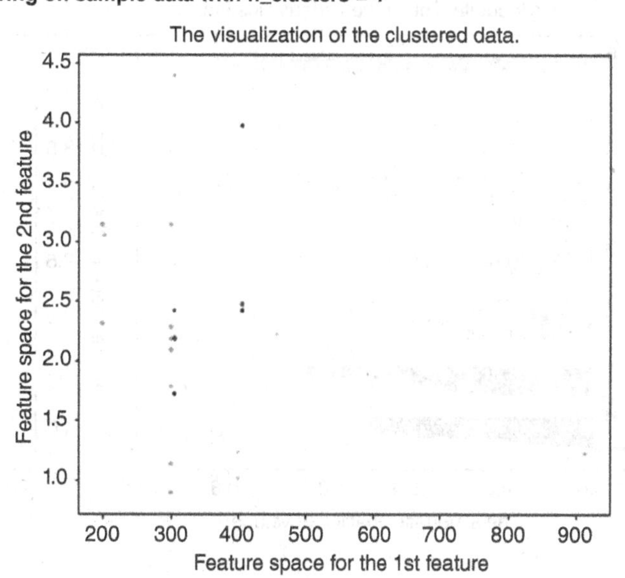

Silhouette analysis for KMeans clustering on sample data with n_clusters = 5

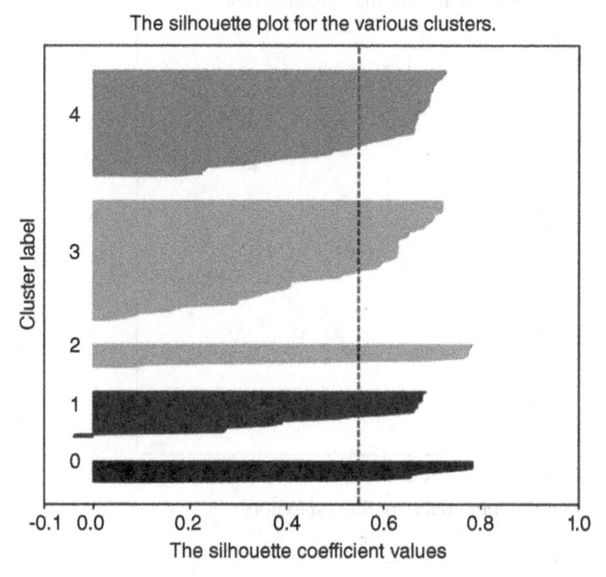

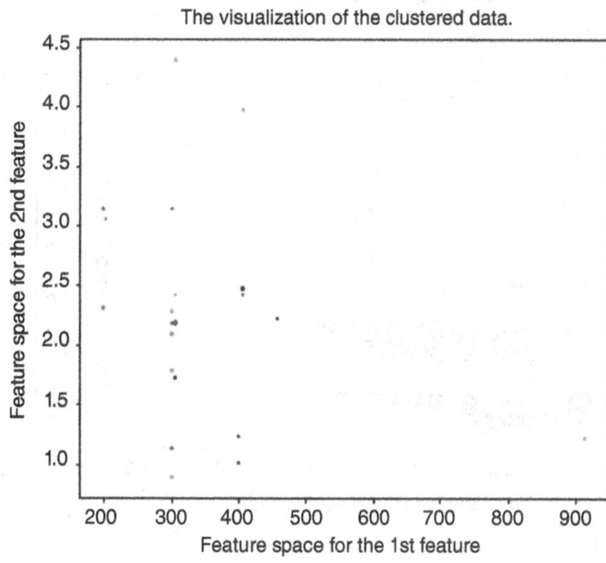

Silhouette analysis for KMeans clustering on sample data with n_clusters = 6

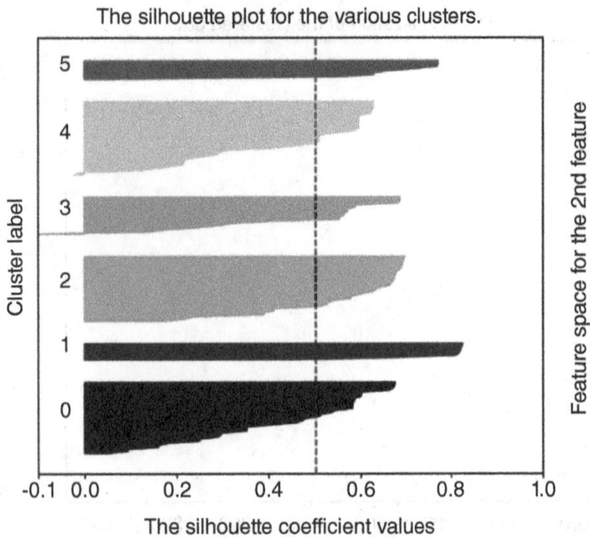

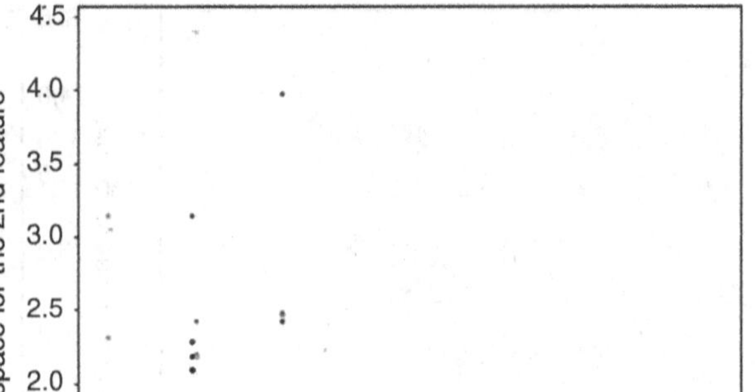

Silhouette analysis for KMeans clustering on sample data with n_clusters = 7

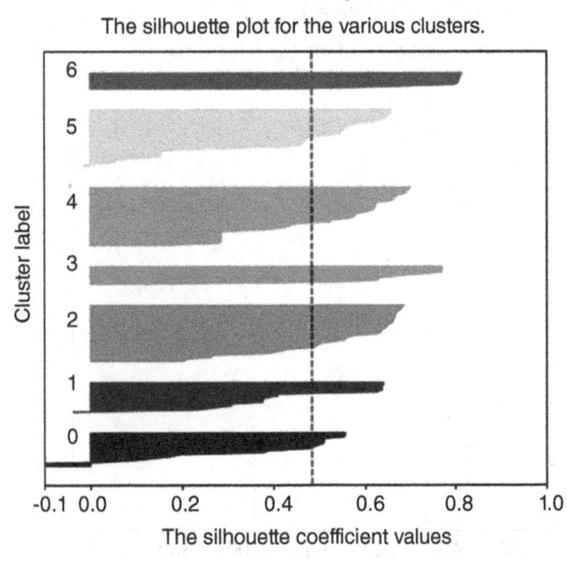

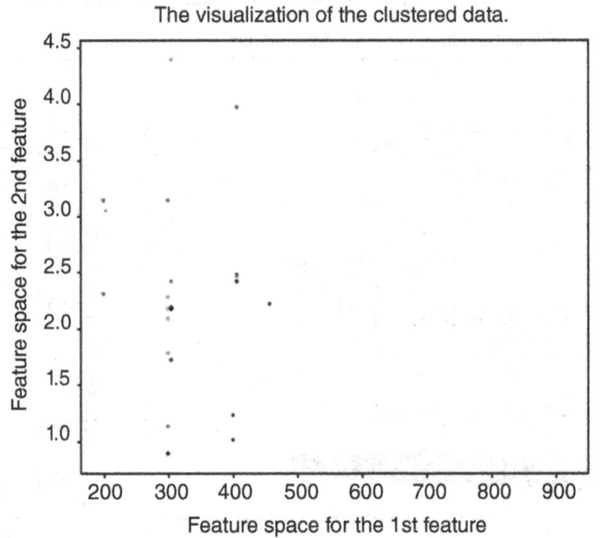

Silhouette analysis for KMeans clustering on sample data with n_clusters = 8

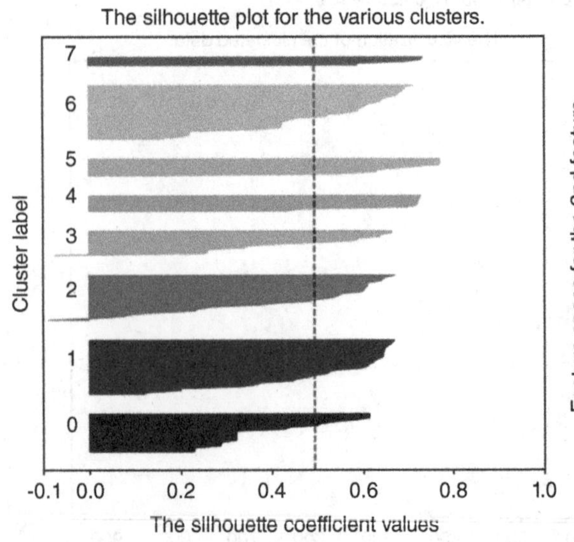

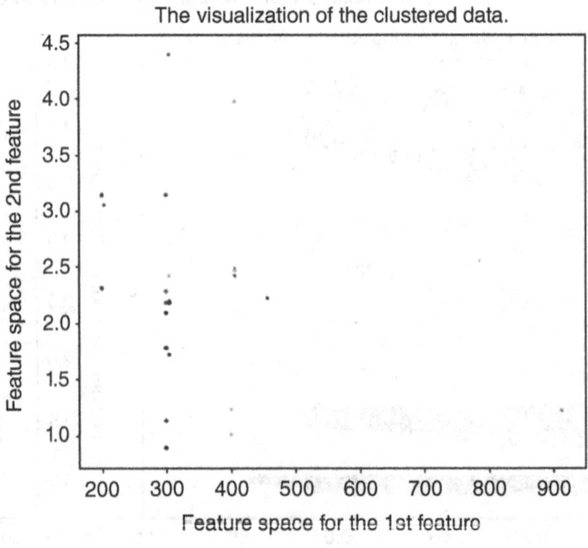

Silhouette analysis for KMeans clustering on sample data with n_clusters = 9

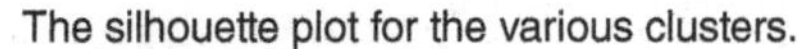

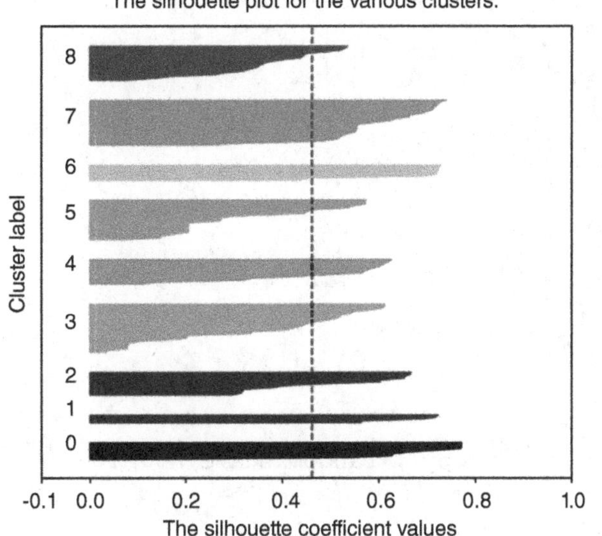

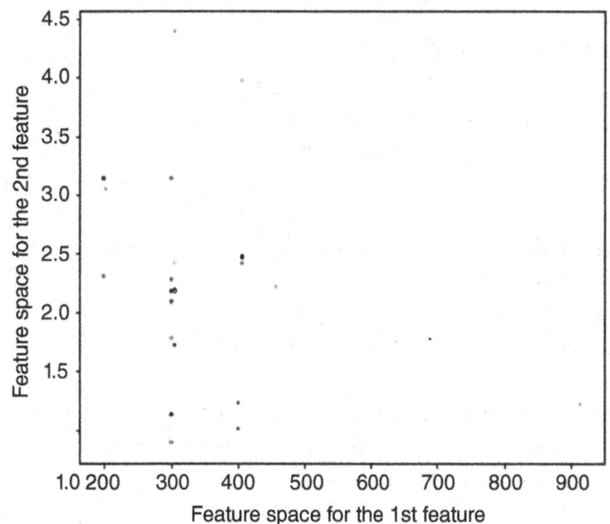

Silhouette analysis for KMeans clustering on sample data with n_clusters = 10

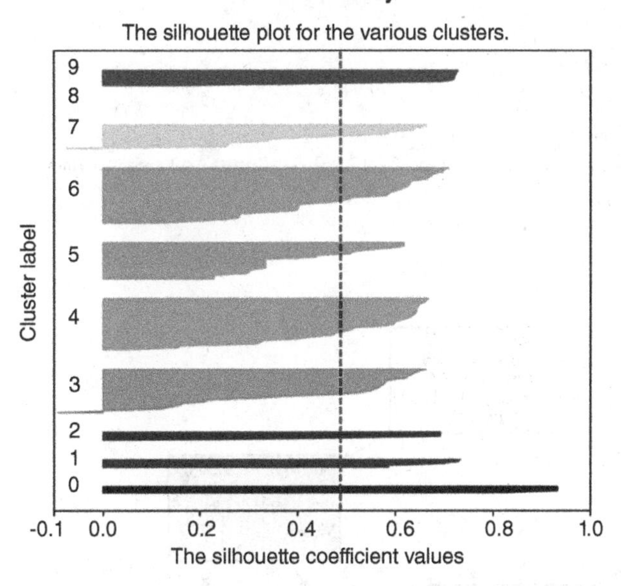

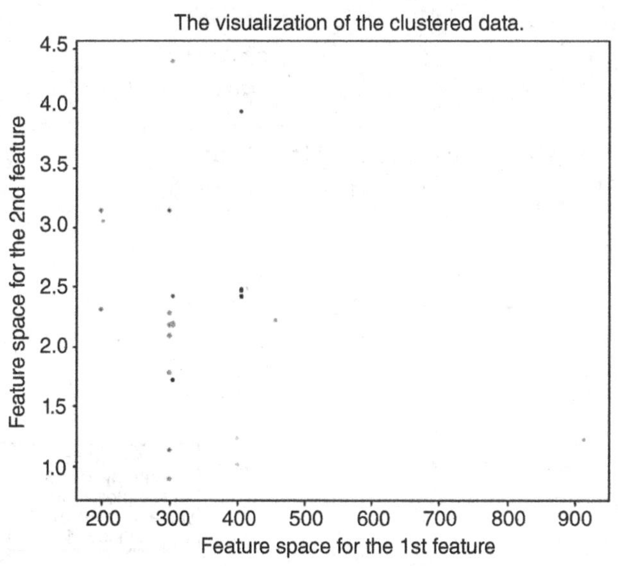


```
fcm = FCM(n_clusters=4)
fcm.fit(X)
fcm_centers = fcm.centers
y = fcm.predict(X)
mapping = {0:1, 1:2, 2:3, 3:4}
y = [mapping[i] for i in y]
'''y2=pd.DataFrame(data=y)
y2.to_excel('y.xlsx', index=False)'''
```

```
import plotly.express as px
clusters=pd.DataFrame(X,columns=dataset.columns)
clusters['cluster']=y
polar=clusters.groupby("cluster").mean().reset_index()
polar=pd.melt(polar,id_vars=["cluster"])
```

```
fig4 = px.line_polar(polar, r="value", theta="variable", color="cluster", line_
close=True,height=600)
fig4.update_layout(
    polar = dict(
      radialaxis_tickfont_size = 21,
      angularaxis = dict(
        tickfont_size=26,
      )))

fig4.update_layout(legend_font_size=26)
fig4.show()

X_normal = dataset.iloc[:, :9].values

from sklearn.tree import DecisionTreeClassifier
classifier = DecisionTreeClassifier(criterion = 'entropy', random_state = 0)
classifier.fit(X_normal, y)
```

```
from sklearn import tree
fig, axes = plt.subplots(nrows = 1,
                        ncols = 1,
                        figsize = (17,11),
                        dpi=100)
tree.plot_tree(classifier,feature_names = dataset.columns,filled = True,class_names
=['1','2','3','4'],fontsize=12.5);
```

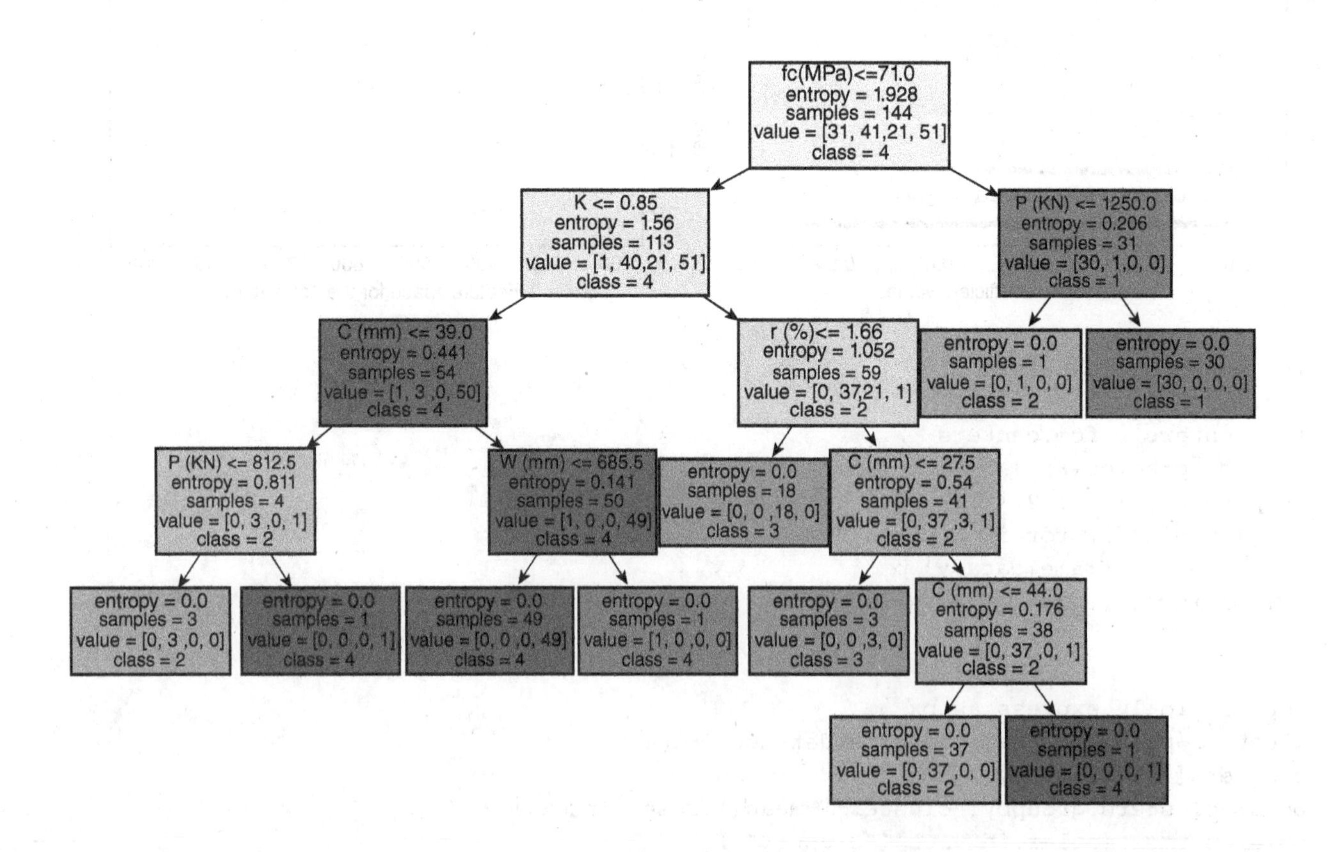

```
text_representation = tree.export_text(classifier)
print(text_representation)
```

```
|--- feature_3 <= 71.00
|   |--- feature_5 <= 0.85
|   |   |--- feature_6 <= 39.00
|   |   |   |--- feature_8 <= 812.50
|   |   |   |   |--- class: 1
|   |   |   |--- feature_8 >  812.50
|   |   |   |   |--- class: 4
|   |   |--- feature_6 >  39.00
|   |   |   |--- feature_0 <= 685.50
|   |   |   |   |--- class: 4
|   |   |   |--- feature_0 >  685.50
|   |   |   |   |--- class: 3
|   |--- feature_5 >  0.85
|   |   |--- feature_1 <= 1.66
|   |   |   |--- class: 2
|   |   |--- feature_1 >  1.66
|   |   |   |--- feature_6 <= 27.50
|   |   |   |   |--- class: 2
|   |   |   |--- feature_6 >  27.50
|   |   |   |   |--- feature_6 <= 44.00
|   |   |   |   |   |--- class: 1
|   |   |   |   |--- feature_6 >  44.00
|   |   |   |   |   |--- class: 4
|--- feature_3 >  71.00
|   |--- feature_8 <= 1250.00
|   |   |--- class: 1
|   |--- feature_8 >  1250.00
|   |   |--- class: 3
```

```
X = dataset.iloc[:, :9].values
y=np.array(y)
In [ ]:
plt.scatter(X[y == 1, 3], X[y == 1, 8], s = 100, c = 'red', label = 'Cluster 1')
plt.scatter(X[y == 2, 3], X[y == 2, 8], s = 100, c = 'blue', label = 'Cluster 2')
plt.scatter(X[y == 3, 3], X[y == 3, 8], s = 100, c = 'green', label = 'Cluster 3')
plt.scatter(X[y == 4, 3], X[y == 4, 8], s = 100, c = 'cyan', label = 'Cluster 4')
plt.title('Clusters of RC columns')
plt.xlabel('fc')
plt.ylabel('P')
plt.legend()
plt.show()
```

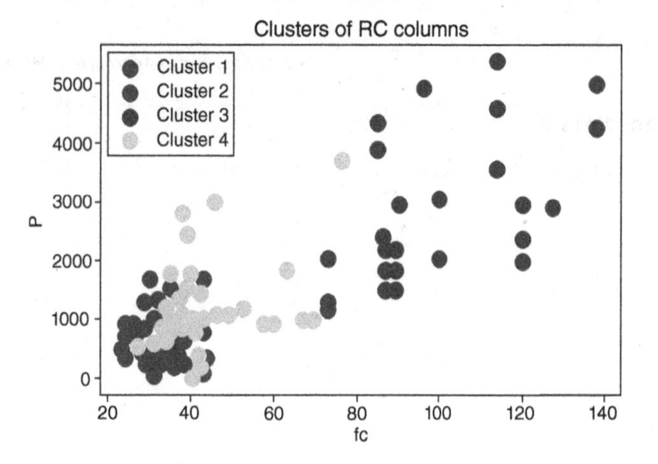


```
#DBSCAN algorithm
import pandas as pd
import numpy as np
import matplotlib.pyplot as plt
from sklearn.preprocessing import StandardScaler
```

```
dataset = pd.read_excel('Fire-exposed RC columns.xlsx')
X = dataset.iloc[:, :9].values
```

```
from sklearn.preprocessing import StandardScaler
sc = StandardScaler()
X = sc.fit_transform(X)

from sklearn.cluster import DBSCAN
dbscan = DBSCAN(eps=2, min_samples=3)
y = dbscan.fit_predict(X)
np.unique(y)
```

```
array([-1,  0,  1,  2,  3])
```

```
mapping = {0:1, 1:2, 2:3, 3:4, -1:'noisy'}
y_r = [mapping[i] for i in y]
'''y2=pd.DataFrame(data=y_r)
y2.to_excel('y.xlsx', index=False)'''
```

```
"y2=pd.DataFrame(data=y_r)\ny2.to_excel('y.xlsx', index=False)"
```

```
import plotly.express as px
clusters=pd.DataFrame(X,columns=dataset.columns)
clusters['cluster']=y_r
polar=clusters.groupby("cluster").mean().reset_index()
polar=pd.melt(polar,id_vars=["cluster"])
fig4 = px.line_polar(polar, r="value", theta="variable", color="cluster", line_
close=True,height=600)
fig4.update_layout(
    polar = dict(
      radialaxis_tickfont_size = 21,
```

```
    angularaxis = dict(
      tickfont_size=26,
    )))

fig4.update_layout(legend_font_size=26)
fig4.show()
```

```
X_normal = dataset.iloc[:, :9].values
```

```
from sklearn.tree import DecisionTreeClassifier
classifier = DecisionTreeClassifier(criterion = 'entropy', random_state = 0)
classifier.fit(X_normal, y_r)
DecisionTreeClassifier(criterion='entropy', random_state=0)
from sklearn import tree
fig, axes = plt.subplots(nrows = 1, ncols = 1, figsize = (17,19), dpi=100)
tree.plot_tree(classifier, feature_names = dataset.columns, filled = True, class_
name s=['1','2','3','4','noisy'],fontsize=13);
```

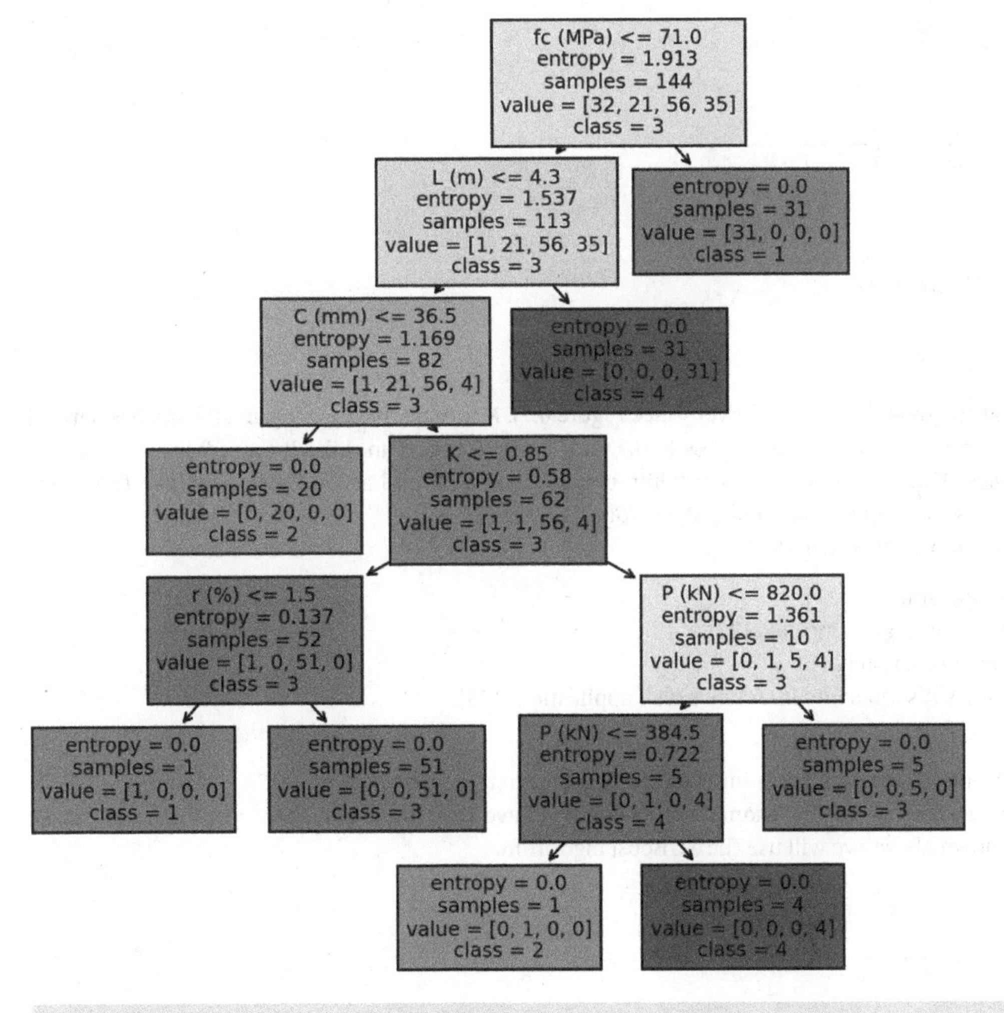


```
X = dataset.iloc[:, :9].values
```

```
plt.scatter(X[y == 0, 3], X[y == 0, 8], s = 100, c = 'red', label = 'Cluster 1')
plt.scatter(X[y == 1, 3], X[y == 1, 8], s = 100, c = 'blue', label = 'Cluster 2')
```

```
plt.scatter(X[y == 2, 3], X[y == 2, 8], s = 100, c = 'green', label = 'Cluster 3')
plt.scatter(X[y == 3, 3], X[y == 3, 8], s = 100, c = 'cyan', label = 'Cluster 4')
#plt.scatter(X[y == 'noisy', 3], X[y == 'noisy', 8], s = 100, c = 'black', label =
'Cluster 5')
plt.title('Clusters of RC columns')
plt.xlabel('fc')
plt.ylabel('P')
plt.legend()
plt.show()
```

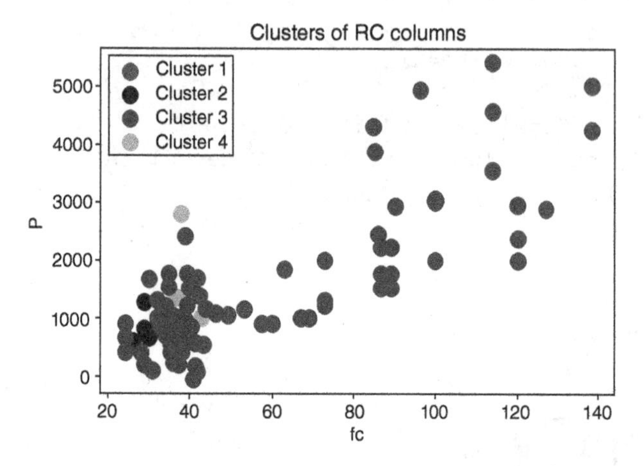

This concludes our example using *Python*.

6.2.2 R

R is the second programming language that we will cover[84] (see Figure 6.7). *R* was created in 1992, and its name is derived from the first letter of the names of its two creators, Ross Ihaka and Robert Gentleman. Like *Python*, *R* is also an open-source programming language. Many of the packages and libraries in *R* can be found at the Comprehensive *R* Archive Network (CRAN).[85] *R* can be used in *Jupyter Notebook* [20] or *RStudio* [21].

Some of the commonly used R packages include:

- *ggplot2*: A package for creating graphics [22].
- *tidyr*: A package for cleaning and organizing data [23].
- *dplyr*: A package for data manipulation [24].
- *Shiny*: A package that converts R scripts into interactive web applications [25].

Example no. 1 [R, Classification]: Predicting fire-induced spalling of concrete mixtures.
We will use the same dataset we used in the first example of *Python*. We have 16 features and one target (with two labels: *no spalling* or *spalling*). As shown above, we will use the XGBoost algorithm.

```
#Installing required packages
>library(modelStudio)
> library(dplyr)
```

84 https://www.r-project.org.
85 https://cran.r-project.org/index.html.

The R Project for Statistical Computing

[Home]

Download

CRAN

R Project

About R
Logo
Contributors
What's New?
Reporting Bugs
Conferences
Search
Get Involved: Mailing Lists
Get Involved: Contributing
Developer Pages
R Blog

R Foundation

Foundation
Board
Members
Donors
Donate

Help With R

Getting Help

Documentation

Manuals
FAQs
The R Journal
Books
Certification
Other

Links

Bioconductor
R-Forge

Getting Started

R is a free software environment for statistical computing and graphics. It compiles and runs on a wide variety of UNIX platforms, Windows and MacOS. To **download R**, please choose your preferred CRAN mirror.

If you have questions about R like how to download and install the software, or what the license terms are, please read our answers to frequently asked questions before you send an email.

News

- **R version 4.2.1 (Funny-Looking Kid)** has been released on 2022-06-23.
- **R version 4.2.0 (Vigorous Calisthenics)** has been released on 2022-04-22.
- **R version 4.1.3 (One Push-Up)** was released on 2022-03-10.
- Thanks to the organisers of useR! 2020 for a successful online conference. Recorded tutorials and talks from the conference are available on the R Consortium YouTube channel.
- You can support the R Foundation with a renewable subscription as a supporting member

News via Twitter

The R Foundation
@_R_Foundation
The #RStats blog has moved to blog.r-project.org
More details from Tomas Kalibera: blog.r-project.org/2022/06/30/mov...
The R Blog
blog.r-project.org
Jun 30, 2022
The R Foundation
@_R_Foundation
New #RStats blog post from Tomas Kalibera: Why to avoid \x in regular expressions.developer.r-project.org/Blog/public/20...

Figure 6.7 Snippet of the *R* website (July 2022).

```
> library(xgboost)
> library(DALEX)
> library(tidyverse)
> library(tidymodels)
> library(caret)
> library(magrittr)
> library(Matrix)
> library(corrplot)
```

```
#Importing data
```

```
> data=read.csv('C:/Users/Classification.csv')
```

```
#Checking data structure
```

```
> str(data)
'data.frame': 646 obs. of 11 variables:
$ Water.binder.ratio…. : num 0.273 0.273 0.3 0.218 0.218 0.3 0.3 0.3 0.273 0.273 …
$ Aggregate.binder.ratio…. : num 2 2 2 1.37 1.37 …
$ Sand.binder.ratio…. : num 1.05 1.05 1.03 1.2 1.2 …
$ Heating.rate…C.min. : num 101 101 101 101 101 101 101 101 101 101 …
$ Maximum.exposure.temperature…C.: int 1034 1034 1034 1034 1034 1034 1034 1034 1034
1034 …
$ Silica.fume.binder.ratio…. : num 0 0 0 0 0 0 0 0 0 0 …
$ Aggregate.size…mm. : num 7 14 20 7 7 14 20 20 7 14 …
```

```
$ GGBS.binder.ratio.... : num 0.25 0.25 0.25 0.25 0.25 0.25 0.25 0.25 0.25 0.25 ...
$ FA.binder.ratio.... : num 0 0 0 0 0 0 0 0 0 0 ...
$ PP.fiber.quantity...kg.m3. : num 0 0 0 0 0 0 0 0 0 0 ...
$ Output : int 1 1 1 1 1 1 1 1 1 1 ...
```

```
#Assiging train-Test split
> smp_size <- floor(0.7 * nrow(data))
```

```
#Data partitioning and setting the seed to make the partitioning reproducible
> set.seed(123)
> train_ind <- sample(seq_len(nrow(data)), size = smp_size)
> trainxg <- data[train_ind, ]
> testxg<- data[-train_ind, ]
#isolate X and Y for train set
> ytrainxg<-trainxg$Output
> xtrainxg<- trainxg %>% select(-Output)
```

```
#Isolating X and Y for test set
> ytestxg<-testxg$Output
> xtestxg<-testxg %>% select(-Output)
```

```
#Plotting histogram for all variables
> names<-names(data)
> classes<-sapply(data,class)
> for(name in names[classes == 'numeric']){
+ dev.new()
+ hist(data[,name])
+ }
```

```
#Plotting histograms for individual variables
> hist(data$Water.binder.ratio....)
```

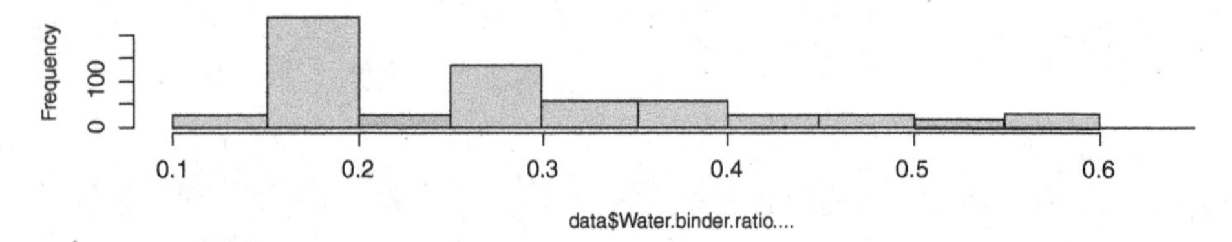


```
> hist(data$Aggregate.binder.ratio....)
```

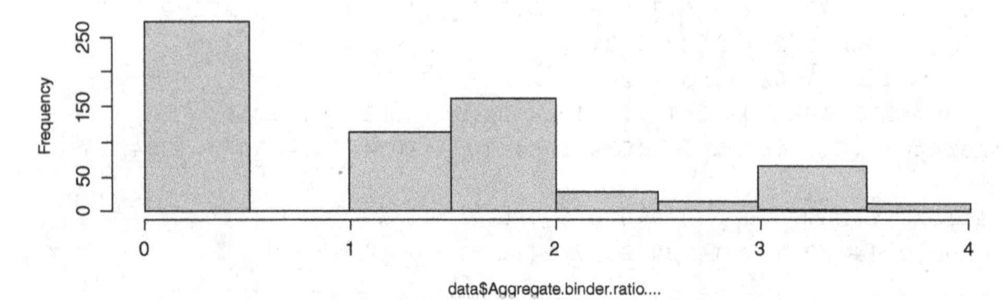

```
> hist(data$Sand.binder.ratio….)
```

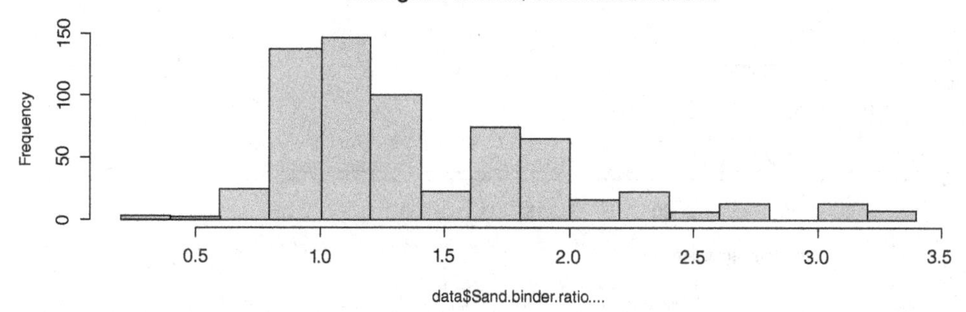


```
> hist(data$Heating.rate…C.min.)
```

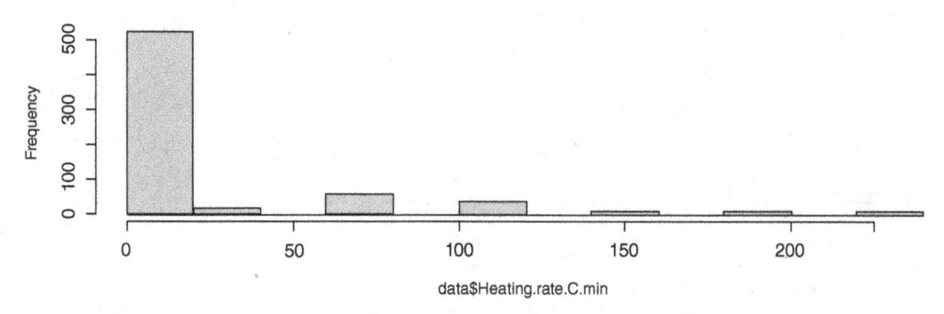


```
> hist(data$Maximum.exposure.temperature…C.)
```

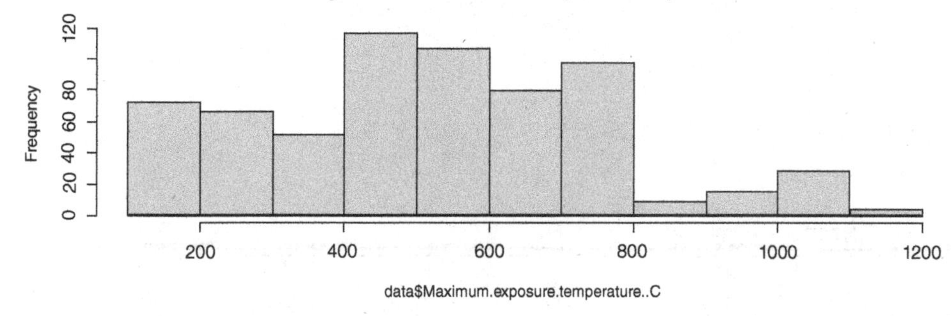


```
> hist(data$Silica.fume.binder.ratio….)
```

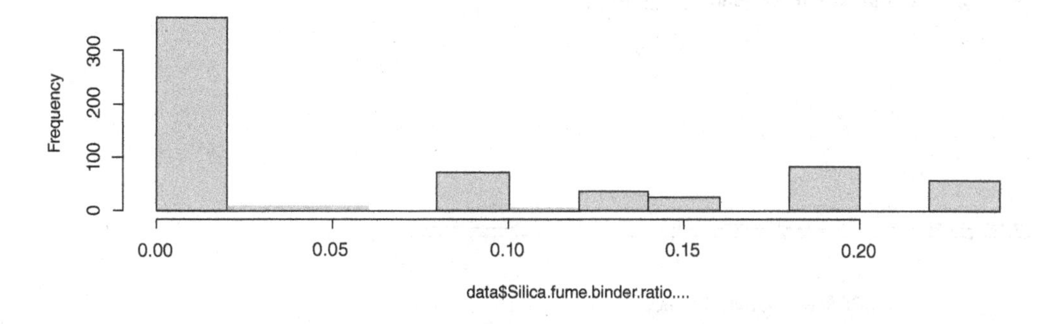

```
> hist(data$Aggregate.size..mm.)
```

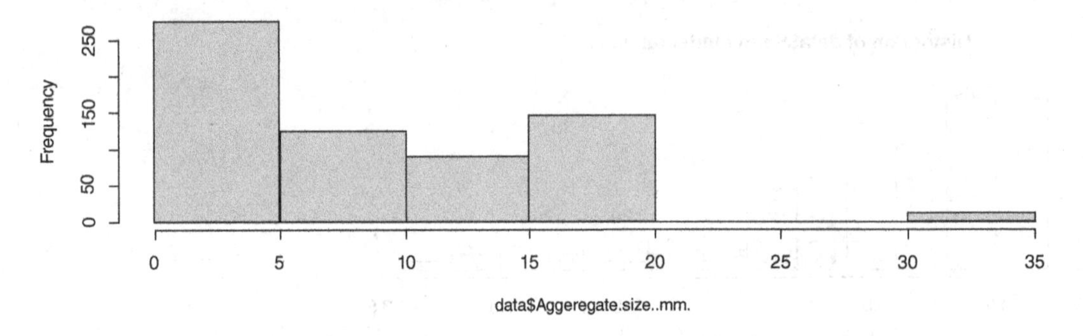


```
> hist(data$GGBS.binder.ratio....)
```

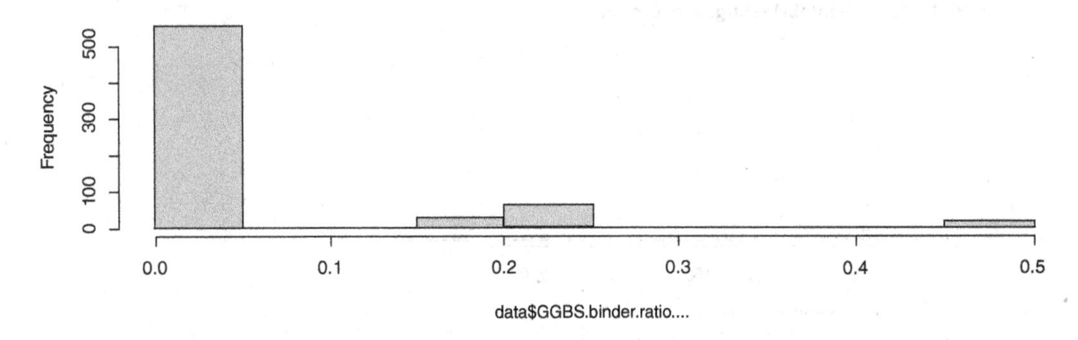


```
> hist(data$FA.binder.ratio....)
```

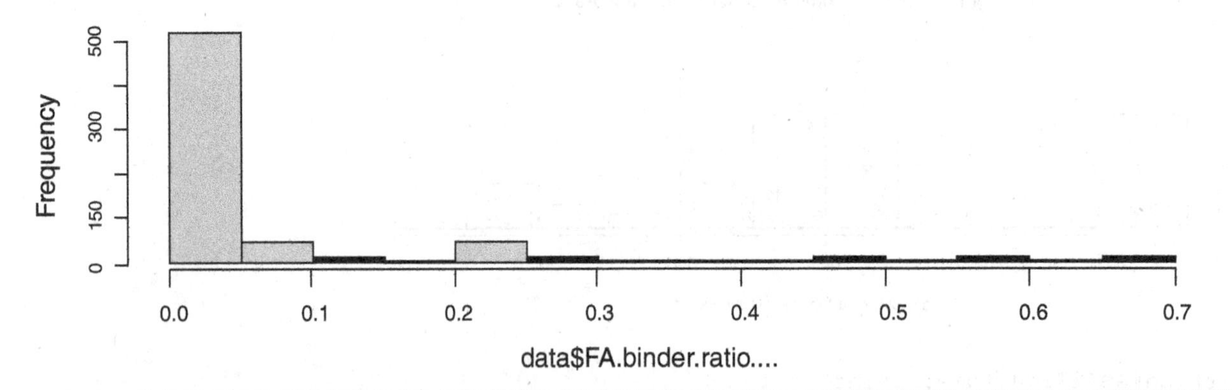


```
> hist(data$PP.fiber.quantity..kg.m3.)
```

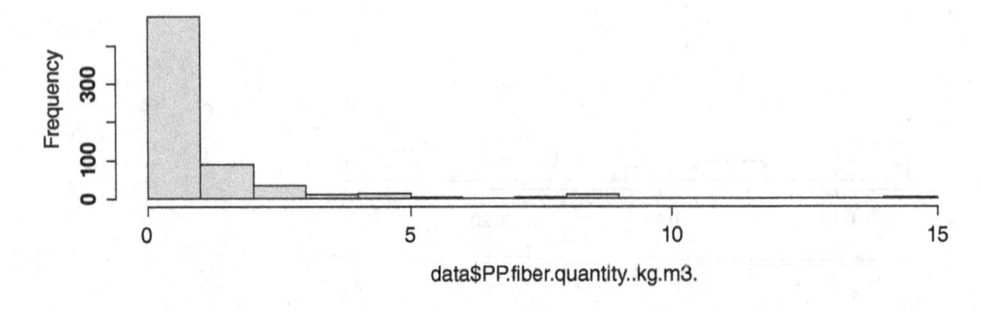

```
> hist(data$Output)
```

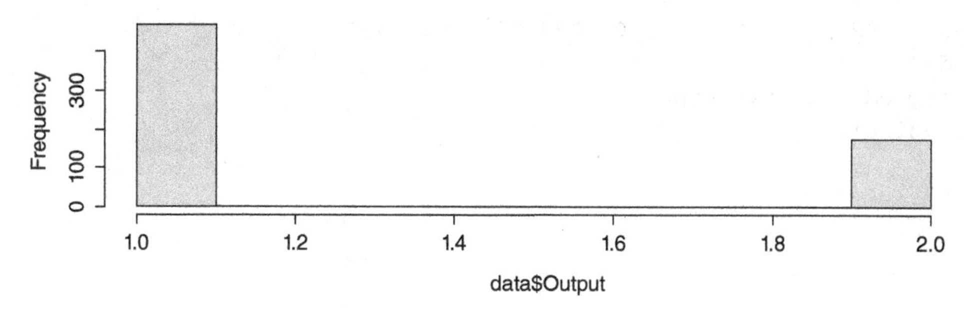

#Setting the parameters for the XG.boost algorithm

```
> params<-list(set.seed=1500,
+ eval_metric="auc",
+ objective="binary:logistic")
```

#Running the XG.boost algorithm

```
> modelxg <-xgboost(data = as.matrix(xtrainxg),
+ label = ytrainxg,
+ params = params,
+ nrounds = 20,
+ verbose = 1)
```

```
[1] train-auc:0.937105
[2] train-auc:0.972061
[3] train-auc:0.970575
[4] train-auc:0.977379
[5] train-auc:0.980787
[6] train-auc:0.986405
[7] train-auc:0.987741
[8] train-auc:0.989051
[9] train-auc:0.989326
[10] train-auc:0.990075
[11] train-auc:0.990762
[12] train-auc:0.990887
[13] train-auc:0.991049
[14] train-auc:0.991149
[15] train-auc:0.991898
[16] train-auc:0.992222
[17] train-auc:0.992422
[18] train-auc:0.992622
[19] train-auc:0.993246
[20] train-auc:0.993046
```

#Checking model accuracy using the confusion matrix metric (sensitivity, specificity, and accuracy)

```
> trained<-model.matrix(trainxg$Output~.-1,data = trainxg)
> trainlabel<-trainxg[,1]
> trainlabel<-as.numeric(trainlable)-1
> tested<-model.matrix(testxg$Output~.-1,data = testxg)
> testlabel<-testxg[,1]
```

```
> testlabel<-as.numeric(testlabel)-1
> pred<-predict(modelxg,tested)
> head(pred)
[1] 0.9829579 0.9880555 0.9820659 0.9829579 0.9829579 0.9820659
> pred<-as.integer(pred>0.5)
> confusionMatrix(table(pred, testxg$Output))
Confusion Matrix and Statistics
```

```
pred 0 1
0 123 8
1 17 46
Accuracy : 0.8711
95% CI : (0.8157, 0.9148)
No Information Rate : 0.7216
P-Value [Acc > NIR] : 4.745e-07
Kappa : 0.6949
Mcnemar's Test P-Value : 0.1096
Sensitivity : 0.8786
Specificity : 0.8519
Pos Pred Value : 0.9389
Neg Pred Value : 0.7302
Prevalence : 0.7216
Detection Rate : 0.6340
Detection Prevalence : 0.6753
Balanced Accuracy : 0.8652
'Positive' Class : 0
```

Example no. 2 [R, Regression]: Predicting axial strength of CFSTs.
We continue the second example of *DataRobot* and *Python* here. We are now trying to predict the axial capacity of concentrically loaded CFSTs. Remember that our dataset has 1260 CFSTs with the following features: column diameter, D (mm), column thickness, t (mm), column length, L (mm), compressive strength of concrete f_c' (MPa), and steel yielding stress, f_y (MPa). The axial compressive strength N (kN) comprises the target for each column. We are still using the Random Forest algorithm.

The *R* code is provided below.

```
#Installing required packages for the Random forest algorithm
library(ggplot2)
library(cowplot)
library(randomForest)
library(rpart)
library(randomForest)
library(rfUtilities)
library(pdp)
library(corrplot)
```

```
#Importing dataset
data=read.csv('C:/Users/(CFST).csv')
```

```
#Checking the structures of the dataset
str(data)
'data.frame':   1260 obs. of 6 variables:
$ D..mm.   : num 169 169 169 169 168 …
$ t..mm.   : num 2.64 2.64 2.62 2.62 3.6 3.6 3.6 5 5 5 …
$ L..mm.   : num 305 305 305 305 305 305 305 305 305 305 …
```

```
$ fy..MPa.  : num 302 302 338 338 288 …
$ fc..Mpa.  : num 18.2 34.7 37.1 34.1 27 33.3 33.3 33.4 33.4 27.9 …
$ N.Test…kN.: num 1326 1219 1308 1330 1557 …
...
```

#Displaying the first six rows of the dataset

```
head(data)
D..mm. t..mm. L..mm. fy..MPa. fc..Mpa. N.Test..kN.
```

```
1 168.8     2.64     305     302.4     18.2     1326
2 168.8     2.64     305     302.4     34.7     1219
3 169.3     2.62     305     338.1     37.1     1308
4 169.3     2.62     305     338.1     34.1     1330
5 168.3     3.60     305     288.4     27.0     1557

6 168.3 3.60  305  288.4  33.3
summary(data)
D..mm.     t..mm.     1432
L..mm.
fy..MPa.
Min. : 44.45   Min. : 0.520     Min. : 152.3     Min. :178.3
1st Qu.: 108.00  1st Qu.: 2.703     1st Qu.: 399.0     1st Qu.:274.9
Median : 129.00  Median : 4.000     Median : 662.0     Median :324.7
Mean : 158.83  Mean : 4.275     Mean :1055.1     Mean :335.7
3rd Qu.: 190.00  3rd Qu.: 5.000     3rd Qu.:1200.0     3rd Qu.:363.3
Max. :1020.00  Max. :16.540     Max. :5560.0     Max. :853.0
fc…Mpa.     N.Test…kN.
Min. : 7.59    Min. : 45.2
1st Qu.: 29.15   1st Qu.: 710.8
Median : 37.40  Median : 1273.0
Mean : 47.50    Mean : 2349.7
3rd Qu.: 51.08  3rd Qu.: 2273.2
Max. :185.94    Max. :46000.0
```

#Selecting a train-test split

```
smp_size <- floor(0.8 * nrow(data))
```

#Partioning data and setting the seeds for reproducibility

```
set.seed(123)
train_ind <- sample(seq_len(nrow(data)), size = smp_size)
trainrf <- data[train_ind, ]
testrf <- data[-train_ind, ]
```

#Fitting the algorithm

```
set.seed(187)
rf.mod<-randomForest(N.Test…kN.~.,ntree=100,mtry=5, data=trainrf)
plot(rf.mod)
```

```
#Setting actual and predicted vectors
```

```
predicted<-rf.mod$predicted
actual<-rf.mod$y
```

```
#Appying error metrics (MAE, RMSE, MSE)
```

```
Metrics::mae(actual,predicted) [1] 221.5164
Metrics::rmse(actual,predicted) [1] 765.9959
Metrics::mse(actual,predicted) [1] 586749.7
```

```
#Selecting functions for model evaluation
```

```
ACC1<-(100-(mean(abs((actual - predicted) / predicted)) * 100))
MAE1<-(mean(abs(actual-predicted)))
MSE1<-(sqrt(mean((actual-predicted)^2)))
ACC1
[1] 91.28963
MAE1
[1] 221.5164
MSE1
[1] 765.9959
```

```
#Plotting histograms for all variables.
```

```
names<-names(data)
classes<-sapply(data,class)
for(name in names[classes == 'numeric']){
+ dev.new()
+ hist(data[,name])
+ }
```

```
#Plotting histograms for individual variables.
```

```
hist(data$N.Test…kN.)
```

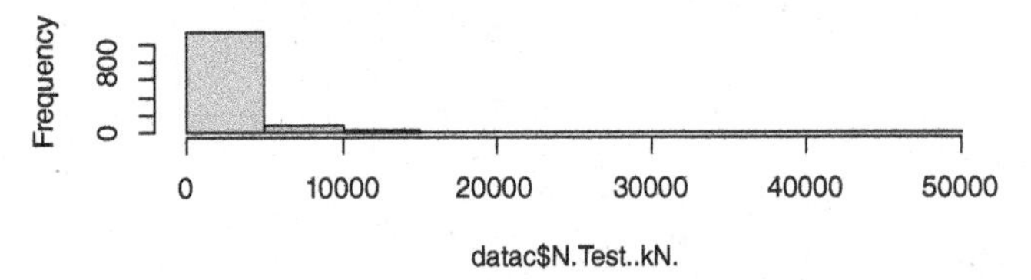

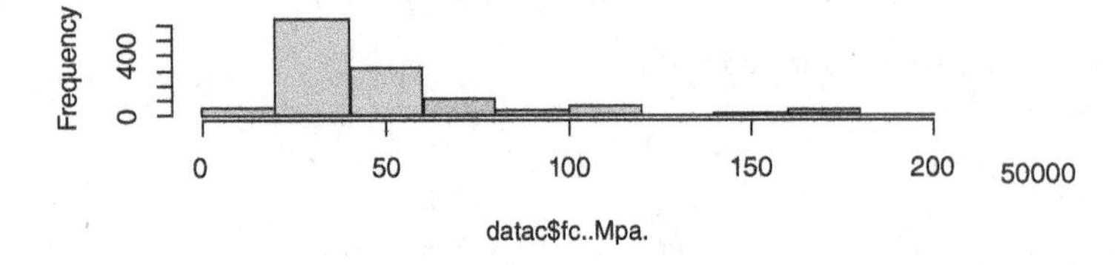

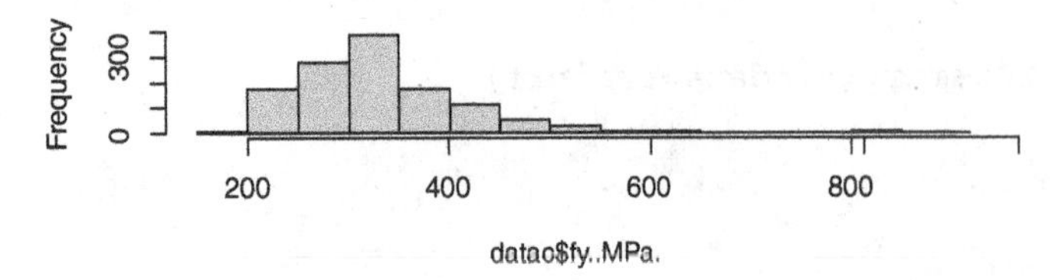

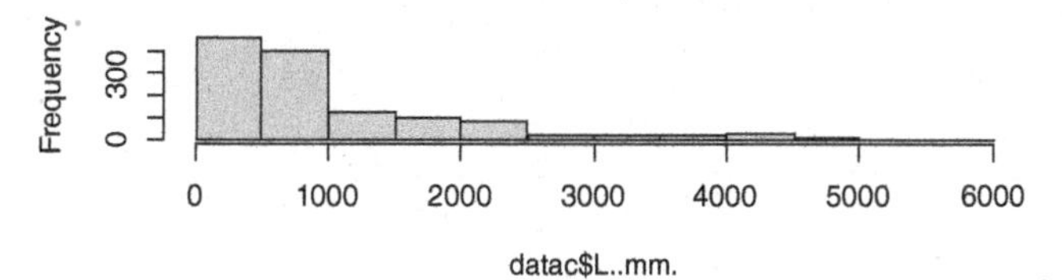

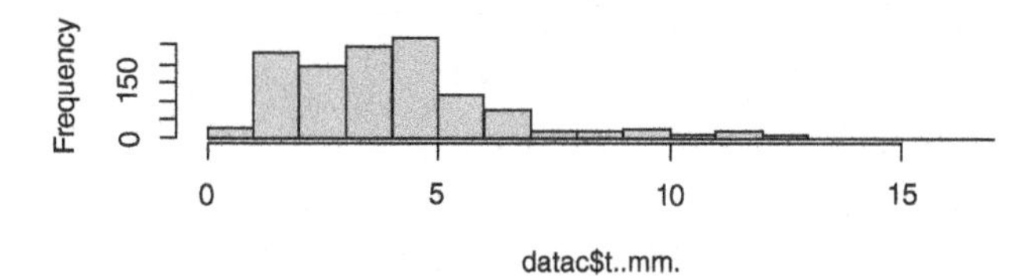

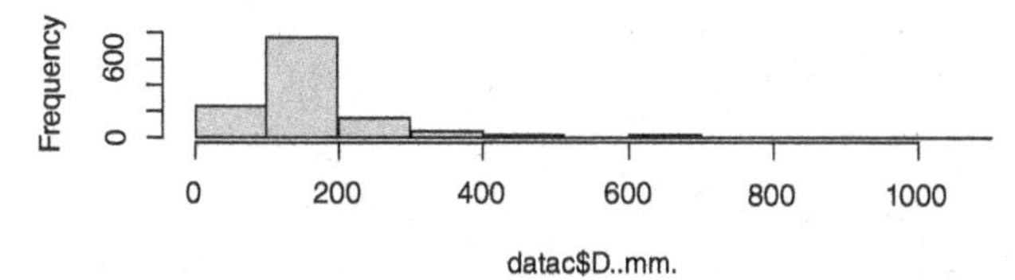


```
#Plotting correlation matrices
#We start with the spearman coefficient86
m=cor(data, method = "spearman")
corrplot(m, method ="circle")
```

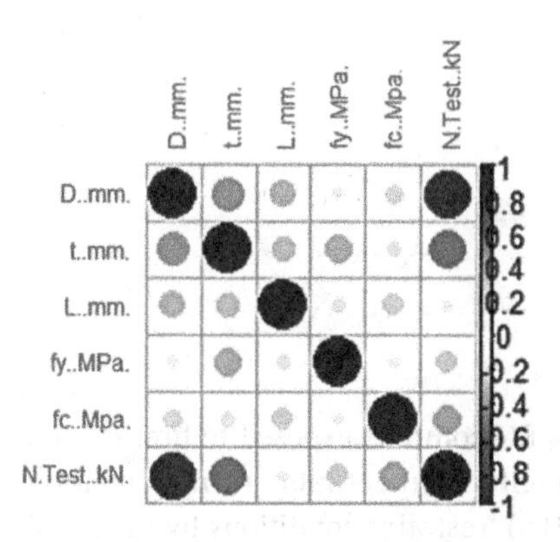


```
#Pearson coefficient
n=cor(data, method ="pearson")
corrplot(n, method ="color",type = "upper",order ="hclust",addCoef.col ="black",
tl.col = "black",tl.srt = 45, diag = FALSE)
```

86As you can see, Arash likes to show different types of charts!

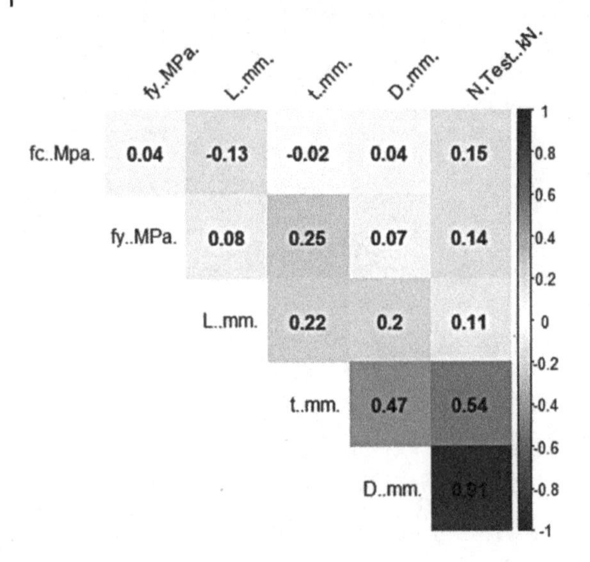


```
#Plotting actuals vs. predictions
plot(x=predicted, y= actual,
+   xlab='Predicted Values',
+   ylab='Actual Values',
+   main='Predicted vs. Actual Values',
+   abline(a=0, b=1))
```

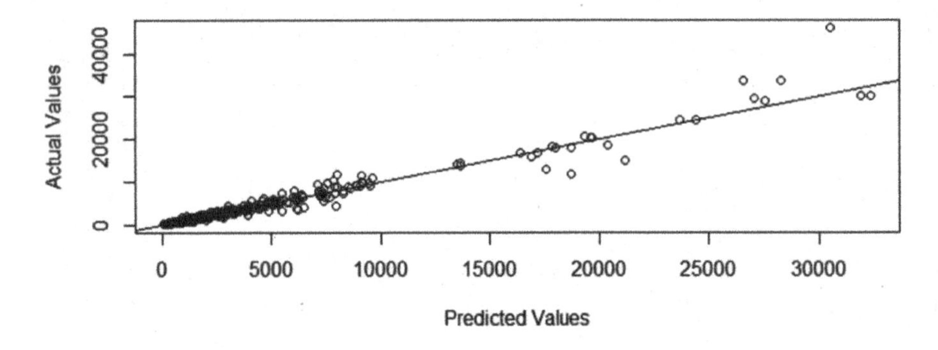

Example no. 3 [*R*, Clustering]: Grouping of fire-exposed RC columns.
The final example of this chapter will re-visit our trusted and tried database of 144 RC columns exposed to fire. As you know by now, the dataset has the following features: column width, W (mm), steel reinforcement ratio, r (mm), column length, L (mm), concrete compressive strength, f_c (MPa), steel yield strength, f_y (MPa), restraint conditions by means of effective length factor, concrete cover to reinforcement, C (mm), eccentricity in applied loading in the main axes, e_x (mm), the magnitude of applied loading, P (kN), and fire resistance, FR (min). We will use the K-Means algorithm in *R*.

The *R* code is provided below.

```
#Reading packages and importing data
install.packages("factoextra")
library(factoextra)
library(ggplot2)
data=read.csv('C:/Fire-exposed RC columns.csv')
data
str(data)
```

```
> data=read.csv('C:/Fire-exposed RC columns.csv')
> str(data)
```

```
'data.frame':144 obs. of  10 variables:
 $ W…mm.                : int  400 300 300 300 300 300 300 300 400 300 …
 $ r….                  : num  1.23 1.13 1.13 0.89 0.89 2.18 2.18 2.18 1.01 1.13 …
 $ L…m.                 : num  3.9 3.9 3.9 3.9 2.1 3.9 2.1 3.9 3.9 2.1 …
 $ fc…MPa.              : int  30 31 33 34 29 29 26 35 30 27 …
 $ fy…MPa.              : int  576 493 493 576 576 576 591 576 576 493 …
 $ K                    : num  1 1 1 1 1 1 1 1 1 1 …
 $ C…mm.                : int  25 25 35 25 25 40 25 25 25 35 …
 $ ex…mm.               : int  20 20 20 0 0 20 0 20 20 0 …
 $ P…kN.                : int  1680 300 283 950 1270 425 878 304 1650 620 …
 $ Fire.Resistance…min. : int  34 60 60 61 63 66 69 69 93 97 …
```

```
> class(data)
[1] "data.frame"
```

```
#Scaling data
```

```
> datascale<-scale(data)
```

```
#Distance
```

```
> data<-dist(data)
```

```
#Calculating number of clusters
```

```
> fviz_nbclust(datascale,kmeans,method = "wss")+
+   labs(subtitle = "Elbow method")
```

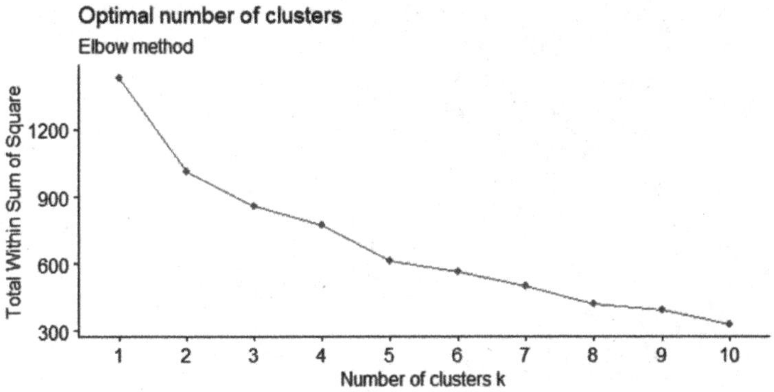

Optimal number of clusters
Elbow method

```
#Applying K-means
```

```
> km.out<-kmeans(datascale,centers = 4, nstart = 100)
> print(km.out)
```

```
K-means clustering with 4 clusters of sizes 41, 52, 21, 30

Cluster means:
        W…mm.       r….       L…m.    fc…MPa.     fy…MPa.          K
1 -0.58816748  0.3632995  1.0970154 -0.6069824  0.17061466  0.8754232
2  0.14500064  0.2378114 -0.5394692 -0.2836764  0.04889508 -0.8493650
3 -0.05081353 -1.5800615 -0.6040576 -0.6289227  1.33564901  1.0005561
4  0.58806392  0.1973272 -0.1413342  1.7614943 -1.25287914 -0.4245683
       C…mm.     ex…mm.      P…kN. Fire.Resistance…min.
1 -0.2627539  0.7231413 -0.6074518           -0.8570342
```

```
2  0.8523465 -0.3422011 -0.1753219            0.7594055
3 -1.7558265 -0.1419463 -0.5347535           -0.6370031
4  0.1107749 -0.2957821  1.5084029            0.3008794

Clustering vector:
  [1] 3 3 3 3 3 1 3 3 3 3 3 3 3 3 3 3 3 3 3 3 2 2 2 2 2 2 2 2 2 2 2 2 2 2 2 2
 [37] 2 2 2 2 1 1 1 1 1 1 1 1 1 1 1 1 1 1 1 1 1 1 1 1 1 1 1 1 1 1 1 1 1 1 1 1
 [73] 1 1 1 1 1 1 1 2 2 2 2 2 2 2 2 2 2 2 2 2 2 2 2 2 2 2 2 2 2 2 2 2 2 2 2 2
[109] 2 3 4 4 4 4 4 4 4 4 4 4 4 1 4 4 4 2 4 4 4 4 4 2 4 2 4 4 4 4 4 4 4 4 4 4

Within cluster sum of squares by cluster:
[1] 198.33392 293.37002  57.86415 119.49425
 (between_SS / total_SS =  53.2 %)

Available components:

[1] "cluster"      "centers"      "totss"        "withinss"
[5] "tot.withinss" "betweenss"    "size"         "iter"
[9] "ifault"
> km.cluster<-km.out$cluster
> fviz_cluster(list(data=datascale,cluster=km.cluster)
```

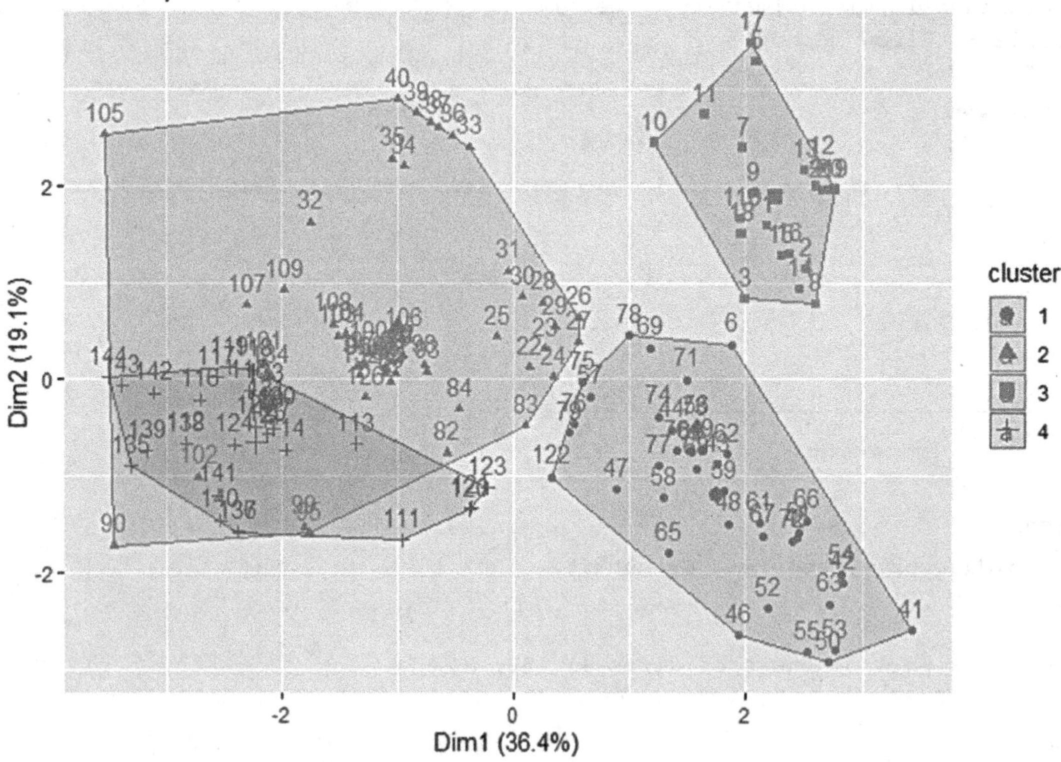

6.3 Conclusions

This chapter covers building and applying ML models using two approaches, a coding-free and a coding-based. In the first approach, the user follows pre-built online or software platforms to input data and build ML solutions pertaining to the phenomenon being investigated. In the second approach, the user reverts to writing scripts of codes in *Python* or *R* programming languages and then executes these scripts to build ML models and examine the outcome of ML analysis. A series of examples were covered on various problems and projects that I hope can be of aid to help practice and visualize the different ML algorithms, platforms, and languages we have covered in this book so far. We will re-visit some of the above listed platforms and languages in future chapters to uncover additional features and options that are likely to further demonstrate the potential of ML and intrigue you about this area.

Definitions

Code library A collection of pre-built modules and codes that can be used repeatedly in different problems.
Coding-based ML A user codes or programs an executable script to run a ML analysis.
Coding-free ML An online portal or software with an intuitive graphical interface used to run a ML analysis.
Relative compactness defined as the volume to surface ratio to that of the most compact shape with the same volume.

Chapter Blueprint

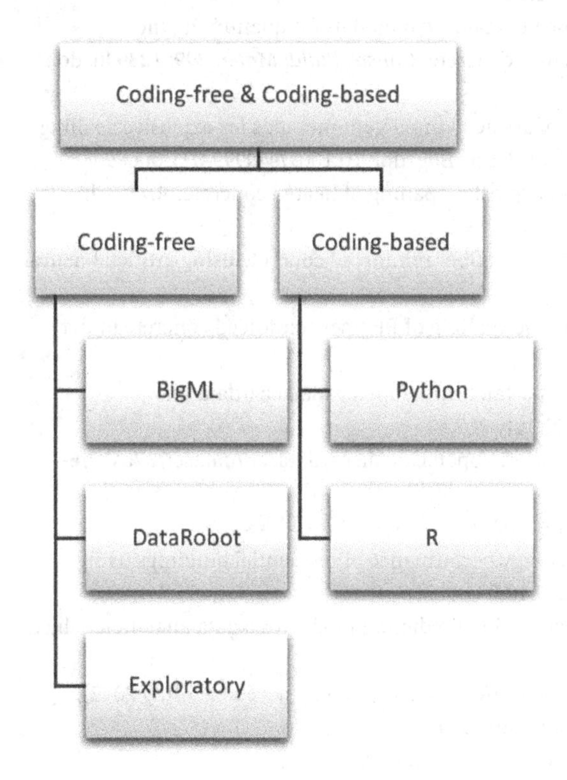

Questions and Problems

6.1 What are advantages and disadvantages to coding-free and coding-based ML.

6.2 Name four programming languages.

6.3 Lists two supervised and two unsupervised algorithms implemented in DataRobot and Exploratory.

6.4 Repeat Example no. 2. [Exploratory, Regression] but for the cooling load as the target. Compare your analysis to that of when the heating load was the target.

6.5 Repeat any example from those covered by the coding-free platforms using all platforms. Compare and comment on the results of your analysis.

6.6 What are the primary differences in the computer vision capabilities between DataRobot and Clarifai?

6.7 Compare fuel consumption between vehicles 1995–1999 and 2022. You may use the datasets provided by Open Canad [7].

6.8 Re-visit Chapter 3 and try out different feature selection methods in Example no. 1 [DataRobot, Classification].

6.9 Given the discussion on manipulating data (noted in Chapter 3), transform Example no. 2. [DataRobot, Regression] into a classification problem.

References

1 BigML. (2022). About. BigML.com. https://bigml.com (accessed 6 July 2022).
2 Liu, J.-C.-C., Huang, L., Tian, Z., and Ye, H. (2021). Knowledge-enhanced data-driven models for quantifying the effectiveness of PP fibers in spalling prevention of ultra-high performance concrete. *Constr. Build. Mater.* 299: 123946. doi: 10.1016/j.conbuildmat.2021.123946.
3 Liu, J.C., Huang, L., Chen, Z., and Ye, H. (2021). A comparative study of artificial intelligent methods for explosive spalling diagnosis of hybrid fiber-reinforced ultra-high-performance concrete. *Int. J. Civ. Eng.* doi: 10.1007/s40999-021-00689-7.
4 Liu, J.C. and Zhang, Z. (2021). A machine learning approach to predict explosive spalling of heated concrete. *Arch. Civ. Mech. Eng.* 20. doi: 10.1007/s43452-020-00135-w.
5 Liu, J.C. and Zhang, Z. (2020). Prediction of explosive spalling of heated steel fiber reinforced concrete using artificial neural networks. *J. Adv. Concr. Technol.* doi: 10.3151/jact.18.227.
6 Liu, J.C. and Zhang, Z. (2020). Neural network models to predict explosive spalling of PP fiber reinforced concrete under heating. *J. Build. Eng.* doi: 10.1016/j.jobe.2020.101472.
7 Open.Canada.ca. (2022). Fuel consumption ratings - open government portal. https://open.canada.ca/data/en/dataset/98f1a129-f628-4ce4-b24d-6f16bf24dd64#wb-auto-6 (accessed 27 July 2022).
8 Open.Canada.ca. (2022). Bridge conditions - open government portal. https://open.canada.ca/data/en/dataset/37a472f6-b7ea-4a41-9d4b-64a0c8e5025a (accessed 27 July 2022).
9 Exploratory. (2022). About. https://exploratory.io/about (accessed 6 July 2022).
10 Tsanas, A. and Xifara, A. (2012). Accurate quantitative estimation of energy performance of residential buildings using statistical machine learning tools. *Energy Build.* doi: 10.1016/j.enbuild.2012.03.003.
11 Sathishkumar, V.E., Shin, C., and Cho, Y. (2021). Efficient energy consumption prediction model for a data analytic-enabled industry building in a smart city. *Build. Res. Inf.* 49 (1): 127–143.
12 Python (2022). Python source releases | python.org. https://www.python.org/downloads/source (accessed 5 July 2022).
13 Pypi. (2022) PyPI · the python package index. https://pypi.org (accessed 5 July 2022).
14 Keras. (2022) Keras: the Python deep learning API. https://keras.io (accessed 5 July 2022).
15 Matplotlib. (2022). Matplotlib — visualization with Python. https://matplotlib.org (accessed 5 July 2022).
16 NumPy. (2022). NumPy. https://numpy.org/install (accessed 5 July 2022).
17 Pandas. (2022). Installation — pandas 1.4.3 documentation. https://pandas.pydata.org/docs/getting_started/install.html#install (accessed 5 July 2022).
18 scikit-learn. (2022). Installing scikit-learn — scikit-learn 1.1.1 documentation. https://scikit-learn.org/stable/install.html#install-official-release (accessed 5 July 2022).
19 TensorFlow. (2022). Install tensorFlow with pip. https://www.tensorflow.org/install/pip (accessed 7 July 2022).
20 Jupyter. (2022). Project jupyter | home. https://jupyter.org (accessed 7 July 2022).
21 RStudio. (2022). RStudio | open source & professional software for data science teams - rstudio. https://www.rstudio.com (accessed 7 July 2022).

22 ggplot2. (2022). Create elegant data visualisations using the grammar of graphics • ggplot2. https://ggplot2.tidyverse.org (accessed 7 July 2022).

23 tidyr. (2022). Tidy messy data • tidyr. https://tidyr.tidyverse.org (accessed 7 July 2022).

24 dplyr. (2022). A grammar of data manipulation • dplyr. https://dplyr.tidyverse.org (accessed 7 July 2022).

25 Shiny. (2022). Shiny. https://shiny.rstudio.com (accessed 7 July 2022).

26 RapidMiner. (2022). RapidMiner for academics - RapidMiner. https://rapidminer.com/platform/educational (accessed 7 July 2022).

27 Weka. (2022). Weka 3 - machine learning software in java. https://www.cs.waikato.ac.nz/ml/weka (accessed 7 July 2022).

28 KNIME. (2022). Downloads | KNIME. https://www.knime.com/downloads (accessed 7 July 2022).

29 Naser, M.Z. (2021). An engineer's guide to eXplainable artificial intelligence and interpretable machine learning: navigating causality, forced goodness, and the false perception of inference. *Autom. Constr* 129: 103821. doi: 10.1016/J.AUTCON.2021.103821.

30 Naser, M.Z. (2020). Machine Learning Assessment of Fiber-Reinforced Polymer-Strengthened and Reinforced Concrete Members. *ACI Struct. J.* doi: 10.14359/51728073.

31 Young, B.A., Hall, A., Pilon, L. et al. (2019). Can the compressive strength of concrete be estimated from knowledge of the mixture proportions?: new insights from statistical analysis and machine learning methods. *Cem. Concr. Res.* 115: 379–388. doi: 10.1016/j.cemconres.2018.09.006.

32 Naser, M.Z. (2022). Digital twin for next gen concretes: on-demand tuning of vulnerable mixtures through explainable and anomalous machine learning. *Cem. Concr. Compos.* 132: 104640. doi: 10.1016/J.CEMCONCOMP.2022.104640.

33 DataRobot. (2020). Anomaly detection: dataRobot docs. https://docs.datarobot.com/en/docs/modeling/special-workflows/unsupervised/anomaly-detection.html (accessed 5 July 2022).

34 Dataiku. (2014). Pronouncing dataiku. https://blog.dataiku.com/2014/08/07/pronouncing-dataiku (accessed 27 July 2022).

35 Sadhwani, J. (2022). GitHub - JatinSadhwani02/water-quality-prediction-using-ML-technique. https://github.com/JatinSadhwani02/Water-Quality-Prediction-using-ML-technique (accessed 16 June 2022).

36 Tsanas, A. and Xifara, A. (2022). UCI machine learning repository: energy efficiency data set. https://archive.ics.uci.edu/ml/datasets/Energy+efficiency# (accessed 6 July 2022).

37 Chandavale, A. (2022). Carbon emission India | kaggle. https://www.kaggle.com/datasets/adityachandavale/carbon-emission-india?resource=download (accessed 6 July 2022).

38 Sathishkumar, V., Lee, M., Lim, J. et al. (2022). Steel industry energy consumption | kaggle. https://www.kaggle.com/datasets/csafrit2/steel-industry-energy-consumption (accessed 6 July 2022).

39 Tilley, A.(2017). Every AI powerhouse wanted this whiz kid. He's taking them on instead. Forbes. https://www.forbes.com/sites/aarontilley/2017/07/13/clarifai-ai-image-recognition/?sh=2b55ef2dfe42 (accessed 9 July 2022).

40 Lutz, M. (2015). *Learning Python*, 5e. O'Reilly Media.

41 Wickham, H. and Grolemund, G. (2011). R for data science. *Interact. Cardiovasc. Thorac. Surg...*

42 Microsoft. (2022). GitHub - microsoft/data-science-for-beginners. https://github.com/microsoft/Data-Science-For-Beginners (accessed 1 August 2022).

43 Python. (2022). General python FAQ — python 3.10.5 documentation. https://docs.python.org/3/faq/general.html (accessed 5 July 2022).

44 Anaconda. (2022). Anaconda | the World's most popular data science platform. https://www.anaconda.com (accessed 5 July 2022).

45 Google. (2022). Welcome to colaboratory - colaboratory. https://colab.research.google.com/?utm_source=scs-index (accessed 7 July 2022).

46 Özyüksel Çiftçioğlu, A. and Naser, M.Z. (2022). Hiding in plain sight: what can interpretable unsupervised machine learning and clustering analysis tell us about the fire behavior of reinforced concrete columns? *Structures* 40: 920–935. doi: 10.1016/J.ISTRUC.2022.04.076.

7

Explainability and Interpretability

Translucent does not even mean invisible – it means semi-transparent.

Billy Butcher (The Boys, 2019)

Synopsis

So far, we have learned about a variety of ML types and algorithms. We have also learned how to build ML models, examine such models, and deploy them. Collectively, we learned all about developing data-driven ML models. What we do not know *yet* is how to explain (or interpret) model predictions.[1] For example, given a set of features, if a model predicts a certain outcome (i.e., a phenomenon will occur[2]), then the following question may arise: why does the model predict that the combination of these features will yield to this particular outcome?

This chapter hopes to shed some light on answers that can help us address the aforementioned question. In this pursuit, this chapter will introduce principles and associated methods for establishing explainability and interpretability[3] from the lens of ML.

7.1 The Need for Explainability

As we know, a ML model maps a group of features to an outcome (e.g., tie concrete mix ingredients to the compressive strength property[4]). In our process, a properly validated model is one that attains a certain level of goodness; and thereby is declared proper and to have a permissible prediction capability.

1 Quick story: I was working with Moe al-Bashiti on a paper in which we built a ML model to predict failure in concrete specimens under fire conditions (I will showcase an analysis in Example no. 3 in this chapter). The developed model was pretty decent as in it was able to predict the failure of specimens with such high accuracy that matched, if not exceeded that available in the literature (at the time of this particular work). From then, our goal has shifted to the following. We were intrigued to know why and how this model was capable of realizing such accuracy. Was there something in the data the model can see that we could not? Was there something about this model that made it very accurate? I hope you are getting at what I am alluding to... If the answer to one or both of the questions is, *yes*, then maybe, just maybe, we could arrive at a glimpse for a possible theory for failure of such specimens! Simply, we needed to know the why and the how. This chapter is about finding answers to these questions from a *data perspective*.

2 Or, for a regression-type problem, the concentration of dissolved oxygen is 12 ppm.

3 At this point in time, there seem to be some differences between explainability and interpretability. Inherently, explainability is tied to systems that fall under AI (i.e., eXpalinable AI (XAI)), while interpretability describes ML models. I would like to avoid such debate and hence will be adopting explainability throughout this chapter. Also, please note that other terminology also exists for explainable models such as whitebox, glass, and transparent models.

4 Or tie road and weather conditions to the expected number of vehicle accidents. Another example would be to tie emission regulations in a particular region to the air quality in that region.

Machine Learning for Civil & Environmental Engineers: A Practical Approach to Data-driven Analysis, Explainability, and Causality, First Edition. M. Z. Naser.

Companion Website: www.wiley.com/go/Naser/MachineLearningforCivilandEnvironmentalEngineers

Let us think about this for a minute. Surely, satisfying selected performance metrics and error indicators is a quantifiable measure that we can attach to the goodness of a model. This is the same procedure we would also follow when deriving mathematical and empirical models. But there seems to be something missing. How do performance metrics/indicators relate to the physics (or mechanics) of the engineering problems we are interested in? The short answer is that they rarely do![5]

If you remember our discussion on PFIEM, then you would also remember that satisfying a group of selected performance metrics does not effectively imply that the ML model captures the physics behind a phenomenon. Rather, this infers the suitability of such a model to predict the outcome as described in our dataset [1].

This brings me to this chapter.Our objective in this chapter is to understand the reasoning behind a ML model's predictions. In other words, what has led[6] the model to arrive at a specific outcome and not some other outcome? One might say, how is knowing the above relevant? A ML model is based on an algorithm/ensemble that is driven by the data. This implies that the data, together with the way the algorithm was trained to tie[7] this data together, are the reasons behind such model predictions.[8] I am hoping the next section may change your mind.

7.1.1 Explainability and Interpretability

As mentioned earlier, explainability and interpretability are often used interchangeably nowadays [2]. However, subtle differences between the two exist. For example, *explainability* is reserved for *"models that are able to summarize the reasons for neural network behavior, gain the trust of users, or produce insights about the causes of their decisions."*[3]. In addition, explainability entails relating the inner mechanisms of a model and their influence on the model's prediction. An explainable model embodies a complex function that may not be understood on its own and hence is likely to require us to augment such function with additional methods/techniques to be able to understand it.

On the other hand, *interpretability* is *"loosely defined as the science of comprehending what a model did (or might have done)"* [3]. In other words, interpretability is humanly *understandable* without aid.[9] The degree of explainability or interpretability depends on a number of items such as user expectations, problem complexity, algorithmic architecture, etc. Table 7.1 compares commonly used ML algorithms in terms of how understandable they can be.

Adopting explainable models or understanding why a model predicts its outcomes can open exciting doors for civil and environmental engineers. Let me revisit our earlier example of tying concrete mix ingredients to the property of compressive strength.[10] If we were able to develop a ML model that can truly capture this property *as well as* explain the inferred hidden relation(s) between the ingredients and the strength property, then it is possible that we would arrive at a new piece of information (aka. knowledge[11]) that we did not have before. The same concept of explainable models could help our engineers discover new mechanics or new hypotheses for some of the common problems they seem to continue to struggle with[12] [5].

The above brings in the important question of *how to infer a model's understanding of a phenomenon?* After all, this chapter hopes to address this question.

Well, there are a number of explainability methods that we can augment ML models with. These methods open the blackbox for us to get to see *how* and *why* the model predicts the way it does. These methods vary in terms of their:

---

5 To be clear, I am thinking about traditional metrics. Special metrics that fall under domain knowledge or physics/mechanics would negate the above answer.

6 Despite its suitability, I am intentionally avoiding the term *cause* as I am preserving this for a later chapter.

7 aka. map the patterns.

8 However, knowing the rationale behind predictions may go further than that.

9 I will be presenting other definitions and philosophical arguments concerning explainability and interpretability in the following section. Others can be found elsewhere [28–31].

10 Here, the property is the outcome and the ingredients are the features. Please feel free to re-format this example to a problem of your choosing.

11 There is more to discovering new knowledge than explainability. I will re-visit these in the causality chapter.

12 This paragraph inspired al-Bashiti's and Tapeh's papers [32, 33].

Table 7.1 Understandability in common ML models [4].

Model type	Is this humanly understandable?	Types of mechanisms used in a model.	Need for explainability methods.
Linear/Logistic Regression	Yes.	Mathematical based.	No.
Tree-like (Decision Tree, Random Forest)	Yes, by displaying a tree formation.	Rule-based that displays how the decisions are taken at each step.	Not necessary.
k-Nearest Neighbors (k-NN)	Yes.	Tackles a large number of variables via mathematical/statistical representations.	Mostly visually explainable.
Genetic Algorithms (GA)	Yes, since they resemble a tree-like structure.	Mathematical representation.	Mostly readable.
General Additive Models (GAM)	Mostly, and especially if interactions between features are smoothened.		No, unless interactions turn complex.
Tree Ensembles (ExGBT, LGBT)	Unlikely.	Blenders of weaker models.	Very complex models that require further analysis via agnostic or specific methods.
Support Vector Machines (SVM)		Categorize data via hyperplanes in N-dimensional space.	
Neural Networks (NNs)		Layers containing neurons and transformation functions.	

- *Nature*
  - Intrinsic [simple models, i.e., trees] vs. Post hoc [complex models, i.e., deep networks]. For example, simple models (i.e., decision trees) comprise a simple structure we can visualize and trace how this algorithm treats the data to realize its predictions. Since we can visualize how the input yields the output, then such as model is intrinsically explainable. On the other hand, a deep learning model contains a complex structure that is hard to visualize and trace. As such, we cannot visualize the mapping between the features and target.
- *Type*
  - Model-agnostic[13] [i.e., can be applied to any ML model] vs. Model-specific[14] [i.e., can only be applied to a particular family of models]. Here, some explainability methods can be applied to any model, while others are very specific to certain type(s) of models.
- *Scale*
  - Local [i.e., explain individual predictions] vs. Global [i.e., explain model behavior as a whole] [6]. The majority of available methods can explain how the feature affects model predictions on average, but only a few can provide us with how each prediction came to be.

7.2 Explainability from a Philosophical Engineering Perspective❄

In engineering,[15] we have an arsenal of methods that we can use for investigation.[16] At a fundamental level, these methods can elegantly express the *why* and *how* an outcome was generated, whether explicitly or in some hypothesized form. For example, if we design a beam to fail in shear and test this beam under shear loading, then such a beam is likely to fail in

13 Including, feature importance, partial dependence plots, feature interactions, SHAP (SHapley Additive exPlanations), surrogates, and local interpretable model-agnostic explanations (LIME).

14 Such as, generalized linear models and tree-based ensemble models [6, 34, 35].

15 As well as other parallel domains.

16 Such as, experimentation, numerical simulation, analytical derivation, etc.

shear.[17] In other words, the beam on hand failed in shear because its configurations and loading set-up have resulted in shear effects that far exceed that of the beam's shear capacity. If we were to investigate the same beam by replicating its configurations and loading set-up in a FE or analytical model, we are likely to get the same outcome.[18] Simply, a method of investigation can infer the relationship between predictor variable(s) and response variable(s) [7, 8].

Now, can we do the same with ML? Given the nature of ML and its reliance on data, the same cannot be articulated as nicely as that by comparing experiments with numerical models.

In a nutshell, we have a great tool (ML) that is good at predicting phenomena, yet we do not know how to explain its predictions nor infer how its predictions came to be. How can we use such a tool? How can we trust such a tool?[19]

What we need to realize is that ML is fundamentally different from the other methods that we are accustomed to. The latter shares a number of qualities, such as 1) they are universal, 2) they have built-in reliability, 3) they are intuitive, 4) they are transparent, and 5) they are often accepted across our community (i.e., committees, codes, and standards) and tend to be regularly updated and disseminated.[20]

Let us compare ML against the above qualities. At the moment, ML can be considered as a universal tool and could be supplemented with reliability measures (i.e., confidence interval, etc.). However, it remains far from being intuitive, accepted, or transparent. These are the same reasons behind the current inertia toward ML,[21] which also explains why the construction industry is often reluctant to adopt new technologies [9, 10].

As we have seen over and over the pages of this book, a typical ML analysis is fundamentally simple as it comprises three stages: collect observations → apply algorithms → build ML models. While the first stage is easy to follow and requires little effort to visualize, the bulk of our engineers are uncomfortable with the second the third stages.

When combined with questions such as 1) why does a model predict the way it does? 2) what to make out of a model's predictions? And, 3) how to trust a model's predictions? The apparent inertia against ML is not only not surprising but, in a way, is also understandable.

ML users[22] need to be able to *trust* their methods and tools. The same is true for civil and environmental engineers since our domain involves the lives and well-being of members of our society and surrounding environment, etc. The concept of explainability strives to build trust between ML and engineers by allowing us to peek into the blackbox.

In the pursuit of trustable models, one must be cautious of the degree of explainability generated by a model, as the degree of *required* explainability may differ from one engineer to another[23] or from one problem to another (i.e., design of a skyscraper vs. scheduling for a public transit system), [11]. This particular idea of founding a *required* degree of explainability is important in communications between engineers of different backgrounds.

Similarly, explainable models make it easy for those with a limited technical background to understand the workings of ML.[24] Furthermore, an explainable model can also be thought of as trustable to stakeholders and government officials since they can easily understand the rationale behind its predictions.

---

17 Here is another example that does not involve beams. Imagine, we conduct a test to investigate how far can flaming embers travel during a wildfire. The result of such an experiment might return a distance of 100 m. If we replicate the same via a Computational Fluid Dynamic (CFD) model, then a proper model is likely to return the same distance (give or take).

18 For instance, a display of shear stresses in a FE model will show stress concentration and intensity that exceed the available shear capacity.

19 Here comes trouble.

20 For instance, an equation, or a chart is transparent. They show us how the parameters interact with each other and with the outcome we happen to be interested in.

21 Intuitiveness takes a bit of time to establish and hence requires practice and education. Transparency can be improved through new methods that open the opaqueness of ML. Acceptance takes place once we are comfortable enough that the method in question is useful to us. This book hopes to address the first and second points. Let us see what the future holds for the third.

22 This book is about engineers. I opted for *users* here as it conveys the general interest of stakeholders using ML.

23 Suppose a bridge collapses, and the city assembles a team of investigators. A structural engineer would be looking to explain the collapse from a structural engineering point of view, while a material scientist would be looking for an explanation on the material side. A transportation engineer will be looking for abnormal travel patterns that may have caused heavy detours. In addition, a sufficient explanation to one individual may not be perceived as sufficient to another. In other words, explainability is fluid and is highly context dependent. A simpler example would be the following, a senior engineer may not require as detailed of an explanation like that to be required by an engineering student. There is a lot that goes into philosophy and engineering!

24 Here is a scenario, imagine a building official is examining the outcome of a ML model. If the model is explainable, then the model can articulate (or more realistically, document) its rationale for its predictions. The building official can go over such rationale and compare it to codal provisions (since at the end of the day, buildings are to comply with such provisions). If the model rationale agrees and follows such provisions, then the building official would be able to follow the outcome of the model with confidence.

Let us talk a bit more about stakeholders. Civil and environmental engineers seek to address societal problems [12]. Unlike adopting ML in physical systems, where the application is straightforward, utilizing ML for societal problems can be hectic. Notably so, in instances where the stakeholders belong to weaker/forgotten sections of our society. This is because, in lieu of satisfying performance metrics and indicators, there is also a need to adhere to social and public expectations to realize a socially just society. Hence, engineers ought to think about means to incorporate such expectations (e.g., *fairness* and *inclusivity*,[25] etc.) into their models.

So far, most of my discussion implicitly assumes that the ML model can yield correct predictions. Thus, when such a model is augmented with explainability measures, then the outcome of the model can be traced back to the rationale behind such a correct prediction.

Now, I have two scenarios to discuss.[26]

The first, say that the model does *not* provide us with accurate nor correct predictions. Then, perhaps looking at the explanations from the model might help us refine our model and see how was can further improve it. The second, say that the model *does* provide us with accurate and correct predictions; however, its explanations are not correct. Then, we would need to investigate what are the possible causes for such behavior.

Please note that the philosophical discussion on explainable models may cover other concepts, including equity, biasness, accessibility, the trade-off between accuracy and explainability, and energy, etc. [11, 13].

7.3 Methods for Explainability and Interpretability

In this section, I will be describing five commonly used methods[27] that can be used to explain a ML model's predictions on the global (whole dataset) or local (a single observation) level. These methods can be applied once the ML model has successfully passed its development stage.[28] We will also explore these techniques in five different examples (the first three covering supervised learning and the last covering unsupervised learning).

7.3.1 Supervised Machine Learning

We will cover five methods herein, namely, feature importance, partial dependence plots, SHaply Additive exPlanations (SHAP), Local Interpretable Model-Agnostic Explanations (LIME), and surrogates.

7.3.1.1 Feature Importance

There are a number of features that can be involved in a ML model.[29] In reality, not all features are created equally, and it is expected that all, if not most, features make a unique and likely quantifiable contribution to the response (prediction) of such a model. This brings a distinctive look into ML models. For instance, if we could trace the contribution of each feature (aka. their influence on model prediction), then it is possible that we can understand[30] model predictions.

Feature importance[31] is a method to measure the extent to which a given feature influences the outcome of a ML model. One common approach to measuring such extent is by evaluating the increase in error as a result of permuting features systematically [14]. In this approach, a feature is deemed *important* if permuting its values increases the model error.[32] Similarly, a feature is deemed *not* important if its permutation does not alter the prediction error significantly. In general, features are ordered in a descending order from the most important to the least important (see Figure 7.1). In this figure, it is clear that Feature 1 is elemental to the model.[33]

---

25 One noteworthy example I came across revolves around building ML evacuation models that account for social stressors associated with low-income and under-representative households.

26 This discussion stems from a recent research area of adversarial attacks on ML.

27 Please review [28, 30, 36] for more information on such methods.

28 I found Christopher Molnar's book [36] to be of great aid. Give it a try!

29 Or, a typical dataset comprises a series of features that are fed in to a ML model.

30 Or, explain.

31 **Feature importance**: the degree to which a ML model relies on a particular feature to arrive at correct predictions.

32 This tells us that the model relies on this particular feature to arrive at a good prediction. Thus, without this feature, the model struggles to arrive at proper predictions.

33 You could argue that Feature 2 is also elemental. I would agree with you. At this point in time, we do not have a standard scale to quantify the importance of features.

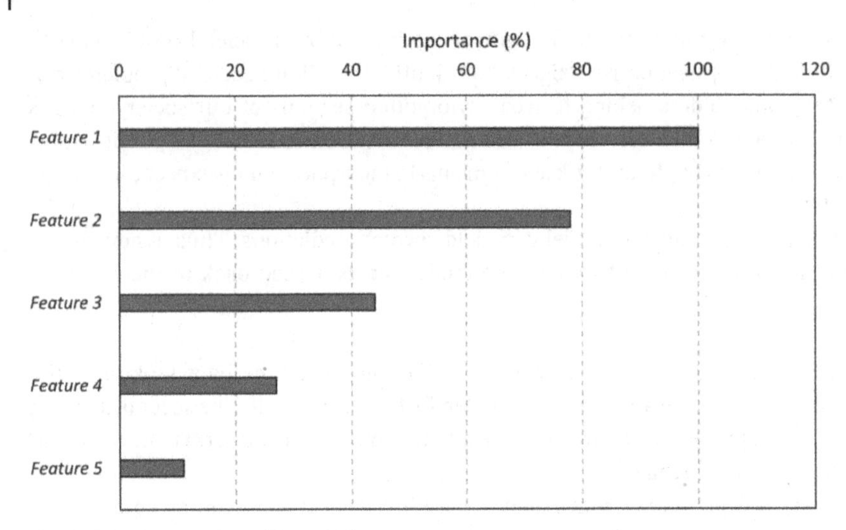

Figure 7.1 Illustration of feature importance.

7.3.1.2 Partial Dependence Plots

The partial dependence plot (PDP)[34] is a plot that infers the type of relationship between a single feature and outcome (response), e.g., linear, nonlinear, or complex. This plot only varied the feature in question while keeping all other features constant [15]. PDPs are built on the assumption that the feature being examined does not correlate with other features.

One can think of PDP as a plot that describes how changing the values of one feature across all observations while keeping the other features unchanged influences the outcome of the phenomenon. Since a PDP examines the average influence of one feature at a time, this plot can determine points of interest in a model's predictive performance. For example, Figure 7.2 shows that lower values of the presented feature tend to push the outcome to have a low value. With an increase in this feature, the outcome increases until it reaches a plateau.[35] The PDP has multiple variations.[36]

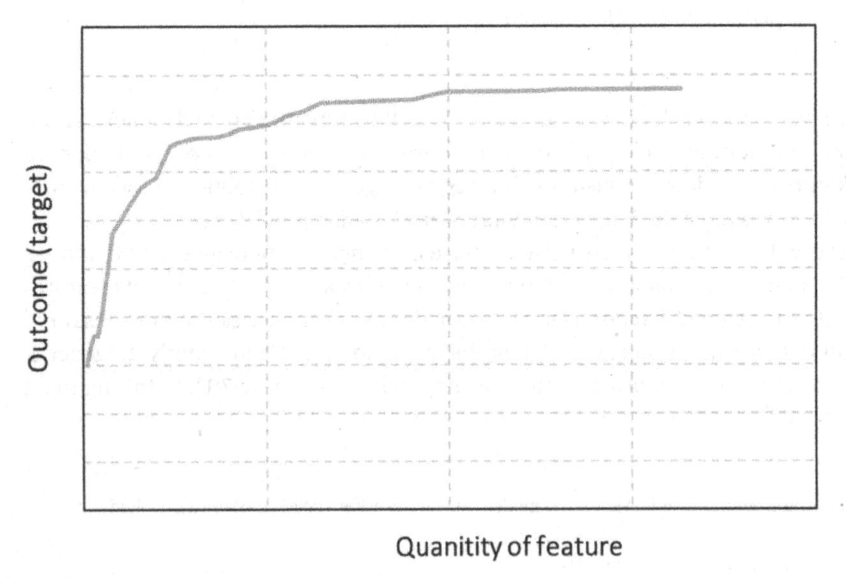

Figure 7.2 Illustration of a PDP.

---

34 **Partial dependence plot**: depicts an individual feature's marginal (average) effect on the response/target variable while holding other features constant.

35 How can you use PDP? Well, imagine that this feature resembles the quantity of an expensive chemical compound that is necessary to distill water. Then, looking at this figure shows us that we may not need to use large quantities of this compound (since the predictions stabilizes beyond the plateau).

36 For example, a PDP can represent more than one feature. In such a variation, both the horizontal and vertical axes would need to be normalized (especially if the units of the features are not of the same magnitude). Another variation is the one described in the next footnote.

Unlike in regression, when a PDP is applied to a classification problem, the PDP gives the probability for an observation to belong to a particular class. One extension to PDP is known as Individual Conditional Expectation (ICE) plot [16], and this plot does not average the relationship between the examined feature and the predicted response.[37]

7.3.1.3 SHAP (SHapley Additive exPlanations)

Similar to feature importance and PDPs, the SHAP method is also regarded as a unified agnostic method that can be applied to explain ML models [17, 18]. SHAP is based on a game theory approach to additively accumulate the contribution of all features involved in a model and dataset. When applied to a ML model, SHAP assigns each feature an *importance value*, and the results of this additive procedure are called *SHAP values*.[38] The SHAP method is mathematically expressed in Equations 7.1 and 7.2. This method iterates over all possible features and their possible combinations to ensure that the model accounts for the interactions between all individual factors. Since the SHAP method accounts for all features and their interactions, this method can be quite computationally exhaustive for large models.

$$\textit{The importance of feature } i = \textit{prediction}(\textit{with feature } i) - \textit{prediction without}(\textit{feature } i) \tag{7.1}$$

And,

$$\varnothing_i = \sum_{S \subseteq N \setminus \{i\}} \frac{|S|!\left(M - |S| - 1\right)!}{M!}\left[f_x\left(S \cup \{i\}\right) - f_x(S)\right] \tag{7.2}$$

where $\varnothing_i$: Shapley value for feature *i*, *S|!*: the number of permutations of feature values that appear before *i*-th feature value. *(|M|-|S|-1)!*: represents the number of permutations of feature values that appear after the *i*-th feature value. The difference term in Equation 7.2 is the marginal contribution of adding the *i*-th feature value to *S*.

The SHAP method can be applied to the global and local scales to explain ML models. On a global scale, SHAP can describe the importance of each individual feature via a *summary plot*. This plot lists a series of horizontal graphics that represent the *contribution of each* feature in a descending order.[39] On the local scale, SHAP can also generate a *force plot* and *waterfall plot* to represent the most critical features influencing the phenomenon on hand and how each factor contributes to the prediction, starting from a *base value* – see Equation 7.3.

$$\textit{Local prediction} = \textit{base value} + \sum \textit{summation of features} \tag{7.3}$$

7.3.1.4 LIME (Local Interpretable Model-Agnostic Explanations)

Local Interpretable Model-Agnostic Explanations (LIME) is another explainability measure that is more suited to the local scale. In LIME, a specific input instance is altered, and the resulting impact of the alteration on this specific prediction is monitored to identify which changes will have the most impact on the prediction (see Equation 7.4). The output of LIME is a summary plot that lists the contribution of each feature

$$\textit{explanation}\,(x) = \arg\,\min_{g \in G} L(f, g, \pi_x) = \Omega(g) \tag{7.4}$$

where *F* denotes the complex model, *g* is the simple model. This simple model *g* comes from a set of interpretable linear models denoted with *G*. π_x defines the local neighborhoods of that data point and is some sort of proximity measure. Ω is a complexity measure aimed at minimizing the complexity.

37 ICE is a PDP for each observation. PDP is the average of all observations. A more common advancement to PDP is the Accumulated Local Effects (ALE) – which can be a bit hard to interpret but work well for correlated features..

38 The average value is referred to as the *base value*.

39 A *summary plot* can be thought of to be somehow similar to that obtained from feature importance. We will see this plot in Example no. 4. In this plot, features of large horizontal spread tend to significantly affect the obtained predictions as opposed to parameters with narrower spread.

7.3.1.5 Surrogates

In this method, a secondary ML model is used to proxy (or simply explain) the predictions of an existing model. The secondary model is often simple in nature, or one that we have a clear understanding of its inner mechanisms (e.g., a linear model); thereby, and once augmented into the complex model, the surrogate may be used to give us insights into why the complex model predicts the way it does.

In practice, the choice of a surrogate model is decoupled from the complex model; as technically, all that a user needs is a model that is simple (such as additive models, genetic algorithms, tree-like models [19–21]) and can capture the predictions of the original model with good accuracy. Note that accuracy, in this context, primarily refers to attaining an adequate performance metrics between the predictions of the surrogate and the predictions from the complex ML model.

7.3.2 Unsupervised Machine Learning

Although unsupervised ML algorithms do not incorporate targets, their predictions can still be traced by means of explanation. Two simple and direct approaches to explaining the predictions of such models are described herein, namely: *explainable trees* and *polar lines plots*.

7.3.2.1 Explainable Trees

In this approach, once the data points have been allocated to the clusters, then the output of a clustering algorithm can be augmented into a tree-like structure (see Figure 7.3a). Each tree contains various nodes, some of which are *split* nodes, and some are *leaf* nodes. Depending on whether the features in the dataset meet certain conditions, a tree structure is formed from the *root* node to the *leaves* (in a top-down approach) [22]. The decision tree divides the data into subsets containing homogeneous (with similar values) samples.

In each tree, *sample* refers to the total number of data points in a particular node, and *entropy* measures the homogeneity of a node. If there is only one class in a node, homogeneity is at its highest, and entropy is zero. In contrast, entropy is highest if classes are evenly distributed within a heterogeneous node. In all cases, each node falls into a coloring scheme depending on its class (and the intensity of color increases as class homogeneity increases (entropy decreases)).

7.3.2.2 Polar Line Plots

A polar line plot represents the vertex comprising polyline marks of each cluster in polar coordinates (along the radial and angular axes). This plot shows the influencing features of all clusters. For example, Figure 7.3b shows the larger influence of Feature 7 and Feature 8 compared to other features.

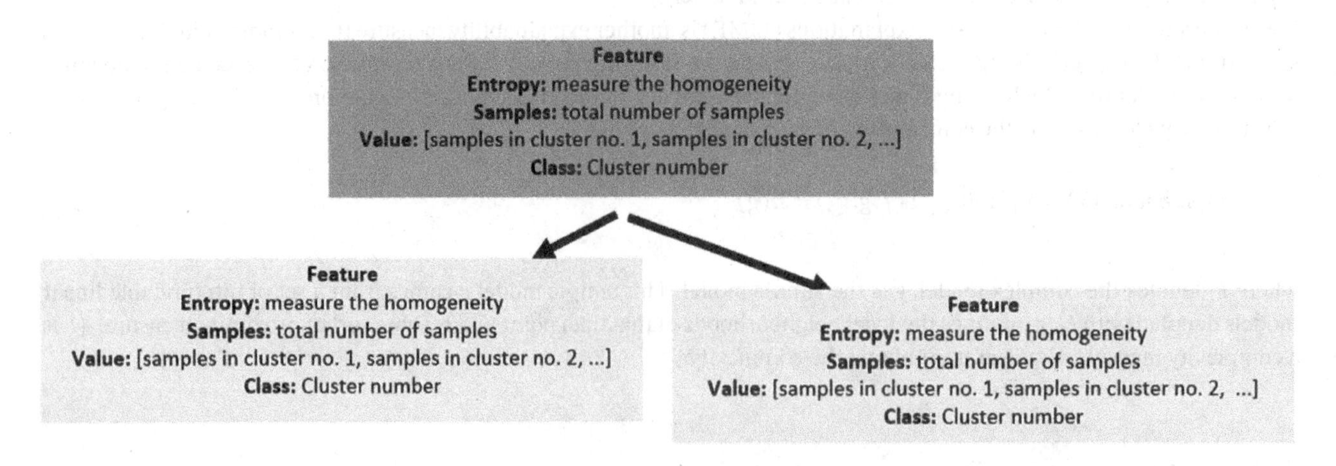

(a) Illustration of an explainable tree [Note: cluster number refers to the largest sample number in the "Value" component]

(Continued)

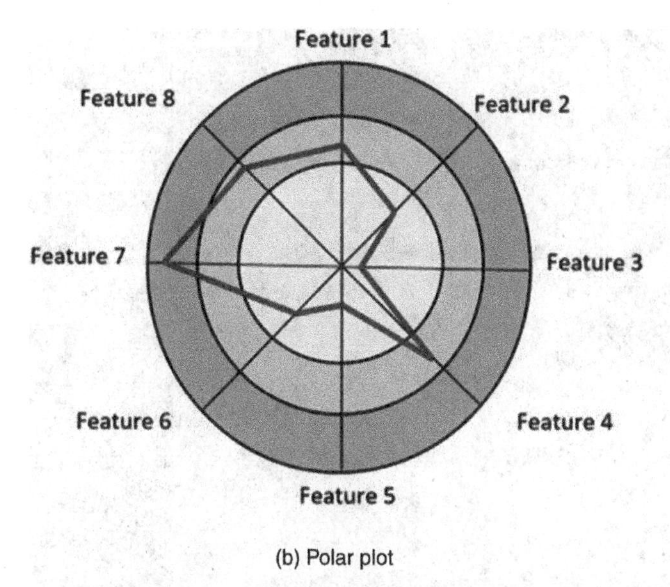

Figure 7.3 (Cont'd) Common explainability methods for unsupervised learning.

7.4 Examples

Let me walk you through five[40] examples that cover the above explainability methods. I will be using the *DataRobot* and *Dataiku* platforms[41] for the first example, and then I will showcase the last three examples with *Python* and *R*.[42]

Example no. 1 (DataRobot platform[43]: Predicting moment capacity and failure mode of RC seams strengthened with fiber reinforced polymer (FRP) composite laminates.

As you remember, we have a dataset[44] comprising data points collected from over 100 experimental tests on RC beams strengthened with FRP systems. This dataset includes compressive strength of concrete (f_c), yield strength of steel reinforcement (f_y), ratio of steel reinforcement (r_s), FRP ratio (r_f), modulus of FRP (E_{fr}), moment arm (a), and strengthening type (T).[45] There are two targets in this dataset, the magnitude of moment capacity (M), and failure mode (FM) (i.e., steel yielding/concrete crushing or failure through the FRP system).[46] Let us carry out this example together.

Step 1: Upload dataset.
Step 2: Select Moment as the *target*.

40 One new example and another three running examples we have seen in earlier chapters (i.e., CFST columns, spalling in concrete mixtures, and clustering of RC columns). I wanted to show Example no. 1 since it combines both regression and classification into one problem! Maybe get to present a few more algorithms within *DataRobot*.

41 The datasets will be provided for you to 1) re-do these examples, and 2) in a programming language of your choosing.

42 Hello Moe al-Bashiti, Arash Tapeh, and Aybike Özyüksel! Thank for preparing these codes. Also note that Example no. 5 is only presented in Python.

43 I have already presented the same example using the *BigML* Platform in a previous chapter. However, I did not cover the explainability component. I told you we will re-visit that soon! See the *Questions and problems* section.

44 Full details on this dataset and data sources can be found in [4, 37].

45 There are traditionally two strengthening types, *plated* (e.g., where plates of FRP are bonded to the surface of the member) and *mounted* (aka. near surface mounted (NSM) where rebars or strips are embedded via grooves cut into the sides of the member).

46 Simply, you could use this dataset in a regression or classification problem.

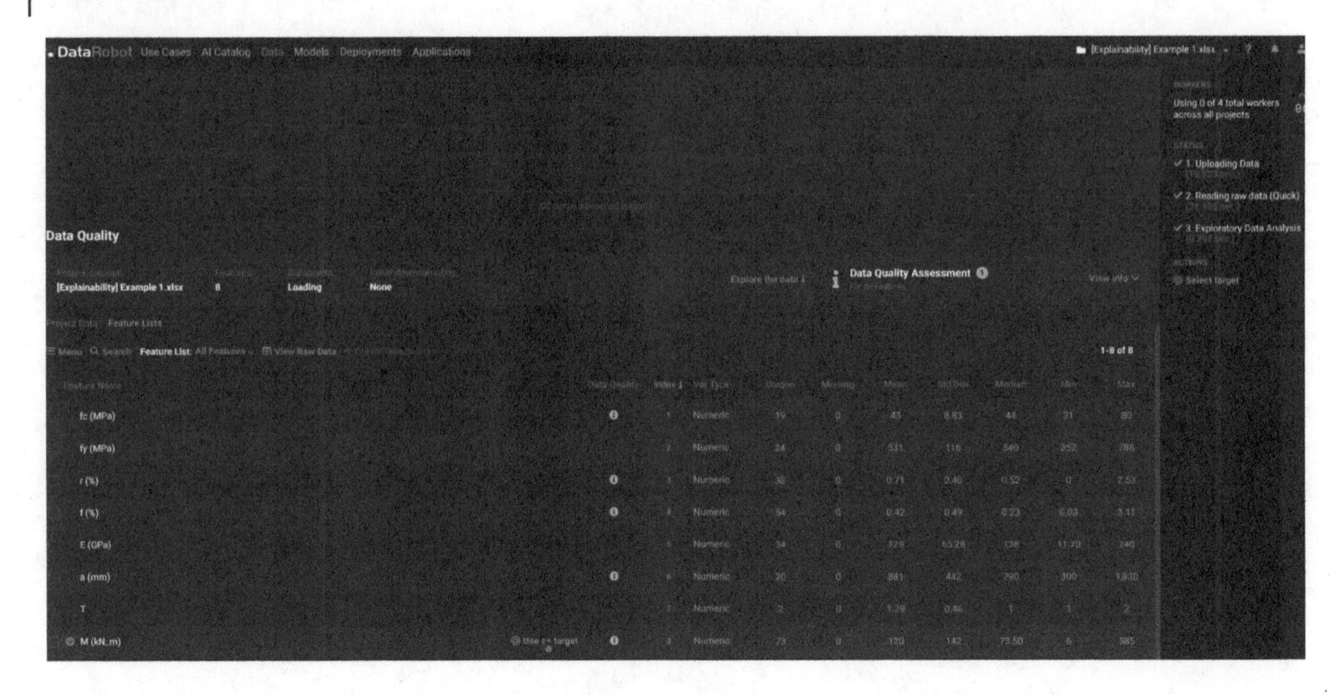

Step 3: Select *Show advanced options.*

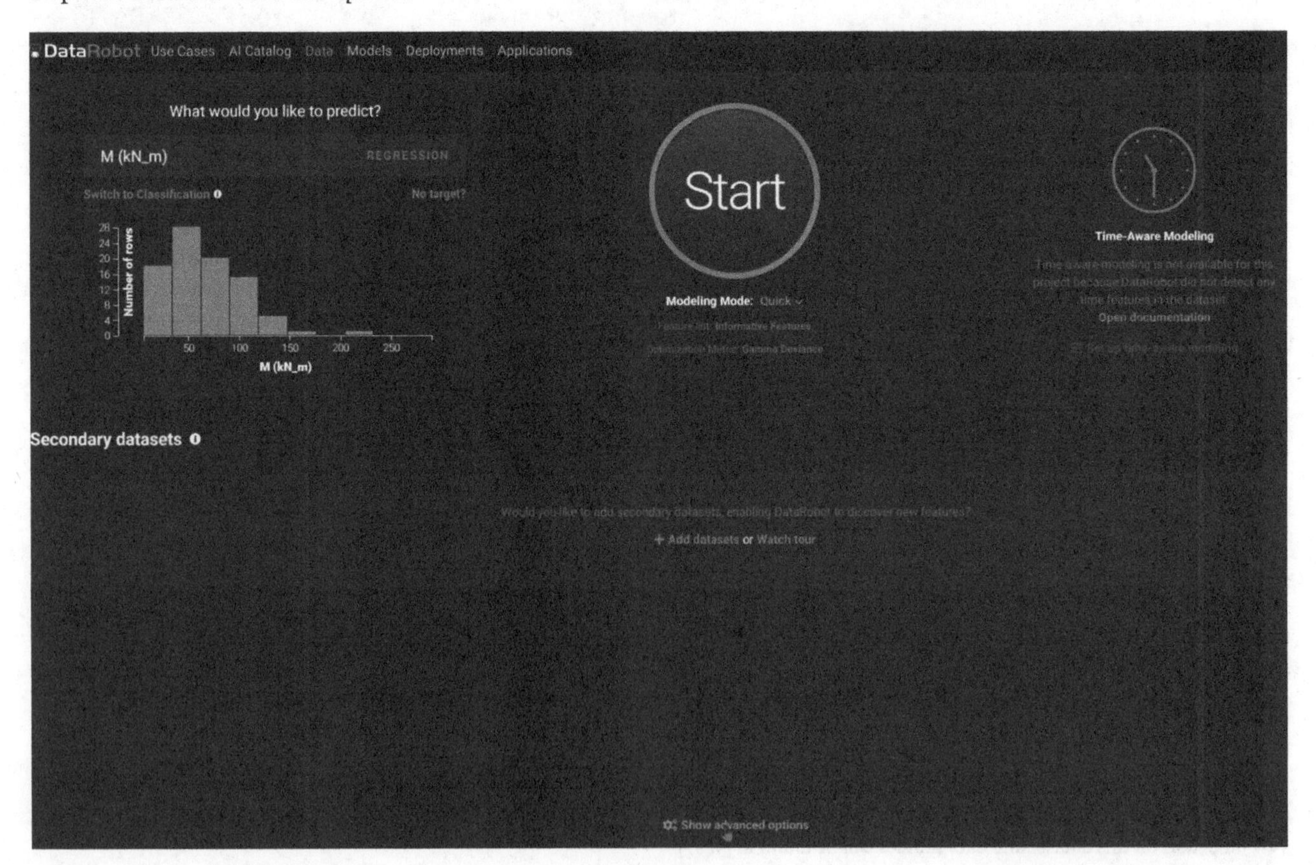

Step 4: Go to *Partioning* to check the default setting [keep as is].

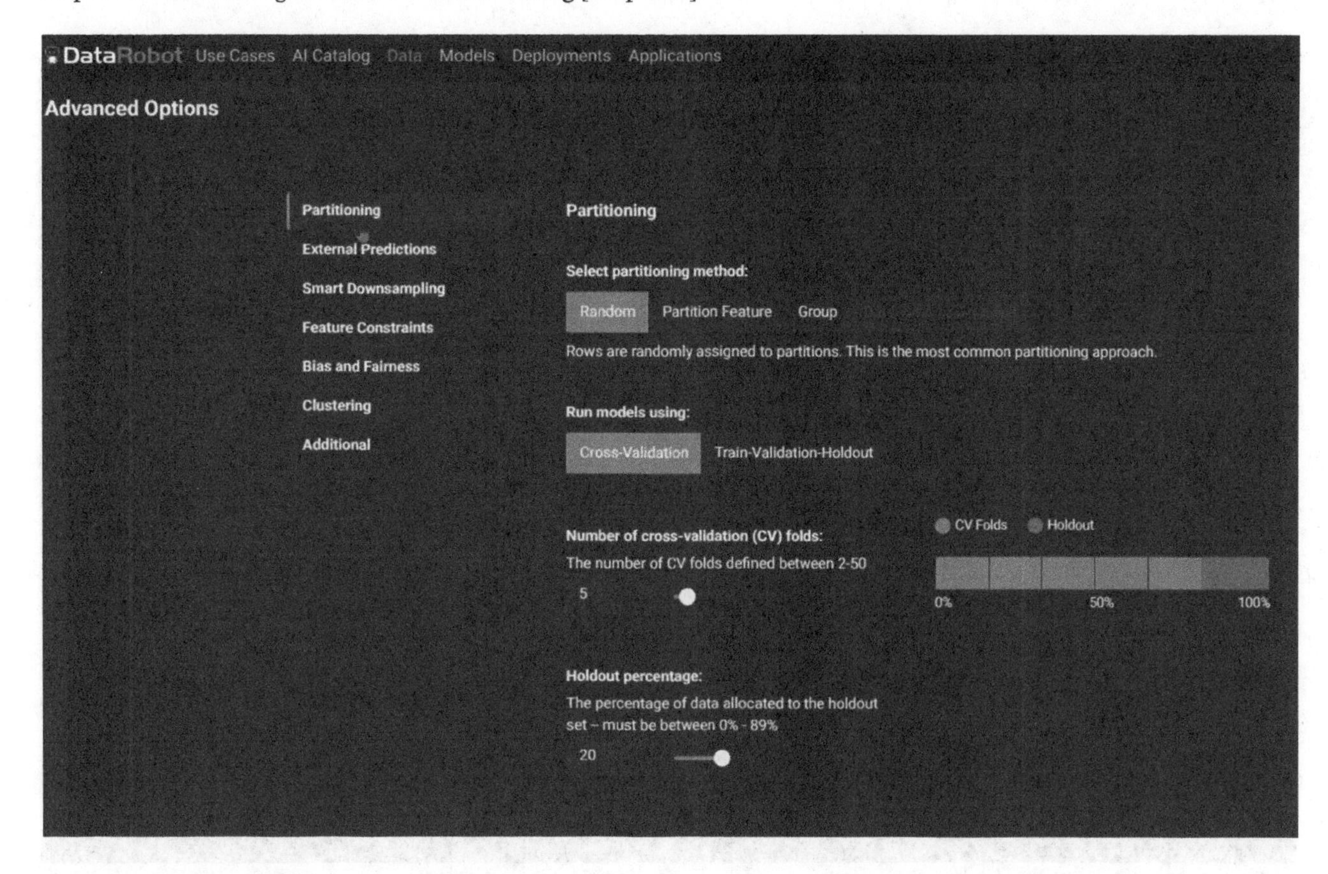

Step 5: Go to *Additional* and then update the settings as shown above. Then, *Hide advanced options.*

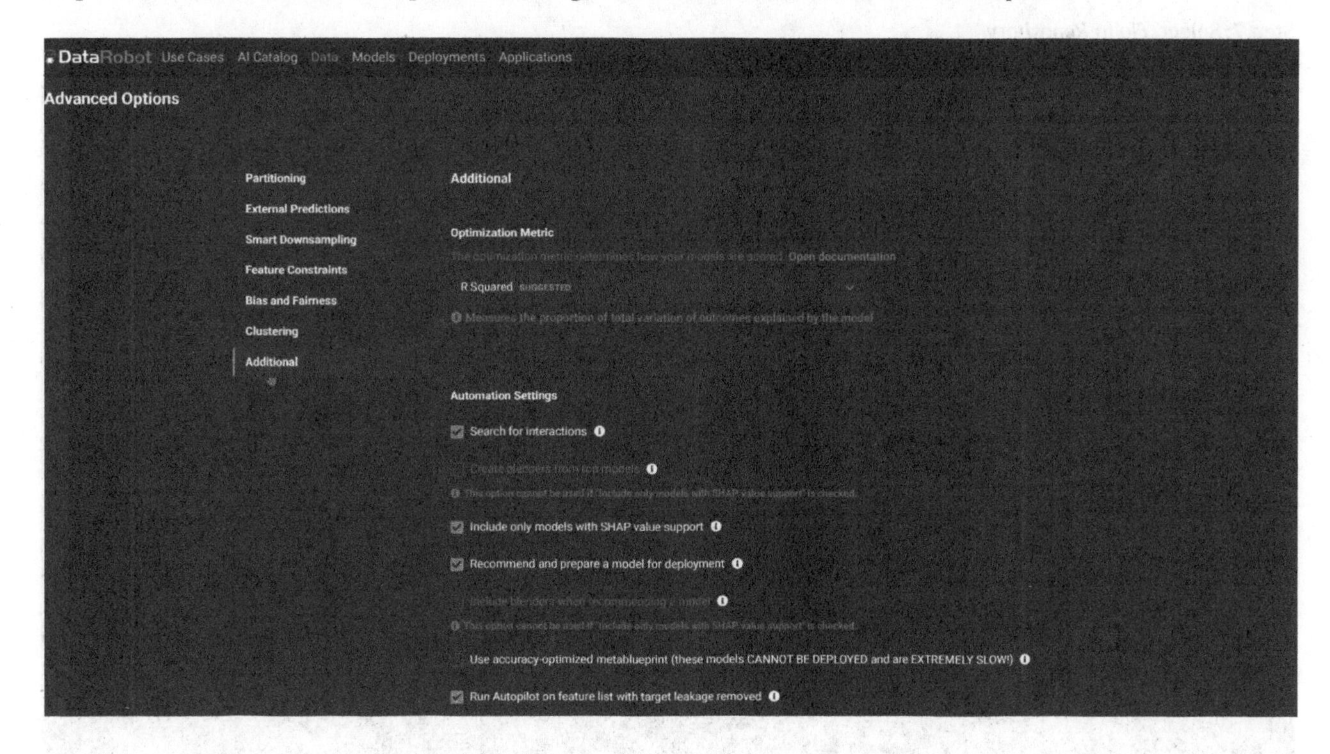

Step 6: Go to *Modeling Mode* and select *Manual*. Then, select, *Start*.

Step 7: Select, *Go to Repository*.

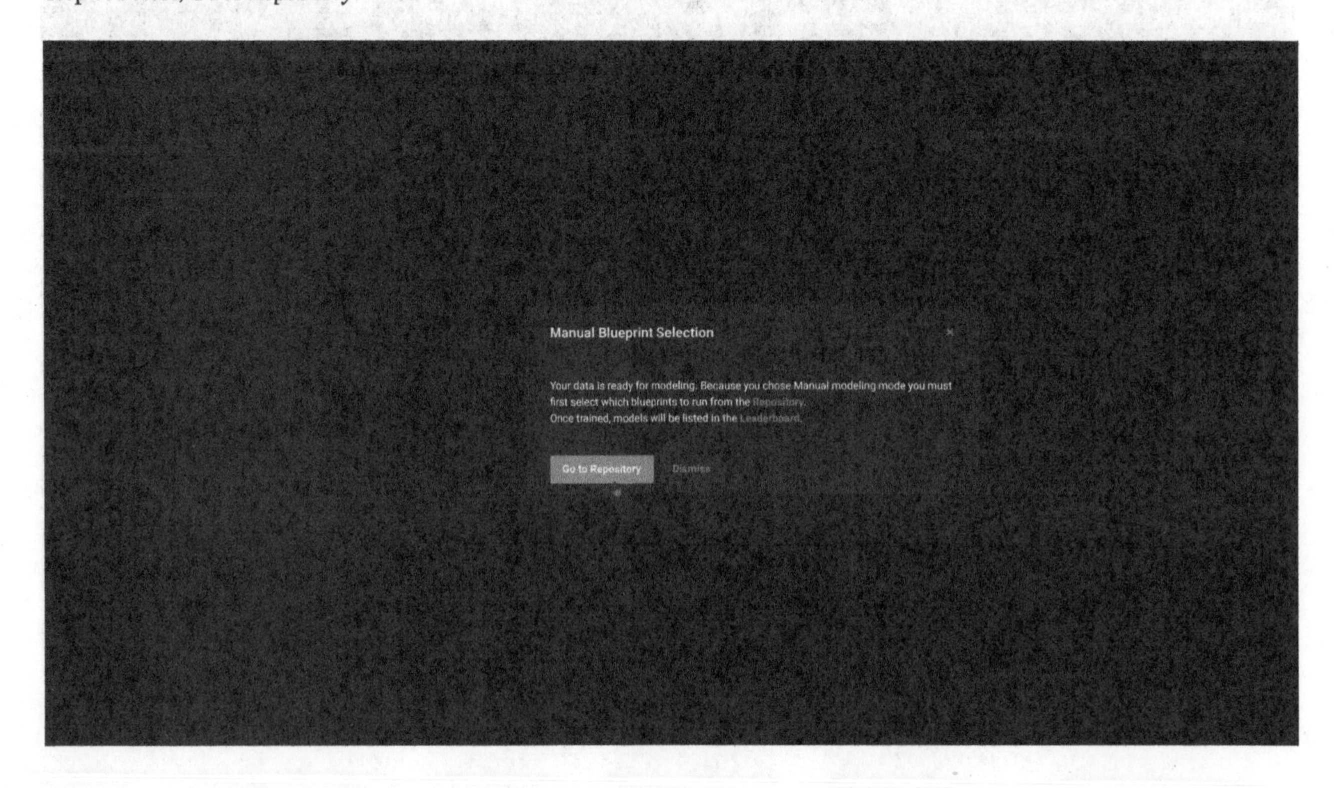

Step 8: Select, *Light Gradient Boosted Trees Regressor (4 leaves)* – BP 32. Then, *Run Task* and move to the *Models* tab.

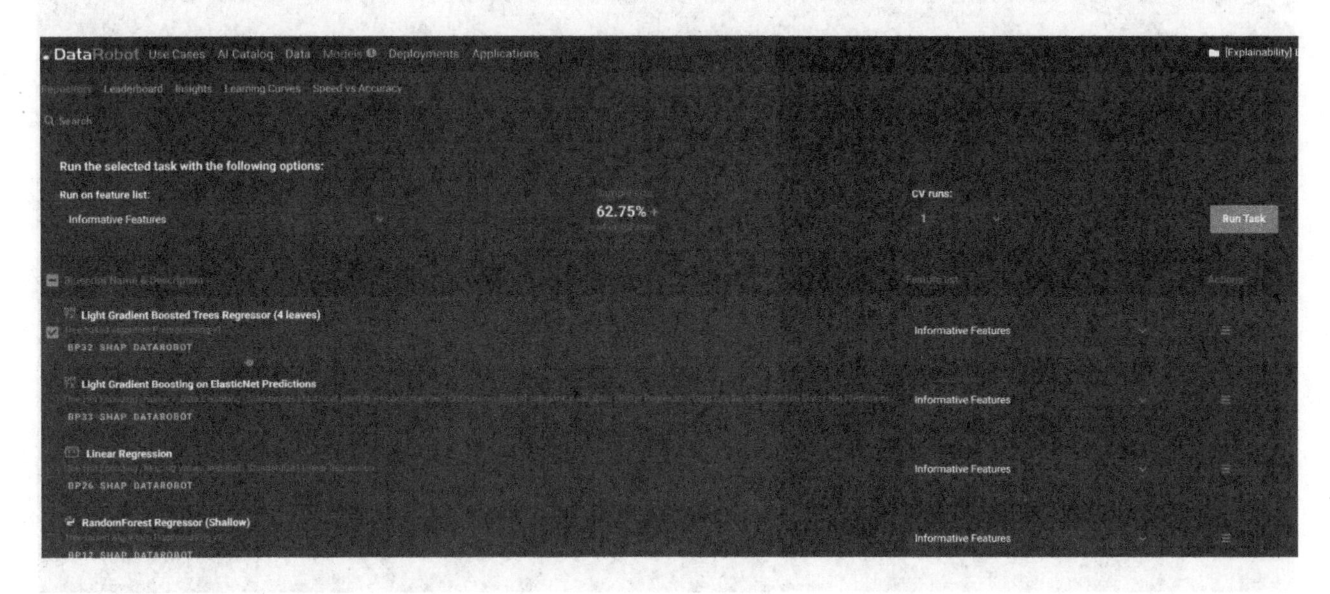

Step 9: While the model is running. Switch to the *Data* tab and check out statistical insights and *Feature Associations*.[47] As you can see, the minimum and maximum compressive and yield strength of the compiled beams range between 31 and 80 MPa and 352 and 788 Mpa, respectively. The steel and FRP ratios have a minimum and maximum range of 0–2.53% and 0.04–2.6%, respectively, typical of that used in RC beams design. The modulus of FRP ranges from low stiffness (11.7 Gpa) to high stiffness (240 Gpa). The *Feature Associations* shows that moment arm has a strong mutual information with the moment and also shows that beams' material properties and geometric features seem to have a medium mutual information with the observed moment at failure.

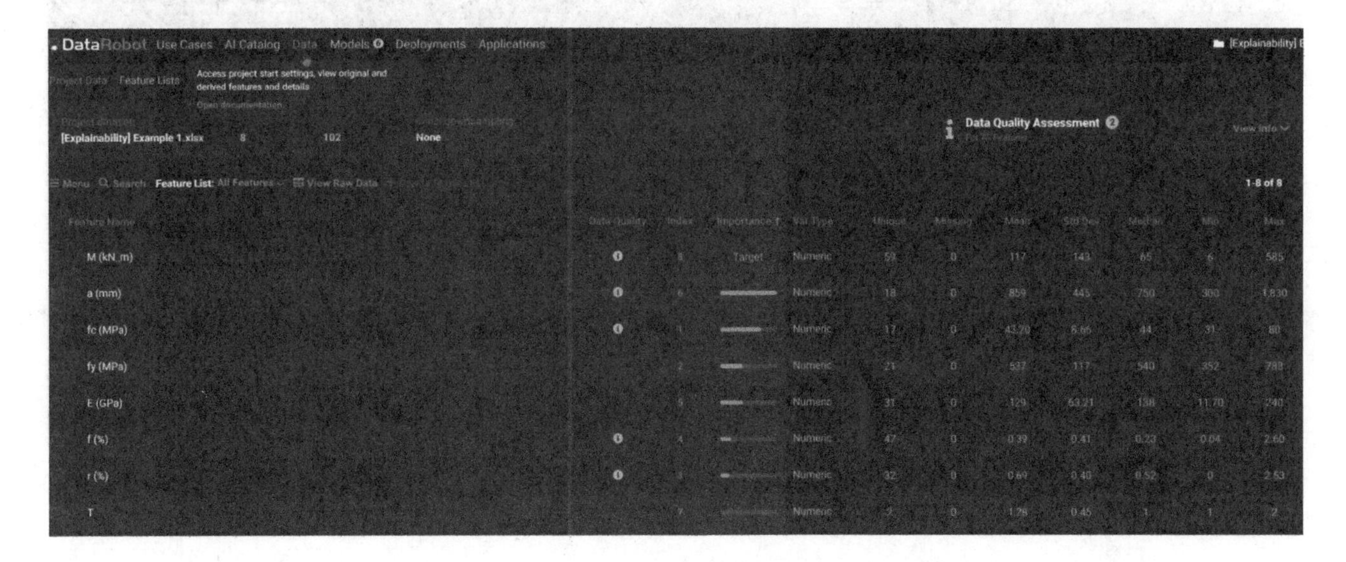

47 In theory, this should be an early step to carry out (as we would like to learn some statistical insights into our data).

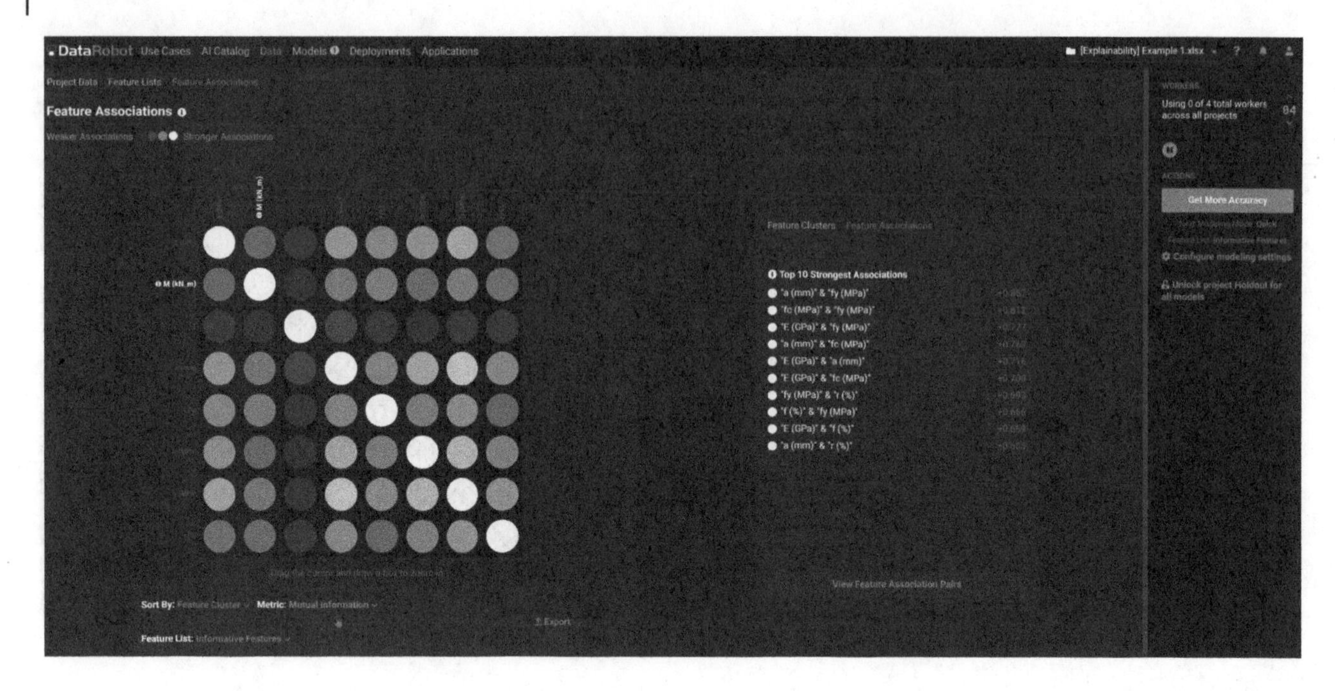

Step 10: You should now be able to see the outcome of the *validation* and *cross validation* process. Let us add two more models from the repository [*Decision Tree Regressor* (BP 25) and *Elastic-Net Regressor* (L2 / Gamma Deviance) (BP 11), *eXtreme Gradient Boosted Trees Regressor* (BP 22), *Generalized Additive Model* (PB2), *Keras Deep Residual Neural Network Regressor using Training Schedule (2 Layers: 512, 512 Units)* (BP4)] and *Run Tasks*.

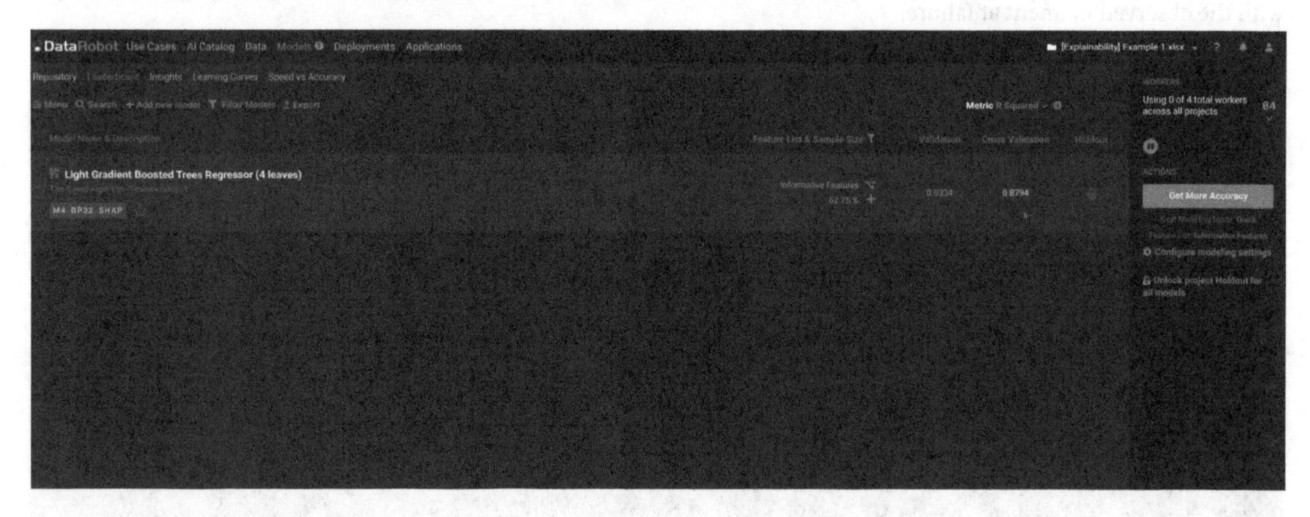

Step 11: You should be able to see the result for all the selected models as per the *Holdout* set. As you can see, the Decision Tree Regressor outperforms the others. Select this model and then navigate to the *Evaluate* tab to examine the performance of the model.

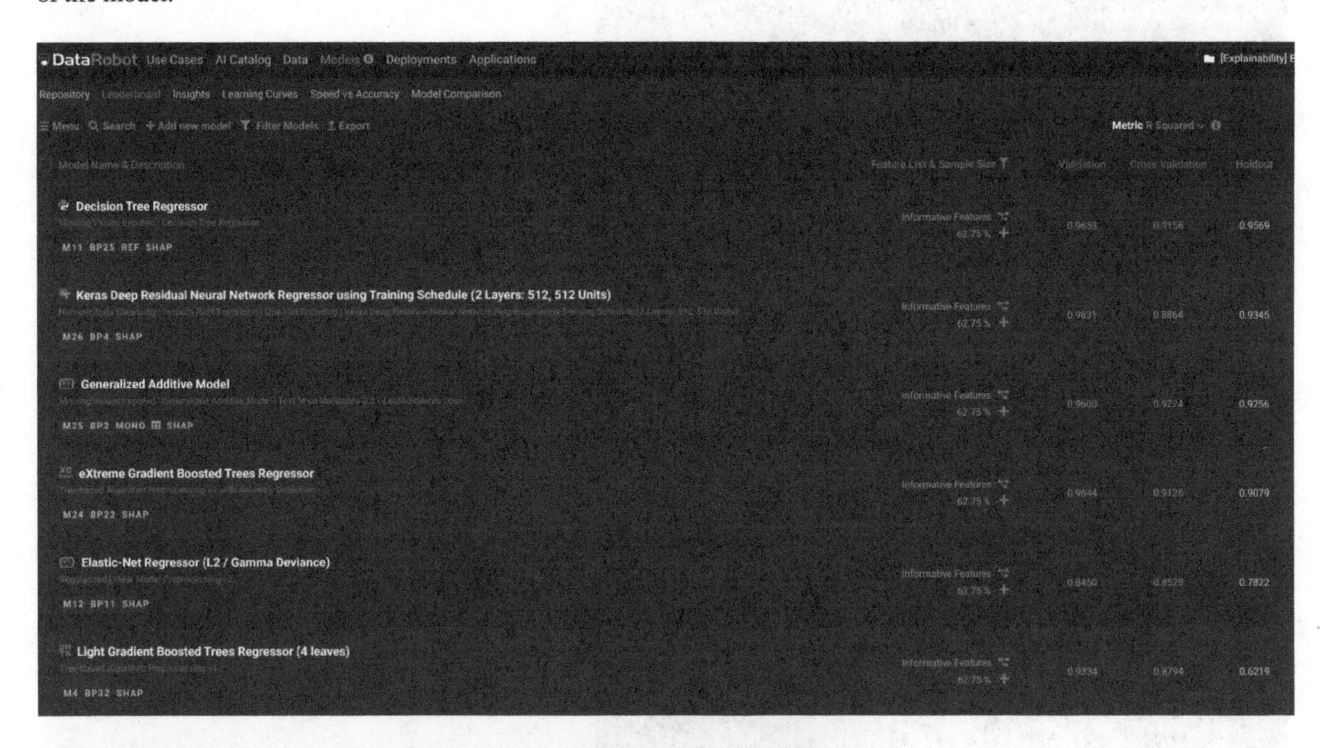

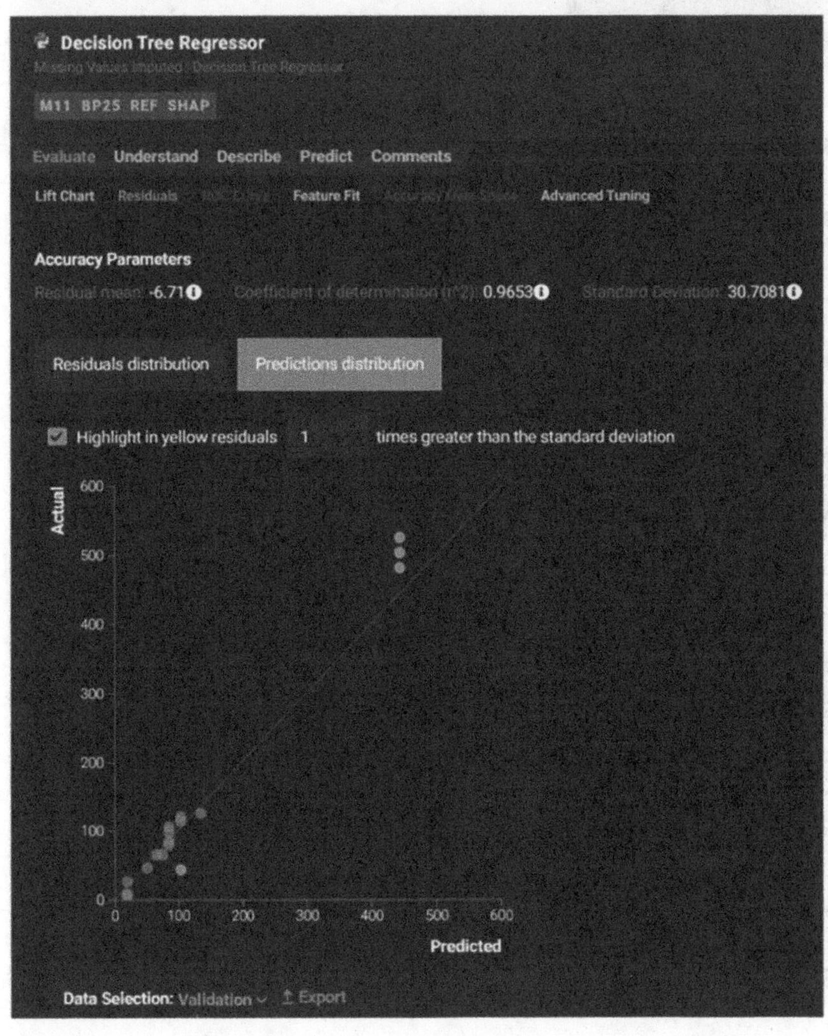

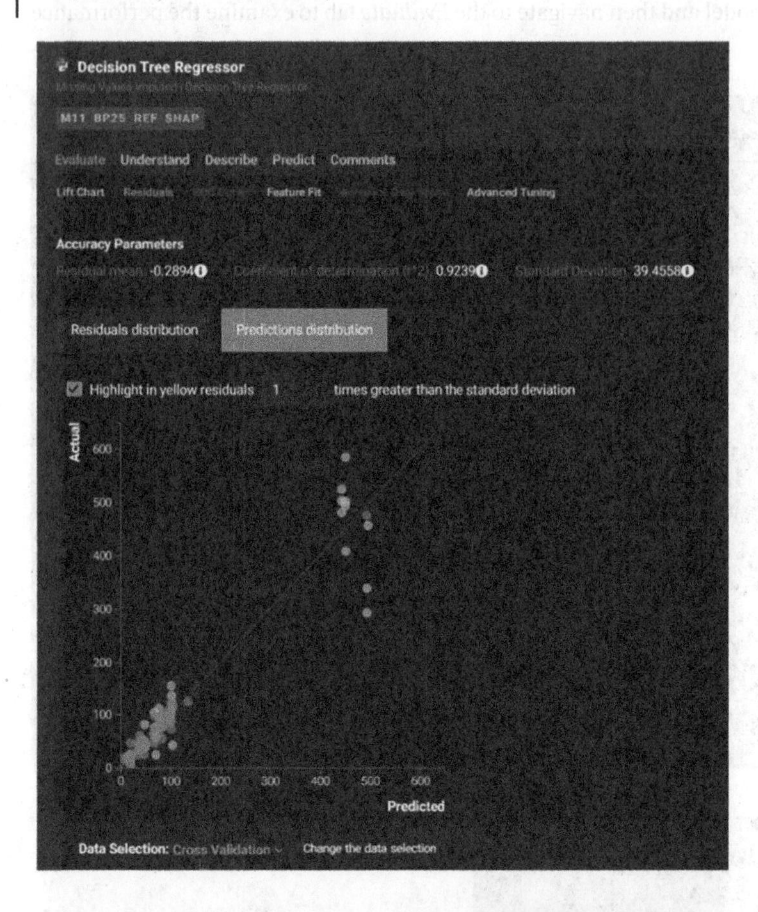
Decision Tree Regressor
M11 BP25 REF SHAP
Evaluate Understand Describe Predict Comments
Lift Chart Residuals Feature Fit Advanced Tuning
Accuracy Parameters
-0.2894
0.9239
Standard Deviation 39.4558
Residuals distribution
Predictions distribution
Highlight in yellow residuals 1 times greater than the standard deviation
Actual
600
500
400
300
200
100
0
0 100 200 300 400 500 600
Predicted
Data Selection: Cross Validation
Change the data selection

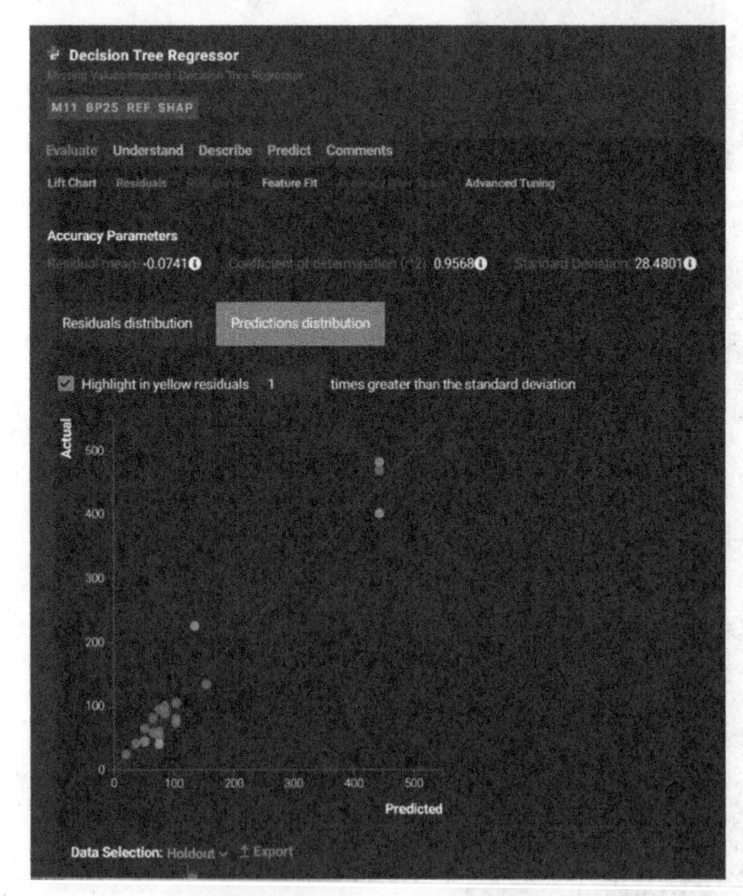
Decision Tree Regressor
M11 BP25 REF SHAP
Evaluate Understand Describe Predict Comments
Lift Chart Residuals Feature Fit Advanced Tuning
Accuracy Parameters
-0.0741
0.9568
Standard Deviation 28.4801
Residuals distribution
Predictions distribution
Highlight in yellow residuals 1 times greater than the standard deviation
Actual
500
400
300
200
100
0
0 100 200 300 400 500
Predicted
Data Selection: Holdout
Export

Step 12: Navigate to the *Understand* tab. Then, compute *Feature Impact* and *Feature Effects*. You should be able to see the following three snippets for *feature importance* and *partial dependence plots*.

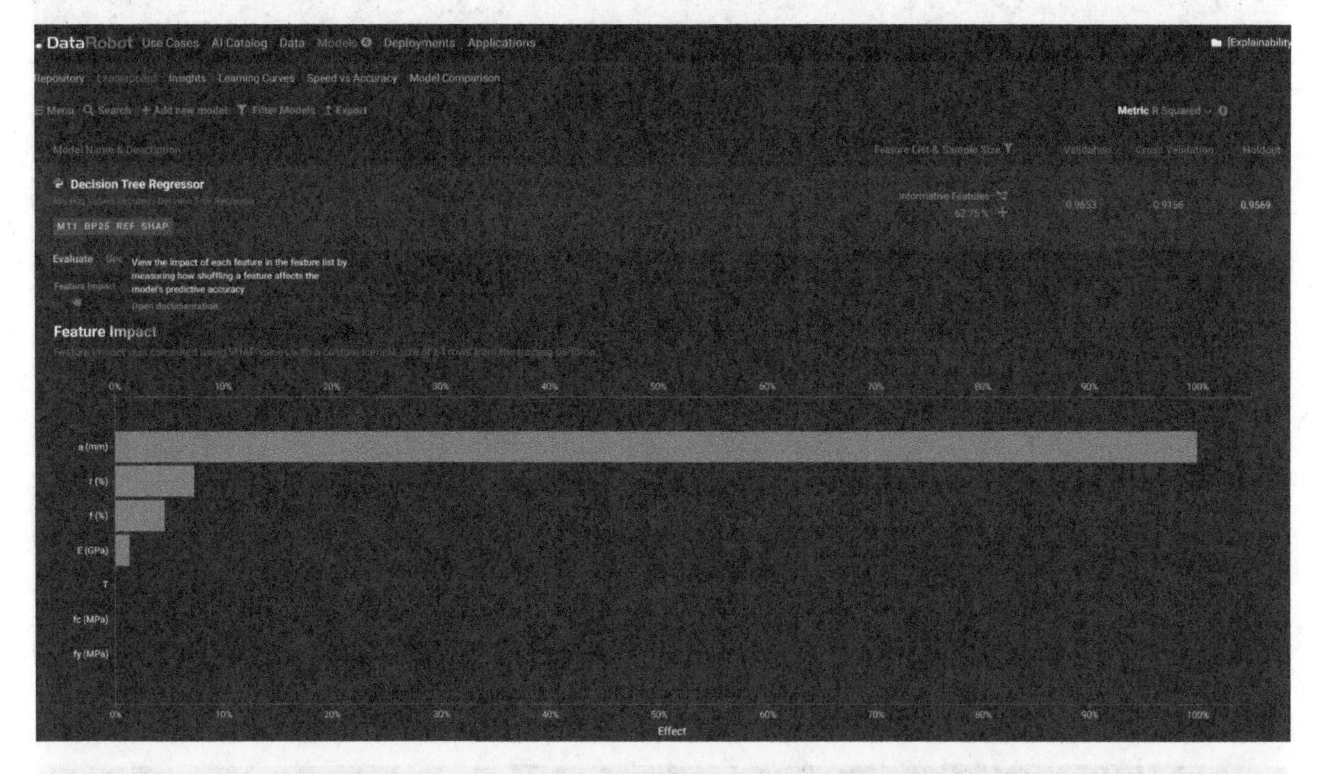

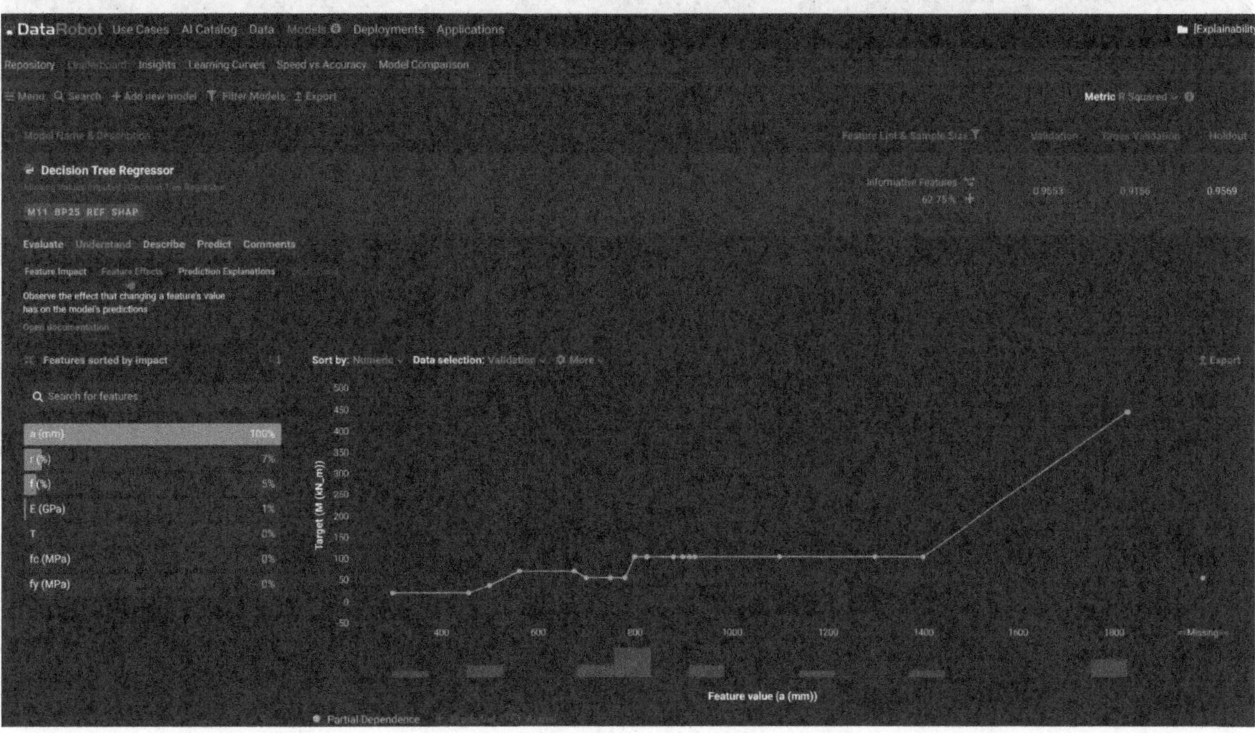

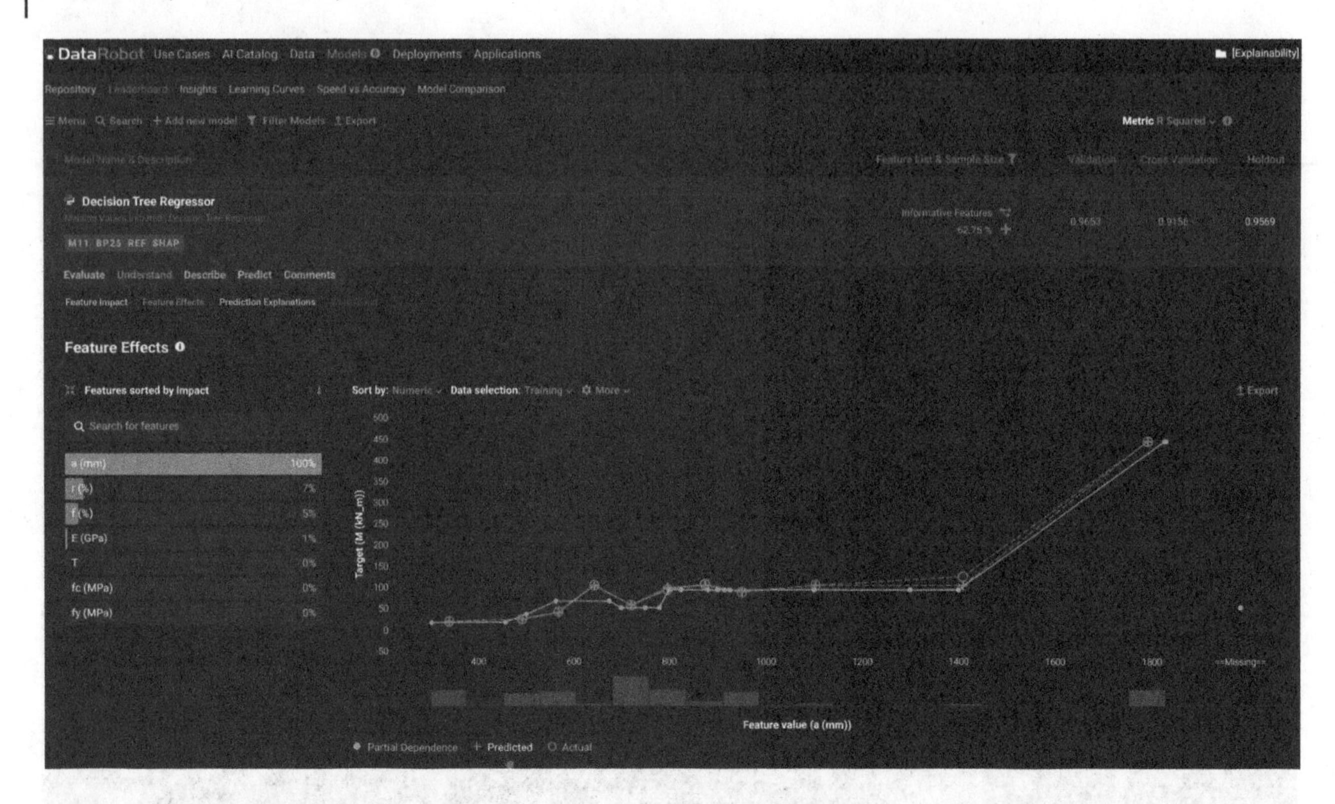

Finally, you can visit the *Describe* tab to check out an overview of the model.

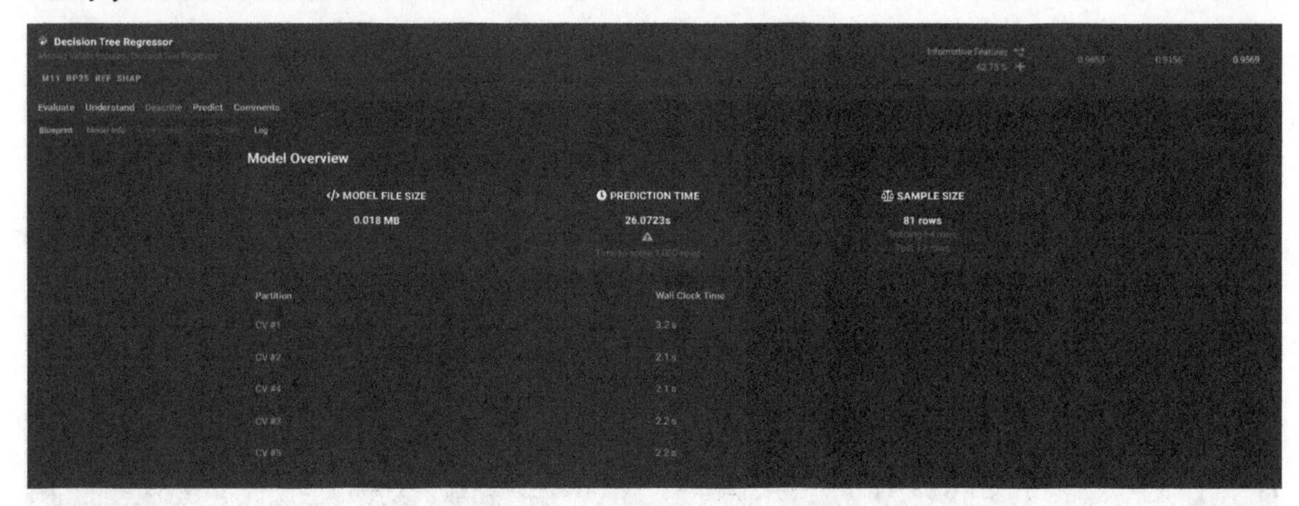

Now, let us compare the results of the Decision Tree Regressor against that of the second and third-ranked models (i.e., *Keras Deep Residual Neural Network Regressor using Training Schedule (2 Layers: 512, 512 Units)* (BP4) and *Generalized Additive Model* (PB2)). Navigate to the *Predict* tab and select *All data* and then *Compute predictions*.

The downloaded data would result in Figure 7.4. As you can see, the three models perform similarly and seem to capture the problem on hand. If we zoom in to the range of 0–200 kN.m, we can see the outperformance of the *Decision Tree Regressor* against the other two models.

Let us now compare the feature importance[48] as calculated by each model. Looking at Figure 7.5 shows that the *Decision Tree Regressor* heavily relied on four features, with the lion's share going to the moment arm (a) and three features scoring

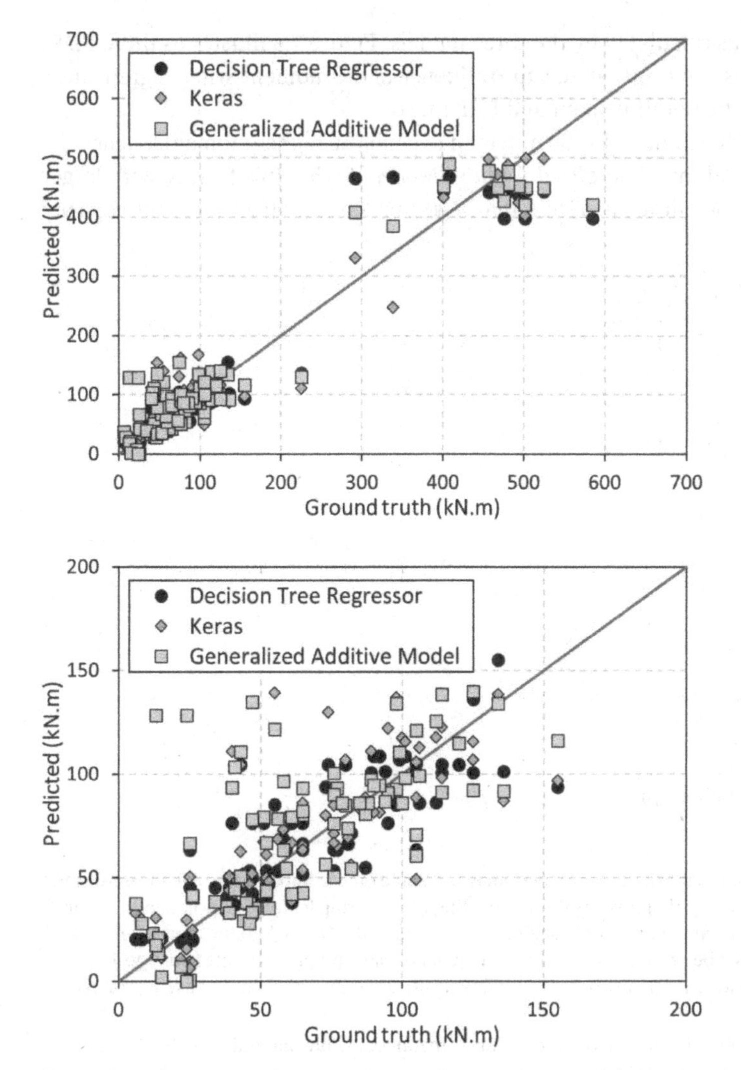

Figure 7.4 Comparison between the top three ranking models in example no. 1 (the bottom sub-figure presents the range between 0 and 200 kN.m).

48 Remember, feature importance showcases measure the extent to which a given feature influences the outcome of a ML model.

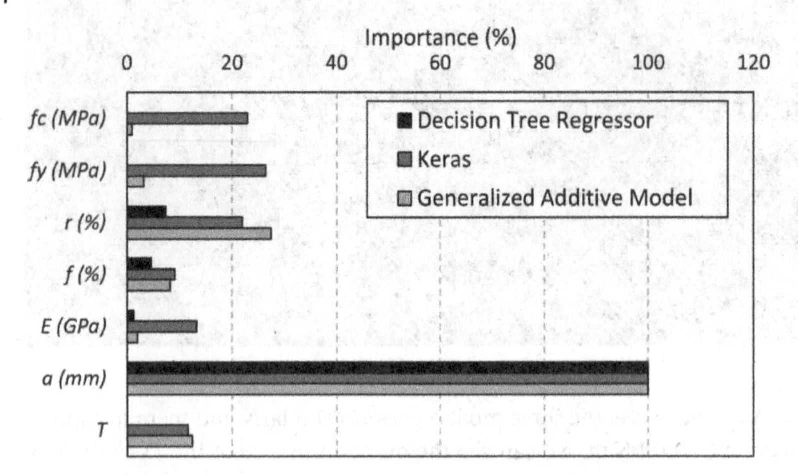

Figure 7.5 Comparison between models in terms of feature importance in regression.

zero (f_c, f_y, and T). The other two models also rely on a, and have noted the importance of all features with varying magnitudes.[49]

Let us now also compare the PDP[50] for the features as calculated by the three models. Figure 7.6 illustrates three PDPs comprising moment arm, reinforcement, and FRP ratios.[51] As you can see, the influence of the moment arm is significant[52] in predicting the moment capacity when compared to the reinforcement and FRP ratios.

For example, larger moment arms beyond 800 mm, 1400 mm, and 1600 mm tend to generate higher moments, hence the possible reasoning for *Keras, Decision Tree Regressor*, and the *Generalized Additive Model of* tying this feature with larger values of moment capacity.[53] Furthermore, smaller moment arms of less than 800 mm do not affect moment capacity

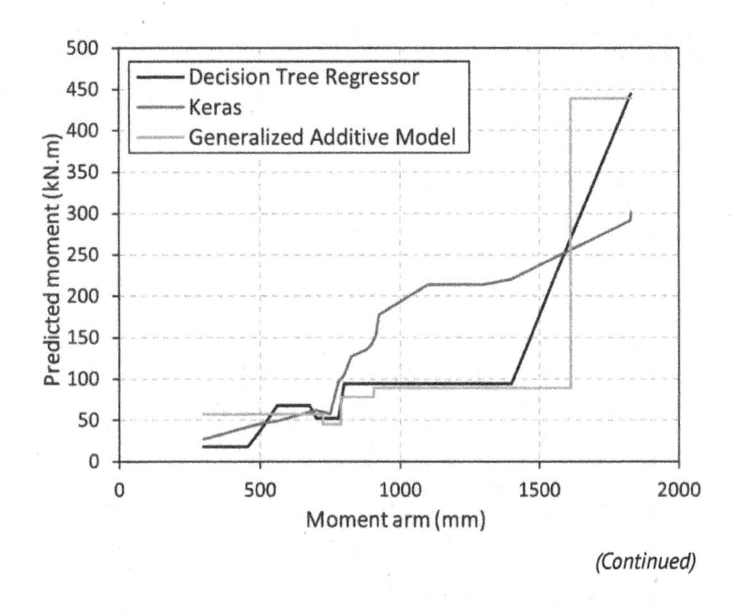

(Continued)

49 What can we learn from such analysis? Well, we can clearly see that ML models may or may not agree on the importance of features (despite having very close performance and goodness)! In addition, I would say that it is worth investigating re-occurring features between models such as a, r, and f. if you happen to be surprised the f_c was not as important as you have thought, here is what to do. Go back to the dataset and see the ranges of this parameter. It is evident that most of the collected beams had ranges that resemble normal strength concrete (average 43.2 MPa) with a few beams of higher strength. In other words, there were not many beams with drastically different f_c that could have warranted a higher importance.

50 Remember that a PDP depicts the marginal effect of an individual feature on the prediction of the model while maintaining all other features constant [15].

51 Calculated as the ratio of the cross sectional area of the FRP to the cross sectional area of concrete.

52 This can be seen by the range of the vertical axis in the PDP for a, as opposed for r or f. Key features tend to have a large span of influence.

53 This also agrees with domain knowledge! The reader is reminded that from a structural engineering perspective, a larger moment arm is expected to generate a higher bending moment (for identical beams and simply supported beams).

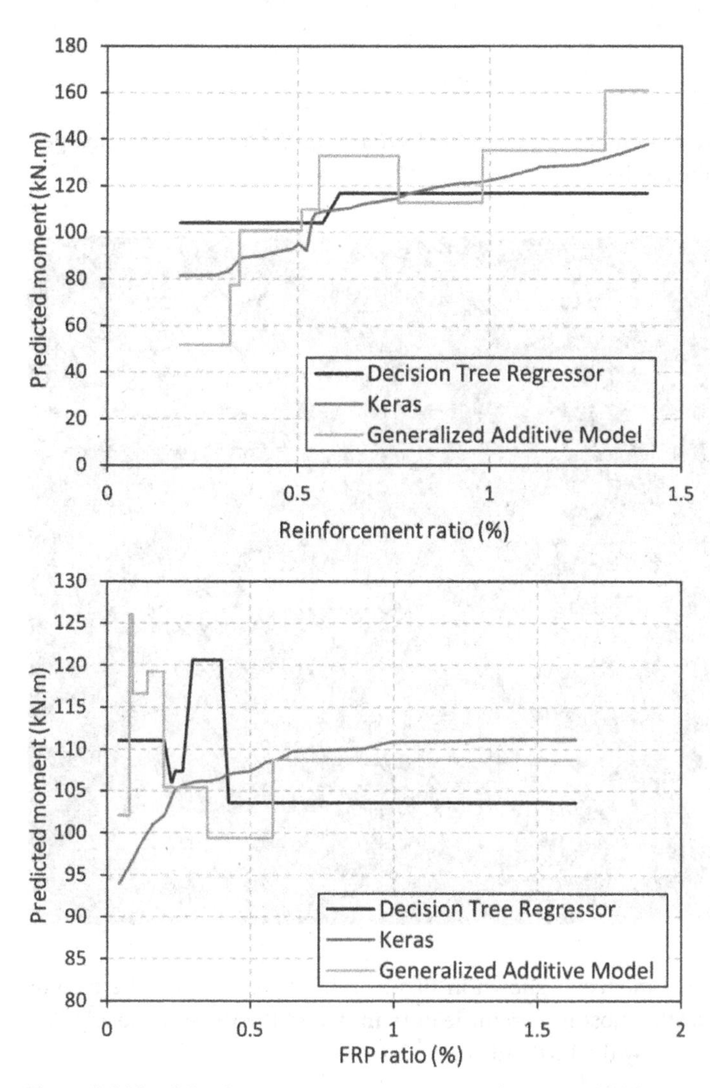

Figure 7.6 (Cont'd) Comparison between models in terms of PDP.

predictions in all three models. Furthermore, this figure shows how the other two features do not seem to influence the model's decision-making process much.

Now, let me repeat the above steps to build similar models but to predict the failure mode of the same beams.[54] I will be using the following three algorithms [*Decision Tree Classifier (Gini)* (BP 17[55]), *Generalized Additive Model* (PB2), *Keras Deep Residual Neural Network Regressor using Training Schedule (2 Layers: 512, 512 Units)* (BP4)].

Once the analysis is completed, you will be able to see the same results shown below.

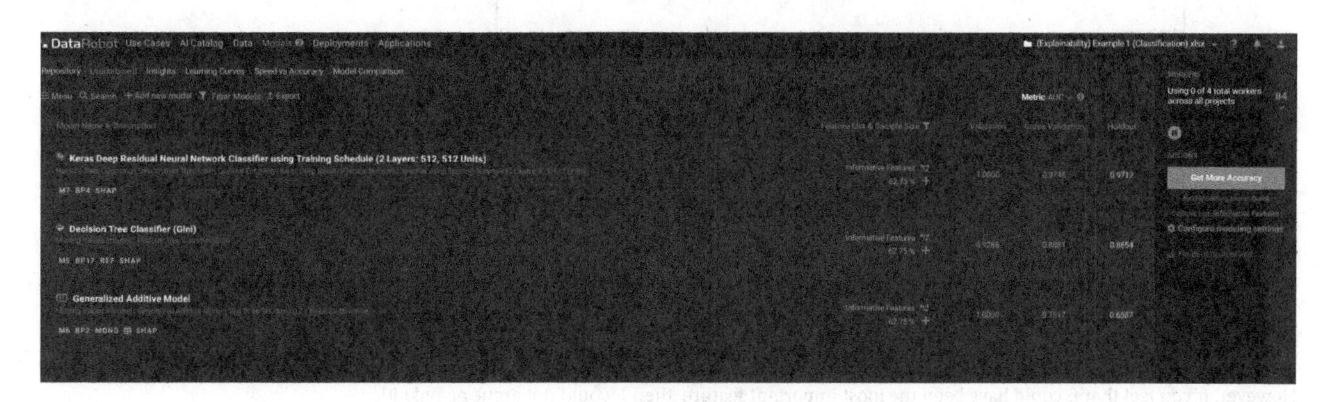

54 To save some pages, I will not be going over this example in a step-by-step manner. We will use the same steps in the regression part, but switching the analysis into a classification by selecting *Failure mode* as our *Target*. Please use *AUC* as the optimization metric.

55 Note: we can use the same decision tree algorithm we used earlier (BP 25) as that one was a regressor.

To dig deeper and evaluate the performance of the highest ranking model, select the *Evaluate* tab. Note that I have changed the *Metric* into *LogLoss* to show you the performance of the model.

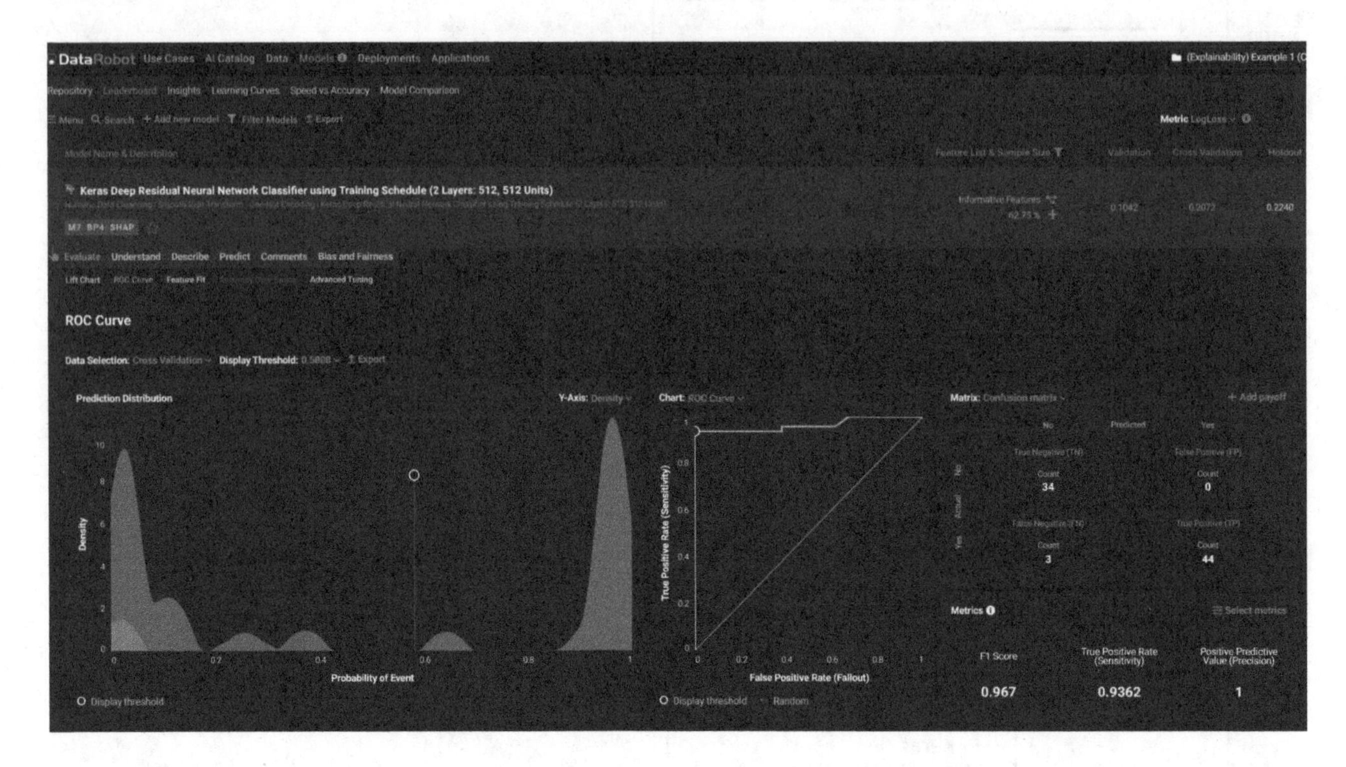

A comparison between the feature importance for all three models is shown in Figure 7.7. As you can see, the type of strengthening system (T) replaces the moment arm (a) as the most important feature in two of the three models.[56] The second feature turns out to be the yield strength (f_y), followed by the FRP ratio (a).[57]

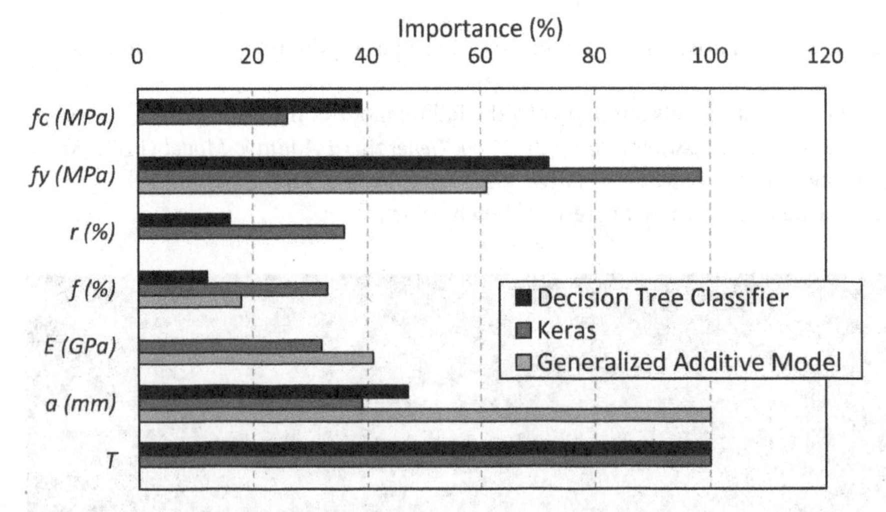

Figure 7.7 Comparison between models in terms of feature importance in classification.

56 If you are wondering why I picked T over f_y (for example, since it ranks well in all models), it is because 100% it is hard to overlook. However, if you feel that f_y could have been the most important feature, then I would not argue against it!

57 Compare those to that obtained from the regression analysis! Makes you think! The reason for this discrepancy is that the two phenomenon (moment capacity at failure) and failure mode could be thought of as dependent. For example, moment capacity is the point where a beam *fails* in bending.

Another comparison is one where we compare the PDPs as calculated by the three models (see Figure 7.8). As you can see, using a non-traditional strengthening method tends to lead to beams failing due to failure in the FRP system. A look into the yield strength shows that beams with a lower yield strength (< 500 Mpa) were noticeably failing due to the failure of the FRP strengthening system.[58] This figure shows that lesser values of (i.e., shorter) moment arm are directly related to an increase of failure through FRP debonding, which agrees with experimental observations wherein shorter moment arms tend to intensify shear effects, and lower strength of reinforcement amplifies the utilization of FRP systems, making it vulnerable to debonding [23].[59]

Before I end this example, I would like to discuss a few items. Let me combine the feature importance from the regression and classification analyses into a new figure first (see Figure 7.9). Intuitively, we can see that the moment arm is the most re-occurring feature in both analyses, which is then followed by the ratio of FRP.

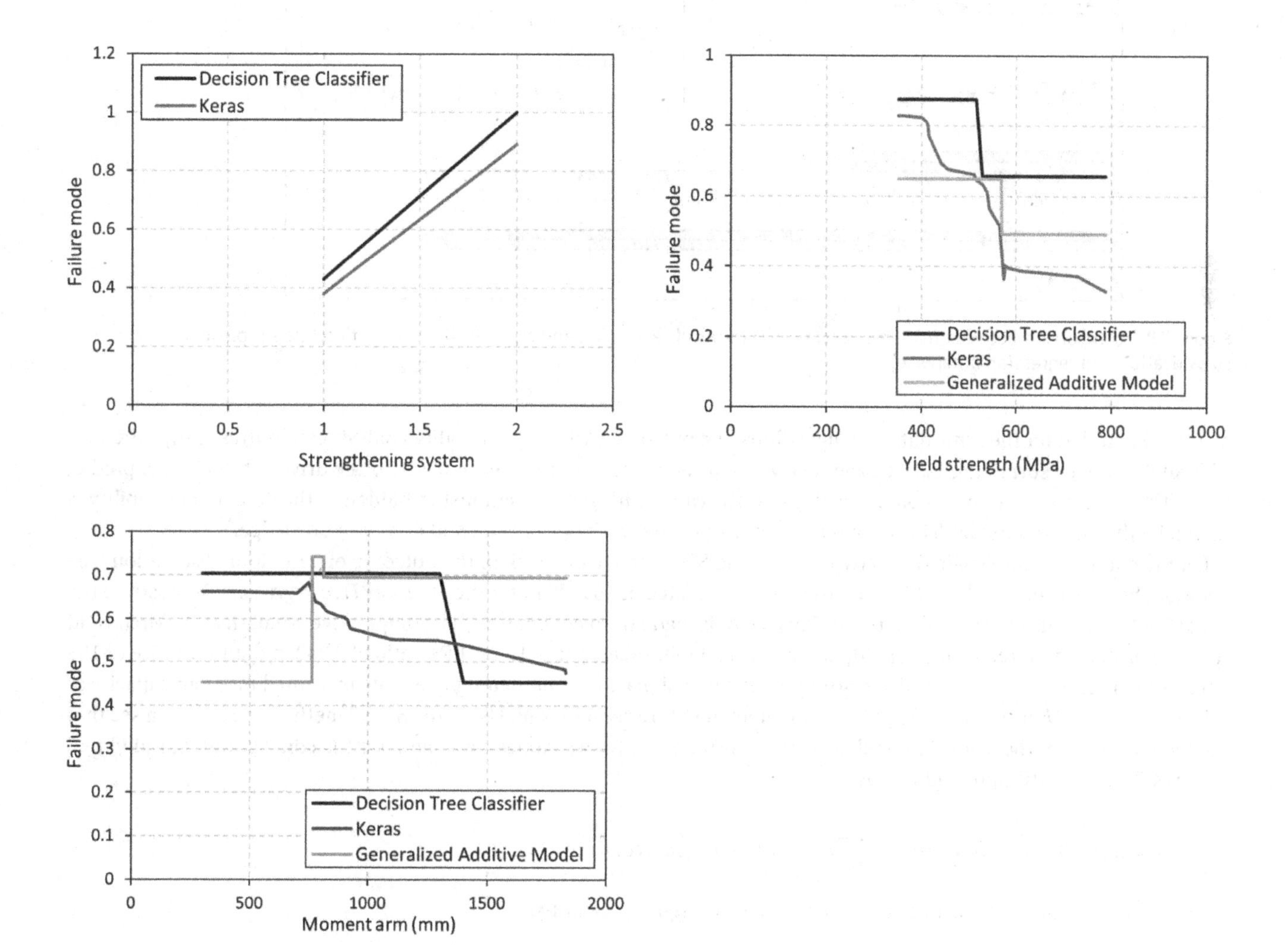

Figure 7.8 Comparison between models in terms of PDP [Note: Failure mode = 1 for beams with FRP system failure and zero otherwise. Also, Strengthening system = 1 for FRP plates and 2 for NSM].

58 Please remember that in classifying a failure mechanism in such a beam, a balance between steel and FRP primarily dictates the expected failure mode (see [38, 39])

59 Please note the PDP from the *Generalized Additive Model* tend to disagree with the other two models here. You are likely to encounter such observation yourself at some point in time!

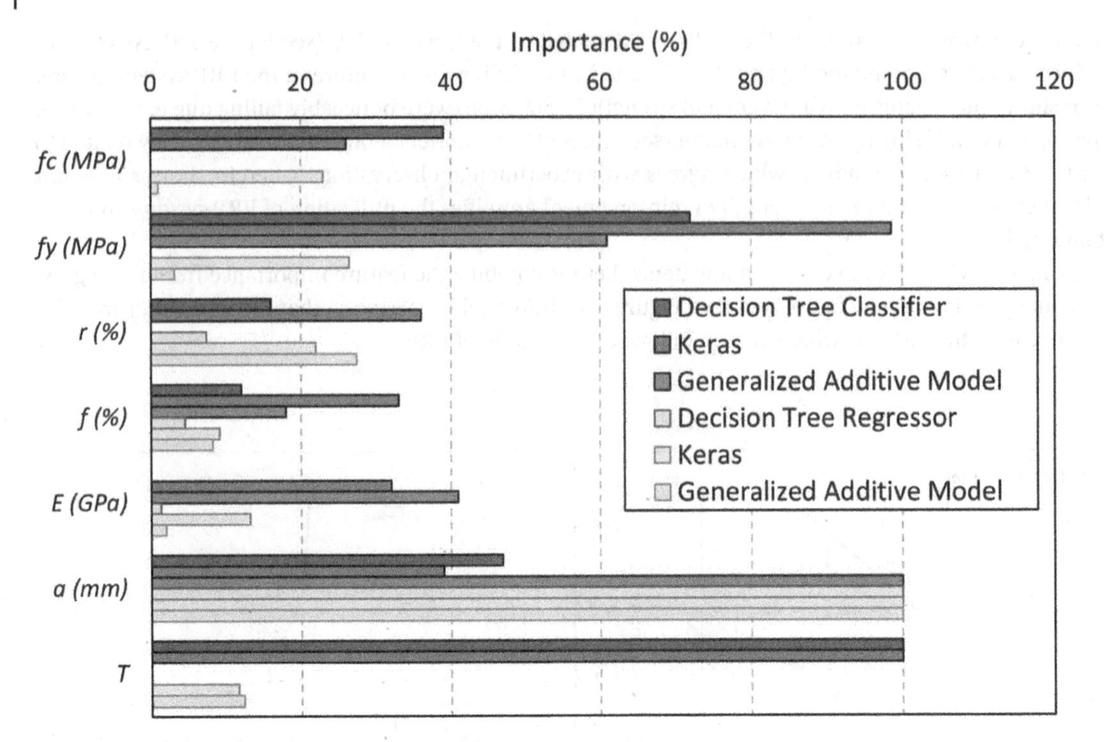

Figure 7.9 Comparison between feature importance in terms of regression and classification [Note: filled bars represent classification and regression otherwise].

It is evident that for the same data and algorithms, the outcome of the explainability analysis can be drastically different. This should not be surprising. As we have reviewed earlier, explainability aims to describe the drive behind model predictions. This does not mean nor equate to arriving at the *cause and effect*[60] relationship hidden in the data! Explainability is driven by the data and model. The same may or may not agree with our physics and domain knowledge.[61]

Let me end this example with a discussion on using SHAP to locally explain the outcome of a random observation (see below). The *base value* for the SHAP[62] method was calculated at 109 kN.m for the *Decision Tree Regressor.* To illustrate the SHAP method, let us take a random beam. This beam is made from concrete with a compressive strength of 31 Mpa, yield strength of steel reinforcement of 414 Mpa, ratio of steel reinforcement of 1.122, FRP ratio of 0.242, modulus of FRP of 183 Gpa moment arm of 1830 mm, and was strengthened with a plated strengthening type. The measured moment capacity of this beam is 476 kN.m. The model predicts a moment at failure as 444 kN.m. Using the SHAP method, we can then see that the model arrives at the 444 kN.m predictions as a result of the SHAP values calculated for the contribution of individual features. These contributions add up as:

$$Local\ prediction = base\ vlaue + \sum summation\ of\ features \tag{7.5}$$

$$Local\ prediction = 109 + (333.65 + 2.01 + 0.2689 - 0.934) = 444\ \text{kN.m}$$

60 I have a whole chapter on this!

61 Please remember that these are important ideas, however some problems may not require us to satisfy these. Understand your problem and what are you expecting from the model.

62 Please note that *DataRobot* does not incorporate LIME at the moment. I will be showing LIME in the coming examples.

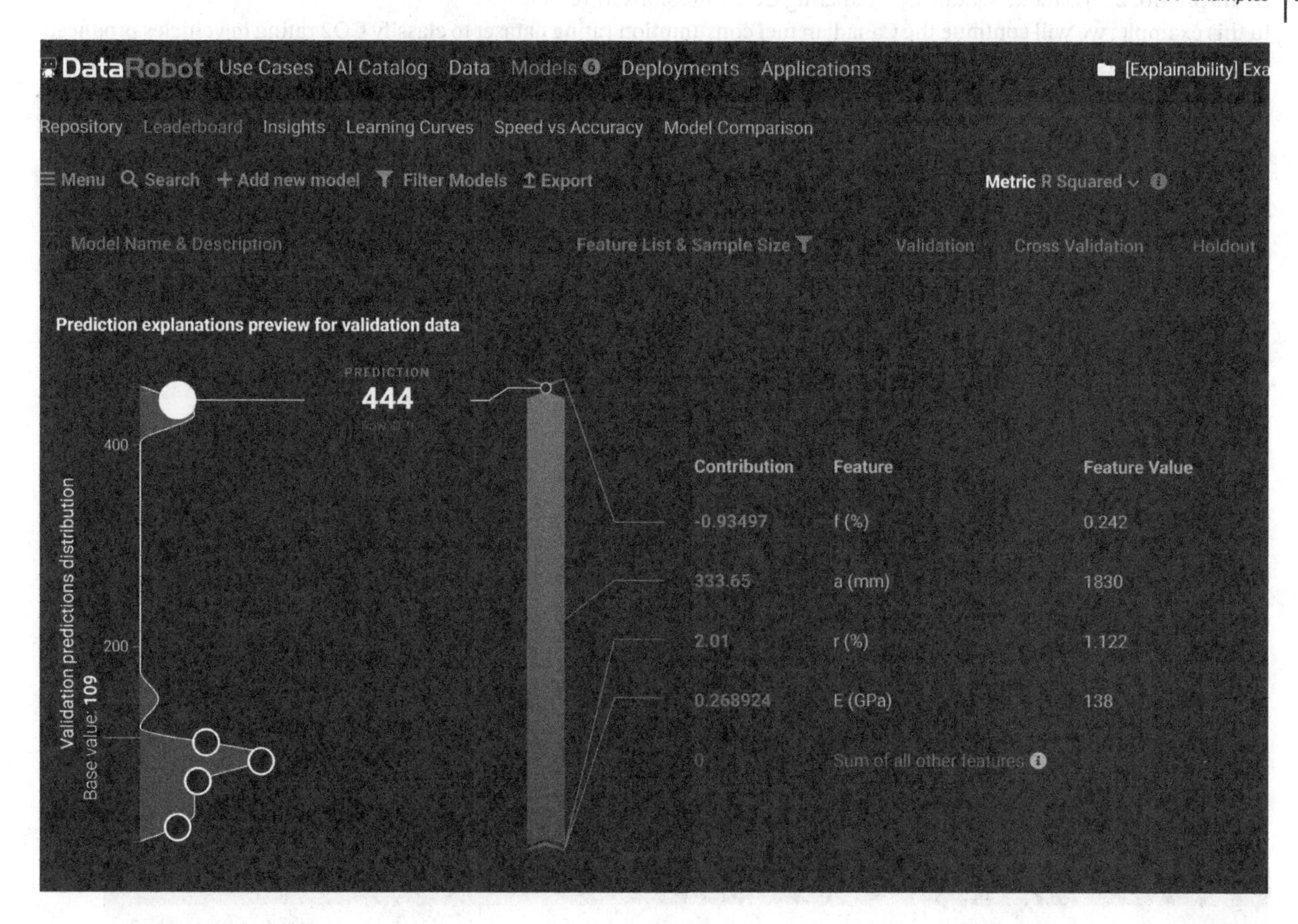

7.4.1 Surrogates*

A surrogate[63] is a simple model that is used to augment the predictions from a more complex model.[64] In this example, I developed two surrogates; one from the generalized additive model (GAM) and another from the logistic regression (LR).

The GAM model can be represented by a linear formula that applies the inverse of the *link* function (i.e., exponential function) being multiplied to the sum of all standardized features multiplied by a computed coefficient which could be accompanied by an intercept (see Equation 7.6). On the other hand, the LR model is another simple model that can come in handy in classification problems (see Equation 7.7).

$$Prediction_{GAM} = link\ function\begin{bmatrix}(63.31\times a)+(16.84\times r)+(-5.83\times f_y)+(-20.44\times f_c)\\+(-1.22\times f_r)+(-2.28\times E_{fr})+(-7.28\times T)+94.10\end{bmatrix} \tag{7.6}$$

$$Prediction_{LR} = logit\begin{bmatrix}(-1.03\times a)+(-1.74\times r)+(0.24\times f_y)+(-0.0002\times f_c)\\+(0.82\times f_r)+(1.06\times E_{fr})+(0.10\times T)+1.39\end{bmatrix} \tag{7.7}$$

Please note that all features in Equations 7.6 and 7.7 are standardized.

63 As I mentioned before, I will be going over surrogates and sharing my opinion on them in an upcoming chapter. For now, you may skip this section and revisit it after you complete that chapter. You may also check out the following paper for more discussion on the developed surrogates herein [4].

64 Say, Model A is complex. What we can do is build a surrogate model that we train using Model A's predictions (not using the ground truth). If you are thinking about the amplification of error, then those exist. If you are thinking about it, why not just apply the simple surrogate to the ground truth data? Then, that is a good question, and the short answer is, yes, you can! Sometimes we cannot know how complex a model can get before we start our analysis!

Example no. 2 [Dataiku platform]: Predicting CO2 Emissions in vehicles

In this example, we will continue the Canadian fuel consumption rating dataset to classify CO2 rating in vehicles produced in 2022 [24]. This dataset contains 945 observations and four features (engine size, number of cylinders, and fuel consumption in the city) and one target, CO2 rating[65] (10 classes). We will add explainability measures.

Step 1: Upload and create your dataset.[66]

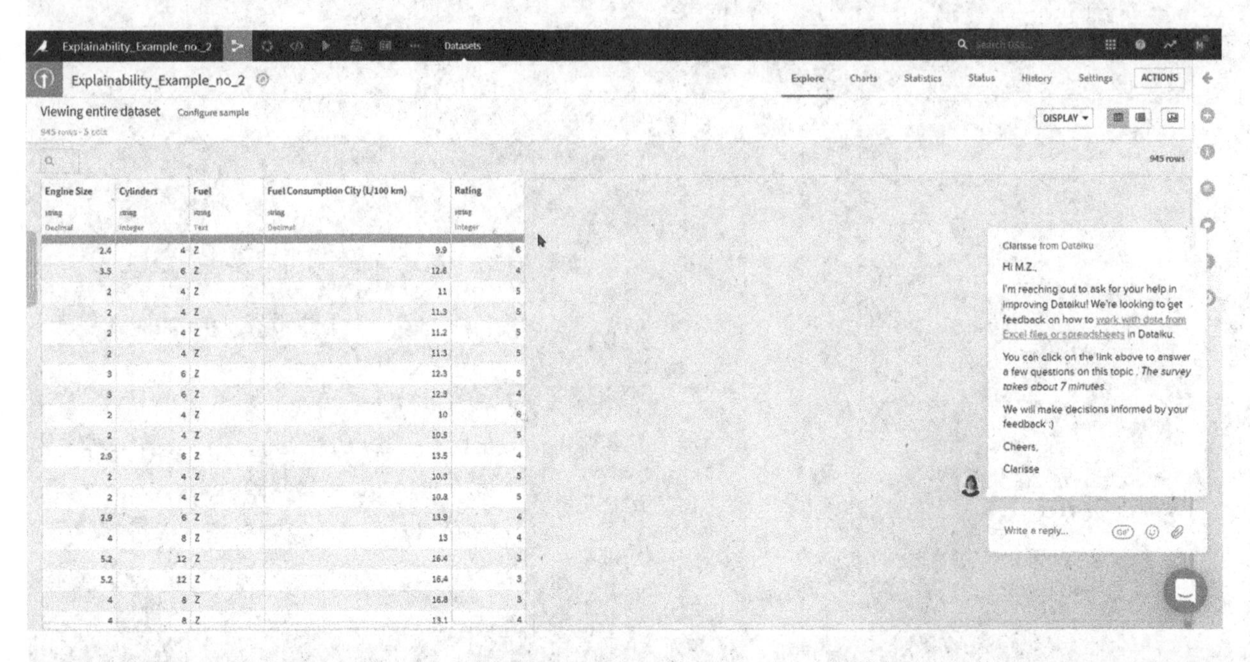

Step 2: Visit the Statistics tab and select a *Multivariate analysis*. Opt to select *Correlation analysis* and then select *Spearman*.

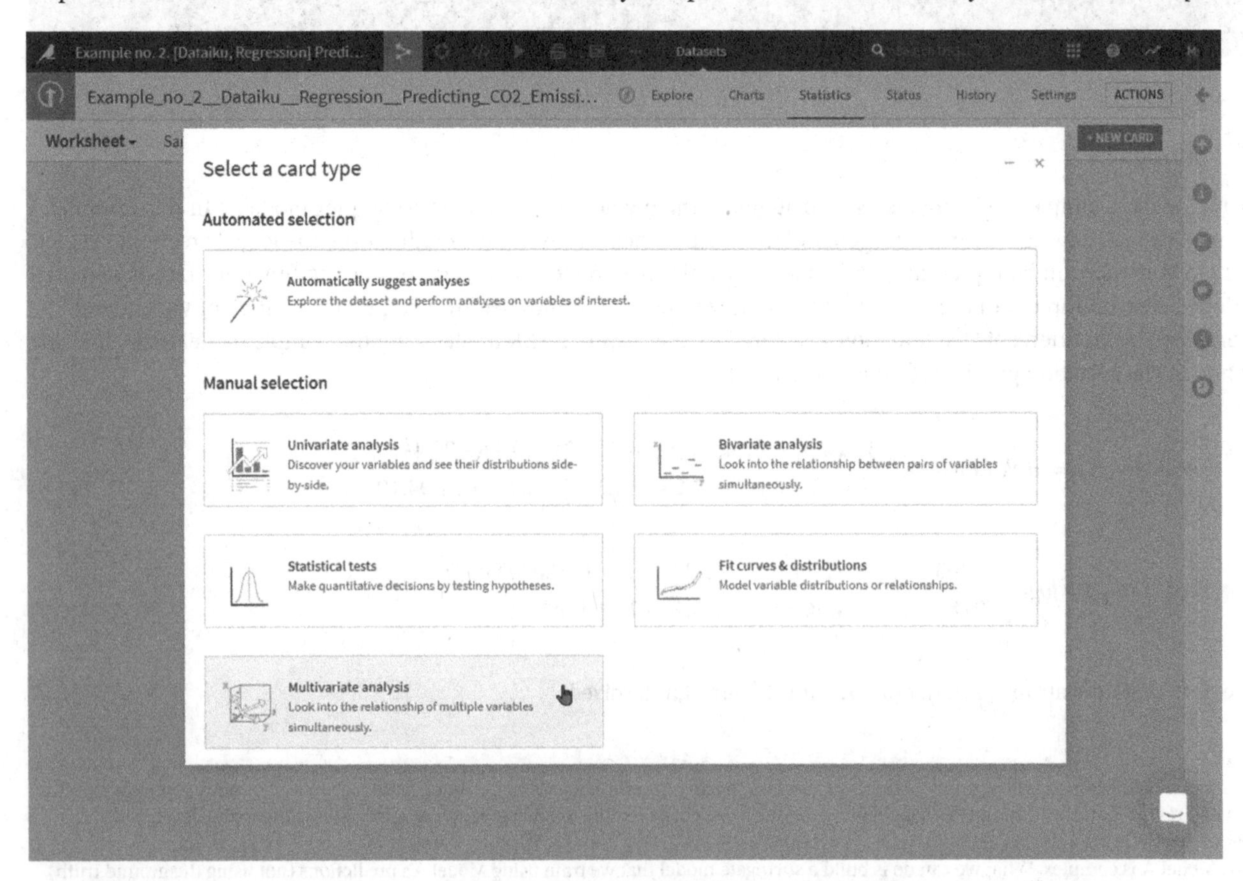

65 The next example will cover the regression counterpart by assigning the CO2 emissions as the target.

66 You may follow Step 1 from the previous example.

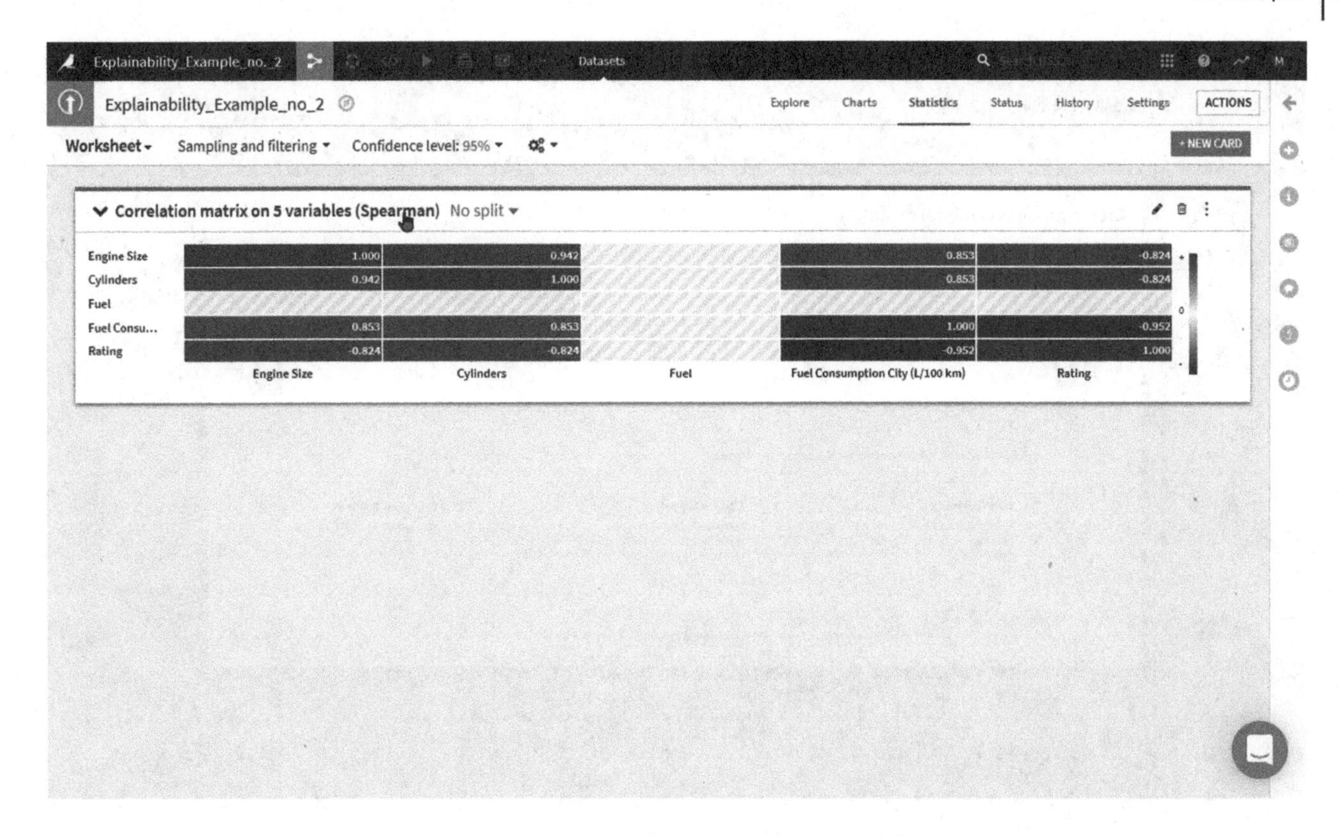

Step 3: Create *Interpretable Models*. Update analysis settings by adding new ML algorithms.

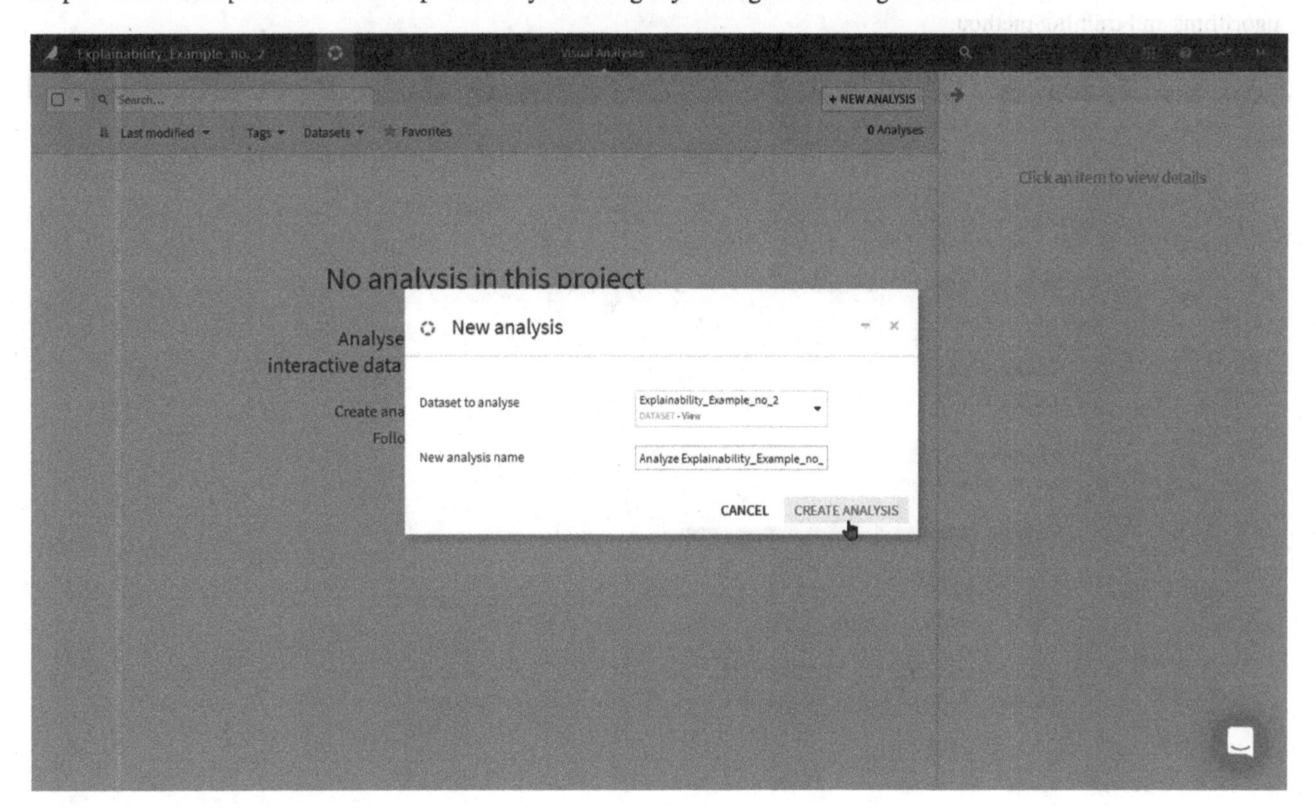

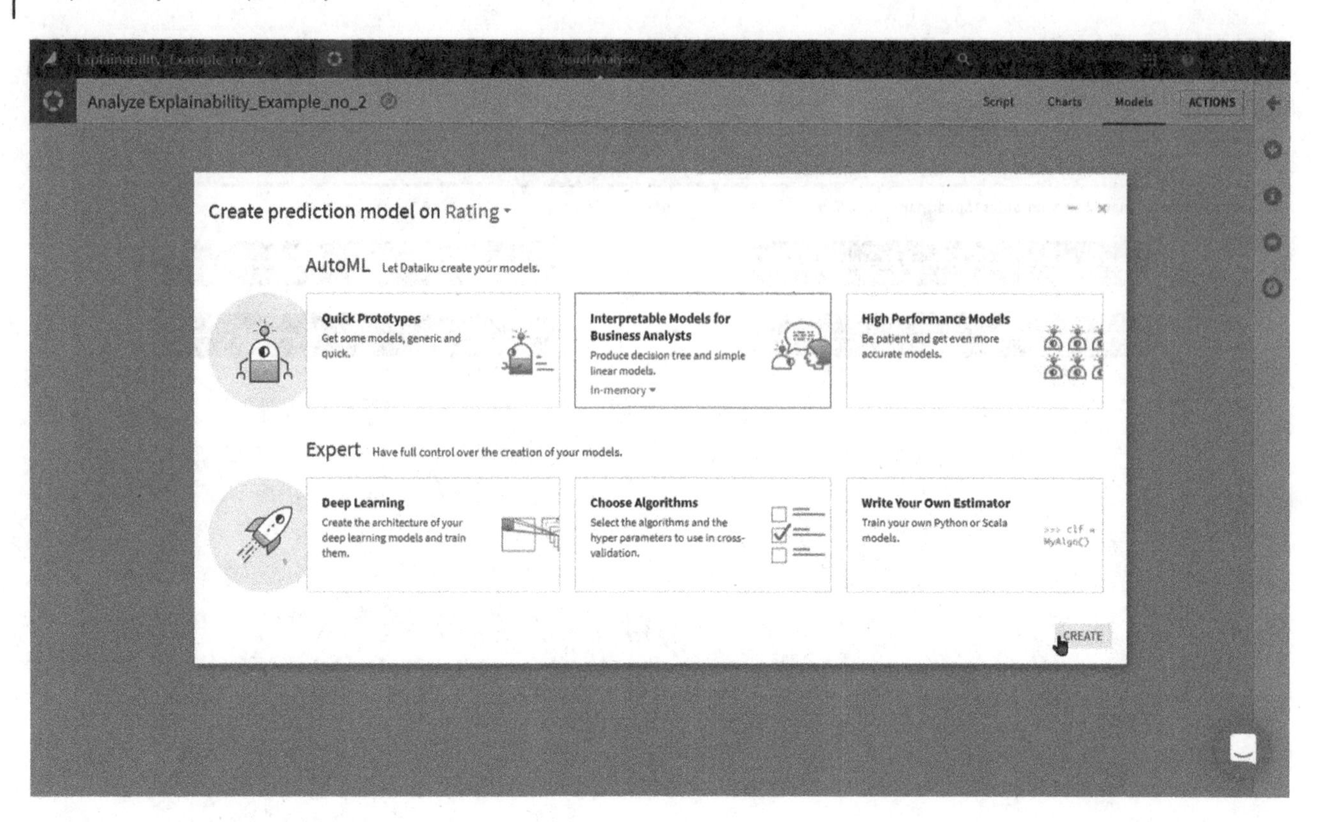

Step 4: Under the *Design* tab, select *Multiclass classification* analysis. Update analysis setting by selecting your choice of algorithms and training method.

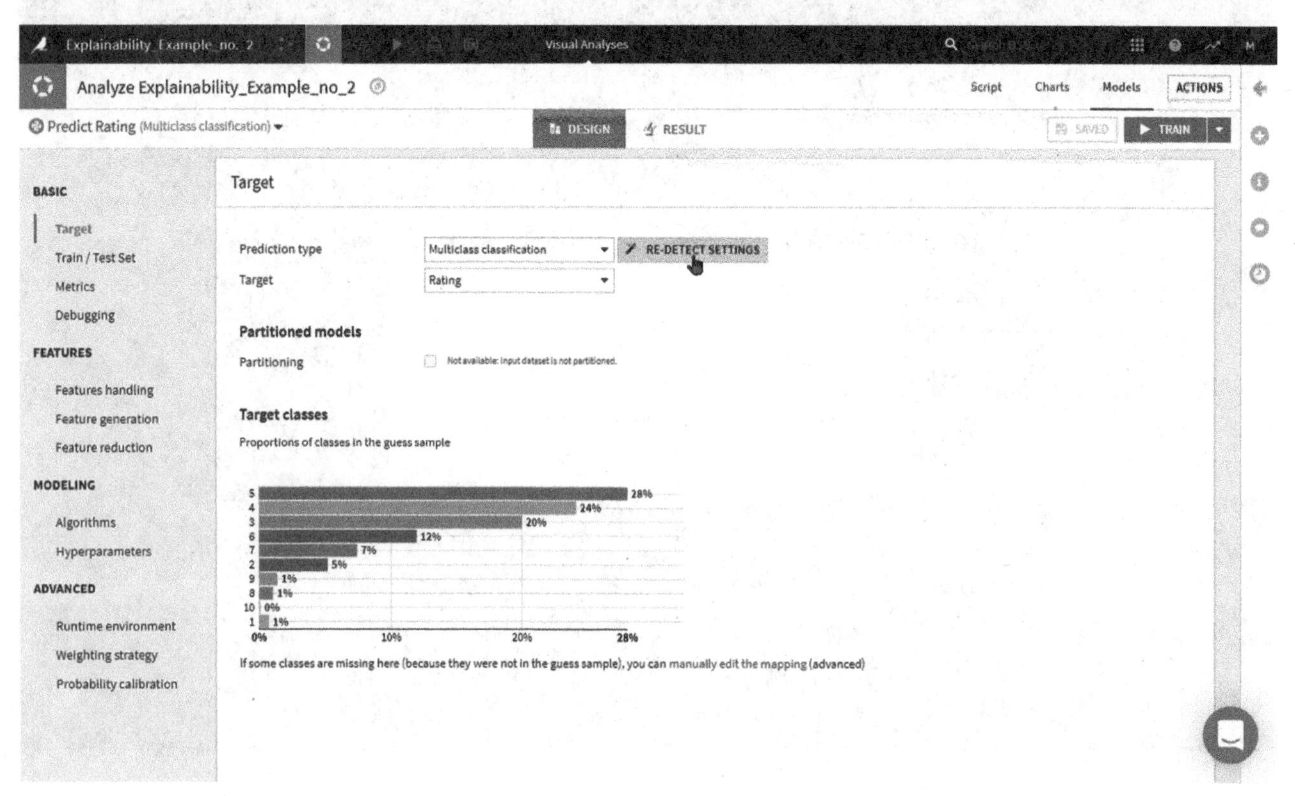

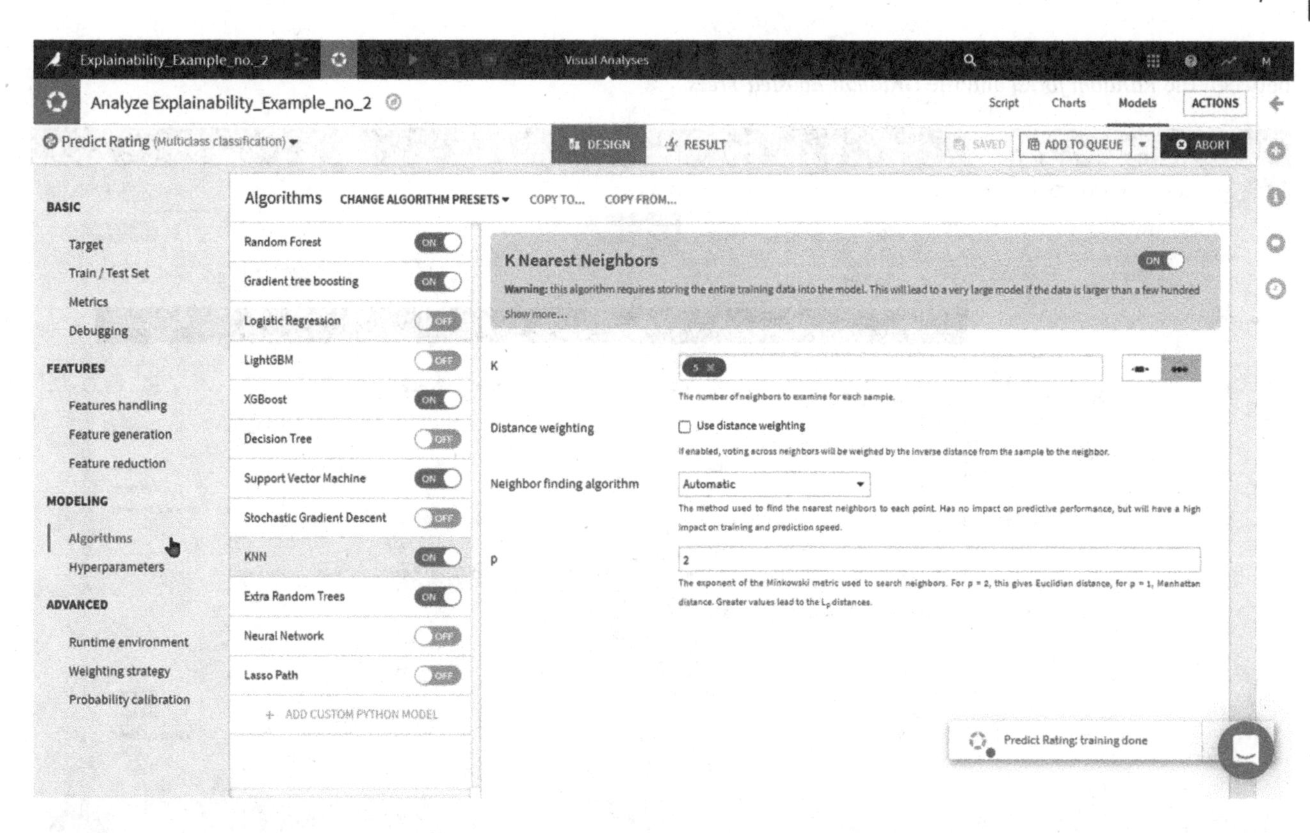

Explainability_Example_no_2
Visual Analyses
Analyze Explainability_Example_no_2
Script
Charts
Models
ACTIONS
Predict Rating (Multiclass classification)
DESIGN
RESULT
SAVED
ADD TO QUEUE
ABORT
BASIC
Target
Train / Test Set
Metrics
Debugging
FEATURES
Features handling
Feature generation
Feature reduction
MODELING
Algorithms
Hyperparameters
ADVANCED
Runtime environment
Weighting strategy
Probability calibration
Algorithms
CHANGE ALGORITHM PRESETS
COPY TO...
COPY FROM...
Random Forest
Gradient tree boosting
Logistic Regression
LightGBM
XGBoost
Decision Tree
Support Vector Machine
Stochastic Gradient Descent
KNN
Extra Random Trees
Neural Network
Lasso Path
ADD CUSTOM PYTHON MODEL
K Nearest Neighbors
Warning: this algorithm requires storing the entire training data into the model. This will lead to a very large model if the data is larger than a few hundred
Show more...
K
The number of neighbors to examine for each sample.
Distance weighting
Use distance weighting
If enabled, voting across neighbors will be weighed by the inverse distance from the sample to the neighbor.
Neighbor finding algorithm
Automatic
The method used to find the nearest neighbors to each point. Has no impact on predictive performance, but will have a high impact on training and prediction speed.
p
2
The exponent of the Minkowski metric used to search neighbors. For p = 2, this gives Euclidian distance, for p = 1, Manhattan distance. Greater values lead to the Lp distances.
Predict Rating: training done

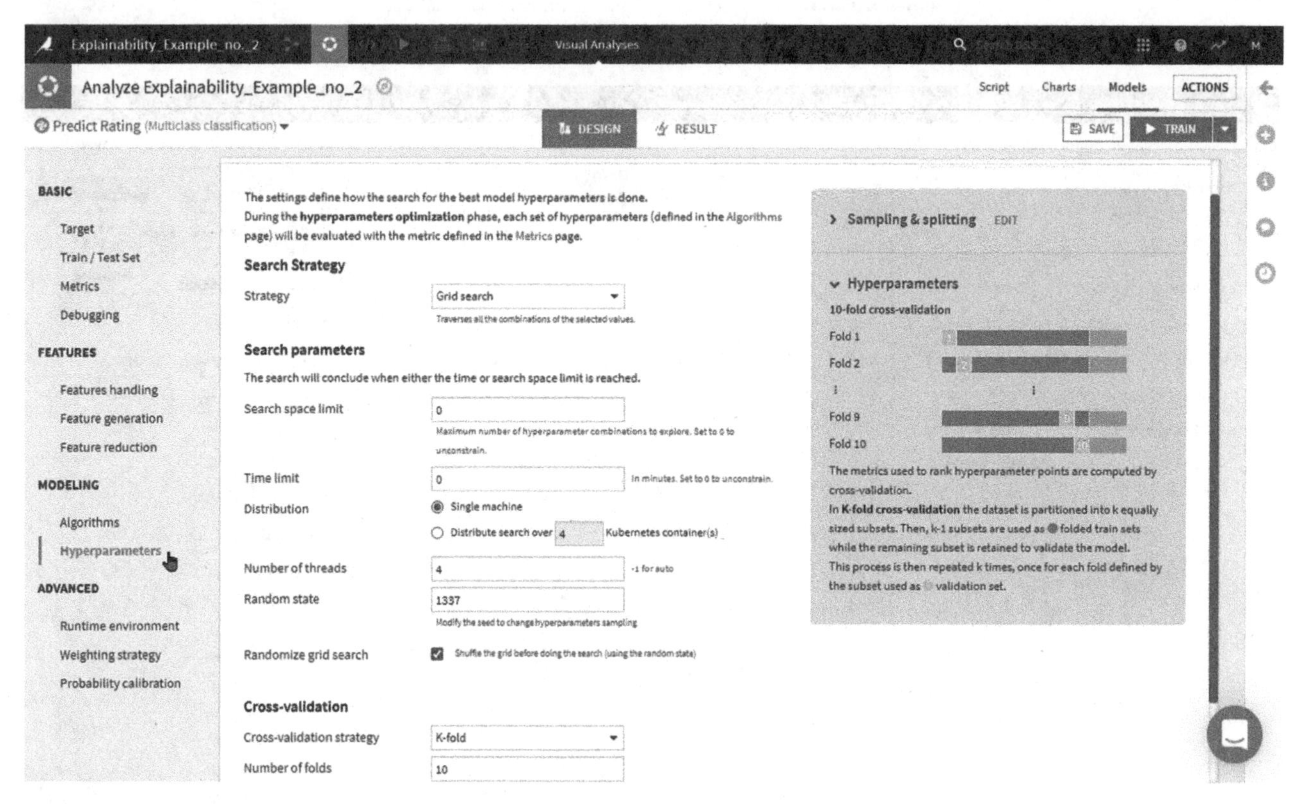

Explainability_Example_no_2
Visual Analyses
Analyze Explainability_Example_no_2
Script
Charts
Models
ACTIONS
Predict Rating (Multiclass classification)
DESIGN
RESULT
SAVE
TRAIN
BASIC
Target
Train / Test Set
Metrics
Debugging
FEATURES
Features handling
Feature generation
Feature reduction
MODELING
Algorithms
Hyperparameters
ADVANCED
Runtime environment
Weighting strategy
Probability calibration
The settings define how the search for the best model hyperparameters is done.
During the hyperparameters optimization phase, each set of hyperparameters (defined in the Algorithms page) will be evaluated with the metric defined in the Metrics page.
Search Strategy
Strategy
Grid search
Traverses all the combinations of the selected values.
Search parameters
The search will conclude when either the time or search space limit is reached.
Search space limit
0
Maximum number of hyperparameter combinations to explore. Set to 0 to unconstrain.
Time limit
0
In minutes. Set to 0 to unconstrain.
Distribution
Single machine
Distribute search over
4
Kubernetes container(s)
Number of threads
4
-1 for auto
Random state
1337
Modify the seed to change hyperparameters sampling
Randomize grid search
Shuffle the grid before doing the search (using the random state)
Cross-validation
Cross-validation strategy
K-fold
Number of folds
10
Sampling & splitting
EDIT
Hyperparameters
10-fold cross-validation
Fold 1
Fold 2
Fold 9
Fold 10
The metrics used to rank hyperparameter points are computed by cross-validation.
In K-fold cross-validation the dataset is partitioned into k equally sized subsets. Then, k-1 subsets are used as folded train sets while the remaining subset is retained to validate the model.
This process is then repeated k times, once for each fold defined by the subset used as validation set.

Step 5: Check out the results of the analysis.[67] Note that the *Random forest* leads the board. Also note the close performance between the *Random forest* and the *Gradient Boosted Trees.*

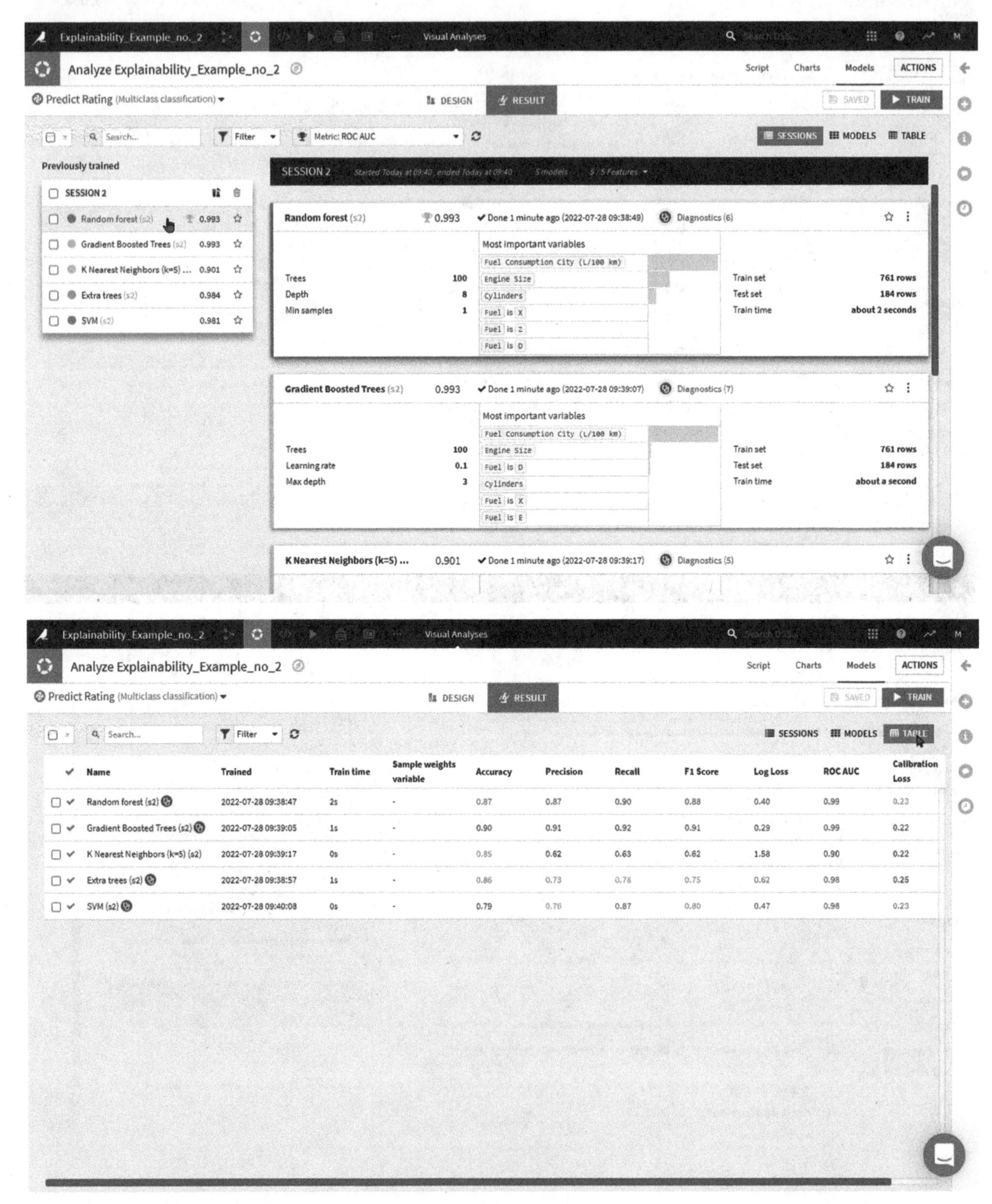

Name	Trained	Train time	Sample weights variable	Accuracy	Precision	Recall	F1 Score	Log Loss	ROC AUC	Calibration Loss
Random forest (s2)	2022-07-28 09:38:47	2s	-	0.87	0.87	0.90	0.88	0.40	0.99	0.23
Gradient Boosted Trees (s2)	2022-07-28 09:39:05	1s	-	0.90	0.91	0.92	0.91	0.29	0.99	0.22
K Nearest Neighbors (k=5) (s2)	2022-07-28 09:39:17	0s	-	0.85	0.62	0.63	0.62	1.58	0.90	0.22
Extra trees (s2)	2022-07-28 09:38:57	1s	-	0.86	0.73	0.76	0.75	0.62	0.98	0.25
SVM (s2)	2022-07-28 09:40:08	0s	-	0.79	0.76	0.87	0.80	0.47	0.98	0.23

67 Note that I have deleted the XGBoost model as it failed (again) in this analysis.

Step 6: While within the *Random forest* model, have a look at other outcomes (i.e., decision trees, variable importance, partial dependence plots, sub-population analysis, SHAP values, interactive scoring, etc.[68]).

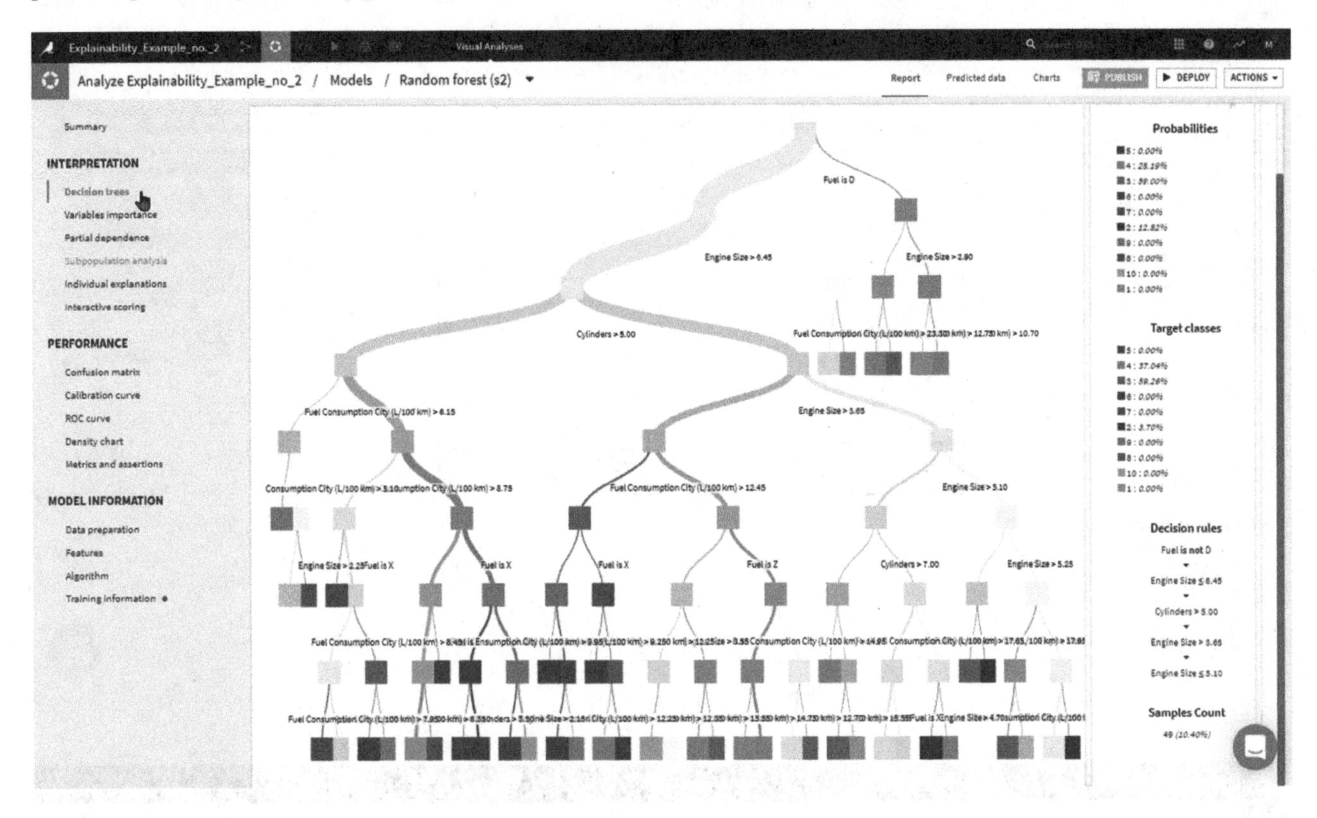

As you can see, the fuel consumption has the largest importance of all features, followed by engine size and number of cylinders.

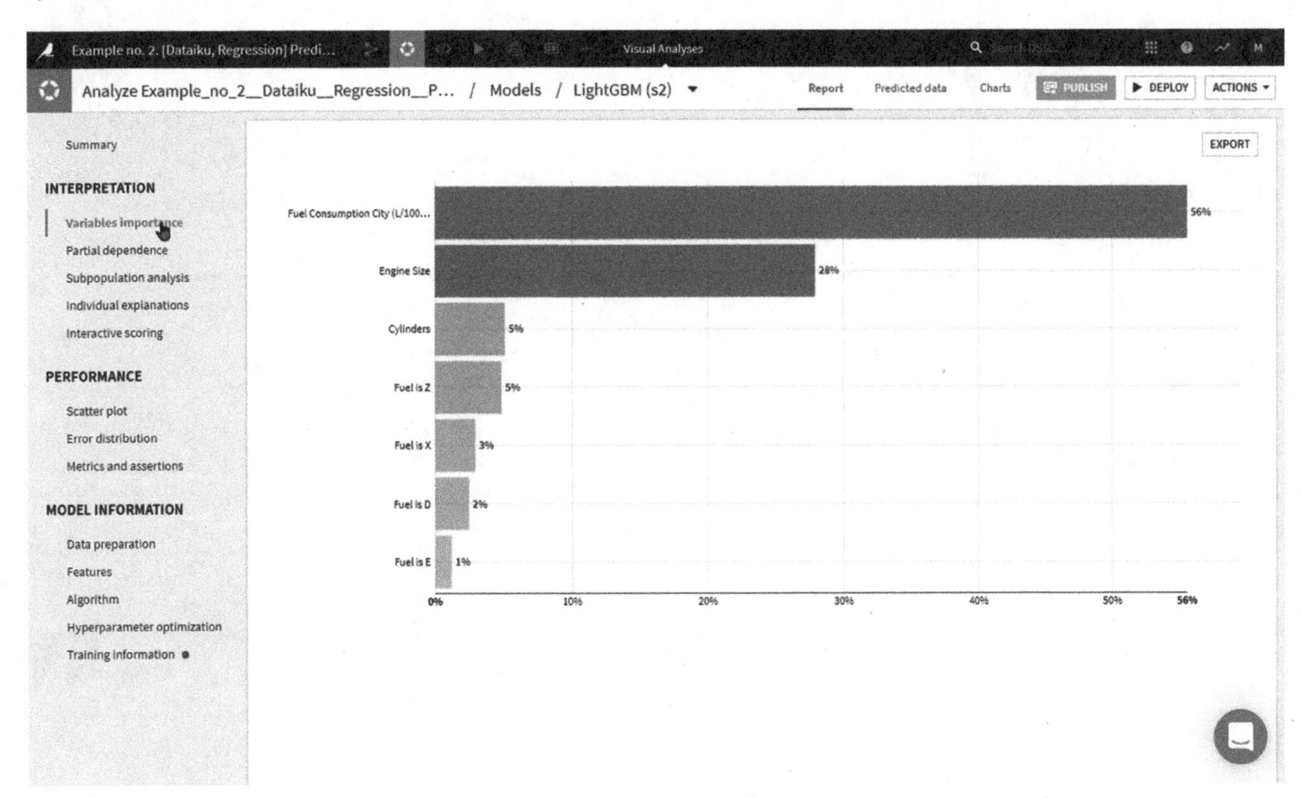

68 The regression coefficients can be seen under the same tab for the *ordinary linear regression* model

Explainability_Example_no_2
Visual Analyses
Analyze Explainability_Example_no_2 / Models / Random forest (s2)
Report
Predicted data
Charts
PUBLISH
DEPLOY
ACTIONS
Summary
INTERPRETATION
Decision trees
Variables importance
Partial dependence
Subpopulation analysis
Individual explanations
Interactive scoring
PERFORMANCE
Confusion matrix
Calibration curve
ROC curve
Density chart
Metrics and assertions
MODEL INFORMATION
Data preparation
Features
Algorithm
Training information
Select your variable
Fuel Consumption City (L/100
COMPUTE
COMPUTE ALL
EXPORT
50 bins for Fuel Consumption City (L/100 km), computed on 184 rows (the full test set)
Distribution
21.7%
0.0%
Partial dependence
Engine Size
Reading tips
A partial dependence plot shows the dependence of the predicted response on a single feature. The x axis displays the value of the selected feature, while the y axis displays the partial dependence.
The value of the partial dependence for a class is by how much the log-odds for this class are higher or lower than those of the average class probability.
Note : the log-odds for a probability p are defined as log(p / (1 - p)) . They are strictly increasing, ie. higher log odds mean higher probability.

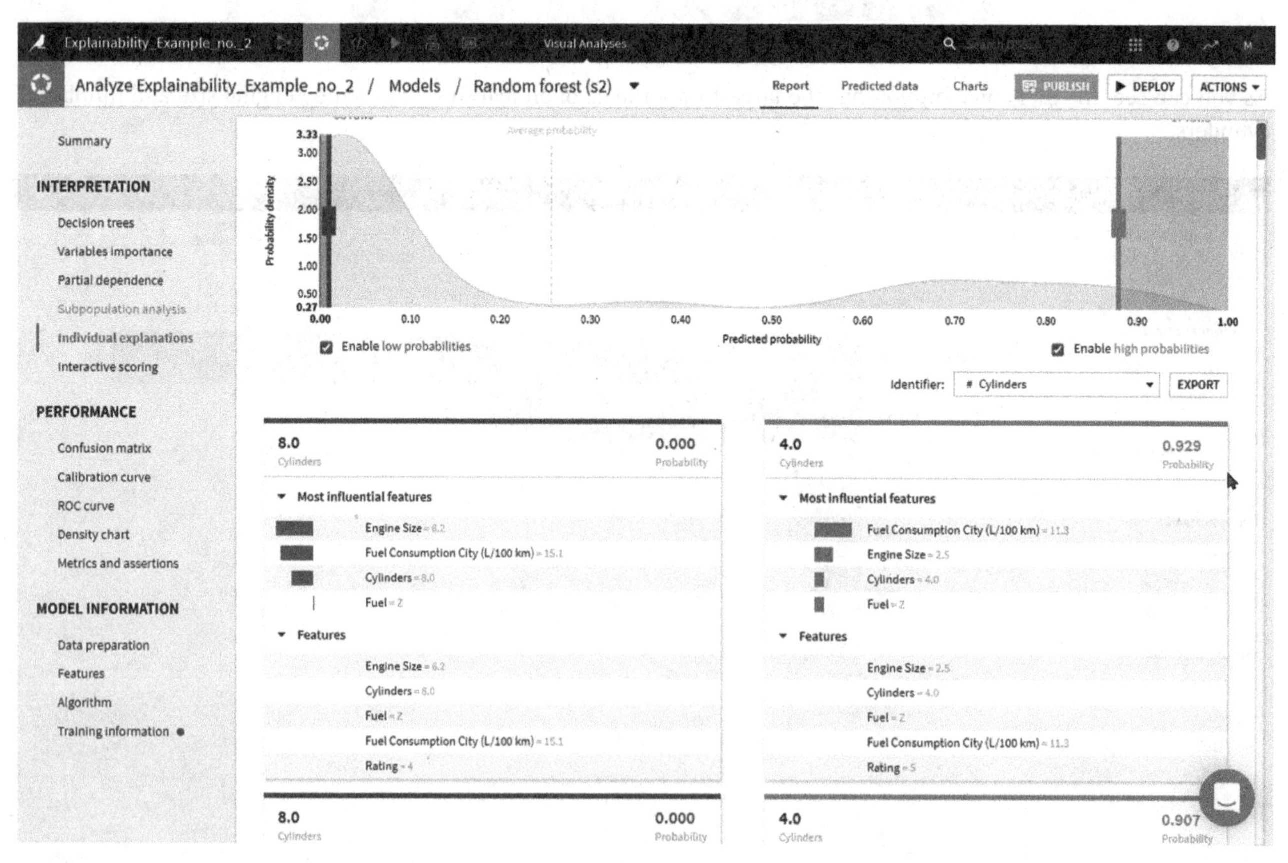
Explainability_Example_no_2
Visual Analyses
Analyze Explainability_Example_no_2 / Models / Random forest (s2)
Report
Predicted data
Charts
PUBLISH
DEPLOY
ACTIONS
Summary
INTERPRETATION
Decision trees
Variables importance
Partial dependence
Subpopulation analysis
Individual explanations
Interactive scoring
PERFORMANCE
Confusion matrix
Calibration curve
ROC curve
Density chart
Metrics and assertions
MODEL INFORMATION
Data preparation
Features
Algorithm
Training information
Probability density
Predicted probability
Enable low probabilities
Enable high probabilities
Identifier:
Cylinders
EXPORT
8.0
Cylinders
0.000
Probability
Most influential features
Engine Size
Fuel Consumption City (L/100 km)
Cylinders
Fuel
Features
Engine Size
Cylinders
Fuel
Fuel Consumption City (L/100 km)
Rating
4.0
Cylinders
0.929
Probability
Most influential features
Fuel Consumption City (L/100 km)
Engine Size
Cylinders
Fuel
Features
Engine Size
Cylinders
Fuel
Fuel Consumption City (L/100 km)
Rating
8.0
Cylinders
0.000
Probability
4.0
Cylinders
0.907
Probability

In this interactive scoring, the applied features identify that this observation is likely to fall under *Rating no. 5* with a 79.99%.

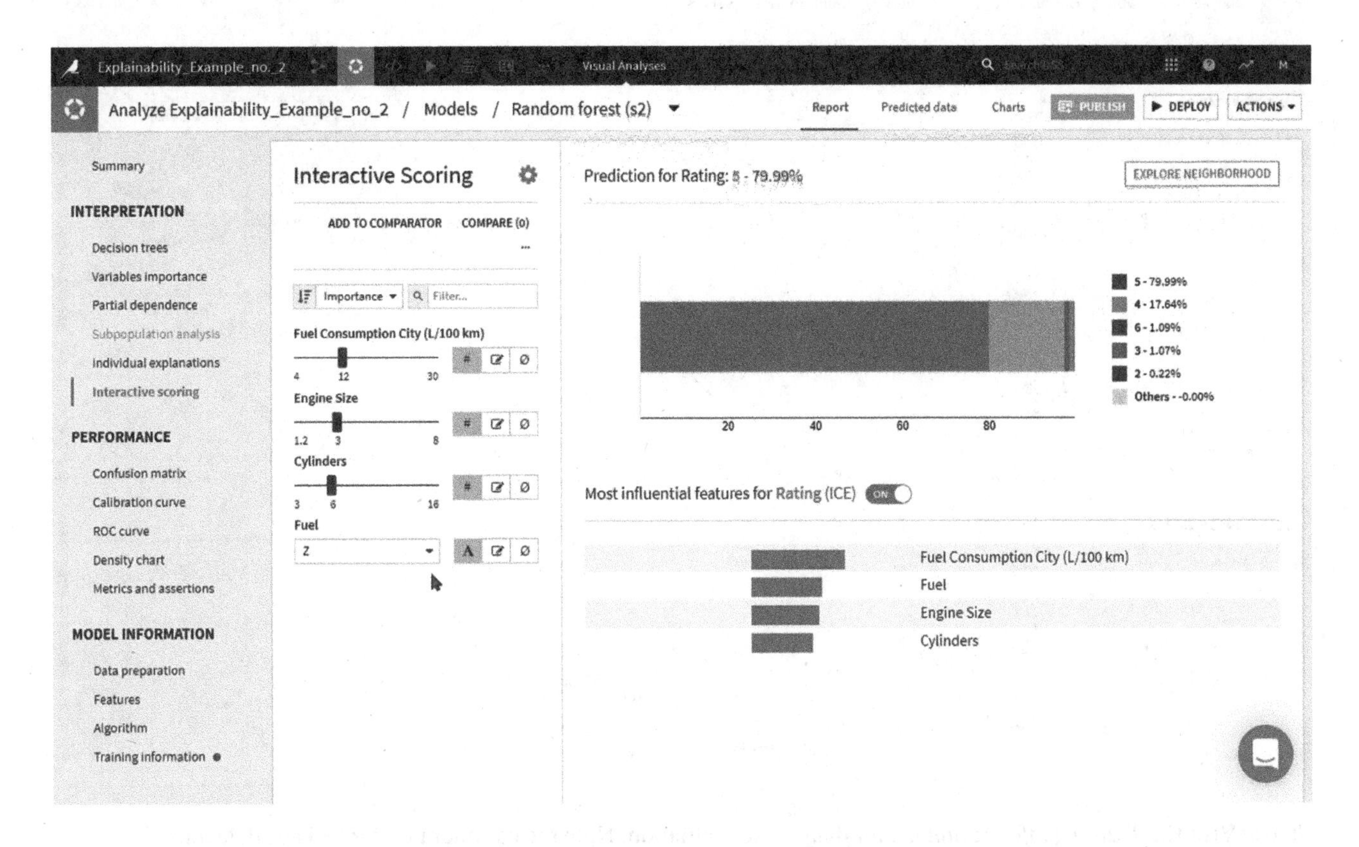

Step 7: Visit the *Performance* tab to visualize model metrics in terms of confusion matrix, and ROC curve.

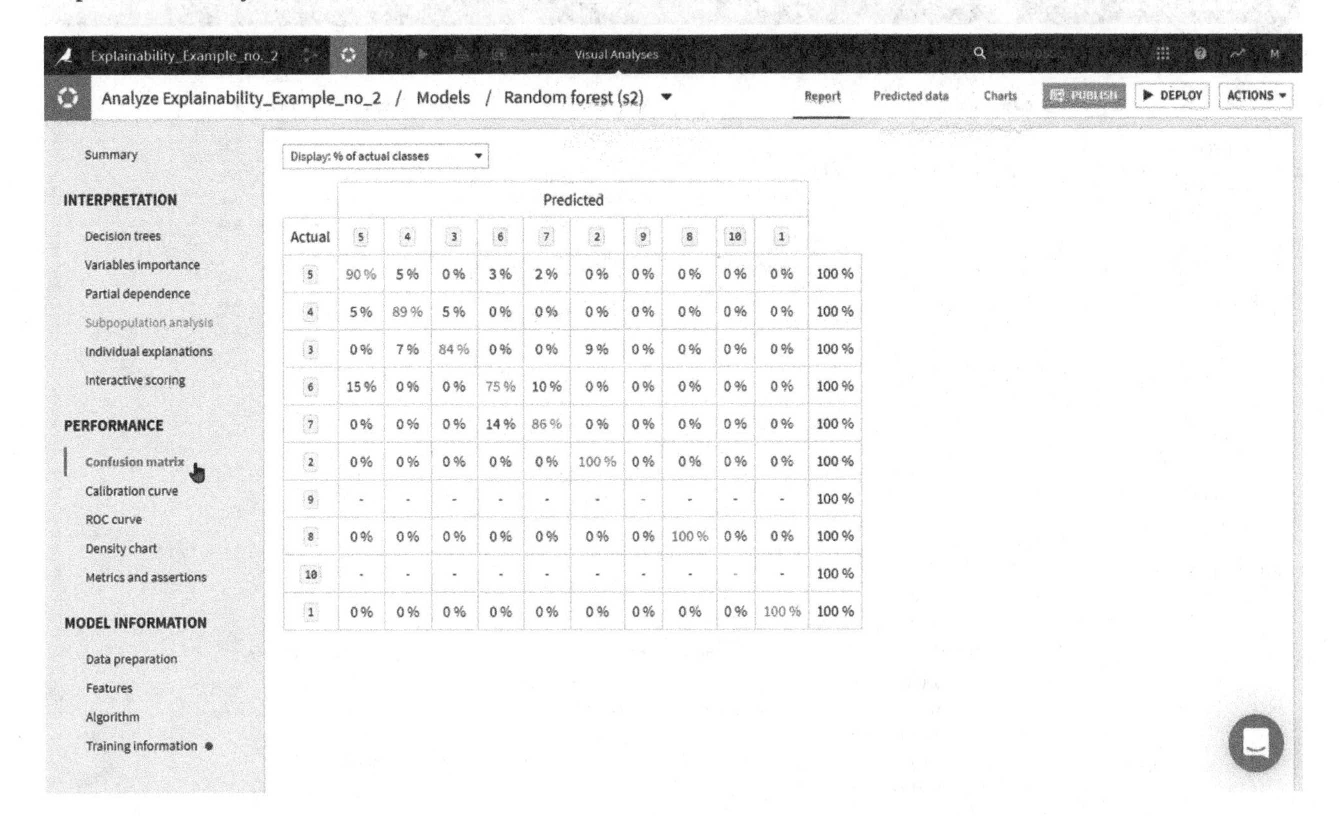

	Predicted										
Actual	5	4	3	6	7	2	9	8	10	1	
5	90 %	5 %	0 %	3 %	2 %	0 %	0 %	0 %	0 %	0 %	100 %
4	5 %	89 %	5 %	0 %	0 %	0 %	0 %	0 %	0 %	0 %	100 %
3	0 %	7 %	84 %	0 %	0 %	9 %	0 %	0 %	0 %	0 %	100 %
6	15 %	0 %	0 %	75 %	10 %	0 %	0 %	0 %	0 %	0 %	100 %
7	0 %	0 %	0 %	14 %	86 %	0 %	0 %	0 %	0 %	0 %	100 %
2	0 %	0 %	0 %	0 %	0 %	100 %	0 %	0 %	0 %	0 %	100 %
9	-	-	-	-	-	-	-	-	-	-	100 %
8	0 %	0 %	0 %	0 %	0 %	0 %	0 %	100 %	0 %	0 %	100 %
10	-	-	-	-	-	-	-	-	-	-	100 %
1	0 %	0 %	0 %	0 %	0 %	0 %	0 %	0 %	0 %	100 %	100 %

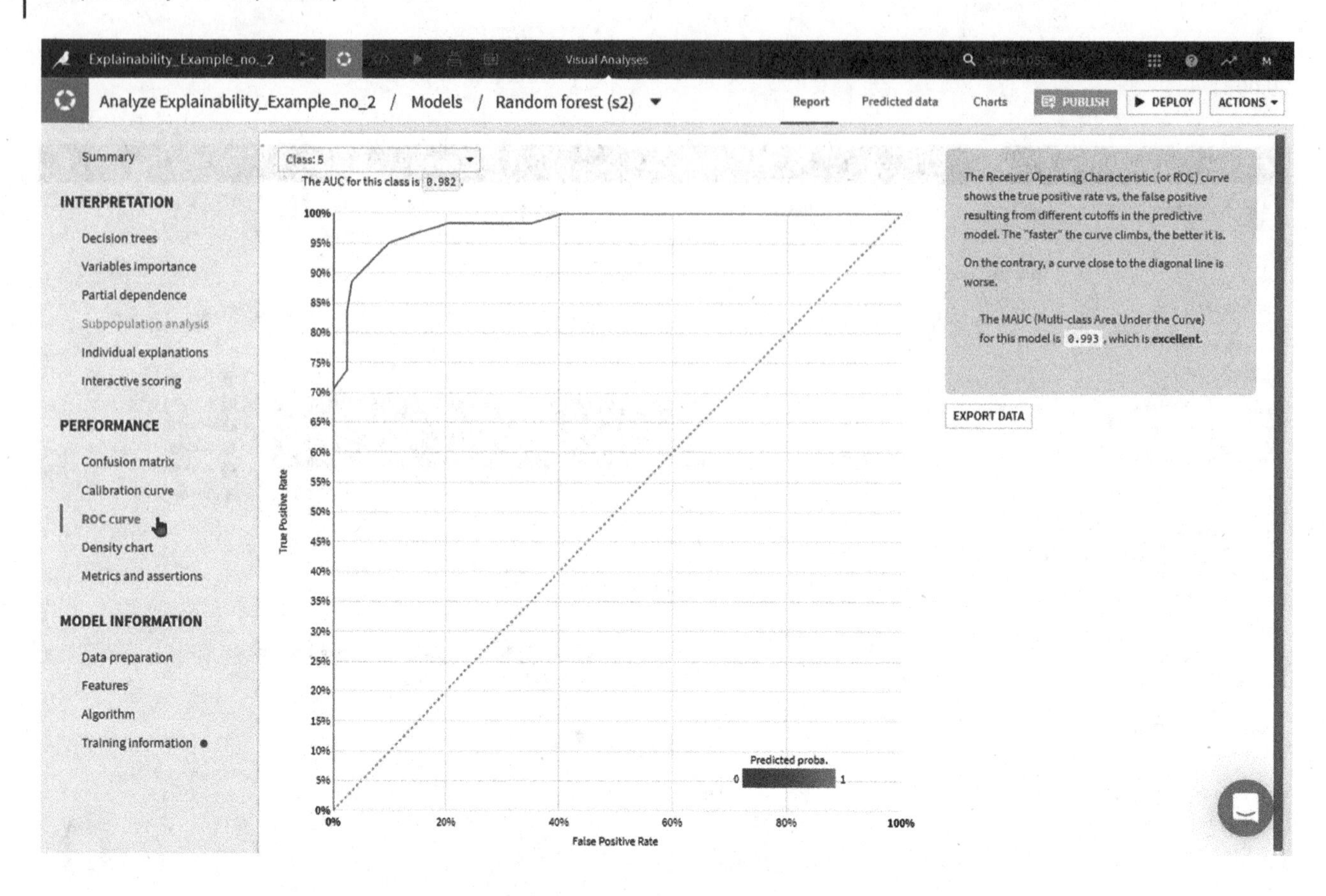

Step 8: Visit the *Training information* tab for diagnostic evaluation. Note the comment under leakage detection!

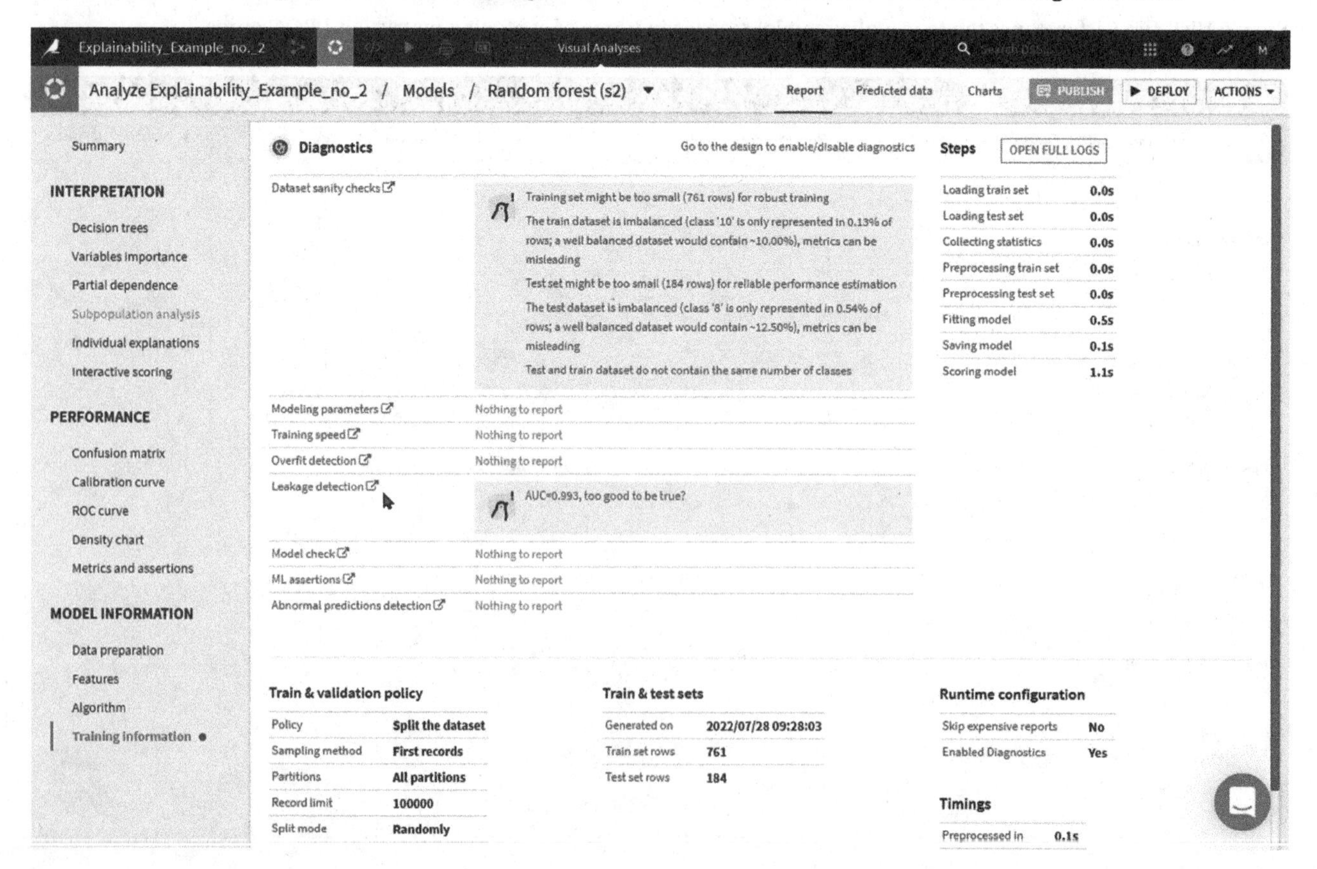

Example no. 3 (Python and R): Predicting fire-induced spalling of concrete mixtures[69]*.
Let us continue our example and augment the blackbox model with explainability measures. I will demonstrate SHAP (on the global and local scales), as well as LIME.

7.4.2 Global Explainability

At the global level, the SHAP method can generate two types of visualizations: *feature importance plot* and *summary plot*. This example continues what was covered earlier.

The *feature importance plot* is presented in Figure 7.10. This plot lists the most significant features in descending order (i.e., the model heavily relies on the PP quantity feature in predicting spalling the most). It should be noted that global importance gives an *average* overview of all parameters and how they contribute to the model. However, Figure 7.10 lacks the direction of impact, e.g., whether a variable has a positive or negative influence on the occurrence of spalling.

In addition to the *feature importance plot*, the *summary plot* is another visualization that can be generated (see Figure 7.11). This plot uses colors to distinguish the influence of the quantity of each feature upon the occurrence of spalling (i.e., red instances represent larger values and affect the model prediction depending on the side on which the red dots lie). For instance, large quantities of PP fibers are likely to mitigate fire-induced spalling, while the presence of high moisture content increases the tendency to spalling.[70] A closer examination of Figure 7.11 shows that some features do not influence spalling. One such feature is the diameter of PP fiber (when SHAP value equals zero).

Let us continue our journey by examining the *partial dependence plots*, another explainability method that can be used at the global scale of XAI models. Such a plot depicts the relationship between a feature and the occurrence of spalling.[71] I will only be showing the PDPs for the top four features listed in Figure 7.12, and I will be referring to the average trend in each PDP.

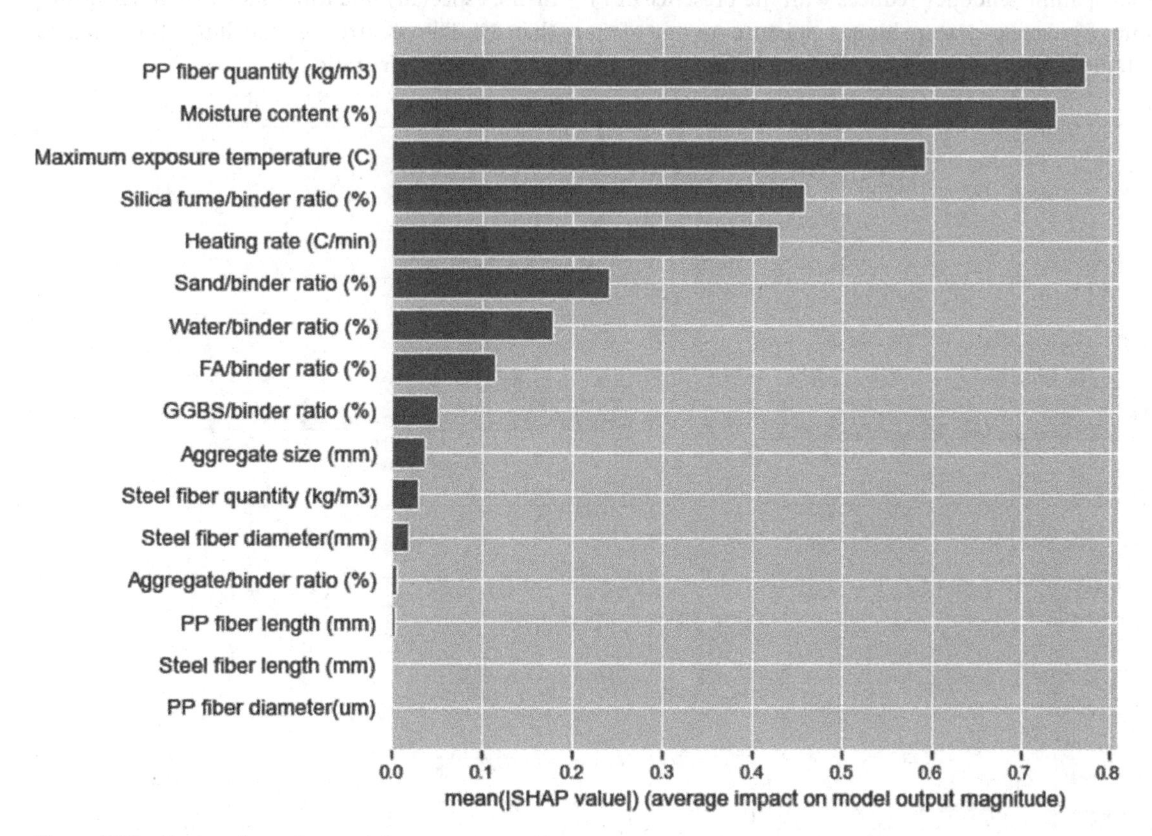

Figure 7.10 Feature importance plot.

69 This is a bit of an advanced example that I am going to snowflake. For a more straight forward example on classification, please check Example no. 1. Personally, I like Example no. 3 as it pushes the envelope a bit. To recap this example, we built an XGBoost classifier to identify concrete mixtures that tend to be vulnerable to fire-induced spalling (which is a failure mode that could occur under elevated temperatures). This example stems from Moe's paper [32].

70 Not surprisingly, these observations agree with our domain knowledge [40, 41].

71 Note that a partial dependence plot for a specific variable assumes that all other parameters remain unchanged (constant).

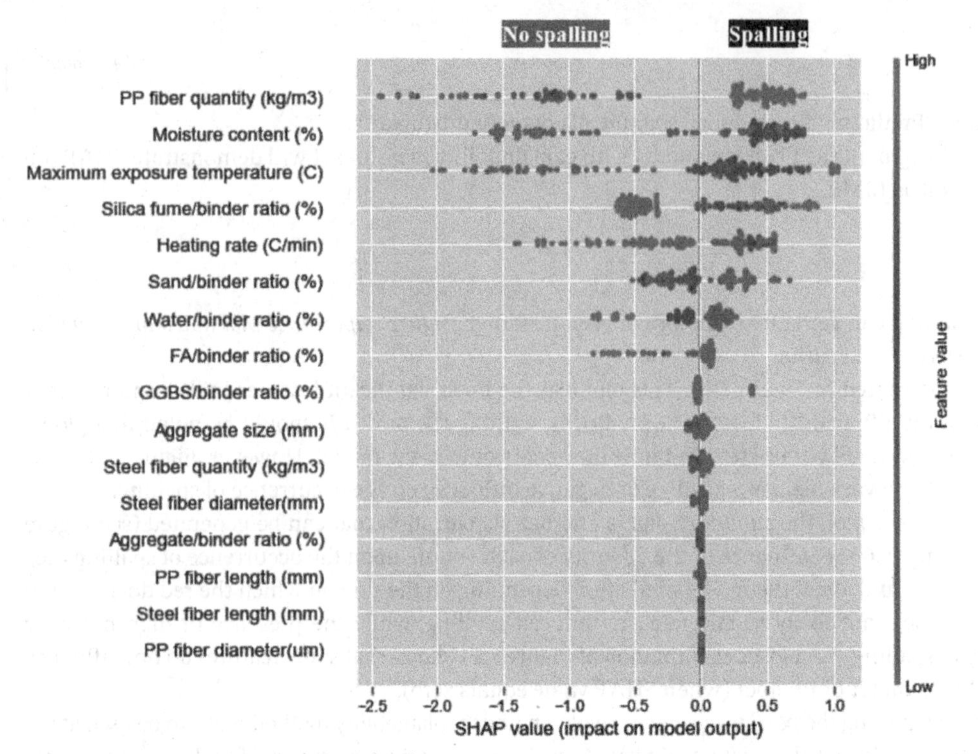

Figure 7.11 Summary plot of SHAP values.

We can see that spalling tendency reduces with the presence of PP[72] fibers; especially quantities larger than 2.5 kg/m^3, and increases once a concrete mixture homes moisture content of more than 3%. The occurrence of spalling also increases when exposure temperature increases and with the larger quantities of silica fume/binder ratio.[73]

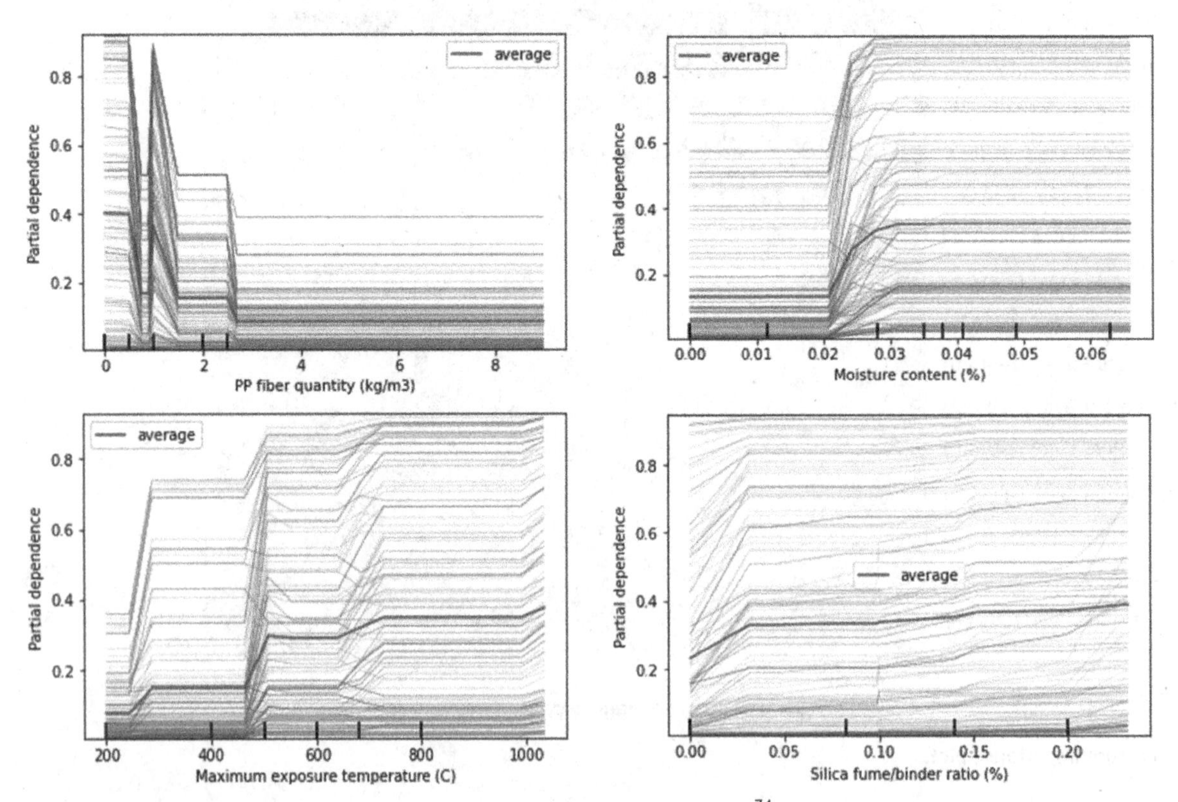

Figure 7.12 Partial dependence plot for the top eight influencers of spalling.[74]

72 In other words, the average PDP drops down. Also, note that the lack of PP fibers pushes the model to 40% of the limit required for spalling to take place.

73 This feature is rising slowly but surely!

74 One thing to keep in mind. A plot like the one shown could be used to hypothesize the exitance of a *possible* theory that revolves around these features and spalling!

7.4.3 Local Explainability

At the local scale, we select an observation and augment model prediction of the outcome of this observation via SHAP and LIME. For instance, Table 7.2 lists the concrete mixture for one specimen that did not spall. Let us apply SHAP and LIME to this specimen.

The *force plot* and *waterfall plot* represent the contribution of each feature to the prediction of our selected observation (see Figure 7.13). These plots are based on a calculated *SHAP base value E[f(X)]* of 0.2363.[75] These plots are intuitive to follow. For example, the larger the bar is, the higher the impact of its corresponding parameter on the predicted outcome. Let us deep-dive[76] into these plots.

Table 7.2 Properties of a typical specimen.

Factor	Value	Factor	Value
Water/binder ratio (%)	0.38	FA/binder ratio (%)	0.0
Aggregate/binder ratio (%)	1.86	PP fibers quantity (kg/m^3)	1.5
Sand/binder ratio (%)	1.86	PP fibers diameter(um)	32
Heating rate (°C/min)	66	PP fibers length (mm)	12
Moisture content (%)	4.1	Steel fibers quantity (kg/m^3)	0.0
Maximum exposure temperature (°C)	678	Steel fibers diameter(mm)	0.0
Silica fume/binder ratio (%)	0.0	Steel fibers length (mm)	0.0
Aggregate size (mm)	10.0		

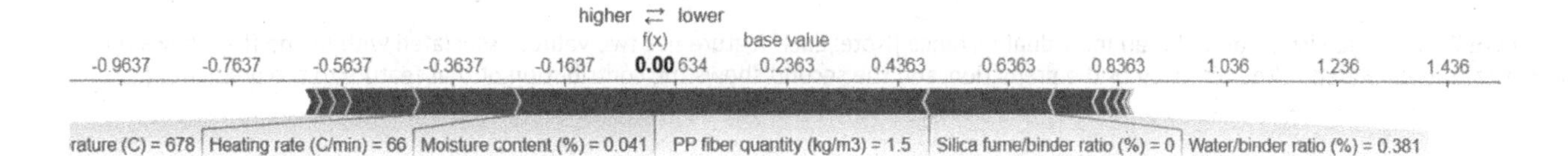

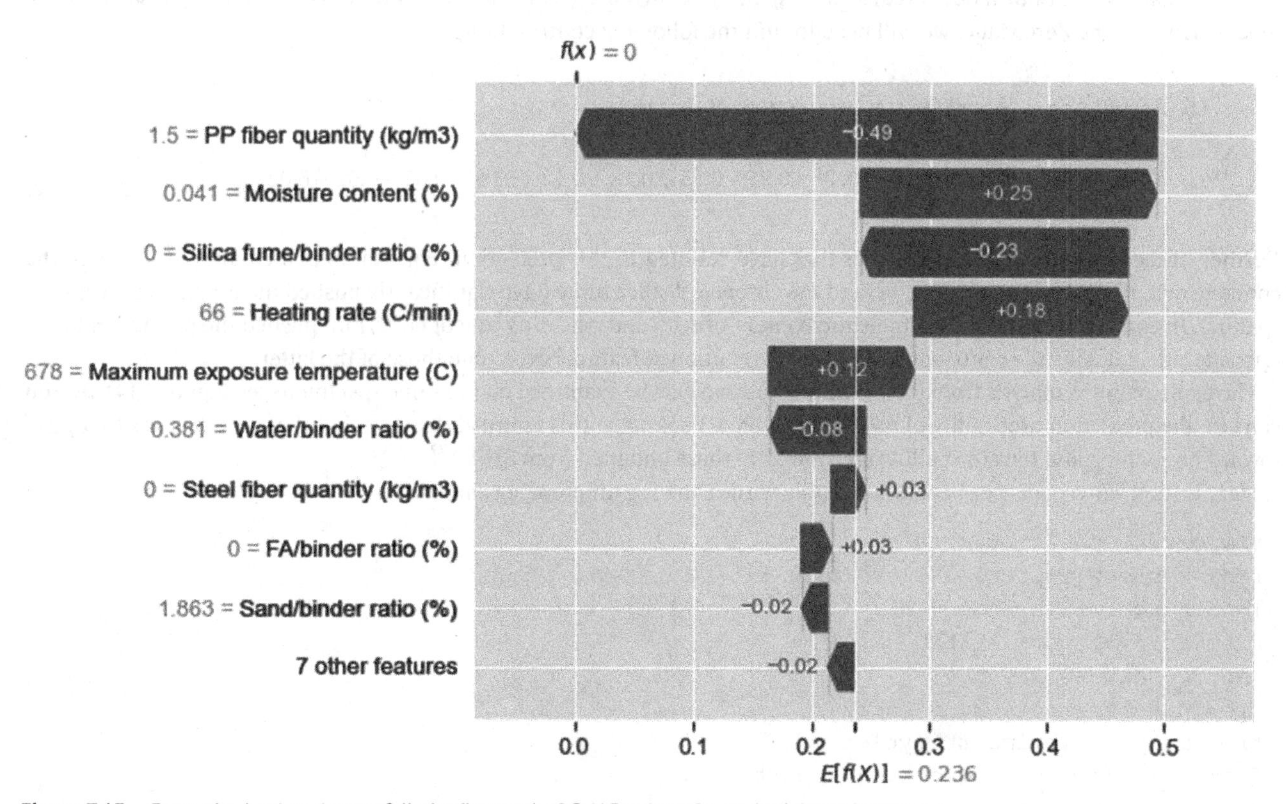

Figure 7.13 *Force plot* (top) and *waterfall plot* (bottom) of SHAP values for an individual instance.

75 Remember that this value is the mean of the model output over the training dataset. More specifically, the bottom of a *waterfall plot* is the expected value from the model, and then each feature shows how the positive (red) or negative (blue) contribution moves the predicted value from the expected model output.

76 TARZ would get this one.

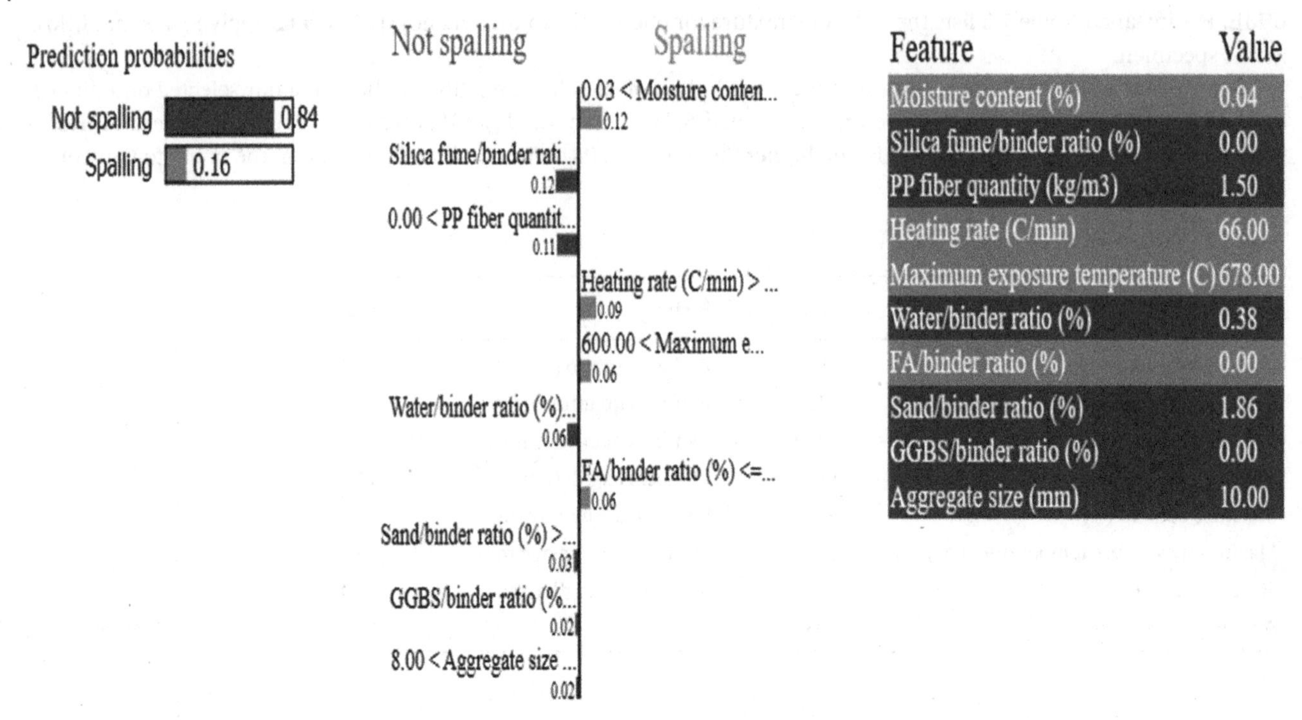

Figure 7.14 LIME visualization for an individual instance [Note: Each feature has two values associated with it. The first shows the critical range related to the direction of the prediction, and the second shows the contribution of that feature to the prediction].

This plot shows that our model predicts a *no spalling* occurrence (which agrees with our observation). To see how the model arrives at the Zero value, we will need to sum the following contributions:

$$Local\ prediction = base\ vlaue + \sum summation\ of\ features$$

$$Local\ prediction = 0.0 = +0.236 - 0.49 + 0.25 - 0.23 + 0.18 + 0.12 - 0.08 + 0.03 - 0.02 - 0.02. \tag{7.8}$$

Further, these plots show the top features that have resulted in this prediction. For example, including PP fibers in the concrete mix of the amount of 1.5 kg/m^3 and the absence of silica fume have significantly pushed the prediction toward no spalling. In contrast, relatively high moisture content of 4.1% and a heating rate of 66°C/min pushed the prediction to the opposite side. Luckily, the contribution of the first-mentioned features overcome those of the latter.

The explainability analysis from the LIME method can also be examined on the same specimen (see Figure 7.14). As you can see, the prediction probability of *no spalling* is 84%. Looking at this figure also shows that the same features that push toward no spalling and toward spalling are similar to those obtained from SHAP.[77]

Here is the *Python* code to repeat this example while showing different specimens.

```
#Importing the libraries
from xgboost import XGBClassifier
import xgboost as xgb
from scipy import stats
import seaborn as sns
from pandas import MultiIndex
from matplotlib import pyplot
from sklearn.model_selection import train_test_split
from sklearn.metrics import roc_curve
import pandas as pd
import numpy as np
```

77 However, this is not always going to be true. Ask Moe al-Bashiti!

```
from sklearn.feature_selection import mutual_info_classif
from sklearn.model_selection import GridSearchCV
from sklearn.metrics import confusion_matrix,classification_report
from sklearn.metrics import accuracy_score
from sklearn.metrics import balanced_accuracy_score, roc_auc_score, make_scorer
from sklearn.metrics import plot_confusion_matrix
from sklearn.model_selection import StratifiedKFold
from sklearn.model_selection import cross_val_score
```

```
#Importing the dataset from your local drive
```

```
fire=pd.read_csv('database.csv')
#fire= fire[fire['Steel fiber quantity (kg/m3)'] > 0]
#A=fire.agg(
   # {
   # "Water/binder ratio (%)":["min", "max", "median", "skew", "std","mean" ],
   # "Aggregate/binder ratio (%)":["min", "max", "median", "skew", "std","mean" ],
       # "Sand/binder ratio (%)":["min", "max", "median", "skew", "std","mean" ],
       # "Heating rate (°C/min)":["min", "max", "median", "skew", "std","mean" ],
       # "Moisture content (%)":["min", "max", "median", "skew", "std","mean" ],
       # "Maximum exposure temperature (°C)":["min", "max", "median", "skew",
"std","mean" ],
       # "Silica fume/binder ratio (%)":["min", "max", "median", "skew",
"std","mean" ],
       # "Aggregate size (mm)":["min", "max", "median", "skew", "std","mean" ],
       # "GGBS/binder ratio (%)":["min", "max", "median", "skew", "std","mean" ],
       # "FA/binder ratio (%)":["min", "max", "median", "skew", "std","mean" ],
       # "PP fiber quantity (kg/m3)":["min", "max", "median", "skew", "std","mean" ],
       # "PP fiber diameter(um)":["min", "max", "median", "skew", "std","mean" ],
       # "PP fiber length (mm)":["min", "max", "median", "skew", "std","mean" ],
       # "Steel fiber quantity (kg/m3)":["min", "max", "median", "skew",
"std","mean" ],
       # "Steel fiber diameter(mm)":["min", "max", "median", "skew", "std","mean" ],
       # "Steel fiber length (mm)":["min", "max", "median", "skew", "std","mean" ],
    # }
#)
fire
```

	Water/binder ratio (%)	Aggregate/binder ratio (%)	Sand/binder ratio (%)	Heating rate (C/min)	Moisture content (%)	Maximum exposure temperature (C)	Silica fume/binder ratio (%)	Aggregate size (mm)	GGBS/binder ratio (%)	FA/binder ratio (%)	PP fiber quantity (kg/m3)	PP fiber diameter(um)
0	0.273	2.000	1.049	101.0	0.049	1034	0.0	7.0	0.25	0.0	0.0	0.0
1	0.273	2.000	1.049	101.0	0.049	1034	0.0	14.0	0.25	0.0	0.0	0.0
2	0.300	2.000	1.033	101.0	0.049	1034	0.0	20.0	0.25	0.0	0.0	0.0
3	0.218	1.374	1.199	101.0	0.047	1034	0.0	7.0	0.25	0.0	0.0	0.0
4	0.218	1.374	1.199	101.0	0.043	1034	0.0	7.0	0.25	0.0	0.0	0.0
...	...	...	...	...	...	...	...	...	...	...	...	...
641	0.200	0.000	1.040	15.0	0.063	800	0.2	0.6	0.00	0.0	1.0	36.0
642	0.200	0.000	1.040	15.0	0.065	800	0.2	0.6	0.00	0.0	1.5	36.0
643	0.200	0.000	1.040	15.0	0.066	800	0.2	0.6	0.00	0.0	2.0	36.0
644	0.200	0.000	1.040	15.0	0.064	800	0.2	0.6	0.00	0.0	2.5	36.0
645	0.200	0.000	1.040	15.0	0.066	800	0.2	0.6	0.00	0.0	3.0	36.0

646 rows × 17 columns

```
x=fire.drop(['Output'],axis=1)
y=fire['Output']
x_train,x_test,y_train,y_test=train_test_split(x,y,test_size=0.30,random_state=111)
#fire['pp fiber length (mm)'] = fire['pp fiber length (mm)'].astype(float)
#fire['maximum exposure temperature'] = fire['maximum exposure temperature'].
astype(float)
#fire['s fiber length (mm)'] = fire['s fiber length (mm)'].astype(float)
#fire['output'] = fire['output'].astype(int)

fire.dtypes
#fire['output'].hist()
```

```
Water/binder ratio (%)                float64
Aggregate/binder ratio (%)            float64
Sand/binder ratio (%)                 float64
Heating rate (C/min)                  float64
Moisture content (%)                  float64
Maximum exposure temperature (C)        int64
Silica fume/binder ratio (%)          float64
Aggregate size (mm)                   float64
GGBS/binder ratio (%)                 float64
FA/binder ratio (%)                   float64
PP fiber quantity (kg/m3)             float64
PP fiber diameter(um)                 float64
PP fiber length (mm)                    int64
Steel fiber quantity (kg/m3)          float64
Steel fiber diameter(mm)              float64
Steel fiber length (mm)                 int64
Output                                  int64
dtype: object
```

```
#Plotting the histogram for the output
wb=fire['Output']

pyplot.hist(wb, edgecolor='black')

pyplot.title('Water/binder ratio')
pyplot.xlabel('Ratio')

pyplot.ylabel('No. of samples')

pyplot.show()
```

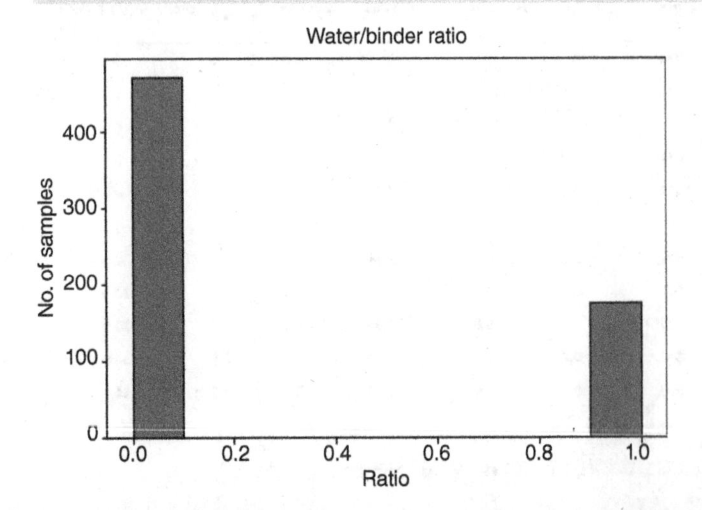

```
#Another way to plot histograms for many features
for i, col in enumerate(fire.columns):
    pyplot.figure(i)
    sns.histplot(fire[col])
```

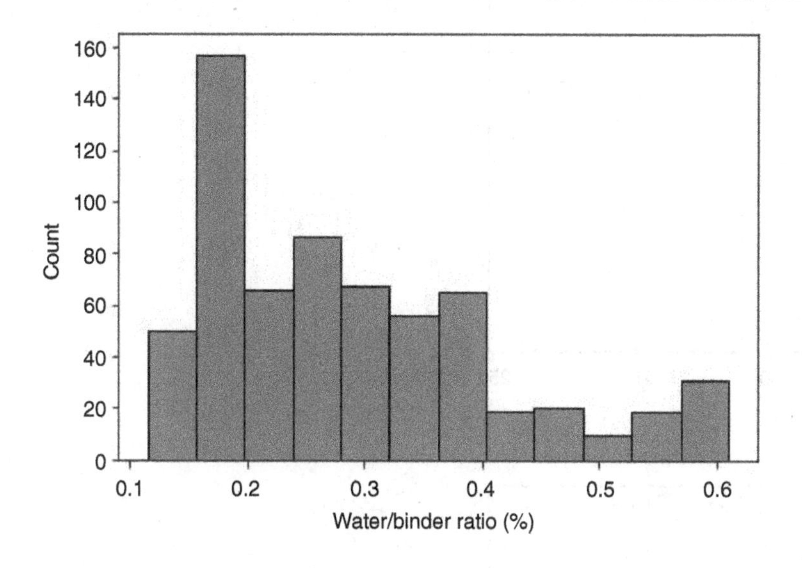

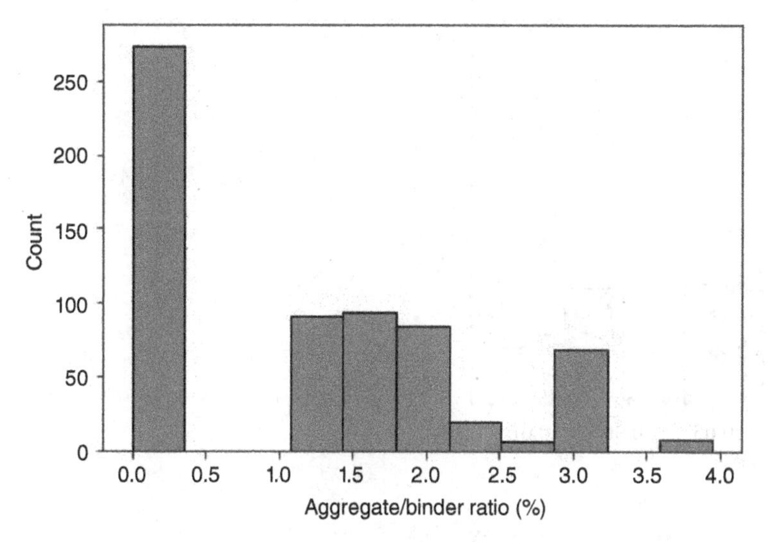

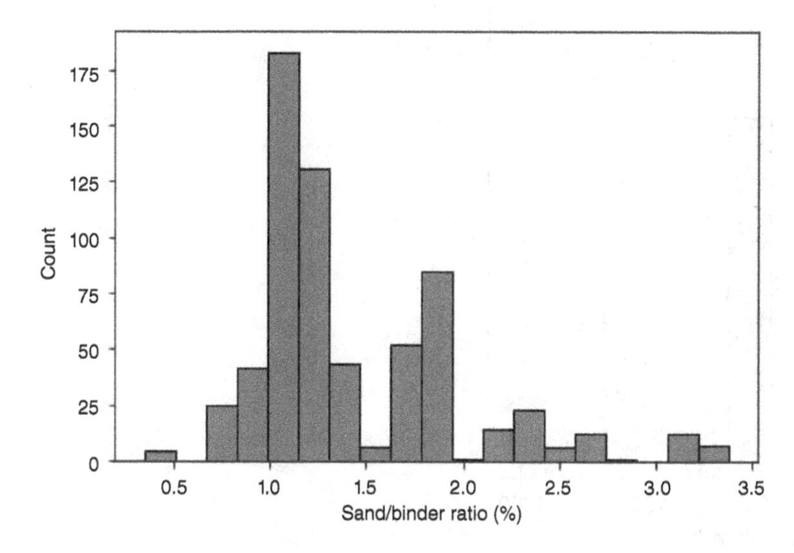

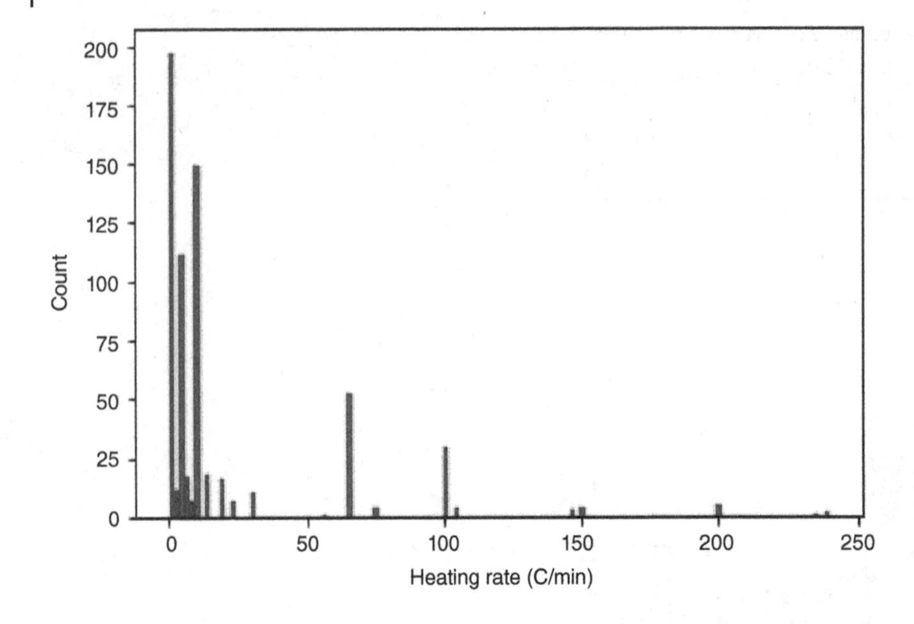
Count
Heating rate (C/min)
200
175
150
125
100
75
50
25
0
0
50
100
150
200
250

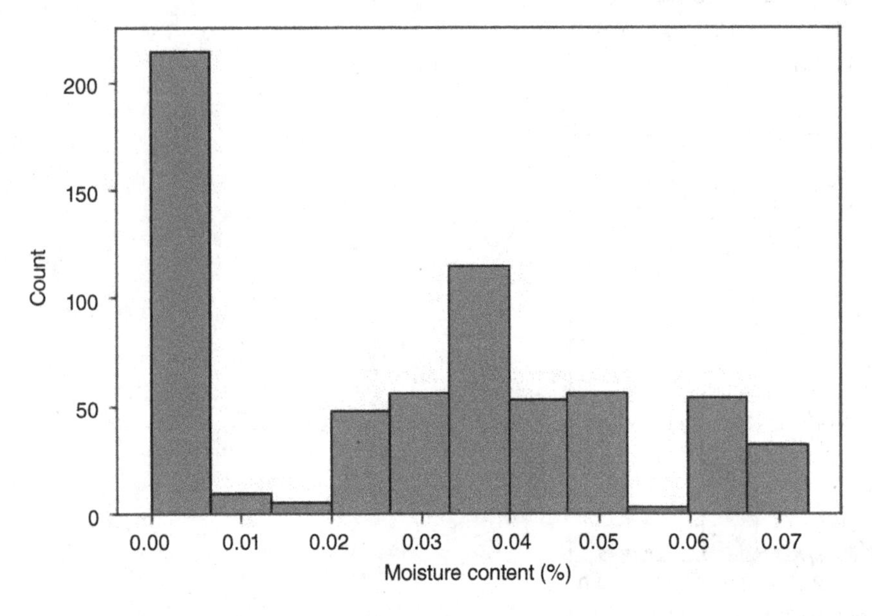
Count
Moisture content (%)
200
150
100
50
0
0.00
0.01
0.02
0.03
0.04
0.05
0.06
0.07

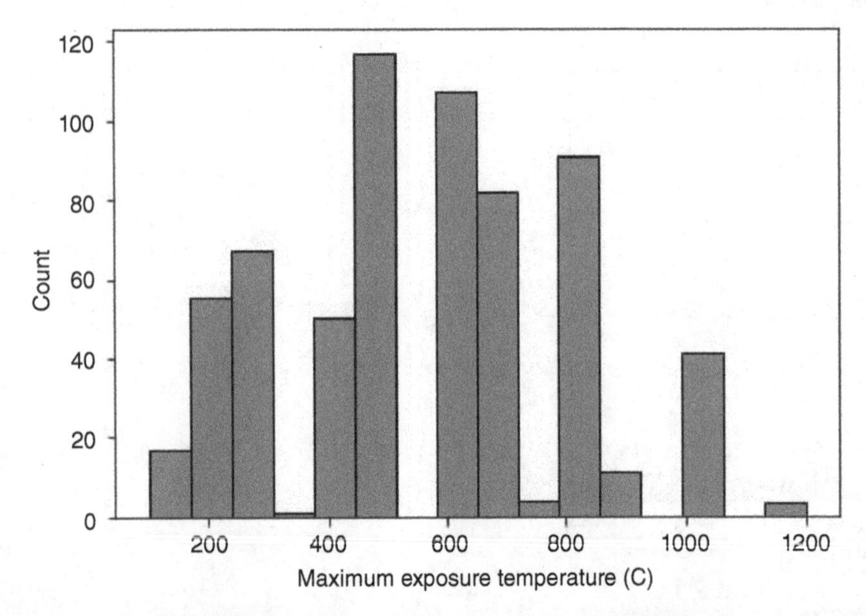
Count
Maximum exposure temperature (C)
120
100
80
60
40
20
0
200
400
600
800
1000
1200

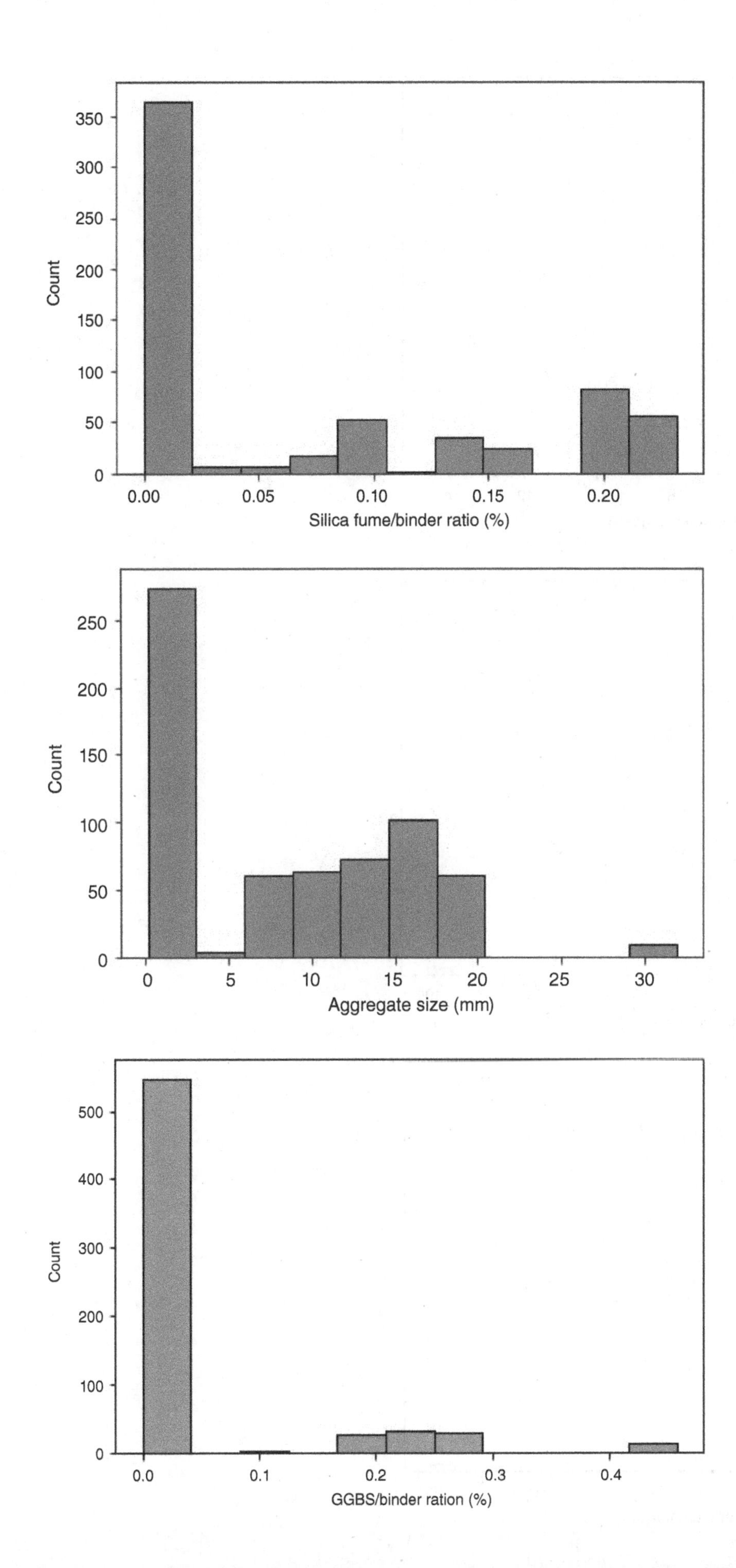
Count
Silica fume/binder ratio (%)
Count
Aggregate size (mm)
Count
GGBS/binder ration (%)

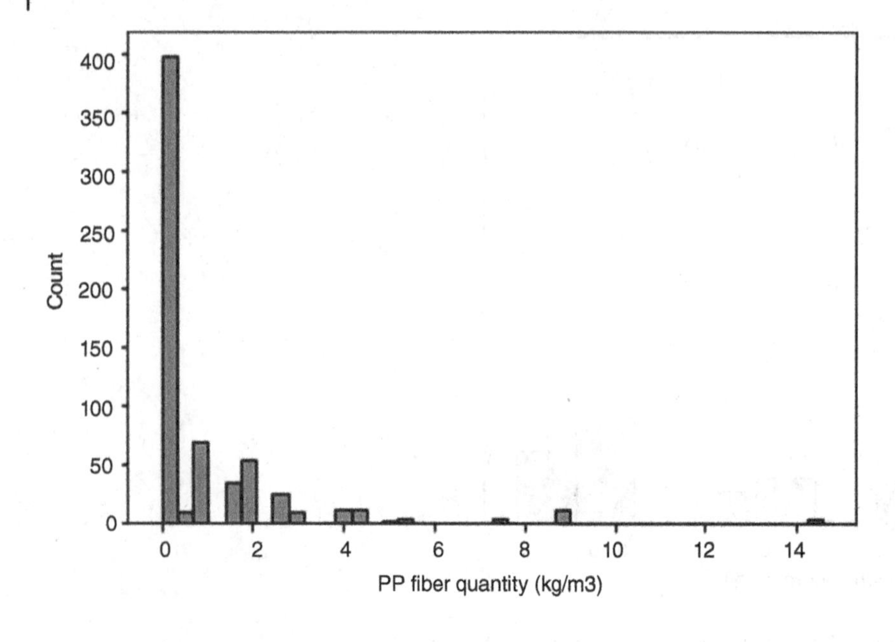

Count
400
350
300
250
200
150
100
50
0
0
2
4
6
8
10
12
14
PP fiber quantity (kg/m3)

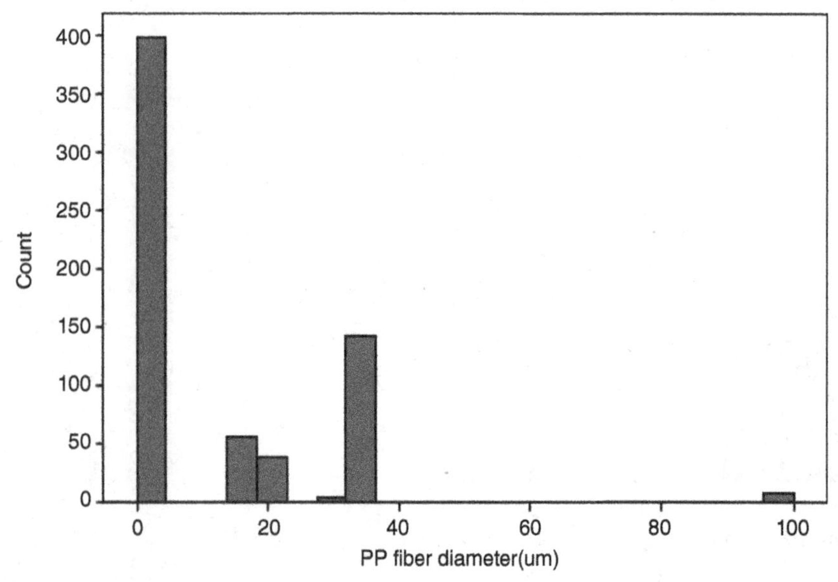

Count
400
350
300
250
200
150
100
50
0
0
20
40
60
80
100
PP fiber diameter(um)

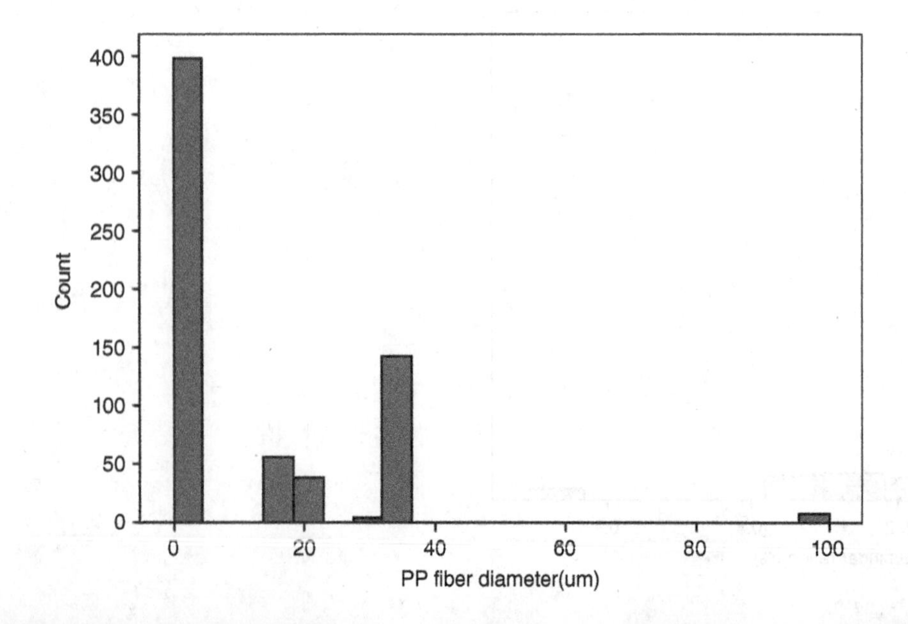

Count
400
350
300
250
200
150
100
50
0
0
20
40
60
80
100
PP fiber diameter(um)

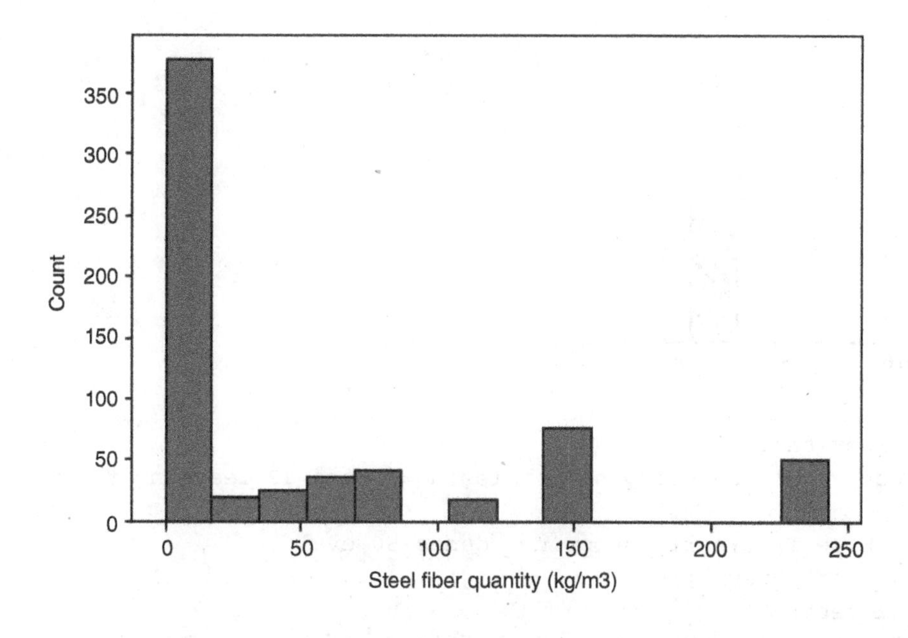
Count
350
300
250
200
150
100
50
0
0
50
100
150
200
250
Steel fiber quantity (kg/m3)

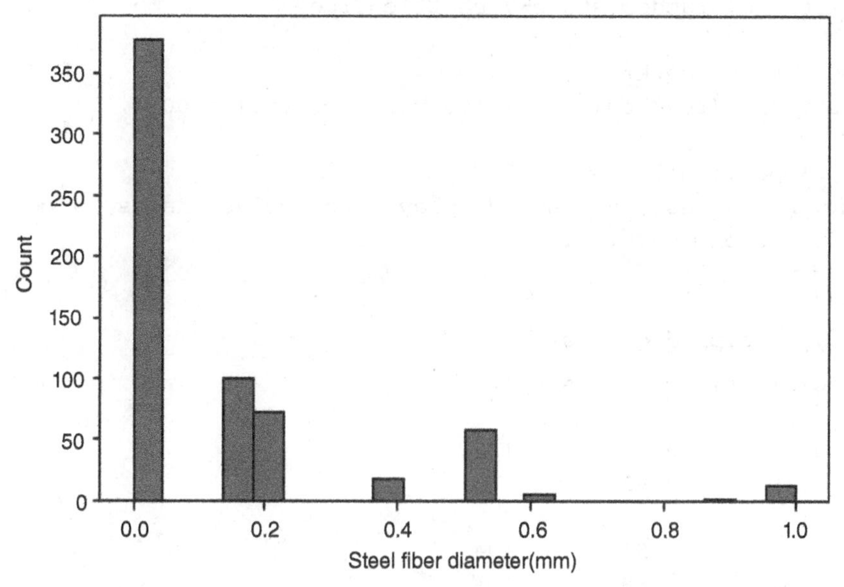
Count
350
300
250
200
150
100
50
0
0.0
0.2
0.4
0.6
0.8
1.0
Steel fiber diameter(mm)

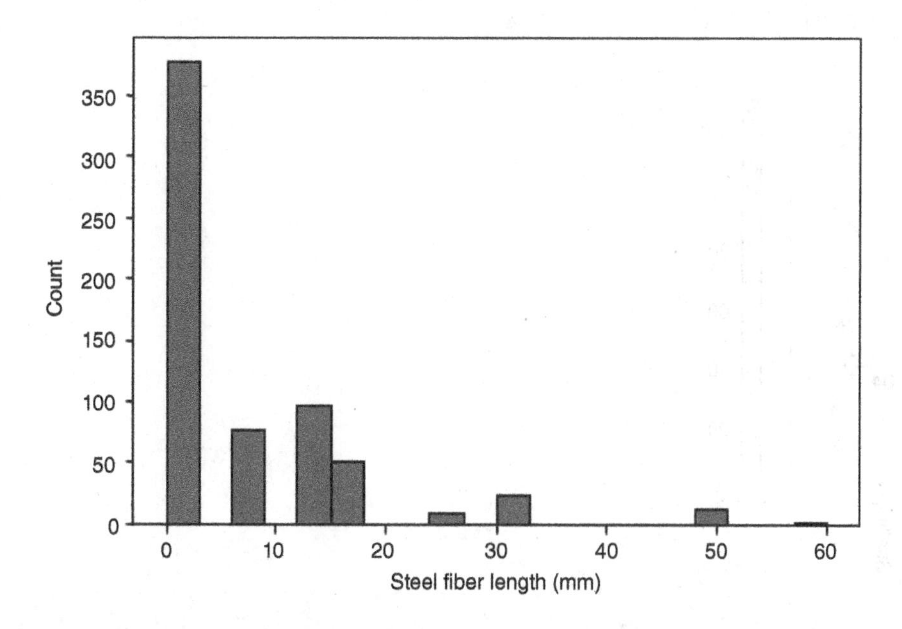
Count
350
300
250
200
150
100
50
0
0
10
20
30
40
50
60
Steel fiber length (mm)

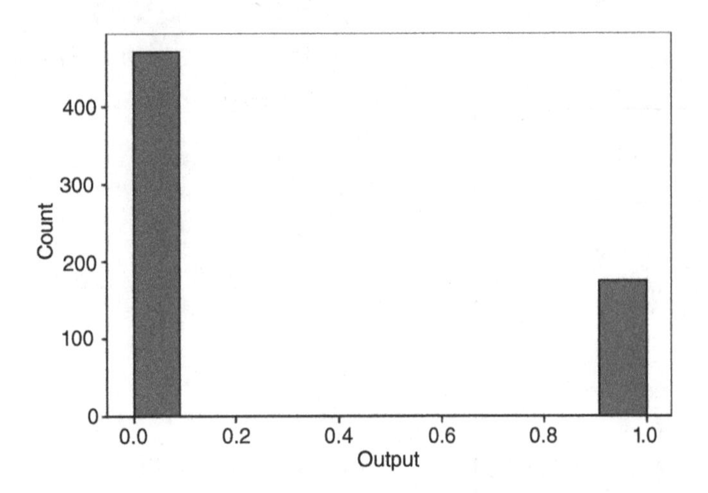


```
#Excuting and fitting the algorithm.
xgbc=xgb.XGBClassifier(objective ='binary:logistic',missing=1,seed=42,learning_rate
= 0.05, max_depth = 3)
xgbc.fit(x_train,y_train,verbose=True,early_stopping_rounds=50,eval_
metric='aucpr',eval_set=[(x_test,y_test)])
predictions = xgbc.predict(x_test)
kfold = StratifiedKFold(n_splits=10, random_state=7,shuffle=True)
)
results = cross_val_score(xgbc, x, y, cv=kfold)
print("Accuracy: %.2f%% (%.2f%%)" % (results.mean()*100, results.std()*100))
```

```
class_names = ['no spalling', 'spalling']
disp = plot_confusion_matrix(xgbc, x_test, y_test, display_labels=class_names,
cmap=pyplot.cm.Blues, xticks_rotation='vertical')
pyplot.grid(b=None)
pyplot.title('Testing set')
print (classification_report(y_test,predictions))
```

```
              precision    recall  f1-score   support

           0       0.90      0.97      0.94       135
           1       0.92      0.76      0.83        59

    accuracy                           0.91       194
   macro avg       0.91      0.87      0.88       194
weighted avg       0.91      0.91      0.90       194
```

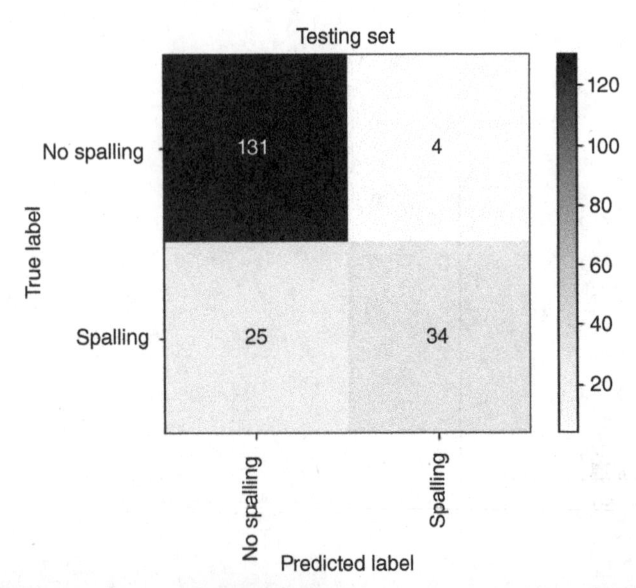

```
class_names = ['no spalling', 'spalling']
disp = plot_confusion_matrix(xgbc, x_train, y_train, display_labels=class_
names,colorbar=True, cmap=pyplot.cm.Blues, xticks_rotation='vertical')
pyplot.grid(b=None)
pyplot.title('Training set')
```

```
Text(0.5, 1.0, 'Training set')
```

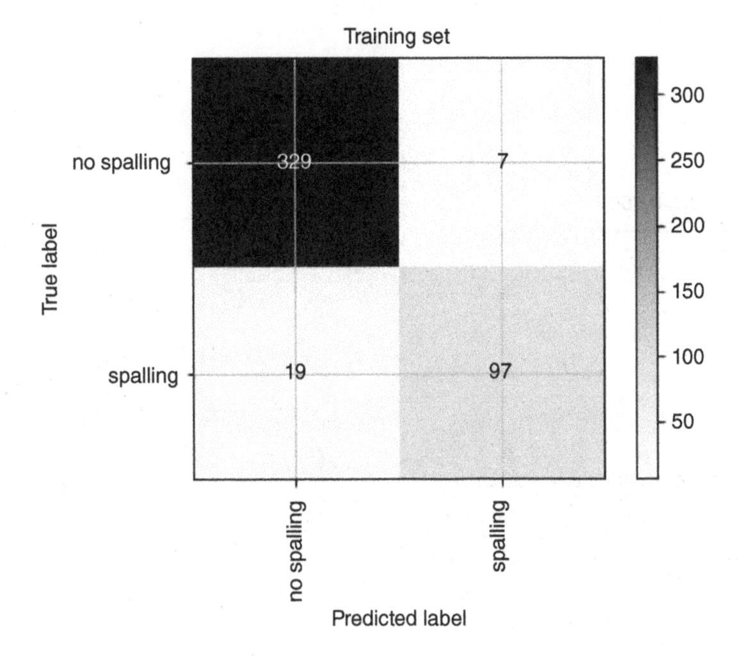


```
r_auc=roc_auc_score(y_train,predictions)
r_auc
```

```
yhat = xgbc.predict_proba(x_test)

# retrieve just the probabilities for the positive class
pos_probs = yhat[:, 1]
# plot no skill roc curve
pyplot.plot([0, 1], [0, 1], linestyle='--')
# calculate roc curve for model
fpr, tpr, _ = roc_curve(y_test, pos_probs)
# plot model roc curve
pyplot.plot(fpr, tpr, marker='.', label='AUC=0.866')
# axis labels
pyplot.xlabel('False Positive Rate')
pyplot.ylabel('True Positive Rate')
# show the legend
pyplot.legend()
pyplot.title('Testing set')
# show the plot
pyplot.show()
```

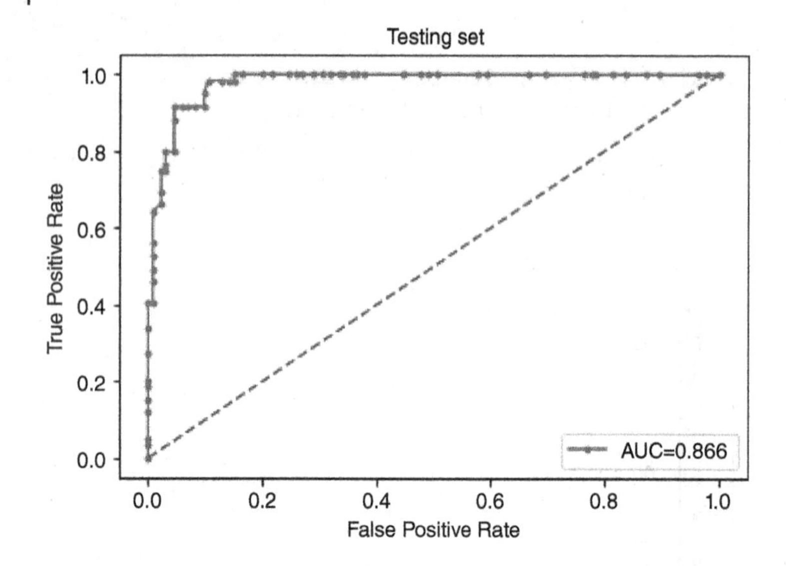


```
yhat = xgbc.predict_proba(x_train)

# retrieve just the probabilities for the positive class
pos_probs = yhat[:, 1]
# plot no skill roc curve
pyplot.plot([0, 1], [0, 1], linestyle='--')
# calculate roc curve for model
fpr, tpr, _ = roc_curve(y_train, pos_probs)
# plot model roc curve
pyplot.plot(fpr, tpr, marker='.', label='AUC=0.907')
# axis labels
pyplot.xlabel('False Positive Rate')
pyplot.ylabel('True Positive Rate')
# show the legend
pyplot.legend()
pyplot.title('Training set')
# show the plot
pyplot.show()
```

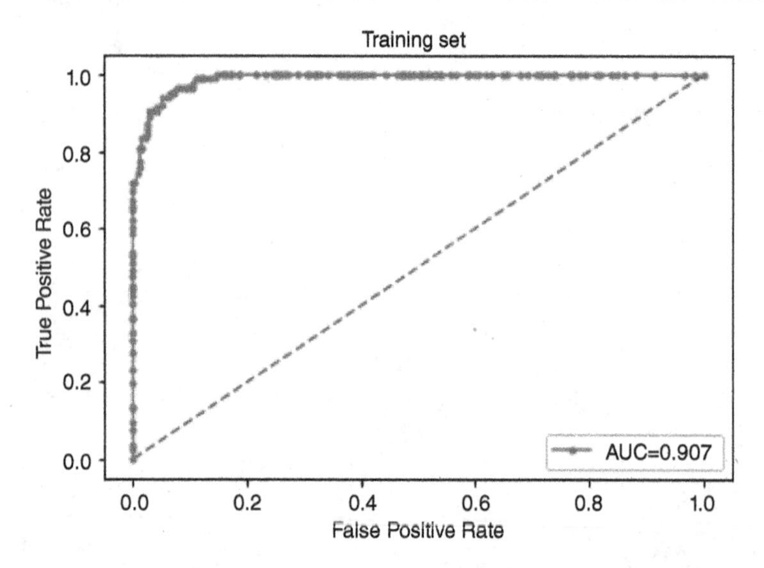

```
importances = xgbc.feature_importances_
indices = np.argsort(importances)
features = x.columns
pyplot.title('Feature Importances')
pyplot.barh(range(len(indices)), importances[indices], color='g', align='center')
pyplot.yticks(range(len(indices)), [features[i] for i in indices])
pyplot.xlabel('Relative Importance')
pyplot.show()
# The problem with global importance is that it gives an average overview of vari-
able contributing to the model
# but it lacks the direction of impact means whether a variable has a positive or
negative influence.
```

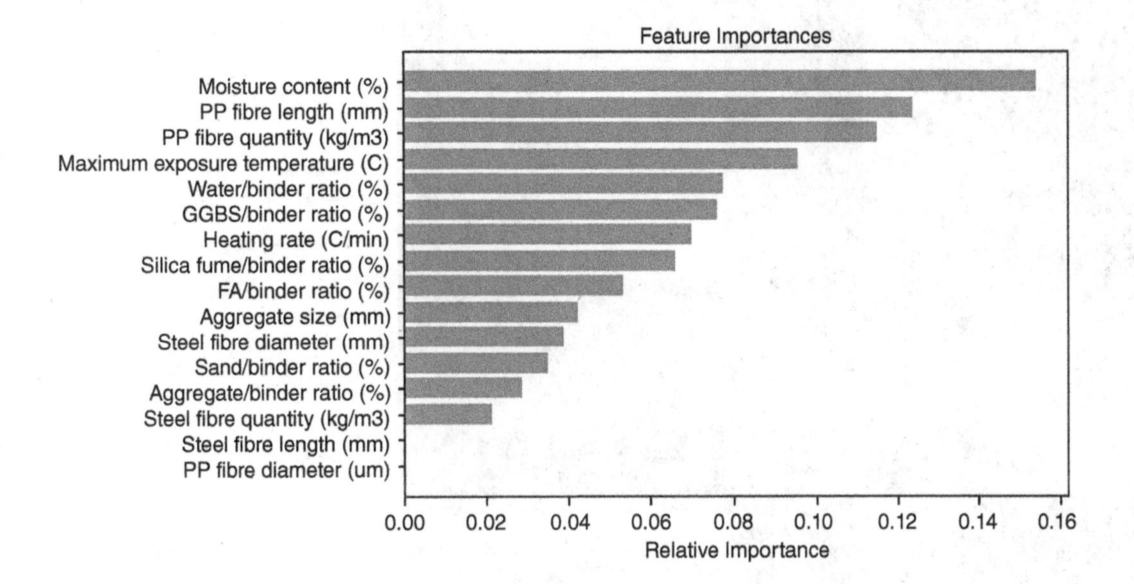


```
#in a more perfect world, I would more prefer the use of correlations attained from
mutual information (shown in the next page) and/or point-biserial correlation to
Pearson or Spearman methods - since our data has binary output.

l=fire.corr()
l
```

```
sns.heatmap(l, annot=True)
sns.set(rc = {'figure.figsize':(4,3)})
pyplot.title("Pearson Correlation")

pyplot.show()
def display_correlation(df):
    r = df.corr(method="spearman")
    pyplot.figure(figsize=(22,12))
    heatmap = sns.heatmap(df.corr(), vmin=-1,
                          vmax=1, annot=True)
    pyplot.title("Spearman Correlation")
    return(r)

r_simple=display_correlation(l)
```

Spearman Correlation

	Water/binder ratio (%)	Aggregate/binder ratio (%)	Sand/binder ratio (%)	Heating rate (C/min)	Moisture content (%)	Maximum exposure temperature (C)	Silica fume/binder ratio (%)	Aggregate size (mm)	GGBS/binder ratio (%)	FA/binder ratio (%)	PP fiber quantity (kg/m3)	PP fiber diameter(um)	PP fiber length (mm)	Steel fiber quantity (kg/m3)	Steel fiber diameter(mm)	Steel fiber length (mm)	Output
Water/binder ratio (%)	1	0.97	0.99	0.39	-0.022	-0.11	-0.89	0.97	-0.47	0.1	-0.76	-0.71	-0.79	-0.78	-0.6	-0.49	-0.22
Aggregate/binder ratio (%)	0.97	1	0.94	0.43	-0.11	-0.14	-0.91	0.99	-0.46	0.18	-0.77	-0.74	-0.79	-0.73	-0.56	-0.46	-0.24
Sand/binder ratio (%)	0.99	0.94	1	0.32	-0.043	-0.15	-0.85	0.93	-0.5	0.046	-0.74	-0.71	-0.77	-0.74	-0.56	-0.45	-0.24
Heating rate (C/min)	0.39	0.43	0.32	1	0.38	0.61	-0.51	0.46	0.19	0.055	-0.45	-0.38	-0.47	-0.64	-0.61	-0.57	0.43
Moisture content (%)	-0.022	-0.11	-0.043	0.38	1	0.61	-0.085	-0.046	0.36	-0.26	0.043	0.21	0.058	-0.39	-0.37	-0.34	0.53
Maximum exposure temperature (C)	-0.11	-0.14	-0.15	0.61	0.61	1	0.025	-0.13	0.46	-0.048	-0.0076	0.14	-0.019	-0.4	-0.43	-0.42	0.68
Silica fume/binder ratio (%)	-0.89	-0.91	-0.85	-0.51	-0.085	0.025	1	-0.91	0.14	-0.19	0.65	0.66	0.7	0.73	0.51	0.41	0.2
Aggregate size (mm)	0.97	0.99	0.93	0.46	-0.046	-0.13	-0.91	1	-0.44	0.12	-0.76	-0.72	-0.78	-0.75	-0.56	-0.46	-0.22
GGBS/binder ratio (%)	-0.47	-0.46	-0.5	0.19	0.36	0.46	0.14	-0.44	1	-0.2	0.41	0.25	0.29	0.16	0.053	-0.008	0.42
FA/binder ratio (%)	0.1	0.18	0.046	0.055	-0.26	-0.048	-0.19	0.12	-0.2	1	-0.15	-0.14	-0.2	-0.19	-0.19	-0.18	-0.24
PP fiber quantity (kg/m3)	-0.76	-0.77	-0.74	-0.45	0.043	-0.0076	0.65	-0.76	0.41	-0.15	1	0.88	0.94	0.6	0.42	0.31	-0.18
PP fiber diameter(um)	-0.71	-0.74	-0.71	-0.38	0.21	0.14	0.66	-0.72	0.25	-0.14	0.88	1	0.95	0.43	0.34	0.28	-0.089
PP fiber length (mm)	-0.79	-0.79	-0.77	-0.47	0.058	-0.019	0.7	-0.78	0.29	-0.2	0.94	0.95	1	0.65	0.54	0.46	-0.17
Steel fiber quantity (kg/m3)	-0.78	-0.73	-0.74	-0.64	-0.39	-0.4	0.73	-0.75	0.16	-0.19	0.6	0.43	0.65	1	0.86	0.76	-0.14
Steel fiber diameter(mm)	-0.6	-0.56	-0.56	-0.61	-0.37	-0.43	0.51	-0.56	0.053	-0.19	0.42	0.34	0.54	0.86	1	0.98	-0.25
Steel fiber length (mm)	-0.49	-0.46	-0.45	-0.57	-0.34	-0.42	0.41	-0.46	-0.008	-0.18	0.31	0.28	0.46	0.76	0.98	1	-0.28
Output	-0.22	-0.24	-0.24	0.43	0.53	0.68	0.2	-0.22	0.42	-0.24	-0.18	-0.089	-0.17	-0.14	-0.25	-0.28	1

1.00 0.75 0.50 0.25 0.00 -0.25 -0.50 -0.75 -1.00

```
mi=mutual_info_classif(x,y)
len(mi)
mi
```

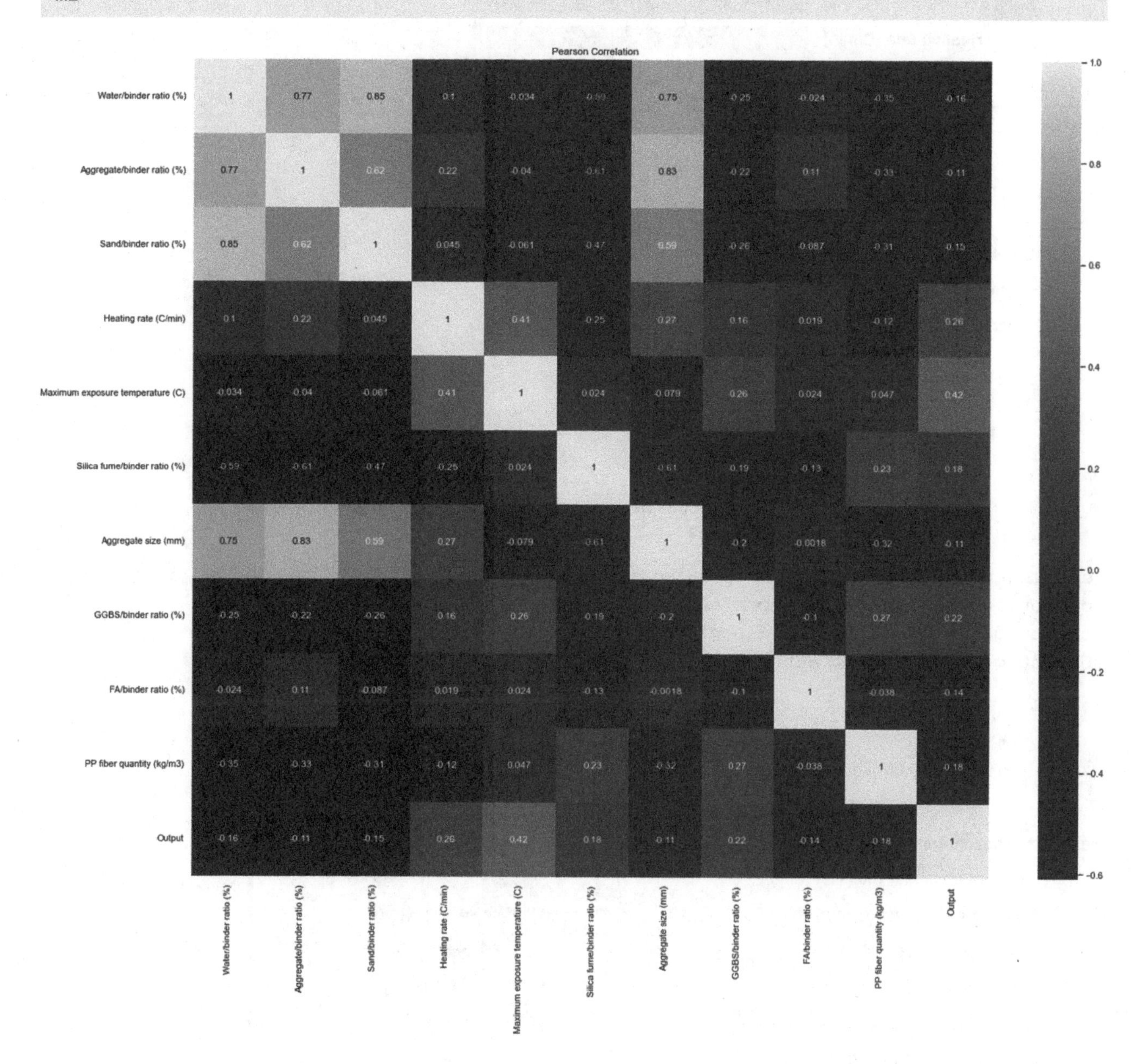


```
import shap

explainer = shap.TreeExplainer(xgbc)
shap_values = explainer.shap_values(x_test)
shap.summary_plot(shap_values, x_test, plot_type="bar")
```

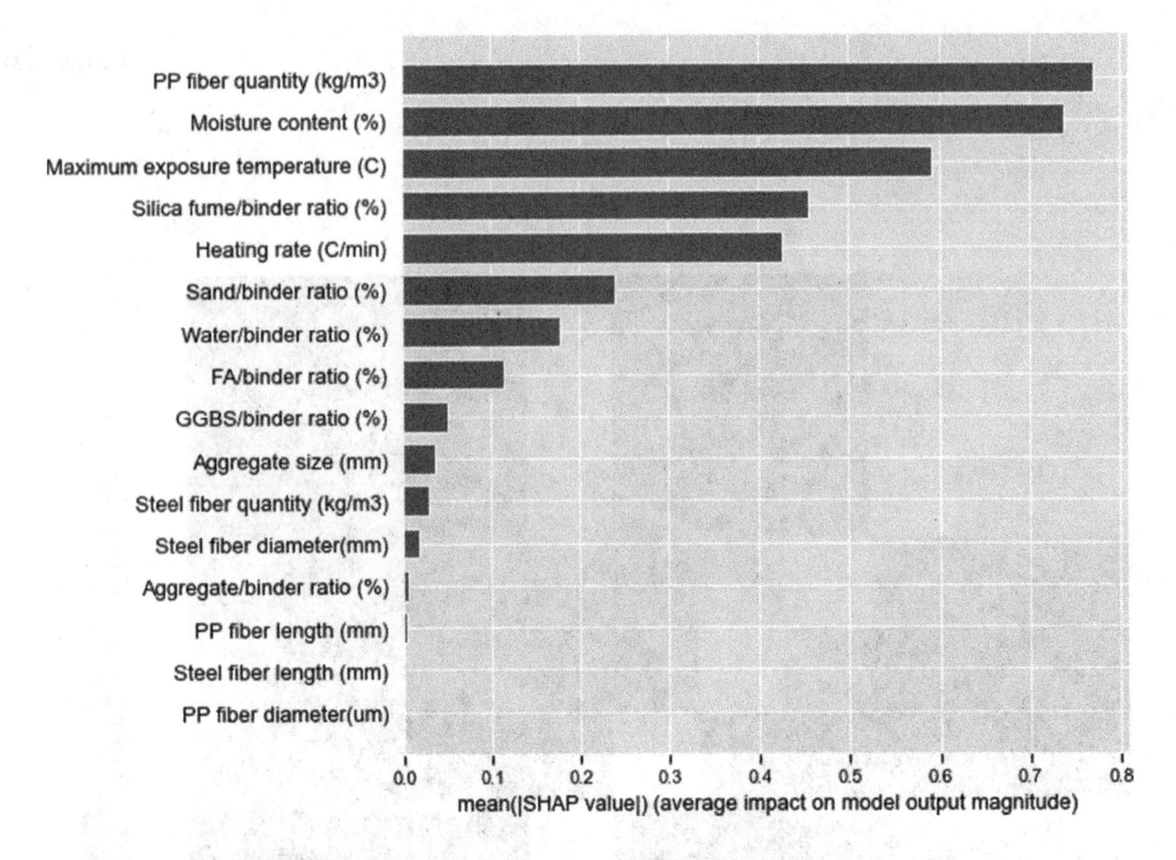


```
shap.summary_plot(shap_values, x_test,show=False)
pyplot.gcf().axes[-1].set_box_aspect(50)
pyplot.gcf().axes[-1].set_aspect(100)
pyplot.gcf().axes[-1].set_box_aspect(100)
```

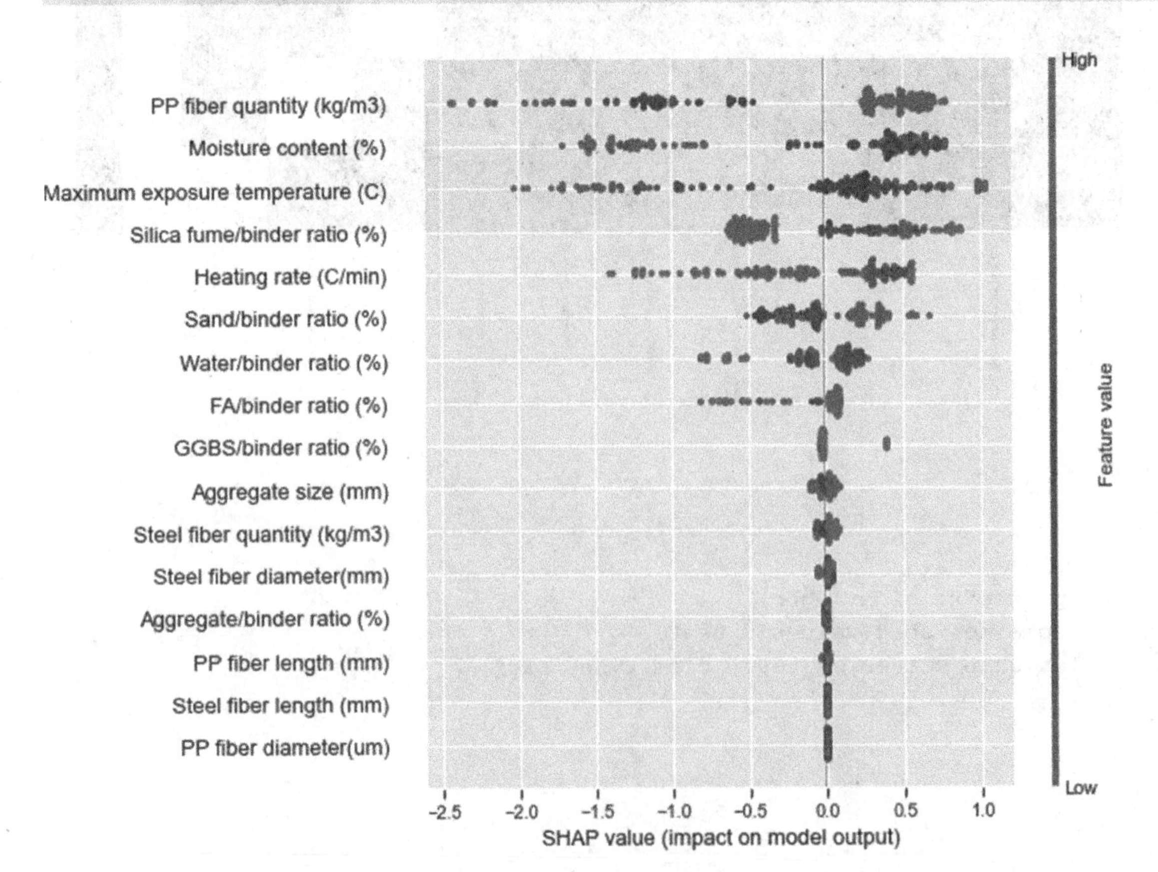

```
shap_interaction_values = shap.TreeExplainer(xgbc).shap_interaction_values(x_test.
iloc[:18,:])

shap.summary_plot(shap_interaction_values, x_test.iloc[:18,:])
```

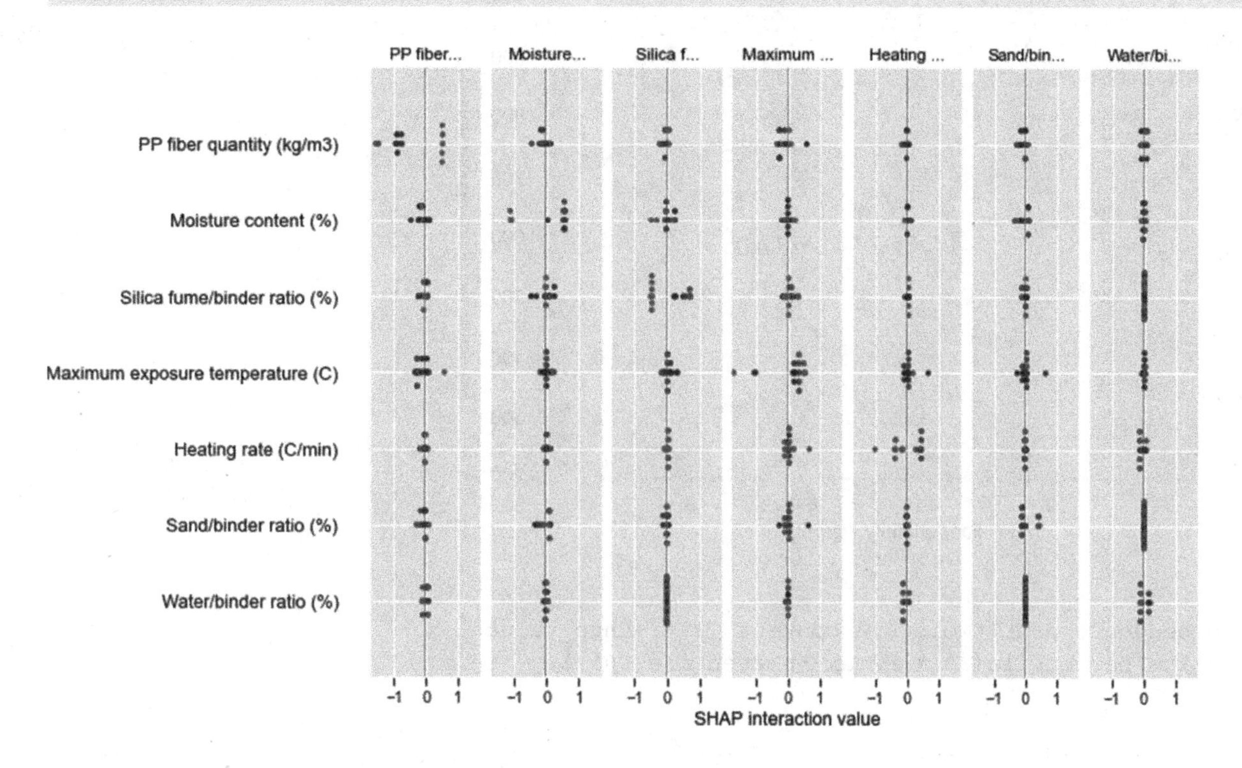


```
shap.dependence_plot("Maximum exposure temperature (C)", shap_values, x_
test,interaction_index=None)
```

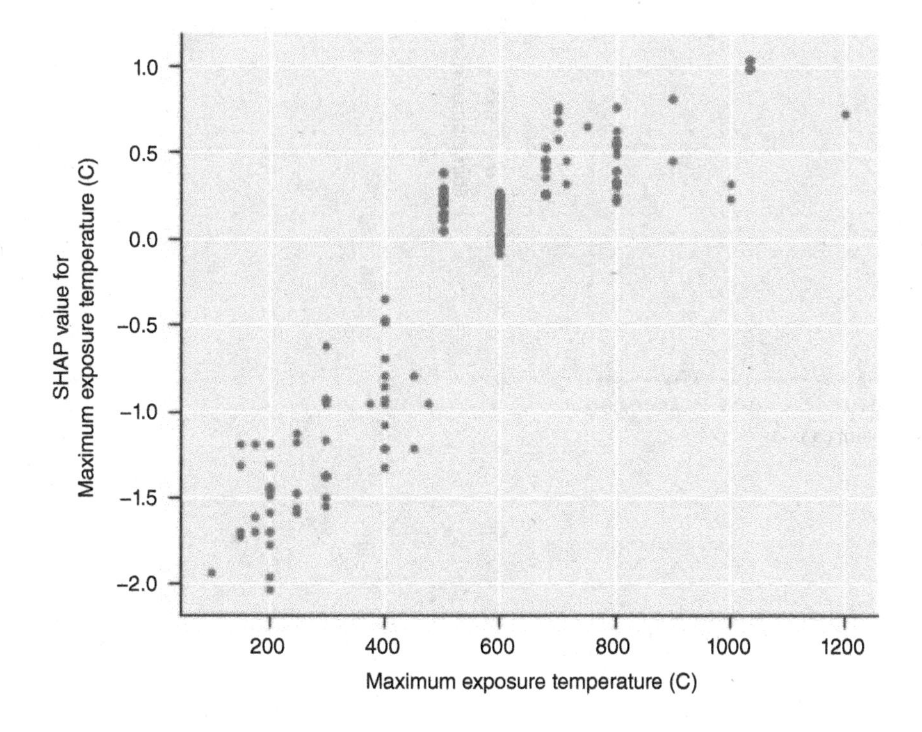

```
shap.dependence_plot("PP fiber quantity (kg/m3)", shap_values, x_test,)
```

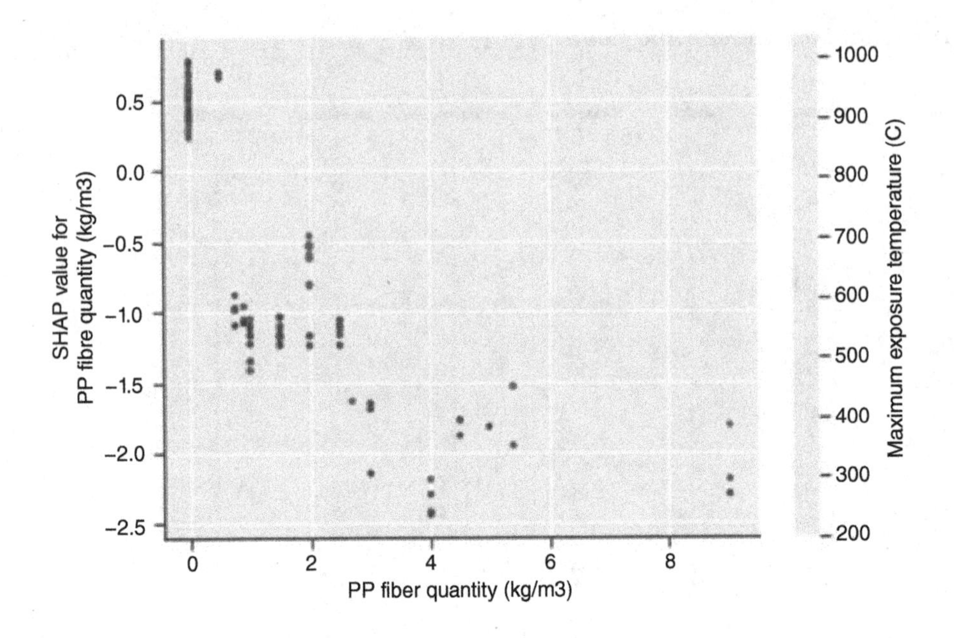


```
shap.dependence_plot("Moisture content (%)", shap_values, x_
test,xmin="percentile(30)", xmax="percentile(90)")
```

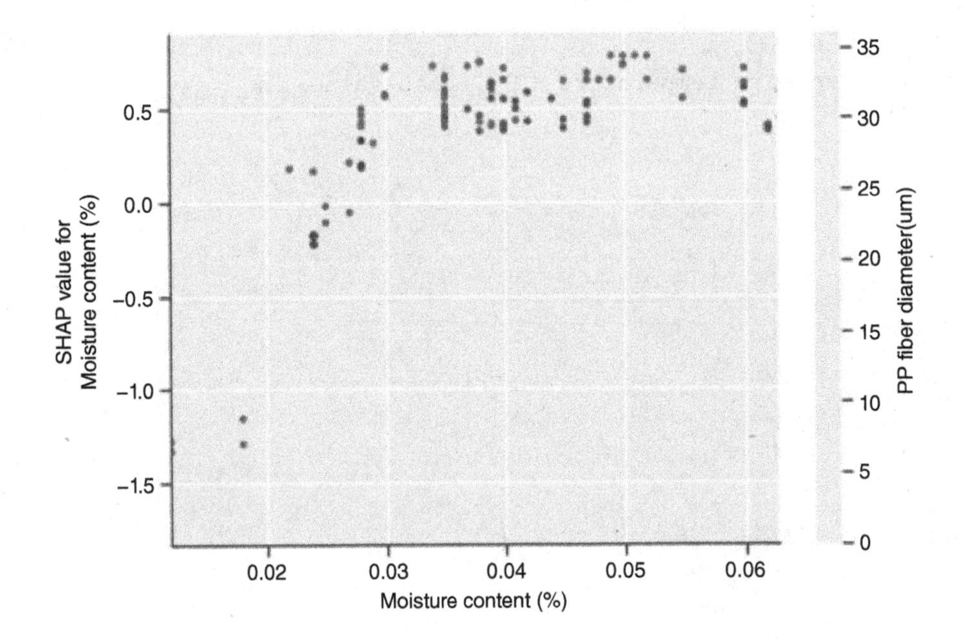

```
shap.dependence_plot(3, shap_values, x_test,xmin="percentile(20)",
xmax="percentile(99)",
                    )
```

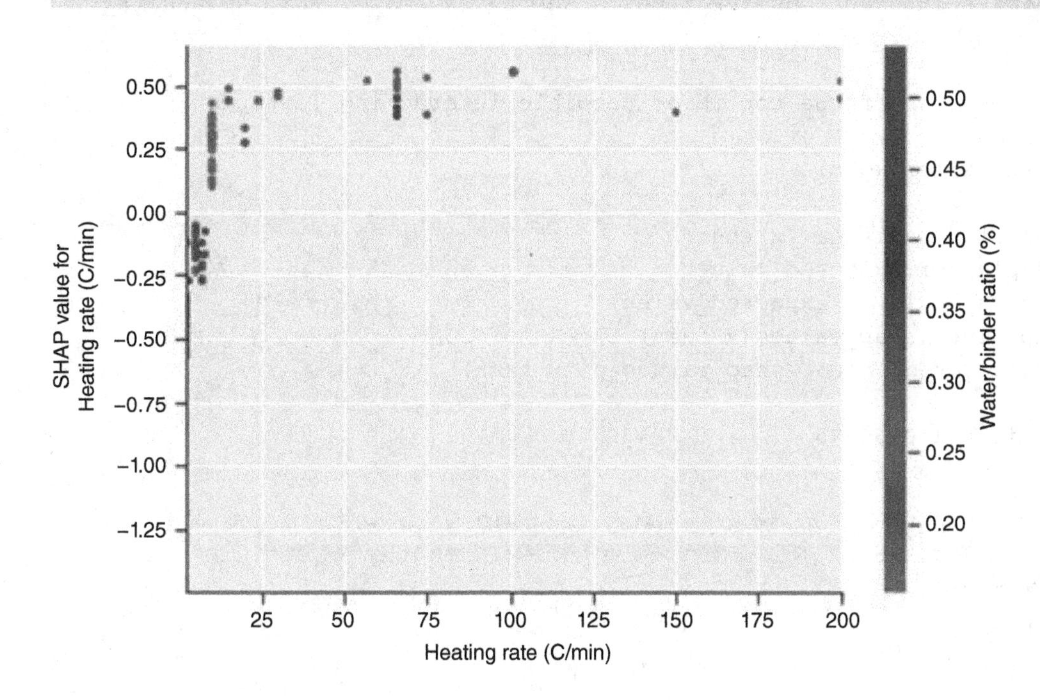


```
shap.dependence_plot(7, shap_values, x_test,xmin="percentile(1)",
xmax="percentile(99)")
```

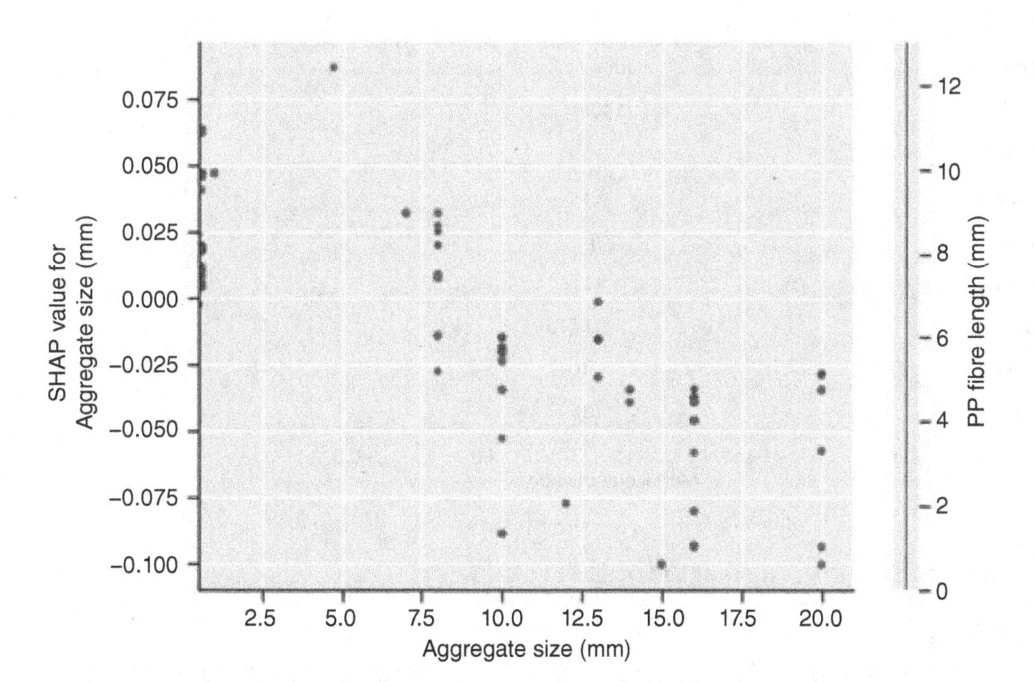

```
#We can use shap.approximate_interactions to guess which features
#May interact with maximum exposure temperature
```

```
inds = shap.approximate_interactions("PP fiber quantity (kg/m3)", shap_values, x_
test)
```

```
#Make plots colored by each of the top three possible interacting features

#Explainer = shap.Explainer(xgbc)
```

```
xgb_binary_shap_values = explainer(x_test)
expected_value = explainer.expected_value
print("The expected value is ", expected_value)
shap_values_p = explainer.shap_values(x_test)[1]
shap.decision_plot(expected_value, shap_values_p, x_test)
```

```
The expected value is  -1.0156575
```

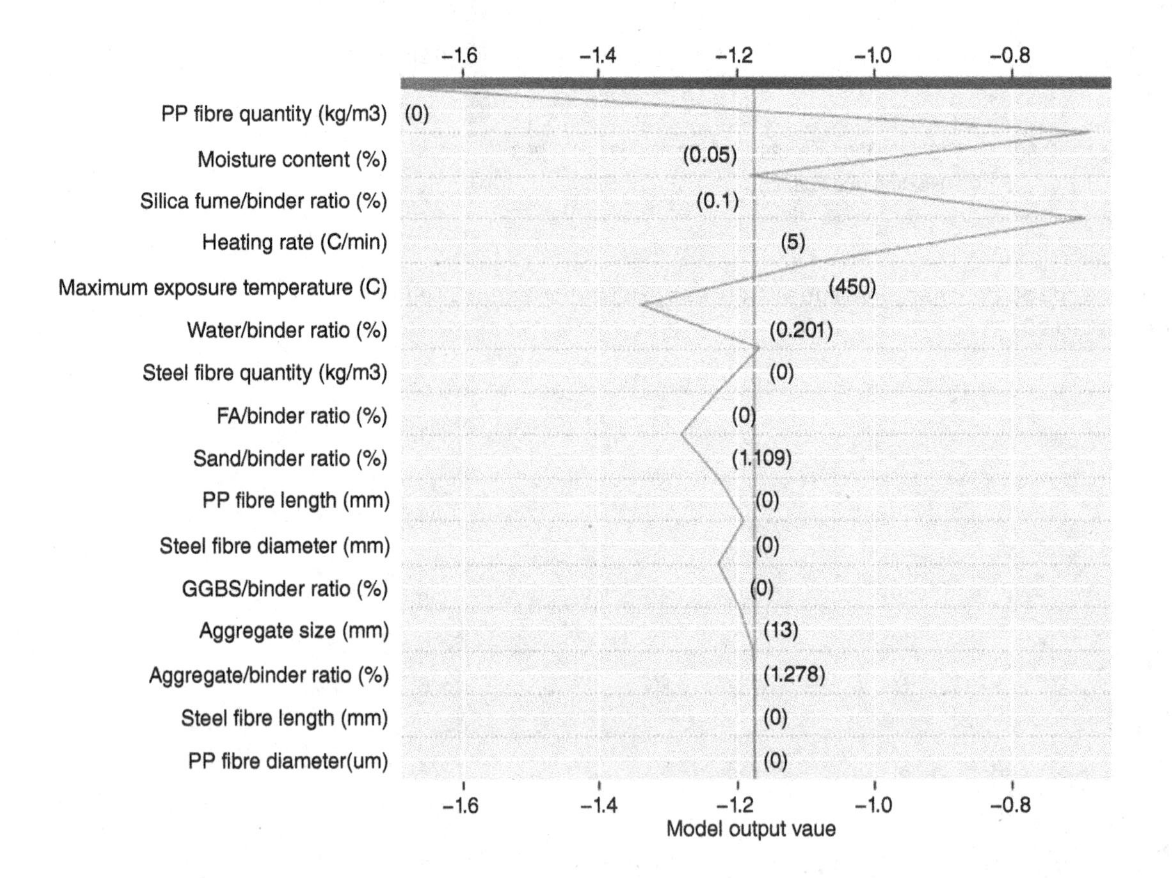


```
test_objects = [x_test.iloc[0:17], x_test.iloc[191:192]]

for float in test_objects:
    print('Probability of class 1 = {:.4f}'.format(xgbc.predict_proba(float)[0][1]))
    print('Formula raw prediction = {:.4f}'.format(xgbc.predict(float)[0]))
    print('\n')
```

```
Probability of class 1 = 0.5410
Formula raw prediction = 1.0000


Probability of class 1 = 0.0275
Formula raw prediction = 0.0000
```

```
shap.initjs()
def p(j):
    explainer
    xgbc_shap_values = explainer.shap_values(x_test)
    return(shap.force_plot(explainer.expected_value, xgbc_shap_values[j,:], x_test.
iloc[j,:], link='logit'))
p(0
)
```

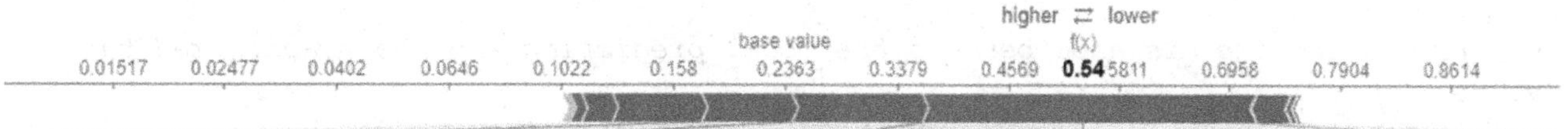


```
#The waterfall is another way of plotting the force plot.
shap.plots.waterfall(xgb_binary_shap_values[0])
```

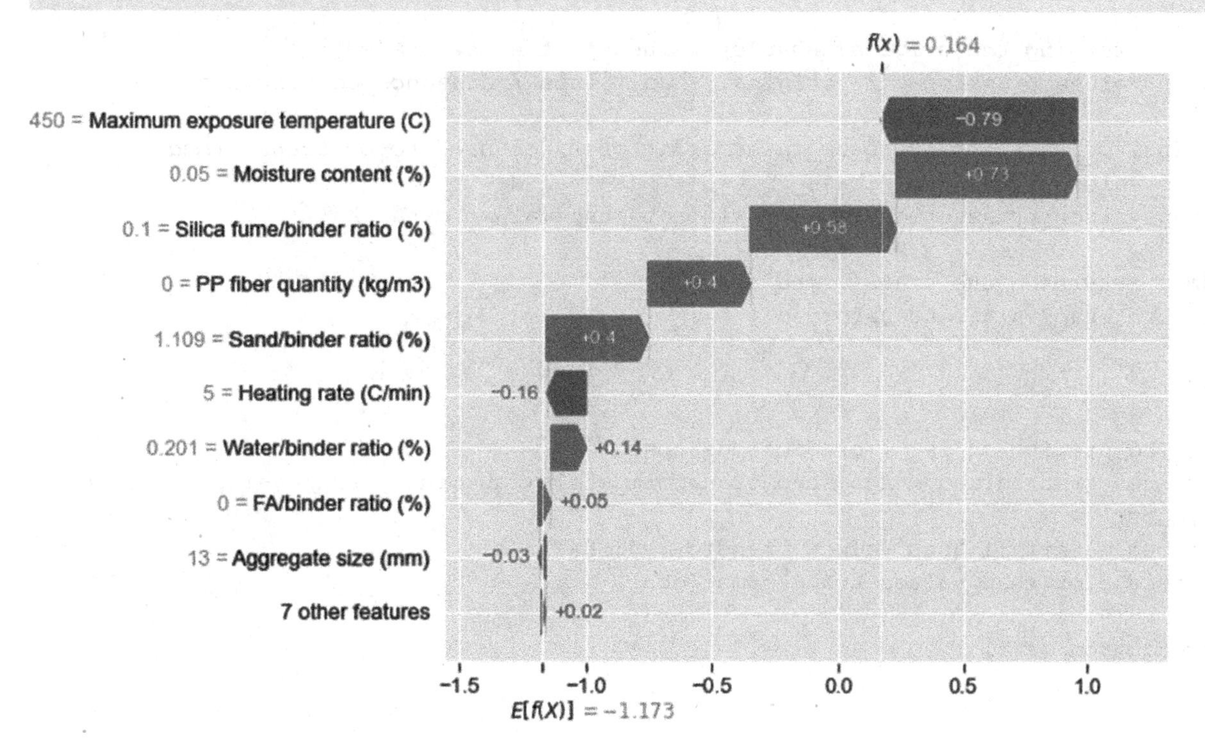

```
def xgb_shap_transform_scale(original_shap_values, Y_pred, which):
    from scipy.special import expit

    #Compute the transformed base value, which consists in applying the logit
function to the base value
    from scipy.special import expit #Importing the logit function for the base value
transformation
    untransformed_base_value = original_shap_values.base_values[-1]

    #Computing the original_explanation_distance to construct the distance_coeffi-
cient later on
    original_explanation_distance = np.sum(original_shap_values.values, axis=1)
[which]

    base_value = expit(untransformed_base_value)  # = 1 / (1+ np.exp(-untransformed_
base_value))

    #Computing the distance between the model_prediction and the transformed base_
value
    distance_to_explain = Y_pred[which] - base_value

    #The distance_coefficient is the ratio between both distances which will be used
later on
    distance_coefficient = original_explanation_distance / distance_to_explain

    #Transforming the original shapley values to the new scale
    shap_values_transformed = original_shap_values / distance_coefficient

    #Finally resetting the base_value as it does not need to be transformed
    shap_values_transformed.base_values = base_value
    shap_values_transformed.data = original_shap_values.data

    #Now returning the transformed array
    return shap_values_transformed

obs =0

print("The prediction is ", predictions[obs])
shap_values_transformed = xgb_shap_transform_scale(xgb_binary_shap_values, predic-
tions, obs)
shap.plots.waterfall(shap_values_transformed[obs])
shap.force_plot(shap_values_transformed[obs])
```

```
The prediction is  1
```

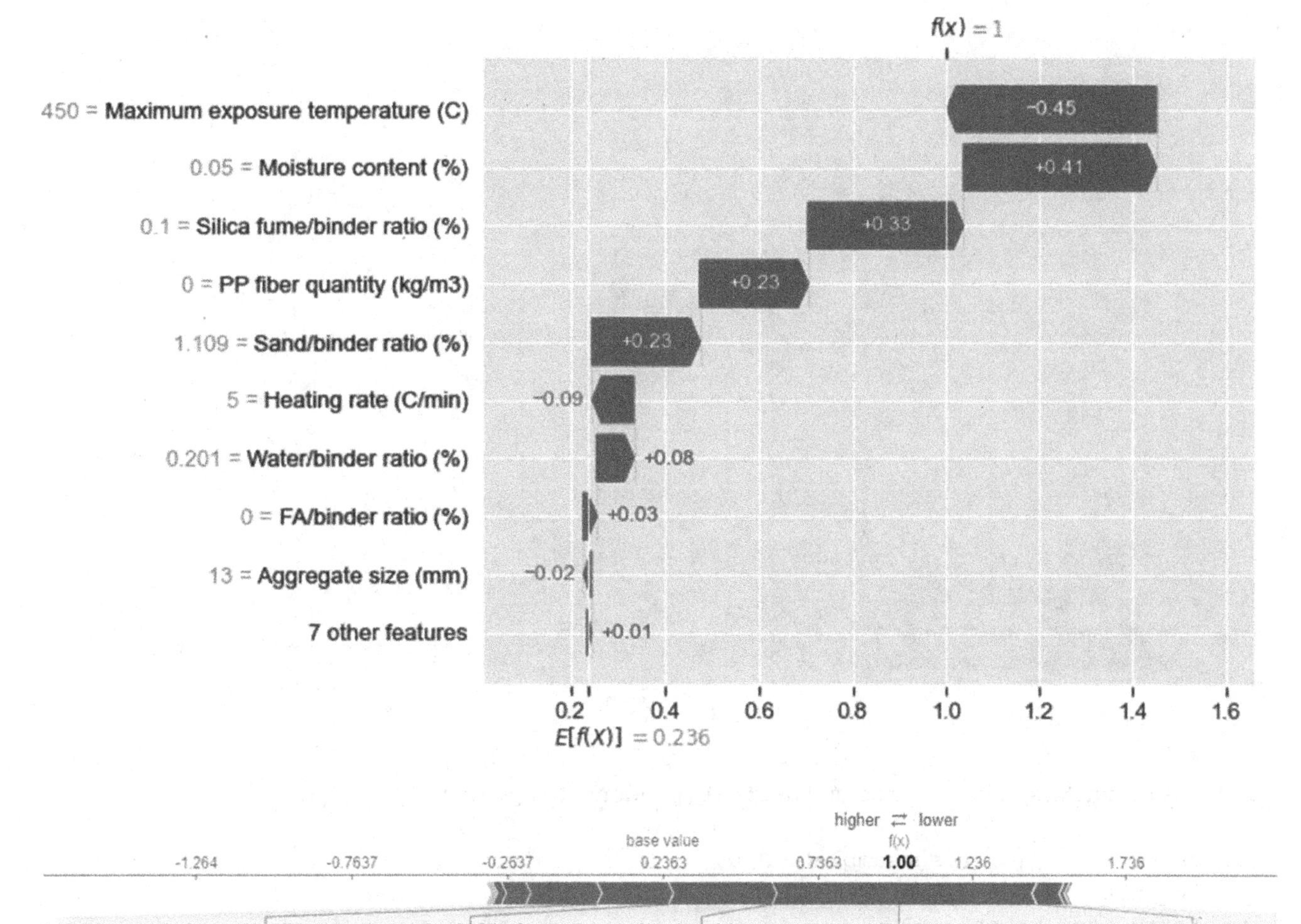


```
import numpy as np
import pandas as pd
from sklearn.datasets import make_hastie_10_2
from sklearn.ensemble import GradientBoostingClassifier
from sklearn.inspection import PartialDependenceDisplay
```

```
features = [#"PP fiber quantity (kg/m3)",
    "Moisture content (%)"#, "Maximum exposure temperature (C)", "Silica fume/binder
ratio (%)",
            #"Heating rate (C/min)","Sand/binder ratio (%)","Water/binder ratio
(%)","FA/binder ratio (%)"
          ]
PartialDependenceDisplay.from_estimator(xgbc, x_train, features, kind='both')

#display.plot(n_cols=5)
```

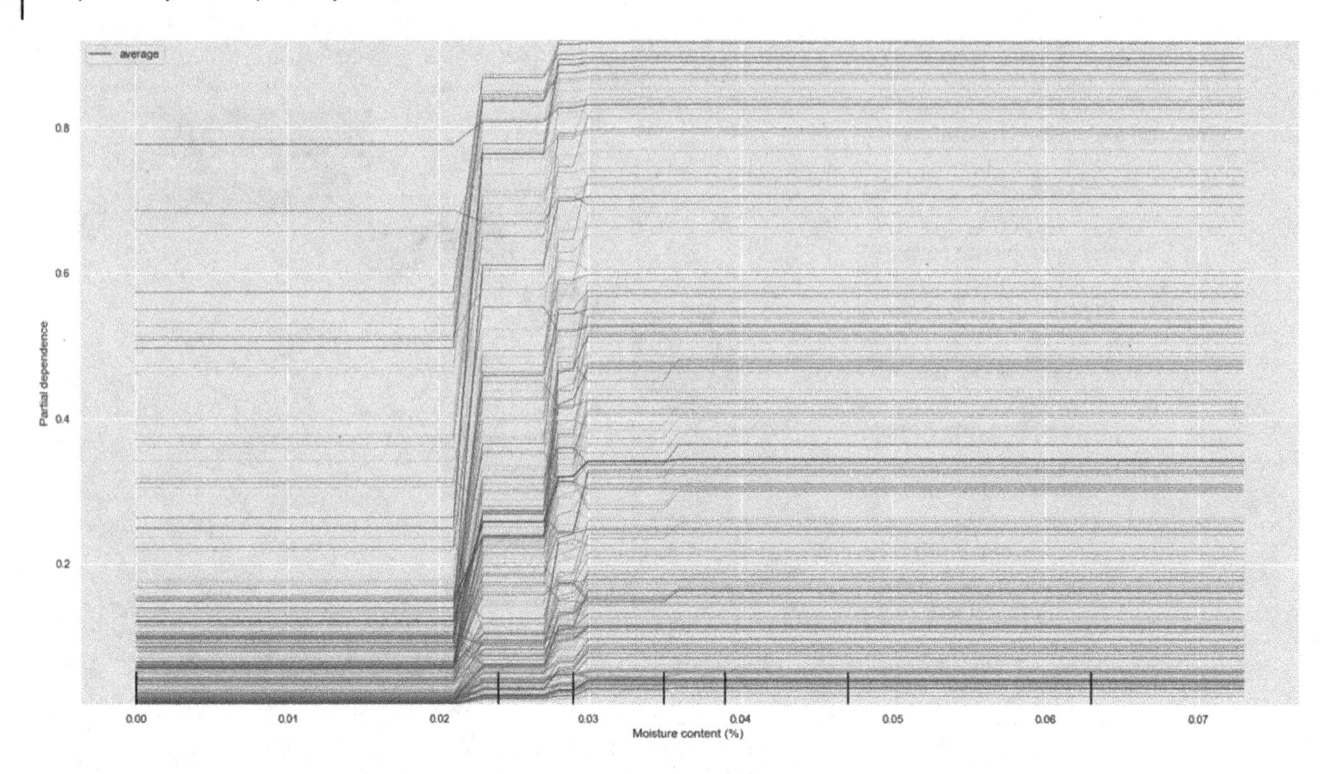


```
from sklearn.inspection import PartialDependenceDisplay

common_params = {    "subsample": 200,
    "n_jobs": 2,
    "grid_resolution": 20,
    "random_state": 0,
#    "subsample": 300,
 #   "n_jobs": 2,
 #   "grid_resolution": 20,
 #   "random_state": 0,
}

print("Computing partial dependence plots…")

display = PartialDependenceDisplay.from_estimator(
    xgbc,
    x_test,
    features=[#"PP fiber quantity (kg/m3)",
        # "Moisture content (%)",
         # "Maximum exposure temperature (C)",
         # "Silica fume/binder ratio (%)",
             # "Heating rate (C/min)",
        #"Sand/binder ratio (%)",
        # "Water/binder ratio (%)",
        "FA/binder ratio (%)"
             ],
    kind="both",
    **common_params,
)
```

```
display.figure_.suptitle(
    "Partial dependence of spalling"
)
#pyplot.legend(['Legend'], loc='upper left',color='red')

display.figure_.subplots_adjust(hspace=0.2)
for i in range(display.lines_.shape[0]):
            display.lines_[0,i,-1].set_color('RED')
            display.axes_[0, i].legend()
```

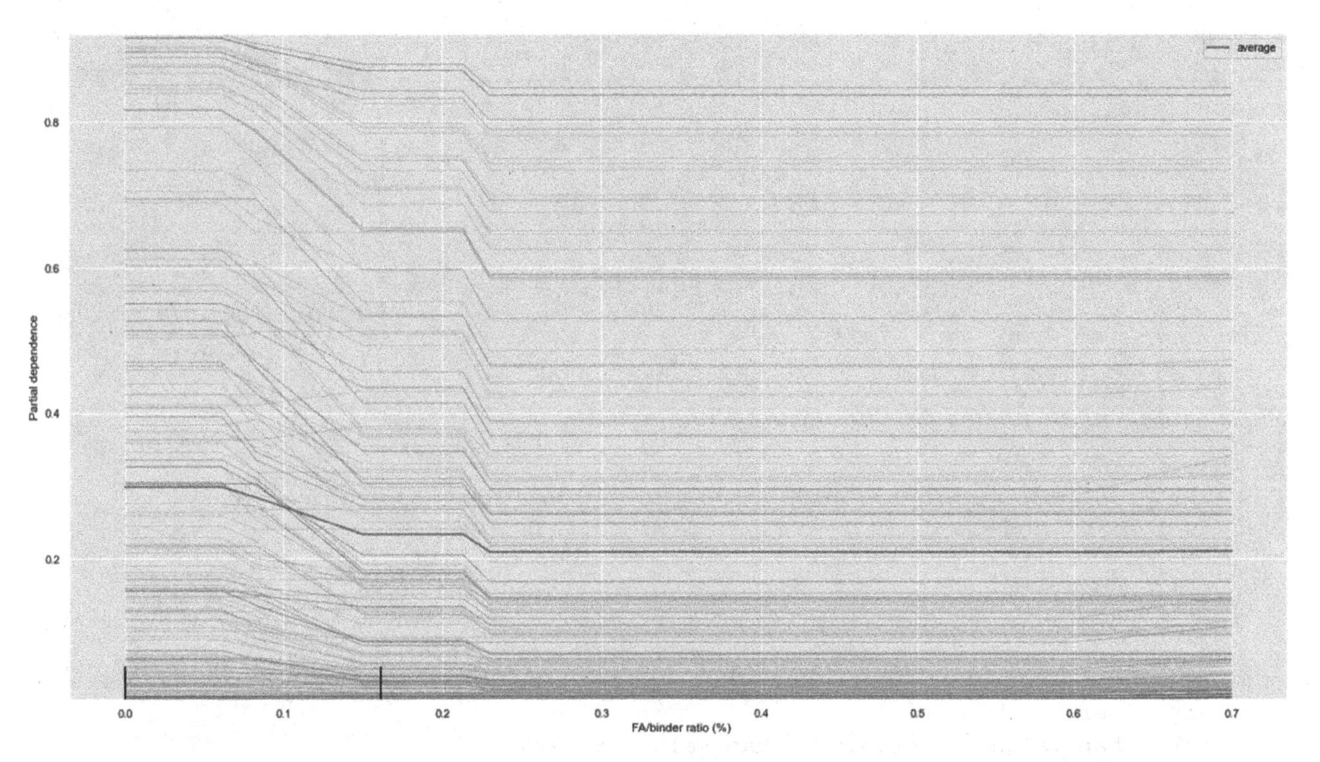


```
import matplotlib.pyplot as plt
from sklearn.inspection import partial_dependence
from sklearn.inspection import PartialDependenceDisplay

features = ['PP fiber quantity (kg/m3)',"Moisture content (%)","Maximum exposure
temperature (C)",
            "Silica fume/binder ratio (%)","Heating rate (C/min)","Sand/binder
ratio (%)",
            "Water/binder ratio (%)",



display = PartialDependenceDisplay.from_estimator(
    xgbc,
```

```
    x_test,
    features,
    kind="both",
    subsample=664,
    n_jobs=10,
    grid_resolution=1200,
    random_state=0,

    )
```

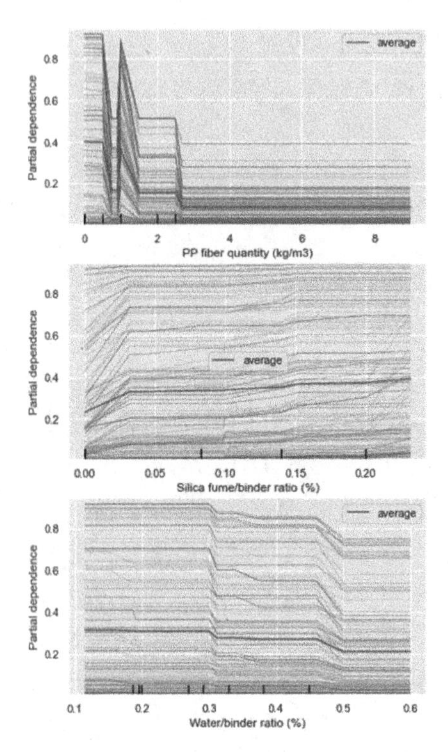

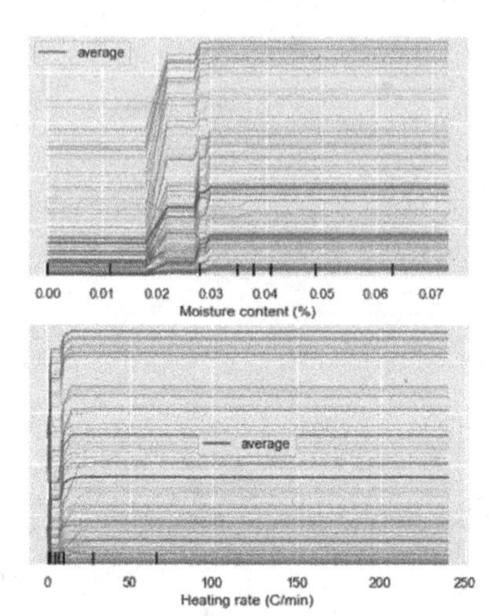

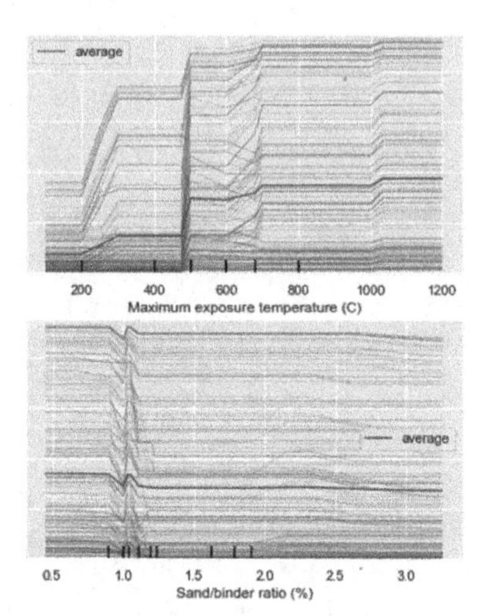


```
def p(j):
    explainer
    xgbc_shap_values = explainer.shap_values(x_test)
    return(shap.force_plot(explainer.expected_value, xgbc_shap_values[j,:], x_test.
iloc[j,:], link='logit'))
p(12
)
```

```
#Here is the code for LIME
import lime
import lime.lime_tabular
from lime import lime_tabular
```

```
explainerL = lime_tabular.LimeTabularExplainer(
    training_data=np.array(x),
    feature_names=x.columns,
    class_names=['Not spalling', 'Spalling'],
    mode='classification')
```

```
exp = explainerL.explain_instance(
    data_row=x.iloc[285],
    predict_fn=xgbc.predict_proba
)
#exp.save_to_file('temp.html')
exp.show_in_notebook(show_table=True)
```

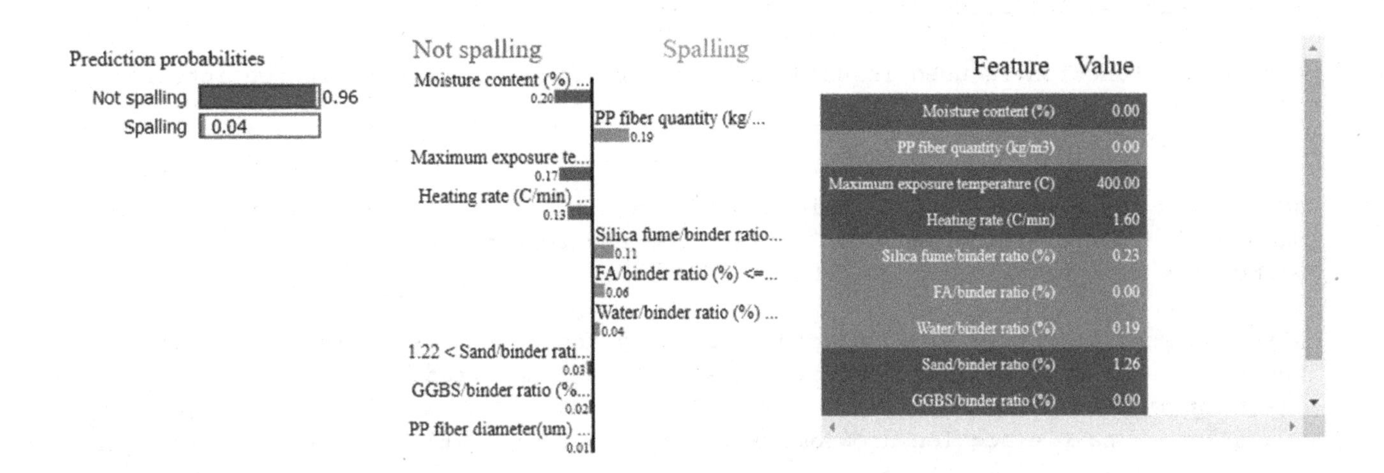

Here is the *R* code to repeat the same example.[78]

```
#Installing necessary and useful packages for training XGboost algorithm in the
selected dataset.
>>install.packages ("modelStudio"): The modelStudio automatically produce explain-
able results from black- box model.
>>install.packages ("dplyr"): An agile tool for working with data frame.
>>install.packages ("xgboost", type = "binary"): is an R package for XGboost imple-
mentation in R environment. This package has ability to perform different tasks
at the same time on a single machine which could be ten times faster than similar
package.
>>install.packages ("DALEX"): DALEX performance is like a transition machine that
transform blackbox model into a box which can demonstrate the structures that lay
behind the blackbox models.
>>install.packages ("tidyverse"): tidyverse is the modeling package that contain
several different packages including rsample, parsnip, and recipes to name but a few
for problems modeling.
>>install.packages ("tidymodels"): tidymodels is the modeling package that contain
several different packages including rsample, parsnip, and recipes to name but a few
for problems modeling.
>>install.packages ("caret"): This package is a function for plot creation and model
creation or classification as well as regression model.
>>install.packages ("Matrix"): A thorough collection of Matrix classes such as
triangular, symmetrical, and diagonal matrices.
>>install.packages ("corrplot"): provides a visual exploratory tool on correlation
matrix that supports automatic variable reordering to help detect hidden patterns
among variables.
```

78 As you will see, some differences might occur between the two languages, and the fact that the *R* example used a 80:20 split to train the model.

#Required packages for Random-forest regression in R programming language to retrieve the installed-packages above.

```
>>Library (modelStudio) Library (dplyr)
Library (xgboost) Library (DALEX) Library (tidyverse) Library (tidymodels) Library (caret) Library (Matrix) Library (corrplot)
```

#Data importation: read.csv() function or Data Input function that reads a file from a local drive in the user's computer in the form of a table.

```
>>data=read.csv('C:/Classification.csv')
```

#The str() function is used instead of a summary function to exhibit the internal structure of dataset.

```
>>str(data)
```

#Splitting the dataset into a training and testing set in this example, 70% of the sample size for training and the 30% left for the testing set.

```
>>smp_size <- floor(0.7 * nrow(data))
```

#Set the seed to ensure a reproducible training and testing set. The sample () function is used to isolate a specified size from sample set.

```
>>set.seed(123)
train_ind <- sample(seq_len(nrow(data)), size = smp_size) trainxg <- data[train_ind, ]
testxg<- data[-train_ind, ]
```

#Isolate X and Y for train set

```
>>ytrainxg<-trainxg$Output
xtrainxg<- trainxg %>% select(-Output)
```

#Isolate X and Y for test set

```
>>ytestxg<-testxg$Output
xtestxg<-testxg %>% select(-Output)
```

#Plot histograms for the features we are interested in. Here, we used for loop to print the histograms for each variable in one function.

```
>>names<-names(data)
>>classes<-sapply(data,class)
for(name in names[classes == 'numeric']){ dev.new()
hist(data[,name])
}
```

#Another approach is to plot histograms for the features using the hist() function can be found as follows:

```
>>hist(data$Water.binder.ratio)
hist(data$Aggregate.binder.ratio )
hist(data$Sand.binder.ratio )
hist(data$Heating.rate..C.min.) hist(data$Maximum.exposure.temperature..C.)
hist(data$Silica.fume.binder.ratio )
hist(data$Aggregate.size..mm.) hist(data$GGBS.binder.ratio )
hist(data$FA.binder.ratio)
hist(data$PP.fiber.quantity..kg.m3.) hist(data$Output)
```

#Setting the parameters for the XG.boost Model using list() function

```
>>params<-list(set.seed=1500, eval_metric="auc", objective="binary:logistic")
```

#Running XG.boost model using xgboost() function a simple and fast way of implementing Extreme Gradient Boosting algorithm.

```
>>modelxg <-xgboost(data = as.matrix(xtrainxg), label = ytrainxg,
params = params, nrounds = 20,
verbose = 1)
```

#Checking the model accuracy using Confusion Matrix (Sensitivity, Specificity, and Accuracy). The confusionMatrix() function can show the result.

```
>>trained<-model.matrix(trainxg$Output~.-1,data = trainxg) trainlable<-trainxg[,1]
trainlable<-as.numeric(trainlable)-1
tested<-model.matrix(testxg$Output~.-1,data = testxg) testlabel<-testxg[,1]
testlabel<-as.numeric(testlabel)-1
pred<-predict(modelxg,tested) head(pred)
pred<-as.integer(pred>0.5) confusionMatrix(table(pred, testxg$Output))
```

#Plotting correlation matrices using cor() function. These plots show the correlation between different variables in the dataset. The first method is spearman that evaluate the monotonic relationship, and the last one is Pearson which evaluates the linear relation between two different continues variables.

#first Method (Spearman)

```
>>m=cor(data, method = "spearman") corrplot(m, type = "full",
method = "circle", tl.cex = 0.5, tl.col = 'black')
corrplot(m, type = "full",
method = "number", tl.cex = 0.55, tl.col = 'black')
corrplot(m, method ="color",type = "upper",order ="hclust",tl.cex = 0.5, tl.col =
"black",tl.srt = 45, diag
= FALSE)
```

#Second method (Pearson)

```
>>n=cor(data, method ="pearson") corrplot(n, type = "full",
method = "circle", tl.cex = 0.5, tl.col = 'black')
corrplot(n, type = "full",
method = "number", tl.cex = 0.55, tl.col = 'black')
corrplot(n, method ="color",type = "upper",order ="hclust",tl.cex = 0.5, tl.col =
"black",tl.srt = 45, diag
= FALSE)
```

#This summary plot is considered one of the most important plots in SHAP. The plot axes are as follows: Vertical location shows what the feature is depicting. The color shows whether that feature was high or low for that row of the dataset. Horizontal location shows whether the effect of that value caused a higher or lower prediction. It can be implemented using xgb.plot.shap.summary() function.

```
>>xgb.plot.shap(data = as.matrix(xtrainxg), model = modelxg,
top_n =5)
>>xgb.plot.shap.summary(data = as.matrix(xtrainxg), model = modelxg,
```

```
top_n = 11)
```

#This code shows the XG.boost model importance for all variables in the dataset.

```
>>imp<-xgb.importance(colnames(ytrainxg),model = modelxg) print(imp)
xgb.plot.importance(imp)
```

#Creating explainer model using boost_tree() function as method to generate a model prepared for further different fitting and DALEX::explain() function as a X-ray to create explainable model for clarification.

```
>>data$Output<-as.factor(data$Output) fit_xgboost<-boost_tree(learn_rate = 0.3)%>%
set_mode("classification")%>%
set_engine("xgboost")%>% fit(data$Output~.-1,data=data)
>>data$Output<-as.numeric(data$Output) explainer<-DALEX::explain(
model= fit_xgboost, data = data, y=data$Output, label = "XGBoost"
)
```

#Calling Model studio to visualize the model which automates the explanatory analysis of machine learning model.

```
modelStudio::modelStudio(explainer = explainer)
```

#Required packages for XGboost classification in R programming language to call the specified packages.

```
library(modelStudio)
library(dplyr)
library(xgboost)
library(DALEX)
library(tidyverse)
library(tidymodels)
library(caret)
library(magrittr)
library(Matrix)
library(corrplot)
```

#Data importation.

```
data=read.csv('C:/Classification.csv')
```

#Checking the structures lay behind the dataset.

```
str(data)
'data.frame':646 obs. of 11 variables:
$ Water.binder.ratio....: num 0.273 0.273 0.3 0.218 0.218 0.3 0.3 0.3 0.273 0.273 …
$ Aggregate.binder.ratio....: num 2 2 2 1.37 1.37 …
$ Sand.binder.ratio....: num 1.05 1.05 1.03 1.2 1.2 …
$ Heating.rate..C.min.: num 101 101 101 101 101 101 101 101 101 101 …
$ Maximum.exposure.temperature..C.: int 1034 1034 1034 1034 1034 1034 1034 1034 1034
1034 …
$ Silica.fume.binder.ratio....: num 0 0 0 0 0 0 0 0 0 0 …
$ Aggregate.size..mm.: num 7 14 20 7 7 14 20 20 7 14 …
$ GGBS.binder.ratio....: num 0.25 0.25 0.25 0.25 0.25 0.25 0.25 0.25 0.25 0.25 …
$ FA.binder.ratio....: num 0 0 0 0 0 0 0 0 0 0 …
$ PP.fiber.quantity..kg.m3.: num 0 0 0 0 0 0 0 0 0 0 …
$ Output: int 1 1 1 1 1 1 1 1 1 1…
```

```
#Train-Test split strategy.
smp_size <- floor(0.7 * nrow(data))

#Data partitioning and setting the seed to make the partitioning reproducible.
set.seed(123)
train_ind <- sample(seq_len(nrow(data)), size = smp_size)
trainxg <- data[train_ind, ]
testxg<- data[-train_ind, ]

#Isolate X and Y for train set.
ytrainxg<-trainxg$Output
xtrainxg<- trainxg %>% select(-Output)

#Isolate X and Y for test set.
ytestxg<-testxg$Output
xtestxg<-testxg %>% select(-Output)

#Plotting histogram for all variables.
names<-names(data)
classes<-sapply(data,class)
for(name in names[classes == 'numeric']){
+ dev.new()
+ hist(data[,name])
+ }

#Plotting histogram for individual variable.
hist(data$Water.binder.ratio)
```

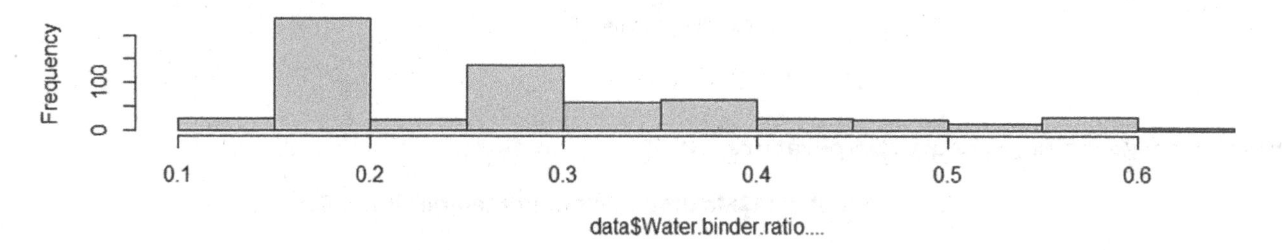


```
hist(data$Aggregate.binder.ratio    )
```

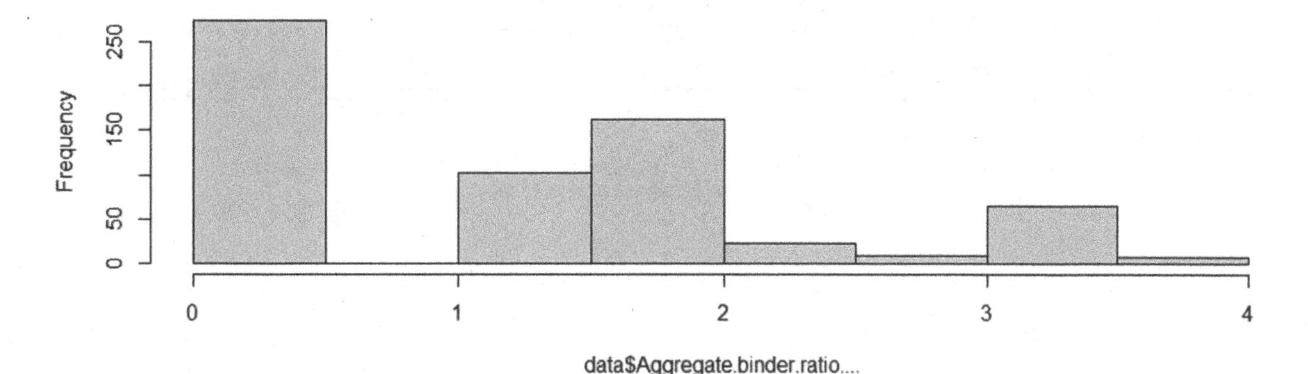

```
hist(data$Sand.binder.ratio    )
```

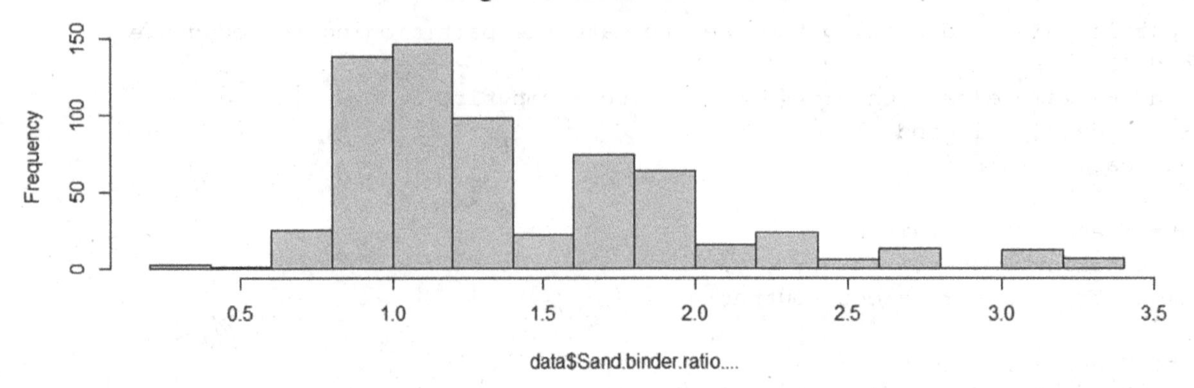


```
hist(data$Heating.rate..C.min.)
```

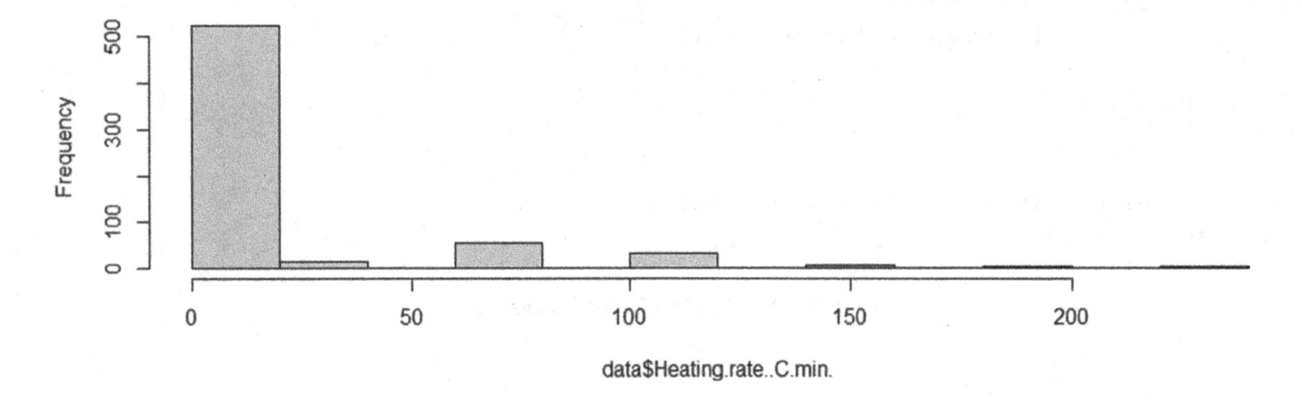


```
hist(data$Maximum.exposure.temperature..C.)
```

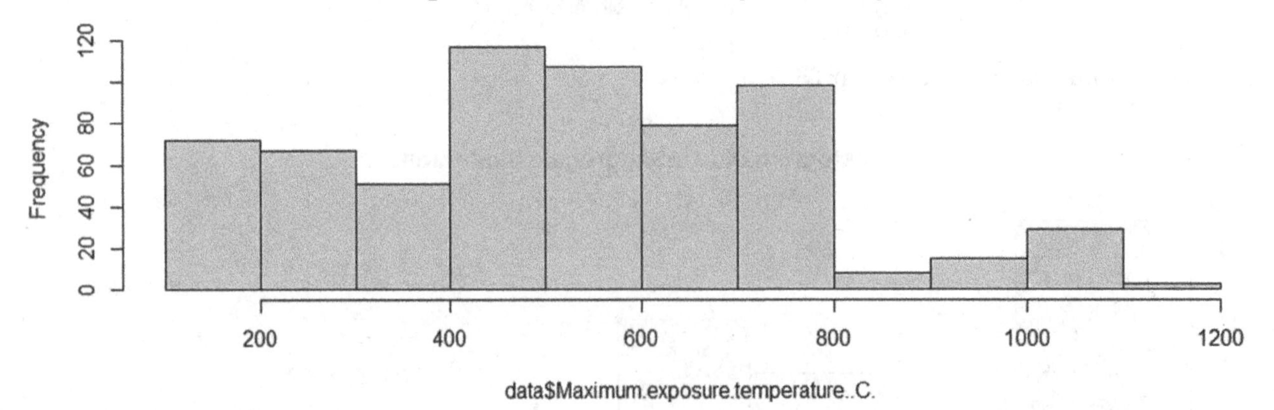

```
hist(data$Silica.fume.binder.ratio    )
```

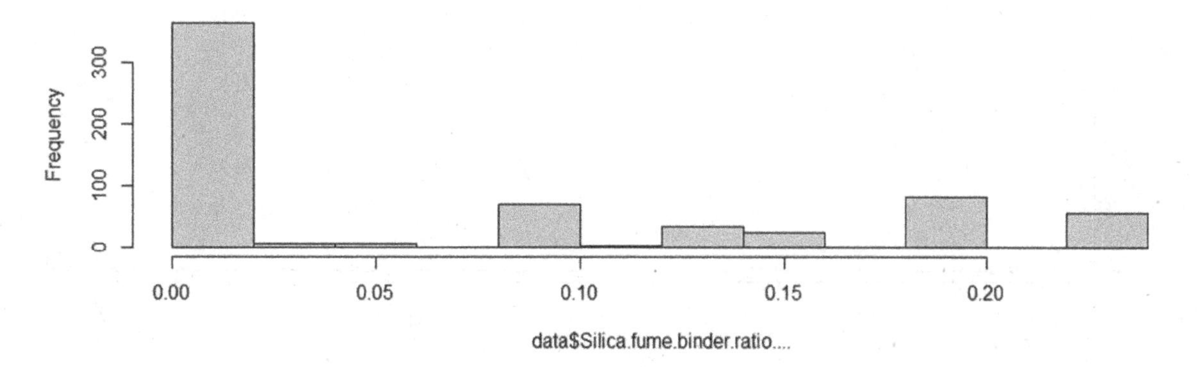


```
hist(data$Aggregate.size..mm.)
```

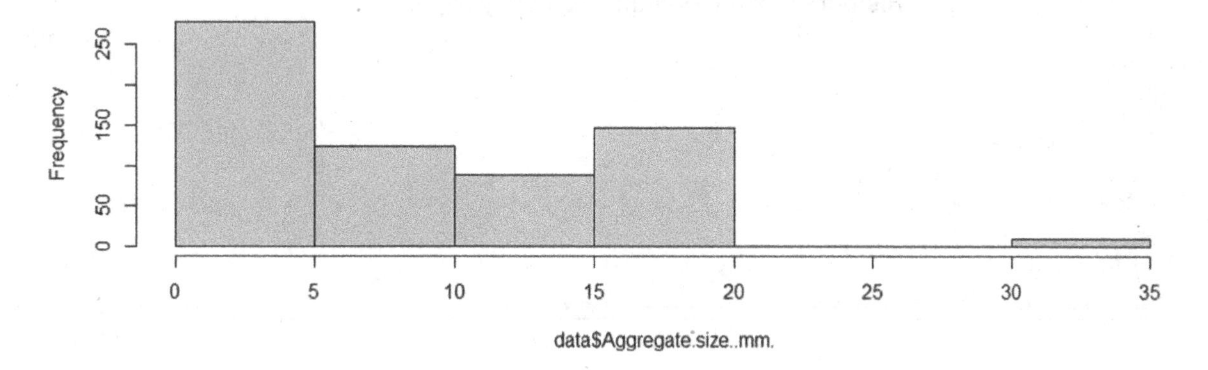


```
hist(data$GGBS.binder.ratio    )
```

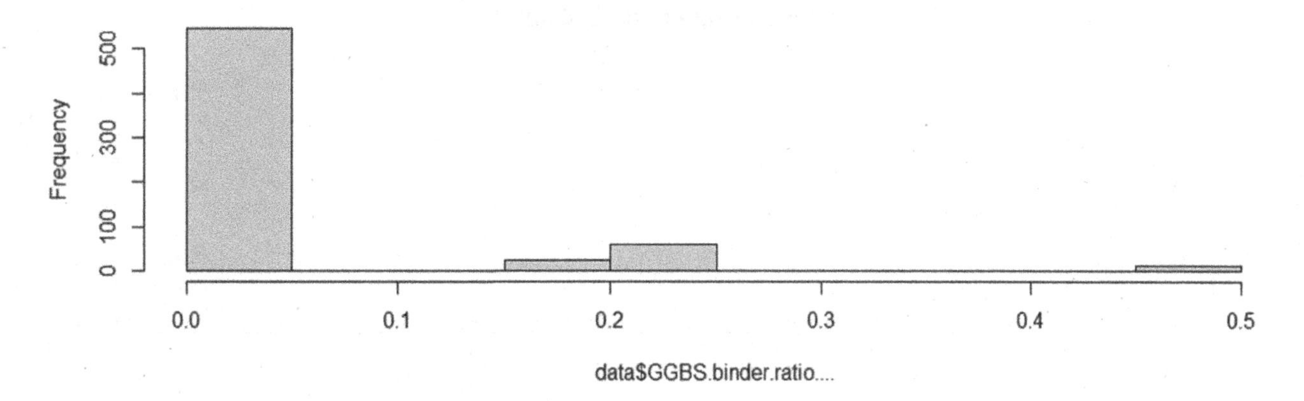

```
hist(data$FA.binder.ratio    )
```

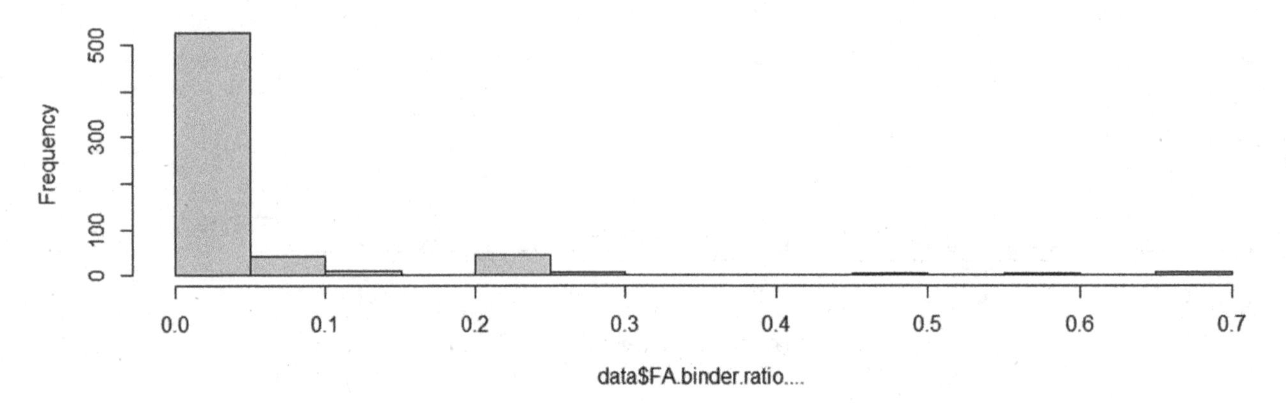


```
hist(data$PP.fiber.quantity..kg.m3.)
```

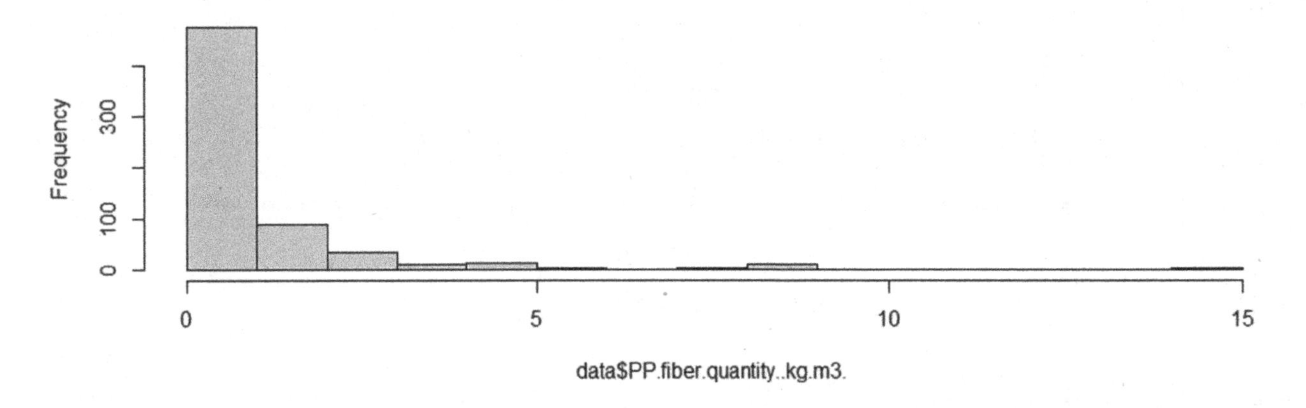


```
hist(data$Output)
```

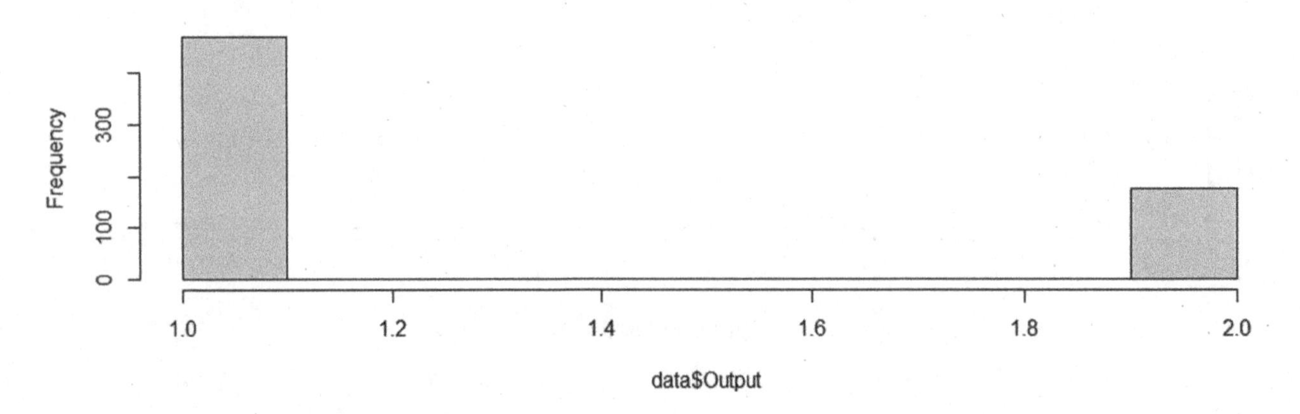

```
#Setting the parameters for the XG.boost Model
params<-list(set.seed=1500,
+   eval_metric="auc",
+            objective="binary:logistic") #Running XG.boost
modelxg <-xgboost(data = as.matrix(xtrainxg),
+   label = ytrainxg,
+   params = params,
+   nrounds = 20,
+   verbose = 1)

[1]   train-auc:0.937105
[2]   train-auc:0.972061
[3]   train-auc:0.970575
[4]   train-auc:0.977379
[5]   train-auc:0.980787
[6]   train-auc:0.986405
[7]   train-auc:0.987741
[8]   train-auc:0.989051
[9]   train-auc:0.989326
[10]  train-auc:0.990075
[11]  train-auc:0.990762
[12]  train-auc:0.990887
[13]  train-auc:0.991049
[14]  train-auc:0.991149
[15]  train-auc:0.991898
[16]  train-auc:0.992222
[17]  train-auc:0.992422
[18]  train-auc:0.992622
[19]  train-auc:0.993246
[20]  train-auc:0.993046
```

```
#Checking the model accuracy using Confusion
#Matrix (Sensitivity, Specificity, and Accuracy)
trained<-model.matrix(trainxg$Output~.-1,data = trainxg)
trainlable<-trainxg[,1]
trainlable<-as.numeric(trainlable)-1
tested<-model.matrix(testxg$Output~.-1,data = testxg)
testlabel<-testxg[,1]
testlabel<-as.numeric(testlabel)-1
pred<-predict(modelxg,tested)
head(pred)
[1] 0.9829579 0.9880555 0.9820659 0.9829579 0.9829579 0.9820659
pred<-as.integer(pred>0.5)
confusionMatrix(table(pred, testxg$Output))
Confusion Matrix and Statistics pred 0 1
0 123 8
1 17 46
Accuracy: 0.8711
95% CI: (0.8157, 0.9148)
No Information Rate: 0.7216
P-Value [Acc > NIR]: 4.745e-07
Kappa: 0.6949
```

```
Mcnemar's Test P-Value: 0.1096
Sensitivity: 0.8786
Specificity: 0.8519
Pos Pred Value: 0.9389
Neg Pred Value: 0.7302
Prevalence: 0.7216
Detection Rate: 0.6340
Detection Prevalence: 0.6753
Balanced Accuracy: 0.8652
'Positive' Class: 0
```

```
#Plotting correlation matrix with two different approaches
#Spearman correlation
#first Method(spearman)
m=cor(data, method = "spearman")
corrplot(m, type = "full",
 +   method = "circle", tl.cex = 0.5, tl.col = 'black')
```

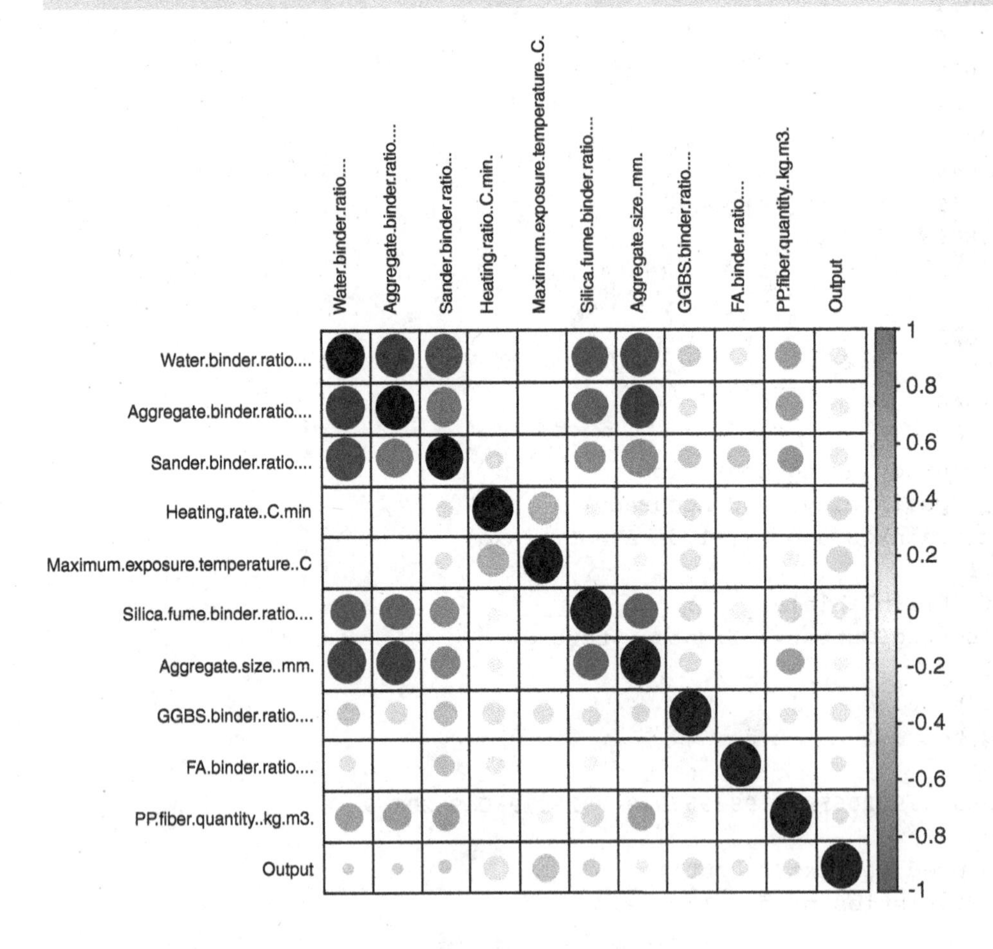

```
corrplot(m, type = "full",
+   method = "number", tl.cex = 0.55, tl.col = 'black')
```

	Water.binder.ratio....	Aggregate.binder.ratio....	Sander.binder.ratio....	Heating.ratio...C.min.	Maximum.exposure.temperature..C.	Silica.fume.binder.ratio....	Aggregate.size..mm.	GGBS.binder.ratio....	FA.binder.ratio....	PP.fiber.quantity..kg.m3.	Output
Water.binder.ratio...	1.00	0.85	0.81	0.04	0.03	-0.65	0.84	-0.27	-0.16	-0.43	0.11
Aggregate.binder.ratio...	0.85	1.00	0.68	0.06		-0.61	0.85	-0.20		-0.43	0.10
Sander.binder.ratio...	0.81	0.68	1.00	0.15	0.66	-0.49	0.61	0.31	0.30	-0.44	0.16
Heating.ratio..C.min.	0.04	0.06	0.15	1.00	0.46	-0.08	0.13	0.26	0.18		0.34
Maximum.exposure.temperature..C.	0.03		0.06	0.46	1.00		-0.04	0.23		0.12	0.40
Silica.fume.binder.ratio....	-0.65	-0.61	-0.49	0.08		1.00	-0.62	-0.18	0.07	0.32	0.17
Aggregate.size..mm.	0.84	0.85	0.61	0.13	0.04	-0.62	1.00	0.18		-0.42	0.09
GGBS.binder.ratio....	-0.27	-0.20	-0.31	0.26	0.23	-0.18	-0.18	1.00		0.08	0.24
FA.binder.ratio....	-0.16		-0.30	0.18		-0.07			1.00	0.05	-0.14
PP.fiber.quantity..kg.m3.	-0.43	-0.43	-0.44		0.12	0.32	-0.42	0.08	0.05	1.00	-0.17
Output	-0.11	-0.10	-0.16	0.34	0.40	0.17		0.24	-0.14	-0.17	1.00

1
0.8
0.6
0.4
0.2
0
-0.2
-0.4
-0.6
-0.8
-1

```
corrplot(m, method ="color",type = "upper",order ="hclust",tl.cex = 0.5, tl.col =
"black",tl.srt = 45, diag
= FALSE)
```

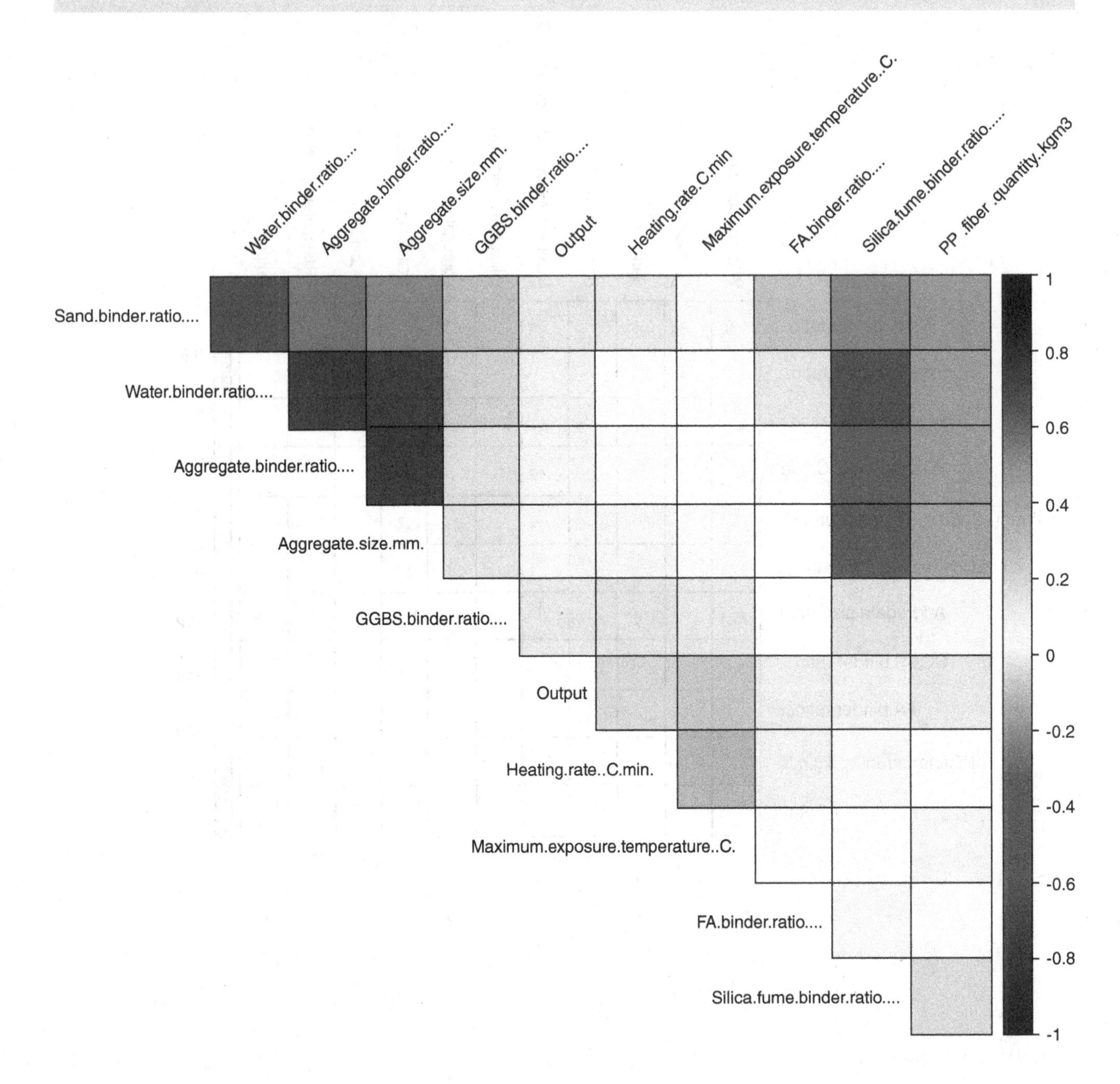

```
#Pearson correlation
n=cor(data, method ="pearson")
corrplot(n, type = "full",
+   method = "circle", tl.cex = 0.5, tl.col = 'black')
```

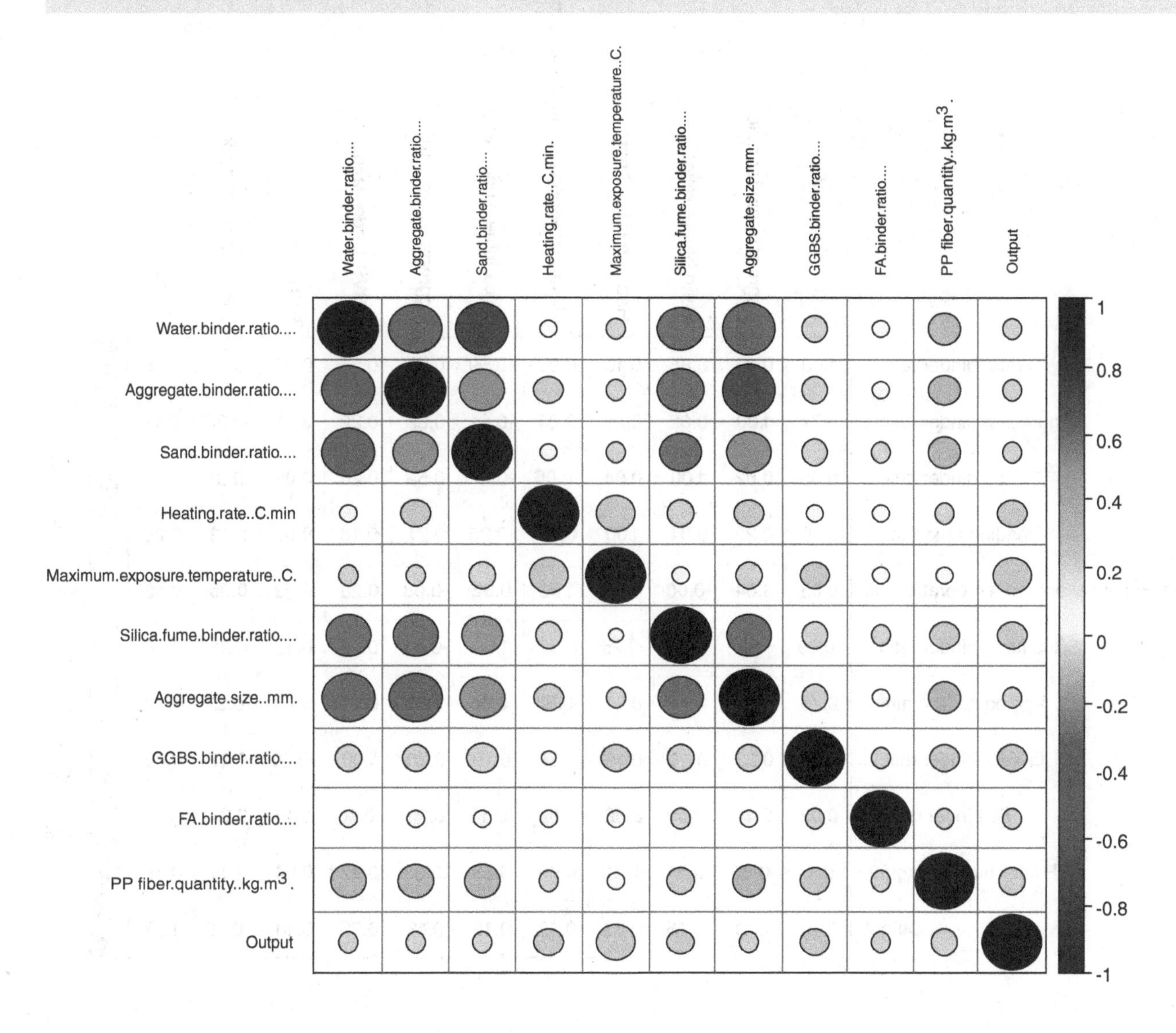

```
corrplot(n, type = "full",
+   method = "number", tl.cex = 0.55, tl.col = 'black')
```

	Water binder ratio...	Aggregate binder ratio...	Sand binder ratio...	Heading ratio C.min.	Maximum exposure temperature...c.	Silica fume binder ratio...	Aggregate size.mm...	GGVS. binder ratio...	FA binder ratio....	PP fiber quantity..kgm3.	Output
Water binder ratio...	1.00	0.77	0.85	0.10	0.03	-0.59	0.75	0.25	0.02	-0.35	0.16
Aggregate binder ratio...	0.77	1.00	0.62	0.22	0.04	-0.61	0.83	0.22	0.11	-0.33	0.11
Sand binder ratio...	0.85	0.62	1.00	0.04	-0.06	-0.47	0.59	0.26	0.09	-0.31	0.15
Heading ratio C.min.	0.10	0.22	0.04	1.00	0.41	-0.25	0.27	0.16	0.00	0.12	0.26
Maximum exposure temperature...c.	0.03	0.04	-0.06	0.41	1.00	0.02	-0.08	0.26	0.02	0.05	0.42
Silica fume binder ratio...	-0.59			-0.25	0.02	1.00	-0.61	0.19	0.13	0.23	0.18
Aggregate size.mm...	0.75			0.27	-0.08	-0.61	1.00	0.20	0.23	-0.32	-0.11
GGVS. binder ratio...	0.25	0.22	0.26	0.16	0.26	0.19	0.20	1.00	-0.10	0.27	0.22
FA binder ratio....	0.02	0.11	-0.09	8.02	0.02	-0.13	0.23	-0.10	1.00	0.04	0.14
PP fiber quantity..kgm3.	-0.35	-0.33	-0.31	-0.12	0.05	0.23	-0.32	0.27	0.04	1.00	0.18
Output	0.16	-0.11	-0.15	0.26	0.42	0.18	0.11	0.22	0.14	0.18	1.00

Color scale: 1, 0.8, 0.6, 0.4, 0.2, 0, -0.2, -0.4, -0.6, -0.8, -1

```
corrplot(n, method ="color",type = "upper",order ="hclust",tl.cex = 0.5, tl.col =
"black",tl.srt = 45, diag
= FALSE)
```

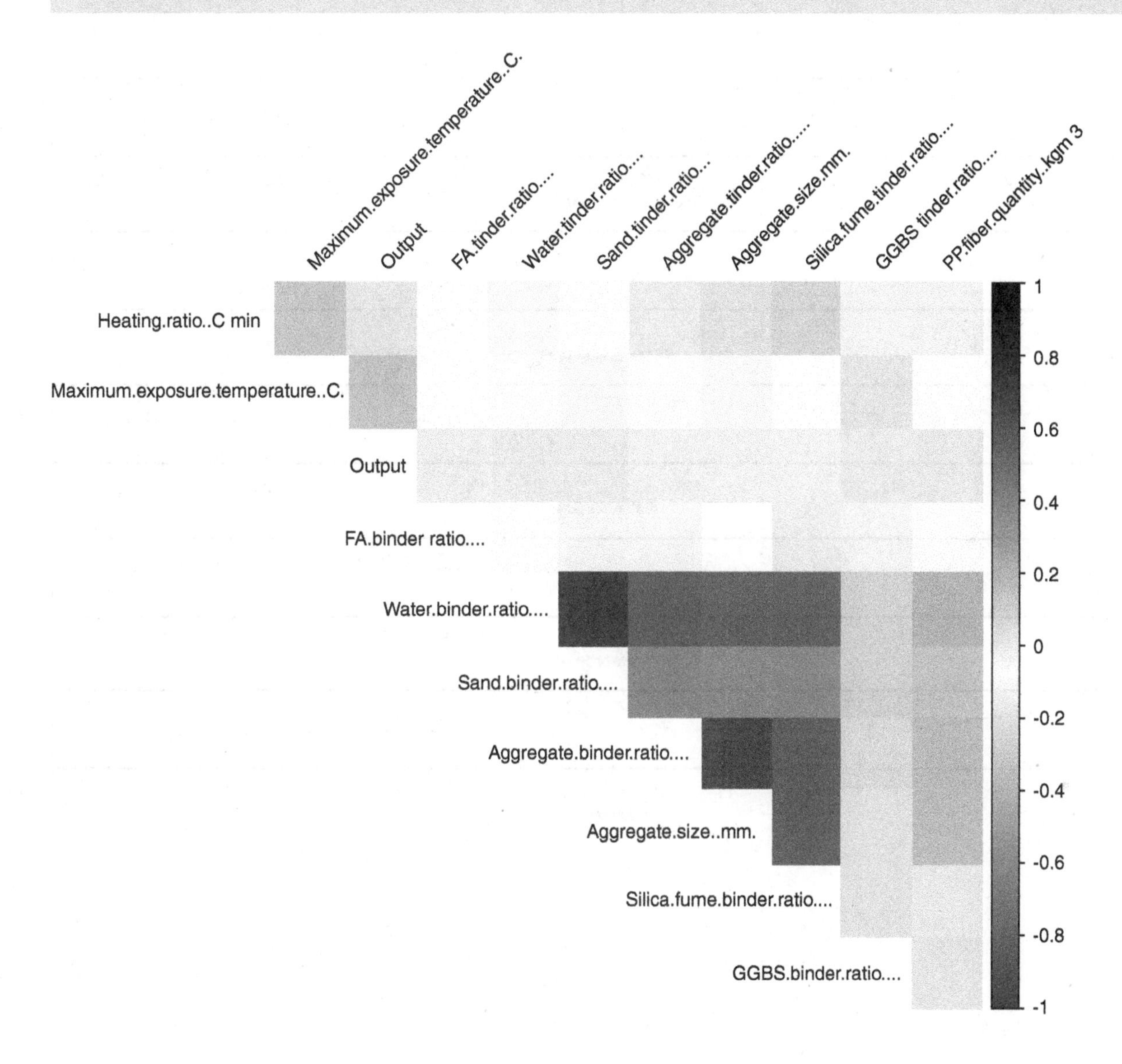

```
#Plotting SHAP summary and SHAP's top 5 features.
xgb.plot.shap(data = as.matrix(xtrainxg),
+   model = modelxg,
+   top_n =5)
```

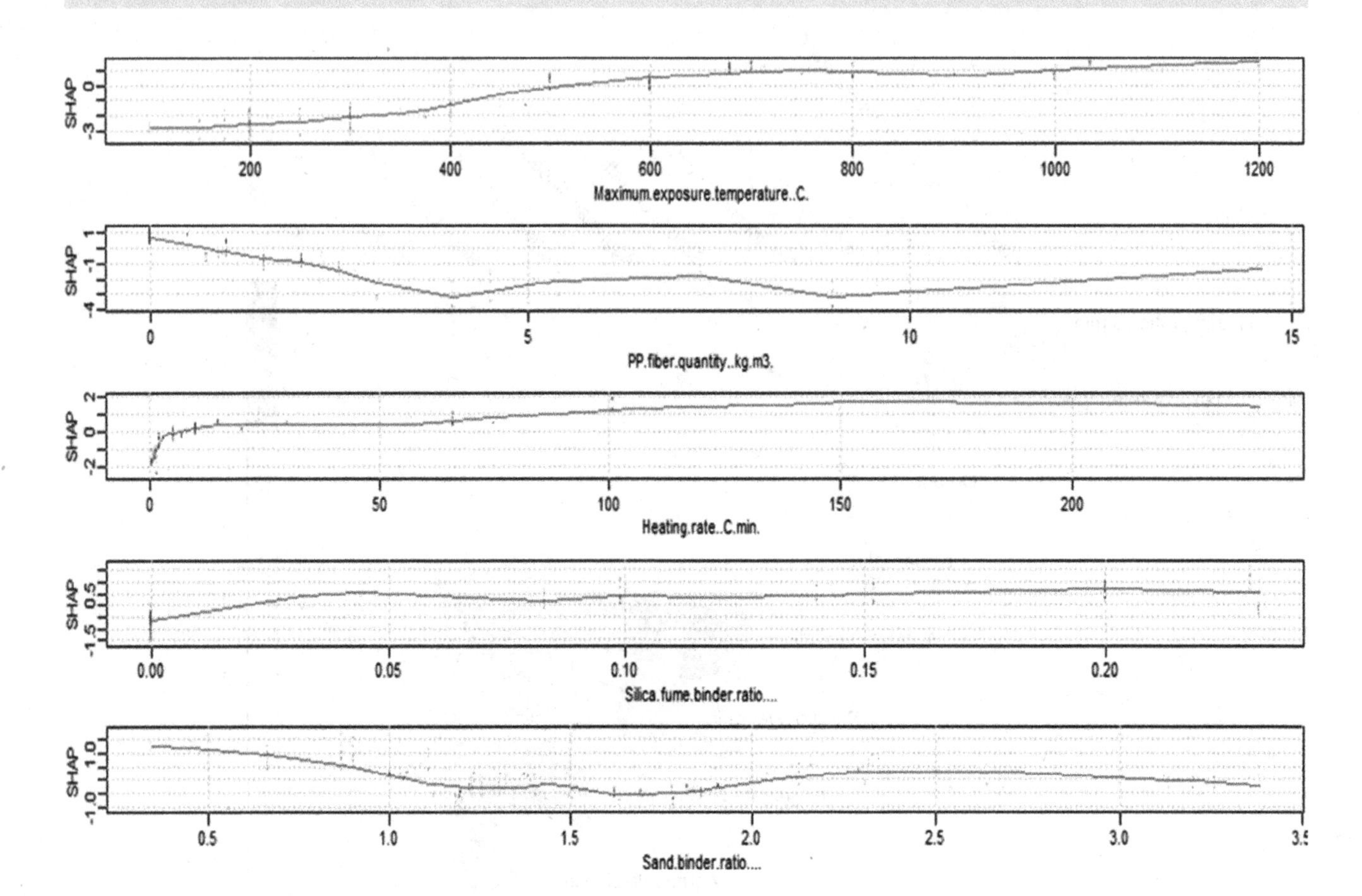

```
xgb.plot.shap.summary(data = as.matrix(xtrainxg),
+   model = modelxg,
+   top_n = 11)
```

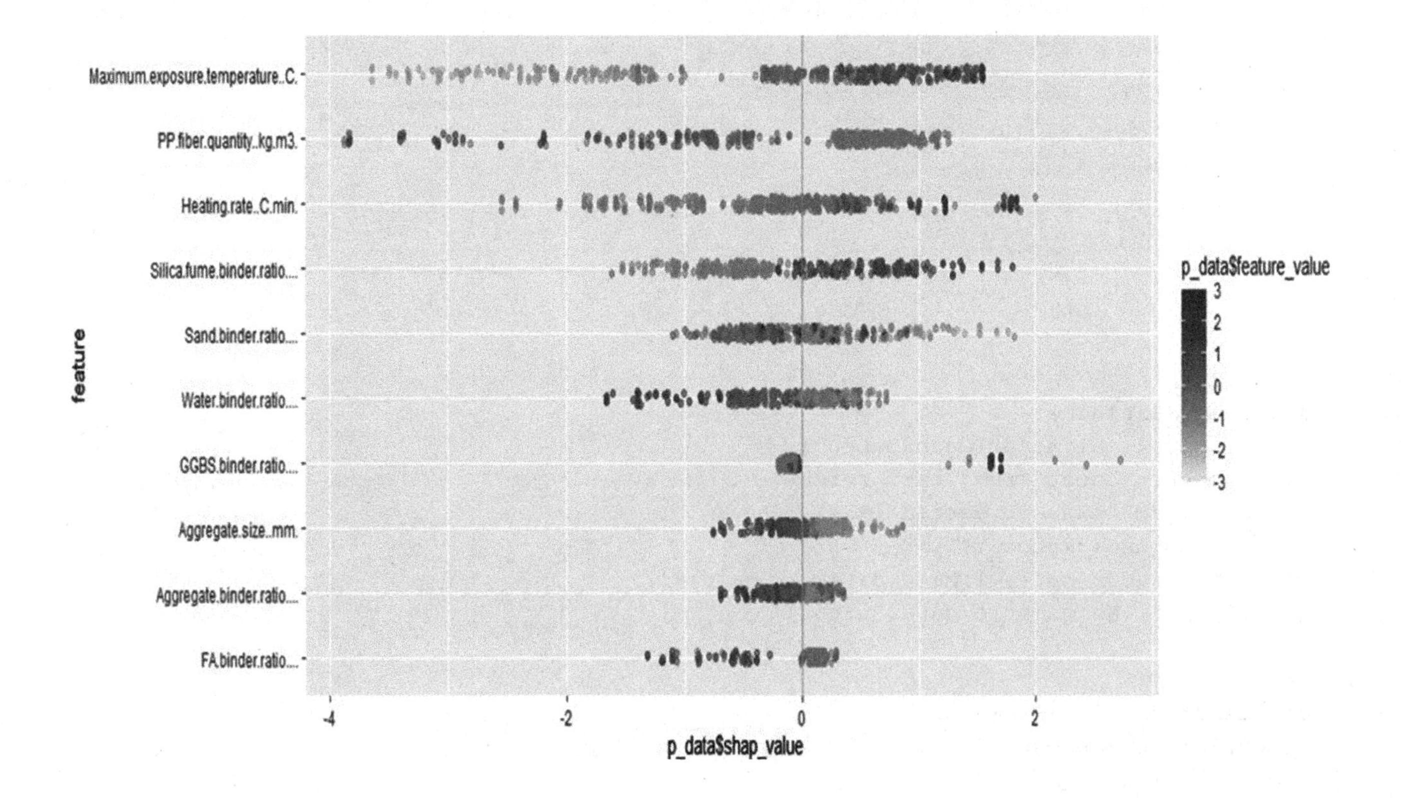


```
#Feature importance
imp<-xgb.importance(colnames(ytrainxg),model = modelxg)
print(imp)
Feature  Gain Cover Frequency
1: Maximum.exposure.temperature..C. 0.23558285 0.20467206 0.12500000
2: PP.fiber.quantity..kg.m3. 0.15665185 0.15931793 0.15878378
3: Heating.rate..C.min. 0.15192102 0.15287276 0.15878378
4: Silica.fume.binder.ratio 0.10385727 0.10755627 0.11486486
5: Sand.binder.ratio 0.08987868 0.11619380 0.13175676
6: Water.binder.ratio 0.08150161 0.08095570 0.13851351
7: GGBS.binder.ratio0.07003093 0.06831878 0.03040541
8: Aggregate.size..mm. 0.03942213 0.04269872 0.05067568
9: Aggregate.binder.ratio 0.03643546 0.03056270 0.05743243
10: FA.binder.ratio 0.03471819 0.03685128 0.03378378
xgb.plot.importance(imp)
```

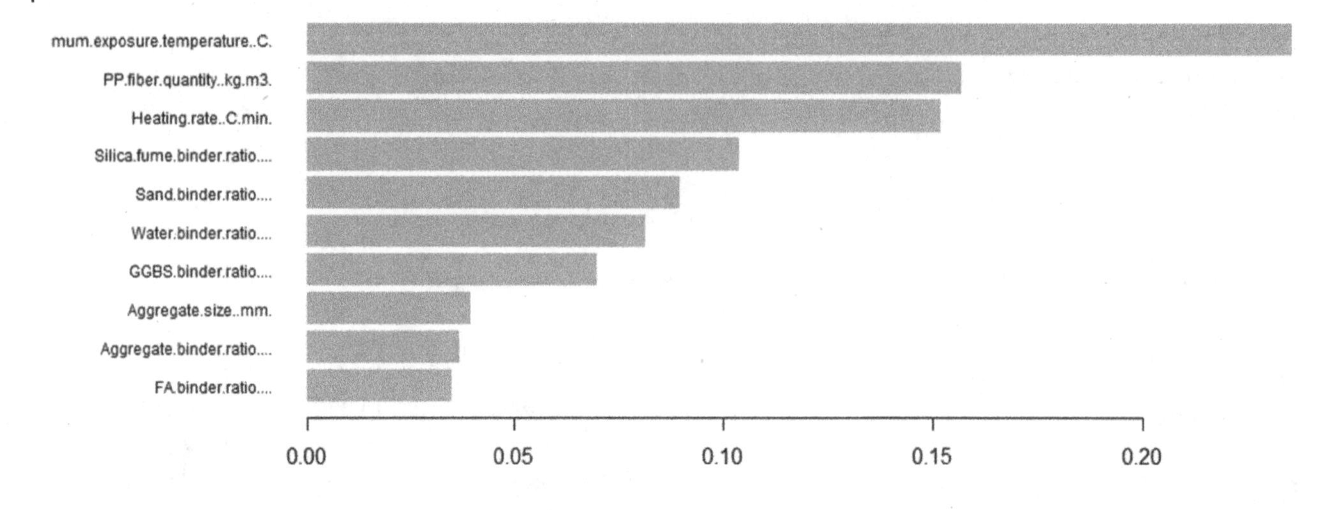


```
#Plotting explainer
data$Output<-as.factor(data$Output)
fit_xgboost<-boost_tree(learn_rate = 0.3)%>%
  + set_mode("classification")%>%
  + set_engine("xgboost")%>%
  + fit(data$Output~.-1,data=data)
data$Output<-as.numeric(data$Output)
```

```
#Explainer
explainer<-DALEX::explain(
+ model= fit_xgboost,
+ data = data,
+ y=data$Output,
+ label = "XGBoost"
+ )
Preparation of a new explainer is initiated
-> model label: XGBoost
-> data: 646 rows 11 cols
-> target variable : 646 values
-> predict function : yhat.model_fit will be used ( default )
-> predicted values : No value for predict function target column. ( default )
-> model_info : package parsnip,  ver. 0.1.7,  task classification ( default )
-> predicted values : numerical, min = 0.008053362,  mean = 0.2773234,  max =
0.9723785
-> residual function : difference between y and yhat ( default )
-> residuals: numerical, min = 0.1408839,  mean = 0.9935744,  max = 1.773495 A new
explainer has been created!
```

```
#Model studio
> modelStudio::modelStudio(explainer = explainer)
`new_observation` argument is NULL. `new_observation_n` observations needed to cal-
culate local explanations are taken from the data.

Calculating …
Calculating ingredients::feature_importance
```

```
Calculating ingredients::partial_dependence (numerical) Calculating
ingredients::accumulated_dependence (numerical) Calculating DALEX::model_performance
Calculating DALEX::model_diagnostics Calculating iBreakDown::local_attributions (1)
Calculating iBreakDown::shap (1)
Calculating ingredients::ceteris_paribus (1)
Calculating iBreakDown::local_attributions (2)
Calculating iBreakDown::shap (2)
Calculating ingredients::ceteris_paribus (2)
Calculating iBreakDown::local_attributions (3)
Calculating iBreakDown::shap (3)
Calculating ingredients::ceteris_paribus (3) Calculating …
Elapsed time: 00:00:22 ETA: 0s >
```

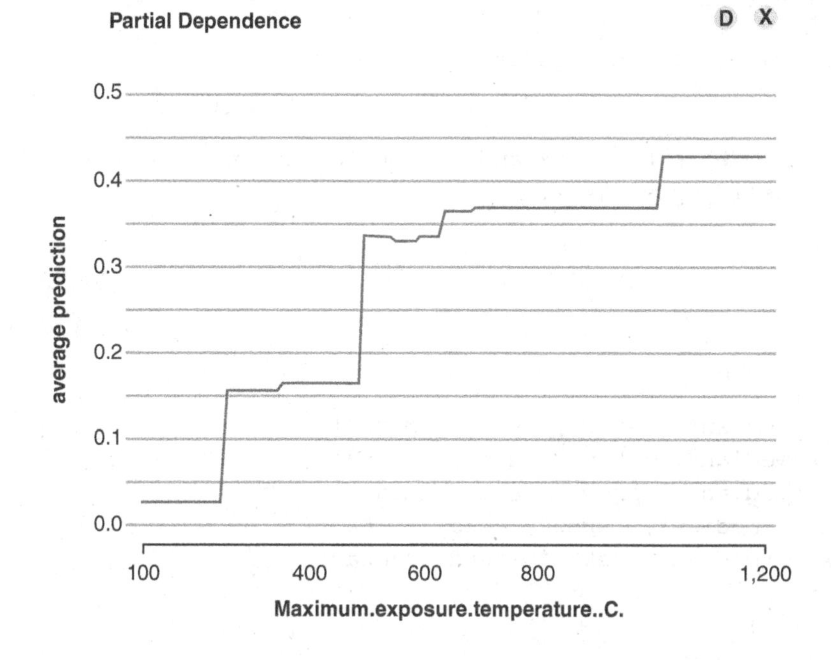

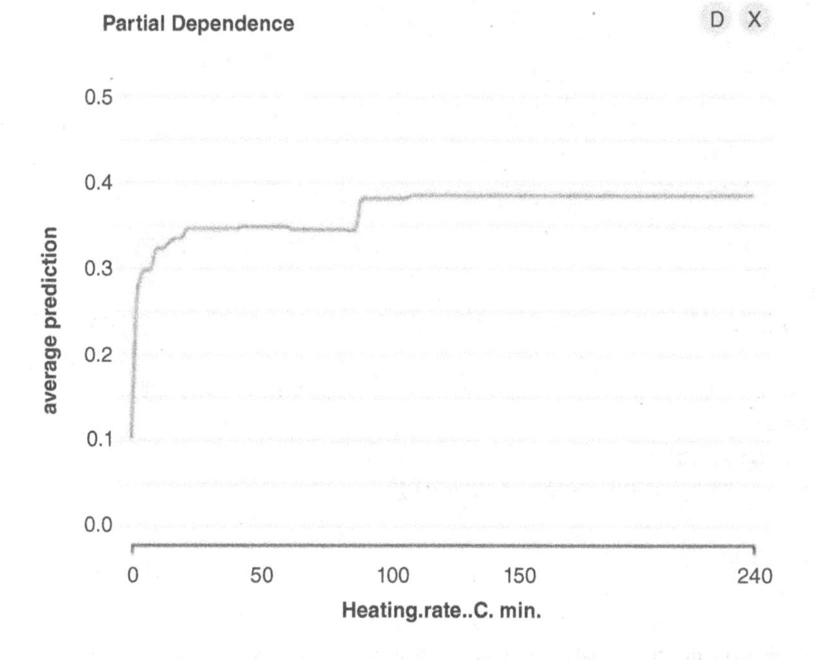

Example no. 4 (Python and R): Predicting axial capacity of concrete-filled steel tubes (CFST) columns.[79]
Let me remind you of the features involved in this example. These are effective length (L_e), tube thickness (t), tube diameter (D), the elastic modulus (E_s), yield stress (f_y) as well as compressive strength (f_c) of in-filled concrete. We will continue the same example but augment it with SHAP and PDPs.[80]

```
#This code continues blackbox model on CFST columns.
    import pandas as pd
    import NumPy as np
    import XGBoost as xgb
    from sklearn.model_selection import train_test_split
    from matplotlib import pyplot
    import seaborn as sns
    from sklearn.metrics import mean_squared_error
    from sklearn.metrics import accuracy_score
    from sklearn.metrics import r2_score
    from sklearn.inspection import PartialDependenceDisplay
    import shap
```

```
#SHAP is a library that explains a model's blackbox and is considered an important
tool for explainable artificial intelligence (XAI).
```

```
    CFST=pd.read_csv('CFST.csv')
    CFST.head(5)
    CFST.dtypes
    Statistical_insights=CFST.agg(
     {
         "D (mm)": ["min", "max", "median", "skew", "std", "mean"],
         "t (mm)": ["min", "max", "median", "skew", "std", "mean"],
         "L (mm)": ["min", "max", "median", "skew", "std", "mean"],
         "fy (MPa)": ["min", "max", "median", "skew", "std", "mean"],
         "fc (Mpa)": ["min", "max", "median", "skew", "std", "mean"]
     }
                                      )

    print (Statistical_insights)
    x=CFST.drop(['N Test (kN)'], axis=1)
    y=CFST['N Test (kN)']
    x_train,x_test,y_train,y_test=train_test_split(x,y,test_size=0.2,

        random_state=200)
    RF=RandomForestRegressor(max_depth = 20, random_state = 0,
                          n_estimators = 10000)
    RF.fit(x_train,y_train)
    predictions = RF.predict(x_test)
    score =RF.score(x_train,y_train)
print ("score : % f" %(Score))
print ('Accuracy',100- (np.mean(np.abs((y_test - Predictions) / y_test)) * 100))
print (r2_score(y_test, Predictions))
print ('MAE',np.mean(abs(y_test-Predictions)))
print ('RMSE',np.sqrt(np.mean(np.square(y_test-Predictions))))

    for i, col in enumerate(CFST.columns):
```

79 Basically, we have 1245 CFST columns that we are hoping to predict their axial capacity under concentric loading. We built a ML model.
80 To keep this chapter from growing, I will not go into full details as I did in Example no. 3.

```
            pyplot.figure(i)
        sns.histplot(CFST[col])
#Printing histograms for all the features and labels we are interested in.
#Here, we used the for loop to print the histograms for each variable in one
function.
```

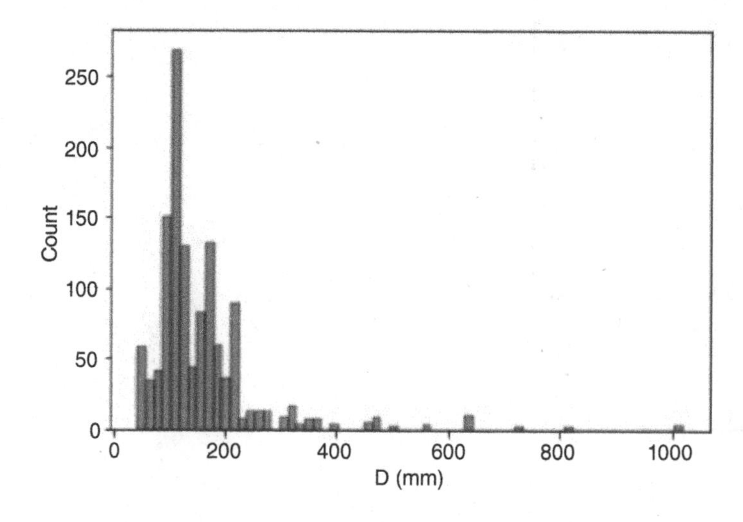

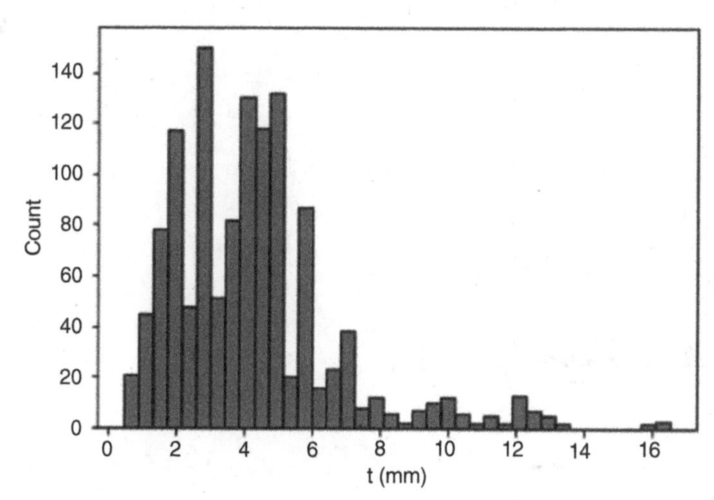

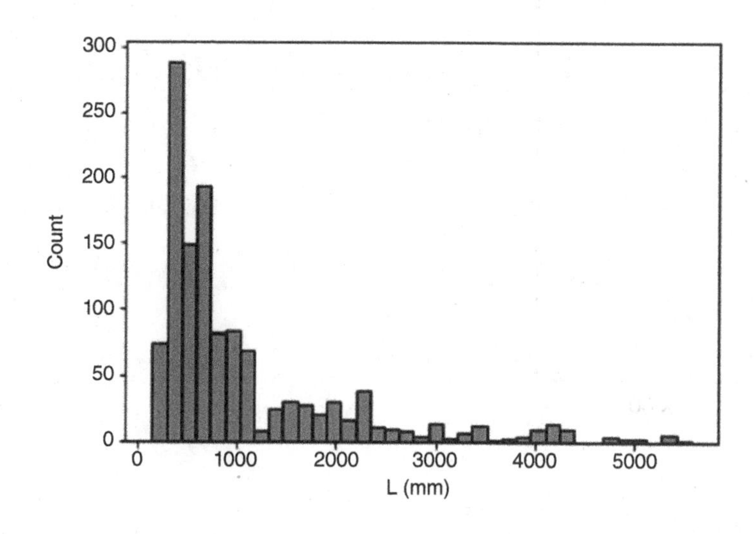

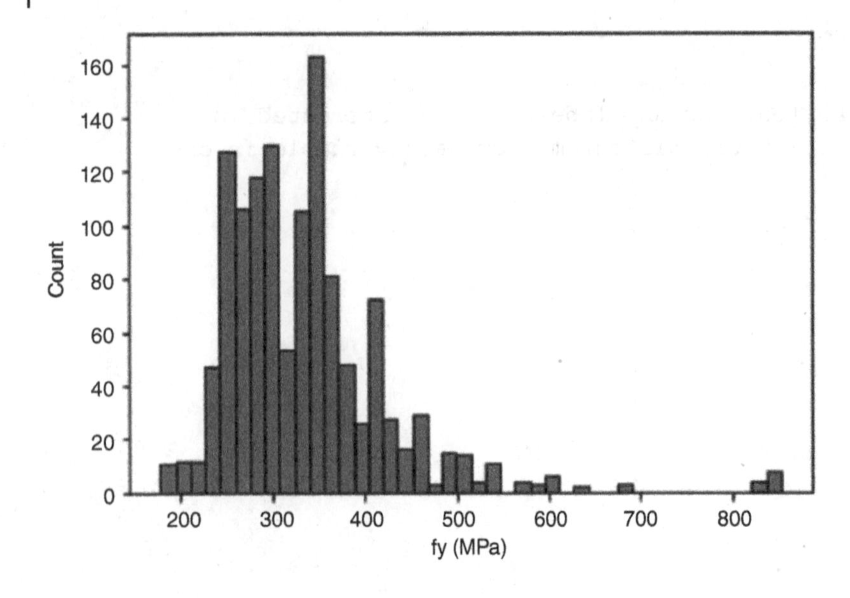
Count
160
140
120
100
80
60
40
20
0
200
300
400
500
600
700
800
fy (MPa)

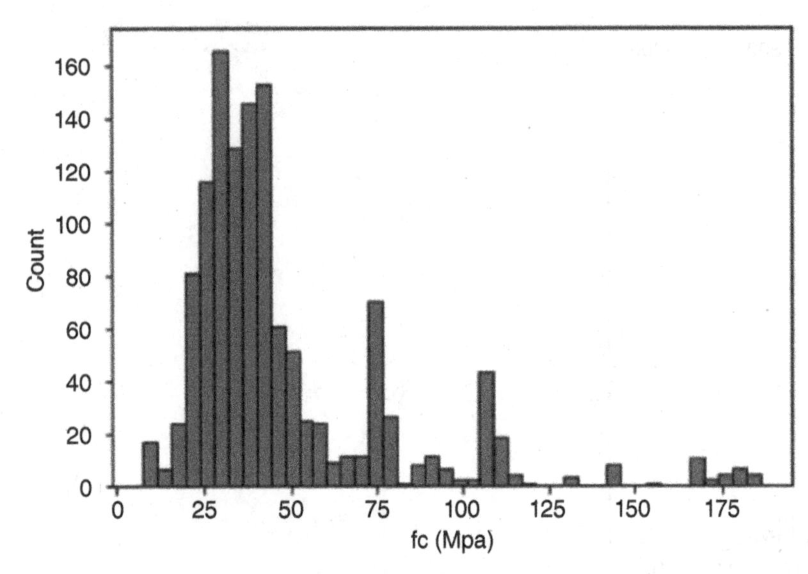
Count
160
140
120
100
80
60
40
20
0
0
25
50
75
100
125
150
175
fc (Mpa)

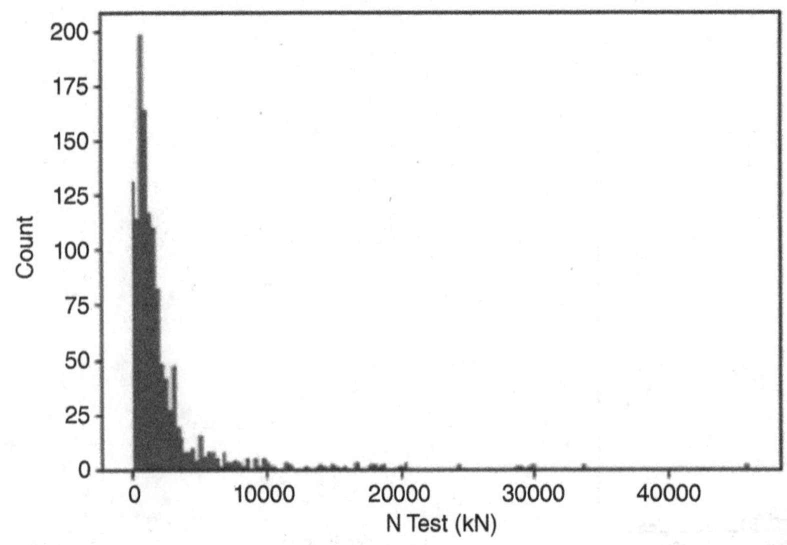
Count
200
175
150
125
100
75
50
25
0
0
10000
20000
30000
40000
N Test (kN)

```
Pearson_HM=CFST.corr()

        sns.heatmap(Pearson_HM, annot=True)
        sns.set(rc = {'figure.figsize':(7,5)}) #figure size command

        pyplot.title("Pearson Correlation") #figure title command
        pyplot.show()
#Here, we used the matplotlib library to visualize the matrix (heatmap), and the
code generates a regression metric called 'Pearson Correlation.'
```

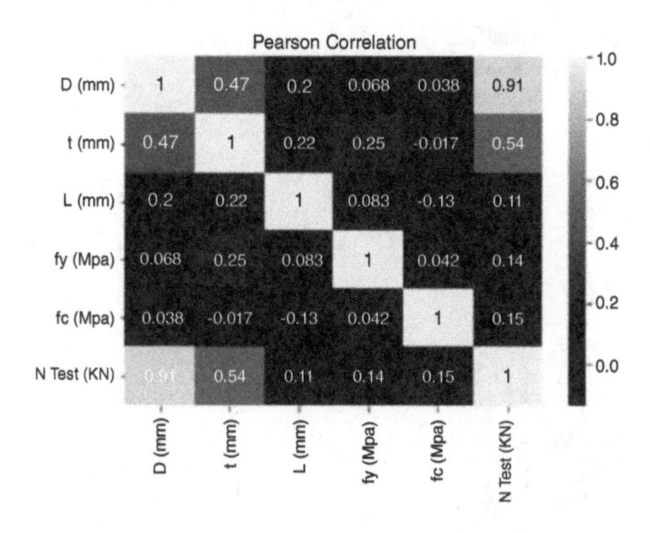


```
def display_correlation(df):
r = df.corr(method="spearman")
    pyplot.figure(figsize=(7,5))
    heatmap = sns.heatmap(df.corr(), vmin=-1,vmax=1, annot=True)
    pyplot.title("Spearman Correlation")
    return(r)
    r_simple=display_correlation(Pearson_HM)
```

```
#Regression metric 'Spearman Correlation'.
```

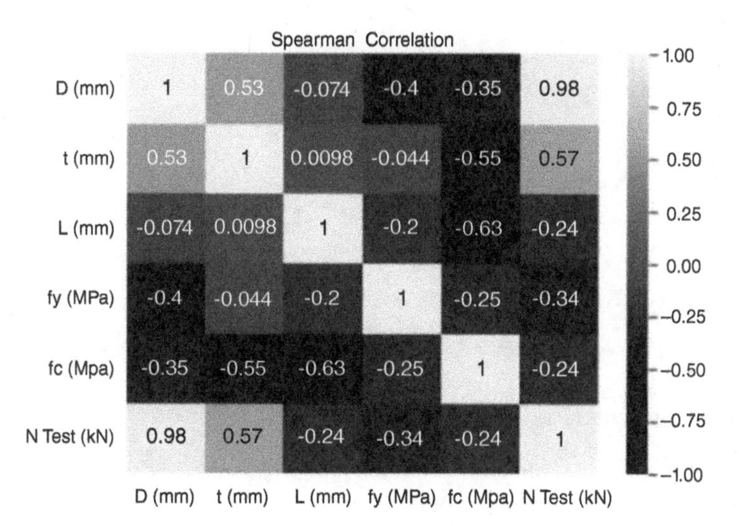

```
ACTUAL_VS_PREDICTED =range(len(y_test))
      pyplot.plot(ACTUAL_VS_PREDICTED,y_test,label='Original')
pyplot.plot(ACTUAL_VS_PREDICTED,Predictions,label='Predicted')
      pyplot.legend()
      pyplot.show()

#Plotting the predicted values and the actual values to check for any outliers.
```

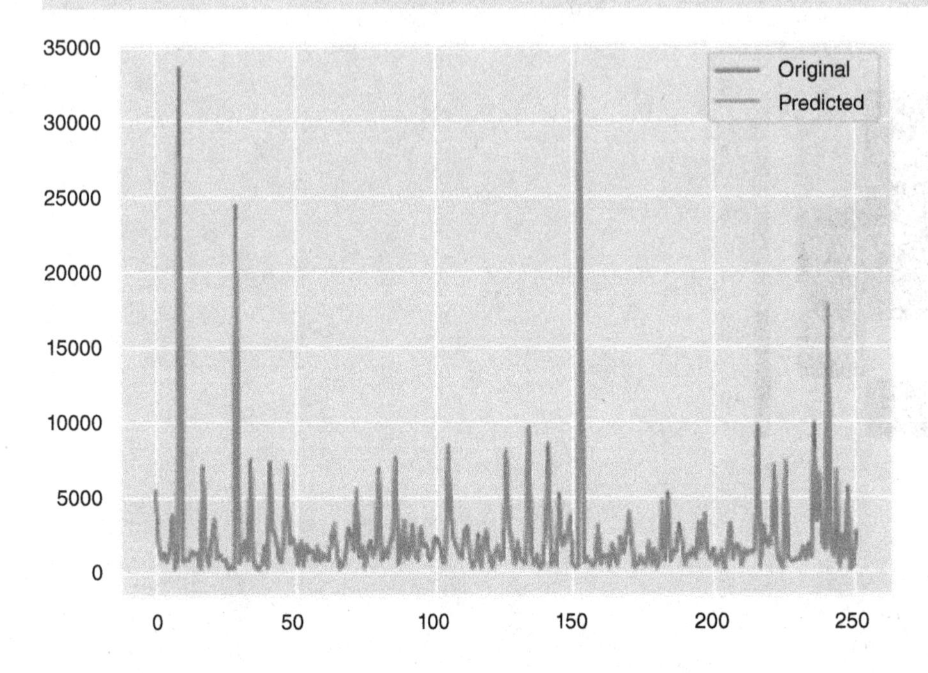


```
feature_importances = pd.Series(RF.feature_importances, index =x.columns) feature_
importances.nlargest(20).plot(kind = 'barh')
```

```
#This plot represents the global feature importance.
```

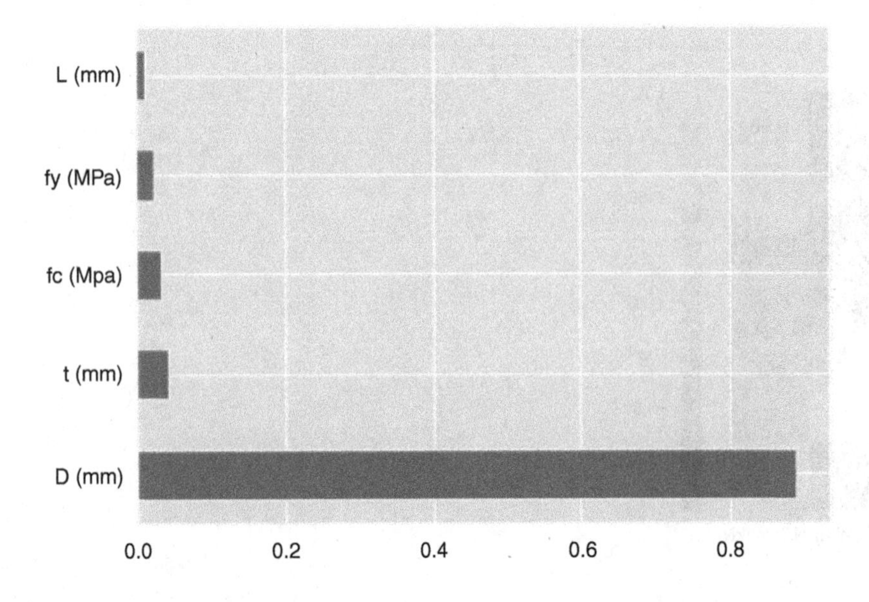

```
explainer = shap.TreeExplainer(RF)
    RF_binary_shap_values = explainer(x_test)
    shap_values = explainer.shap_values(x_test)
    shap.summary_plot(shap_values, x_test, plot_type="bar")
```

#Similar to the feature importance as previously stated in the code but generated using SHAP. Note that both plots do not need to be identical. Also, the figure shows the most significant variables in descending order.

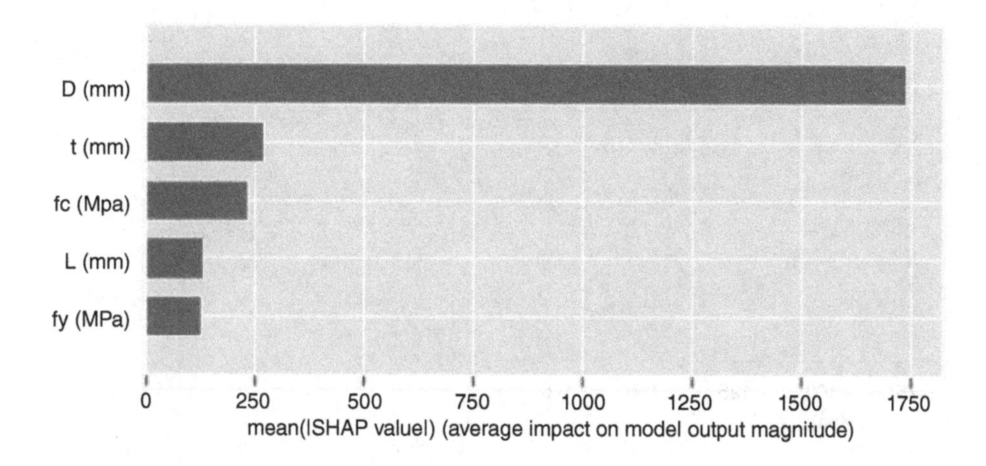


```
shap.summary_plot(shap_values, x_test)
```

#This summary plot is considered one of the most important plots in SHAP.
#The coloring scheme shows whether that feature was high or low for that row of the dataset.
#Horizontal positions show whether the effect of that value yields a higher or lower prediction.

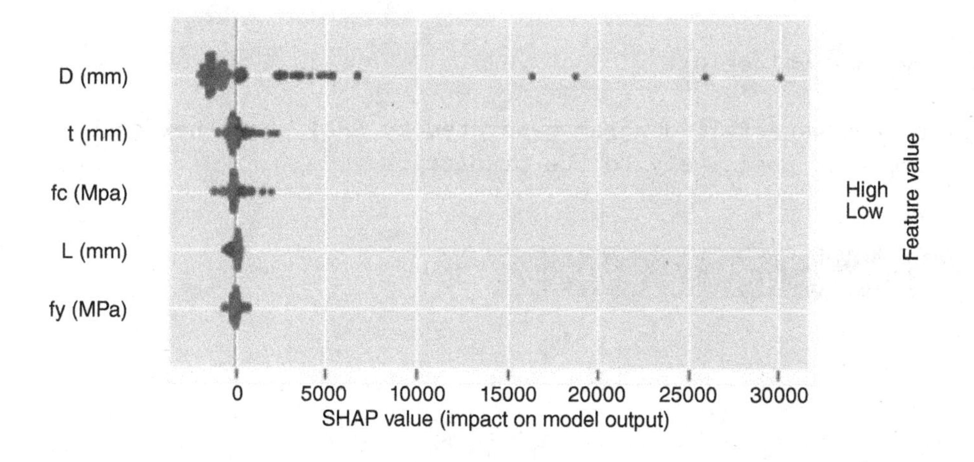


```
shap.dependence_plot("fc (Mpa)", shap_values, x_test,
    interaction_index=None )
```

```
#SHAP dependence plots demonstrate the relationships between any chosen feature and
the target variable.
```

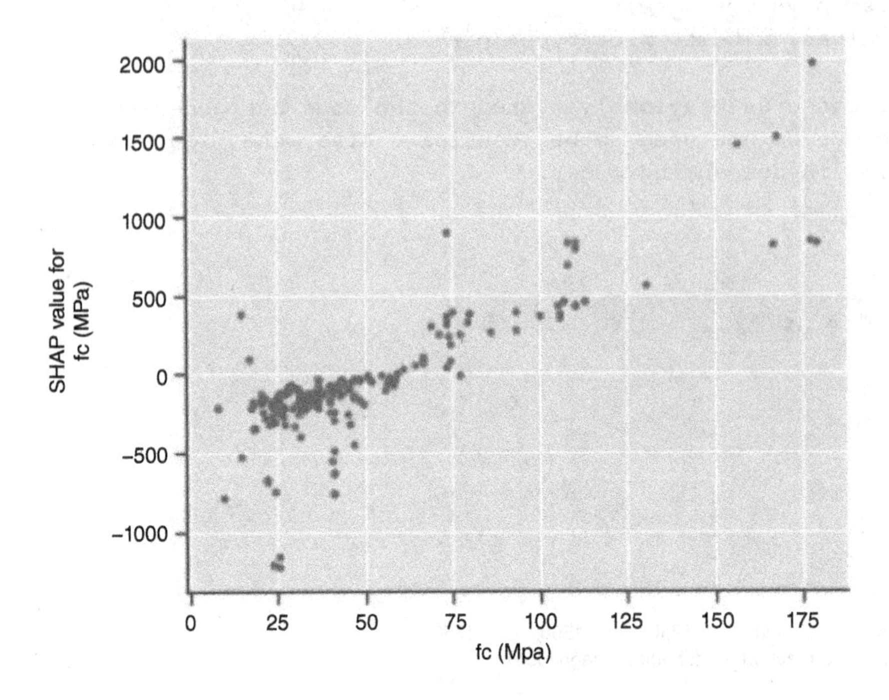


```
shap.initjs()
  def p(j):
          explainer = shap.TreeExplainer(RF)
          RF_shap_values = explainer.shap_values(x_test)
    return  (shap.force_plot(explainer.expected_value, RF_shap_values[j,:],
             x_test.iloc[j,:]))
    p (0)
#Force plot generated by SHAP shows how each feature contributes negatively or pos-
itively to the prediction.
```

```
shap.plots.bar(RF_binary_shap_values[0])
```

```
#Similar to the force plot, the waterfall diagram generated by SHAP shows how each
feature contributes negatively or positively to the prediction.
```

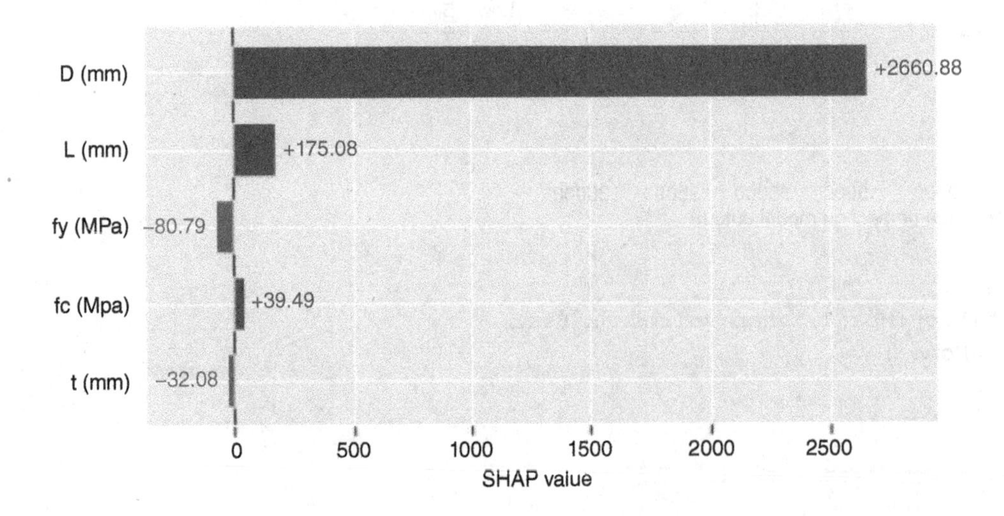

```
features = [0]
        PartialDependenceDisplay.from_estimator(RF, x_train, features, kind='both')
```

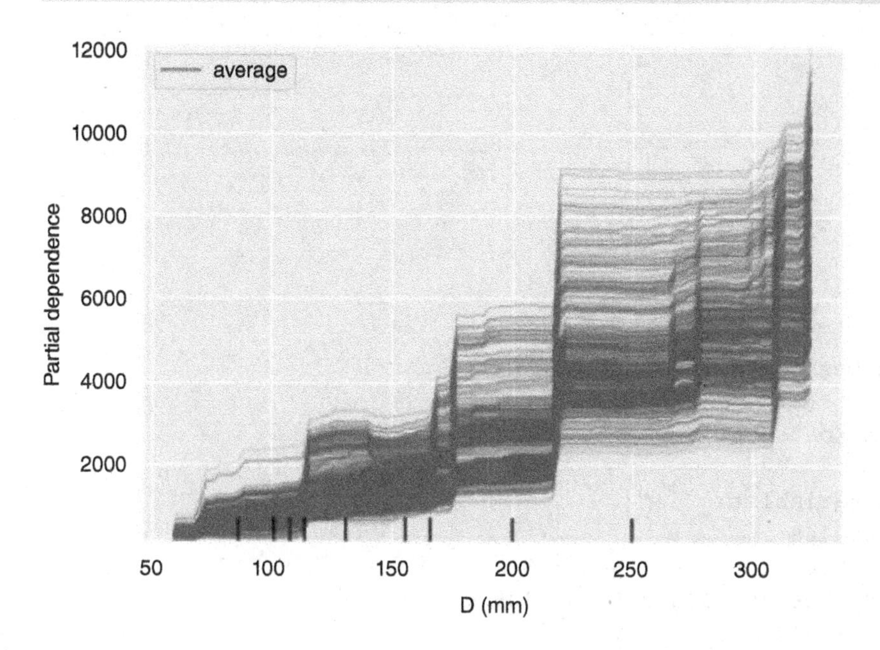


```
#The Partial dependence plots are another model agnostic method that can be used
for outlining the global explainability of XAI models. Note that '0' represents the
first feature in the data set, and the 'for' loop can be used to show a partial
dependence plot for the other features or by changing the '0' to 1,2, or 3 to show
the plots separately.
features = [1]
        PartialDependenceDisplay.from_estimator(RF, x_train, features, kind='both')
```

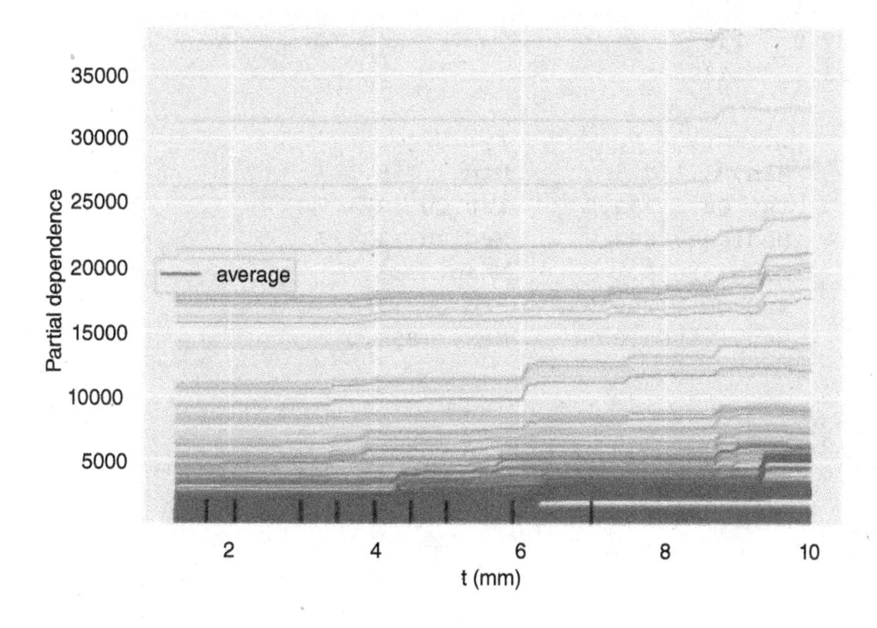

Here is the *R* version of the same example.

```
#Install required packages for Random-forest regression in R programming language to
call the specified packages.
library(ggplot2)
library(cowplot)
library(randomForest)
library(rpart)
library(randomForest)
library(rfUtilities)
library(pdp)
library(corrplot)

#Data importation
data=read.csv('C:/Users/Arash/Desktop/mz naser/CFST/(CFST).csv')

#Checking the structures of the dataset.
str(data)
'data.frame':  1260 obs. of 6 variables:
$ D..mm.   : num 169 169 169 169 168 …
$ t..mm.   : num 2.64 2.64 2.62 2.62 3.6 3.6 3.6 5 5 5 …
$ L..mm.   : num 305 305 305 305 305 305 305 305 305 305 …
$ fy..MPa. : num 302 302 338 338 288 …
$ fc..Mpa. : num 18.2 34.7 37.1 34.1 27 33.3 33.3 33.4 33.4 27.9 …
$ N.Test..kN.: num 1326 1219 1308 1330 1557 …

#Displaying the 6th first row of dataset.
head(data)
D..mm. t..mm. L..mm. fy..MPa. fc..Mpa. N.Test..kN.

1 168.8   2.64   305   302.4   18.2   1326
2 168.8   2.64   305   302.4   34.7   1219
3 169.3   2.62   305   338.1   37.1   1308
4 169.3   2.62   305   338.1   34.1   1330
5 168.3   3.60   305   288.4   27.0   1557
6 168.3   3.60   305   288.4   33.3   1432
summary(data)
D..mm.       t..mm.    1432            L..mm.            fy..MPa.
Min.  : 44.45   Min.  : 0.520       Min.  : 152.3      Min.  :178.3
1st Qu.: 108.00  1st Qu.: 2.703     1st Qu.: 399.0     1st Qu.:274.9
Median : 129.00  Median : 4.000     Median : 662.0     Median :324.7
Mean  : 158.83   Mean  : 4.275      Mean  :1055.1      Mean  :335.7
3rd Qu.: 190.00  3rd Qu.: 5.000     3rd Qu.:1200.0     3rd Qu.:363.3
Max.  :1020.00   Max.  :16.540      Max.  :5560.0      Max.  :853.0
fc..Mpa.        N.Test..kN.
Min.  : 7.59      Min.  : 45.2
1st Qu.: 29.15     1st Qu.: 710.8
Median : 37.40     Median : 1273.0
Mean  : 47.50      Mean  : 2349.7
3rd Qu.: 51.08     3rd Qu.: 2273.2
Max.  :185.94      Max.  :46000.0

#Train-Test split strategy.
smp_size <- floor(0.8 * nrow(data))
```

```
#Data partitioning and setting the seeds to make the partitioning reproducible.
set.seed(123)
train_ind <- sample(seq_len(nrow(data)), size = smp_size)
trainrf <- data[train_ind, ]
testrf <- data[-train_ind, ]

#Fitting the Random forest model.
set.seed(187)
rf.mod<-randomForest(N.Test..kN.~.,ntree=100,mtry=5, data=trainrf)
plot(rf.mod)

#Actual and predicted vectors.
predicted<-rf.mod$predicted
actual<-rf.mod$y

#Metrics for checking model accuracy.
Metrics::mae(actual,predicted) [1] 221.5164
Metrics::rmse(actual,predicted) [1] 765.9959
Metrics::mse(actual,predicted) [1] 586749.7

#Functions for model evaluation.
ACC1<-(100-(mean(abs((actual - predicted) / predicted)) * 100))
MAE1<-(mean(abs(actual-predicted)))
MSE1<-(sqrt(mean((actual-predicted)^2)))
ACC1
[1] 91.28963
MAE1
[1] 221.5164
MSE1
[1] 765.9959

#Plotting histogram for all variables.
names<-names(data)
classes<-sapply(data,class)
for(name in names[classes == 'numeric']){
+ dev.new()
+ hist(data[,name])
+ }

#Plotting histogram for individual variables.
hist(data$N.Test..kN.)
```

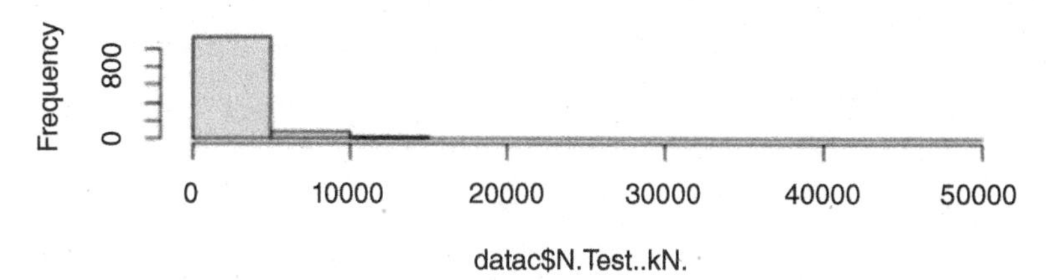

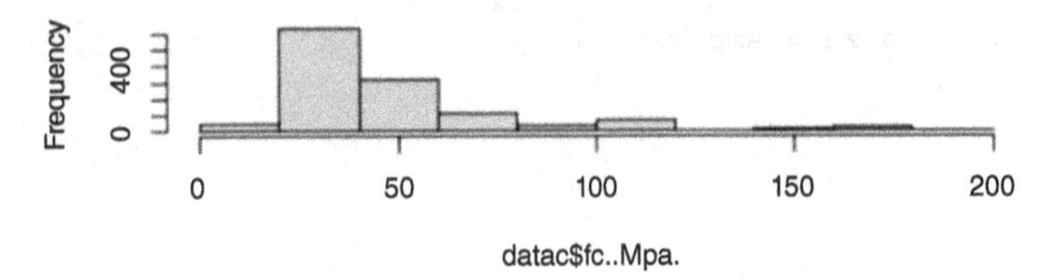
Histogram of datac$fc..Mpa.
Frequency
0
400
0
50
100
150
200
datac$fc..Mpa.

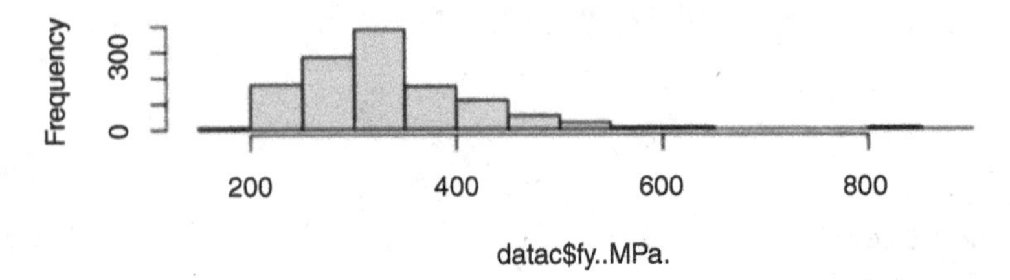
Histogram of datac$fy..MPa.
Frequency
0
300
200
400
600
800
datac$fy..MPa.

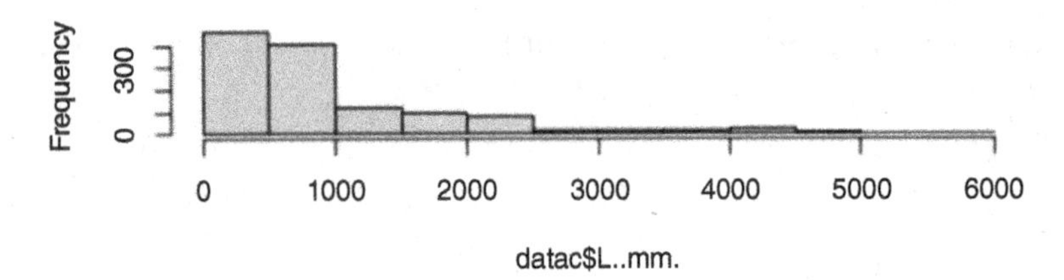
Histogram of datac$L..mm.
Frequency
0
300
0
1000
2000
3000
4000
5000
6000
datac$L..mm.

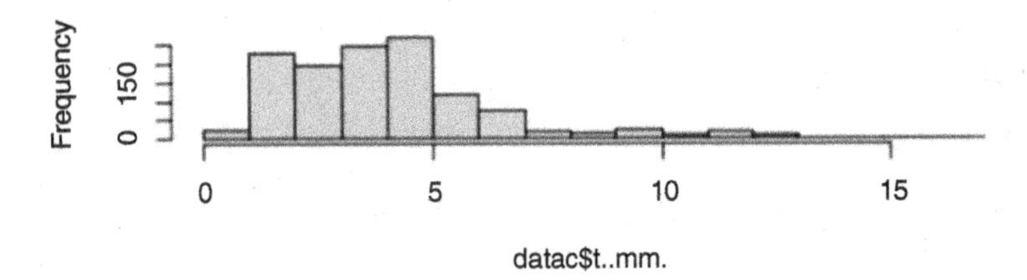
Histogram of datac$t..mm.
Frequency
0
150
0
5
10
15
datac$t..mm.

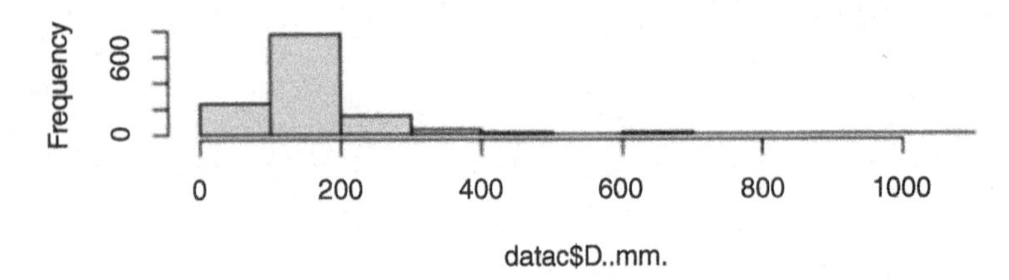
Histogram of datac$D..mm.
Frequency
0
600
0
200
400
600
800
1000
datac$D..mm.

```
#Plotting correlation matrices.
#Spearman methods:81
m=cor(data, method = "spearman")
corrplot(m, method ="circle")
```

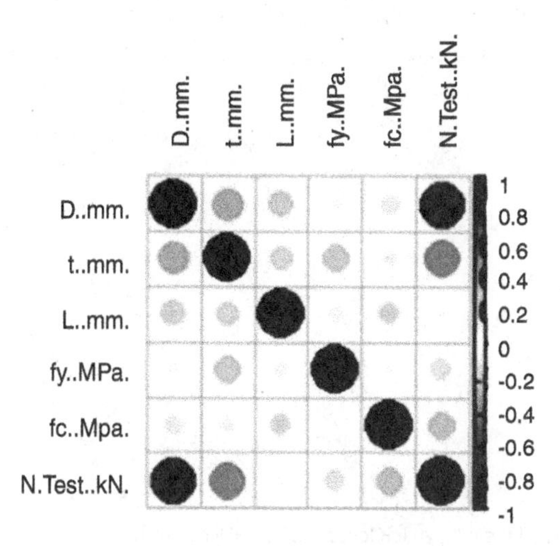


```
corrplot(m, method ="pie")
```

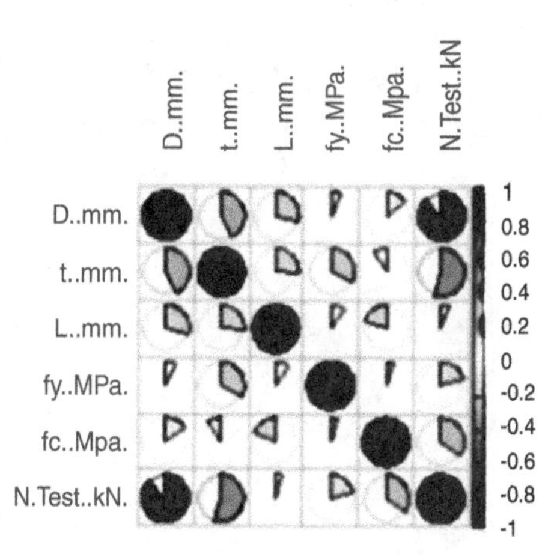


```
corrplot(m, method ="number")
```

	D..mm.	t..mm.	L..mm.	fy..MPa.	fc..MPa.	N.Test..N.
D..mm.	1.00	0.43	0.29	0.07	0.15	0.88
t..mm.	0.43	1.00	0.26	0.32	0.10	0.55
L..mm.	0.29	0.26	1.00	0.10	0.20	0.06
fy..MPa.	0.07	0.32	0.10	1.00	0.08	0.20
fc..MPa.	0.15	0.10	0.20	0.08	1.00	0.34
N.Test..kN.	0.88	0.55	0.06	0.20	0.34	1.00

1 0.8 0.6 0.4 0.2 0 -0.2 -0.4 -0.6 -0.8 -1

81 As you can see, Arash likes to show different types of charts!

```
corrplot(m, method = "square")
```

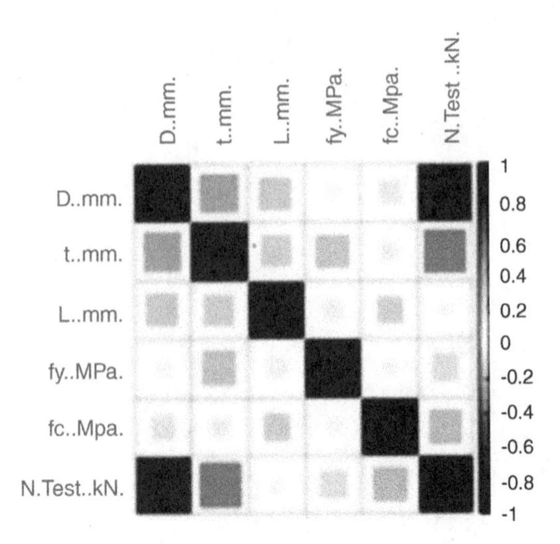


```
corrplot(m, method ="color",type = "upper",order ="hclust",addCoef.col ="black",
tl.col = "black",tl.srt
= 45, diag = FALSE)
```

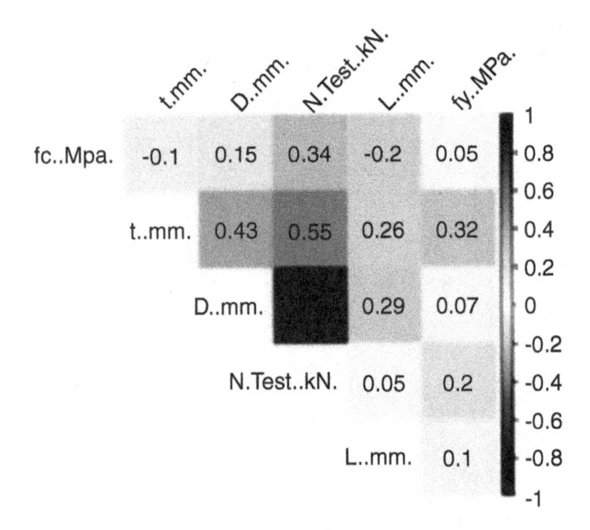


```
#Pearson correlation.
n=cor(data, method ="pearson")
corrplot(n, method ="color",type = "upper",order ="hclust",addCoef.col ="black",
tl.col = "black",tl.srt = 45, diag = FALSE)
```

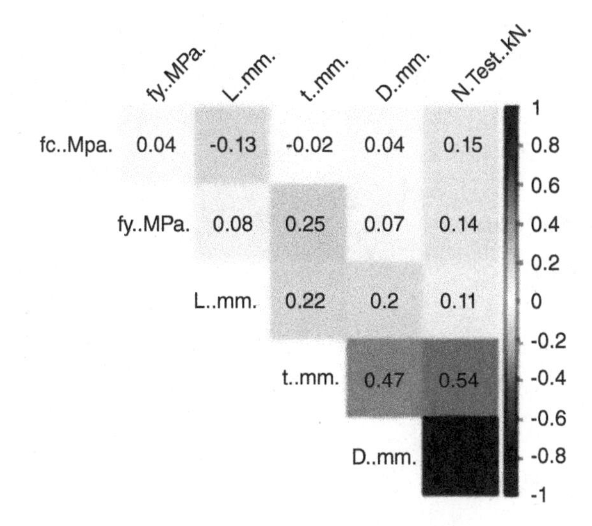


```
#Plotting actuals vs. predictions.
plot(x=predicted, y= actual,
+  xlab='Predicted Values',
+  ylab='Actual Values',
+  main='Predicted vs. Actual Values',
+  abline(a=0, b=1))
```

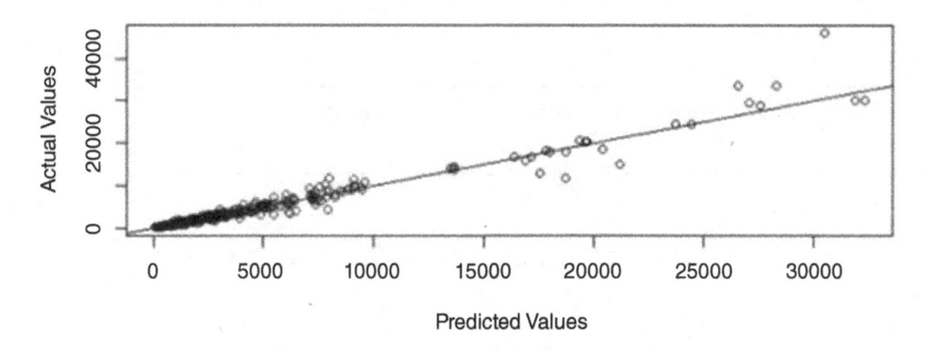


```
#Plotting SHAP summary and SHAP top 5 features.
xgb.plot.shap.summary(data = as.matrix(trainrf),
+        model = modelxg,
+        top_n = 11)
```

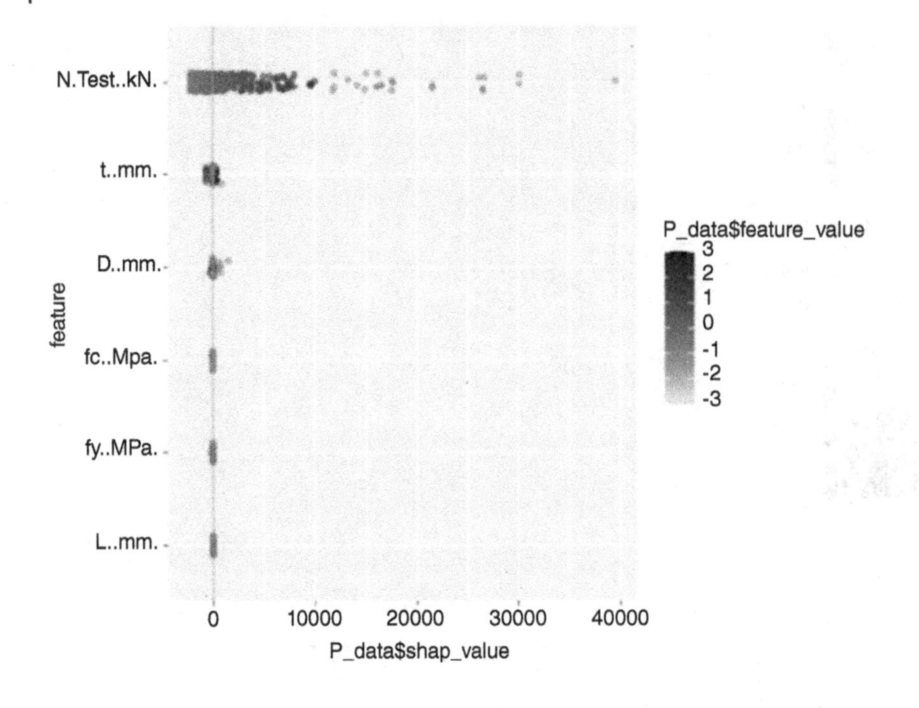


```
xgb.plot.shap(data = as.matrix(trainrf),
+     model = modelxg,
+     top_n = 5)
```

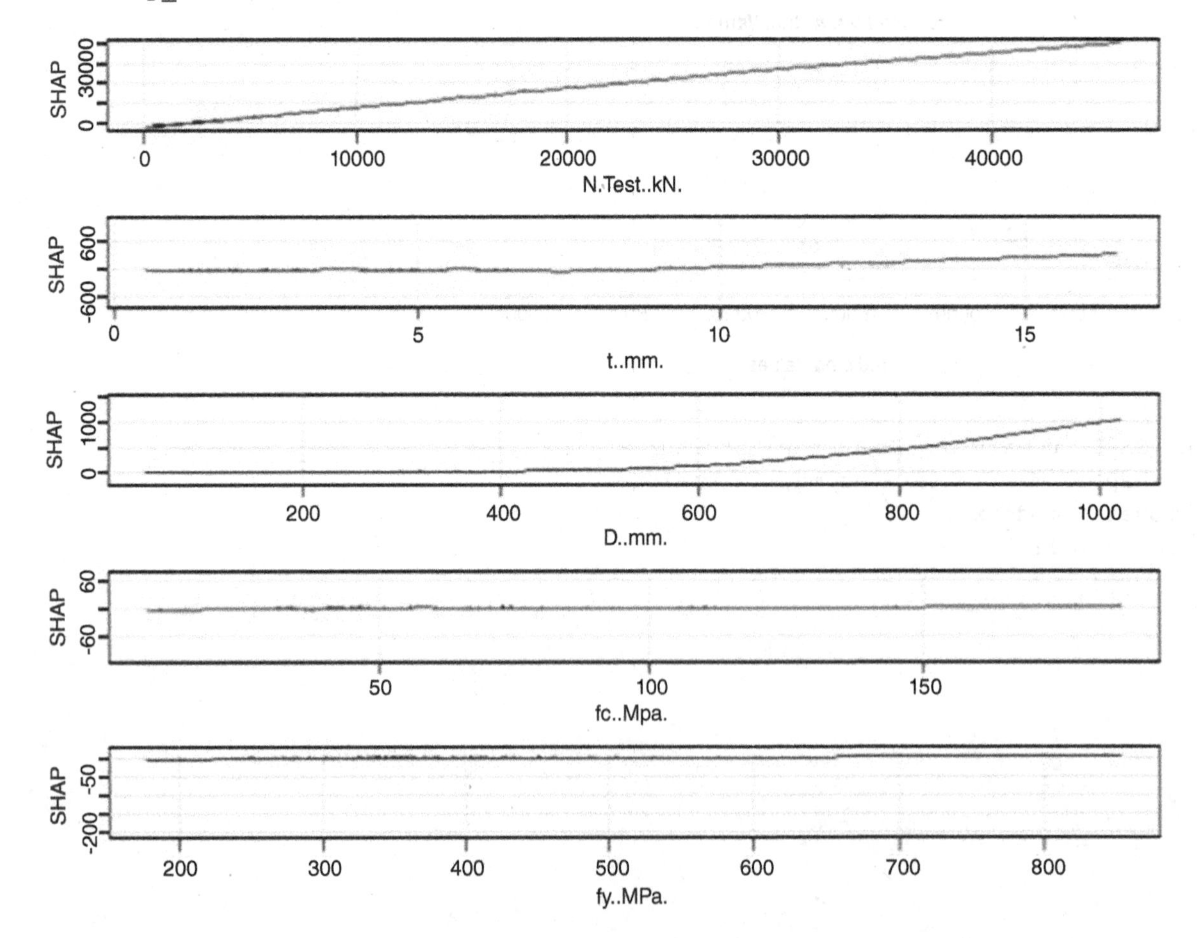

```
#Plotting Explainer using XG.boost modeling.
fit_xgboost<-boost_tree(learn_rate = 0.3)%>%
+    set_mode("regression")%>%
+    set_engine("xgboost")%>%
+    fit(data$N.Test..kN.~.,data=data)

#Explainer
explainer<-DALEX::explain(
+ model= rf.mod,
+ data = data,
+ y=data$N.Test..kN.,
+ label = "RandomForest"
+ )
Preparation of a new explainer is initiated
-> model label: RandomForest
-> data: 1260 rows 6 cols
-> target variable: 1260 values
-> predict function: yhat.randomForest will be used ( default )
-> predicted values: No value for predict function target column. ( default )
-> model_info: package randomForest,  ver. 4.7.1.1,  task regression ( default )
-> predicted values: numerical, min = 76.89726,  mean = 2348.331,  max = 37481.5
-> residual function: difference between y and yhat ( default )
-> residuals: numerical, min = -3136.158,  mean = 1.386442,  max = 8518.503 A new
explainer has been created!

#Model studio
modelStudio::modelStudio(explainer = explainer)
`new_observation` argument is NULL. `new_observation_n` observations needed to calcu-
late local explanations are taken from the data.

Calculating …
Calculating ingredients::feature_importance
Calculating ingredients::partial_dependence (numerical) Calculating
ingredients::accumulated_dependence (numerical) Calculating DALEX::model_performance
Calculating DALEX::model_diagnostics Calculating iBreakDown::local_attributions (1)
Calculating iBreakDown::shap (1)
Calculating ingredients::ceteris_paribus (1)
Calculating iBreakDown::local_attributions (2)
Calculating iBreakDown::shap (2)
Calculating ingredients::ceteris_paribus (2)
Calculating iBreakDown::local_attributions (3)
Calculating iBreakDown::shap (3)
Calculating ingredients::ceteris_paribus (3) Calculating …
Elapsed time: 00:00:23 ETA: 0s
```

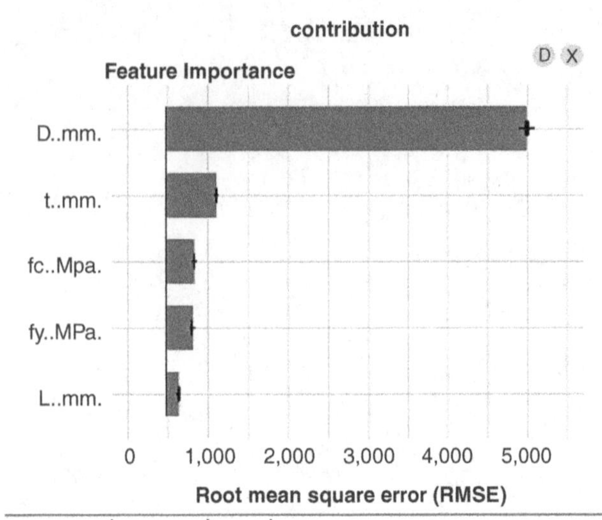
contribution
D X
Feature Importance
D..mm.
t..mm.
fc..Mpa.
fy..MPa.
L..mm.
0
1,000
2,000
3,000
4,000
5,000
Root mean square error (RMSE)
mse: 234796.779
rmse: 484.558
r2: 0.985
mad: 38.157

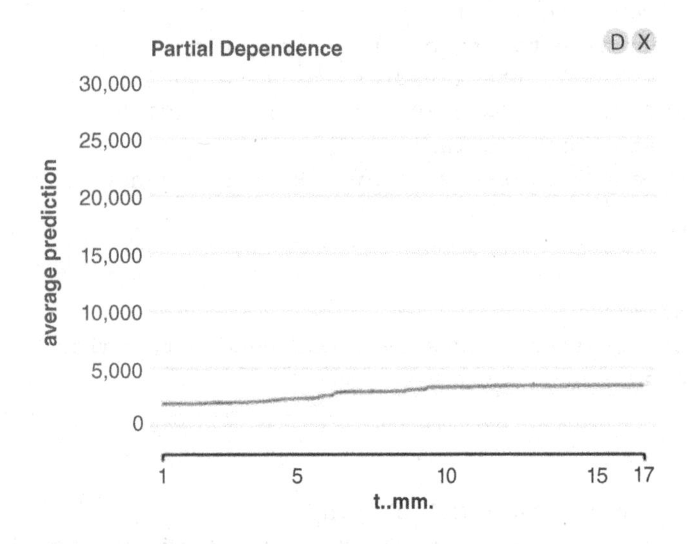
Partial Dependence
D X
average prediction
30,000
25,000
20,000
15,000
10,000
5,000
0
1
5
10
15
17
t..mm.

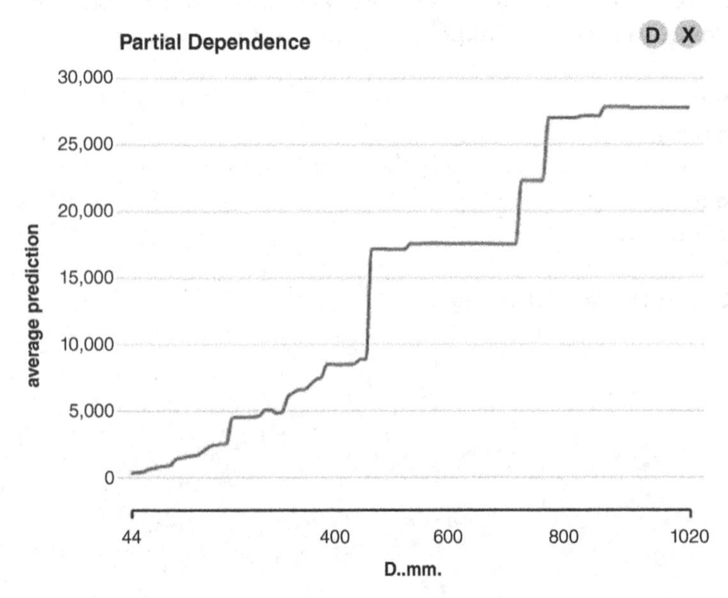
Partial Dependence
D X
average prediction
30,000
25,000
20,000
15,000
10,000
5,000
0
44
400
600
800
1020
D..mm.

Example no. 5 (Python): Clustering fire resistance of RC columns.
I will be continuing the same example we covered in our discussion on clustering. This would be a good time to review that particular example.[82]

We will apply the explainable trees approach to understand the reasoning behind such clustering analysis. Table 7.3 shows the first and second splits taken in each algorithm, along with the number of splits and leaves.

I will now go over two explainable trees obtained from the K-Means and Hierarchical clustering models (plotted in Figures 7.15 and 7.16).[83] Overall, we can also see that the entropy reduces with each split (as reflected by the higher intensity in color). The first split from the K-Means model takes place at the compressive strength of 71 MPa. From there, columns of $f_c > 71$ MPa are moved into cluster no. 1, and the rest are further analyzed. The next split occurs in columns with a length (L) longer than 4.3 m joining cluster no. 4, and the remaining columns continue the analysis. The third split occurs at the concrete cover (C) with columns of a smaller cover thickness (less than 36.5 mm) joining cluster no. 2. Columns with pin-pined conditions are examined against the level of applied loading, and columns with higher rigidity ($K \leq 0.85$) are examined against the steel reinforcement ratio.

Table 7.3 Splits of algorithms.

Split	K-Means	Hierarchical	Fuzzy-C-Means	DBSCAN
First Split	f_c	C	f_c	L
Second Split	L	C, f_c	K, P	f_c, e_x
Number of split node	7	11	9	12
Number of leaf node	8	12	10	13

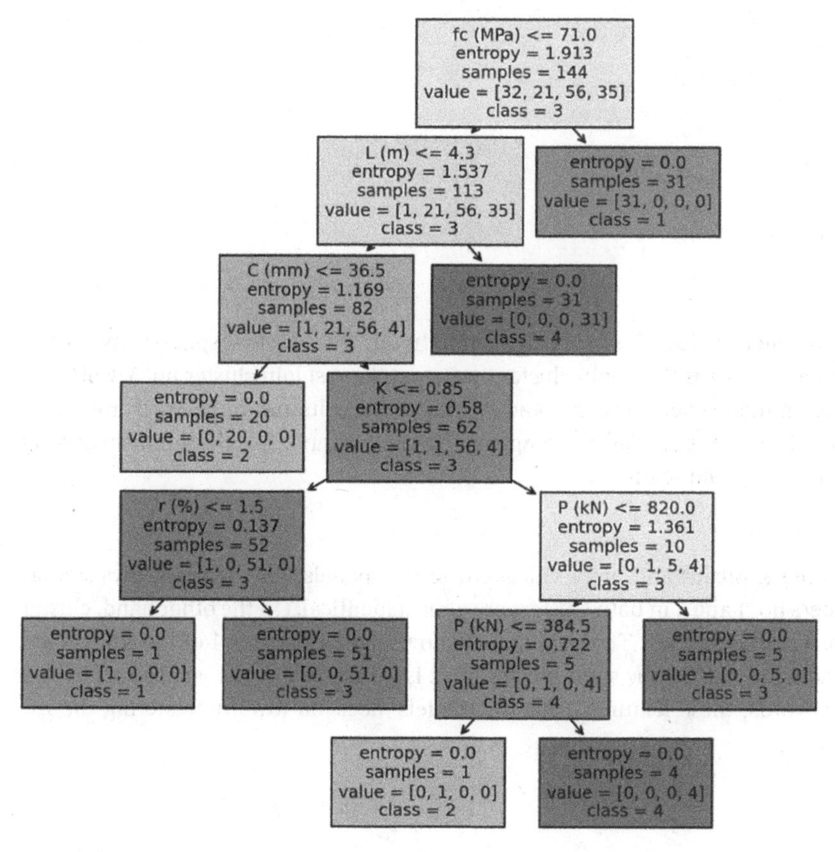

Figure 7.15 Explainable tree for K-Means.

82 Let me recap what we did. We have used four clustering algorithms to cluster over 140 RC columns. Our multi-algorithm analysis has noted that the examined RC columns fall under four clusters.

83 And will leave the other two trees for you to try and compare your analysis to that shown at the end of this example. You have seen such trees earlier, but we did not discuss them then. I promised we will. Here we go!

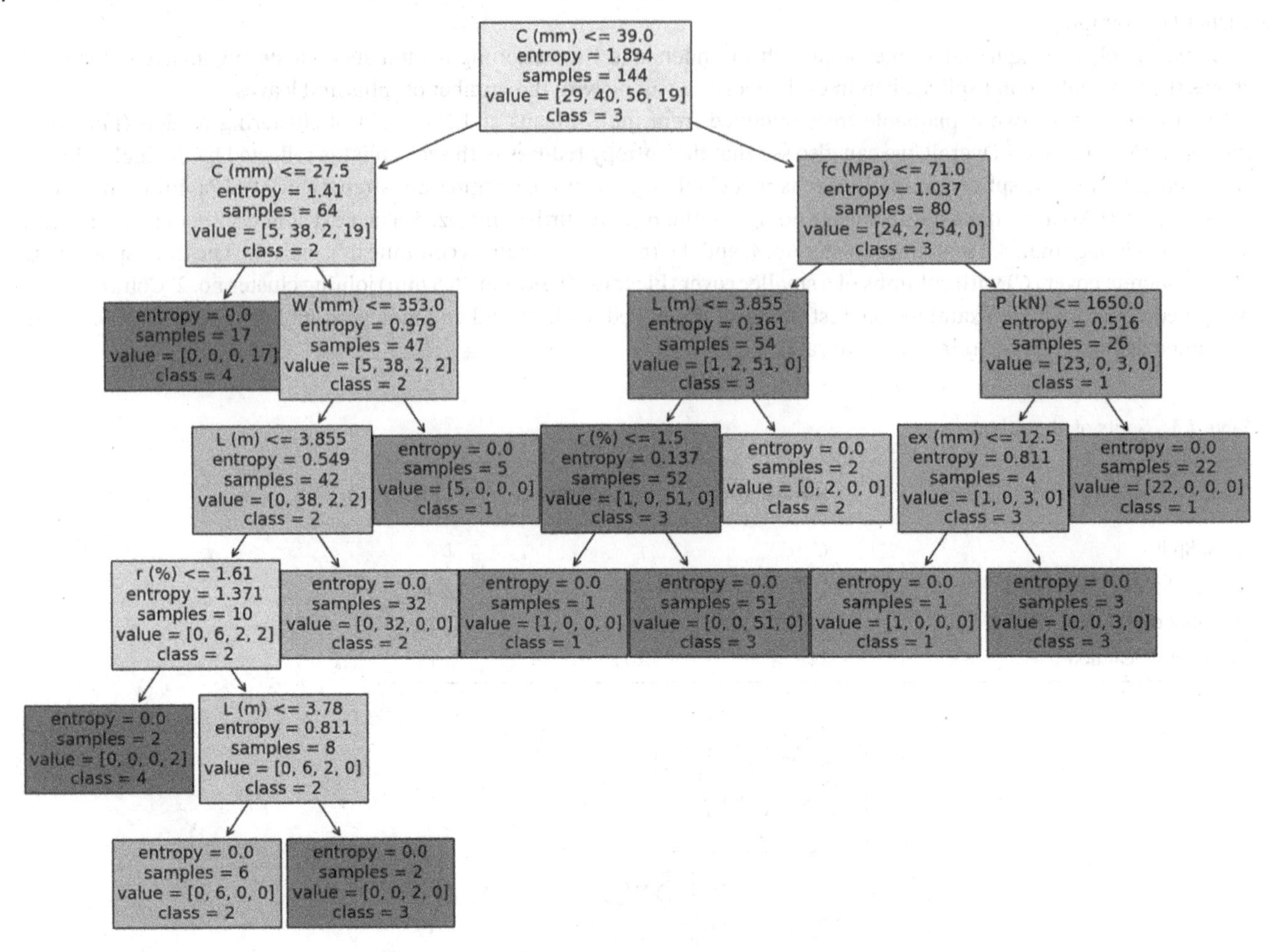

Figure 7.16 Explainable tree for hierarchical clustering.

Figure 7.16 shows the explainable tree for the hierarchical clustering algorithm. In this tree, the first split occurs for columns with a smaller cover thickness (less than 39 mm) as these join cluster no. 2, and the rest join cluster no. 3. Columns in cluster no. 2 continue to be analyzed based on their cover thickness, and the rest of the columns in this track are examined against their other features. In cluster no. 3, columns of 3.86 m in length move to cluster no. 3, and columns receiving loads of less than 1650 kN are moved to cluster no. 1, and so on.

7.4.3.1 Polar Line Plots

Figure 7.17 shows the polar line plots from the K-Means and hierarchical clustering models. This figure shows similar trends between clusters.[84] For example, clusters no. 1 and 3 in both clusters are almost identical. On the other hand, cluster no. 2 in K-Means is equivalent to cluster no. 4, and vice versa. The same figure shows that cluster no. 1 of the K-Means is primarily influenced by strength (f_c) and loading (P), while the same cluster no. 2 is influenced by the yield strength and stability/restraint conditions (K).[85] In other words, these features drive the models' decision toward clustering the RC columns.

84 Despite having similar trends, the numbering of the clusters is not identical.

85 I trust that you can get the explanations for the other clusters! For example, cluster no. 4 in both models I governed by the collected features about equally – with a bit of sensitivity toward concrete cover.

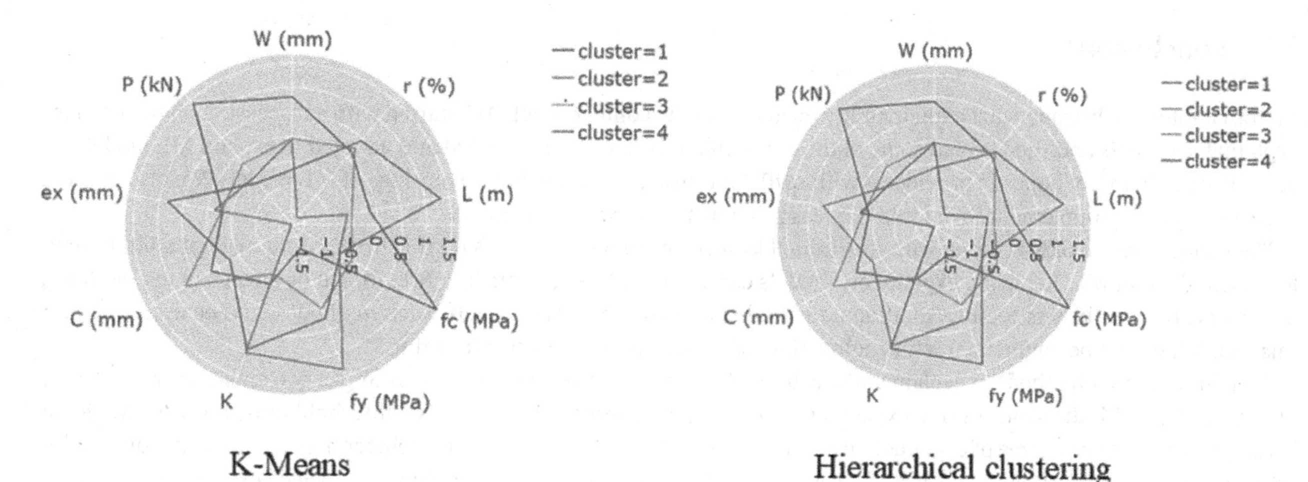

Figure 7.17 Comparison of polar line plots.

Here is the code to get the above trees and polar line plots.

```
#Adding Explainable trees
X_normal = dataset.iloc[:, :9].values

from sklearn.tree import DecisionTreeClassifier
classifier = DecisionTreeClassifier(criterion = 'entropy', random_state = 0)
classifier.fit(X_normal, y)

from sklearn import tree
fig, axes = plt.subplots(nrows = 1, ncols = 1, figsize = (8,8), dpi=200)
tree.plot_tree(classifier, feature_names = dataset.columns,filled = True, class_name
s=['1','2','3','4'],fontsize=9)

text_representation = tree.export_text(classifier)
print(text_representation)

#Visualizing clusters via polar line plots
X = dataset.iloc[:, :9].values
y=np.array(y)

plt.scatter(X[y == 1, 3], X[y == 1, 8], s = 100, c = 'red', label = 'Cluster 1')
plt.scatter(X[y == 2, 3], X[y == 2, 8], s = 100, c = 'green', label = 'Cluster 2')
plt.scatter(X[y == 3, 3], X[y == 3, 8], s = 100, c = 'blue', label = 'Cluster 3')
plt.scatter(X[y == 4, 3], X[y == 4, 8], s = 100, c = 'cyan', label = 'Cluster 4')
#plt.scatter(X[y_kmeans == 4, 0], X[y_kmeans == 4, 1], s = 100, c = 'magenta', label
= 'Cluster 5')
#plt.scatter(kmeans.cluster_centers_[:, 0], kmeans.cluster_centers_[:, 1], s = 300,
c = 'yellow', label = 'Centroids')
plt.title('Clusters of RC columns')
plt.xlabel('fc')
plt.ylabel('P')
plt.legend()
plt.show()
```

7.5 Conclusions

In this chapter, we learned what explainability means from the context of ML. We started with a discussion on the need for explainable models and then visited some of the main methods that can be applied and adopted to explain ML models. In general, explainability hopes to provide us with sufficient insights into the inner workings of ML models. We also covered a few examples to supplement to dig into the reasoning behind model predictions.

The opaque nature of ML, along with the limited knowledge of our engineers on this front, could make it a bit difficult to trust and adopt ML. Realizing explainable models can help ease[86] engineers into ML. In lieu of the presented methods, engineers are also likely to require additional methods to explain their ML models. Such methods could be best suited for this task if they can be augmented with engineering principles or engineering reasoning.[87]

Keep in mind that methods of explainability can satisfy users to varying degrees. For example, two engineers may or may not be satisfied with the same level of explainability [25–27]. It is possible that there is a threshold or potentially a trade-off between an acceptable degree of explainability to the expected level of satisfaction an engineer may require depending on an engineer's background, sophistication, and goals (i.e., need to comply with codal provisions or societal expectations, etc.).

Definitions

Explainability Entails relating the inner mechanisms of a model and their influence on the model's prediction. Also, assume that a model is capable of summarizing or producing insights to their decisions.
Feature importance The degree to which a ML model relies on a particular feature to arrive at correct predictions.
Interpretability The science of understanding what a model did (or might have done).
Partial dependence Depicts an individual feature's marginal (average) effect on the response/target variable while holding other features constant.
Summary plot A plot that summarizes information similar to that obtained from feature importance.

Questions and Problems

7.1 What are the qualities of the traditional methods of investigation?

7.2 Contrast explainability and interpretability.

7.3 Discuss the main differences between model predictions and model explanations.

7.4 Write one paragraph to describe why explainable ML is important to your area of civil and environmental engineering. Supplement your write-up with a practical example.

7.5 Redo Example no. 1 in *DataRobot*.

7.6 Repeat Example no. 1 in *Dataiku*.

7.7 Build a regression model and a classifier in a programming language of your choice and compare your analysis to that provided in Example no. 1.

86 Keep in mind that explainability is one way to do so. There are many dimensions to get us there.
87 Versus looking into explanations from a purely data-driven perspective. The data can say so much to the phenomenon, but the available data may not cover all the possible mechanisms to the phenomenon on hand.

7.8 Compare SHAP and LIME predictions for local explainability of three observations from Example no. 3. Use the provided dataset.

7.9 Extend Example no. 5 by using FCM and DBSCAN algorithms.

Chapter Blueprint

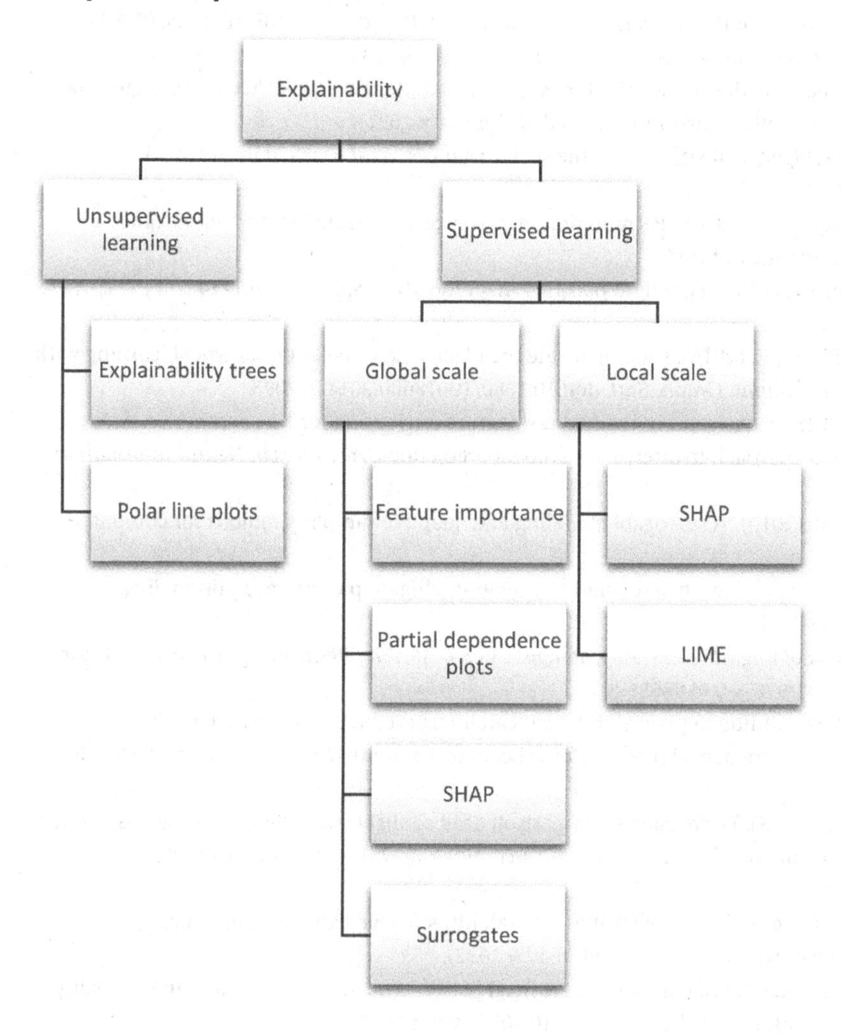

References

1 Adadi, A. and Berrada, M. (2018). Peeking inside the black-box: a survey on Explainable Artificial Intelligence (XAI). *IEEE Access*. doi: 10.1109/ACCESS.2018.2870052.

2 Dosilovic, F.K., Brcic, M., and Hlupic, N. (2018). Explainable artificial intelligence: a survey. *2018 41st International Convention on Information and Communication Technology, Electronics and Microelectronics (MIPRO)*. doi: 10.23919/MIPRO.2018.8400040.

3 Gilpin, L.H., Bau, D., Yuan, B.Z. et al. (2019). Explaining explanations: an overview of interpretability of machine learning. *2018 IEEE 5th International Conference on Data Science and Advanced Analytics (DSAA)*. doi: 10.1109/DSAA.2018.00018.

4 Naser, M.Z. (2021). An engineer's guide to eXplainable Artificial Intelligence and Interpretable Machine Learning: navigating causality, forced goodness, and the false perception of inference. *Autom. Constr.* 129: 103821. doi: 10.1016/J.AUTCON.2021.103821.

5 Morgan, D. and Jacobs, R. (2020). Opportunities and challenges for machine learning in materials science. *Annu. Rev. Mater. Res.* doi: 10.1146/annurev-matsci-070218-010015.

6 Du, M., Liu, N., and Hu, X. (2020). Techniques for interpretable machine learning. *Commun. ACM.* doi: 10.1145/3359786.

7 Rutherford, J. (2002). Practical experiment designs for engineers and scientists. doi: 10.1198/tech.2002.s79.

8 Antony, J. (2014). *Design of Experiments for Engineers and Scientists*, 2e. doi: 10.1016/C2012-0-03558-2.

9 Economist (2017). Can we fix it? – the construction industry's productivity problem | Leaders | The Economist. https://www.economist.com/leaders/2017/08/17/the-construction-industrys-productivity-problem (accessed 27 August 2020).

10 Bogue, R. (2018). What are the prospects for robots in the construction industry? *Ind. Rob.* doi: 10.1108/IR-11-2017-0194.

11 Lipton, Z.C. (2018). The mythos of model interpretability. *Commun. ACM.* doi: 10.1145/3233231.

12 Cohen, B. (2018). The rise of engineering's social justice. James G. Martin Cent. Acad. Renew. https://www.jamesgmartin.center/2018/11/the-rise-of-engineerings-social-justice-warriors (accessed 22 February 2021).

13 Miller, T. (2019). Explanation in artificial intelligence: insights from the social sciences. *Artif. Intell.* doi: 10.1016/j.artint.2018.07.007.

14 Altmann, A., Toloşi, L., Sander, O., and Lengauer, T. (2010). Permutation importance: a corrected feature importance measure. *Bioinformatics.* doi: 10.1093/bioinformatics/btq134.

15 Friedman, J.H. (2001). Greedy function approximation: a gradient boosting machine. *Ann. Stat.* doi: 10.1214/aos/1013203451.

16 Goldstein, A., Kapelner, A., Bleich, J., and Pitkin, E. (2015). Peeking inside the black box: visualizing statistical learning with plots of individual conditional expectation. *J. Comput. Graph. Stat.* doi: 10.1080/10618600.2014.907095.

17 Lundberg, S.M., Erion, G.G., and Lee, S.I. (2018). Consistent individualized feature attribution for tree ensembles. ArXiv.

18 Lundberg, S.M. and Lee, S.I. (2017). A unified approach to interpreting model predictions. *Advances in Neural Information Processing Systems 30 (NIPS 2017).*

19 Gorissen, D., Couckuyt, I., Demeester, P. et al. (2010). A surrogate modeling and adaptive sampling toolbox for computer based design. *J. Mach. Learn. Res.* 11.

20 Naser, M.Z. and Seitllari, A. (2020). Concrete under fire: an assessment through intelligent pattern recognition. *Eng. Comput.* 36: 1–14. doi: 10.1007/s00366-019-00805-1.

21 Naser, M.Z. and Corbett, G. (2022). *Handbook of Cognitive and Autonomous Systems for Fire Resilient Infrastructures.* Cham: Springer International Publishing. doi: 10.1007/978-3-030-98685-8.

22 Özyüksel Çiftçioğlu, A. and Naser, M.Z. (2022). Hiding in plain sight: what can interpretable unsupervised machine learning and clustering analysis tell us about the fire behavior of reinforced concrete columns? *Structures* 40: 920–935. doi: 10.1016/J.ISTRUC.2022.04.076.

23 Zhang, H.Y., Lv, H.R., Kodur, V., and Qi, S.L. (2018). Performance comparison of fiber sheet strengthened RC beams bonded with geopolymer and epoxy resin under ambient and fire conditions. *J. Struct. Fire Eng.* 9: 174–188. doi: 10.1108/JSFE-01-2017-0023.

24 Open.Canada.ca (2022). Fuel consumption ratings – Open Government Portal. https://open.canada.ca/data/en/dataset/98f1a129-f628-4ce4-b24d-6f16bf24dd64#wb-auto-6 (accessed 27 July 2022).

25 Emmert-Streib, F., Yli-Harja, O., and Dehmer, M. (2020). Explainable artificial intelligence and machine learning: a reality rooted perspective. *Wiley Interdiscip. Rev. Data Min. Knowl. Discov.* doi: 10.1002/widm.1368.

26 Hoffman, R.R., Klein, G., and Mueller, S.T. (2018). Explaining explanation for "Explainable AI. *Proceedings of the Human Factors and Ergonomics Society Annual Meeting.* doi: 10.1177/1541931218621047.

27 Hase, P. and Bansal, M. (2020). Evaluating explainable AI: which algorithmic explanations help users predict model behavior? doi: 10.18653/v1/2020.acl-main.491.

28 Rudin, C. (2019). Stop explaining black box machine learning models for high stakes decisions and use interpretable models instead. *Nat. Mach. Intell.* doi: 10.1038/s42256-019-0048-x.

29 Gunning, D., Stefik, M., Choi, J. et al. (2019). XAI-Explainable artificial intelligence. *Sci. Robot.* doi: 10.1126/scirobotics.aay7120.

30 Boehmke, B., Greenwell, B., Boehmke, B., and Greenwell, B. (2019). Interpretable machine learning. In: *Hands-On Machine Learning with R.* doi: 10.1201/9780367816377-16.

31 T. Royal Society. (2021). Explainable AI: the basics.

32 al-Bashiti, M.K. and Naser, M.Z. (2022). Verifying domain knowledge and theories on Fire-induced spalling of concrete through eXplainable artificial intelligence. *Constr. Build. Mater.* 348: 128648. doi: 10.1016/J.CONBUILDMAT.2022.128648.

33 Tapeh, A. and Naser, M.Z. (2022). Discovering graphical heuristics on fire-induced spalling of concrete through eXplainable artificial intelligence. *Fire Technol.* doi: 10.1007/s10694-022-01290-7.

34 Hastie, T.J. and Pregibon, D. (2017). Generalized linear models. In: *Statistical Models in S*. doi: 10.1201/9780203738535.

35 Chen, T. and Guestrin, C. (2016). XGBoost: a scalable tree boosting system. *KDD '16: Proceedings of the 22nd ACM SIGKDD International Conference on Knowledge Discovery and Data Mining*. doi: 10.1145/2939672.2939785.

36 Molnar, C. (2019). Interpretable machine learning. In: *A Guide for Making Black Box Models Explainable*. Book.

37 Naser, M.Z. (2020). Machine learning assessment of fiber-reinforced polymer-strengthened and reinforced concrete members. *ACI Struct. J.* doi: 10.14359/51728073.

38 Al-Tamimi, A.K., Hawileh, R., Abdalla, J., and Rasheed, H.A. (2011). Effects of ratio of CFRP plate length to shear span and end anchorage on flexural behavior of SCC RC beams. *J. Compos. Constr.* 15: 908–919. doi: 10.1061/(ASCE)CC.1943-5614.0000221.

39 De Lorenzis, L. and Teng, J.G. (2007). Near-surface mounted FRP reinforcement: an emerging technique for strengthening structures. *Compos. Part B Eng.* doi: 10.1016/j.compositesb.2006.08.003.

40 Hertz, K.D.D. (2003). Limits of spalling of fire-exposed concrete. *Fire Saf. J.* 38: 103–116. doi: 10.1016/S0379-7112(02)00051-6.

41 Kodur, V.K.R. and Phan, L. (2007). Critical factors governing the fire performance of high strength concrete systems. *Fire Saf. J.* 42: 482–488. doi: 10.1016/j.firesaf.2006.10.006.

8

Causal Discovery and Causal Inference

There is no result in nature without a cause. Understand the cause, and you will have no need for experiments.
Leonardo da Vinci

Synopsis

The vast majority of this textbook is devoted to exploring, creating, applying, and assessing machine learning models. While this chapter continues this trend, yet unlike the rest, this chapter dives into the data in the hope of discovering the hidden mechanism(s) that generate the observations we are interested in.[1] My goal in this chapter is to divert your attention, for a bit, from the linear process of collecting data → run algorithms → arrive at a blackbox/whitebox ML model and toward collecting data → identify the data generating process[2] (DGP) → arrive at a causal ML → infer the effects of manipulation(s)/treatment(s). By now, you might be wondering. What is a DGP? What is a causal model? What are manipulations and treatments?

For this, allow me to first introduce some of the big ideas behind causality, its principles, and how we can integrate those into a ML paradigm.

8.1 Big Ideas Behind This Chapter

A fundamental pursuit for us is to understand how a phenomenon comes to be. As engineers, we design and conduct experiments to uncover cause(s) and effect(s)[3] behind a phenomenon we happen to be interested in. Our thinking process is simple, manipulate the system in one way to observe how the system responds. The outcome of such manipulation can then be thought of as the result that reflects the manipulation. Overall, our goal is to create a model to predict and infer the response of the system without having to resort to physical testing (or numerical simulation, etc.). Simply

1 *Quick stories*: Unlike the other chapters, I will be briefly going over two quick stories.
- I have two nieces. They are four and three years old. Recently, they started to ask, *why?*, in response to many of our conversations, requests, and questions. We love their curiosity! As you know, the practice of questioning is very common to kids, and is, indeed, a process in which they learn! From this view, knowing the *why* is elementary! From this context, this chapter hopes to present a case for causality and causal ML.
- In the previous chapter, I have italicized the *data perspective* in the last sentence of the first footnote. This chapter is about going beyond the data-driven nature of ML to identify causes and effects. I have debated the inclusion/de-inclusion this chapter from this textbook multiple times. Evidently, I *feel* that including this chapter would be more beneficial than excluding it.

2 The **data generating process (DGP)** can be simply defined as the process that generates the data. For example, loaded beams tend to deflect. A possible question targeting DGP would be, what causes a load beam to deflect? Thus, in a journey aimed at answering such a question, our goal is to identify or retrieve the underlying structure of DGP. Please refer to my review in [36] from which this chapter stems.
3 Say what causes a beam to fail in a certain way (i.e., flexure) as opposed to another (e.g., shear). Also, note that I am confining this discussion to physical testing. The same discussion could possibly be extended to FE simulations and analytical investigations.

Machine Learning for Civil & Environmental Engineers: A Practical Approach to Data-driven Analysis, Explainability, and Causality, First Edition. M. Z. Naser.

Companion Website: www.wiley.com/go/Naser/MachineLearningforCivilandEnvironmentalEngineers

put, manipulation causes the system to change. The effect of such change allows us to peek into a specific direction as to how the phenomenon occurs.

Causal[4] knowledge depicts how a phenomenon, *Y*, comes to be by answering a specific set of questions such as, what causes *Y*? How, when, and why does *Y* occur? What would happen if we were to influence *Y* or one of its factors, etc.? [1]. Arriving at precious answers to such questions is an exciting and ambitious quest. All of which makes causal knowledge difficult to obtain [2].

Let us be optimistic for a minute[5] and assume that we were able to arrive at the true DGP for testing beams under fire conditions – as in, we uncovered the DGP for how beams deflect when exposed to fire. With that in mind, it is then plausible that this knowledge will be tremendous since we will not need to rely heavily on expensive and complex fire tests.[6] One might say, instead of testing, we could build FE models or a digital twin (DT) model – which happen to be more affordable than testing. This is a good suggestion! However, these models would still need to be validated, preferably per case basis, against a real fire test. To add a second shade of gray to the above, we continue to lack a standardized procedure to build and validate such models. In some instances, we cannot truly model some testing conditions. We are still constrained by the set of many assumptions and idealizations our models (numerical, analytical, physical, ML, etc.) have to obey. I, hence, believe that pursuing causal models is of high[7] merit.

8.2 Re-visiting Experiments

As engineers, we devise an experiment (or a series of experiments) in hope of comprehending a phenomenon. In general, experiments are driven by a hypothesis. A hypothesis targets a specific direction to uncover the DGP (i.e., the cause(s) leading to the effect). We carry out our experiments and observe their outcome. Such outcome is then used to establish a set of theories. A theory postulates the workings of a DGP and gives us the recipe to predict the phenomenon with a level of certainty [3]. In other words, experiments draw a causal path (i.e., cause(s) → effect), for, without the cause(s), the phenomenon would not have been generated[8] [4].

In the above procedure, we design controlled experiments. A benchmark specimen is kept as a reference line, which is also accompanied by a series of altered (manipulated) specimens. Each manipulated specimen is modified by changing a single parameter while holding all other parameters identical to the benchmark. Allowing only one parameter (say, *X*) to vary at a time gives us insights into how such a parameter affects the observed response (e.g., the difference between the outcome of the altered specimen and that of the benchmark specimen is *directly*[9] *caused* by the alteration of *X*) [4].

Let us think of the following simple example. Say we are trying to understand the concept of flexural capacity (bending) in simply supported steel beams. First, we would fabricate a *Beam A* with certain geometrical and material features.[10] Then, we load this beam in a manner that enables us to capture its flexural response. We load the beam[11] and report that this beam fails once the level of the applied moment reaches 361.6 kN.m.

In such an experiment, we associate the reported failure moment with the geometrical and material features as well as the loading configuration of *Beam A*. This link draws from the following observation: applying a bending moment of 361.6 kN.m to *Beam A* is followed by the failure of the beam. A series of questions may arise: what caused *Beam A* to fail? was failure triggered by the geometric configurations or material properties of *Beam A*? How do boundary and loading configurations affect such failure? These are causal questions.

---

4 Seeking causal knowledge is a foundational pursuit with branching philosophical, epistemological, and ontological ties. For a more cohesive look at causality from the intersection of philosophy, epistemology, and ontology, please refer to [48, 49]. For completion, a philosophical and historical discussion and a more in-depth review on causality can also be found elsewhere [10, 14] and [50], respectively. A side note, one of the earliest papers that I came across to discuss causality from an engineering perspective is that of Claudius [51] (published in 1932).

5 Do your best to remain optimistic – all the time. Jasmine has a mug that says, *always optimistic!*

6 Let us not forget the fact that such tests require specialized equipment, facilities, and personnel.

7 If not, highest.

8 A broader look into causality can be found herein [10, 52, 53] as well as in my reviews [36, 74] and Henry Burton's look at causality in earthquake engineering [75]. We were chatting in a recent meeting, and it turned out that both of us were working on this area around the same time!

9 In most cases, we avoid altering two or more parameters simultaneously to minimize any ambiguity of establishing *indirect causation*.

10 Or simply select a standard shape (i.e., W16 × 36) that is made from Grade A992 structural steel.

11 Say with one point load at mid-span.

Another set of questions that could also arise includes: other things constant [*Ceteris Paribus*] would *Beam A* have failed if it was not for the applied loading? would this beam fail at 361.6 kN.m had it been made of a W18 × 40 section or Grade A36? These are a special type of causal question, referred to as counterfactuals. Both causal and counterfactual questions need to be answered *causally*.[12]

Building on what a civil and environmental engineer might learn early in their second year of school (in a mechanics course), we know that moment capacity (or simply, the resistance) of a W-shaped steel beam under bending is a function of its geometric features; conveniently lumped into the plastic modulus, *Z*, as well as its material features; represented by the yield strength of steel, f_y. Both *Z* and f_y are tied into Equation 8.1. Equation 8.1 conveys that both *Z* and f_y contribute to the moment capacity in a complimentary (equal) manner.

$$Resistance = Z \times f_y \tag{8.1}$$

Not surprisingly, the same equation allows us to answer all the above questions intuitively.

For instance:

1) What has caused the failure of *Beam A*?
   Ans. *The effect of a bending moment that exceeds the level of moment capacity* – and hence the equality.
2) Was failure triggered by the geometric configurations or material properties of *Beam A*?
   Ans. *Building on Equation 8.1, both Z and f_y have contributed equally.*
3) *Ceteris Paribus*, could Beam A have failed if it was not for the presence of the bending moment?
   Ans. *No. Since if it was not for such loading, the capacity (resistance) of the beam would not have been exceeded.*
4) *Ceteris Paribus*, would *Beam A* have failed if it was not for the applied loading? Would this beam fail at 361.6 kN.m if it had been a W18 × 40? Or had it been made from Grade A36?
   Ans. *No. Without the loading, there would not have been a bending moment. Also, W18 × 40 has a larger plastic modulus than W16 × 36, and Grade 36 has lower yield strength than Grade A992.*

As you can see, my answers to these four questions are somewhat trivial.[13] This simplicity stems from the narrow space of parameters involved in the phenomenon of flexural failure in W-shaped steel beams. More directly, the answers seem trivial because we can visualize [14] and imagine each answer as the outcome of an experiment.

Let us extend the same exercise to beams of complex shapes, made from hybrid material grades or those supported by mixed restraints. Repeating the same exercise is still doable but might require us to use a bit more horsepower[15] to get to an answer.

However, what if we now extend the same exercise to a phenomenon that we do not have a working model for, one that we do not have an equation for, or one that we cannot as easily visualize.[16] It is clear that the subject of causality exponentially grows with the addition of higher dimensions and/or problems we lack domain knowledge of.

8.3 Re-visiting Statistics and ML

The majority of the time, the outcomes of our investigations are examined through statistical methods. In such examination, we may implicitly *assume* that statistical methods are applicable to our data/observations. In reality, preset assumptions and requirements with regard to data size, distribution, quality, etc., must be met[17] [5]. If one or more assumptions

12 At this point of my writing, I felt that this discussion can use a proper example. You may skip the coming few paragraphs if you feel comfortable.

13 A new question can also be formulated: how could we have prevented failure in Beam A? *Ans. Failure can be prevented by switching Beam A with a larger section, and/or using a higher Grade of structural steel.* On a more fundamental and philosophical notion, *failure could have been prevented by reducing the level of bending moment, or possibly eliminating the existence of such bending moment.* The two former solutions stem from our engineering knowledge, while the latter stems from a causal perspective to our engineering understanding of the flexural failure problem.

14 E.g., we can visualize that W18 × 40 is a larger section than W16 × 36 and hence it has a larger moment capacity.

15 Shout out to my friend, Atif Al Khatib, for this analogy.

16 Or perhaps we cannot design an experiment for!

17 This is a good time to re-visit Chapter 3!

are breached, then the application of statistical methods is poor, and the resulting statistical inference could be weak or violated.

In parallel, let us take a step back to go over what a statistical model does. Such a model is designed to *fit* parameters into a functional form (whether linear, nonlinear, etc.) that allows the prediction of a given phenomenon [6]. Knowing that real data may not confer to commonly adopted statistical requirements, especially when the number/relationship of parameters grows/becomes complex [7], then our reliance on statistical methods, while it may present us with the tools to chart approximate solutions, may not allow us to unlock the DGP of our phenomena.[18]

To prevail over the limitations of statistics, nonparametric methods such as ML become attractive [8]. Unlike statistical methods, ML directly learns from data to search for patterns that tie the input parameters to a phenomenon's output (hence, becoming applicable in scenarios where data do not conform to specific distributions or comprise highly non-linear/dimensional relations) [9]. Yet, as you know by now, traditional ML models rarely look for causality.

Further, many of the existing ML algorithms are primarily blackboxes with limited potential to establish causality [10]. Such models can be comfortably used in an *application* manner to predict the outcome of a given problem from a data-driven perspective.

You might now be remembering our discussion on explainability. Here are my two cents. Explainable models do have the capability to display the influence of each input parameter on the predicted outcome. In a way, understanding how each parameter influences model predictions *may* allow us to uncover possible hidden links.[19] Such links can turn into descriptive relationships (i.e., *mapping functions*) that could be extended to map a perceived cause and effect. However, fundamentally speaking, these models cannot identify cause and effect functional relations simply because they are not created to follow causal principles. Rather, they are built to leverage association and high-level correlations [11, 12].

8.4 Causality

8.4.1 Definition and a Brief History

The pursuit of causality is a domain-independent journey. On the other hand, establishing causality is likely to be a domain-dependent effort as we are likely to leverage our domain knowledge to answer causal questions (see Table 8.1).

Hume [14] defines causality as a *philosophical* relation and a *natural* relation. In the former, a cause is "*an object precedent and contiguous to another, and where all objects resembling the former are placed in like relations of precedency and contiguity to those objects that resemble the latter.*" Alternatively, a cause as a natural relation is "*... an object precedent and contiguous to another, and so united with it that the idea of the one determines the mind to form the idea of the other, and the impression of the one to form a more lively idea of the other.*" A more straightforward definition, also attributed to Hume, is "*We may define a cause to be an object, followed by another, [...] where, if the first object <u>had not</u> been, the second <u>had never</u> existed.*" The underlined portion of this definition describes *counterfactuals*, a key companion concept to *causality*.

In terms of counterfactuals, Lewis [15] defines causality as, "*We think of a cause as something that makes a difference, and the difference it makes must be a difference from what would have happened without it. Had it been absent, its effects – some*

Table 8.1 Causality based on Brady [13].

	Neo-Humean regularity	Counterfactuals	Manipulation via experimentation	Mechanisms & capacities
Approach	Observation of constant conjunction and correlation that requires atemporal precedence	Truth in otherwise similar worlds of "*if the cause occurs, then so does the effect*" and vice versa	Recipe that regularly produces the effect from the cause and requires an observation of the manipulation	Consideration of whether there is a mechanism or capacity that leads from the cause to the effect
Emphasis	Causes of effects (e.g., focus on the dependent variable in regressions)	Effects of causes (e.g., focus on treatment's effects in experiments)	Effects of causes (e.g., focus on treatment's effects in experiments)	Causes of effects (e.g., focus on the mechanism that creates effects)

18 In a way, statistical inference is rarely equivalent to causal inference [50].

19 Especially, when supplemented with engineering domain knowledge.

of them, at least, and usually all – would have been absent as well." For its similarity[20] to Hume's yet evident simplicity, I tend to favor Lewis' definition.

Let us reinforce the above definitions with examples. One may say the following; a given *Beam A* would deflect upon loading. This implies that the application of loading *causes Beam A* to deflect. Similarly, one may also then *infer* the counterfactual that if it was not for the application of loading, *Beam A* would not have deflected.

Here is another example, there was a 30-minute delay in traffic on the I-85 highway as a result of a car crash. The implied *cause* is: the car crash, and the inferred *counterfactual* is: if it was not for the car crash, traffic would not have been delayed.[21]

Now that we know what causality is, a question must arise. *How can we establish causality?*

Generally, there are two approaches. The first is referred to as the *Potential Outcomes (PO)* approach [16] and the second as the *Directed Acyclic Graph (DAG)* approach [17].

The PO approach analyzes data[22] to estimate *population-level* causal effects. This approach applies a form of difference between two states, the intervened upon/treated and control, to quantify average causal effects (see Equation 8.2):

$$\textit{Average causal effect} = \left|\textit{observed response of the treated units} - \textit{observed response of the control units}\right| \tag{8.2}$$

The DAG approach adopts graphical methods to identify possible links (→) between causes and effects. For example, a look at Figure 8.1 shows a sample graphical model with links between X_1 and X_2, X_1 and Y, and X_2 and Y. These links denote that X_1 causes X_2 and Y. Similarly, X_2 also causes Y. In the DAG approach, causality is a three-leveled hierarchy[23] that begins with:

- *Association* [purely statistical relationships as defined by the data, i.e., how does a unit change in *X* affect the observed response?]
- *Intervention* [how does changing one parameter to a specific value (say, *X* to *2X*) affect the observed response?]
- *Counterfactuals* [what would have happened if a parameter or more were altered?] [10, 18].

Recent works on causality stem from investigations primarily carried out by non-engineering domains (e.g., social sciences and medicine). The need for causality rises due to the fact that problems in such domains are essentially different from those in engineering. For example, a medical phenomenon (say, the suitability of a treatment (i.e., drug or a

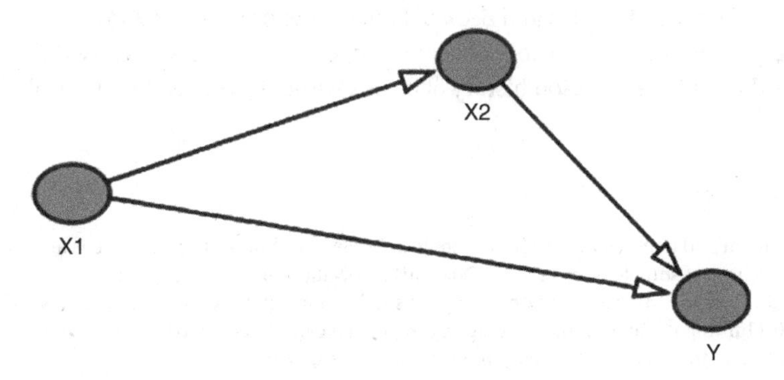

Figure 8.1 Sample DAG.

20 There are many, many other definitions that describe causality and counterfactuals in literature [31, 54]. I would advise you to read a bit on these. "Google is your friend" – shout out to Prof. Clark Glymour. You may also read my recent review [24].

21 Here is a third example. The concentration of chemical compound *X* has risen in the past 5 years due to the improper management of chemical waste generated in Factory A. Thus, the cause is: improper management of chemical waste, and the counterfactual is: if it was not for the improper management of chemical waste, concentration of compound *X* would not have risen. In all of these examples, as well as others, remember that we are assuming that the event/phenomenon occurs due to a single cause (in reality, *Beam A* could have deflected due to settlement effects, traffic on I-85 could have been delayed due to shutting down a few lanes to pave of a new layer of asphalt, etc.). This noted simplicity is warranted in this introductory section on causality.

22 Collected from results of randomized experiments or observational studies.

23 We will dive into this hierarchy in a few sections from now.

procedure) for individuals with certain pre-existing conditions) involves a series of parameters (e.g., health history, age, sex, parental background, etc.) – some of which may not be observable or measurable. In addition, the same problems are likely to require randomizations to normalize the effect of bias and confounders.[24] If an individual is treated with a specific treatment, then it is impossible to observe how the same individual would have been treated with another treatment. This impossibility complicates arriving at causal effects and is the reason that "*causal inference is inherently a missing data problem*" [19].

On the other hand, it is more likely for civil engineers to compare the state of similar units in a typical experimental setting.[25] As articulated earlier, a typical testing program contains a benchmark specimen and altered specimens (that varies one parameter at a time). After each test, the response of an altered specimen is compared against that of the benchmark. Simply, the systematic nature of our tests allows us to arrive at causal conclusions in a structured manner.

This chapter started with a quote often attributed to Leonardo da Vinci, "*There is no result in nature without a cause. Understand the cause, and you will have no need for experiments.*" From an engineering perspective, this quote stresses the need for understanding the relationship between a cause and a result (i.e., response/effect). The same quote highlights that in order to observe a result, an experiment (e.g., manipulation[26]) is required.

Let me re-visit my favorite example so far! Civil engineers are familiar with beams.[27] For brevity, let us assume that a Grade 50 ksi W-shaped beam, *Beam A*, can only fail in bending as soon as its moment capacity, M_{50ksi}, is exceeded. Now, say that we are interested in knowing if an identical beam, *Beam B*, would have failed under a similar moment level if this beam had been made[28] from Grade 36 ksi. To answer the above, *Beam B* is fabricated to be identical to *Beam A* (but from Grade 36 ksi) and tested.[29] Not surprisingly, we report that *Beam B* fails once its capacity reaches M_{36ksi} (which is naturally smaller than M_{50ksi}). This ends our investigation and answers our counterfactual question. A second question arises, did we *really* have to fabricate and test *Beam B*?

From an empirical view and *ceteris paribus*, beams made from stronger construction materials (in this case, Grade 50 ksi steel) tend to have larger moment capacities than identical beams made from inferior materials. In parallel, Equation 8.1 also states that since the yield strength of Grade 50 ksi is larger than the yield strength of Grade 36 ksi steel, then it is only logical that $M_{36ksi} < M_{50ksi}$. Now, the difference between $|M_{50ksi}$ and $M_{36ksi}|$ is the *direct causal effect* of altering the material grade on the moment capacity between the above two beams.

Let us extend the above example to investigate the difference in the load-deformation history of *Beams A* and *B*. Then, instead of relying on Equation 8.1 as a basis for comparison, we will need to either test these two beams or model them via a FE software (since we do not have a simple equation that enables us to plot the load-deformation history of beams directly).

Similar to the above example, the load-deformation history we record while testing *Beams A* and *B* also encode causality (i.e., represent *direct causal result* to *Beam A* being made from Grade 50 ksi and *Beam B* being made from Grade 36 ksi – see Figure 8.2). Thus, if the *causal* mechanism behind the load-deformation history of beams is decoded, then it is possible that we would be able to derive an expression that predicts the load-deformation history of beams without the need for physical tests or numerical simulations.[30]

---

24 A **confounder** is a variable that influences both the independent and dependent variables. Here is a simple and short example. There is a positive association between ice cream consumption and the purchase of sun blocking creams. Does this association mean that ice cream consumption causes individuals to buy sun blocking creams? No. Does the same association mean that sun blocking creams causes individuals to buy ice cream? Also, no. This likely means that there is a third lurking variable, summer temperature, which causes people to buy ice cream and sun blocking creams. The next few pages will re-visit this concept from a civil and environmental engineering perspective.

25 Let me dig a little deeper into this. Do you still remember our beams example (i.e., W16 × 36 vs. W18 × 40)? Well, here is a story. Let us assume that they both have the same grade of steel (A992) and come from the same mill. These two beams are pretty much identical, except for their size. This makes establishing causality a bit easier since our space of parameters is narrow. Now, let us assume that one beam comes from mill A and the other comes from mill B. In this instance, it is possible that the same beams might have differences in material properties (since each mill has its own material recipe). However, such differences are small enough to ignore. At the end of the day, a steel grade A992 is a steel grade A992. Unlike the simple comparison of the above beams, a comparison between two individuals comprises a much larger space of parameters that we cannot simply ignore but rather we need to acknowledge.

26 Please note that others may disagree with the notion of "need for manipulation" to realize causality [55]. I will leave this debate here. You may find such debate interesting!

27 Remember that **beams** are load bearing members designed to primarily resist lateral actions.

28 Side note: the terminology of "would have" and "had been made" implies the need for a counterfactual causal investigation.

29 Tommy Cousins and Brandon Ross would be very happy to be part of this testing!

30 I cannot promise you that we will be able to do so in this chapter, but it is possible that you will see a few papers on this soon!

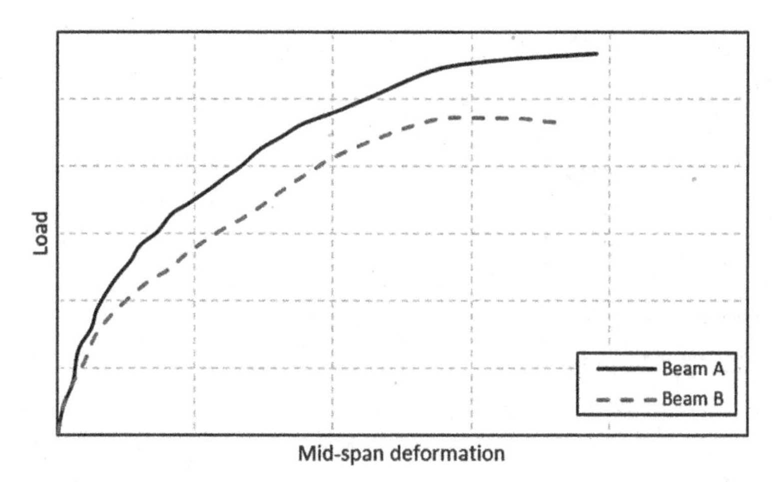

Figure 8.2 Samples of load-deformation curves for *Beam A* (solid) and *Beam B* (dashed).

8.4.2 Correlation and Causation

We have mentioned earlier that causality seeks to identify how phenomena come to be (or are caused by their causes). Unfortunately, more often than not, causality is tied to correlation.[31] It is critical to remember that correlation alone does *not* imply causation.[32]

A prime example of spurious[33] correlation can be seen in Figure 8.3a. This figure shows the result of a historical analysis of the number of failed buildings in the US between 1991 and 1996 [20, 21]. The same figure also shows the number of received PhDs in civil engineering (as per the US National Science Foundation [22]). As one can see, Figure 8.3a shows a strong positive correlation (+84.9%) between the above two events![34]

Let us now look at a larger picture between the above two events. Let us extend Figure 8.3a to a much larger timeframe. Plotting the number of failed buildings in the US against the number of received PhDs in civil engineering between 1989 and 2000 yields a weak correlation of + 35.5% (see Figure 8.3b).[35] Figure 8.3 is a prime example of spurious correlation, especially if one mistakenly implies a causal link between the number of failed buildings and the uprise in civil engineering PhD graduates. In reality, there could be many, many reasons that can explain the uprise and decline of both building failures and PhD graduates.

The above is but one example of why correlation alone does not necessitate causation. Put in other words, statistical relations do not exclusively define causal relations.[36] Figure 8.3 represents *marginal* associations as opposed to *conditional* associations[37] (and hence are unlikely to have a definite causal effect). While Figure 8.3 demonstrates shocking claims (yet, supported by authentic data), let us look at a more practical example.

Figure 8.4 presents another set of data aimed at reinforcing the difference between marginal associations and conditional associations from a practical civil engineering perspective by examining the relationship between the applied loads on fire-exposed reinforced concrete (RC) columns against the compressive strength of concrete in each corresponding column. Take a look at Figure 8.4a first.

---

31 Even so in our laboratory experiments!

32 However, causation may imply some form of correlation [1].

33 **Spurious correlation** occurs when two parameters appear causally related, but in reality they are not.

34 Does this mean that an uptake in PhDs leads to more failure? No, it does not.

35 This plot may be perceived to support the notion that more PhD graduates in civil engineering do not correlate with a rise in building failures. This interpretation is <u>also</u> not correct.

36 Tyler Vigen [https://www.tylervigen.com/spurious-correlations] has collected a dense of list of illustrations on spurious correlations.

37 **Marginal association**: two parameters are independent while ignoring a third. **Conditional association**: two parameters are independent given a third.

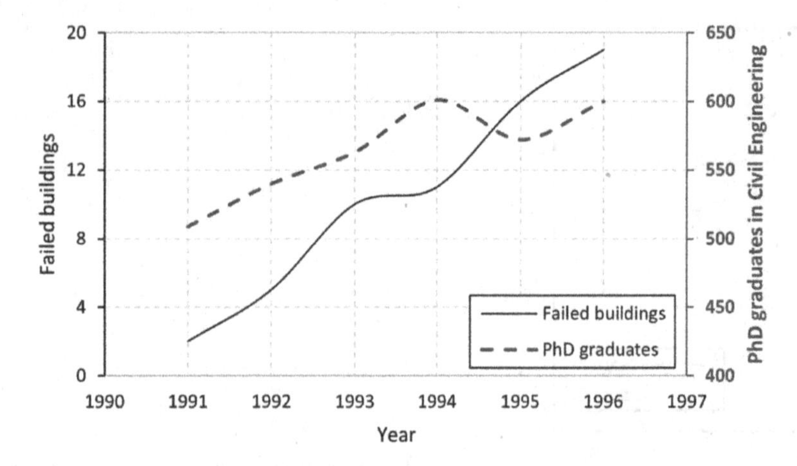

(a) Correlation between failed buildings in the US and the number of received PhDs in civil engineering within 1991--1996 (R = + 84.9%)

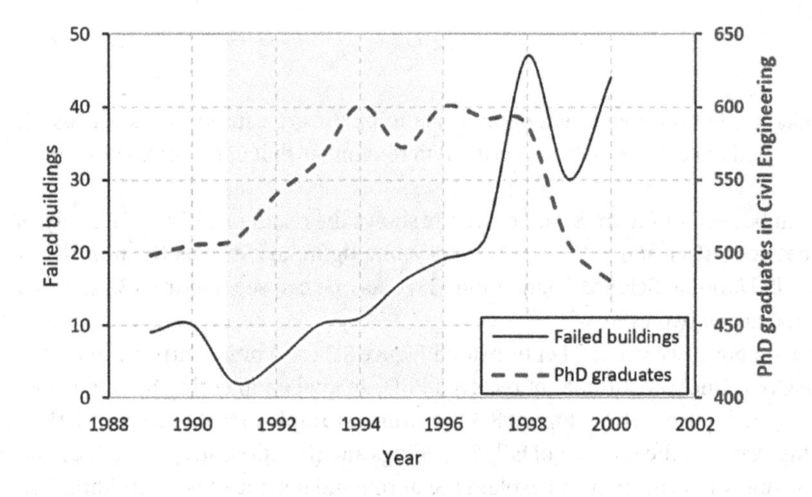

(b) Correlation between failed buildings in the US and the number of received PhDs in civil engineering between 1989 and 2000 (R = + 35.6%) – the shaded area is the one examined in Figure 8.3a

Figure 8.3 Examples of spurious correlations.

Figure 8.4a shows the *marginal* dependence[38] between the aforementioned two variables. As you can see, columns subjected to relatively lower loads (between 250 and 1500 kN) are also made from lower strength concrete (i.e., a reduction in loading (*X*) co-occurs with smaller compressive strength (*Y*)). The same figure also shows that columns made from high strength concrete (i.e., > 60 MPa) are also likely to be subjected to higher loads (>1500 kN). The above two observations imply that the depicted marginal dependence is positively linear.

However, let us now take these two variables and *condition* them on a third variable, *Z*. This variable represents the occurrence of fire induced spalling[39] – see Figure 8.3b. Now, you can clearly see that spalling seems to occur regardless of any marginal association between the applied loading and the strength of concrete. In other words, *X* is independent (⫫) of *Y* given the occurrence of spalling (*Z*).

38 Simply, marginal independence implies that knowledge of *Y*'s value does not affect our belief in the value of *X*. Also, conditional independence implies that knowledge of *Y*'s value does not affect our belief in the value of *X*, given a value of *Z*.

39 Please do me a favor and draw a mental horizontal line at 60 MPa. Now, one may be tempted to focus on the top half of this figure to notice the high occurrences of spalling in columns of high strength concrete. Please do not forget to also focus on the bottom half of the same figure to see the large number of specimens that spalled.

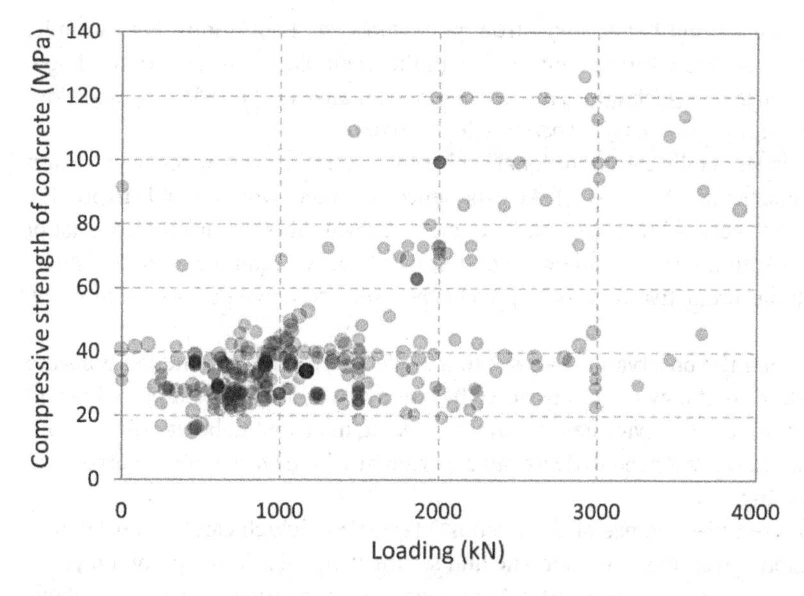

(a) Marginal dependence

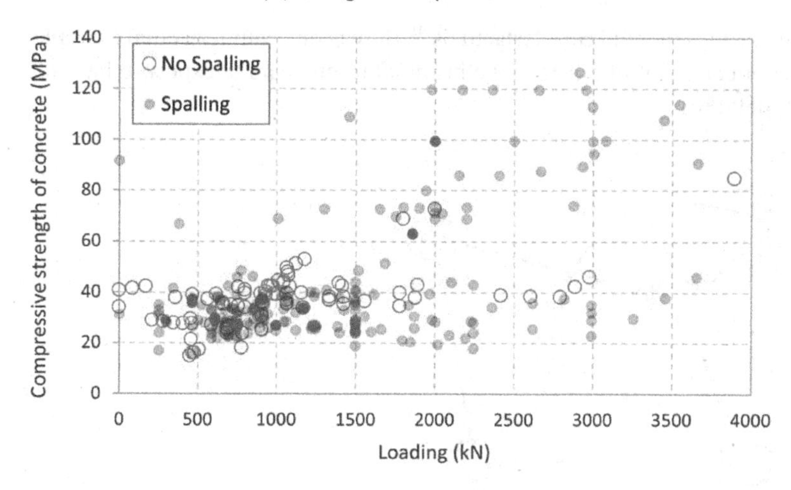

(b) Conditional independence

Figure 8.4 Marginal vs. conditional association between the compressive strength of concrete and loading in fire-exposed RC columns.

8.4.3 The Causal Rungs

At the time of drafting this book, most supervised ML models are applied in a similar manner to that shown in Figure 8.5a, where the independent variables (X_1, X_2, X_3, X_4) are used to predict Y. In contrast, Y is likely to happen via a DGP (see Figure 8.5b for a possible DGP). Thus, to accurately predict Y, we need to identify the correct DGP.

A good start to causality is to discuss how this approach differs from statistical and empirical methods.[40] I will supplement this contrast with a discussion on the ladder of causation as pioneered by Pearl[41] [17]. In this system, Pearl [17] identifies three levels (or rungs) to causality.

---

40 Here is what is happening at this moment. I know this book is about ML in civil and environmental engineering, however, this particular section of the book could be a bit tricky to visualize and hence I am confining my discussion and illustrations to examples covering beams (again! My bad – think of it this way. Beams are there in statics, mechanics, etc. not just in structural engineering courses). That does not mean that this discussion does not apply to other examples/sub-fields – trust me, it does! This simply means that I found these examples to be eas-*ier* to visualize.Also, I will also limit the amount of mathematical background behind causality to maintain a reasonable chapter size since this chapter is likely to push the limit I had in set I mind. As always, please feel free to navigate the following resources on mathematical conditions and background required for establishing causality [10, 16, 56–58]. A discussion on such items (including *do*-calculus, conditional and interventional distributions, etc.) can simply be too advanced for one chapter on causality.

41 In my eyes, graphs are easier to get the point across. For a discussion on causality from the lens of the potential outcomes (PO) approach, please refer to the following [16, 18]. For brevity, we will discuss the PO approach in a later section.

The first rung builds upon the observations[42] we *see* and hence stems from pure *statistical* associations. For example, uncoated metallic beams experience some level of corrosion during their service. In this example, our observations of how metallic beams are likely to corrode are from a sense of association. This association can answer the following question: *What is the expected corrosion level an uncoated metallic beam could experience in five years?*

The above question can be mathematically represented by the conditional probability, P, of the occurrence of corrosion (y), given that we observed (|) an uncoated metallic beam (x), or $P(y \mid x)$. An experienced engineer would be able to provide us with a possibly accurate estimation of the expected level of corrosion if we are to provide this engineer with a list of information about the grade and type of metal, surrounding environmental conditions, etc. As you can see, *statistical* inference, empirical analysis, and ML can help us build predictive models to predict the expected degree of corrosion solely from data.[43]

The second rung of causality extends further than the observations we see to include *interventions*.[44] Interventions are tasks that we *do* to answer a separate set of questions that may not be answered by our observation. For example, what will happen to the metallic beam *if* we change one or more of its parameters (say, from being uncoated to being painted with an anti-corrosion coating)? To answer this question, we will need to devise an experiment where we actually intervene on this beam and paint it with an anti-corrosion coating.

Mathematically, this interventional question resembles the use of *do*-calculus[45] $P(y|do(x))$, which can be translated as the probability of event y (occurrence of corrosion) given that we intervene and set the value of x (addition of a layer of anti-corrosion coating) and observe the same event (five years of service).[46] As one can see, an accurate answer to the above question cannot be answered from data alone.[47]

The third and final rung goes beyond the first two rungs to build counterfactuals.[48] Here is one counterfactual example, what would have happened to the expected level of corrosion of our metallic beam after ten years of exposure had this particular girder been fabricated from a different metal B?

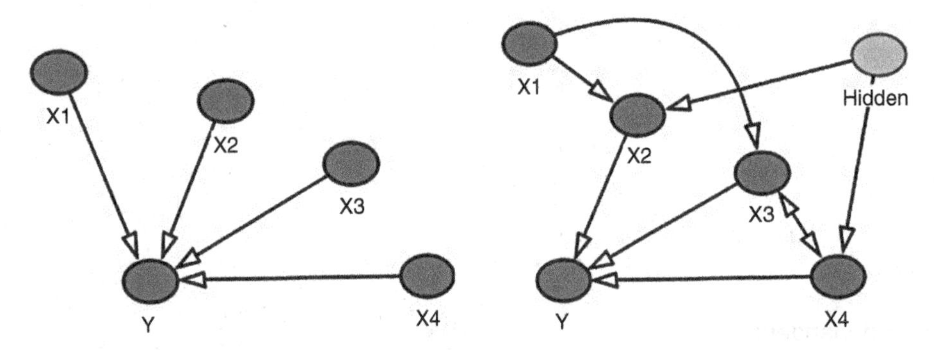

Figure 8.5 Difference between a traditional ML model (left) and a causal model (right).

---

42 **Observations**: Things we see.

43 Just as many machine learning models do!

44 **Interventions**: Things we can do.

45 **Do-calculus**: an axiomatic mathematical system that simulates physical interventions by deleting certain functions from a model and replacing them with a constant while keeping the rest of the model unchanged [59]. This can be used to replace probability formulas containing the *do* operator with ordinary conditional probabilities [60].

46 There is a subtle difference between $P(y \mid x)$ and $P(y|do(x))$ wherein the former describes which value Y would likely take on when X happened to be x (i.e., observational distribution), while the latter describes the values of y when X is set to x (i.e., interventional distribution). These are not equal since the coated girder is likely to experience minimal corrosion.

47 Unless we have previous results from an earlier experiment or, from a new experiment in which we test (i.e., intervene) on an identical beam to the one in question.

48 **Counterfactuals**: Things we can imagine.

Mathematically, this question could be translated to $P(y_x|x', y')$; the probability that event $Y = y$ would be observed had X been x, given that we observed X to be x' and Y to be y'. This question is not empirically testable since we cannot test the *exact same metallic beam* twice, especially if it is fabricated from two different materials and examined in two different service durations.[49] So, how can we answer this question?

A plausible solution would be to carry out experiments on identical metallic beams (where one is fabricated from metal A and the other from metal B). Such an experiment is likely to give us a realistic answer to our counterfactual question. Fundamentally, this may answer our counterfactual question.[50] Yet, one can think of many examples where running an experiment may not be possible or ethical (e.g., What would have happened to the occupants of skyrise, *X*, had the earthquake been twice as strong? What would have happened to the residents of a community, *T*, had its evacuation halted for two more hours? What would have happened to air quality in city, *C*, had the wildfire lasted two more weeks? etc.). These examples, among many other that could have been drawn,[51] emphasize the need for a causal approach.

As you can see, our domain, and just like many others, incorporates all three rungs. Yet, we tend to heavily prioritize the first rung and a bit from the second rung (i.e., testing and simulation). In all cases, little attention has been paid to identifying DGPs from a graphical or PO point of view. I cannot, nor wish to, speculate on the above. However, let me share these two thoughts with you. It is quite possible that: 1) identifying DGPs is not only complex but is ambitious as well, and 2) realizing DGPs may not be necessary since methods spanning from the first two rungs are suitable to provide us with appropriate solutions that simply work and overcome our problems.[52]

8.4.4 Regression and Causation[53]

Ask yourself the following question, why are we interested in identifying a DGP? If looking at Figure 8.4 does not answer this question, then allow me to answer it. We are interested in identifying DGP because a DGP can be thought of as equivalent to a model, tool, chart, or equation that represents how the governing factors interact to generate the phenomenon of interest. This brings a question of how a causal model differs from a typical statistical (i.e., regression) model.

First, and unlike a regression-based predictive tool/equation, a causal model goes beyond statistical inference and correlation to establish evidence for how causes generate effects [23, 24]. Second, regression is used to estimate the *influence* of dependent variables on an independent variable, *Y*. Regression is a "purely statistical form of explanation." [25].

By defining one variable as *X* and another as a *Y*, then the direction of the relationship is explicitly stated. However, the direction of each relationship should be justifiable and defensible.[54] Remember that regression can be comfortably used to *predict* an outcome of interest, *Y*, given a set of parameters (or regressors). A prediction in this manner does not imply nor indicate that the regressors, *X's*, are causes of *Y*.[55]

---

49 I hope by now you have noticed that counterfactuals answer individual-unit questions (while interventions answer population-level questions [61]).

50 Philosophically speaking, this may not help us solve our query since we did not actually test the exact same beam under different conditions. Let us leave this here, for now.

51 Feel free to re-read this sentence a couple of times!

52 Evidently, you can stop reading this chapter here! I would still encourage you to continue to finish this chapter since we are more than half-way through.

53 Please visit the Appendix for a short companion discussion.

54 In a typical ML model, say from any of the previous chapters, we know *Y* beforehand as *Y* is a response or an outcome we are interested in. The trouble comes in assigning the *X*'s. What are these *X*'s? How do we know we have them? How do we know we have all of them? Or, some of them? What about if we cannot collect data on an *X*? Or, worse, what if we do not know that a particular *X* exists? Do not panic yet! Here is, and just like any other investigation method, where working assumptions become valuable! In hindsight, assigning a series of regressors, X_1, X_2, X_3, to predict *Y* does not imply that these regressors are causes of *Y*. In the event that a thorough domain knowledge exists on such regressors and that these regressors satisfy causal principles, then a regression analysis can indeed turn causal. Say that an engineer notices that heavily loaded columns (*Y*) are often built from materials of high strength (*X*). Then, in order for this engineer to predict the magnitude of loading a column can carry, this engineer can simply regress the loading on the material strength. The resulting equation ($Y = f(X)$) can probably estimate the expected loading a column of some material grade can carry with good accuracy – especially if supplemented with enough data points. Let us remember two things, this equation does not imply that higher loading capacity is *caused* by the presence of high material strength nor that the presence of higher material strength is *caused* by higher loading! This equation simply presents a statistical explanation. Remember that it is quite logical to say that having higher material strength is one reason of having higher loading capacity.

55 Even if such assignment may arrive from domain knowledge. Further, assigning such regressors is hardly associated with checking of confounders or common causes of the *X's* either.

Let us reminisce a bit about Chapter 3. In linear regression, a unit change in Y can be represented via a coefficient that is tied to a corresponding X (while keeping all other variables constant). Take a look at Equation 8.3:

$$Y = 3X + 2 \tag{8.3}$$

In this equation, 2 is the y-intercept and refers to the estimated value of Y when X is zero. The coefficient 3 is the regression coefficient and estimates the increase in Y per one unit increase in X.[56]

It is expected that Equation 8.3 is bound to be flawed if there is some unmeasured common cause between the X's themselves and/or with Y. A traditional solution to remedy the above is to incorporate additional dependent variables. This solution has been shown to cause instability (since the newly added variables may be confounders, may have common causes, or may, in fact, be an effect as opposed to a cause of Y) [26]. This is an elemental difference between regression and causation.

Another key difference pertains to the exclusive nature of the time hierarchy. From this lens, some variables may naturally occur before, together, or after the occurrence of others. For example, let us talk about loading and cracking.

The application of loading on a new beam may cause it to crack. Similarly, poorly maintained load bearing elements, which already suffer from aging and cracks, can further crack upon an increase in loading. Finally, creep effects induced by long-term loading can accelerate settling effects which can cause cracking. Thus, incorporating this natural process of time hierarchy is elemental to why causal models can be better than regression models in the above examples.

If you would like to *remember* one thing about this section, then please remember that regression projects the values of Y given its X's, while causality aims to discover/infer the relationship between any two individual variables.

8.4.5 Causal Discovery and Causal Inference

How can we establish causality? Well, causality can be established through causal models.[57] Causal models are models derived via causal mechanisms. There are two primary causal mechanisms, *causal discovery*, and *causal inference*.

Causal discovery[58] hopes to discover the underlying structure of causal relations between variables pertaining to a DGP. Let us re-visit *Beam A*. Say that this beam is simply supported, has a depth, D, a width, W, a span, L, and deflects by a magnitude of X due to loading, P. If we would like to observe how the deflection changes in response to interventions on D, W, L, and P, then we can design and carry out experiments to collect a series of observations pertaining specifically to each of our interventions.

Analyzing such observations allows us to develop a tool (say, an expression/equation) to predict the deflection[59] of *Beam A* without having to carry out future experiments. The same tool can help us identify a suitable combination of interventions that limit the deformation of *Beam A* to a predefined limit.[60]

Causal inference, on the other hand, is a process where we draw conclusion(s) that a particular intervention caused an effect (or outcome) [27]. Simply put, causal inference infers how interventions, treatments, and manipulations systematically alter the observations. For example, one of our interventions in *Beam A* is to shorten L and observe what happens to X (while keeping all the other parameters unchanged). At the end of this experiment, we observe that $X_{\textit{Post intervening on L}} < X_{\textit{Pre-intervening on L}}$, then we can safely infer that shortening L has caused the deformation level of *Beam A* to be smaller.

A *causal structure* can be represented mathematically by causal models. A causal model spells out the probabilistic (in) dependence of variables and the effects of interventions [28]. Such a model can be represented by a set of equations (namely, structural equation models[61] (SEMs)) and/or graphs (i.e., Directed Acyclic Graphs[62] (DAGs)). An SEM represents causal relationships between the involved parameters in an equational[63] format, while a DAG visually represents the causal structure of the DGP. I will focus on DAGs.

---

56 More formally, $Y = \beta_0 + \beta_1 X + \varepsilon$, where β_0 is the y-intercept, β_1 is the regression coefficient, and ε is the random error component. This is a good time to re-visit Chapter 3!

57 **Causal models**: models that describe (mathematically or otherwise) the possible DGP responsible for producing the phenomenon.

58 Can also be referred to as *causal structure search*.

59 aka. The GDP of deflection in *Beam A*.

60 You can think of this as the *maximum allowable deflection* (or *drift* in case we substitute the example of *Beam A* to *Column A*).

61 **Structural equation models**: a set of techniques used to measure and analyze the relationships of observed and latent variables while accounting for measurement error. Unlike regression, SEM can examine a set of relationships between one or more continuous or discrete independent variables and one or more continuous or discrete dependent variables [62].

62 A graph is a is a mathematical object that consists of nodes and edges. A DAG is a graph with directed edges. An example is shown in Figure 8.5.

63 In a parametric or non-parametric form.

A DAG contains edged arrows that flow in one direction (e.g., $A \rightarrow B$, meaning A causes B) and does not allow for circular causation (see Figure 8.6). In Figure 8.6, A is the direct cause of D. A also is the direct cause of B, and C. Similarly, B is a direct cause of C. In addition, A also has an indirect effect on C, given its role on B.

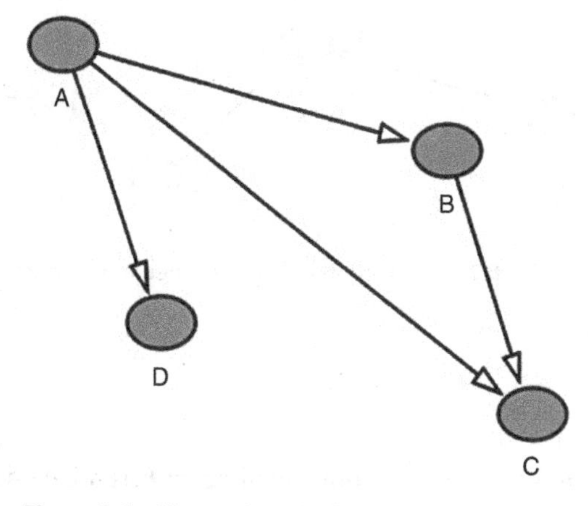

Figure 8.6 Illustration of a DAG.

A node, say B, is called a descendent of A (and so on). In the Figure 8.6 DAG, $A \rightarrow B \rightarrow C$ is called a path. DAGs are interpreted causally at the interventional and counterfactual levels, while our associational levels are used to describe conditional independencies between variables.[64] Other than direct causes and indirect causes, there are several ways two nodes can be tied together. Three main relationships are of interest to causal analysis: *mediators*, *moderators*, and *colliders*.[65]

A mediator[66] is a pattern in the form of $X \rightarrow Z \rightarrow Y$; effectively implying that X causally affects Y *through* Z.[67] Here is an example, fire (X) generates heat (Z), which in turn degrades the sectional capacity of a load bearing member (Y).

Moderators can change the *size* (or *direction*) of the relationship between other variables. For example, the magnitude of deformation (Y) in a given load bearing member can be amplified during fire conditions. In this example, the adverse influence of fire (Z) can lead a beam to undergo larger deformations under the same level of loading (X). Since the thermally-induced deformations are not observed under ambient conditions from X, then Z *amplifies* the deformation (Y) of a load bearing member.

Reversing the arrow from Z to X changes a moderator to a *confounder*.[68] In this formation, $X \leftarrow Z \rightarrow Y$, Z causally affects X and Y; and hence Z is a confounder that stirs *non-causal* association between X and Y. As mentioned earlier, confounders are not easy to investigate since they can be variables that we do not observe, do not often consider, do not know that it exists, nor measure. Here is one possible example of a confounder: seismically active regions (Z) witness low structural failures (X) than other regions despite having a large number of structures (Y).[69] In this instance, the confounder Z statistically correlates X and Y despite the lack of a direct causal link between X and Y.[70]

A *collider* pattern can be identified by flipping the arrows of moderators toward Z. Now, Z is no longer a shared cause of X and Y but rather a *common effect*. For example, harsh environmental conditions (X) and poor maintenance (Y) can lead to structural issues (Z) such as cracking and corrosion; thus, harsh environmental conditions (X) $\rightarrow$ structural issues (Z) $\leftarrow$ poor maintenance (Y). If we are to condition on Z,[71] then the association between X and Y in the mediator and confounder is blocked. However, doing the same on a collider induces non-causal association.

---

64 There is more to DAGs (and SEMs). Please review [57, 63].

65 From here on, you might notice a shift in my use of symbols as I depart from using A, B, C, etc. I only do this to remain true to the terminology used in causality; hence, X denotes a cause, Y is an outcome (target), and Z is a third variable that could be a mediator, moderator, or a collider.

66 Also called, *Chain*.

67 Similarly, isolating $A \rightarrow B \rightarrow C$ from Figure 8.6 into a new DAG can be considered a chain.

68 aka. *Common cause* or *Fork*.

69 low structural failures (X) $\leftarrow$ seismically active regions (Z) $\rightarrow$ high number of structures (Y).

70 An explanation would be that a seismically active region might home a metropolitan. On the one hand, such a metropolitan would indeed house a large population which requires the development of many structures. On the other hand, structures of seismically active regions are often required to comply with codal provisions that enforce additional detailing and requirement to mitigate damage. The result is, a metropolitan houses large number of structures with enforced additional provisions to mitigate seismic hazards and hence this metropolitan does not suffer from major structural failures!

71 Building on the above discussion and referring to Figure 8.7 while doing so may raise the question of how to distinguish a confounder from a collider (from the data alone or when domain knowledge is limited). This task can be completed via the *back-door* and *front-door adjustments* [16]. In the confounding formation, $X \leftarrow Z \rightarrow Y$, the back-door adjustment states that 1) no node in Z is a descendent of X, and 2) any path between X and Y that begins with an arrow into X (known as a back-door path) is blocked by Z, then controlling for Z blocks all non-causal paths between X and Y. If the confounder is *unobservable* or *hypothetical*, then the front-door adjustment can be applied. In this process, we add a new variable that we may assume is not caused directly by the confounder. Now, we can use the aforementioned back-door adjustment to estimate the effect of the new parameter on the outcome. For an exhaustive description of the above two adjustments, please refer to [64, 65].

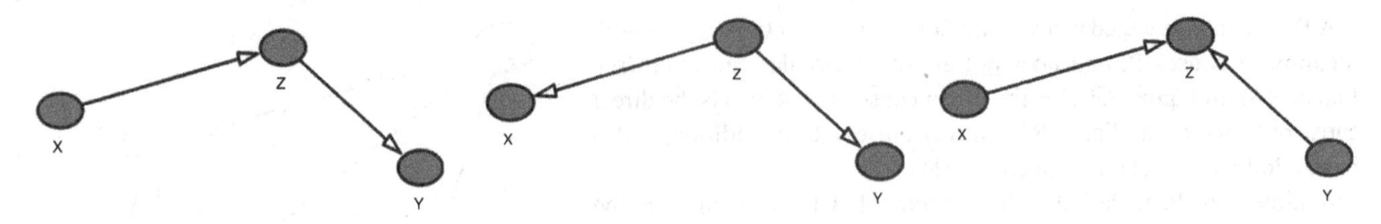

Figure 8.7 Illustration of DAGs for mediators, confounders, and colliders (from left to right).

8.4.6 Assumptions Required to Establish Causality

There are several fundamental assumptions that are required to establish causal discovery and causal inference. Allow me to briefly describe these here.[72]

- The *causal Markov assumption*. This assumption states that a variable, X, is independent of each other variable (except X's effects) conditional on its direct causes. Take a look at the left DAG in Figure 8.7. In this figure, X is independent of Y conditional on Z.
  - The *d-separation* criterion is a close companion to the Markovian assumption [10]. This criterion establishes whether X is independent of Y, given Z (i.e., $X \perp\!\!\!\perp Y \mid Z$), by associating *independence* with the *separation* of variables in a causal graph. For example, a path is d-separated if it contains 1) a chain or a fork such that the middle node is not in Z, or 2) a collider such that the middle node is not in Z, nor its descendants are in Z.
- The *causal faithfulness assumption* states that any population produced by a causal graph has the independence relations obtained by applying d-separation criterion, and hence eliminates all cases of unfaithfulness.[73] The causal Markov and faithfulness assumptions suggest that d-separation relationships in a causal DAG have a one-to-one correspondence with conditional independencies in the distribution of that particular graph [29].
- The *causal sufficiency* assumption[74] refers to the absence of hidden or latent parameters, i.e., our data contains measurements on all of the common causes of the selected variables [30].

8.4.7 Causal Graphs and Graphical Methods

So far, we have discussed the principles and the assumptions needed to identify causality and DGPs. We also know that a causal model has a causal structure to represent the DGP. While we may not know the true structure of a true DGP, we are likely to have some domain knowledge[75] and data that can get us close enough to a possible causal structure for a DGP[76].

If you ask me, this sounds like a good spot to move into the graphical methods that can help us arrive at causal graphs and, by extension, causal models.[77]

Graphical methods span a variety of granularities – see Figure 8.8. For example, a *directed graph (DG)* is a graph without arrows and hence allows loop feedback. A *partially directed graph (PDG)* is a type of graph that may contain both directed and undirected edges. On the other hand, a DAG assumes acyclicity.

---

72 A rich example and corresponding mathematical details can be found elsewhere [25, 28, 30, 66].

73 Independences that are not a consequence of the causal Markov condition or d-separation. An example of unfaithfulness is where X is a cause of Y and Y is a cause of Z, but X and Z are independent of each other [52].

74 Other assumptions also exist such as Gaussianity of the noise distribution, one or several experimental settings, and linearity/nonlinearity, acyclicity – see [29].

75 **Domain knowledge**: information and concepts accepted by a domain.

76 Such a DGP may in fact turn out to be the ground truth or one that we are comfortable with labeling as a causal structure pending a series of assumptions arising from domain knowledge.

77 A thorough review on graphical models can be found in [52].

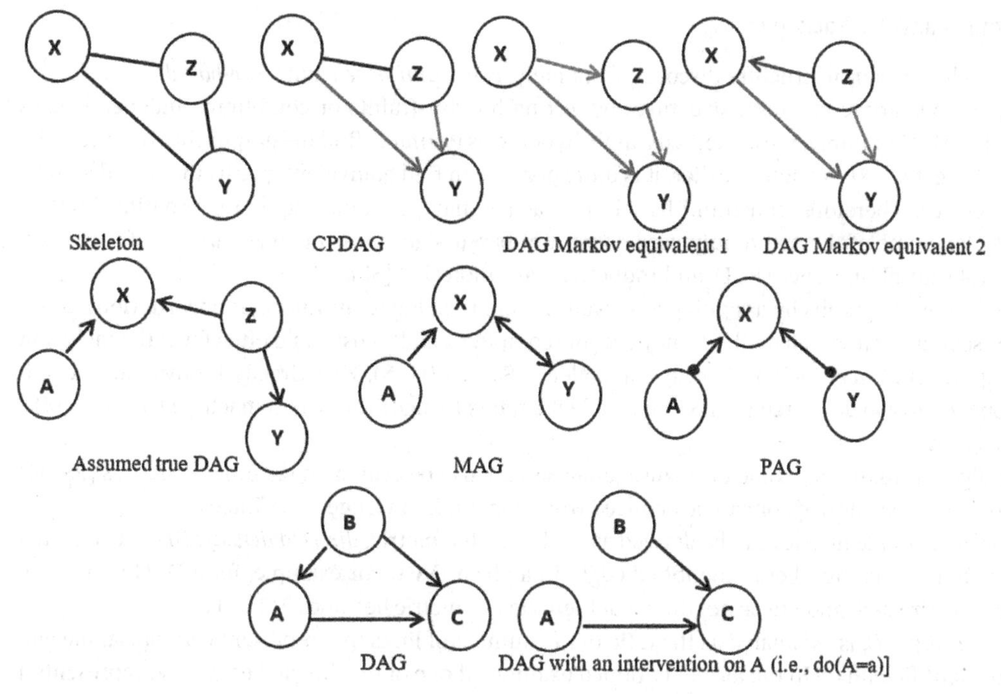

Figure 8.8 Graphs.

Truthfully, a causal DAG may not be easily, if at all, identifiable, but rather a causal search method returns a set of equivalent DAGs. Such graphs are referred to as Markov equivalence.[78] For example, $X \rightarrow Z \rightarrow Y$, $X \leftarrow Z \rightarrow Y$, and $X \leftarrow Z \leftarrow Y$ home the same conditional independence (i.e., $X \perp\!\!\!\perp Y \mid Z$). A *Completed Partially Directed Acyclic Graph (CPDAG)* can be used to represent the above set of Markov equivalent graphs. A CPDAG has a directed edge if there is only one graph in the Markov equivalence class with an edge in that direction.[79]

Unobserved variables or confounders can be represented with bidirected edges[80] in an *Acyclic Directed Mixed Graph (ADMG). Maximal Ancestral Graph (MAG)* has the capability to represent the same features of ADMGs and selection bias.[81] Figure 8.8 provides a visual illustration of the aforenoted types of graphs.

The presented graphs show a number of links (i.e., relationships). Some of the primary relationships are described below.

$X \rightarrow Y$ where X is a cause of Y

$X - Y$ where X is a cause of Y or vice versa

$X \leftrightarrow Y$ where there is an unmeasured confounder

$X \circ\!\!\rightarrow Y$ either X is a cause of Y, or there is a confounder

$X \circ\!\!\rightarrow Y$ either X is a cause of Y, and/or vice versa, and/or there exists a confounder

78 Encode the same set of d-separation relationships.

79 If there is uncertainty about the direction, this particular edge is left undirected.

80 $X \leftrightarrow Y$; meaning there exists a possible confounder between X and Y.

81 Preferential omission of data points [67].

8.4.8 Causal Search Methods and ML Packages

There are two primary approaches to causal structure discovery,[82] namely, *constraint-based* and *score-based.*[83]

Constraint-based search methods arrive at causal structures by testing for constraints or conditional independencies while constructing a graph (i.e., DAG). Simply, these methods apply hypothesis testing to find the graphs that represent the independence relationships correctly [28]. As noted earlier, it is quite possible to find equivalent graphs that fulfill a given set of conditional independencies. Therefore, constraint-based approaches may generate graphs representing Markov equivalence (CDPAGs or PAGs). Some of the commonly used constraint-based algorithms include variants of the Peter-Clark (PC) algorithm [31], Fast Causal Inference (FCI), and Inductive Causation (IC) [32].

Score-based search methods identify graphs by assigning a relevance score (i.e., Bayesian Information Criterion). Since each candidate graph is to be scored, such search methods may become computationally costly and often favor the inclusion of greedy heuristics. Such approaches may include Greedy Equivalence Search (GES), Fast Greedy Equivalence Search (FGES), Greedy interventional equivalence search (GIES), etc. Table 8.2 refers to algorithmic and machine learning (ML) packages for various causal discovery tools.

Similar to traditional ML algorithms, the outcome of a causal analysis can also be evaluated via unique metrics [33, 34] – especially if the ground truth DGP is known[84] or can be deduced from domain knowledge – see Table 8.3.

In the latter, two additional and simple metrics can be derived as well. The first metric, *directed density, DD,* is calculated as the ratio of directed edges to the total number of identified edges in a given DAG. For example, for a DAG with three identified links wherein two are directed, and one is not directed, then the *DD* metric becomes $2/3 = 0.67$.

The second metric, *causal density, CD,* is calculated as the ratio of the number of links that represent true causal mechanisms to the total number of identified links. In the aforementioned example, if one of the directed edges also represents a true causal mechanism, then the *CD* metric also turns 0.33.

It should be noted that, unlike the *DD*, the *CD* metric is not only applicable to directed edges.

8.4.9 Causal Inference and ML Packages

To be able to infer, one must have a *causal structure* that describes the phenomenon we are interested in. Going back to our example of the uncoated metallic beam. Say that we were able to discover the DGP responsible for the corrosion of this beam. Then, using this DGP, we can answer questions about how altering the features of this beam would influence the magnitude of expected corrosion.[85]

Unlike causal discovery, we can also causally infer aspects of a certain phenomenon for which we do not know its DGP. For this to work, we will need data (i.e., observations). If we examine this data via a series of causal assumptions, then we can causally infer the corresponding causal effects. These assumptions and rules are discussed herein.[86]

In the Potential Outcome (PO) approach, pioneered by Rubin [16], causal effects can be estimated by comparing the *potential outcomes* of a given intervention. In reality, only one outcome can be observed since we cannot intervene on a unit and then un-intervene on the same unit at the same time.[87] In the PO approach, the potential outcomes are noted with Y. Y_0 notes the potential outcome for the untreated/benchmarked unit(s), and Y_1 notes the potential outcome for the treated (or intervened upon unit(s)). It is common to denote the intervention or treatment with a T.

Let us go over an example. Say, we are interested in causally inferring the benefits on flexural capacity of fabricating RC beams using ultra-high performance concrete (UHPC) instead of normal strength concrete (NSC). One way to carry out this investigation is by casting two groups of beams. These beams have identical features, except that *Group A* $[A_{0,1}, A_{0,2}, A_{0,3}]$ is made from NSC (Y_0), and *Group B* $[B_{0,1}, B_{0,2}, B_{0,3}]$ is made from UHPC (Y_1). We test all beams under bending and collect our observations. This would be a good time to causally examine our observations to infer an answer(s) to our question.

---

82 Vowels et al. [68] and Heinze-Deml [29] identify additional approaches such as causal association rules, causal forest, and causal networks.

83 I am only covering key algorithms on each front. Others can be found elsewhere [29, 68, 69].

84 When the ground truth DGP is not known, the performance of the causal discovery algorithm is evaluated on how well the causal structure predicts the phenomenon on hand.

85 More interestingly, we hope to infer the influence of a variable that we have not explored in an experiment!

86 I promised to use PO. Here we will see how we can use this approach. Please note that the PO and Pearlian approaches present equivalent concepts [18].

87 I hope you still remember that causal inference is *a missing data problem*, for which we can only observe one/some but not all of the outcomes.

Table 8.2 Algorithmic and ML packages for causal discovery (full description of each package and incorporate algorithms can be found in its original source).

Packages for causal discovery	
CausalNex	[https://causalnex.readthedocs.io/en/latest/index.html] A Python library that combines ML and domain expertise for causal reasoning.
Pcalg	[https://cran.r-project.org/web/packages/pcalg/index.html] An R package for causal discovery that includes: *Constraint-based structure learning algorithms*: • PC • FCI • Really fast causal inference (RFCI) • Greedy interventional equivalence search (GIES) *Score-based structure learning algorithms*: • GES • Greedy interventional equivalence search (GIES) *Hybrid structure learning algorithms*:Adaptively restricted GES (ARGES)
bnlearn	[https://www.bnlearn.com] • A Python package for causal discovery that includes: *Constraint-based structure learning algorithms*: • PC (the *stable* version) • Grow-Shrink (GS) • Incremental Association Markov Blanket (IAMB) • Fast Incremental Association (Fast-IAMB) *Score-based structure learning algorithms*: • Hill Climbing (HC) • Tabu Search (Tabu) *Hybrid structure learning algorithms*: • Max-Min Hill Climbing (MMHC) • Hybrid HPC (H2PC) • General 2-Phase Restricted Maximization (RSMAX2) *Local discovery algorithms*: • Chow-Liu • ARACNE *Bayesian network classifiers*: • Naive Bayes • Tree-Augmented naive Bayes (TAN)
Tetrad ***Py-causal*** ***R-causal***	[https://www.ccd.pitt.edu/data-science] Software, Python, and R packages created by the Center for Causal Discovery at Pittsburgh University.
Causal discovery toolbox	[https://fentechsolutions.github.io/CausalDiscoveryToolbox/html/data.html] Python and R packages for causal discovery.
TIGRAMITE	[https://github.com/jakobrunge/tigramite] A Python package for causal discovery for time series datasets.

The simplest approach is to average the results of the flexural capacities of *Group A* and do the same for *Group B* and compare the average (or expectation ($\mathbb{E}$)) from each group. Mathematically, this results in the following:

The difference in average (expectation) between

$$\text{Group A and Group B} = \mathbb{E}\ [Y_1 - Y_0] \tag{8.4}$$

Similarly, one can use the Pearlian approach (i.e., *do*-calculus) by applying the do-operator (see Equation 8.5).[88]

$$[\,Y \mid do\,(T = 1)] - \mathbb{E}\ [\,Y \mid do(\,T = 0)\,] \tag{8.5}$$

Both Equations 8.4 and 8.5 estimate the Average Treatment Effect (ATE).

---

88 Note that $p(Y = y \mid do(X = x)) = p_m(Y = y \mid X = x) = \sum(Y = y \mid X = x)p(Z = z)$, where p_m is the manipulated distribution arising from our intervention. A review on do-calculus can be found in [54], and a sample solved example is shown in Example no. 5.

Table 8.3 Metrics used for causal discovery analysis [28, 29].

Metric	Description
Missing or extra edges	Number of edges that are missing or present in the original model but not in the generated one.
Incorrect adjacencies	Number of undirected edges that are present in the generated model but not in the original one.
Incorrect and correct directed edges	Number of directed edges that are missing are present in the generated model that were correctly directed.
Structural hamming distance	Sum of missing edges, extra edges, and incorrectly directed edges.
Adjacency precision	Correctly predicted adjacencies/predicted adjacencies.
Adjacency recall	Correctly predicted adjacencies/true adjacencies.
Arrowhead precision	Correctly predicted arrowheads/predicted arrowheads.
Arrowhead recall	Correctly predicted arrowheads/true arrowheads.
Area-above-curve	AOC = 1 − AUC, where AUC (area-under-curve) measures the area below the graph of false positive rate and true positive rate.

That is not the only quantity that we can infer. In fact, there are additional causal effects, such as the conditional average treatment effect (CATE)[89] and the Average Treatment Effect on Treated (ATT),[90] as well as the Individual Treatment Effect of unit (ITE),[91] etc.

To find the ATT, it is only logical to only examine the beams in *Group B*.

To continue our example in order to estimate ATT, we would want to go back to *Group B* only. Then, ATT would be the $\mathbb{E}\left[Y_1 - Y_0\right]T = 1$[92].

There are a number of ML packages that can be used in causal inference investigations. These are listed in Table 8.4.

8.4.10 Causal Approach

In the following pages, I will be presenting an approach to causal discovery and causal inference that I will be adopting in the following examples pertaining to this chapter.

Let us start by presenting the approach.[93]

It is only natural for the proposed causal approach to comprise three steps. Causal discovery [to uncover the underlying structure through a causal graph, say a DAG[94]], causal inference [to infer how would the observed phenomenon/output change by intervening[95]], and, when possible, examine causal findings against a companion or independent analysis (say, from ML, statistical analysis, domain knowledge, etc.). Figure 8.9 demonstrates the proposed approach in more detail.

89 Simply, an ATE conditioned on a subset of the population.

90 Which answers the following question, what is the expected causal effect of the treatment for individuals in the treatment group?

91 Which answers the following question, what is the expected causal effect of the treatment on the outcome of a specific unit? It is obvious that ITE may not be easily nor possibly estimated from this example since we can only test each exact beam once. A workaround would be to estimate the difference in expectation between each identical beam to get a close estimate of ITE.

92 The PO approach extends to cover a wide range of applications and conditions (i.e., propensity scoring, confoundness, missing data, matching, etc.). Please feel free to review [19].

93 Or, mental map!

94 We will leverage domain knowledge too!

95 Remember that an intervention equates to *setting* $X = x$ as opposed to *observing* $X = x$. The former relates to causation (what is the fire resistance of a RC slab if its thickness is *increased* to 200 mm?) and the latter to prediction (what is the fire resistance of the same slab *given* it has a width of 200 mm?).

Table 8.4 Algorithmic and ML packages for causal inference (full description of each package and incorporate algorithms can be found in its original source).

Packages for causal inference	
DoWhy	[https://microsoft.github.io/dowhy] A Python package for causal thinking and analysis that provides a systematic four-step interface for causal inference for modeling and validating causal assumptions.
EconML	[https://www.microsoft.com/en-us/research/project/econml] A Python package that can be used to estimate individualized causal responses from observational or experimental data.
CausalGAM	[https://cran.r-project.org/web/packages/CausalGAM/index.html] An R package that implements estimators for average treatment effects and a standard regression estimator through generalized additive models.
FLAME	[https://cran.r-project.org/web/packages/FLAME/index.html] An R package for causal inference.
CausalNex	[https://causalnex.readthedocs.io/en/latest] A Python package for causal reasoning (discovery and inference).
Causal ML	[https://causalml.readthedocs.io/en/latest/about.html] A Python package for causal inference.

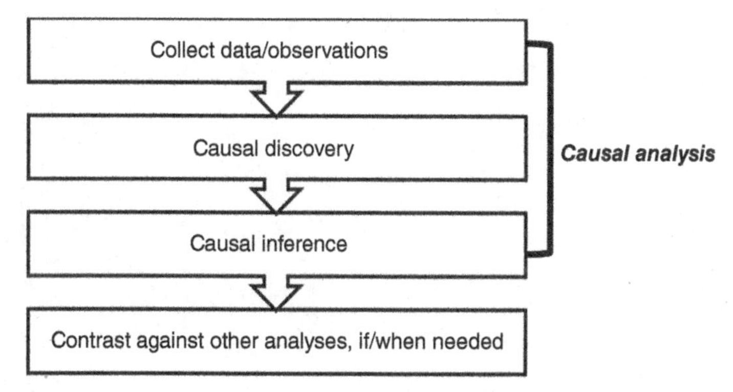

Figure 8.9 Flowchart of the proposed approach [35, 36].

8.5 Examples

This section describes five examples[96] to be highlighted in this work. The first three revolve around causal discovery (via constraint-based and score-based algorithms),[97] and the last two cover causal inference.

8.5.1 Causal Discovery

The first two examples leverage the graphical interface of *Tetrad* to discover possible causal structures for a known and unknown DGP. The third example is presented through *Python*.

Example no. 1 discovering causal structure for a known DGP (Tetrad)
The first case study is simple and hopes to verify our domain knowledge of a known DGP. The DGP of interest pertains to the moment capacity, M, of W-shaped steel beams. Looking at Equation 8.1, the moment capacity is estimated by multi-

96 These examples stem from my research and papers [24, 35, 36].

97 The used algorithms are the PC and FGES in their default settings (given the illustrative nature of this study). For more information on these algorithms, please refer to their original sources [29, 70].

plying the plastic modulus, Z, with the yield strength of steel, f_y. In other words, M is the effect of Z and f_y and hence labels M as a collider, $Z \rightarrow M \leftarrow f_y$.

To execute this analysis, a database was created from all W-shaped structural steel sections as provided by the American Institute of Steel Construction (AISC) manual[98] [37]. This database contains two independent features (Z and f_y) for each beam, as well as M (as a dependent variable). Simply, since we know the DGP (i.e., Equation 8.1), I multiplied the plastic modulus of each steel section by two steel Grades (A992 and A36).

The compiled database was used in an analysis that I carried out through the *Tetrad*[99] software and using two algorithms (namely, the PC algorithm, a *constraint-based*, and the FGES algorithm, a *score-based*). The procedure to use *Tetrad*[100] and the results of this analysis are shown below. As one can see, the generated DAG resembles that of our domain knowledge and is identical to the PC and FGES algorithms. This is both expected and comforting. As mentioned, only score-based algorithms are fitted with the capability to calculate a score-of-fitness to the developed DAG (estimated at −19,236.4 based on the default Bayesian Information Criterion (BIC)[101]).

Here are the steps to replicating my analysis:

Step 1: Select *Data*

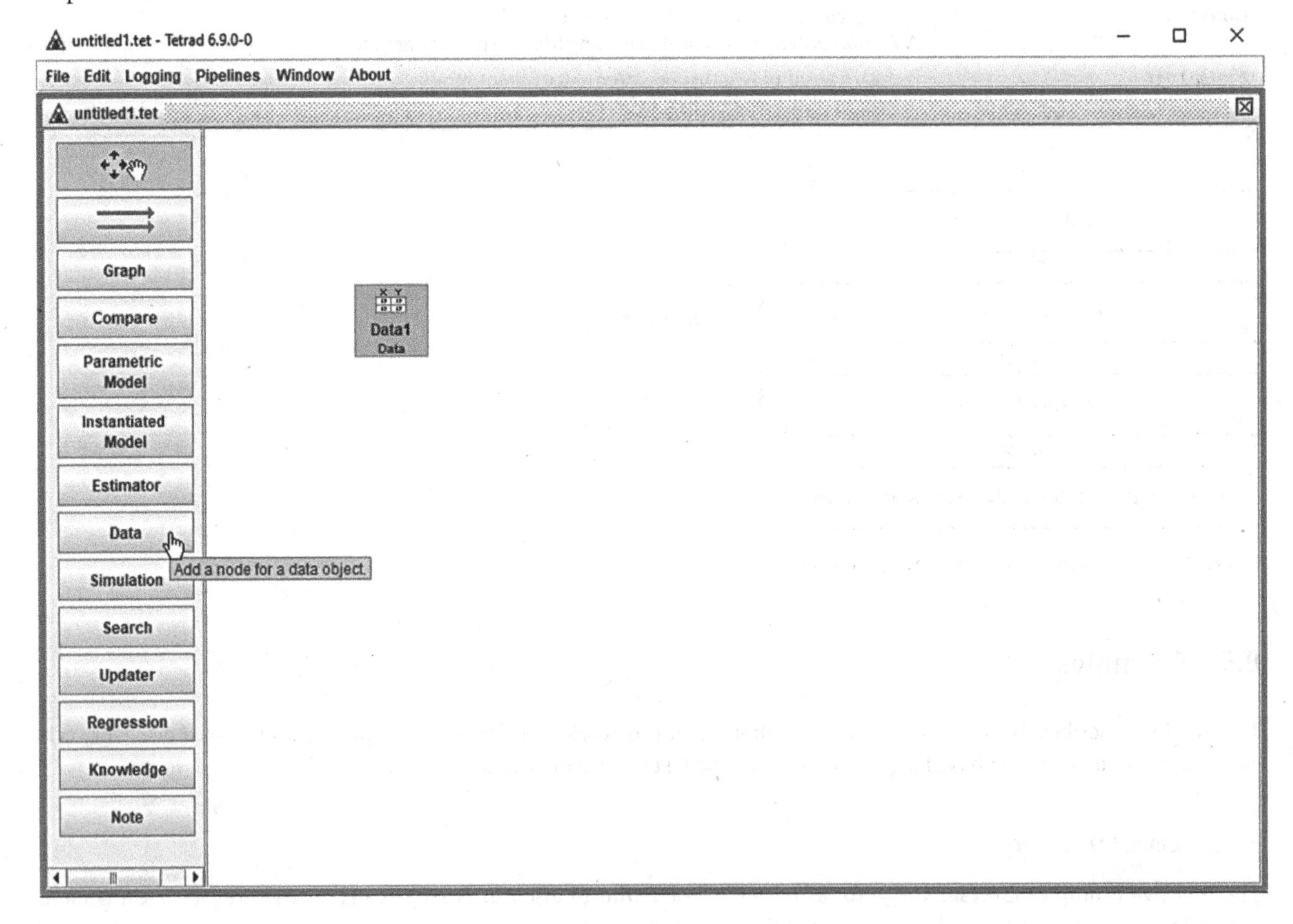

98 If you did not know, the AISC provides many useful (and free) tools such as an Excel spreadsheet with all standard steel sections. As always, I will be providing the data used on my website.

99 A link to download *Tetrad* is provided in Table 8.2. Let me take a minute to also acknowledge the Center for Causal Discovery, supported by grant U54HG008540.

100 Version 6.9.0 [Last modified: on *Mon May 03 20:18:40 EDT 2021*] and can be found at [71].

101 BIC score is calculated as chi-square statistics - degrees of freedom × log(sample size). Low values of BIC imply the goodness of the model.

Step 2: *Load* your dataset

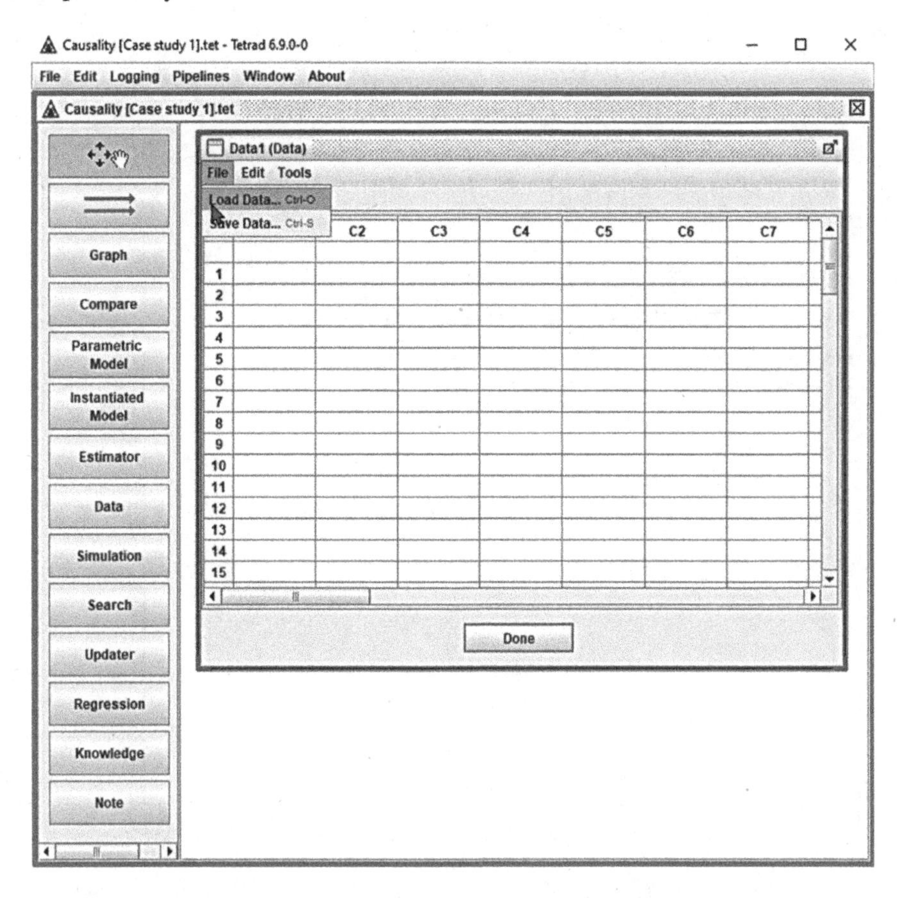

Step 3: Locate and check the needed setting (you may follow the same settings I used) and *Validate* and *Load* your data.

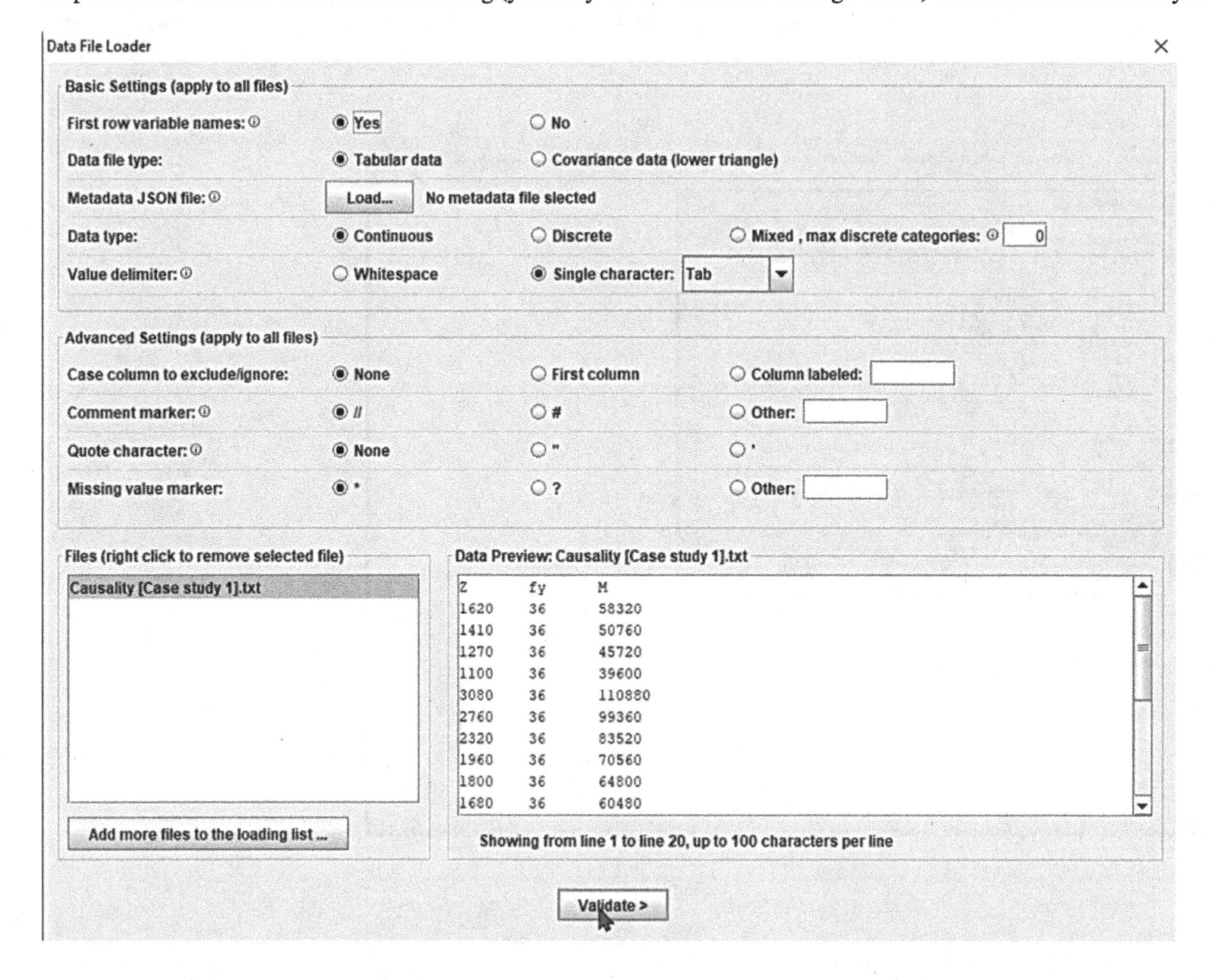

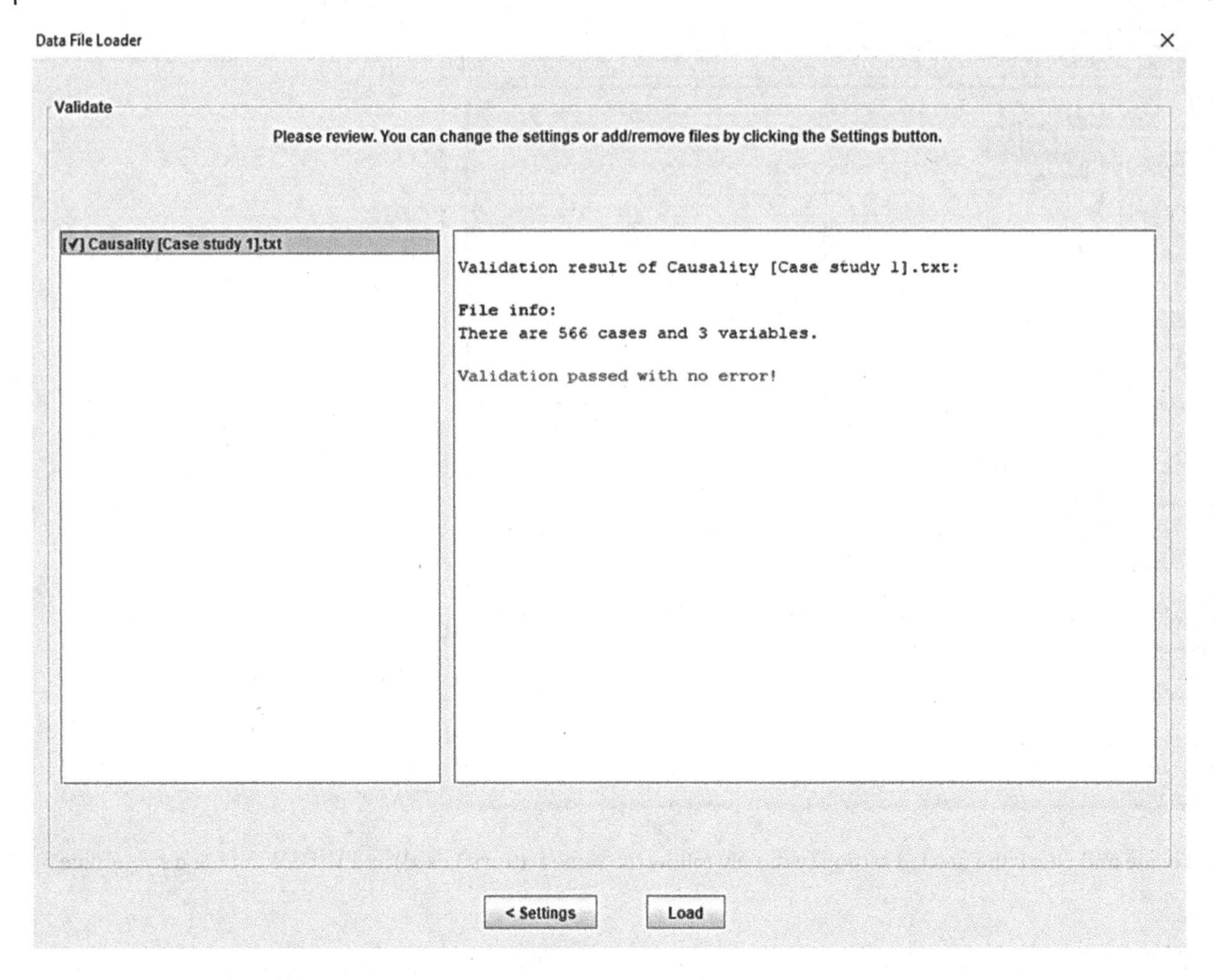

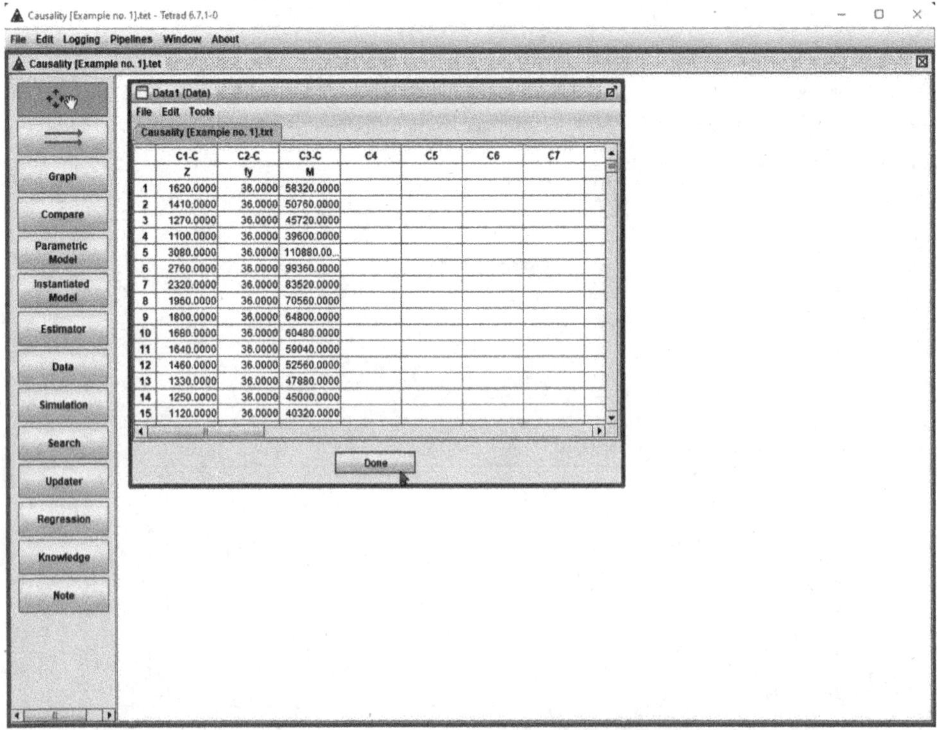

	C1-C	C2-C	C3-C	C4	C5	C6	C7
	Z	fy	M				
1	1620.0000	36.0000	58320.0000				
2	1410.0000	36.0000	50760.0000				
3	1270.0000	36.0000	45720.0000				
4	1100.0000	36.0000	39600.0000				
5	3080.0000	36.0000	110880.00...				
6	2760.0000	36.0000	99360.0000				
7	2320.0000	36.0000	83520.0000				
8	1960.0000	36.0000	70560.0000				
9	1800.0000	36.0000	64800.0000				
10	1680.0000	36.0000	60480.0000				
11	1640.0000	36.0000	59040.0000				
12	1460.0000	36.0000	52560.0000				
13	1330.0000	36.0000	47880.0000				
14	1250.0000	36.0000	45000.0000				
15	1120.0000	36.0000	40320.0000				

Step 4: Select *Search* and link it to the data by dragging an *Arrow* from *Data* to *Search*.

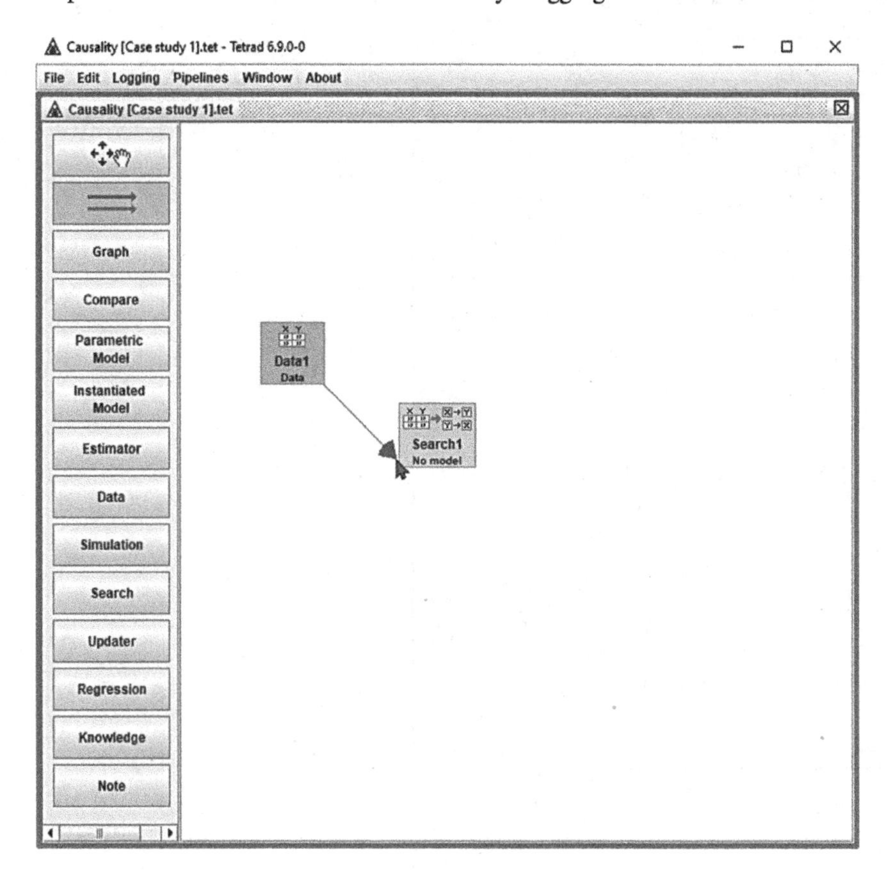

Step 5: Double click on the *Search* and select *PC Variants*. Update the setting for colliders to 1.0. Click *Run Search & Generate Graph*.

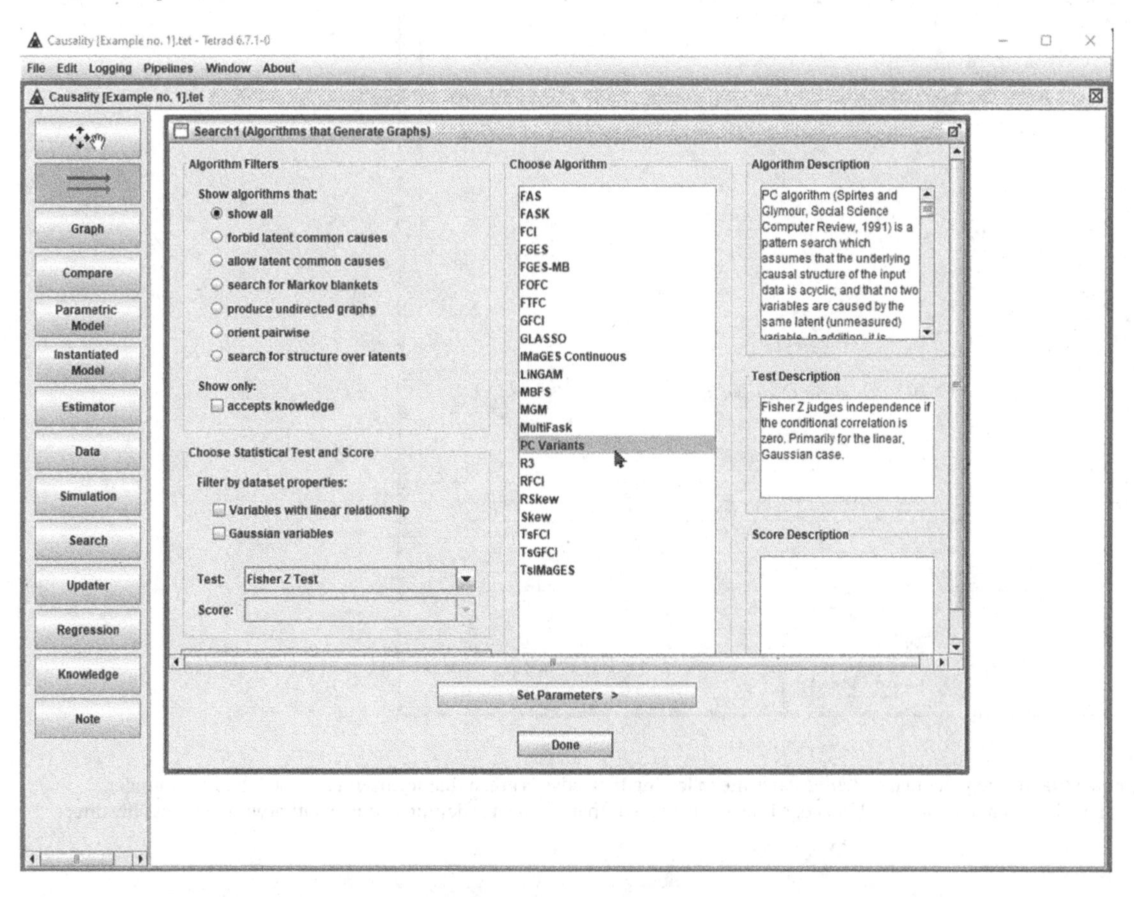

Search1 (Algorithms that Generate Graphs)

PC Variants

Collider discovery: 1 = Lookup from adjacency sepsets, 2 = Conservative (CPC), 3 = Max-P: 1

Collider conflicts: 1 = Overwrite, 2 = Orient bidirected, 3 = Prioritize existing colliders: 3

Maximum size of conditioning set (unlimited = -1): -1

Test ordering: 1 = PC-1 1, 2 = PC-2, 3 = PC-3: 1

Maximum path length for the unshielded collider heuristic for max P (min = 0): 3

Yes if a concurrent FAS should be done: Yes / No

Yes if the 'stable' FAS should be done: Yes / No

Yes if the heuristic for orienting unshielded colliders for max P should be used: Yes / No

Yes if verbose output should be printed or logged: Yes / No

Fisher Z Test

Cutoff for p values (alpha) (min = 0.0): 0.01

Bootstrapping

The number of bootstraps/resampling iterations (min = 0): 0

The percentage of resample size (min = 0.1): 90

Ensemble method: Preserved (0), Highest (1), Majority (2): 0

Yes, if adding an original dataset as another bootstrapping: Yes / No

Yes, if sampling with replacement (bootstrapping): Yes / No

< Choose Algorithm | Run Search & Generate Graph >

Done

Step 6: Assess the graph.[102]

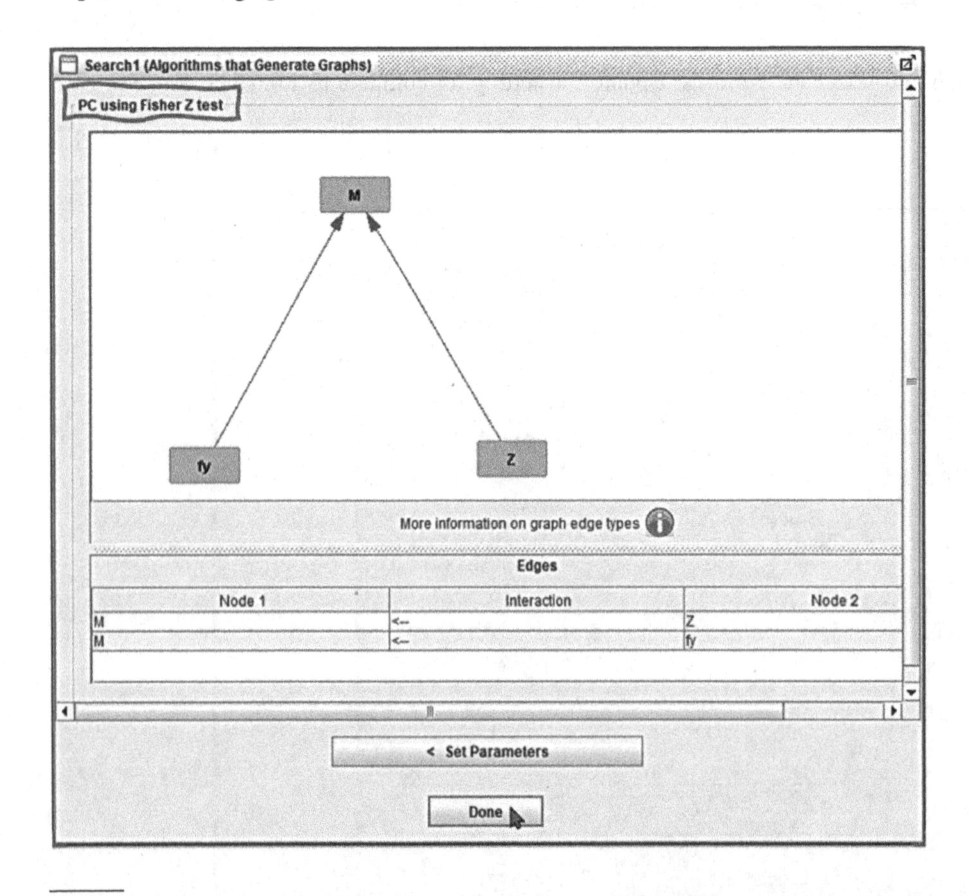

102 It should be noted that the help manual of *Tetrad* states the following,If an edge is green that means there is no latent confounder. Otherwise, there is possibly latent confounder. If an edge is bold (thickened) that means it is definitely direct. Otherwise, it is possibly direct.

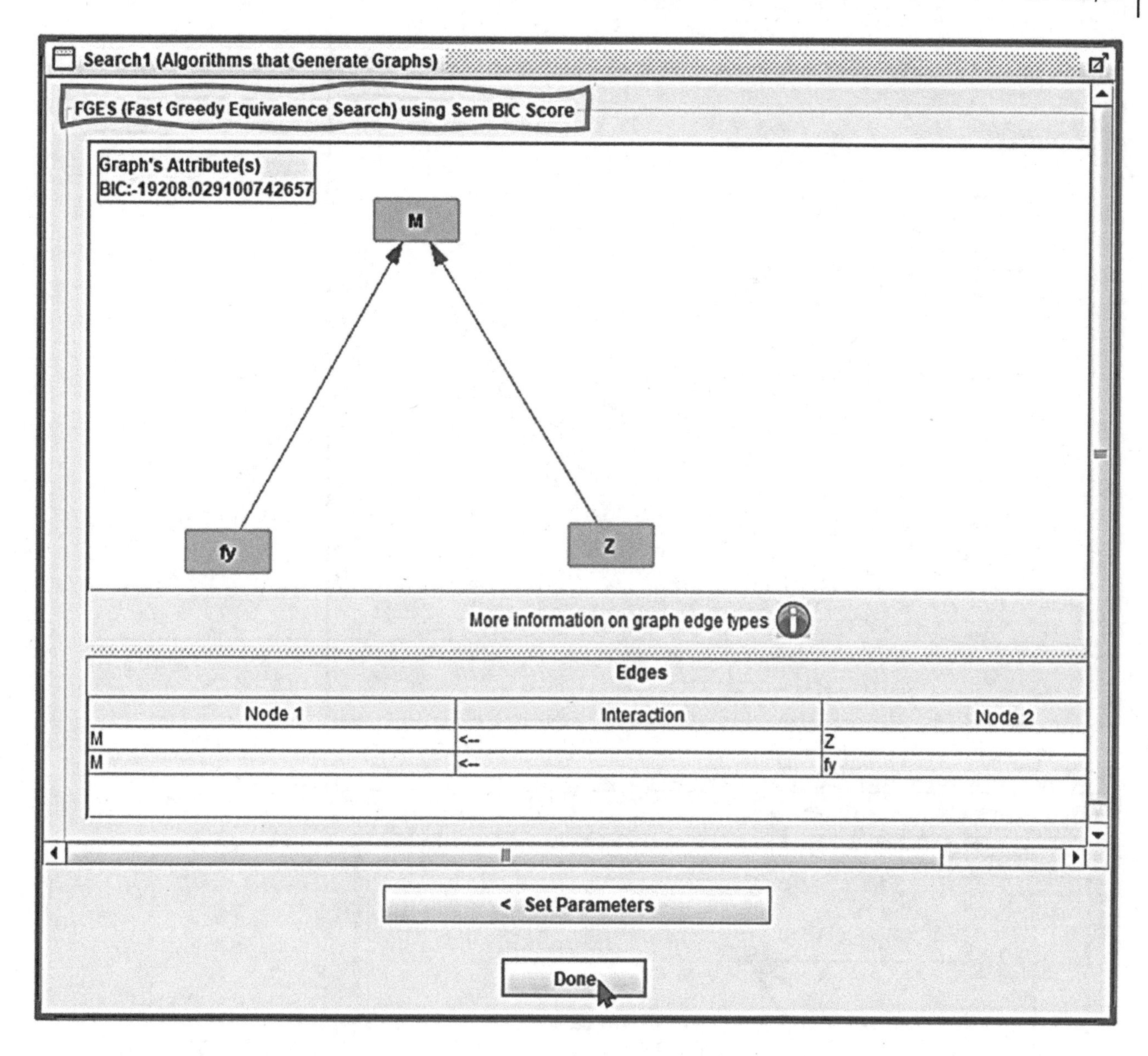

Example no. 2 discovering causal structure for an unknown DGP (Tetrad)
In the second case study, I am particularly interested in uncovering the DGP of how RC columns spall under fire conditions causally as opposed to predicting this phenomenon as we did in an earlier chapter. For the data, I will use the same data we used earlier. As you remember, our database includes the following variables: column width and depth, *b*, steel reinforcement ratio, *r*, concrete compressive strength, *f*, concrete cover to reinforcement, *C*, the magnitude of applied loading, *P*, and Spalling, *SP* (Yes, or No).

Similar to the first case study, the constraint-based PC algorithm and the score-based FGES algorithm were applied to identify the causal structure of the DGP. The outcome of this analysis is shown below. This figure shows that both algorithms were not able to establish spalling (*SP*) as the mechanism of interest.[103] Similarly, the algorithms also show results that do not agree with our domain knowledge (i.e., *SP* → concrete cover (*C*)).

103 This is not surprising. In most instances, we are obligated to identify the target variable.

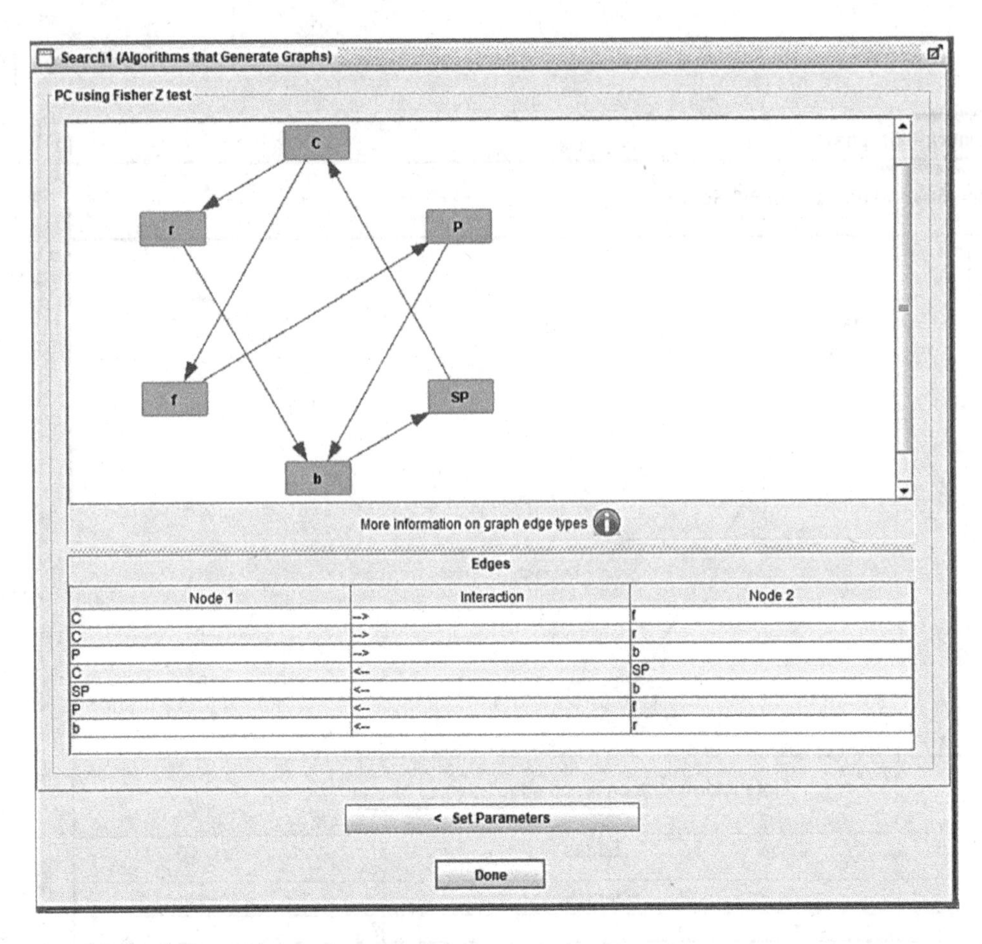
Search1 (Algorithms that Generate Graphs)
PC using Fisher Z test
C
r
P
f
SP
b
More information on graph edge types
Edges
Node 1 | Interaction | Node 2
C --> f
C --> r
P --> b
C <-- SP
SP <-- b
P <-- f
b <-- r
< Set Parameters
Done

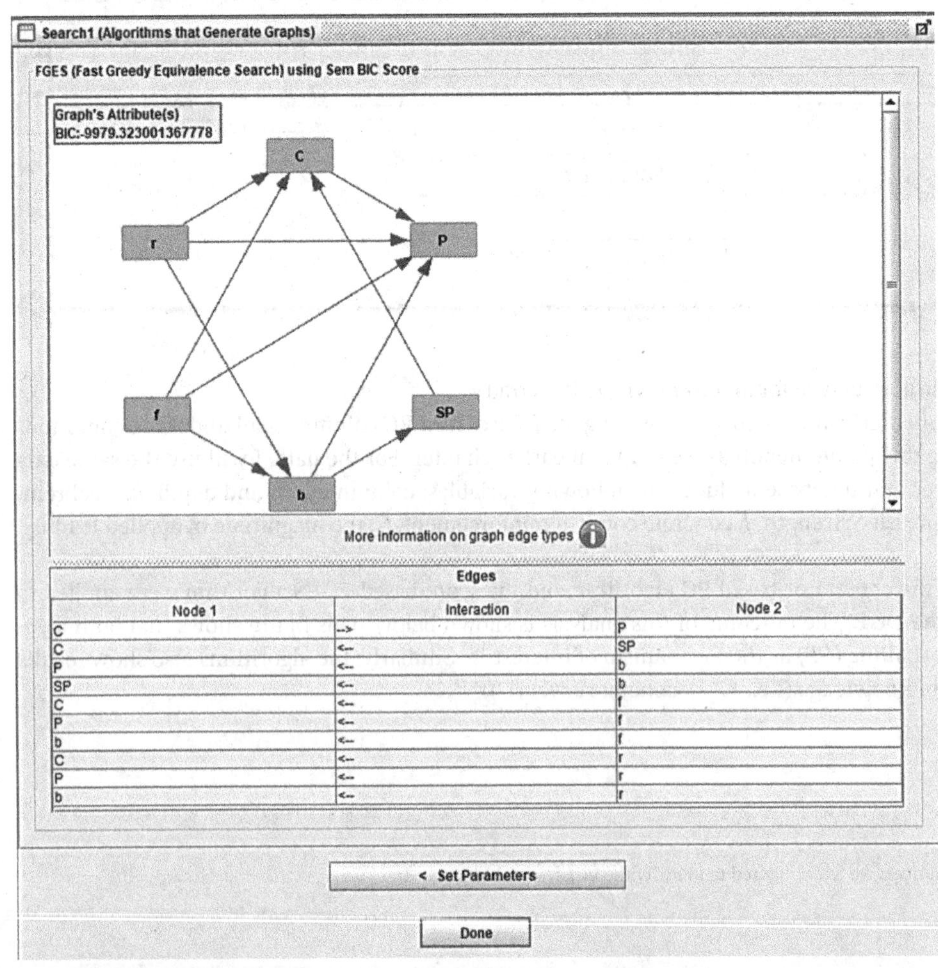
Search1 (Algorithms that Generate Graphs)
FGES (Fast Greedy Equivalence Search) using Sem BIC Score
Graph's Attribute(s)
BIC:-9979.323001367778
C
r
P
f
SP
b
More information on graph edge types
Edges
Node 1 | Interaction | Node 2
C --> P
C <-- SP
P <-- b
SP <-- b
C <-- f
P <-- f
b <-- f
C <-- r
P <-- r
b <-- r
< Set Parameters
Done

This is a good time to put our domain knowledge into good use! Luckily, and as we have seen before, *Tetrad* can incorporate domain knowledge into a causal analysis. The procedure to incorporate domain knowledge into *Tetrad* and the result of this new analysis is demonstrated below.

A look into this figure shows that both algorithms do not agree on the DGP. In fact, this figure notes that spalling is caused by *C* and *b* (and only the FGES algorithm noted a third variable, *f*, as well). It is clear that both of these figures do not provide us with sufficient information to uncover a likely DGP.[104] It is worth noting that both DAG agree with our domain knowledge as obtained from notable studies on spalling that qualitatively paint a picture of the possible DGP behind spalling [38–41]. This causal analysis may shed some gray over the suitability of the previously presented ML analysis (as that particular analysis cannot be assumed to uncover the causality of spalling but is rather a simple prediction tool). Hence, a more in-depth analysis, and possibly data, is needed to uncover the DGP of spalling.[105]
Follow these steps to incorporate domain knowledge.

Step 1: Once you have loaded and validated the data. Add a *Knowledge* box

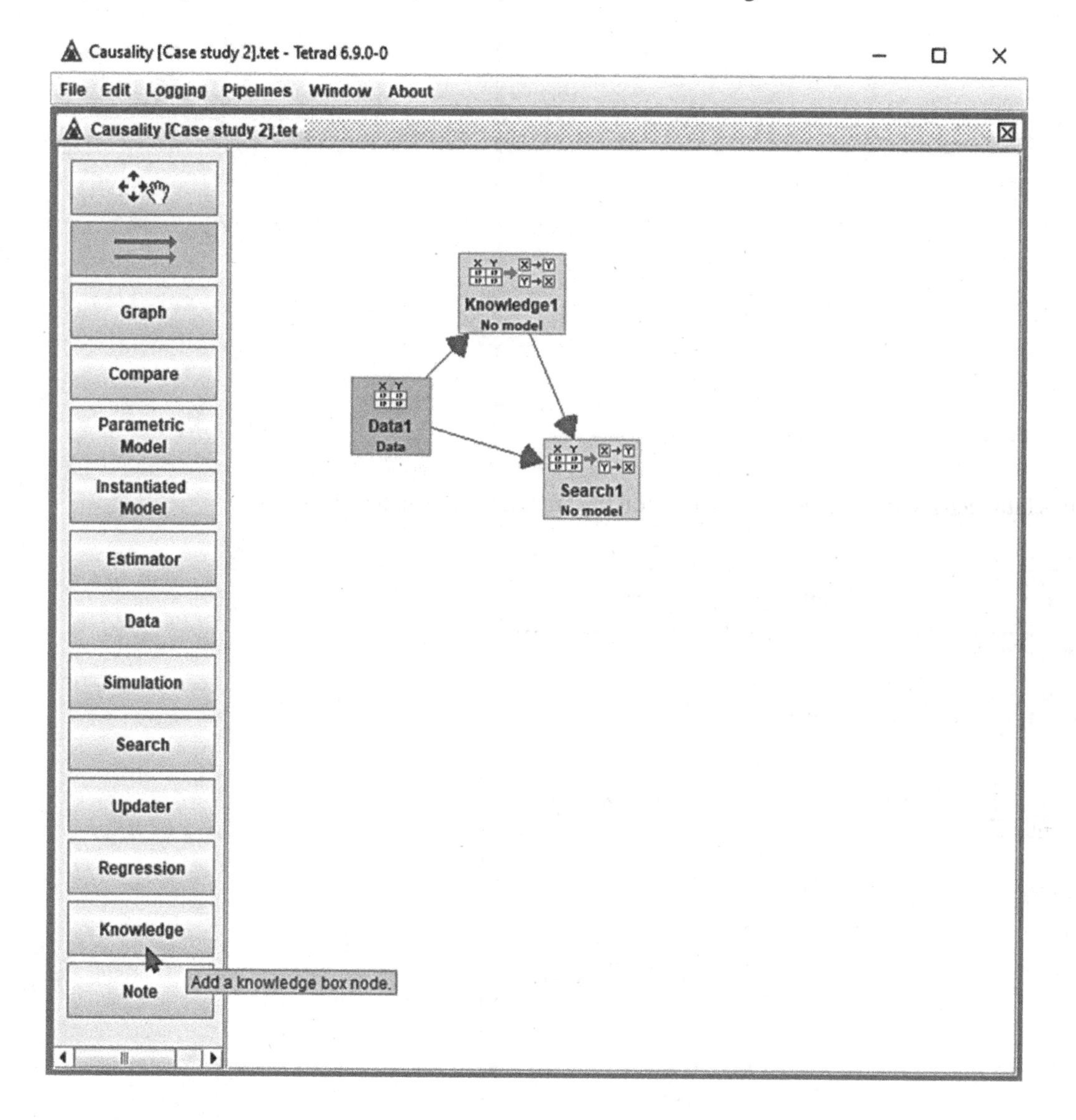

104 Do you remember how we used the same database to predict spalling through ML with high accuracy! Makes you wonder!

105 Another reason to not delve into the DGP in this study is that preliminary analysis conducted using other algorithms (i.e., RFCI and FAS) identified the presence of confounders. Tackling such confounders is hectic and is best left for a future work.

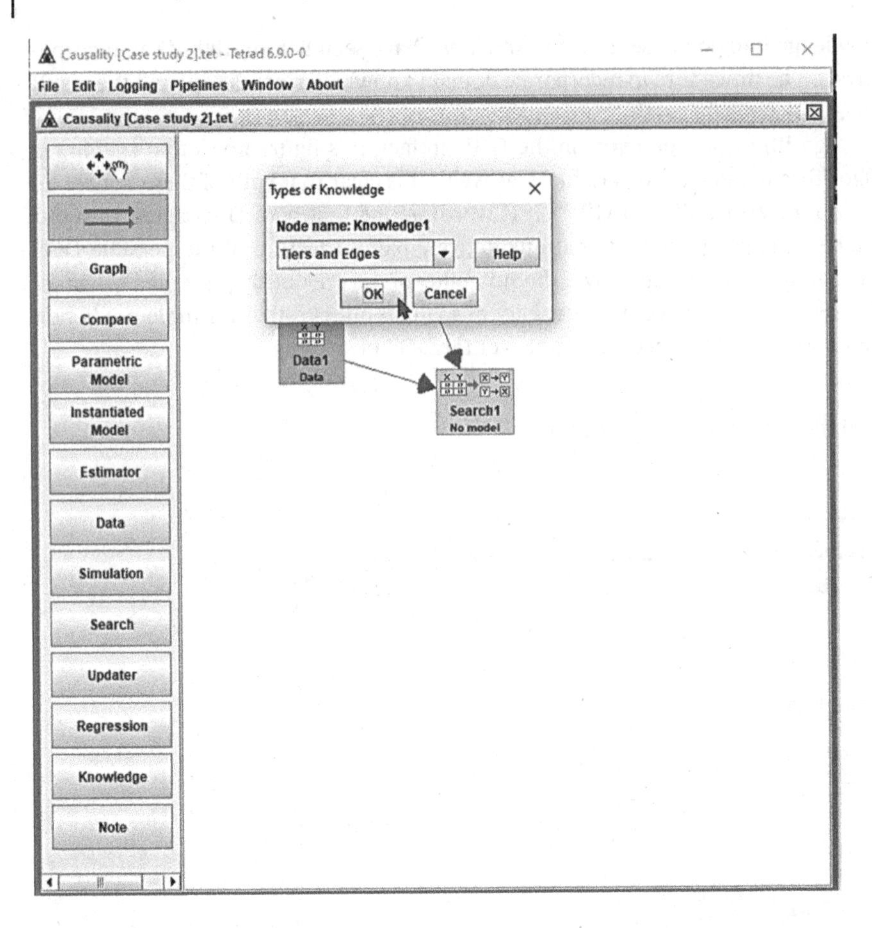

Step 2: Add the features into Tier 1 and the target into Tier 2. Run the analysis and visualize the results.

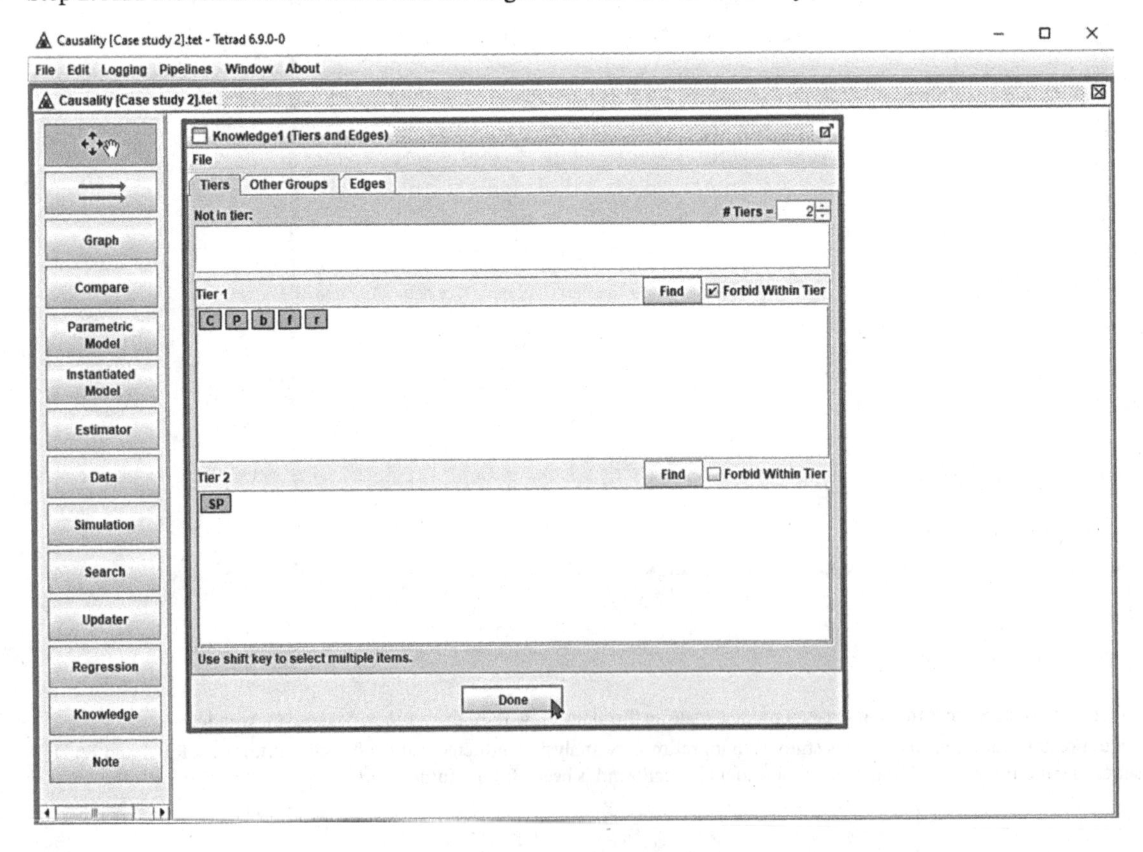

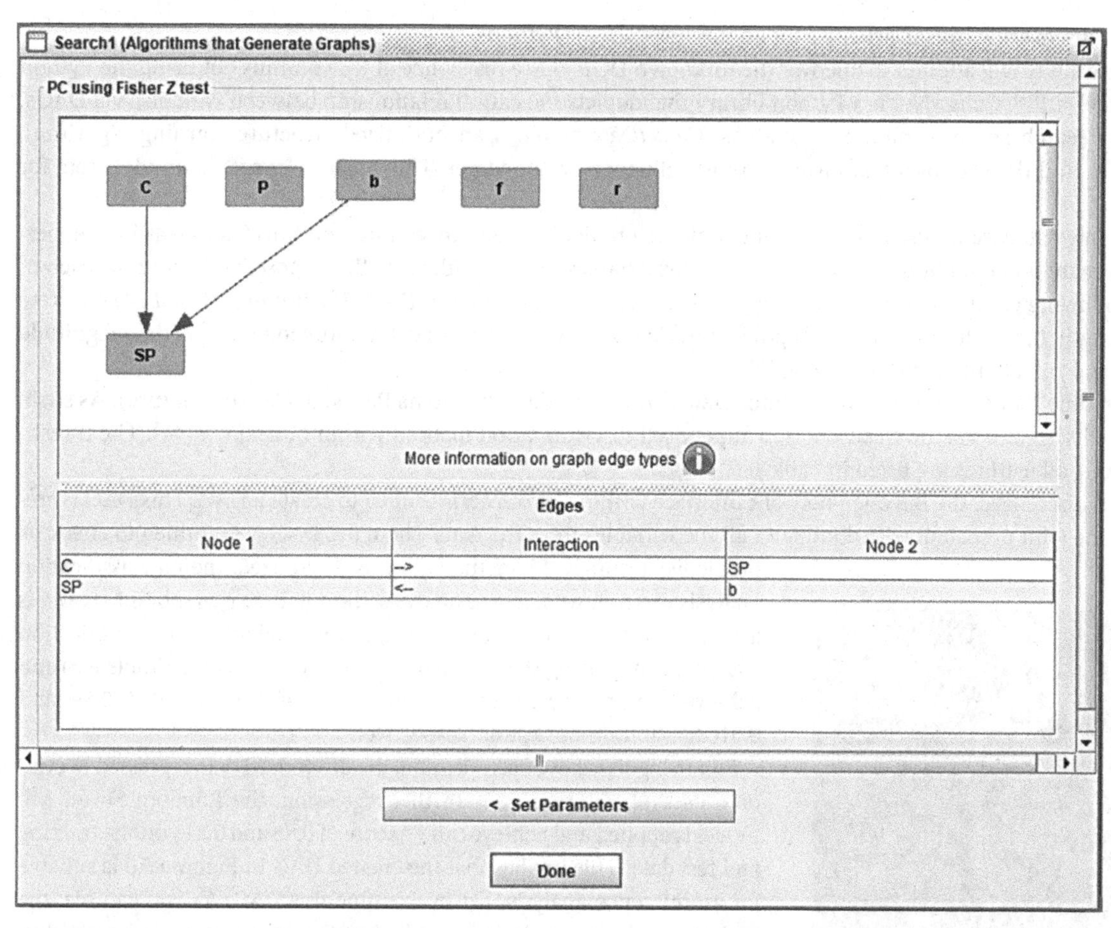
Search1 (Algorithms that Generate Graphs)
PC using Fisher Z test
C
p
b
f
r
SP
More information on graph edge types
Edges
Node 1
Interaction
Node 2
C
-->
SP
SP
<--
b
< Set Parameters
Done

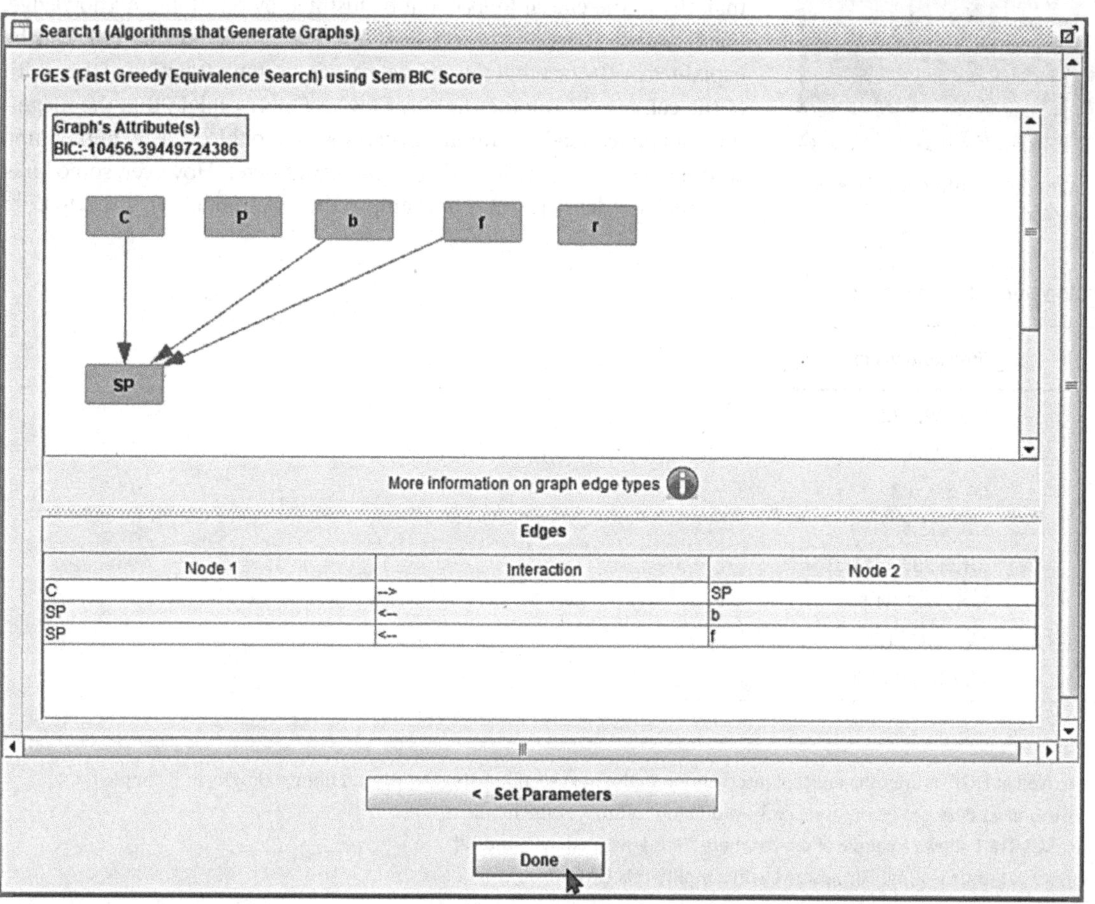
Search1 (Algorithms that Generate Graphs)
FGES (Fast Greedy Equivalence Search) using Sem BIC Score
Graph's Attribute(s)
BIC:-10456.39449724386
C
p
b
f
r
SP
More information on graph edge types
Edges
Node 1
Interaction
Node 2
C
-->
SP
SP
<--
b
SP
<--
f
< Set Parameters
Done

Example no. 3 discovering causal structure for an unknown DGP (Python)

The following case study will attempt to uncover the unknown DGP of fire resistance of RC columns but using the Python library, *CausalNex* [42][106]. *CausalNex* is a Python library that depicts the causal relationship between variables via DAGs. To overcome the search space of multiple variables, *CausalNex* employs an optimized structure learning algorithm, *NOTEARS* [43], to find the variables that have the most influence on the target. This makes *CausalNex* an ideal tool for causal discovery.

I will use the same database we used in an earlier chapter to predict fire resistance of RC columns (as well as in unsupervised/clustering analysis) herein too. As you have seen, this database has 144 full scale fire-exposed RC columns, each of which has the following parameters: 1) column width, *W*, 2) steel reinforcement ratio, *r*, 3) column length, *L*, 4) concrete compressive strength, f_c, 5) column effective length factor, *K*, 6) concrete cover to steel reinforcement, *C*, 7) the magnitude of applied loading, *P*, and 8) fire resistance time, *FR*.

At this point in time, *CausalNex* does not accommodate continuous data (due to its Bayesian Network nature). As such, the database was discretized via the Decision Tree Supervised discrimination method within *CausalNex* [44]. The thresholds for discretizing all features are listed in Table 8.5.

Once the data is discretized, the Bayesian network interface within *CausalNex* is trained to create a DAG. This DAG is created by learning the joint probability distribution of all the variables from the data. Then, the DAG is examined to check its predictive accuracy. Since the data was discretized, then a classification analysis was followed. In this analysis, the database was split into two sets as 70/30 for the training and testing of the model[107]. The classification was performed using the Bayesian Network Classifier (see Table 8.6) and achieved an accuracy, precision, and recall of 0.86/0.86, 0.90/0.89, and 0.910.86 for training/testing, respectively.

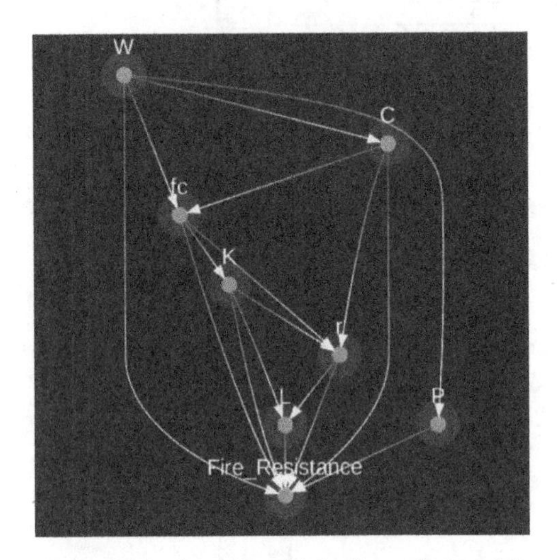

Figure 8.10 Causal model for fire resistance RC columns [from *CausalNex*].

Following the above classification, the discretized data regressed to evaluate its explanatory power. In this regression, the Random Forest ML model is applied and achieves an R^2 score of 0.85 and 0.811 on the training and test data. This implies that the created DAG in Figure 8.10 is suitable for an inference analysis. Let us examine this DAG. As you can see, the majority of the causal links could be justified by our domain knowledge. For instance, there is a direct link from *W* to *C*, and this link can be explained by the fact that the concrete cover is often a function of the width of the column.[108] Another link from *W* to *P* may explain that the magnitude of applied load a column can carry is tied to the capacity of the same column, which also include *W* as a main parameter. However, some links are much harder to justify. For example, the link from *C* to f_c and f_c to *K*.[109]

Table 8.5 Threshold values of each feature.

Variables	Threshold values
C (mm)	27.5, 39, 44.5
K	0.59, 0.85
L (m)	2.4, 3.9, 4.3
P (kN)	538, 975.5, 1174
W (mm)	201.5, 302.5, 431.5
f_c (MPa)	26.5, 33.5, 34.5
R (%)	0.95, 1.18, 1.9
FR (min)	60, 120, and 180

106 We will use the arrived at DGP to infer the effect of intervention in the next section using the Python library *DoWhy* in Example no. 4.
107 Please note that *CausalNex* does not incorporate cross-validation training at the moment.
108 A quick look into ACI 216.1 shows a couple of tables where *FR* is governed by *C* and *W*!
109 What to do? Well, we continue testing! For now, we will simplify this DAG.

Table 8.6 Characteristics of the classification algorithm.

Classification Method	Bayesian Estimator
Bayes prior	K2
Discretiser method	fixed
Discretiser max depth	1
Discretiser random state	2020

Allow me now to provide you with the full *Python* code[110] to carry out the above analysis. In this code, I am showcasing more analysis than that shown above since this codes also cover Example no. 4.[111]

```
#Import libraries
import numpy as np
import pandas as pd
!pip install dowhy
import dowhy
from dowhy import CausalModel
from IPython.display import Image, display

from sklearn.preprocessing import LabelEncoder
!pip install causalnex
import causalnex
from causalnex.discretiser import Discretiser
from IPython.display import Image
from causalnex.plots import plot_structure, NODE_STYLE, EDGE_STYLE

#Import dataset
df = pd.read_excel('Causality_Example_no._3.xlsx')
df=pd.DataFrame(df)

#Inputs discretization
from causalnex.discretiser.discretiser_strategy import (
    DecisionTreeSupervisedDiscretiserMethod,
)
features = list(df.columns.difference(['Fire_Resistance']))
tree_discretiser = DecisionTreeSupervisedDiscretiserMethod(
    mode="single",
    tree_params={"max_depth": 2, "random_state": 2021},
)
tree_discretiser.fit(
    feat_names=features,
    dataframe=df,
    target_continuous=True,
    target="Fire_Resistance",
for col in features:
    df[col] = tree_discretiser.transform(df[[col]])
tree_discretiser.map_thresholds

#Visualization
!apt install libgraphviz-dev
```

110 Thanks to Aybike Özyüksel for building this code. The actual codes can be accessed from my website.

111 It felt more intuitive to have the full code example in one spot.

```
!pip install pygraphviz
from causalnex.structure.notears import from_pandas
sm = from_pandas(df)
sm.remove_edges_below_threshold(0.3)
sm = sm.get_largest_subgraph()
viz = plot_structure(
    sm,
    graph_attributes={"scale": "0.5"},
    all_node_attributes=NODE_STYLE.WEAK,
    all_edge_attributes=EDGE_STYLE.WEAK,
)
Image(viz.draw(format='png'))
viz = plot_structure(sm, prog="dot")
Image(viz.draw(format="jpg"),width=500)

#Output discretization
from causalnex.network import BayesianNetwork
bn = BayesianNetwork(sm)
discretised_data = df.copy()
discretised_data['Fire_Resistance'] = Discretiser(
    method="fixed",
    numeric_split_points = [120,180],
).transform(discretised_data["Fire_Resistance"].values)
Fire_Resistance_map = {0: "Class_1", 1: "Class_2", 2: "Class_3"}
discretised_data['Fire_Resistance'] = (
    discretised_data['Fire_Resistance'].map(Fire_Resistance_map)
)
bn = bn.fit_node_states(discretised_data)
bn = bn.fit_cpds(
    discretised_data,
    method="BayesianEstimator",
    bayes_prior="K2",
)
discretised_data

#Classification
#Train data classification_report
from sklearn.model_selection import train_test_split
train, test = train_test_split(discretised_data, train_size=0.7, test_size=0.3,
random_state=7)

from causalnex.evaluation import classification_report
classification_report(bn, test, "Fire_Resistance")

#Test data classification_report
from causalnex.evaluation import classification_report
classification_report(bn, train, "Fire_Resistance")
predicts = bn.predict(test, "Fire_Resistance")

from sklearn.metrics import classification_report, confusion_matrix
report = classification_report(test["Fire_Resistance"], predicts)
matrix = confusion_matrix(test["Fire_Resistance"], predicts)
print(report)
print(matrix)

#XGBOOST regression
y = df["Fire_Resistance"]
```

```
X = df.drop(["Fire_Resistance"], axis=1)
from sklearn.model_selection import train_test_split
X_train, X_test, y_train, y_test = train_test_split(X, y, test_size = 0.2, random_state = 0)
from xgboost import XGBRegressor
regressor = XGBRegressor()
regressor.fit(X_train, y_train)
y_pred=regressor.predict(X_test)
from sklearn.metrics import r2_score
r2_score(y_test, y_pred)
y_pred = regressor.predict(X_train)
from sklearn.metrics import r2_score
r2_score(y_train, y_pred)
#RANDOM_FOREST regression
y = df["Fire_Resistance"]
X = df.drop(["Fire_Resistance"], axis=1)
from sklearn.model_selection import train_test_split
X_train, X_test, y_train, y_test = train_test_split(X, y, test_size = 0.2, random_
state = 0)
from sklearn.ensemble import RandomForestRegressor
regressor = RandomForestRegressor(n_estimators= 100, random_state=0)
regressor.fit(X_train, y_train)
y_pred = regressor.predict(X_test)
from sklearn.metrics import r2_score
r2_score(y_test, y_pred)
y_pred = regressor.predict(X_train)
from sklearn.metrics import r2_score
r2_score(y_train, y_pred)
#DoWhy analysis (intervention on the width, w)[112]
#Import libraries
import os, sys
import random
sys.path.append(os.path.abspath("../../../"))
import numpy as np
import pandas as pd
!pip install dowhy
import dowhy
from dowhy import CausalModel
from IPython.display import Image, display
#Correlation matrix
dataset = pd.read_excel('Causality_Example_no._3_w.xlsx')
import matplotlib.pyplot as plt
import seaborn as sb
X = dataset
X.head(1)
```

112 Start of interventions and Example no. 4.

```
#Correlation matrix
import seaborn as sns
plt.figure(figsize=(15,13))
sns.heatmap(X.corr(),
            #annot = True,
            fmt = '.1g',
            center = 0,
            #cmap='GnBu')
            cmap = sb.diverging_palette(0, 230, 90, 60, as_cmap=True))
            #cmap = sns.diverging_palette(250, 10, as_cmap=True))
            #linewidths = 1)
            #linecolor = 'black')
plt.xticks(rotation=45,fontsize=18)
plt.yticks(rotation=45,fontsize=18)

#Import dataset
dataset = pd.read_excel('Causality_Example_no._3_w.xlsx')
T = dataset.iloc[:, 7].values
y = dataset.iloc[:, -1].values
W=dataset.iloc[:, 0].values
r=dataset.iloc[:, 1].values
L=dataset.iloc[:, 2].values
fc=dataset.iloc[:, 3].values
K=dataset.iloc[:, 4].values
C=dataset.iloc[:, 5].values
P=dataset.iloc[:, 6].values

df = pd.DataFrame({'W': W, 'r': r, 'L': L, 'fc': fc, 'K': K, 'C': C, 'P': P,
                   'T': T,
                   'Y': y})
#Model:Causal Model
causal_graph = """digraph {
W->T;
T->Y;
W->P; W->C; W->fc;
C->fc; C->r; K->Y;
fc->r; fc->K; fc->Y;
r->L; r->Y;
K->L; K->r; L->Y;
P->Y; C->Y;
}"""

from dowhy import CausalModel
model=CausalModel(
        data = df,
        treatment='T',
        outcome='Y',
      graph=causal_graph.replace("\n", " "),
        )
model.view_model()
from IPython.display import Image, display
display(Image(filename="causal_model.png"))

#Identify effect
```

```
identified_estimand = model.identify_effect(proceed_when_unidentifiable=True)
print(identified_estimand)

#Estimate using Linear Regression
#Estimate the causal effect and compare it with Average Treatment Effect
estimate = model.estimate_effect(identified_estimand,
        method_name="backdoor.linear_regression", test_significance=True
)
print(estimate)
#Refute the outcome
#Random Common Cause - add an independent random variable as a common cause to the
dataset; If the assumption was correct, the estimation should not change.
refute_results=model.refute_estimate(identified_estimand, estimate,
        method_name="random_common_cause")
print(refute_results)

#Data Subset Refuter - replace the given dataset with a randomly selected subset;
If the assumption was correct, the estimation should not change that much.

res_subset=model.refute_estimate(identified_estimand, estimate,
        method_name="data_subset_refuter", subset_fraction=0.9)
print(res_subset)

#Placebo Treatment - replace the true treatment variable with an independent random
variable; If the assumption was correct, the estimate should go close to zero.
res_placebo=model.refute_estimate(identified_estimand, estimate,
        method_name="placebo_treatment_refuter", placebo_type="permute")
print(res_placebo)

#Model:Modifying Causal model
causal_graph = """digraph {
K->T;
T->Y;
C->Y;
W->C;
fc->r;  W->Y;
r->Y;
L->Y;
K->L;
P->Y;
fc->Y;
P->W;
P->fc;
P->r;
}"""

from dowhy import CausalModel
model=CausalModel(
        data = df,
        treatment='T',
        outcome='Y',
#common_causes=['K'],
      graph=causal_graph.replace("\n", " "),
#instruments=['C'],
#effect_modifiers=['W','r', 'L','fc','C','P', 'K']
```

```
    )
model.view_model()
from IPython.display import Image, display
display(Image(filename="causal_model.png"))

#Identify
identified_estimand = model.identify_effect(proceed_when_unidentifiable=True)
print(identified_estimand)

#Estimate using Linear Regression
#Estimate the causal effect and compare it with Average Treatment Effect
estimate = model.estimate_effect(identified_estimand,
        method_name="backdoor.linear_regression", test_significance=True
)
print(estimate)

#Refute
#Random Common Cause – add an independent random variable as a common cause to the
dataset; If the assumption was correct, the estimation should not change.

refute_results=model.refute_estimate(identified_estimand, estimate,
        method_name="random_common_cause")
print(refute_results)

#Data Subset Refuter – replace the given dataset with a randomly selected subset;
If the assumption was correct, the estimation should not change that much
res_subset=model.refute_estimate(identified_estimand, estimate,
        method_name="data_subset_refuter", subset_fraction=0.9)
print(res_subset)

#Placebo Treatment– replace the true treatment variable with an independent random
variable; If the assumption was correct, the estimate should go close to zero.
res_placebo=model.refute_estimate(identified_estimand, estimate,
        method_name="placebo_treatment_refuter", placebo_type="permute")
print(res_placebo)

#Model:Hypothetical DAG
causal_graph = """digraph {
fc->T;
T->Y;
C->Y;
W->Y;
K->Y;
L->Y;
r->Y;
P->Y
}"""
from dowhy import CausalModel
model=CausalModel(
        data = df,
        treatment='T',
        outcome='Y',
#common_causes=['K'],
      graph=causal_graph.replace("\n", " "),
#instruments=['W','r', 'L','fc','C','P', 'K'],
```

```
#effect_modifiers=['W','r', 'L','fc','C','P', 'K']
        )
model.view_model()
from IPython.display import Image, display
display(Image(filename="causal_model.png"))
#Identify
identified_estimand = model.identify_effect(proceed_when_unidentifiable=True)
print(identified_estimand)

#Estimate using Linear Regression
#Estimate the causal effect and compare it with Average Treatment Effect
estimate = model.estimate_effect(identified_estimand,
        method_name="backdoor.linear_regression", test_significance=True
)

print(estimate)

#Refute
#Random Common Cause – add an independent random variable as a common cause to the
dataset; If the assumption was correct, the estimation should not change.

refute_results=model.refute_estimate(identified_estimand, estimate,
        method_name="random_common_cause")
print(refute_results)

#Data Subset Refuter – replace the given dataset with a randomly selected subset;
If the assumption was correct, the estimation should not change that much.**
res_subset=model.refute_estimate(identified_estimand, estimate,
        method_name="data_subset_refuter", subset_fraction=0.9)
print(res_subset)

#Placebo Treatment – replace the true treatment variable with an independent random
variable; If the assumption was correct, the estimate should go close to zero.
res_placebo=model.refute_estimate(identified_estimand, estimate,
        method_name="placebo_treatment_refuter", placebo_type="permute")
print(res_placebo)

#Model:Isolated Model
causal_graph = """digraph {
W->T;
T->Y;
}"""

from dowhy import CausalModel
model=CausalModel(
        data = df,
        treatment='T',
        outcome='Y',
#common_causes=['K'],
      graph=causal_graph.replace("\n", " "),
#instruments=['C'],
#effect_modifiers=['W','r', 'L','fc','C','P', 'K']
        )
model.view_model()
from IPython.display import Image, display
display(Image(filename="causal_model.png"))
```

```
#Identify
identified_estimand = model.identify_effect(proceed_when_unidentifiable=True)
print(identified_estimand)

#Estimate using Linear Regression
#Estimate the causal effect and compare it with Average Treatment Effect
estimate = model.estimate_effect(identified_estimand,
        method_name="backdoor.linear_regression", test_significance=True
)
print(estimate)

#Refute
#Random Common Cause – add an independent random variable as a common cause to the
dataset; If the assumption was correct, the estimation should not change.
refute_results=model.refute_estimate(identified_estimand, estimate,
        method_name="random_common_cause")
print(refute_results)

#Data Subset Refuter – replace the given dataset with a randomly selected subset;
If the assumption was correct, the estimation should not change that much.
res_subset=model.refute_estimate(identified_estimand, estimate,
        method_name="data_subset_refuter", subset_fraction=0.9)
print(res_subset)

#Placebo Treatment – replace the true treatment variable with an independent random
variable; If the assumption was correct, the estimate should go close to zero.
res_placebo=model.refute_estimate(identified_estimand, estimate,
        method_name="placebo_treatment_refuter", placebo_type="permute")
print(res_placebo)
```

8.5.2 Causal Inference

I will be presenting two examples herein. Example no. 4 is a companion to Example no. 3, and then Example no. 5 (to present a simple case on *do*-calculus).

Example no. 4 inferring the effect of interventions[113]
The *DoWhy* [45] *Python* library is a causal inference library that can be used to estimate the average causal effect[114] of one variable on another or upon a target/response. There are four steps to conducting a causal analysis via *DoWhy*: 1) modeling the underlying causes of a problem, 2) identifying trends/relationships between variables, 3) estimating parameters based on previous models or observations, and 4) refuting the obtained estimates by testing the robustness of model's assumptions (through three sub-models, the *Random common cause, Data subset Refuter*, and *Placebo Treatment Refuter*).[115] For example, if our assumptions are valid, then estimations from the Random common cause and Data subset refuter models should be close enough. However, if our assumptions are invalid, then the estimate should be near zero (per the Placebo treatment refutation model). This allows for greater accuracy and efficiency in research [46].

Now, let us continue the analysis in Example no. 3. Once the causal structure was arrived at from *CausalNex*, we turn toward *DoWhy* to infer the outcome of possible interventions upon the fire resistance of RC columns. I will present three

113 This analysis is a companion to Example no. 3.

114 Or simply ACT.

115 **Random Common Cause**: Adds randomly drawn variables to the database and re-runs the analysis to see if the causal estimate changes or not. The causal estimate shouldn't change by much due to a random variable. **Data Subset Refuter**: Creates subsets of the data and checks whether the causal estimates vary across subsets. In order to effectively measure causation, there should not be large variances in the estimates. **Placebo Treatment Refuter**: Randomly assigns a variable as a treatment and re-runs the analysis. If a causal relationship exists, then the causal estimate will move toward zero.

sub-analyses on interventions: DAG from *CausalNex*, augmenting CausalNex's DAG with domain knowledge, and by simplifying *CausalNex*'s DAG with domain knowledge. The results from each sub-analysis are presented below.

8.5.3 DAG from CausalNex

We import the *CausalNex*-developed DAG model shown in Figure 8.11 to *DoWhy*. For this, we build seven distinct models wherein each model examines the effect of intervention/treatment (T) on each variable, and then we observe the associated change to FR. The same figure demonstrates that all of the variables, together with the treatment values, have an effect on FR (and, as discussed above, some of the variables influence each other).

Table 8.7 shows the value of FR estimates and refutes for each variable. As you can see, the FR reduces by 39.6 min when $T = 1$ for the steel reinforcement ratio variable.[116] Furthermore, we can also observe that our treatment has a negative impact on *FR* for L, f_c, and K and a positive impact on the other variables. In addition, the refute analysis confirms the reliability of these results.

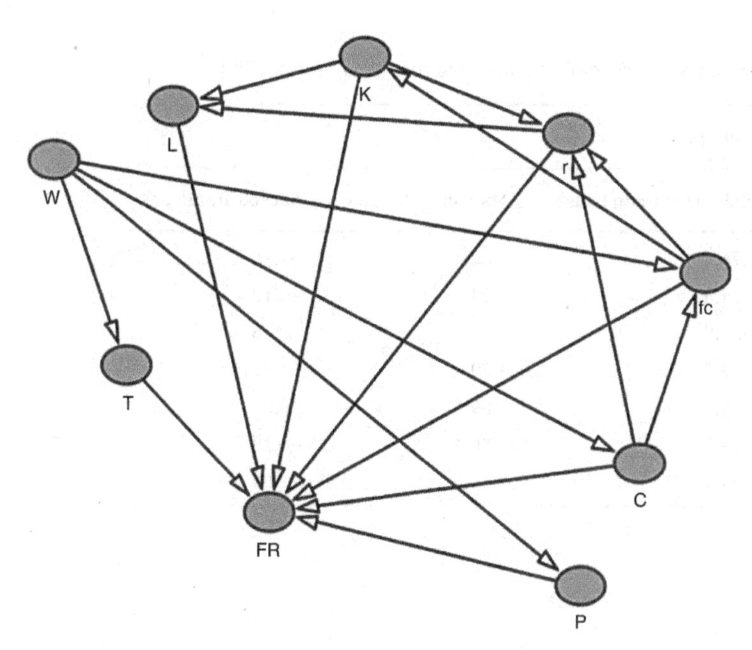

Figure 8.11 CausalNex model [Note: *T*: intervention/treatment, *FR*: fire resistance].

8.5.4 Modifying CausalNex's DAG with Domain Knowledge

Given that some of the identified causal links in the DAG by *CausalNex* do not align with our domain knowledge, an attempt was carried out herein to augment this DAG. Thus, the original DAG was trimmed into a new DAG (see Figure 8.12). In creating this DAG, I am assuming the following coarse assumptions and that this DAG is representable of the FR's DGP:

- The boundary conditions of a particular column affect its length, L, via K.
- Knowing the applied level of loading, P, influences the design of columns in terms of f_c, W, and r. In turn, f_c also affects r.
- Knowing W, influences the size of the concrete cover, C.

As shown in Table 8.8, when the effects of all variables on each other and on *FR* are considered together with the effect of $T = 1$ on *FR*, then W, L, f_c, and K negatively affect *FR*, whereas the opposite is true for r, C, and P. When the refute models are cross checked, it is clear that the estimations seem trustworthy.

116 I took the treatment/intervention as the average of each particular variable. For example, how FR would causally differ if r for a typical column changes from its initial value to the average value of the steel reinforcement ratio. To maintain consistency with the discussion on causal inference, I am referring to treatments/interventions with a T.

Table 8.7 Results of analysis for fire resistance (min) [bolded *p*-values imply significance that is larger than 5%].

	Estimate		Refute (min)		
Variable	Mean value (min)	*p*-value	Random Common Cause	Data Subset Refuter	Placebo Treatment
W	−252.0	0.260	−249.8	−275.3	−1.8
r	39.6	**0.001**	39.61	39.7	1.4
L	−97.6	**0.019**	−97.4	−96.8	0.6
f_c	−100.7	0.186	−100.8	−101.8	1.2
K	−60.3	0.688	−60.5	−49.3	1.5
C	89.4	**3.9e-08**	89.0	90.2	0.7
P	38.2	**0.0001**	38.4	37.7	2.1

Table 8.8 Results of analysis for fire resistance (min) [bolded *p*-values imply significance that is larger than 5%].

	Estimate		Refute (min)		
Variable	Mean value (min)	*p*-value	Random Common Cause	Data Subset Refuter	Placebo Treatment
W	−276.2	0.18	−276.6	−282.4	−1.88
r	25.0	0.78	24.8	24.2	−2.12
L	−86.9	0.78	−86.9	−86.5	2.36
f_c	−73.1	0.54	−73.3	−79.0	−4.71
K	−69.6	**1.9e-05**	−69.6	−69.6	0.003
C	79.9	**5.2e-09**	79.7	80.4	−0.13
P	44.4	**0.01**	44.1	43.6	−2.69

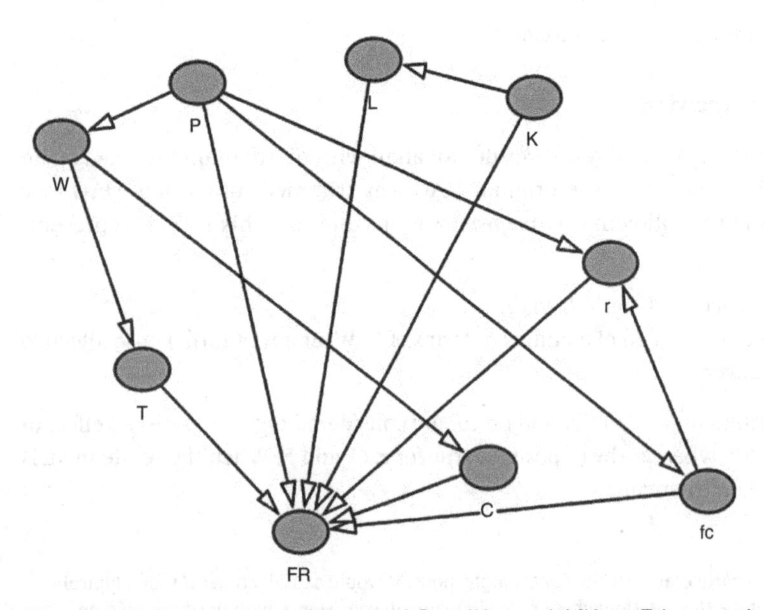

Figure 8.12 Domain-knowledge revised *CausalNex* model [Note: *T*: intervention/treatment, *FR*: fire resistance].

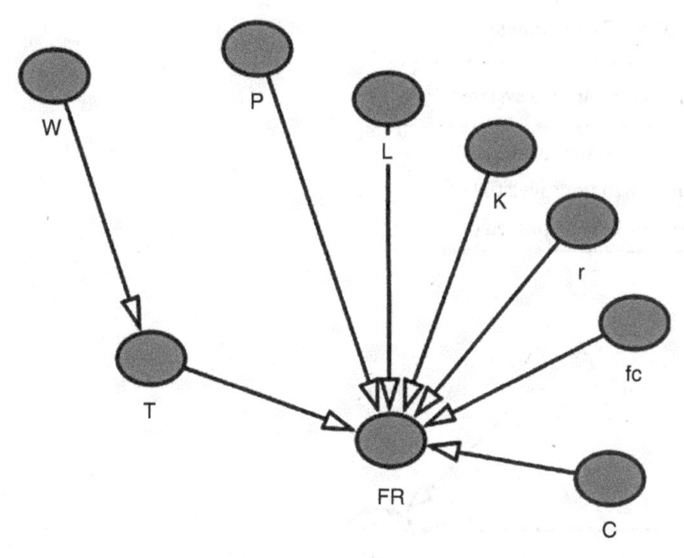

Figure 8.13 Simplified model [Note: *T*: intervention/treatment, *FR*: fire resistance].

Table 8.9 Results of analysis [bolded *p*-values imply significance that is larger than 5%].

	Estimate		Refute (min)		
Variable	Mean value (min)	*p*-value	Random Common Cause	Data Subset Refuter	Placebo Treatment
W	−245.0	0.11	−247.3	−225.7	−2.1
r	19.0	0.76	19.2	18.4	0.27
L	−82.0	0.97	−81.7	−78.0	−2.1
f_c	40.9	0.85	42.2	41.3	0.28
K	−81.1	**0.02**	−81.1	−80.5	1.73
C	87.8	**5.8e-9**	87.9	87.6	−0.07
P	36.3	**0.004**	36.4	36.3	0.99

8.5.5 A DAG Similar to a Regression Model

In this analysis, I am disregarding the inter-variable effects and only assuming that each variable affects *FR* directly (see Figure 8.13). The result of this analysis is listed in Table 8.9. This table shows that when all variables are assessed for their impacts on FR, positive interventions negatively influence FR for *W*, *L*, and *K*, whereas they positively influence FR for *r*, f_c, *C*, and *P*. The analysis from this DAG also seems to satisfy all refuting models. It is interesting to note that results from this sub-analysis match well (with the exception of f_c) with that from the DAG that was augmented with domain knowledge.

Example no. 5 inferring the benefits of alternative systems
The following is a case study about inferring the causal effects of whether the adoption of a new water treatment system[117] (i.e., $T = 1$) is more beneficial than an existing water filtering system (e.g., $T = 0$) in reducing the concentration of toxic compounds (*C*) in homes. Data were collected from two branches of a national laboratory (Branch A and Branch B) to be analyzed to decide if this national laboratory could adopt the new system across the country (see Table 8.10). In hindsight, we are trying to estimate the causal effect of the treatment (*T*) on the concentration of toxic compounds (*C*) measured in homes.

117 Remember when I told you I will do my best to borrow examples from a variety of civil & environmental engineering fields. Here is one!

Table 8.10 Data compiled by Branch A and Branch B (total samples = 7160 homes).

Element	New water treatment system ($T = 1$)	Existing water filtering system ($T = 0$)
Branch A	810 out of 870 improved (93%)	2430 out of 2800 improved (87%)
Branch B	1900 out of 2620 improved (73%)	600 out of 870 improved (69%)
Total	2710 out of 3490 improved (78%)	3030 out of 3670 improved (83%)

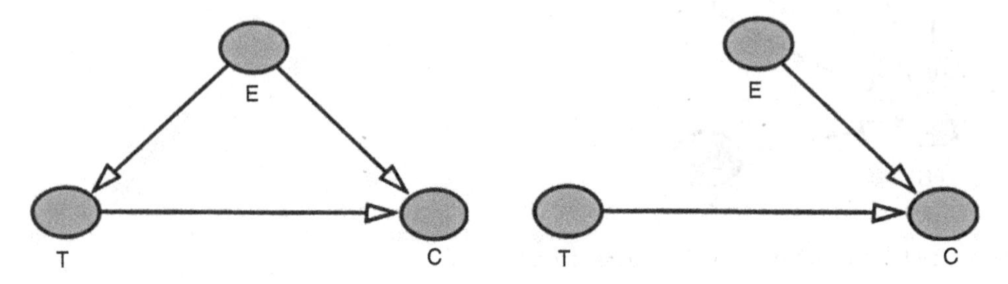

Figure 8.14 DAGs for example no. 5.

A quick look at Table 8.10 shows that the new water treatment system outperforms the existing water filtering system in both branches (93% vs. 87% and 73% vs. 69%, respectively). Surprisingly, a glance at the Totals at the bottom of this table reveals that the new water treatment system underperforms (78% vs. 83% for Branch A and Branch B, respectively).[118]

Let us continue the narrative of this example. The newly elected chief environmental engineer (E) is an advocate of the new system. Thus, being the chief engineer has an effect on adopting the new system ($T = 1$), which in turn has an effect on the concentration of toxic compounds (C). This can be translated into the DAG to the left in Figure 8.14.

If we take a step back to think about this example, it is evident that this national laboratory is interested in the probability of improvement in filtering toxic compounds if both branches are compelled to adopt (or not adopt) the new water treatment system. In other words, the difference between these two probabilities is indeed the *average causal effect* in the population of the data listed in Table 8.10.

Simply, we will need to calculate the $p(C = 1 | \text{do}(T = 1))$ and $p(C = 1 | \text{do}(T = 0))$ which can be arrived at from the interventional DAG (see Figure 8.14 Right). For completion, the required *do-calculus* calculations are presented below.

$$p(C=1|\ do(T=1)) = p(C=1|T=1,E=0)p(E=0) + p(C=1|T=1,E=1)p(E=1)$$
$$= 810/870 \times (870+2800)/7160 + 1900/2602 \times (2620+807)/7160 = 0.83$$

And,

$$p(C=1|\ do(T=0)) = p(C=1|T=0,E=0)p(E=0) + p(C=1|T=0,E=1)p(E=1)$$
$$= 2430/2800 \times (870+2800)/7160 + 600/870 \times (262+870)/7160 = 0.78$$

Then, the expectation (ACE) is:

$$[Y\ do(T=1)] - \mathbb{E}\,[Y\ do\ (T=0)] = 0.83 - 0.78 = +0.05$$

118 This intriguing data is an example of *Simpson's Paradox* [72, 73] and hence warrants a proper causal inference investigation to estimate the true causal effects.

Please note that this estimate (+5%) infers that, on average, 5% more homes would benefit from using the new water treatment system. Surprisingly, this is opposite to the initial conclusion we drew from the last row of Table 8.11.

8.6 A Note on Causality and ML

I hope by now you were able to realize that traditional ML can be a synonym for predictive modeling. When a ML model predicts a value for an outcome, this prediction is likely to arise from a non-causal procedure since the ML model, in principle, was not developed using the three rungs of causal principles. Two more items to remember, a typical ML model assumes that the data points are *independent and identically distributed* (iid)[119] and rarely checks for confoundness and selection bias.

Unlike traditional ML, a causal model seeks to estimate a causal effect, emphasizes the role of the input variables, and seeks evidence (domain knowledge and judgment[120]) to identify such a role in establishing causality, minimizing/eliminating confoundness, selection bias, etc. An important note to remember is that a causal model is interpretable by nature as it conveys the DGP; however, a ML model may not be.[121]

While this chapter is a proponent of causal ML, please remember that we may not need to truly establish causality and build causal models in every single problem we face! In some problems, a blackbox model suffices, while in others, we may be in need of an explainable model. In civil and environmental engineering, we may have solid domain knowledge that can overcome the need to search for causal models! In all cases, it is vital to keep in mind that realizing causality is challenging as we may lack the proper data or tools.[122]

8.7 Conclusions

This chapter presents big ideas and concepts behind causality, causal discovery, and causal inference from a civil and environmental engineering lens. To sum up this chapter, causality aims to establish the causes and their effects. There are two main approaches to causality (the *Pearlian* approach and the *Potential Outcome* approach). We often start with causal discovery to uncover a possible structure to a DGP. We can also leverage domain knowledge to help finetune the discovered structure (as well as associated DAGs). Yet, we may opt to infer the influence of interventions as per the discovered model (with/out a DAG). Remember that inferring is different from predicting – the focus of many of our discussions in earlier chapters.

The following also help summarizes the main ideas behind this chapter:

- Causality is an exciting and uncharted area of research, and integrating causal principles is expected to further accelerate knowledge discovery. Thus, our domain is expected to thrive significantly from future success in this area.
- Identifying data generating mechanisms (DGPs) is a natural pursuit for engineers and their quest to discover new knowledge.
- There is a dire need to integrate causal principles into traditional ML methods since, unlike traditional ML, the causal analysis provides us with the most realistic predictions given its capability to accommodate interventions (without needing new tests or experiments) vs. pure statistical associational predictions provided by traditional ML.
- Causal graphs and search algorithms may, and sometimes may not, agree with our domain knowledge, and hence incorporating such knowledge into causal analysis is beneficial.

---

119 Each data point was generated without reference to others, and the underlying distributions in the DGP are the same for all of the points.

120 For transparency, some ML models also apply domain knowledge to justifying the rationale behind selecting or neglecting some features – this still does not equate to a causal model.

121 There could be a debate with regard to explainable ML. However, as we have seen in earlier chapters, explainability thrives on reasoning a prediction from a data perspective (and not with regard to DGP).

122 Are you looking for a research problem? Look no farther.

Definitions

Causal discovery A process to discover the underlying structure of causal relations between variables pertaining to a data generating mechanism.
Causal inference A process where we draw a conclusion that a particular intervention caused an effect (or outcome).
Causal models Models that describe (mathematically or otherwise) the possible DGP responsible for producing the phenomenon.
Causality The science of cause and effect.
Conditional association Two parameters are independent given a third.
Confounder Is a variable that influences both the independent and dependent variables.
Counterfactuals Things we can imagine.
Data generating process Defined as the process that generates the data.
Do-calculus An axiomatic mathematical system that simulates physical interventions by deleting certain functions from a model and replacing them with a constant while keeping the rest of the model unchanged.
Domain knowledge Information and concepts accepted by a domain.
Interventions Things we can do.
Marginal association Two parameters are independent while ignoring a third.
Observations (from a causal perspective) Things we see.
Spurious correlation Occurs when two parameters appear causally related, but in reality they are not.
Structural equation models A set of statistical techniques used to measure and analyze the relationships of observed and latent variables while accounting for measurement error.

Questions and Problems

8.1 How does causal discovery differ from causal inference?

8.2 How does causality differ from traditional ML models?

8.3 Comment on regression vs. causality.

8.4 What are the main differences between the PO and Pearlian approaches?

8.5 What are the advantages and disadvantages of score-based and constraint-based causal methods?

8.6 Repeat the covered examples and compare the results of your analysis to that discussed in each example.

8.7 Following Example no. 1, use *Tetrad* to arrive at a DAG representing the shear capacity[123] phenomenon in W-shaped steel beams. Complete this problem using the following database *"Causality [Problem 6].txt."*

8.8 Select a ML package of your choosing (i.e., *Pcalg*, or *bnlearn*, or any of those listed in Table 8.2) to explore how the discovered DAGs differ from those uncovered in examples 2 and 3.

8.9 Redo Example no. 5 using a ML package of your choosing (i.e., *DoWhy*).

8.10 Evaluate Example no. 1 on FRP-strengthened RC beams (from the *Explainability and Interpretability* chapter) using *Tetrad.*

8.11 Read, Sir Austin Bradford Hill [47] criteria for establishing causality. Summarize these criteria.

---

123 Shear capacity $(V) = 0.6 f_y \times A_w$, and A_w is the web area $= d \times t_w$, where d is the depth of section and t_w is the thickness of web.

Chapter Blueprint

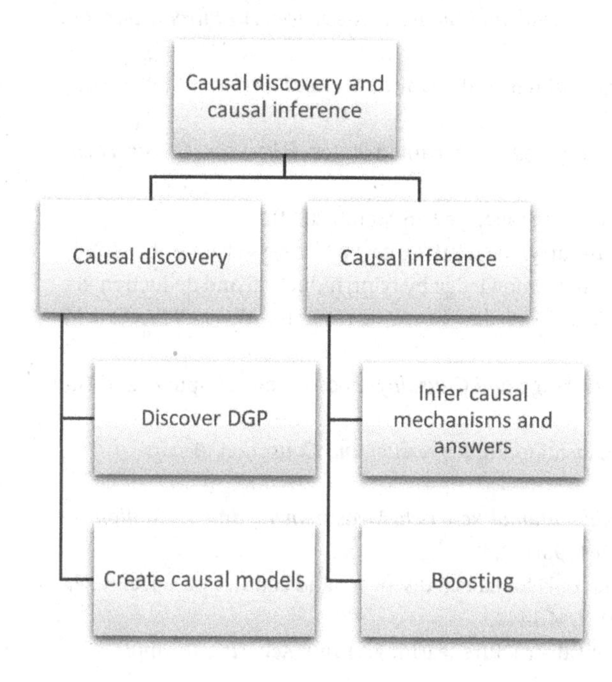

References

1 Bunge, M. (1979). *Causality and Modern Science*. doi: 10.4324/9781315081656.
2 Schölkopf, B. (2022). *Causality for Machine Learning*. doi: 10.1145/3501714.3501755.
3 Pearson, K. (1892). *The Grammar of Science*. doi: 10.1038/046247b0.
4 Chambliss, D.F. and Schutt, R.K. (2013). Causation and experimental design. In: *Making Sense of the Social World: Methods of Investigation*. https://digitalcommons.hamilton.edu/books/30.
5 Ziegel, E.R. (2003). *The Elements of Statistical Learning, Technometrics*. doi: 10.1198/tech.2003.s770.
6 Christodoulou, E., Ma, J., Collins, G.S. et al. (2019). A systematic review shows no performance benefit of machine learning over logistic regression for clinical prediction models. *J. Clin. Epidemiol.* doi: 10.1016/j.jclinepi.2019.02.004.
7 Bzdok, D., Altman, N., and Krzywinski, M. (2018). Statistics versus machine learning. *Nat. Methods.* doi: 10.1038/nmeth.4642.
8 Sun, H., Burton, H.V., and Huang, H. (2021). Machine learning applications for building structural design and performance assessment: state-of-the-art review. *J. Build. Eng.* 33: 101816. doi: 10.1016/j.jobe.2020.101816.
9 Hand, D.J. (2013). Probability for statistics and machine learning: fundamentals and advanced topics by Anirban DasGupta. *Int. Stat. Rev.* doi: 10.1111/insr.12011_5.
10 Pearl, J. (2009). Causal inference in statistics: an overview. *Stat. Surv.* doi: 10.1214/09-SS057.
11 Rudin, C. (2019). Stop explaining black box machine learning models for high stakes decisions and use interpretable models instead. *Nat. Mach. Intell.* doi: 10.1038/s42256-019-0048-x.
12 Naser, M.Z. (2021). An engineer's guide to eXplainable artificial intelligence and interpretable machine learning: navigating causality, forced goodness, and the false perception of inference. *Autom. Constr.* 129: 103821. doi: 10.1016/J.AUTCON.2021.103821.
13 Brady, H.E. (2009). Causation and explanation in social science. In: *The Oxford Handbook of Political Methodology* (ed. J.M. Box-Steffensmeier, H.E. Brady and D. Collier). doi: 10.1093/oxfordhb/9780199286546.003.0010.
14 Klemme, H.F. (2020). Hume, David: a treatise of human nature. In: *Kindlers Literatur Lexikon (KLL)* (ed. H.L. Arnold). doi: 10.1007/978-3-476-05728-0_9644-1.
15 Lewis, D. (1973). Causation. *J. Philos.* 70: 556. doi: 10.2307/2025310.
16 Rubin, D.B. (2005). Causal inference using potential outcomes. *J. Am. Stat. Assoc.* doi: 10.1198/016214504000001880.
17 Pearl, J. and Mackenzie, D. (2018). *The Book of Why: The New Science of Cause and Effect-Basic Books*. Basic Books.

18 Imbens, G.W. (2020). Potential outcome and directed acyclic graph approaches to causality: relevance for empirical practice in economics. *J. Econ. Lit.* doi: 10.1257/JEL.20191597.

19 Imbens, G.W. and Rubin, D.B. (2015). *Causal Inference: For Statistics, Social, and Biomedical Sciences: An Introduction*. doi: 10.1017/CBO9781139025751.

20 Wardhana, K. and Hadipriono, F.C. (2003). Analysis of recent bridge failures in the United States. *J. Perform. Constr. Facil.* 17: 144–150. doi: 10.1061/(ASCE)0887-3828(2003)17:3(144).

21 Wardhana, K. and Hadipriono, F.C. (2003). Study of recent building failures in the United States. *J. Perform. Constr. Facil.* 17: 151–158. doi: 10.1061/(ASCE)0887-3828(2003)17:3(151).

22 Surveys | NCSES | NSF (2022). https://www.nsf.gov/statistics/surveys.cfm (accessed 26 March 2022).

23 Wasserman, L. (2021). Causal inference. https://www.stat.cmu.edu/~larry/=stat401/Causal.pdf accessed 3 April 2022).

24 Naser, M.Z. (2022). Causality in structural engineering: discovering new knowledge by tying induction and deduction via mapping functions and eXplainable artificial intelligence. *AI Civ. Eng.* 1(1):6. doi: 10.1007/s43503-022-00005-9 (accessed 7 September 2022).

25 Huntington-Klein, N. (2021). *The Effect: An Introduction to Research Design and Causality*. Boca Raton: Chapman and Hall/CRC. doi: 10.1201/9781003226055.

26 Glymour, C., Schemes, R., Spirtes, P. et al. (1994). Regression and causation. In: *Regression and Causation. Report CMU-PHIL-60*.

27 Frey, B.B. (2018). Causal inference. In: *The SAGE Encyclopedia of Educational Research, Measurement, and Evaluation* (ed. B.B. Frey). USA: University of Kansas. doi: 10.4135/9781506326139.n102.

28 Nogueira, A.R., Pugnana, A., Ruggieri, S. et al. (2022). Methods and tools for causal discovery and causal inference. *Wiley Interdiscip. Rev. Data Min. Knowl. Discov.* 12: e1449. doi: 10.1002/WIDM.1449.

29 Heinze-Deml, C., Maathuis, M.H., and Meinshausen, N. (2017). Causal structure learning. *Annu. Rev. Stat. Its Appl.* doi:10.1146/annurev-statistics-031017-100630.

30 Scheines, R. (n.d.). An Introduction to Causal Inference. Carnegie Mellon University.

31 Spirtes, P., Glymour, C., and Scheines, R. (2000). Causation, prediction, and search (Springer lecture notes in statistics). *Lect. Notes Stat.* 554.

32 Yu, K., Li, J., and Liu, L. (2016). *A Review on Algorithms for Constraint-based Causal Discovery*. doi: 10.48550/arxiv.1611.03977.

33 Hossin, M. and Sulaiman, M.N. (2015). A review on evaluation metrics for data classification evaluations. *Int. J. Data Min. Knowl. Manag. Process.* doi: 10.5121/ijdkp.2015.5201.

34 Naser, M.Z. and Alavi, A.H. (2021). Error metrics and performance fitness indicators for Artificial Intelligence and machine learning in engineering and sciences. *Archit. Struct. Constr.* 1: 1–19. doi:10.1007/S44150-021-00015-8.

35 Naser, M.Z. and Ciftcioglu, A.O. (2022). *Causal Discovery and Causal Learning for Fire Resistance Evaluation: Incorporating Domain Knowledge*. doi: 10.48550/arxiv.2204.05311.

36 Naser, M.Z. (2022). Causality, causal discovery, and causal inference in structural engineering. *Under Rev.* doi: 10.48550/arxiv.2204.01543.

37 AISC (2022). AISC shapes database v15.0H. Am. Inst. Steel Constr. Database. https://www.aisc.org/globalassets/aisc/manual/v15.0-shapes-database/aisc-shapes-database-v15.0.xlsx (accessed 30 March 2022).

38 Hertz, K.D. (2003). Limits of spalling of fire-exposed concrete. *Fire Saf. J.* 38: 103–116. doi: 10.1016/S0379-7112(02)00051-6.

39 Khoury, G.A. (2001). Effect of fire on concrete and concrete structures. *Prog. Struct. Eng. Mater.* 2: 429–447. doi: 10.1002/pse.51.

40 Kodur, V.K.R. (2000). Spalling in high strength concrete exposed to fire: concerns, causes, critical parameters and cures. In: *Advanced Technology in Structural Engineering* (ed. M. Elgaaly, 1–9. Reston, VA: American Society of Civil Engineers. doi: 10.1061/40492(2000)180.

41 Sanjayan, G. and Stocks, L.J. (1993). Spalling of high-strength silica fume concrete in fire. *ACI Mater. J.* doi:10.14359/4015.

42 Beaumont, P., Horsburgh, B., Pilgerstorfer, P. et al. (2021). CausalNex.

43 Zheng, X., Aragam, B., Ravikumar, P., and Xing, E.P. (2018). Dags with no tears: continuous optimization for structure learning. In: *Part of Advances in Neural Information Processing Systems 31 (NeurIPS 2018)*. https://papers.nips.cc/paper/2018/hash/e347c51419ffb23ca3fd5050202f9c3d-Abstract.html.

44 Pedregosa, F., Varoquaux, G., Gramfort, A. et al. (2011). Scikit-learn: machine learning in python. *J. Mach. Learn. Res.* 12: 2825–2830.

45 Sharma, A. and Kiciman, E. (2020). DoWhy: an end-to-end library for causal inference. doi: 10.48550/arxiv.2011.04216.

46 Sharma, A. and Kiciman, E. (2019). DoWhy: a Python package for causal inference.
47 Hill, A.B. (1965). The environment and disease: association or causation? *J. R. Soc. Med.* doi: 10.1177/003591576505800503.
48 Michotte, A. (1963). *The Perception of Causality*. doi;10.4324/9781315519050.
49 Salmon, W.C. (1998). *Causality and Explanation*. doi:10.1093/0195108647.001.0001.
50 Holland, P.W. (1985). Statistics and Causal Inference. *J. Am. Stat. Assoc.* doi: 10.1080/01621459.1986.10478354.
51 Claudius, R.H. (1932). Causality from the point of view of the engineer. *Philos. Rev.* doi: 10.2307/2179802.
52 Allen, G.I. (2020). Handbook of graphical models. *J. Am. Stat. Assoc.* doi: 10.1080/01621459.2020.1801279.
53 Hoerl, C., McCormack, T., and Beck, S.R. (2011). *Understanding Counterfactuals, Understanding Causation: Issues in Philosophy and Psychology*. doi: 10.1093/acprof:oso/9780199590698.001.0001.
54 Pearl, J. (2009). *CAUSALITY*, 2e. Cambridge University Press. http://bayes.cs.ucla.edu/BOOK-2K (accessed 28 March 2022).
55 Woodward, J. (2008). Causation and manipulability. *Stanford Encycl. Philos.*.
56 Glymour, C., Zhang, K., and Spirtes, P. (2019). Review of causal discovery methods based on graphical models. *Front. Genet.* doi: 10.3389/fgene.2019.00524.
57 Forney, A. and Mueller, S. (2021). Causal inference in AI education: a primer. *J. Causal Inference.* (n.d).
58 Spirtes, P. and Zhang, K. (2016). Causal discovery and inference: concepts and recent methodological advances. *Appl. Informatics.* doi:10.1186/s40535-016-0018-x.
59 Pearl, J. (2012). The do-calculus revisited, in: uncertain. *UAI'12: Proceedings of the Twenty-Eighth Conference on Uncertainty in Artificial Intelligence.*
60 Causal Models (Stanford Encyclopedia of Philosophy) (n.d.) https://plato.stanford.edu/entries/causal-models/index.html (accessed 2 June 2022).
61 Dablander, F. (n.d.) An Introduction to Causal Inference. doi: 10.31234/OSF.IO/B3FKW.
62 Ullman, J.B. and Bentler, P.M. (2012). Structural equation modeling. In: *Handbook of Psychology*, 2e (ed. I.B. Weiner). Tampa, FL. doi: 10.1002/9781118133880.HOP202023.
63 Bollen, K.A. and Pearl, J. (2013). Eight myths about causality and structural equation models. In: *Handbook of Causal Analysis for Social Research* (ed. S.L. Morgan). doi:10.1007/978-94-007-6094-3_15.
64 Glynn, A.N. and Kashin, K. (2018). Front-door versus back-door adjustment with unmeasured confounding: bias formulas for front-door and hybrid adjustments with application to a job training program. *J. Am. Stat. Assoc.* doi: 10.1080/01621459.2017.1398657.
65 Correa, J.D. and Bareinboim, E.. (2017). Causal effect identification by adjustment under confounding and selection biases. *Thirty-First AAAI Conference on Artificial Intelligence.* doi: 10.1609/aaai.v31i1.11060.
66 Pearl, J. (2013). Causal diagrams and the identification of causal effects. In: *Causality* (ed. J. Pearl). doi: 10.1017/cbo9780511803161.005.
67 Bareinboim, E., Tian, J., and Pearl, J. (2022). Recovering from selection bias in causal and statistical inference. *Probabilistic and Causal Inference: The Works of Judea Pearl* doi: 10.1145/3501714.3501740.
68 Vowels, M.J., Camgoz, N.C., and Bowden, R. (2021). *D'ya like DAGs? A Survey on Structure Learning and Causal Discovery.* doi: 10.48550/arxiv.2103.02582.
69 Nogueira, A.R., Gama, J., and Ferreira, C.A. (2021). Causal discovery in machine learning: theories and applications. *J. Dyn. Games.* doi: 10.3934/jdg.2021008.
70 Ramsey, J., Glymour, M., Sanchez-Romero, R., and Glymour, C. (2017). A million variables and more: the fast greedy equivalence search algorithm for learning high-dimensional graphical causal models, with an application to functional magnetic resonance images. *Int. J. Data Sci. Anal.* doi: 10.1007/s41060-016-0032-z.
71 Tools – Center for Causal Discovery (2022). https://www.ccd.pitt.edu/tools (accessed 6 August 2022).
72 Wagner, C.H. (1982). Simpson's Paradox in Real Life. *Am. Stat.* doi: 10.1080/00031305.1982.10482778.
73 Blyth, C.R. (1972). On Simpson's Paradox and the Sure-Thing Principle. *J. Am. Stat. Assoc.* doi: 10.1080/01621459.1972.10482387.
74 Naser, M.Z. (2022). Simplifying Causality: A Brief Review of Philosophical Views and Definitions with Examples from Economics, Education, Medicine, Policy, Physics and Engineering. arXiv preprint arXiv:2212.13537. https://doi.org/10.48550/arXiv.2212.13537 (accessed 27 December 2022).
75 Burton, H. (2022). Causal inference on observational data: opportunities and challenges in earthquake engineering. Earthquake Spectra: 87552930221125492.

9

Advanced Topics (Synthetic and Augmented Data, Green ML, Symbolic Regression, *Mapping Functions*, Ensembles, and AutoML)

On the other hand, omitting things makes the model simpler. But the real world is complex.

Nick Huntington-Klein[1]

Synopsis

The chapter will cover some of the *other* great ML ideas we did not get to cover in previous chapters. It is in this chapter that an engineer can truly appreciate the powerfulness of ML. I will be presenting techniques and ideas[2] behind the following concepts, generating synthetic and augmented data, Green ML,[3] symbolic regression, *Mapping Functions*,[4] and ensembles and AutoML.[5] Personally,[6] I had these lumped into one chapter as I believe that these techniques can, and are likely, to be handy in practical or research problems for some senior/graduate students or practicing engineers.[7]

9.1 Synthetic and Augmented Data

In reality, and given a healthy dataset, the majority of the accepted ML algorithms are likely to work well. These algorithms have been verified against many, many datasets and applications.[8] When these algorithms are applied to different problems, they could use a bit of tuning to arrive at decent performance. Beyond that, an algorithm can only get you as good of a performance as the data can get![9] As we have noticed, our data can make or break ML models.

---

1 From Chapter 6 in *The Effect: An Introduction to Research Design and Causality* [51].

2 With plenty of examples. Given the nature of this chapter and it's make up, you will notice a different organization to this chapter wherein each section starts with a sub-section on *Big Ideas* and ends with a *A note* sub-section. You can think of these sections as mini *synopses* and *conclusions*.

3 **Green ML**: A machine learning approach to energy-cognizant models.

4 ***Mapping Functions***: a mathematical expression that intelligently ties features to outcomes. If you are wondering why this term is italicized, then this is a good question. The answer is that the term was also italicized when I published my first paper on *Mapping Functions*.

5 **AutoML**: Automated Machine Learning.

6 Hence, the absence of a snowflake. As a side note, if you happen to be an educator teaching a course on ML, I may allocate one lecture at the end of the course to skim through the presented ideas and techniques. I would not expect a *deep dive* into these – however, it would be pretty *cool* if you do!

7 *Quick story*: The story for this chapter stems from the chapter's quote as well as this supporting quote by Hubert M. Blalock Jr. "*Put simply, the basic dilemma faced in all sciences is that of how much to oversimplify reality*." from *Causal Inferences in Nonexperimental Research* [52]. While it is common to attempt to simplify reality through idealizations and assumptions to create workable models, it is also important to have the knowledge and tools to minimize such simplifications and overcome their limitations. From this context, this chapter hopes to present you with some knowledge and techniques that go beyond those traditionally adopted in ML.

8 Finding bugs or errors in traditional models can be hard. That does not stop us from trying to improve such models!

9 This is also why some algorithms are assumed to work well in some problems, while others are not. I think that there might be some truth to that wherein the architecture of the algorithm works better for some data vs. others. But I still subscribe to the *no-free lunch* theorem. I would love to do my own series of tests on this! You too, please try to do that as well.

Machine Learning for Civil & Environmental Engineers: A Practical Approach to Data-driven Analysis, Explainability, and Causality, First Edition. M. Z. Naser.

Companion Website: www.wiley.com/go/Naser/MachineLearningforCivilandEnvironmentalEngineers

9.1.1 Big Ideas

Simply, to affordably improve model performance, we can either modify the structure of the algorithm or work on our data. However, if you remember our discussion on Equation 2.2 in Chapter 2,[10] then you will realize that improving the data is likely to have a more favorable response on model performance. The first portion of this book was regarding the former; here, we tackle the latter. This brings me to the idea of synthetic and augmented data.

In many instances, our data could be limited or expensive to collect. For example, conducting full-scale experiments is hectic and costly. Sometimes, researchers may not have access to testing facilities or equipment, which further complicates data collection, or our data may contain heavy noise/outliers which can equally complicate data collection. In some instances where we collect historical data,[11] some aspects of the data collection process may not be accessible to us, nor can we verify its measurements or adequateness.[12]

So, what can we do? Let us think about this for a minute. In the majority of cases, we do tend to have *some* data that we are confident of. Can we use this data pool to extend our dataset? Here is where generating synthetic[13] and augmented[14] data can be of interest to us.

We can start with the strong assumption that the data we have so far is comprehensive and representative enough[15] (see Figure 9.1). Then, we can try to apply a few techniques capable of drawing additional data points from the observations we have. I will cover three techniques here, namely conservative interpolation [1], synthetic minority over-sampling technique (SMOTE) [2], and generative adversarial networks (GANs) [3].

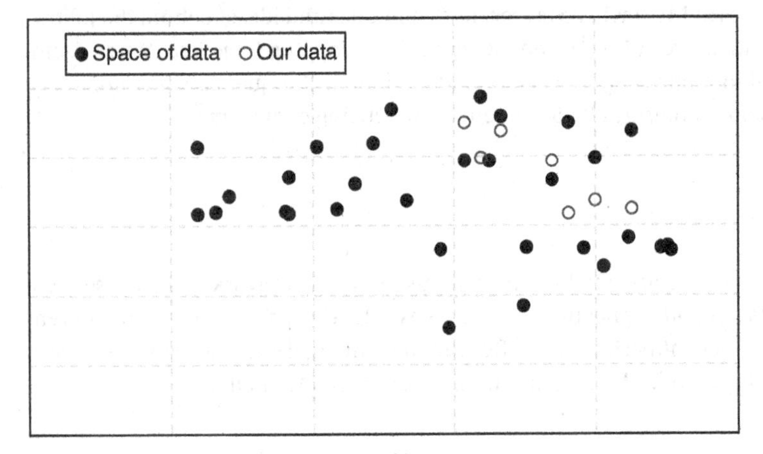

Figure 9.1 Our data vs. space of data.[16]

9.1.2 Conservative Interpolation

While this first technique is coarse and raw, it may help double our dataset quickly and conservatively, especially in a classification problem. Here is what we can do. We cluster observations with similar features in pairs. We then look at the paired observations. We make the *assumption* that if the paired observations have the same outcome, then a third observation can be generated with averaged features of the two (real) parents that will have a similar outcome to the parents. In case that the paired observations have a different outcome, then the synthetic observation is assumed to have the worst case scenario of the two parents (for conservativeness).

---

10 *Machine Learning* = *Algorithms* + *Data*

11 Such as data published in the literature.

12 Sensors used 20–30 years ago may not have the same reliability/sensitivity as those available to us now. Similarly, let us say an engineer is trying to build a ML model to predict the strength of concrete mixtures. This engineer may opt to add mixture data from an experimental program that was conducted in 1990. This effectively implies that the engineer is assuming that such historical data is adequate to predict the properties of concretes we use in 2022! A side note, always be cognizant and transparent of your data sources.

13 **Synthetic data**: Data generated by the manipulation of real observations.

14 **Augmented data**: Data generated by means of simulations or modeling.

15 Hence, the *strong* assumption!

16 Perhaps the term *space* better implies the use of a 3D figure.

Now, here is an illustration.

Suppose we clustered two fire-exposed RC columns with similar features into a pair. If both columns have spalled (or not spalled) under fire, then a third synthetic column can be generated with features equal to the averaged features of the parents. In other words, if the parent columns have spalled, then the synthetic column is also likely to spall since it has averaged features to its parents.[17] In contrast, if one of the columns spalled while the other did not spall, then the third (synthetic column) is assumed to spall. Table 9.1 lists the statistical properties of the running example of fire-exposed RC columns.[18] As you can see, the statistical insights from synthetic and real data match well.

Table 9.1 Comparison between real and synthetic data as obtained from the fire-exposed RC columns dataset.

	Statistic	Width (mm)	Steel ratio (%)	Compressive strength of concrete (MPa)	Concrete cover (mm)	Loading (kN)
Real observations	Minimum	152.0	0.3	15.0	13.0	0.0
	Maximum	514.0	11.7	126.5	64.0	5373.0
	Average	326.1	2.3	42.1	33.5	1342.0
	Standard deviation	71.8	1.6	24.3	7.7	1001.3
	Skewness	0.8	2.5	1.9	−0.4	1.6
Synthetic observations	Minimum	152.0	0.5	16.1	13.0	84.5
	Maximum	511.0	10.8	119.7	51.0	4970.0
	Average	325.6	2.3	42.0	33.5	1338.4
	Standard deviation	66.0	1.3	23.1	7.3	911.8
	Skewness	0.8	2.5	1.9	−0.7	1.6
All observations	Minimum	152.0	0.7	16.0	25.0	0.0
	Maximum	514.0	4.9	126.5	64.0	5373.0
	Average	325.3	2.5	54.3	37.6	1556.9
	Standard deviation	69.4	0.8	27.9	4.4	1109.1
	Skewness	0.7	1.0	1.1	0.6	1.4

9.1.3 Synthetic Minority Over-sampling Technique (SMOTE)

The synthetic minority over-sampling technique (SMOTE), together with its derivates, creates new observations by examining how the nearest two observations of the *smaller* class are joined by a form of distance (i.e., straight line) and then multiplies such distance by a random number between zero and unity, and places the newly generated synthetic observation at this new distance from one of the real observations belonging to the smaller class used for distance calculation. Figure 9.2 illustrates the basic idea behind SMOTE.

9.1.4 Generative Adversarial Networks (GANs) and Triplet-based Variational Autoencoder (TVAE)

As we learned before, GANs is a deep learning technique [4]. If you think about it, since GANs follow an unsupervised approach, then they could be used to learn the structure of the patterns in our limited data. Thus, GANs can also be used to generate synthetic data that could have been possibly drawn from the space of our dataset[19]! These generated synthetic data are statistically indistinguishable from actual data [4]. A good library to use GANs, as well as other advanced techniques (such as Triplet-based Variational Autoencoder (TVAE)) is the Synthetic Data Vault (SDV) [5].

Allow me now to go over the following short example. I will extend one of the running examples we have covered earlier on RC columns exposed to fire since having limited data is a prime problem within the structural fire engineering area. As you might

17 Here is the thing, to be conservative, we are assuming a conservative approach and hence it is our assumption that if a parent spalled, then the newly generated data *might* spall as well. In other words, this is our strong assumption!

18 Dataset is proved for you to repeat this analysis.

19 I have covered how GANs work at an earlier chapter. Please review our discussion.

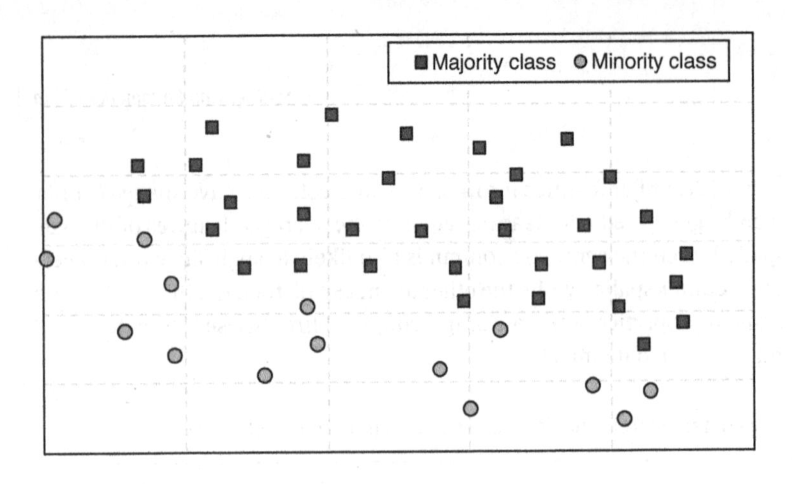

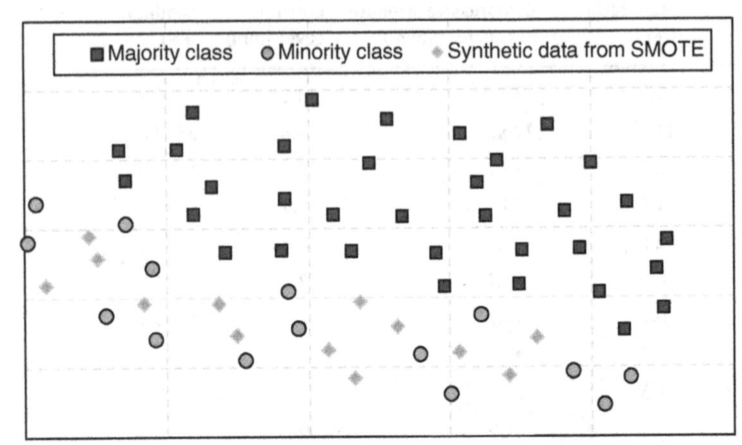

Figure 9.2 Illustration of the SMOTE technique (Top: real dataset, Bottom: dataset with synthetic data).

have remembered, we have about 140 RC columns exposed to fire, and we are trying to predict the fire resistance of these columns. We are hoping to expand this dataset via the TVAE method.[20] I am shooting to create 10,000 synthetic (or fake) data points.

Here is our *Python* code.

```
#Let's generate synthetic (fake) data using TVAE.
#Import libraries and data.

!pip install sdv
from sdv.tabular import TVAE
import numpy as np
import pandas as pd
data = pd.read_excel('RC.xlsx')

#Create model.
model = TVAE(epochs=5000)
model.fit(data)

#Generate synthetic (fake) data.
new_data =model.sample(10000)

#Evaluate synthetic (fake) data.
!pip install table-evaluator
real_data=data
synthetic_data=new_data
from table_evaluator import TableEvaluator
table_evaluator = TableEvaluator(real_data, synthetic_data)
table_evaluator.evaluate(target_col='Fire_Resistance', notebook=True)

table_evaluator.visual_evaluation()
```

20 This would be a good HW problem! Another good HW problem is to create a ML model to predict fire resistance using the real data and the synthetic. I have a code that can help do so!

Absolute Log Mean and STDs of numeric data

Means of real and fake data

Stds of real and fake data

fake data mean (log)

real data mean (log)

fake data std (log)

real data std (log)

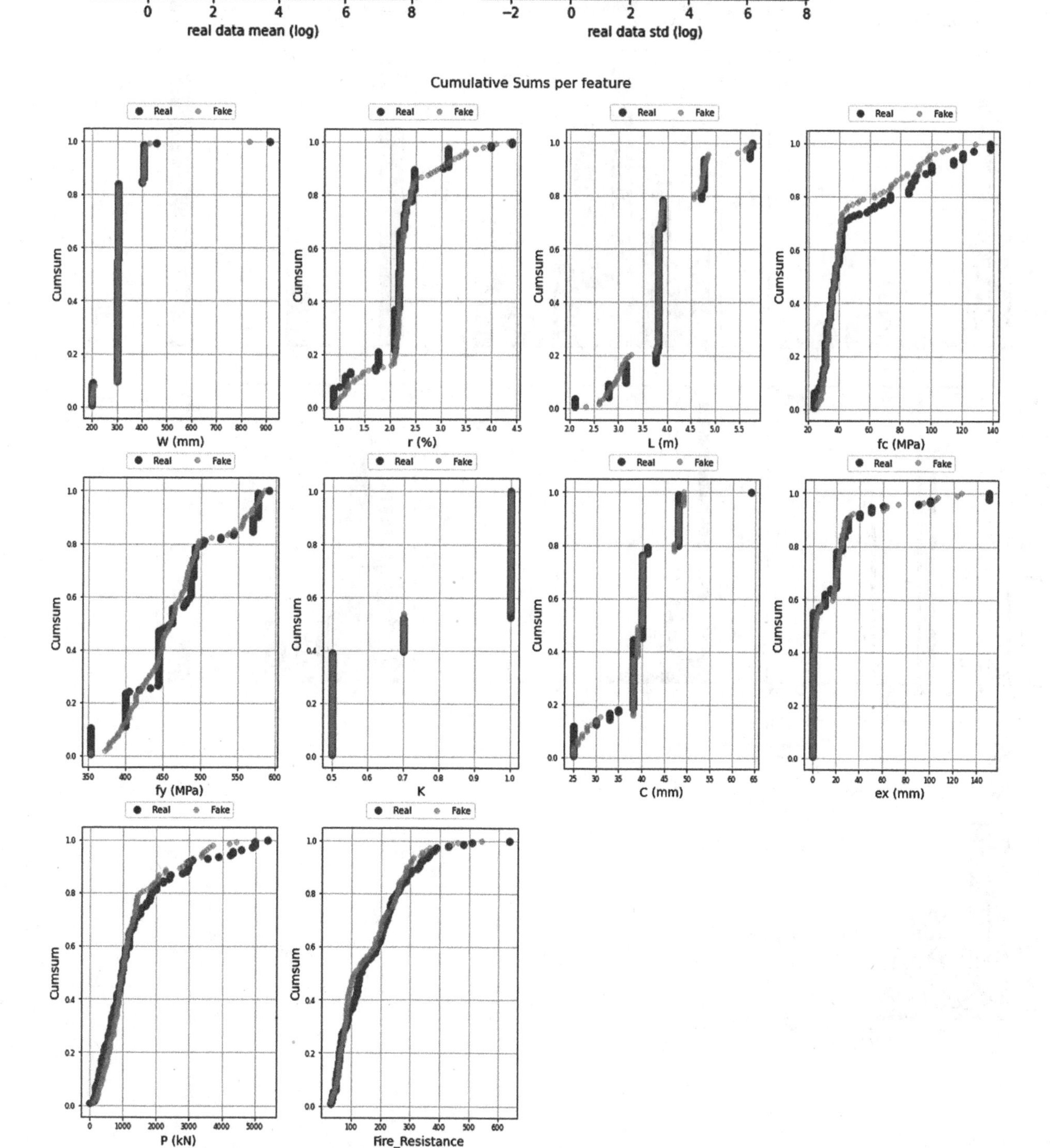

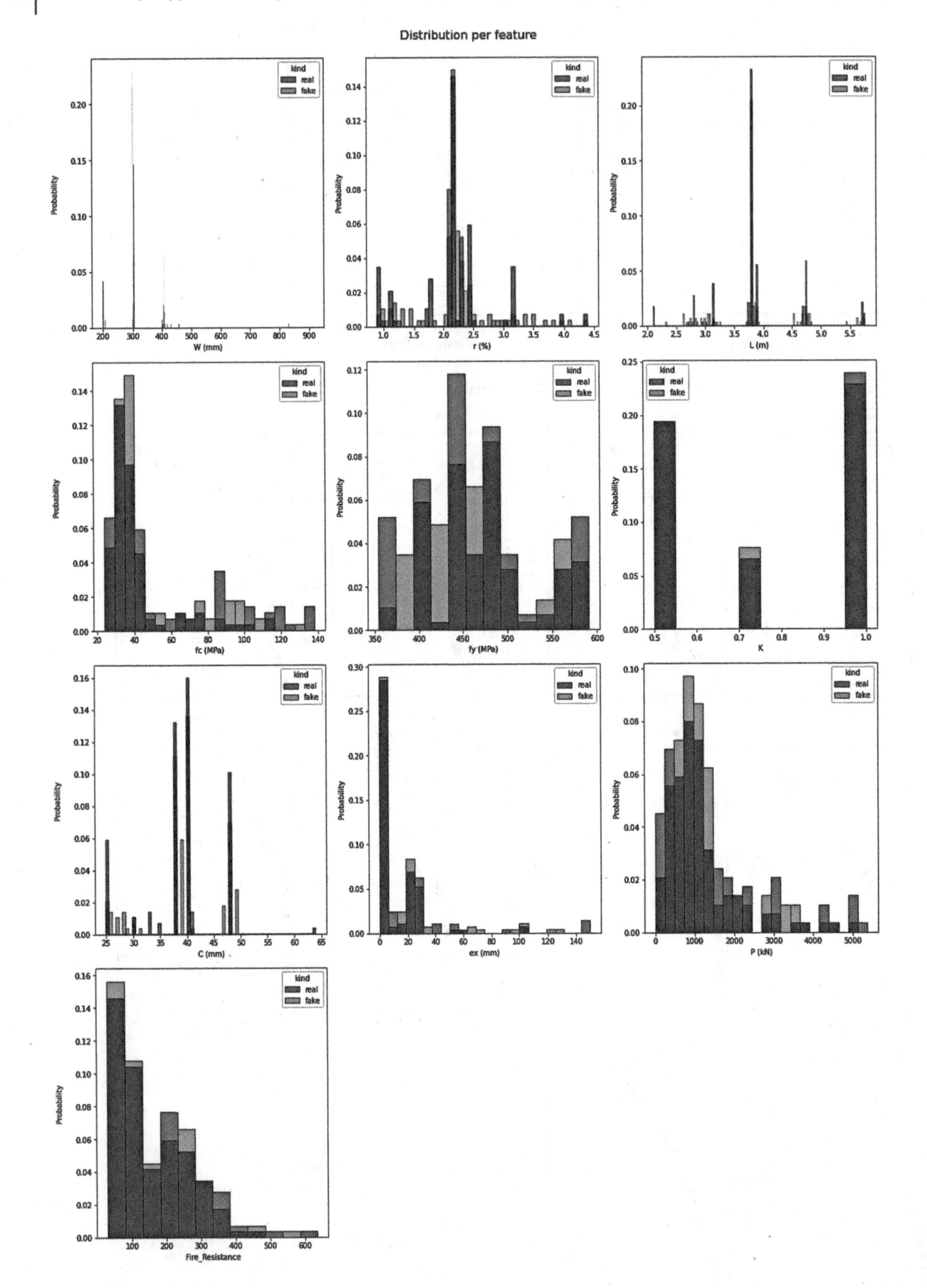
Distribution per feature
kind
real
fake
Probability
W (mm)
r (%)
L (m)
fc (MPa)
fy (MPa)
K
C (mm)
ex (mm)
P (kN)
Fire_Resistance

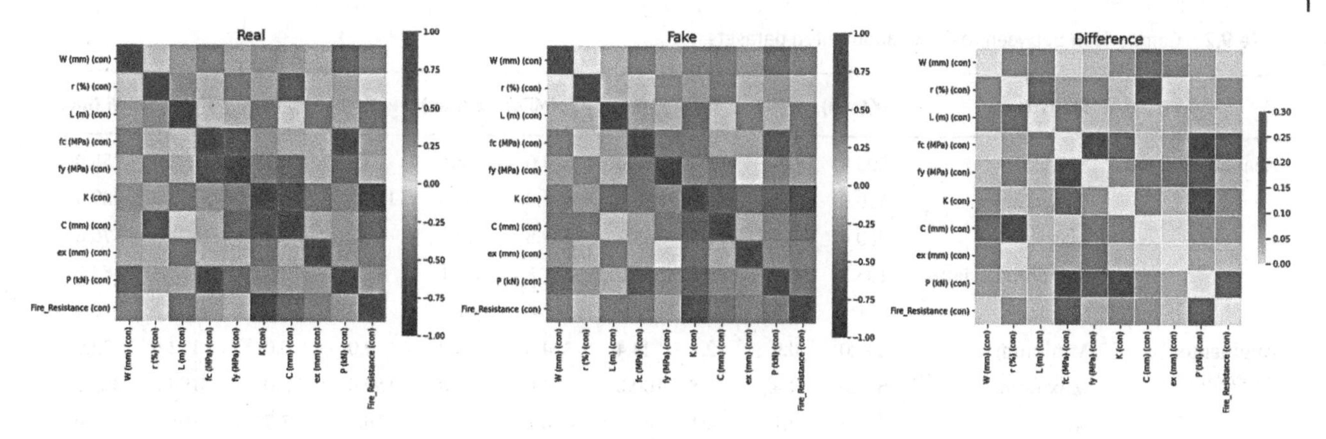


```
#The properties of the generated synthetic data match that of our real (smaller)
dataset.

new_data.describe()
```

	W (mm)	r (%)	L (m)	fc (MPa)	fy (MPa)	K	C (mm)	ex (mm)	P (kN)	Fire_Resistance
count	10,000.0000	10,000.0000	10,000.0000	10,000.0000	10,000.0000	10,000.0000	10,000.0000	10,000.0000	10,000.0000	10,000.0000
mean	308.1444	2.1648	3.8855	47.7973	462.5226	0.7840	38.9143	15.0718	1,243.6419	155.2673
std	46.2391	0.5327	0.6707	25.4207	58.0016	0.2280	6.4090	24.9852	1,029.8042	100.4984
min	200.0000	0.8900	2.1000	24.0000	354.0000	0.5000	25.0000	0.0000	0.0000	31.0000
25%	300.0000	2.1200	3.8000	32.0000	422.0000	0.5000	38.0000	0.0000	583.7500	73.0000
50%	301.0000	2.1900	3.8100	36.0000	463.0000	1.0000	39.0000	2.0000	908.0000	118.0000
75%	304.0000	2.3000	3.8800	46.0000	489.0000	1.0000	40.0000	22.0000	1,453.0000	227.0000
max	914.0000	4.3900	5.7600	138.0000	591.0000	1.0000	64.0000	150.0000	5,373.0000	636.0000

```
#Export data.
new_data.to_excel('X_ synthetic_data.xlsx', index=False)
```

9.1.5 Augmented Data[21]

The last technique that can be used to enlarge the size of a database is to use augmented observations by means of advanced analysis methods such as validated FE or analytical simulations [1]. Simply, a FE/CFD model is validated against a real test. This model is then used to run a parametric study or additional cases to generate data that can be compiled into a dataset. This hybrid dataset (containing real and augmented data) is fed into the ML model for training and testing. Table 9.2 shows statistical insights as collected from two datasets on RC columns aimed at predicting the fire resistance of such columns. The real data comprised 144 RC columns, and the augmented dataset comprised 166 columns that were simulated via FE models.[22]

21 The term augmented images can also be used in computer vision problems. In such problems, the original images are augmented with rotation, grayscale, etc. to yield new images that can be added to a dataset. I will be providing a dataset of images collected from FE models that could be used to build a small deep learning model that can identify buckling in beams (see [53]).

22 I wanted to acknowledge my good friends, Nikhil Raut [25] and Thomas Gernay [54], who modeled many, many of these columns in their papers.

Table 9.2 Comparison between real and augmented datasets.

	Statistic	W (mm)	r (%)	L (m)	f_c (MPa)	f_y (MPa)	C (mm)	e_x (mm)	e_y (mm)	P (kN)	FR (min)
Real observations	Minimum	203.0	0.9	2.1	24.0	354.0	25.0	0.0	0.0	0.0	55.0
	Maximum	610.0	4.4	5.7	138.0	591.0	64.0	150.0	75.0	5373.0	389.0
	Average	350.4	2.1	3.9	55.7	439.3	42.4	12.8	3.2	1501.8	176.6
	Standard deviation	105.3	0.5	0.5	33.0	61.8	7.1	23.3	12.9	1168.6	82.0
	Skewness	1.1	1.0	−0.5	0.9	1.0	−1.0	3.0	4.0	1.3	0.4
Augmented observations	Minimum	200.0	0.9	2.1	24.4	354.0	23.0	0.0	0.0	122.0	22.0
	Maximum	610.0	4.4	5.8	100.0	576.0	50.0	150.0	75.0	3190.0	391.0
	Average	333.9	2.1	4.1	48.1	436.3	41.2	16.1	3.7	1108.1	155.9
	Standard deviation	117.2	0.5	0.6	26.4	52.2	9.9	30.2	13.6	874.5	81.8
	Skewness	1.2	0.9	0.5	1.3	1.2	−0.8	2.7	3.7	1.3	0.2
All observations	Minimum	200.0	0.9	2.1	24.0	354.0	23.0	0.0	0.0	0.0	22.0
	Maximum	914.0	4.4	5.8	138.0	591.0	64.0	150.0	75.0	5373.0	636.0
	Average	324.3	2.1	4.0	49.3	449.4	40.2	15.8	2.0	1204.8	161.0
	Standard deviation	99.2	0.6	0.7	28.1	60.1	8.7	29.7	10.1	1031.6	97.6
	Skewness	1.9	0.6	0.3	1.4	0.7	−0.6	2.9	5.3	1.7	0.9

9.1.6 A Note

I would like to take the next few lines to point out a few items pertaining to our discussion on synthetic and augmented data. First, we need to remember that this newly added data is not real data – as this data was not generated from actual tests or scenarios. While this data *may* resemble many of the same characteristics of the parent data, we still do not know if such fake data actually exist or is replicable.[23]

Second, augmented data might suffer from what I call *idealization syndrome*, which arises from some of the inherent assumptions used in numerical/analytical models. For example, finite elements have assumptions associated with them.[24] The input material properties are also idealized and hence could miss some of the complex phenomena that may occur. In some unique cases, some phenomena cannot be easily captured in such simulations (i.e., charring of wood or spalling in concrete), and hence, augmented data may not produce similar results to that expected in real tests.[25]

9.2 Green ML

Back in the Performance Fitness Indicators and Error Metrics chapter, I presented a small section on energy-based functional indicators.[26] Now, I have a bit more leeway in this chapter to cover the dimension of energy from a more practical view.

23 I can see how results from validated FE models do not fall into this statement.

24 Another example, and this may be a bit specific to structural fire engineering, using augmented results from researchers belonging to a specific region who are likely to apply their region's norm/recommendations to their FE models may lead to some bias (i.e., for example, European researchers are likely to apply Eurocode material models to their FE models, and North American researchers are likely to use ASCE material models in their FE models – such models could be too specific to be assumed to generalize well).

25 These pointers are good to have in mind. You may be thinking about other items. Please send me your thoughts.

26 And promised you to take a more in-depth look into ML and energy in this chapter. Here we go!

9.2.1 Big Ideas

In this section, I argue that there is an ethical component to adopting ML responsibly, and thus ML users (such as civil and environmental engineers) are to be cognizant of the hidden costs of energy consumption and subsequent carbon emissions arising from ML in our domain. In this section, I will show you how much the variation in algorithms, their architecture, and language can influence model predictivity and energy consumption. I will do so via an example on a relatively large dataset (~8000 observations).

Before I do so, allow me to present the following few paragraphs, which I hope you will find informative for our discussion.[27]

With the growing accumulation of big data and the need for larger and more accurate ML models, our engineers will likely end up creating models with intricate architectures/topologies. Not surprisingly, complex models will require tremendous resources during training, development, storage, and deployment.[28] Well, how much resources can we expect?

A rough answer can be found in two recent works that have estimated the monetary and environmental[29] cost of training various ML models. These studies reported a wide range of estimates between $41.0 and $3.2 million and between 6.0 and 150,000 kg of carbon emissions per model[30] [6, 7]. If we take a step back to look at global carbon emissions, we will find that 2.3% of such emissions are generated by the information and communication technology sector[31] [8].

Fundamentally, creating new (or complex) models hopes to realize improved performance. However, the improved performance associated with complex or exotic[32] ML models may or may not positively reflect upon the consumed resources to attain such improvements. For instance, the notable ML model *ResNeXt* consumed 35% more energy to train than its predecessors, *ResNet*, to achieve a 0.5% improvement in accuracy [9]! While improvement in accuracy[33] does come in handy, in a variety of applications, minimal improvement in accuracy may not justify the consumed energy needed to realize such accuracy. Let me illustrate a few examples.

- Suppose we are building a ML model to predict the compressive strength property of concrete. In this example, if the model predicts a strength of 38 MPa ∓ 10% may not *significantly* alter the practical application of this concrete mixture since both extreme ranges remain within the range of normal strength concrete.
  - Building on the first point. Creating an intricate ML model (say, an ensemble or deep learning model) to predict the same property may not be warranted – especially if a simpler model can provide us with a comparable[34] prediction.[35]
- On the other hand, building a ML model to predict the level of toxins in drinking water is likely to be of much higher accuracy and reliability than that of the first example. Such a model is then expected to justify the need for additional computational resources.
  - The above, however, does not rationalize our lack of trying to build a leaner model that is both accurate and of low computational needs.

27 I have debated the addition of this section for a while. However, at the end of the day, this textbook is hoping to establish a foundation for ML education in our domain. While a discussion on fundamentals and presentation of examples is of utmost merit, the continued advances in ML are exponential, with new algorithms and methods being developed on almost a daily basis. In my eyes, I felt it is important to cover some bit-out-of-the-norm concepts, such as Green ML, as I feel it is warranted to learn. If we engrave such concepts early into our journey, this will make such concepts a bit formal. Other big ideas may also include, Ethical ML, Equitable ML, Trustworthy ML, etc.

28 Think of it this way, a 3D FE model is more complex than a simple 1D model!

29 In terms of carbon emissions.

30 Please note that a human adult generates, on average, about 5000 kg of carbon emissions annually. Also, note that 150,000 kg of carbon emissions is equivalent to that emitted through the lifetime of five gasoline-based vehicles [6].

31 For comparison purposes, the cement and aviation industry generates carbon emissions of about 5% and 1.7%, respectively [55, 56]. I hope you can now agree with the need to use ML responsibly.

32 **Complex models** are those that join/ensemble different algorithms/models into one and **exotic models** are those that use one algorithm to optimize or finetune the parent algorithm/algorithm. Please see [69] as much of this discussion stems from there.

33 As well as other aspects of ML models.

34 I am intentionally not defining what is a comparable performance. This could be related to the accuracy, training time, energy indicator, etc.

35 The same example could be extended to a ML model designed to predict the number of commuters passing through a given intersection, or an expected rough budget for a project.

In lieu of the above, the next few years will initiate a ML-based race in our domain. Thus, being aware of the hidden costs of ML will enable us to be efficient users of ML. Investments in new hardware by engineering firms are often expected to last for a few years/decades, and hence embracing a green approach to ML may help extend the lifespan of such hardware and upgrade costs.

9.2.2 Example

Let us now explore ML from an energy perspective. The idea behind this example is simple. I will compare several algorithms in terms of their predictive performance, model size, energy consumption, and storage needs. In addition, I will also be examining the influence of algorithm architecture and programming language.[36] Here are the algorithms: XGBoost, Light Gradient Boosting Machine (LGBM), Keras Deep Residual Neural Network (KDNN)), TensorFlow Deep Learning (TDL), Vowpal Wabbit (VW), and Random Forest (RF). I will compare the performance of these algorithms in classifying if a given concrete mixture can successfully attain its specified design strength at in-situ conditions.[37]

I will use a large dataset comprising 8000 observations compiled by [10, 11].[38] In this dataset, the following features were collected: water, cement, fly ash contents, water-reducing admixture, air-entraining admixture contents, coarse, and fine aggregate contents. The outcome of this dataset is the measured compressive strength.[39] For simplicity, I will be using the *DataRobot* platform's default settings.[40]

To better illustrate this example, I will first be showcasing how all of the selected models achieve comparable predictivity performance yet significantly vary in terms of their energy expenditure.

9.2.2.1 Predictivity

The selected ML models were evaluated from a predictivity point of view by plotting the performance of all models in terms of accuracy (ACC), Area under the ROC curve (AUC), and Log Loss Error (LLE). Figure 9.3 and Table 9.3 show that most, if not all, models seem to predict the target variable adequately. A look into the same figure also shows that any arising variations are minor (as per the scale of the horizontal axis). For example, all models (except KDNN of one layer of 64 units) scored 98% or more in the ACC and AUC metrics. Similarly, most ML models scored within 0.05 of the LLE.[41] It is evident that no particular model scored the highest in all metrics, hence the benefit of applying the predictivity indicator we discussed earlier (also shown in Figure 9.3).

9.2.2.2 Programming Language

When it comes to the effect of programming languages, we can compare the two versions of the RF (*Python* and *R*) and GBT (*Python* and *R*) algorithms. Table 9.3 and Figure 9.3 show that the *R* version of both models performed better than the *Python* version on this dataset.

---

36 Given the wide possibilities and approaches of tuning algorithms (especially those of different topologies), I find it necessary to establish that in order to enable the reproduction of this analysis, all of the examined algorithms were applied in their default settings. Please remember that the goal of this example is <u>*not*</u> to identify the most *suited* algorithm to solve our phenomenon, but rather to compare the selected algorithms from a predictivity and energy point of view.

37 Attaining a specific strength in controlled conditions (i.e., laboratory) is more likely than in site conditions (where conditions and materials are less controlled). As such, ensuring that a concrete mixture attains its specified strength in-situ conditions is important.

38 Yes, I may be a bit biased toward structural engineering and material science. However, I could not resist the temptation of showcasing an example on concrete and ML from an energy and emission perspective – the irony! On a more practical front, I am providing you with a transportation-based, *very* large dataset that you can do and estimate ML models' energy consumption. Thanks to Paul-Stefan Popescu and Cojocaru Ion for sharing this dataset on Kaggle [50].

39 In this example, if the attained strength was equal to or exceeded the design strength, then the mixture is deemed *satisfactory*. If not, then the mixture is deemed *under-designed*. In this dataset, there were 6647 and 1099 observations labeled as *satisfactory*, *under-designed*, respectively.

40 Here are the settings per algorithm used to repeat this analysis: XGBoost: BP65, LGBM: BP70, RF (*R*): BP46, RF (*Python*): BP39, GBT (*Python*): BP40, GBT (*R*): BR63, TFDL: BP25, VW: PB61, LR: BP42, EN: BP12, Keras: BP01, Keras: BP04, Keras BP61, and Keras BP62.

41 I hope you still remember that a score of 0.0 implies a perfect model.

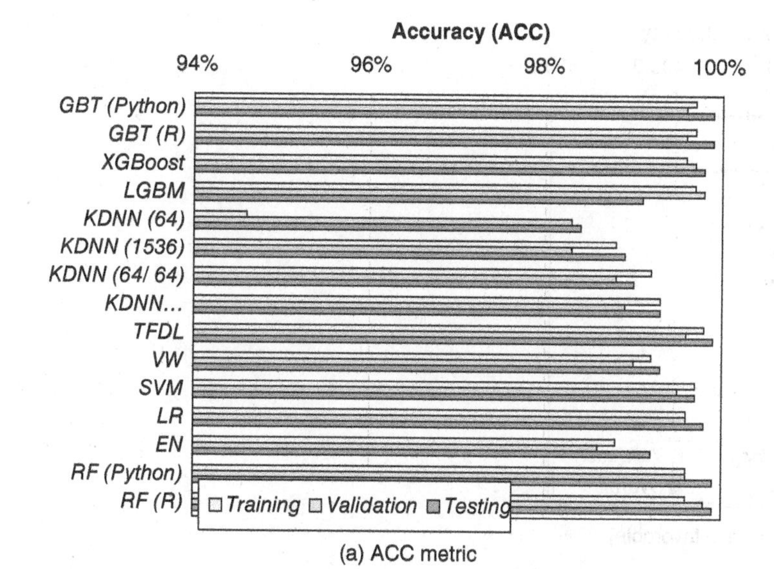

(a) ACC metric

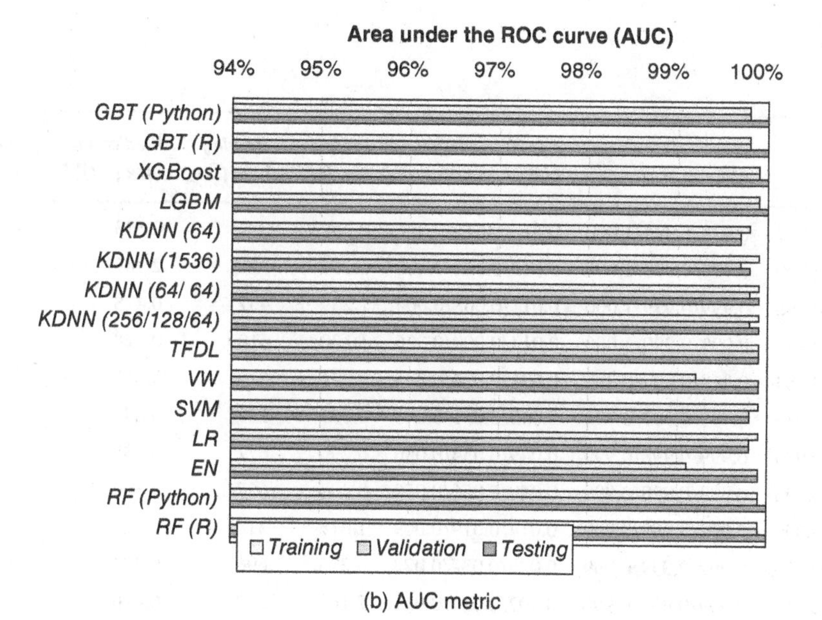

(b) AUC metric

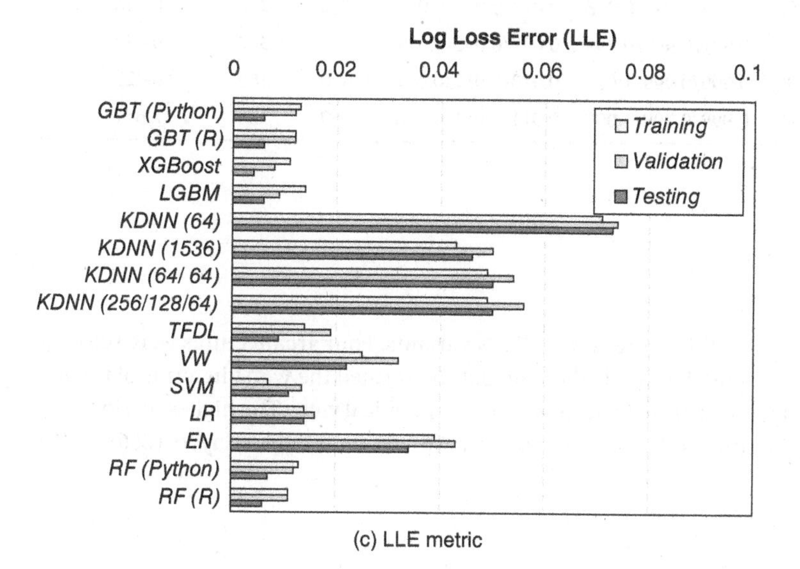

(c) LLE metric

(Continued)

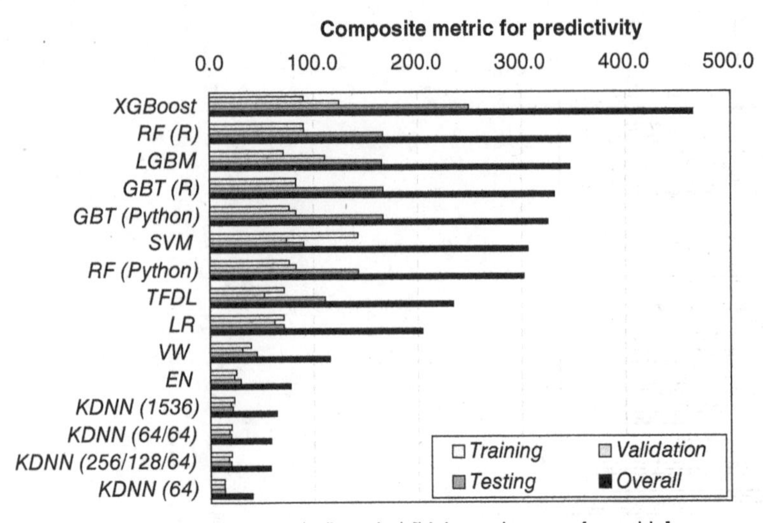

(d) Composite metric (in order) [higher values are favorable]

Figure 9.3 (Cont'd) Comparison of metrics.

Table 9.3 Evaluation of ML models.

Model		DataRobot BP	ACC**	AUC	LLE	Model size (MB)	Training time (Sec)	Prediction time (sec)* **
GBT	Python	BP40	0.997/0.996/0.999	0.998/0.998/1.000	0.013/0.012/0.006	0.337	389	0.523
	R	BP63	0.997/0.996/0.999	0.998/0.998/1.000	0.012/0.012/0.006	0.259	391	2.567
XGBoost		BP65	0.996/0.997/0.998	0.999/0.999/1.000	0.011/0.008/0.004	1.239	437	0.478
LGBM		BP70	0.997/0.998/0.991	0.999/0.999/1.000	0.014/0.009/0.006	1.130	610	0.428
KDNN	64*	BP62	0.946/0.983/0.984	0.998/0.997/0.997	0.071/0.074/0.073	0.271	677	0.851
	1536	BP01	0.988/0.983/0.989	0.999/0.997/0.998	0.043/0.051/0.046	4.757	197	0.895
	64/64	BP04	0.992/0.988/0.990	0.999/0.998/0.999	0.049/0.054/0.050	0.290	677	0.835
	256/128/64	BP04	0.993/0.989/0.993	0.999/0.998/0.999	0.049/0.056/0.050	0.446	677	1.300
TFDL		PB25	0.998/0.996/0.999	0.999/0.999/0.999	0.014/0.019/0.009	1.818	318	0.778
VW		PB61	0.992/0.990/0.993	0.992/0.999/0.999	0.025/0.032/0.022	0.223	696	0.404
SVM		PB14	0.997/0.995/0.997	0.999/0.998/0.998	0.007/0.014/0.011	2.162	233	0.490
LR		BP42	0.996/0.996/0.998	0.999/0.998/0.998	0.014/0.016/0.014	0.152	259	0.430
EN		BP12	0.988/0.986/0.992	0.991/0.999/0.999	0.039/0.043/0.034	0.223	322	0.489
RF	Python	BP39	0.996/0.996/0.999	0.999/0.999/1.000	0.013/0.012/0.007	0.626	366	0.427
	R	BP46	0.996/0.998/0.999	0.999/0.999/1.000	0.011/0.011/0.006	1.460	386	2.657

*Training/validation/testing.
**Number of units in each layer.
***To score 1000 observations.

9.2.2.3 Model Architecture

The best candidate to examine the influence of model architecture is the KDNN variants. Four architectures were selected by changing the number of units and/or layers in each model. While these models performed the worst in terms of predictivity, I would still like to remind you that these models performed *adequately* from a practical view. The highest performing model of the KDNNs is that of one layer with 1536 units, followed by two layers (64/64 units), three layers (256/128/64 units), and finally, one layer of 64 units.

9.2.3 Energy Perspective

Now that we have explored the predictivity of ML models and showed that they have comparable performance, it is time to delve into how much energy each model used. Let us use the energy-based functional indicator.

As you remember, this indicator comprises model size (in MB), training time (in seconds), and prediction time to score 1000 rows (in seconds). These three metrics can be found in *DataRobot*'s platform (see Figure 9.4). Figure 9.4 and Table 9.3 list such metrics for all models. As one can see, the largest and smallest model size belongs to KDNN (with 1536 units[42]) and LR, respectively. Overall, all ML models completed training within 197–696 seconds and had prediction times between 0.404 and 2.657 seconds.

The above metrics are then combined into a functional indicator:[43]

$$Estimated\ Energy\ (E) = MS \times (TT + PT) \tag{9.1}$$

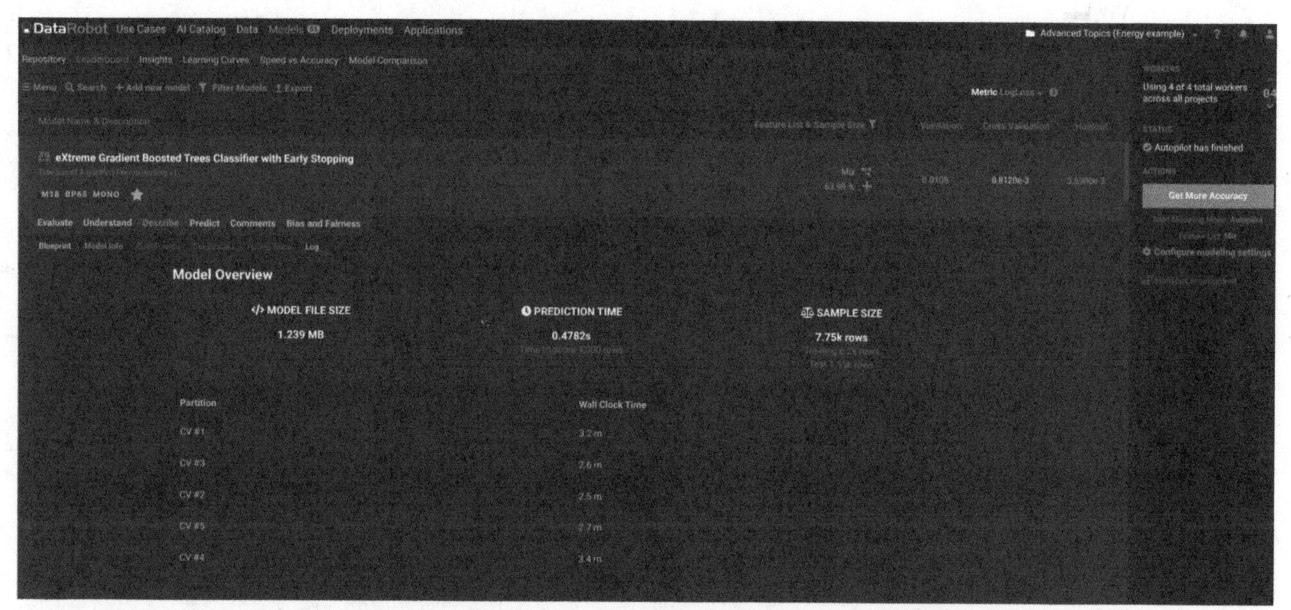

(a) Model overview from *DataRobot* (Model size and Prediction time)

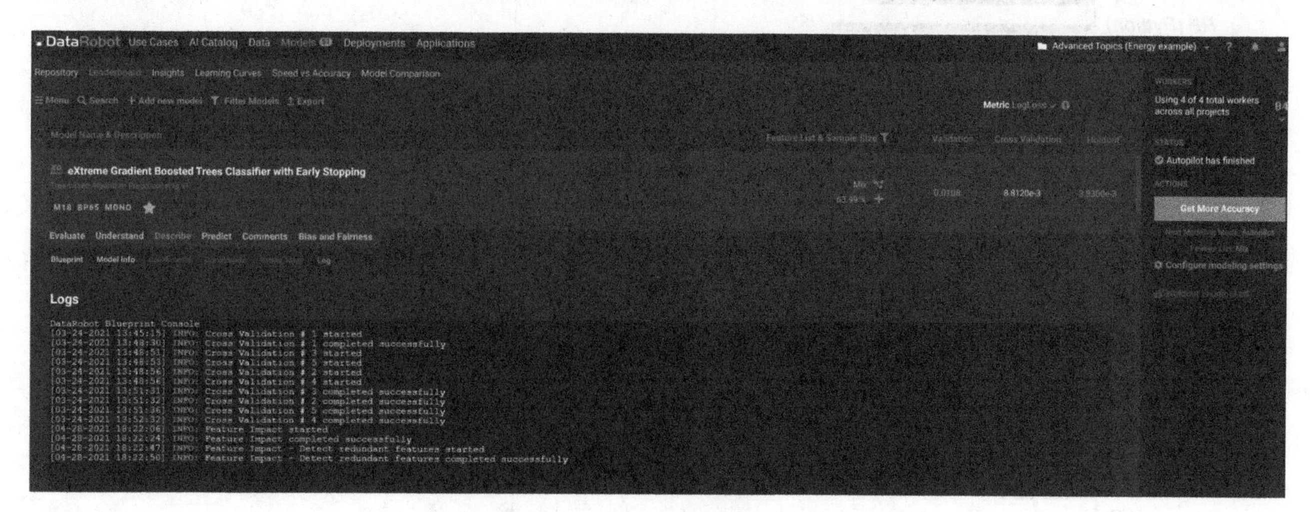

(b) Model overview from *DataRobot* (Training time)

Figure 9.4 Insights into energy metrics from *DataRobot*.

42 This model also happens to be the fastest training model.

43 Please remember that smaller values are favorable.

where *MS*: model size, *TT*: training time, and *PT*: prediction time.

Figure 9.5 shows the results from the functional indicator. It is clear that the most energy efficient model is that of LR, followed by EN and the *R* version of GBT, and the most expensive model is the KDNN (with 1536 units).[44] As you can see, these metrics show that despite the models having similar predictivity, these do not have similar energy expenditure.

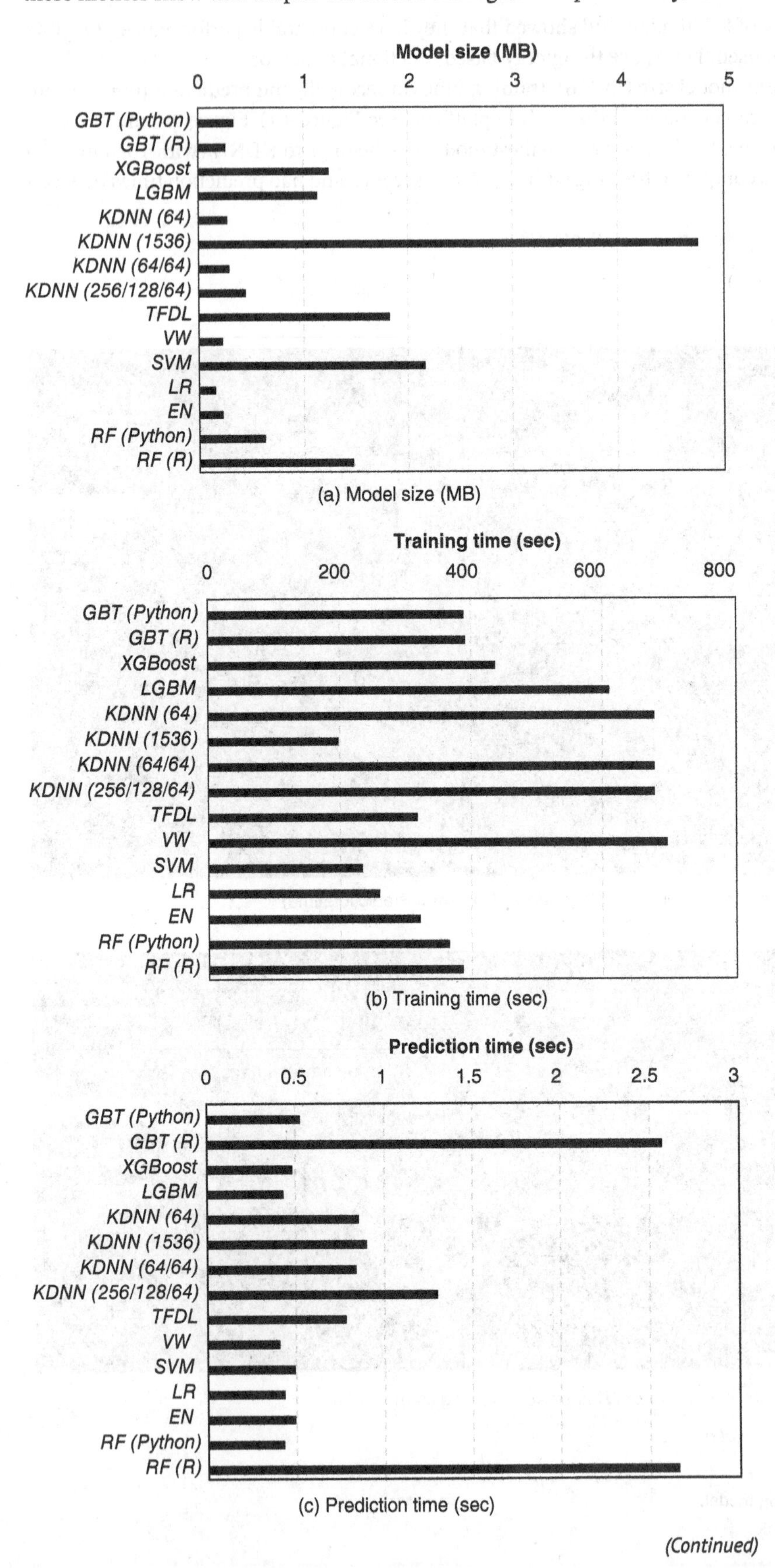

(a) Model size (MB)

(b) Training time (sec)

(c) Prediction time (sec)

(Continued)

44 In this example, the LR model is estimated to consume energy at 4.0% of that of the KDNN (with 1536 units).

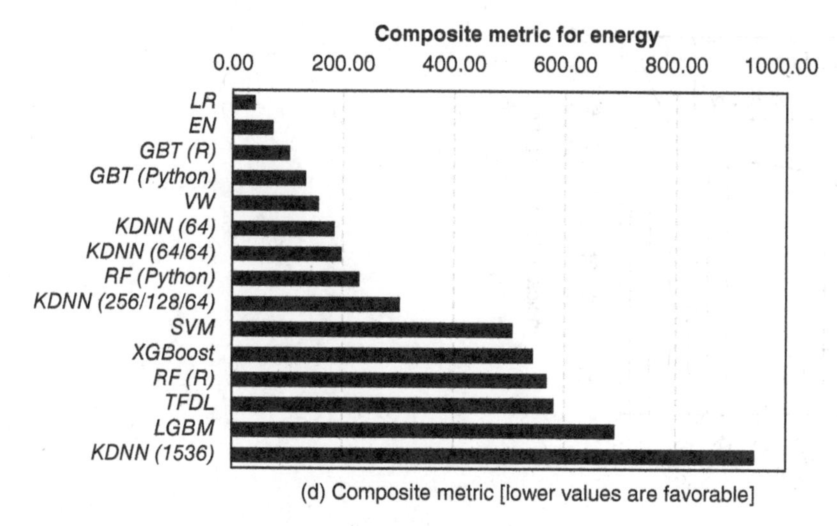

Figure 9.5 (Cont'd) Comparison of metrics.

Let us now revisit the influence of programming language and model architecture from an energy perspective.

9.2.3.1 Programming Language

Generally speaking, the influence of programming language is apparent. For instance, the RF (*R*) variant consumed about 64% more energy than its *Python* variant (and the GBT (*Python*) variant consumed 48% more energy than its *R* variant).

9.2.3.2 Model Architecture

As expected, ML models with simple architectures (i.e., LR) consume less energy as opposed to those of a more complex nature. In terms of KDNN, it is clear that the number of units/layers correlates well with the increase in energy consumption.

9.2.3.3 Additional Insights

Figure 9.6 ties the predictivity of ML models to their energy consumption. Comparatively speaking, it seems that the KDNNs have a similar performance in terms of predictivity and energy consumption (expect KDNN 1536). Similarly, the last tree-like models note substantial predictivity with an accompanying rise in energy. As you can see, the TFDL presents a case for a balance model, and the GBTs[45] have positioned themselves to be at the frontier of this analysis.

Let us dig a little deeper and try to link the above energy expenditure to carbon emissions.[46] Here is a list of assumptions:

- The US Energy Information Administration [12] reports that 0.42 kg of CO_2 is emitted per 1 kWh.
- Lawrence Berkeley National Laboratory, USA [13], estimates the unit annual energy consumption (AEC) for desktops to be 194 kWh/yr, with 20% of desktops consuming more than 300 kWh/yr and high-end computing desktops reaching 600 kWh/yr.

Most engineers are likely to use above-average desktops in their modeling (or a cloud computing service) for their work/modeling. Given the above assumptions, a typical desktop is expected to generate $0.42\text{kg of } CO_2 / \text{kWh} \times 300 \text{ kWh} / \text{yr} = 126\text{kg}$ of CO_2 annually.

45 You could make a case for LR as well.

46 I would like you to know that we do not, yet, have a straightforward method to relate the energy expenditure of ML models to carbon emissions as model development is dependent upon the combination of software/hardware, geographical location of workstations (which may advocate for carbon neutrality by imposing a "carbon tax"), and resources used in generating electricity (renewable resources vs. nonrenewable resources), etc [6, 57, 58]. Setting this aside, the above can safely be assumed as a hypothetical, and exciting, case study.

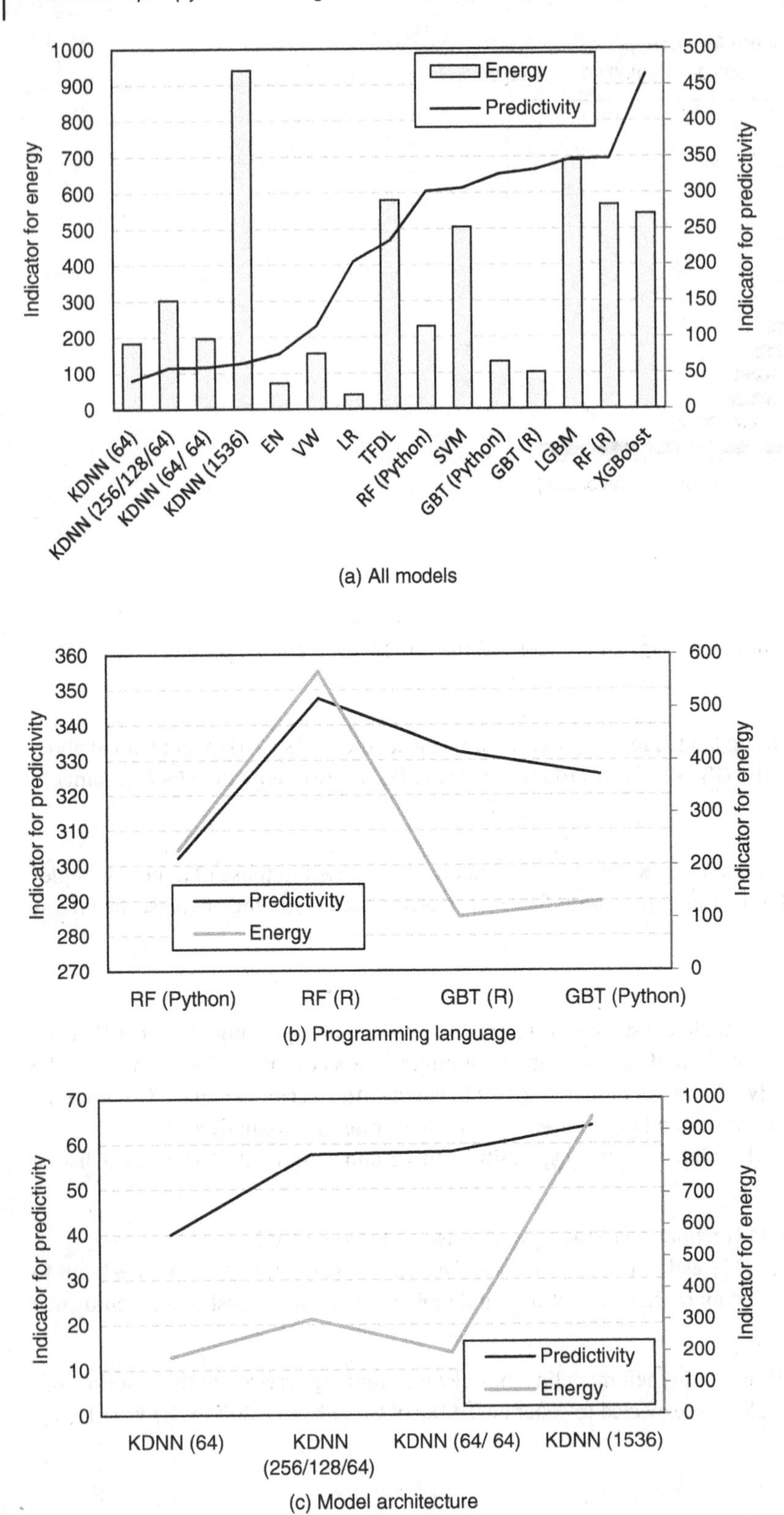

Figure 9.6 Comparison between predictivity and energy for the examined ML models.

Now, the American Society of Civil Engineers estimates that there are about 150,000 civil engineers in 177 countries [14]. Assuming 1%[47] of these engineers are dedicated to ML, this yields us:

$$0.01 \times 150{,}000 \times 126 = 189{,}000\,\text{kg of } CO_2 \text{ annually} \tag{9.2}$$

This is equivalent to emissions generated from 41.1 passenger vehicles driven for one year (about 509,619.7 km) as per the published estimations made by the United States Environmental Protection Agency (EPA) [15].

Since Table 9.3 clearly shows that all selected models can be used to accurately predict if a concrete mixture can develop its design strength under in-situ conditions, the use of energy-hungry models may not be warranted in such a problem.[48]

A question might arise, *how do the presented models compare against each other?*

A straightforward answer would be to normalize the models by their respective highest benchmarks (i.e., XGBoost for predictivity and KDNN (1536) for energy consumption). This comparison is shown in Figure 9.7 and infers that developing simple models such as LR or GBT (R), can lead to a 96% or 89% reduction in energy (and possible CO_2) as opposed to KDNN (1536); effectively implying the merit in favoring simple models. My point from the above lengthy example is to advocate for responsible and Green ML early into the journey of adopting ML to maximize the benefits of ML and negate unresponsible energy consumption.[49]

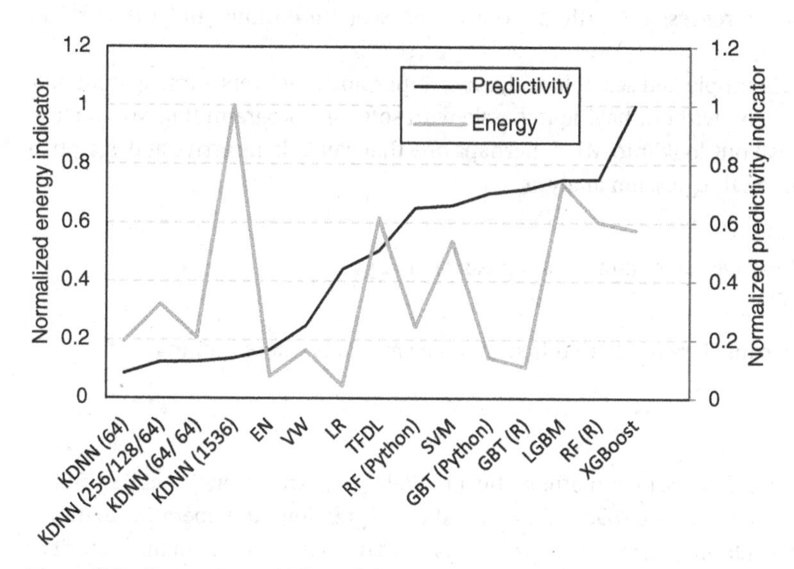

Figure 9.7 Comparison of ML models.

9.2.4 A Note

Now that we have established a benchmark for energy consumption in common ML models, can we extend the same to traditional methods (such as FE modeling)?

While both approaches can potentially use similar computing stations, traditionally, FE models require not only large computational capacities but also demand large storage units – especially for FE models that utilize finer mesh and nonlinear effects.

Here is an example,

A few years ago,[50] I developed a three-dimensional nonlinear FE model to predict the deformation history of RC beams under fire conditions [16]. This model was composed of 50,125 elements and of a size of 64.58 MB, and the results file was

47 Seems reasonable!

48 Which may also extend to other problems as well.

49 There are many other dimensions that I did not get to cover such as exploring pre-trained models and transfer learning techniques.

50 This was my first every journal article!

700 MB in size. In a more recent work [17], I have developed a ML model to predict the same phenomenon. This ML model has a size of 305 KB.[51]

The American Council for an Energy-Efficient Economy estimates 5.12 kWh of electricity per gigabyte[52] of transferred data. This implies that:

One GB emits

$$0.42\text{kg of } CO_2 / \text{kWh} \times 5.12\text{kWh} = 2.15\text{kg of } CO_2 \quad (9.3)$$

Thus, carbon emissions arising solely from *storing* the FE model (without the result file), excluding those arising from model development and simulation time, are 215 times higher than that of the ML mode!

9.3 Symbolic Regression

When we covered supervised ML, we spent quite some time going over the principles of regression. This was needed since, in essence, regression ML models carry out regression operations to tie the features to the outcome (target) of the phenomenon we are interested in. During this time, we focused our attention on linear regression[53] since it is the most fundamental. As you remember, linear regression suffers from a series of limitations and the *curse of linearity*.[54]

Despite that, linear regression is a great tool. It is simple and scalable and, most importantly, delivers a transparent outcome – an expression that an engineer can use easily (without having to hassle with software/programming ML models, etc.).[55] This brings an exciting opportunity to revisit our look into ML to perhaps one that can help us arrive at descriptive expressions that represent the outcome of a traditional regression analysis.

Simply, can we:

1) use ML to derive expressions that resemble those we can obtain from regression analysis?
2) convert a ML model's logic into an expression?[56]

The answer to the above two questions is, yes, we can! Here is where this section becomes a bit more exciting.

9.3.1 Big Ideas

Symbolic regression[57] is a technique that allows users to discover mathematical models (i.e., expressions) that best fit the data on hand. In other words, symbolic regression searches the space of mathematical expressions and operators to derive an expression or a series of expressions. Our goal in adopting symbolic regression is to arrive at an expression that describes how the features, when combined, yield the observed outcome.

Such an expression can be used to understand the interaction between the features, as well as the interaction between the features and outcome. The expression can also be used to trace the logic behind the ML model, i.e., explain model/expression predictivity. Given its mathematical form, the expression can be a stepping stone toward a new hypothesis or

51 While both models were developed using different working stations and approximately 10 years apart, the truth of the matter is that both models still require to be stored.

52 1 GB = 1000 MB.

53 And a bit on nonlinear regression.

54 I would like to take a minute here to mention that every method/tool is applicable for specific conditions. When such conditions are not met, then the application of the tool is voided. This does not mean that this method is poor or inadequate. All this means is that this particular tool is perhaps not the most suited for this particular application. The same is also the reason why we have other tools, and we continue to develop some more tools!

55 If you are thinking if there is a way to covert a ML model's logic into an expression, then you are already on a great start to this section!

56 If this sounds similar to a surrogate model, then you are correct. It does.

57 **Symbolic regression**: an approach to find suitable mathematical expressions to build model(s) to describe data [59].

theory. While the outcome of symbolic regression is one such expression, this expression can be mistakenly assumed to reflect the physics behind the phenomenon. In reality, other expressions may also *fit* the data,[58] while none may necessarily describe the physics behind the data.[59]

Let us walk through how symbolic regression works before we jump into some examples. Symbolic regression can be thought of as an evolutionary process that mimics the natural selection process as conveyed by principles of the Darwinian evolution.[60] In this process, we seek to identify patterns. Once a viable pattern is identified, this pattern directs the search within the data space. In this model, a string of characters is expressed in terms of expression trees, a mathematical equation comprising a tree-like data structure, where each node represents an expression.

We start with a random population of individuals often referred to as "trees" to represent a number of possible solutions through the structural ordering of mathematical symbols. For example, a possible solution is a ranked tree consisting of *functions* and *terminals*. A function (*F*) may contain basic mathematical operations (+, ×, −, etc.), power functions (^, log, exp), trigonometric functions (sin, cos, tan, etc.), step, and logistic functions, etc. On the other hand, the terminal (*T*) comprises arguments and numerical constants/variables.

Hence, a developed model has a tree-like configuration in which branches can extend from a function and end in a terminal (see Figure 9.8). This tree can be converted into a *Karva-expression*. The first position in this expression denotes the root of the tree, and the transformation process starts from the root and reads through the string layer by layer). A solution is declared fit once such a solution satisfies our goodness. Symbolic regression can be carried out via genetic algorithms or programming [18]. More recently, deep learning and hybrid ML systems were also developed to carry out symbolic regression tasks [19, 20].

9.3.2 Examples

Let me walk you through a few software capable of performing symbolic regression. I will be covering three examples; a simple example[61] and two of a more complex nature.

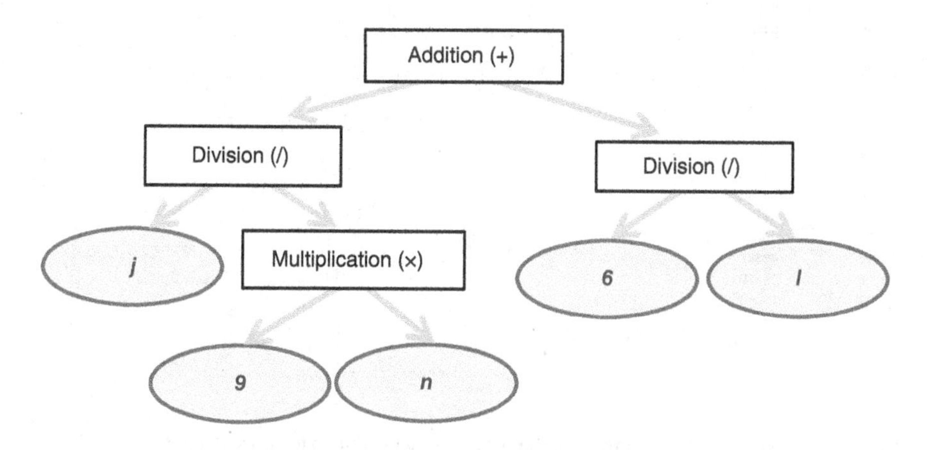

Figure 9.8 Karva-expression for a tree-like representation of $\frac{j}{9n} + \frac{6}{l}$.

58 Hence my emphasis on using the term *descriptive* expressions. In the next section, we will go over *mapping functions* which are one step higher than descriptive expressions.

59 This misconception becomes an issue in problems where the data is limited or hard to collect – ask me, I know!

60 There are also a series of genetic operations that can be undertaken such as selection, crossover, and mutation.

61 Which we will also re-visit in the next section and have seen in the causality chapter.

Example no. 1: Deriving an expression for moment capacity of W-shaped steel beams.
For this example, I will use the same dataset we used in the causality chapter.[62] I will run the same dataset in four software (namely, *Eureqa*,[63] *TuringBot*,[64] *HeuristicLab*,[65] and *GeneXproTools*[66]),[67] an online graphical interface (*MetaDemoLab*), and *Python*. I will cover the steps associated with each software in detail[68] and provide you with the saved files for your reference.

9.3.3 Eureqa

Eureqa[69] was introduced in 2009 [21] and then acquired by *DataRobot*[70] in 2017. *Eureqa* has a clever graphical interface and is intuitive to use. To use, *Eureqa*, open a new *Eureqa* file.

Step 1: Copy the dataset and then paste it into the *Enter Data* sheet within *Eureqa* as shown. Then, navigate to the *Prepare Data* sheet.

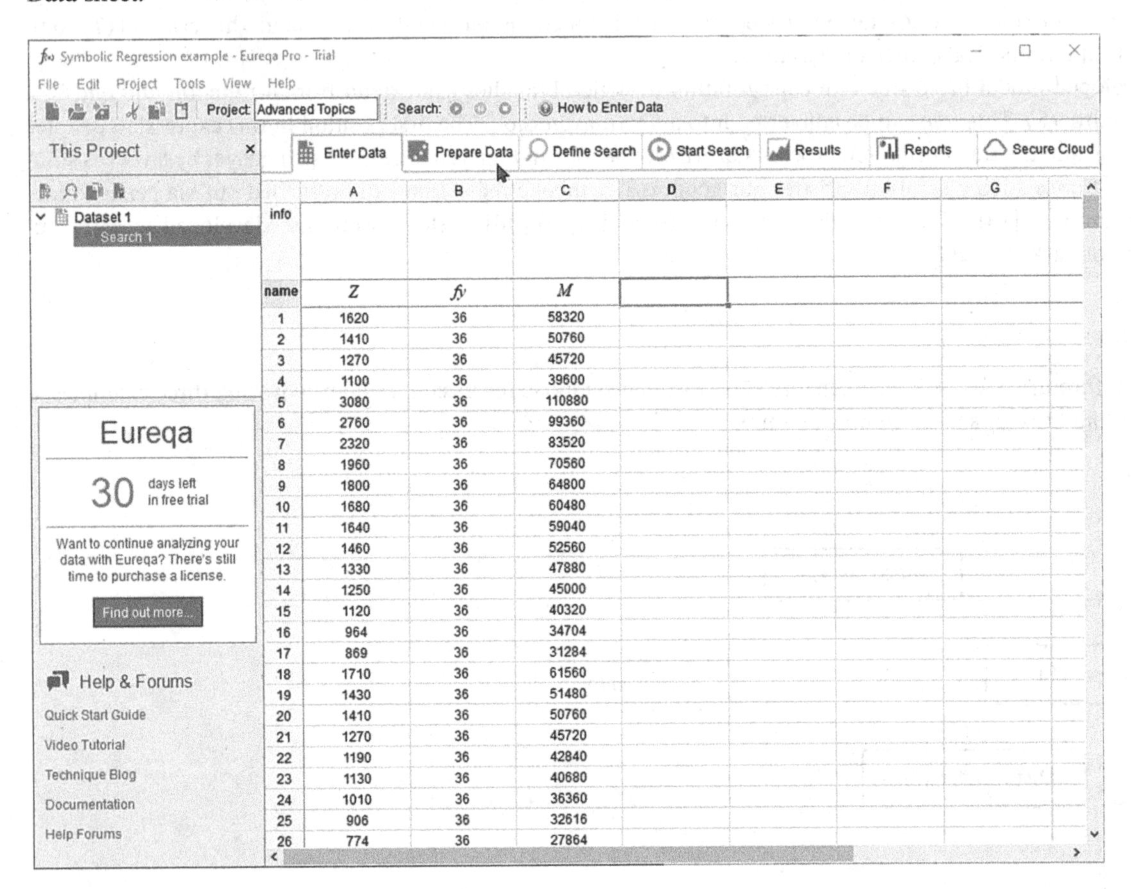

62 As you might remember, this dataset was compiled of W-shaped steel beams from the AISC manual with two features, Z and f_y.

63 Version 1.24.0.

64 Version 1.80.

65 Version 3.3.16.

66 Version 5.0.0. This software is a special case that we will briefly go over.

67 You can obtain these software from their developers – often free for use by students and educators. Also, other software worthy of mentioning include: *GPTIPS* code [60, 61] and *Discipulus* [62].

68 For the sake of page limitation, and the fact that we cannot cover many examples or problems that civil and environmental engineers might face, I will not be going into much detail for each software. I would advise you to explore these and go through their embedded tutorials and help manuals. I will not be thoroughly comparing these software against each other as I will be using their default setting. Saying this, such a comparision does sound like a good benchmarking paper.

69 Fun fact. This is the first symbolic regression that I have ever used.

70 I will show you an example on how to use *Eureqa* from the *DataRobot* platform in the next section.

Step 2: You may follow up with the bolded (suggested) steps, or move to the next sheet, *Define Search*.

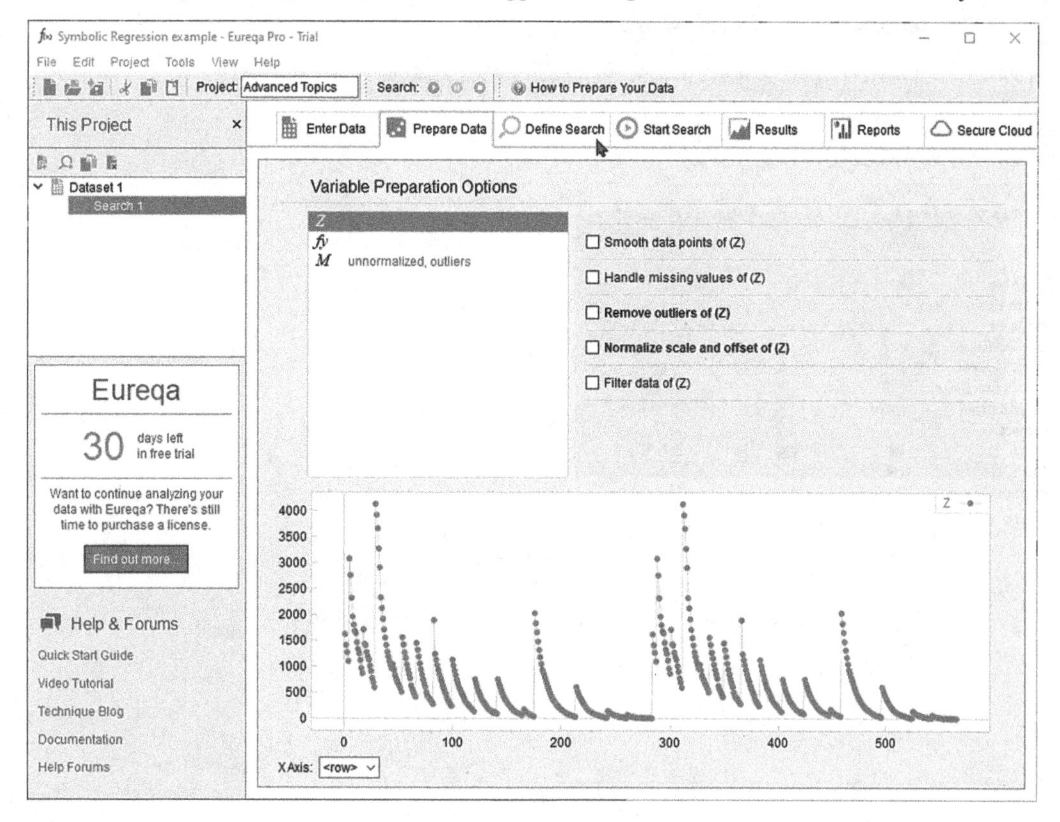

Step 3: Check out the *Formula building-blocks*, *Error metric*, and *Data Splitting*. You may adjust these settings to the provided ones.[71]

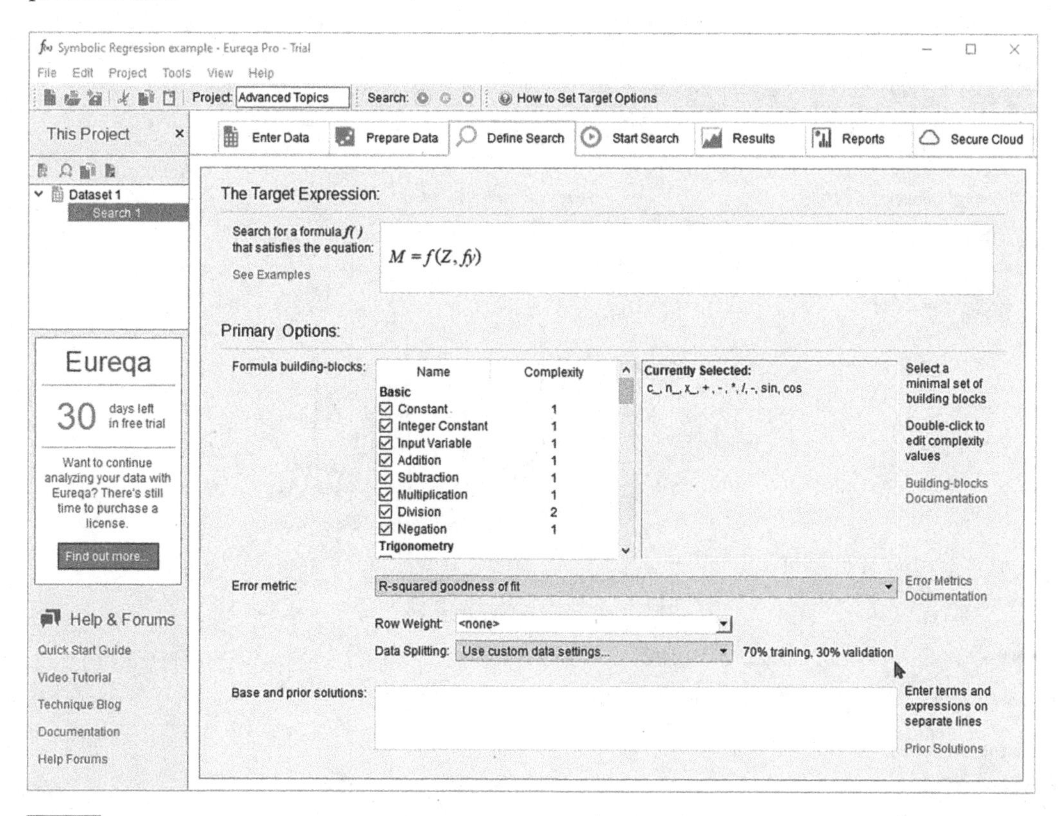

71 In some instances, *Eureqa* may drop a few features if they have low importance. A work around to ensure that a certain feature remains (i.e., forced into the expression) is to apply the following: *Target* = *f*1()*feature1 + *f*3()*feature2... + *fn*(feature1, feature1, ...)

Step 4: *Run Search.*

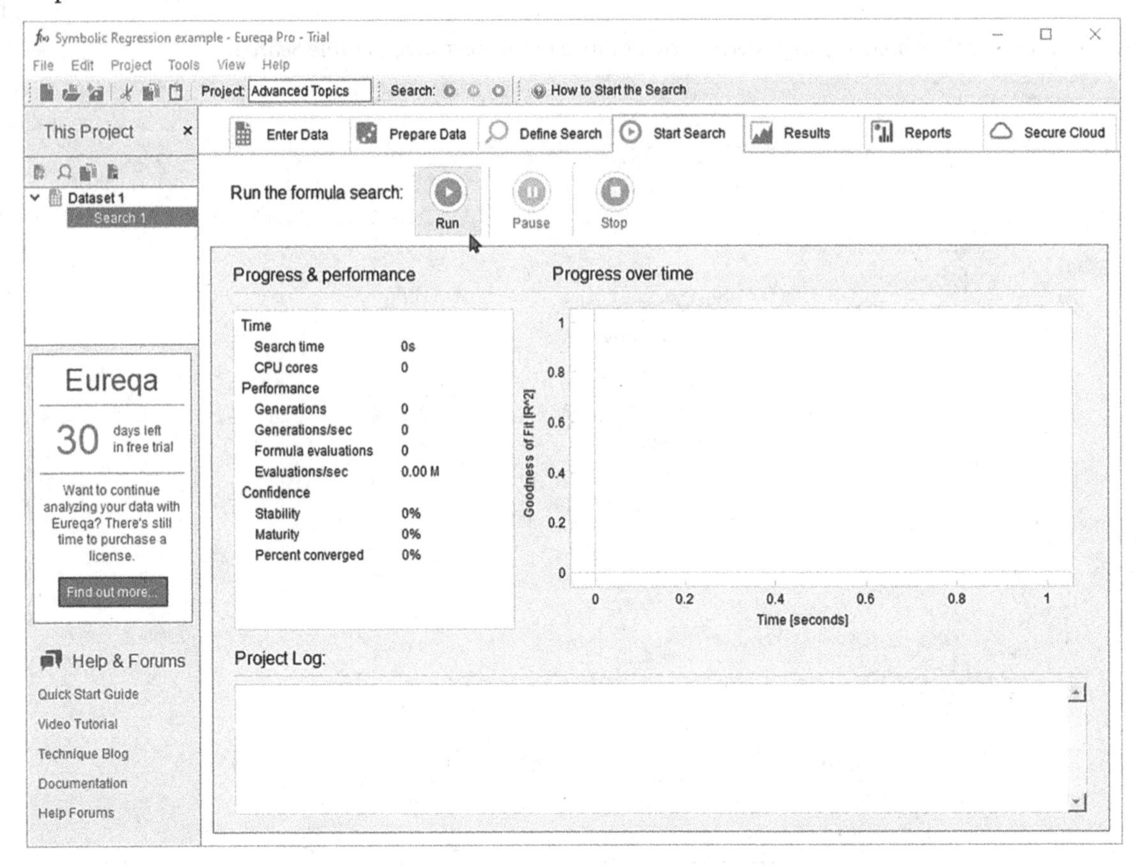

Step 5: As you can see, *Eureqa* was able to arrive at the correct functional form immediately. You can change *Plot Type* into *Observed vs. Predicted Plot*.

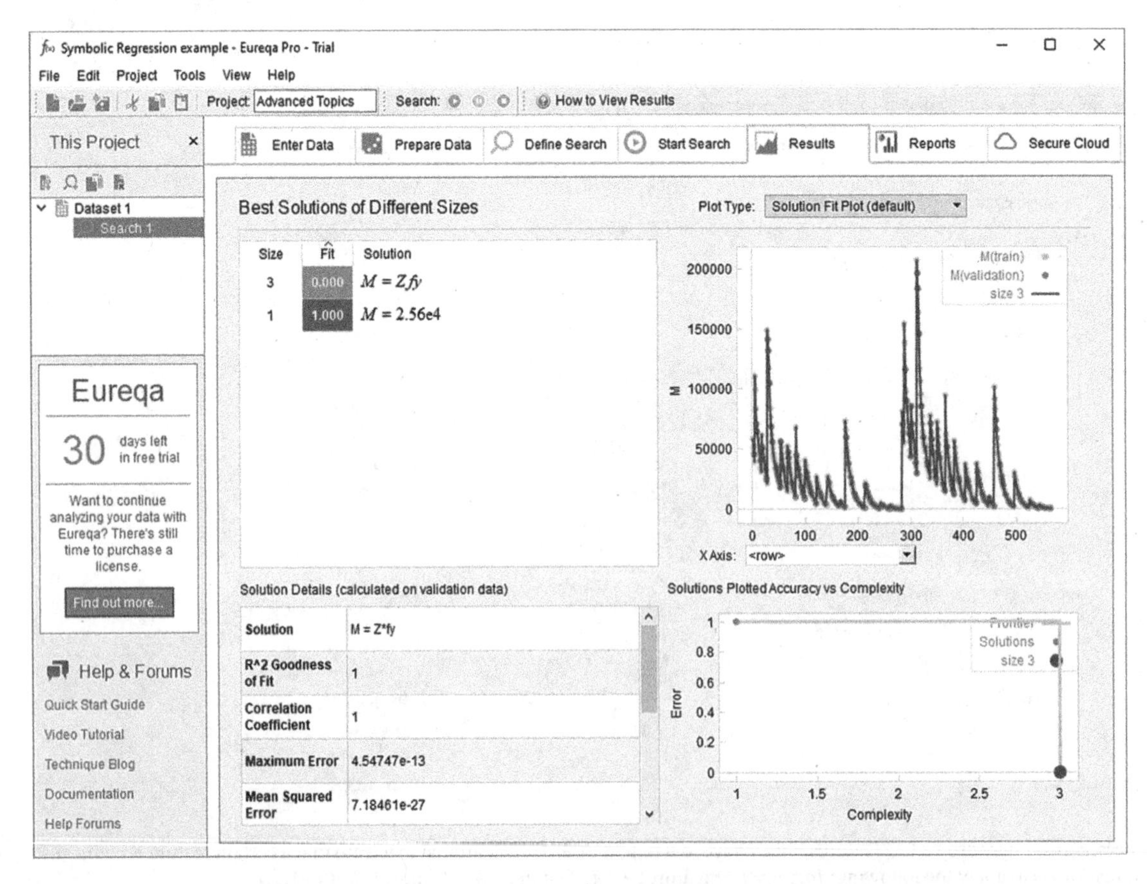

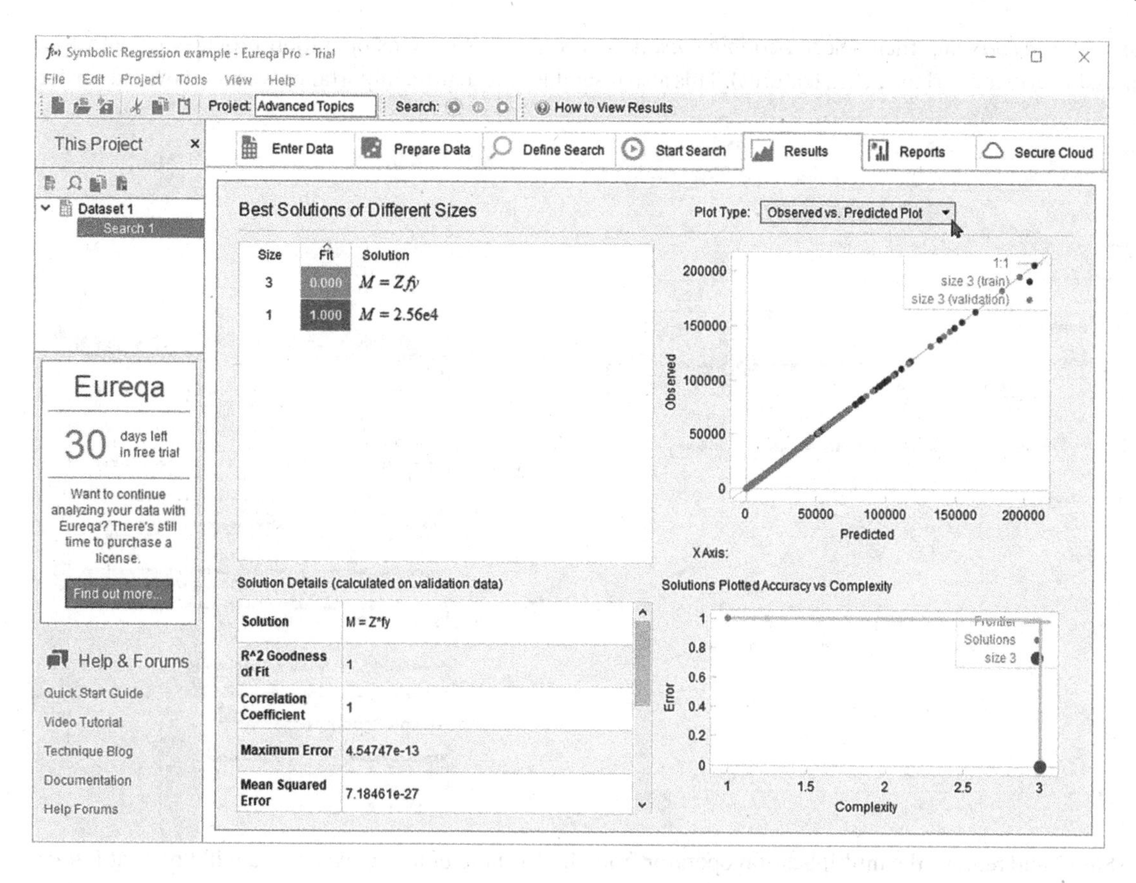

Step 6: Copy the arrived at expression. Go to the *Reports* tab. Select *calculate model statistics on a dataset* and *Run*.

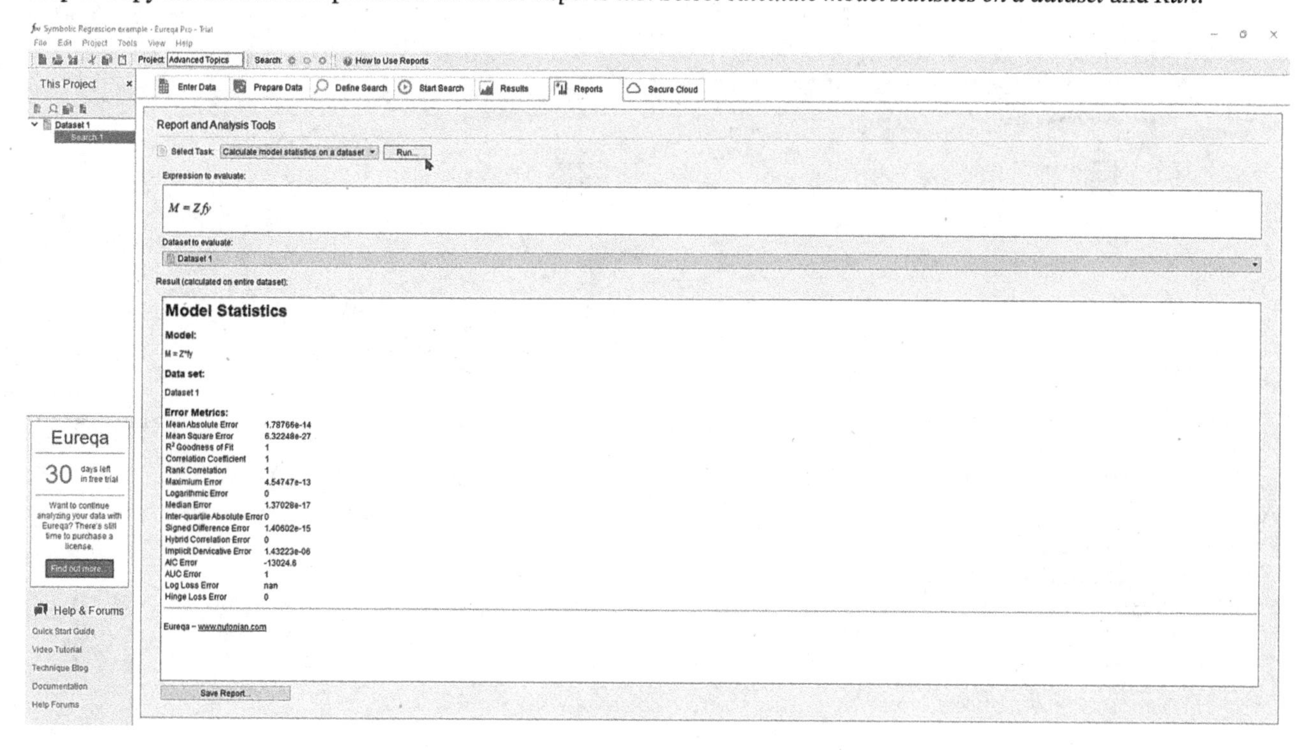

Step 7: Navigate to the *Reports* tab, then select *Variable sensitivity report* and *Run*. This operation provides us with how sensitive the expression is to the variables (i.e., features). This also resembles the feature importance we have covered earlier.

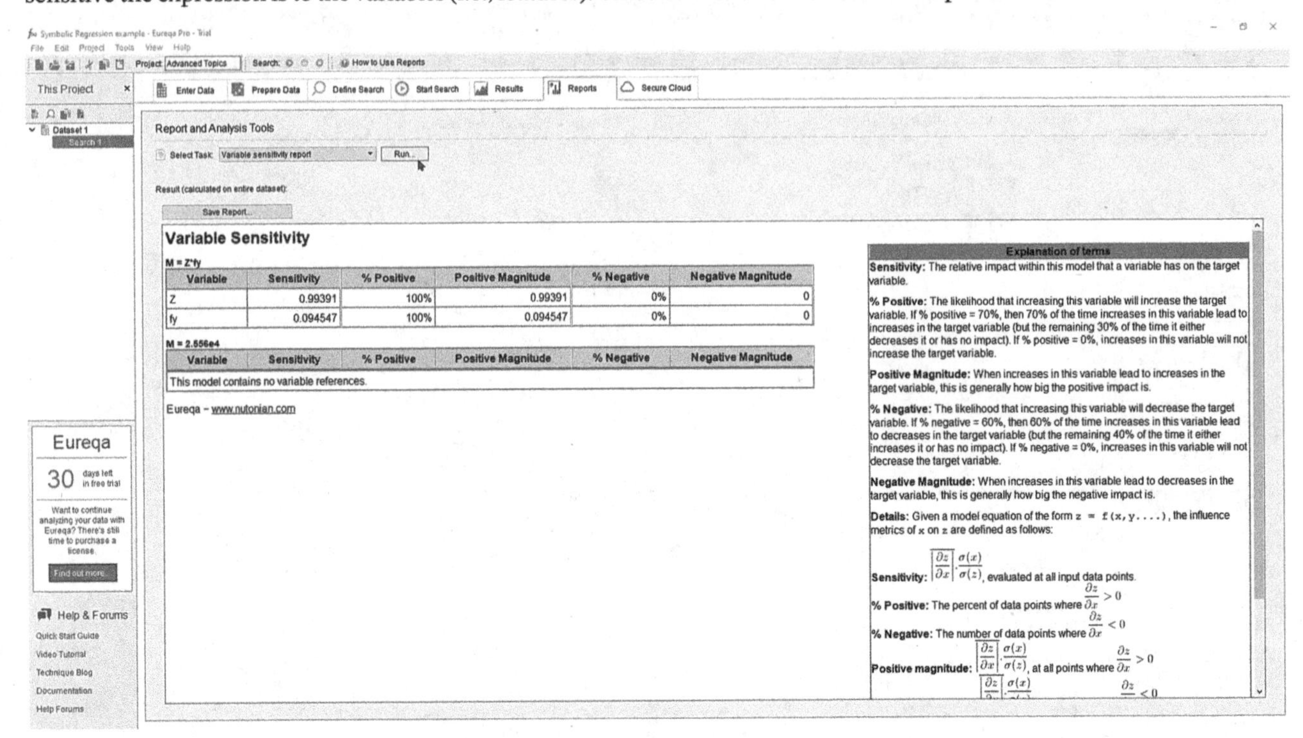

If you go back to Step 3 and remove the multiplication operator from the *Formula building-blocks*, you will find that *Eureqa* struggles to find an expression to the problem on hand[72]!

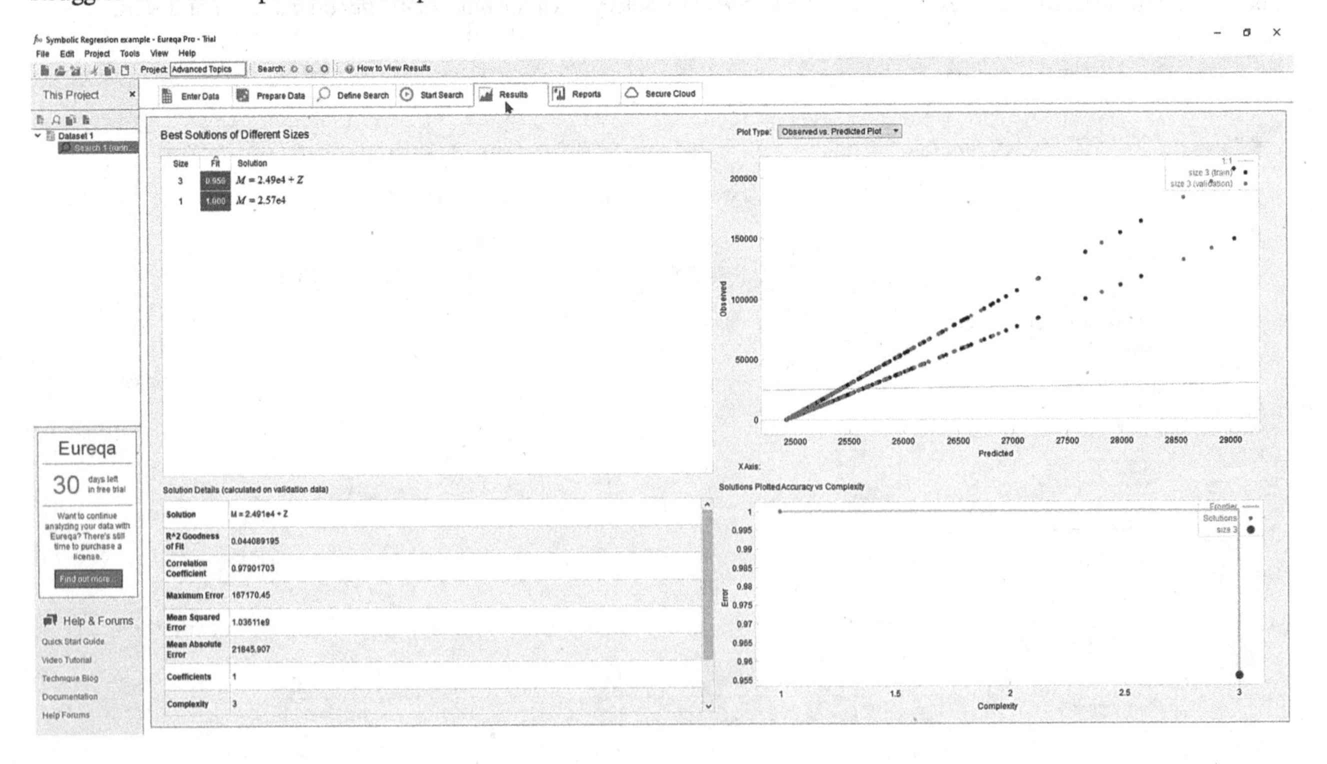

72 Not surprisingly!

9.3.4 TurningBot

TuringBot is another symbolic regression software. This software is named after great mathematician and computation pioneer Alan Turing [22]. *TuringBot* uses a novel algorithm based on simulated annealing[73] to arrive at mathematical expressions and has a lean interface. Let us repeat the above example using *TuringBot*.[74]

Step 1. Input your dataset file[75] and identify the *input* and *target* variables. Set up the *search options* and operators as shown below. Then, *start* the search.

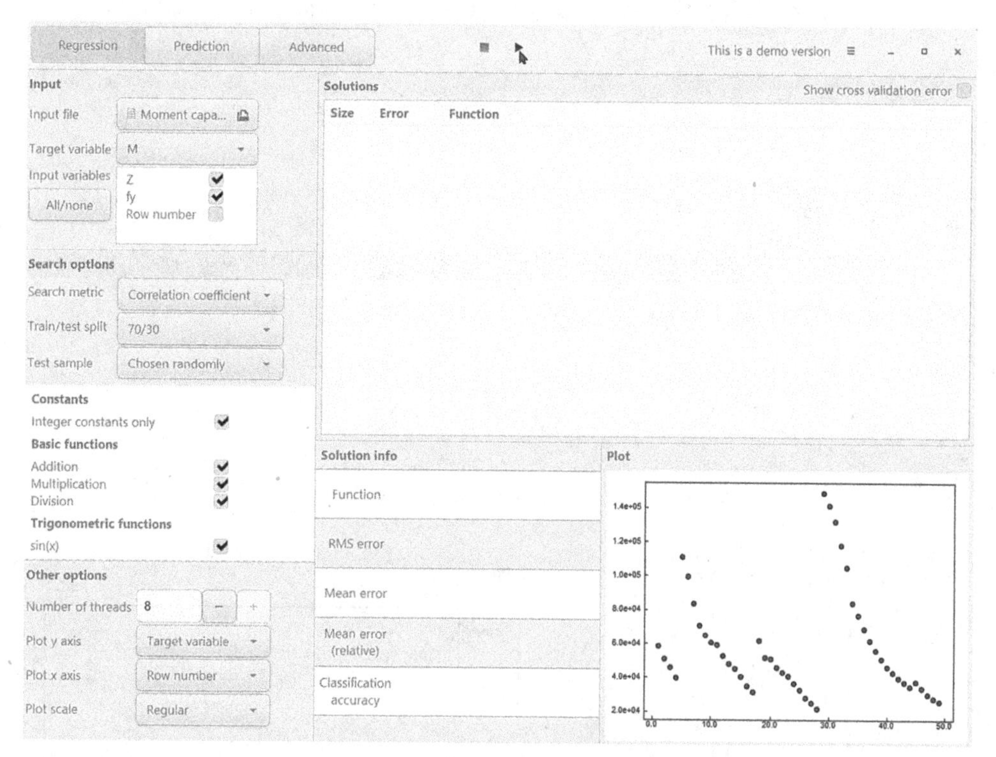

Step 2: Check the outcome of the analysis.

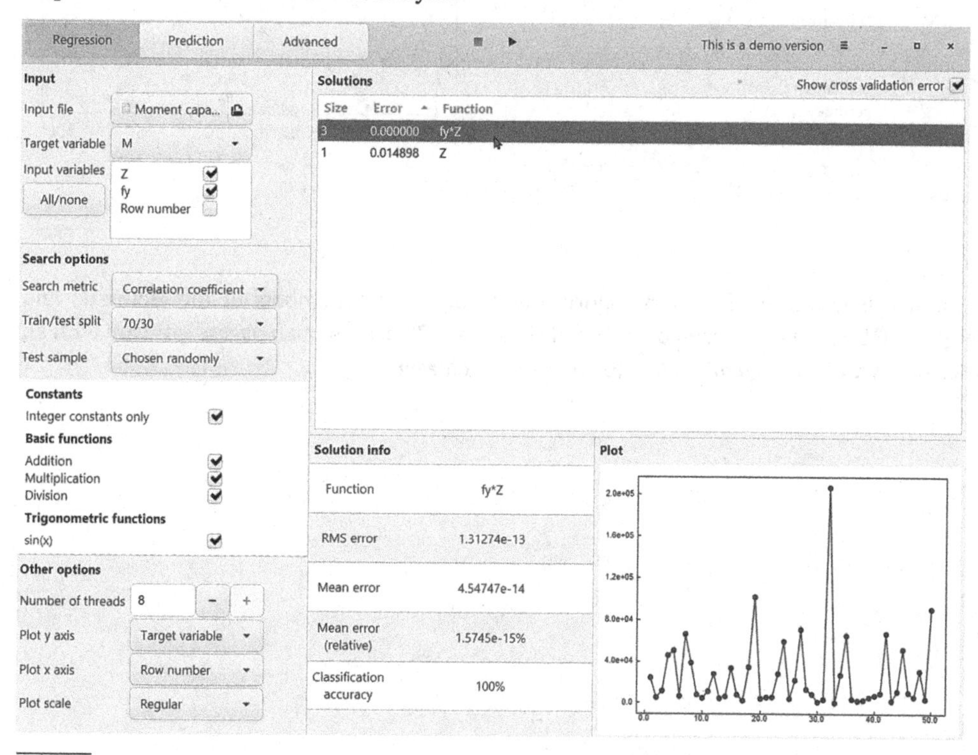

73 A probabilistic search *optimization* algorithm that approximates the global optimum of a given function.
74 At this point in time, *TuringBot* can only allow the user to input 50 observations and no more than three features into the demo version. As such, I will be inputting two random samples comprising the two different grades of structural steel.
75 As .txt or .csv.

If we repeat the same exercise and remove the multiplication operator, then we can see *TuringBot* struggles as well. Just like in the above software, the approach of symbolic regression can get close forms to the ground truth. However, the fact that the key operator is not allowed in the analysis complicates this issue.

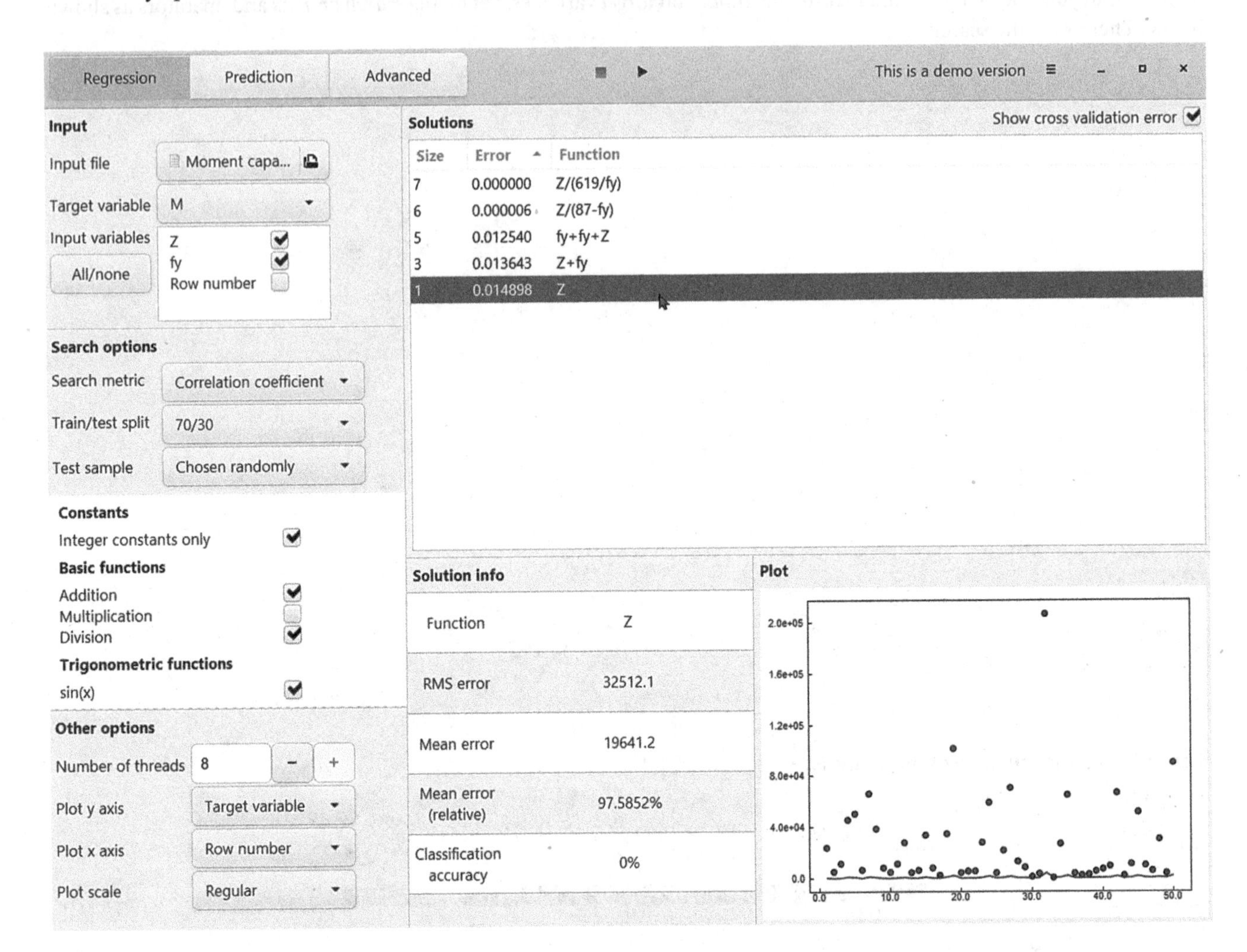

9.3.5 HeuristicLab

HeuristicLab is a framework for heuristic and evolutionary algorithms developed by members of the Heuristic and Evolutionary Algorithms Laboratory (HEAL) and has been operational since 2002. This is an open-access software with an active online presence. Let us see how we can use *HeuristicLab* to analyze our dataset.

Step 1: Open *HeuristicLab Optimizer*. Then, *Start*. Load the dataset. Then, *Finish*.

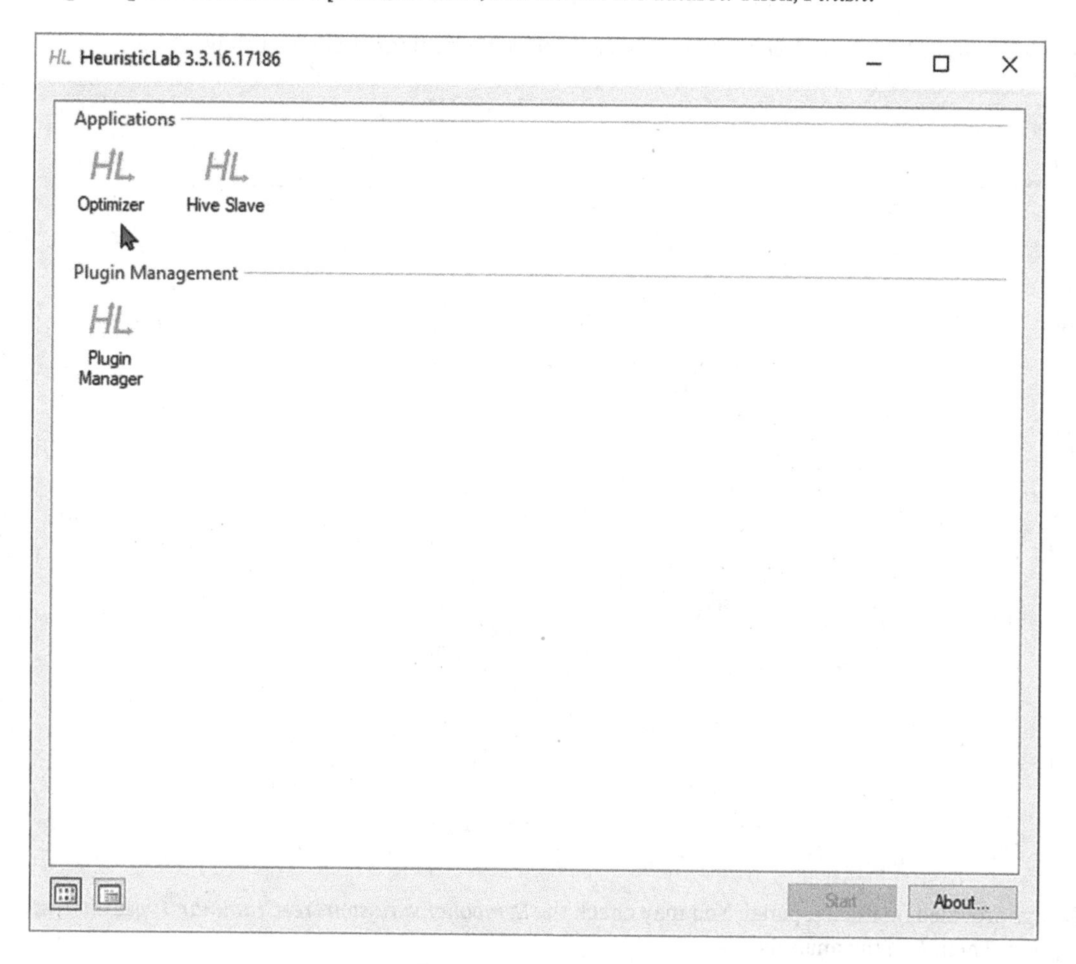

Step 2: Select *Genetic Programming – Symbolic Regression*. Then, double click.

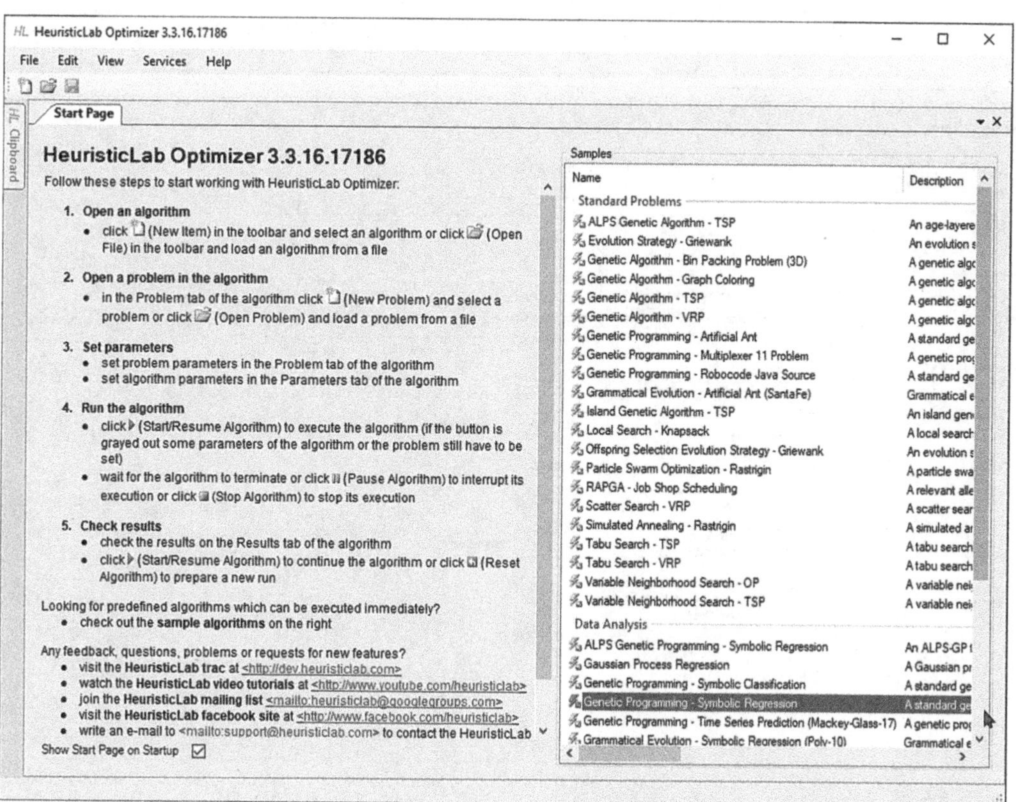

Step 3: Input our *Moment capacity of W-shaped steel beams.csv* dataset. Check the settings as shown below.

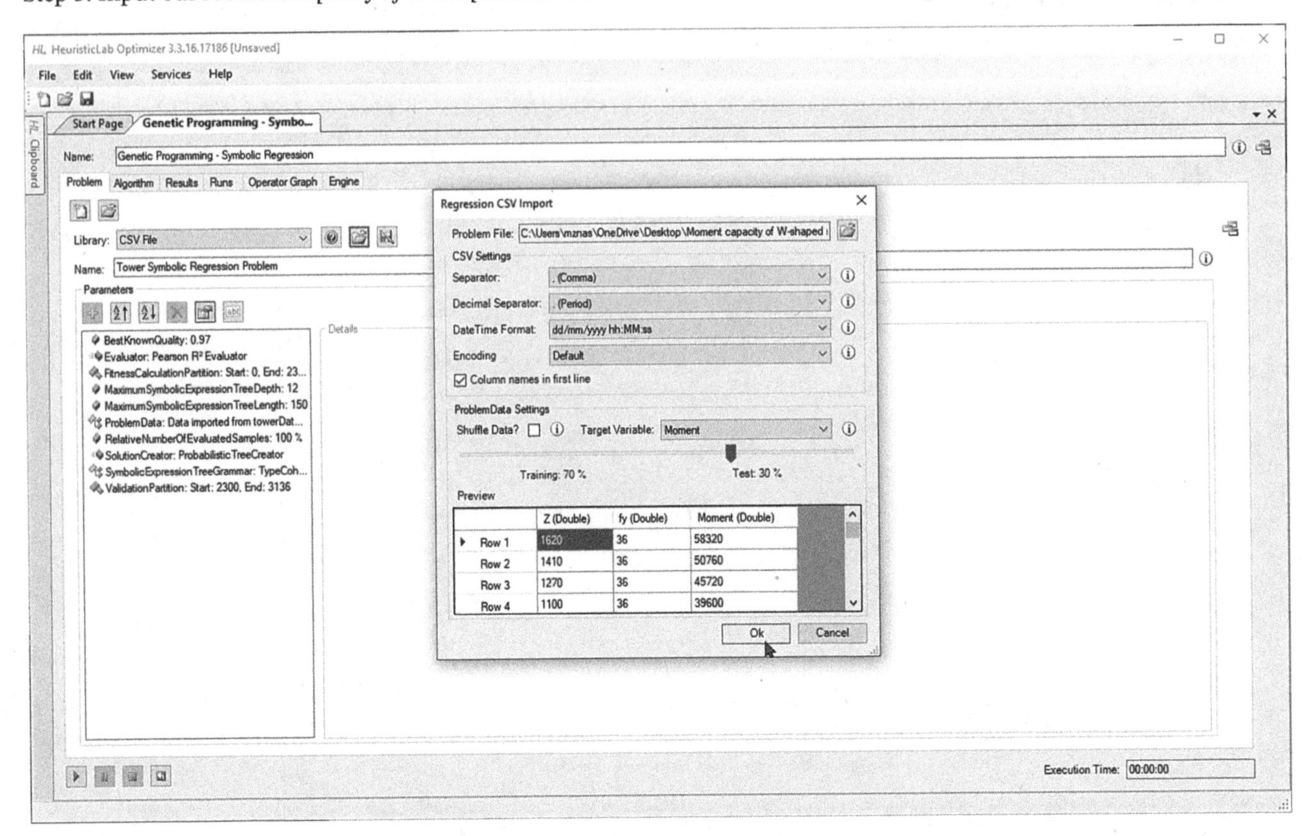

Step 4: Go over the settings provided on the left panel. You may check the *SymbolicExpressionTreeGrammar:TypeCoherent ExpressionGrammar* option. Then, *Run* the analysis.

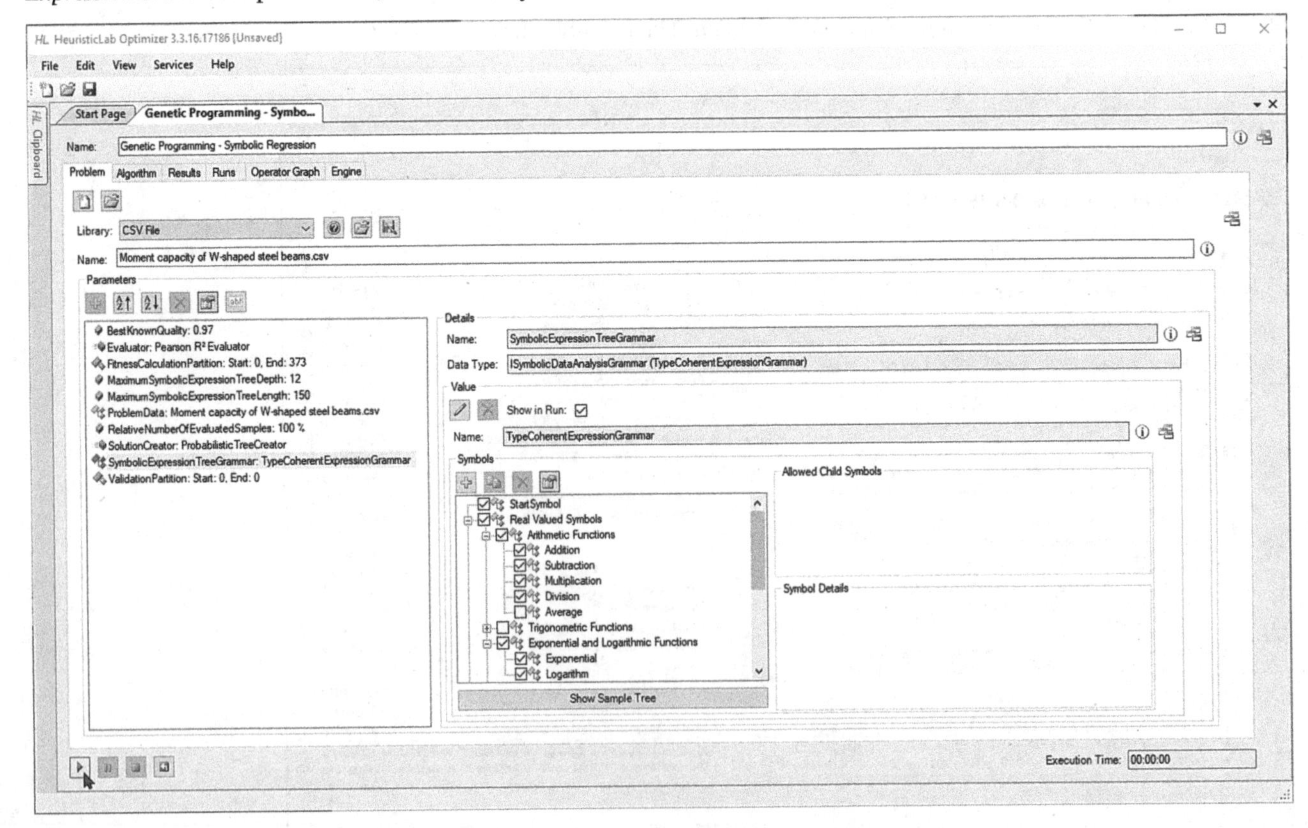

Step 5: Check the outcome of the analysis. Switch the view into *Mathematical Representation.*[76]

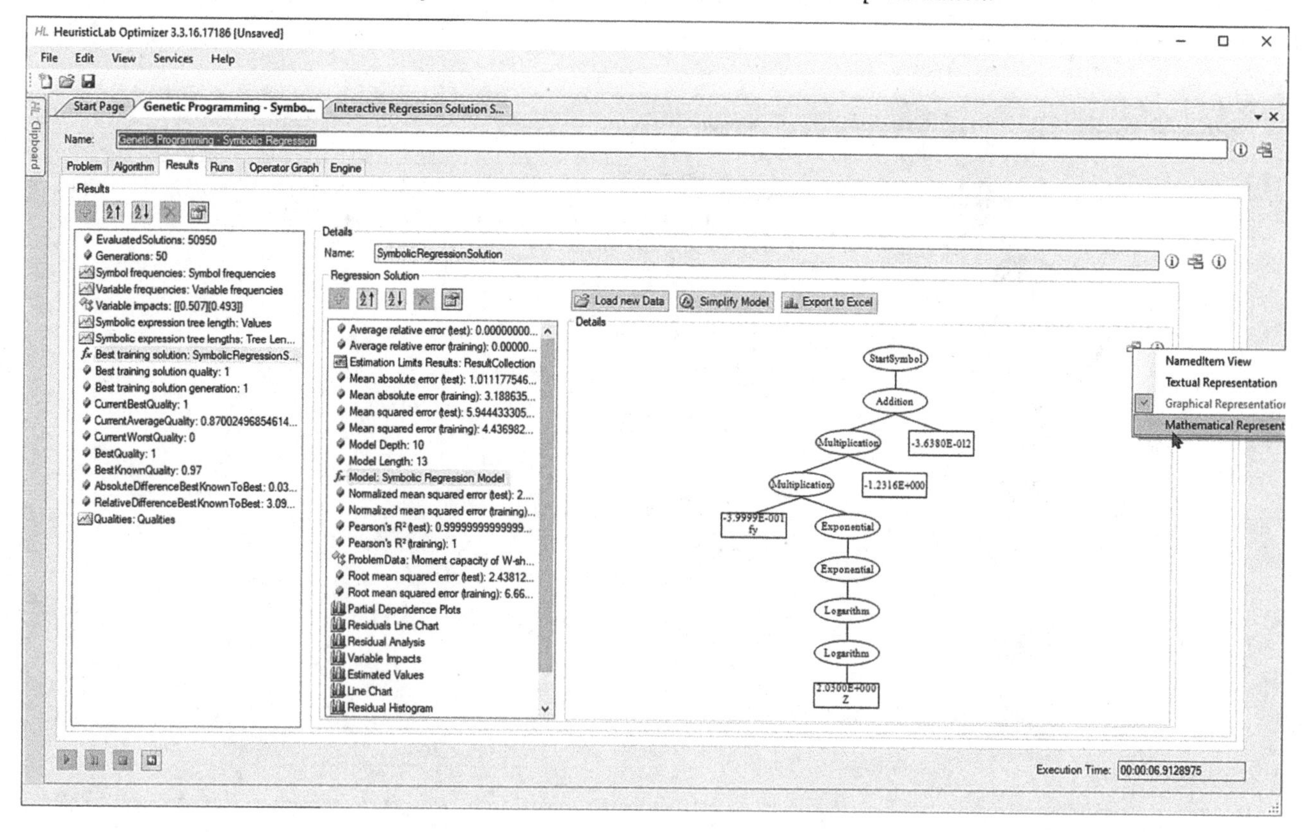

Step 6: As you can see. This is a much more complex form of the ground truth. To further simplify, click on *Simply Model.*

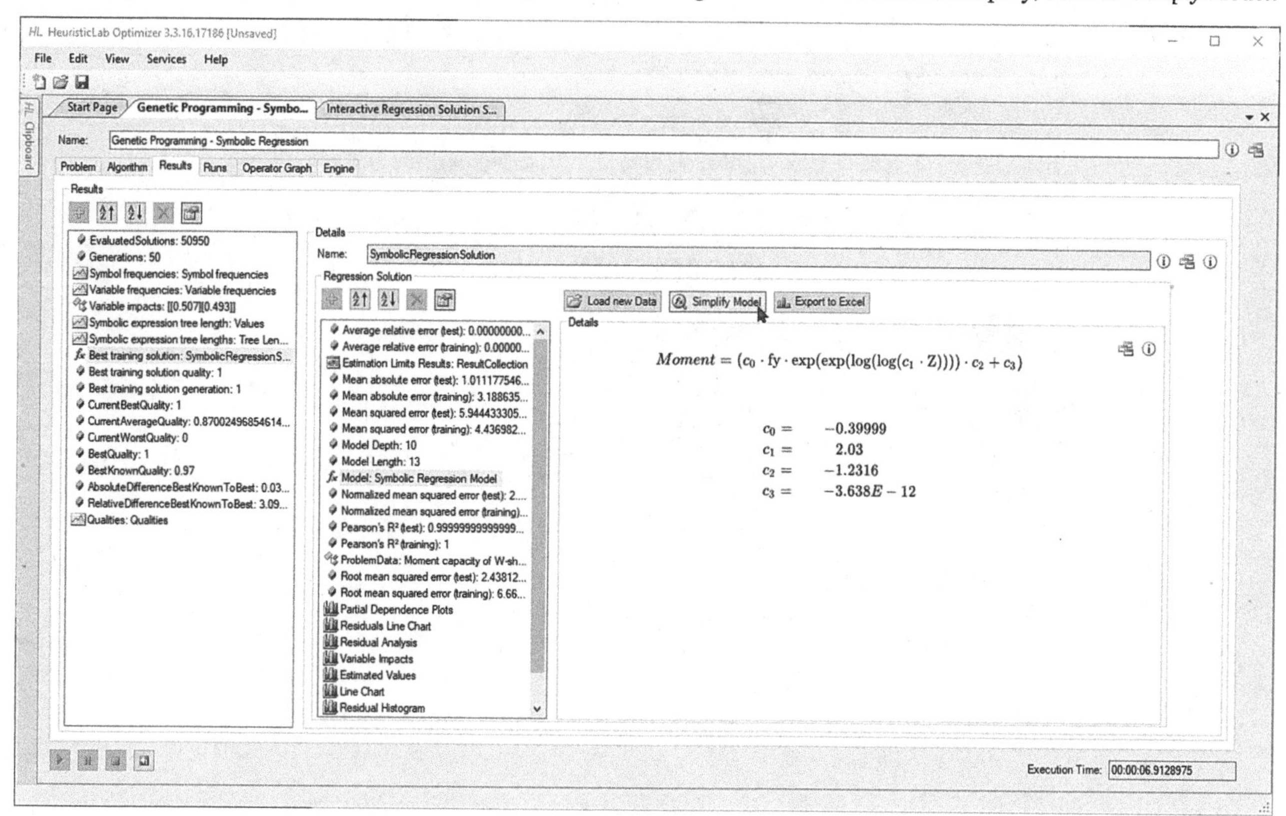

76 A right click on the right panel should do the trick.

Step 7: Further *simplify*. As you can see, the coefficient C_0 is practically zero and we have arrived at the ground truth!

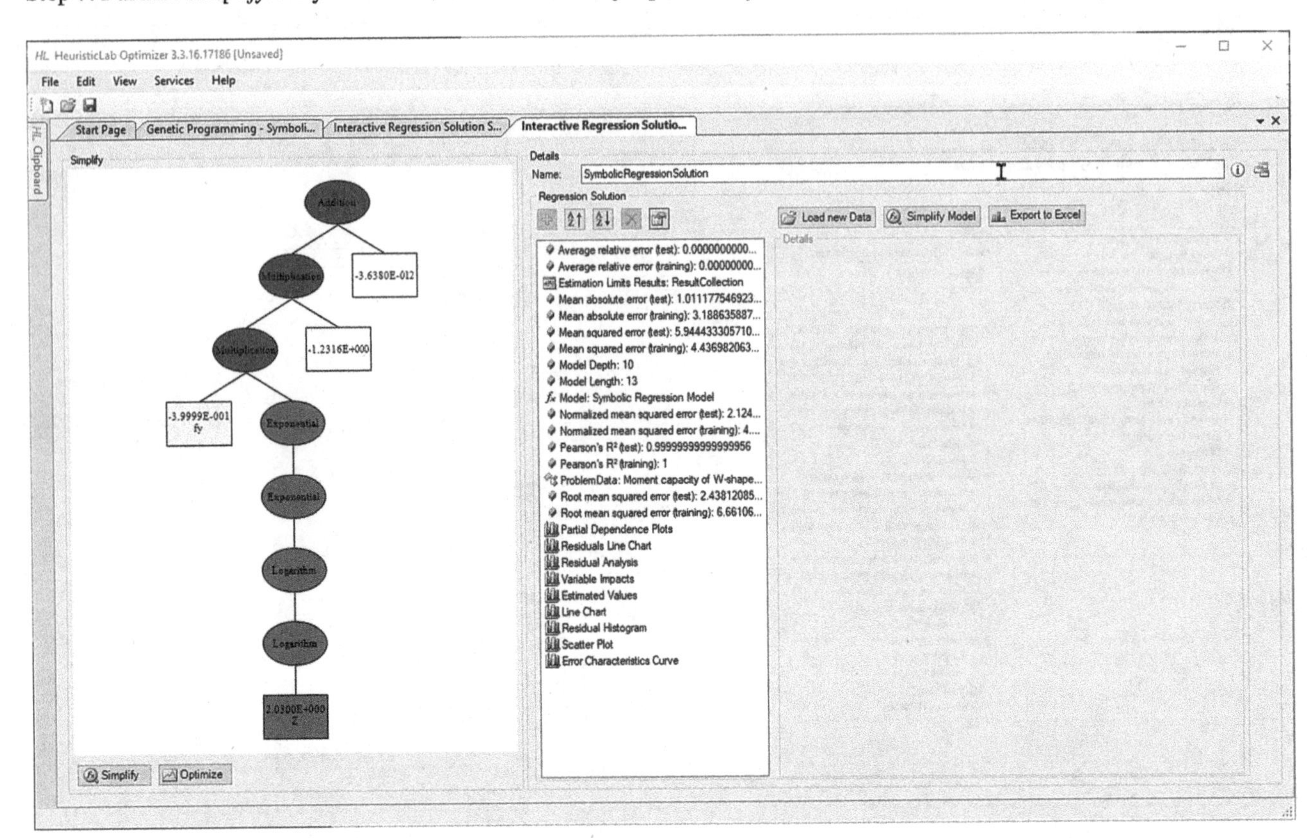

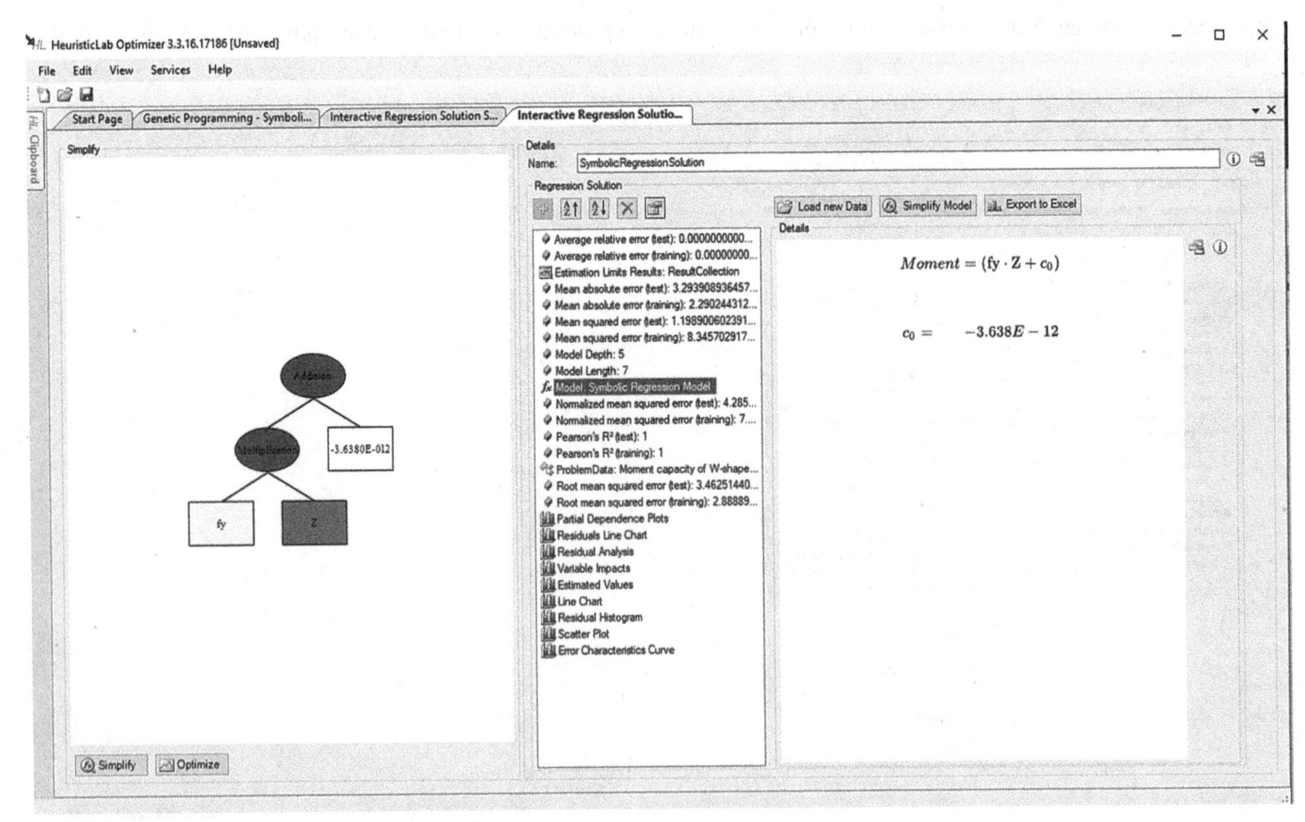

9.3.6 GeneXproTools*

GeneXproTools is a predictive modeling software that utilizes Gene Expression Programming (GEP) technique as invented by Dr. Candida Ferreira [23]. One feature of *GeneXproTools* is its ability to export the outcome of its analysis into a ready-to-use programming code.[77] Let us start our analysis.

Step 1: Start a new model.

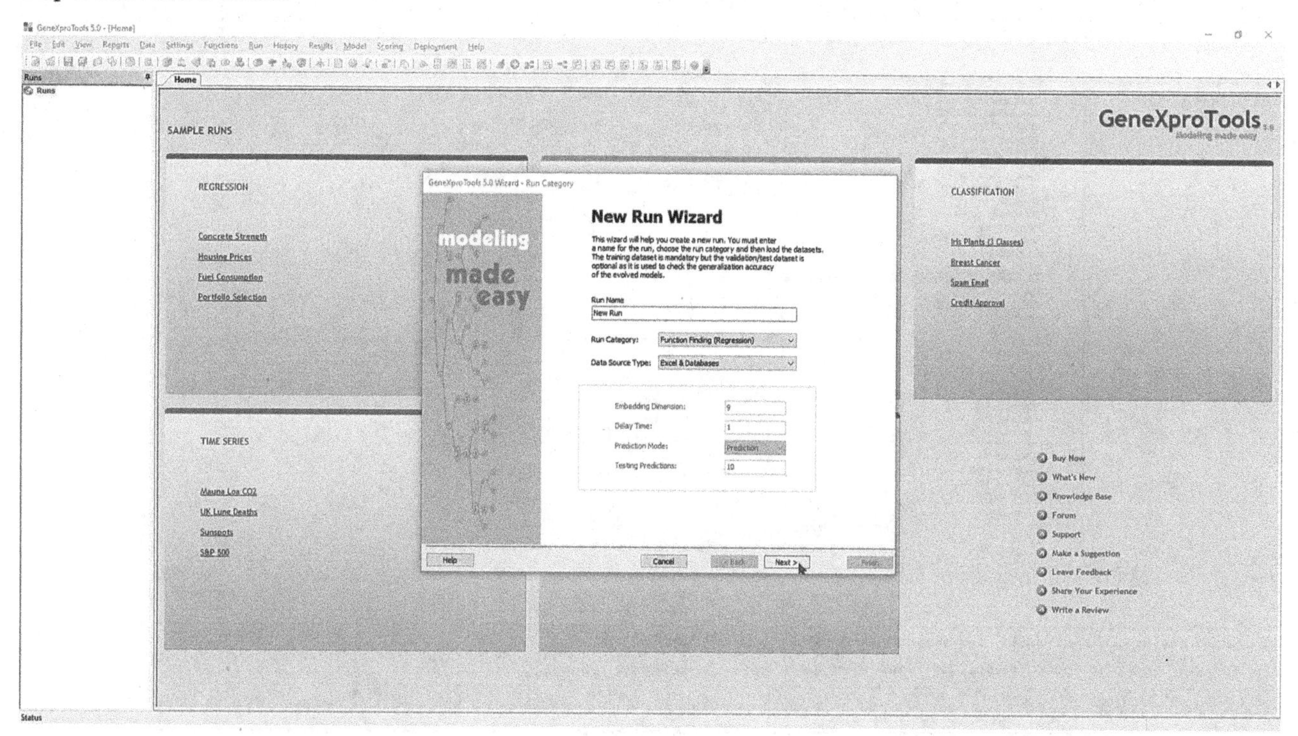

Step 2: Enter the analysis panel.

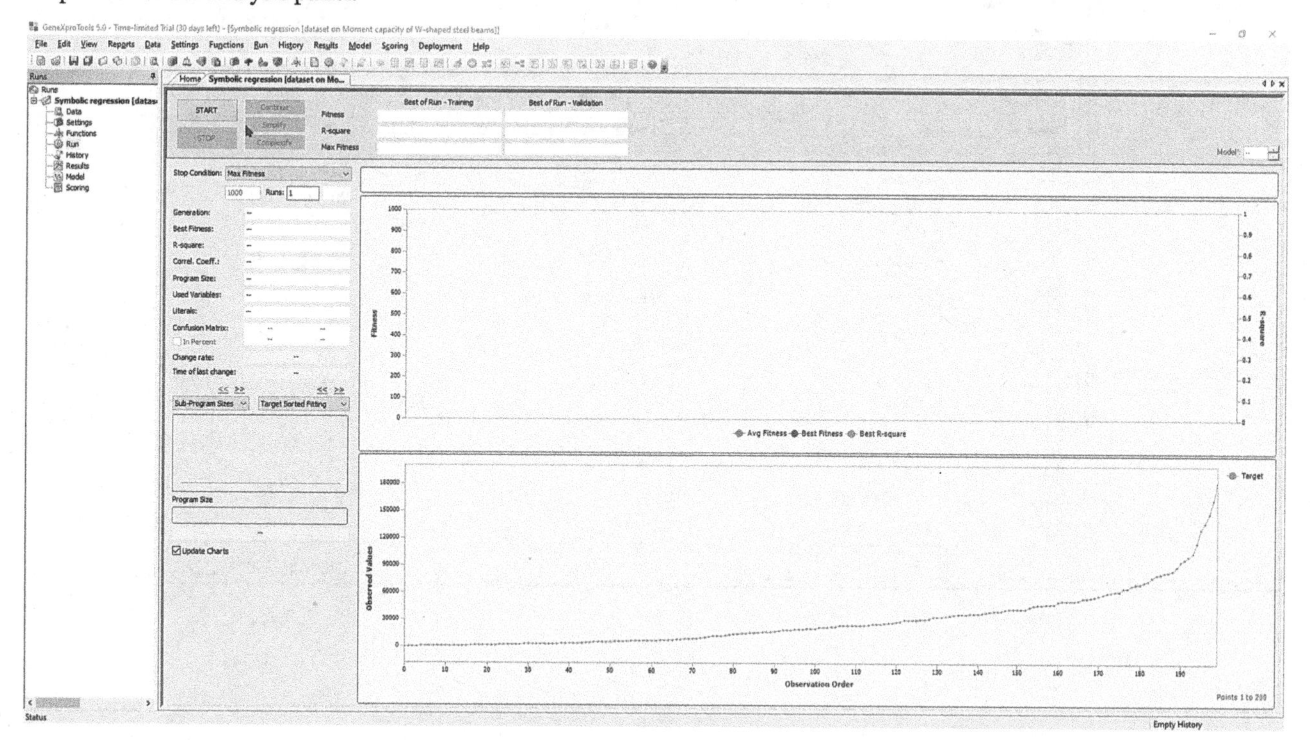

---

77 Including Ada, C, C++, C#, Fortran, Java, Java Script, Matlab, Pascal, Perl, PHP, Python, Visual Basic, VB.Net, Verilog, and VHDL. Since the output is slightly different than that of other software, I thought of snowflake in this section.

Step 3: Check out *General Settings* under *Settings*. Also, check *Functions* from the left panel.

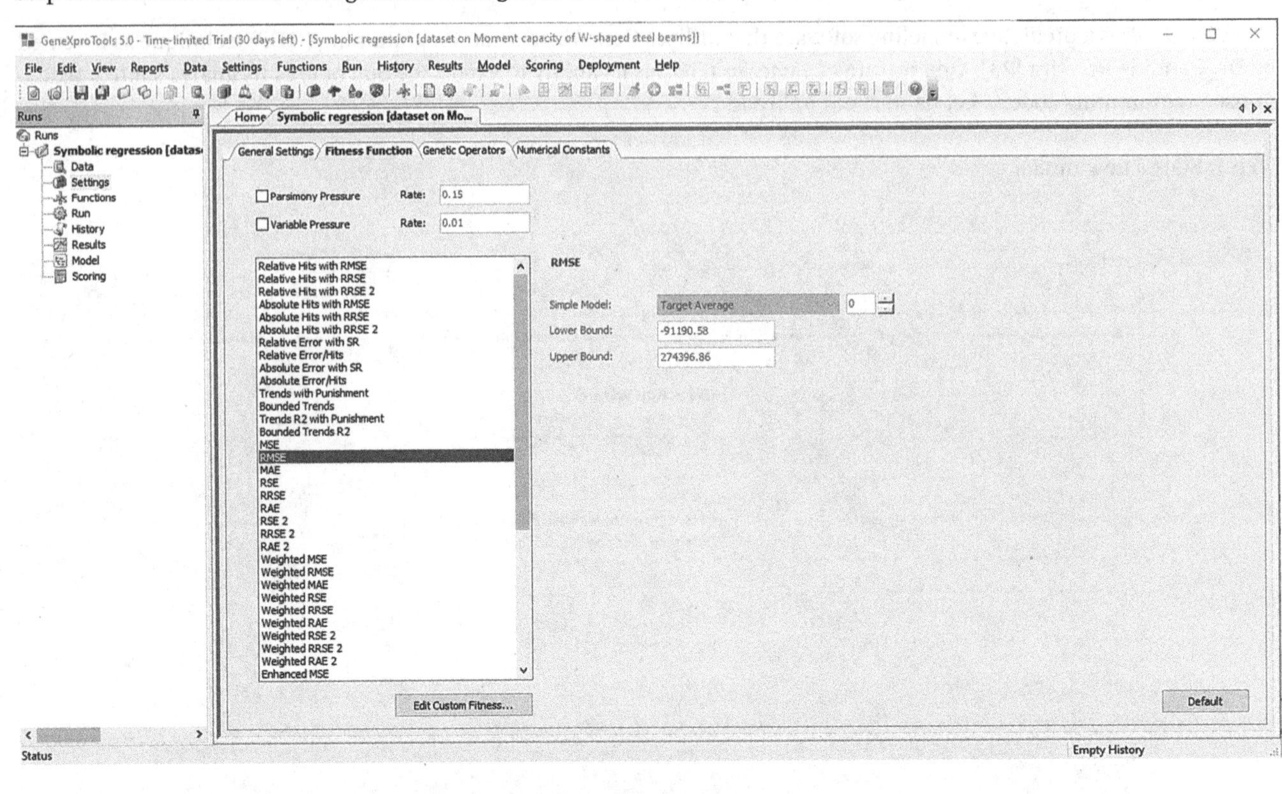

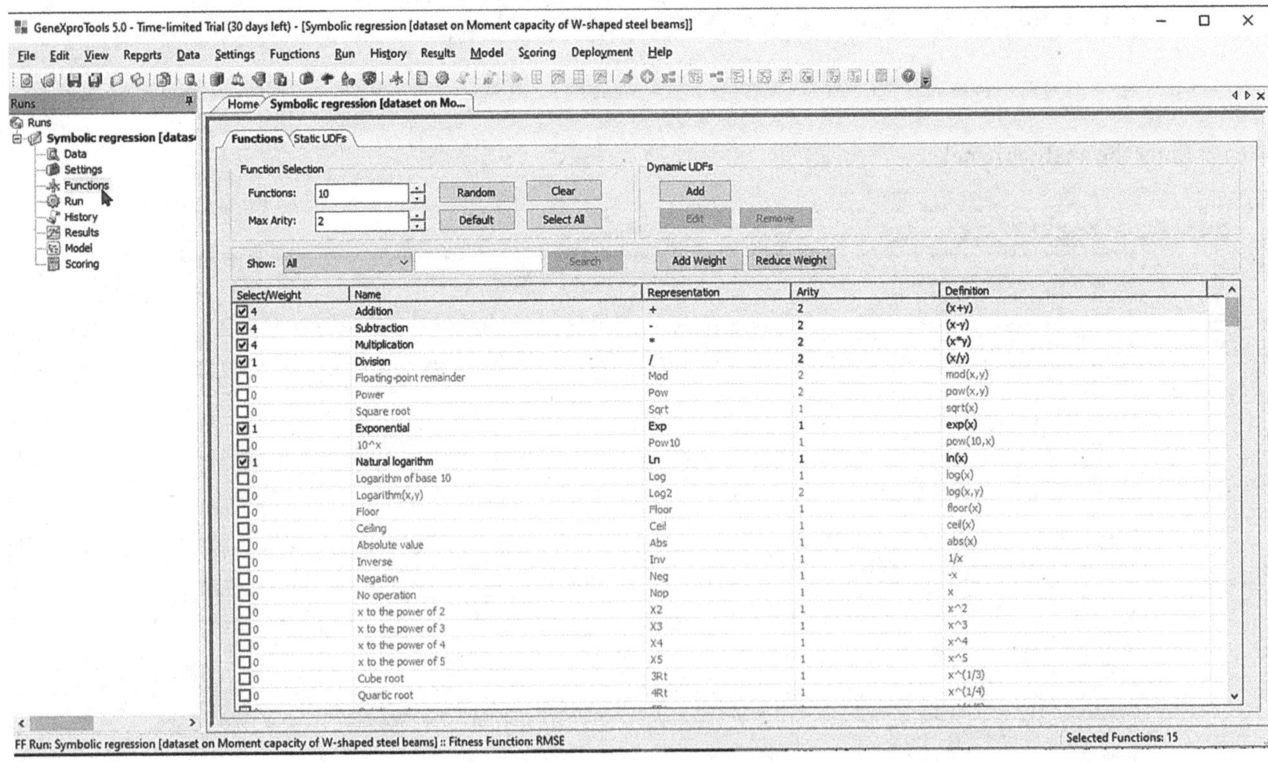

Step 4: You will notice that the *Stop condition* can be controlled. Select *R-square Threshold*. Then, *Run* the analysis.

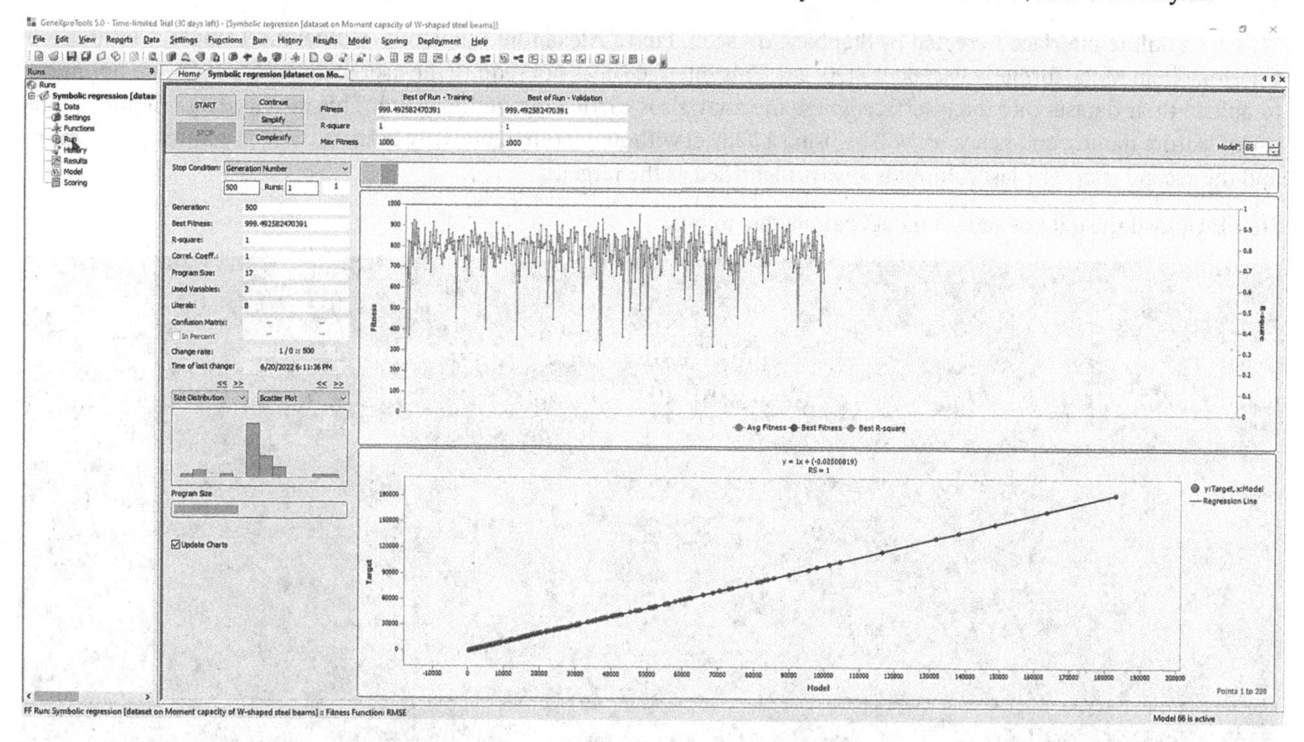

Step 5: Export the outcome into a language of your choice (*Python* is shown).

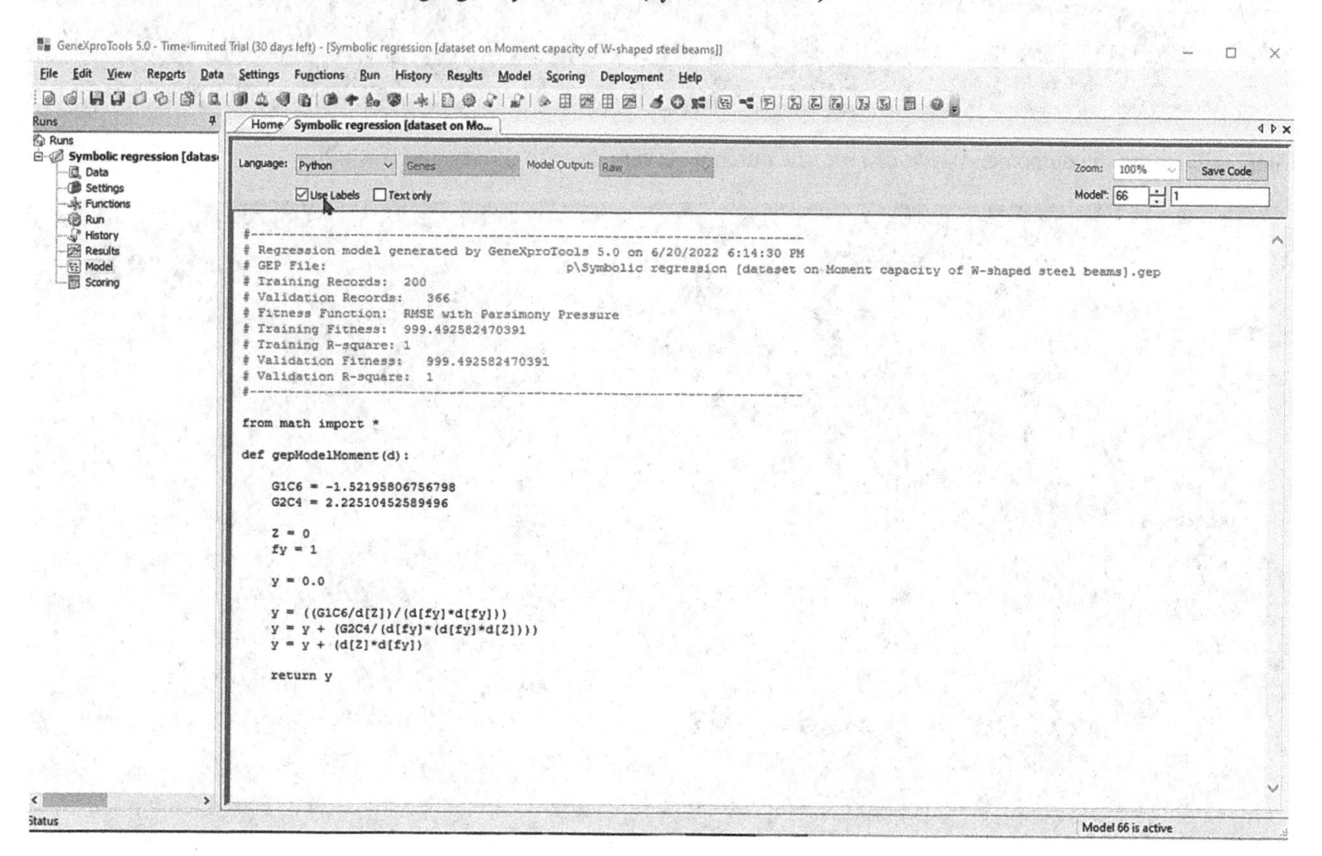

9.3.7 Online Interface by MetaDemoLab

This is an online interface[78] created by Stéphane d'Ascoli, Pierre-Alexandre Kamienny, Guillaume Lample, and François Charton (from Meta, formerly Facebook) [20, 24]. This interface does not require the user to download software but rather to upload their dataset into the interface where the analysis is carried out on the cloud. This interface does not allow the use of feature names, and hence we will be using a dataset without feature names. By default, the first column is named x_0, and the second is x_1. The last column is always identified as the target.[79]

Step 1: Upload the dataset and set model parameters to two.

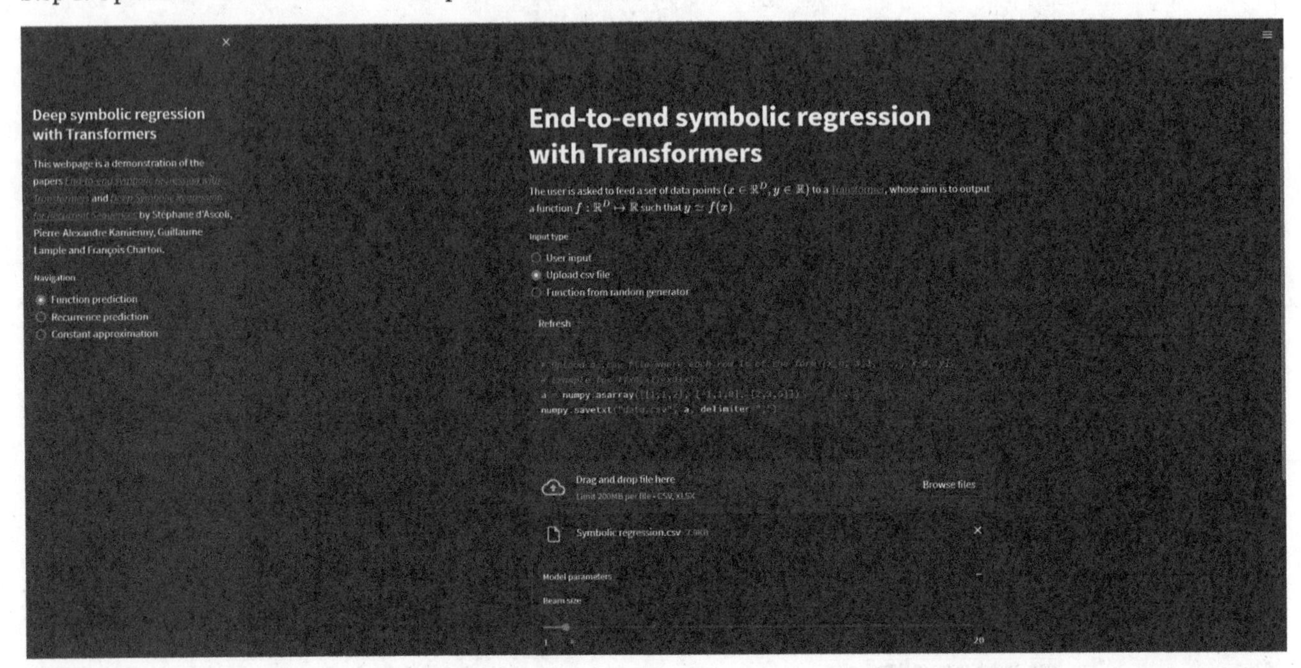

Step 2: Examine the outcome. As you can see, the outcome is very close to the ground truth.[80]

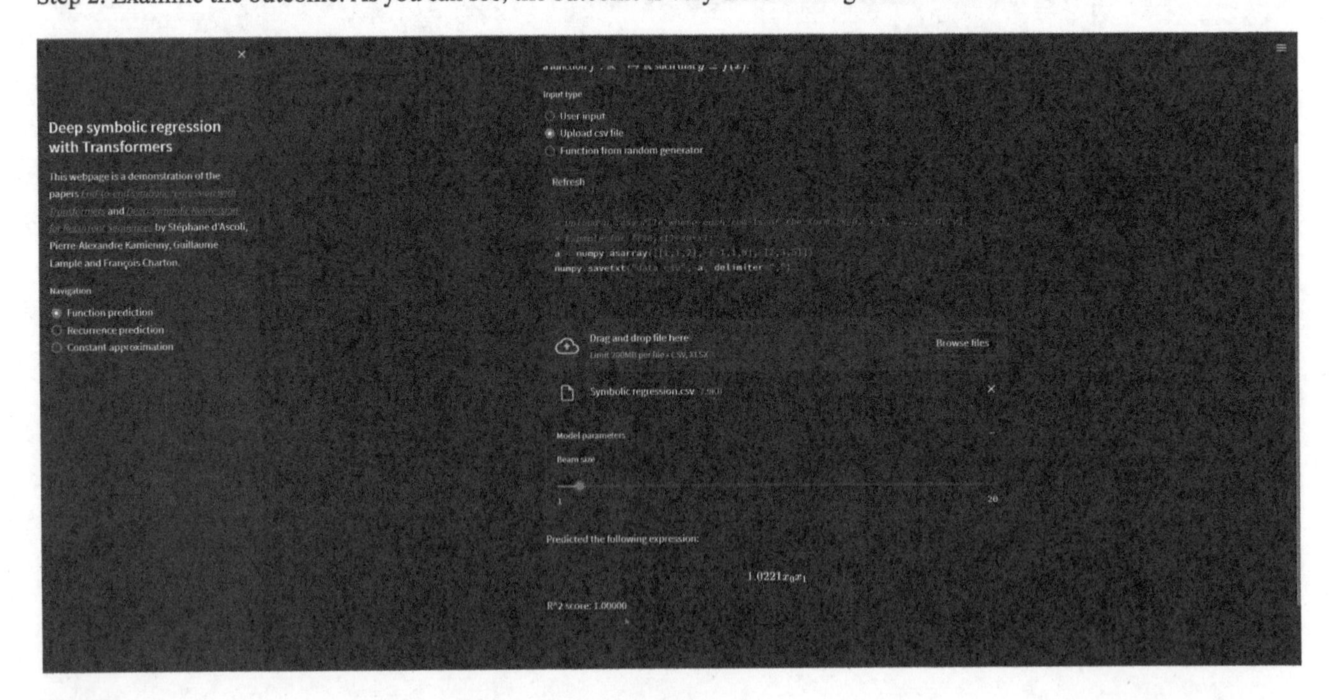

78 The interface can be accessed at: https://symbolicregression.metademolab.com

79 Make sure you save your dataset in CSV (Comma delimited) [*.csv] format. This interface may not accommodate classification at the moment.

80 For transparency, $1.0221x_o x_1 = 1.0221 f_y Z$.

9.3.8 Python

Here is the *Python* code using the *gplearn* library.

```
from gplearn.genetic import SymbolicRegressor
from sklearn.model_selection import train_test_split
import pandas as pd

Moment=pd.read_excel('Moment.xlsx')
x= Moment.drop(['M'],axis=1)
y= Moment['M']
x_train,x_test,y_train,y_test=train_test_split(x,y,test_size=0.30,random_state=111)

est_gp = SymbolicRegressor(population_size=1260,
                           generations=20, stopping_criteria=0.01,
                           p_crossover=0.7, p_subtree_mutation=0.1,
                           p_hoist_mutation=0.05, p_point_mutation=0.1,
                           max_samples=0.9, verbose=1,
                           parsimony_coefficient=0.01, random_state=0)

#Fitting dataset
est_gp.fit(x_train, y_train)

#The symbolic regressor predictions
y_gp = est_gp.predict(x_test)
```

```
    |   Population Average    |             Best Individual              |
---- ------------------------- ------------------------------------------ ----------
 Gen   Length          Fitness   Length          Fitness      OOB Fitness  Time Left
   0    38.00      4.27846e+25        3                0                0     18.74s
```

```
#Model score
score=est_gp.score(x_test,y_test)
score
```

```
1.0
```

```
converter = {
    'mul': lambda x, y : x*y,
}

from sympy import *
N = sympify(str(est_gp._program), locals=converter)
N
```

X_0X_1

Where X_0 is f_y and X_1 is Z.

Example no. 2: Deriving an expression for axial capacity of CFST columns.
Let me now take a bit of a larger phenomenon with multiple features. How about we explore the CFST columns[81] and see if we can derive a descriptive function for the axial capacity of these columns using symbolic regression? Let's even go one extra step and compare predictions from this new expression against that of available building codes. This will be the case in the second example.

For this example, I will be using *Eureqa*, *MetaDemoLab*, and *Python*. I am not going to go through each step. Rather, I will share some screenshots of each software's final expression. Then, I will plot how these predictions match that from codal expressions in American Institute of Steel Construction (AISC), Eurocode 4, and the Australian Standard (AS) 2327. Our features include tube diameter (D), tube thickness (t), tube length (L), yield strength of tube (f_y), and compressive strength of concrete (f_c).

81 Hello Tai Thai and Son Thai [63, 64]!

9.3.9 Eureqa

$$Axial\ capacity\ of\ Circular\ CFSTs = 0.004748 \times D^2 + 0.00396 \times D \times t \times f_y + 0.0007937 \times f_c \times D^2 - 36.94 - 1.055e-7 \times D \times L \times f_y \times f_c \tag{9.4}$$[82]

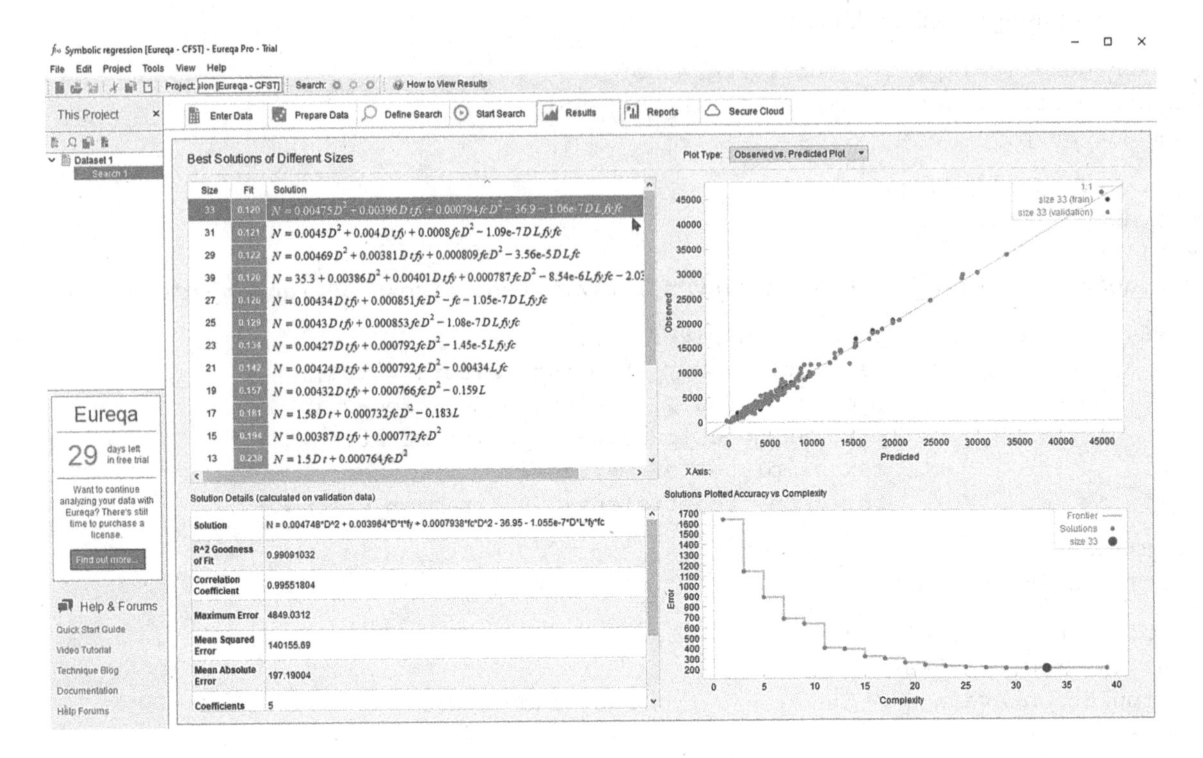

9.3.10 MetaDemoLab

$$Axial\ capacity\ of\ Circular\ CFSTs = (0.009x_0 - 0.543)(2.453x_1 + 0.026x_3 + 0.469x_4 - 13.345)\left(0.079x_0 - 0.008x_2 - 0.184x_4 - 3.433\cos(37.388x_1 - 162.125) + 74.607 + \frac{0.021}{0.007x_3 - 3.422}\right) \tag{9.5}$$

where $x_0 = D$, $x_1 = t$, $x_2 = L$, $x_3 = f_y$, and $x_4 = f_c$.

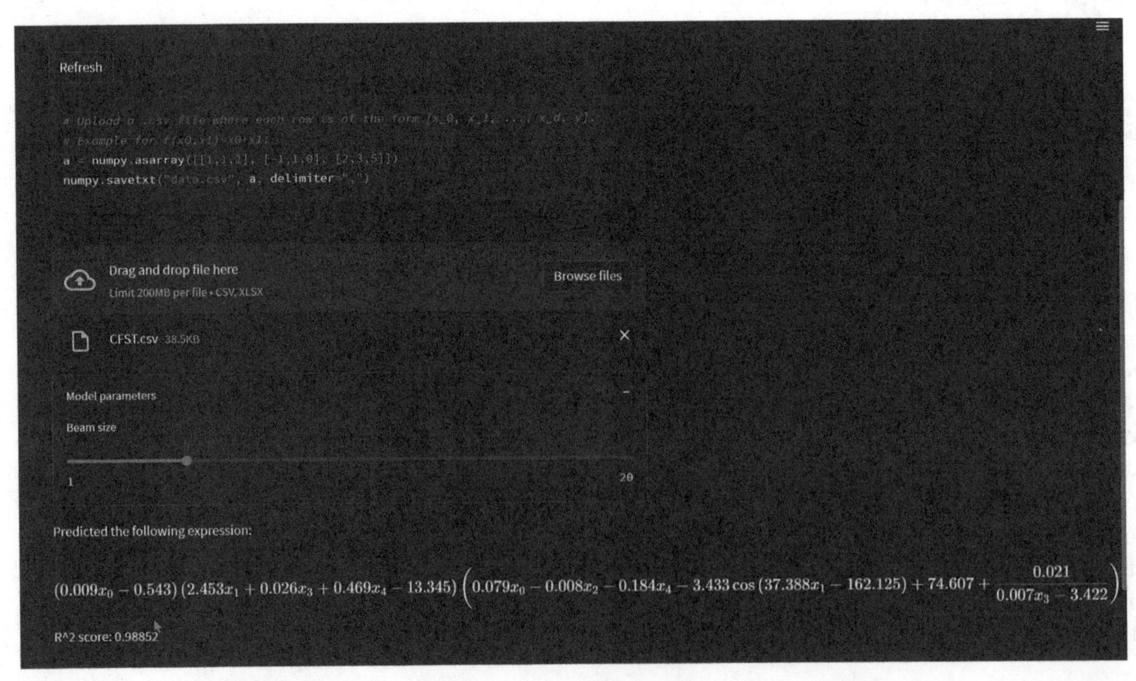

82 I used N in *Eureqa* to describe the axial capacity of Circular CFSTs. Hello Eurocode!

9.3.11 Python

Here is the *Python* code using the *gplearn* library.

```
#Python Data Analysis Library, it helps by creating a dataframe
#from CSV or TSV or XLSX files
import pandas as pd
from pandas import MultiIndex, Int64Index

#A library of multidimensional array objects and a collection of #routines for
#processing of array
import numpy as np

#A ML algorithm that can be used for regression and classificati
#on models

import xgboost as xgb
#A ML algorithm that can be used for regression and classificati
#on models
from sklearn.ensemble import RandomForestRegressor

#A package from sklearn that helps with splitting
#the data into training and testing datasets
from sklearn.model_selection import train_test_split

#Matplotlib and sns is a library for creating static,
#animated, and interactive visualizations in Python
from matplotlib import pyplot
import seaborn as sns
#Metrics from the sklearn library to evaluate the model
from sklearn.metrics import mean_squared_error
from sklearn.metrics import accuracy_score
from sklearn.metrics import r2_score

#Calling the data function
CFST=pd.read_csv('CFSTc.csv')

#Printing the first 5 rows of the data to check for errors
CFST.head(5)
CFST
```

	D	t	L	fy	fc	N
0	168.8	2.6	305.0	302.4	18.2	1326.0
1	168.8	2.6	305.0	302.4	34.7	1219.0
2	169.3	2.6	305.0	338.1	37.1	1308.0
3	169.3	2.6	305.0	338.1	34.1	1330.0
4	168.3	3.6	305.0	288.4	27.0	1557.0
...	...	...	...	...	...	...
1255	131.0	4.0	393.0	306.0	36.5	1542.0
1256	133.0	5.0	399.0	306.0	36.5	1659.0
1257	133.0	5.0	399.0	306.0	36.5	1673.0
1258	133.0	5.0	399.0	306.0	36.5	1712.0
1259	133.0	5.0	399.0	306.0	36.5	1774.0

```
1260 rows × 6 columns

#checking how python reads the dataset
#(float64 = numirical numbers)
CFST.dtypes

D     float64
t     float64
L     float64
fy    float64
fc    float64
N     float64
dtype: object
                                                                        In [4]:

Statistical_insights=CFST.agg(
    {
"D": ["min", "max", "median", "skew", "std", "mean"],
"t": ["min", "max", "median", "skew", "std", "mean"],
"L": ["min", "max", "median", "skew", "std", "mean"],
"fy": ["min", "max", "median", "skew", "std", "mean"],
"fc": ["min", "max", "median", "skew", "std", "mean"]
    }
        )

#Agg is a function that is used for aggregating the data using #one or more operations
#The output of this function will demonstrate a summary
#of statistical insights for the parameters of the dataset

print(Statistical_insights)
#This dataset is used for regression model
#which means you are trying to train the model
#to predict a numerical value (compressive strength in this exam#ple)

                  D          t            L           fy          fc
min       44.500000   0.500000   152.300000   178.300000    7.600000
max     1020.000000  16.500000  5560.000000   853.000000  185.900000
median   129.000000   4.000000   662.000000   324.700000   37.400000
skew       3.720751   1.584916     2.004010     2.167371    2.150750
std      104.833942   2.455295  1000.293248    90.955122   30.665316
mean     158.835238   4.276746  1055.142222   335.676508   47.503492

#Assigning the 'X' variable to use the features
#as a data frame for X
#Dropping the target label
#(it will be reassigned for y variable as a target)

x=CFST.drop(['N'], axis=1)
y=CFST['N']

#Assigning the target label

x_train,x_test,y_train,y_test=train_test_split(x,y
                                               , test_size=0.2
                                               , random_state=200
                                               )

#Splitting the dataset into 80% training set and 20% as a
#testing set
RF=RandomForestRegressor(max_depth = 20,
```

```
                          random_state = 0,
                          n_estimators = 10000
                         )
#Machine learning algorithm
RF.fit(x_train,y_train)

#Fitting our split data into training and testing datasets
RandomForestRegressor(max_depth=20, n_estimators=10000, random_state=0)
Predictions = RF.predict(x_test)

#The prediction functions
Score =RF.score(x_train,y_train)

#Prediction score for the training set
print ("score : % f" %(Score))
print ('Accuracy',100-(np.mean(np.abs((y_test - Predictions) / y_test)) * 100))
print (r2_score(y_test, Predictions))
print ('MAE',np.mean(abs(y_test-Predictions)))
print ('RMSE',np.sqrt(np.mean(np.square(y_test-Predictions))))

#Above are metrics to evaluate the performance of the model
#and elevate the confidence that the model
#is accurately predicting the outcome of the model 'N test'
score:  0.994501
Accuracy 91.61711747850501
0.9766233358294548
MAE 190.32685368078842
RMSE 555.1837875375517

#Importing the SymbolicRegressor library to initiate the
#algorithm
from gplearn.genetic import SymbolicRegressor
est_gp = SymbolicRegressor(population_size=1260,
                           generations=20, stopping_criteria=0.01,
                           p_crossover=0.7, p_subtree_mutation=0.1,
                           p_hoist_mutation=0.05, p_point_mutation=0.1,
                           max_samples=0.9, verbose=1,
                           parsimony_coefficient=0.01, random_state=0)

#Initiating the algorithm using the above settings, which
#increased the accuracy of the model
est_gp.fit(x_train, y_train)

#Fitting the training dataset

y_gp = est_gp.predict(x_test)

#Symobilic regressor predictions

    |   Population Average    |             Best Individual              |
---- ------------------------- ------------------------------------------ ----------
 Gen   Length          Fitness   Length          Fitness      OOB Fitness  Time Left
   0    29.54      1.33284e+21       15          1173.65          1756.85     16.44s
   1    11.15      1.85592e+08       13          899.473          723.376     10.53s
   2    10.77      2.72142e+06       13          828.562          1360.17     10.60s
   3    13.74           493065       17          725.416          590.774      9.98s
   4    16.42          95669.6       27          681.176          981.357     10.14s
   5    20.54      3.24654e+06       37           661.54          844.249     10.23s
   6    23.47      3.03561e+07       25          664.564          882.518      9.76s
   7    24.23      5.48202e+07       27          654.208          1017.65      8.96s
```

```
   8    29.96      7.49274e+06       35       643.528       987.419     9.06s
   9    35.43      2.61246e+07       37       635.438       979.855     9.16s
  10    39.70           184599       47       632.776       968.959     8.46s
  11    43.96           103712       63       622.238       1080.15     7.73s
  12    49.71           226647       85       626.494       1037.15     7.30s
  13    53.44           381808       21       625.923       1092.07     6.49s
  14    62.79      1.63537e+10       73       631.003       743.992     5.95s
  15    69.80       8.3692e+07       85       615.331       1026.55     5.10s
  16    79.79           125911       69       608.597       1061.46     4.88s
  17    89.14      2.29499e+08      109       587.532       1112.72     3.07s
  18    93.87           127462       75       590.906       1004.49     1.59s
  19   101.48           267719       99       594.219       872.978     0.00s

score=est_gp.score(x_test,y_test)
print('R2:',z)

#Model score
R2: 0.891905456818484

#The equation that the symbolic regressor was able to generate,
#but needs alot of simplificaation  which will be provided in
#the next couple codes
print(est_gp._program)
add(add(div(add(mul(X2, X4), div(X2, 0.920)), X2), X0), add(div(sub(add(X0,
sub(add(X0, X0), sub(div(div(sub(add(X0, sub(add(X0, X1), sub(div(sub(X3, X4),
sub(add(div(sub(X2, X0), sub(X2, X0)), X0), add(X0, add(X0, add(sub(mul(0.095, X2),
X0), X0)))))), -0.911))), sub(mul(0.095, X2), -0.911)), div(0.920, X1)), mul(0.095,
add(X0, sub(add(X0, div(X0, X2)), sub(div(add(mul(0.095, X2), X2), mul(0.095, X2)),
-0.911)))))), -0.911))), sub(mul(0.095, X2), -0.911)), div(0.920, X1)), div(X0, X2)))

#Simplify the above equation
converter = {
    'add': lambda x, y : x + y,
    'sub': lambda x, y : x - y,
    'mul': lambda x, y : x*y,
    'div': lambda x, y : x/y,
    'sqrt': lambda x : x**0.5,
    'log': lambda x : log(x),
    'abs': lambda x : abs(x),
    'neg': lambda x : -x,
    'inv': lambda x : 1/x,
    'max': lambda x, y : max(x, y),
    'min': lambda x, y : min(x, y),
    'sin': lambda x : sin(x),
    'cos': lambda x : cos(x),
    'pow': lambda x, y : x**y,
}
from sympy import *
N = sympify(str(est_gp._program), locals=converter)
```

$$N = X_0 + \frac{X_0}{X_2} + 1.086X_1\left(3X_0 - \frac{1.0869X_1\left(2X_0 + X_1 - 0.095X_2 - \frac{X_3 - X_4}{-X_0 - 0.095X_2 + 1} - 1.822\right)}{0.19X_0 + \frac{0.095X_0}{X_2} - 1.1815} - 0.095X_2 - 1.822\right) + \frac{X_2X_4 + 1.0869X_2}{X_2}$$

```
save.N
```

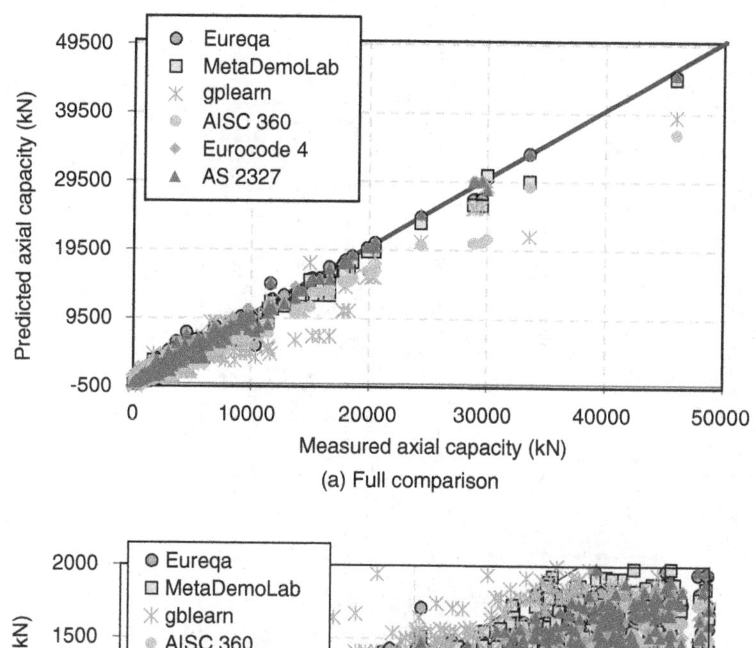

(a) Full comparison

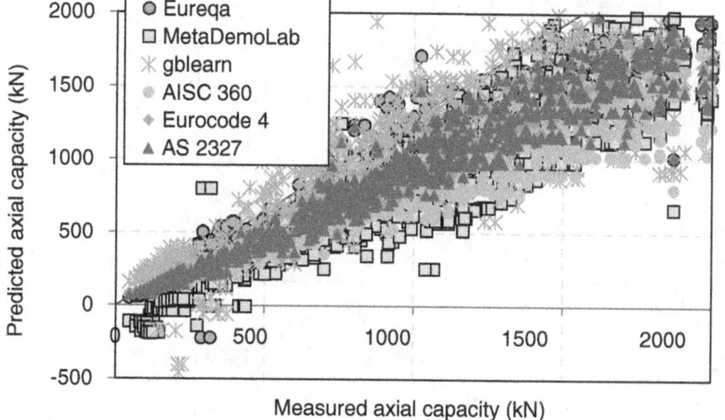

(b) Comparison (spot analysis [up to 1000 N])

Figure 9.9 Comparison between descriptive and codal expressions. [Note the existence of zero and negative predictions].

As you can see, there is a good match between the descriptive and codal expressions.[83] However, it is worth noting to show that some predictions from the descriptive expression turned zero and negative (see Figure 9.9). This is one of the common issues with descriptive expressions (since they rely on the data vs. the mechanics of the problem). An easy fix would be to add absolutes to the derived expressions to negate the negatives. Unfortunately, the zero and unrealistic small predictions may not be as easily fixed. The engineer is then tasked with selecting different/more appropriate expressions that do not suffer from such issues.[84] On the other hand, you can also see that some codal expressions do not also seem to capture the axial capacity of CFST columns – especially those with high capacity (exceeding 20,000 kN[85]).

9.3.12 GeneXproTools*

For completion, I have also developed a script from *GeneXproTools*. You can find it here and the solution file can be downloaded from my website.

83 Both R and R^2 ranges between 0.99 and 1.00 for most cases (expect those from *gplearn* which comes to be 0.96 and 0.92).

84 In my examples, I intentionally showed such issues to give you a glimpse on some of the dangers of blindly accepting ML outputs, as well as relying on global metrics without checking for *spot analysis (local stability)*! If you do run a reliability analysis on predictions from descriptive expression, they may unveil *other* issues!

85 In reality, codal expressions can be often limited to a certain range of applicability and inherently designed to be conservative.

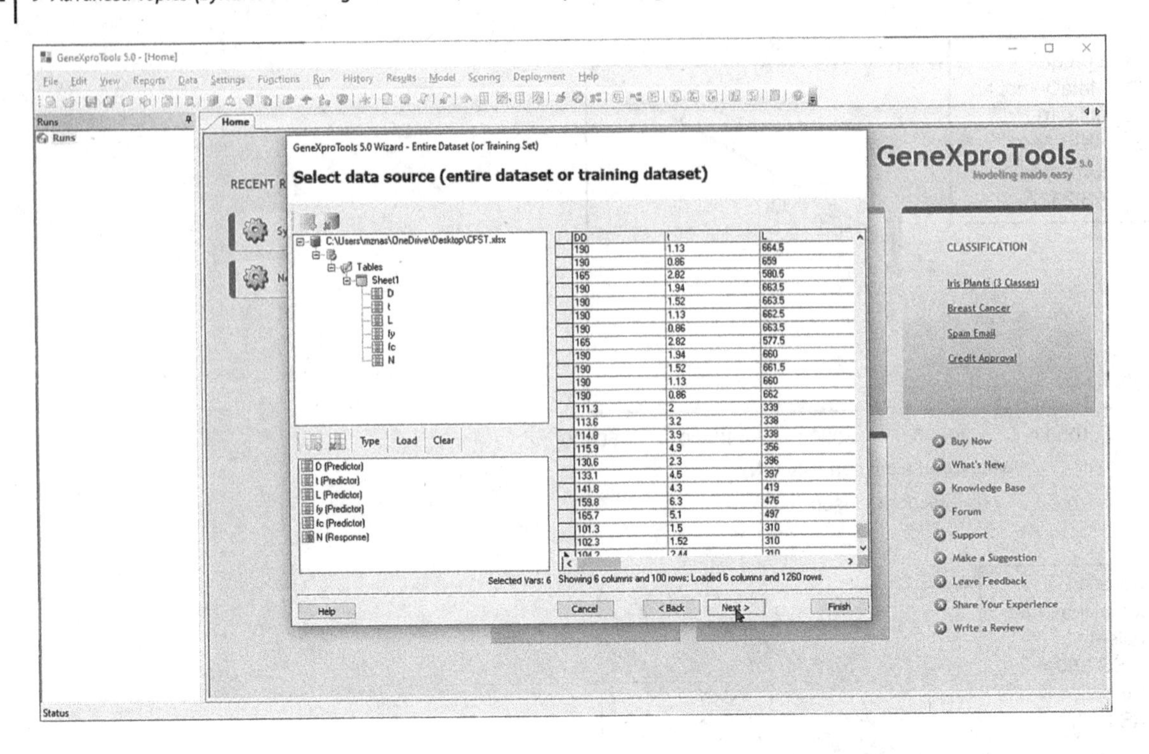

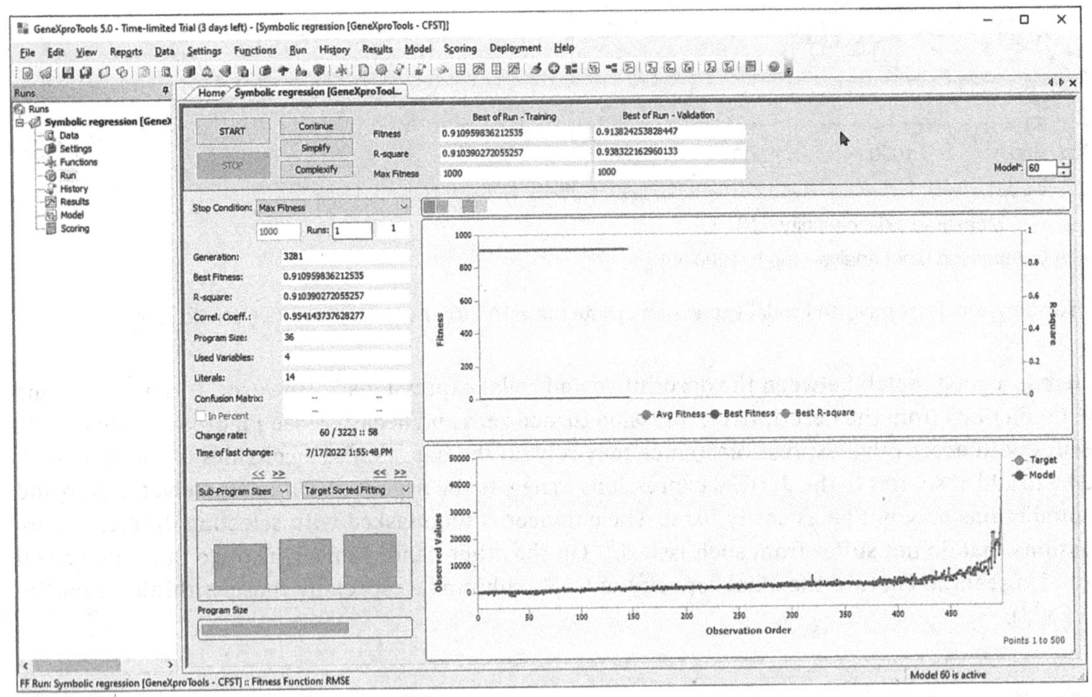

Solution via *Python*:

```
#---------------------------------------------------------------------
# Regression model generated by GeneXproTools 5.0 on 7/17/2022
#1:56:32 PM
# GEP File: C:\Symbolic regression [GeneXproTools - CFST].gep
# Training Records:  500
# Validation Records:   760
# Fitness Function:  RMSE
# Training Fitness:  0.910959836212535
# Training R-square: 0.910390272055257
```

```
# Validation Fitness:   0.913824253828447
# Validation R-square:  0.938322162960133
#-------------------------------------------------------------------------
from math import *

def gepModel(d):

    G1C7 = 6.49150059511094
    G1C3 = 9.63011566515091
    G2C9 = -2.13608203375347
    G2C8 = 5.58519241920225

    DD = 0
    t = 1
    L = 2
    fc = 4

    y = 0.0

    y = ((G1C7*(1.0-G1C3))*log(((d[DD]*d[DD])*d[t])))
    y = y + ((d[t]+((gep3Rt(d[DD])-G2C9)+((d[t]+G2C8)/2.0)))*d[DD])
    y = y + (((((d[fc]+d[fc])-d[L])+(d[DD]+d[DD]))+(d[DD]*d[t]))/2.0)
    return y

def gep3Rt(x):
    if (x < 0.0):
        return -pow(-x,(1.0/3.0))
    else:
        return pow(x,(1.0/3.0))
```

Solution via *R*:

```
#-------------------------------------------------------------------------
# Regression model generated by GeneXproTools 5.0 on 7/17/2022
#1:58:53 PM
# GEP File: C:\Users\mznas\OneDrive\Desktop\Symbolic regression [GeneXproTools -
CFST].gep
# Training Records:  500
# Validation Records:   760
# Fitness Function:  RMSE
# Training Fitness:  0.910959836212535
# Training R-square: 0.910390272055257
# Validation Fitness:   0.913824253828447
# Validation R-square:  0.938322162960133
#-------------------------------------------------------------------------

gepModel <- function(d)
{

    G1C7 <- 6.49150059511094
    G1C3 <- 9.63011566515091
    G2C9 <- -2.13608203375347
    G2C8 <- 5.58519241920225

    DD <- 1
    t <- 2
    L <- 3
    fc <- 5

    y <- 0.0
```

```
    y <- ((G1C7*(1.0-G1C3))*log(((d[DD]*d[DD])*d[t])))
    y <- y + ((d[t]+((gep3Rt(d[DD])-G2C9)+((d[t]+G2C8)/2.0)))*d[DD])
    y <- y + (((((d[fc]+d[fc])-d[L])+(d[DD]+d[DD]))+(d[DD]*d[t]))/2.0)
    return (y)
}
gep3Rt <- function(x)
{
    return (if (x < 0.0) (-((-x) ^ (1.0/3.0))) else (x ^ (1.0/3.0)))
}
```

Example no. 3: Descriptive expression of a continuous phenomenon.
Unlike the above two examples, where our focus was directed at deriving descriptive expressions for single outcomes, this example will extend our approach to continuous phenomena. In such a phenomenon, the output we are interested in continues as a function of the event (i.e., increase/decrease in applied loading,[86] presence/absence of stimulus, etc.).

Allow me to revisit one of the most suited problems here[87] as we will be looking to derive an expression that can predict how the temperature rises in RC beams under fire conditions. There are three folds to this example, 1) we do not have such an expression at the moment,[88] 2) this example can mimic many of the continuous problems that fall under our domain, and 3) I would like to incorporate the concept of physics-guided and domain knowledge feature selection.[89]

Let us look at our dataset. We have a dataset from 20 normal strength and uninsulated reinforced concrete (RC) columns tested under standard fire conditions that were collected in an earlier work [17] with the following features: duration of exposure under standard fire (t), the width of column (W), concrete cover (C), aggregate type (A[90]) and temperature rise (T) as a function of fire exposure time.

As you can see, these features all agree with our domain knowledge and physics principles.[91] As you can also see, this is a continuous phenomenon since we are tracing the temperature rise history (and not just the temperature in cross section at a specific point in time), and this history changes as a function of fire exposure time.

We will be using the following two software: *Eureqa*, *MetaDemoLab*, and *HeuristicLab*.

9.3.13 Eureqa

$$T = 9.55 + 0.122 \times A \times t + 0.113 \times t \times C - 0.00337 \times t \times W - 0.00811 \times t^2 \quad (9.6)$$

9.3.14 MetaDemoLab

$$T = (0.061x_0 + 0.147)(-0.124x_0 - 0.602\tan(1.378x_1 + 0.294x_2 + 17076x_3 - 178.621) + 73.246 \quad (9.7)$$

$x_0 = t$, $x_1 = C$, $x_2 = W$, and $x_3 = A$

9.3.15 HeuristicLab

As you can see, the descriptive expression from HeuristicLab is a bit complex,[92] and hence I will not be discussing it here.

86 Which could be mechanical or thermal, etc.

87 That we have covered in an earlier chapter.

88 Aside from those published in my previous papers [17, 65]. In other words, if an engineer would like to predict the deformation of a structural member, this engineer is likely to conduct a new test or build a nonlinear FE model.

89 **Physics-guided and domain knowledge feature selection**: selecting features in a ML model based on our domain knowledge and physics principles.

90 This is a binary and dummy feature (1 for carbonate, 2 for siliceous aggregates).

91 I would like to note a few things: 1) all columns are tested under similar conditions (standard fires and simply supported), 2) all columns are homogenous and made from RC. These two items narrow our feature space. In other words, if we were to include different fire testing regimes or insulation materials, then our feature space will grow. If we do not change our feature space, then the selected features do not align with our domain knowledge and physics principles.

92 I tend to favor compact and simple expressions. Saying that, there may not be nothing wrong with a bit of complexity!

Let us now plot predictions from the two expressions in Figure 9.10. It is clear that there is good convergence between the two expressions.[93] When we apply these expressions to predict temperature rise in RC beams, Figure 9.11 shows that these expressions can realize decent predictions.[94]

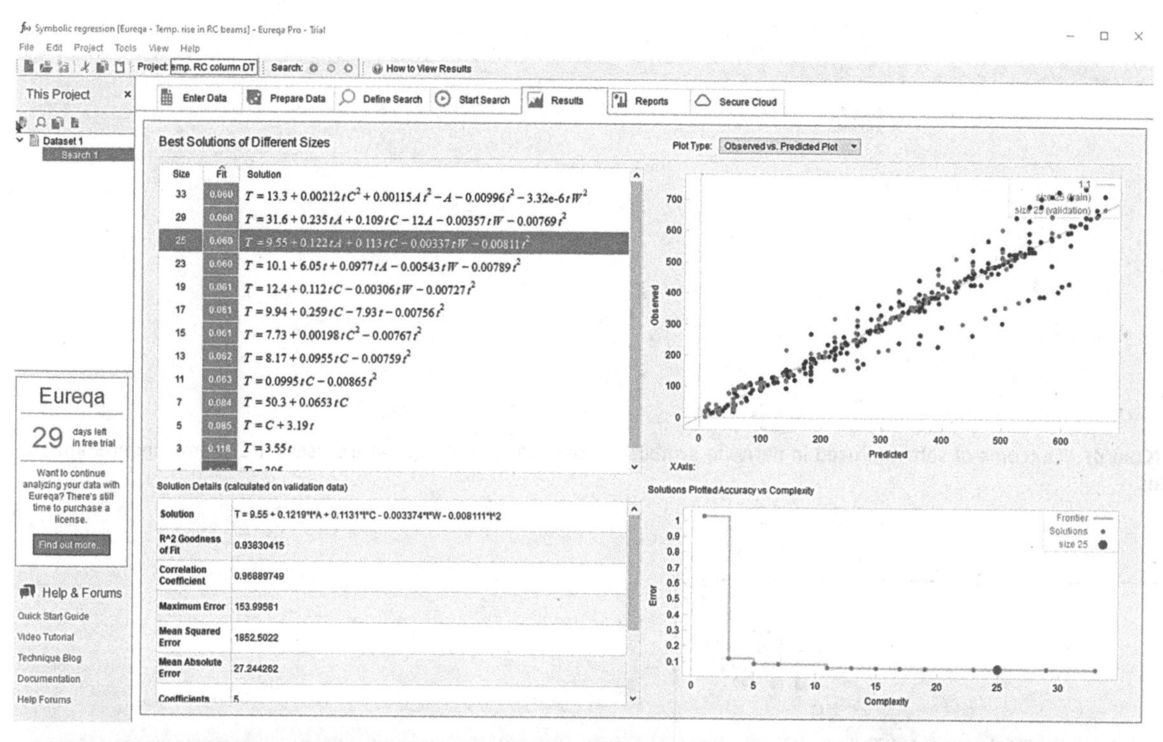

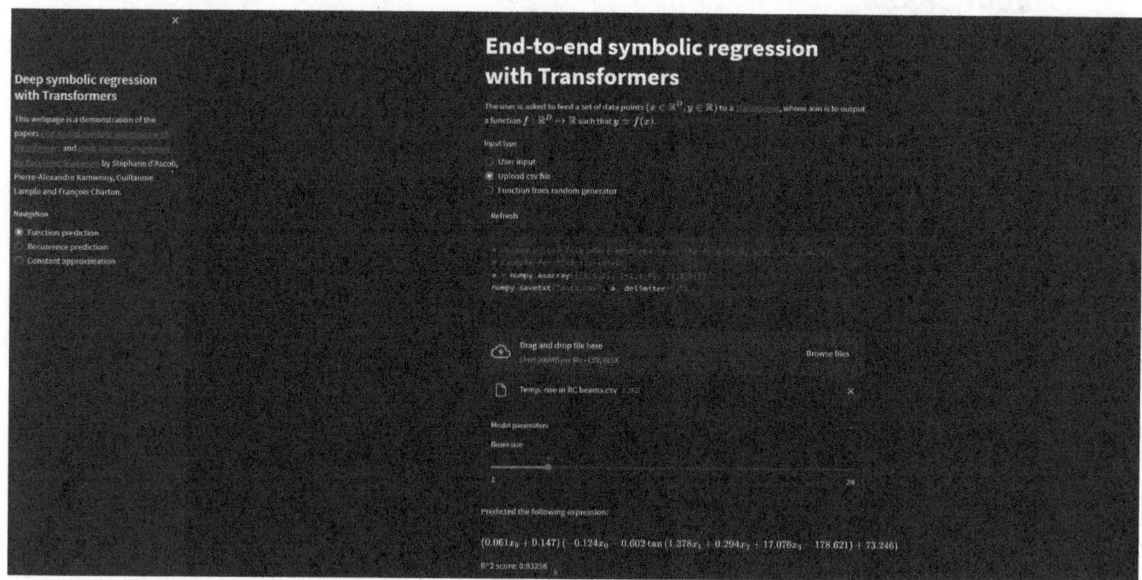

(Continued)

93 In fact, R and R^2 for predictions from these two expressions are unity.

94 I would like to make very clear that these expressions, while seeming to be able to capture the phenomenon on hand, they cannot be thought of as causal expressions (as in, they may not articulate the cause → effect relationships).

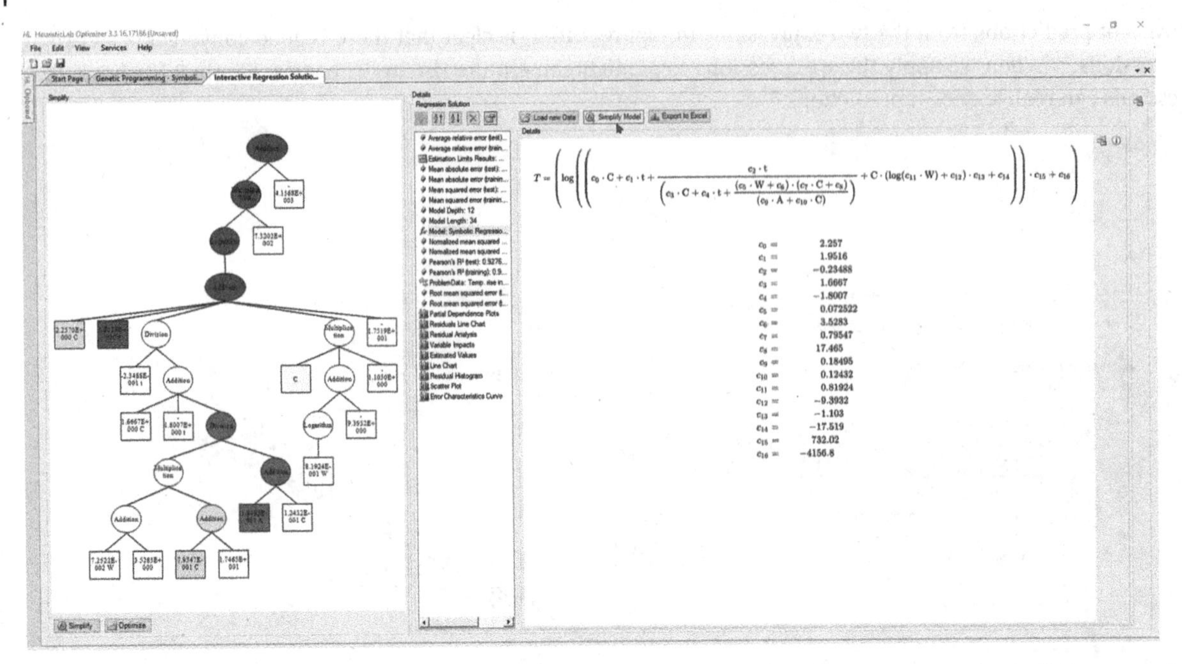

Figure 9.10 (Cont'd) Outcome of software used in deriving symbolic expressions for temperature rise in RC beams under standard fire conditions.

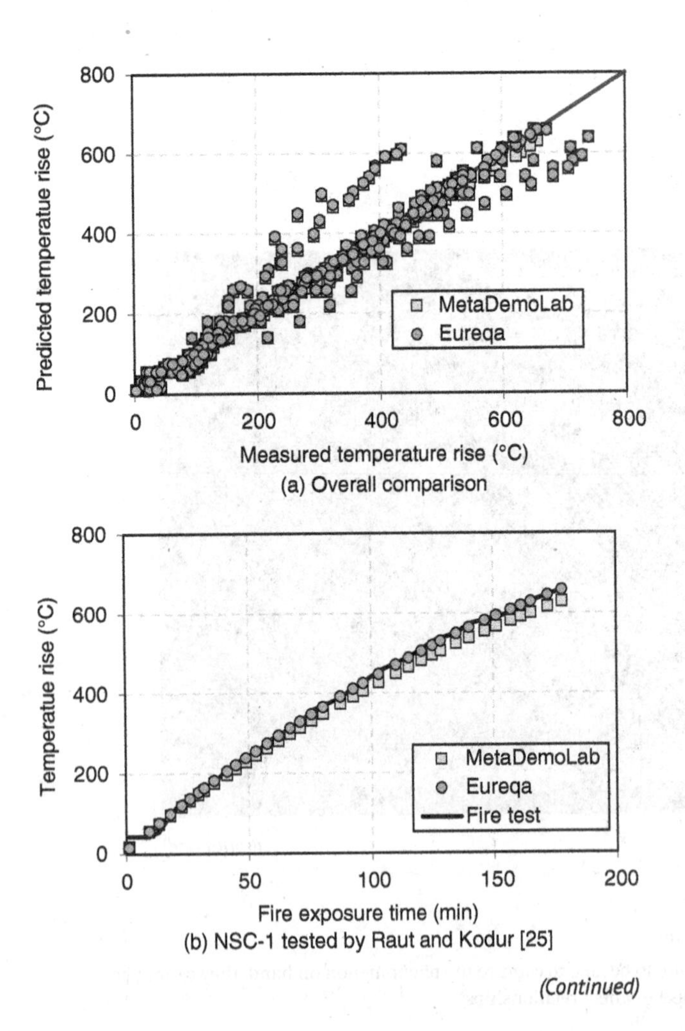

(Continued)

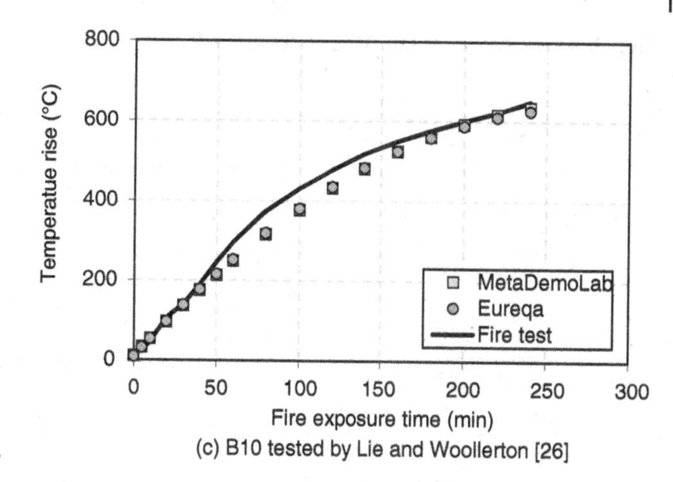

Figure 9.11 (Cont'd) Comparision between descriptive expressions and observed temperature rise under fire conditions.

Example no. 4: Descriptive expression of a classification problem.

The last example will tackle a classification problem with regard to transportation infrastructure. Say that we have a collection of bridges, would we be able to derive an expression that can identify which bridges are vulnerable to collapse?[95]

This dataset comprises the following features, structural systems (*S*), construction materials (*M*), Span (*P*), Age (*A*), and the number of lanes (*N*).[96] The outcome in this dataset is binary (collapse, no collapse).

To carry out a classification analysis, a few tweaks are to be applied to *Eureqa*. These are shown in the following snippets as well as in the provided *Eureqa* file. The outcome of this analysis is this Step function, which returns a value of unity if a bridge is vulnerable to collapse.

$$Collapse = step(564.74 \times M - 1278.08 - 0.81 \times P - 4.72 \times A - 14.61 \times N - 44.19 \times S - 632.72 \times or(0.85 \times M - 2.70 \times P - 116.91)) \tag{9.8}$$

Step 1: Assign the proper *Formula building-blocks* (you may use the *Step* or *Logistic* functions).

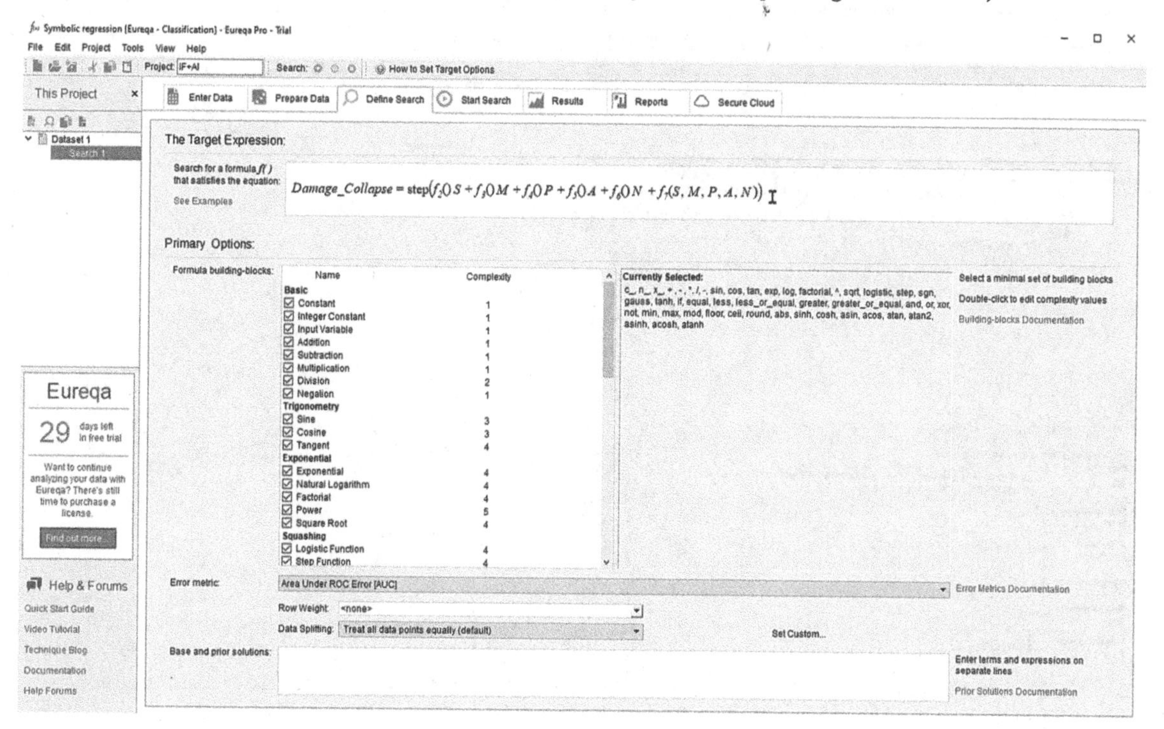

95 Based on historical data of failed bridges.
96 This example is based on a more complex dataset that can be found elsewhere [66].

Step 2: Keep an eye on the AUC ROC Error.[97]

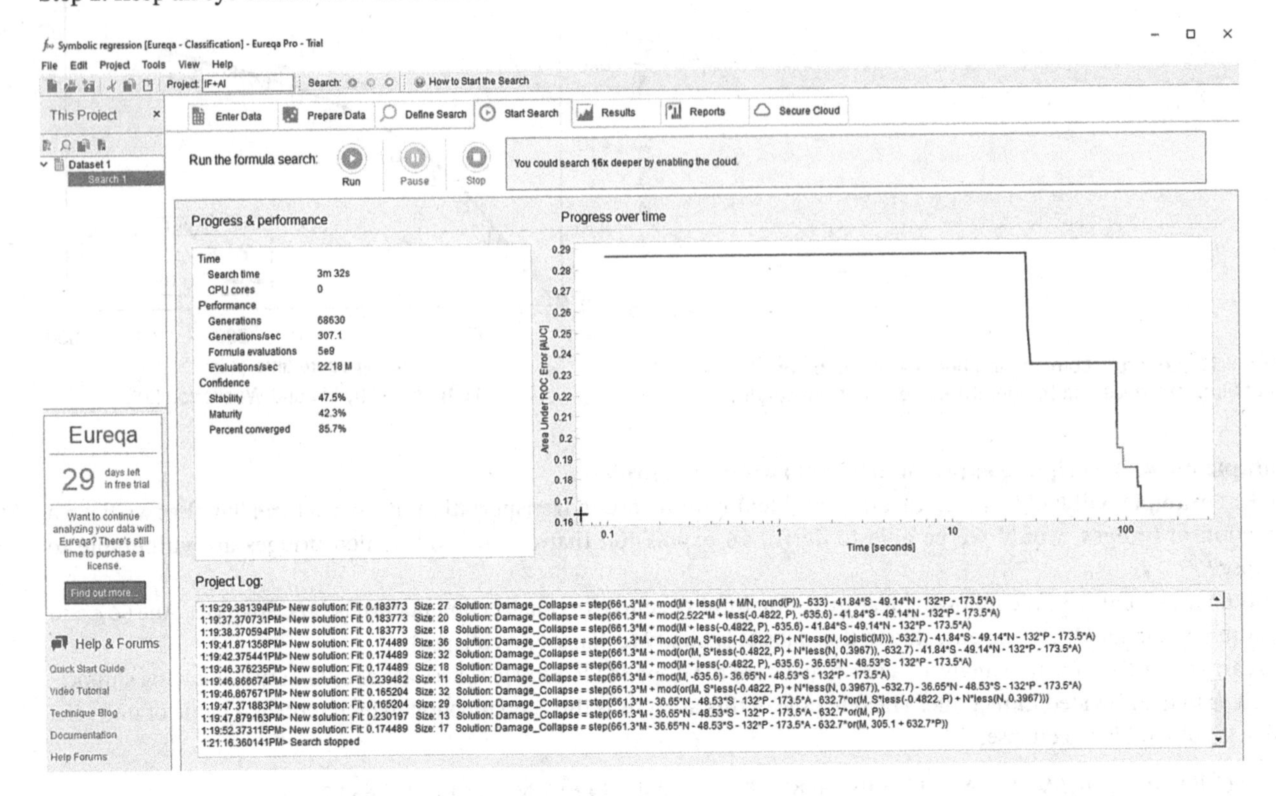

Step 3: Arrive at a possible descriptive expression.

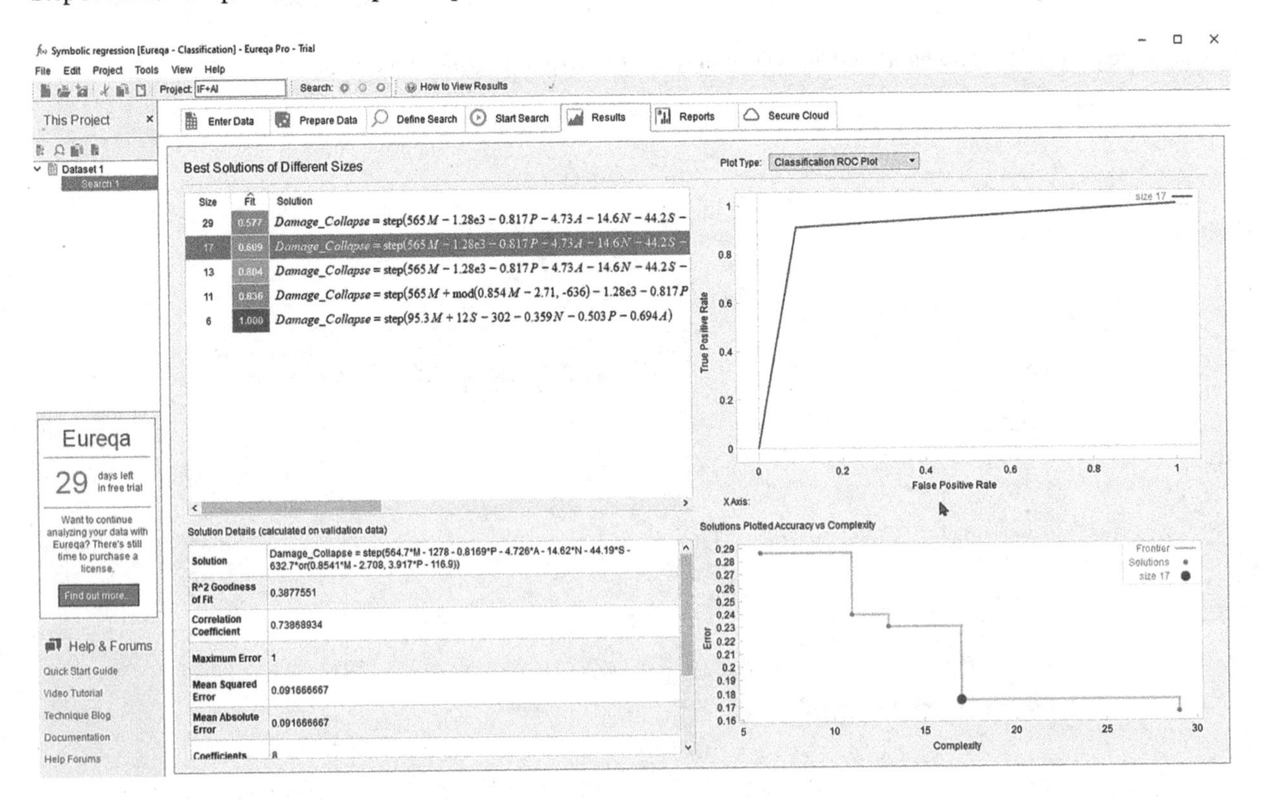

97 We usually calculate the AUC ROC. *Eureqa* calculates the error (i.e., 1-AUC ROC) which implies that smaller error is favorable.

9.3.16 A Note

This section was all about the following concept. Can we use ML to arrive at a simple model that represents a formula? If so, how? As you can see, we can derive such expressions simply with many of the available software that we have access to. We can also do the same via programming scripts.

What I would like to remember is that for some problems, arriving at such descriptive expressions can be of merit as an engineer will be substituted into a formula as opposed to a ML model.[98] The same can also be extended to the following question, could you just update codal equations via symbolic regression into one equation of much higher performance? This is a good question. My answer is that while symbolic regression searches for symbols, the goal of codal expressions is not just to predict a phenomenon with high accuracy. These expressions go beyond that to incorporate reliability and confidence measures. The same expressions often arise from first principles vs. a data-driven analysis.[99]

Now, what if we can supplement symbolic regression with reliability and confidence measures, while at the same time incorporating domain knowledge and physics and causal principles? This is a good question that I hope we can address one day.

9.4 *Mapping Functions*

The previous discussion on symbolic regression opens the door to this section on *mapping functions*. While symbolic regression ties a set of features together[100] without considering any insights in our data, *mapping functions*, on the other hand, ensure that features are properly identified, selected, and tied by incorporating physics principles, domain knowledge, and insights from our data.[101] The overarching goal of a *mapping function*[102] is to mimic a ML model as much as possible to allow the engineer to adopt and apply the principles of ML without having to hassle with building models or programming. One of the great strengths of a *mapping function*, and given its mathematical representation, is its ability to form the basis for a hypothesis which could then be tested and explored.

9.4.1 Big Ideas

Let me start with the following. Deriving a *mapping function* requires an understanding of the phenomenon we happen to be interested in. At a minimum, we need observations that depict some form of consistent response regarding the problem on hand. It is because of this consistency that it could be possible to encode the response we see into a mathematical expression that can be repeatedly used with confidence. To ensure its wide acceptance and ease of use, a *mapping function* is preferred to be compact[103] and reliable.

To demonstrate the concept of *mapping functions*, allow me to revert back to some visual examples.[104]

9.4.1.1 Sectional Phenomenon

In a general sense, structural engineering phenomena that is related to the *sectional* level (i.e., moment) are primarily governed by two components: geometric features and material properties comprising this structural member.

To reinforce this discussion, I will be referring to Equation 9.1, which we can arrive at from mechanics. This equation illustrates the functional relation between the *geometry* and *material properties* for the moment capacity of a typical rectangular reinforced concrete (RC) beam with tensile reinforcement. Thus, at a sectional level, the *governing parameters* are expected to belong to the class of geometric features and material properties.

98 Which requires access to the code, as well as knowledge of programming and a computing device/cloud.

99 Remember that the data can take us so far. That does not mean the data can always unlock the physics behind our problems.

100 Whether collected from literature, or identified by domain/expert knowledge, or based on their availability.

101 And in some cases, a *mapping function* can also answer causal questions.

102 A *mapping function* maps the features to the target. There is a lot that we still do not know on *mapping functions* and they could tie to structural causal models. Perhaps the next edition will house a more rich discussion and examples/codes.

103 Keep in mind that complex phenomena may sometimes result in complex functional forms.

104 The concept behind mapping functions could be generalized to other problems. I will stick to my comfort zone here and will advise to explore this concept for your own problems.

$$M = A_s \times f_y \left(d - \frac{a}{2} \right) \quad (9.9)$$

$$\text{and, } a = \frac{A_s \times f_y}{0.85 f_c \times b}$$

where A_s: Area of the tensile steel reinforcement, d: depth of the concrete beam, and b: width of the concrete beam, f_c: compressive strength of the concrete, and f_y: yield strength of the tensile steel reinforcement.

9.4.1.2 Behavioral Phenomenon

On the other hand, *behavioral* structural engineering phenomena that occur at the element level, such as load-deformation history, require the addition of more governing features in lieu of the geometric features and material properties. Most notably, such features will arise from the type of *applied loading* and/or *boundary conditions*. Intuitively, we can infer that 1) complex phenomena are likely to require a higher degree of features, and 2) there is mutual independence between the classes of features.

Let us walk through two examples.

Let us revisit one of the first examples we covered in this book. Beams are load bearing members designed to satisfy strength and serviceability criteria [27]. A primary interest to structural engineers is to trace the load-deformation history[105] of beams. From a practical perspective, the deformation a beam undergoes can be measured in tests or calculated following mechanics principles (such as the moment-curvature method) or simulated via FE models. Given that the magnitude of deformation a beam undergoes primarily reflects the degree of degradation in its load bearing capacity arising from the applied loading, then it is quite possible to associate these two events together [17].

Obtaining the load-deformation history experimentally is an involved process that requires building and testing a series of beam specimens.[106] Similarly, to model such history, an engineer is tasked with developing an analytical or a FE model that captures the interaction between the applied loads, beam geometry, materials properties, etc. In both cases, the mentioned methods follow and obey mechanics and physics principles [28].

Similarly, if you happen to witness or visualize such a test, then domain knowledge informs us that the deformation history of a loaded beam in bending results in a parabolic-like or curved-like history plot. This plot insinuates that the observed deformations, which result from stresses generated from a continually applied load, P, reflect the degradation of the beam's sectional capacity[107] [17].

If we confine our discussion to RC beams,[108] then such a plot is likely to have distinct points that reflect distinct phenomena (e.g., concrete cracking, steel yielding, concrete crushing, and at the point of failure). For example, Figure 9.12 illustrates the deformation history of two idealized beams, *Beam 1* and *Beam 2*. The only difference between these two beams is that *Beam 2* is shallower than *Beam 1*, and hence under identical loading conditions,[109] *Beam 2* will fail before *Beam 1*.

Arriving at the above conclusion is trivial since we are only considering one feature (*height of beam*) and only this feature is varied between the two beams. Now, if other parameters were to vary as well, then the problem substantially grows as our perception of physics principles and domain knowledge becomes hard to bridge.

Let us extend the Figure 9.12 example by trying to trace the deformation-history of beams once exposed to fire conditions.[110]

As we have seen above, the load bearing capacity and degree of deformation in a beam can be calculated following mechanics principles or FE simulations. However, under certain conditions (say, under fire effects), arriving at load bearing capacity and deformation history turns into a multifaceted problem [29]. Unlike tracing the load-deformation of beam at ambient conditions, the magnitude of deformation a beam undergoes under fire primarily reflects the degree of

105 The load-deformation history of beam is a chart that plots the load and mid-span deformation at the vertical and horizontal axes, respectively. It is a *history* since it shows the propagation of deformation as a function of the applied loading. In a way, this plot reflects how the features of the beam responds to the applied loading.

106 This turns into a more complex endeavor at extreme events, which we will visit in the next example.

107 A function of geometric features, material properties, restraints, etc.

108 Re-visit this discussion by referring to steel or timber beams.

109 I hope you can read between the lines and identify the little bit of causal language embedded here.

110 In this instance, the addition of fire adds a new layer of complexity that you are going to see in a bit.

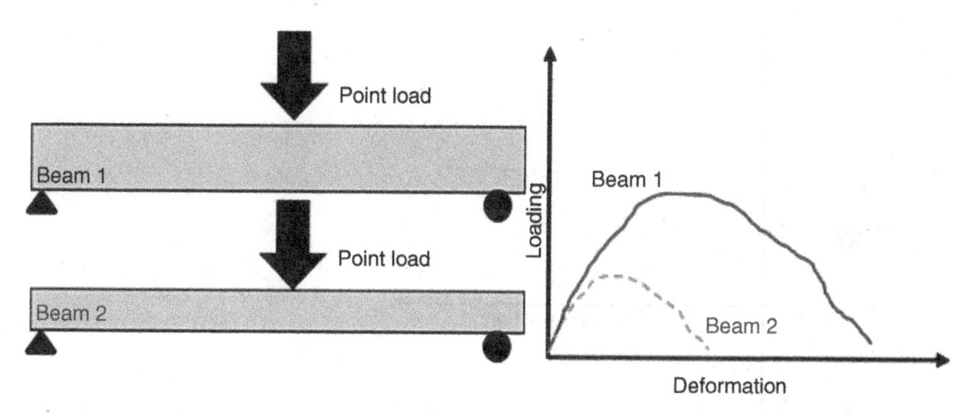

Figure 9.12 Illustration of load-deformation history.

degradation within its load bearing capacity due to the temperature-induced material properties.[111] Thus, it is quite possible to associate these two phenomena[112] [17].

Physics principles show that the mid-span deformation resulting in a fire-exposed beam is generated from applied loading, P, and degradation to the beam's sectional capacity[113] [17]. Since P remains virtually constant during fire testing,[114] then the extent of deformation reflects the cumulative degradation in material properties and any possible losses in cross-section size.

Let us look at Figure 9.13. Now, for two identical beams,[115] *Beam 1* and *Beam 2*, where *Beam 1* is loaded with P_1 and *Beam 2* is loaded with P_2 (and $P_1 > P_2$). Since $P_1 > P_2$, then *Beam 1* is expected to undergo higher levels of deformation under fire. Arriving at this notion is trivial since only one feature (P) is varied between the two beams. However, if other features were to be varied as well, then, and similar to the example at ambient temperature, the problem on hand substantially grows.

9.4.2 Concept of *Mapping Functions*

Based on the above examples, one could argue that the sectional capacity equation and load-deformation history plots contain the DNA of beams as they distill the collective interaction between the geometrical and material features, as well as loading and restraint conditions.

For example, this load-deformation history is what we seek to obtain when we design our tests and remains a primary outcome of our investigations.[116] The same history gives us valuable insights into a beam's response (i.e., cracking, yielding, etc.), performance (ductile, brittle, etc.), and failure (via bending, shear, etc.). Thus, it is essential to be able to obtain such a history especially via a modern approach that minimizes the need for tests or simulations.

Let us form a hypothesis; "*in order to be able to understand or predict a phenomenon, all that is needed is to identify the primary features and the governing relationship that ties these features to patterns belonging to the so-called phenomenon.*" Such a governing relationship can be a *mapping function*. A *mapping function* does not require complex simulations; but instead, it stems from reverse-engineering the observations that already hold the hidden knowledge we seek into an elegant formula that can be applied in one step.

---

111 The rise in temperature softens construction materials. In other words, the strength of building material under 500°C can be a fraction of that under ambient conditions.

112 Frankly speaking, obtaining the history experimentally is a costly and complex process due to the harsh nature of fire tests and the need for specialized equipment and experienced personnel [67].

113 A function of temperature rise, geometric features, material properties, restraints, etc.

114 One of the rules of fire testing is that we maintain the same level of loading. This is different than testing at ambient conditions, where the level of applied loading is consistently being increased.

115 In case you are wondering, Beam 2 in this particular example is not the same as that we discussed at ambient temperature. Hence, the difference in colors.

116 Whether at ambient or extreme conditions. An engineer could develop a FE model that captures the interaction between the environment, beam features, and materials properties, as well as applied loads. Both testing and FE approaches are well accepted and have been proven effective, yet continue to suffer on a few fronts (e.g., cost/time associated with setting up fire tests, need for dedicated software/workstations, etc.) [28]. Thus, an opportunity to develop a new approach to tracing phenomena (i.e., deformation history of beams or any structural elements for that matter) presents itself.

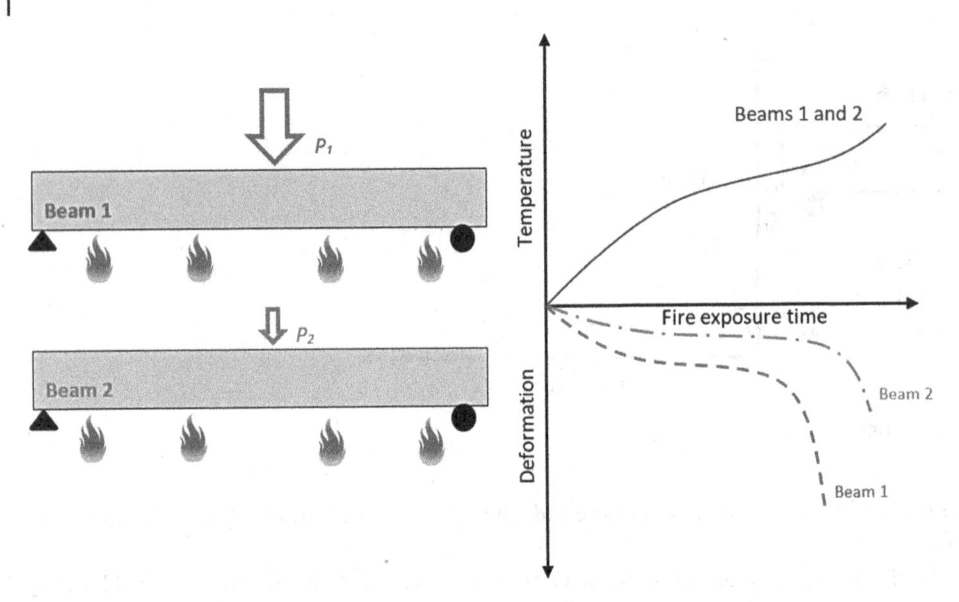

Figure 9.13 Typical response of beams under fire conditions.

Mapping functions can be a bit different than those we often develop from our experiments. For example, in our experiments, specimens are designed to satisfy constraints of testing equipment[117] and funds allocated for testing.[118] Therefore, researchers pick a handful of features[119] and build specimens to examine such parameters using the available equipment they have. Then, researchers adopt traditional statistical methods (i.e., linear regression) to analyze such data. While this leads to good and proper design expressions, given the above limitations, it may also not properly account for all parameters.[120]

On the other hand, a *mapping function* aims to narrow the primary limitation of limited tests by combining data from different experiments and then identifying the key parameters (from a data point of view) to derive a generalized function (equation[121]).[122] Therefore, a *mapping function* is expected to be more encompassing of the examined phenomenon. But that is not all for *mapping functions*. An additional feature of *mapping functions* is their ability to infer explainable and causal relationships.[123]

Let me address the identification problem, and then I will come back to go over the explainability and causal component.

How do we know if a parameter/feature has high fidelity or not? Frankly speaking, recent reviews [30–32] infer that: 1) we lack guidance on setting mathematical standard scores for feature fidelity, 2) we do not know the influence of algorithmic topology/formulation or data quality on such scores, and 3) this is a highly-specific domain problem.

Thus, fundamentally identifying what high fidelity means for our domain by combining physics principles, domain knowledge, and pure data perspective is critical and is one to be handled on a problem-basis.

To arrive at an inclusive view of fidelity, we can: 1) leverage findings of classical approaches (i.e., examine the parameters within formulas adopted in codal provisions or derived by researchers), and 2) seek experts' guidance. In addition, we can also leverage ML! However, instead of applying one algorithm, as commonly accepted, we apply a series of algorithms with different topologies (e.g., thus giving us distinctly various methods to process our data). Then, high fidelity features will be those that satisfy a set of *problem-based conditions*.

For example, the following two conditions can be tried (for now): attaining an importance score of 50% or higher across 75% of the adopted algorithms. Simply put, a feature *X* with an importance score of 50% in algorithm A, 57% in algorithm B, 40% in algorithm C, and 70% in algorithm D, will satisfy the above set of conditions.

117 In a way, we do not have labs that contain all equipment to test all combinations of parameters in a single campaign.

118 Thus, it is rare to test all parameters in a given campaign.

119 In this instance, the use of *parameters* is more adequate. Also, these parameters are primarily identified by expert judgment and knowledge domain.

120 Which, as we have seen before, is one of the reasons why data-driven models seem to outperform codal expressions.

121 Note that there is no guarantee that the *mapping function* will have a similar form to one obtained in the traditional manner.

122 Which overcomes the limitations of traditional statistical methods.

123 Which symbolic regression does not account for.

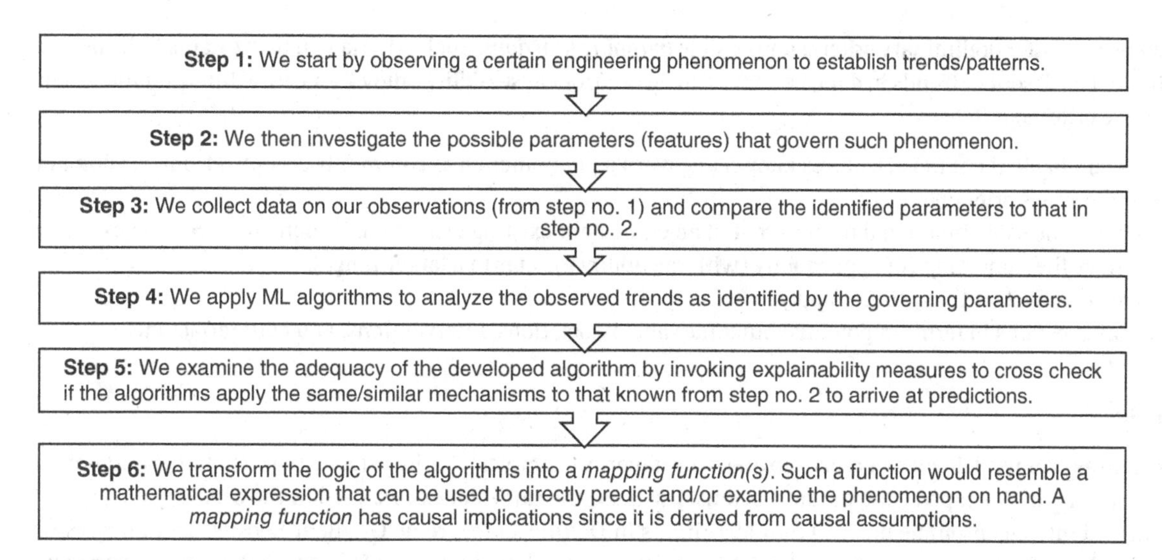

Figure 9.14 Outline of the proposed approach.

The rationale behind this approach is that it is unlikely that the majority of the tested algorithms (given their different processing mechanisms) to constantly rank features at higher scores unless they prove to be elemental to algorithmic predictions [33–35]. The above two arbitrary conditions were selected after a series of preliminary tests and serve as a stepping-stone for this discussion.[124]

In contrast to *mapping functions*, existing explainability methods such as SHAP or LIME aim to explain model predictions solely based on the available data for that particular model, which: 1) are likely to be limited, and 2) do not have a physical or domain knowledge basis. On the other hand, *mapping functions* draw causal effects from physics principles, domain knowledge, and data-driven perspectives and hence are more appropriate for engineering problems. Unlike existing explainability methods, *mapping functions* is not just about the data.

9.4.3 Approach to *Mapping Functions*

Figure 9.14 presents a flowchart for arriving at *mapping functions*. The first three steps are self-explanatory as they relate to our observations and data collection. In this step 4, we apply ML algorithms[125] to examine the collected features and evaluate the performance of models. In step no. 5, we explore explainability measures that indicate the *importance* of each feature on the models/ensemble's predictions (i.e., how often such a feature was used to arrive at a correct prediction) and the *partial effect* of each parameter when varied across the range of the outcome while keeping other parameters fixed.[126] In step no. 6, the ML models/ensemble is converted into a *mapping function* which, given its 1) transparency, 2) resemblance of design expressions engineers are familiar with, and 3) independence from the need for ML coding/software can have room in our domain.[127]

124 We have also seen this in the problems we covered in previous sections. Side note: we still need more testing!

125 You could build an ensemble too and do the same!

126 Knowing the importance of a feature allows us to visualize its significance from a data-driven perspective. Similarly, knowing how a feature changes throughout the range of a phenomenon sheds light on the role such a parameter plays in predicting the phenomenon.For example, think of how the cracking strength of concrete is expected to play a dominant role during the early stages of a load-deformation history in RC beams (prior and up to cracking) than that in the post cracking regime (since the load-deformation plot post cracking is primarily governed by the behavior of tensile reinforcement and compressive strength of concrete [68]). Hence, a proper explainability assessment would yield results that match our knowledge domain expectations. In other words, we are trying to parallel our knowledge domain (as well as physics principles) to that obtained from ML, and thereby causal capability for *mapping functions*.

127 In case you happen to come across some of my papers [70], you may find a 7th step that I did not outline above. In this step, the derived *mapping function* is tested to explore causality from the perspectives of the *Potential Outcomes* and *Directed Acyclic Graph* approaches. For example, a *mapping function* can be assessed by exploring how the response of two identical elements (say, beams) would vary by altering a particular parameter, *X*, in one beam ($Beam_{Alt}$) over a benchmark beam ($Beam_{Bench}$). The same *mapping function* can also be examined in terms of the Pearlian principles outlined by the *Directed Acyclic Graph* approach (e.g., by exploring *association* → *intervention* → *counterfactuals*). For instance, the *mapping function* could be used to explore how varying feature *X* by one unit is *associated* with varying the observed response. Similarly, the same function could be used to predict a new response when feature *X* is *intervened* upon. Finally, the same function can also be used to answer hypothetical questions, such as how to vary *X* to realize a desired response?

The above steps help infer preliminary criteria for possible *mapping functions*. Such criteria will help establish the degree of describe-ability (i.e., describes trends in data, i.e., data-driven) and/or cause-ability (allows for causal interpretation). For now, there are five criteria:

1) Preferable to contain all, if not most, features known to govern a phenomenon as commonly accepted from our domain knowledge and physics principles.
2) Preferable to be in the same functional form as that of an equivalent existing expression if such an expression exists.
3) Preferable to have the same range of applicability (with the ability to cater to a larger range).
4) Preferable to be compact and easy to use/apply.
5) Answers causal questions in terms of potential outcomes and associations, interventions, and counterfactuals.

9.4.4 Example[128]

In this section, and as promised, I will revert back to one of my favorite examples[129] to investigate the *sectional* phenomenon of the moment capacity of W-shaped compact steel beams. I will use the same dataset we used in the causality chapter.[130] This dataset was fed into an ensemble made of four algorithms in *DataRobot* (eXtreme Gradient Boosted Trees Regressor (BP89), Light Gradient Boosting on ElastiNet Predictions (BP86), Random Forest (BP87), and Elastic-Net Regressors (BP83)). We will build the models and an ENET Blender ensemble and then compare their predictions, feature importance, and partial dependence plots. Predictions from the ensemble are then compared against Equation 9.6 (ground truth). Figure 9.15 and Table 9.4 present a look in various performance metrics.

$$M = f_y \times Z \tag{9.10}$$

Let us now evaluate the feature importance of the features (see Figure 9.16). As you can see, we have a great match between all models, as well as the models and the ensemble.[131] What we can also see is that none of the models (including the ensemble which came the closest) were able to attain a perfect score. I am highlighting this here not because we are chasing goodness but rather because the phenomenon on hand is comparatively simple. It is a multiplication form of two features.[132] In other words, we have a *simple-enough* problem that is examined with four ML models and one super model (ensemble), yet we were not able to capture the ground truth.[133]

Since we only have two features, let us use both in a *mapping function*. This problem is simple-enough so let us apply a form of symbolic regression. Since *DataRobot* homes Eureqa models, then these sound like an attractive solution to try (see Figure 9.16).[134] If you run Eureqa models BP5, BP32, BP33 as well as BP34 as BP42 you will get the following three expressions:

Table 9.4 Performance of the ensemble and *mapping functions* for training/validation/testing.

Model	MAE (k.in)			RMSE (k.in)			R^2		
BP83	2784.9987	3262.9209	3268.0417	3428.2588	4654.9674	4523.3411	0.9736	0.9756	0.9762
BP86	250.2709	457.5039	282.1472	761.6212	1283.9251	904.9877	0.9987	0.9982	0.9991
BP87	606.6892	1284.7378	961.9335	1214.6767	3741.1688	2858.9634	0.9967	0.9840	0.9905
BP89	288.4201	654.8781	350.2897	401.0465	2065.3350	724.4136	0.9996	0.9949	0.9994
Ensemble	220.7427	525.4814	276.1248	327.2889	1471.0975	515.5873	0.9998	0.9974	0.9997
MF (BP32)	0.0000	0.0000	0.0000	0.0000	0.0000	0.0000	1.0000	1.0000	1.0000

128 This is perhaps one of the very few examples where I will be using traditional units. Remember: 1 in = 25.4 mm and 1 ksi = 6.89 MPa.
129 That we are still exploring to realize causal interpretations for.
130 As you might remember, this dataset was compiled of W-shaped steel beams from the AISC manual with two features, Z and f_y.
131 Some variation is apparent in the PDP plot of f_y (which can parallel that from the slight variation in the PDP for Z. All in, the trends match).
132 Another reminder of the difference between prediction and causation.
133 This may not always be the case! Different phenomenon, different stories.
134 I would normally opt for genetic algorithms, however, going for *Eureqa* models makes this example a bit concise.

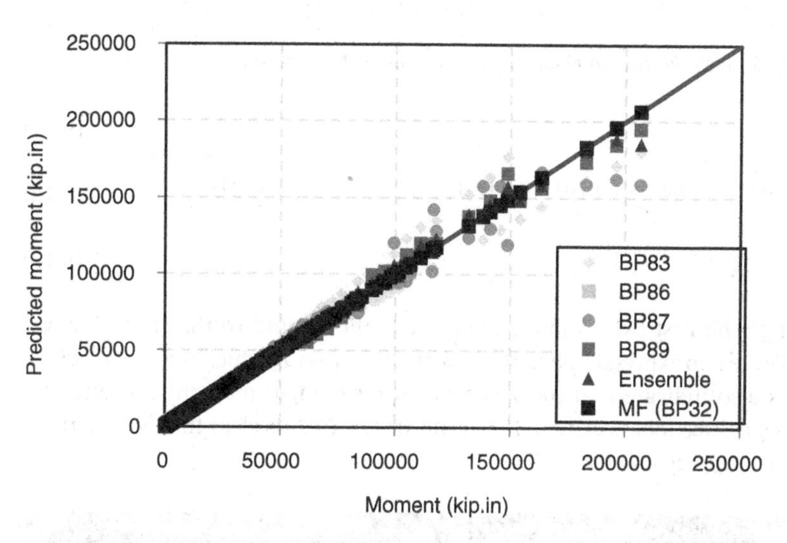

Figure 9.15 Validation.

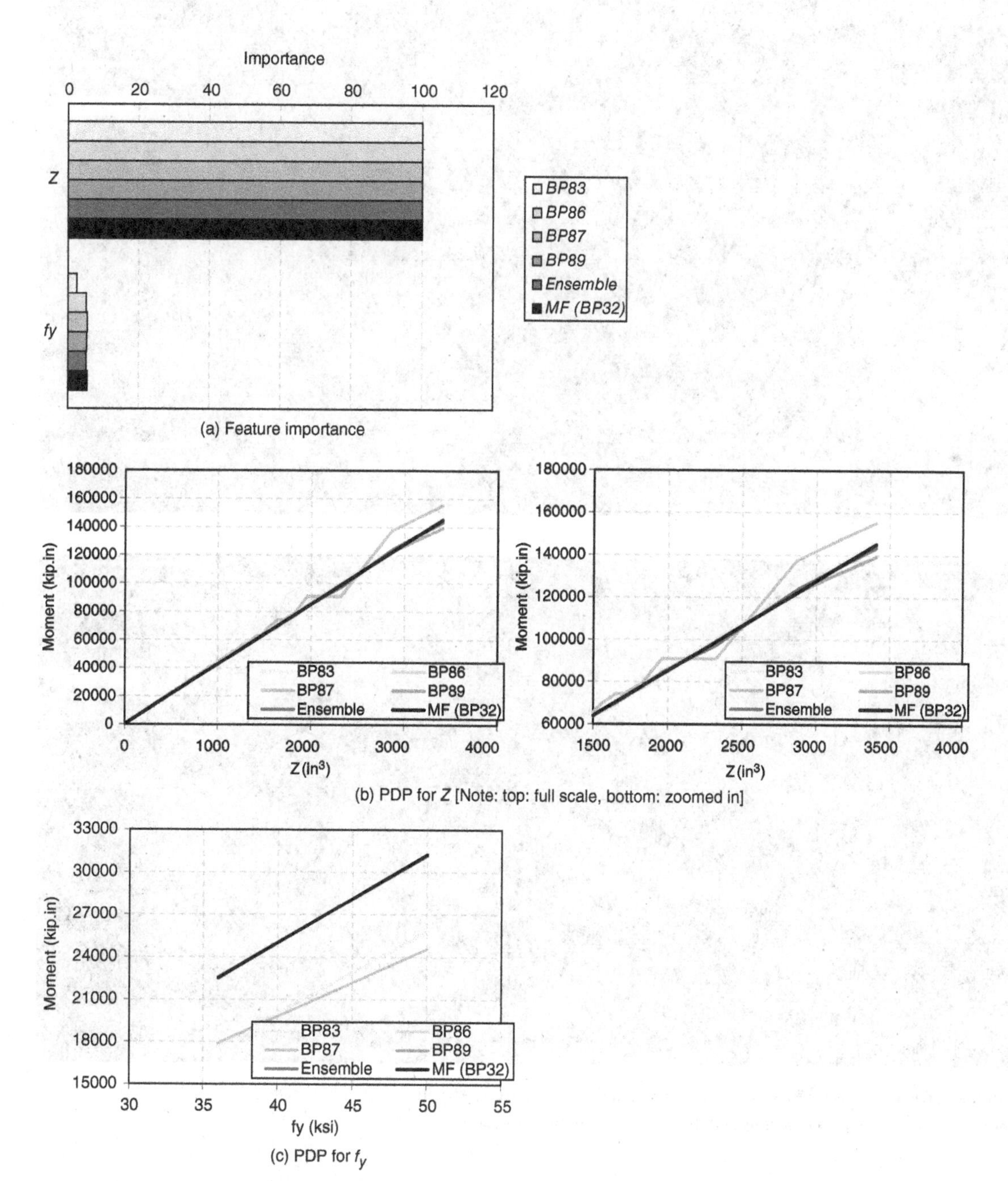

Figure 9.16 Comparison between models.

$$M_{BP5,PB32,BP33} = f_y \times Z \tag{9.11}$$

$$M_{BP34} = 2512.61 + 0.95 \times Z \times f_y - 1318.58 \times less(Z - 86.14, 25.76) - 0.94 \times f_y \times if(-43.07, Z, 25.76) \tag{9.12}$$

$$M_{BP42} = f_y \times step(Z) \times min(Z, 3589.45) - 18.22 \tag{9.13}$$

Looking at these expressions infers the following. The first expression is identical to the ground truth. The other two expressions seem to be precursors to the first.[135] These two expressions show a heavy reliance on Z, which is expected given the importance of this feature. These two forms also indicate that there is a taken for granted moment capacity of a constant value (i.e., -18.22 k.in) even for fictitious steel sections (i.e., with $Z = 0$ and/or $f_y = 0$). It is clear that both of these forms void some of the earlier criteria for selecting *mapping functions*.[136]

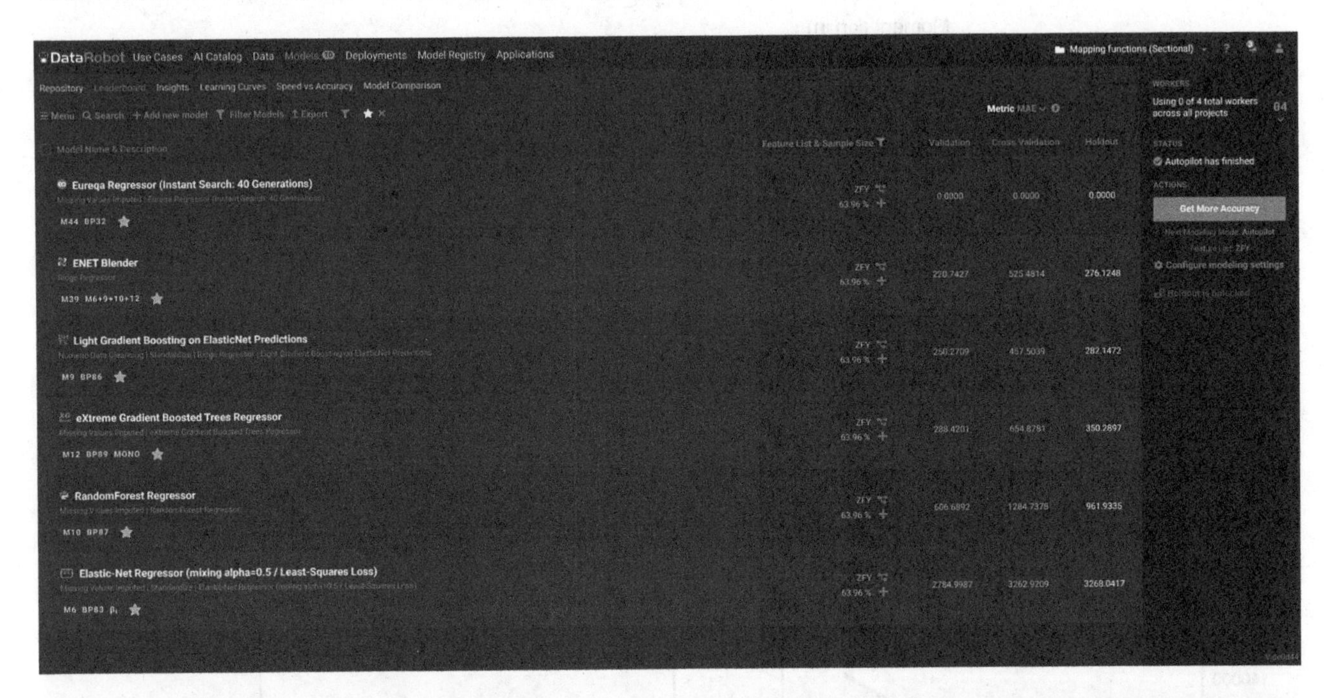

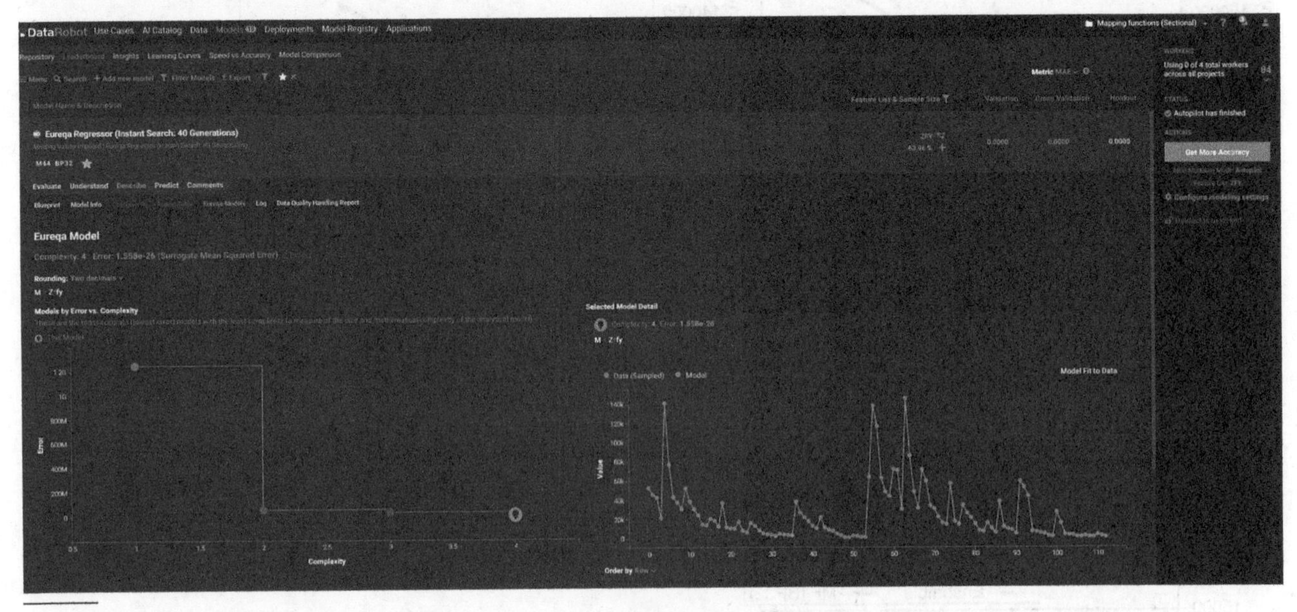

135 Adjusting the number of generations for BP34 as BP42 will return the ground truth – try it!

136 Hence the difference between *mapping functions* and those from symbolic regression.

Table 9.5 Comparison against the Pearlian approach.

Eq./Question	Association	Intervention	Counterfactuals
	How does varying parameter *X* by one unit is *associated* with varying the observed response (*M*)?	**How does changing parameter *X* to a specific value *ceteris paribus* affect the observed response (*M*)?**	**Which parameter could we have changed to realize a predefined response (maximum an increase in *M*)?**
$f_y = 50$ ksi $Z = 114$ in^3 $M = f_y \times Z = 5700$ k.in	1 unit increase in f_y or Z leads to increase in M.	When, $f_y = 51$ ksi $\rightarrow M = 5814$ k.in (+2%) $Z = 115$ in$^3 \rightarrow M = 5750$ k.in (+0.8%)	Practically, a change in f_y from 36 ksi to 50 ksi can maximize the increase in M.
$f_y = 50$ ksi $d = 18$ in $b_f = 7.5$ in $t_f = 0.57$ in $t_w = 0.36$ in $M = f_y \times (b_f t_f (d - t_f) + 0.25 t_w (d - 2t_f)^2) = 5707$ k.in	1 unit increase in any feature increases M.	When, $f_y = 51$ ksi $\rightarrow M = 5104.9$ k.in (+2%) $d = 19$ in $\rightarrow M = 5374.8$ k.in (+7%) $b_f = 8.5$ in $\rightarrow M = 5501.1$ k.in (+10%) $t_f = 1.36$ in $\rightarrow M = 10{,}666.8$ k.in (+213%) $t_w = 1.57$ in $\rightarrow M = 5646.5$ k.in +70%)	Increase d.
$f_y = 50$ ksi $d = 18$ in $b_f = 7.5$ in $t_f = 0.57$ in $t_w = 0.36$ in $M = f_y \times (b_f t_f (d - t_f) + 0.25 t_w (d - 2t_f)^2) = 5707$ k.in	1 unit increase in any feature increases M.	When, $f_y = 51$ ksi $\rightarrow M = 5104.9$ k.in (+2%) $d = 19$ in $\rightarrow M = 5374.8$ k.in (+7%) $b_f = 8.5$ in $\rightarrow M = 5501.1$ k.in (+10%) $t_f = 1.36$ in $\rightarrow M = 10{,}666.8$ k.in (+213%) $t_w = 1.57$ in $\rightarrow M = 5646.5$ k.in +70%)	Increase d.

Since the first functional form matches the ground truth, we could further explore answers to causal principles outlined by the Pearlian approach (e.g., *association* → *intervention* → *counterfactuals*). Table 9.5 lists such questions and answers.

You could argue that the above example and derived function could be a reflection of the simplicity of the sectional capacity phenomenon. As in, there were only two features Z and f_y, wherein f_y is reduced to two re-occurring values (36 ksi and 50 ksi). I would agree with you. In fact, having this example serves two purposes: 1) a good introductory problem, and 2) brings the practical aspect where W-shaped structural steel sections are primarily fabricated using two grades of steel – for the most part (which influences how feature importance looks down on f_y[137]). I will be offering a problem at the end of this chapter to extend this example into multi-steel grades to inflate the number of data points. I feel that this would be a good homework problem to showcase the influence of data size and domain accepted practices on ML models/*mapping function* predictions.

To complement this case study, an effort is introduced to add a new layer of complexity by breaking down the plastic modulus Z into its basic components, i.e., depth, d, flange width, b_f, flange thickness, f_t, and web thickness, t_w. Our goal is to repeat the above analysis to see if we can arrive at the ground truth relationship between Z and its components (shown in Equation 9.7). Thus, the same dataset is extended by substituting Z with its basic components. We finally arrive at Equation 9.8.[138]

$$M = f_y \times (b_f t_f (d - t_f) + 0.25 t_w (d - 2t_f)^2) \tag{9.14}$$

$$M = 6.24 d t_f f_y + 0.494 d b_f t_w f_y + 0.0194 b_f t_f f_y d^2 - 0.063 t_w t_f f_y d^2 \tag{9.15}$$

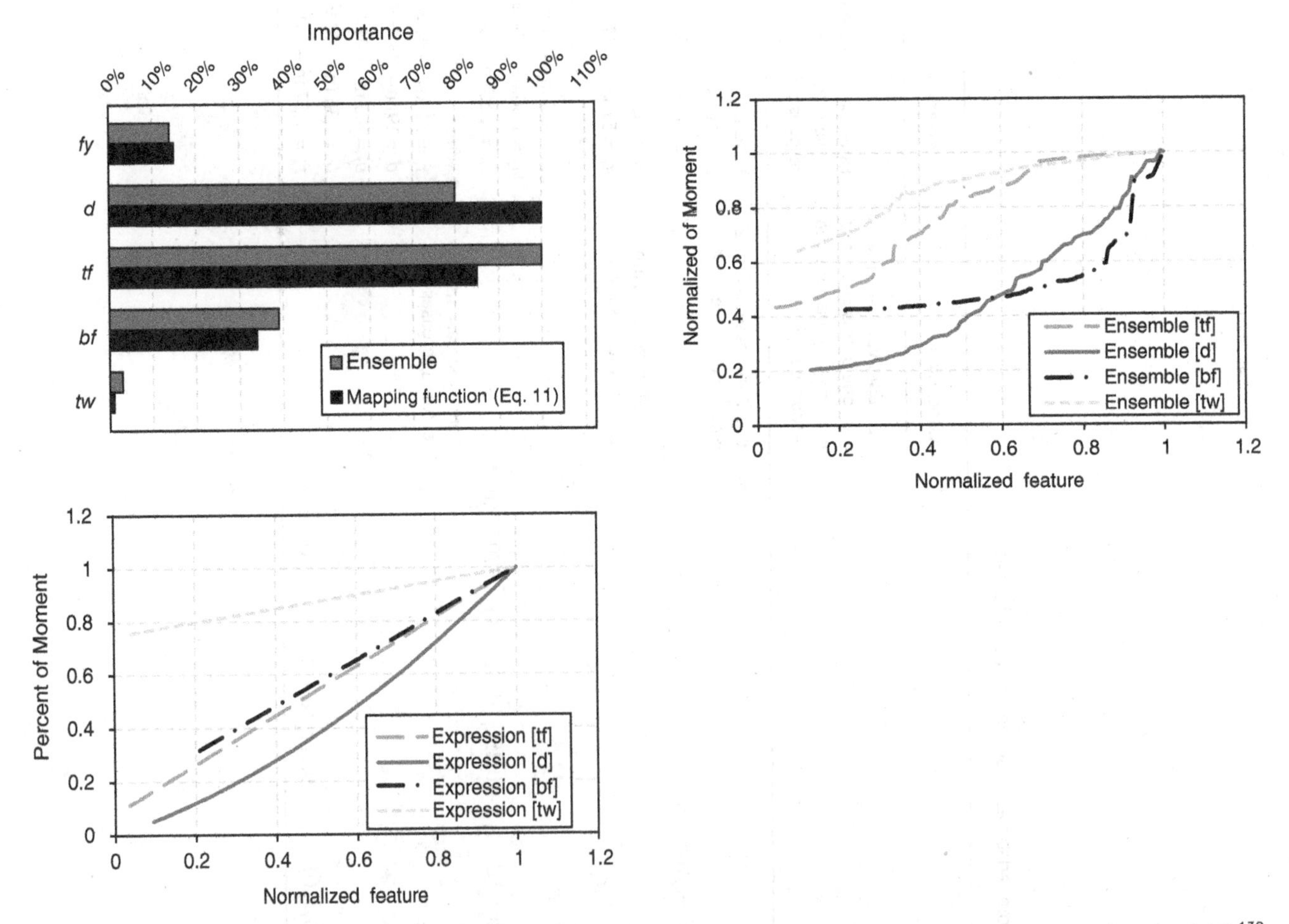

Figure 9.17 Feature importance (top), and PD plots for Z (middle) and f_y (bottom) from ensemble vs. *Mapping function* (Equation 9.11).[139]

137 Remember this is an explainability tool that is based on data! Not domain knowledge!

138 We are a few generations away from getting a close match to the ground truth! Another good homework problem.

139 The axes were normalized since each feature has a different scale.

For completion, Figure 9.17 showcases feature importance (top) and PDP for the newly examined features. An agreement between the two is evident.

9.4.5 A Note

Mapping functions capitalize on the fact that many of our problems are closed-loop systems. These problems are bound by practical considerations[140] and construction practices,[141] etc. Thus, when the above is combined with physical facts (e.g., members deflect when loaded, congestion takes place in bottlenecks, etc.) or domain knowledge, this effectively narrows our space to search for expressions that can be used to describe[142] our problems.

Another direction that we need to acknowledge is that of ML components. For example, we need to better understand the new tool we are adopting. For instance, what is the influence of algorithmic families, optimization functions, choice of performance metrics, etc. on the derived functions? A question I find most interesting is how to turn a descriptive function (i.e., from symbolic regression) into a *mapping function* (with causal and counterfactual expressibility).

9.5 Ensembles

We have discussed the *No Free Lunch Theorem* in a former chapter. This theory stipulates that all optimization algorithms perform equally well when their performance is averaged across all possible problems [36]. In other words, there is not a single best algorithm. We have also seen the same across many of our examples and as recently as that in the Green ML section. This theory sets the stage for the present section on ensembles and Automated ML (AutoML).

9.5.1 Big Ideas

If the above theory is true, then utilizing a particular[143] algorithm to examine a phenomenon does not guarantee arriving at an ideal, let alone an optimal model to solve the problem on hand. The same implicitly invites us to explore multiple algorithms.

But, what about combining ML algorithms or models together?

Say, we have two ML models which happen to yield decent predictivity. Would combining both into one model improve such predictivity? The answer is yes, more likely than not. Combining models together is what ensembling is.

An ensemble capitalizes on exploiting multi-algorithmic searches of two (or more) algorithms into a ML model of higher accuracy, less vulnerability to overfitting, and better handling of missing or imbalanced data[144] [37, 38].

In a regression-based problem, an ensemble may take the predictions of sole models and perform a *statistical analysis* to arrive at a combination of its own prediction. Such analysis can be in the form of averaging[145] the sole predictions. In some instances, the user may opt to favor conservative predictions, and hence the lowest sole prediction can be selected, and so on.

In a classification-based problem, an ensemble can apply a *voting* mechanism to arrive at a final prediction from each sole prediction. In such a mechanism, the largest number of votes for a given prediction (or class) are tallied and then assigned to the ensemble.[146] There are a number of ways voting can take place, majority[147] voting (selects the ensemble's prediction as the class with more than half the votes), weighted voting (selects the final prediction by weighing each sole prediction in some fashion[148]), etc.

---

140 For example, load bearing members have practical size ranges wherein very small or very large members are rarely used. Highway lanes are restricted to specific sizes.

141 There is a handful of building material families that can be used in construction.

142 Causally or otherwise.

143 If you take a minute to ask, how do we select an algorithm to apply to a problem? Well, this section can be random or predetermined (i.e., stemming from our familiarity with an algorithm, suggestions received from publications/experts). In a way, this is the same process as selecting a software to use! Some might opt to use ANSYS, while others may opt to adopt ABAQUS, etc.

144 You can think of ensembles as a way to harness the **wisdom of crowds**, where the aggregate knowledge of a diverse, independent, decentralized body of individuals often exceeds the knowledge of single individuals/experts.

145 Or Median. Say, five sole models predict the following outcomes: 12, 34, 25, 7, and 60. Then, the median would be 25 (e.g., 7, 12, 25, 34, and 60) and the average is 27.6.

146 To avoid ties, the use of an odd number of models becomes an intuitive solution.

147 Unanimous voting is a special case of majority voting.

148 One of my favorite weighting techniques is one that weighs each prediction proportional to the confidence each model has in its own prediction. Another, and simple, technique would be to weigh each prediction in a proportional manner to the performance (e.g., accuracy) of each model.

There are other techniques than those listed above. For example, stacking, blending, bagging, and boosting are three such common techniques. *Stacking*[149] combines predictions from various models. These predictions are stacked together and used as input to train the ensemble. *Blending*[150] is similar to stacking but uses a holdout set from the training set to make predictions. *Bagging*[151] adopts bootstrapping[152] to generate data that can be input to train parallel models. Each model then learns the error predicted by previous models. *Boosting*, on the other hand, converts weaker models into a strong one.[153] Boosting is a process to correct arising errors from previous iterations by adjusting emphasis (i.e., weights) according to the degree of errors.[154] Boosting is applied sequentially and does not allow parallel operation (unlike bagging[155] – see Figure 9.18).

9.5.2 Examples

Let us build ensembles using the intuitive platform of *DataRobot* and *BigML*. I will be using the dataset on water potability by Jatin Sadhwani [39]. We will first build traditional ML models and compare these against each other. Then, we will build two ensembles from these individual models. First, I will be using the following *DataRobot* blueprints (BP) BP58, BP63, BP64, BP65, and BP66[156] to predict the quality of water (i.e., if the water is safe to drink or not) given a set of features for 3276 observations. These features include pH value, hardness, total dissolved solids, chloramine, sulfates, conductivity, total organic carbon, trihalomethanes, and turbidity.

Let us input our dataset into *DataRobot*. Follow my steps.

Step 1: Select *Potability* as *Target*.

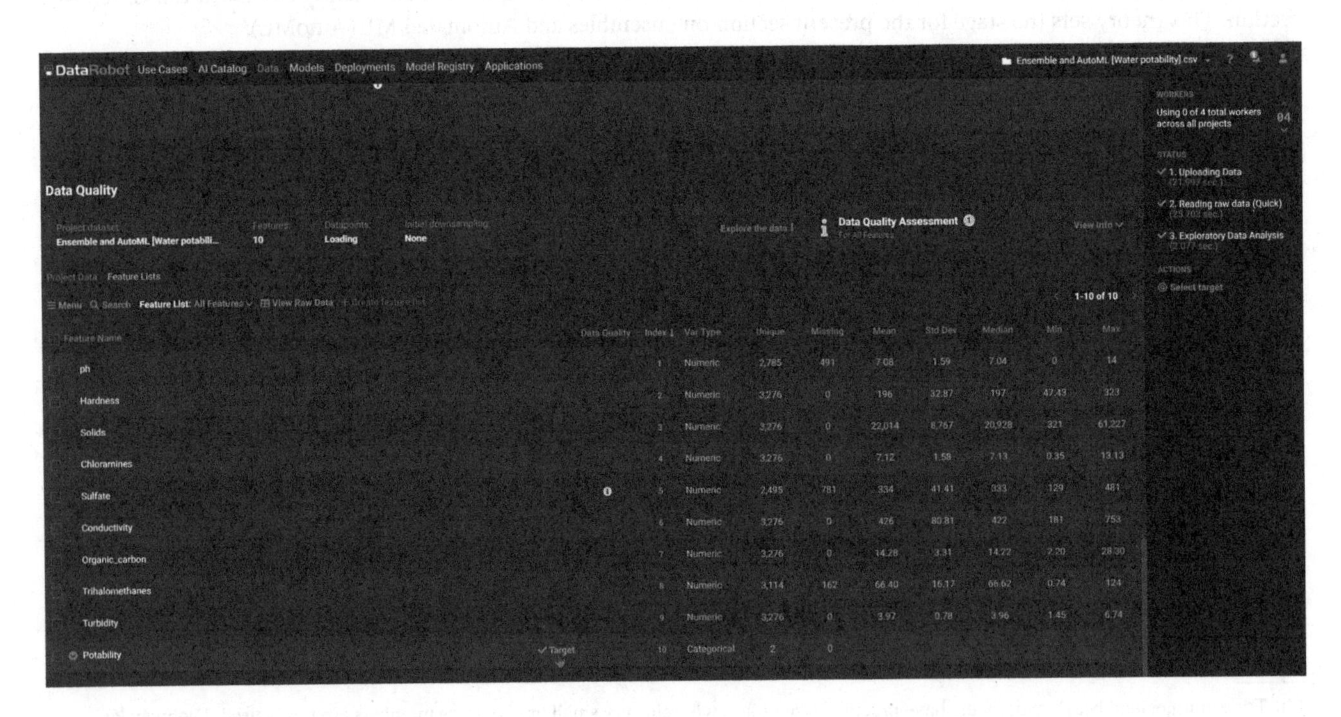

149 Also called *stacked generalization*. Stacking can be applied to *heterogenous* models.

150 Blending is simpler than stacking and prevents leakage. However, blending uses less data and may lead to overfitting.

151 Also called *Bootstrap Aggregating*.

152 **Bootstrapping**: A process of creating multiple sets (bags) of the original training data with replacement. In this process, some data may only appear in some bags, may appear more than once, or may not appear at all. This process reduces variance and minimizes overfitting.

153 Hence, the term *boost*!

154 Or simply, focuses on the observations that are hard to predict (which drives the error up).

155 On the other hand, bagging and boosting can be *primarily* applied to homogenous models.

156 These blueprints are: Keras Slim Residual Neural Network Classifier using Training Schedule (1 Layer: 64 Units), Light Gradient Boosting on ElasticNet Predictions, RandomForest Classifier (Gini), RuleFit Classifier and eXtreme Gradient Boosted Trees Classifier, respectively.

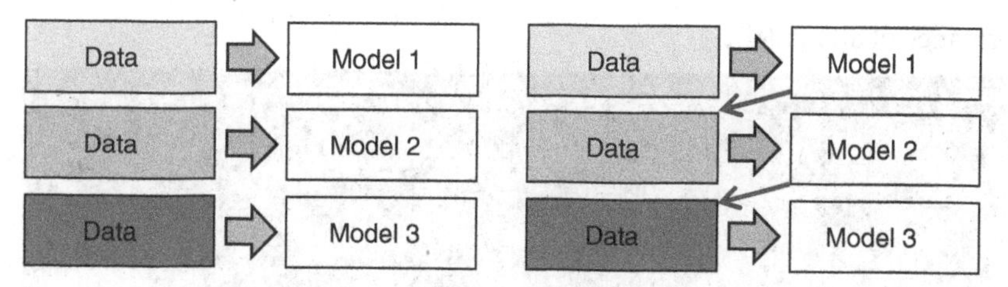

Figure 9.18 Comparison between bagging (Left) and boosting (Right).

Step 2: Under *Additional*. Select *Create blenders from top models* and *Search for interactions*.

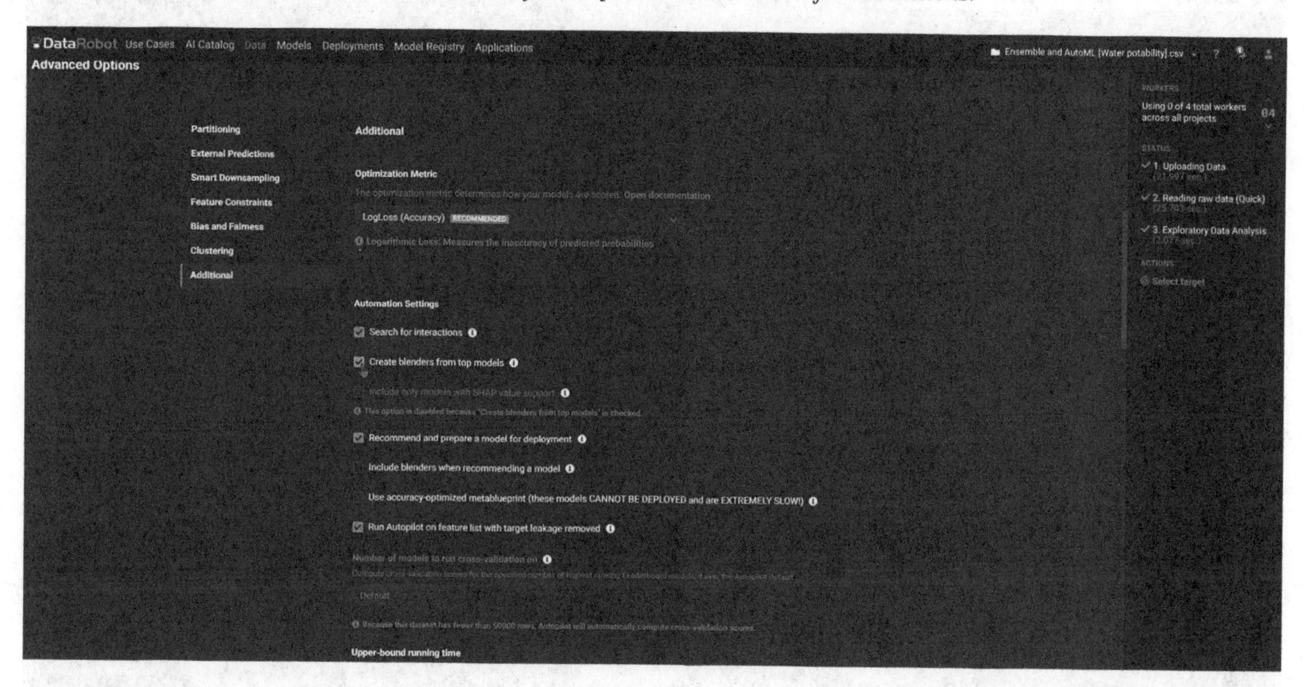

Step 3: Select *Quick* under *Modeling Mode*.

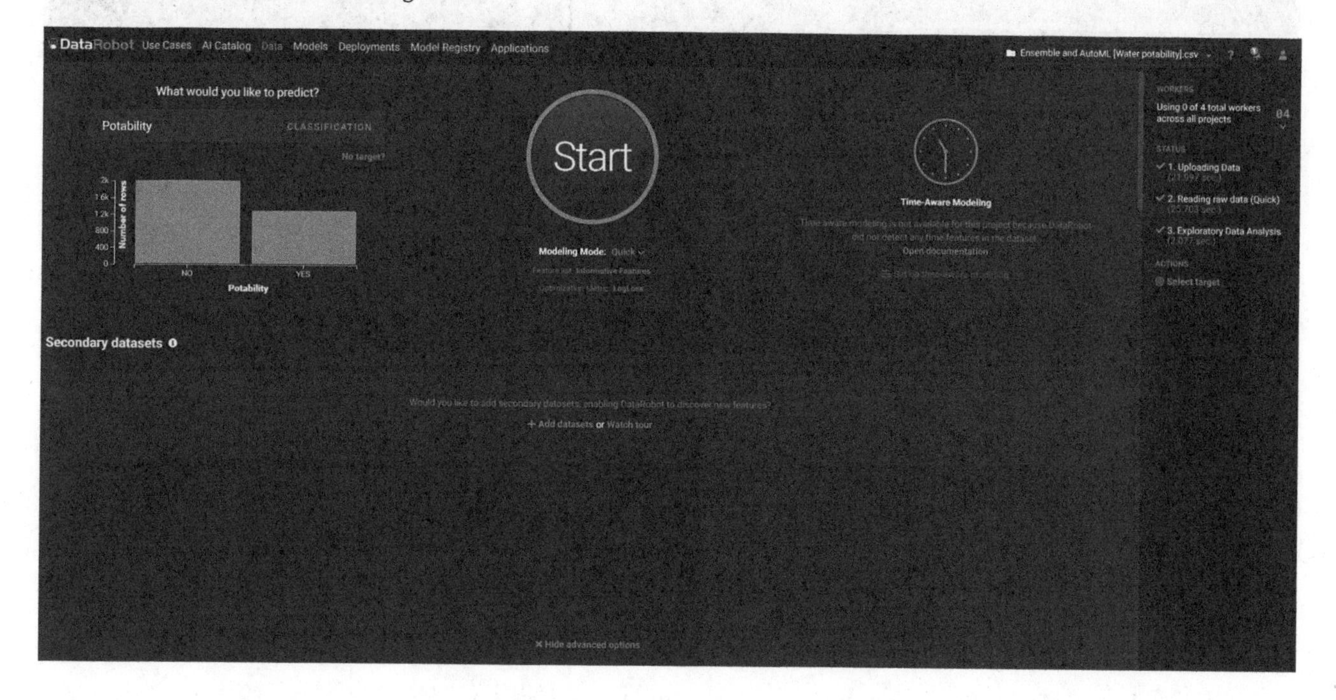

Step 4: Under the *Models* tab, check the performance of all models.

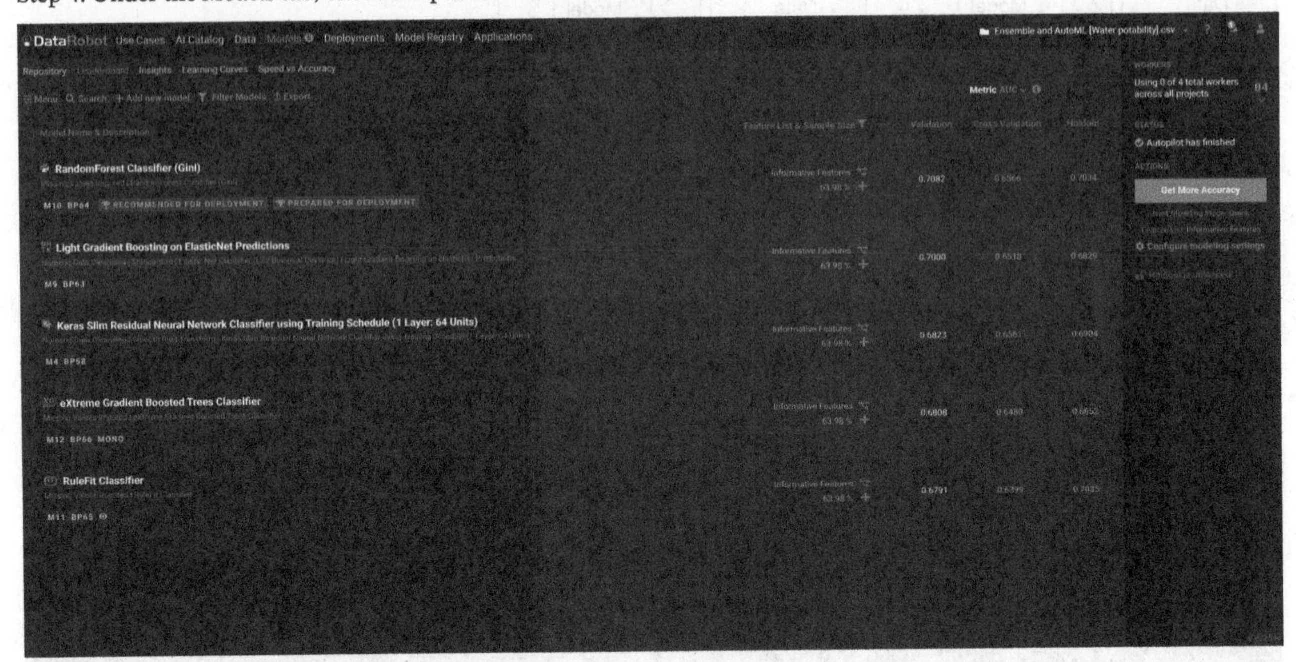

Step 5: Select the *Models* tab, and then go to *Blending* and select *Average Blend*.[157] Repeat the above to select *Median Blend*.[158]

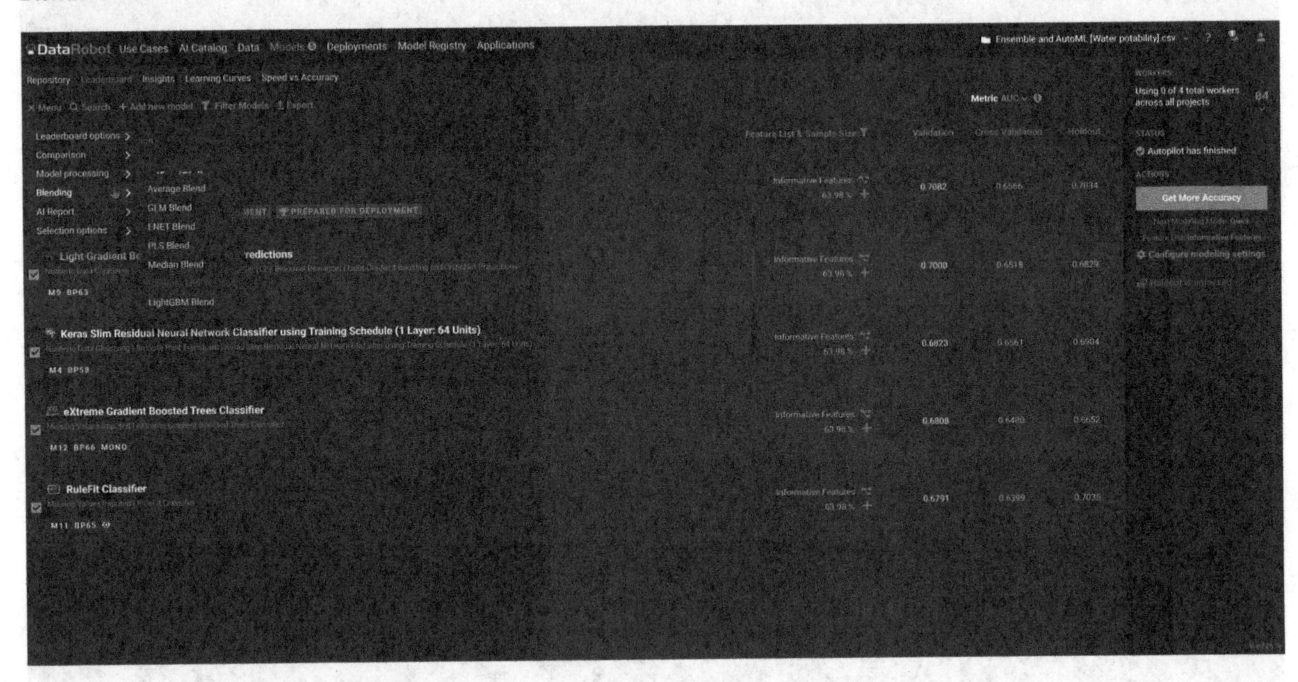

157 To average prediction between different models.
158 To get the median of prediction between different models. Also, please try a few other blending methods as well.

Step 6: Check the final performance of all models and ensembles and note how both ensembles perform better than all other models.

As you can see, the *DataRobot* platform makes it easy to build ensembles. In reality, programming ensembles can be a bit hectic. I would rather preserve some trees and direct you to some references that can provide you with additional information on programming ensembles [40, 41].

Let us now repeat the above example using *BigML*. Follow my steps and *upload* our dataset.

Step 1: Under the *Datasets* tab, build a *Model*, *Logistic regression* model, and a *Deepnet model*.

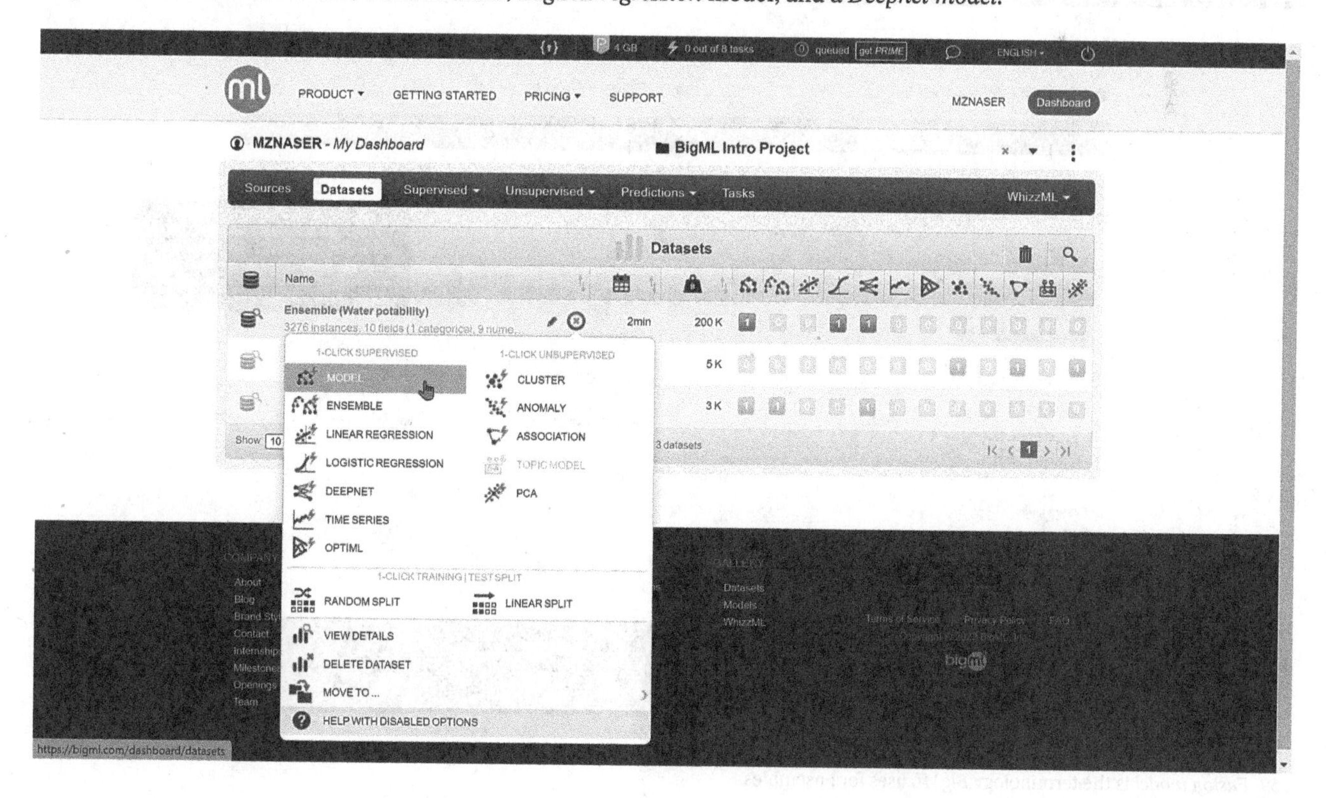

Step 2: Check your results against those shown here.

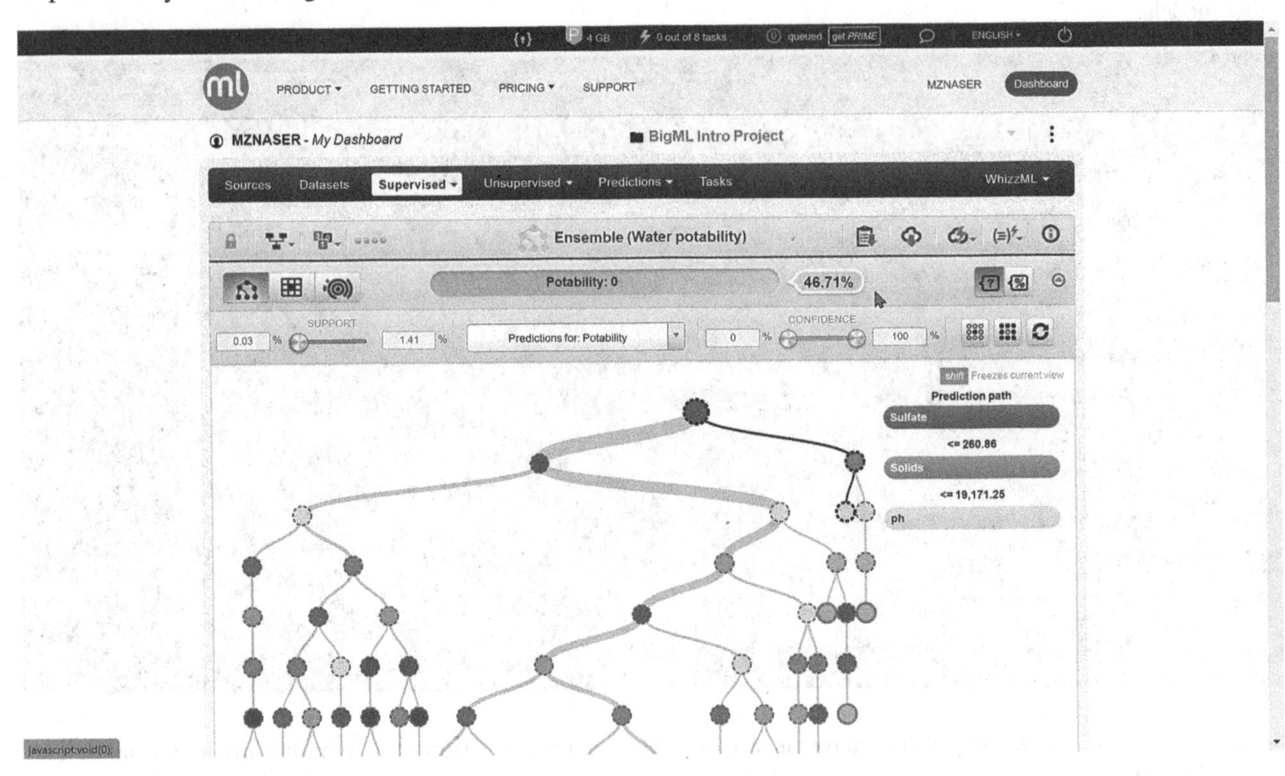

Then, navigate to the *Supervised* tab and build a *fusion model*.[159] Type the first few letters of this project to select compatible models. Then, select *Create fusion*.

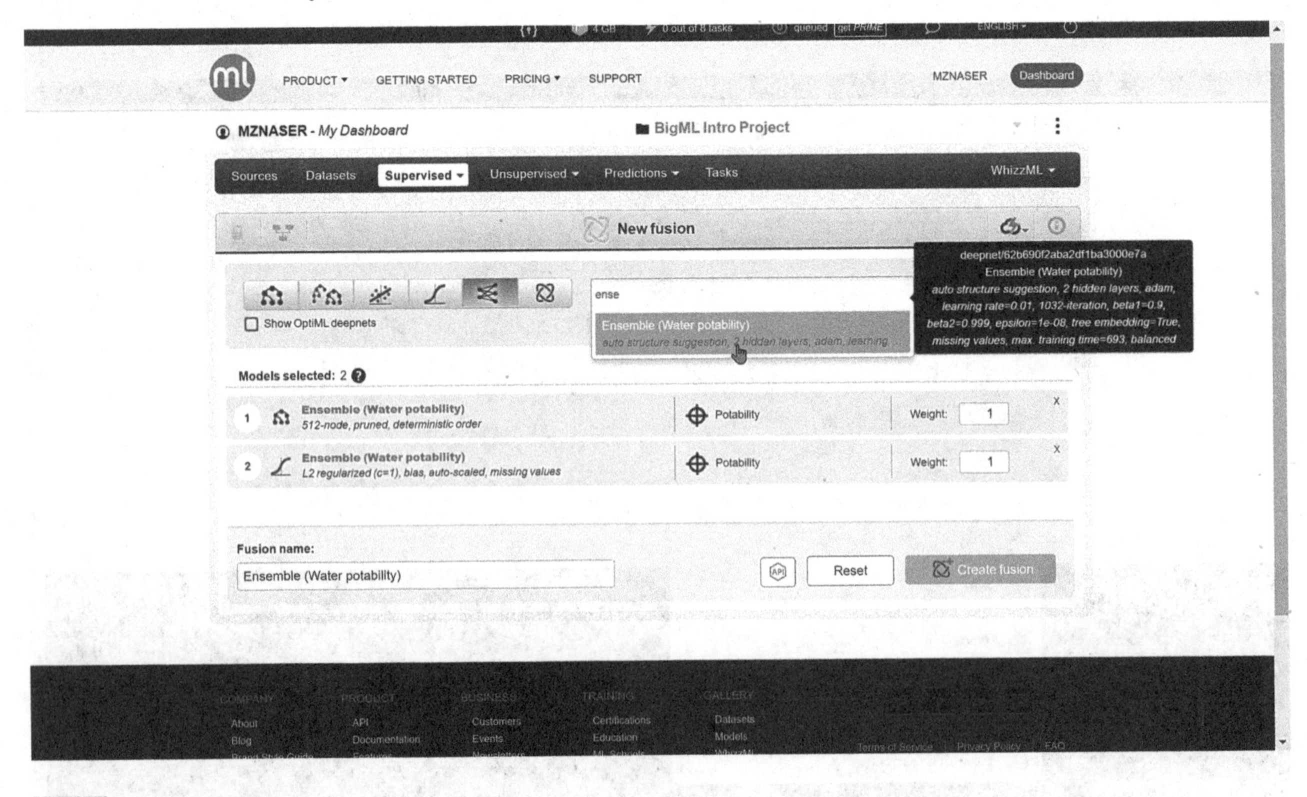

159 *Fusion model* is the terminology *BigML* uses for ensembles.

Step 3: Once the analysis completes, visit the *Fusion Summary Report*. Have a look at the *Evaluation* tab (to visit the *confusion matrix*[160] and the *area under the ROC*).

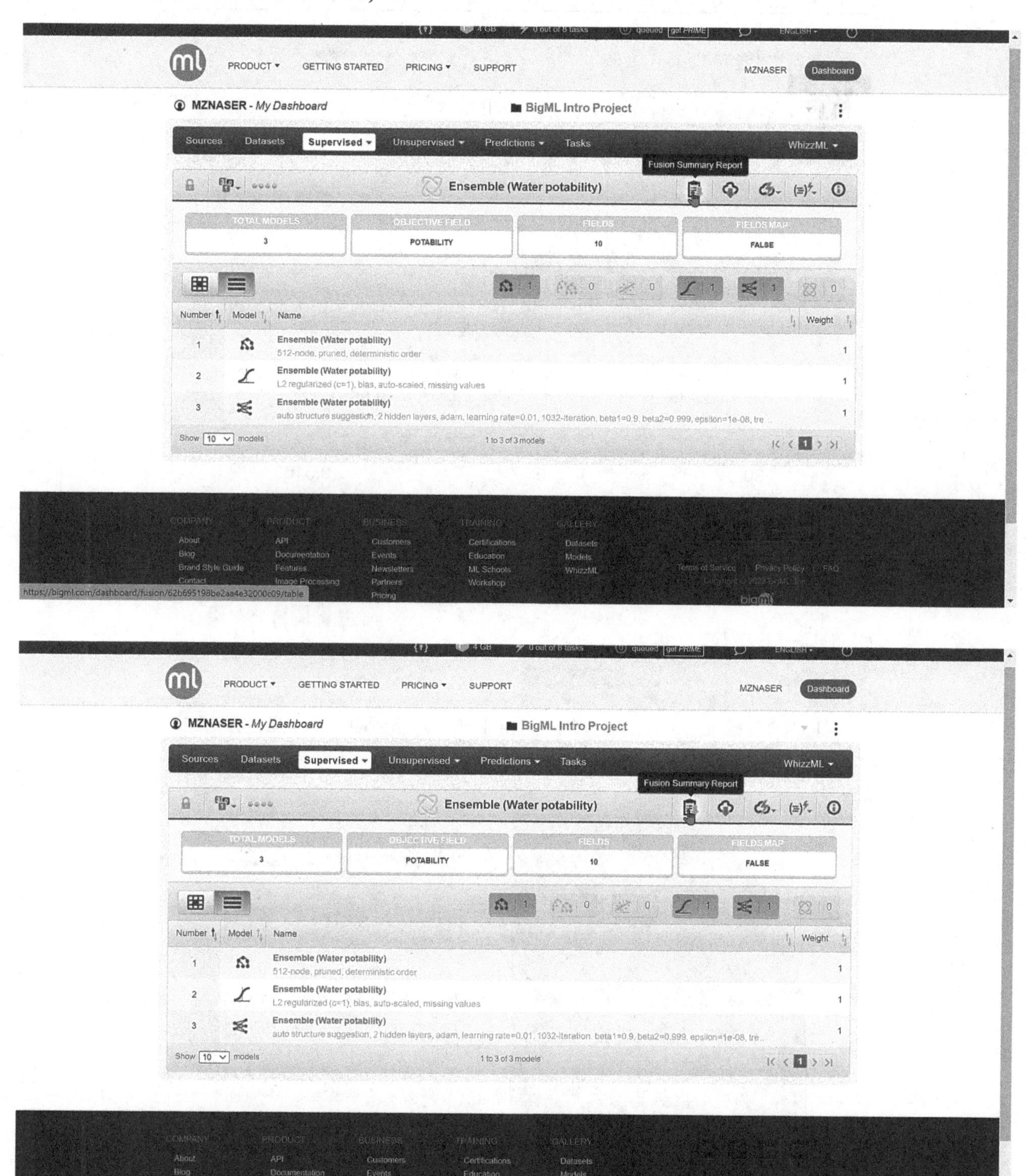

160 Note that this can be downloaded into an Excel-ready file.

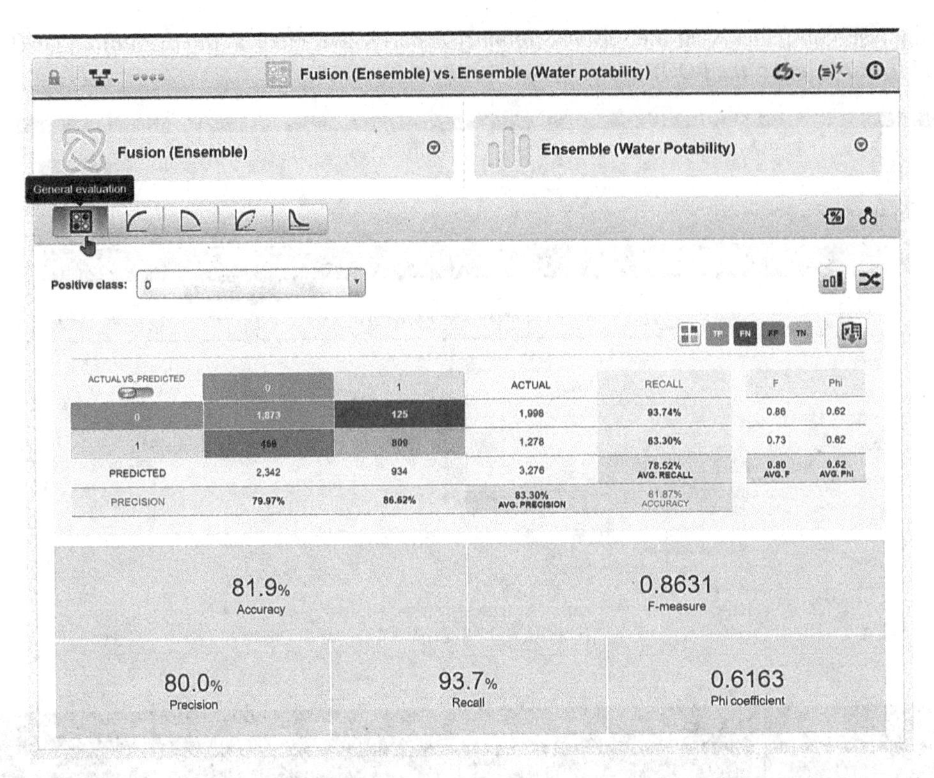
Fusion (Ensemble) vs. Ensemble (Water potability)
Fusion (Ensemble)
Ensemble (Water Potability)
General evaluation
Positive class: 0
ACTUAL VS. PREDICTED
ACTUAL
RECALL
F
Phi
PREDICTED
PRECISION
81.9% Accuracy
0.8631 F-measure
80.0% Precision
93.7% Recall
0.6163 Phi coefficient

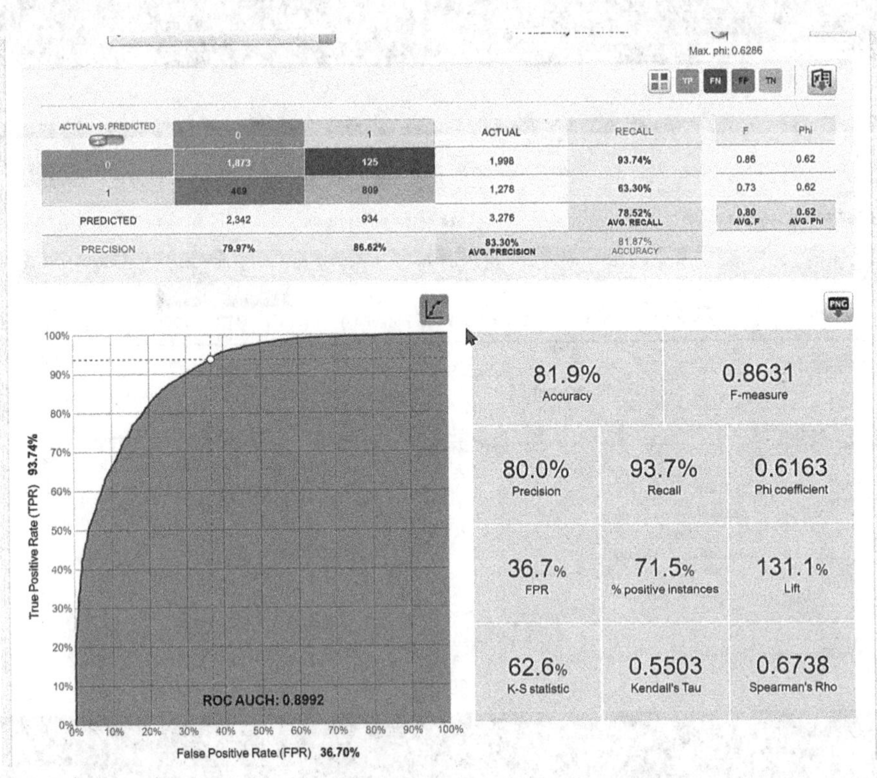
Max. phi: 0.6286
ACTUAL VS. PREDICTED
ACTUAL
RECALL
PREDICTED
PRECISION
True Positive Rate (TPR) 93.74%
False Positive Rate (FPR) 36.70%
ROC AUCH: 0.8992
81.9% Accuracy
0.8631 F-measure
80.0% Precision
93.7% Recall
0.6163 Phi coefficient
36.7% FPR
71.5% % positive instances
131.1% Lift
62.6% K-S statistic
0.5503 Kendall's Tau
0.6738 Spearman's Rho

Step 4: Visit the *Field importance* plot.

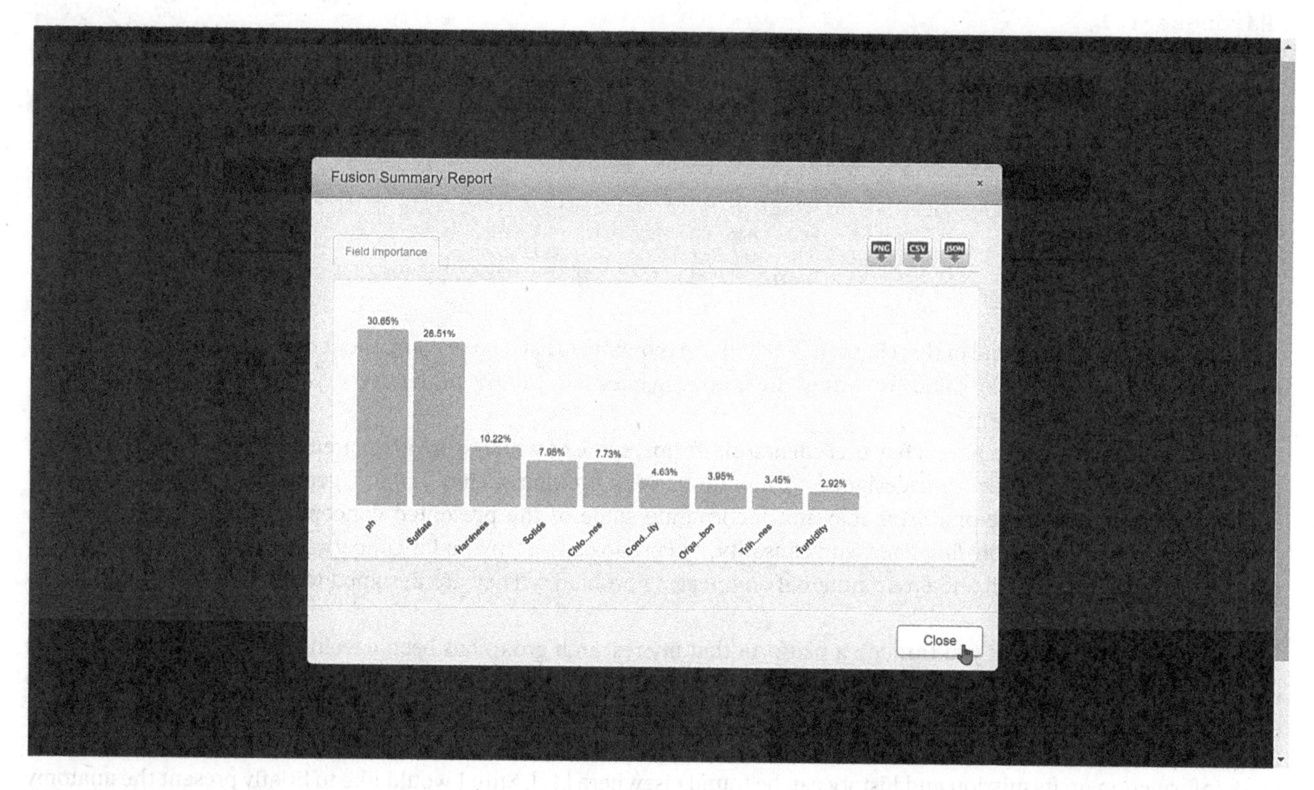

Step 5: As always, you may create and download the *Python* code for this fusion model.

Python code:

```
# Requires BigML Python bindings
#
# Install via: pip install bigml
#
# or clone it:
# git clone https://github.com/bigmlcom/python.git
from bigml.fusion import Fusion
from bigml.api import BigML
# Downloads and generates a local version of the FUSION,
# if it hasn't been downloaded previously.
fusion = Fusion('fusion/62b69e7b8be2aa494b000c93',
api=BigML("mznaser",
"33dcd88020b98f21a24f68ce3b9c943e6ab4e4b6",
domain="bigml.io"))
# To make predictions fill the desired input_data
# in next line.
input_data = {
"Organic_carbon": 1,
"Trihalomethanes": 1,
"Sulfate": 1,
"Conductivity": 1,
"Solids": 1,
"Chloramines": 1,
```

```
"ph": 1,
"Hardness": 1,
"Turbidity": 1
}
fusion.predict(input_data)
```

9.6 AutoML

9.6.1 Big Ideas

As we are approaching the end of this chapter,[161] it helps to remember that a large component of ML revolves around automation to better our lives. The same also brings the following question, *can we automate everything we learned in this textbook into one pipeline/platform?*

The answer is *yes*! We also have a few precedents/platforms, some of which were presented in this book too. However, unlike these platforms which are designed to handle various problems, they remain general and not application/domain-oriented. The same platforms may not incorporate some of the presented concepts, such as synthetic data, symbolic regression, *mapping functions*, and causality.[162] Perhaps one of my major concerns is that these platforms are not designed to address civil and environmental engineering problems. They are designed to address ML and data pertaining to any domain.

In this section, I will walk you through a platform that my research group has been developing. This platform aims to automate and optimize ML for civil and environmental engineering problems. This AutoML virtual assistant[163] (VA) enables engineers to carry ac**CeLE**rated, si**M**ulation-free, tran**S**parent, reduced-**O**rder, and infere**N**ce-based fire resistance analysis with ease and hence is called CLEMSON. At the moment, CLEMSON is in the final stages of development, and more information on its mission and history can be found elsewhere [42]. Still, I would like to briefly present the anatomy behind this VA for two reasons, 1) it sums up most of the notable concepts presented in this book, and 2) it may be of interest to some readers.

9.6.2 The Rationale and Anatomy of CLEMSON

In many examples of this book, we have seen that traditional (or simple) ML models perform well. As per one of our earlier conversations, the choice behind selecting a ML algorithm often stems from a personal preference. As such, and while the selected model may yield good predictions, this practice does not allow us to explore other models. In an extreme scenario, the developed model may have associated *hidden* costs with it (i.e., energy-wise, accessibility-wise, etc.). However, and realistically speaking, such models may not be able to adequately attain good training/validation performance (or when deployed) or may not be able to correctly uncover the underlying mechanisms of a given problem.

One way to solve the above problem is to utilize multi-search algorithms and exotic or ensemble models. The idea behind this utilization is simple. The engineer is advised to explore a variety of ML algorithms (especially those incorporating a mixture of architectures, topologies, etc.) in a tournament style. Once fully trained, all models can be ranked according to pre-specified metrics/indicators in a *leaderboard*.

A key notion to ML belongs to realizing the reasons behind model predictions – especially in our domain. Unlike other traditional methods[164] which obey physics and mechanics principles, ML continues to be data-driven. To allow our engineers to peek into the reasoning behind model predictions, there is a need to incorporate explainability.[165]

---

161 As well as textbook!

162 Not that they cannot be extended and improved.

163 We are hoping for this to be a virtual assistant vs. a traditional platform.

164 FE or CFD methods.

165 As well as causality. However, we are working on CLEMSON to add causal principles to it. Also, we hope to support data handling and feature engineering tools for model monitoring, and support for computer vision and unsupervised learning.

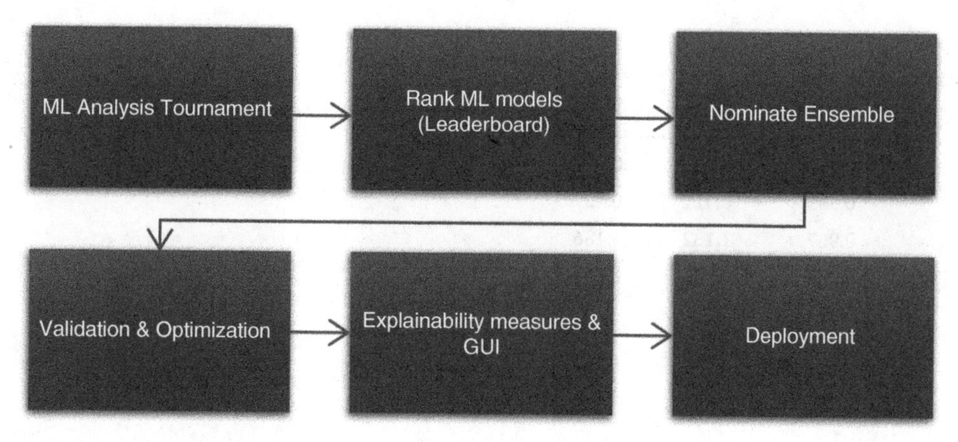

Figure 9.19 Flowchart of CLEMSON.

One of the persistent key limitations to adopting ML is the need for programming/coding.[166] Thus, removing this hurdle can be completed by adding a graphical user interface that allows a user (as opposed to the model's developer) to modify its architecture or update its settings.

CLEMSON builds upon the success of ML-based platforms that enables the automated development of ML models to overcome the above issues. Simply, CLEMSON develops a series of ML models to explore phenomena. Identifies the best scoring models and blends them into an ensemble (if necessary). Augment such ensemble with explainability measures and then deploy such ensemble via a graphical user interface (GUI) – see Figure 9.19. When needed, the VA also surrogates the created model/ensemble in a mathematical expression (*mapping function*) to negate the need for the parent model/ensemble.

CLEMSON follows the same training/validation/testing procedure we covered earlier. In this procedure, a dataset is split into unequal proportions.[167] The largest two are reserved for training and validating the model. Then, the model is independently tested against the last set. The above process also incorporates a k-fold cross-validation procedure[168] that can also be used. Finally, the predictivity of the recommended model/ensemble is cross-checked against traditional[169] and functional/domain-specific[170] metrics/indicators.

9.6.3 Example

Let us now go over a compound example that starts with a classification component and ends with a regression component. In this example, CLEMSON will be applied to assess the fire resistance rating[171] of concrete-filled steel tubular (CFST) columns and then predict temperature-time history at the core of columns and axial deformation-time history of the same columns.

In total, we have 102 CFST columns that were tested under fire conditions. These columns have the following features. Column diameter, D, column slenderness, n, tube thickness, t, concrete compressive strength, f_c, yield strength of steel, f_y, steel reinforcement ratio, r, restraint conditions, K, eccentricity, e, and level of applied loading, P, and time of exposure.[172]

166 Thank you *DataRobot* and *Exploratory*!

167 Say, 70/20/10 or 60/25/15.

168 Where $k = 5$ or 10, etc.

169 For classification (Area under the ROC curve (AUC), Balanced Accuracy (BA), Log Loss Error (LLE)) and regression (Mean Absolute Error (MAE), Symmetric Mean Absolute Percentage Error (SMAPE), Root Mean Squared Error (RMSE), Coefficient of Determination (R^2)).

170 For classification (Clemson Metric (CM), Cumulative Clemson Metric (CCM)) and for regression (Maximum Response Deviation (MRD)).

171 Also known as FRR.

172 As you will see, FRR was predicted via P, f_c, D, f_y, k, n, t, r, and e, while temperature-time history used D and t, and deformation-time history applied exposure time, P, D, t, and k. If you are asking why different features were used to predict different phenomena, it is because domain knowledge tell us that each phenomenon is governed by specific features! For example, since all selected examined CFST columns were exposed to standard fires, made of hot-rolled tubes and plain normal strength concrete, then the effect of concrete type, boundary conditions, etc., can be normalized.

Table 9.6 List of selected performance metrics.

Fire Resistance Rating	T	V	S
AUC	0.906	0.781	0.817
Balanced Accuracy	0.505	0.421	0.427
Log Loss (LL) Error	0.967	1.132	1.186
Cumulative Clemson Metric (CCM)	0.732	0.800	0.671
Temperature-time history			
Mean Absolute Error (MAE)	17.563	16.296	17.498
Symmetric Mean Absolute Percentage Error (SMAPE)	14.547	13.510	12.705
Root Mean Squared Error (RMSE)	27.271	23.567	25.593
Coefficient of Determination (R^2)	0.954	0.958	0.957
Deformation-time history			
Mean Absolute Error (MAE)	1.184	1.793	2.543
Symmetric Mean Absolute Percentage Error (SMAPE)	36.247	35.37	44.60
Root Mean Squared Error (RMSE)	2.008	3.603	8.182
Coefficient of Determination (R^2)	0.964	0.903	0.712

[Note – T: Training sample, V: Validation sample, S: Testing sample].

The developed ensemble is built from the following five algorithms: extreme gradient boosted trees [43, 44], light gradient boosted trees [45], deep neural networks [46], random forest [47], and TensorFlow [48].

Table 9.6 lists the outcome of the classification and regression analyses. As one can see from the results listed in Table 9.6, the developed ensemble seems to perform well. In addition, Figure 9.20 also shows that predictions from ML for one sample column fall within the MDR bounds as an additional means to assess ML predictions at the micro-level (specific intervals of 10 min each).

Let us look at the ensemble's feature importance and PDPs.[173] As you can see, Figure 9.21 shows that ensemble predictions matche that of each individual model for the most part. The same figure. also shows PDPs. For instance, there is a positive linear relationship between exposure time and temperature rise in CFST columns and a negative relationship between tube diameter size and temperature rise.[174] This figure also shows that the first 15–20% of fire exposure is associated with deformation expansion followed by a contraction due to strength degradation.

Finally, CLEMSON exports the model into a spreadsheet with a GUI. This spreadsheet is presented in Figure 9.22 and can provide engineers with a ML-approach to predict FRR, temperature-time history, and deformation-time history without having to hassle with ML programming.

9.6.4 A Note

Machine learning automates our lives on a daily basis. AutoML is our way of automating ML to carry out our routine operations pertaining to our domain. A fully functional AutoML solution can be a powerful tool in our arsenal of methods of

173 A PDP reveals the type of relationship a feature has on model predictions.

174 Larger columns tend to heat at a slower rate (since there is more mass to concrete than in smaller columns).

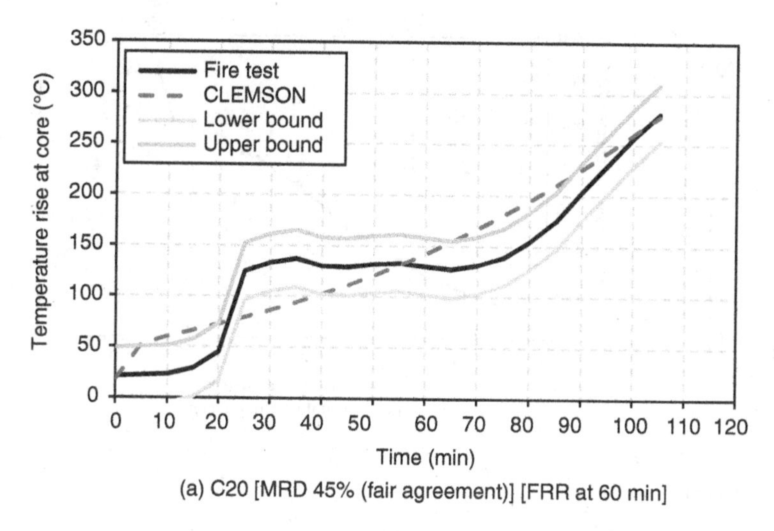

(a) C20 [MRD 45% (fair agreement)] [FRR at 60 min]

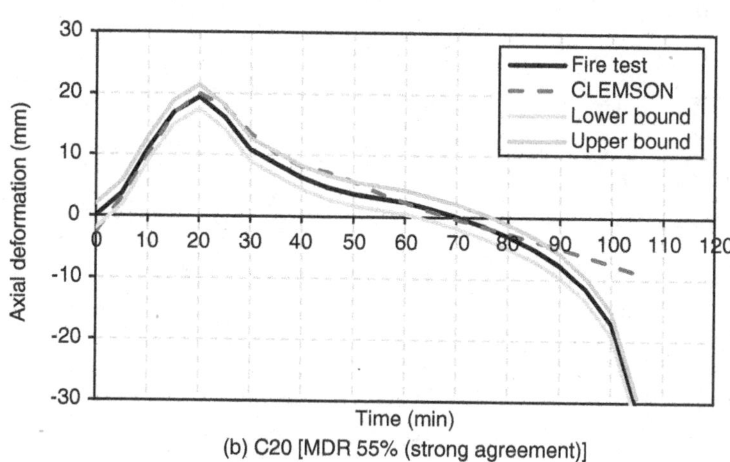

(b) C20 [MDR 55% (strong agreement)]

Figure 9.20 Comparison via MDR.

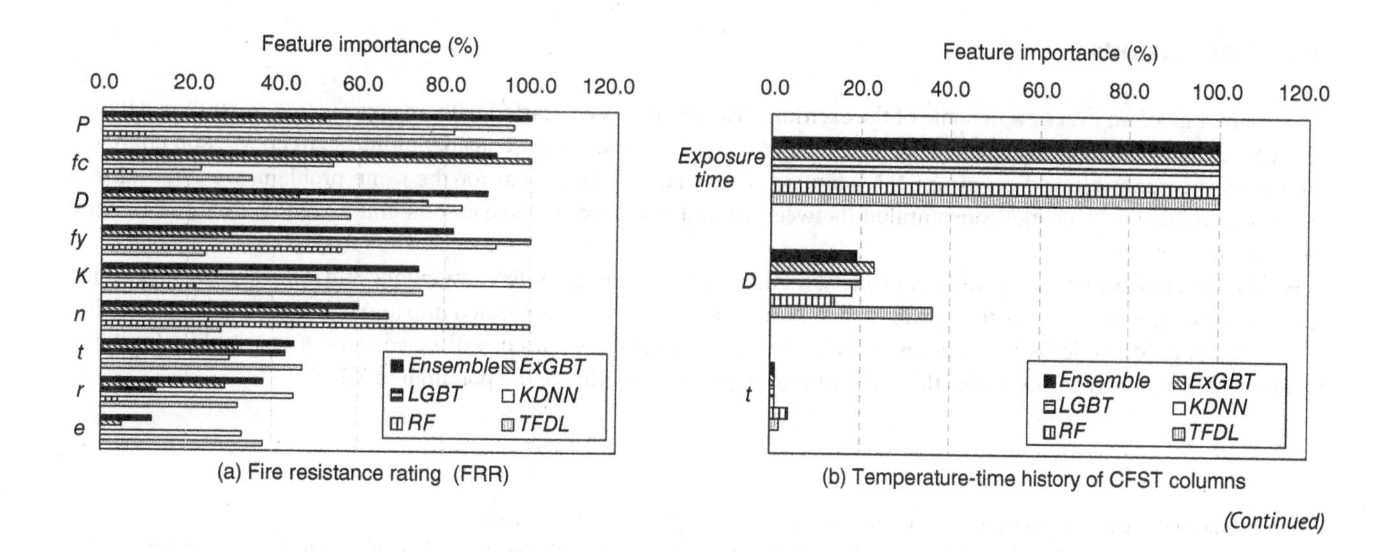

(a) Fire resistance rating (FRR)

(b) Temperature-time history of CFST columns

(Continued)

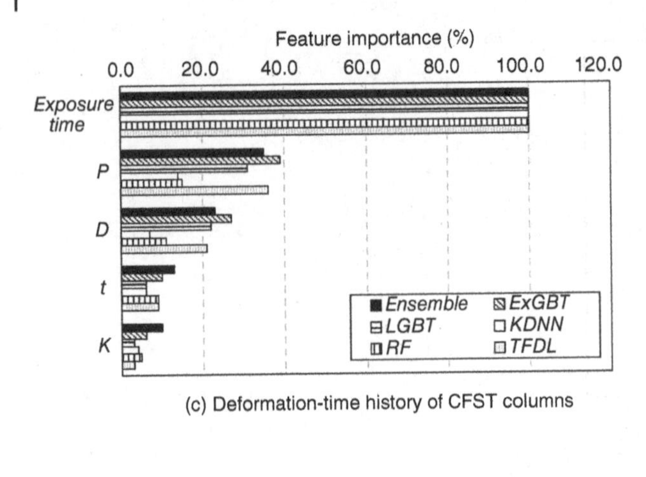

(c) Deformation-time history of CFST columns

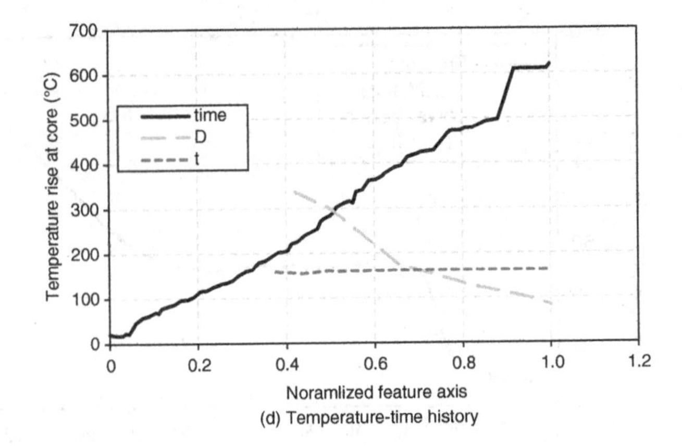

(d) Temperature-time history

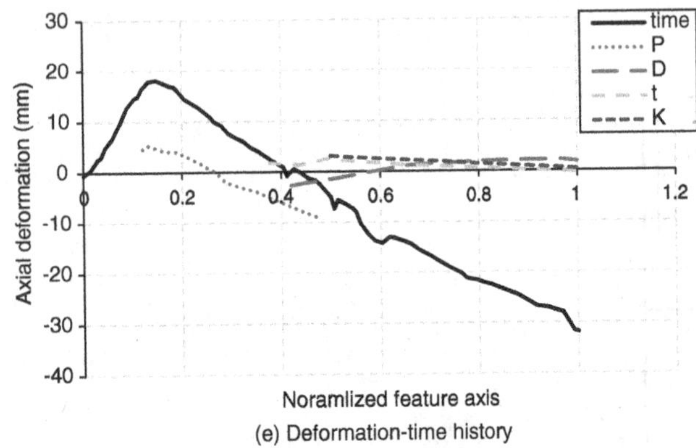

(e) Deformation-time history

Figure 9.21 (Con'd) Insights from explainability measures.

investigation. With the rise of the big data era, such a tool will prove warranted and needed soon,[175] especially if it can accommodate multi-ML areas (unsupervised, supervised, and semi-supervised learning) and can be supplemented with physics, domain knowledge, and causal principles.

9.7 Conclusions

This chapter gives an overview of some of the exciting ideas in ML.[176] We started with an approach to expanding datasets via synthetic and augmented means. We then dived into exploring the energy considerations of Green ML and charted a path toward responsible and cognizant ML adoption. Our analysis shows that for the same problem, we can realize a significant reduction in energy consumption (between 23 and 99%) and possible carbon emissions. This emphasizes the need to embrace Green ML.

We also covered examples of readily available ML software that can help us derive symbolic and *mapping functions*. Such functions resemble mathematical formulas that are ML-based and can be substituted into and hence may aid in a situation where access to ML codes and computing stations is limited. Finally, we discussed the concept of ensembling (building combined ML models) and extended this concept into AutoML to maximize the potential of ML.

175 Think about similar tools and platforms offered by major companies. These are there for a reason!

176 There are much, much more ideas and concepts that I did not get to discuss. Partially due to page limitations and because they may not be directly applicable to our domain. In all cases, please take the initiative to learn more about ML. There is more to ML than that housed within the pages of this book.

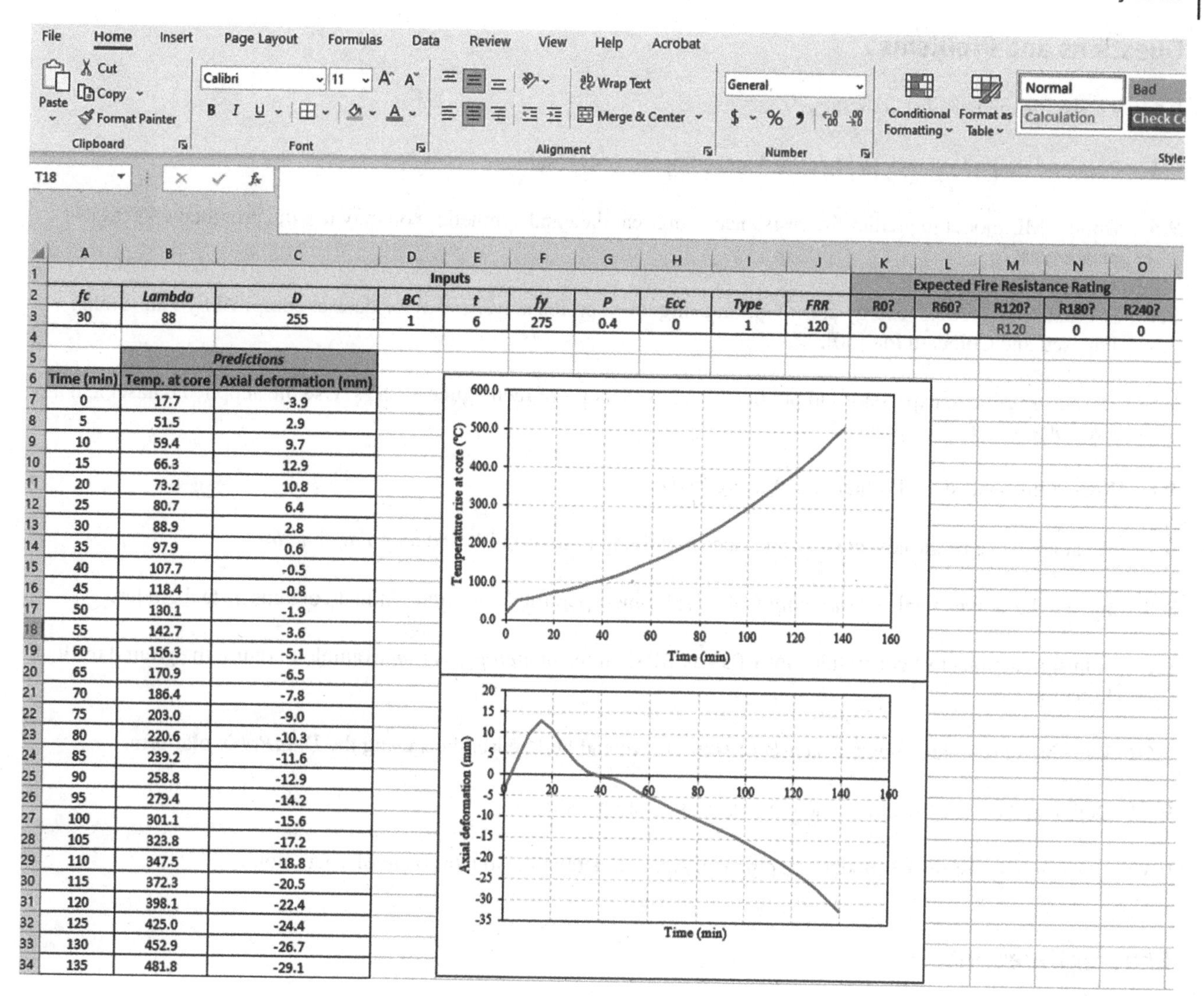

fc	Lambda	D	BC	t	fy	P	Ecc	Type	FRR
30	88	255	1	6	275	0.4	0	1	120

R0?	R60?	R120?	R180?	R240?
0	0	R120	0	0

Time (min)	Temp. at core	Axial deformation (mm)
0	17.7	-3.9
5	51.5	2.9
10	59.4	9.7
15	66.3	12.9
20	73.2	10.8
25	80.7	6.4
30	88.9	2.8
35	97.9	0.6
40	107.7	-0.5
45	118.4	-0.8
50	130.1	-1.9
55	142.7	-3.6
60	156.3	-5.1
65	170.9	-6.5
70	186.4	-7.8
75	203.0	-9.0
80	220.6	-10.3
85	239.2	-11.6
90	258.8	-12.9
95	279.4	-14.2
100	301.1	-15.6
105	323.8	-17.2
110	347.5	-18.8
115	372.3	-20.5
120	398.1	-22.4
125	425.0	-24.4
130	452.9	-26.7
135	481.8	-29.1

Figure 9.22 GUI of the developed Excel spreadsheet.

Definitions

Augmented data Data generated by means of simulations or modeling.

AutoML Automated Machine Learning.

Bagging A method that adopts bootstrapping to generate data that can be input to train parallel models.

Bootstrapping A process of creating multiple sets (bags) of the original training data with replacement.

Complex models ML models that join/ensemble different algorithms/models into one model.

Ensemble A collection of models.

Exotic models ML model that use one algorithm to optimize or finetune the parent algorithm/model.

Green ML A machine learning approach to energy-cognizant models.

Mapping Functions A mathematical expression that intelligently ties features to outcomes.

Physics-guided and domain knowledge feature selection Selecting features in a ML model based on our domain knowledge and physics principles.

Stacking A method that combines predictions from various models to be used as input to train the ensemble.

Symbolic regression An approach to find suitable mathematical model(s) to describe data.

Synthetic data Data generated by the manipulation of real observations.

Wisdom of crowds The aggregate knowledge of a diverse, independent, decentralized body of individuals often exceeds the knowledge of single individuals/experts.

Questions and Problems

9.1 Repeat the analysis listed in Table 9.1.

9.2 Redo the example shown in Figure 9.3 and compare your results.

9.3 Build a ML model to predict fire resistance using real data and synthetic. You may use the provided code as a benchmark.

9.4 Estimate ML models' energy consumption using the supplied *driver behavior* dataset compiled by Paul-Stefan Popescu and Cojocaru Ion [50].

9.5 Derive descriptive expressions to predict the axial capacity of rectangular CFSTs. Use the supplied dataset and any two software of your choice.

9.6 Repeat the analysis in Problem no. 4 using *Python*.

9.7 What are some of the advantages and issues pertaining to descriptive and *mapping functions*?

9.8 Repeat the first example on *mapping functions* by incorporating a new dataset that contains 5 steel grades.

9.9 Adjust the number of generations for BP34 as BP42 in the *mapping function* example to return the ground truth of Equation 9.6.

9.10 What are ensembles? Select a dataset of your choice and build ensembles using the *DataRobot* platform.

9.11 Define and contrast stacking, blending, bagging, and boosting.

9.12 Class project. Build a 3-model ML pipeline. Apply this pipeline to a problem of your choice.

Chapter Blueprint

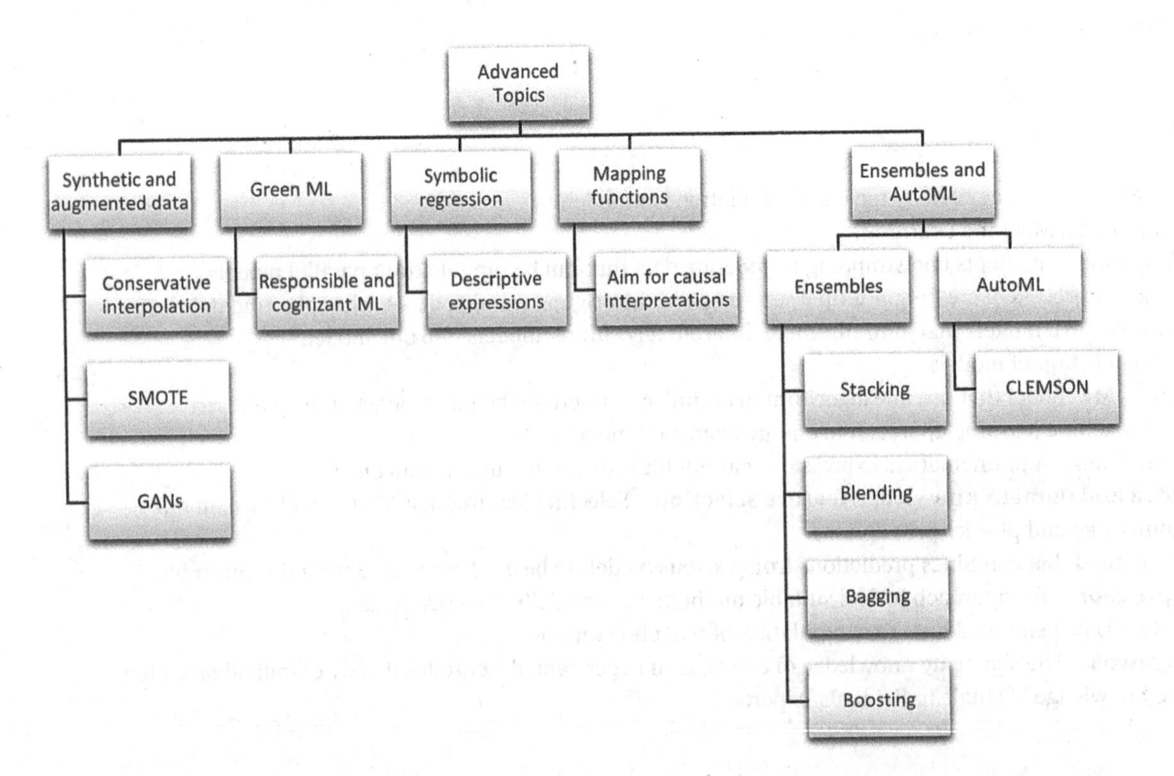

References

1 Naser, M.Z. and Kodur, V.K. (2022). Explainable machine learning using real, synthetic and augmented fire tests to predict fire resistance and spalling of RC columnsI. *Eng. Struct.* 253: 113824. doi: 10.1016/J.ENGSTRUCT.2021.113824.

2 Chawla, N.V., Bowyer, K.W., Hall, L.O., and Kegelmeyer, W.P. (2002). SMOTE: Synthetic Minority Over-sampling Technique. *J. Artif. Intell. Res.* doi: 10.1613/jair.953.

3 Xu, L., Skoularidou, M., Cuesta-Infante, A., and Veeramachaneni, K. (2019). Modeling tabular data using conditional GAN. *33rd Conference on Neural Information Processing Systems (NeurIPS 2019)*, Vancouver, Canada.

4 Goodfellow, I., Pouget-Abadie, J., Mirza, M. et al. (2020). Generative adversarial networks. *Commun. ACM.* 63. doi: 10.1145/3422622.

5 Synthetic Data Vault (SDV). (2022). https://sdv.dev/SDV/index.html (accessed 15 June 2022).

6 Strubell, E., Ganesh, A., and McCallum, A. (2020). Energy and policy considerations for modern deep learning research. Proc. AAAI Conf. Artif. Intell. doi: 10.1609/aaai.v34i09.7123.

7 Jackson, R. (2019). A roadmap to reducing greenhouse gas emissions 50 percent by 2030. https://earth.stanford.edu/news/roadmap-reducing-greenhouse-gas-emissions-50-percent-2030#gs.zij3zf. July 2022.

8 GeSI SMARTer. (2020). The role of ICT in driving a sustainable future GeSI SMARTer 2020 2 the role of ICT in driving a sustainable future independent review by, n.d.

9 Lu, Y., Zhang, A., Li, Q., and Dong, B. (2018). Beyond finite layer neural networks: bridging deep architectures and numerical differential equations | OpenReview. *6th International Conference on Learning Representations* https://openreview.net/forum?id=ryZ283gAZ (accessed 23 April 2021).

10 Young, B.A., Hall, A., Pilon, L. et al. (2019). Can the compressive strength of concrete be estimated from knowledge of the mixture proportions?: new insights from statistical analysis and machine learning methods. *Cem. Concr. Res.* 115: 379–388. doi: 10.1016/j.cemconres.2018.09.006.

11 Yu, A. (2021). hyperparameter_tune/Clean_data.csv at master · huiziy/hyperparameter_tune · GitHub https://github.com/huiziy/hyperparameter_tune/blob/master/Clean_data.csv (accessed 18 March 2021).

12 EIA. (2021). Frequently Asked Questions (FAQs) – How much carbon dioxide is produced per kilowatthour of U.S. electricity generation? *U.S. Energy Inf. Adm.* https://www.eia.gov/tools/faqs/faq.php?id=74&t=11 (accessed 28 April 2021).

13 Desroches, L., Fuch, H., Greenblatt, J. et al. (2014). Computer usage and national energy consumption: results from a field-metering study. https://www.osti.gov/servlets/purl/1166988 (accessed 28 April 2021).

14 ASCE. (n.d.). About ASCE | ASCE. https://www.asce.org/about-asce (accessed 16 June 2022).

15 EPA. (2021). Greenhouse Gas Equivalencies Calculator, EPA. https://www.epa.gov/energy/greenhouse-gas-equivalencies-calculator (accessed 28 April 2021).

16 Hawileh, R.A., Naser, M., and Rasheed, H.A. (2011). Thermal-Stress Finite Element Analysis of CFRP Strengthened Concrete Beam Exposed to Top Surface Fire Loading. *Mech. Adv. Mater. Struct.* 18: 172–180. doi: 10.1080/15376494.2010.499019.

17 Naser, M.Z. (2019). AI-based cognitive framework for evaluating response of concrete structures in extreme conditions. *Eng. Appl. Artif. Intell.* 81: 437–449. https://www.sciencedirect.com/science/article/pii/S0952197619300466 (accessed 1 April 2019).

18 Koza, J.R. (1992). A genetic approach to finding a controller to back up a tractor-trailer truck. *Proceedings 1992 American Control Conference (ACC)*

19 Petersen, B.K., Larma, M.L., Mundhenk, T.N. et al. (n.d.). Deep symbolic regression: Recovering mathematical expressions from data via risk-seeking policy gradients.

20 d'Ascoli, S., Kamienny, P.-A., Lample, G., and Charton, F. (2022). Deep Symbolic Regression for Recurrent Sequences. doi: 10.48550/arxiv.2201.04600.

21 Schmidt, M. and Lipson, H. (2009). Distilling Free-Form Natural Laws from Experimental Data. *Science (80-)* 324: doi: 10.1126/science.1165893.

22 TuringBot. (2022). Documentation - TuringBot: symbolic regression software. https://turingbotsoftware.com/documentation.html#introduction (accessed 20 June 2022).

23 Ferreira, C. (2001). Gene Expression Programming: A New Adaptive Algorithm for Solving Problems. *Gene Expr. Program. a New Adapt. Algorithm Solving Probl. Complex Syst.* 13. https://www.semanticscholar.org/paper/Gene-Expression-Programming%3A-a-New-Adaptive-for-Ferreira/3232b2a24c2584ca8e81cb5bf6f55aef34f0aefe (accessed 16 March 2019).

24 Kamienny, P.-A., d'Ascoli, S., Lample, G., and Charton, F. (2022). End-to-end symbolic regression with transformers. doi: 10.48550/arxiv.2204.10532.

25 Raut, N. and Kodur, V. (2011). Response of Reinforced Concrete Columns under Fire-Induced Biaxial Bending. *ACI Struct. J.* 108: doi: 10.14359/51683218.

26 Lie, T. and Woollerton, J. (1988). Fire resistance of reinforced concrete columns – NRC Publications Archive – National Research Council Canada. https://doi.org/http://doi.org/10.4224/20386656. July 2022.

27 ASCE (2016). *Minimum Design Loads For Buildings And Other Structures*. American Society of Civil Engineers. doi: 10.1061/9780872629042.

28 Kodur, V.K.R., Garlock, M., and Iwankiw, N. (2012). Structures in fire: state-of-the-art, research and training needs. *Fire Technol.* 48: 825–839. doi: 10.1007/s10694-011-0247-4.

29 Fellinger, J.H.H. and Twilt, L. (n.d.). Fire Behaviour of Long Span Composite Floor. http://iafss.org/publications/fss/5/1093/view/fss_5-1093.pdf (accessed 8 February 2019).

30 Iguyon, I. and Elisseeff, A. (2003). An introduction to variable and feature selection. *J. Mach. Learn. Res.* 3.

31 Wehenkel, M., Sutera, A., Bastin, C. et al. (2018). Random forests based group importance scores and their statistical interpretation: application for Alzheimer's disease. *Front. Neurosci.* doi: 10.3389/fnins.2018.00411.

32 Williamson, B.D., Gilbert, P.B., Simon, N.R., and Carone, M. (2020). A General Framework for Inference on Algorithm-Agnostic Variable Importance. *ArXiv.*

33 Molnar, C. (2019). *Interpretable Machine Learning. A Guide for Making Black Box Models Explainable*. Learnpub.

34 Boehmke, B., Greenwell, B., Boehmke, B., and Greenwell, B. (2019). Interpretable machine learning. In: *Hands-On Machine Learning with R*, 38. Taylor & Francis. doi: 10.1201/9780367816377-16.

35 Lundberg, S.M. and Lee, S.I. (2017). A Unified Approach to Interpreting Model Predictions. *NIPS'17: Proceedings of the 31st International Conference on Neural Information Processing Systems* December 2017 4768–4777.

36 Wolpert, D.H. and Macready, W.G. (1997). No free lunch theorems for optimization. *IEEE Trans. Evol. Comput.* 1. doi: 10.1109/4235.585893.

37 Polikar, R. (2009). Ensemble learning. *Scholarpedia*. 4. doi: 10.4249/scholarpedia.2776.

38 Chou, J.S. and Pham, A.D. (2013). Enhanced artificial intelligence for ensemble approach to predicting high performance concrete compressive strength. *Constr. Build. Mater.* 49. doi: 10.1016/j.conbuildmat.2013.08.078.

39 Sadhwani, J. (2022). GitHub - JatinSadhwani02/Water-Quality-Prediction-using-ML-technique. https://github.com/JatinSadhwani02/Water-Quality-Prediction-using-ML-technique (accessed 16 June 2022).

40 Asteris, P.G., Skentou, A.D., Bardhan, A. et al. (2021). Predicting concrete compressive strength using hybrid ensembling of surrogate machine learning models. *Cem. Concr. Res.* 145. doi: 10.1016/j.cemconres.2021.106449.

41 Fernández-Delgado, M., Sirsat, M.S., Cernadas, E. et al. (2019). An extensive experimental survey of regression methods. *Neural Networks.* 111. doi: 10.1016/j.neunet.2018.12.010.

42 Naser, M.Z. (2022). CLEMSON: an automated machine learning virtual assistant for accelerated, simulation-Free, transparent, reduced-Order and inference-Based reconstruction of fire response of structural members. *ASCE Jounral Struct. Eng.* 148. doi: 10.1061/(ASCE)ST.1943-541X.0003399.

43 Scikit. (2020). sklearn.ensemble.GradientBoostingRegressor — scikit-learn 0.24.1 documentation. https://scikit-learn.org/stable/modules/generated/sklearn.ensemble.GradientBoostingRegressor.html (accessed 9 February 2021).

44 XGBoost Python Package. (2020). Python Package Introduction — xgboost 1.4.0-SNAPSHOT documentation. https://xgboost.readthedocs.io/en/latest/python/python_intro.html#early-stopping (accessed 10 February 2021).

45 LightGBM. (2020). Welcome to LightGBM's documentation! — lightGBM 3.1.1.99 documentation. https://lightgbm.readthedocs.io/en/latest (accessed 9 February 2021).

46 Keras. (2020). GitHub - keras-team/keras: deep Learning for humans. https://github.com/keras-team/keras (accessed 9 February 2021).

47 Scikit. (2020). sklearn.ensemble.RandomForestClassifier — scikit-learn 0.24.1 documentation. https://scikit-learn.org/stable/modules/generated/sklearn.ensemble.RandomForestClassifier.html (accessed 9 February 2021).

48 TensorFlow. (2020). GitHub - tensorflow/tensorflow: an open source machine learning framework for everyone. https://github.com/tensorflow/tensorflow (accessed 9 February 2021).

49 Lie, T.T. and Chabot, M. (1992). Experimental studies on the fire resistance of hollow steel columns filled with plain concrete. *Natl. Res. Counc. Canada.* 44.

50 Popescu, P. and Ion, C. (n.d.). Driving Behavior. Kaggle. https://www.kaggle.com/datasets/outofskills/driving-behavior/metadata (accessed 15 June 2022).

51 Huntington-Klein, N. (2021). *The Effect : An Introduction to Research Design and Causality*. Boca Raton: Chapman and Hall/CRC. doi: 10.1201/9781003226055.

52 Blalock, H.M., Jr. (1964). *Causal Inferences in Nonexperimental Research*. UNC Press Books.

53 Naser, M.Z. (2020). Enabling cognitive and autonomous infrastructure in extreme events through computer vision. *Innov. Infrastruct. Solut.* 5: 99. doi: 10.1007/s41062-020-00351-6.

54 Gernay, T. (2019). Fire resistance and burnout resistance of reinforced concrete columns. *Fire Saf. J.* 104. doi: 10.1016/j.firesaf.2019.01.007.

55 Farfan, J., Fasihi, M., and Breyer, C. (2019). Trends in the global cement industry and opportunities for long-term sustainable CCU potential for Power-to-X. *J. Clean. Prod.* 217. doi: 10.1016/j.jclepro.2019.01.226.

56 Ritchie, H. and Roser, M. (2021). Emissions by sector. Our World in Data. https://ourworldindata.org/emissions-by-sector (accessed 28 April 2021).

57 Brownlee, A.E.I., Adair, J., Haraldsson, S.O., and Jabbo, J. (n.d.). Exploring the accuracy-energy trade-off in machine learning. https://geneticimprovementofsoftware.com/paper_pdfs/gi2021icse/brownlee_gi-icse_2021.pdf (accessed 28 April 2021).

58 García-Martín, E., Rodrigues, C.F., Riley, G., and Grahn, H. (2019). Estimation of energy consumption in machine learning. *J. Parallel Distrib. Comput.* 134. doi: 10.1016/j.jpdc.2019.07.007.

59 Augusto, D.A. and Barbosa, H.J.C. (2000). Symbolic regression via genetic programming. *Proceedings - Sixth Brazilian Symposium on Neural Networks*. doi: 10.1109/SBRN.2000.889734.

60 Searson, D. and Searson, D. (2009). GPTIPS genetic programming & symbolic regression for MATLAB user guide. http://gpbib.cs.ucl.ac.uk/gp-html/GPTIPSGuide.html (accessed 22 January 2019).

61 GPTIPS. (2010). GPTIPS: Genetic Programming and symbolic regression for MATLAB. http://gptips.sourceforge.net (accessed 22 January 2019).

62 Discipulus (n.d.). Discipulus professional. G6G directory of omics and intelligent software. https://www.g6g-softwaredirectory.com/ai/genetic-programming/20047RMLTechnolDiscipulusProfess.php (accessed 23 January 2019).

63 Thai, S., Thai, H.T., Uy, B., and Ngo, T. (2019). Concrete-filled steel tubular columns: test database, design and calibration. *J. Constr. Steel Res.* 157. doi: 10.1016/j.jcsr.2019.02.024.

64 Naser, M.Z., Thai, S., and Thai, H.-T.H.-T. (2021). Evaluating structural response of concrete-filled steel tubular columns through machine learning. *J. Build. Eng.* 34: 101888. doi: 10.1016/j.jobe.2020.101888.

65 Naser, M.Z. (2021). Mapping functions: a physics-guided, data-driven and algorithm-agnostic machine learning approach to discover causal and descriptive expressions of engineering phenomena. *Measurement.* 185: 110098. doi: 10.1016/J.MEASUREMENT.2021.110098.

66 Abedi, M., Naser, M.Z., and Rapid, R.A.I. (2021). Autonomous and Intelligent machine learning approach to identify fire-vulnerable bridges. *Appl. Soft Comput.* 113. doi: 10.1016/j.asoc.2021.107896.

67 Wiesner, F. and Bisby, L. (2019). The structural capacity of laminated timber compression elements in fire: a meta-analysis. *Fire Saf. J.* 107. doi: 10.1016/j.firesaf.2018.04.009.

68 Chern, J.-C., You, C.-M., and Bazant, Z.P. (1992). Deformation of progressively cracking partially prestressed concrete beams. *PCI J.* 37. doi: 10.15554/pcij.01011992.74.85.

69 Naser, M.Z. (2023). Do We Need Exotic Models? Engineering Metrics to Enable Green Machine Learning from Tackling Accuracy-Energy Trade-offs. *J. Clean. Prod*. 382:135334. https://doi.org/10.1016/j.jclepro.2022.135334 (accessed 1 January 2023).

70 Naser, M.Z. (2022). Causality in structural engineering: discovering new knowledge by tying induction and deduction via mapping functions and eXplainable artificial intelligence. *AI Civ. Eng*. 1(1):6. doi: 10.1007/s43503-022-00005-9 (accessed 7 September 2022).

10

Recommendations, Suggestions, and Best Practices

What would a mistake here be like?

Ludwig Wittgenstein

Synopsis

Now that we are inching toward the end of this textbook, I find it timely to list and talk about some *recommendations, suggestions*, and *best practices* that can come in handy in a ML-based investigation. These are not meant to be followed to the teeth but rather are presented herein as I have found them to be of aid in some, if not most, of my group's work, as well as those I had the pleasure of meeting/reading their works.[1]

Before we dive into this chapter, let us start by distinguishing between what this chapter refers to as *recommendations, suggestions*, and *best practices*:[2]

- A *recommendation* is an advice that has been duly noted in the open literature. Often, a recommendation comprises a general statement.
- A *suggestion* is a proposition of opinion (and may largely stem from personal experiences).
- A *best practice* is a trend that has been commonly accepted by publications in our domain. The origin of a best practice could have been arrived at systemically, such as through a sensitivity study or empirical analysis or via heuristics (rules of thumb). Best practices tend to vary across timeframes and groups.

10.1 Recommendations

The following is a list of recommendations that could be of aid to civil and environmental engineers interested in adopting ML.

10.1.1 Continue to Learn

My favorite recommendation revolves around learning. At a fundamental level, we are public servants wherein our role is to maintain the well-being of our society. To stay up to date with the latest advancement, we ought to continue to learn. In a way, we have a duty for lifelong education – especially one that stretches beyond the award of our degrees. This continued

1 Also, I must acknowledge the many new things I learned from the anonymous reviewers and journal editors too. There are many *quick stories* that I remember that could fit into the theme of this chapter. On one occasion, one reviewer requested the addition of a short section toward the end of an article that outlines the limitations, and range of applicability of a developed model that my co-authors and I built. This chapter is inspired by such a request.

2 Some of the presented items builds up some of my recently co-authored publications [37, 38] while others were adopted from Google's Rules of Machine Learning [39].

Machine Learning for Civil & Environmental Engineers: A Practical Approach to Data-driven Analysis, Explainability, and Causality,
First Edition. M. Z. Naser.

Companion Website: www.wiley.com/go/Naser/MachineLearningforCivilandEnvironmentalEngineers

learning ensures that we are aware of new methods, building code provisions, and the latest advancements in our areas. The above is notably true given that a civil and environmental engineering course plan does not and cannot cover every facet an engineer is expected to know and learn.

The rise of digitalization and automation[3] has opened the door to developing new methods and procedures that could benefit us. While some might say that the rise of such modern technologies is disturbing the calmness of our domain, I tend to disagree. I view these as an opportunity to modernize and transform civil and environmental engineering. Think of it this way. If a form of technology already exists (say, drones) and has been shown to be more affordable and productive than existing (surveying) methods, then engineers will use it whether it is covered in a curriculum or not. By comparison, the notion that because ML is not an integrated component of current course plans, therefore, we do not need to learn ML, is simply *flawed*.

10.1.2 Understand the Difference between Statistics and ML

Up to this point, most of the previous pages covered ideas and concepts of ML, and it has been a while since we visited Chapter 3, so let me switch gears and talk a bit about statistics. *Statistical methods* are "mathematical formulas, models, and techniques that are used in statistical analysis of raw research data" [1]. The primary goal of applying such methods is to extract insights from data and/or provide means to assess the robustness of research outputs. In a statistical analysis, a statistical model often assumes a *pre-determined* distribution to fit the input parameters to the result of a phenomenon [2]. In the event that the number of involved parameters increases or the relationship between them is evidently intricate, or the data does not satisfy assumptions/conditions tied with statistical methods, such methods become less effective – and possibly inapplicable [3].

Unlike statistical methods, ML makes minimal assumptions about the type and quality of data. ML looks for hidden patterns within data points and hence becomes effective when applied to unstandardized data or those with highly nonlinear interactions. A unique distinction between statistical methods and ML is that the former aims to satisfy a set of rules and assumptions, while the latter pursues to satisfy a penalization function [4]. Both statistical methods and ML can be used in predictions. However, it will take a much deeper analysis to explain predictions or arrive at causal effects.

10.1.3 Know the Difference between Prediction via ML and Carrying Out Tests and Numerical Simulations

The goal of carrying out an experiment is to investigate a hypothesis or research question (say, what is the influence of X on Y?) [5, 6]. It is unfortunate that testing is a costly endeavor and continues to limit our capacity to fully explore phenomena.

So, we resort to a more affordable approach by augmenting our tests via numerical models [7]. First, a model is developed. This model is created based on first principles (i.e., those impeded into FE software) and then validated against the outcome of benchmarked tests [8, 9]. Since such a model satisfies first principles as well as matches well with our tests, then we can declare it valid and use it to *understand* the cause and effect governing the phenomenon on hand [10].

However, let us not forget that a FE model is a shell skeleton. Say, a model for a rectangular beam contains the geometrical model that describes the geometrical component and a material model that conveys the behavior of the material(s) within this beam. Thus, for the same geometry, a modeler can observe different responses (say, deflection) if the material model is varied between concrete, steel, and timber. In a way, the FE model does not recognize that we are trying to simulate a beam, but rather it solves the developed skeleton for deflection. While the same is also true for ML models, FE models solve our problems by first solving principles. This is not the case for the majority of ML models.

The procedure for creating traditional ML models comprises a three-step linear procedure: 1) collect data on a phenomenon, 2) apply an algorithm, and 3) declare that such an algorithm can predict the phenomenon on hand via fitness tests. As mentioned earlier, a traditional ML model is a Blackbox that we do not know why it predicts the way it does, nor how it ties the parameters to the outcome of the phenomenon[4]! However, we know that the data drive the decision of

3 Look at how drones transformed the inspection, construction, and surveying.

4 Unless we use some of the covered explainability, *mapping functions*, or causality tools.

such an algorithm, and we know that despite the model's fitness and robustness of data, a traditional ML model may not still agree with physics principles.[5] This is *critical* to remember.

10.1.4 Ask if You Need ML to Address the Phenomenon on Hand

It can be tempting to follow the latest trends in scholarly works. Trying new trends can be an exciting learning experience. However, blindly assuming such trends to be applicable to a phenomenon you happen to be interested in can be detrimental. The existence of a technology (in this case, ML) does not justify the need to favor it over other existing and well-established methods. We ought to learn, investigate, and explore where and how ML can help us.[6]

Another view to leveraging ML is to answer the following question: *did we fully examine the data in our tests and simulations?* It is quite plausible that we did not – especially since we know that traditional methods are constrained to certain assumptions and hence suffer from limitations [11, 12]. It is due to these assumptions and limitations that we may need to explore other means of investigation. This can be apparent in the evidence of many works that showcased how predictions from ML outperform those obtained from building code provisions and statistical methods [13, 14].

By noting how many of our problems resemble a closed-loop system wherein a number of parameters are associated with a response. Then, one can form a hypothesis that if we are to collect enough data on parameters and responses, we could potentially better understand our problems and create/optimize novel solutions to overcome such problems.[7]

10.1.5 Establish a Crystal Clear Understanding of Model Assumptions, Outcomes, and Limitations

Data-driven approaches can be tremendously helpful for cases where we have plenty of data, such as the compressive strength of concrete as a function of its mixture proportions, water quality as a function of its chemical characteristics, etc.

A key *assumption* of data-driven ML is the availability of comprehensive and healthy datasets. This is a valid assumption and one that is commonly adopted whenever statistical methods are applied.[8] Despite being blackboxes, ML models have been shown to perform well. However, we do not know how such models were able to arrive at such good predictions [15]. So, this leaves us with a tool that seems to work well on our assumed-to-be comprehensive datasets. Again, we may not know how this tool works. How can we trust it?[9]

Here is where one draws a line. A decision is to be made as to how the ML model is expected to be used. Are you looking for a blackbox prediction model? An explainable model? Or a causal model? While many of our problems will greatly benefit from any of the above three models, we are tasked to align our interpretations of the model outcomes to the model assumptions and limitations.[10]

10.1.6 Remember that an Explainable Model Is Not a Causal Model

Poorly prepared databases (i.e., with incomplete/loose/confounding features) can deceive explainability tools into mistakenly labeling poor features as important [16]. This may also mislead the engineer into thinking that such features are

5 Just like in other methods of investigation, you ought to check and verify your data and assumptions.

6 I would like to take the following lines to reminisce a bit on this recommendation. As engineers, we tend to favor some methods over others. Such preference stems from an engineer's competence, training, past experience, and established professional practices. As such, opting to adopt ML to solve a given problem could be a result of inherent favoritism toward ML. I cannot remember how many times I reviewed manuscripts where the authors developed a highly complex ML model to solve a simple problem that could have been easily solved via reviewing mechanics principles. I also fell into the same trap as well. I do not think that this is a good practice, just like I do not believe that we need to conduct experiments for every problem we have. We have a toolbox (shout out to Brandon Ross as I am borrowing his analogy) of methods and we can pick and choose which method is likely to be more suitable than others.

7 Please note that I am carefully using the terms "could potentially", as well as "better understand our problems and create/optimize novel solutions." These should not be confused with establishing causality.

8 Please refer to [40] for a detailed discussion on some of the limitations of statistical methods.

9 The same can also be said to trusting an unknown version of a codal equation or an untested FE model.

10 I tend to favor properly developed explainable and casual models. My argument is simple, once we identify the driving mechanism(s) governing a phenomenon, then we know all the ins and outs of such a phenomenon and unlock valuable knowledge. Yet, I would also like to re-iterate the fact that there is a time and place for each and all ML approaches.

important, especially for engineers with limited experience. This could trigger a domino effect and lead to poor decision making. Thus, augmenting ML models with explainability[11] measures, while admirable, is still constrained to understanding the model's behavior across its dataset, and hence may or may not agree with prior engineering knowledge. Remember that explainability methods may not reveal the data generating process (DGP) as neatly as causal ML models, if at all [17]. Still, augmenting blackbox models with such methods is an improvement over that of pure blackbox models. Let us not forget that explainability tools may unveil or pinpoint new information that could form the foundation for new knowledge or identify a new research direction, or at the very least verify existing domain knowledge.[12]

10.1.7 Master Performance Metrics and Avoid the Perception of False Goodness

A ML algorithm is but a script of a few lines of code (i.e., a set of rules). Once the data is fed and trained, such an algorithm becomes a model. Performance metrics describe how well this ML model predicts the phenomenon on hand with respect to the supplied dataset. A common fallacy associated with ML models is that they are only regarded fit if they score highly in terms of performance metrics [18, 19]. This fallacy has caused an unjustified race to excessively tune ML models.[13]

Take a step back to remember that performance metrics are mathematical/logical constructs that describe the difference between ML predictions and those known to be the ground truth. Thus, achieving high metrics indicates that the model on hand captures the phenomenon as described by the provided dataset (and not the phenomenon as a whole or in general or the, as you might know by now, the physics of such a phenomenon).

False goodness occurs when inappropriate or dependent metrics are applied or when their outcome is misinterpreted[14] [20]. Thus, the user has the responsibility to identify the most suitable metrics for the given phenomenon. The user also has the responsibility to understand the assumptions and limitations of such metrics. I would like to put emphasis on the merit of adopting a combination of metrics (say, those that convey the overall performance of a ML model across a dataset), as well as those that test the performance of the same model in local regions (especially, for end regions of continuous phenomena where limited and extreme values may reside).

10.1.8 Acknowledge that Your Model Is Likely to Be Biased

Overall, data can be expensive to generate and collect.[15] In many instances, individuals may not have sufficient resources to conduct and process hundreds or thousands of data points. As I will discuss in a later section, the average of a recent survey on datasets shows that the average size of adopted datasets is about 247 data points.[16] To appreciate this piece of information, imagine a phenomenon you happen to be interested in (say, the axial capacity of RC columns). Would you say that having data on the axial capacity of 247 RC columns would be enough to create a data-driven model capable of predicting the axial capacity of RC columns?[17] The chances are that it is unlikely for such a database to cover all configurations of columns. However, if such a database is restricted to a particular column type with specific features, then the generalizability of such a model is more plausible.

The biasness of models, whether ML or otherwise, becomes a critical issue when the data is extremely limited or when the data is not representative of our population. For instance, datasets on urban water distribution, electric grids, or mass

11 Simply put, when a blackbox model is augmented with a measure(s) that allow us to peek through its internal mechanisms to understand how it works, such a model is referred to as an explainable and/or interpretable model (note: other terms may include *eXplainable AI (XAI)*, *glassbox*, *whitebox*, or *transparent AI* – see [41, 42]). From a civil engineering perspective, I would prefer ML models that can articulate how a prediction is arrived at, and what are the mechanisms that the model uses to tie the features to a prediction. A suitable ML model would have the properties of an equation or a chart. We know how an equation is derived, we know the assumptions used to derive such an equation, and we also know how the features interact without each other to arrive at a prediction. Note that the above argument on the resemblance of ML models to an equation can be extended to FE models.

12 Ask Arash and Moe.

13 This is unlike the race to develop "better," "new," or "improved" generalizing models.

14 Or, when metrics' assumptions are nullified [20, 43].

15 Sometimes it is not.

16 There were instances where the number of data points fell below 70 points and even 50 points. I am not citing such resources intentionally.

17 This is a rhetorical question.

evacuation in relation to resident income, ethnicity, or race may not reflect the entire population[18] and are likely to leave marginalized groups out [21].

10.1.9 Consult with Experts and Re-visit Domain Knowledge to Identify Suitable Features

Features refer to the inputs, or independent variables, fed into a ML model. At a minimum, an engineer should re-visit the open literature to identify the key features known to govern a phenomenon. Similarly, experts can be consulted, or the engineer may opt to leverage some of the unsupervised and/or explainability tools to evaluate the importance of features.

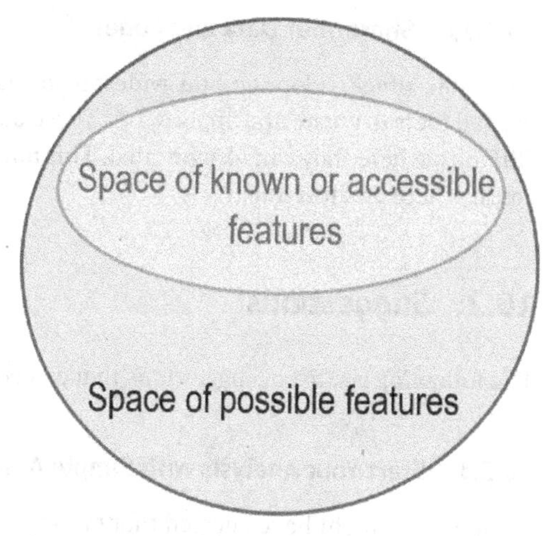

Figure 10.1 Space of features.

Figure 10.1 demonstrates a typical space of features governing a phenomenon. As you can see, the space of *possible features* is large, out of which there is a smaller space for *known* or *accessible features* (i.e., those features that we could measure or compile). This brings an interesting view to ML models. This tells us that a fully validated ML model is one capable of predicting the phenomenon on hand by tying the collected features to the outcome of the phenomenon.[19] This also tells us that depending on the used features, such a model can have varying degrees of usefulness.

10.1.10 Carefully Navigate the Trade-offs

We remain constrained by the availability of computational resources.[20] For a decent-sized ML model, we may not afford to run lengthy or complex trails. Thus, we ought to be aware of some of the notable trade-offs associated with the ML investigations. Some of these trade-offs include *bias-variance, accuracy-explainability, accuracy-computational resources*, to name a few (see Figure 10.2). At this moment, I can only suggest aiming for a *balance* between the trade-offs. For example, I favor simple and interpretable models over complex blackboxes, on the cost of accuracy. Practically speaking, I find little admiration for models that provide fractional increases or those that provide 1–5% improvement over other models.[21]

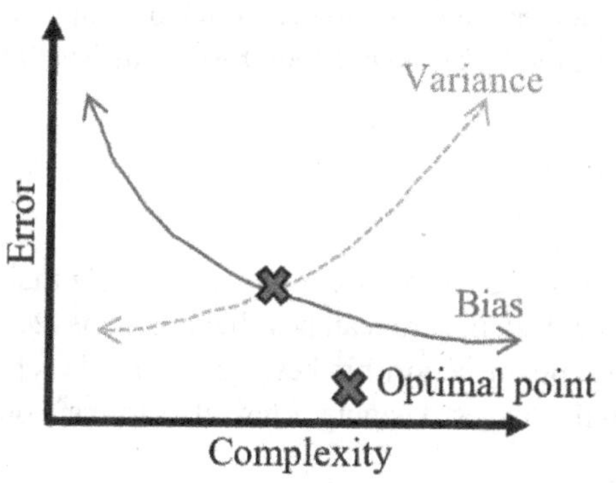

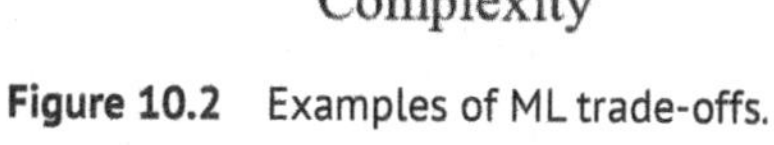

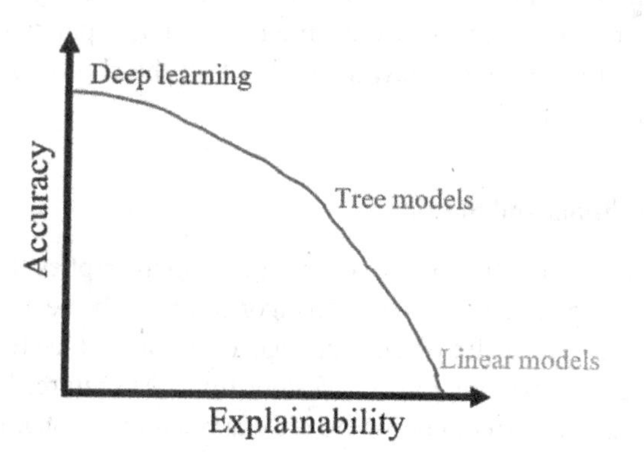

Figure 10.2 Examples of ML trade-offs.

18 Remember, civil engineering is all about our community.

19 A side note. A causal model goes beyond the space of known features into the space of possible features (and can identify the presence of latent features) to articulate how the features interact with each other to yield the phenomenon on hand.

20 One can think of a numerical model where dense mesh is used vs. low quality mesh. A trade-off on the front of accuracy-computation resources is expected to exist.

21 I am taking a big leap of faith by presenting this range – but I had to present some values as the original statement was a bit vague. I would not be too hooked on these values. Think, would this small increase obtained on a dataset reflect well on the generalizability of the model? *Ceteris paribus*, would you use a blackbox with $R^2 = 97\%$ over a explainable model with $R^2 = 94\%$? Hint: there is not a right or wrong answer.

10.1.11 Share Your Data and Codes

To enable timely adoption and widespread integration of ML, we ought to provide our readers and students with the needed tools to create and apply ML.[22] These databases and codes can be hosted freely on data repositories (i.e., Mendeley, GitHub), where they can also be cited. This allows us to keep track of models' histories/versions and ensures that we give credit where credit is due.

10.2 Suggestions

The following is a list of suggestions that can come in handy in the journey of ML.

10.2.1 Start Your Analysis with Simple Algorithms

By now, you might have guessed that my approach to learning ML parallels that of learning civil and environmental engineering material.[23] For example, new engineering students (say, in the structures track) are first introduced to *statics → mechanics → theory of structures → structural design*. During this journey, every course is structured similarly: start with simple concepts and then dive into complex concepts (i.e., statics starts by introducing the *equilibrium of a point* before moving to more advanced problems (e.g., friction and moment of inertia)). The same approach can also be taken in a ML investigation.

Let us acknowledge that the majority of ML algorithms are created by mathematicians and computer scientists. These have been extensively tested across multiple problems. Due to the above, ML models are likely to perform well even in their default settings [22]. Thus, remember to satisfy the Law of Parsimony of *Occam's Razor*, which implies that simplicity is a goal in itself [23]. There is genius in simplicity. This is the core of this suggestion.

If the goal is to arrive at an answer to a problem (say, build a classifier), then you would like to think of what would be the least path of resistance? This path will likely be a simple algorithm that *can do the job right!* Simple models are flexible, easy to use, tune, explain, and require little resources to run. On the other hand, if the goal is to dig deeper into a problem, then my advice is still the same. Start with a simple model, and see what you can learn. You will be surprised at how many challenging problems can be solved using simple models. If such models do not prove helpful, then start to add layers of complexity to see if complex, exotic, or causal models can add new insights.

Nowadays, the flow of information on advancements in ML is rapid, which makes it a hassle to keep track of simple or fundamental models. At the end of the day, remember that the choice of a ML model/strategy depends on your level of experience, dataset, and objective(s).

10.2.2 Explore Algorithms and Metrics

This suggestion stems from the "*No Free Lunch*" theory, which implies that some algorithms are not more sensible than others, but rather the averaged performance of all algorithms can be declared equivalent across all possible problems [24]. In addition, this book covers a wealth of learning algorithms, and since there is no one algorithm best suited for all problems, it becomes of merit to explore a series of algorithms. Therefore, I would suggest keeping a few algorithms[24] on hand.[25] These algorithms can then be contrasted once applied to a problem.

---

22 At the moment, very few journals in our area *require* the submission of data and codes.

23 During my FE modeling days, I remember that I tended to immediately build highly nonlinear three-dimensional models. I enjoyed building these complex models even though I knew that simple two-dimensional models are easier to build, run and, of course, diagnose. Luckily, I did not make the same mistake, to that extent, when I started my ML journey.

24 Preferably, ones that do not share the same origin or architecture.

25 I am writing this chapter while updating my chapter on eccentric connection design in my advanced steel design course. Guess what, there are also a couple of methods to design such connections!

10.2.3 Be Conscious of Data Origin

Real data (i.e., collected from experiments) may contain randomness effects and measurement errors, which could be attributed to equipment sensitivity or outliers/noise or may, in fact, be part of a sub-phenomenon that we are unaware of.[26] Unlike real data, data obtained by synthetic means (drawn from space of observations) or augmented (i.e., via numerical or FE models) can be free of the aforenoted effects due to the assumptions used in drawing such data or idealization associated numerical models. In a way, the origin of data has an inherent, often forgotten, influence on model predictions and any insights we may extract from such models. Therefore, I suggest that ML users carry out data analytics health examinations to ensure the generalizability of datasets.

10.2.4 Emphasize Model Testing

A ML investigation is likely to be more affordable and has a shorter lifespan than that carried out via experiments and numerical simulations.[27] This allows us some extra time to further test ML models, and hence it is suggested to examine ML models with scrutiny in terms of data transformation methods or validation techniques (i.e., *k*-fold validation, train/validate/test splits, out of the bag testing, the inclusion of confidence intervals, benchmark testing,[28] tailoring data subsets,[29] etc.), ensembling procedures, or sensitivity and reliability studies can be worthy of investigation.

10.2.5 Think Beyond Training and Validation

A ML model is often trained and validated using a representative dataset. During the training process, we hope to uncover how a certain phenomenon occurs (with an aim to be able to accurately predict[30] such phenomenon). Such a dataset is much more likely to be *modest* compared to the possible observations that one could potentially collect. If we take a step back to think about the above, we will realize that we have a model that is likely to perform well during its development stage [25]. However, this does not mean this model will perform as well when it is deployed in real-time and/or real-world data points. Be careful of misinterpreting the performance of a model to justify untested generalizations or inferences.

10.2.6 Trace Your Model Beyond Deployment

About 90% of the developed ML models are never deployed [26]. I do believe that the above statistic is alarming.[31] My suggestion is to aim to deploy your models (even if locally) and monitor their progress with the addition of new data. Recent works have shown that a model performance tends to degrade with future data [27, 28]. Update your model regularly with new data and techniques.

10.2.7 Convert Your ML Models into Web and Downloadable Applications

In many of the published works, such models are shown as prototypes with limited access or shareability. This begs the question of whether such models are really necessary. I do not believe that every model should be converted into an application, nor do I wish for it to be sharable.[32] Nowadays, ML codes can be converted into simple applications with a neat GUI

---

26 One can argue that real data from tests may not be representative of real world observations (given the limitation of testing set-ups). This is a valid philosophical argument.

27 Relatively speaking, a testing program may take a few months to design and complete (even much longer if one is exploring durability issues [44]). A numerical program may take a few days/weeks to complete. A ML program might take a few hours/days (for application-based) to a few weeks/months (creation).

28 Benchmarked databases and algorithms on key structural engineering problems have been recently published (see [22]).

29 Say, in case of concrete mixtures. One can tailor a data sub-set that contains outcomes of tests conducted from multiple groups/countries, or spans a number of decades.

30 Creating a predictive ML model is possible. However, uncovering why a phenomenon occurs is a causal pursuit.

31 This could be due to the ML hype phase. This may also imply the need for some form of a regulatory body/initiative to keep track of such codes and applications.

32 Some projects/funding agencies may have restrictions as to when/if developed models and data can be sharable.

that can be hosted online. These applications allow users to access ML models online and without the need to re-run the ML code and hence have the potential to expedite the adoption of ML (in industry or education settings). Such software can also be copyrighted and marketed for subscriptions which bring a new revenue stream to model creators.

10.2.8 Whenever Possible, Include Physics Principles in ML Models

Instilling prior knowledge of first principles or knowledge domain is valuable to regularize and improve the robustness of the transformation functions (often hidden within algorithmic topologies). In addition, incorporating such principles into a ML model is likely to improve model prediction capability.

10.3 Best Practices

Here is a list of best practices that are likely to be helpful.

10.3.1 Avoid the Use of "Small" and Low Quality Data

Rules of thumb for a minimum size for datasets vary between 10 observations per predictor [29], 23 observations per predictor [30], and 100 observations per predictor [31]. Another criterion recommends maintaining a ratio between 3 and 5 between the number of observations and input parameters [32]. These rules of thumb assume the presence of quality data points and hence should be regarded as rough guides.

There is a lot more to datasets than their size. For example, if a dataset contains a large number of points (>1000), but such points only cover a narrow range of values or contain many repetitions, etc., then such a dataset may not be as helpful as one with a smaller number of unique and diverse data points. Strive to have a representative dataset. A good practice is to check for an acceptable dataset size from recently published works in your area and/or to seek an expert opinion. Remember that more data may not always correspond to better models, but more high quality data is likely beneficial. Also, remember that there is *little room* to finetune the setting of algorithms in search of improved accuracy, yet there is significant room to improve accuracy by enhancing data quality.[33]

10.3.2 Be Aware of the Most Commonly Favored ML Algorithms[34]

Figure 10.3 presents the most frequently used algorithms in civil engineering as reported in publications published between 2011 and 2021. It is apparent that the top five algorithms comprise ANN, GA, GP, SVM, and PCA. Of these algorithms, ANN and GA ranked the highest, reaching more than 50% of all publications. This preference for ANN and GA might have arisen from the fact that these two algorithms found a home in civil engineering a few decades ago. This could also explain the familiarity of our engineers with these algorithms.

10.3.3 Follow the Most Favored Model Development Procedures

At the moment, the majority of the examined scholarly works primarily adopt two model development procedures: *train/test splits* or *k-fold cross-validation* (see Figure 10.4). Historically, earlier works favor the former, wherein about 85% applied splits ranging between 70 and 80% in training the ML model and 30 and 20% in validating and testing the model. However, there is a recent trend toward utilizing variations of the *k*-fold cross validation method due to its benefits in terms of minimizing biasness over the train/test splitage.

In line with the data in Figure 10.4, the top five performance metrics include MAE, MSE, R, R^2, RMSE, and MAPE – with R being the most used metric (almost in every study reviewed), followed by MSE and RMSE (see Figure 10.5). The use of these three metrics is understandable as the same metrics are also used in regression problems as well as in evaluating the outcome of experimental tests. Very few researchers created dedicated objective functions [33, 34] or new indicators specific to specialized problems [35]. For classification-based studies, the prominently used metrics were those related to the confusion matrix and AUC.

33 Please refer to [45] for a good discussion on the role of data in ML modeling.

34 The next three best practices were arrived at via a literature survey led by Arash Tapeh.

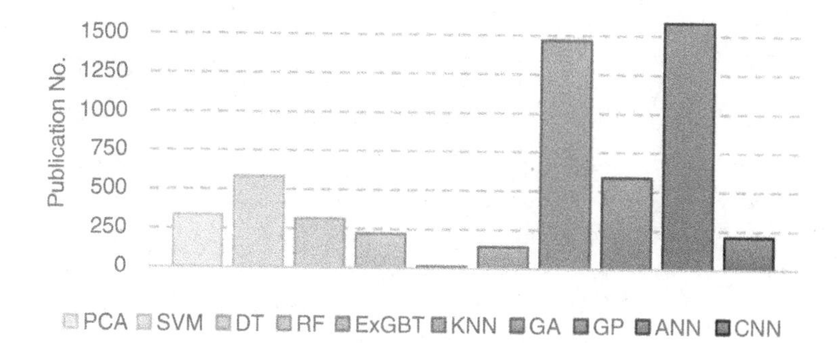

(a) No. of publications per surveyed algorithm

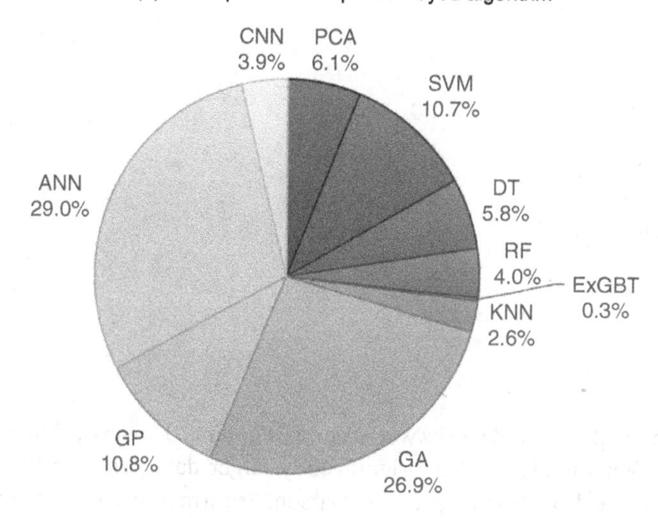

(b) Breakout of used algorithms

Figure 10.3 Insights into frequently used algorithms in civil engineering.

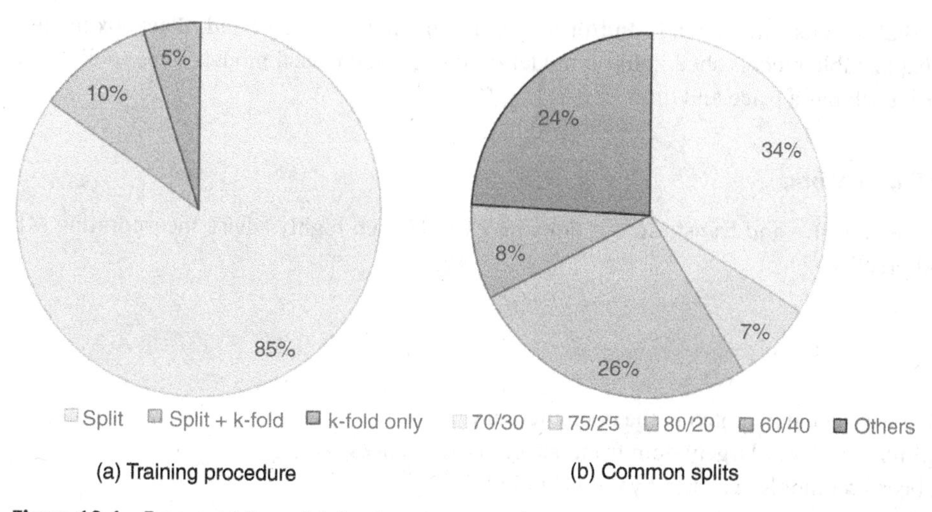

(a) Training procedure (b) Common splits

Figure 10.4 Frequent ML model development procedures.

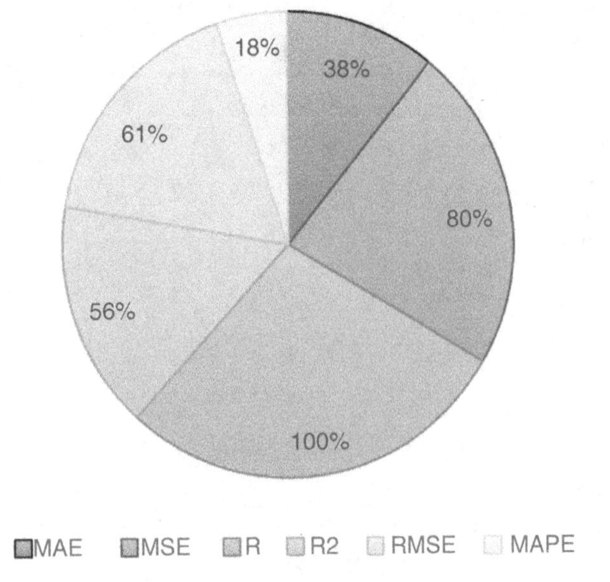

Figure 10.5 Frequently used performance metrics in ML models.

10.3.4 Report Statistics on Your Dataset

As mentioned a few pages ago, the average dataset size from all reviewed works was around 247 points. However, datasets used in deep learning models and computer vision problems tended to have significantly larger data points (>1000). Remember to report details of your dataset via basic statistical treatments such as Pearson, Spearman correlation and mutual information matrices, and frequency plots. In addition, report information with regard to data range (i.e., max, min, average) and distribution and treatment of outliers/noise.

10.3.5 Avoid Blackbox Models in Favor of Explainable and Causal Models (Unless the Goal Is to Create a Blackbox Model)

Traditionally, we create solutions that help us "understand" and/or predict the phenomenon on hand. Blackbox models can help us predict phenomena, explainable models shed light on model rationales, and causal models hope to discover why a phenomenon occurs [10, 36]. Each has a place and time.

10.3.6 Integrate ML into Your Future Works

Machine learning methods continue to evolve and transform our domain. I would then highly advise incorporating ML into our curriculum, research, and practice.[35]

Definitions

Recommendation An advice that has been duly noted in the open literature.
Suggestion A proposition of opinion (and may largely stem from personal experiences).
Best practice A trend that has been commonly accepted by publications.

---

35 For some odd reason, this reminds of the following conversation from Interstellar (2014)."It's not possible." – Case "*No, it's necessary.*" – Cooper.

Questions and Problems

10.1 What are the main differences between *recommendations*, *suggestions*, and *best practices*?

10.2 How does a causal model differ from an explainable model?

10.3 Discuss the effect of data origin on model generality.

10.4 What is the rationale behind favoring real/experimental data over simulated or synthetic data?

10.5 When would a blackbox model be sufficient?

10.6 Debate the need to adopt simple models before diving into complex and exotic models.

10.7 List venues that allow open access/free share of data and codes.

References

1 Nature. (2022). Statistical methods. Nature. https://www.nature.com/subjects/statistical-methods (accessed 24 May 2022).

2 Christodoulou, E., Ma, J., Collins, G.S. et al. (2019). A systematic review shows no performance benefit of machine learning over logistic regression for clinical prediction models. *J. Clin. Epidemiol.* 110. doi: 10.1016/j.jclinepi.2019.02.004.

3 Bzdok, D., Altman, N., and Krzywinski, M. (2018). Statistics versus machine learning. *Nat. Methods* 15. doi: 10.1038/nmeth.4642.

4 Ivezic, Ž., Connolly, A.J., VanderPlas, J.T., and Gray, A. (2014). Statistics, data mining, and machine learning in astronomy. doi: 10.23943/princeton/9780691151687.001.0001.

5 Christensen, B.T. and Schunn, C.D. (2007). The relationship of analogical distance to analogical function and preinventive structure: the case of engineering design. *Mem. Cogn.* 35. doi: 10.3758/BF03195939.

6 Biot, M.A. (1943). Analytical and experimental methods in engineering seismology. *Trans. Am. Soc. Civ. Eng.* 108. doi: 10.1061/taceat.0005571.

7 Zienkiewicz, O.C. and Taylor, R.L. (2000). *The Finite Element Method Volume 1 : The Basis, Methods.* Wiley.

8 Tekkaya, A.E. (2005). A guide for validation of FE-simulations in bulk metal forming. *Arab. J. Sci. Eng.* 30.

9 Ferreira, J., Gernay, T., Franssen, J., and Vassant, O. (2018). Discussion on a systematic approach to validation of software for structures in fire. *SiF 2018*– The 10th International Conference on Structures in Fire FireSERT, Ulster University, Belfast, UK, June 6-8, 2018. http://hdl.handle.net/2268/223208 (accessed 1 April 2020).

10 Mackenzie, D. and Pearl, J. (2018). *The Book of Why: The New Science of Cause and Effect.* Basic Books.

11 Tran-Ngoc, H., Khatir, S., Le-Xuan, T. et al. (2020). A novel machine-learning based on the global search techniques using vectorized data for damage detection in structures. *Int. J. Eng. Sci.* 157. doi: 10.1016/j.ijengsci.2020.103376.

12 Rahmanpanah, H., Mouloodi, S., Burvill, C. et al. (2020). Prediction of load-displacement curve in a complex structure using artificial neural networks: A study on a long bone. *Int. J. Eng. Sci.* 154. doi: 10.1016/j.ijengsci.2020.103319.

13 Zarringol, M. and Thai, H.T. (2022). Prediction of the load-shortening curve of CFST columns using ANN-based models. *J. Build. Eng.* 51. doi: 10.1016/j.jobe.2022.104279.

14 Naser, M.Z. (2020). Machine learning assessment of fiber-reinforced polymer-strengthened and reinforced concrete members. *ACI Struct. J.* 117. doi: 10.14359/51728073.

15 Rudin, C. (2019). Stop explaining black box machine learning models for high stakes decisions and use interpretable models instead. *Nat. Mach. Intell.* 1. doi: 10.1038/s42256-019-0048-x.

16 Papadimitriou, P. and Garcia-Molina, H. (2010). Data leakage detection. *IEEE Trans. Knowl. Data Eng.* 23. doi: 10.1109/TKDE.2010.100.

17 Pearl, J. (2009). Causal inference in statistics: an overview. *Stat. Surv.* 3. doi: 10.1214/09-SS057.

18 Corbett-Davies, S. and Goel, S. (2018). The measure and mismeasure of fairness: a critical review of fair machine learning. *ArXiv*. doi: 10.48550/arXiv.1808.00023.

19 Naser, M.Z. and Alavi, A.H. (2021). Error metrics and performance fitness indicators for artificial intelligence and machine learning in engineering and sciences. *Archit. Struct. Constr.* 1. doi: 10.1007/s44150-021-00015-8.

20 Arp, D., Quiring, E., Pendlebury, F. et al. (2021). *Dos and Don'ts of Machine Learning in Computer Security*. Usenix Security Symposium (USENIX), USENIX.

21 Mehrabi, N., Morstatter, F., Saxena, N. et al. (2019). A survey on bias and fairness in machine learning. *ACM Comput. Surv.* 54.

22 Naser, M.Z., Kodur, V., Thai, H.-T. et al. (2021). StructuresNet and FireNet: benchmarking databases and machine learning algorithms in structural and fire engineering domains. *J. Build. Eng.* 44: 102977. doi: 10.1016/J.JOBE.2021.102977.

23 Hildebrand, C.D.W. (1938). Ockham, studies and selections by stephen chak tornay. LaSalle, Ill.: open court publishing company, 1938. viii, 207 pages.$1.75. *Church Hist.* doi: 10.2307/3160457.

24 Wolpert, D.H. and Macready, W.G. (1997). No free lunch theorems for optimization. *IEEE Trans. Evol. Comput.* 1. doi: 10.1109/4235.585893.

25 Baylor, D., Breck, E., Cheng, H.T. et al. (2017). TFX: a tensorflow-based production-scale machine learning platform. *Proc. ACM SIGKDD Int. Conf. Knowl. Discov. Data Min.* doi: 10.1145/3097983.3098021.

26 Joury, A. (2020). Why 90 percent of all machine learning models never make it into production. towards data science https://towardsdatascience.com/why-90-percent-of-all-machine-learning-models-never-make-it-into-production-ce7e250d5a4a (accessed 25 May 2022).

27 Adam, G.A., Chang, C.-H.K., Haibe-Kains, B., and Goldenberg, A. (2020). Hidden risks of machine learning applied to healthcare: unintended feedback loops between models and future data causing model degradation. *Proceedings of the 5th Machine Learning for Healthcare Conference* PMLR 126: 710–731.

28 Young, Z. and Steele, R. (2022). Empirical evaluation of performance degradation of machine learning-based predictive models – a case study in healthcare information systems. *Int. J. Inf. Manag. Data Insights.* 2. doi: 10.1016/j.jjimei.2022.100070.

29 van Smeden, M., Moons, K.G., de Groot, J.A. et al. (2018). *Sample size for binary logistic prediction models: beyond events per variable criteria*. 28. 2455–2474. doi: 10.1177/0962280218784726.

30 Riley, R.D., Snell, K.I.E., Ensor, J. et al. (2019). Minimum sample size for developing a multivariable prediction model: PART II - binary and time-to-event outcomes. *Stat. Med.* 38. doi: 10.1002/sim.7992.

31 Spirtes, P., Glymour, C., and Scheines, R. (2000). Causation, prediction, and search (Springer lecture notes in statistics). *Lect. Notes Stat.*

32 Frank, I. and Todeschini, R. (1994). *The Data Analysis Handbook*. https://books.google.com/books?hl=en&id=SXEpB0H6L3YC&oi=fnd&pg=PP1&ots=zfmIRO_XO5&sig=dSX6KJdkuav5zRNxaUdcftGSn2k (accessed June 21, 2019).

33 Naser, M.Z. and Alavi, A.H. (2022 July). Error metrics and performance fitness indicators for artificial intelligence and machine learning in engineering and sciences. *Archit. Struct. Constr.* 1: 1–19. doi: 10.1007/S44150-021-00015-8.

34 Amlashi, A.T., Golafshani, E.M., Ebrahimi, S.A., and Behnood, A. (2022). Estimation of the compressive strength of green concretes containing rice husk ash: a comparison of different machine learning approaches. *Eur J Environ Civ Engrg.* 1–23. doi: 10.1080/19648189.2022.2068657.

35 Naser, M.Z. (2022). CLEMSON: an automated machine learning (AutoML) virtual assistant for accelerated, simulation-free, transparent, reduced-order and inference-based reconstruction of fire response of structural members. *ASCE J. Struct. Eng.* 148. doi: 10.1061/(ASCE)ST.1943-541X.0003399.

36 Naser, M.Z. (2021). Mapping functions: a physics-guided, data-driven and algorithm-agnostic machine learning approach to discover causal and descriptive expressions of engineering phenomena. *Measurement* 185: 110098. doi: 10.1016/J.MEASUREMENT.2021.110098.

37 Naser, M.Z. and Ross, B. (2022). An opinion piece on the dos and don'ts of artificial intelligence in civil engineering and charting a path from data-driven analysis to causal knowledge discovery. doi: 10.1080/10286608.2022.2049257.

38 Naser, M.Z. (2021). Demystifying ten big ideas and rules every fire scientist & engineer should know about blackbox, whitebox & causal artificial intelligence. https://arxiv.org/abs/2111.13756v1 (accessed 26 January 2022).

39 Zinkevich, M. (2022). Rules of machine learning. google developers. https://developers.google.com/machine-learning/guides/rules-of-ml (accessed 4 August 2022).

40 Bzdok, D., Altman, N., and Krzywinski, M. (2018). Points of Significance: statistics versus machine learning. *Nat. Methods.* 15. doi: 10.1038/nmeth.4642.

41 Adadi, A. and Berrada, M. (2018). Peeking inside the black-box: a survey on explainable artificial intelligence (XAI). *IEEE Access* 6. doi: 10.1109/ACCESS.2018.2870052.

42 Barredo Arrieta, A., Díaz-Rodríguez, N., Del Ser, J. et al. (2020). Explainable artificial intelligence (XAI): concepts, taxonomies, opportunities and challenges toward responsible AI. *Inf. Fusion.* 58. doi: 10.1016/j.inffus.2019.12.012.

43 Botchkarev, A. (2019). A new typology design of performance metrics to measure errors in machine learning regression algorithms. *Interdiscip. J. Information, Knowledge, Manag.* 14: 045–076. doi: 10.28945/4184.

44 Mohammed, T.U., Hamada, H., and Yamaji, T. (2004). Concrete After 30 Years of Exposure - Part 1: mineralogy, microstructure, and interfaces. *ACI Mater. J.* 101. doi: 10.14359/12982.

45 Lones, M.A. (2021). How to avoid machine learning pitfalls: a guide for academic researchers. https://doi.org/10.48550/arXiv.2108.02497 (accessed 10 December 2021).

11

Final Thoughts and Future Directions

Just because something works does not mean that it cannot be improved.

Shuri (Black Panther)

Synopsis

This chapter concludes the body of our textbook with a look into some of the future directions and research needs. Some of the discussed ideas stem directly from the limitations and challenges we have seen over the previous chapters, while others build upon those noted as identified in the open literature.

11.1 Now

Let us look at some of the key points we learned throughout this book. There are plenty of ML algorithms, and these primarily fall under three main types of learning (supervised, unsupervised, and semi-supervised). The collection of algorithms, as well as learning strategies, can find a home in many problems and areas that belong to the civil and environmental engineering discipline. The above has been common knowledge for quite some time (+20 years) [1, 2]. The adoption of machine learning (ML) has just started to prosper.

This could be due to the fact that ML is fundamentally different from the other methods we are familiar with. For example, to use or apply ML, our engineers have to learn the principles of ML. The majority of such principles remain absent from our education, and some revolve around coding and programming – an exercise that we rarely cover in our curriculum. This is further darkened by the opaque nature of most blackbox ML methods.

As you can see, we are already standing at a disadvantage. This hurdle is likely to diffuse with time and strategic/well-thought out exposure to ML.[1] There is rising support in favor of ML, whether from our academia or industry.[2] The path we select now will be elemental to seeing how soon, or far, ML becomes synonymous with civil and environmental engineering.[3]

11.2 Tomorrow

The look for tomorrow could be summarized through the following open-ended discussions.

1 I hope so!

2 Including start-ups and entrepreneurship.

3 If you pause for a minute. Our discipline includes massive built infrastructure, natural and human resources. At the moment, we are collecting tremendous amounts of data (and of data of various types/formats) and could possibly be one of the most attractive disciplines to apply (as well as create) ML tools.

Machine Learning for Civil & Environmental Engineers: A Practical Approach to Data-driven Analysis, Explainability, and Causality, First Edition. M. Z. Naser.

Companion Website: www.wiley.com/go/Naser/MachineLearningforCivilandEnvironmentalEngineers

11.2.1 Big, Small, and Imbalanced Data

A key assumption in a given ML analysis is the one where we deem our datasets to be comprehensive, representative, and balanced. In reality, this is somehow a fair assumption[4] and one that is needed to jump-start a ML analysis.

However, the real data can be small and messy (imbalanced). This data is what we train our ML models on and hence such models, despite their scoring performance, already include some messiness[5] in them. Would it be possible to devise a ML strategy to overcome the above? Further, in some cases, we may not be able to collect quantitative data and opt to use qualitative data. How can qualitative-based ML models be used in a quantitative manner? Given some of the previous chapters, it sure looks like many of the key limitations to ML reside in the data and strategies to ML, as opposed to algorithmic or modeling aspects.

On a more positive front, the above has been gaining interest, and it is likely that we will see new solutions to some of the mentioned problems soon with the emergence of data-centric ML.

11.2.2 Learning ML

Engineers on the verge of applying ML are advised, if not required, to get hands-on experience[6] before doing so. With the rise of coding-free approaches to ML, using ML is becoming more accessible and affordable. The automation provided by such platforms could be, intentionally or unintentionally, misused and, more alarmingly, abused. A critical future direction can revolve around delivering educational methods that best fit our needs and, quite frankly, backgrounds.[7]

11.2.3 Benchmarking ML

Building on my above discussion on standardization, it is also possible to seek to develop our own ML algorithms and practices that are applicable to our discipline. Creating and benchmarking our own models would require a collaborative community effort to collect suitable and international datasets that strives to document our experimental and numerical observations. ML algorithms can then be trained on such comprehensive and well-accepted datasets to maximize the potential of such models and ensure their generalizability.

11.2.4 Standardizing ML

Recent history has shown that new technologies are only adopted once harmony, convergence, and understanding between academia and industry are realized.[8] This convergence is often translated by the development and publication of standards, codes, manuals of practice, etc. Therefore, one possible path of least resistance to embracing ML is through pursuing standardization, as it is via standardization that a committee can be comfortable enough to share their ideas and beliefs in a matter that is representable to the engineering community. This effort is likely to limit the intentional or unintentional cherry-picking in the ML pipeline and could yield developing approved sets of ML-based *virtual assistants*[9] with experiences spanning decades[10] of research/practice.

11.2.5 Unboxing ML

While blackbox ML has its merit and place, it could be plausible to move past such models and toward those of white and causal nature. Whitebox and causal models have been noted to arrive at similar prediction capability as those of their blackbox counterparts, yet with the potential to provide us with the means to understand the logic behind ML predictions. This

4 Especially, if we clearly acknowledge our, and data's, limitations.

5 Think about how sensitive strain data can be in a dynamic test! Also, think about how accurate a sensor measurement can be under 800°C! Real and reliable data is not easy to collect, and can be very expensive to pursue.

6 Learning ML is similar to learning a new tool, method, or software. You've got to study and practice. Practice makes perfect!

7 In the meantime, I imagine it would be beneficial to form complementary relationships with data scientists.

8 At the frontier of a particular technology.

9 For example, a building code assists an engineer in their design process. A similar concept can be created wherein ML aids our engineers.

10 Imagine being able to distill years of experiences into ML. Imagine being able to use decades of blueprints and designs into a ML to create a ML assistant, or a ML engineer!

can be of significant help to our engineers and researchers (especially in deriving new theories) as this provides a layer of confidence against bias and unexplainability.

11.2.6 Popularizing ML

I must admit that much of the presented discussion on ML in this book is dominated by the soft side[11] of ML. However, there is an equally exciting component to ML that I did not discuss, nor most of the published works in our discipline. This component is with regard to the hardware of ML where ML is augmented into robots or equipment that fall under civil and environmental engineering. In the past few years, we have seen the construction sector pushing new heights and depths! Those modern constructions are pushing our abilities to climb thousands of feet high skyscrapers.[12] Construction robots armed with ML could be the future of our industry.[13]

11.2.7 Engineering ML

Machine learning is a method and philosophy. I often think of ML as a language that is likely to evolve[14] with time. As such, we can create our own version of ML.[15] In a way, we can engineer ML into one that best suits us, our problems, and our domain. In this process, we can break free from the norm of being ML appliers and become ML creators.

11.3 Possible Ideas to Tackle

I often receive emails and requests to suggest a few problems suitable for research projects (arguably for those that may fit into an MSc or a PhD thesis). Here are some such generic[16] suggestions:

- Find a problem in your area with well-established solutions.[17] Ask yourself, can ML provide you with new but intuitive, affordable, or faster solutions?
- Find a problem in your area that lacks a solution. Ask yourself, could ML provide us with a possible solution?

---

11 As in software/virtual/program-like.

12 The tallest building in the world (Burj Khalifa) right now stands at 2722 ft (829 m). The height of the most likely to be the tallest building in the world (Burj Al Meel, a.k.a. The Mile Tower) is speculated to be twice as high as that of Burj Khalifa! For completion, the China Jinping Underground Laboratory is perhaps the deepest of its kind in the world as it is located at the depth of 2400 m.

13 Not to mention that these robotic workers are also needed to fabricate and build extraterrestrial construction for space exploration missions (perhaps even more so than are needed here on Earth since they will need to be able to not only carry out construction tasks, but to also characterize in-situ materials, and be fully independent given the lack of communication and poor relays with Earth [3]).

14 Vocabulary evolves. Terms evolve. Styles and accents evolve, etc.

15 This brings Al-Mutanabbi's [4] famous lines to mind:

(1) مَا كلُّ ما يَتَمَنّى المَرْءُ يُدْرِكُهُ تجرِي الرّياحُ بمَا لا تَشتَهي السّفُنُ

And, the recent response to these lines:

(2) تَجْرِي الرِّيَاحُ كَمَا تَجْرِي سَفِينَتُنَا نَحْنُ الرِّيَاحُ وَنَحْنُ الْبَحْرُ وَالسُّفُنُ

(3) إن الذي يرتجي شيئاً بهمّتهِ يلقاهُ لو حاربَتْهُ الانسُ والجن

(4) (فاقصد الى قمم الاشياءِ تدركها تجري الرّياح كما رادت لها السفن

[I feel that translating these lines from Arabic dilutes their essence and flow. However, let me briefly go over why I have cited these here. The first line is by Al-Mutanabbi (965 AD) and is probably a lil more than a thousand years old. This is, by far, one of the most memorable lines of his poems. The line simply asks the reader to be content with their reality. The next few lines, on the other hand, have surfaced a few years ago on social media by an *unknown* modern poet who skillfully uses most of the terms, essence, and flow of the first line to ask the reader to create their own version of reality wherein challenges can be overcome with persistence (line 2). Lines 3 and 4 re-affirm the notion of line 2].

16 These are generic indeed, as I feel that research projects should have a personal component that stems from one's interest/intrigue as opposed to being *told* what to research, or *invited* to research. I would highly suggest to sit down and think, as in really think, to read and explore our, as well as other domains'literature in search of a problem that you would like to tackle. In hindsight, I usually end up recommending what *not* to research instead! PS. Solving problems is easy. You already know what the problem is. All what you need now is to find solutions (or just a solution) to this problem. Coming up with a problem worthy of your time is one of the hardest decisions you could make. Not giving up on solving such a problem, despite all arising challenges, is, possibly, the second hardest.

17 Problems that have/are built with empirical solutions might be a good candidate to this suggestion.

- Find a phenomenon[18] with unknown/contradicting theories on its mechanisms. Explore such a phenomenon via explainable and causal ML in the hope of uncovering/discovering possible mechanisms.
- Find a problem where you can compare traditional solutions to those of ML nature. A goal is to explore if ML can extend the range of applicability of traditional methods.
- Create a framework to teach ML the physics of your area. Do the same to incorporate domain knowledge.
- There seem to be distinct practices for identical tasks.[19] Will ML be able to harmonize such tasks?
- Think of ML frameworks where similar problems can be combined into one.[20]
- Develop approaches to reproduce results of ML analysis and ensure its consistency and reliability.
- Create an open-source ML bot or software.[21]
- Work on ideas to combine data of different origins and types.[22]
- Think about ways and means to convince stakeholders of the merit of adopting ML.
- Fundamentally, most of our engineers are interested in the solution[23] component that a method investigation yields. The same is also true in the case of ML. Thus, you may think of approaches that can convert ML codes/programs into solutions that we are familiar with. This may accelerate our trust in ML.
- As of late, I have been finding myself thinking about the education side[24] of ML. Perhaps something on this front can be interesting to tackle.

11.4 Conclusions

This chapter concludes the main body of our book. This chapter also offers a perspective into our present and a possible future of our discipline from the context of ML.[25]

References

1 Adeli, H. (2001). Neural Networks in Civil Engineering: 1989–2000. *Comput. Civ. Infrastruct. Eng.* 16. doi: 10.1111/0885-9507.00219.

2 Watts, J.M. (1987). Expert systems. *Fire Technol* 23: 1–2. doi: 10.1007/BF01038360.

3 Naser, M.Z. (2019). Extraterrestrial construction materials. *Prog. Mater. Sci.* 105. doi: 10.1016/j.pmatsci.2019.100577.

4 Wikipedia (2022). Al-Mutanabbi.https://en.wikipedia.org/wiki/Al-Mutanabbi (accessed 15 July 2022).

18 For example, fire-induced spalling is one such problem where our body of literature (i.e., fire tests) has generated a decent amount of data but research papers published on this front seem to not agree on a specific mechanism for it. Similar phenomena exist in each area.

19 As in, individuals belonging to a specific region/country/school of thought, etc. carry out tasks in their own special way (i.e., perhaps by following the standard/body of authority under their jurisdiction). For example, testing properties of construction materials are bound by ISO standards (e.g., Europe) or ASTM standards (US). In this example, the same property can be tested using the ISO standard or the ASTM standard. However, both standards may or may not agree in the procedure, and hence the reported measurement (of the same property) can be drastically different.

20 How many papers are published annually with the theme of: we used ML to predict property X of construction material Y? Many of these are simple problems that, more or less, adopt a standard methodology (collect data → apply ML → create a model). Unfortunately, a good portion of such papers reapply already tired algorithms and approaches and provide minor adjustments and tweaks to chase accuracy metrics. Perhaps by building frameworks, the need for such papers can be minimized (since the framework can potentially host thousands of data and hundreds of models with the ability to self-evolve and improve) and can help move our focus toward other worthy problems.

21 This can be tailored toward a specific problem such as crash forecasting, inspection issues detection, blueprint extenders, etc.

22 Say that we collect data from 20 different experiments carried out by 20 different researchers in 20 different labs. How can ML help us harmonize this data?

23 Much of our engineering revolves around applications! In an application setting, engineers tend to be more interested in applying a method. For example, we carry tests to understand why something happens and once we know, we derive a solution (i.e., an equation) that conveys this understanding so that other engineers can use such solutions to overcome the problem on hand!

24 How to best deliver ML-based education to civil and environmental engineers? When to introduce our engineers to ML? Shall we go with coding-free, or do we need to maintain the traditional coding-based approach to ML?

25 In case you are wondering, it has been seven weeks since I started working on this book. What a journey... Stay in touch!

Index

Note: *Italic* page numbers indicate illustrations, **bold** page numbers indicate tables.

Machine Learning for Civil & Environmental Engineers: A Practical Approach to Data-driven Analysis, Explainability, and Causality, First Edition. M. Z. Naser.

Companion Website: www.wiley.com/go/Naser/MachineLearningforCivilandEnvironmentalEngineers

e

n

o

p

t